THE COMPLETE OFFICIAL TRIUMPH TR7

Model Years 1975-1981

Selected Titles From Bentley Publishers

Engineering

Robert Bentley's Repair Manual for British Cars *John Organ*
ISBN 978-0-8376-0041-3

Vintage Racing British Sports Cars *Terry Jackson*
ISBN 978-0-8376-0153-3

Scientific Design of Exhaust & Intake Systems *Philip H. Smith and John C. Morrison*
ISBN 978-0-8376-0309-4

The Design and Tuning of Competition Engines *Philip H. Smith* ISBN 978-0-8376-0140-3

New Directions in Suspension Design: Making the Car Faster
Colin Campbell
ISBN 978-0-8376-0150-2

Racing Car Design and Development *Len Terry with Alan Baker* ISBN 978-0-8376-0080-2

Austin-Healey

Austin-Healey 100/6 NS 3000 Workshop Manual: 1956-1968
British Leyland Motors
ISBN 978-0-8376-1743-5
Hardcover, 416 pages

The Complete Official 1275 cc Sprite/Midget: 1967-1974
British Leyland Motors
ISBN 978-0-8376-1748-0
Hardcover, 400 pages

The Complete Official 948 cc & 1098 cc Sprite/Midget
British Leyland Motors
ISBN 978-0-8376-1747-3
Hardcover, 256 pages

MG

The Complete Official MGB: 1962-1974 *British Leyland Motors*
ISBN 978-0-8376-1761-9
Hardcover, 480 pages
Includes: Driver's Handbook, Workshop Manual, Special Tuning Manual

The Complete Official MGB: 1975-1980 *British Leyland Motors*
ISBN 978-0-8376-1760-2
Hardcover, 304 pages

The Complete Official MG Midget 1500: 1975-1979
British Leyland Motors
ISBN 978-0-8376-1744-2
Hardcover, 384 pages

The MG Workshop Manual: 1929-1955 *W.E. Blower*
ISBN 978-0-8376-1746-6
Hardcover, 602 pages

Triumph

The Complete Official Triumph GT6, GT6+ & GT6 MK III: 1967-1973 *British Leyland Motors*
ISBN 978-0-8376-1749-7
Hardcover, 480 pages
Includes: Driver's Handbook, Workshop Manual

The Complete Official Triumph TR6 & TR250: 1967-1976
British Leyland Motors
ISBN 978-0-8376-1762-6
Hardcover, 608 pages
Includes: Driver's Handbook, Workshop Manual

The Complete Official Triumph Spitfire MK III, MK IV & 1500: 1968-1974 *British Leyland Motors*
ISBN 978-0-8376-1763-3
Hardcover, 480 pages
Includes: Driver's Handbook, Workshop Manual

The Complete Official Triumph Spitfire 1500: 1975-1980
British Leyland Motors
ISBN 978-0-8376-1745-9
Hardcover, 536 pages
Includes: Driver's Handbook, Workshop Manual

The Complete Official Triumph TR7: 1975-1981
British Leyland Motors
ISBN 978-0-8376-1742-8
Hardcover, 544 pages
Includes: Driving Manual, Workshop Manual

The Complete Official Triumph TR4 & TR4A: 1961-1968
British Leyland Motors
ISBN 978-0-8376-1741-1
Hardcover, 404 pages
Includes: Driving Manual, Workshop Manual, Competition Preparation Manual by R.W. "Kas" Kastner

The Complete Official Triumph TR2 & TR3: 1953-1961
British Leyland Motors
ISBN 978-0-8376-1750-3
Hardcover, 456 pages
Includes: Driver's Instruction Book, Service Instruction Manual

Jaguar

The Complete Official Jaguar "E" Series 1 & Series 2: 1961-1971
British Leyland Motors
ISBN 978-0-8376-1764-0
Hardcover, 790 pages
Includes: Driver's Handbook, Workshop Manual, Special Tuning Manual

B BentleyPublishers®
Automotive Reference

Bentley Publishers has published service manuals and automobile books since 1950. For more information, please contact Bentley Publishers at 1734 Massachusetts Avenue, Cambridge, MA 02138 USA, or visit our web site at **BentleyPublishers.com**

THE COMPLETE OFFICIAL TRIUMPH TR7 1975-1981

Comprising the official
Driver's Handbook
Repair Operation Manual

BentleyPublishers®
.com

BENTLEY PUBLISHERS® | Automotive Reference™

Bentley Publishers, a division of Robert Bentley, Inc.
1734 Massachusetts Avenue
Cambridge, MA 02138 USA
800-423-4595 / 617-547-4170

Information that makes the difference®

BentleyPublishers®.com

Technical contact information

We welcome your feedback. Please submit corrections and additions to our technical discussion forum at:

http://www.BentleyPublishers.com

Updates and corrections

We will evaluate submissions and post appropriate editorial changes online as updates or tech discussion. Appropriate updates and corrections will be added to the book in future printings. Check for updates and corrections for this book before beginning work on your vehicle. See the following web address for additional information:

http://www.BentleyPublishers.com/updates/

WARNING—Important Safety Notice

In this book we have attempted to describe repairs, modifications and accessories which may be used with Triumph vehicles, using examples and instructions which we believe to be accurate. However, the examples, instructions, and other information are intended solely as illustrations and should be used in any particular application only by personnel who are experienced in the repair and modification of Triumph vehicles and who have independently evaluated the repair, modification or accessory. Implementation of a modification or attachment of an accessory described in this book may render the vehicle, attachment, or accessory unsafe for use in certain circumstances. **REPAIR AND MODIFICATION OF MOTOR VEHICLES CAN BE DANGEROUS UNLESS UNDERTAKEN WITH FULL KNOWLEDGE OF THE CONSEQUENCES.**

Do not use this book unless you are familiar with basic automotive repair procedures and safe workshop practices. This book illustrates procedures required for some service and modification work; it is not a substitute for full and up-to-date information from the vehicle manufacturer or aftermarket supplier, or for proper training as an automotive technician. Note that it is not possible for us to anticipate all of the ways or conditions under which vehicles may be serviced or modified or to provide cautions as to all of the possible hazards that may result.

The vehicle manufacturer and aftermarket suppliers will continue to issue service information updates and parts retrofits after the editorial closing of this book. Some of these updates and retrofits will apply to procedures and specifications in this book. We regret that we cannot supply updates to purchasers of this book.

We have endeavored to ensure the accuracy of the information in this book. Please note, however, that considering the vast quantity and the complexity of the information involved, we cannot warrant the accuracy or completeness of the information contained in this book.

FOR THESE REASONS, NEITHER THE PUBLISHER NOR THE AUTHOR MAKES ANY WARRANTIES, EXPRESS OR IMPLIED, THAT THE EXAMPLES, INSTRUCTIONS OR OTHER INFORMATION IN THIS BOOK ARE FREE OF ERRORS OR OMISSIONS, ARE CONSISTENT WITH INDUSTRY STANDARDS, OR THAT THEY WILL MEET THE REQUIREMENTS FOR A PARTICULAR APPLICATION, AND WE EXPRESSLY DISCLAIM THE IMPLIED WARRANTIES OF MERCHANTABILITY AND OF FITNESS FOR A PARTICULAR PURPOSE, EVEN IF THE PUBLISHER OR AUTHOR HAVE BEEN ADVISED OF A PARTICULAR PURPOSE, AND EVEN IF A PARTICULAR PURPOSE IS INDICATED IN THE BOOK. THE PUBLISHER AND AUTHOR ALSO DISCLAIM ALL LIABILITY FOR DIRECT, INDIRECT, INCIDENTAL OR CONSEQUENTIAL DAMAGES THAT RESULT FROM ANY USE OF THE EXAMPLES, INSTRUCTIONS OR OTHER INFORMATION IN THIS BOOK. IN NO EVENT SHALL OUR LIABILITY WHETHER IN TORT, CONTRACT OR OTHERWISE EXCEED THE COST OF THIS BOOK.

Your common sense and good judgment are crucial to safe and successful automotive work. Read procedures through before starting them. Think about how alert you are feeling, and whether the condition of your vehicle, your level of mechanical skill or your level of reading comprehension might result in or contribute in some way to an occurrence which might cause you injury, damage your vehicle,or result in an unsafe repair or modification. If you have doubts for these or other reasons about your ability to perform safe work on yourvehicle, have the work done at an authorized Triumph dealer or other qualified shop.

Before attempting any work on your Triumph, read the Warnings and Cautions on page vi and any warning or caution that accompanies a procedure or description in the book. Review the Warnings and Cautions each time you prepare to work on your Triumph.

ACKNOWLEDGEMENT

Special thanks are due H.C. Taylor, National Service Training Manager, and Gary Kolesaire, Parts Publication and Sales Promotion Supervisor, British Leyland Motors, Inc.

This book is prepared, published and distributed by Bentley Publishers, 1734 Massachusetts Avenue, Cambridge, Massachusetts 02138 USA.

The Complete Official Triumph TR7: 1975-1981

ISBN 978-0-8376-1742-8
Job code: X128-07

Library of Congress Catalog Card No. 81-68901

The paper used in this publication is acid free and meets the requirements of the National Standard for Information Sciences-Permanence of Paper for Printed Library Materials. ∞

Manufactured in the United States of America.

Foreword

This Manual has been compiled in order to provide complete, accurate, and comprehensive maintenance and repair data to both car owners and professional mechanics. The *Driver's Handbook*, which comprises the first part of this Manual, is similar to the handbook provided with every new Triumph sports car. The *Repair Operation Manual*, which comprises the second — and largest — portion of this Manual, is the official factory manual and was originally intended for use by dealer service departments. The Publisher has added new material to the factory data so that the present Manual is suitable both for persons of limited mechanical experience and for those trained mechanics who may not be altogether familiar with Triumph cars.

Home Maintenance and Repair

Though professional mechanics will find this Manual indispensable as a source of specifications, testing procedures, and precision fitting data, such mechanics usually carry out basic assembly and disassembly work guided mainly by their years of training and experience. For example, professional mechanics know without being told that engine, transmission, and rear axle parts must be properly pre-lubricated when they are reassembled following an overhaul. Therefore, such information is not commonly included in factory workshop manuals. On the other hand, non-professionals often need to be reminded of such things since mechanical work may not be a part of their daily routine. Wherever possible, the Publisher has included *Notes* and *Cautions* to alert the non-professional mechanic to those additional considerations that a professional mechanic would take into account as a matter of course.

Though simple maintenance procedures — such as those described in the *Driver's Handbook* portion of this Manual — require only a minimum of tools, the car owner will sooner or later find it desirable to obtain a fairly complete set. The basic tools include an assortment of open- and box-end wrenches and a selection of screwdrivers, pliers, one or two adjustable wrenches and other miscellaneous items such as feeler gauges, a spark plug socket, etc. A socket wrench set — preferably ⅜-in. drive — will speed up many common jobs and may be indispensable to others. A quality set of tools can be a good investment; they will last a lifetime if not abused. Cheaply made tools will not last long without wear and breakage and often damage bolts and fittings which cost more to replace than the money saved in buying inferior tools.

Liberal reference is made in this Manual to the special service tools which are recommended by the factory. Such special tools are not supplied by British Leyland Motors but can be ordered from Messrs. V. L. Churchill & Company Limited, P. O. Box No. 3, London Road, Daventry, Northants, England — information that may prove of little value to a car owner in the U.S. who must repair his car by the next weekend! Fortunately, generic equivalents for most of these special tools are available from American tool manufacturers that supply professional mechanics. Do not, however, overlook the possibility of obtaining the needed tool from a tool rental agency. In many U.S. cities, there are also do-it-yourself garages that rent or loan special tools to those who patronize the shop.

This Manual describes certain precision reconditioning jobs which even Authorized Dealers commonly farm out — either to an automative machine shop or to a specialty shop. These jobs include cylinder reboring, crankshaft reconditioning, valve grinding, starter rebuilding, sheet metal repair, etc. Instead of delivering the entire car to a professional mechanic, who may take the parts to an outside machine shop, you can save considerable money by removing the worn or damaged components and taking them to the machine shop yourself.

Advice for Professionals and Car Owners Alike

Assembly work on small, highly-tuned machines such as the Triumph sports cars must be carried out with greater precision than is commonly practiced on large American cars. Particular emphasis must be given to the proper use of torque wrenches and to the strict adherance to tightening torque specifications that are given in this Manual. Though all of the torque specifications may fall within the range of a 0–100 ft. lb. (0–14 mkg) torque wrench, torque wrenches operate most efficiently when used in the middle two quarters of their scales. Thus, a small torque wrench with a range of 0–50 ft. lb. (0–7 mkg) is recommended in addition to a larger torque wrench with a 150 ft. lb. (21 mkg) capacity. A fastener that is too tight can be worse than one that is too loose — especially on a lightweight sports car. Stretched or broken bolts and distorted parts, which result from overtightening by muscle-bound mechanics, become a serious concern where the precision fitting of light alloy and thin-wall iron castings is involved.

Lastly, the importance of cleanliness cannot be overemphasized. Under no circumstances should an engine or gearbox be repaired on the ground or on a garage floor. Thoroughly clean the exteriors of major components prior to disassembly in order to keep road dirt and other grime out of the working parts. No more than a pinch of abrasive dust in a gearbox can cause rapid failure of the synchronizers and bearings.

During the final assembly of an engine or gearbox, the cleaned parts should be laid out on a clean workbench which has been covered with clean sheets of new carboard or wrapping paper. The engine or gearbox itself, if not mounted on a special stand, should likewise be placed on a clean workbench. Sandpapering, valve grinding, or the use of bench grinders should not be permitted near the area where final assembly is taking place. If assembly cannot be completed in a day, enclose the partially-assembled engine or gearbox in a large plastic bag — such as a new trash bag or a dry cleaner's garment return bag — so that dust and dirt will be excluded until assembly work resumes.

David N. Wenner
Automotive Editor
Robert Bentley, Inc.

Please read these warnings and cautions before proceeding with maintenance and repair work.

WARNING—

• Do not re-use any fasteners that are worn or deformed in normal use. Many fasteners are designed to be used only once and become unreliable and may fail when used a second time. This includes, but is not limited to, nuts, bolts, washers, self-locking nuts or bolts, circlips, cotter pins. For replacements always use new parts.

• Never work under a car unless it is solidly supported on stands designed for the purpose. Do not support a car on cinder blocks, hollow tiles, or other props that may crumble under continuous load. Do not work under a car that is supported solely by a jack.

• If you are going to work under a car on the ground, make sure that the ground is level. Block the wheels to keep the car from rolling. Never work under the car while the engine is running. Disconnect the battery ground strap to prevent others from starting the car while you are under it.

• Never run the engine unless the work area is well ventilated. Carbon monoxide kills.

• Friction materials such as brake or clutch discs may contain asbestos fibers. Do not create dust by grinding, sanding, or by cleaning with compressed air. Avoid breathing asbestos fibers and asbestos dust. Breathing asbestos can cause serious diseases such as asbestosis or cancer, and may result in death.

• Tie long hair behind your head. Do not wear a necktie, a scarf, loose clothing, or a necklace when you work near machine tools or running engines. If your hair, clothing or jewelry were to get caught in the machinery, severe injury could result.

• Disconnect the battery ground strap whenever you work on the fuel system or the electrical system. When you work around fuel, do not smoke or work near heaters or other fire hazards. Keep an approved fire extinguisher handy.

• Illuminate your work area adequately but safely. Use a portable safety light for working inside or under the car. Make sure the bulb is enclosed by a wire cage. The hot filament of an accidentally broken bulb can ignite spilled fuel or oil.

• Catch draining fuel, oil or brake fluid in suitable containers. Do not use food or beverage containers that might mislead someone into drinking from them. Store flammable fluids away from fire hazards. Wipe up spills at once, but do not store the oily rags, which can ignite and burn spontaneously.

• Finger rings should be removed so that they cannot cause electrical shorts, get caught in running machinery or be crushed by heavy parts.

• Keep sparks, lighted matches and open flame away from the top of the battery. If hydrogen gas escaping from the cap vents is ignited, it will ignite gas trapped in the cells and cause the battery to explode.

• Always observe good workshop practices. Wear goggles when you operate machine tools or work with battery acid. Gloves or other protective clothing should be worn whenever the job requires it.

• Do not attempt to work on your car if you do not feel well. You increase the danger of injury to yourself and others if you are tired, upset or have taken medicine or any other substance that may impair you from being fully alert.

• Some aerosol tire inflators are highly flammable. Use extreme caution when repairing a tire that may have been inflated using an aerosol tire inflator. Keep sparks, open flame or other sources of ignition away from the tire repair area. Inflate and deflate the tire at least four times before breaking the bead from the rim. Completely remove the tire from the rim before attempting any repair.

• Greases, lubricants and other automotive chemicals contain toxic substances, many of which are absorbed directly through the skin. Read manufacturer's instructions and warnings carefully. Use hand and eye protection. Avoid direct skin contact.

CAUTION—

• If you lack the skills, tools and equipment, or a suitable workshop for any procedure described in this Manual, we suggest you leave such repairs to an authorized dealer or other qualified shop. We especially urge you to consult an Authorized Dealer before beginning repairs on any car that may still be covered wholly or in part by the new-car warranty.

• British Leyland Motors was constantly improving its cars and sometimes these changes, both in parts and specifications, were made applicable to earlier models. Therefore, part numbers listed in this manual are for reference only. Always check with your Authorized Parts Department for the latest information.

• Before starting a job, make certain that you have all the necessary tools and parts on hand. Read all the instructions thoroughly, do not attempt shortcuts. Use tools appropriate to the work and use only replacement parts meeting specifications. Makeshift tools, parts and procedures will not make good repairs.

• Use pneumatic and electric tools only to loosen threaded parts and fasteners. Never use these tools to tighten fasteners, especially on light alloy parts.

• Be mindful of the environment and ecology. Before you drain the crankcase, find out the proper way to dispose of the oil. Do not pour oil onto the ground, down a drain, or into a stream, pond or lake. Consult local ordinances that govern the disposal of wastes.

Contents

2 OFFICIAL TRIUMPH TR7 REPAIR OPERATION MANUAL

TRIUMPH TR7
DRIVER'S HANDBOOK

PART 1

SERVICE RECORDS

A 'Passport to Service' is included in the literature package for your new car. The 'Passport to Service' not only contains a record of vital information, but also contains information about Warranty, Jaguar Rover Triumph Inc., car clubs and tyre manufacturers. The passport also contains a voucher for the 1000 mile inspection.

WARRANTIES

The 'Passport to Service' furnished with your new TR7 contains detailed information about warranties applicable to it, which include:

—The limited vehicle warranty.

—The emission control system warranty and covered parts list.

Tyres are not covered by Jaguar Rover Triumph Inc., but by the specific manufacturer of the tyres on your vehicle. Details of tyre warranties are included in the new vehicle literature pack. The 'Passport to Service' is a convenient document to record all maintenance services on the vehicle wherever performed. Replacement, blank 'Passports to Service' may be obtained by writing to Consumer Relations Department, Jaguar Rover Triumph Inc., 600 Willow Tree Road, Leonia, New Jersey. 07605.

In all communications relating to Service or Spares, please quote the Commission Number (Chassis Number.) paint and trim numbers. These are stamped on a plate attached to the left-hand door rear closing face. The Commission Number is also stamped on a plate attached to the left-hand windscreen pillar and is visible through the screen.

NOTE: When L.H. or R.H. is used in the text, this refers to the Left-hand or Right-hand of the vehicle, viewed from the driver's seat.

The Manufacturers reserve the right to vary their specifications with or without notice, and at such times and in such manner as they think fit. Major as well as minor changes may be involved in accordance with the Manufacturer's policy of constant product improvement.

Whilst every effort is made to ensure the accuracy of particulars contained in this Handbook, neither The Manufacturer nor the Dealer, by whom this Handbook is supplied, shall in any circumstances be held liable for any inaccuracy or the consequences thereof.

CONTENTS

IMPORTANT
CATALYTIC CONVERTOR SPECIFICATION VEHICLES ONLY
(FEDERAL INCLUDING CALIFORNIA)

The TR7 (Catalytic Convertor Specification) is designed to operate on UNLEADED fuels of 91 octane.

It is essential that only unleaded fuels are used in this vehicle otherwise serious damage can be caused to the Catalytic Converter.

WARNING: Do not use oxygenated fuels such as blends of methanol/gasoline or ethanol/gasoline (e.g. GASAHOL).

CANADA SPECIFICATION

The TR7-CANADA which is not fitted with a Catalytic Converter, performs efficiently on fuels of 91 octane (Regular Fuel).

NOTE: The engine is designed to use either leaded or unleaded fuel.

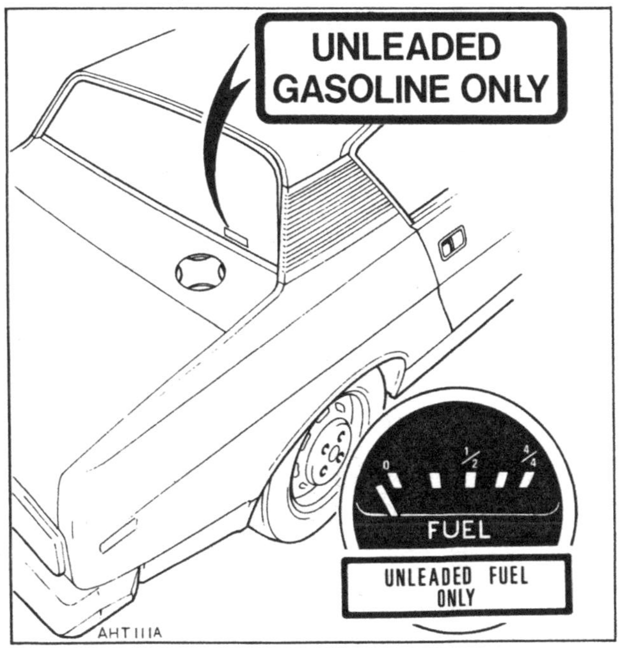

INTRODUCTION

Upon taking delivery of your new car it is in your interest to read the contents of this Handbook and so become familiar with the controls and other features which are provided for your comfort and safety.

A combined steering-column lock and ignition switch fitted to the vehicle is of particular importance and, if used properly, will greatly reduce the possibility of the car being stolen.

The lock key number appears only on a tag attached to the larger key of your key-set and does not appear on the lock or key. Enter the key number in your diary NOW and store the key tag in a safe place at home. This will enable you to order the correct key if required and prevent duplication of the key by unauthorized persons. Instructions for using the switch are given in the following pages under the heading 'Controls, Instruments and Indicators'.

Servicing

Before receiving your new car it will have undergone an inspection to ensure that all systems are working and the vehicle is complete to specification.

After completing 1000 miles the vehicle should be returned to the selling Dealer who will carry out the first important service and make whatever adjustments are required, free of charge except for cost of lubricants.

Subsequent to the free service, maintenance and servicing is the responsibility of the owner who is advised to make full use of the planned maintenance operated by all Rover-Triumph franchise holders. The operations carried out by your Dealer will be in accordance with current recommendations and may be subject to revision from time to time.

The 'Passport to Service' which includes the 'Owners Service Statement' has provision for the Dealer to certify that the work has been performed in accordance with Rover-Triumph recommendations. Regular use of the 'Passport to Service' at the specified intervals is the best safeguard against the possibility of abnormal repair bills at a later date and would constitute proof of regular servicing thus enhancing the value of your car to a prospective buyer.

Rover-Triumph Dealers are provided with the latest information concerning special tools and workshop techniques enabling them to undertake your service and repairs in the most efficient and economic manner.

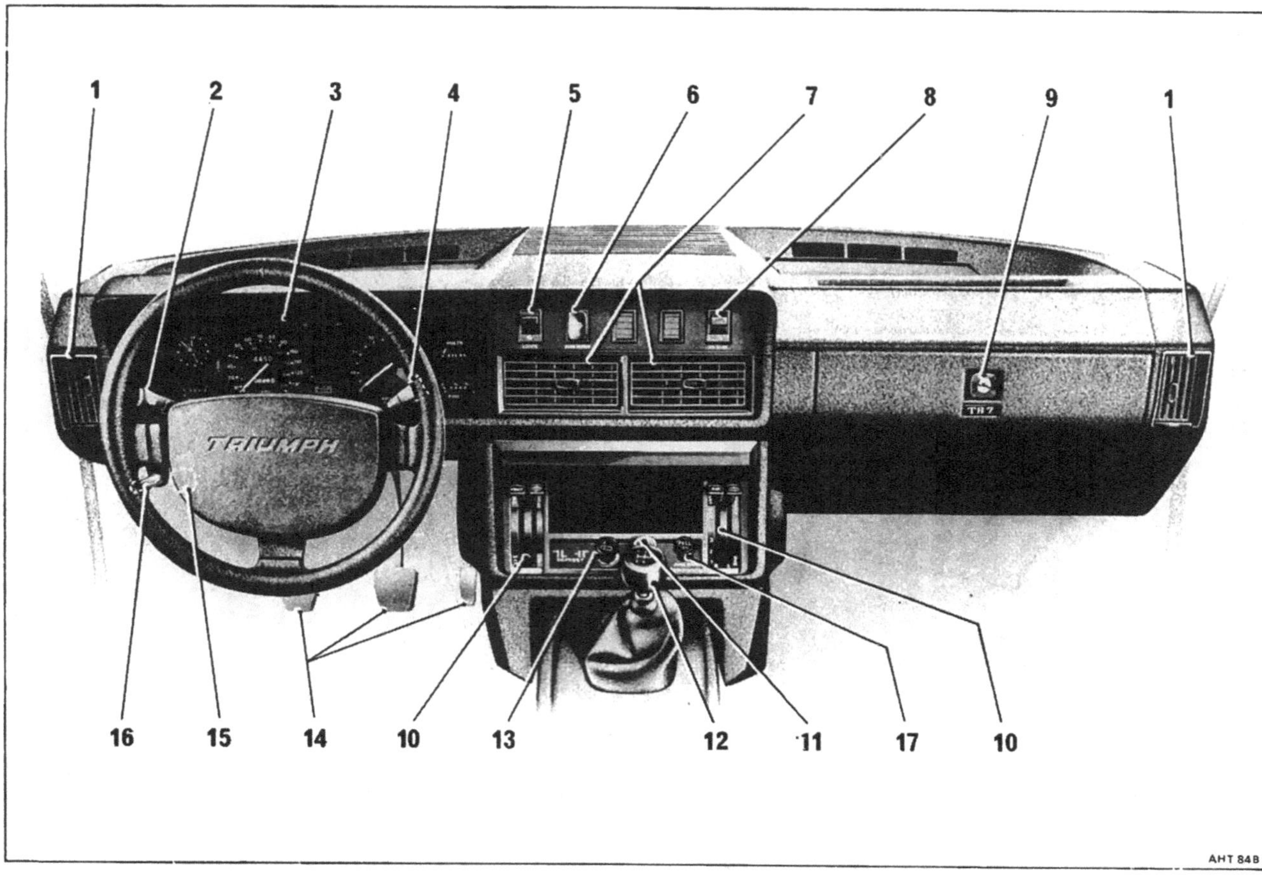

CONTROLS, INSTRUMENTS AND INDICATORS

Item	Description	Page Ref	Item	Description	Page Ref
1	Facia side vent	25	10	Air conditioning/heater controls	23-24
2	Combination switch	11	11	Instrument illumination rheostat	16
3	Instruments and warning indicators	9	12	Gear selector lever	13
4	Windscreen wiper and washer control	11	13	Cigar lighter	16
5	Master lighting switch	12	14	Pedals	13
6	Backlight heater switch	12	15	Choke control (Canada only)	26
7	Fascia central air vent	25	16	Bonnet release control	15
8	Hazard warning switch	12	17	Air conditioning cut-out switch [1976 and later models]	23
9	Glove box lock	16			

Location of instruments and controls

Ignition/Starter Switch and Steering Lock

0. All electrical circuits switched off: mechanical lock pre-set when key is removed.
1. Auxiliary position.
2. Ignition.
3. Start.

Unlocking sequence (clockwise): Insert the key and turn it to position '1'. If difficulty is encountered, turn the steering wheel slightly to relieve the load on the lock bolt. The steering wheel should now be free to turn.

NOTE: This operation should be performed only after the following requirements have been carried out in the correct sequence:- 1. OCCUPANTS SEATED; 2. DOORS CLOSED; 3. SEAT BELTS FASTENED; 4. GEARS NEUTRAL. Refer to 'Running Instructions'.

To switch on the ignition, turn the key to position '2'. Further movement to position '3' operates the starter motor. The key will return to position '2' when released.

Locking sequence (anti-clockwise) Turning the key from position '2' to '1' will switch off the ignition. Push the key fully towards the lock and turn it to position '0'. Withdraw the key and move the steering wheel slightly to ensure that the lock bolt is fully engaged.

5

WARNING. If for any reason the ignition is switched off while the car is in motion DO NOT ATTEMPT TO DEPRESS OR TURN THE KEY TO POSITION 'O', as this constitutes part of the steering lock sequence. The designed operating sequence prevents the engine being started with the steering LOCKED. Serious consequences can result from alterations or substitution of the ignition start switch which would permit the engine to be started with the LOCK ENGAGED. Under no circumstances must the ignition start function be separated from the steering lock.

NOTE: If the engine fails to start [on a 1977 or later car], the key must be returned to position '1' before you can again operate the starter.

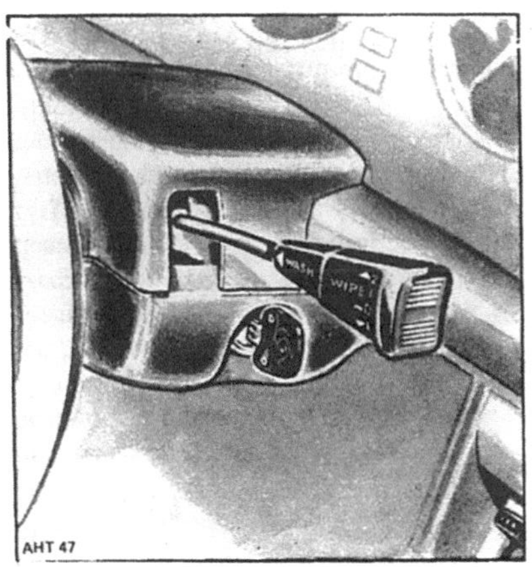

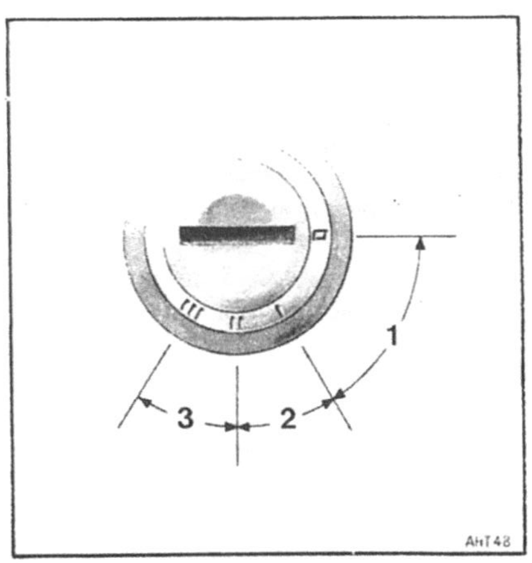

Speedometer

The speedometer indicates the road speed of the vehicle in miles and kilometres per hour. It also combines the following indicators:

(a) *Trip odometer* — The figures within the aperture above the centre of the dial may be used to record the distance of each journey provided that the figures are pre-set to zero by turning clockwise the knob that extends downwards from behind the instruments.

(b) *Odometer* — The figure within the aperture below the centre of the dial records the total mileage of the vehicle and may be used as a guide to periodic lubrication and maintenance.

Tachometer

The tachometer indicates the engine speed in revolutions per minute and is calibrated in increments of 500, extending to 7,000 r.p.m.

Recommendations concerning the speed range are given under 'Running Instructions'.

Clock

The hands of the electrically operated clock can be reset by turning the knob located below the instrument and extending from the lower facia rail.

Fuel Gauge

The gauge indicates the contents of the fuel tank. Allow thirty seconds for the gauge needle to reach a steady reading after switching on the ignition.

When the ignition is switched off the needle should return to zero.

Water Temperature Indicator

The gauge is marked 'C' (cold) and 'H' (hot), indicating the temperature of the coolant as it leaves the cylinder head. It is effective when the ignition is switched on.

When the ignition is switched off the needle should return to zero.

Voltmeter

The voltmeter registers the battery's state of charge. Under all normal running conditions the needle should register approximately midway along the scale. If the needle registers within the area at either extremity of the scale, the battery or charging system needs attention.

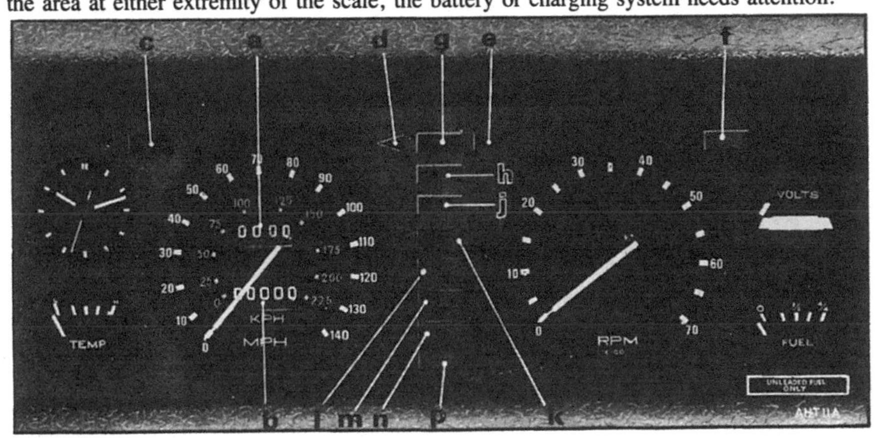

Warning Lights

Main beam (c) [on 1975–1979 models; *n* on 1980 and later models] — The 'Beam' indicator glows blue when the main beams are selected and is extinguished during dipped beam operation.

Direction indicator monitors (d) and (e) — The arrowed indicator (d) glows green intermittently when the steering column combination switch is moved to signal a left-hand turn. The (e) indicator glows similarly when the switch is moved to signal a right-hand turn.

Demist (f) — [on 1975–1979 models] — The 'Demist' indicator glows when the backlight heater element is switched on.

Oxygen sensor (f) — [on 1980 and later models] — The 'Oxygen Sensor' indicator glows when the sensors require renewal. This will occur at the mileage interval stated in the Maintenance Handbook. Consult your dealer as soon as convenient after the indicator illuminates.

Ignition (g) — The 'Ignition' indicator glows red when the ignition is switched on and is extinguished when the engine is running.

Oil pressure (h) — The 'Oil' indicator glows when the ignition is switched on and is extinguished when the engine is running.

Brakes (j) — Should failure of the front or rear hydraulic brake lines occur, or if the handbrake is left 'on' the indicator will glow. If the lamp glows, indicating failure of the hydraulic system, the cause must be rectified immediately.

The indicator should illuminate when starter motor is operated; failure to do so indicates bulb failure.

Belts (k) — The 'Seat Belt' warning light will glow when the ignition is switched on and a vehicle occupant has not fastened a seat belt. This will be accompanied by an audible alarm. Refer to 'Running Instructions'.

Park brake (l) [on 1980 and later models] — The 'Park Brake' indicator glows when the handbrake is applied and the ignition is switched on.

Choke (l) (Canada only) — The 'Choke' indicator glows when the 'cold start' control is pulled and is extinguished when the control is pushed fully home.

Fuel (m) [on 1975–1979 models; *(p)* on 1980 and later models] — The 'Fuel' indicator glows red when the quantity of fuel in the tank falls below 6.8 litres (1.5 gallons). Fuel surge may cause the indicator to glow intermittently before the fuel is reduced to this amount.

Coolant [*(m)* on 1980 and later models] — The 'Coolant' indicator glows when the coolant level in the header tank is too low. The fault should be rectified immediately by topping up the header tank, refer to 'Cooling System'.

A prudent watch should be kept on the coolant temperature gauge to ensure that the engine does not overheat.

The indicator glows for three seconds when the ignition is switched on, failure to do so indicates bulb failure.

Fog (n) (optional fitment) — If the vehicle is equipped with rear fog lamps, the indicator will glow when the circuit is operative.

6

Combination Switch

The steering-column combination switch serves the following purposes:

Horns (A) — To operate the horns, press the end of the control.

Direction indicators (L and R) — To indicate a left-hand turn, move the lever anti-clockwise. Move it clockwise for right-hand turns. This control is effective only when the ignition is switched on.

Head lamp flash (C) [1980 and later] — Pull the lever towards the steering wheel against spring pressure to flash the lamps. This control is effective regardless of the position of any other switch.

Windscreen Wiper and Washer Control

The control is effective only when the ignition is switched on.

For continuous operation of the two-speed wiper, move the control upwards to position '1' for slow-speed operation, or fully upwards to position '2' for high-speed operation. [For 1975–1977 models,] for brief use of the wiper, pull the control towards the steering wheel, against spring pressure. Wiping will cease when the control is released.

[1980 and later]: For intermittent use of the wipers, move the control downwards against the spring pressure to position '3'. Repeat this movement to cancel the function.

To operate the windscreen washer, press the knob on the end of the control.

CAUTION: Should the ignition be switched off with the wiper switch in an operating position during freezing conditions, the wiper blades can freeze to the windscreen. When the ignition is switched on again, the wiper motor may be damaged by overload. **Always switch off the wiper motor before switching off the ignition.**

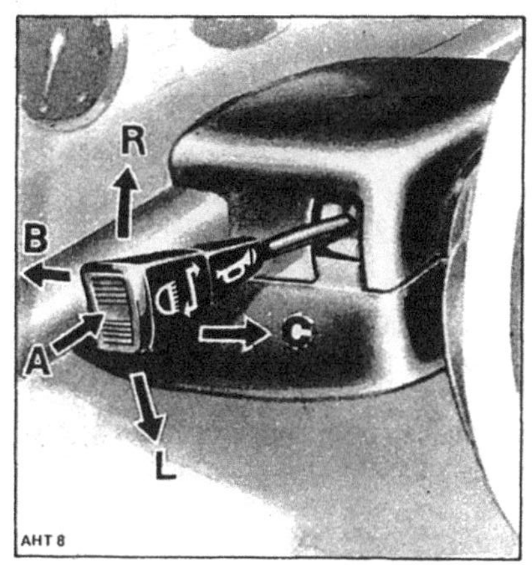

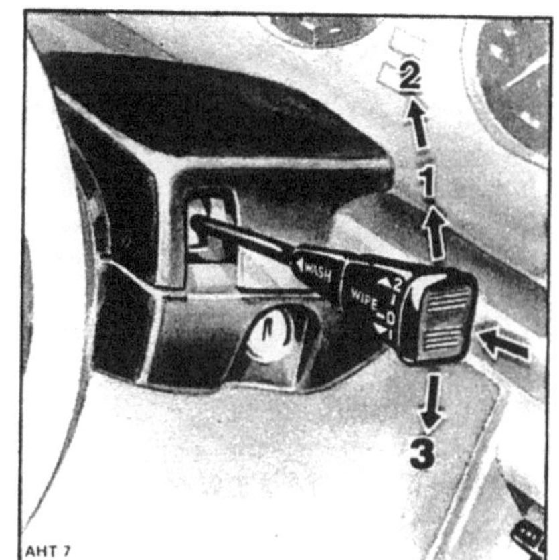

Master Lighting Switch (1)

To operate the three-position lighting switch:

- Move the switch bar to the upper position to extinguish all external an l fascia lamps.

A - Move the switch bar to the mid position to illuminate front and rear parking and marker lamps and the fascia lamps.

A - Move the switch bar to the lower position to raise and illuminate the headlamps and illuminate the parking, marker and fascia lamps.

NOTE: If the headlamp lift mechanism fails to raise the headlamps, refer to 'Electrical System' — Headlamps.

Back Light Heater Switch (2)

Depress the switch to operate the back-light demister heater which will operate only when the ignition is switched on. A lamp incorporated in the instrument panel will illuminate while the heater is operating. Do not leave switched on for longer than is necessary because of heavy current consumption. To switch off, depress the switch again.

The heater should not be used to remove snow or ice, but may be used periodically to remove condensation or to prevent ice formation.

Fog Lamp Switch [1980 and later] (3)

Move the switch bar (3) to the mid position to illuminate the rear fog guard lamps only.
Move the switch bar (3) to the lower position to illuminate the front and rear fog lamps.
Move the switch bar (3) to the upper position to extinguish the lamps.

A warning lamp incorporated in the switch will illuminate while the lamp circuit is switched 'ON'.

This circuit is operative only when the side/tail lamps are illuminated.

7

Hazard Warning Switch (4)

If the vehicle is immobilized and constitutes a hazard to other road users, warning may be given by using the 'hazard warning system'. To operate move the switch bar to the lower position.

Instrument Illumination Rheostat

From the 'off' position (fully anti-clockwise), turn the switch knob clockwise to illuminate the instruments at high intensity, and further clockwise to diminish the intensity.

See 'Master Lighting Switch'.

Gear Selector Lever

The gear positions are indicated on the lever knob. Engage reverse only when the vehicle is stationary.

Automatic transmission selector details are given in section — 'Borg-Warner Automatic Transmission'.

For optimum fuel economy the recommended gear change speeds are:

4 speed	5 speed
1st/2nd - 15 mph	1st/2nd - 15 mph
2nd/3rd - 25 mph	2nd/3rd - 25 mph
3rd/4th - 40 mph	3rd/4th - 35 mph
	4th/5th - 45 mph

Handbrake Lever

To apply the rear wheel brakes, pull the handbrake lever upwards; it is retained by a ratchet and pawl. Release the hand brake by pulling it slightly upwards as the button is depressed to free the pawl; then allow the lever to move downwards to the off position.

Refer to 'Warning Lights' — page 10

Pedals

The pedals are arranged in the conventional positions. The brake pedal operates the brake hydraulic system to apply the brakes on all four wheels. Warning lights at the rear of the vehicle function automatically when the brake pedal is pressed while the ignition is switched on. To avoid needless wear of the clutch withdrawal mechanism, do not rest the foot on the clutch pedal while driving.

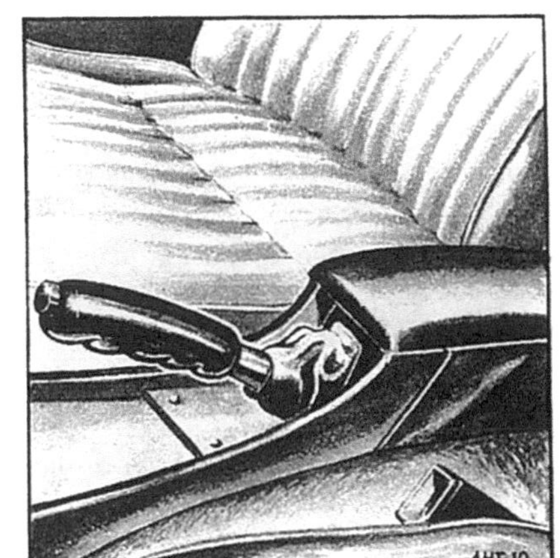

BODY FITTINGS

Keys

Two keys and a duplicate set are provided. The large key is used for operating the ignition/steering-column lock, door locks and the fuel filler cap; the small key operates the glovebox and luggage compartment locks.

CAUTION: The ignition and steering column lock key is supplied complete with an identification disc. This must not be mislaid since it provides the only record of the key. In the event of the loss of keys the disc will have to be submitted to your selling dealer to obtain duplicates.

Door Locks

Key locking — With button 'A' raised, close the door, insert the key into the lock and turn it a quarter turn towards the front of the vehicle. Return the key to the vertical position to withdraw it.

To unlock — Insert the key, turn it a quarter turn rearwards and return it to the vertical position to withdraw it.

Keyless locking — Either door may be locked from the inside by pressing the button 'A' downwards. This position is cancelled when the door is unlocked from the outside or when button 'A' is lifted and the lever 'B' is pulled from the inside.

To lock either door from the outside (as an alternative to using the key), hold the external release lever in the open position and depress button 'A' before closing the door.

Do not leave the keys in the car.

IMPORTANT: Do not force down button 'A' when the door is open.

Door windows

Turn the handle (C) to lower or raise the door windows.

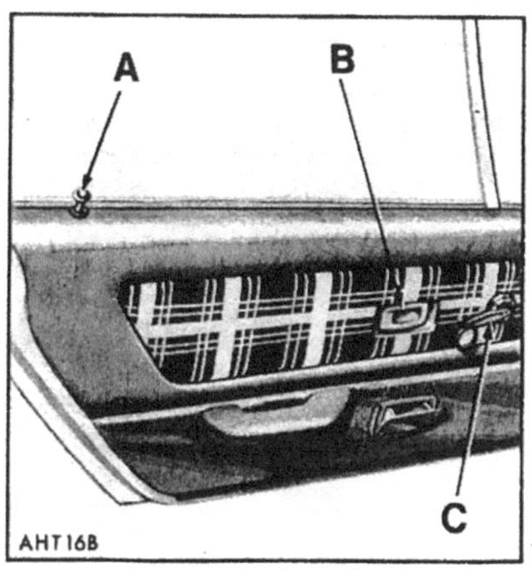

Ashtrays

Swivel the top of the tray outwards for use and press closed when not required. To remove an ashtray, open, depress the spring clip to release the top edge and pull the bottom edge clear of the surround.

Bonnet Release Control

To open the bonnet, pull the control knob located below the fascia at the left-hand side of the car. This disengages the locking plate and allows the bonnet to spring open sufficiently to insert fingers and release the safety catch. When fully opened the bonnet is held by a support stay.

To close, disengage the stay and lower the bonnet, firmly pressing its catch into positive engagement with the locking plate.

Driving Mirror

The fully adjustable interior mirror is mounted on a break-away support for personal safety. If, through impact, the stem is dislodged from its mounting, it can be refitted to its original position.

The driving mirror is adjustable for direction, and may be 'dipped' temporarily to prevent reflected glare from the headlamps of following vehicles. To 'dip' the mirror move the protruding tab rearward; to regain normal adjustment move the tab forward.

Door Mirror

The driver's door mirror should be adjusted to provide an area of vision slightly overlapping that of the interior rear view mirror.

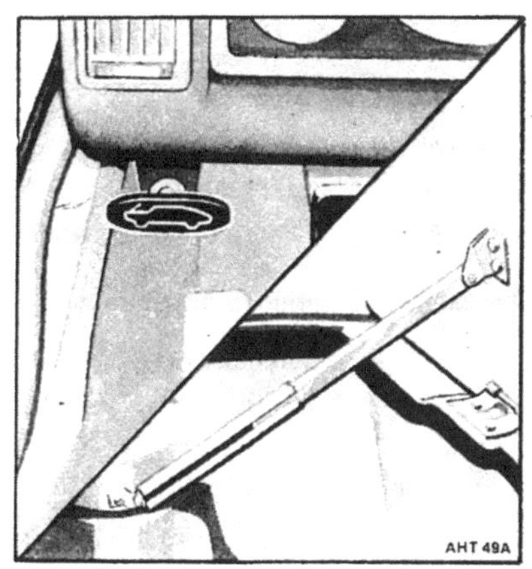

Sun visors

Two adjustable sun visors (that on the passenger's side having a vanity mirror) are padded to reduce the risk of impact injury. Either visor can be swung downwards or can be unclipped from the driving mirror bracket and swung sideways to eliminate sun glare.

The drivers sun visor incorporates a ticket pocket.

Glovebox locker

To open, turn the knob and lower the hinged lid. The lid is retained in the closed position by a spring catch.

Map Light [1977 and later models only]

The map light is clipped into a holder mounted between the seats.

The light is switched on automatically when the lamp is removed from the holder.

To switch off and replace the lamp, locate the pegs in the holder recesses and pivot the lamp body upwards until it is retained by the clip.

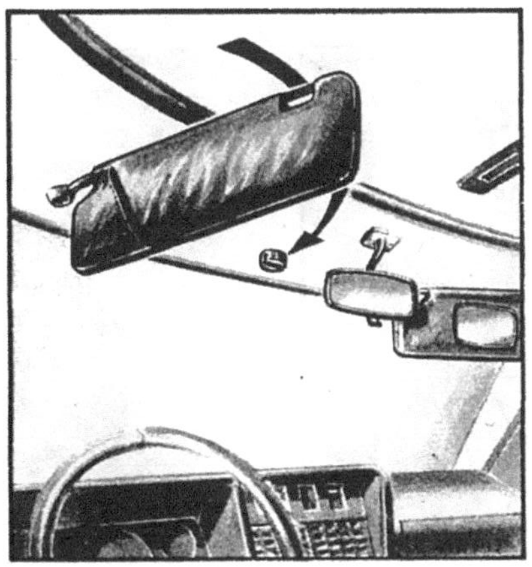

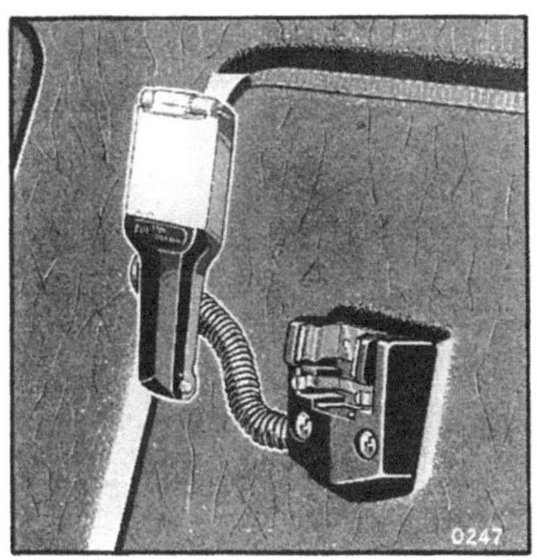

Cigar-lighter

The cigar-lighter is operated by pushing its knob inwards to heat the element. When a predetermined temperature is reached, the element is ejected from the 'heat' position from whence it may be withdrawn for use. A small pilot lamp is incorporated within the element carrier to facilitate replacement of the element during darkness.

Radio Facility

A manually operated radio aerial is fitted in the rear wing panel and is raised by lifting the protruding lip. Ensure that the aerial is lowered before entering an automatic car wash.

Loudspeaker units are fitted in the door panels.

Heated back-light

The heated back-light window has the heating element on the inside surface of the glass and with reasonable care will last indefinitely. The following practices will damage the element and must be avoided:

1. Scratching off labels and advertising stickers.
2. Wiping the glass with the back of a ringed hand.
3. Stowing hard and metal objects so that they abrade the glass.
4. Cleaning with harsh abrasives.

Courtesy Lamp

[1980 and later]: A three position lamp is fitted to each door. With the lens flush (2) the lamp is 'OFF'. With the lens tilted up (3) the lamp is switched 'ON'. With the lens tilted down (1) the lamps operate when a door is opened.

[1978–1979]: The door lamps are operated by a switch on each door pillar.

1975–1977: The roof lamp is operated by a switch on each door pillar. The lamp can be illuminated when the doors are closed by operating a switch located at the edge of the lens.

[1975–1977 models]

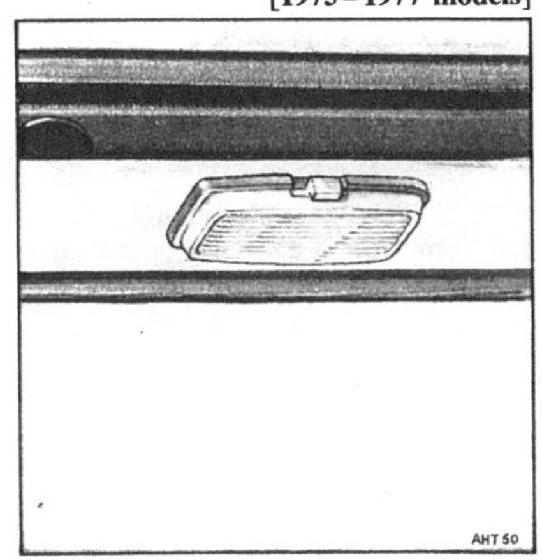

[1978 and later models]

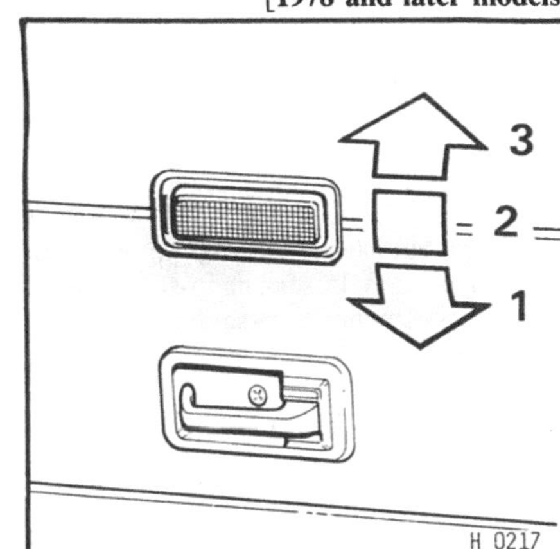

Opening Roof (Optional)

To open the roof panel, pull the handle downwards, turn a half turn anti-clockwise and pull the panel rearwards. To close, reverse the procedure.

The roof panel may be locked in any position but the locking mechanism is not designed to hold against firm hand pressure.

Luggage Compartment [1975 models]

The luggage compartment lid is spring-loaded to permit easy lifting.

To open, turn the unlocked handle clockwise and allow the lid to rise gently.

Luggage Compartment [1976 models]

To open, turn the unlocked handle clockwise, raise the lid to its limit and lower the lid onto its telescopic support.

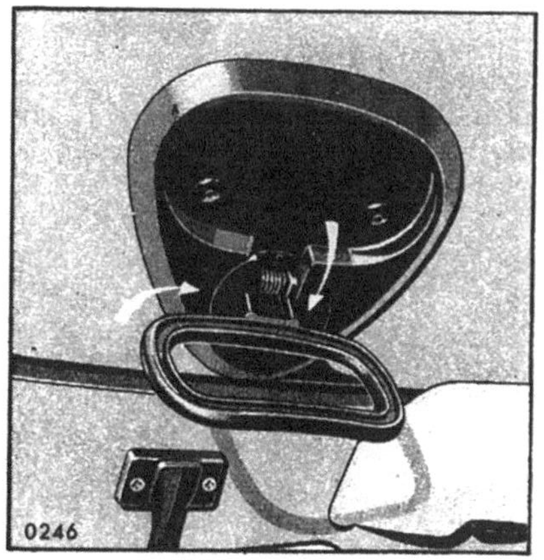

[1975 models]

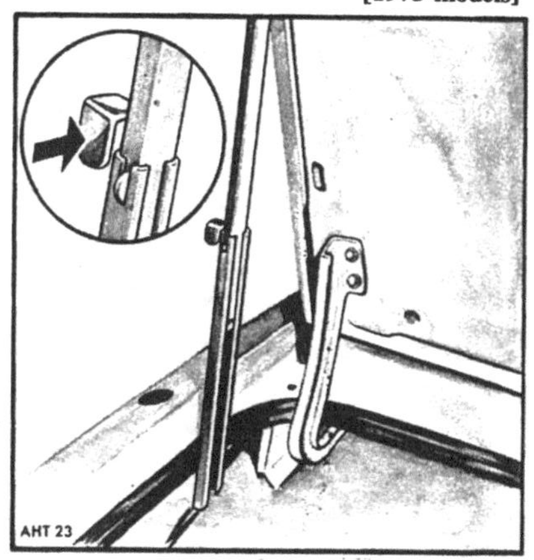

[1976 – 1979 models]

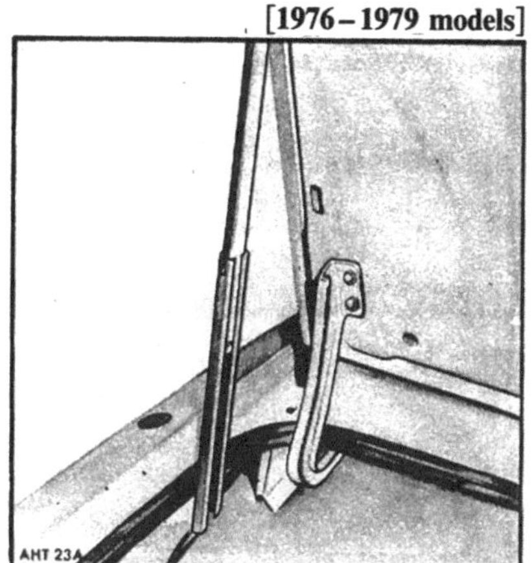

Luggage Compartment [1977 – 1979 models]

The luggage compartment lid is spring-loaded to permit easy lifting.

To open, turn the unlocked handle clockwise and allow the lid to rise gently.

Luggage Compartment [1980 and later models]

To open, insert the key, turn it anti-clockwise and depress the button. Assistance in raising and retention in the open position is provided by gas-filled telescopic struts. To close, lower the lid and when nearly closed, gently push the lid down to engage the lock catch.

To lock, insert the key, turn clockwise and remove key.

Interior Stowage Space

In addition to the glovebox, further stowage space is provided for small packages, maps, etc. in the centre console armrest, in rear quarter panel pockets and on the rear shelf behind the seats. It is dangerous practice to place heavy or sharp objects on the rear shelf as these can become lethal projectiles in the event of an impact.

Fuel Filler Cap

To open, turn the cap a quarter turn anti-clockwise and lift off.

[1980 and later models]

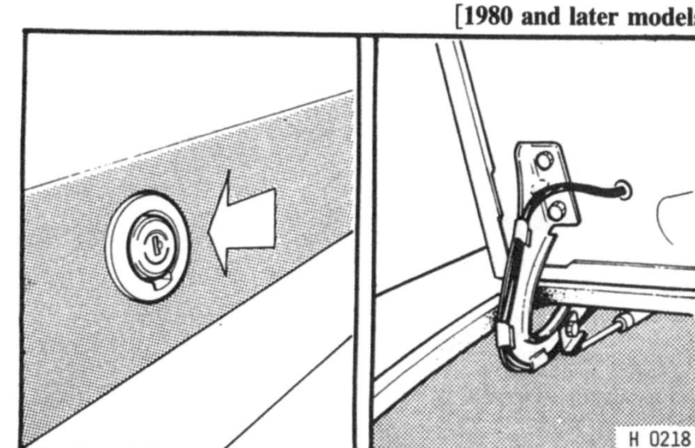

Soft Top

It is most important that the instructions for raising, lowering, and folding the hood are carried out in the sequence given. Do not apply pressure to the frame-members other than the header rail; undue force is not necessary and should be avoided. Do not fold or store the hood when it is wet or damp.

Lowering Soft Top

Release the header rail from the windscreen frame by turning the two catch levers towards the inside of the car and push the header rail upwards (Fig. A). From outside the car, release the eight snap fasteners, four per side, which secure the hood to the body.

Fig. A

Fig. B

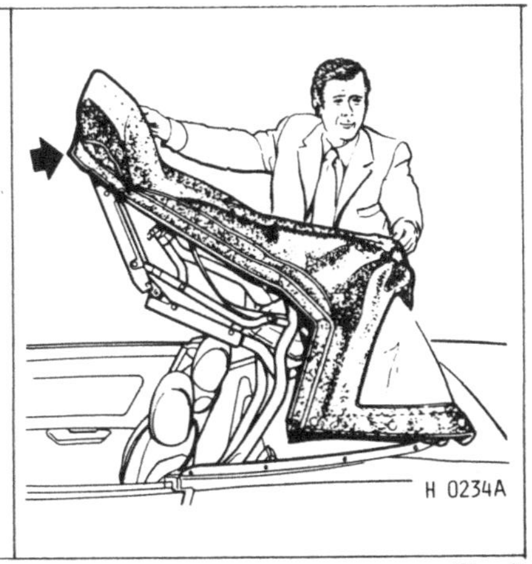

Fig. C

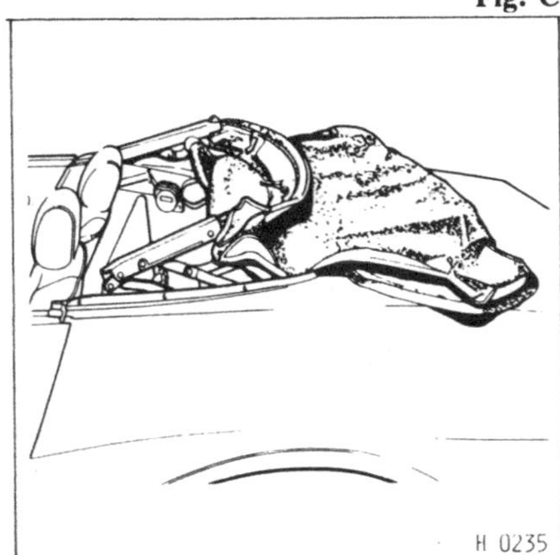

Fig. D

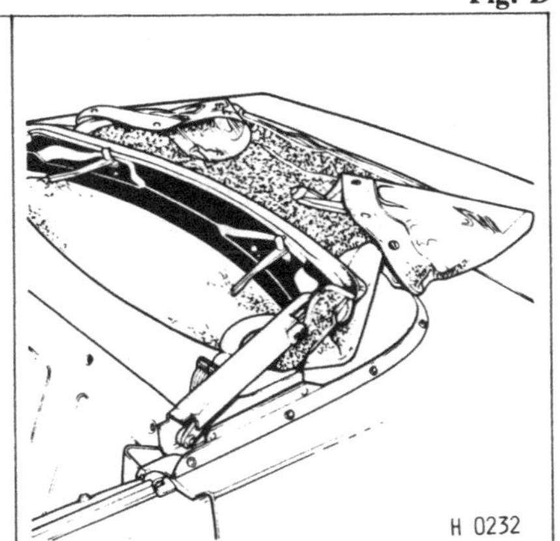

Lift the header rail upwards and push it rearwards. At the same time pull the hood cover rearwards by holding the flap at the top of the rear window (Fig. B) so that the rear window and quarter lights rest flat on the boot lid and are not trapped under the header rail (Fig. C).

Ensure the hood is flat, then fold the quarter lights inwards on top of the cover (Fig. D). Then fold the rear window forward onto the hood frame behind the seats (Fig. E).

Soft Top in the Down Position (Fig. F)

Lay the cover over the hood and secure it with the twelve snap fasteners to the body, the three on the rear trim pad behind the seats and the one on each door pillar.

Raising the Soft Top

Unfasten and remove the hood cover. Fold the hood fabric backwards over the boot lid and fold out the rear quarter lights. Lifting the soft top header rail, raise the assembly to allow the fabric to lie evenly over the frame and position the header rail onto the windscreen frame. Position the hood draught excluder arrowed in Fig. B between the inner and outer door pillar seals. Secure the header rail catches by turning the levers outwards. Fasten the four snap fasteners on each side of the hood to the body.

Opening Rear Window

Pull the zip-fastener around the window and roll the window downwards.

Fig. E

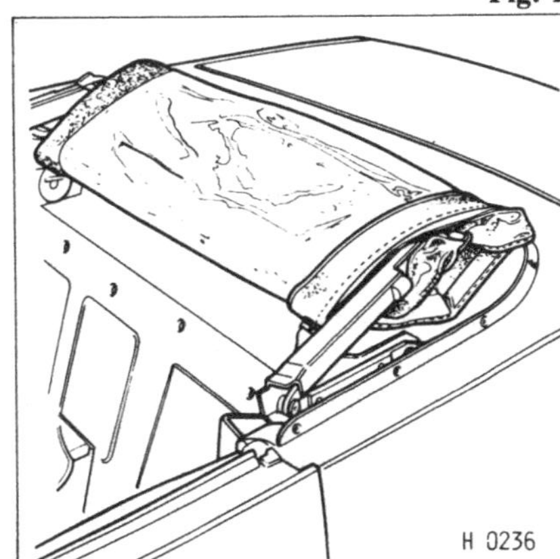

Fig. F

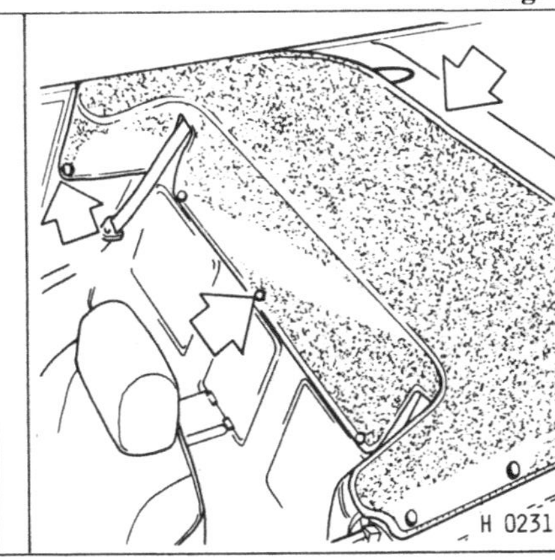

SEATS AND SAFETY HARNESS

Seat Adjustment

Either front seat can be adjusted for leg reach while the occupant is seated by lifting a bar against spring pressure to release the seat runner. The seat can then be moved to the desired position before allowing the bar to engage in its nearest locking notch. Some additional seat movement may be necessary to achieve this.

The angle of the seat back-rest can be altered to provide a reclining position. To achieve this, raise and hold the lever (3) and, using controlled body pressure, move the back-rest to the desired position. Release the lever to lock the back-rest at the chosen angle. The seat back, if unrestrained, will return to the upright position when the locking lever is raised.

Head Restraints

The head restraints should be adjusted to support the back of the head and not the neck. This can be achieved by lifting or lowering the device as required.

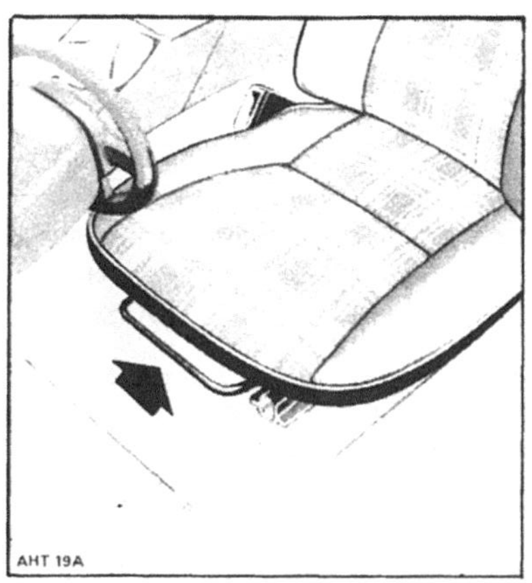

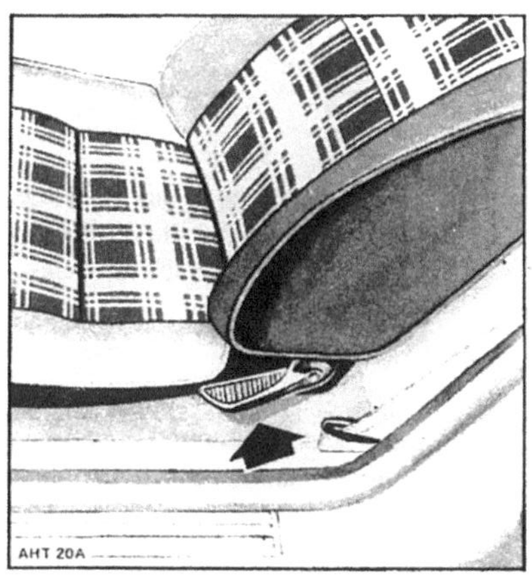

Description of Inertia Reel Type Harness

One end of the harness (A) is attached to a self-retracting inertia reel mounted behind the passenger compartment. The harness passes through a 'running eye' (B) and the buckle (C) before passing to a fixed point (D) on the seat frame.

The harness is prevented from unreeling when retardation of the vehicle causes the displacement of a heavy steel ball. This tilts a pivoting plate which applies a positive lock to the reel mechanism.

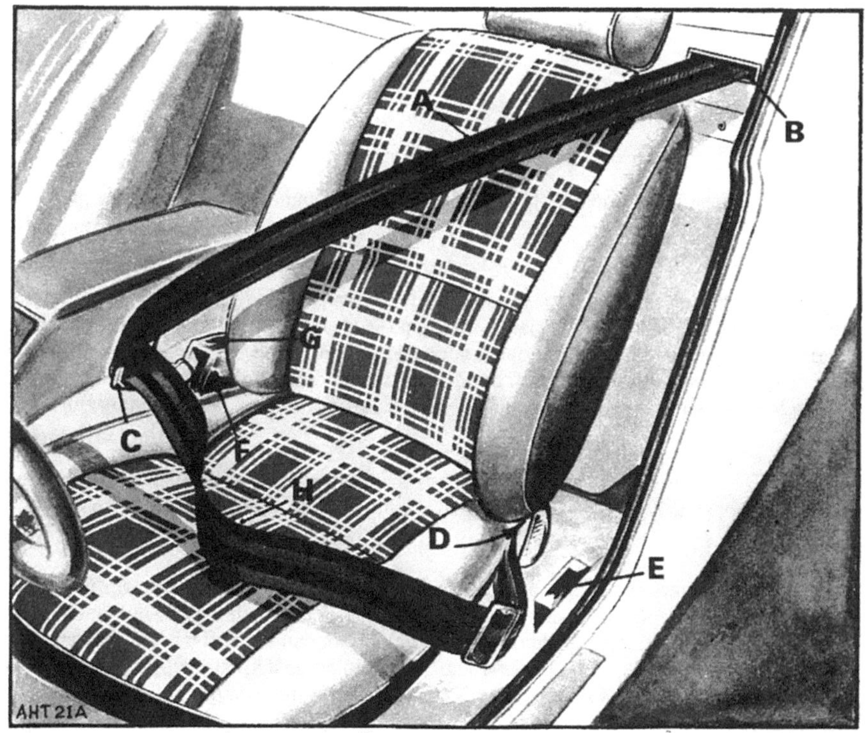

Using the Harness

While seated remove the buckle (C) from the stowage unit (E), pull the strap (A) away from the 'running eye' (B) until a loop is formed and pass the arm nearer the door through the loop. Continue pulling the harness over the hips until the buckle (C) can be inserted into the buckle unit (F) mounted on the transmission tunnel. A positive 'click' ensures that the harness is safely locked.

To release the harness, depress the release button (G), when the harness tongue will automatically disengage from the buckle unit and retract towards the door pillar.

To obtain the maximum designed protection from the safety harness it is essential that it is properly fitted over the body. The lap strap (H) should be placed so that it rests securely over the bony part of the hips. The shoulder strap (A) should pass over the shoulder and diagonally downwards across the body. No adjustment is required as the automatic retraction of the reel retains the harness at the correct tension.

In its normal fitted position the reel will allow the harness to extend and retract to permit normal body movement without locking but will lock automatically in the event of emergency braking.

To provide the users with maximum freedom during normal driving conditions the seat belts are of the inertia reel type. Hard braking or fast cornering locks the belts immediately.

13

Inertia Reel Mechanism Check

Inertia Lock Test (vehicle stationary)

Whilst seated, fasten the belt and grip the shoulder belt at approximately shoulder level with the opposite hand. Pull the belt sharply in a downwards direction, the belt should lock.

Inertia Lock Test (vehicle in motion)

The following road test must be carried out only under maximum safe road conditions, i.e., on a dry, straight, traffic-free road.

With the safety harness fitted to the driver and passenger as previously described, drive the car at 8 km/h (5 m.p.h.); ensuring that it is safe to do so, brake sharply.

The safety harness should lock automatically, holding both driver and passenger securely in position.

It is important when braking that the reaction of both driver and passenger is normal, i.e., the body must not be thrown forward in anticipation, thus causing a 'snatching' action of the reel which will not operate the locking mechanism. The harness is locked by retardation of the car, not by body movement.

If the belt fails to lock on either test, consult your Dealer.

Harness Stowage

The harness is self retracting but to provide an easily accessible position of the locking buckle (C) always replace it in the stowage unit (E).

Cleaning

An occasional wipe with a warm soapy sponge will keep the harness clean. Do not use bleach or dyes otherwise the efficiency of the harness may be affected.

AIR CONDITIONING (Optional)

The system is controlled by four levers mounted in the central console.

Vent — Lever 'A' controls the volume of dehumidified unheated air flowing through the fascia central air vents. The vent control flap may be closed or opened by moving the lever to its upper or lower position respectively.

Temperature — Lever 'B' controls the temperature of dehumidified air discharged into the vehicle. Air discharged from the footwell or windscreen vents can be heated or cooled. Air discharged from the fascia central and side vents can be cooled, but not heated.

Air — Lever 'C' controls the respective volumes of fresh or recirculated air flowing through the unit.

With the lever in its topmost position airflow through the unit is negligible.

Moving the lever to the first detent (Max) provides maximum boosted flow of recirculated air. As the lever is moved downwards recirculated air is supplemented by increasing amounts of fresh air.

Positions 3, 2 and 1 operate a three speed blower fan providing fresh air only.

The air control also operates the condenser fan motors which will run independently of the ignition circuits and may continue to run after switching off the ignition until the condenser has cooled.

Distribution — Lever 'D' controls the distribution of air between the windscreen and footwells.

Air Conditioning Cut-Out Switch — This is a fuel saving device which provides manual over-ride of the air conditioning refrigeration system. [The 1975 models do not have this control.]

The switch should be in the off position when 'defrosting' operations are carried out.

The switch should be in the 'ON' position when demisting operations are carried out, do not put the switch to the 'OFF' position in wet or humid conditions.

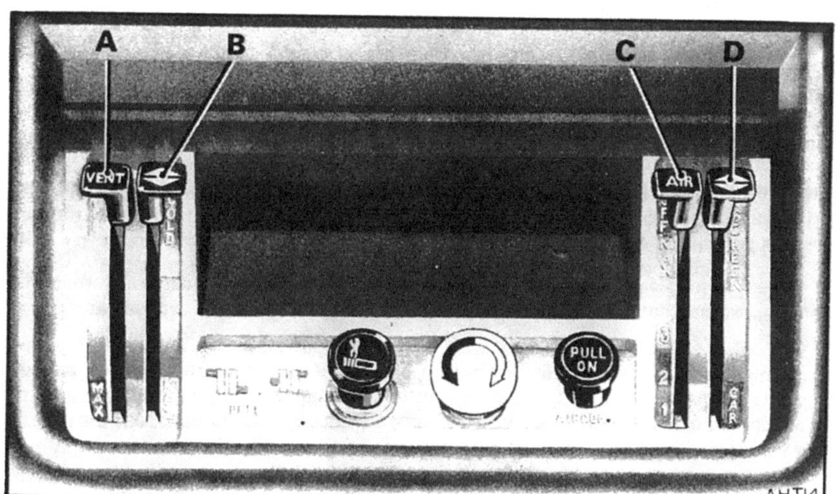

Rapid 'Defrost'

Maximum windscreen defrosting is achieved by setting the control levers as follows:

Vent — Off — fully up
Temperature — Hot — fully down
Air — Maximum boost — first position down from 'OFF'
Distribution — Screen — fully up
Fascia Control and Side Vents — Closed

The air conditioning cut-out switch should be in the 'OFF' position when defrosting operations are carried out.

Rapid Demist

Maximum windscreen demisting is achieved by setting the control levers as follows:

Vent — Off — fully up
Temperature — Hot — fully down
Air — Position 3 — fresh air
Distribution — Screen — fully up
Fascia Control and Side Vents — Closed

The air conditioning cut-out switch should be in the 'ON' position when demisting operations are carried out.

HEATING AND VENTILATION SYSTEM

The system is controlled by four levers mounted in the central console.

Vent — Lever 'A' controls the volume of unheated air flowing through the fascia central air vents.

The vent control flap may be closed or opened by moving the lever to the upper or lower position respectively.

Temperature — Lever 'B' controls the temperature of air flowing through the footwell or windscreen vents.

Air — Lever 'C' controls the volume of fresh air flowing into the vehicle. Moving the lever to its upper position closes a flap valve and prevents air flow. Moving the lever downwards increases the air flow which can be boosted at low road speeds by further movement of the lever to positions 1, 2 and 3 which operates a blower fan at low or high speeds.

Distribution — Lever 'D' controls the distribution of heated air between the windscreen and footwells.

Rapid 'Defrost'

Maximum windscreen defrosting is achieved by setting the control levers as follows:-

Vent — Off - fully up.

Temperature — Hot - fully down.

Air — Maximum boost - fully down.

Distribution — Screen - fully up.

Facia Control and Side Vents — Closed

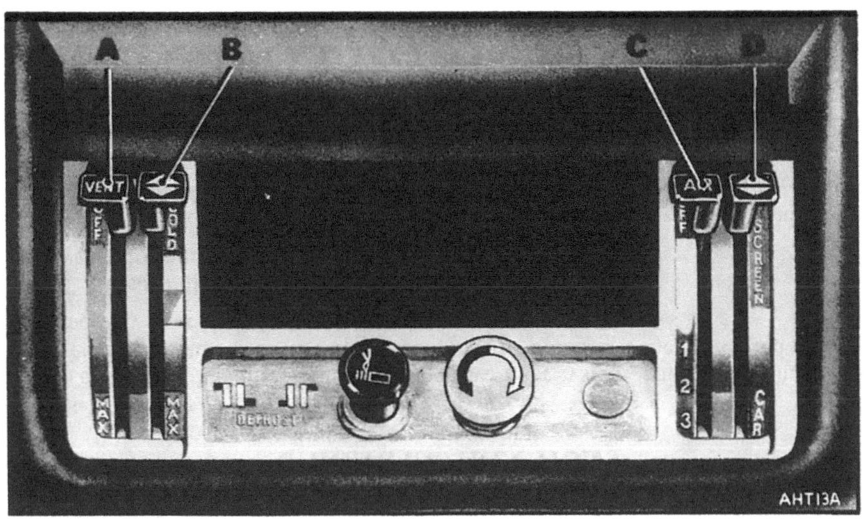

AHT13A

Fascia side vents

A swivelling air vent located at either end of the fascia is provided with a valve that can be opened or closed by turning the control knob below the vent.

This admits unheated air which can be directed as required by adjusting the vanes.

Fascia central air vents

By suitable adjustment of the vent control lever, unheated air can be directed through the central air vents to the car interior. The flow may be directed as required by adjusting the vanes.

Air extraction

Whilst the vehicle is in motion, stale air from the interior flows via slots located at the base of the rear window into the luggage compartment, from where it is extracted by a low pressure air stream through the grills in the roof quarter panels.

Footwell ventilation

Cold air may be admitted to the footwells by pushing open a flap valve located in the outboard side panel of each footwell.

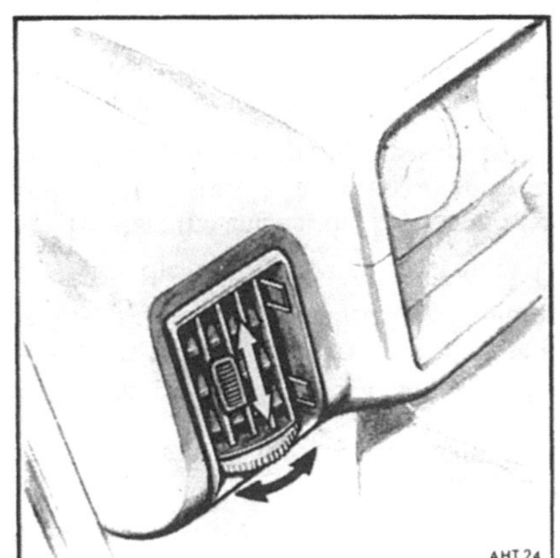

AHT 24

AHT 25

RUNNING INSTRUCTIONS

Seat Belt Warning System

Your Triumph is fitted with a seat belt warning system which will, when used correctly, be of major benefit and minimum inconvenience to the driver and passenger and may save their lives.

The basis of the system is to warn you both audibly and visually to wear your seat belt whilst driving the car.

A seat belt warning light located on the facia, and an electric buzzer will operate for a maximum period of eight seconds after the ignition is switched on and the driver's seat belt is not coupled.

Immediately the belt is coupled the buzzer will cease to operate but the warning light will remain illuminated for the complete eight seconds.

Audible Alarm System

The Audible Alarm System functions when the ignition key is positioned in the switch and the driver's door is open. The "warning" denoted by a continuous buzzing sound will terminate when the driver's door is closed or the ignition key is completely removed. The audible alarm is also used in conjunction with the seat belt interlock system.

Use of the Choke Control — Not Catalytic Convertor Specification

Under cold ambient temperatures — Pull the choke control out fully to start the engine. As soon as the engine starts push the choke control in until the engine maintains an idling speed of approximately 1100 r.p.m. The choke control should be pushed in fully as soon as the engine will maintain normal running without its use.

Under warm ambient temperatures — Use full choke to start the engine, but immediately after the engine fires return the choke control to the fast idle position where a definite notch will be felt on the cable pull. After approximately one minute driving return the control knob to the fully in position.

Starting a hot engine — When re-starting a hot engine, depress the throttle pedal to about one-third of its travel before operating the starter switch. The choke control should not be used.

1975-1977 Starting Procedure — Catalytic Convertor Specification only

Depress the throttle pedal fully and release. Crank the engine until it starts; DO NOT touch the throttle during this operation.

As soon as the engine is running, blip the throttle in order to set the auto-choke in its correct running position.

1978 and later Starting Procedure — U.S.A. Specification only

Hot or Cold Starts — Without touching the accelerator, crank the engine until it starts. If, however, the cranking period is more than 2—3 seconds, depress the accelerator pedal slightly and pump it until the engine fires and runs.

Preparing to Drive

When seated in the car with the seat belts fastened it is good practice to switch on the ignition and check the gauges and warning lights before driving off.

To Disengage Steering Lock and Start Engine

Insert the key and turn in clockwise direction. If difficulty is experienced in turning the key, this can be rectified by simultaneous movement of the steering wheel.

Turn the key clockwise to symbol II (Ignition) the ignition will be switched on.

To start the engine, the key should be turned a little more against spring pressure to symbol III ("start"), as soon as the engine fires release the key which will return automatically under spring pressure to the ignition position (symbol II).

Do not operate the starter for longer than five to six seconds and wait until the engine has stopped before re-using the starter. If after a few attempts the engine fails to start, switch off the ignition and investigate the cause. Continued use of the starter will not only discharge the battery but may also damage the starter.

To Stop Engine and Engage Steering Lock

Turn the key in an anti-clockwise direction from the "ignition" position (symbol II) to the "lock" position (symbol 0). This action stops the engine.

Removal of the key in this position automatically actuates the steering lock mechanism. (See Audible Alarm System). Slight movement of the steering wheel may be necessary to ensure the lock bolt is fully engaged.

WARNING: If, for any reason, the (ignition) engine is switched off while the car is in motion DO NOT ATTEMPT TO TURN THE KEY INTO THE "LOCK" POSITION O, as this constitutes part of the locking sequence.

Running-in

The importance of correct running-in cannot be too strongly emphasized, for during the first few thousand miles of motoring, all working surfaces of the vehicle are 'bedding down'.

Avoid placing heavy loads upon the engine, such as using full throttle at low speeds or when the engine is cold. Running-in should be progressive and no harm will result from the engine being allowed to 'rev' fairly fast for short periods provided that it is thoroughly warm and not pulling hard.

Always select a lower gear if necessary to relieve the engine of load. Full power should not be used until at least 1,600 km (1,000 miles) have been covered, and even then it should be used only for short periods at a time. These periods can be extended as the engine becomes more responsive.

Maximum Engine Speeds

After running-in has been completed, drivers are advised to restrict engine speeds as follows:

Maximum recommended engine speed (continuous) 6000 rev/min.
Maximum recommended engine speed (intermittent) 6500 rev/min.

Towing

For recovery the car should be towed with the key in the ignition/steering lock at position '1'. For tow-starting the key should be at position '2'.

If towing a model fitted with a manual gearbox, the propeller shaft must be disconnected or removed.

See special instructions for Automatic Transmission (where fitted).

Two towing brackets are fitted beneath the front bumper. It is recommended that when towing, ropes are attached to either, but not both brackets.

Roof Rack

Bulky rather than heavy loads **no greater than 50 kg (110 lb.) may be carried on a roof rack.** Any load on the roof may affect the handling of the car, especially in a cross wind or when cornering.

TOWING A TRAILER

The following information is of a general nature and should be used in the light of regulations prevailing in the territories concerned for the type of trailer being towed. Laws concerning towing vary in different countries and it is advisable to ensure that the car and the vehicle to be towed comply with existing regulations.

This information applies to travelling on paved or hard surface roads and not on dirt roads or across country.

Safety, stability, and comfort are the essentials that must be satisfied when towing is being considered.

The unladen weight of the trailer will be included in the manufacturer's catalogue or handbook together with the gross laden weight, that is, the unladen weight plus the weight of personal effects and equipment.

Owners new to towing should bear in mind the effect on the car's performance when deciding the weight of trailer to be towed. Owners who constantly tow loads should have their car serviced more frequently.

Tow Bar

It is important that only an approved tow bar is fitted.

Nose Weight (The weight applied to the tow bar when stationary)

Whilst the nose weight on most trailers (such as horse boxes) remains fairly constant, it can vary considerably with trailers according to the weight and distribution of the load. All trailer manufacturers recommend a maximum nose weight and in the interest of stability, it is unwise to exceed this figure. Generally, the recommended weight is between 45 to 55 kg (99 to 121 lbs) which allows a sufficient margin for individual preference and trimming to give the best towing performance.

Most trailer accessory stores sell nose weight gauges which are simple to use and enable frequent checks to be made. Another method is to place a stout piece of timber between the coupling and ordinary bathroom scales with the jockey wheel clear of the ground. Ensure that the parking brake is applied for safety.

Tow Ball Hitch Height

An important factor in towing stability is the height from the ground to the centre of the tow ball. The approved tow bar kit will ensure that with both vehicles on level ground and carrying their full permitted loads, the hitch height will, subject to the trailer nose weight being correct, provide a level or preferably slightly nose-down attitude.

Whilst most trailer manufacturers conform to this standard, some recommend a lower hitch height. In this case it will be necessary to fit a 'drop plate' which is a flat plate that bolts on the tow bar in place of the tow ball and the ball may then be fitted to give the correct height. Drop plates are available with holes at various distances apart to suit individual requirements.

Loading the Car

The maximum permissible loading of the vehicle (passengers and luggage) is the difference between the 'Unladen Weight' (kerb weight) and the 'Gross Vehicle Weight', which must not be exceeded. When towing with this load, luggage to the equivalent nose weight of the trailer must be removed from the car.

Towing Mirrors

It is a legal requirement when towing in some countries to fit an extra mirror to the left side of the car. In the interest of safety, however, it is suggested that one is also fitted to the right side.

Taking to the Road

Assuming that all the maintenance operations have been carried out on the car and vehicle to be towed, make these checks before starting a journey.

(1) The ball coupling spring safety device is engaged.
(2) Electrical connections are made to the car/trailer.
(3) Reversing mechanism is disengaged.
(4) Jockey wheel is raised and properly secured.
(5) Trailer corner steadies are fully raised.
(6) Trailer doors, windows and ventilators are closed.
(7) Gas bottles are secure.
(8) All lights and indicators are working.
(9) Driving mirrors are adjusted.
(10) Trailer parking brake is released.
(11) Tyre pressures on the car are adjusted for the additional load.

The following advice may be helpful to owners new to towing.

(1) Since a trailer may be wider than the car, care should be taken when negotiating bends particularly when there is a kerb.
(2) Remember the extra length when overtaking and cornering. On freeways in particular, it is very easy to forget that one is towing.
(3) Take extra care when passing cyclists.
(4) Make constant use of mirrors.
(5) Always brake gently and in good time.

Recommended Maximum Trailer and Gross Train Weights

Braked Trailers

Towing vehicle load conditions — 2 up
Restart gradient at sea level — 1 in 8 (12½%)
Trailer weight — 1016 kgs (2240 lbs)

With additional payload, such as luggage, the equivalent extra weight should be deducted from the maximum recommended trailer weight.

With extra care, braked trailers in excess of the above weight may be towed.

The Gross Train Weight should not exceed: 2680 kgs (5908 lbs).
Restart gradient at sea level is then 1 in 10 (10%).

Gross Train Weight is the Gross Vehicle Weight plus maximum trailer weight.

When towing a trailer, the towing vehicle maximum rear axle load must not be exceeded. The recommended static vertical nose load (A) at trailer hitch is 50 ±5 kgs (110 ±11 lbs).

Recommended Maximum Trailer and Gross Train Weights

Unbraked Trailers

Towing vehicle load conditions — 1 up
Trailer weight — 495 kgs (1091 lbs)
The Gross Train Weight should not exceed: 1655 kgs (3649 lbs).

Gross Train Weight is the Gross Vehicle Weight plus maximum trailer weight.
When towing a trailer, the towing vehicle maximum rear axle load must not be exceeded.
The recommended static vertical nose load (A) at trailer hitch is 50 ±5 kgs (110 ±11 lbs).

Vehicle Weights

Unladen minimum (kerb weight) — 1092 kgs (2407 lbs).
Gross Vehicle Weight — 1340 kgs (2954 lbs)
Maximum rear axle load — 665 kgs (1466 lbs)

Gross Vehicle Weight (G.V.W.) is the maximum design weight the vehicle can carry and includes the weight of the vehicle. It must not under any circumstances be exceeded.

2145

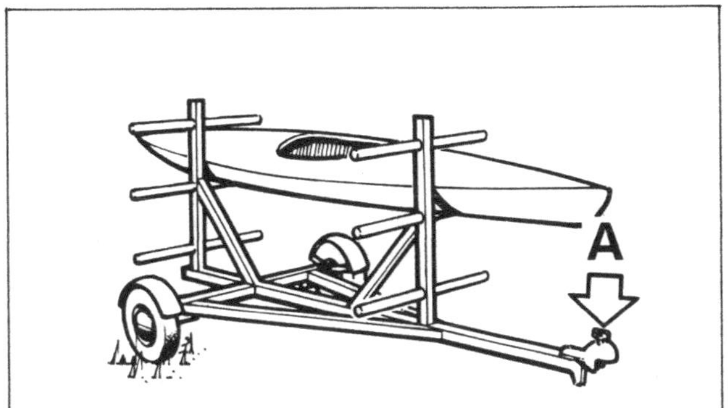

2144

AUTOMATIC TRANSMISSION

Introduction

The transmission consists of two main parts — a fluid driven torque converter capable of varying torque-multiplication, and a hydraulically controlled epicyclic gearbox, featuring planetary gears which provide reverse and three forward ratios.

Selector lever positions

A selector lever, is mounted on the transmission tunnel. The upper face of the quadrant is marked with the symbols 'P', 'R', 'N', 'D', '2', '1', which indicate the following functions:

'P' for park.
'R' for reverse.
'N' for neutral.
'D' for automatic forward drive.
'2' for first and second gear only.
'1' for first gear hold.

A safety lock prevents inadvertent selection of 'R', 'P', 'N' or '1', by baulking movement of the selector lever into these positions until a button on top of the lever is depressed. The button must also be depressed to allow disengagement of 'P'.

Free movement of the lever is permitted to allow unhindered selection between 'D' and '2' during normal driving and also of 'N' from 'R'.

AHT 233

Starting the engine

The starter switch will operate only when the selector lever is placed either in 'Park' or 'N'. This safety feature prevents possible damage resulting from vehicle movement when starting the engine in a confined space.

For normal starting, first APPLY THE HAND BRAKE and move the selector lever to 'N', then start the engine as described under 'Running Instructions'. Should the vehicle be on a steep incline, start the engine without moving the lever from 'Park' position (see 'Park').

Release the hand brake and use the accelerator pedal to accelerate the car from rest. The transmission will start in first gear, automatically selecting second and third at varying speeds in accordance with load and throttle opening. Smaller throttle openings will result in gear shifts at lower speeds than will occur with a fully open throttle.

Normal driving

When the engine is started without the use of the choke (Canadian Specification only), release the accelerator pedal and allow the engine to revert to idling. Move the selector lever to 'D', where it may be left for all normal driving.

NOTE: Under fast-idle conditions when using the choke control for cold starting, movement of the selector lever from 'N' or 'Park' will produce a cushioned thump; this is not detrimental to the car or transmission.

20

Manual control

Manual selection of first, second and third gears can be made by engaging '1', '2' and 'D' in sequence.

'1' Engagement of '1' from rest holds the transmission in first gear and prevents automatic up-shift to second gear.

If '1' is selected during second gear operation the transmission will remain in second gear until the road speed is reduced; first gear will then be engaged and held.

'2' Engagement of '2' renders third gear inoperative and permits automatic shifts between first and second gears only.

If position '2' is selected during third gear operation an immediate down-shift to second gear will result; third gear will remain inoperative until 'D' is re-selected.

'D'

See 'Normal Driving'.

'Park'

Selection of 'Park' operates a mechanical device which locks the transmission, perventing the car from moving forward or backward. This position must be selected ONLY when the vehicle is at rest.

When parking on an incline, APPLY THE HAND BRAKE AS AN ADDITIONAL PRECAUTION. When about to move off, move the selector to the appropriate driving position before releasing the hand brake.

Maximum acceleration

Maximum acceleration is achieved by pushing the accelerator pedal to the full throttle position, i.e. about seven-eighths of available pedal movement. Further downward movement, overcoming built-in resistance, will bring the pedal to the kick-down position.

Automatic down-shifts

Down-shifts will occur in relation to throttle position and load, or when the accelerator is depressed to the kick-down position. To protect the engine, down-change speeds are pre-set.

Halt/manoeuvring/creep

To halt the vehicle in traffic, release the accelerator pedal and apply the footbrake. It will be noted that there is a slight 'creep' when the engine is idling and the selector is at '1', '2' or 'D'. This feature can be used to advantage in traffic, or to hill-hold on a slight incline, when the vehicle will be prevented from rolling backward. Use can be made of the tendency to creep when manoeuvering the car in a confined space, or when moving away on icy roads.

NOTE: To avoid overheating when '1', '2', 'D' or 'R' is selected and the vehicle is stationary, do not run the engine above idling speed, except when using the choke (Canadian Specification only). Select 'Park' or 'N' for prolonged periods of idling.

Emergency braking procedure

FIRST GEAR MUST BE ENGAGED AND RETAINED before descending long steep gradients, such as those encountered in Alpine regions. To achieve this, adopt either of the following procedures:

1. Stop the vehicle, move and retain the selector in position '1'.
2. Move and retain the selector in position '1' and reduce the road speed below the 2 — 1 Shift speed given in the 'General Specification'.

Towing

When ascending long gradients steeper than 1 in 15 whilst towing a caravan, etc., engage position '1'.

Emergency starting

An emergency start cannot be made by towing or pushing.

Vehicle recovery

For local recovery, the vehicle can be towed a maximum distance of **32 km (20 miles)** at a speed NOT EXCEEDING **48 km/h (30 m.p.h.)** provided that the transmission is undamaged, the oil level is correct, the selector lever is at 'N', and the ignition key is turned to position '1'. An extra 3.6 U.S. pts. (3 pints) of the recommended fluid must be added to the gearbox through the filler tube.

After towing, the oil level should be corrected before the vehicle is driven under its own power.

If the transmission unit is damaged, or if the towing distance will exceed **32 km (20 miles)** the propeller shaft must first be completely removed. Alternatively, the car can be towed while the rear wheels are lifted.

Maintenance

Refer to 'Periodic Checks'

WHEELS AND TYRES

General

Tyres of the correct type and dimensions, at the correct cold inflation pressures, are an integral part of the vehicle's design and regular maintenance of tyres contributes not only to safety but to the designed functioning of the vehicle. Road-holding, steering and braking are especially vulnerable to incorrectly pressurized, badly fitted or worn tyres.

Tyres of the same size and type but of different make have widely varying characteristics. It is therefore recommended that tyres of the same make are fitted to all wheels.

Size, Type, Pressures

The pressures recommended (see 'General Specification') provide optimum ride and handling characteristics for all normal operating conditions. The pressures should be checked, and adjusted if necessary, once per week. This should be done with the tyres cold. Tyre temperatures and pressures increase when running. Bleeding a warm tyre to the recommended pressure will result in under-inflation which may be dangerous. A slight natural pressure loss occurs with time. If this exceeds 0.14 kg/cm^2 (2 lb/in^2) per week, the cause should be investigated and rectified.

It is dangerous and in some countries illegal to drive a vehicle with tyres inflated to incorrect pressures. Drivers are advised to familiarise themselves with local legislation.

Wear

All tyres fitted as original equipment include tread wear indicators in their tread pattern. When the tread has worn to a remaining depth of 1.5 mm (0.06 in.) the indicators appear at the surface as bars which connect the tread pattern across the full width of the tyre. It is illegal in certain states to continue to use tyres after the tread has worn to less than this depth.

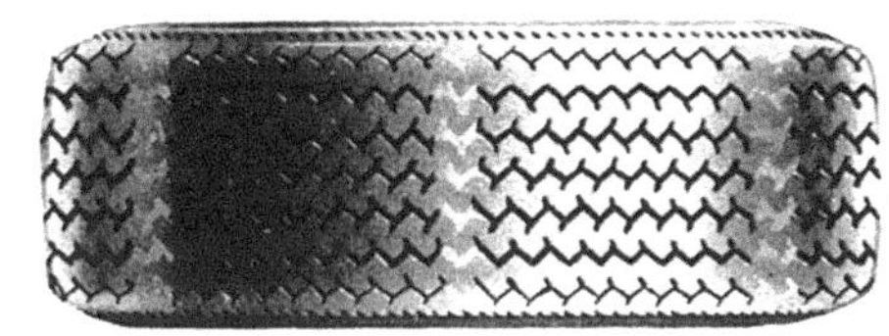

AHT 52

It should be noted that the properties of many tyres alter progressively with wear. In particular the 'wet grip' and aquaplaning resistance are gradually but substantially reduced. Extra care and speed restriction should therefore be exercised on wet roads as the effective tread depth diminishes.

Incorrect wheel alignment will accelerate tyre wear. Fins on the inside or outside edges of the tread pattern are caused by excessive toe-in or toe-out respectively. As fins may also be caused by high cornering speeds or road camber it is better to have the cause ascertained by having the wheel alignment checked. See 'General Specification' page

Damage

Excessive local distortion can cause the casing of a tyre to fracture and may lead to premature tyre failure. Tyres should be examined especially for cracked walls, exposed cords, etc. Flints and other sharp objects should be removed from the tyre tread; if neglected they may work through the cover. Any oil or grease which may get onto the tyres should be cleaned off using a suitable cleaner. Do not use paraffin (kerosene), which has a detrimental effect on rubber.

Repairs

Tubeless tyres A temporary repair can be made to tubeless tyres, using a special kit, provided the puncturing hole is small and confined to the central tread area. The following precautions, must, however, be observed.

1. Do not use more than one plug in each hole.
2. Do not use the tyre for high speeds.
3. Ensure that a permanent 'cold patch' or vulcanized repair is made at the earliest opportunity.

Valves

Whenever a new tubeless tyre is fitted, the Schrader snap-in type valve must also be renewed. To facilitate fitting, lubricate the valve with soap solution before using a special tool to snap the valve squarely into an airtight position in the rim hole.

Spare wheel and tools

The spare wheel and tools are housed below the luggage compartment floorboards. Inflate the tyre to the highest pressure specified in the 'General Specification' section and readjust to the correct pressure when required.

22

Lifting jack

Two jacking points are provided at each side of the car under the sills, one forward of the rear wheel and the other rearward of the front wheel. The jacking points are conical pins which accommodate the top plate of the lifting jack.

Position the head of the jack under the relevant point, ensuring that it locates correctly over the pin. Engage the hooked end of the handle with the eye of the screwed rod and wind the handle to take the weight of the car.

WARNING. It is unsafe to work under the car using only the jack to support it. Always use stands or other suitable supports under the rear axle or front chassis members to provide adequate support and safety.

Neglect of the jack may lead to difficulty in a roadside emergency. Examine the jack occasionally, clean and grease the thread to prevent the formation of rust.

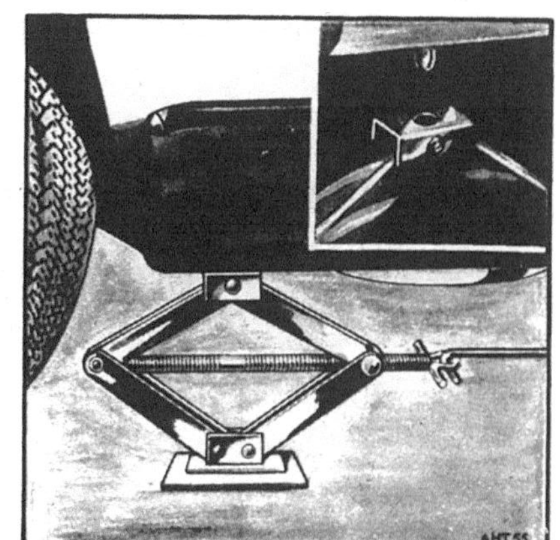

Road wheel and securing nuts

Should it be necessary to lift the vehicle when on an inclined road surface, exercise the greatest care. Firmly apply the handbrake and chock one of the wheels not being lifted.

Remove the spare wheel from the luggage compartment and make sure that its pressure is correct.

Place the jack in position, slacken the wheel nuts and lift the wheel clear of the ground.

Completely remove the wheel nuts, remove the hub trim, exchange the road wheels and replace the hub trim (ensuring that the anti-rattle springs are secure on their location pegs) and wheel nuts.

Lower the jack and securely tighten the wheel nuts.

CAUTION: The optional cast alloy road wheels combine lightness with great structural strength but are susceptible to localised damage that could result from the careless use of hammers, levers or spanners etc. Therefore to avoid damage take great care to ensure that the wheel nuts are fully seated before finally tightening alternate nuts.

[1975 & 1976 models]

[1977 and later models]

COOLING SYSTEM

Expansion Tank

The pressurized 'No loss' system incorporates an expansion tank (1) which collects excess coolant spilled from the radiator as the coolant expands when heated: Displaced coolant returns to the radiator as the system cools. Maintain the level of coolant in the expansion tank at half-full when cold.

[On 1980 and later models, the expansion tank is on the left fender.]

NOTE: As the transfer of coolant to and from the expansion tank takes place only when the cooling system is completely water and air-tight, the level within the expansion tank must not be accepted as a true indication of a full cooling system where there is evidence of coolant leakage. Such a situation requires that a check be made by removing the filler plug and, where coolant loss is confirmed the cause rectified immediately.

Pressure Cap (1)

A pressure cap on the expansion tank permits a build-up of pressure within the system as engine temperature increases enabling the system to operate at temperatures above the normal boiling-point of coolant under atmospheric pressure.

It is extremely dangerous to remove the pressure cap or the filler plug from a hot engine as the coolant will boil immediately when pressure is released.

[1975–1979 models]

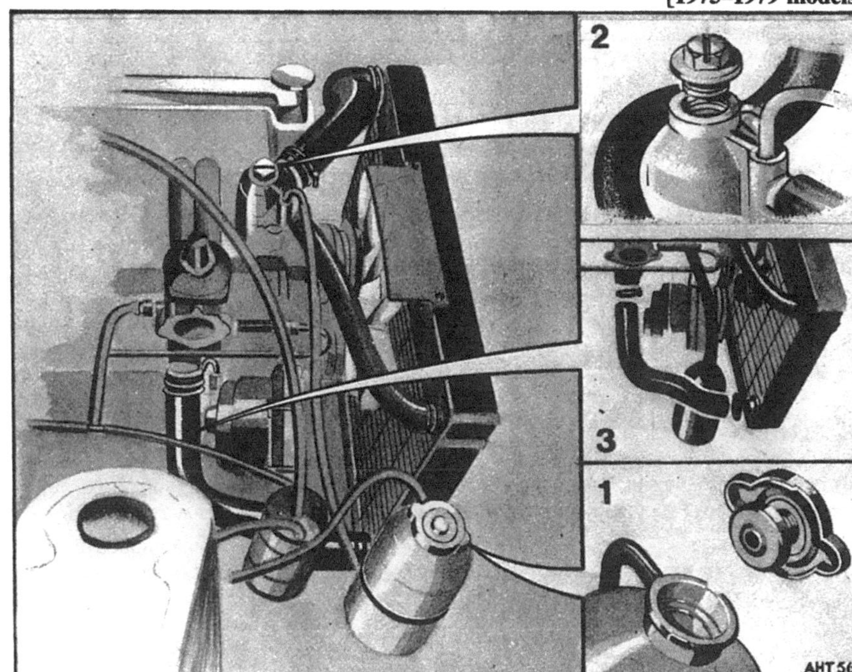

WARNING: When it is necessary to remove the pressure cap from a hot engine, exercise great care by protecting the hands against escaping steam. Slowly turn the pressure cap anti-clockwise until resistance of the safety stop is felt. Leave the cap in this position until all pressure is released. Press the cap downwards against the spring to clear the safety stops, and continue turning until it can be lifted off.

Filler Plug (2) [not on 1980 and later models]

Under no circumstances should the filler plug be removed before releasing the pressure cap.

Draining

When the engine is cold, remove the pressure cap (1), the filler plug (2) and the cylinder block drain plug (4).

Slacken the bottom hose clip (3) and disconnect the hose from the radiator.

Filling

Refit the bottom hose (3), ensuring that all hoses are in good condition and the hose connections are tight.

Refit the cylinder block drain plug (4).

Set the heater controls to the maximum heat position.

[1975–1979]: Fill the system through the filler orifice (2) and refit the sealing plug. Half fill the expansion tank and refit the pressure cap (1).

24

Warm the engine by running at approximately 1,200 r.p.m. for three minutes. Stop the engine, remove the pressure cap (1) and filler plug (2), completely fill the system with coolant and refit the plug and cap.

[1980 and later]: Remove the pressure cap (B) from the header tank (A), and fill the system until the coolant level is 25mm (1 in.) below the neck of the header tank.

Refit the pressure cap and run the engine at approximately 1500 rev/min. until the temperature rises sufficiently to open the thermostat.

Stop the engine and carefully remove the pressure cap. Top up the coolant level in the header tank to within 25mm (1 in.) of the neck of the tank, and refit the pressure cap.

[1980 and later models]

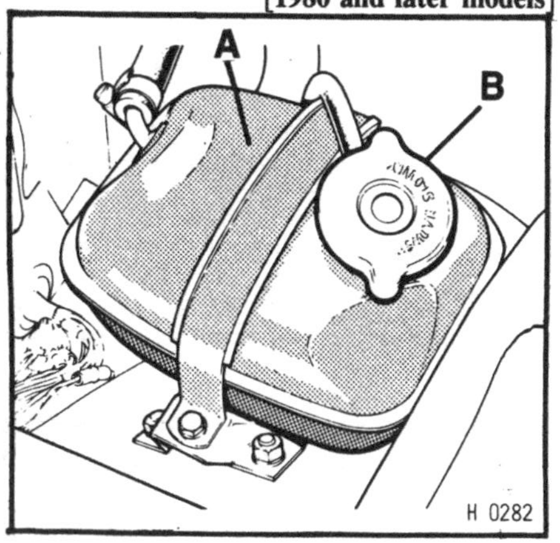

Frost Precautions

Water expands when it freezes, and if precautions are not taken there is considerable risk of bursting the radiator, cylinder block or heater. The heater unit cannot be drained with the cooling system; it is therefore essential to use anti-freeze in the cooling system in freezing conditions.

After filling with anti-freeze solution, attach a warning label in a prominent position on the car stating the type of anti-freeze contained in the cooling system to ensure that the correct type is used for topping-up.

Anti-freeze can remain in the cooling system for two years provided that the specific gravity of the coolant is checked periodically and anti-freeze added as necessary. The specific gravity check should be carried out by an authorized Distributor or Dealer. After the second year the system should be drained and flushed by inserting a hose in the filling orifice, removing the drain plugs, and allowing water to flow through until clean. Make sure that the cooling system is water-tight, examine all joints and replace any defective hose with a new one. Refill with the appropriate anti-freeze solution and add 0.25 pints (0.34 U.S. pints, 0.15 litre) of neat anti-freeze to the expansion tank.

The recommended quantities are given below.

Do not use radiator anti-freeze solution in the windscreen-washing equipment. Use the correct washer solvent, which will not damage the paintwork.

[1977 and later models]

Solution	Amount of anti-freeze			Commences Freezing		Frozen Solid	
%	Pts	U.S. pts	Litres	°C	°F	°C	°F
25	3.5	4.2	2.0	—13	9	—26	—15
33⅓	4.7	5.8	2.6	—19	—2	—36	—33
50	7	8.7	4.0	—36	—33	—48	—53

[1975 & 1976 models]

ANTI-FREEZE CONCENTRATION		25%	30%	35%	50%
SPECIFIC GRAVITY OF COOLANT AT 15·5°C (60°F)		1.039	1.048	1.054	1.076
ANTI-FREEZE QUANTITY	PINTS (U.S.)	3·9	4·7	5·5	7·8
DEGREE OF PROTECTION	**Complete** Car may be driven away immediately from cold	—12°C 10°F	—16°C 3°F	—20°C — 4°F	—36°C —33°F
DEGREE OF PROTECTION	**Safe Limit** Coolant in a mushy state. Engine may be started and driven away after short warm-up period	—18°C 0°F	—22°C — 8°F	—28°C —18°F	—41°C —42°F
DEGREE OF PROTECTION	**Lower Protection** Prevents frost damage to cylinder head, block and radiator. Thaw out before starting the engine.	—26°C —14°F	—32°C —25°F	—37°C —35°F	—47°C —53°F

ELECTRICAL SYSTEM

Alternator

To prevent damage to the alternator, do not run the engine while the battery or any of the charging circuit cables are disconnected. Ensure that all electrical connections in the charging circuit are maintained tight at all times. Should it be necessary to run the engine while the charging circuit is incomplete, disconnect the electrical cables from the alternator.

IMPORTANT. The alternator has polarity-sensitive components that may be irreparably damaged if subjected to incorrect polarity. Ensure that the battery earth lead is always connected to the battery negative terminal.

Battery

Ensure that the battery top and terminals remain clean and dry. Coat the terminals with petroleum jelly (Vaseline) to prevent corrosion.

Check the electrolyte level monthly and, if required, replenish with distilled water as detailed in 'Periodic Checks'. If electrolyte has been spilled, clean the affected area with a cloth moistened with ammonia to neutralize the acid and prevent acid corrosion.

Ensure that the battery is always firmly clamped in position by the retaining assembly. When fitting battery leads do not hammer the terminals to the terminal posts. Such action may damage the battery.

The battery will deteriorate rapidly if left in a discharged condition. If the unit is reduced to a low state of charge it should be recharged at the first opportunity.

Fuses

A fusebox is mounted on a relay plate and is accessible after removal of the cover panel at the front end of the glovebox.

Failure of a particular fuse is indicated when all circuits protected by it become inoperative. Before renewing a blown fuse inspect the wiring of the circuits that have failed for evidence of a short-circuit or other fault. If a new fuse blows immediately and the cause of the trouble cannot be found, have the circuit checked by your Rover-Triumph Distributor or Dealer.

The fuses in the fuse box protect the following circuits:

1 (top)	Air conditioning
2	Cigar lighter, courtesy lamp, horn, hazard warning system, clock
3 & 4	Air conditioning control, windscreen wiper, reverse lamp, windscreen washer, seat belt warning system, stop lamp, heater motor, direction indicator

In-line fuses:

Fuse next to fuse box	Cigar lighter illumination, instrument illumination, heater control illumination, plate illumination, front parking lamp, front marker lamp, rear marker lamp, tail lamp
Fuse in right-hand side of passenger footwell	Heated backlight

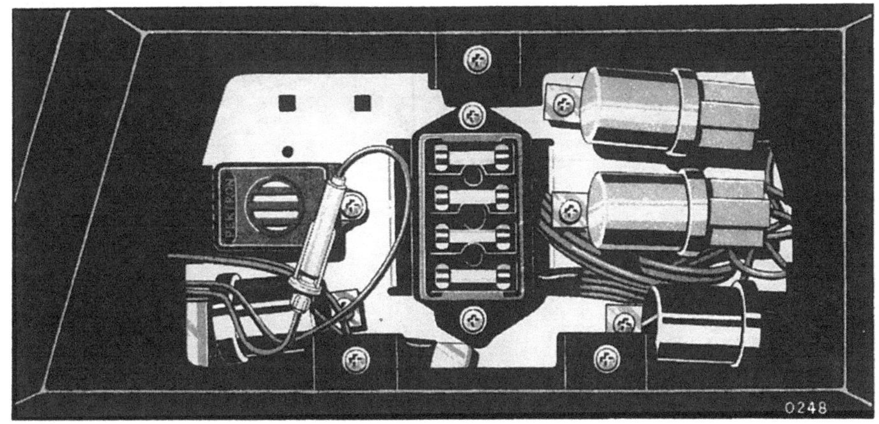

Circuits (cars with 4-fuse box)

The fuses in the fuse box protect the following circuits:

1 (top left)	10 amp	Left low beam
2	25 amp	Turn signals, windshield wipers and washers, choke heater, brake lights, reverse lights
3	10 amp	Left high beam
4	35 amp	Main air conditioner relay, heater/AC blower
5	10 amp	Right high beam
6	10 amp	Right low beam
7 (left bottom)	35 amp	Right condenser fan
8	15 amp	Rear demister
9	50 amp	Hazard lights, horn relay, seat belt warning, fog lamps, boot lamp, cigar lighter, courtesy lights
10	10 amp	Right side marker lamps, tail light
11	3 amp	Left side marker lamps, tail light
12	35 amp	Left condenser fan

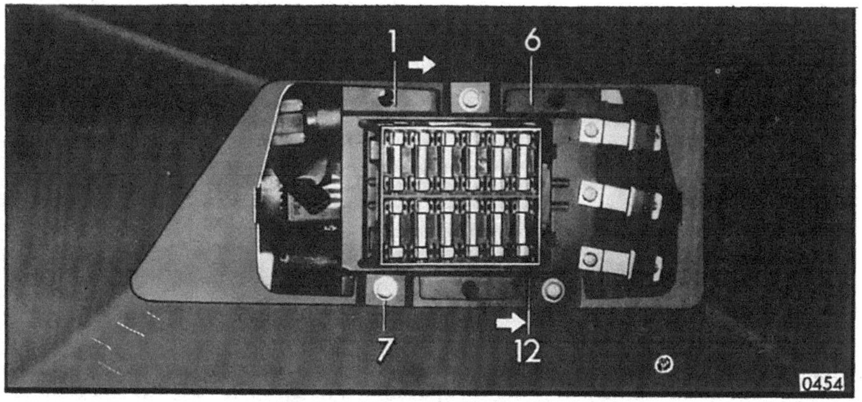

Circuits (cars with 12-fuse box)

Headlamps

In the event of electrical failure, the headlamps can be raised manually as follows: disconnect the plug in the engine bay (refer to underbonnet label), and locate the knob situated below the headlamp motor and turn it only in the direction indicated in the illustration, until the headlamp is raised fully. To lower, continue to turn the knob in the same direction until the headlamp is lowered.

DO NOT TURN THE KNOB IN THE WRONG DIRECTION.

Filament failure — Raise the affected headlamp manually. Take out four screws (1) and remove the cowl.

Remove three screws (1, 2 and 3) to release the retaining rim and light unit. Pull the connector from the light unit or bulb. Renew faulty item and reassemble.

Beam Aiming — Adjust screw 'A' to move the beam in a vertical plane; adjust screw 'B' to move the beam in a horizontal plane.

NOTE: To comply with legal requirements, beam checking and resetting should be entrusted to a Leyland Cars Distributor or Dealer who has special equipment for this purpose.

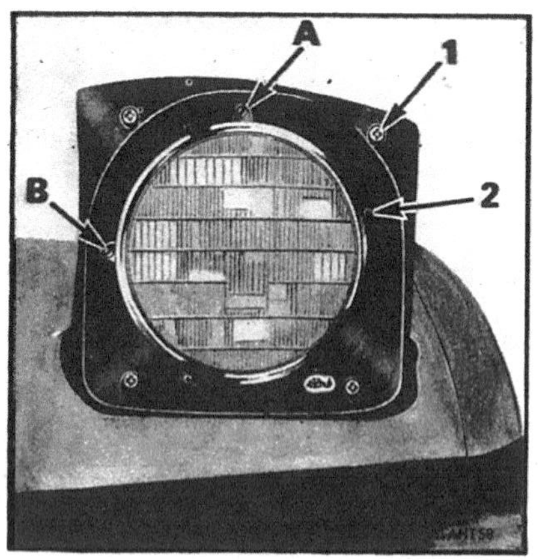

42

Front Parking and Direction Indicator Lamps

Remove two screws (1) and withdraw the lamp lens (2) to gain access to the parking bulb (3) and direction indicator bulb (4).

Rear Direction Indicator, Stop/Park and Reverse Lamps.

Open the luggage compartment lid. Remove the screws and turn back the trim panel to gain access to the lamp cluster.

Twist the appropriate bulb holder from the lamp cluster. Renew the bulb and re-assemble.

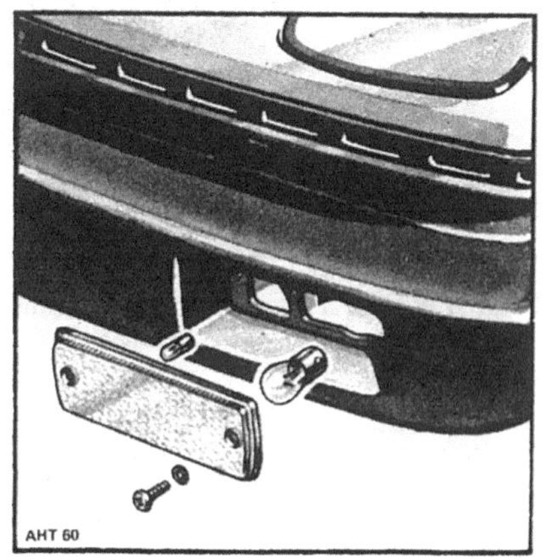

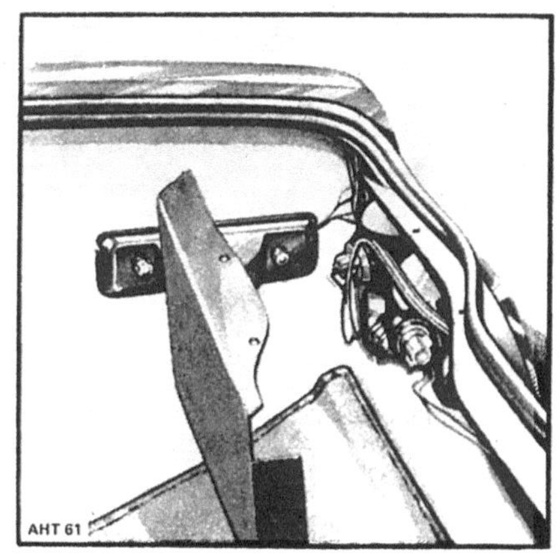

43

Side Marker Lamps

Unscrew the two nuts and washers, detach the cover and pull out the bulbholder. Renew the bulb and re-assemble.

Plate Illumination Lamp

Remove two screws and detach the cover to gain access to the bulb(s).

Luggage Compartment Lamp

Take out two screws and withdraw the lamp and lens from the panel. Replace the bulb and re-assemble.

Courtesy Lamp(s)

[On earlier models], disconnect the battery and lift off the lens by gently squeezing the longest sides. Carefully pry out the bulb and insert a new one. Refit the lens by engaging the top clips and squeezing the lens into position. Reconnect the battery.

[On later models, remove lens and lamp housing by carefully prising the rearward edge of the lamp outward to free it from the door trim pad.]

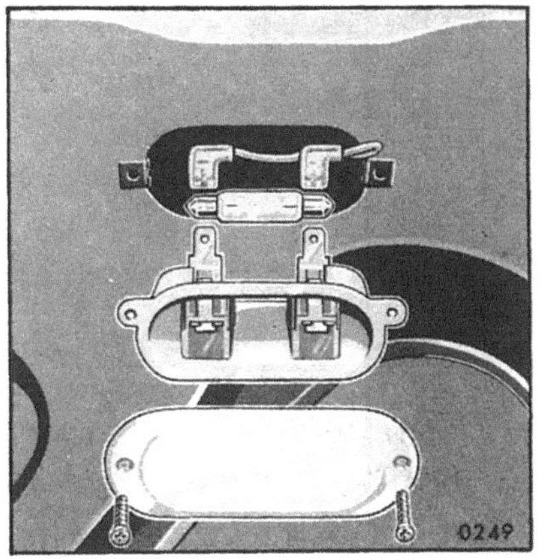

[Luggage Compartment Lamp]

[Side Marker Lamp]

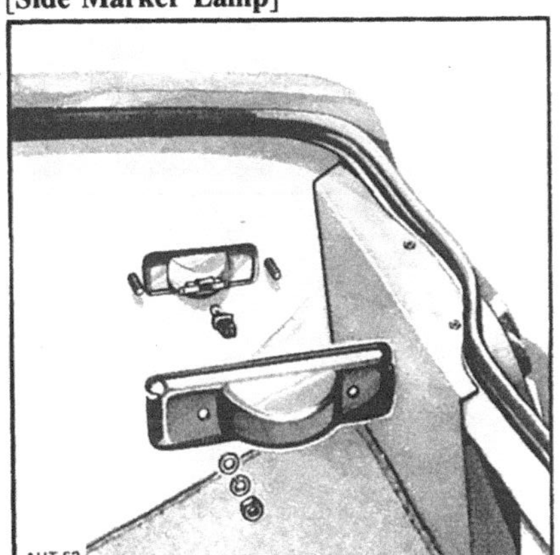

[Plate Illumination Lamp]

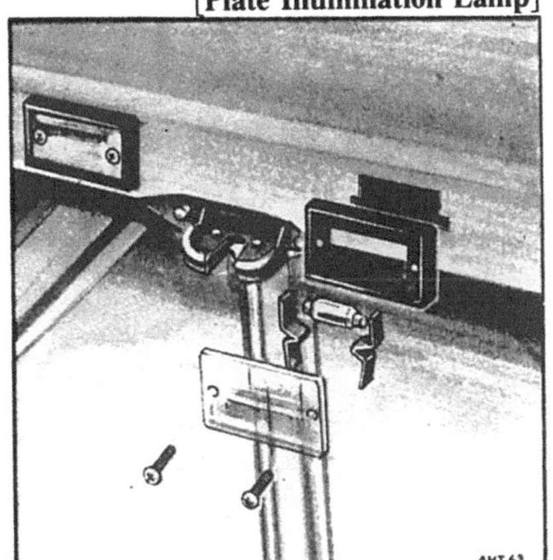

Windscreen Wipers

Service position of driver's wiper arm and blade — Wet the windscreen. Switch on ignition and wipers. Stop the wiper assembly in a vertical position by switching off the ignition at an appropriate moment. Lift the wiper arm and blade from the screen so that it falls into its service positon.

Do not switch on the ignition until the arm is returned to its normal position on the screen. If this is done the pantograph arm will jam, the motor will stall and the appropriate fuse will 'blow' to prevent damage to the arm or motor.

Renew driver's wiper blade — Depress the clip and withdraw the wiper blade from the pivot block.

Renew passenger's wiper blade — Depress the clip and withdraw the wiper blade from the arm.

Renew passenger's wiper arm and blade — Lift the wiper arm and blade from the screen so it falls into its service position. Position screwdriver as shown and impart a twisting action to lift the clip from the spindle groove. The assembly may now be removed by hand.

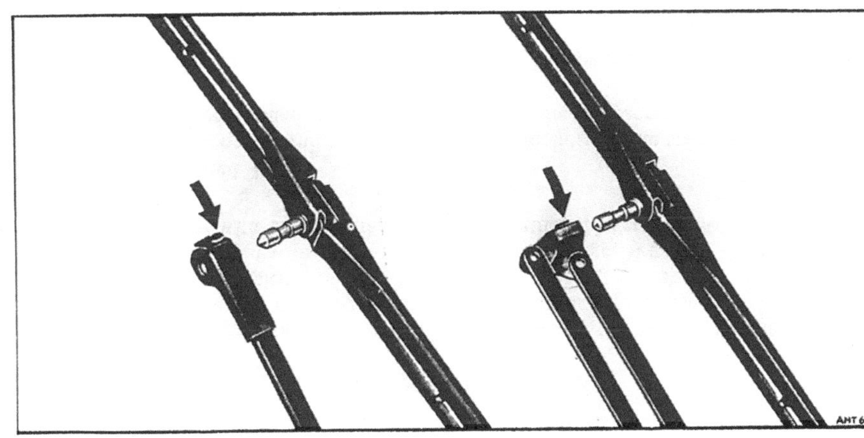

BULB CHART

Lamp	Watts	Lucas Part No.	Unipart No.	Triumph Part No.
Headlamp				
L.H. Dip Normal	75/60		UKC2423	
R.H. Dip — U.S.A.	50/40			514578
R.H. Dip — France	45/40	411	GLB 411	510219
R.H. Dip — Normal	60/50	54523079	GLU 114	215735
Front parking lamp	4			
Front flasher lamp	21	382	GLB 382	502379
Front marker lamp	3	504	GLB 504	518414
Rear marker lamp	3	504	GLB 504	518414
Tail lamp	5	207	GLB 207	57591
Stop lamp	21	382	GLB 382	502379
Rear flasher lamp	21	382	GLB 382	502379
Reverse lamps	21	382	GLB 382	502379
Plate illumination lamp	5			
Instrument panel illumination . .	2·2			
Courtesy lamp	6			586438
Warning lights	1·2	286	GLB 286	UKC2412

CARE OF BODYWORK

Washing

Avoid using a dry cloth to wipe dust from the paintwork and chromium surfaces. Dust is an abrasive and if removed in this way it will scratch the polished surfaces. Wash the car frequently with plenty of running water and a clean, soft sponge. Soften, and if possible remove the mud with water before using the sponge. When all dirt is removed, sponge off and dry with a clean, damp chamois-leather. Never wash or polish the car under a hot sun.

Removing Grease and Tar

Remove grease or tar with methylated spirits (alcohol). White spirit is also effective, but this must not be applied to rubber, particularly the windscreen wiper blades.

Glass Surfaces

Glass is easily scratched. This can be avoided by always using a damp chamois-leather which is especially reserved for use on glass only. If silicone polishes have been used on the body, take care that the polish does not come into contact with the glass. It is extremely difficult to remove and causes the windscreen to smear.

Rear Window (soft top)

Extreme care must be used when cleaning the plastic window in the convertible top as it is easily scratched. Use only a soft cloth with water or a suitable plastic cleaner.

Chromium Plating and Stainless Steel

Frequent washing and thorough drying is recommended, especially during the winter months when there is likelihood of corrosion through contamination with road salt.

Polishing

After a period of use, the formation of traffic film will cause the paintwork to lose some of its lustre, even though the car has been carefully and regularly washed. The original brilliance may be restored after washing by using a reputable non-abrasive cleaner and polish.

Being the most durable, wax preparations are preferable, but where these are used regularly the old wax must first be removed with a cleaner before further application of new wax. The frequency at which polishing is necessary will depend upon local conditions of air pollution.

Interior of Car and Soft Top

Brush and clean the inside of your car each time and wash and polish the outside. Use a vacuum cleaner where possible and ensure complete removal of all dust from the interior and trim.

Wash the upholstery and soft top with luke-warm, non-caustic, soapy water. Do not use detergents or household cleaners as these may cause damage.

Remove all traces of suds with a clean, damp cloth and thoroughly dry the upholstery with a dry duster or towel.

Wipe the fascia and instrument panel with a damp cloth only. Wax or other polishes should not be used inside the car.

Opening Roof

Periodically, clean the sliding roof side rail rebate and, if sticking or jamming occurs, lubricate very sparingly using a silicone base aerosol penetrating fluid.

Inflammability

The car interior conforms to State and Federal laws on flammability. To preserve this condition do not clean interior other than as described above.

PERIODIC CHECKS

Lubrication

The lubrication systems of your new car are filled with high-quality oils. Continue to use high-quality oils of the correct viscosity range for the engine, gearbox and rear axle during subsequent maintenance or when topping up. Adhere to the servicing intervals stated in the 'Maintenance Summary'.

The use of oils that are not to the recommended specification can lead to high oil and fuel consumption, excessive wear, and may result in premature damage to engine, gearbox or rear axle components.

Oils to the recommended specification marketed by reputable oil companies contain additives which disperse corrosive acids formed by combustion. The additives also assist in preventing the formation of sludge which can restrict the oil circulation.

Do not use additional additives.

Engine Oil

When a new car is delivered, the engine sump contains an adequate quantity of special running-in oil. Provided that the level is not allowed to fall below the 'low' mark on the dipstick during the running-in period, topping-up should be unnecessary.

At the completion of the first 1,000 miles (1,600 km) it is important that the special oil is drained and the engine sump replenished to the 'high' mark on the dipstick with a recommended grade of oil.

Once a week, or prior to starting out on a long run, or every 250 miles (500 km), check the level of oil in the engine sump, first making sure that the car is standing on level ground. If the engine has been running, wait a few minutes to allow the oil to drain back into the sump.

Withdraw the dipstick (1), wipe it clean and push it fully home before again withdrawing it for reading. Add oil via the filler cap (2) until the level reaches the 'high' mark on the dipstick. DO NOT OVERFILL, and ensure that the filler cap and dipstick are replaced.

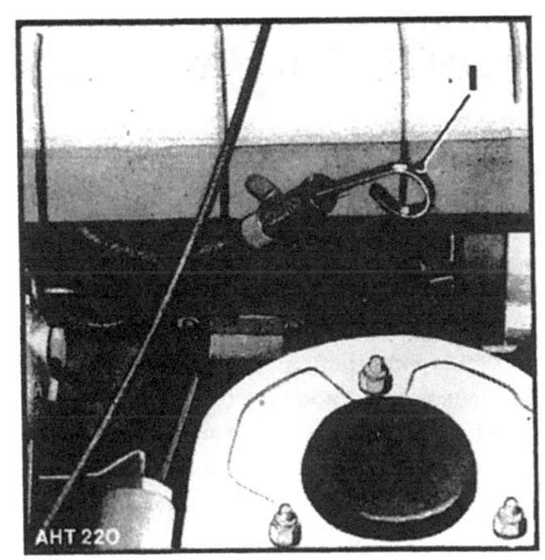

Engine Oil Change

Drain the sump while the engine is warm by unscrewing the drain plug (1) and directing the oil stream into a suitable receptacle.

When the oil has drained completely, refit the plug and fill the engine to the high mark on the dipstick with new oil; see 'Lubrication Recommendations'.

Owners are strongly recommended to change the engine oil every 5,000 kms (3,000 miles) or 3 months if the car is frequently driven for long periods at high speeds approaching the operating maximum. Engine oil deteriorates with increased temperature and if high speeds are maintained, the oil reaches a very high temperature. This instruction does not apply if an oil cooler is fitted.

Oil Filter Element

To renew the filter element unscrew the securing bolt (4), remove the container (5) and discard the element (6). Wash out the container and insert a new element.

Renew the sealing ring (7) ensuring that it is correctly located in the cylinder block and re-attach the filter assembly by tightening the bolt (4) sufficiently to ensure an oil-tight joint.

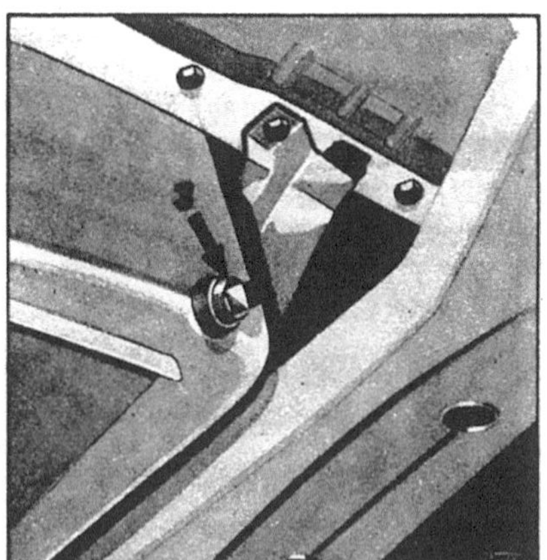

Manual Gearbox

With the vehicle standing on level ground, remove the oil level plug (arrowed) and, using a suitable dispenser such as a pump-type oil can with flexible nozzle filled with an extreme pressure (Hypoid) lubricant, top up the gearbox until the oil is level with the bottom of the filler plug threads.

Allow surplus oil to drain away before refitting the level plug and wiping clean.

Rear Axle

Remove the oil level plug (shown arrowed), and top-up the rear axle until the oil is level with the bottom of the filler plug threads. Allow surplus oil to drain before fitting the plug and wiping clean.

5 speed

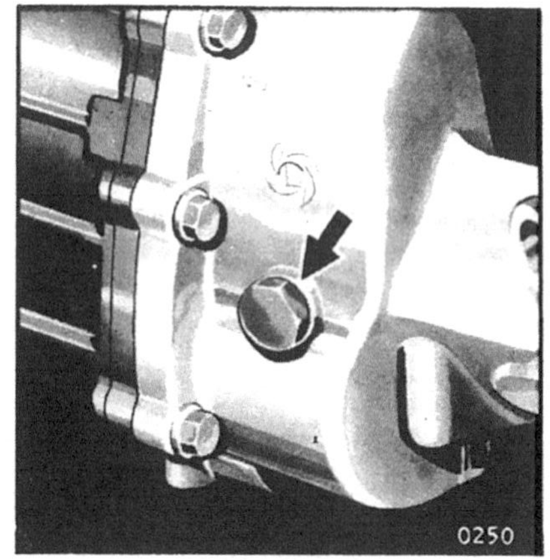

5 speed

4 speed

4 speed and automatic

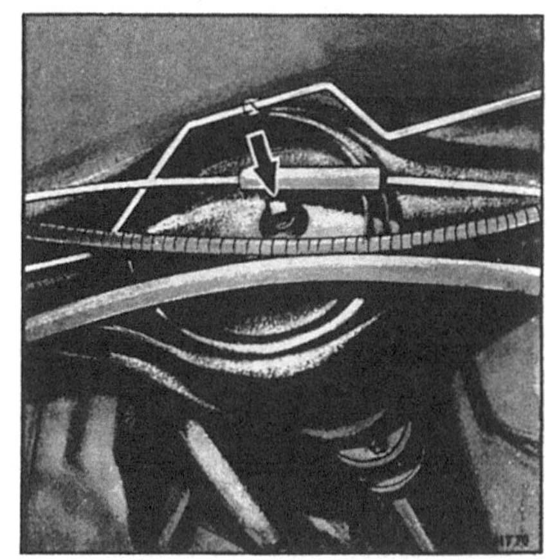

Automatic Transmission Unit

Check the fluid level as follows:

1. Stand the car on level ground and apply the handbrake firmly. Start the engine from cold and, with the footbrake firmly applied, run the engine at idle speed for 2 to 3 minutes, passing the selector lever through the complete range of positions to ensure that the transmission is primed.
2. Select the 'P' (Park) position and apply the handbrake. Leave the engine running at idle speed.
3. Remove the transmission dipstick and wipe it with a clean, non-fluffy cloth.
4. Replace the dipstick, ensuring that it is pushed fully into the tube and withdraw it immediately for reading.
5. Check the fluid level on the side of the dipstick marked 'COLD' and if necessary, add fluid; see Lubrication Recommendations.
6. Repeat instructions 1 to 5 until the fluid level is correct.

DO NOT OVERFILL THE TRANMISSION.

Where the reverse side of the dipstick carries marks denoted 'HOT', the fluid level check may be carried out with the transmission at normal operating temperature. The procedure is as described above except that the vehicle must be driven for 25-30 km (15-20 miles) to warm the transmission. The check is then carried out using the 'HOT' side of the dipstick.

NOTE: When operating at high ambient temperatures and on unmetalled roads, periodically inspect and remove dust and mud deposits from the slots and screen on the underside of the torque converter housing, and from the under side of the transmission oil sump these deposits can adversely affect proper cooling of the unit.

Coolant

CAUTION. To avoid injury from scalding when it is necessary to remove the pressure cap from a hot engine, exercise great care by protecting the hands against escaping steam. Slowly turn the pressure cap (8) anti-clockwise until resistance of the safety stop is felt. Leave the cap in this position until all pressure is released. Press the cap downwards against the spring to clear the safety stops, and continue turning until it can be lifted off.

Maintain the level of coolant in the expansion tank at half-full by topping-up as necessary when the system is cool. Refer to 'Cooling System' for anti-freeze protection.

Screen Washer

Check the level of water in the translucent windscreen washer container (9) and, if required, replenish the container with clean soft water. The addition of a small amount of mild detergent will help prevent smearing.

During freezing conditions it is beneficial to fill the container with a mixture of one part methylated spirits (wood alcohol) to two parts of water. This will assist in the dispersal of snow and ice from the screen.

Do not add anti-freeze solutions to the container as this will discolour the paint work and damage wiper blades and sealing rubber.

Should a screen washer jet become obstructed, a strand of fine wire not greater than 0.7 mm (0.030 in) diameter can be used to clear the jet orifice.

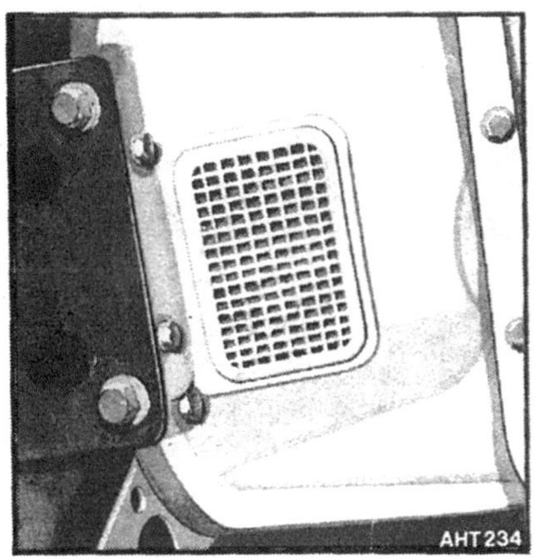

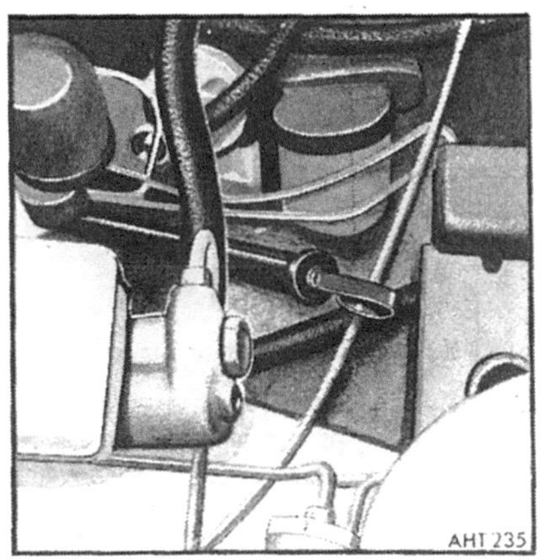

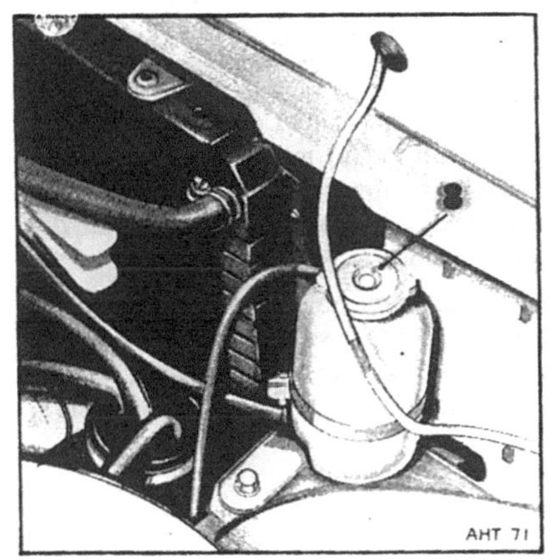

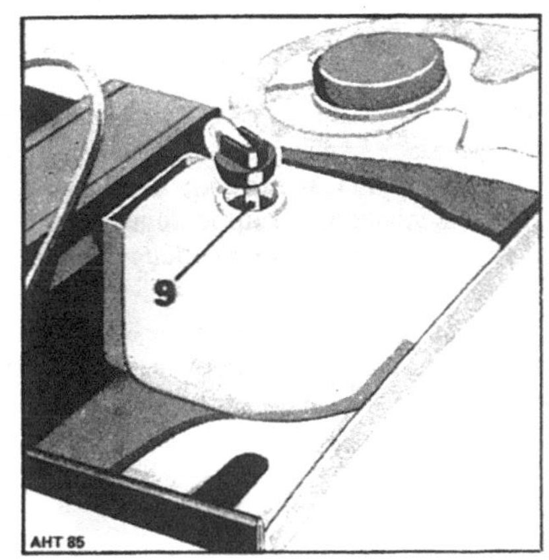

Battery

At periods not exceeding four weeks, examine the level of the electrolyte in the cells, and top up with distilled water if necessary. The electrolyte level should just cover the separators. Check more often during hot weather and before each long journey.

WARNING. Never use a naked flame to examine a battery as gases given off by it are dangerously explosive.

Clutch Fluid

Check the level of fluid monthly by wiping clean and removing the cap from the clutch master cylinder reservoir. If required, top up with new brake fluid of the recommended type. Re-seal the brake fluid tin and refit the reservoir cap immediately.

Brake Fluid

Check the level of fluid weekly by examining the fluid reservoir. Top up when required with new fluid of the type recommended. Do not allow the fluid level to drop below the danger mark on the reservoir — a sudden appreciable drop in fluid level must be investigated and the cause rectified immediately. Do not take off the cap (0) unless you need to; re-seal the brake fluid tin and refit the reservoir cap immediately.

Tyre Pressures

Check pressures weekly and examine the tyres for cuts, tread depth and defects. Adjust the pressures while the tyres are cold. See 'General Specification' for correct pressures.

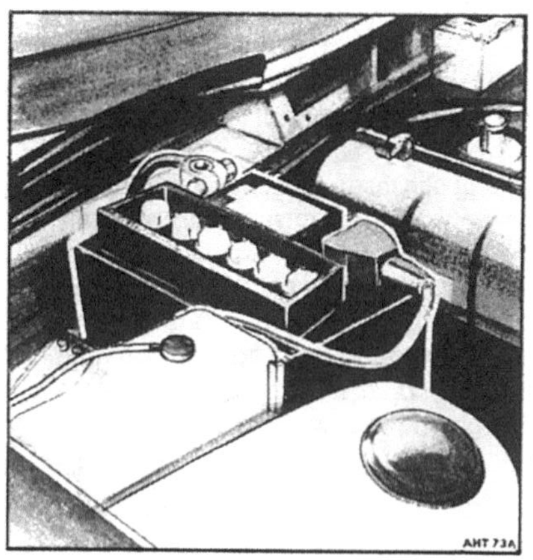

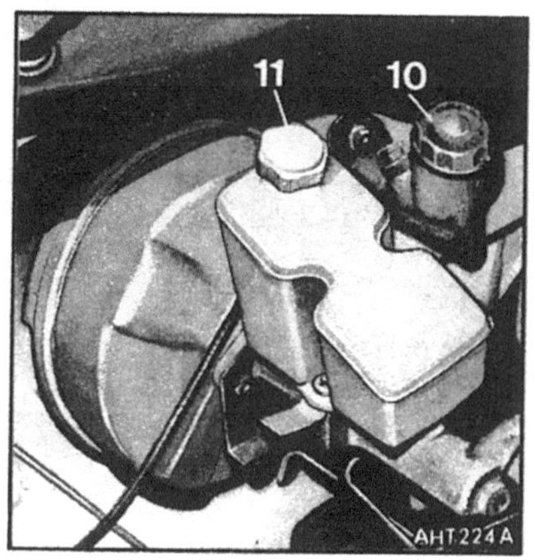

ROUTINE SERVICING

1,000 Miles (1,600 km.) Free Service

At the completion of 1,000 miles (1,600 km.) or as near to this figure as possible, it is important that the car is returned to the selling Dealer, who will complete, free of charge except for the cost of oil and grease, a free service.

The importance of carrying out this free service cannot be over emphasised, as it corresponds with the end of the running-in period when many of the components will have settled down, thus necessitating re-adjustment. Furthermore, the special running-in engine oil, used for the initial fill, is unsuitable for normal motoring and must, therefore, be replaced by the correct grade of oil.

Scheduled Maintenance

The 'Maintenance Summary' lists all the maintenance operations and the frequency at which they are required. Because of the need to use specialized equipment and techniques, especially on operations concerned with 'Emission and Evaporative Control Systems' and in compliance with the terms of the 'Emission Warranty', it is strongly recommended that all the 'scheduled services' are entrusted to your Distributor or Dealer.

Preventive Maintenance

In addition to the recommended periodical inspection of brake components it is advisable as the car ages, and as a precaution against the effects of wear and deterioration, to make a more searching inspection and renew parts as necessary.

It is recommended that:

(1) Disc brake pads, drum brake linings, hoses and pipes should be examined at intervals no greater than those laid down in the Maintenance Summary of the Handbook.

(2) Brake fluid should be changed completely every 18 months or 19,000 miles (30,000 km) whichever is the sooner.

(3) All fluid seals in the hydraulic system should be renewed, and all flexible hoses should be examined and renewed if necessary every 3 years or 37,500 miles (60,000 km) whichever is the sooner. At the same time the working surface of the piston and of the bores of the master cylinder, wheel cylinders, and other slave cylinders should be examined and new parts fitted where necessary.

Care must be taken always to observe the following points:

(a) At all times use the recommended brake fluid.
(b) Never leave fluid in unsealed containers. It absorbs moisture quickly and can be dangerous if used in your braking system in this condition.
(c) Fluid drained from the system or used for bleeding is best discarded.
(d) The necessity for absolute cleanliness throughout cannot be over-emphasized.

Replacing Brake-shoes

When it becomes necessary to renew the brake-shoes, it is essential that only genuine shoes, with the correct grade of lining, are used. Always fit new shoes as complete axle sets, never individually or as a single wheel set. Serious consequences could result from out-of-balance braking due to mixing of linings.

Replacement brake-shoes are obtainable from your Distributor or Dealer.

EMISSION AND EVAPORATIVE CONTROL SYSTEMS
U.S.A. SPECIFICATION

[**NOTE:** This section contains general information. For more detailed information, consult the Repair Operation Manual section of this book (Part 2).]

All Triumph models entering the U.S.A. markets incorporate efficient emission control systems. These systems enable the vehicles to conform with all current Federal and California State Regulations governing the emission of hydrocarbons, carbon monoxide, nitric oxide and the emission of fuel, by evaporation, from the fuel delivery system.

Fuel

IT IS ESSENTIAL THAT UNLEADED FUELS ARE USED IN THIS VEHICLE OTHERWISE SERIOUS DAMAGE CAN BE CAUSED TO THE CATALYTIC CONVERTER.

EMISSION CONTROL SYSTEMS — SERVICING

The importance of servicing at the correct intervals cannot be over-stressed as improvements in design and manufacturing techniques count for nothing if the servicing standards are not upheld.

Routine servicing, carried out at the mileage intervals quoted the Maintenance Summary helps to prevent deterioration of the systems.

CARBURETTER

The emission carburetters are prime components of the emission system and great care is exercised during the manufacture and initial adjustment of these instruments. Adjustment of the carburetters necessitates the use of specialised equipment and it is recommended that this is carried out by a British Leyland Dealer or other established and qualified service facility.

Adjustments for Altitude

To ensure that the vehicle meets stringent exhaust emission standards at different altitudes, the carburetter settings must be adjusted at the altitude at which principle use is intended. (Areas above 4000ft. are considered 'high' altitude and below 4000ft. 'low' altitude).

If a vehicle is purchased in an area of 'high' altitude, it will be adjusted prior to delivery.

If the area of principle use is changed, then the vehicle should be taken to an Authorised Dealer and the necessary carburetter adjustments made; they will only take a few minutes.

CALIFORNIA — Vehicles sold in California have not been designed or tested at 'high' altitude, no adjustments should, therefore be made for this specification.

Carburetter Controls

The throttle rod linkage will not require adjustments during normal operation. To ensure complete throttle closure a degree of 'lost motion' or slackness is incorporated into the linkage; no attempt must be made to eliminate this.

Carburetter Servicing Schedules

To maintain the carburetter at peak efficiency, regular servicing as detailed in the 'Maintenance Summary' is essential. The appropriate servicing operations, checking and adjusting the idle speed, CO at idle, topping up the carburetter piston dampers and checking and adjusting the deceleration by-pass valve, should be performed by authorised dealers, who are trained in the use of the special equipment needed.

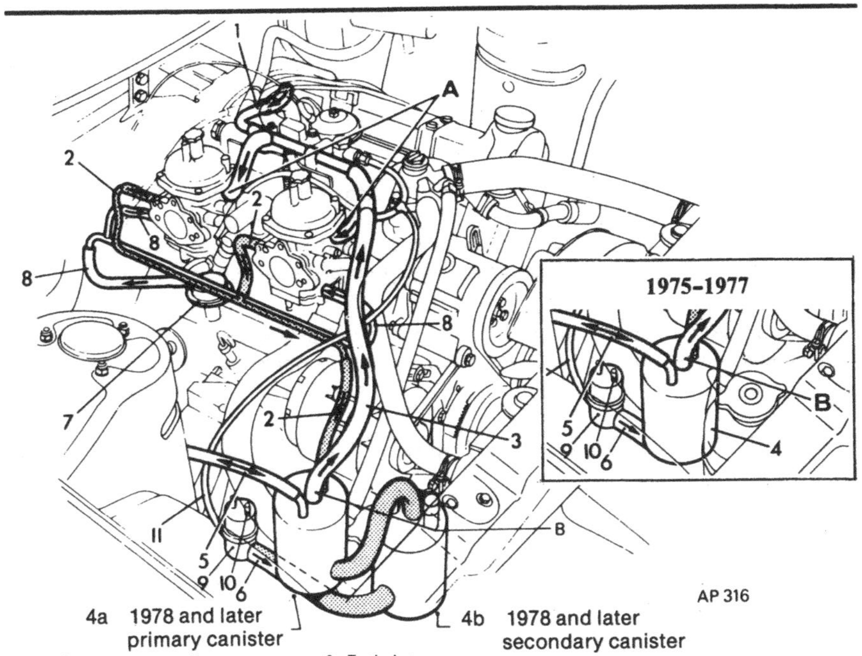

1. Crankcase purge line
2. Carb. float chamber vent pipe
3. Canister purge pipe
4. Charcoal canister
5. Fuel tank vent pipe
6. Purge air to canister
7. Fuel pump
8. Fuel pipe
9. Anti run on valve (California only)
10. Electrical connections for run on valve
11. Manifold vacuum line*

A. .280–.300" Restrictors
B. 3/32" restrictor

*[Routed differently, to the rear of the engine and into the crankcase purge line, on 1975 and 1976 models; this is shown later in the diagram for Canadian models.]

Fig. 1 - TR7 Crankcase and Canister Purge System

Crankcase Breathing (Fig. 1)

Crankcase breathing and evacuation of 'blow by' gases is achieved by utilizing the characteristic partial vacuum in the constant depression carburetters. By this method crankcase emissions are burned in the engine combustion process (Fig. 1). A wire gauze strainer in the engine top cover acts as an oil separator/flame trap.

Running-on Control — (Anti Run-on Valve) (Fig. 1)

NOTE: The running-on control valve is fitted to California Specified vehicles only.

This valve prevents the 'running-on' of the engine after the ignition is switched off when, due to the heat of the engine, a condition of compression ignition is set up.

The method of achieving a cut off is by applying a slight 'vacuum' to the float chamber of the carburetters when the ignition is switched off.

System Decription

With the ignition off, a solenoid is activated which operates a valve that seals off the inlet to the botton of the canister. With the inlet sealed a connection to the intake manifold applies a partial vacuum to the canister and consequently to the float chamber via an interconnecting pipe. When the engine has stopped and the oil pressure drops to zero, the solenoid is deactivated and the engine is thus ready again for operation.

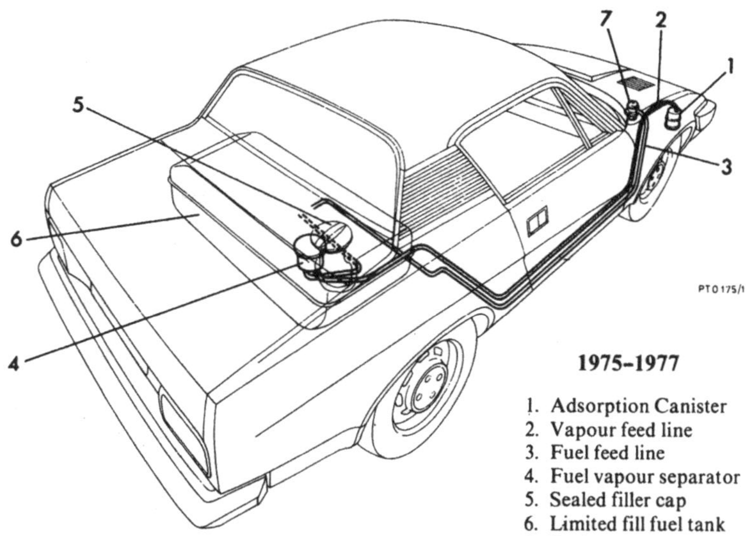

1975-1977

1. Adsorption Canister
2. Vapour feed line
3. Fuel feed line
4. Fuel vapour separator
5. Sealed filler cap
6. Limited fill fuel tank
7. Fuel pump

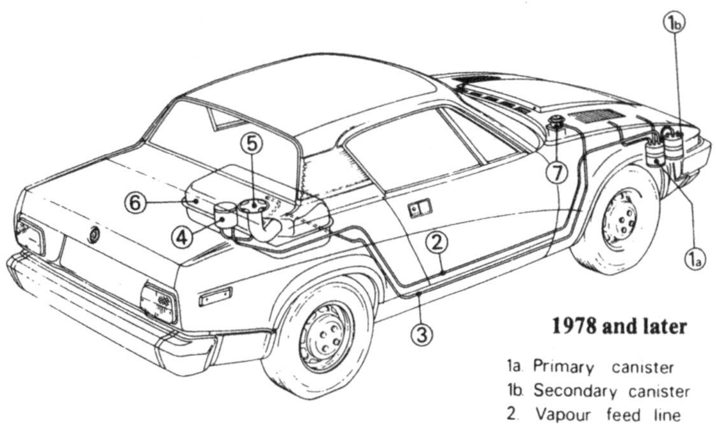

1978 and later

1a Primary canister
1b Secondary canister
2 Vapour feed line
3 Fuel feed line
4 Fuel vapour separator
5 Sealed filler cap
6 Limited fill fuel tank
7 Fuel pump

Fig. 2 Evaporative Control Sytsem

Function Checks

If the valve is not working then it will be apparant by engine running-on. A system check can be made by applying current to the solenoid which, if working correctly, will stop the engine.

EVAPORATIVE CONTROL SYSTEM

The evaporative control system uses an activated adsorption canister through which the fuel tank is vented. The following are features of the evaporative control system (see Fig. 2):

(1) The carburetter float chamber is vented to the engine during open throttle conditions to the adsorption canister at closed throttle.
(2) The constant depression of the carburetter is used to induce a purge condition through the adsorption canister. Via the running-on control valve - (California only). The crankcase breathing is also linked into the system.
(3) A separator tank prevents fuel surges from reaching the canister and thus saturating the system.
(4) The fuel filler cap is sealed to prevent evaporative losses.
(5) The fuel tank filler tube extends into the tank to prevent complete filling and so allow for expansion of fuel in hot weather.

WARNING: The use of compressed air to clean an adsorption canister or clear a blockage in the evaporative system is very dangerous. An explosive gas present in a partly saturated canister may be ignited by the heat generated when compressed air passes through the canister.

Servicing

Minimal servicing is required on the evaporative system apart from renewing the adsorption canister at the specified intervals and checking visually the security of piping on the system.

EXHAUST GAS RECIRCULATION (E.G.R.) SYSTEM (Fig. 3)

System Description

To reduce the nitric oxides content in the exhaust, the peak combustion temperatures are lowered by recirculating a controlled quantity of the exhaust gases through the combustion process.

The E.G.R. valve is mounted on the inlet manifold. A control signal, taken from a throttle edge tapping in the carburetter, gives no recirculation at idle or full load, but does allow an amount of recirculation, dependent on the vacuum signal and a metering profile on the valve, Fig. 3 under part load conditions.

Servicing

After the completion of the initial 1000 miles check the security of the E.G.R. valve operating lines. On the completion of 50,000 mile periods, the following checks are necessary:

(1) Security of the E.G.R. valve operating lines.
(2) Operation of the E.G.R. valve.
(3) Clean the valve and pipework.

NOTE: It is recommended that this work is carried out by a British Leyland Dealer.

An E.G.R. Service Repair Kit can be purchased from a British Leyland Dealer. Complete details on the servicing of the E.G.R. system are detailed in Repair Operation Manual Pt. No. RTC 9200/77A obtainable from British Leyland Dealers.

1. E.G.R. valve
2. Inlet manifold
3. Exhaust port
4. Choke cam
5. E.G.R. cut-out valve
6. Petrol trap
7. Vacuum source — throttle edge tapping

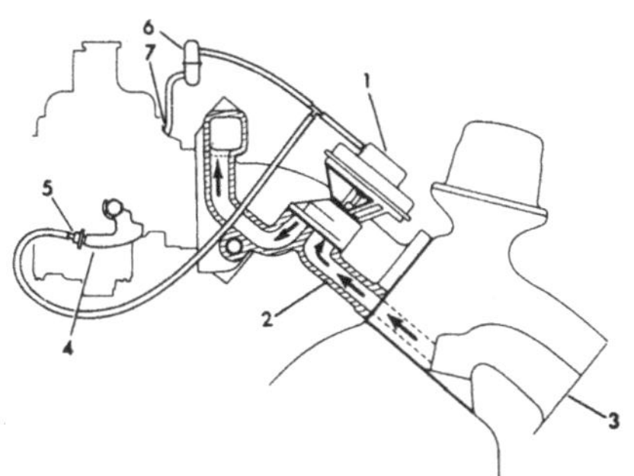

[1975 & 1976 models]

1. E.G.R. valve
2. E.G.R. pipe
3. Vacuum source — throttle edge tapping
4. Petrol trap
5. Inlet manifold
6. Exhaust port

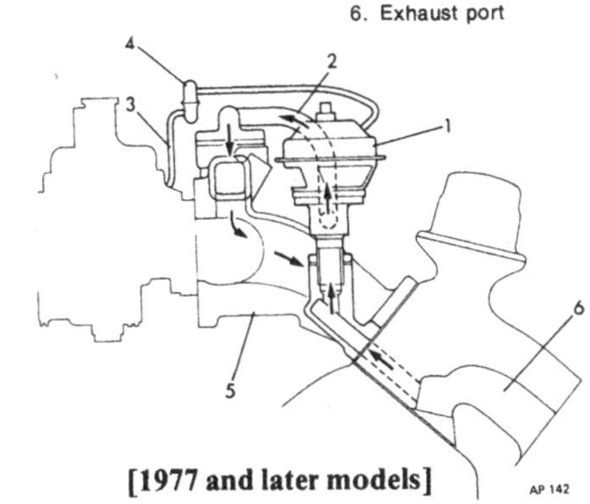

[1977 and later models]

Fig. 3 Exhaust Gas Recirculation System

Function Checks

Warm the engine to normal running temperature. Ensure that the idle speed returns to normal. Blip the throttle and observe the valve, which should open and close as the engine speed changes. If the valve is not operating, remove the E.G.R. valve and check the valve operation using a vacuum test gauge. Fit a new E.G.R. valve if it is found to be defective.

AIR INJECTION SYSTEM

The air pump, driven by the engine, delivers air into a single entry point in the exhaust manifold. A one-way valve prevents damage to the pump should back-fire or belt breakage occur. A relief valve, located in the air-pump, is required to dump part of the air at high speeds, to prevent pump damage.

Servicing

Minimal servicing is required on the system apart from inspecting the hoses and pipes at specified intervals.

CATALYTIC CONVERTERS

A catalytic converter is fitted into the exhaust system in order to reduce carbon monoxide and hydrocarbon emissions.

The following precautions apply to catalytic devices:

1. In order to maintain the efficiency of the emission control system it is essential to use UNLEADED gasoline only; this fuel minimises spark plug fouling thereby sustaining engine performance.
2. DO NOT tamper with the engine settings; they have been established to ensure that the vehicle will comply with stringent exhaust emision regulations. Incorrect engine settings could cause unusually high catalytic converter temperatures and thus result in damage to the converter and vehicle. If adjustment to the settings is considered necessary, this should be performed by a British Leyland Dealer or other qualified facility.
3. A correctly tuned engine optimises exhaust emissions performance and fuel economy and it is recommended that the vehicle is maintained as outlined under the MAINTENANCE SUMMARY of this handbook.
4. DO NOT continue to operate the vehicle if any engine malfunction is evident; malfunctions should be rectified immediately. For instance, misfire, loss of engine performance or engine run-on may lead to unusually high catalytic converter temperatures and may result in damage to the converter and car.
5. NEVER leave the vehicle unattended with the engine running.
6. The use of a catalytic converter increases exhaust system temperatures, (particularly under engine malfunction), therefore do not operate or park the vehicle in areas where combustible materials such as dry grass or leaves may come into contact with the exhaust system.
7. The vehicle is designed for normal road use. Below are examples of abuse which could damage the catalytic converter and car and may lead to a dangerous condition due to excessively high catalytic converter temperatures:

   (a) Competition use
   (b) Off roadway use
   (c) Excessive engine revolutions
   (d) Overloading the vehicle
   (e) Excessive towing loads
   (f) Switching off the engine and coasting in gear

8. DO NOT run the engine with either a spark plug lead disconnected or a spark plug removed. DO NOT use any device that requires an insert into a spark plug hole in order to generate air pressure, (e.g. tyre pump, paint spray attachment, etc.), as this could also result in catalytic converter damage.
9. DO NOT push or tow the vehicle to start it, this could damage the catalytic converter. It is recommended that jumper leads are used.
10. Heavy impact on the converter casing must be avoided as it contains ceramic material which is easily damaged.

MAINTENANCE SUMMARY

USA

The following items should be checked weekly by the driver:
Engine oil level
Brake fluid level
Radiator coolant level
Battery electrolyte level
Windshield washer reservoir level
All tyre pressures
Operation of all lights
Operation of horn
Operation of all washers and wipers

MAINTENANCE INTERVALS

Service	Miles x 1000	Monthly Intervals
A	1	1
B	3, 9, 16, 22, 34, 41, 47	3
C	6, 19, 31, 44	6
D	12·5, 37·5	12
E	25, 50	24

NOTE: The service schedules are based on an annual mileage of approximately 12,500 miles or 20,000 Kms. Should the vehicle complete substantially less miles than this per annum, it is recommended that a 'C' service is completed at six month intervals and a 'D' service at twelve month intervals.

'A' SERVICE

Engine
Check for oil leaks
Renew engine oil
Top up carburetter piston damper(s)
Check/adjust carburetter idle settings
Check cooling/heater systems for leaks and hoses for security and condition
Check/top up cooling system
Check/adjust operation of all washers and top up reservoirs
Check driving belts, adjust or renew
Lubricate accelerator control linkage and pedal pivot — check operation
Check/adjust deceleration by-pass valve
Check/Adjust torque of cylinder head nuts/bolts
Check security of engine mountings
Check security of EGR valve operating lines
Check crankcase breathing and evaporative loss systems hoses for security

'A' SERVICE — *continued*

Ignition
Check/adjust ignition timing using electronic equipment
Check security of distributor vacuum unitline and operation of vacuum unit [when fitted]
Check ignition wiring for fraying, chafing and deterioration

Exhaust and Fuel Pipes
Check fuel system for leaks, pipes and unions for chafing and corrosion
Check exhaust system for leaks and security

Transmission
Check for oil leaks
Check/top up gearbox oil
Check/top up rear axle/final drive oil.
Check/top up automatic gearbox fluid
Check/top up clutch fluid reservoir
Check clutch pipes for leaks and chafing

Steering and Suspension
Check condition and security of steering unit joints and gaiters
Check security of suspension fixings
Check/adjust front (and rear) wheel alignment
Check steering rack/gear for oil/fluid leaks.
Check shock absorbers for fluid leaks

Brakes
Check visually, hydraulic pipes and unions for chafing, leaks and corrosion
Check/top up brake fluid reservoir(s)
Check handbrake operation; adjust to manufacturers instructions
Check footbrake operation; adjust to manufacturers instructions (Self adjusting)
Check brake servo hose(s) for security and condition

Wheels and Tyres
Check/adjust tyre pressures including spare
Check tightness of road wheel fastenings
Check tyres for external cuts in tyre fabric, exposure of ply or cord structure, lumps or bulges

IMPORTANT: If tyres do not conform with legal requirements, report to owner.

Body
Lubricate all locks and hinges (Not steering lock)
Check condition and security of seats and seat belts
Ensure cleanliness of controls, door handles, steering wheel
Check operation of all door, bonnet and boot locks
Check operation of window controls
Check operation of seat belt inertia reel mechanism.
Check tightness of sub-frame/body mountings

'A' SERVICE — *continued*

Electrical
Check function of original equipment i.e. interior and exterior lamps, horns, wipers and warning indicators
Check/top up battery electrolyte
Check/adjust headlamp alignment
Check output of charging system

General
Road/roller test and check function of all instrumentation

'B' SERVICE

Engine
Check for oil leaks
Check/top up engine oil
Check cooling/heater systems for leaks and hoses for security and condition
Check/top up cooling system
Check/adjust operation of all washers and top up reservoirs

Exhaust and Fuel Pipes
Check fuel system for leaks, pipes and unions for chafing and corrosion
Check exhaust system for leaks and security

Transmission
Check for oil leaks
Check/top up clutch fluid reservoir
Check clutch pipes for leaks and chafing

Steering and Suspension
Check condition and security of steering unit joints and gaiters
Check steering rack/gear for oil/fluid leaks.
Check shock absorbers for fluid leaks

Brakes
Check visually, hydraulic pipes and unions for chafing, leaks and corrosion
Check/top up brake fluid reservoir(s)
Inspect brake pads for wear and discs for condition
Check brake servo hose(s) for security and condition

Wheels and Tyres
Check that tyres comply with manufacturers specification
Check/adjust tyre pressures including spare
Check tyres for tread depth and visually for external cuts in fabric, exposure of ply or cord structure, lumps or bulges
Check tightness of road wheel fastenings
IMPORTANT: If tyres do not conform with legal requirements, report to owner.

'B' SERVICE — *continued*

Body
Check condition and security of seats and seat belts
Check rear view mirror(s) for cracks.
Ensure cleanliness of controls, door handles, steering wheel
Check operation of seat belt warning system
Check operation of seat belt inertia reel mechanism.

Electrical
Check function of original equipment i.e. interior and exterior lamps, horns, wipers and warning indicators
Check/top up battery electrolyte
Check/adjust headlamp alignment
Check, if necessary renew wiper blades

General
Report additional work required

'C' SERVICE

Engine
Check for oil leaks
Renew engine oil filter
Renew engine oil
Check cooling/heater systems for leaks and hoses for security and condition
Check/top up cooling system
Check/adjust operation of all washers and top up reservoirs
Lubricate accelerator control linkage and pedal pivot — check operation

Exhaust and Fuel Pipes
Check fuel system for leaks, pipes and unions for chafing and corrosion
Check exhaust system for leaks and security

Transmission
Check for oil leaks
Check/top up gearbox oil
Check/top up rear axle/final drive oil.
Check/top up automatic gearbox fluid
Check/top up clutch fluid reservoir
Lubricate automatic gearbox exposed selector linkage
Check clutch pipes for leaks and chafing

'C' SERVICE — *continued*

Steering and Suspension
Lubricate steering rack and pinion
Check condition and security of steering unit joints and gaiters
Check/adjust front (and rear) wheel alignment
Check steering rack/gear for oil/fluid leaks.
Check shock absorbers for fluid leaks

Brakes
Check visually, hydraulic pipes and unions for chafing, leaks and corrosion
Check/top up brake fluid reservoir(s)
Check handbrake operation; adjust to manufacturers instructions
Check footbrake operation; adjust to manufacturers instructions (Self adjusting)
Inspect brake linings/pads for wear, drums/discs for condition
Lubricate handbrake mechanical linkage and cable guides (lever pivot).
Check brake servo hose(s) for security and condition

Wheels and Tyres
Check that tyres comply with manufacturers specification
Check/adjust tyre pressures including spare
Check tyres for tread depth and visually for external cuts in fabric, exposure of ply or cord structure, lumps or bulges.
Check tightness of road wheel fastenings
IMPORTANT: If tyres do not conform with legal requirements, report to owner.

Body
Lubricate all locks and hinges (Not steering lock)
Check condition and security of seats and seat belts
Check rear view mirror(s) for cracks.
Ensure cleanliness of controls, door handles, steering wheel
Check operation of seat belt warning system
Check operation of seat belt inertia reel mechanism.

Electrical
Check function of original equipment i.e. interior and exterior lamps, horns, wipers and warning indicators
Check/top up battery electrolyte
Clean and grease battery connections
Check/adjust headlamp alignment
Check, if necessary renew wiper blades
Check output of charging system

General
Road/roller test and check function of all instrumentation
Report additional work required
For further maintenance regarding brakes, refer to page (55)

'D' SERVICE

Engine
Check for oil leaks
Renew engine oil filter
Renew engine oil
Renew carburetter/air-intake air cleaner element(s)
Top up carburetter piston damper(s)
Check/adjust carburetter idle settings
Check cooling/heater systems for leaks and hoses for security and condition
Check/top up cooling system
Check/adjust operation of all washers and top up reservoirs
Check driving belts, adjust or renew
Lubricate accelerator control linkage and pedal pivot — check operation
Check/top up air-conditioning compressor fluid
Check crankcase breathing and evaporating loss systems. Check hoses/pipes and restrictors for blockage, security and condition.
Check air-intake temperature control system.
Check/adjust deceleration by-pass valve
Check air injection system hoses/pipes for security.

Ignition
Renew spark plugs.
Lubricate distributor
Check/adjust ignition timing using electronic equipment
Check ignition wiring for fraying, chafing and deterioration
Clean distributor cap, check for cracks and tracking
Check coil performance on oscilloscope
Check security of distributor vacuum unitline and operation of vacuum unit (California only)

Exhaust and Fuel Pipes
Check fuel system for leaks, pipes and unions for chafing and corrosion
Check exhaust system for leaks and security
Check condition of fuel filler cap seal

Transmission
Check for oil leaks
Check/top up gearbox oil
Check/top up rear axle/final drive oil.
Check/top up automatic gearbox fluid
Check/top up clutch fluid reservoir
Lubricate automatic gearbox exposed selector linkage
Check clutch pipes for leaks and chafing
Check tightness of propeller shaft coupling bolts
Lubricate clutch pedal pivots

'D' SERVICE — *continued*

Steering and Suspension
Lubricate steering rack and pinion
Check condition and security of steering unit joints and gaiters
Check security of suspension fixings
Adjust front hub bearing end-float
Check/adjust front (and rear) wheel alignment
Check steering rack/gear for oil/fluid leaks.
Check shock adsorbers for fluid leaks

Brakes
Check visually, hydraulic pipes and unions for chafing, leaks and corrosion
Check/top up brake fluid reservoir(s)
Check handbrake operation; adjust to manufacturers instructions
Check footbrake operation; adjust to manufacturers instructions (Self adjusting)
Inspect brake linings/pads for wear, drums/discs for condition
Lubricate handbrake mechanical linkage and cable guides (lever pivot).
Lubricate brake pedal pivot(s)
Check brake servo hose(s) for security and condition
At 37,500 miles, (60,000 kms)/3 year intervals, whichever is sooner:- Renew air filter in brake servo unit.

Wheels and Tyres
Check that tyres comply with manufacturers specification
Check/adjust tyre pressures including spare
Check tyres for tread depth and visually for external cuts in fabric, exposure of ply or cord structure, lumps or bulges
Check tightness of road wheel fastenings
IMPORTANT: If tyres do not conform with legal requirements, report to owner.

Body
Lubricate all locks and hinges (Not steering lock)
Check condition and security of seats and seat belts
Check rear view mirror(s) for cracks.
Ensure cleanliness of controls, door handles, steering wheel
Check operation of all door, bonnet and boot locks
Check operation of window controls
Check operation of seat belt warning system
Check operation of seat belt inertia reel mechanism.

Electrical
Check function of original equipment i.e. interior and exterior lamps, horns, wipers and warning indicators
Check/top up battery electrolyte
Clean and grease battery connections
Check/adjust headlamp alignment
Check, if necessary renew wiper blades
Check output of charging system

'D' SERVICE — *continued*

General
Road/roller test and check function of all instrumentation
Report additional work required
For further maintenance regarding brakes, refer to page (55)

'E' SERVICE

This service comprises the 'D' Service Operations, except that in addition the following operations should be performed at 50,000 miles only:
Check E.G.R. system.
Renew adsorption canister
Renew catalytic converter (25,000 California only).

EMISSION AND EVAPORATIVE CONTROL SYSTEMS

Triumph TR7 — Canadian Specification Only

All Triumph models entering the Canadian market incorporate efficient emission control systems. These systems enable the vehicles to conform with all current regulations governing the emission of hydrocarbons, carbon monoxide, nitric oxide and the emission of fuel, by evaporation, from the fuel delivery system.

Fuel

The TR7 (Canada) performs efficiently on fuels of 91 octane (Regular fuel).

NOTE: THE ENGINE IS DESIGNED TO USE EITHER LEADED OR UNLEADED FUEL.

EMISSION CONTROL SYSTEMS — SERVICING

The importance of servicing at the correct intervals cannot be overstressed, improvements in design and manufacturing techniques count for nothing if the servicing standards are not upheld.

Routine servicing, carried out at the mileage intervals quoted in the Maintenance Summary helps to prevent deterioration of the systems.

40

CARBURETTERS

The engine is fitted with two ZENITH CDSEV carburetters which are prime components of the emission system and great care is exercised during the manufacture and initial adjustment of these instruments. Adjustment of the carburetters necessitates the use of specialised equipment and must be carried out by a British Leyland Dealer or other established and qualified service facility.

Carburetter Controls

The throttle rod linkage will not require adjustments during normal operation. To ensure complete throttle closure a degree of 'lost motion' or slackness is incorporated into the linkage; no attempt must be made to eliminate this.

Carburetter Servicing Schedules

To maintain the carburetter at peak efficiency, regular servicing as detailed in the 'Maintenance Summary' is essential. The appropriate servicing operations should be performed by authorized dealers, who are trained in the use of the special equipment needed.

Crankcase Breathing (Fig. 1)

Crankcase breathing and evacuation of 'blow by' gases is achieved by utilizing the characteristic partial vacuum in the constant depression carburetter. By this method crankcase emissions are burned in the engine combustion process (Fig. 1). A wire gauze strainer in the engine top cover acts as an oil separator/flame trap.

Anti Run-On Valve (Fig. 1)

This valve prevents the 'running-on' of the engine after the ignition is switched off when, due to the heat of the engine, a condition of compression ignition is set up.

The method of achieving a cut-off is by applying a slight 'vacuum' to the float chamber of the carburetter when the ignition is switched off.

System Description

When the ignition is switched off a solenoid is activated which operates a valve that seals off the inlet to the bottom of the intake canister. With the inlet sealed a connection to the inlet manifold applies a partial vacuum to the canister and consequently to the float chamber via an interconnecting pipe. The vacuum thus applied is sufficient to prevent fuel being drawn into the engine. When the engine has stopped and the oil pressure drops to zero the solenoid is deactivated and the engine is thus ready again for operation.

Servicing

The system requires no servicing other than checking for deterioration and leak free connection to the system piping.

Function Checks

If the system is not working then it will be apparent by engine running-on. A system check can be made by applying current to the solenoid which, if working correctly, will stop the engine.

EVAPORATIVE CONTROL SYSTEM

The evaporative control system uses an activated adsorption canister through which the fuel tank is vented. The following are features of the evaporative control system (see Fig. 2).

(1) Both carburetter float chambers are vented to the engine during open throttle conditions and to the adsoption canister at closed throttle.
(2) The constant depression of the carburetters is used to induce a purge condition through the adsorption canister via the running-on valve. The crankcase breathing is also linked into the system.
(3) A separater tank prevents fuel surges from reaching the canister and thus saturating the system.
(4) The fuel filler cap is sealed to prevent evaporative losses.
(5) The fuel tank filler tube extends into the tank to prevent complete filling and so allow for expansion of fuel in hot weather.

WARNING: The use of compressed air to clean an adsorption canister or clear a blockage in the evaporative system is very dangerous. An explosive gas present in a partly saturated canister may be ignited by the heat generated when compressed air passes through the canister.

Servicing

Minimal servicing is required on the evaporation control system apart from renewing the adsorption canister at the specified interval, and checking visually the security of piping on the system.

1. Crankcase purge line
2. Carburetter float chamber vent pipe
3. Canister purge line
4. Charcoal canister
5. Fuel tank vent pipe
6. Anti run-on valve
7. Manifold vacuum line
8. Electrical connections for anti run-on valve
9. Purge air to canister
10. Flame arrestor

A. $^{3}/_{32}$in. restrictor
B. $^{5}/_{32}$in. restrictor

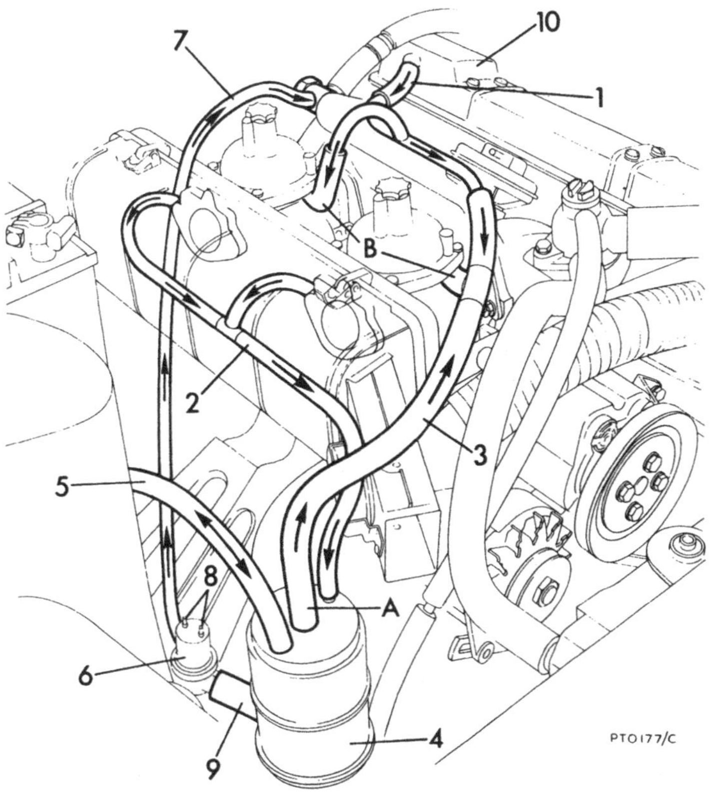

Fig. 1 Crankcase and Canister Purge System

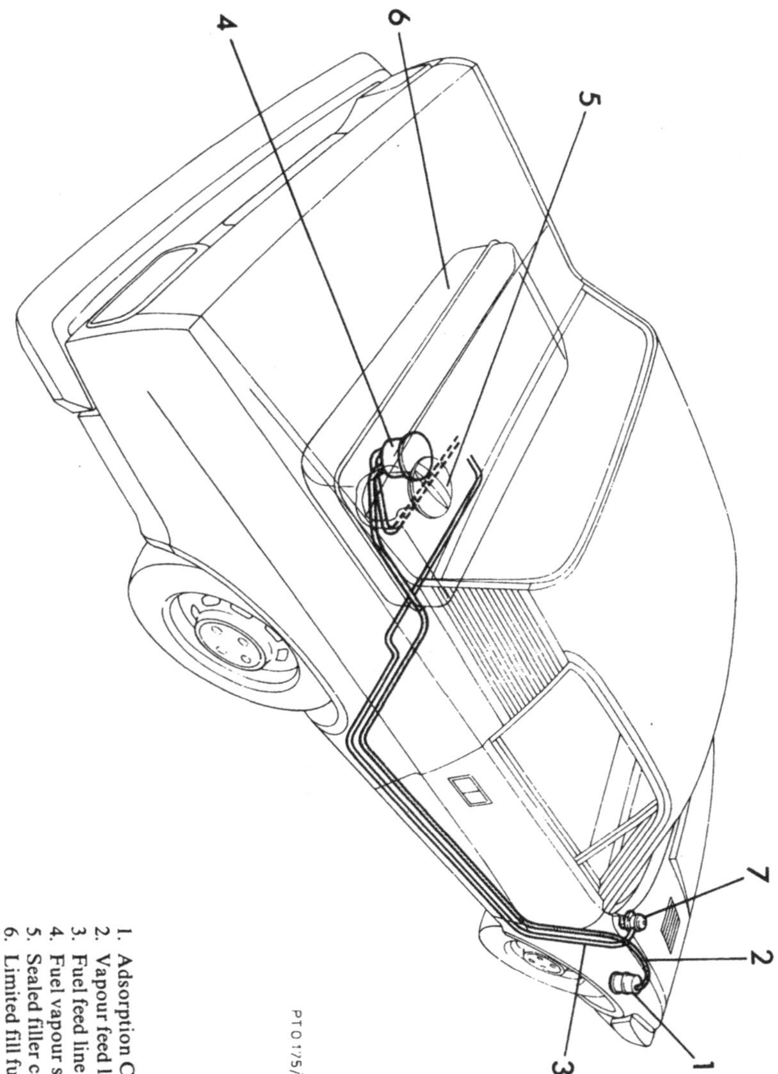

Fig. 2 Evaporative Emission Control System

1. Adsorption Canister
2. Vapour feed line
3. Fuel feed line
4. Fuel vapour separator
5. Sealed filler cap
6. Limited fill fuel tank
7. Fuel pump

EXHAUST GAS RECIRCULATION (E.G.R.) SYSTEM

System Description

To reduce the nitric oxides content in the exhaust, the peak combustion temperatures are lowered by recirculating a controlled quantity of the exhaust gases through the combustion process.

The E.G.R. valve is mounted on the inlet manifold. A control signal, taken from a throttle edge tapping in the carburetter, gives no recirculation at idle or full load, but does allow an amount of recirculation, dependent on the vacuum signal and a metering profile on the valve, Fig. 3 under part load conditions.

Servicing

After the completion of the initial 1000 miles check the security of the E.G.R. valve operating lines. On the completion of subsequent 12,500 mile periods, the following checks are necessary:

(1) Security of the E.G.R. valve operating lines.
(2) Operation of the E.G.R. valve
(3) Clean the valve and pipework

NOTE: It is recommended that this work is carried out by a British Leyland Dealer. As E.G.R. Service Repair Kit can be purchased from a British Leyland Dealer. Complete details on the servicing of the E.G.R. system are detailed in Repair Operations Manual Pt. No. RTC 9200/77 obtainable from British Leyland Dealers.

42

1. E.G.R. valve
2. Inlet manifold
3. Exhaust port
4. Choke cam
5. E.G.R. cut-out valve
6. Petrol trap
7. Vacuum source — throttle edge tapping

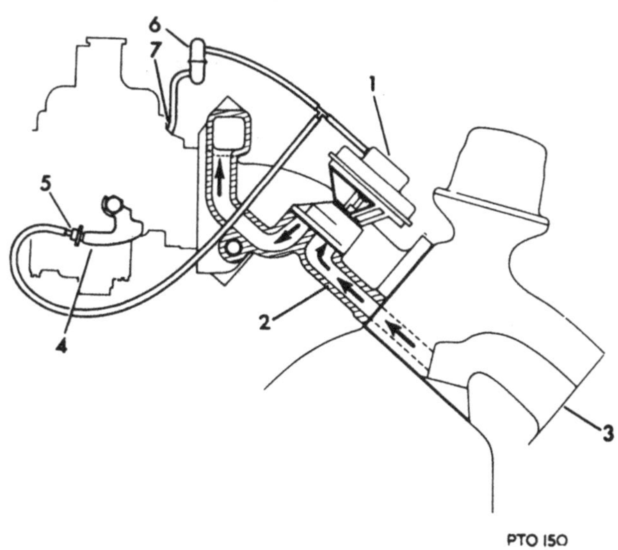

Fig. 3 Exhaust Gas Recirculation System

MAINTENANCE SUMMARY

TR7 — CANADA — 1977

The following items should be checked weekly by the driver:

Engine oil level
Brake fluid level
Radiator coolant level
Battery electrolyte level
Windshield washer reservoir level
All tyre pressures
Operation of all lights
Operation of horn
Operation of all washers and wipers

MAINTENANCE INTERVALS

Service	Miles x 1000	Monthly Intervals
A	1	1
B	3, 9, 16, 22, 34, 41, 47	3
C	6, 19, 31, 44	6
D	12·5, 37·5	12
E	25, 50	24

NOTE: The service schedules are based on an annual mileage of approximately 12,500 miles or 20,000 Kms. Should the vehicle complete substantially less miles than this per annum, it is recommended that a 'C' service is completed at six month intervals and a 'D' service at twelve month intervals.

'A' SERVICE

Engine
Check for oil leaks
Renew engine oil
Top up carburetter piston damper(s)
Check/adjust carburetter idle settings
Check cooling/heater systems for leaks and hoses for security and condition
Check/top up cooling system
Check/adjust operation of all washers and top up reservoirs.
Check driving belts, adjust or renew
Lubricate accelerator control linkage and pedal pivot — check operation
Check/adjust choke settings (Manual choke)
Check/adjust deceleration by-pass valve
Check/Adjust torque of cylinder head nuts/bolts
Check security of engine mountings
Check security of EGR valve operating lines
Check crankcase breathing and evaporative loss systems hoses for security

'A' SERVICE — *continued*

Ignition
Check/adjust ignition timing using electronic equipment
Check ignition wiring for fraying, chafing and deterioration
Check security of distributor vacuum unitline and operation of vacuum unit

Exhaust and Fuel Pipes
Check fuel system for leaks, pipes and unions for chafing and corrosion
Check exhaust system for leaks and security

Transmission
Check for oil leaks
Check/top up gearbox oil
Check/top up rear axle/final drive oil.
Check/top up automatic gearbox fluid
Check/top up clutch fluid reservoir
Check clutch pipes for leaks and chafing

Steering and Suspension
Check condition and security of steering unit joints and gaiters
Check security of suspension fixings
Check/adjust front (and rear) wheel alignment
Check steering rack/gear for oil/fluid leaks.
Check shock absorbers for fluid leaks

Brakes
Check visually, hydraulic pipes and unions for chafing, leaks and corrosion
Check/top up brake fluid reservoir(s)
Check handbrake operation; adjust to manufacturers instructions
Check footbrake operation; adjust to manufacturers instructions (Self adjusting)
Check brake servo hose(s) for security and condition

Wheels and Tyres
Check/adjust tyre pressures including spare
Check tightness of road wheel fastenings
Check tyres for external cuts in tyre fabric, exposure of ply or cord structure, lumps or bulges

IMPORTANT: If tyres do not conform with legal requirements, report to owner.

Body
Lubricate all locks and hinges (Not steering lock)
Check condition and security of seats and seat belts
Ensure cleanliness of controls, door handles, steering wheel
Check operation of all door, bonnet and boot locks
Check operation of window controls
Check operation of seat belt inertia reel mechanism.
Check tightness of sub-frame/body mountings

'A' SERVICE — *continued*

Electrical
Check function of original equipment i.e. interior and exterior lamps, horns, wipers and warning indicators
Check/top up battery electrolyte
Check/adjust headlamp alignment
Check output of charging system

General
Road/roller test and check function of all instrumentation

'B' SERVICE

Engine
Check for oil leaks
Check/top up engine oil
Check cooling/heater systems for leaks and hoses for security and condition
Check/top up cooling system
Check/adjust operation of all washers and top up reservoirs

Exhaust and Fuel Pipes
Check fuel system for leaks, pipes and unions for chafing and corrosion
Check exhaust system for leaks and security

Transmission
Check for oil leaks
Check/top up clutch fluid reservoir
Check clutch pipes for leaks and chafing

Steering and Suspension
Check condition and security of steering unit joints and gaiters
Check steering rack/gear for oil/fluid leaks.
Check shock absorbers for fluid leaks

Brakes
Check visually, hydraulic pipes and unions for chafing, leaks and corrosion
Check/top up brake fluid reservoir(s)
Inspect brake pads for wear and discs for condition
Check brake servo hose(s) for security and condition

Wheels and Tyres
Check that tyres comply with manufacturers specification
Check/adjust tyre pressures including spare
Check tyres for tread depth and visually for external cuts in fabric, exposure of ply or cord structure, lumps or bulges
Check tightness of road wheel fastenings

IMPORTANT: If tyres do not conform with legal requirements, report to owner.

'B' SERVICE — *continued*

Body
Check condition and security of seats and seat belts
Check rear view mirror(s) for cracks.
Ensure cleanliness of controls, door handles, steering wheel
Check operation of seat belt warning system
Check operation of seat belt inertia reel mechanism.

Electrical
Check function of original equipment i.e. interior and exterior lamps, horns, wipers and warning indicators
Check/top up battery electrolyte
Check/adjust headlamp alignment
Check, if necessary renew wiper blades

General
Report additional work required

'C' SERVICE

Engine
Check for oil leaks
Renew engine oil filter
Renew engine oil
Check cooling/heater systems for leaks and hoses for security and condition
Check/top up cooling system
Check/adjust operation of all washers and top up reservoirs
Lubricate accelerator control linkage and pedal pivot — check operation

Exhaust and Fuel Pipes
Check fuel system for leaks, pipes and unions for chafing and corrosion
Check exhaust system for leaks and security

Transmission
Check for oil leaks
Check/top up gearbox oil
Check/top up rear axle/final drive oil.
Check/top up automatic gearbox fluid
Check/top up clutch fluid reservoir
Lubricate automatic gearbox exposed selector linkage
Check clutch pipes for leaks and chafing

'C' SERVICE — *continued*

Steering and Suspension
Lubricate steering rack and pinion
Check condition and security of steering unit joints and gaiters
Check/adjust front (and rear) wheel alignment
Check steering rack/gear for oil/fluid leaks.
Check shock absorbers for fluid leaks

Brakes
Check visually, hydraulic pipes and unions for chafing, leaks and corrosion
Check/top up brake fluid reservoir(s)
Check handbrake operation; adjust to manufacturers instructions
Check footbrake operation; adjust to manufacturers instructions (Self adjusting).
Inspect brake linings/pads for wear, drums/discs for condition
Lubricate handbrake mechanical linkage and cable guides (lever pivot).
Check brake servo hose(s) for security and condition

Wheels and Tyres
Check that tyres comply with manufacturers specification
Check/adjust tyre pressures including spare
Check tyres for tread depth and visually for external cuts in fabric, exposure of ply or cord structure, lumps or bulges
Check tightness of road wheel fastenings
IMPORTANT: If tyres do not conform with legal requirements, report to owner.

Body
Lubricate all locks and hinges (Not steering lock)
Check condition and security of seats and seat belts
Check rear view mirror(s) for cracks.
Ensure cleanliness of controls, door handles, steering wheel
Check operation of seat belt warning system
Check operation of seat belt inertia reel mechanism.

Electrical
Check function of original equipment i.e. interior and exterior lamps, horns, wipers and warning indicators
Check/top up battery electrolyte
Clean and grease battery connections
Check/adjust headlamp alignment
Check, if necessary renew wiper blades
Check output of charging system

General
Road/roller test and check function of all instrumentation
Report additional work required
For further maintenance regarding brakes, refer to page (55)

'D' SERVICE

Engine
Check for oil leaks
Renew engine oil filter
Renew engine oil
Renew carburetter/air-intake air cleaner element(s)
Top up carburetter piston damper(s)
Check/adjust carburetter idle settings
Check cooling/heater systems for leaks and hoses for security and condition
Check/top up cooling system
Check/adjust operation of all washers and top up reservoirs
Check driving belts, adjust or renew
Lubricate accelerator control linkage and pedal pivot — check operation
Check/adjust choke settings (Manual choke)
Check/top up air-conditioning compressor fluid
Check crankcase breathing and evaporating loss systems. Check hoses/pipes and restrictors for blockage, security and condition
Check air-intake temperature control system.
Renew adsorption canister
Check/adjust deceleration by-pass valve
Check air injection system hoses/pipes for security.

Ignition
Clean/adjust spark plugs
Lubricate distributor
Check/adjust ignition timing using electronic equipment
Check ignition wiring for fraying, chafing and deterioration
Clean distributor cap, check for cracks and tracking
Check coil performance on oscilloscope
Check security of distributor vacuum unitline and operation of vacuum unit

Exhaust and Fuel Pipes
Check fuel system for leaks, pipes and unions for chafing and corrosion
Check exhaust system for leaks and security
Check condition of fuel filler cap seal

Transmission
Check for oil leaks
Check/top up gearbox oil
Check/top up rear axle/final drive oil.
Check/top up automatic gearbox fluid
Check/top up clutch fluid reservoir
Lubricate automatic gearbox exposed selector linkage
Check clutch pipes for leaks and chafing
Check tightness of propeller shaft coupling bolts
Lubricate clutch pedal pivots

'D' SERVICE — *continued*

Steering and Suspension
Lubricate steering rack and pinion
Check condition and security of steering unit joints and gaiters
Check security of suspension fixings
Adjust front hub bearing end-float
Check/adjust front (and rear) wheel alignment
Check steering rack/gear for oil/fluid leaks.
Check shock absorbers for fluid leaks

Brakes
Check visually, hydraulic pipes and unions for chafing, leaks and corrosion
Check/top up brake fluid reservoir(s)
Check handbrake operation; adjust to manufacturers instructions
Check footbrake operation; adjust to manufacturers instructions (Self adjusting)
Inspect brake linings/pads for wear, drums/discs for condition
Lubricate handbrake mechanical linkage and cable guides (lever pivot).
Lubricate brake pedal pivot(s)
Check brake servo hose(s) for security and condition
At 37,500 miles, (60,000 kms)/3 year intervals, whichever is sooner:- Renew air filter in brake servo unit.

Wheels and Tyres
Check that tyres comply with manufacturers specification
Check/adjust tyre pressures including spare
Check tyres for tread depth and visually for external cuts in fabric, exposure of ply or cord structure, lumps or bulges
Check tightness of road wheel fastenings
IMPORTANT: If tyres do not conform with legal requirements, report to owner.

Body
Lubricate all locks and hinges (Not steering lock)
Check condition and security of seats and seat belts
Check rear view mirror(s) for cracks.
Ensure cleanliness of controls, door handles, steering wheel
Check operation of all door, bonnet and boot locks
Check operation of window controls
Check operation of seat belt warning system
Check operation of seat belt inertia reel mechanism.

Electrical
Check function of original equipment i.e. interior and exterior lamps, horns, wipers and warning indicators
Check/top up battery electrolyte
Clean and grease battery connections
Check/adjust headlamp alignment
Check, if necessary renew wiper blades
Check output of charging system

'D' SERVICE — *continued*

General

Road/roller test and check function of all instrumentation

Report additional work required

For further maintenance regarding brakes, refer to page (55)

'E' SERVICE

This service comprises the 'D' Service Operations except that in addition the following operations should be performed at 50,000 miles only:

Renew adsorption canister

APPROVED AUTOMATIC TRANSMISSION FLUIDS

(The products listed are not in order of preference)

Supplier	Product
B.P.	B.P. Autran B
Esso Petroleum Co.	Esso Glide
Shell	Donax T7 S.7995
Mobil Oil Co.	Mobil ATF 210
Castrol	Castrol TQF
Regent/Texaco/Caltex	Texamatic 6991, Type F
Alex Duckham & Co.	Duckhams Q Matic
Petrofina (UK) Ltd.	Fina Purifimatic 33F
Filtrate Limited	Filtrate auto. trans. fluid F
Amoco (UK) Ltd.	Amoco ATF Type F
Sun Oil Co.	Sunamatic 126
Antar	Transantar 33
Mitsubishi Oil Co.	Diamond ATF Special F
Idemitsu Kosan Co. Ltd.	Appolloil Mission F
Automatic Transmission Services	Automatic Transmission Fluid
Total	Total ATF 33
Veedol GmbH	Veedol ATF F
Gulf Oil (GB) Ltd.	Gulf ATF, Type F
Rudolf Fuchs Mineraloelwerk KG	Fuchs Automatic TF
Tiona Petroleum Co.	Flo-Matic ATF Type F
Valvoline Oil Co.	Valvomatic ATF Type FA
Aral Aktiengesellschaft	Aral Getriebeol ATF 33
Motul S.A.	Motul Omatic 126F
Elf Union	Elfmatic F
Wenzel & Weidmann GmbH	Ecubsol ATF 33-F
Pennzoil Company	Pennzoil Hydra Flo
Century Oil Co.	Transfluid F

RECOMMENDED LUBRICANTS AND ANTI-FREEZE SOLUTIONS OVERSEAS

(4 speed transmission)

	Service Classification	SAE Viscosity	Ambient Temperature °C (-30 -20 -10 0 +10 +20)
Engine	API - SE	5W/20	◄────
		5W/30	◄────
		5W/40	◄────
		10W/30	────►
		10W/40	────►
		10W/50	────
		15W/40	────►
		15W/50	────►
		20W/40	────►
		20W/50	────►
Final Drive	API - GL5	Hypoid 90	────►
	MIL-L-2105B	Hypoid 80W	◄────
Manual Gearbox (Top-up only)‡	API - GL5	Hypoid 90	────►
	MIL-L-2105B	Hypoid 80W	◄────
Automatic Gearbox	ATF Type F [1975–1978] ATF Type G [1979 and later]		◄────►
Steering Rack Hubs & Chassis Grease Points	NLGI - 2 Multipurpose Grease		
Brake & Clutch Fluid	Castrol-Girling Universal brake fluid, Lockheed 329S brake fluid, Unipart 550 brake fluid (or other brake fluids DOT.3 or SAE J1703d specifications).		
Antifreeze*	Permanent type ethylene glycol base with suitable inhibitor for mixed metal engines		
Windshield Washer	Windshield Washer Antifreeze Fluid (Proprietary Brands)		

‡ See service instructions for refill if unit is drained completely.

* For anti-freeze proportions refer to page 39.

47

RECOMMENDED LUBRICANTS AND ANTI-FREEZE SOLUTIONS OVERSEAS

(5 speed transmission)

	Service Classification	SAE Viscosity	Ambient Temperature °C (-30 -20 -10 0 +10 +20)
Engine	API - SE	5W/20	◄────
		5W/30	◄────
		5W/40	◄────
		10W/30	────►
		10W/40	────►
		10W/50	────►
		15W/40	────►
		15W/50	────►
		20W/40	────►
		20W/50	────►
Final Drive &	API - GL5	Hypoid 90	────►
	MIL-L-2105B	Hypoid 80W	◄────
Manual Gearbox ‡	API - GL4	EP 75W	◄────►
	MIL-L-2105B	[1980 and later] use ATF Type G (M2C33G)	
Automatic Gearbox	ATF Type F [1975–1978] ATF Type G [1979 and later]		◄────►
Hubs & Chassis Grease Points Steering Rack	NLGI - 2 Multipurpose Grease		
Brake & Clutch Fluid	Brake Fluids must have a minimum boiling point of 260°C (500°F) and comply with FMV SS 116, DOT 3		
Antifreeze*	Permanent type ethylene glycol base with suitable inhibitor for mixed metal engines		
Windshield Washer	Windshield Washer Antifreeze Fluid (Proprietary Brands)		

* For anti-freeze proportions refer to page 39.

GENERAL SPECIFICATION

Engine

Number of cylinders	4 in line — inclined 45°
Bore of cylinders	3·56 in (90·3 mm)
Stroke of crankshaft	3·07 in (78 mm)
Engine capacity	122 in³ (1998 cm³)

Cooling System

Type	Pressurized 'No loss' system incorporating a separate expansion tank.
Circulation	Impellor type pump. Thermostatically controlled flow.
Fan — number of blades	13 blades
— diameter	14 in (356 mm)
Pressure cap	15 lb/in² (1·05 Kgs/cm²)

Lubrication System

Pump	High capacity rotor pump
Oil filter	'Full-flow' type with replaceable element
Oil filler	Sealed cap on camshaft cover

Fuel System

Pump (carburetter models)	Mechanically operated diaphragm type
(fuel injection models)	Electrical
Carburetter [CANADA and U.S.A. models without catalytic converter — 1975–1976]	Twin side-draught Stromberg 175 CDS(E)V
[1977 models]	Twin side-draught Stromberg 175 CDFEVX
[1978 and later FEDERAL and CALIFORNIA models]	Twin side-draught Stromberg 175 CDFVX—Auto-choke
[1975–1976 models with catalytic converter specifications only]	Single side-draught Stromberg 175 CD4VT
Air Cleaner	Combined air cleaner and silencer with replaceable paper element. Temperature-controlled inlet air flow
Fuel Injection	Bosch, Analog, Electronic

Ignition System

Coil	6 volt with ballast resistor incorporated in the wiring harness: Type 15C6
Distributor [1977 and later FEDERAL models]	Lucas electronic, with centrifugal advance controls
Distributor [1977 and later CALIFORNIA models and 1975 & 1976 FEDERAL models]	Lucas electronic, with centrifugal advance and vacuum retard controls
Distributor air-gap	0·014 in - 0·016 in (0·4 mm)
Rotation	Anti-clockwise-viewed from top of rotor arm
Firing order	Numbered from front of engine, i.e. timing chain end — 1:3:4:2

Spark plug type [1975 models]	N-11Y (Champion)
Spark plug type [1976 and later models] [can also be used in 1975 engines]	N-12Y (Champion)
Spark plug gap	0·025 in (0·63 mm)
Ignition timing - FEDERAL (static)	10° BTDC
- FEDERAL (dynamic - 800±100 r.p.m.).	10° BTDC
Ignition timing - CALIFORNIA (static)	10° BTDC
- CALIFORNIA (dynamic - 800±100 r.p.m.)	2° ATDC

Transmission (Manual)

Clutch	Single dry plate diaphragm type
— diameter	8.5in (216mm)
Gearbox	Synchromesh on all forward gears
Final drive ratio	3·63:1

	Top	3rd	2nd	1st	Rev
Gearbox ratios to 1	1·00	1·25	1·78	2·65	3·05
Overall ratios to 1	3·63	4·56	6·47	9·62	11·08
Engine r.p.m. at 10 m.p.h.	554	696	988	1471	
Engine r.p.m. at 10 Km.p.h.	344	432	614	913	
Road speed at 1000 engine r.p.m.					
m.p.h.	18				
Km.p.h.	29				
Maximum recommended road speed in intermediate gears corresponding to engine speed of 6500 rpm					
m.p.h.		93	66	44	
Km.p.h.		150	106	71	

Transmission (Optional, 5 speed).

Gearbox (5 speed)	Synchromesh on all forward gears
Final drive ratio	3.9:1 [1981 models: 3.45:1]

	5th	4th	3rd	2nd	1st	Rev.
Gearbox ratios to 1	0.83	1.00	1.40	2.09	3.32	3.43
Overall ratios to 1	3.24	3.90	5.46	8.15	12.95	13.38
Engine r.p.m. at 10 m.p.h.	481	579	810	1209	1921	
Engine r.p.m. at 10 Km.p.h.	299	359	503	751	1193	
Road speed at 1000 engine r.p.m.						
m.p.h.	17					
Km.p.h.	28					
Maximum recommended road speed in intermediate gears corresponding to engine speed of 6500 rpm						
m.p.h.			80	54	34	
Km.p.h.			129	87	54	

Gearbox — Automatic —
Borg Warner Type 65 [1975–1980 models]*

	3rd	2nd	1st	Rev
Transmission conversion range	1.00	1.45	2.39	2.09
Overall ratios to 1	3.27	4.74	7.83	6.85
Road speed corresponding to 1000 engine r.p.m.				
m.p.h.	20.0			
Km.p.h.	32.0			

AUTOMATIC TRANSMISSION SHIFT SPEEDS

Throttle Position	*Zero Throttle*	*Light Throttle*		*Part Throttle*
Selector	1	2	D	D
Shift	2—1	1—2	2—3	3—2
Road speed m.p.h. / Km.p.h.	30-38 48-61	8-12 13-19	12-16 23-26	30-40 48-64

Throttle Position	*Kickdown*						
Selector	D	D	D	1	D	2	2
Shift	1—2	2—3	3—2	2—1	3—1	1—2	2—1
Road speed m.p.h. / Km.p.h.	38-44 61-71	65-71 105-114	58-66 93-106	42-50 68-81	30-38 48-61	38-44 61-71	30-38 48-61

*[**NOTE:** 1981 models use Borg Warner Type 66 Automatic Gearbox. See Section 04 in Repair Operations Manual (Part 2 of this book) for data.]

49

Electrical System

Voltage	12
Polarity	Negative ground
Fuse box — fuses	3 x 50 amp. 2 x 35 amp. 1 x 15 amp.
Battery — voltage	12 Volt
Amps. capacity at 20hr. discharge rate	40
Alternator [1975 models]	Lucas 16 ACR or 20 ACR
Alternator [1976 models]	Lucas 17 ACR, 23 ACR, or 25 ACR
Alternator [1977 & later models]	Lucas 17 ACR or 20 ACR
Starter motor	Lucas 2M100PE

Brake System

Front — type	Caliper disc
Diameter	9.75in (248mm)
Rear — type	Drum — self adjusting shoes
Diameter/width	8 x 1.5in (204 x 38mm) — 4 speed and Auto. 9 x 1.75in (299 x 44.5mm) — 5 speed
Brake servo — type	Direct acting
Nominal boost ratio	2·2:1

Chassis Data

Wheelbase	85 in (2160 mm)
Track — front	55·5 in (1409 mm)
Track — rear	55·3 in (1404 mm)
Front wheel alignment (Kerb condition)	Toe-in 0-0·0625 in (0-1·6 mm)
Turning circle — between kerbs	29·0 ft (8·8 metres)
Turning circle — between walls	31·5 ft (9·5 metres)
Ground clearance	3.5 to 4.5 in (90 to 114 mm), depending on market, model year, and load condition.

Overall Dimensions

Length	164.3 in (4173 mm) [1981: 165.4 in (4203 mm)]
Width	66.2 in (1681 mm)
Height unladen	49.4 to 49.9 in (1255 to 1268 mm), depending on market, model year, and load condition

Weights (approximate ranges)

*Shipping	2158 lbs (979 kgs)	2372 lbs (1076 kgs)
*Kerb	2249 lbs (1020 kgs)	2511 lbs (1139 kgs)
*Gross vehicle weight	2820 lbs (1279 kgs)	2954 lbs (1340 kgs)†
Roof rack capacity	112 lbs (51 kgs)	
Max. axle load—front	1499 lbs (680 kgs)	1576 lbs (715 kgs)†
Max. axle load—rear	1349 lbs (612 kgs)	1466 lbs (665 kgs)†
Towing capacity braked trailer	1680 lbs (762 kgs)	2240 lbs (1016 kgs)†

*[**NOTE:** These figures must be taken as a guide only and vary according to market, model year and equipment fitted (except where specified).]
†[1981 models]

Capacities (nominal)	**U.S.A.**	**Imperial**	**Metric**
Fuel tank (U.S.A.)	14·4 gals	12·0 gals	54·5 litres
Engine sump and filter	9·5 pts	8·0 pts	4·5 litres
Engine sump — drain/refill	8·4 pts	7·0 pts	4·0 litres
Gearbox from dry	2·4 pts	2·0 pts	1·1 litres
Gearbox from dry (5 speed)	3.3 pts	2·75 pts	1.5 litres
Rear axle from dry	2·7 pts	2·25 pts	1·3 litres
Rear axle from dry (5 speed)	2·4 pts	2.0 pts	1.1 litres
Cooling system, including reservoir and heater or air cond.	15·6 pts	13 pts	7·4 litres
Automatic transmission with oil cooler	11·4 pts	9·5 pts	5·4 litres

Wheels

Diameter	13 in
Rim section	5½J

Tyres
(4 speed gearbox)

Inflation Pressures							
Size	Loading Condition	Front			Rear		
		Bars	kg/cm^2	lb/in^2	Bars	kg/cm^2	lb/in^2
Radial ply — tubeless 175/70 SR 13	All	1·7	24	1·6	2·0	28	1·9

(5 speed gearbox)

Inflation Pressures							
Size	Loading Condition	Front			Rear		
		Bars	kg/cm^2	lb/in^2	Bars	kg/cm^2	lb/in^2
Radial ply — tubeless 175/70 SR 13	All	1·7	24	1·6	2·0	28	1·9

TRIUMPH TR7
REPAIR OPERATION MANUAL

PART 2

SPECIFICATION

Purchasers are advised that the specification details set out in this Manual apply to a range of vehicles and not to any one. For the specification of a particular vehicle, purchasers should consult their Distributor or Dealer.

The Manufacturers reserve the right to vary their specifications with or without notice, and at such times and in such manner as they think fit. Major as well as minor changes may be involved in accordance with the Manufacturer's policy of constant product improvement.

Whilst every effort is made to ensure the accuracy of the particulars contained in this Manual, neither the Manufacturer nor the Distributor or Dealer, by whom this Manual is supplied, shall in any circumstances be held liable for any inaccuracy or the consequences thereof.

[**NOTE:** Any material added to this *Repair Operation Manual* by Robert Bentley, Inc. has been indicated by the use of brackets ([ ]). Any of this material that has been taken directly from British Leyland publications has been indicated by the insertion of a source reference at the bottom of the appropriate page.]

CONTENTS

[**NOTE:** From page 53 to page 533 of this Manual, when a reference on one page is made to information found on another page, numbers (usually six) are used. These numbers are found in the lower outside margin of each page.

For example, on page 151 (numbered 12.65.50 in the lower right-hand margin), we are referred in step 1 to 12.65.01. This number reference is found in the lower right-hand margin of page 145.]

INTRODUCTION

The purpose of this manual is to assist skilled mechanics in the efficient repair and maintenance of Triumph TR7 vehicles. Using the appropriate service tools and carrying out the procedures as detailed will enable the operations to be completed in the time stated in the 'Repair Operations Times'.

Indexing

The contents pages list the titles and reference numbers of the divisions in alphabetical order.

Operation Numbering

A master index of numbered operations has been compiled for universal application to all vehicles manufactured by the British Leyland Motor Corporation and, therefore, because of the different specifications of various models, continuity of the numbering sequence cannot be maintained throughout this manual.

Each operation described in this manual is allocated a number from the master index and cross-refers with an identical number in the 'Repair Operations Times'. The number consists of six digits arranged in three pairs.

Each instruction within an operation has a sequence number and, to complete the operation in the minimum time, it is essential that the instructions are performed in numerical sequence commencing at 1 unless otherwise stated. Where applicable, the sequence numbers identify the relevant components in the appropriate illustration.

Service Tools

Where performance of an operation requires the use of a service tool, the tool number is quoted under the operation heading and is repeated in, or following, the instruction involving its use. An illustrated list of all necessary tools is included in Division 99.

References

References to the left- or right-hand side in the manual are made when viewing from the rear. With the engine and gearbox assembly removed, the 'timing cover' end of the engine is referred to as the front. A key to abbreviations and symbols is given on page 01-9.

REPAIRS AND REPLACEMENTS

When service parts are required it is essential that only genuine Triumph or British Leyland Unipart replacements are used.

Attention is particularly drawn to the following points concerning repairs and the fitting of replacement parts and accessories.

Safety features embodied in the car may be impaired if other than genuine parts are fitted. In certain territories, legislation prohibits the fitting of parts not to the vehicle manufacturer's specification. Torque wrench setting figures given in the Repair Operation Manual must be strictly adhered to. Locking devices, where specified, must be fitted. If the efficiency of a locking device is impaired during removal it must be renewed. Owners purchasing accessories while travelling abroad should ensure that the accessory and its fitted location on the car conform to mandatory requirements in their country of origin.

The car warranty may be invalidated by the fitting of other than genuine British Leyland parts. All British Leyland Unipart replacements have the full backing of the factory warranty.

British Leyland Distributors and Dealers are obliged to supply only genuine service parts.

UNIPART

ABBREVIATIONS AND SYMBOLS

Across flats (bolt size)	A.F.
After bottom dead centre	A.B.D.C.
After top dead centre	A.T.D.C.
Alternating current	a.c.
Amperes	amp
Ampere-hour	Ah
Atmospheres	Atm
Before bottom dead centre	B.B.D.C.
Before top dead centre	B.T.D.C.
Bottom dead centre	B.D.C.
Brake horse-power	b.h.p.
Brake mean effective pressure	b.m.e.p.
British Standards	B.S.
Carbon monoxide	CO
Centigrade (Celsius)	C
Centimetres	cm
Cubic centimetres	cm^3
Cubic inches	in^3
Cycles per minute	c/min
Degree (angle)	deg. or °
Degree (temperature)	deg. or °
Diameter	dia.
Direct current	d.c.
Fahrenheit	F
Feet	ft
Feet per minuteft/min	ft/min
Fifth	5th
Fig. (illustration)	Fig.
First	1st
Fourth	4th
Gallons (Imperial)	gal.
Gallons (U.S.)	U.S. gal
Grammes (force)	gf
Grammes (mass)	g
High compression	h.c.
High tension (electrical)	h.t.
Horse-power	hp
Hundredweight	cwt
Inches	in
Inches of mercury	inHg
Independent front suspension	i.f.s.
Internal diameter	i.dia.
Kilogrammes (force)	kgf
Kilogrammes (mass)	kg
Kilogramme centimetre	kgf cm
Kilogramme metres	kgf m
Kilogrammes per square centimetre	kg/cm^2
Kilometres	km
Kilometres per hour	km/h
Kilovolts	kV
King pin inclination	k.p.i.
Left-hand	L.H.
Left-hand steering	L.H. Stg.
Left-hand thread	L.H. Thd.
Low compression	l.c.
Low tension	l.t.
Maximum	max.
Metres	m
Microfarad	mfd
Midget Edison Screw	MES
Miles per gallon	m.p.g.
Miles per hour	m.p.h.
Millimetres	mm
Millimetres of mercury	mmHg
Minimum	min.
Minus (of tolerance)	–
Minute (of angle)	'
Negative (electrical)	–
Newton Metres	Nm
Number	No.
Ohms	ohm
Ounces (force)	ozf
Ounces (mass)	oz
Ounce inch (torque)	ozf in
Outside diameter	o.dia.
Overdrive	O/D
Paragraphs	para.
Part Number	Part No.
Percentage	%
Pints (Imperial)	pt
Pints (U.S.)	U.S. pt
Plus or minus	±
Plus (tolerance)	+
Positive (electrical)	+
Pounds (force)	lbf
Pounds (mass)	lb
Pounds feet (torque)	lbf ft
Pounds inches (torque)	lbf in
Pounds per square inch	lb/in^2
Radius	r
Ratio	:
Reference	ref.
Revolutions per minute	rev/min
Right-hand	R.H.
Right-hand steering	R.H.Stg.
Second (angle)	"
Second (numerical order)	2nd
Single carburetter	SC
Society of Automobile Engineers	S.A.E.
Specific gravity	sp. gr.
Square centimetres	cm^2
Square inches	in^2
Standard	std.
Standard wire gauge	s.w.g.
Synchronizer/synchromesh	synchro
Third	3rd
Twin carburetters	TC
United Kingdom	UK
Volts	V
Watts	W
Screw threads	
American Standard Taper Pipe	N.P.T.F.
British Association	B.A.
British Standard Fine	B.S.F.
British Standard Pipe	B.S.P.
British Standard Whitworth	B.S.W.
Unified Coarse	U.N.C.
Unified Fine	U.N.F.

GENERAL SPECIFICATION DATA

ENGINE

Type	In line — inclined at 45 degrees
Number of cylinders	4
Bore	3.56 in (90,3 mm)
Stroke	3.07 in (78,0 mm)
Capacity	122 in^3 (1998 cm^3)
Valve operation	Single overhead camshaft

Crankshaft

Main journal diameter	2.1260 to 2.1265 in (54,000 to 54,013 mm)
Minimum regrind diameter	2.0860 to 2.0865 in (52,984 to 52,997 mm)
Crankpin journal diameter	1.7500 to 1.7505 in (44,450 to 44,463 mm)
Minimum regrind diameter	1.7100 to 1.7105 in (43,434 to 43,447 mm)
Crankshaft end thrust	Taken on thrust washers at centre main bearing
Crankshaft end float	0.003 to 0.011 in (0,07 to 0,28 mm)

Main bearings

Number	5
Diametrical clearance	0.0012 to 0.0022 in (0,030 to 0,058 mm)
Undersizes	0.010 in 0.020 in 0.030 in 0.040 in (0,254 mm 0,508 mm 0,762 mm 1,016 mm)

Connecting rods

Small end bush fitted internal diameter	0.9377 to 0.9380 in (23,818 to 23,825 mm)
Bush external diameter	1.0015 to 1.0025 in (25,438 to 25,464 mm)
Small end diameter less bush	0.9995 to 1.0005 in (25,387 to 25,413 mm)
Big end diameter less shells	1.8955 to 1.8960 in (48,146 to 48,158 mm)
Length between centres	5.123 to 5.127 in (130,12 to 130,23 mm)
Maximum bend	0.0015 in (0,038 mm) for length of the gudgeon pin
Maximum twist	0.0015 in (0,038 mm) per inch length of gudgeon pin

Big end bearings

Diametrical clearance	0.0008 to 0.0023 in (0,020 to 0,058 mm)
End float on crankpin	0.006 to 0.013 in (0,15 to 0,33 mm)
Undersizes	0.010 in 0.020 in 0.030 in 0.040 in (0,254 mm 0,508 mm 0,762 mm 1,016 mm)

Gudgeon pins

Length	2.626 to 2.638 in (66,70 to 67,00 mm)
Diameter	0.9374 to 0.9376 in (23,811 to 23,815 mm)
Clearance in con rod	0.0001 to 0.0006 in (0,002 to 0,015 mm)
Clearance in piston	Zero to 0.0004 in (0,010 mm)

Pistons

Clearance in bore measured at bottom of skirt at right angles to gudgeon pin	0.0005 to 0.0015 in (0,013 to 0,039 mm)
Top compression ring groove width	0.0705 to 0.0713 in (1,790 to 1,810 mm)
Second compression ring groove width	0.0700 to 0.0709 in (1,780 to 1,800 mm)
Oil control ring groove width	0.1579 to 0.1587 in (4,010 to 4,030 mm)

Piston rings

Number one (top) compression ring	Chrome periphery
Gap in bore	0.015 to 0.025 in (0,40 to 0,65 mm)
Clearance in groove	0.0019 to 0.0039 (0,050 to 0,082 mm)
Number two compression ring	Upper face marked 'TOP'
Gap in bore	0.015 to 0.025 in (0,40 to 0,65 mm)
Clearance in groove	0.0015 to 0.0025 in (0,040 to 0,065 mm)
Oil control – rails	
Gap in bore	0.015 to 0.055 in (0,40 to 1,40 mm)
Oil control – expander	
Gap in bore	Ends to abut

Camshaft

Location	Overhead
Bearings	Non-serviceable
Timing chain	0.375 in (9,52 mm) pitch x 106 pitches

Oil pump

Outer ring end-float	0.004 in (0.1 mm)
Inner ring end-float	0.004 in (0.1 mm)
Outer ring to pump body diametrical clearance	0.008 in (0.2 mm)
Rotor lobe clearance	0.010 in (0.25 mm)
Relief valve spring free length	1.70 in (43.18 mm)

Valves

Inlet	
Overall length	4.259 to 4.269 in (108,17 to 108,43 mm)
Head diameter	1.560 in (39,62 mm)
Angle of face	45°
Stem diameter	0.3107 to 0.3113 in (7,881 to 7,907 mm)
Stem to guide clearance	0.0017 to 0.0023 in (0,043 to 0,058 mm)
Exhaust	
Overall length	4.254 to 4.258 in (108,05 to 108,15 mm)
Head diameter	1.280 in (32,51 mm)
Angle of face	45°
Stem diameter	0.3100 to 0.3106 in (7,87 to 7,89 mm)
Stem to guide clearance	0.0014 to 0.0030 in (0,035 to 0,076 mm)

Valve seats

Outside diameter: Inlet	1.6695 to 1.6705 in (42,405 to 42,430 mm)
Exhaust	1.3335 to 1.3345 in (33,870 to 33,896 mm)
Angle of seat	**42½°**

Valve springs

Free length	1.600 in (40,40 mm)
Fitted length	1.440 in (36,57 mm)
Number of working coils	3¾

Valve timing

Inlet opens	16° B.T.D.C.
Inlet closes	56° A.B.D.C.
Inlet duration	252°
Inlet peak	110° A.T.D.C.
Exhaust opens	56° B.B.D.C.
Exhaust closes	16° A.T.D.C.
Exhaust duration	252°
Exhaust peak	70° A.B.D.C.

continued

Lubrication

System	Wet sump, pressure fed
Oil warning light extinguishes at	3 to 5 lbf/in² (0,21 to 0,35 kgf/cm²)
Oil filter	Full-flow, replaceable element
Oil pump type	Hobourn-Eaton
Oil pressure relief valve	
Type	Non-adjustable
Relief valve spring free length	1.700 in (43,18 mm)

FUEL SYSTEM

Carburetter	**Refer to 'ENGINE TUNING DATA'**
Fuel pump	**Mechanically-operated, diaphragm-type**
Operating pressure	**2.5 to 3.5 lbf/in² (0,17 to 0,24 kgf/cm²)**
Fuel injection	**Bosch, analogue, electronic**
Fuel pump	**Electrical**
Operating pressure	**Approximately 36 lbf/in² (2.5 kgf/cm²)**

COOLING SYSTEM

Type	**Pressurized spill return with thermostat control, pump and fan-assisted**
Type of pump	**Centrifugal**
Thermostat	
Up to Engine No. CL 1468	**82°C (180°F)**
From Engine No. CL 1469	**88°C (190°F)**
Pressure cap	**15 lbf/in² (1,05 kgf/cm²)**

CLUTCH

Make/type	Borg & Beck, diaphragm-type
Clutch plate diameter	8.5 in (215,9 mm)
Facing material	2124F
Number of damper springs	6
Damper spring colours	Red and Violet
Clutch release bearing	Ball journal
Clutch fluid	Refer to Division 09

GEARBOX – 4 speed

Type	Single helical, constant mesh
Speeds	4 forward, 1 reverse
Synchromesh	All forward speeds
Ratios: Fourth (Top)	1.00 : 1
Third	1.25 : 1
Second	1.78 : 1
First	2.65 : 1
Reverse	3.98 : 1 Earlier Models
	3.05 : 1 Later Models
Overall ratios: Fourth (Top)	3.63 : 1
Third	4.56 : 1
[Second	6.47 : 1]
First	9.62 : 1
Reverse	10.95 : 1 Earlier Models
	11.08 : 1 Later Models

GEARBOX – 5 speed

Type	Single helical, constant mesh		
Speeds	5 forward, 1 reverse		
Synchromesh	All forward speeds		
Ratios: Fifth (Top)	0.83 : 1 [1981: 0.79 : 1]	Overall ratios: Fifth (Top)	3.24 : 1
Fourth	1.00 : 1	Fourth	3.90 : 1
Third	1.40 : 1	Third	5.46 : 1
Second	2.09 : 1	Second	8.15 : 1
First	3.32 : 1	First	12.95 : 1
Reverse	3.43 : 1	Reverse	13.38 : 1

GEARBOX – AUTOMATIC

Type [1975–1980 uses Borg Warner 65, 1981 uses Borg Warner 66]

	3rd	2nd	1st	Rev.
Transmission conversion range—type 65	1.00–1.91	1.45–2.77	2.39–4.57	2.09–3.99
—type 66	1.00	1.45	2.39	2.09
Overall ratios to 1 —type 65	3.27–6.25	4.74–9.05	7.82–14.95	6.85–13.08
—type 66	3.08	4.47	7.36	6.44
Road speed corresponding to 1000 engine r.p.m. —type 65	20.0–10.5 m.p.h.		32.0–16.8 km.p.h.	
—type 66	21.9 m.p.h.		35.2 km.p.h.	

AUTOMATIC TRANSMISSION SHIFT SPEEDS

Throttle Position	Zero throttle	Light throttle		Part throttle
Selector—type 65	1	2	D	D
—type 66	1	D	D	D
Shift	2–1	1–2	2–3	3–2
Road speed m.p.h. type 65	30–38	8–12	12–16	30–40
km.p.h. type 65	48–61	13–19	23–26	48–64
m.p.h. type 66	28–39	7–13	12–19	27–43
km.p.h. type 66	45–63	11–21	19–31	43–69

Throttle Position	Kickdown						
Selector	D	D	D	1	D	2	2
Shift	1–2	2–3	3–2	2–1	3–1	1–2	2–1
Road speed m.p.h. type 65	38–44	65–71	58–66	42–50	30–38	38–44	30–38
km.p.h. type 65	61–71	105–114	93–106	68–81	48–61	61–71	48–61
m.p.h. type 66	38–48	70–80	60–72	39–50	25–38	none	none
km.p.h. type 66	61–77	112–128	96–116	63–80	40–61		

PROPELLER SHAFT

Type	One piece straight tube
Universal joints	Constant velocity at both ends [1977 and later models; Hookes joint at rear 1975 & 1976 models]
Overall length, face to face, fully compressed	936 mm (36.85 in)

FINAL DRIVE – Used with 4 speed gearbox

Type	Hypoid
Ratio	3.63 : 1

FINAL DRIVE — from VIN 402047

Type	Hypoid
Ratios : Used with 5 speed gearbox	
U.K.	3.90 : 1
Europe, USA and Canada	3.45 : 1
1981 (North American)	3.45 : 1
Used with Automatic gearbox	
Ratio—1975–1978	3.27 : 1
—1979–1980	3.63 : 1
—1981	3.08 : 1

SUSPENSION – UK & Europe, [U.S.A. 1975 & 1976 models]

Front	Independent. McPherson strut telescopic damper units with co-axial coil springs and anti-roll bar
Coil springs	
Number of working coils	5.5
Free length	13.855 in (352,0 mm)
Rate	93.7 lb/in (16.4 N/mm)
Identification	Red paint stripe
Rear	Four link system. Lower trailing arms, upper trailing radius rods with telescopic damper units, coil springs and anti-roll bar
Coil springs	
Number of working coils	5
Free length	10.71 in (272,0 mm)
Rate	165 lb/in (28.9 ± 5% N/mm)
Identification	White paint stripe

SUSPENSION – U.S.A. Market Specifications [1977 and later models]

Front	**Independent. McPherson strut telescopic damper units with co-axial coil springs and anti-roll bar**

Coil springs

Vehicles with Manual or Automatic Transmission and Heater

Up to Commission Nos. ACG13000/ACW7000 and from Commission Nos. ACG25001/ ACW30001

Wire Diameter	**0.448 in (11.38 mm)**
Number of working coils	**5.25**
Free length	**13.74 in (349 mm)**
Rate	**88 lbf/in (15.4 N/mm)**
Identification	**One green stripe**

From Commission Nos. ACG13001/ACW7001 to Commission Nos. ACG25000/ ACW30000

Wire diameter	**0.448 in (11.38 mm)**
Number of working coils	**5.25**
Free length	**13.30 in (338 mm)**
Rate	**88 lbf/in (15.4 N/mm)**
Identification	**One white stripe and one blue stripe.**

Vehicles with Manual or Automatic Transmission and Air Conditioning

Up to Commission Nos. ACG13000/ACW7000 and from Commission Nos. ACG25001/ ACW30001

Wire diameter	**0.46 in (11.68 mm)**
Number or working coils	**5.5**
Free length	**13.85 in (352 mm)**
Rate	**93.7 lbf/in (16.4 N/mm)**
Identification	**One red stripe**

From Commission Nos. ACG13001/ACW7001 to Commission Nos. ACG25000/ ACW30000

Wire diameter	**0.46 (11.68 mm)**
Number of working coils	**5.5**
Free length	**13.39 in (340 mm)**
Rate	**93.7 lbf/in (16.4 N/mm)**
Identification	**One white stripe and one yellow stripe**
Rear	**Four link system. Lower trailing arms, upper trailing radius rods with telescopic damper units, coil springs and anti-roll bar**

Coil springs

Up to Commission Nos. ACG13000/ACW7000

Wire diameter	**0.463 in (11.76 mm)**
Number of working coils	**5**
Free length	**10.71 in (272 mm)**
Rate	**165 lbf/in (28.9 N/mm)**
Identification	**One white stripe**

From Commission Nos. ACG13001/ACW7001

Wire diameter	**0.463 in (11.76 mm)**
Number of working coils	**5**
Free length	**10.27 in (261 mm)**
Rate	**165 lbf/in (28.9 N/mm)**
Identification	**One white stripe and one red stripe**

STEERING

Make/type	Alford & Alder, rack and pinion
Steering wheel diameter	14.5 in (368 mm)
Steering wheel turns, lock-to-lock [1975–1980]	3.875
[1981]	3
Steering angles and dimensions:	
Front wheel alignment	Zero to 0.062 in (1,5 mm) toe-in
Camber angle	¼° negative ± 1° (Check with the vehicle in the kerb condition)
Castor angle	3½° positive ± 1° (Check with the vehicle in the kerb condition)
King pin inclination	11¼° ± 1°

Triumph TR7 Manual. Part No. RTC 9200. Issue 1.
British Leyland UK press release. January, 1977.
AKM3079B
AKM3079/1

BRAKES

Foot brake

Type	Divided hydraulic system with discs at front and drums at rear	
Operation	Hydraulic, servo-assisted, boost ratio 2.3 : 1 nominal	
Front brake		
Type	Hub-mounted discs with two pistons	
Disc diameter	9.75 in (247,6 mm)	
Disc thickness	0.375 in (9,5 mm)	
Front pad area	16.6 in² (107 cm²)	
Front swept area	183.5 in² (1182 cm²)	
[Pad material 1975 & 1976 models	DON 227]	
Pad material [1977 & later models]	FERODO 2441-FG	
Rear brake		
Type	Drums, service brake, actuated, self adjusting.	
	4 speed specification	5 speed specification
Drum internal diameter	7.995 to 8.000 in (203,20 to 203,33 mm)	8.995 to 9.000 in
Drum maximum worn dia.	8.050 in (204,47 mm)	9.050 in
Shoe arrangement	One leading & one trailing	One leading & one trailing
Shoe lining width	1.50 in (38,1 mm)	1.75 in (44.5 mm)
Lining material	DON 202 GG	
Rear lining area	46 in² (297 cm²)	60.4 in² (390 cm²)
Rear swept area	75.4 in² (486 cm²)	98.9 in² (638 cm²)
Hand brake		
Type	Mechanical, operating rear shoes	

WHEELS

Size/type	5½J x 13 in dia pressed steel rims

TYRES

Size/type	4 speed specification	5 speed specification
	175/70 HR 13	185/70 SR 13
	Radial ply-tubeless	Radial ply-tubeless
Pressures: Front	24 lbs/in² 1.7 bar	24 lbs/in² 1.7 bar
Rear	28 lbs/in² 1.9 bar	28 lbs/in² 1.9 bar

ELECTRICAL EQUIPMENT

System	12 volt, negative earth
Fuses	See page 86.70.02
[Battery capacity—1975 & 1976 models	50 amp-hour at 20 hr rate]
Battery capacity [1977 & later models]	40 amp-hour at 20 hr rate
[Alternator—1975 models	Lucas 16 ACR (20 ACR when air conditioning is fitted)
—1976 models	Lucas 17 ACR or 23 ACR (25 ACR when air conditioning is fitted)
—1977 & later models	Lucas 17 ACR or 23 ACR (20 ACR or 25 ACR when air conditioning is fitted)]

ELECTRICAL EQUIPMENT (continued)

Starter motor	Lucas 2M 100PE
Wiper motor	Lucas 16 W
Windscreen washer pump	Lucas 103 J
Headlamp actuator	Lucas 15 W
Distributor	See Engine Tuning Data
Bulb chart	See page 86.00.01

GENERAL DIMENSIONS (APPROXIMATE)

Overall length [1975–1980]	160.0 in (4065 mm) UK & Europe
	164.3 in (4173 mm) U.S.A.
[1981]	165.4 in (4203 mm) U.S.A.
Overall width	66.2 in (1681 mm)
Overall height [1975–1980]	49.9 in (1268 mm)
[1981]	49.5 in (1258 mm)
Wheelbase	85.0 in (2160 mm)
Track, front	55.5 in (1409 mm)
Track, rear	55.3 in (1404 mm)
Ground clearance	4.5 in (114 mm)
Turning circle: Between curbs	29.0 feet (8.8 m)
Between walls	31.5 feet (9.5 m)

WEIGHTS [1975–1980]

	4 speed		5 speed	
Vehicle dry weight	2158 lbs	979 kgs	2264 lbs	1027 kgs
Kerbside - basic	2249 lbs	1020 kgs	2355 lbs	1068 kgs
Kerbside - with options	2403 lbs	1090 kgs	2463 lbs	1117 kgs
Vehicle capacity	571 lbs	259 kgs	551 lbs	250 kgs
Gross vehicle weight	2820 lbs	1279 kgs	2906 lbs	1318 kgs
Maximum axle load: Front	1499 lbs	680 kgs	1499 lbs	680 kgs
Rear	1349 lbs	612 kgs	1442 lbs	654 kgs
Towing capacity (braked trailer)	1684 lbs	764 kgs	1684 lbs	764 kgs
Roof rack maximum load	112 lbs	51 kgs	112 lbs	51 kgs

WEIGHTS [1981] (approximate)

Coupe	Manual gearbox		Automatic gearbox	
Showroom—minimum	2317 lbs	1051 kgs	2310 lbs	1048 kgs
Unladen—maximum	2511 lbs	1139 kgs	2504 lbs	1136 kgs
Gross vehicle	2954 lbs	1340 kgs	2954 lbs	1340 kgs
Maximum axle load—front	1576 lbs	715 kgs	1576 lbs	715 kgs
—rear	1466 lbs	665 kgs	1466 lbs	665 kgs
Towing capacity—braked trailer	2240 lbs	1016 kgs	2240 lbs	1016 kgs

Convertible	Manual gearbox		Automatic gearbox	
Showroom—minimum	2335 lbs	1059 kgs	2328 lbs	1056 kgs
Unladen—maximum	2502 lbs	1135 kgs	2496 lbs	1132 kgs
Gross vehicle	2954 lbs	1340 kgs	2954 lbs	1340 kgs
Maximum axle load—front	1576 lbs	715 kgs	1576 lbs	715 kgs
—rear	1466 lbs	665 kgs	1466 lbs	665 kgs
Towing capacity—braked trailer	2240 lbs	1016 kgs	2240 lbs	1016 kgs

PAINT AND TRIM CODING SYSTEM

The Commission number plate bears symbols for identification of the vehicle's exterior and trim colours.

Colour Code

The basic colours are allocated a letter as shown in the table. Shades of these colours are allocated two suffix letters. For example, a commission number plate stamped 'Paint JAA. Trim PAA, denotes that the vehicle is painted 'French Blue' and trimmed 'Black'.

Basic Colour	Basic Colour Letter	Paint Trim Code					
Brown/Buff	A	AAA Beige (trim)	AAB Sand Fleck (trim)	AAC Maple	AAD Chestnut (trim)	AAE Russet Brown	
Bronze	B						
Red/Maroon	C	CAA Carmine	CAB Pimento	CAE Vermilion			
Pink	D						
Orange	E	EAA Topaz					
Yellow	F	FAA Mimosa	FAB Inca Yellow				
Gold	G						
Green	H	HAA British Racing Green	HAD Tara Green	HAE Brooklands Green			
Blue	J	JAA French Blue	JAB Delft	JAC Ice Blue	JAD Shadow Blue (trim)	JAF Astral Blue	JAG Pageant Blue
Mauve/Purple/Violet	K						
Grey	L	LAA Birch Grey (trim)	LAB Grey (trim)				
Silver/Aluminium	M						
White/Ivory/Cream	N	NAA White (trim)	NAB White	NAC Sebring White (trim)	NAD Honeysuckle	NAF Leyland White	
Black	P	PAA Black (trim)	PAB Black (matt)	PAC Black (gloss)	PAD Black (rally)		
Multi-coloured	R	RAA Red/Black	RAB Green/Black				

ENGINE TUNING DATA – U.K & Europe

ENGINE	
Type	O.H.V. in line 4 cylinder
Capacity	122 in^3 (1998 cm^3)
Compression ratio	9.25 : 1
Firing order	1 – 3 – 4 – 2
Number one cylinder	Front
Idle speed	650 to 850 rev/min
Fast idle speed (engine hot)	1500 to 1700 rev/min
Ignition timing: Static	10° B.T.D.C.
Dynamic (with vacuum pipe connected)	10° B.T.D.C. at 650 to 850 rev/min
Location of ignition timing marks	Scale on front cover – notch on pulley
Valve clearance (engine cold): Inlet	0.008 in (0,2 mm)
Exhaust	0.018 in (0,5 mm)
Valve clearance adjustment	Pallets between valve and cam follower
Valve timing	See Division 04
IGNITION DISTRIBUTOR	
Make/type	A.C. Delco D302
Rotation, viewed on rotor	Anti-clockwise
Contact breaker gap	0.014 to 0.016 in (0,34 to 0,41 mm)
Dwell angle	39° ± 1°
Centrifugal advance	See page 86.35.00
Vacuum advance	See page 86.35.00
SPARK PLUGS	
Make	Champion
Type	N12Y
Gap	0.024 to 0.026 in (0,61 to 0,66 mm)
IGNITION COIL	
Make/type	Lucas 15C6
Primary resistance at 20°C (68°F)	1.3 to 1.5 ohms
CARBURETTERS	
Make/type	Twin – S.U. HS6
Needle size	BDM
Jet size	100

ENGINE TUNING DATA – U.S.A. Markets

ENGINE	
Type	O.H.V. in line 4 cylinder
Capacity	122 in^3 (1998 cm^3)
Compression ratio	8.0 : 1
Firing order	1 – 3 – 4 – 2
Number on cylinder	Front
Idle speed	800 ± 100 rev/min
Fast idle speed (engine hot)	[1500 to 1700 rev/min]

ENGINE TUNING DATA — U. S. A. Markets (continued)

	[1977 & later Federal]	[1977 & later California and all 1975 & 1976 models]
Ignition timing:		
Static	10° B.T.D.C.	10° B.T.D.C.
Dynamic at idle	10° B.T.D.C.	2° A.T.D.C.
Location of ignition timing marks	Scale on front cover – notch on pulley	
Valve clearance (engine cold)		
Inlet	0.008 in (0,2 mm)	
Exhaust	0.018 in (0,5 mm)	
Valve clearance adjustment	Pallets between valve and cam follower	
Valve timing	See Division 04	

IGNITION DISTRIBUTOR – FEDERAL Market Specification	
Make/type	[Lucas 47 DE4. Opus electronic in built]
Rotation viewed on rotor	Anti-clockwise
Pick up air gap	0.014 to 0.016 in (0.34 to 0.41 mm)
Drive resistor	Lucas 10 ohms ± 5%]
Centrifugal advance	See 86.35.00

IGNITION DISTRIBUTOR – CALIFORNIA Market Specification
[**NOTE:** All other specifications are the same as those found under FEDERAL Market Specifications, above.]

Vacuum retard	See 86.35.00
SPARK PLUGS	
Make	Champion
Type	N12Y [(can be used in 1975 cars, which originally had N-11Y spark plugs)]
Gap	0.024 to 0.026 in
IGNITION COIL	
Make/type	Lucas 15C6
Primary resistance at 20°C (68°F)	1.3 to 1.5 ohms
[CARBURETTERS	
Quantity and make	Two Zenith Strombergs
Type (1975 & 1976 U.S.A. and all Canada without catalytic converter)	175 CD SEV Manual chokes
(1975 & 1976 U.S.A. with catalytic converter)	175 CD 4VT Automatic chokes
(1977 U. S. A.)	175 CD FEVX Automatic chokes
(1978–1980 U. S. A.)	175 CD FVX Automatic chokes
Choke size	1.625 in (41.275 mm)
Metering needle (1975 & 1976 with catalytic converter)	45C
(1977–1979 Federal)	B1DH
(1980 and later Federal)	B1EP
(1977 and later California)	B1EP
Orifice	0.100 in
Piston return spring	Red]
FUEL INJECTION	
Make/Type	Bosch, analogue, electronic

Triumph TR7 Manual. Part No. RTC 9200. Issue 1.
Triumph TR7 Handbook. Part No. RTC.9209/75. Edition 2.
Triumph TR7 Handbook. Part No. RTC.9209/77. Edition 1.
Service Division Dealer Training Aid # S1034.
AKM3079R

TORQUE WRENCH SETTINGS

Operation	Description	Specified torque Nm	lbf ft	kgf m
ENGINE				
Air Cleaner to Carburetter	5/16" UNC Setscrew	11	8	1,1
Air Pump Adjusting Link and Mounting Bracket to Block	5/16" UNF Bolt	27	20	2,7
Air Pump Bracket and Alternator Adjusting Link to Block	5/16" UNF Bolt	27	20	2,7
Air Pump Adjusting Link Attachment	5/16" UNF Bolt	27	20	2,7
Air Pump to Mounting Bracket	5/16" UNF Bolt	27	20	2,7
Alternator to Mounting Bracket	5/16" UNF Bolt	27	20	2,7
Alternator Mounting Bracket to Bracing Link	5/16" UNF Bolt	27	20	2,7
Alternator Adjusting Link and Timing Cover to Block	5/16" UNF Bolt	27	20	2,7
Bearing Caps to Block	7/16" UNF Bolt	88	65	8,9
Camshaft Chain Wheel Attachment	¼" UNF Setscrew	14	10	1,38
Camshaft Cover Attachment	¼" UNC Pan Hd. Setscrew	7	5	0,69
Camshaft Bearing Cap Attachment	5/16" UNF Stud	19	14	1,9
Camshaft Bearing Cap and Cover Attachment	5/16" UNF Stud	3	2	0,27
Carburetter Flexible Mounting to Inlet Manifold	5/16" UNC Setscrew	19	14	1,9
Carburetter to Flexible Mounting	5/16" UNF Stud	19	14	1,9
Clutch to Flywheel	5/16" UNF Setscrew	30	22	3,0
Clutch Housing to Rear Engine Plate	3/8" UNF Dowel Bolt	46	34	4,7
Clutch Housing to Rear Engine Plate	5/16" UNF Bolt	27	20	2,7
Clutch Housing and Front Engine Support Bracket to Rear Engine Plate	5/16" UNF Bolt	27	20	2,7
Clutch Housing to Block	5/16" UNF Bolt	27	20	2,7
Connecting Rod Bolt	3/8" Bolt	61	45	6,2
Crankshaft Pulley Attachment	5/8" UNF Bolt	160	120	16,5
Cylinder Head to Block	7/16" UNC Bolt	75	50	6,9
Cylinder Head to Block	7/16" UNC Stud	75	50	6,9
Distributor to Block	¼" UNF Setscrew	12	9	1,2
Diverter Valve to Mounting Bracket	¼" UNF Setscrew	14	10	1,38
Drive Plate to Crank	3/8" UNF Bolt	61	45	6,2
Engine Stay Bracket to Block	5/16" UNF Bolt	27	20	2,7
Engine Stay Attachment	3/8"UNF Jam Nut	46	34	4,7
Engine Support to Block (R.H.)	5/16" UNF Setscrew	27	20	2,7
Evaporative Loss Canister Mounting Bracket to Body	6 mm Setscrew	10	7	0,9
Fan to Holset Coupling	¼" UNF Bolt	12	9	1,2
Fan Pulley Bearing Housing to Timing Cover	5/16" UNC Setscrew	19	14	1,9
Flywheel to Crankshaft	3/8" UNF Bolt	61	45	6,2
Front Lifting Eye to Block	3/8" UNF Setscrew	46	34	4,7
Fuel Pump to Block	5/16" UNF Setscrew	19	14	1,9
Gearbox Adaptor Attachment	5/16" UNF Stud	27	20	2,7
Heater By-Pass Pipe to Inlet Manifold	11/16" Union Nut	19	14	1,9
Hot Air Manifold to Exhaust Manifold	5/16" UNF Setscrew	27	20	2,7
Inlet Manifold to Carburetter Flex Mounting	5/16" UNC Setscrew	27	20	2,7
Inlet Manifold Attachment	5/16" UNC Setscrew	27	20	2,7
Idler Shaft Chainwheel Attachment	3/8" UNF Setscrew	51	38	5,2
Oil Pump to Cylinder Block	5/16" UNF Bolt	27	20	2,7
Oil Seal Housing Attachment	¼" UNF Setscrew	12	9	1,2
Oil Seal Housing Attachment	¼" Setscrew	12	9	1,2
Oil Sump Drain Plug	3/8" Dryseal Plug	34	25	3,4
Oil Sump to Block	5/16" UNF Bolt	27	20	2,7
Oil Sump to Block	5/16" UNF Screw	27	20	2,7
Oil Sump to Timing Cover	5/16" UNF Screw	27	20	2,7
Oil Suction Pipe to Block	¼" UNF Bolt	12	9	1,2
Oil Suction Pipe to Block	¼" UNF Setscrew	12	9	1,2
Oil Transfer Adaptor to Block	3/8" UNF Bolt	46	34	4,7
Rear Engine Mounting to Crossmember	8 mm Setscrew	28	21	2,9
Rear Engine Mounting to Bracket	12 mm Setscrew	80	59	8,1
Rear Engine Mounting Bracket to Gearbox	8 mm Setscrew	28	21	2,9
Rear Engine Mounting Crossmember to Body	8 mm Setscrew	28	21	2,9
Rear Engine Plate to Block	5/16" UNF Setscrew	27	20	2,7
Right Hand Front Engine Mounting Bracket to Body	8 mm Setscrew	28	21	2,9
Right Hand Front Engine Mounting to Engine Bracket	8 mm Setscrew	28	21	2,9
Right Hand and Left Hand Engine Mounting Attachment	10 mm	50	37	5,1
Starter Motor to Clutch Housing	3/8" Bolt	46	34	4,7
Throttle Linkage and Mounting Bracket Assembly to Inlet Manifold	¼" UNC Setscrew	10	7	0,9
Timing Chain Tensioner to Block	¼" UNF Bolt	12	9	1,2
Timing Chain Support Bracket and Guides to Block	5/16" UNF Setscrew	27	20	2,7
Timing Cover to Cylinder Block	5/16" UNF Bolt	27	20	2,7
Timing Cover to Cylinder Head	5/16" UNF Setscrew	27	20	2,7
Water Outlet Elbow to Inlet Manifold	5/16" UNC Setscrew	27	20	2,7
Water Pump Cover to Block	5/16" UNF Setscrew	27	20	2,7
Water Pump Impeller Retaining	5/16" UNC Bolt	19	14	1,9
Water Transfer Housing to Cylinder Head	5/16" UNC Setscrew	27	20	2,7
FUEL INJECTION SYSTEM				
Air meter support plate/bracket	8 mm Stud/Setscrew	28	21	2,9
Auxiliary air valve to water pump cover	6 mm × 12 mm Setscrew	7	5	0,7
Fuel rail to injectors	14 mm/13 mm dia.	0.68	6 lbf in	0,07
Injector clamp plates to manifold	6 mm × 16 mm Setscrew	10	7	0,9
Lambda (oxygen) sensor to exhaust manifold	18 mm	60	44	6,0
Pressure regulator to fuel rail	14 mm	28	21	2,9
Temperature sensor to manifold	12 mm	15	11	1,4
Thermo time switch to manifold	14 mm	28	21	2,9

Operation	Description	Nm	lbf ft	kgf m
EXHAUST SYSTEM				
Exhaust Manifold Attachment	3/8" UNF Setscrew	46	34	4,7
Exhaust Pipe to Manifold	10 mm Stud	50	37	5,1
Exhaust Pipe to Brackets on Gearbox and Clutch Housing	8 mm Setscrew	28	21	2,9
Exhaust Silencer to Front Exhaust Pipe	8 mm Bolt	10	7	0,9
Exhaust Silencer to Rear Silencer and Tail Pipe	8 mm Bolt	10	7	0,9
Exhaust Support Bracket to Pipe	8 mm Setscrew	28	21	2,9
Rear Silencer Strap to Support Bracket	8 mm Bolt	28	21	2,9
Rear Silencer Support Bracket to Floor	8 mm on Mounting Rubber	28	21	2,9
CLUTCH, 4 SPEED GEARBOX AND PROPELLER SHAFT				
Boss to Clutch Housing	5/16" UNF Setscrew	27	20	2,7
Clutch and Brake Pedal Mounting Bracket Attachment	8 mm Setscrew	28	21	2,9
Clutch Housing to Gearbox Case	3/8" UNF Wedgelock Bolt	43	32	4,4
Clutch Housing to Gearbox Case	3/8" UNF Setscrew	43	32	4,4
Clutch Housing to Rear Engine Plate	3/8" UNF Dowel Bolt	43	32	4,4
Clutch Housing to Rear Engine Plate	5/16" UNF Bolt	27	20	2,7
Clutch Housing and Front Exhaust Support Bracket to Rear Engine Plate	5/16" UNF Bolt	27	20	2,7
Clutch Housing to Cylinder Block	5/16" UNF Bolt	27	20	2,7
Clutch Master Cylinder to Dash	8 mm Setscrew	28	21	2,9
Cover to Gearbox Case	¼" UNF Setscrew	12	9	1,2
Cover Plate to Top Cover	¼" UNF Setscrew	12	9	1,2
Exhaust Pipe Support Bracket Attachment	5/16" UNF Setscrew	27	20	2,7
Flange to Mainshaft	5/8" UNF Nut	163	120	16,5
Magnetic Drain Plug	3/8" Dryseal Plug	34	25	3,4
Oil Filler Plug	3/8" NP Taper Plug	34	25	3,4
Oil Sump Coupling Plate to Clutch Housing	10 mm Setscrew	50	37	5,1
Overdrive Adaptor Fixings	5/16" UNF Setscrew	27	20	2,7
Pipe, Clutch Master Cylinder to Slave Cylinder	12 mm Tube Nut	12	9	1,2
Prop Shaft to Gearbox and Rear Axle	3/8" UNF Bolt	46	34	4,7
Prop Shaft Safety Strap Attachment	8 mm Bolt	28	21	2,9
Reverse Idler Spindle Locating Screw	5/16" UNF Pointed Setscrew	19	14	1,9
Selector Shaft to Forks	5/16" UNF Taper Setscrew	13	10	1.4
Slave Cylinder to Boss	5/16" UNF Bolt	12	9	1,2
Slave Cylinder to Clutch Housing	8 mm Bolt	28	21	2,9
Slave Cylinder Attachment	5/16" UNF Setscrew	27	20	2,7
Sump Coupling Plate to Clutch Housing	10 mm Setscrew	50	37	5,1
Top Cover to Gearbox Case	¼" UNF Bolt	13	10	1,3
Top Cover to Gearbox Case	¼" UNF Setscrew	13	10	1,3

Operation	Description	Nm	lbf ft	kgf m
GEARBOX – 5 speed				
Driving flange to mainshaft	18 mm nyloc nut	200	150	20,4
Dustcap assembly to extension housing	6 mm x 12 mm Setscrew	10	7	0,9
Extension and centre plate to main case	8 mm x 55 mm bolt	28	21	2,9
Fifth gear selector fork pivot bracket to centre plate	8 mm x 25 mm	28	21	2,9
Front cover to maincase	8 mm x 25 mm Setscrew	28	21	2,9
Interlock spool retainer to gearbox case	6 mm x 16 mm Setscrew	10	7	0,9
'J' coupling pin to main selector shaft	8 mm on pin	20	15	1,8
Mounting bracket upper fixing	8 mm x 20 mm Setscrew	28	21	2,9
Mounting bracket lower fixing	8 mm x 30 mm Bolt	28	21	2,9
Magnetic drain plug	16 mm Plug	35	26	3,5
Oil pump body to rear extension	6 mm x 20 mm screw	10	7	0,9
Oil inlet access hole blanking	8 mm x 8 mm socket Locwel Setscrew	20	15	1,8
Reverse lever mounting pin to centre plate	10 mm on pin	28	21	2,9
Reverse baulk plate to gearbox extension	6 mm x 40 mm Bolt	10	7	0,9
Remote control housing to gearcase rear extension	8 mm x 40 mm Setscrew	20	15	1,8
Speedo cable clip to gearbox	6 mm x 20 mm Setscrew	10	7	0,9
Torsion spring brackets to gearbox extension	6 mm x 20 mm Setscrew	10	7	0,9
Torsion spring adjuster screw locking	8 mm x 25 mm Setscrew	20	15	1,8
REAR AXLE – 4 SPEED GEARBOX				
Crown Wheel Attachment	3/8" UNF Bolt	62	46	6,3
Drain Plug	3/8" Plug	34	25	3,4
Hub to Axle Shaft	5/8" UNF on Axle	160	120	16,5
Hypoid Housing to Axle Casing	5/16" UNF Bolt	27	20	2,7
Hypoid Housing Bearing Cap Retention	3/8" UNF Bolt	52	38	5,2
Hypoid Flange to Pinion	5/8" on Pinion	Tighten to obtain correct pre load as Specified in Division 51.		
Wheel Attachment	12 mm Stud	100	75	10,3
Alloy Wheel Attachment	12 mm Stud	120	88	12,2
REAR AXLE – 5 SPEED GEARBOX				
Rear cover to axle case	8 mm Setscrew	28	21	2,9
Pinion oil seal housing to axle case	10 mm Setscrew	50	37	4,0
Differential unit bearing caps to axle case		100	75	10,2
Crown wheel to differential unit (with Loctite)	12 mm Bolt	122	90	12,2
Pinion Flange to pinion	16 mm Myloc nut	160	120	16,5
Axle shafts/backplates to axle casing	10 mm Setscrew	50	37	5,1

Operation	Description	Specified torque Nm	lbf ft	kgf m
FRONT SUSPENSION				
Anti-Roll Bar Clamp Bolts	10 mm Bolt	40	30	4,0
Anti Roll Bar to Crossmember	10 mm Bolt	50	37	5,1
Anti Roll Bar to Lower Link	12 mm on Bar	80	59	8,1
Sub-frame Attachment Front Fixing	12 mm Bolt	80	59	8,1
Sub-frame Attachment Rear Fixing	12 mm Bolt	80	59	8,1
Damper Unit Closure Nut	1.781" x 24 UNF	100	74	10,2
Hub to Stub-Axle	9/16" UNF on Axle	Tighten to 5 lbf ft (0,69 kgf m) unscrew one flat and insert pin		
Lower Link to Crossmember	12 mm Bolt	80	59	8,1
Lower Link to Strut Assembly	½" UNF on Ball Joint	61	45	6,2
Strut Mounting to Body	8 mm on Mounting Studs	28	21	2,9
Strut to Mounting	12 mm on Cartridge	60	44	6,0
Tie Rod Lever to Stub Axle	12 mm Bolt	100	74	10,2
Wheel Attachment	12 mm Stud	100	74	10,2
Alloy Wheel Attachment	12 mm Stud	120	88	12,2
REAR SUSPENSION				
Anti Roll Bar to Lower Link	10 mm x 30 Bolt	50	37	5,1
	Clamp Nut	27	20	2,7
Damper to Trailing Arm	3/8" UNF on Damper Lock Nut	19	14	1,9
	Clamp Nut	27	20	2,7
Lower Link to Body and Axle	7/16" UNF Bolt	65	48	6,6
Upper Link to Body and Axle	7/16" UNF Bolt	65	48	6,6
STEERING				
Rack (Pinion Side) to Crossmember	10 mm Bolt	40	30	4,0
Rack to Crossmember	10 mm Bolt	40	30	4,0
Steering Column Clamp to Upper Column	8 mm Bolt	10	7	0,9
Steering Column Lock Shear Head Bolt	Shear Bolt		Tighten to shear	
Steering Column to Body	Shear Bolt		Tighten to shear	
Steering Column Clamp Locating Plate Attachment	7/16" UNF Grub Screw	20	15	2,0
Steering Column Clamp Locating Plate Attachment	7/16" UNF Jam Nut	50	37	5,1
Steering Column Clamp to Column	¼" Bolt	12	9	1,2
Steering Column Rail to 'A' Post	6 mm Screw	10	7	0,9
Steering Column Support Rail to Tunnel Bracket	6 mm Screw	10	7	0,9
Steering Wheel to Column	11/16" Steering Wheel Nut	50	37	5,1
Tie Rod End Ball Joint Assembly	12 mm on Ball Joint	50	37	5,1
Tie-Rod End Ball Joint to Tie-Rod	14 mm Locknut	50	37	5,1
Tube to Support Rail	6 mm Screw	10	7	0,9
Tube to Dash and Pedal Bracket	6 mm Screw	10	7	0,9
Universal Joint Assembly to Upper and Lower Columns and Rack	8 mm Bolt	28	21	2,9

Operation	Description	Specified torque Nm	lbf ft	kgf m
BRAKES				
Brake and Clutch Pedal Mounting Bracket Attachment	8 mm Setscrew	28	21	2,9
Brake Master Cylinder Attachment	8 mm Stud	14	10	1,3
Caliper Assembly Attachment	12 mm Bolt	100	74	10,2
Disc to Hub	3/8" UNC Bolt	43	32	4,4
Disc Shield Assembly to Stub Axle	6 mm Setscrew	10	7	0,9
Front Hose to Caliper	10 mm Nut	15	11	1,4
Hand Brake Cable to Lever on Backplate	5/16" UNF Fork End	10	7	0,9
Hand Brake Compensator Lever to Abutment Boss	6 mm Setscrew	10	7	0,9
Hand Brake Fulcrum Compensator	8 mm Nut	28	21	2,9
Hand Brake Mounting Bracket and Clamp Plate to Floor	6 mm Setscrew	10	7	0,9
Master Cylinder to Pressure Reducing Valve	10 mm Male tube nut	12	9	1,2
Pipe, Pressure Reducing Valve to Front Hoses	10 mm Female pipe nut 10 mm Male tube nut	12	9	1,2
Pipe, Pressure Reducing Valve to Rear Hoses	10 mm Female pipe nut 10 mm Male tube nut	12	6	1,2
Pipe, Pressure Reducing Valve and Bracket to Turret	8 mm Bolt	28	21	2,9
Rear Brake Backplate Attachment	5/16" UNF Bolt	19	14	1,9
Rear Brake Drum to Hub	¼" UNF Setscrew	10	7	0,9
Rear Hose to R.H. Rear Wheel Cylinder	10 mm Female pipe nut 10 mm Male tube nut	12	9	1,2
R.H. Rear Wheel Cylinder to L.H. Rear Wheel Cylinder	10 mm Female pipe nut 10 mm Male tube not	12	9	1,2
BODY				
Accelerator Pedal to Mounting Bracket	6 mm Setscrew	10	7	0,9
Accelerator Mounting Bracket to Dash	6 mm Setscrew	10	7	0,9
Bonnet Buffer Fixing	8 mm Jam Nut	12	9	1,2
Bonnet Hinge to Front Panel Assembly	6 mm Setscrew	10	7	0,9
Bonnet Lock Striker to Bonnet	8 mm Setscrew	28	21	2,9
Bonnet Lock to Body	8 mm Setscrew	28	21	2,9
Bonnet Release Cable to Fascia Support Rail Bracket	½" UNF Jam Nut	12	9	1,2
Body Side to Front End Assembly	6 mm Setscrew	10	7	0,9
Bonnet Striker Pin	3/8" UNF on striker pin	19	14	1,9
Dash Assembly	6 mm Weld Stud	8	6	0,8
Deflector to Re-inforced Steering Column Support Rail	6 mm Setscrew	10	7	0,9
Door Division Channel to Door	6 mm Setscrew	10	7	0,9
Door Glass Run Channel to Door	6 mm Setscrew	10	7	0,9
Door Hinge to Door	8 mm Stud on Hinge Assembly	28	21	2,9
Door Lock Attachment	6 mm Setscrew Csk.	10	7	0,9
Door Striker to 'B' Post	6 mm Setscrew	10	7	0,9
Fascia Attachment Brackets (Upper) to Dash Top	6 mm Setscrew	10	7	0,9

continued

Operation	Description	Specified torque Nm	lbf ft	kgf m
Fascia Support Bracket (Lower) to Support Rail	6 mm Setscrew	6	4.5	0,6
Fascia to Dash	6 mm Setscrew	10	7	0,9
Front Bumper Cover Assembly to Front Bumper Bar	6 mm Setscrew	3	2.2	0,3
Front Bumper Assembly to Body	8 mm Weld Bolt	14	10	1,4
Front Tie Down Bracket to Front Crossmember Panel	8 mm Setscrew	28	21	2,9
Front Upper Assembly to Front Fenders	6 mm Setscrew	10	7	0,9
Fuel Tank Attachment	10 mm Stud	50	37	5,1
Mirror Mounting Bracket to Door Bracket	6 mm Setscrew	10	7	0,9
Mirror Mounting Bracket to Door Inner Panel	6 mm Setscrew	10	7	0,9
Plenum Panel Fixing	6 mm Pan Head Screw	6	4.5	0,6
Radiator Crossmember to Longitudinal Members	8 mm Setscrew	28	21	2,9
Radiator Fan Guard to Radiator Attachment	8 mm Setscrew	28	21	2,9
Radiator Mounting Bracket to Bonnet Hinge Mounting Panel	8 mm Weld Bolt	14	10	1,4
Rear Bumper to Body	8 mm Weld Bolt	20	15	2,0
Rear Bumper Assembly to Body	8 mm Weld Bolt	14	10	1,4
Rear Bumper Corner Struts to Body	8 mm Setscrew	28	21	2,9
Rear Bumper Cover to Rear Bumper	6 mm setscrew	3	2.2	0,3
Rear Bumper Corner Struts to Rear Bumper	8 mm Weld Bolt	20	15	2,0
Relay Plate to Bracket on Dash Front	6 mm Setscrew	10	7	0,9
Seat Belt Buckle Assembly to Seat Slide	7/16" UNF Setscrew	43	32	4,4
Seat Belt (Inner to Seat Slide)	7/16" UNF Setscrew	43	32	4,4
Seat Belt to Wheel Arch	7/16" UNF Setscrew	43	32	4,4
Seat Belt to Mounting Bracket	7/16" UNF Setscrew	43	32	4,4
Seat Belt Warning Switch to Gearbox Extension	3/8" UNC Switch	20	15	2,0
Seat Slides to Floor	8 mm Cap Screw	28	21	2,9
Seat Slides to Seat Frame	6 mm Cap Screw	10	7	0,9
Side Lamp and Head Lamp Cables to Radiator closing Panel	6 mm Setscrew	10	7	0,9
Trunk Lid Hinges to Trunk Lid	8 mm Setscrew	28	21	2,9
Trunk Lid Hinges to Body	6 mm Setscrew	10	7	0,9
Trunk Lid Support to Support Bracket Tonneau	6 mm Setscrew	10	7	0,9
Trunk Lock to Lid	6 mm Setscrew	10	7	0,9
Trunk Lock Striker to Body	¾" UNF Setscrew	12	9	1,2
Window Regulator Attachment	¼" UNF Setscrew	12	9	1,2

Operation	Description	Specified torque Nm	lbf ft	kgf m
ELECTRICAL				
NOTE: Refer to Section 86 for additional torque settings on proprietary items				
Battery Attachment	¼" on Fixing Rod	3	2	0,27
Battery Earth to Body	6 mm Setscrew	10	7	0,9
Headlamp Box to Hinge Arm	6 mm Weld Bolt	6	4.5	0,6
Headlamp Hinge Arm to Pivot Bracket	10 mm	40	30	4,0
Headlamp Hinge Arm to Bracket Assemblies	8 mm Setscrew	20	15	2,0
Headlamp Hinge Arm to Pivot Pin	8 mm	20	15	2,0
Headlamp Mechanism to Body	8 mm Weld Bolt	14	10	1,4
Headlamp Mechanism Crank Arm to Gearbox	6 mm Nut	15	11	1,4
Headlamp Raising Link Adjustment	6 mm Jam Nut	6	4.5	0,6
Headlamp Tension Spring Attachment	6 mm Shouldered Bolt	8	6	0,8
Heater Mountings	6 mm Setscrew	10	7	0,9
Horn Attachment	6 mm Setscrew	10	7	0,9
Ignition Coil to Mounting Plate	6 mm Setscrew	10	7	0,9
AIR CONDITIONING				
Compressor Mounting Bracket to Cylinder Head	5/16" UNF Bolt	27	20	2,7
Compressor Mounting Bracket to Cylinder Block	5/16" UNF Bolt	27	20	2,7
Compressor Mounting Bracket to Cylinder Block	5/16" UNF Setscrew	27	20	2,7
Compressor Mounting Bracket to Compressor	3/8" UNC Setscrew	40	30	4,0
Compressor Mounting Bracket to Compressor	3/8" UNC Stud	32	24	3,2
Compressor Belt Tension Adjustment	3/8" UNF Locknut	19	14	1,9
Condenser Fan Motor Attachment	5 mm Setscrew	3	2	0,27
Hoses — Flare Type				
Compressor to Condenser	¾" Crimp Back Nut	40	30	4,0
Condenser to Receiver Drier	⅝" Crimp Back Nut	28	21	2,9
Evaporator to Compressor	⅞" Crimp Back Nut	50	37	5,1
Evaporator to Receiver Drier	⅝" Crimp Back Nut	28	21	2,9
Ranco Valve to Receiver Drier	7/16" Flare Nut	12	9	1,2
Rotalock Valve to Compressor	1" UNS	50	37	5,1
Hoses — 'O' Ring Type				
Compressor to Condenser				
Compressor end	1" UNS Gasket Nut	27	20	2,7
Condenser end	¾" Gasket Nut	27	20	2,7
Hoses – Evaporator to Compressor				
Compressor end	1" UNS Gasket Nut	27	20	2,7
Evaporator end	⅞" UNF Nut	37	27	3,7
Condenser to Receiver Drier	⅝" UNF Gasket Nut	18	13	1,8
Receiver Drier to Evaporator	⅝" UNF Nut	18	13	1,8
Ranco Valve to Receiver Drier	7/16" UNF Gasket Nut	9	7	0,9

RECOMMENDED LUBRICANTS & ANTI-FREEZE SOLUTIONS – BRITISH ISLES – ALL SEASONS

(THE PRODUCTS RECOMMENDED ARE NOT LISTED IN ORDER OF PREFERENCE)

COMPONENT	BP	CASTROL	DUCKHAMS	ESSO	MOBIL	PETROFINA	TEXACO	SHELL
Engine Carburetter Dashpots and Oil can Oils must meet Leyland Cars Specification BLS.OL.02	BP Super Visco-Static 20–50	Castrol GTX	Duckhams Q Motor Oil	Esso Uniflo 15W/50	Mobil Super 15W–50	Fina Super grade Motor Oil S.A.E. 20W–50	Havoline Motor Oil 20W–50	Shell Super Multigrade 20W–50
Manual Gearbox – 4 Speed*	BP Gear Oil S.A.E. 90 EP	Castrol Hypoy	Duckhams Hypoid 90	Esso Gear Oil GX 85W/140	Mobilube HD 90	Fina Pontonic MP S.A.E. 90	Texaco Multigear Lubricant EP 90	Shell Spirax 90 EP
Manual Gearbox – 5 Speed†	BP Gear Oil S.A.E. 80 EP	Castrol Hypoy B 75	Duckhams Hypoid 75	Esso Gear Oil GP 80W	Mobilube HD 80	Fina Pontonic MP SAE 80	Texaco Multigear Lubricant EP 80	Shell Spirax HD 75/80W Shell Donax TF
Rear Axle	BP Gear Oil S.A.E. 90 EP	Castrol Hypoy	Duckhams Hypoid 90	Esso Gear Oil GX 85W/140	Mobilube HD 90	Fina Pontonic MP S.A.E. 90	Texaco Multigear Lubricant EP 90	Shell Spirax 90 EP
Rear Hubs, Brake Cables, Grease Gun‡	BP Energrease L 2	Castrol LM Grease	Duckhams LD 10 Grease	Esso Multi-Purpose Grease H	Mobilgrease MP or MS	Fina Marson HTL 2 Grease	Marfak All Purpose Grease	Shell Retinax A
Automatic Transmission and Power Steering	BP Autran G	Castrol TQF	Duckhams 'Q'-Matic	Esso Glide (Type 'G')	Mobil ATF 210	Fina Purfimatic 33F	Texamatic Type 'G'	Shell Donax TF
Clutch and Brake Reservoirs	Unipart Universal Brake Fluid or other brake fluids having a minimum boiling point of 260°C (500°F) and complying with FMVSS DOT 3 or SAE J1703 specification.							
Approved Anti-Freeze	Unipart Universal Anti-Freeze.							
	* Manual gearbox, if drained, MUST be refilled with Hypoid 75W gear oil. † Manual gearbox, if drained, MUST be refilled with Hypoid 75W gear oil. For service top-up use Hypoid 75W gear oil, or where this oil is not available, the Hypoid 80W oils shown above may be used.							
	‡ Front hub grease – Mobilgrease MS or Super.							

RECOMMENDED LUBRICANTS AND ANTI-FREEZE SOLUTIONS – OVERSEAS MARKETS

	SERVICE CLASSIFICATION	AMBIENT TEMPERATURE °C (–30 –20 –10 0 +10 +20)
ENGINE	Oils must meet Leyland Cars' Specification BLS.OL.02 and/or the European Motor Manufacturers' requirements and/or A.P.I.–SE Performance Level	5W/20 5W/30 5W/40 10W/30 10W/40 10W/50 15W/40 15W/50 20W/40 20W/50
FINAL DRIVE	A.P.I. – GL5	HYPOID 90
	MIL-L-2105B	HYPOID 80W
MANUAL GEARBOX* –4 SPEED	A.P.I.– GL4	HYPOID 90
	MIL-L-2105	HYPOID 80
MANUAL GEARBOX† – 5 SPEED	A.P.I. – GL4	HYPOID 75W
	MIL-L-2105	
AUTOMATIC GEARBOX POWER STEERING	ATF TYPE F [through 1978] ATF TYPE G [1979 and later]	
HUBS‡ AND CHASSIS GREASE POINTS	N.L.G.I.–2 Multi-purpose Grease	
BRAKE & CLUTCH RESERVOIRS – USA	LOCKHEED 3295 UNIVERSAL BRAKE FLUID or other brake fluids having a minimum boiling point of 260°C (500°F) and complying with FMVSS 116 DOT3 specifications.	
– except USA	UNIPART UNIVERSAL BRAKE FLUID or other brake fluids having a minimum boiling point of 260°C (500°F) and complying with FMVSS 116 DOT3 or S.A.E. J1703d specifications.	
ANTI-FREEZE	Permanent type ethylene glycol base with suitable inhibitor for MIXED METAL ENGINES.	
WINDSHIELD WASHER	Windshield Washer Anti-Freeze Fluid (Proprietary Brands).	

* Manual gearbox, if drained, MUST be refilled with Hypoid 75W gear oil.
† Manual gearbox, if drained, MUST be refilled with Hypoid 75W gear oil. For Service top-up use Hypoid 75W gear oil, or where this oil is not available, Hypoid 80W oils may be used.
‡ Front hub grease = N.L.G.I. No. 2 high melting-point grease.

ANTI-FREEZE SOLUTIONS

ANTI-FREEZE CONCENTRATION			25%	30%	35%	50%
SPECIFIC GRAVITY OF COOLANT AT 15.5° C (60° F)			1.039	1.048	1.054	1.076
ANTI-FREEZE QUANTITY EXPANSION TANK SYSTEM		US pints	3.875	4.65	5.425	7.75
		UK pints	3.25	3.9	4.55	6.5
		Litres	1.85	2.22	2.59	3.7
ANTI-FREEZE QUANTITY HEADER TANK SYSTEM		US pints	4.02	4.82	5.628	8.04
		UK pints	3.35	4.02	4.69	6.7
		Litres	1.9	2.28	2.66	3.8
DEGREE OF PROTECTION	**Complete** Car may be driven away immediately from cold		−12° C 10° F	−16° C 3° F	−20° C −4° F	−36° C −33° F
DEGREE OF PROTECTION	**Safe Limit** Coolant in mushy state. Engine may be started and driven away after short warm-up period		−18° C 0° F	−22° C −8° F	−28° C −18° F	−41° C −42° F
DEGREE OF PROTECTION	**Lower Protection** Prevents frost damage to cylinder head, block and radiator. Thaw out before starting engine.		−26° C −15° F	−32° C −26° F	−37° C −35° F	−47° C −53° F

CAPACITIES

	UK	METRIC	USA
Fuel tank [except California & Federal]	13.8 gal.	62.8 litre	16.6 gal.
[Fuel tank—California & Federal	12.0 gal.	54.5 litre	14.4 gal.]
Engine sump and oil filter	8 pint	4.5 litre	9.5 pint
Engine sump (drain and refill)	7 pint	4.0 litre	8.4 pint
Gearbox (from dry)			
4 speed	2.0 pint	1.1 litre	2.4 pint
5 speed option	2.70 pint	1.53 litre	2.3 pint
Rear axle (from dry)			
4 speed	2.25 pint	1.3 litre	2.7 pint
5 speed option	1.60 pint	0.9 litre	1.9 pint
Automatic transmission with oil cooler	9.5 pint	5.4 litre	11.4 pint

LUBRICATION – Note. Figures in italics refer to U.S.A. market specification vehicles

NOTE: Ensure that the vehicle is standing on a level surface when checking the oil levels.

Weekly or before a long journey
1 Check/top up cooling system level.
13 Check engine oil level.

Every 3,000 Miles (5,000 km) – *3, 9, 16, 22, 28, 34, 41, 47 miles* x *1000*
1 Check/top up cooling system level.
13 Check/top up engine oil level.
12 Check/top up brake fluid reservoir.
11 Check/top up clutch fluid reservoir.

Every 6,000 Miles (10,000 km) – *6, 19, 31, 44 miles* x *1000*
1 Check/top up cooling system.
12 Check/top up brake fluid reservoir.
11 Check/top up clutch fluid reservoir.
13 Drain engine oil and refill.
3 Renew oil filter.
10 Lubricate accelerator linkage/pedal fulcrum.
4 Clean and grease battery connections.
6 Check/top up level of gearbox oil.
8 Check/top up level of final drive oil.
14 Lubricate steering rack and pinion.
7 Lubricate handbrake linkage and cable guides.
9 Lubricate brake and clutch pedal pivots.
Lubricate all locks, door hinges, strikers and bonnet release.

Every 12,000 Miles (20,000 km) – *12.5, 25, 37.5, 50 miles* x *1000*
1 Check/top up cooling system.
12 Check/top up brake fluid reservoir.
11 Check/top up clutch fluid reservoir.
13 Drain engine oil and refill.
3 Renew oil filter.
10 Lubricate accelerator linkage/pedal fulcrum.
4 Clean and grease battery connections.
6 Check/top up level of gearbox oil.
8 Check/top up level of final drive oil.
14 Lubricate steering rack and pinion.
7 Lubricate handbrake linkage and cable guides.
9 Lubricate brake and clutch pedal pivots.
Lubricate all locks, door hinges, strikers and bonnet release.
5 Lubricate distributor.
2 Check/top up carburetter piston dampers.

[NOTE: For fuel injection vehicles, refer to Maintenance Summary at the end of Section 10.]

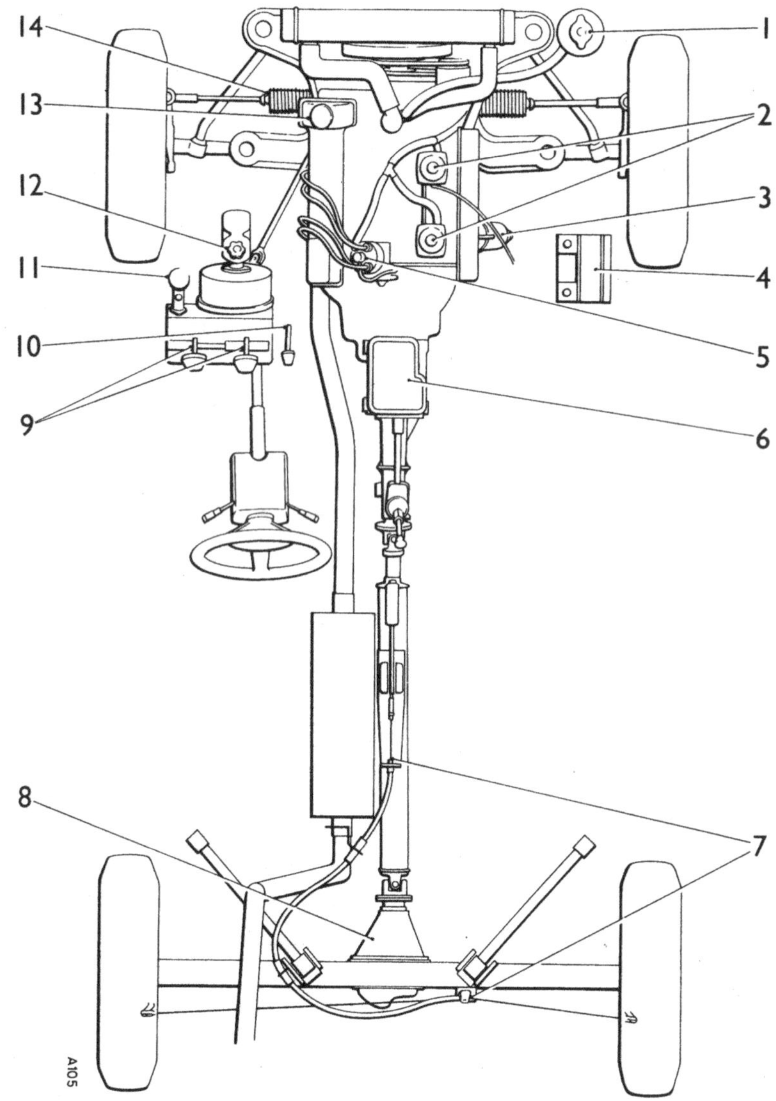

MAINTENANCE SUMMARY, U.K & EUROPE

KEY: MILEAGE x 1,000 MILES

OPERATION No.	MILEAGE	SERVICE
10.10.03	1	
10.10.06	3, 9, 15, 21, 27, 33, 39, 45	C
10.10.12	6, 18, 30, 42	A
10.10.24	12, 24, 36, 48	B

SERVICE / Operation Description	1000 Miles	C	A	B
ENGINE				
1 Check/top up engine oil		x		
2 Check/top up cooling system	x	x	x	x
3 Check/adjust operation of all washers and top up reservoirs	x	x	x	x
4 Renew engine oil	x		x	x
5 Renew engine oil filter			x	x
6 Lubricate accelerator control linkage (and pedal pivot) – check operation	x		x	x
7 Check cooling and heater system for leaks and hoses for security and condition	x	x	x	x
8 Check for oil leaks	x	x	x	x
9 Check/adjust torque of cylinder head nuts/bolts	x			
10 Check driving belts, adjust or renew	x	x	x	x
11 Check security of engine mountings	x			
12 Check/adjust carburetter idle settings	x		x	x
13 Top up carburetter piston dampers	x		x	x
14 Renew carburetter air cleaner element				x
15 Clean fuel pump filter	x			x
IGNITION				
16 Clean/adjust spark plugs			x	
17 Renew spark plugs				x
18 Check distributor points: adjust or renew			x	x
19 Lubricate distributor	x		x	x
20 Check/adjust ignition timing using electronic equipment	x		x	x

OPERATION No.	1000 Miles 1600 km	C	A	B
TRANSMISSION				
21 Check for oil leaks	x	x	x	x
22 Check/top up gearbox oil	x		x	x
23 Check/top up rear axle/final drive oil	x		x	x
24 Check tightness of propeller shaft coupling bolts				x
25 Check clutch pipes for leaks and chafing	x	x	x	x
26 Check/top up clutch fluid reservoir	x	x	x	x
27 Lubricate clutch pedal pivots				x
28 Check/top up automatic transmission fluid			x	x
29 Lubricate automatic gearbox exposed selector Linkage			x	x
STEERING AND SUSPENSION				
30 Check steering rack/gear for oil/fluid leaks	x	x	x	x
31 Check security of suspension fixings	x			x
32 Check condition and security of steering unit, joints and gaiters	x	x	x	x
33 Check/adjust front wheel alignment	x		x	x
34 Adjust front hub bearing end float				x
35 Lubricate steering rack and pinion			x	x
36 Check shock absorbers for fluid leaks	x	x	x	x
BRAKES				
37 Inspect brake pads for wear and discs for condition		x		
38 Inspect brake linings/pads for wear and drums/discs for condition			x	x
39 Check/top up brake fluid reservoir(s)	x	x	x	x
40 Check foot-brake operation: adjust to manufacturer's instructions (Self-adjusting)	x		x	x
41 Check handbrake operation, adjust to manufacturer's instructions	x		x	x
42 Lubricate brake pedal pivot(s)				x
43 Lubricate handbrake mechanical linkage and cable guides (lever pivot)			x	x
44 Check visually hydraulic hoses, pipes and unions for chafing, cracks, leaks and corrosion	x	x	x	x
45 Check brake servo hose(s) for security and condition	x	x	x	x
ELECTRICAL				
46 Check function of original equipment, i.e. interior and exterior lamps, horns, wipers and warning indicators	x	x	x	x
47 Check/top up battery electrolyte	x	x	x	x
48 Clean and grease battery connections			x	x
49 Check/adjust headlamps alignment	x	x	x	x
50 Check, if necessary renew, wiper blades		x	x	x

OPERATION No.	1000 Miles 1600 km	10.10.06	10.10.12	10.10.24
EXHAUST AND FUEL PIPES				
51 Check exhaust system for leaks and security	x	x	x	x
52 Check fuel system for leaks, pipes and unions for chafing and corrosion	x	x	x	x
WHEELS AND TYRES				
53 Check/adjust type pressures including spare	x	x	x	x
54 Check that tyres comply with manufacturer's specification		x	x	x
55 Check tightness of road wheel fastenings	x	x	x	x
56 Check tyres for external cuts in tyre fabric, exposure of ply or cord structure, lumps or bulges	x			
57 Check tyres for tread depth and visually for external cuts in tyre fabric, exposure of ply or cord structure, lumps or bulges		x	x	x
Important: If tyres do not conform with legal requirements report to owner				
BODY				
58 Lubricate all locks, hinges and door check mechanism (not steering lock)	x		x	x
59 Check condition and security of seats and seat belts	x	x	x	x
60 Check rear view mirrors for cracks and crazing		x	x	x
61 Check operation of all door, bonnet and boot locks	x			x
62 Check operation of seat belt warning system			x	x
63 Check operation of seat belt inertia reel mechanism	x	x	x	x
64 Check operation of window controls	x			x
65 Check tightness of sub frame/body mountings	x			
66 Ensure cleanliness of controls, door handles and steering wheel	x	x	x	x
ROAD TEST				
67 Road/roller test and check function of all instrumentation	x		x	x
68 Report additional work required		x	x	x

At 36,000 miles (60,000 km) or 3 years (whichever is the sooner).

Renew the air filter in the brake servo unit.

PREVENTIVE MAINTENANCE

In addition to the recommended periodical inspection of brake components it is advisable as the car ages and as a precaution against the effects of wear and deterioration to make a more searching inspection and renew parts as necessary.

It is recommended that:

1 Disc brake pads, drum brake linings, hoses and pipes should be examined at intervals no greater than those laid down in the Maintenance Summary Chart.
2 Brake fluid should be changed completely every 18,000 miles (30,000 km) or 18 months whichever is the sooner.
3 All fluid seals in the hydraulic system should be renewed and all flexible hoses should be examined and renewed if necessary every 36,000 miles (60,000 km) or 3 years whichever is the sooner. At the same time the working surfaces of the pistons and bores of the master cylinder, wheel cylinders and other slave cylinders should be examined and new parts fitted where necessary.

Care must be taken always to observe the following points:

a At all times use the recommended brake fluid.
b Never leave fluid in unsealed containers, it absorbs moisture quickly and can be dangerous if used in the brake system in this condition.
c Fluid drained from the system or used for bleeding should be discarded.
d The necessity for absolute cleanliness throughout cannot be over-emphasized.

Replacing brake shoes.

When it becomes necessary to renew the brake-shoes it is essential that only genuine shoes, with the correct grade of lining, are used. Always fit new shoes as complete sets, never individually or as a single wheel set. Serious consequences could result from out-of-balance braking due to the mixing of lining.

MAINTENANCE SUMMARY – UK & Europe

The Maintenance Summary list on pages 10–2 and 10–3 gives details of mileage intervals for the following operations. The figure in parenthesis to the left of each heading refers to the item number on the summary list.

ENGINE

(1) Check/top up engine oil level
NOTE: Allow time for the oil to drain back into the sump after running the engine.

Stand the vehicle on level ground.
1 Withdraw the dipstick, wipe it clean and push fully home again before withdrawing it for reading.
2 Add oil via the filler cap until the level reaches the 'High' mark on the dipstick.
DO NOT OVERFILL and ensure that the dipstick and filler cap are replaced.

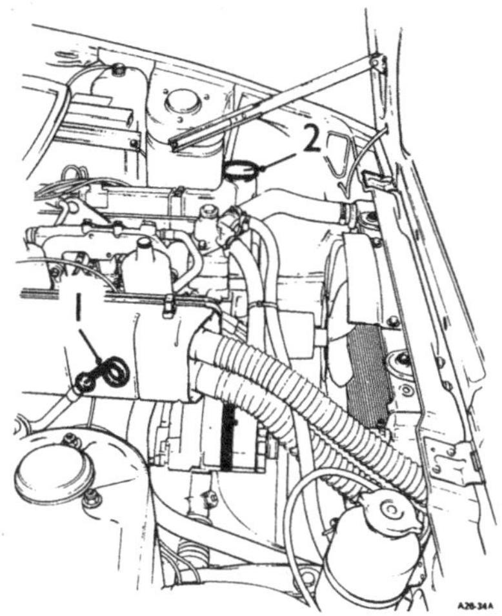

(2) Check/top up cooling system
WARNING: Do NOT remove cooling system filler caps or plugs when engine is hot.
1 Slowly turn the pressure cap anti-clockwise until the resistance of the safety stop is felt. Leave the cap in this position until all pressure is released. Press the cap downwards against the spring to clear the safety stops and continue turning until it can be lifted off.
2 Maintain the level of the coolant in the expansion tank at half full by topping up as necessary whilst the system is cool.

NOTE: Ensure that the specific gravity of the coolant is maintained, see Section 09.

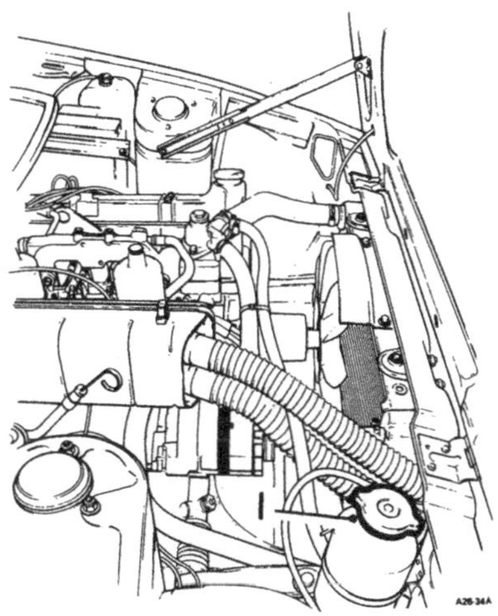

(3) Check/adjust operation of all washers and top up reservoirs
1 Replenish the container with clean soft water. The addition of a small amount of mild detergent will prevent smearing on the windscreen.
2 During freezing conditions it is beneficial to fill the container with a mixture of one part methylated spirits (wood alcohol) to two parts of water. This will assist in the dispersal of snow and ice from the screen.
3 Do not add anti-freeze solutions to the container as this will discolour the paintwork and damage wiper blades and sealing rubber.
4 Should a screen washer jet become obstructed, a strand of fine wire not greater than 0.7 mm (0.030 in) diameter can be used to clear the jet orifice.

(4) Renew engine oil
NOTE: This operation is best carried out when the engine is warm and with the vehicle standing level on a ramp or over a pit.
1 To drain the sump, unscrew the drain plug three complete turns to direct the oil stream into a receptacle while the engine is warm. When the rate of flow lessens remove the plug completely. Refit the plug and refill the sump with the appropriate grade of engine oil. The use of additives is unnecessary.

(5) Renew engine oil filter
See 12.60.01 and 12.60.02.

(6) Lubricate accelerator linkage/pedal fulcrum and check operation

Lubricate
Use an oil can, lubricate the accelerator linkage on the carburetter and the accelerator pedal pivot. Wipe away any surplus oil to avoid drips or any possibility of staining the carpet inside the car.

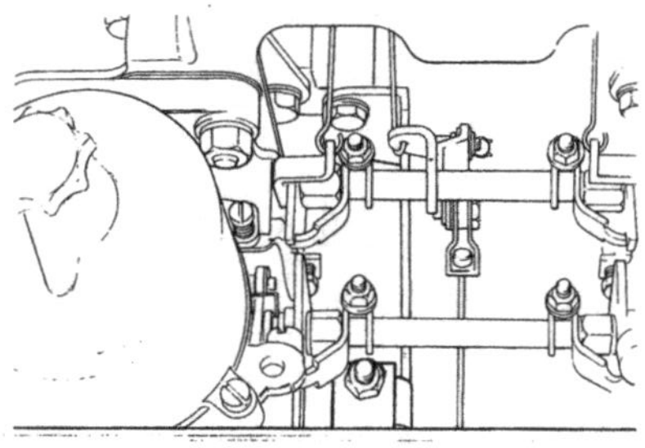

A19-121

Check operation
Check carburetter throttle response to initial movement of the accelerator pedal. If required, adjust the throttle cable at the carburetter.
Check carburetter throttle position with accelerator pedal fully depressed, by observing that the carburetter lever(s) move to a positive stop.
If the throttle movement is not satisfactory, investigate the cause and rectify as necessary, proceeding in the following order.
Check that the throttle pedal movement is not restricted by floor mats or carpet etc.
Check the carburetter lever(s) for correct position and settings, see 19.15.01 or 19.15.02, as applicable.
Check the throttle cable location and condition, see 19.20.06.

(7) Check cooling and heater systems for leaks and hoses for security and condition
1 Check for leaks from engine and radiator drain taps/plugs (where fitted).
2 Check for leaks from the water hose joints.
3 Check for leaks from the water hoses through damage or porosity.
4 Check for leaks from the water pump, thermostat housing, radiator and heater unit.
5 Report any leaks found.

(8) Check/report oil/fuel/fluid leaks
1 Check for oil leaks from the engine and transmission.
2 Check for fuel leaks from the pump, carburetter pipes, joints and unions.
3 Check for fluid leaks from the brake master cylinder, pipes, joints and unions.
4 Check for fluid leaks from the clutch master cylinder, pipes, joints and unions.
5 Report any leaks found.

(9) Check/adjust torque of cylinder head nuts/bolts
See 12.29.27.

(10) Check driving belt adjust or renew
Check and adjust alternator drive belt, see 86.10.05 if visibly worn or damaged renew.

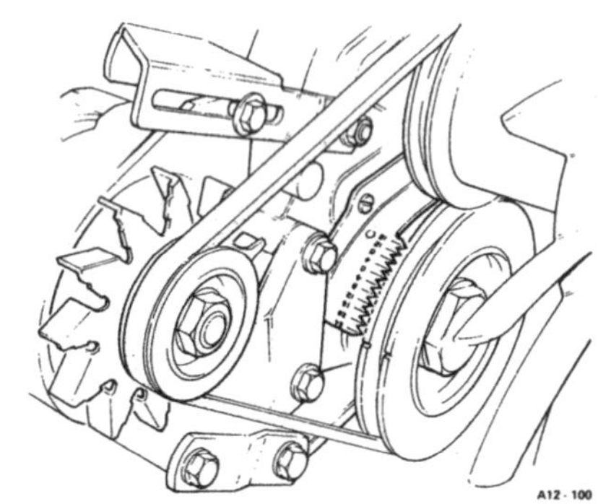

A12-100

(11) Check security of engine mountings
– Using the torque figures shown in Section 06.

(12) Check/adjust carburetter idle speed
See 19.15.01 or 19.15.02 where applicable.

(13) Top up carburetter piston dampers

1 Unscrew and remove the damper assembly from the top of the carburetter.
2 Top up with clean engine oil to bring the level 13 mm (½ in) above the top of the hollow piston rod.
3 Push the damper assembly back into position and screw the cap firmly into the reservoir.

NOTE: Under no circumstances should a heavy-bodied lubricant be used.

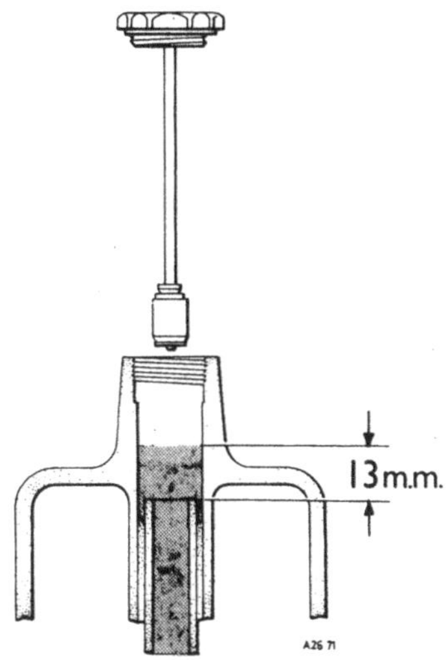

(14) Renew carburetter air cleaner element

See 19.10.08.

(15) Clean fuel pump filter

1 Remove the centre bolt from the fuel pump top cover.
2 Withdraw the filter gauge and wash it in clean fuel.
3 Using a small screwdriver, loosen any sediment in the chamber.
4 Blow out the dirt using a low pressure air line.
5 Renew the gasket if necessary and refit the filter gauge and cover.

NOTE: Leakage of air at the gasket or centre bolt will prevent the pump from functioning correctly.

IGNITION

(16) Clean/adjust spark plugs – for each spark plug in turn

1 Remove the ignition high tension lead from the plug.
2 Unscrew the plug from the engine using a special plug spanner or a box type spanner.
3 Wipe clean the ceramic body of the plug.
4 Visually check the plug body for cracks and renew the plug if any cracks are present.
5 Unscrew the end terminal cap from the plug.
6 Clean the plug terminal threads with a wire brush.
7 Clean the cap threads using a low pressure air line.
8 Screw the end terminal cap firmly into position on the plug.
9 Clean the electrode area and the plug threads with a wire brush or sand blasting machine.
10 Visually check the electrode for damage and renew the plug if there are any signs of damage.
11 Check the electrode gap, which if correct will just allow 0.025 in (0.64 mm) feeler gauge to slide slowly between the electrodes under light pressure.

If adjustment is necessary

12 a Using a suitable tool, carefully move the side electrode.
b Check the gap, repeat this procedure until the gap is correct.
13 Refit the plug to the engine and tighten to the correct torque.
14 Refit the high tension lead to the plug.

(17) Renew spark plugs – for each plug in turn

1 Remove the ignition high tension lead from the plug.
2 Unscrew the plug from the engine using a special plug spanner or a suitable box spanner.
3 Discard the plug.
4 Visually check the new plug for damage to the body and electrodes, discard the plug if there are any signs of damage.
5 Check the electrode gap on the new plug which when correct will just allow a 0.025 in (0.64 mm) feeler gauge to slide slowly between the electrodes under light pressure.
If adjustment is necessary, using a suitable tool carefully move the side electrode. Re-check the gap and repeat this procedure until the gap is correct.
6 Fit the new plug to the engine.
7 Tighten the plug to the correct torque.
8 Refit the high tension lead to the plug.

(18) Check distributor points: Adjust or renew

See 86.35.14.

(19) Lubricate distributor

See 86.35.18

(20) Check/adjust ignition timing using electronic equipment

See 86.35.00 or 86.35.15.

TRANSMISSION

(21) Check for oil leaks
See Maintenance Operation No. 8.

(22) Check/top up gearbox oil
1 With the vehicle standing in a level position, remove the oil level/filler plug and using a suitable dispenser filled with the correct grade oil, see Section 09, top up the gearbox until the oil is level with the bottom of the filler hole threads.
2 Allow surplus oil to drain away, then refit the filler plug and wipe clean.

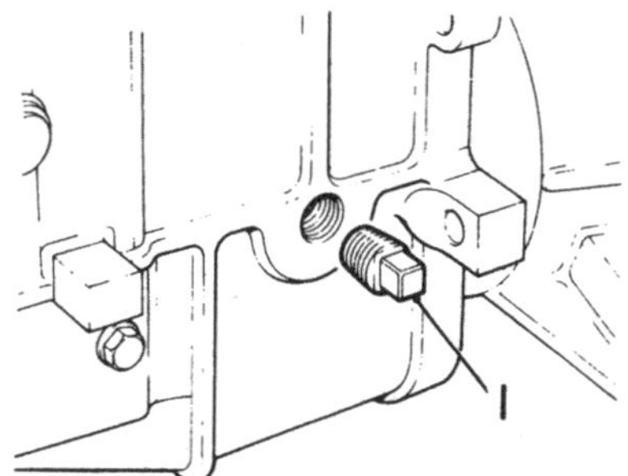

A10 18

(23) Check/top up level of rear axle/final drive oil
1 With the vehicle standing in a level position, remove the oil lever/filler plug and top up the final drive unit with the correct grade oil, see Section 09 until the oil is level with the bottom of the filler hole threads.
2 Allow surplus oil to drain away, then refit the filler plug and wipe clean.

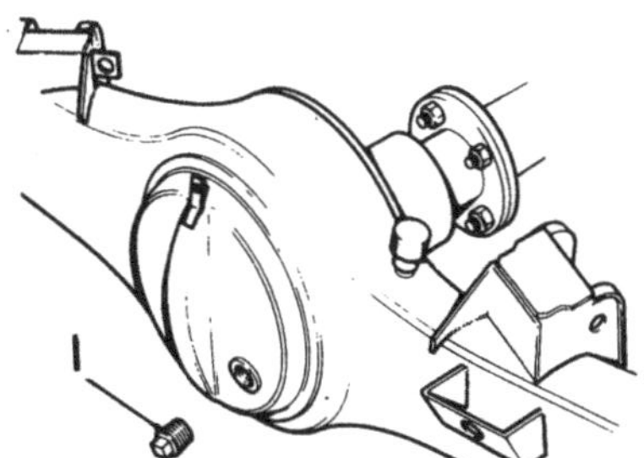

A10 19

(24) Check tightness of propeller shaft coupling bolts
Refer to Section 06 for torque settings.

(25) Check clutch pipes for leaks and chafing, check visually
1 Clutch pipes, hoses and unions for chafing, leaks and corrosion.
2 Report any defects found.

(26) Check/top up clutch fluid reservoir
Top up when required with new fluid of the type recommended, see Section 09.

(27) Lubricate clutch pedal pivots
1 Using an oil can lubricate the clutch pedal pivot.
2 Wipe away the surplus oil to prevent staining the carpet.

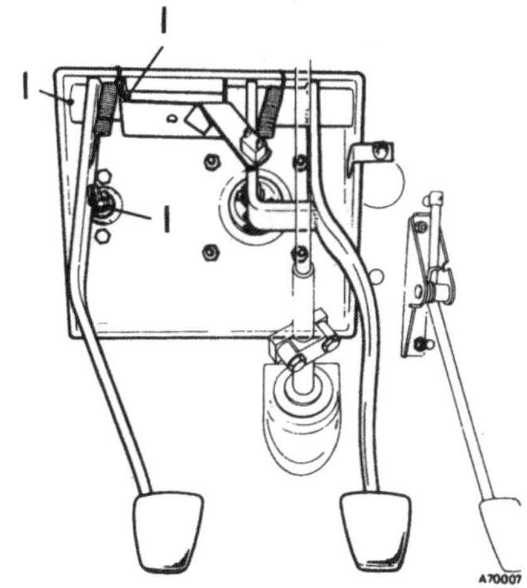

A70007

(28) Check/top-up automatic gearbox fluid
Check the fluid level as follows:
1 Stand the car on level ground and apply the handbrake firmly. Start the engine from cold and, with the footbrake firmly applied, run the engine at idle speed for 2 to 3 minutes, passing the selector lever through the complete range of positions to ensure that the transmission is primed.
2 Select the 'P' (Park) position and apply the handbrake. Leave the engine running at idle speed.
3 Remove the transmission dipstick and wipe it with a clean, non-fluffy cloth.
4 Replace the dipstick, ensuring that it is pushed fully into the tube and withdraw it immediately for reading.
5 Check the fluid level on the side of the dipstick marked 'COLD' and if necessary, add fluid; see Lubrication Recommendations (Section 09).
6 Repeat instructions 1 to 5 until the fluid level is correct.

CAUTION: Do not overfill the transmission.
Where the reverse side of the dipstick carries marks denoted 'HOT', the fluid level check may be carried out with the transmission at normal operating temperature. The procedure is as described above except that the vehicle must be driven for 25–30 km (15 miles) to warm the transmission. The check is then carried out using the 'HOT' side of the dipstick.

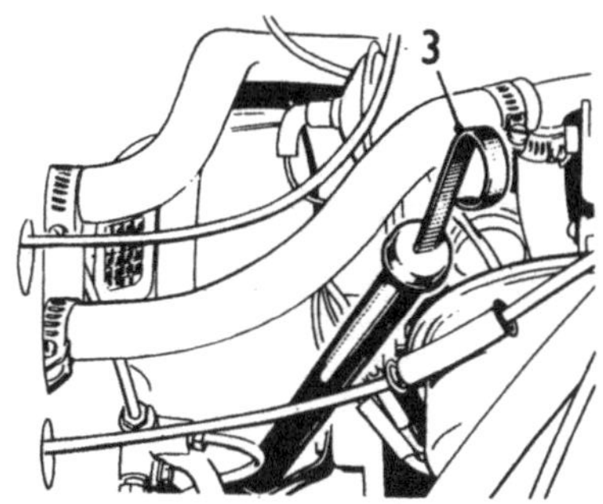

1673

IMPORTANT: When operating at high ambient temperatures and on unmetalled roads, periodically inspect and remove dust and mud deposits from the slots and screen on the underside of the torque converter housing, and from the underside of the transmission oil sump. These deposits can adversely affect proper cooling of the unit.

(29) Lubricate automatic gearbox exposed selector linkage

STEERING AND SUSPENSION

(30) Check steering rack/gear for oil fluid leaks
See Maintenance Operation No.8.

(31) Check security of suspension fixings
Refer to Section 06 for torque settings.

(32) Check condition and security of steering unit, joints and gaiters
1 Check security of steering unit mounting and steering joints – refer to Section 06 for torque settings.
2 Check steering for backlash.
3 Check condition of steering gaiters.
4 Report any defects found.

(33) Check/adjust front wheel alignment
See 57.65.01.

(34) Adjust front hub bearing end-float
1 Remove the road wheel.
2 Check the hub for end-float.
3 If any adjustment is required, remove the grease cap, split pin, and nut retaining cap.
4 Tighten the slotted nut as required to eliminate the end-float.

NOTE: A torque of 5 lbf ft (0.691 kgf m) must not be exceeded or damage may be caused to the bearings and the bearing tracks.
5 Refit the nut retaining cap.
6 Insert and lock the split pin.
7 Clean grease cap and refit.
8 Fit the road wheel.

(35) Lubricate steering rack and pinion
1 Wipe clean the plug and surrounding area.
2 Remove the plug.
3 Fit a suitable grease nipple in place of the plug.
4 Apply a grease gun, filled with the correct grade of grease, see Section 09, to the grease nipple and give five strokes only.

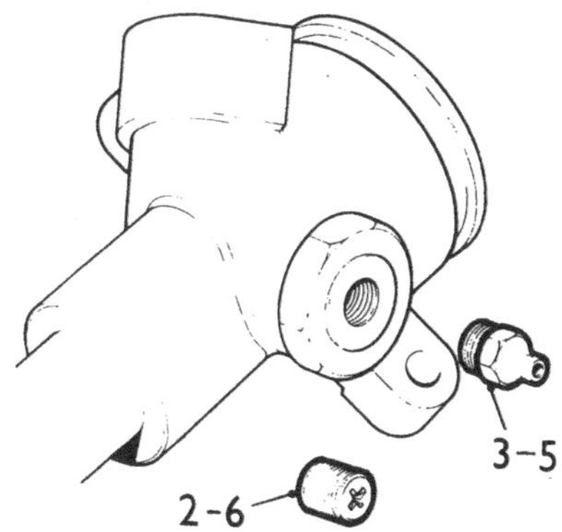

CAUTION: Over greasing can cause damage to the protective bellows.
5 Remove the grease nipple.
6 Refit the plug.
7 Wipe away any surplus grease.

(36) Check shock absorbers for fluid leaks
Report any fluid leaks found.

BRAKES

(37) Inspect brake pads for wear and discs for condition
1 Jack up the front of the car and place safely onto stands before removing the disc brake pads (see 70.40.02).

CAUTION: Do NOT depress the brake pedal while the pads are removed.
2 Report pad condition if the friction lining has been reduced to 0.125 in (3 mm) or if there is not sufficient material to provide a thickness of 0.125 in (3 mm) at the completion of a further 3,000 miles (5,000 km) motoring.
3 Check the brake discs for excessive scoring and run out and report this if present.
4 Refit the pads, road wheels and lower the car.
5 Firmly depress the footbrake several times to correctly locate the friction pads.

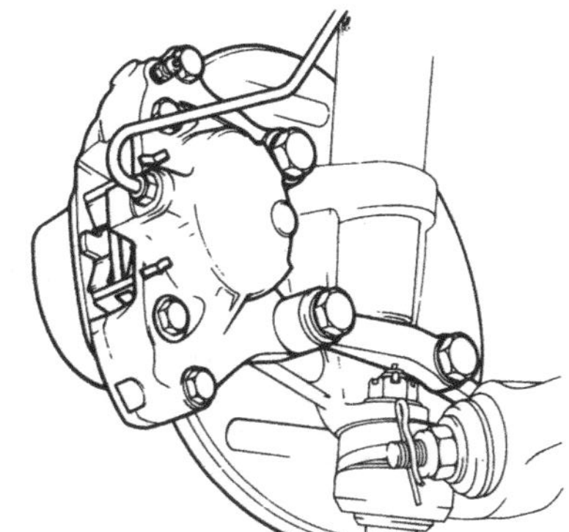

(38) Inspect brake linings/pads for wear and drums/discs for condition
1 Jack up the rear of the car and support the body on stands.
2 Remove the rear road wheels.
3 Remove the rear brake drum, see 70.10.03.
4 Check the brake linings for wear, report if they are excessively worn, damaged or contaminated by oil or grease. Remove any surplus oil or grease and dust linings and drums.
5 Remove the brake drum, see 70.10.03.
6 Fit the rear road wheels and lower the car.
7 See Maintenance No. 35 for the front brake pads and discs inspection.

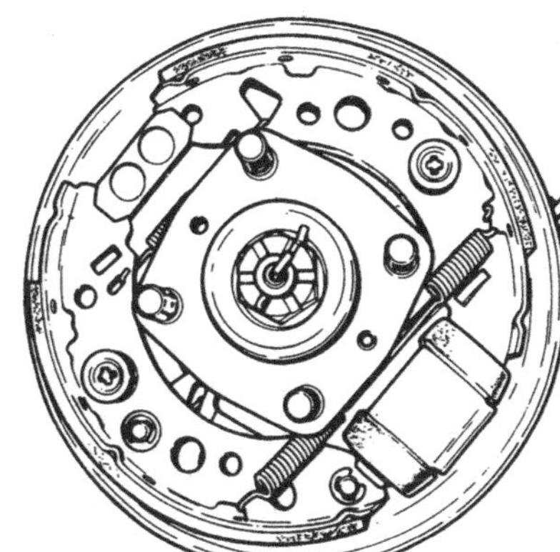

(39) Check/top up brake fluid reservoir(s)
1 Wipe clean the reservoir cap and surrounding area.
2 Remove the reservoir cap.
3 Check the fluid level against the mark on the side of the reservoir.
4 If necessary, add fluid to bring the level up to the mark on the side of the reservoir.

WARNING: Use only new fluid of the correct specification, see Section 09. DO NOT use fluid of unknown origin, or fluid that has been exposed to the atmosphere, or fluid that has been discharged during bleeding operations.
5 Replace the reservoir cap.
6 Remove any spilled fluid with a clean cloth.

CAUTION: Paintwork can be damaged by direct contact with brake fluid.

(40) Check footbrake operation, adjust to manufacturer's instructions (Self-adjusting)
1 With the handbrake off, check the brake pedal for spongy operation.
2 If the pedal has spongy operation, bleed and adjust the brakes, see 70.25.01.

(41) Check handbrake operation, adjust to manufacturer's instructions
1 If the handbrake travel is excessive, adjust the handbrake, see 70.35.10.

(42) Lubricate brake pedal pivot(s)
1 Using an oil can, lubricate the brake pedal pivot.
2 Wipe away the surplus oil to prevent staining the carpet.

(43) Lubricate handbrake mechanical linkage and cable guides (lever pivot)
1 Lubricate the handbrake pivot.
2 Smear grease around the handbrake lever cable connection, working it well into the clevis.
3 Smear grease around the brake drum cable connections, working it well into the clevis pin.
4 Grease the exposed section of the inner cable to combat corrosion.

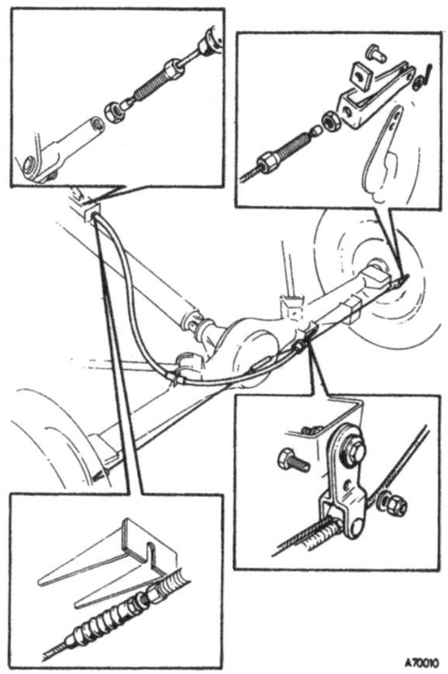

A7000

(44) Check visually hydraulic pipes and unions for chafing, leaks and corrosion
Report any defects found.

(45) Check brake servo hose(s) for security and condition

ELECTRICAL

(46) Check function of original equipment i.e. interior and exterior lamps, horns, wipers and warning indicators
In sequence, check operation of:
1 Side, tail and headlamps (including dip/main beam and 'Flash' control).
2 Instrument panel illuminations.
3 Interior light.
4 Horns.
5 Auxiliary lights.
With ignition circuits energised, check operation of:
6 All warning lights:
7 Fuel lever indicator.
8 Heater blower motor.
9 Windscreen washers.
10 Windscreen wipers
11 Direction indicators.
12 Brake lights.
13 Reversing lights.
14 Start engine and note that the oil pressure warning light has extinguished.
Check operation of:
15 Charging system warning light in relation to engine speed.
16 Temperature indicator.
17 Switch off the engine and return the ignition switch to the auxiliary position, then re-check the function of any fitted accessories e.g. a radio, that are supplied with power from this switch position.
18 Report any defects found.

(47) Check/top up battery electrolyte
Using DISTILLED WATER ONLY top up; the electrolyte level is correct when it just covers the separators.

(48) Clean and grease battery connections
1 Check the battery and surrounding area for corrosion from battery chemicals.
2 Clean off any corrosion found.
3 Check visually for cracks in the battery case and report on any case cracks found.
4 Check security of terminal connections.

(49) Check/adjust headlamp alignment
See 86.40.17, Headlamp beam aiming.

(50) Check, if necessary renew, wiper blades
1 Examine each wiper blade in turn for damage.
2 With the wiper blades in position and the windscreen wet, operate the wiper motor.
3 Check the wiper blade operation for smearing and adequate removal of dirt.
4 Stop the wiper motor.
5 If the checks in procedures 1 and 3 are not satisfactory replace the wiper blades as necessary.
See 84.15.06 or 84.15.07.

EXHAUST AND FUEL PIPES

(51) Check exhaust system for leaks and security

1 Check the security of the exhaust pipe to the manifold nuts, correct tightening torque is 37 lbf ft (5.1 kgf m).
2 Check the security of the exhaust pipe joint clips.
3 Check the security of the exhaust system mounting bolts.
4 Using a second operator, run the engine at fast idle speed.
5 Check the exhaust system joints for leaks.
6 Check the exhaust pipes for leaks arising from damage or deterioration.
7 Check the exhaust silencers for leaks arising from damage or deterioration.
8 Stop the engine.
9 Report any defects found.

(52) Check fuel system for leaks, pipes and unions for chafing and corrosion

Report any defects found.

WHEELS AND TYRES

(53) Check/adjust tyre pressures including spare

Refer to Section 04 for pressure settings.

WARNING: It can be dangerous :

a To use a car fitted with tyres in a damaged condition.
b To mix cross ply and radial ply tyres on the same axle or to fit radial ply tyres to the front wheels only.
c To use a car fitted with tyres that have a tread depth of less than 0.039 in (1,0 mm) over three-quarters of the tread width for the entire circumference of the tyre.
d To use a car with the tyres inflated to a pressure that is not suitable for the use to which the vehicle is put.

If the tyres do not conform with legal requirements report to the owner.

(54) Check that tyres comply with manufacturer's specifications

See Section 04.

Report any deviations that may influence the car's performance or the accuracy of the speedometer.

(55) Check tightness of road wheel fastenings

Refer to Section 06 for the torque settings.

(56) Check tyres for external cuts in the fabric, exposure of ply or cord structure, lumps or bulges.

(57) Check tyres for tread depth and visually for external cuts in tyre fabric, exposure of ply or cord structure, lumps or bulges

BODY

(58) Lubricate all locks and hinges (not steering lock)

Using an oil can sparingly, lubricate all locks, door hinges, strikers and bonnet release. Wipe away any surplus oil to avoid staining paintwork or interior fittings.

(59) Check condition and security of seat and seat belts

(60) Check rear view mirror for cracks and crazing

(61) Check operation of all door, bonnet and boot locks

(62) Check operation of seat belt warning system

(63) Check operation of seat belt inertia reel mechanism

(64) Check operation of window controls

(65) Check tightness of sub-frame/body mountings

Refer to Section 06 for torque settings.

(66) Ensure cleanliness of controls, door handles and steering wheel

ROAD TEST

(67) Road/roller test and check function of all instrumentation

(68) Report additional work required

At 36,000 miles (60,000 km) or 3 years, whichever is the sooner. Renew the air filter in the brake servo unit.

FAULT FINDING CHART – NORTH AMERICA FEDERAL SPECIFICATION

This chart indicates the possible areas of fault causes. Progressively work through the 'possible causes' cross referring to the key.

Possible causes shown in brackets only relate to the condition bracketed in the fault column.

[**NOTE:** For Fuel Injection models, also see Section 19.70.00]

FAULT	POSSIBLE CAUSES IN ORDER OF CHECKING
Will not start	1, 2, 17, 18, 9, 10, 11, 12, 22
Poor or erratic idle	17, 30, 21, 26, 21, 11, 12, 13, 36, 38, 39, 40, 41, 6, 4, 20, 18, 31, 19, 23, 33, 34, 35, 24, 22, 29, 25, 27, 43, 42, 32, 45, 14, 47, 51, 52
Hesitation or flat spot (cold engine)	17, 18, 28, 50, 26, 33, 5, 23, 24, 25, 11, 15, 12, 35, 4, 6, 20, 27, 36, 38, 39, 40, 41, 43, 47, 49, 52, 51, 50, 48, 22, 29, 3, 14, 42
Excessive fuel consumption	19, 31, 24, 5, 26, 25, 22, 35, 4, 11, 13, 15, 12, 47, 49, 52, 51, 50, 48, 43, 6, 20, 27, 36, 38, 39, 40, 41, 29, 3, 14, 42
Lack of engine braking or high idle speed	33, 30, 32, 12, 26, 5, 22, 24, 25
Lack of engine power	17, 18, 34, 5, 35, 11, 13, 15, 12, 26, 24, 25, 22, 4, 47, 49, 52, 51, 50, 48, 43, 6, 20, 27, 36, 38, 39, 40, 41, 29, 14, 3, 42
Engine overheating	7, 12, 51, 50, 48, 26, 25, 8, 14, 42
Engine misfires	17, 18, 11, 13, 15, 12, 35, 4, 26, 24, 25, 6, 20, 27, 36, 38, 39, 40, 41, 49, 14, 3
Fuel smells	19, 31, 24, 39, 37, 38, 40, 41, 26, 25, 35
Engine "runs on"	17, 30, 33, 32, 38, 7, 12, 51, 50, 50, 48, 26, 25, 8, 14, 42
Engine knock or "pinking"	17, 12, 7, 51, 50, 48, 26, 25, 14, 8, 42
Arcing at plugs	13, 15
Lean running (low Co)	26, 17, 18, 24, 25, 6, 20, 27, 36, 38, 39, 40, 41, 43, 49
Rich running (Excess Co)	26, 40, 24, 25, 52, 51, 50, 48, 44, 46, 45
Backfiring in exhaust	17, 18, 35, 4, 6, 12, 20, 36, 43, 27, 32
Noisy air injection pump	44, 46, 45
Engine cuts out or stalls (at idle)	17, 18, (30), 35, 4, 11, 13, 15, 12, 26, 24, 25, 6, 20, 27, 36, 38, 39, 40, 41, 43, 49, 22, 23, 29, 52, 51, 50, 48, 14, 42, 3

KEY TO POSSIBLE CAUSES

Basic Engine Checks

1 Low battery condition or poor connections.
2 Starter system deficient.
3 Poor compressions.
4 Exhaust system leaking or blocked.
5 Faults on areas of the vehicle other than engine.
6 Air leaks at inlet manifold.
7 Cooling system blocked or leaking.
8 Cylinder head gasket leaking.

Ignition System Checks

9 H.T. circuit faults.
10 L.T. power faults.
11 L.T. switching faults.
12 Ignition timing incorrect.
13 System deterioration.
14 Advance mechanism faults.
15 Spark plug faults.
16 Distributor failure.

Fuel System Checks

17 Insufficient or incorrect fuel.
18 Fuel starvation.
19 Fuel leaks.

Carburetter checks

20 Air leak at carburetter/manifold joints.
21 Diaphragm uncorrectly located or damaged.
22 Air valve sticking.
23 Obstructed float chamber or diaphragm vent holes.
24 Incorrect fuel level in float chamber.
25 Metering needle faults.
26 Mixture incorrectly set
27 Leakage at throttle spindles.
28 Piston damper inoperative.
29 Air valve spring missing, or incorrect.
30 Idle speed incorrectly set.
31 Leakage from fuel connection joints or seals.
32 Incorrectly set or faulty by-pass valve.
33 Sticking throttle.
34 Throttle linkage inhibited or incorrectly set.
35 Dirty or blocked air cleaner.

Evaporative & Crankcase Ventilation System Checks

36 Engine oil filler cap loose or leaking.
37 Fuel filler cap defective.
38 Restrictors missing or blocked.
39 Hoses blocked or leaking.
40 Adsorption canister restricted or blocked.
41 Vapour separator blocked.
42 E.G.R. valve malfunction.
43 Leaks at E.G.R. vacuum control lines.
44 Incorrectly tensioned air pump driving belt.
45 Relief valve malfunction or insufficient + pump pressure.
46 Check valve sticking.
47 Hot air inlet hose loose, adrift or blocked.
48 Flap valve jammed.
49 Vacuum pipes disconnected or leaking.
50 One way (low leak) valve faults.
51 Temperature sensor faulty, leaking or jammed.
52 Flap valve diaphragm leaking.

POSSIBLE CAUSE	CHECK AND REMEDIAL ACTION
1 Low battery condition or poor connections.	Check battery condition with hydrometer. Recharge, clean and secure terminals or renew as necessary. (if battery is serviceable but discharged, trace and rectify cause of flat battery e.g. short circuit or insufficient charge from alternator).
2 Starter system deficient.	If starter fails to turn engine briskly, check engagement circuit and connections. Check and clean main starter circuit and connections.
3 Poor compressions.	Check compressions with proprietary tester. If compressions are low or uneven, remove cylinder head for further examination and rectification.
4 Exhaust system leaking or	Check and rectify as necessary.
5 Faults on areas of vehicle other than engine.	Check for binding brakes, slipping clutch etc.
6 Air leaks at inlet manifold.	Check inlet manifold/cylinder head joint. Remake with new gasket if necessary. Check manifold tappings for leaks-seal as necessary.
7 Cooling system blocked or leaking.	Flush system and check for blockage. Check hoses and connections for security and leakage. Renew as necessary. Check thermostat and renew if faulty.
8 Cylinder head gasket leaking.	Check cylinder block/head joint for signs of leakage. Renew gasket if necessary.
9 H.T. circuit faults.	Disconnect king lead at distributor and position the end approximately 3/16″ (5 mm) from earthed metal. Switch on ignition, disconnect white/blue lead and check for spark at king lead end each time white/blue lead is disconnected. If spark is non-existent or weak, renew king lead and retest. If spark is satisfactory, check H.T. leads for fraying, deterioration and security, distributor cap for cracks, tracking, dirt or condensation, distributor rotor for deterioration and spark plugs. Renew leads, cap, rotor or plugs as necessary.
10 L.T power faults.	a) Ensure ignition switch on and check supply voltage. If less than 11 volts, check ignition switch, wiring and connections. If more than 11 volts, check voltage at coil "+ ve" terminal. If this is in the range 4–8 volts, proceed to (b) below. If it is not 4–8 volts renew the ballast resistance wire. b) Check voltage at coil "– ve" terminal. If less than 2 volts, proceed to (c) below. If more than 2 volts, check the drive resistor (8–11 ohms) and renew if necessary. c) Disconnect the white/blue lead and recheck the coil "– ve" terminal voltage. If this is more than 9 volts, check the coil by substitution. If less than 9 volts, disconnect coil "– ve" lead and again check voltage on coil "– ve" terminal. If the voltage is then 9 volts, check the coil by substitution, but, if more than 9 volts, refer to 16.
11 L.T switching faults.	Ensure ignition off and check pick-up gap in distributor. This should to 0.014 to 0.016 in (0.36 to 0.41 mm). Adjust if necessary.
12 Ignition timing incorrect.	Check ignition timing and adjust as necessary.
13 System deterioration.	Check ignition wiring for fraying, chafing and deterioration. Check distributor cap for cracks and tracking and rotor for condition. Renew leads, cap or rotor as necessary.
14 Advance mechanism faults.	Check operation of advance mechanism against figures using a stroboscopic timing light. Lubricate or renew as necessary.
15 Spark plug faults.	Remove plugs, clean, reset gap and test on proprietary spark plug testing machine. Renew if in doubt.
16 Distributor failure.	a) Ensure that the distributor is earthed. b) With the distributor cap removed, crank the engine and check that the distributor shaft rotates. If not, investigate and rectify. c) Recheck as detailed in 9, 10, 11, and 15. d) As the last resort, renew the distributor complete.
17 Insuficient or incorrect fuel.	Ensure that the fuel tank has an adequate amount of the correct type and grade of fuel. **NOTE**: It is essential that unleaded fuels, are used in this vehicle otherwise serious damage can be caused to the catalytic converter.
18 Fuel starvation.	Insert a pressure gauge into the pump to carburetter fuel line and check the pressure with the engine running. If not

POSSIBLE CAUSE	CHECK AND REMEDIAL ACTION
	satisfactory, check fuel feed and breather pipes for leaks or blockage. Renew connectors if damaged or deteriorated. If contamination of fuel is discovered, flush fuel system and clean or renew the tank filter. If necessary, renew the pump to rectify low pressure.
19 Fuel leaks.	Check fuel system for leaks and rectify as necessary. Renew any doubtful connectors.
20 Air leak at carburetter/ manifold joints.	Check joints. Remake with new gaskets if necessary. Check CO.
21 Diaphragm incorrectly located or damaged.	Remove air valve cover and check location of diaphragm – piston depression holes should be in line with and face towards the throttle spindle. Renew diaphragm if damaged or deteriorated.
22 Air valve sticking.	Clean air valve and guide and reassemble. Check free movement by hand-unit should move freely and return, to carburetter bridge with an audible 'click'.
23 Obstructed float chamber, or diaphragm vent holes.	23 Obstructed float chamber or Ensure that gaskets, piping or dirt are not blocking the holes.
24 Incorrect fuel level in float chamber.	Reset float height and clean or renew needle valve and or float as necessary.
25 Metering needle faults.	Check that needle is correct type. Ensure that shoulder of needle is flush with face of air valve and that needle bias is correct.
26 Mixture incorrectly set.	Check/adjust CO at idle.
27 (After considerable service) leakage at throttle spindles.	Check spindles and seals – renew as necessary.
28 Piston damper inoperative.	Check/top up damper oil level. Check damper operation by raising piston by hand. Resistance should be apparent.
29 Air valve spring missing, or incorrect.	Check that spring is correct (colour code) and refit/renew as necessary.
30 Idle speed incorrectly set.	Check/adjust idle speed.
31 Leakage from fuel connection, joints or seals.	Check fuel inlet connection – renew connector if necessary. Inspect float chamber joint and sealing plug 'O' ring seal – renew if necessary.
32 Incorrectly set or faulty	Reset or renew by-pass valve as a unit.
33 Sticking throttle.	Check throttle operation – free off and reset as necessary.
34 Throttle linkage inhibited or incorrectly set	Check that throttle linkage and accelerator pedal are not inhibited by carpets, mats, sound insulation pads etc. and that full throttle is obtainable. Lubricate and reset if necessary.
35 Dirty or blocked air cleaner.	Inspect air cleaner element. Fit new element of correct type if necessary.

POSSIBLE CAUSE	CHECK AND REMEDIAL ACTION
36 Engine oil filler cap loose or leaking.	Check cap for security. Renew cap if seal is deteriorated.
37 Fuel filler cap defective.	Check seal for condition – renew if deteriorated. Check filler cap for security – rectify or renew as necessary.
38 Restrictors missing or	Check and clear or renew as necessary.
39 Hoses blocked or leaking.	Check and clear as necessary. Renew any deteriorated hoses.
40 Adsorption canister restricted	Inspect and renew if necessary.
41 Vapour separator blocked.	Check and clear or renew as necessary.
42 E.G.R valve malfunction.	Check the function of the E.G.R valve on the vehicle. If not satisfactory, remove and clean valve and pipework. Renew valve if spring is broken, diaphragm ruptured or other fault obvious.
43 Leaks at E.G.Rvacuum control lines.	Check security and condition of control lines. Renew connectors or lines as necessary.
44 Incorrectly tensioned air pump driving belt.	Check/adjust belt tension. Renew belt if necessary.
45 Relief valve malfunction or insufficient + pump pressure	Check that the relief valve operates at the correct pressure by using a pressure guage between the relief valve and outlet pipe to measure the point at which air pressure ceases to rise with engine speed. If any deviation is apparent, renew the pump/valve assembly. If the pump does not produce enough pressure to operate the relief valve check 44. and renew pump if necessary.
46 Check valve sticking.	Disconnect air hose at check and blow through valve. If the valve obstructs flow from the pump to the manifold or allows flow from the manifold towards the pump, renew the valve after checking hoses for blockage.
47 Hot air inlet hose loose, adrift or blocked.	Check hot air inlet hose for condition and security. Renew if necessary.
48 Flap valve jammed.	Check operation of flap valve. Renew flap valve if performance is unsatisfactory.
49 Vacuum pipes disconnected or leaking.	Check vacuum pipes for security of connections and deterioration. Renew if necessary.
50 One way (low leak) valve faults.	Blow through valve to check "one-way" action. If the valve leaks, fit a new valve.
51 Temperature sensor faulty leaking or jammed.	Check and renew if necessary.
52 Flap valve diaphragm leaking.	Check with a distributor vacuum test unit. If leakage is apparent, renew servo motor and cover unit.

FAULT FINDING CHART – AMERICA CALIFORNIA SPECIFICATION

This chart indicates the possible areas of fault causes. Progressively work through the 'possible causes' cross referring to the key.

Possible causes shown in brackets only relate to the condition bracketed in the fault column.

[**NOTE:** For Fuel Injection models, also see Section 19.70.00.]

FAULT	POSSIBLE CAUSES IN ORDER OF CHECKING
Will not start (warm engine)	1, 2, 18, 19, 37, (38), 9, 10, 11, 12, 23
Poor or erratic idle	18, 31, 22, 27, 16, 10, 11, 12, 15, 13, 39, 42, 43, 44, 45, 6, 4, 21, 19, 32, 20, 24, 34, 35, 36, 25, 38, 23, 30, 26, 24, 28, 47, 46, 33, 49, 14, 51, 55, 56, 52, 8, 3
Hesitation or flat spot (cold engine)	18, 19, 29, (37, 54), 27, 34, 5, 24, 25, 30, 26, 11, 13, 16, 12, 15, 36, 4, 6, 21, 28, 39, 42, 43, 44, 45, 47, 51, 53, 56, 55, 54, 52, 23, 30, 3, 14, 46
Excessive fuel consumption	20, 32, 25, 5, 27, 38, 26, 23, 36, 4, 11, 13, 16, 12, 15, 51, 53, 56, 55, 54, 52, 47, 6, 21, 28, 39, 42, 43, 44, 45, 30, 3, 14, 46
Lack of engine braking or high idle speed	34, 31, 15, 33, 12, 27, 5, 23, 25, 38, 26
Lack of engine power	18, 19, 35, 5, 36, 11, 13, 16, 12, 27, 25, 38, 26, 23, 4, 51, 53, 56, 55, 54, 52, 47, 6, 21, 15, 28, 39, 42, 43, 44, 47, 30, 14, 3, 46
Engine overheating	7, 12, 15, 55, 54, 52, 27, 26, 8, 14, 46
Engine misfires	18, 19, 11, 13, 16, 12, 36, 4, 27, 25, 38, 26, 6, 21, 15, 28, 39, 42, 43, 44, 45, 53, 14, 3
Fuel smells	20, 32, 25, 43, 40, 42, 44, 45, 38, 27, 26, 36
Engine 'runs on'	18, 31, 34, 33, 42, 41, 7, 12, 27, 55, 54, 52, 27, 26, 38, 8, 14, 46
Engine knock or 'pinking'	18, 12, 15, 7, 55, 54, 52, 27, 26, 14, 8, 46
Arcing at plugs Lean running (low CO)	13, 16 27, 18, 19, 25, 26, 6, 21, 15, 28, 39, 42, 43, 44, 45, 47, 53
Rich running (excess CO)	27, 44, 25, 38, 26, 56, 55, 54, 52, 48, 50, 49
Backfiring in exhaust	18, 19, 36, 4, 6, 12, 15, 21, 39, 47, 28, 33
Noisy air injection pump	48, 50, 49
Engine cuts out or stalls (at idle)	18, 19, (31), (37), 36, 4, 11, 13, 16, 12, 38, 27, 25, 26, 6, 21, 15, 28, 39, 42, 43, 44, 45, 47, 53, 23, 24, 30, 56, 55, 54, 52, 14, 46, 3

KEY TO POSSIBLE CAUSES

Engine checks
1 Low battery condition or poor connections.
2 Starter system deficient.
3 Poor compressions.
4 Exhaust system leaking or blocked.
5 Faults on areas of vehicle other than engine.
6 Air leaks at inlet manifold.
7 Cooling system blocked or leaking.
8 Cylinder head gasket leaking.

Ignition System Checks
9 H.T. circuit faults.
10 L.T. power faults.
11 L.T. switching faults.
12 Ignition timing incorrect.
13 System deterioration.
14 Advance mechanism faults.
15 Vacuum system faults.
16 Spark plug faults.
17 Distributor failure.

Fuel System Checks
18 Insufficient or incorrect fuel.
19 Fuel starvation.
20 Fuel leaks.

Carburetter Checks
21 Air leak at carburetter/manifold joints.
22 Diaphragm incorrectly located or damaged.
23 Air valve sticking.
24 Obstructed float chamber or diaphragm vent holes.
25 Incorrect fuel level in float chamber.
26 Metering needle faults.
27 Mixture incorrectly set.
28 Leakage at throttle spindles (after considerable service).
29 Piston dampers inoperative.
30 Air valve spring missing or incorrect.
31 Idle speed incorrectly set.
32 Leakage from fuel connection joints or seals.
33 Incorrectly set or faulty by-pass valve.
34 Sticking throttle.
35 Throttle linkage inhibited or incorrectly set.
36 Dirty or blocked air cleaner.
37 Choke inoperative.
38 Choke sticking on.

Evaporative & Crankcase Ventilation System Checks.
39 Engine oil filler cap loose or leaking.
40 Fuel Filler cap defective.
41 Anti-run on valve inoperative.
42 Restrictors missing or blocked.
43 Hoses blocked or leaking.
44 Adsorption canister restricted or blocked.
45 Vapour separator blocked.

E.G.R. System Checks
46 E.G.R. valve malfunction.
47 Leaks at E.G.R. vacuum control lines.

Air Injection System Checks
48 Incorrectly tensioned air pump driving belt.
49 Relief valve malfunction or insufficient + pump pressure.
50 Check valve sticking.

Miscellaneous
51 Hot air inlet hose loose, adrift or blocked.
52 Flap valve jammed.
53 Vacuum piped disconnected or leaking.
54 One way (low leak) valve faults.
55 Temperature sensor faulty, leaking or jammed.
56 Flap valve diaphragm leaking.

POSSIBLE CAUSE	CHECK AND REMEDIAL ACTION
1 Low battery condition or poor connections.	Check battery condition with hydrometer. REcharge, clean and secure terminals or renew as necessary. (if battery is serviceable but discharged, trace and rectify cause of flat battery e.g. short circuit or insufficient charge from alternator).
2 Starter system deficient.	If starter fails to turn engine briskly, check engagement circuit and connections. Check and clean main starter circuit and connections.
3 Poor compressions.	Check compressions with proprietary tester. If compressions are low or uneven, remove cylinder head for further examination and rectification.
4 Exhaust system leaking or blocked.	Check and rectify as necessary.
5 Faults on areas of vehicle other than engine.	Check for binding brakes, slipping clutch etc.
6 Air leaks at inlet manifold.	Check inlet manifold/cylinder head joint. Remake with new gasket if necessary. Check manifold tappings for leaks-seal as necessary.
7 Cooling system blocked or leaking.	Flush system and check for blockage. Check hoses and connections for security and leakage. Renew as necessary. Check thermostat and renew if faulty.
8 Cylinder head gasket leaking.	Check cylinder block/head joint for signs of leakage. Renew gasket if necessary.
9 H.T. circuit faults.	Disconnect king lead at distributor and position the end approximately 3/16" (5 mm) from earthed metal. Switch on ignition, disconnect white/blue lead and check for spark at king lead end each time white/blue lead is disconnected. If spark is non-existent or weak, renew king lead and retest. If spark is satisfactory, check H.T. leads for fraying, deterioration and security, distributor cap for cracks, tracking, dirt or condensation, distributor rotor for deterioration and spark plugs (16). Renew leads, cap, rotor or plugs as necessary.
10 L.T. power faults.	a) Ensure ignition switch on and check supply voltage. If less than 11 volts, check ignition switch, wiring and connections. If more than 11 volts, check voltage at coil "+ ve" terminal. If this is in the range 4–8 volts, proceed to (b) below. If it is not 4–8 volts renew the ballast resistance wire. b) Check voltage at coil "– ve" terminal. If less than 2 volts, proceed to (c) below. If more than 2 volts, check the drive resistor (8–11 ohms) and renew if necessary. c) Disconnect the white/blue lead and recheck the coil "– ve" terminal voltage. If this is more than 9 volts, check the coil by substitution. If less than 9 volts, disconnect coil "– ve" lead and again check voltage on coil "– ve" terminal. If the voltage is then 9 volts, check the coil by substitution, but, if more than 9 volts, refer to 17.
11 L.T. switching faults.	Ensure ignition off and check pick-up gap in distributor. This should to 0.014 to 0.016 in (0.36 to 0.41 mm). Adjust if necessary.
12 Ignition timing incorrect.	Check ignition timing and adjust as necessary
13 System deterioration.	Check ignition wiring for fraying, chafing and deterioration. Check distributor cap for cracks and tracking and rotor for condition. Renew leads, cap or rotor as necessary.
14 Advance mechanism faults.	Check operation of advance mechanism against figures using a stroboscopic timing light. Lubricate or renew as necessary.
16 Spark plug faults.	Remove plugs, clean, reset gap and test on proprietary spark plug testing machine. Renew if in doubt.
17 Distributor failure.	a) Ensure that the distributor is earthed. b) With the distributor cap removed, crank the engine and check that the distributor shaft rotates. If not, investigate and rectify. c) Recheck as detailed in 9, 10, 11, and 16. d) As the last resort, renew the distributor complete.

POSSIBLE CAUSE	CHECK AND REMEDIAL ACTION
18 Insuficient or incorrect fuel.	Ensure that the fuel tank has an adequate amount of the correct type and grade of fuel. **NOTE**: It is essential that unleaded fuels, are used in this vehicle otherwise serious damage can be caused to the catalytic converter.
19 Fuel starvation.	Insert a pressure gauge into the pump to carburetter fuel line and check the pressure with the engine running. If not satisfactory, check fuel feed and breather pipes for leaks or blockage (see also 32 and 24). Renew connectors if damaged or deteriorated. If contamination of fuel is discovered, flush fuel system and clean or renew the tank filter. If necessary, renew the pump to rectify low pressure.
20 Fuel leaks.	Check fuel system for leaks and rectify as necessary. Renew any doubtful connectors.
21 Air leak at carburetter/manifold joints.	Check joints. Remake with new gaskets if necessary. Check CO.
22 Diaphragm incorrectly located or damaged.	Remove air valve cover and check location of diaphragm – piston depression holes should be in line with and face towards the throttle spindle. Renew diaphragm if damaged or deteriorated.
23 Air valve sticking.	Clean air valve and guide and reassemble. Check free movement by hand-unit should move freely and return, to carburetter bridge with an audible 'click'.
24 Obstructed float chamber or diaphragm vent holes.	Ensure that gaskets, piping or dirt are not blocking the holes.
25 Incorrect fuel level in float chamber.	Reset float height and clean or renew needle valve and or float as necessary.
26 Metering needle faults.	Check that needle is correct type. Ensure that shoulder of needle is flush with face of air valve and that needle bias is correct.
27 Mixture incorrectly set.	Check/adjust CO at idle.
28 (After considerable service) leakage at throttle spindles.	Check spindles and seals – renew as necessary.
29 Piston damper inoperative.	Check/top up damper oil level – Check damper operation by raising piston by hand. Resistance should be apparent.
30 Air valve spring missing, or incorrect.	Check that spring is correct (colour code) and refit/renew as necessary.
31 Idle speed incorrectly set.	Check/adjust idle speed.
32 Leakage from fuel connection, joints or seals.	Check fuel inlet connection – renew connector if necessary. Inspect float chamber joint and sealing plug 'O' ring seal – renew if necessary.
33 Incorrectly set or faulty by-pass valve.	Reset or renew by-pass valve as a unit.
34 Sticking throttle.	Check throttle operation – free off and reset as necessary.
35 Throttle linkage inhibited or incorrectly set.	Check that throttle linkage and accelerator pedal are not inhibited by carpets, mats, sound insulation pads etc. and that full throttle is obtainable. Lubricate and reset if necessary.
36 Dirty or blocked air cleaner.	Inspect air cleaner element. Fit new element of correct type if necessary.
37 Choke inoperative.	Check connections and seals of mixture feed pipe at inlet manifold and autochoke. Check connections of water feed pipes and bleed autochoke system. Check electrical connections, at cut-out switch on inlet manifold, and autochoke. If still unsatisfactory renew autochoke.
38 Choke sticking on.	Bleed autochoke.
39 Engine oil filler cap loose or leaking.	Check cap for security. Renew cap if seal is deteriorated.
40 Fuel filler cap defective.	Check seal for condition – renew if deteriorated. Check filler cap for security- – rectify or renew as necessary.
41 Anti-run on valve inoperative	Apply current to the run on valve solenoid with the engine running. If the engine does not stop, the valve is suspect – renew valve and retest. If the engine does not stop, check wiring between valve, control (oil pressure) switch and the ignition switch.
42 Restrictors missing or blocked.	Check and clear or renew as necessary.
43 Hoses blocked or leaking.	Check and clear as necessary. Renew any deteriorated hoses.
44 Adsorption canister restricted or blocked.	Inspect and renew if necessary.
45 Vapour separator blocked.	Check and clear or renew as necessary.
46 E.G.R. valve malfunction.	Check the function of the E.G.R. valve on the vehicle. If not satisfactory, remove and clean valve and pipework. Renew valve if spring is broken, diaphragm ruptured or other fault obvious.
47 Leaks at E.G.R. vacuum control lines.	Check security and condition of control lines. Renew connectors or lines as necessary.
48 Incorrectly tensioned air pump driving belt.	Check/adjust belt tension. Renew belt if necessary.
49 Relief valve malfunction. or insufficient + pump pressure.	Check that the relief valve operates at the correct pressure by using a pressure guage between the relief valve and outlet pipe to measure the point at which air pressure ceases to rise with engine speed. If any deviation is apparent, renew the pump/valve assembly. If the pump does not produce enough pressure to operate the relief valve check 48 and renew pump if necessary.

POSSIBLE CAUSE	CHECK AND REMEDIAL ACTION
50 Check valve sticking.	Disconnect air hose at check and blow through valve. If the valve obstructs flow from the pump to the manifold or allows flow from the manifold towards the pump, renew the valve after checking hoses for blockage.
51 Hot air inlet hose loose, adrift or blocked.	Check hot air inlet hose for condition and security. Renew if necessary.
52 Flap valve jammed.	Check operation of flap valve. Renew flap valve if performance is unsatisfactory.
53 Vacuum pipes disconnected or leaking.	Check vacuum pipes for security of connections and deterioration. Renew if necessary.
54 One way (low leak) valve faults.	Blow through valve to check "one-way" action. If the valve leaks, fit a new valve.
55 Temperature sensor faulty, leaking or jammed.	Check and renew if necessary.
56 Flap valve diaphragm leaking.	Check and renew if necessary.
	Check with a distributor vacuum test unit. If leakage is apparent, renew servo motor and cover unit.

MAINTENANCE SUMMARY, USA Market Specification
Carburetter engined vehicles
KEY: MILEAGE *x* 1,000 MILES

OPERATION No.	MILEAGE
10.10.03	1
10.10.14	3, 9, 16, 22, 28, 34, 41, 47
10.10.26	6, 19, 31, 44
10.10.28	12.5, 37.5
10.10.50	25, 50

Key to operation numbers and mileage	10.10.03	10.10.14	10.10.26	10.10.28	10.10.50
Operation Description					
ENGINE					
1 Check/top up engine oil		X			
2 Check/top up cooling system	X	X	X	X	X
3 Check/adjust operation of all washers and top up reservoir	X	X	X	X	X
4 Renew engine oil	X		X	X	X
5 Renew engine oil filter			X	X	X
6 Lubricate accelerator control linkage and pedal pivot – check operation	X		X	X	X
7 Check cooling/heater systems for leaks and hoses for security and condition	X	X	X	X	X
8 Check for oil leaks	X	X	X	X	X
9 Check/adjust torque of cylinder head nuts/bolts	X				
10 Check driving belts, adjust or renew	X			X	X
11 Check security of engine mountings	X				
12 Check/adjust carburetter idle settings	X			X	X
13 Top up carburetter, piston damper(s)	X			X	X
14 Renew carburetter/air intake, air cleaner element				X	X
15 Check/adjust deceleration by-pass valve	X			X	X
16 Check security of E.G.R. valve operating lines	X				
17 Check E.G.R. system					X
18 Check air-intake temperature control system				X	X

Key to operation numbers and mileage intervals	10. 10. 03	10. 10. 14	10. 10. 26	10. 10. 28	10. 10. 50
19 Check crankcase breathing and evaporate loss system hoses for security	X				
20 Check crankcase breathing and evaporative loss systems. Check hoses/pipes and restrictors for blockage, security and condition				X	X
21 Check air injection system hoses/pipes for security and condition				X	X
22 Check/top up air conditioning compressor fluid				X	X
23 Renew adsorption canister					50 only
24 Renew catalytic converter					50 only
IGNITION					
25 Check security of distributor vacuum unit and operation of vacuum unit (California Market only)	X			X	X
26 Lubricate distributor				X	X
27 Check/adjust ignition timing using electronic equipment	X			X	X
28 Check ignition wiring for fraying, chafing and deterioration	X			X	X
29 Clean distributor cap, check for cracks and tracking				X	X
30 Renew spark plugs				X	X
31 Check coil performance on oscilloscope				X	X
TRANSMISSION					
32 Check/top up gearbox oil	X		X	X	X
33 Check/top up rear axle/final drive oil	X		X	X	X
34 Check for oil leaks	X	X	X	X	X
35 Check tightness of propeller shaft coupling bolts				X	X
36 Check/top up automatic gearbox fluid	X		X	X	X
37 Lubricate automatic gearbox exposed selector linkage			X	X	X

Key to operation numbers and mileage	10. 10. 03	10. 10. 14	10. 10. 26	10. 10. 28	10. 10. 50
38 Check clutch pipes for leaks and chafing	X	X	X	X	X
39 Check/top up clutch fluid reservoir	X	X	X	X	X
40 Lubricate clutch pedal pivots				X	X
STEERING AND SUSPENSION					
41 Check steering rack for oil leaks	X	X	X	X	X
42 Check security of suspension fixings	X			X	X
43 Check condition and security of steering unit joints and gaiters	X	X	X	X	X
44 Check/adjust front wheel alignment	X		X	X	X
45 Adjust front hub bearing end float				X	X
46 Lubricate steering rack and pinion			X	X	X
47 Check shock absorber for fluid leaks	X	X	X	X	X
BRAKES					
48 Inspect brake pads for wear and discs for condition		X			
49 Inspect brake linings/pads for wear, and drums/discs for condition			X	X	X
50 Check/top up brake fluid reservoir	X	X	X	X	X
51 Check footbrake operation; adjust to manufacturer's instructions (self-adjusting)	X		X	X	X
52 Check handbrake operation; adjust to manufacturer's instructions	X		X	X	X
53 Lubricate brake pedal pivot				X	X
54 Lubricate handbrake mechanical linkage and cable guides			X	X	X
55 Check visually, hydraulic pipes and unions for chafing leaks and corrosion	X	X	X	X	X
56 Check brake servo hoses for security and condition	X	X	X	X	X
57 Renew **Brake** servo filter unit	**37,500 miles**				

OPERATION No.	10. 10. 03	10. 10. 14	10. 10. 26	10. 10. 28	10. 10. 50
ELECTRICAL					
58 Check function of original equipment, i.e. interior and exterior, lamps, horns, wipers and warning indicators	X	X	X	X	X
59 Check/top up battery electrolyte	X	X	X	X	X
60 Clean and grease battery connections			X	X	X
61 Check/adjust headlamp alignment	X	X	X	X	X
62 Check, if necessary renew wiper blades		X	X	X	X
63 Check output of charging system	X		X	X	X
64 Check brake system warning light	X	X	X	X	X
EXHAUST AND FUEL PIPES					
65 Check exhaust system for leaks and security	X	X	X	X	X
66 Check fuel system for leaks, pipes and unions for chafing and corrosion	X	X	X	X	X
67 Check condition of fuel filler cap seal				X	X
WHEELS AND TYRES					
68 Check/adjust tyre pressures including spare	X	X	X	X	X
69 Check that tyres comply with manufacturer's specification		X	X	X	X
70 Check tightness of road wheel fastenings	X	X	X	X	X
71 Check tyres for external cuts in tyre fabric, exposure of ply or cord structure, lumps or bulges	X				
72 Check tyres for tread depth and visually for external cuts in fabric, exposure of ply or cord structure, lumps or bulges **IMPORTANT**: If tyres do not conform with legal requirements report to the owner		X	X	X	X

OPERATION No.	10. 10. 03	10. 10. 14	10. 10. 26	10. 10. 28	10. 10. 50
BODY					
73 Lubricate all locks, hinges and door check mechanism (not steering lock)	X		X	X	X
74 Check condition and security of seats and seat belts	X	X	X	X	X
75 Check rear view mirrors for security, cracks and crazing		X	X	X	X
76 Check operation of all door, bonnet and boot locks	X			X	X
77 Check operation of seat belt warning system		X	X	X	X
78 Check operation of seat belt inertia real mechanism	X	X	X	X	X
79 Check operation of window controls	X			X	X
80 Check tightness of sub frame/body mountings	X				
81 Ensure cleanliness of controls, door handles, steering wheel	X	X	X	X	X
GENERAL					
82 Road/roller test and check function of all instrumentation	X		X	X	X
83 Report additional work required		X	X	X	X

At 37,500 miles or 3 yrs whichever is the sooner – Renew air filter in brake servo unit

ADDITIONAL WORK FOR CANADIAN MARKET SPECIFICATION VEHICLES

1 Check/adjust choke settings (manual choke) – operation Nos. 10.10.03, 10.10.28, 10.10.50.
2 Check operation of distributor vacuum unit – operation Nos. 10.10.28, 10.10.50.

(1) Check/adjust choke settings (manual choke)
1 Maintain the engine at normal running temperature.
2 Check that the mixture control cam lever on both carburetters returns to its stop.
3 Check and if necessary adjust the mixture control cable so that there is free movement.
4 Pull-out the mixture control cable knob on the control cowl approximately ¼ in until the fast idle cams are correctly engaged with the ball locators.
5 Slaken the fast idle screw lock-nut on both carburetters, start the engine and rotate the screw head (on both carburetters) against the cam until the engine revolutions reach a steady 1600 rev/min.
6 Use the air flow meter to check that the carburetters are in balance at 1600 rev/min. if necessary adjust the fast idle screws as required.
7 Tighten the fast idle screw lock-nut on both carburetters and push the mixture control knob fully home.

NOTE: An engine set to 1600 rev/min. while hot is equivalent to a fast idle speed of approximately 1300 rev/min. when cold.

(2) Check operation of distributor vacuum unit.
1 Check security of distributor vacuum unit operating line.
2 Start the engine and warm to normal running temperature.
3 When the engine is idling steadily, disconnect the vacuum pipe at the distributor.
4 A noticeable rise in engine speed (approximately 500 rev/min) should be apparent if the vacuum unit is functioning, otherwise overhaul or renew the distributor.
5 When satisfied that the vacuum unit is operating correctly, reconnect the vacuum unit pipe, ensuring a secure connection.

MAINTENANCE SUMMARY
North American Specification

The Maintenance summary on pages 10–16 to 10–18 give details of mileage intervals for the following operations. The figure in parenthesis to the left of each heading refers to the item number on the summary list.

ENGINE
(1) Check/top up engine oil
NOTE: Allow time for oil to drain back into sump after running engine.
Stand vehicle on level ground.
1 Withdraw the dipstick, wipe it clean and push fully home again before withdrawing it for reading.
2 Add oil via the filler cap until the level reaches the 'High' mark on the dipstick.
DO NOT OVERFILL and ensure that the dipstick and filler cap are replaced.

(2) Check/top up cooling system
WARNING: Do NOT remove cooling system filler caps or plugs when engine is hot.

[1975–1978]

1 Slowly turn the pressure cap anti-clockwise until the resistance of the safety stop is felt. Leave the cap in this position until all pressure is released. Press the cap downwards against the spring to clear the safety stops and continue turning until it can be lifted off.
2 Maintain the level of the coolant in the expansion tank at half full by topping up as necessary whilst the system is cool.
NOTE: Ensure that the specific gravity of the coolant is maintained – see 09–3.

[1979 and later]

Top up the coolant level until it is 25 mm (1 inch) below the header tank filler neck. Run the engine at approximately 1,500 rev/min until the thermostat opens. Stop the engine, and observing the following warning, check coolant level, and top up to 25 mm (1 inch) below the header filler neck.
WARNING: When it is necessary to remove the pressure/filler cap from a hot engine, exercise great care by protecting the hands against escaping steam. Slowly turn the pressure cap anti-clockwise until resistance of the safety stop is felt. Leave the cap in this position until all pressure is released. Press the cap downwards against the spring to clear the safety stops, and continue turning until it can be lifted off.

(3) Check/adjust operation of all washers & top up reservoirs.
1 Replenish the container with clean soft water. The addition of a small smount of mild detergent will prevent smearing on the windscreen.
2 During freezing conditions it is beneficial to fill the container with a mixture of one part methylated spirits (wood alcohol) to two parts of water. This will assist in the dispersal of snow and ice from the screen.
3 Do not add anti-freeze solutions to the container as this will discoloru the paintwork and damage wiper blades and sealing rubber.

(4) Renew engine oil
NOTE: This operation is best carried out when the engine is warm and with the vehicle standing level on a ramp or over a pit.

To drain the sump, unscrew the drain plug three complete turns to direct the oil stream into a receptacle while the engine is warm. When the rate of flow lessens, remove the plug completely. Refit the plug and refill the sump with the appropriate grade of engine oil. The use ofadditives is unnecessary.

(5) Renew oil filter element
See 12.60.02.

(6) Lubricate accelerator linkage/pedal pivot – check operation
Lubricate
Using an oil can, lubricate the accelerator linkage on the carburetter and the accelerator pedal pivot. Wipe away any surplus oil to avoid drips or any possibility of staining the carpet inside the car.

Check operation
Check carburetter throttle response to initial movement of the accelerator pedal. If required, adjust the throttle cable at the carburetter.
Check carburetter throttle position with accelerator pedal fully depressed, by observing that the carburetter lever(s) move to a positive stop.

If the throttle movement is not satisfactory, investigate the cause and rectify as necessary, proceeding in the following order.
Check that the throttle pedal movement is not restricted by floor mats or carpet etc.
Check the carburetter lever(s) for correct position and settings – see 19.15.01 or 19.15.02, as applicable.
Check the throttle cable location and condition – see 19.20.06.

(7) Check cooling/heater systems for leaks and hoses for security and condition
1 Check for leaks from engine and radiator drain taps/plugs (where fitted).
2 Check for leaks from water hose joints.
3 Check for leaks from water hoses through damage or porosity.
4 Check for leaks from the water pump thermostat housing, radiator and heater unit.
5 Report any leaks found.

(8) Check for oil leaks
Report any found.

(9) Check/adjust torque of cylinder head nuts/bolts
See 12.29.27.

(10) Check driving belts, adjust or renew
Check condition where belt is visibly worn or damaged. Fit new belt if necessary.
Check and adjust:
a Alternator drive belt – see 86.10.05.
b Air pump drive belt – see 17.25.13.
c Compressor drive belt (where applicable) – see 82.10.01.

(11) Check security of engine mountings
Check the security of the following engine fixings using the torque figures shown in 06.
1 Inlet manifold and exhaust manifold.
2 Cam cover.
3 Air cleaner and carburetters.
4 Air pump.
5 Sump.
6 Timing cover.
7 E.G.R. valve.
8 Engine and gearbox mountings.
9 Alternator.

(12) Check/adjust carburetter idle settings
See – 19.15.01 or 19.15.02, as applicable.

(13) Top up carburetter piston damper(s)
1 Release the two toggle clips securing the outer section of the air cleaner housing.
2 Pivot the outer section about the hot air intake hose to gain access to the air cleaner element. Take care that the air intake control vacuum pipe is not disturbed.
3 Remove the carburetter damper assembly from the carburetter by unscrewing the hexagonal plug in top of the carburetter.
4 Raise the piston by inserting a finger into the carburetter air intake hole. With the piston raised top up the hollow damper guide with a recommended engine oil, until the oil level is ¼" (6 mm) below the top of the guide.
5 Release the piston and refit the damper assembly, screwing down the plastic plug.
6 To ensure correct location of the oil retaining cup in the damper guide, again raise and lower the piston.

NOTE: A certain amount of pressure will be left when lifting the piston, but it is essential that the piston is lifted to its maximum height to ensure correct location of the oil retaining cup.

(14) Renew carburetter/air intake air cleaner element
See 19.10.08.

(15) Check/adjust deceleration by-pass valve
See – 19.15.01 or 19.15.02, as applicable.

(16) Check security of E.G.R. valve operating lines
Check E.G.R. valve operating pipe lines for security of push fit connections at the following units:
a E.G.R. valve.
b Cut-off valve.
c Tee-piece (Canada only).
d Fuel trap.
e Carburetter.
Renew any pipes that show signs of deterioration.

(17) Check E.G.R. system
1 Disconnect the vacuum pipe from the top of the E.G.R. valve.
2 Remove the two bolts securing the E.G.R. valve to the inlet manifold and withdraw the valve.
3 Clean the base of the valve with a wire brush.
4 Use a standard spark plug machine to clean the valve seat and pintle. Insert the valve opening into the machine and lift the diaphragm evenly by using two fingers. Blast the valve for approximately 30 seconds, remove the inspect. If necessary, repeat until all carbon deposits are removed. Use compressed air to remove all traces of carbon grit from the valve.
5 Examine the E.G.R. ports in the manifold. Light deposits are not detrimental to system function and should not be disturbed. Heavy deposits must be cleared by removing the inlet manifold and cleaning the ports.

CAUTION: Do not attempt to clean the ports with the manifold in position.
6 Using a new gasket, refit the E.G.R. valve and two securing bolts.
7 Reconnect the vacuum pipe to the top of the E.G.R. valve.
8 Check E.G.R. valve operating lines for security of push fit connections at E.G.R. valve, cut-off valve, tee-piece, fuel trap and carburetter. Renew any pipes that show signs of deterioration.
9 Check function of E.G.R. valve as follows:
Start and run the engine until normal operating temperature is attained. Ensure that the choke control knob is pushed fully in. Open and close the throttle several times and observe or feel the E.G.R. valve, which should open and close with the changes in engine speed. The valve should close instantly when the throttle is closed.
10 If the operation of the valve does not appear completley satisfactory, check the valve by connecting the vacuum pipe of a distributor vacuum test unit to the valve. Ensure that the valve is actuated, held, and that there is no leak of vacuum, otherwise, fit a new E.G.R. valve complete.

(18) Check air intake temperature control system, excluded on catalytic converter vehicles
See 17.30.01.

(19) Check crankcase breathing and evaporative loss system hoses for security
1 Visually check the following hoses for security, and rectify as necessary.
2 Crankcase purge line.
3 Carburetter float chamber vent pipe.
4 Adsorption canister purge line.
5 Fuel tank vent pipe.
6 Manifold vacuum line.

(20) Check breathing and evaporative loss systems
Check hoses and restrictors for blockage, security and condition.
1 Inspect all hoses illustrated, items 2 to 6, for condition and renew any hoses that show signs of deterioration, including slackness or cracking.
2 Completely disconnect the crankcase purge line and the adsorption canister purge line, illustration items 2 and 4, then using a low pressure air supply, blow through these hoses to check for blockage. Investigate and clear any blockage, paying particular attention to the restrictors fitted at points 'A' and 'B'.
3 Refit the purge lines, ensuring that all connections are secure and tight. Renew any doubtful hose.

(21) Check air injection system, hoses pipes for security and condition

(22) Check/top up air conditioning compresser fluid

(23) Renew adsorption canister
See 17.15.13.

(24) Renew Catalytic converter Excluded on Non Catalytic Converter Vehicles
See 17.50.01.
Reset Catalytic Converter service interval indicator, using special key.

IGNITION
(25) Check security of distributor vacuum unit line and operation of vacuum unit (California and Canadian Market only)
1 Check security of distributor vacuum unit operating line connections at carburetter, fuel trap and distributor.
2 Start the engine and warm it to normal running temperature.
3s When the engine is idling steadily, disconnect the vacuum unit pipe at the distributor.
4 A noticeable rise in engine speed (approximately 500 rev/min) should be apparent if the vacuum unit is functioning.
5 When satisfied that the vacuum unit is operating correctly, reconnect the vacuum unit pipe to the distributor, ensuring a secure connection.

(26) Lubricate distributor
1 Remove the distributor cap and rotor arm. Using an oil can, apply three drops of a clean, recommended engine oil into the reservoir in the rotor carrier.
2 Remove flash-over shield and lubricate pick-up centre bearing plate with a drop of the same oil in each of the two holes provided. DO NOT disturb the screw securing the base plate. Refit the flash-over shield.

(27) Check/adjust ig ition timing using electronic equipment
See 86.35.15.

(28) Check ignition wiring for fraying, chafing and deterioration

Low tension circuit

1 Check connections of balst resistor wire, drive resistor, distributor, coil and ignition switch.
2 Check ignition coil connections.
3 Check wiring between coil and distributor.
4 Check distributor external connections.
5 Remove distributor cap and check internal wiring.
6 Check distributor internal connections.
7 Refit distributor cap.

High tension circuit.

8 Check lead between coil and distributor.
9 For each spark plug in turn: Check lead between plug and distributor.
10 Check high tension lead connections.
11 Report wiring condition.

(29) Clean distributor cap, check for cracks and tracking

1 Using a nap-free cloth, wipe the distributor cap and rotor arm clean. Inspect the cap and rotor, internally and externally, for cracks and any trace of tracking.
2 Refit the rotor arm and cap if serviceable, otherwise, fit new components as necessary.

(30) Renew spark plugs

For each plug in turn

1 Withdraw ignition high tension lead from plug.
2 Unscrew plug from engine using a special plug spanner or a suitable tube spanner.
3 Discard the plug.
4 Visually check new plug for damage to body and electrodes, discard plug if damaged.
5 Check electrode gap on new plug, which when correct, will just allow a 0.0025 in (0,64 mm) feeler gauge to slide slowly between the electrodes under light pressure.
6 If adjustment is necessary, using a suitable tool, carefully move the side electrode. Recheck the gap and repeat this procedure until the gap is correct.
7 Check sealing washer for cracks and distortion, and renew washer if necessary.
8 Fit new spark plug to engine.
9 Tighten plug to 20 lbf ft (2,8 kgf m).
10 Refit the high tension lead to the plug.

(31) Check coil performance on oscilloscope

1 Using proprietary electronic testing equipment, in accordance with the equipment instructions, check the resistance of the ignition coil primary winding, which must be 1.2 to 1.5 ohms. If necessary, fit a new ignition coil.

TRANSMISSION

(32) Check/top up gearbox oil

1 With the vehicle standing in a level position, remove the oil level/filler plug and, using a suitable dispenser filled with the correct grade oil, see 09–2, top up the gearbox until the oil is level with the bottom of the filler hole threads.
2 Allow surplus oil to drain away, then refit the filler plug and wipe clean.

(33) Check/top up rear axle/final drive oil

1 With the vehicle standing in a level position, remove the oil level/filler plug and top up the final drive unit with the correct grade oil, see 09–2, until the oil is level with the bottom of the filler hole threads. Allow surplus oil to drain away, then refit the filler plug and wipe clean.

(34) Check for oil leaks

Report any found.

(35) Check tightness of propeller shaft coupling bolts

Refer to Division 06 for torque settings.

(36) Check/top up automatic gearbox fluid

1 Stand the car on level ground and apply the handbrake firmly. Start the engine from cold and, with the footbrake firmly applied, run the engine at idle speed for 2 to 3 minutes, passing the selector lever through the complete range of positions to ensure that the transmission is primed.
2 Select the 'P' (Park) position and apply the handbrake. Leave the engine running at idle speed.
3 Remove the transmission dipstick and wipe it with a clean, non-fluffy cloth.
4 Replace the dipstick, ensuring that it is pushed fully into the tube and withdraw it immediately for reading.
5 Check the fluid level on the side of the dipstick marked 'COLD' and if necessary, add fluid; see Lubrication Recommendations, Section 09.

CAUTION: Do not overfill the transmission.

NOTE: Where the reverse side of the dipstick carries marks denoted 'HOT', the fluid level check may be carried out with the transmission at normal operating temperature. The procedure is as described above except that the vehicle must be driven for 25–30 km (15 miles) to warm the transmission. The check is then carried out using the 'HOT' side of the dipstick.

IMPORTANT: When operating at high ambient temperatures and on unmetalled roads, periodically inspect and remove dust and mud deposits from the slots and screen on the underside of the torque converter housing, and from the underside of the transmission oil sump. These deposits can adversely affect proper cooling of the unit.

(37) Lubricate automatic gearbox exposed selector linkage

(38) Check clutch pipes for leaks and chafing

Report any found.

(39) Check/top up clutch fluid reservoir

(40) Lubricate clutch pedal pivots

STEERING AND SUSPENSION

(41) Check steering rack/for oil/leaks

Report any found.

(42) Check security of suspension fixings

Refer to Division 06 for torque settings.

(43) Check condition and security of steering unit joints and gaiters

Refer to Division 06 for torque settings, report any defects found.

(44) Check/adjust front wheel alignment

See 57.65.01.

(45) Adjust front hub bearing end float

1 Remove road wheel.
2 Check hub for end float.
3 If adjustment is required, remove hub cap and split pin.
4 Tighten slotted nut as required to eliminate end float. A torque of 5 lbf ft (0,691 kgf m) must not be exceeded or damage may be caused to bearings and bearing tracks. Slacken nut to permit entry of split pin.
5 Insert and lock split pin.
6 Clean hub cap and refit.
7 Fit road wheel.

(46) Lubricate steering rack and pinion

1 Wipe clean the plug and surrounding area.
2 Remove the plug.
3 Fit a suitable grease nipple in place of the plug.
4 Apply a grease gun, filled with the correct grade of grease, see 09–2, to the grease nipple and give five strokes only.

CAUTION: Over greasing can cause damage to the protective bellows.

5 Remove the grease nipple.
6 Refit the plug.
7 Wipe away any surplus grease.

(47) Check shock absorbers for fluid leaks

Report any found.

BRAKES

(48) Inspect brake pads for wear and discs for condition

1 Jack up the front of the car and place safely onto stands before removing the disc pads (see 70.40.02).

CAUTION: Do NOT depress the brake pedal while the pads are removed.

2 Report pad condition if the friction lining has been reduced to 0.125 in (3 mm) or if there is not sufficient material to provide a thickness of 0.125 in (3 mm) at the completion of a further 3,000 miles (5,000 km) motoring.

3 Check the brake discs for excessive scoring and run out and report this if present.

(49) Inspect brake lining/pads for wear, drums/discs for condition

1 Jack up the car and place safely onto stands before removing the road wheel (held by 4 nuts) and the brake drum (see 70.10.03).

2 Check the brake linings for wear and report if they are excessively worn, damaged or contaminated by oil or grease. Remove surplus oil or grease and any brake lining dust before replacing the brake drum.

3 For discs see Maintenance Operation 48 for front pads and disc inspection.

(50) Check/top up brake fluid reservoir

Top up when required with new fluid of the correct type recommended – see **'LUBRICANTS & ANTI-FREEZE SOLUTIONS.'** Do not allow the fluid level to drop below the danger mark on the reservoir.

WARNING: Use only new brake fluid of the correct specification. DO NOT use fluid of unknown origin, or fluid that has been exposed to the atmosphere, or discharged during bleeding operations. If significant topping is required, check the hydraulic system for leaks.

CAUTION: Paintwork can be damaged by direct contact with brake fluid.

(51) Check footbrake operation, adjust to manufacturers instructions (self adjusting)

1 With the handbrake off, check the brake pedal for spongy operation.

2 If the pedal has spongy operation, bleed and adjust the brakes, see 70.25.02 and 70.25.03.

(52) Check handbrake operation; adjust to manufacturers instructions

1s If the handbrake travel is excessive, adjust the handbrake, see 70.35.10.

(53) Lubricate brake pedal pivot

1 Using an oil can, lubricate the brake pedal pivot.

2 Wipe away the surplus oil to prevent staining the carpet.

(54) Lubricate handbrake mechanical linkage and cable guides

1 Smear grease around the handbrake compensator, working it well into the clevis pins.

2 Smear grease around the rear brake-drum clevis pins.

3 Grease all exposed sections of the inner cable to resist corrosion.

(55) Check visually, hydraulic pipes and unions for chafing, leaks and corrosion

(56) Check brake servo hoses for security and condition.

(57) Renew brake servo filter unit

1 Remove the brake stop light switch.

2 Remove the split pin, plain washer and clevis pin securing the servo rod to the brake pedal.

3 Remove the rubber boot from the push-rod.

4 Withdraw the filter.

5 Reverse instructions 1 to 4.

ELECTRICAL

(58) Check function of original equipment, i.e. interior and exterior lamps, horns, wipers and warning indicators

In sequence, check operation of:

1 Side, tail and headlamps (including dip/main beam and 'Flash' control) also number plate illumination lamp.

2 Instrument panel illumination.

3 Interior light(s).

4 Horns.

5 Auxiliary lights.

With ignition circuits energised check operation of:

6. All warning lights.

7. Fuel level indicator.

8. Heater blower motor.

9. Windscreen washers.

10. Windscreen wipers.

11. Direction indicators.

12. Brake lights.

13 Reversing lights.

14 Start engine and note that the oil pressure warning has extinguished.

Check operation of:

15 Charging system warning light in relation to engine speed.

16 Temperature indicator.

17 Switch off the engine and return the ignition switch to the auxiliary position, then re-check the function of any fitted accessories e.g. a radio, that are supplied with power from this switch position.

18 Report any defects found.

(59) Check/top up battery electrolyte

Using DISTILLED WATER ONLY top up; the electrolyte level is correct when it just covers the separators.

(60) Clean and grease battery connections

1 Check the battery and surrounding area for corrosion from battery chemicals.

2 Clean off any corrosion found.

3 Check visually for any cracks in the battery case and report on any case cracks found.

4 Check security of terminal connections.

5 Coat terminals with petroleum jelly.

(61) Check/adjust headlamp alignment

See 86.40.17.

(62) Check, if necessary renew wiper blades

1 Examine each wiper blade in turn for damage before wetting the windscreen and operating the wiper motor control.

2 Replace wiper blades if they are damaged or if the screen is smeared.

(63) Check output of charging system

(64) Check brake system warning light

EXHAUST AND FUEL PIPES

(65) Check exhaust system for leaks and security

1 Check security of exhaust pipe to manifold nuts, correct tightening torque is 30 to 37 lbf ft (4,2 to 5,1 kgf m).

2 Check security of exhaust pipe joint clips.

3 Check security of exhaust system mounting bolts.

4 Using a second operator, run engine at fast idle speed.

5 Check exhaust system joints for leaks. Check exhaust pipes for leaks arising from damage or deterioration.

6 Check exhaust silencers for leaks arising from damage or deterioration.

7 Stop engine.

8 Report any defects found.

(66) Check fuel system for leaks, pipes and unions for chafing and corrosion

Report any defects found.

(67) Check condition of fuel filler cap seal

1 Make a visual check of the seal for the fuel tank filler cap.

2 Renew the seal if its condition is doubtful.

WHEELS AND TYRES

(68) Check/adjust tyre pressures including spare

Refer to Driver's Handbook for pressure settings.

(69) Check that tyres comply with manufacturers specification

See Driver's Handbook.
Report any deviations that may influence the car's performance or the accuracy of the speedometer.

(70) Check tightness of road wheel rastenings

Refer to 'TORQUE WRENCH SETTINGS' for data.

(71) Check tyres for external cuts in tyre fabric, exposure of ply or cord structure, lumps or bulges

(72) Check tyres for tread depth and visually for external cuts in fabric, exposure or ply or cord structure, lumps or bulges.

WARNING
It can be dangerous

a To use a car fitted with tyres in a damaged condition.
b To mix cross ply and radial ply tyres on the same axle or to fit radial ply tyres to the front wheels only.
c To use a car fitted with tyres that have a tread depth of less than 0.039 in (1.0 mm) over three-quarters of the tread width for the entire circumference of the tyre.
d To use a car with the tyres inflated to a pressure that is not suitable for the use to which the vehicle is put.

If the tyres to do not conform with legal requirements report to the owner.

BODY

(73) Lubricate all locks and hinges (not steering lock)

(74) Check condition and security of seats and seat belts

(75) Check rear view mirror for cracks and crazing

(76) Check operation of all door, bonnet and boot locks

(77) Check operation of seat belt warning system

(78) Check operation of seat belt inertial reel mechanism

(79) Check operation of window controls

(80) Check tightness of subframe/body mountings

Refer to 'TORQUE WRENCH SETTINGS' for data.

(81) Ensure cleanliness of controls, door handles, steering wheel

GENERAL

(82) Road/roller test and check function of all instrumentation

(83) Report additional work required

PREVENTIVE MAINTENANCE

In addition to the recommended periodical inspection of brake components it is advisable as the car ages and as a precaution against the effects of wear and deterioration to make a more searching inspection and renew parts as necessary.
It is recommended that:

1 Disc brake pads, drum brake linings, hoses and pipes should be examined at intervals no greater than those laid down in the Maintenance Summary Chart.
2 Brake fluid should be changed every 18 months or 19,000 miles whichever is the sooner.
3 All fluid seals in the hydraulic system should be renewed and all flexible hoses should be examined and renewed if necessary every 3 years or 37,500 miles whichever is the sooner. At the same time the working surface of the piston and of the bores of the master cylinder, wheel cylinders and other slave cylinders should be examined and new parts fitted where necessary.

Care must be taken always to observe the following points:

a At all times use the recommended brake fluid.
b Never leave fluid in unsealed containers. It absorbs moisture quickly and can be dangerous if used in your braking system in this condition.
c Fluid drained from the system or used for bleeding should be discarded.
d The necessity for absolute cleanliness throughout cannot be overemphasized.

Replacing brake-shoes
When it becomes necessary to renew the brake-shoes, it is essential that only genuine shoes, with the correct grade of lining, are used. Always fit new shoes as complete sets, never individually or as a single wheel set. Serious consequences could result from out-of-tolerance braking due to the mixing of lining.

MAINTENANCE SUMMARY — USA Market Specification — Fuel Injection Engined Vehicles

Operation No.	Miles × 1000	Kilometres × 1000	Service
10.10.03	1	1-6	A
10.10.15	7.5, 22.5, 37.5, 57.5, 72.5, 87.5, 102.5, 117.5, 132.5	12, 36, 60, 92, 116, 140, 164, 188, 212	B
10.10.30	15, 45, 65, 95, 125	24, 72, 104, 152, 200	C
10.10.60	30, 80, 110, 140	48, 128, 176, 224	D
10.10.99	50, 100	80, 160	E

To meet USA legislative requirements
Items marked thus † are applicable up to 50,000 miles only
Items marked thus †† are applicable at 50,000 miles and onwards

Operation / Service	A	B	C	D	E
ENGINE					
Renew engine oil	X	X	X	X	X
Check for oil leaks	X	X	X	X	X
Renew engine oil filter		X	X	X	X
Lubricate accelerator control linkage (and pedal pivot); check operation			X	X	X
Renew fuel filter(s)				††	††
Renew air intake air cleaner element				X	X
Check/adjust all driving belts	†	†	†	†	
Check/adjust all driving belts; renew as necessary		††	††	††	††
Check cooling and heater systems for leaks and hoses for security and condition	X	X	X	X	X
Check/adjust torque of cylinder head nuts/bolts	X				
Check/adjust operation of all washers and top up reservoirs	X	X	X	X	X
Check crankcase breathing and evaporative loss control systems					††
Check hoses/pipes and restrictors for blockage, security and condition			X	X	X
Check/adjust engine idle speed	†			††	††
Check/top up cooling system	X	X	X	X	X

Operation / Service	A	B	C	D	E
IGNITION					
Lubricate distributor (Lucas only)				X	X
Check/adjust ignition timing using electronic equipment			††	††	††
Clean/adjust spark plugs; renew as necessary			††	††	††
Renew spark plugs				†	
Check security of distributor vacuum line and operation of vacuum unit			††	††	††
Check ignition wiring (including electric fuel pump wiring) for security, fraying, chafing and deterioration				†	
Check ignition wiring (including electric fuel pump wiring) for security, fraying, chafing and deterioration			††	††	††
TRANSMISSION					
Check for oil leaks	X	X	X	X	X
Check/top-up gearbox oil	X	X	X	X	X
Check/top-up automatic gearbox oil	X	X	X		
Renew automatic gearbox oil and filter				X	X
Check clutch pipes for cracks, chafing, leaks and corrosion	X	X	X	X	X
Check/top-up rear axle/final drive oil	X	X	X	X	X
Check/top-up clutch fluid reservoir	X	X	X	X	X
Check tightness of propeller shaft coupling bolts	X		X	X	X
STEERING AND SUSPENSION					
Check condition and security of steering unit, joints and gaiters		X	X	X	X
Check steering rack/gear for oil/fluid leaks	X	X	X	X	X
Check/top-up fluid in power steering reservoir	X	X	X	X	X
Check/adjust front wheel alignment	X				
Check shock absorbers for fluid leaks	X	X	X	X	X
Check power steering system for leaks, hydraulic pipes and unions for chafing and corrosion	X	X	X	X	X
Check/adjust front hub bearing end float	X		X	X	X
Lubricate steering rack and pinion			X	X	X

Operation	Service	A	B	C	D	E
BRAKES						
Check visually hydraulic pipes and unions for cracks, chafing, leaks and corrosion		X	X	X	X	X
Check/top-up brake fluid reservoir(s)		X	X	X	X	X
Inspect brake pads/linings for wear, discs/drums for condition; adjust brakes as necessary			X	X	X	X
Check brake servo hose(s) for security and condition		X	X	X	X	X
Renew hydraulic brake fluid		22,500 miles (36,000 km) or 18 months whichever is the sooner.				
Renew air filter in brake servo unit		37,500 miles (60,000 km) or 3 years whichever is the sooner.				
ELECTRICAL						
Check function of original equipment, i.e. interior and exterior lamps, horns, wipers and all warning indicators		X	X	X	X	X
Check/adjust headlamp alignment			X	X	X	X
Check, if necessary renew, wiper blades			X	X	X	X
Check/top up battery electrolyte		X	X	X	X	X
Clean and grease battery connections				X	X	X
FUEL AND EXHAUST SYSTEMS						
Check exhaust system for leaks and security		X	X	X	X	X
Check fuel system for leaks, pipes and unions for chafing and corrosion		X	X	X	X	X
Check condition of fuel filler cap seal				X	X	X
Renew Lambda (oxygen) sensor and reset service interval counter					X	X
WHEELS AND TYRES						
Check/adjust tyre pressures including spare		X	X	X	X	X
Check tightness of road wheel fastenings		X	X	X	X	X
Check that tyres comply with manufacturer's specification		X	X	X	X	X
Check tyres for tread depth and visually for external cuts in fabric, exposure of ply or cord structure, lumps or bulges		X	X	X	X	X

Operation	Service	A	B	C	D	E
BODY						
Lubricate all locks, hinges and door check mechanisms (not steering lock)		X	X	X	X	X
Check operation of all door, bonnet luggage compartment and steering column locks		X	X	X	X	X
Check condition, security and operation of seats and seat belts		X	X	X	X	X
Check operation of window controls		X	X	X	X	X
GENERAL						
Road/roller test. Check brake operation and function of all instrumentation		X	X	X	X	X
Report additional work required		X	X	X	X	X

ADDITIONAL RECOMMENDATIONS

In addition to the recommended periodical inspection of brake components it is advisable as the car ages and as a precaution against the effects of wear and deterioration to make a more searching inspection and renew parts as necessary.

It is recommended that:

1 Disc brake pads, drum brake linings, hoses and pipes should be examined at intervals no greater than those laid down in the Maintenance Summary.

2 Brake fluid should be changed completely every 18 months or 22,500 miles (36 000 km) whichever is the sooner.

3 All fluid seals in the hydraulic system and all flexible hoses should be examined and renewed if necessary every three years or 37,500 miles (60 000 km) whichever is the sooner. At the same time the working surface of the pistons and of the bores of the master cylinder, wheel cylinders and other slave cylinders should be examined and new parts fitted where necessary.

Care must be taken always to observe the following points:

a At all times use the recommended brake fluid.

b Never leave fluid in unsealed containers; it absorbs moisture quickly and can be dangerous if used in the braking system in this condition.

c Fluid drained from the system or used for bleeding should be discarded.

d The necessity for absolute cleanliness throughout cannot be over-emphasized.

ENGINE

Renew engine oil

NOTE: This operation is best carried out when the engine is warm and with the vehicle standing level on a ramp or over a pit.

1 To drain the sump, unscrew the drain plug three complete turns to direct the oil stream into a receptacle while the engine is warm. When the rate of flow lessens remove the plug completely. Refit the plug and refill the sump with the appropriate grade of engine oil. The use of additives is unnecessary.

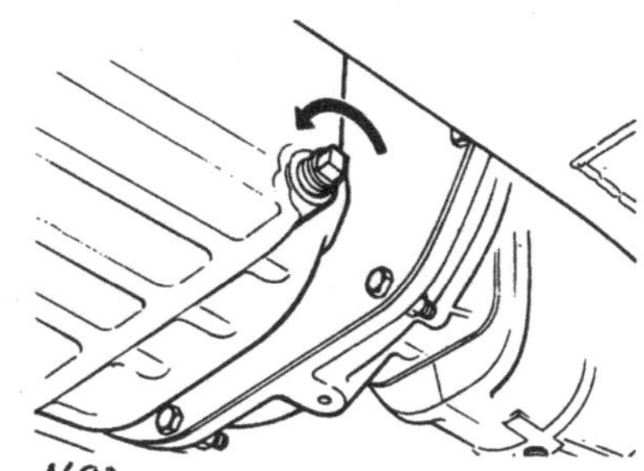

1693

Check for oil leaks

Renew engine oil filter

See 12.60.02.

Lubricate accelerator control linkage (and pedal pivot); check operation

Renew fuel filter(s)

A single or twin filters may be fitted. They are located at the right-hand rear of the vehicle, mounted beneath the floor, forward of the rear axle. A cover may be fitted on later models, see 19.25.03.
The fuel system must be depressurized before commencing this operation, see 19.50.02.

1 Disconnect the fuel pipes from each end of the filter(s).
2 Slacken the clip securing the filter(s).
3 Withdraw and discard the filter(s).
4 Fit the new filter with the end marked 'IN' connected to the fuel supply pipe. (Alternative arrow markings denote fuel flow direction.)
5 Tighten the securing clip and fuel connections.

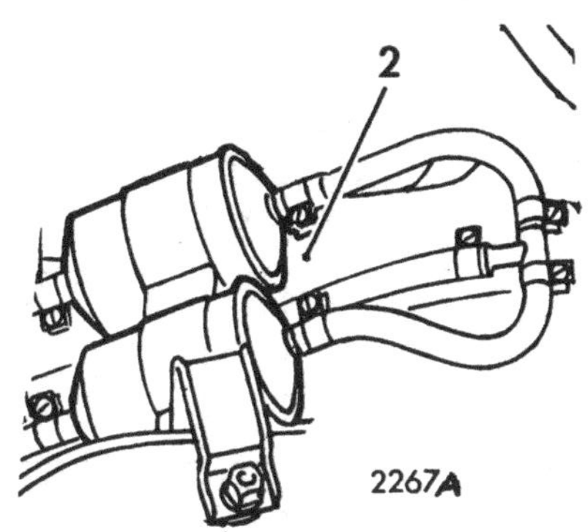

2267A

Renew air intake air cleaner element

Refer to Operation 19.10.08.

Check/adjust all driving belts

Check/adjust all driving belts; renew if necessary

Check condition where belt is visibly worn or damaged. Fit new belt if necessary.
Check and adjust:
a Alternator drive belt, see 86.10.05.
b Compressor drive belt (where applicable), see 82.10.01.

Check cooling/heating systems for leaks and hoses for security and condition

1 Check for leaks from engine and radiator drain taps/plugs (where fitted).
2 Check for leaks from water hose joints.
3 Check for leaks from water hoses through damage or porosity.
4 Check for leaks from the water pump thermostat housing, radiator and heater unit.
5 Report any leaks found.

Check/adjust torque of cylinder head nuts/bolts

See 12.29.27.

Check/adjust operation of all washers and top-up reservoirs

1 Replenish the container with clean soft water. The addition of a small amount of mild detergent will prevent smearing on the windscreen.
2 During freezing conditions it is beneficial to fill the container with a mixture of one part methylated spirits (wood alcohol) to two parts of water. This will assist in the dispersal of snow and ice from the screen.
3 Do not add anti-freeze solutions to the container as this will discolour the paintwork and damage wiper blades and sealing rubber.

Check crankcase breathing and evaporative loss control systems

Check hoses and restrictors for blockage, security and condition

1 Inspect all hoses for condition and renew any hoses that show signs of deterioration, including slackness or cracking.
2 Completely disconnect the crankcase purge line and the adsorption canister purge line, then using a low pressure air supply, blow through these hoses to check for blockage. Investigate and clear any blockage, paying particular attention to the restrictors.
3 Refit the purge lines, ensuring that all connections are secure and tight. Renew any doubtful hose.

Check/adjust engine idle speeds

See 19.20.18.

Check/top-up cooling system

WARNING: When it is necessary to remove the pressure/filler cap from a hot engine, exercise great care by protecting the hands against escaping steam. Slowly turn the pressure cap anti-clockwise until resistance of the safety stop is felt. Leave the cap in this position until all pressure is released. Press the cap downwards against the spring to clear the safety stops, and continue turning until it can be lifted off.

The pressurized cooling system incorporates a header tank which provides a single point for coolant filling and level checking.

The coolant level should be maintained at 1 inch (25 mm) below the neck of the header tank.

If the cooling system has been drained the procedure for refilling the system is as follows:

1. Remove the header tank filler cap.
2. Set the interior heater controls to the maximum heat position.
3. Fill the system until the coolant level is 1 inch (25 mm) below the neck of the header tank.
4. Refit the header tank filler cap and run the engine at approximately 1500 rev/min until the coolant temperature rises sufficiently to open the thermostat.
5. Stop the engine and, observing the caution above, remove the header tank filler cap.
6. Top-up the coolant level as necessary until it is 1 inch (25.4 mm) below the header tank filler neck.
7. Refit the filler cap.

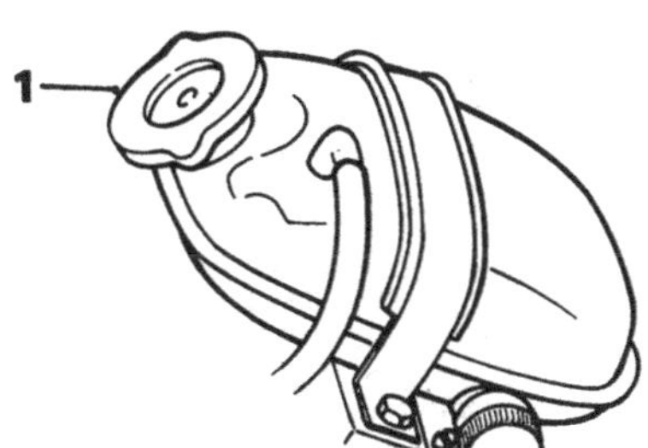

2138

IGNITION

Lubricate distributor (Lucas only)

1. Remove the distributor cap and rotor arm. Using an oil can, apply three drops of a clean, recommended engine oil into the reservoir in the rotor carrier.
2. Remove flash-over shield and lubricate pick-up centre bearing plate with a drop of the same oil in each of the two holes provided. DO NOT disturb the screw securing the base plate. Refit the flash-over shield.

Check/adjust ignition timing using electronic equipment

See 86.35.15.

Clean/adjust spark plugs

For each spark plug in turn:

1. Remove the ignition high tension lead from the plug.
2. Unscrew the plug from the engine using a special plug spanner or a box type spanner.
3. Wipe clean the ceramic body of the plug.
4. Visually check the plug body for cracks and renew the plug if any cracks are present.
5. Unscrew the end terminal cap from the plug.
6. Clean the plug terminal threads with a wire brush.
7. Clean the cap threads using a low pressure air line.

Renew spark plugs

For each plug in turn:

1. Remove the ignition high tension lead from the plug.
2. Unscrew the plug from the engine using a special plug spanner or a suitable box spanner.
3. Discard the plug.
4. Visually check the new plug for damage to the body and electrodes, discard the plug if there are any signs of damage.
5. Check the electrode gap on the new plug which when correct will just allow a 0.025 in (0.64 mm) feeler gauge to slide slowly between the electrodes under light pressure.
   If adjustment is necessary, using a suitable tool carefully move the side electrode. Recheck the gap and repeat this procedure until the gap is correct.
6. Fit the new plug to the engine.
7. Tighten the plug to the correct torque.
8. Refit the high tension lead to the plug.

Check security of distributor vacuum line and operation of vacuum unit

1. Check security of distributor vacuum unit operating line connections.
2. Start the engine and warm it to normal running temperature.
3. When the engine is idling steadily, disconnect the vacuum unit pipe at the distributor.
4. A noticeable rise in engine speed (approximately 500 rev/min) should be apparent if the vacuum unit is functioning.
5. When satisfied that the vacuum unit is operating correctly, reconnect the vacuum unit pipe to the distributor ensuring a secure connection.

Check ignition wiring (including electric fuel pump wiring) for security, fraying, chafing and deterioration

TRANSMISSION

Check for oil leaks

Check/top-up gearbox oil

With the vehicle standing on level ground:

1. Remove the oil level plug.
2. Using a suitable dispenser such as a pump-type oil can with flexible nozzle filled with a recommended lubricant, top up the gearbox until the oil is level with the bottom of the filler plug threads.
3. Allow surplus oil to drain away before refitting the level plug and wiping clean.

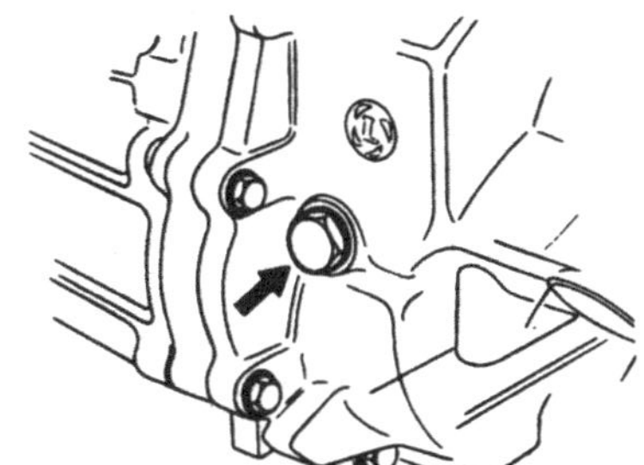

AHT 313

Check/top-up automatic gearbox oil

1 Stand the car on level ground and apply the handbrake firmly. Start the engine from cold and, with the footbrake firmly applied, run the engine at idle speed for 2 to 3 minutes, passing the selector lever through the complete range of positions to ensure that the transmission is primed.
2 Select the 'P' (Park) position and apply the handbrake. Leave the engine running at idle speed.
3 Remove the transmission dipstick and wipe it with a clean, non-fluffy cloth.
4 Replace the dipstick, ensuring that it is pushed fully into the tube and withdraw it immediately for reading.
5 Check the fluid level on the side of the dipstick marked 'COLD' and, if necessary, add fluid; see 'Lubrication Recommendations'.
6 Repeat instructions 1 to 5 until the fluid level is correct. DO NOT OVERFILL THE TRANSMISSION.

Where the reverse side of the dipstick carries marks denoted 'HOT', the fluid level check may be carried out with the transmission at normal operating temperature. The procedure is as described above except that the vehicle must be driven for 15 to 20 miles (25 to 30 km) to warm the transmission. The check is then carried out using the 'HOT' side of the dipstick.

Renew automatic gearbox oil and filter

1 Remove the automatic gearbox sump, see 44.24.04.
2 Remove and discard the filter.
3 Fit a new filter.
4 Replace the sump, see 44.24.04.
5 Refill the gearbox with new oil.
6 Run the engine, prime the automatic gearbox and check for oil leaks.

Check clutch pipes for cracks, chafing, leaks and corrosion

Check/top-up rear axle/final drive oil

1 Remove the oil level plug.
2 Top-up the rear axle until the oil is level with the bottom of the filler plug thread.
3 Allow surplus oil to drain before fitting the plug and wiping clean.

Check/top-up clutch fluid reservoir

1 Wipe clean the area around the reservoir cap.
2 Unscrew the cap.
3 If necessary, top-up the reservoir to within ¼ in (6 mm) of the top with new brake fluid taken from a sealed container.
4 Replace the reservoir cap.

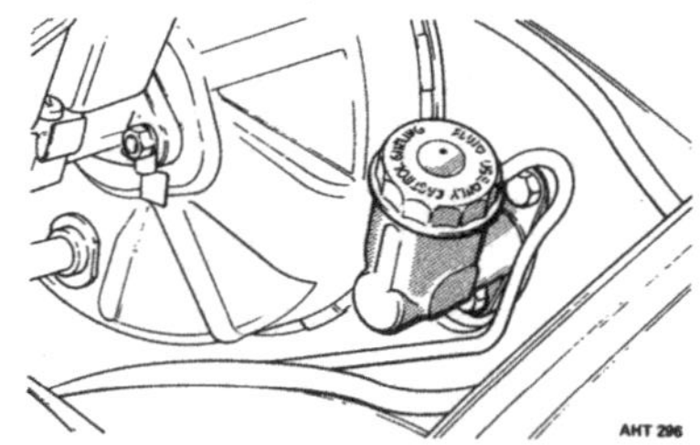

Check tightness of propeller shaft coupling bolts

Check the following fastenings for tightness:
Propeller shaft to gearbox and rear axle:
10 mm bolt 50 Nm (37 lbf ft).

STEERING AND SUSPENSION

Check condition and security of steering unit, joints and gaiters

Refer to Section 06 as applicable.

Check steering rack/gear for oil/fluid leaks

Check/top-up fluid in power steering reservoir

1 Wipe clean the reservoir cap and surrounding area.
2 Unscrew the reservoir cap and dipstick.
3 Wipe the dipstick clean and replace it in position.
4 Withdraw the dipstick again and note the fluid level.

If topping-up is necessary:
5 Add recommended fluid via the filler cap to bring the level just below the high mark on the dipstick. **Do not overfill.**
6 Replace the reservoir cap.

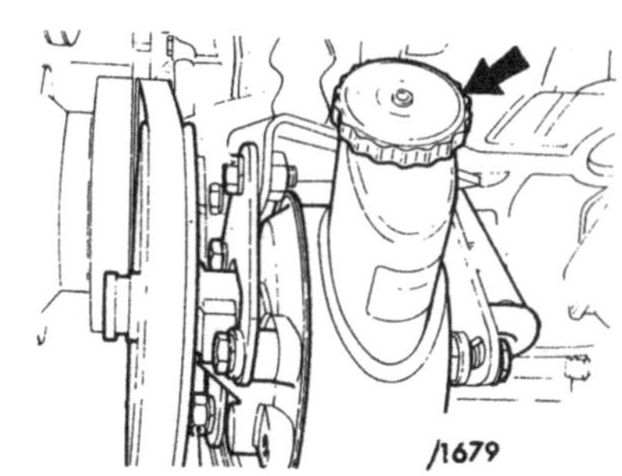

Check/adjust front wheel alignment

Refer to Operation 57.65.01.

Check shock absorbers for fluid leaks

Check power steering system for leaks, hydraulic pipes and unions for chafing and corrosion

Check/adjust front hub bearing end float

Refer to Operation 60.25.13.

Lubricate steering rack and pinion

Using a recommended grease, lubricate the steering rack and pinion as follows:

1 Wipe clean the plug and surrounding area.
2 Remove the grease nipple plug, taking care not to disturb the larger damper plug.
3 Fit a suitable grease nipple in place of the plug.
4 Turn the steering wheel to full right-hand lock.
5 Apply a grease gun to the grease nipple and give five strokes only.
CAUTION: Overgreasing can cause damage to the protective gaiters and/or seals.
6 Remove the grease nipple and refit the plug.
7 Wipe away any surplus grease.

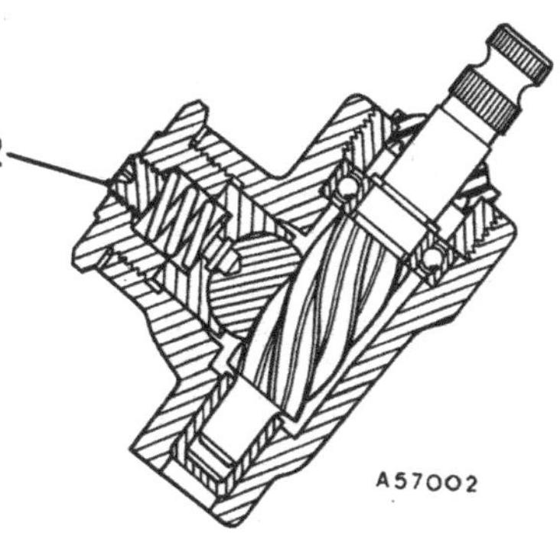

A57002

BRAKES

Check visually hydraulic pipes and unions for cracks, chafing, leaks and corrosion

Check/top-up brake fluid reservoir

Top-up when required with new fluid of the correct type recommended — see **'LUBRICANTS & ANTI-FREEZE SOLUTIONS'.** Do not allow the fluid level to drop below the danger mark on the reservoir.

WARNING: Use only new brake fluid of the correct specification. **DO NOT** use fluid of unknown origin, or fluid that has been exposed to the atmosphere, or discharged during bleeding operations. If significant topping is required, check the hydraulic system for leaks.

CAUTION: Paintwork can be damaged by direct contact with brake fluid.

Inspect brake pads/linings for wear, discs/drums for condition, adjust brakes as necessary

Front brakes

1 Remove the brake pads, see 70.40.02.
2 Renew the pads if the lining thickness is likely to be below 3 mm (1/8 in) before the next service.

Rear brakes

1 Remove the brake drum, see 70.10.03.
2 Renew the brake linings if they are likely to have worn near the rivets or be less than 1.5 mm (1/16 in) thick before the next service.

Brake-shoe renewal

When it becomes necessary to renew the brake-shoes, it is essential to use only genuine shoes, with the correct grade of lining. Always fit new shoes as complete sets, never individually or as a single wheel set. Serious consequences could result from out-of-balance braking due to the mixing of linings.

Adjustment

The front and rear brakes are self-adjusting, but the handbrake may be adjusted following instructions in Operation 70.35.10.

Check brake servo hose(s) for security and condition

Renew hydraulic brake fluid

Refer to notes following 'Check/top-up brake fluid reservoir'.

Renew air filter in brake servo unit

Refer to Operation 70.50.25.

ELECTRICAL

Check function of original equipment i.e. interior and exterior lamps, horns, wipers and all warning indicators

Check/adjust headlamp alignment

Refer to Operation 86.40.17.

Check, if necessary renew, wiper blades

Refer to Operation 84.15.06/07.

Check/top-up battery electrolyte

1 Lift the cover.
2 Check the electrolyte level, which if correct should just cover the separators.
3 Add distilled water as necessary.

1646

Clean and grease battery connections

FUEL AND EXHAUST SYSTEMS

Check exhaust system for leaks and security

1 Place the car on a ramp or over a pit.
2 Check the exhaust system fixings for security, paying particular attention to heat shields, flexible mounting plates and clamps.
3 Using a second operator, run the engine at fast idle speed.
4 Check exhaust system joints for leaks.
5 Check exhaust pipes for leaks arising from damage or deterioration.
6 Check exhaust silencers for leaks arising from damage or deterioration.
7 Stop the engine.
8 Report any defects found. Silencers or pipes found to be leaking or badly corroded should be renewed.
9 Fit new parts as necessary.

Check fuel system for leaks, pipes and unions for chafing and corrosion

Visually check the fuel feed system for leaks as follows:
1 Check for leaks from fuel system connections.
2 Check the fuel pipes for chafing, corrosion and damage.
3 Check for leaks from the fuel and expansion tanks, pump, rails and injectors.
4 Renew any items which show signs of deterioration.

Check condition of fuel filler cap seal

Renew Lambda (oxygen) sensor and reset service interval counter

1 Disconnect the electrical lead from the sensor.
2 Unscrew the sensor from the exhaust manifold, taking care not to strain the exhaust system.
3 Lubricate the threads of the new sensor and fit it to the exhaust manifold.
Tighten the sensor sufficient to make a gas-tight seal, but do not overtighten.
4 Reconnect the electrical leads to the sensor.
5 Reset the service interval counter using the special tool necessary for this purpose.

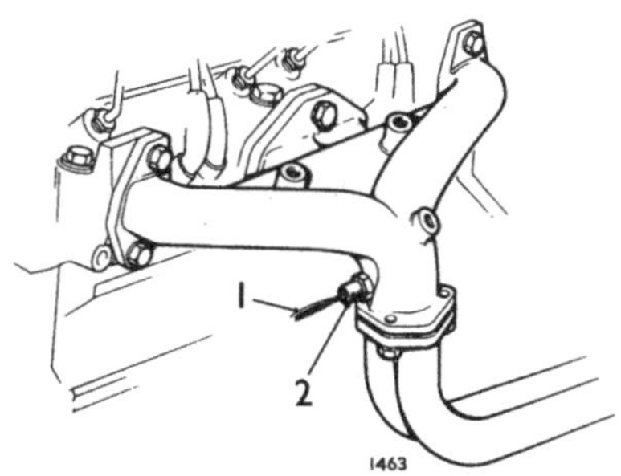

WHEELS AND TYRES

Check/adjust tyre pressures, including spare

Refer to Section 04.

Check tightness of road wheel fastenings

If correct, these will be tightened to a torque of 75 lbf ft (10.3 kgf m).

Check that tyres comply with manufacturer's specification

Refer to Section 04.

Check tyres for tread depth and visually for external cuts in fabric, exposure of ply or cord structure, lumps or bulges

BODY

Lubricate all locks, hinges and door check mechanisms (not steering lock)

Check operation of all door, bonnet, tailgate and steering column locks

Check operation of window controls

Check condition, security and operation of seats and seat belt

The seat belt inertia reel mechanism may be checked using the following procedure:
1 **IMPORTANT:** The following road test must be carried out only under maximum safe road conditions, i.e. on a dry, straight, traffic-free road.
With the safety harness fitted to the driver and front seat passenger drive the car at 5 mph (8 km/h); ensuring that it is safe to do so, brake sharply. The safety harness should lock automatically, holding both driver and passenger securely in position.
It is important when braking that the reactions of both driver and passenger are normal, i.e. the body must not be thrown forward in anticipation, thus causing a 'snatching' action of the belt which would operate the locking mechanism.
2 *Snatch test:* Whilst seated, fasten the seat belt and grip the shoulder belt at approximately shoulder level with the opposite hand. Pull the belt sharply in a downwards direction, the belt should lock.
If the belt fails to lock on either test, the seat belt should be replaced, see 76.73.10.

GENERAL

Road/roller test. Check brake operation and function of all instrumentation

Brake operation
The operation of the footbrake and handbrake independently should be sufficient to stop the vehicle without pull to one side, within a distance required by any local territory legislation.
In addition the footbrakes and/or the handbrake must prevent vehicle movement on a steep incline. Again reference should be made to any applicable local territory legislation.

Report additional work required

ENGINE

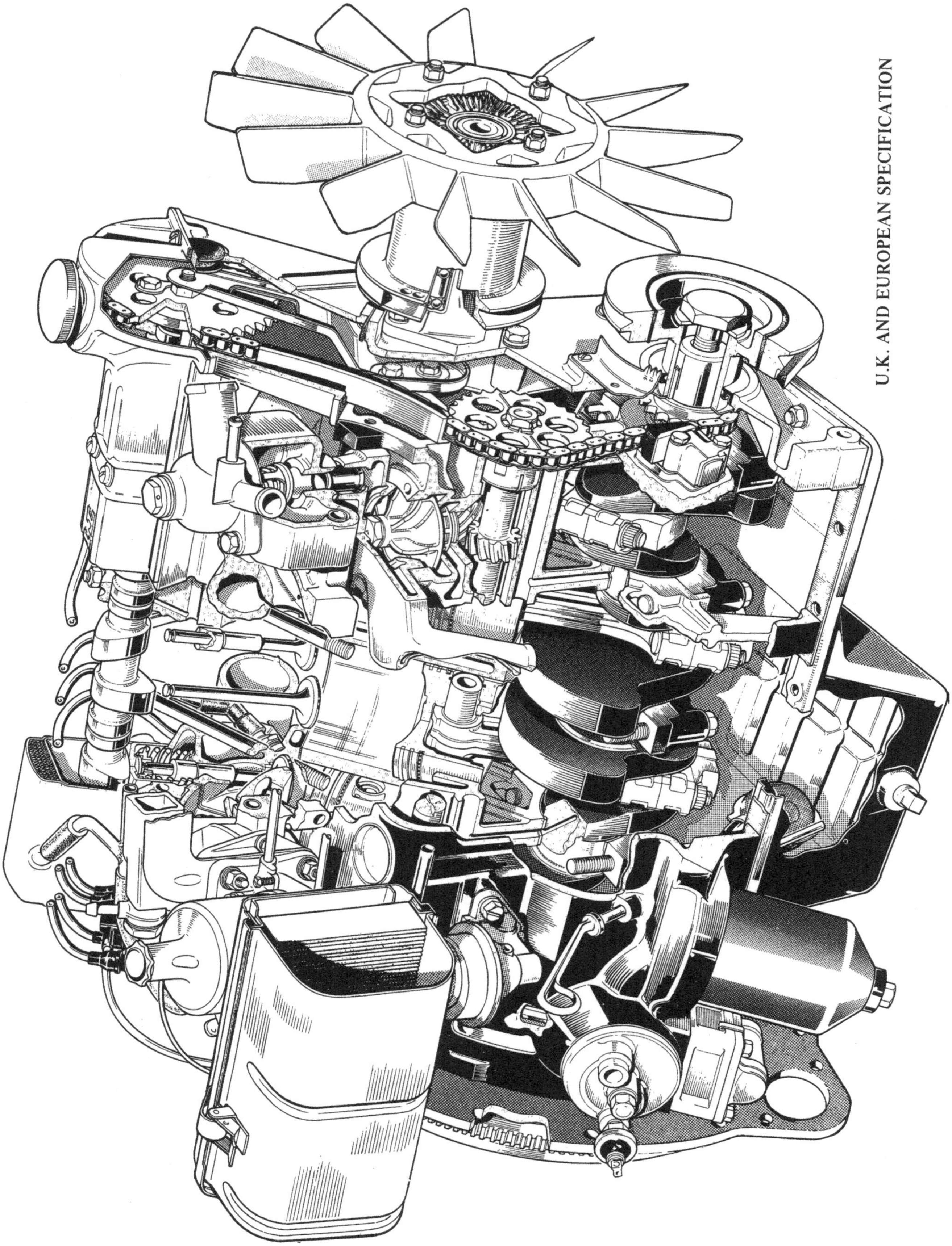

U.K. AND EUROPEAN SPECIFICATION

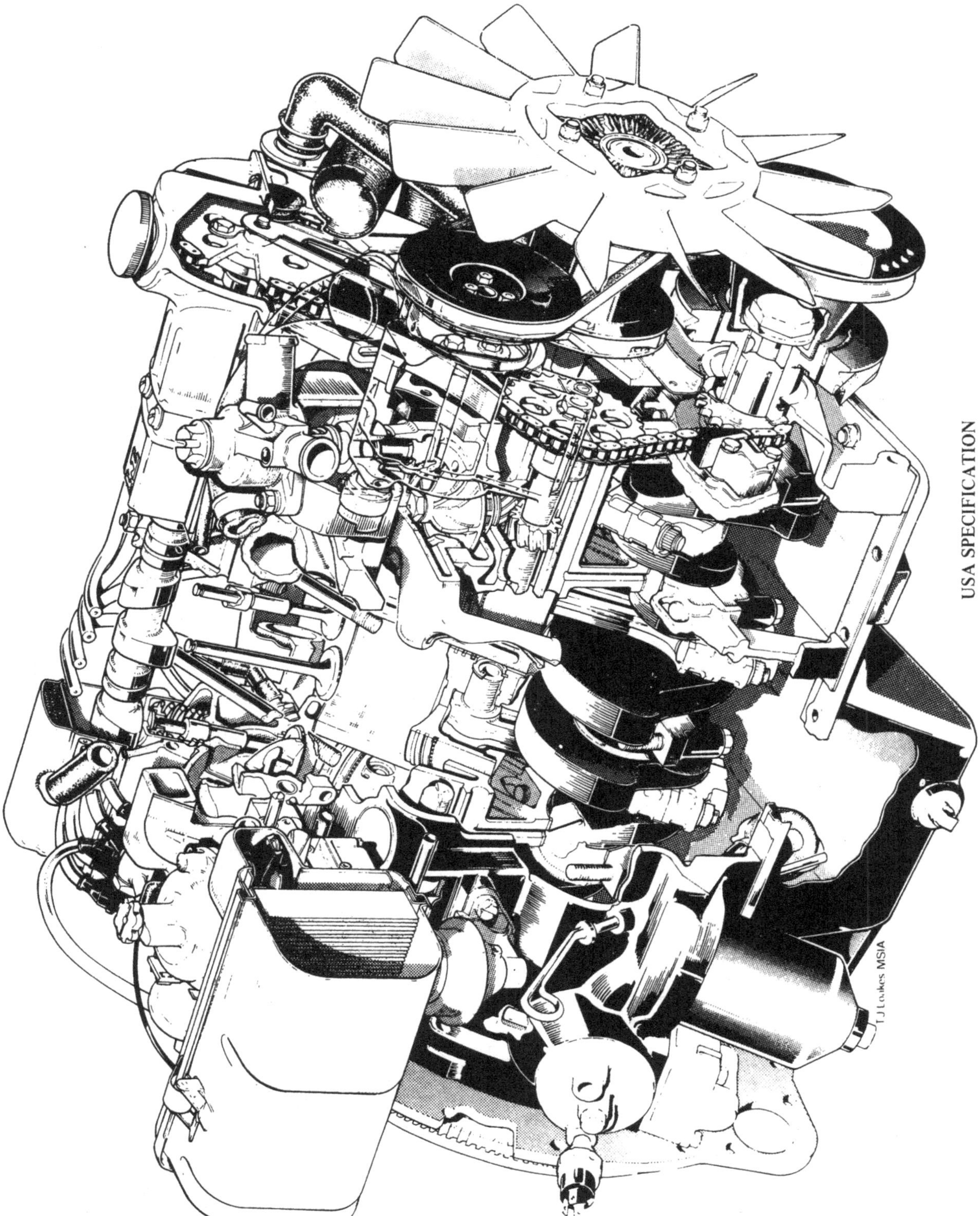

USA SPECIFICATION

JACKSHAFT

– Remove and refit **12.10.14**

Removing

1 Disconnect the battery.
2 Remove the fresh air duct. 80.15.31.
3 Remove the radiator. 26.40.01.
4 Remove the air conditioning condenser 82.15.10 (if fitted).
5 Remove the timing cover. 12.65.01.

[**NOTE:** To prevent abrasive dirt from entering the engine, thoroughly clean the timing cover and the camshaft cover before you remove them.]

6 Remove the inlet manifold complete with carburetters. 19.15.15.
7 Remove the water pump cover. 26.50.01.
8 Remove the impeller. 26.50.01.
9 Remove the petrol pump. 19.48.08.
10 Remove the camshaft cover. 12.29.42.
11 Turn the engine over so that the timing mark on the camshaft flange is in line with the groove on the camshaft front bearing cap.
12 Remove the distributor cap and check that the rotor arm points to the last manifold bolt hole in the cylinder head thus indicating that the engine is at T.D.C. number one cylinder firing.
13 Remove the distributor.
14 Remove the hydraulic timing chain tensioner. 12.65.28.

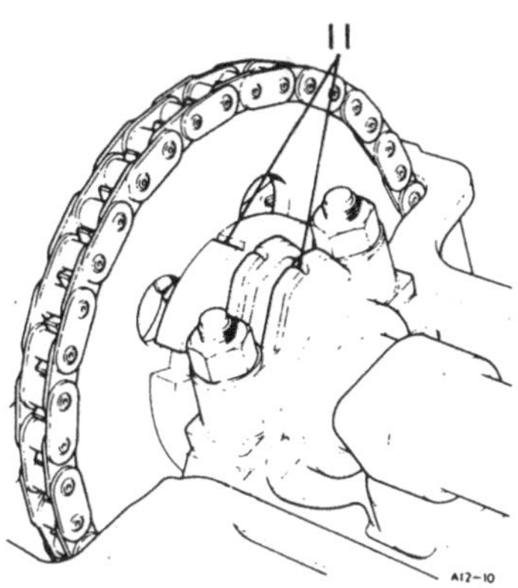

15 Remove the adjustable timing chain guide. 12.65.50.
16 Remove the two Allen screws and withdraw the jackshaft keeper plate.
17 Lift the timing chain clear of the sprocket and withdraw the jackshaft complete with the sprocket.
18 Hold the jackshaft in a vice and remove the sprocket retaining bolt, tab washer and sprocket.

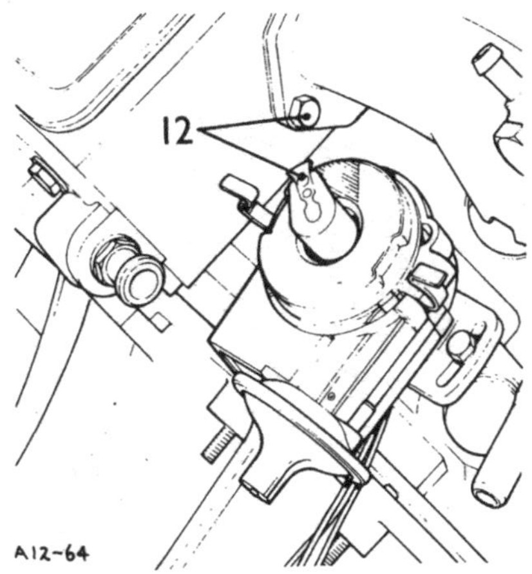

USA SPECIFICATION

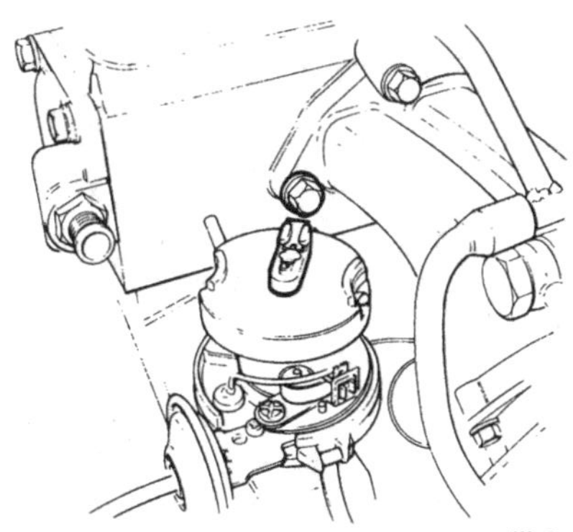

UK & EUROPE SPECIFICATION

Refitting

19 Fit the sprocket to the jackshaft using a new tab washer.
20 Insert the jackshaft into the cylinder block and secure with the keeper plate and two Allen screws.
21 Turn the jackshaft so that the scribed line across the sprocket is equidistant between the lower bolt securing the camshaft sprocket support bracket and the timing cover centre retaining bolt hole. Fit the chain whilst holding the jackshaft in this position.
22 Fit the adjustable chain guide and leave the bolt slack.
23 Fit the hydraulic timing chain tensioner 12.65.28 instructions 7 to 19.
24 Fit the petrol pump ensuring that the lever rides on top of the jackshaft cam.
25 Fit the water pump impeller and cover – instructions 9 to 14. 26.50.01.
26 Fit the distributor and line-up the rotor as in instruction 12 and check the timing. 86.35.15.
27 Fit the timing chain cover. 12.65.01.
28 Fit the camshaft cover.

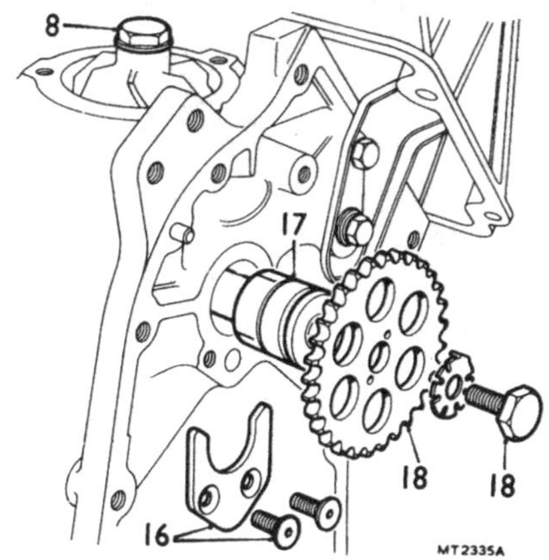

29 Fit the air conditioning condenser 82.15.10.
30 Fit the radiator. 26.40.01.
31 Fit the inlet manifold complete with carburetters. 19.15.15.
32 Fit the fresh air duct. 80.15.31.
33 Check that the cooling system has been filled and connect the battery.

[**NOTE:** To prevent undue wear when the engine is first started, lubricate the chain, the sprockets, and the jackshaft bearing with engine oil during installation. If the engine is to be stored for some time following repair, lightly lubricate these parts with assembly lubricant (available from automotive supply stores) or with a thin coating of multipurpose grease.]

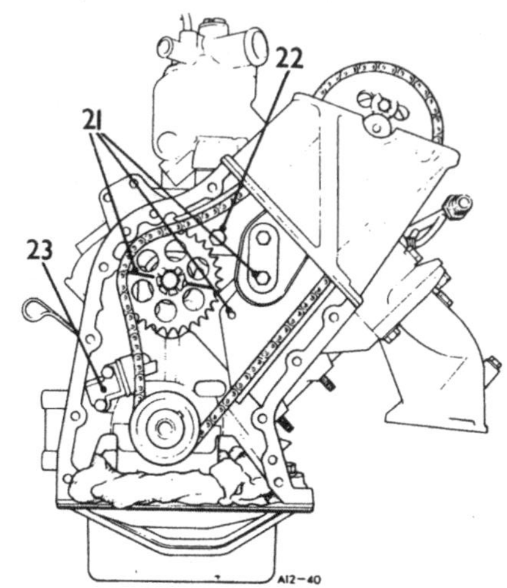

CAMSHAFT

– Remove and refit 12.13.01

Removing

1 Disconnect the battery.
2 Remove the fresh air duct. 80.15.31.
3 Remove the camshaft cover. 12.29.42.

[**NOTE:** To prevent abrasive dirt from entering the engine, thoroughly clean the camshaft cover before you remove it.]

4 Turn the engine over so that the timing mark on the camshaft flange is 180° distant from the groove in the camshaft front bearing cap to gain access to the lower bolt.
5 Anchor the camshaft sprocket to the support bracket using a 'slave' nut.
6 Unlock and remove the exposed camshaft sprocket lower retaining bolt.
7 Turn the engine over so that the timing mark on the camshaft flange is exactly in line with the groove in the camshaft front bearing cap.
8 Unlock and remove the remaining sprocket retaining bolt and lock washer.
9 Evenly slacken-off the ten camshaft bearing cap nuts and remove them complete with plain washers.
10 Check that the bearing caps are numbered for identification i.e. starting at the front of the engine, number one is recognized by the timing grooves and the remainder are numbered 2 to 5.
11 Withdraw the caps and remove the camshaft.

Refitting

12 Clean and lubricate the camshaft journals and corresponding bearings in the cylinder head and fit the camshaft.
13 Lubricate and fit the bearing caps in their correct order.

[**NOTE:** If you remove the tappets, be careful not to interchange them. During camshaft installation, it is best to use assembly lubricant (available from automotive supply stores) or a thin coating of multipurpose grease to lubricate the journals, the cam lobes, the tappets, and the bearing caps, rather than ordinary engine oil.]

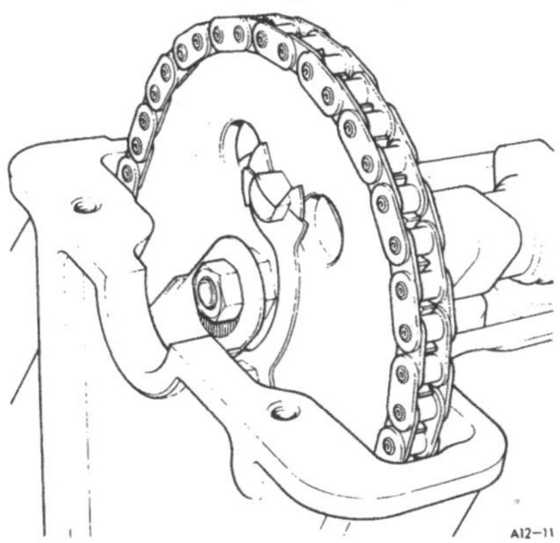

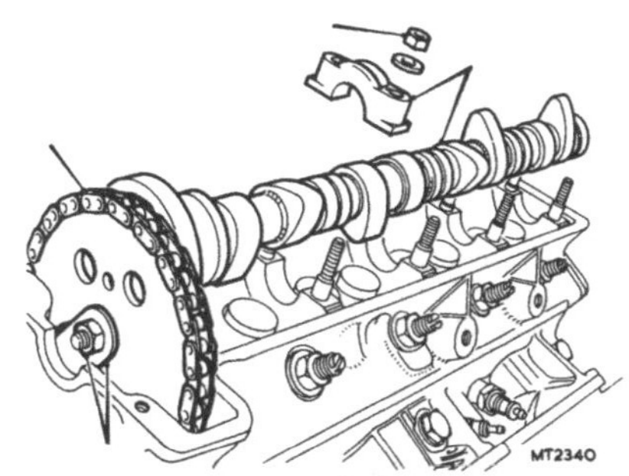

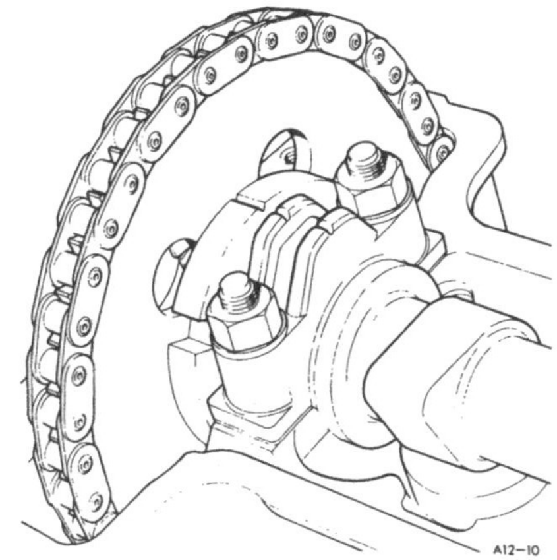

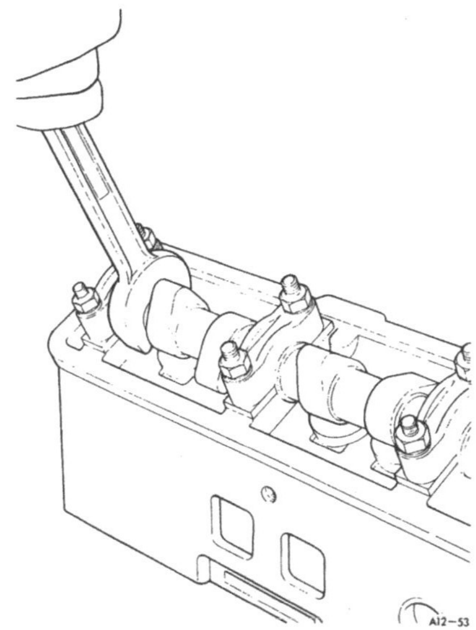

CAUTION: Before commencing instruction 14 ensure that the timing mark on the camshaft flange is approximately in line with the groove in the front bearing cap to avoid damage being caused to the pistons by the valve heads.

14 Fit and tighten the bearing cap nuts and washers evenly to the correct torque figure – see section 06.

NOTE: At this stage in the operation the valve clearances may be checked if desired – see 12.29.48 but it is vital that before doing so the crankshaft is turned 90° so that the pistons are not at T.D.C. and therefore not in danger of being damaged by the valve heads.

15 Finally adjust the position of the camshaft timing mark in relation to the groove in the front bearing cap by turning the shaft by means of a spanner on the hexagon at the rear of the shaft.
16 Secure the camshaft sprocket to the camshaft using a new lock washer – do not at this stage fully tighten or lock the bolt.
17 Turn the engine over so that the remaining bolt may be fitted.
18 Tighten both bolts to the correct torque figure – see section 06 and bend over the lock tabs.
19 Remove the 'slave' nut holding the sprocket to the support bracket.
20 Fit the semi-circular sealing rubber to the cylinder head.
21 Fit the camshaft cover. 12.29.42.
22 Fit the fresh air duct. 80.15.31.
23 Reconnect the battery.

CONNECTING RODS AND PISTONS

– **Remove and refit** **12.17.01**

Special tool: 38 U3.

CAUTION: Do not mix any components during this operation.

Removing

1 Drive the car onto a ramp and disconnect the battery.
2 Remove the inlet manifold complete with carburetters. 19.15.15.
3 Remove the cylinder head. 12.29.10.
4 Remove the sump. 12.60.44.
5 Check the identification numbers on the connecting rods and caps.
6 Turn the crankshaft to bring numbers 1 to 4 pistons to T.D.C. and remove the nuts of numbers 2 and 3 connecting rods.
7 Withdraw the bearing caps and lower shells and fit plastic or rubber sleeves over the connecting rod bolts to prevent damage to the crankpins.
8 Push the piston and connecting rod assemblies upwards and withdraw through the top of the bores and remove the upper shells
9 Turn the crankshaft to bring numbers 1 and 4 connecting rod bolts to an accessible position and remove the nuts and withdraw the bearing caps and lower shells.
10 Push the piston and connecting rod assemblies upwards and withdraw as instruction 8.

[**NOTE:** Lubricate the pistons, the piston rings, and the cylinders with engine oil only. Other lubricants may interfere with the proper seating of new piston rings. If the engine is to be stored for some time following repairs—or if you are installing new bearing shells—it is advisable to use assembly lubricant (available from automotive supply stores) or a light coating of multipurpose grease on the crankpins rather than ordinary engine oil.]

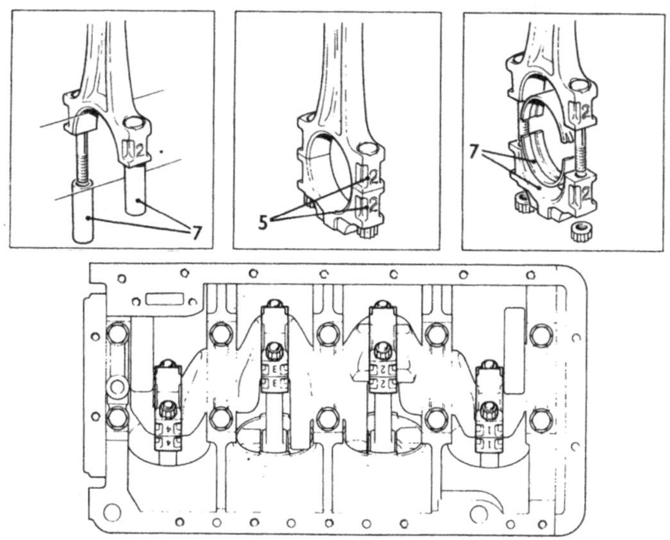

Refitting

11 Stagger the piston ring gaps.
12 Lubricate the pistons and rings and compress the rings with special tool number 38 U3.
13 Insert the connecting rod and piston assemblies into their respective bores ensuring that the raised flat part of the piston crown and the valve head recesses are towards the right-hand side of the engine (as from the driver's position).

NOTE: Some pistons may have arrows stamped on both sides of the skirt, on the gudgeon pin bore side, to indicate the direction of the gudgeon pin off-set. Ensure that when fitting the piston assemblies to the bores that these arrows point to the right-hand side of the engine also. Alternatively some pistons may have an arrow on the crown and the piston assemblies must be fitted with the arrow pointing to the front of the engine.

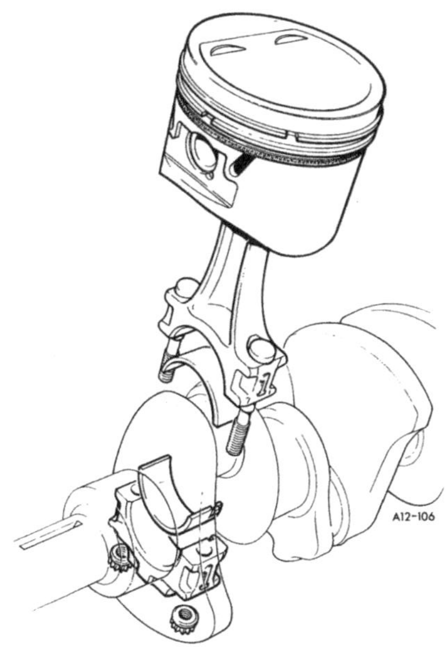

14 Fit the upper bearing shells to the connecting rods ensuring that the keeper tags locate correctly in the connecting rod recesses.
15 Fit the lower bearing shells to the caps ensuring that the keeper tags locate in the recesses.
16 **Pull** the connecting rods onto the crankpins and fit the bearing caps to their respective connecting rods making sure that the identification numbers coincide and are adjacent. Note also that the bearing keeper recesses in the connecting rods and the caps are on the same side.
17 Secure the bearing caps with NEW nuts and tighten evenly to the correct torque – see section 06.
18 Fit the oil sump. 12.60.44.
19 Fit the cylinder head. 12.29.10.
20 Fit the inlet manifold complete with carburetters. 30.15.02.
21 Refill the sump with the recommended grade of oil to the high mark on the dip-stick.
22 Refill the cooling system. 26.10.01
23 Reconnect the battery.

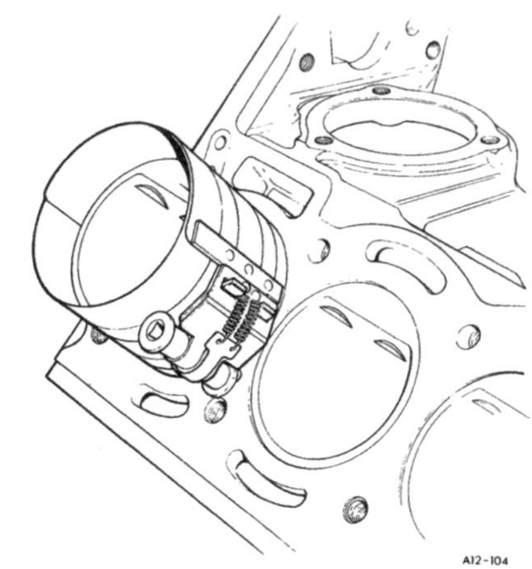

CONNECTING RODS AND PISTONS

– Overhaul 12.17.10

Gudgeon pin bush – each –
– remove and refit 12.17.13

Special tool: 335

CAUTION: Do not mix the components during this operation.

Removing

1 Remove the connecting-rods and pistons. 12.17.01.

Dismantling

[**NOTE:** If wear has left ridges of metal around the tops of the cylinders (known as top cylinder ridges), you should not replace the piston rings unless these ridges are first removed with a special ridge reaming tool. If unremoved, the ridges may break the new top rings.]

2 Remove the two gudgeon pin retaining circlips.
3 Push out the gudgeon pin and separate the piston from the connecting-rod.
4 Remove the top, scraper and oil control rings.
5 Repeat instructions 2 to 4 on the remaining pistons and connecting-rod assemblies.
6 De-grease all components and remove carbon deposits from the pistons.

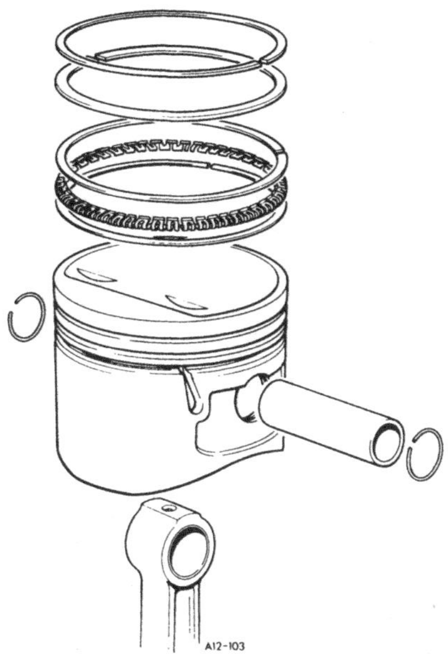

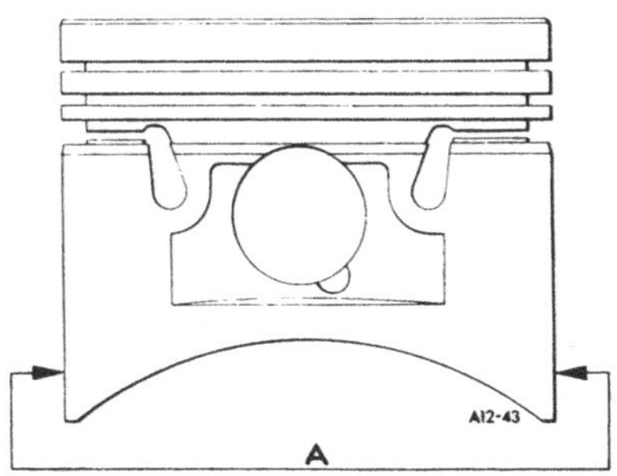

Examination – Pistons and Gudgeon Pins

[**CAUTION:** Every check and fitting operation described here is essential. If you lack the skills or the proper equipment, please turn this work over to your Authorized Dealer or to a qualified automotive machine shop.]

7 Examine the pistons for damage, scoring and cracks.
8 Determine the maximum clearance that exists between the pistons and their respective bores as follows:
   a Measure the piston across the skirt – dimension A – at right angles to the gudgeon pin.
   b Using a cylinder bore measuring gauge determine the maximum cylinder bore wear.
   c Subtract (a) from (b) and compare the result with the clearance given in data for a new engine.
9 Check the dimension of the piston ring grooves and the gap between the piston ring and groove – see data.
10 Examine the gudgeon pin for scores and pitting. Check for wear – see data. The gudgeon pin should be a thumb push fit in the piston at a temperature of 68°F.
11 Check the top and scraper piston ring gaps when inserted squarely into the bores – see data.

NOTE: Two grades of standard piston designated F and G are fitted to new engines built on production. The cylinder block is stamped on the L.H. side as illustrated to indicate the individual bore grade. The grade of the corresponding piston is stamped on the crown.
Whilst these pistons are not supplied for service purposes, a single standard piston 0.001 in (0,0254 mm) oversize is available. Therefore should it be necessary to fit a new piston to a standard bore, the bore must be honed to accommodate the piston with the specified clearance – see data.

continued

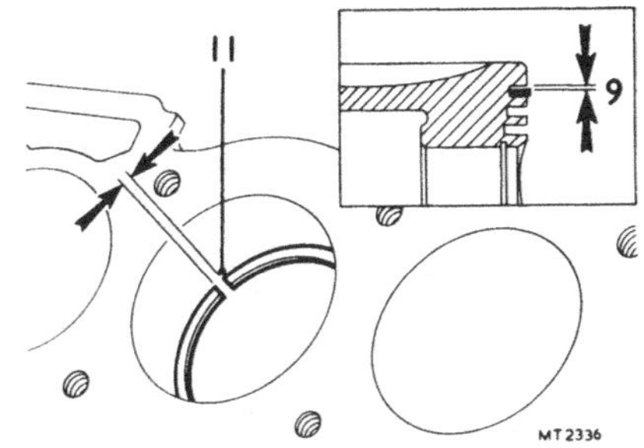

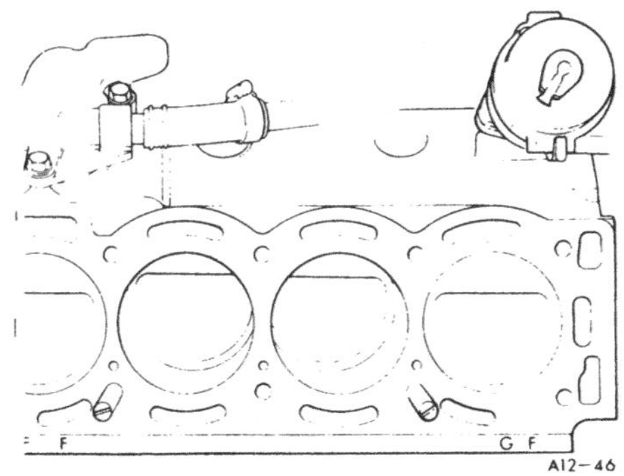

USA SPECIFICATION

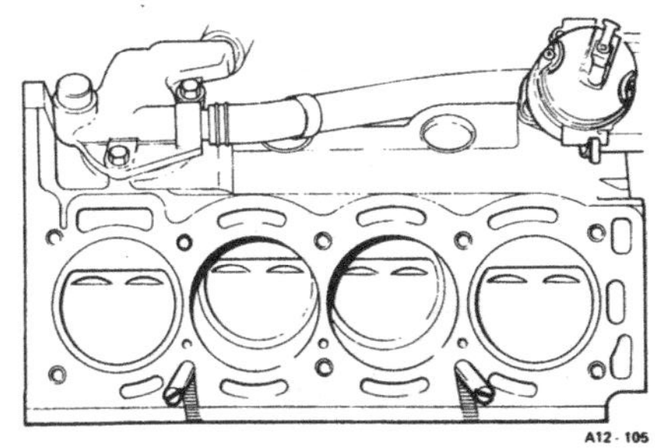

U.K. AND EUROPE SPECIFICATION

Examination – Connecting-rods

12 Using special tool 335 check the connecting-rods for:
A. Bend
B Twist.
Rods that exceed the tolerances in data should be realigned or renewed.

13 Check the gudgeon pin bush for wear and if necessary remove the old bush and fit a new one using a suitable press. Ensure that the oil hole in the bush corresponds exactly with the hole in the connecting-rod. Ream the new bush to size – see data.

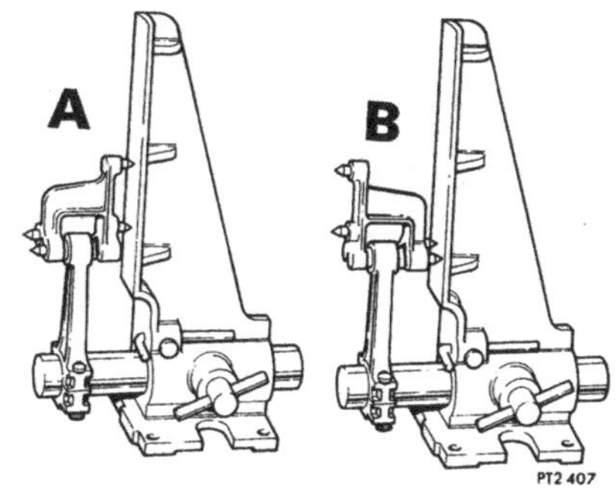

Reassembling

14 Fit the piston rings in the following order:

NOTE: The oil control ring comprises three parts, A, B and C, namely the centre expander rail, flanked by two identical chrome rails.

A. Fit the expander rail into the bottom groove ensuring that the ends butt, not overlap.
B. Fit the bottom chrome rail to the bottom groove.
C. Fit the top chrome rail to the bottom groove.
D. Fit the scraper ring to the centre groove in the piston with the word 'TOP' uppermost.
E. Fit the top compression chrome ring to the top groove.

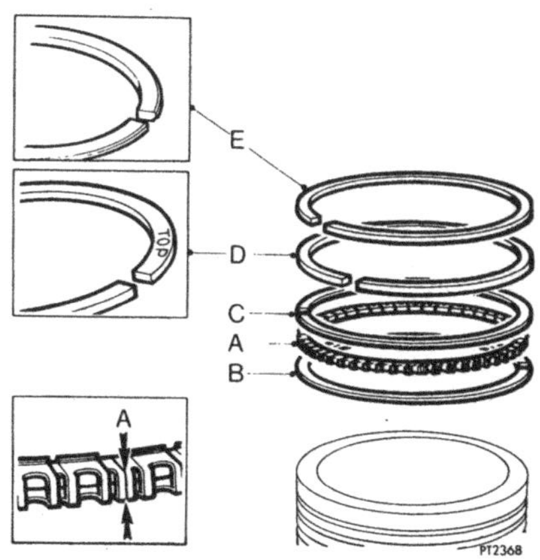

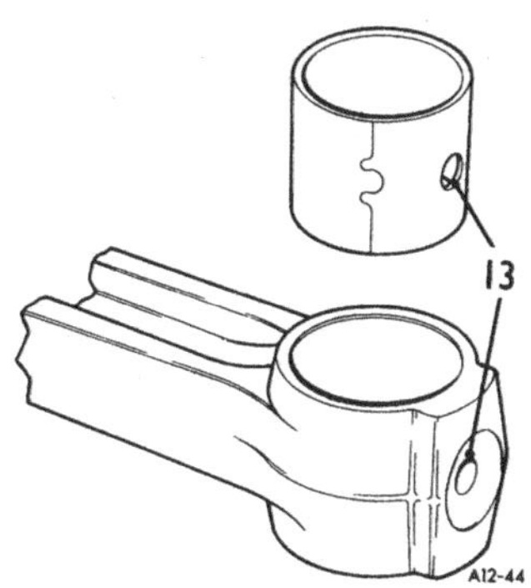

15 Refit the pistons to the connecting-rods so that the identification numbers and the shell bearing keeper recesses are on the opposite side to the raised part of the piston crown.

16 Locate the gudgeon pin with the two circlips ensuring that they fit properly in the grooves.

17 Fit the connecting-rods and pistons to the engine – instructions 10 to 21 12.17.01 ensuring that the sump is refilled with oil to the high mark on the dipstick.

18 Refill the cooling system before connecting the battery.

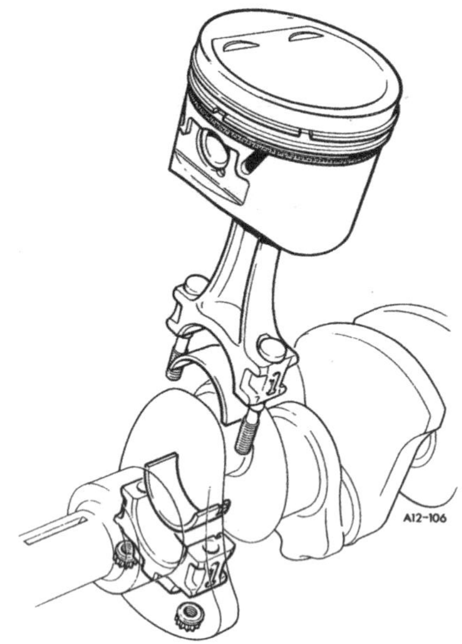

DATA

Cylinder bore size	3.56 in (90,3 mm)
Piston clearance in bore measured at bottom of skirt at right angles to the gudgeon pin	0.0005 to 0.0015 in (0,013 to 0,039 mm)
Pistons	
Top compression ring groove width	0.0705 to 0.0713 in (1,790 to 1,810 mm)
Second compression ring groove width	0.0700 to 0.0709 in (1,780 to 1,800 mm)
Oil control ring groove width	0.1579 to 0.1587 in (4,010 to 4,020 mm)
Piston rings	
Top compression – gap in bore	0.015 to 0.025 in (0,40 to 0,65 mm)
Top compression – clearance in groove	0.0019 to 0.0039 in (0,050 to 0,082 mm)
Second compression ring – gap in bore	0.015 to 0.025 in (0,40 to 0,65 mm)
Second compression ring – clearance in groove	0.0015 to 0.0025 in (0,040 to 0,065 mm)
Oil control rails – gap in bore	0.015 to 0.055 in (0,40 to 1,40 mm)
Oil control expander – gap in bore	ends to butt
Oversize rings	0.010 to 0.020 in (0,254 to 0,508 mm)
Connecting rods	
Small end bush fitted internal diameter	0.9377 to 0.9380 in (23,818 to 23,825 mm)
Bush external diameter	1.0015 to 1.0025 in (25,438 to 25,464 mm)

CONNECTING ROD BEARINGS

– **Remove and refit – set** 12.17.16
  – **one** 12.17.17
  – **extra each** 12.17.18

Removing

1 Drive the car onto a ramp.
2 Disconnect the battery.
3 Drain the sump oil, and remove the dipstick.
4 Remove the sump. 12.60.44.
5 Turn the engine until numbers 1 and 4 big-end bearings are in an accessible position.
6 Check that the connecting-rods and caps are numbered correctly.

CAUTION: Do not mix components whilst carrying out the following instructions.

7 Remove the two special nuts securing each big-end cap and withdraw the caps complete with lower shells.
8 Push the connecting rod upwards sufficiently to enable the upper shell bearing to be removed.
9 Fit rubber or plastic sleeves over the big-end bolts to prevent damage being caused to the crankpins.
10 Turn the crankshaft sufficiently to bring numbers 3 and 4 big-end bearings to an accessible position.
11 Repeat instructions 6 to 9 on numbers 2 and 3 big-ends.

[**NOTE:** If the engine is to be stored for some time following repairs, it is advisable to use assembly lubricant (available from automotive supply stores) or a light coating of multipurpose grease on the crankpins rather than ordinary engine oil. Many experienced mechanics choose to use assembly lubricant whenever new bearing shells are installed.]

Refitting

12 Clean numbers 2 and 3 crankpins and corresponding bearings and caps.

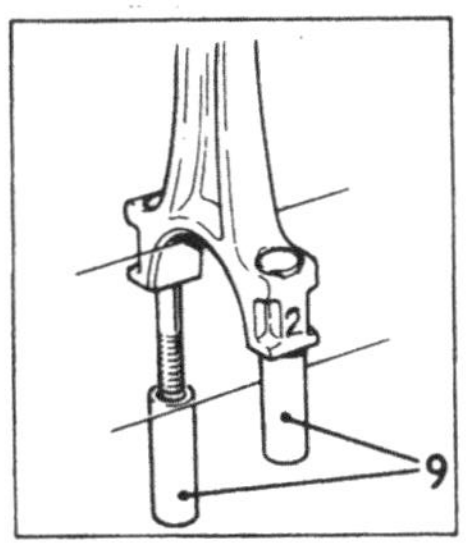

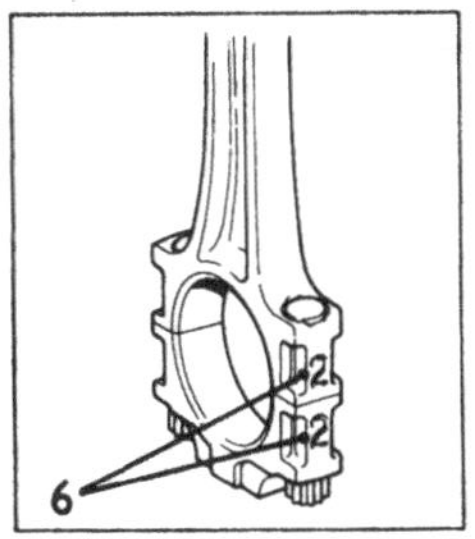

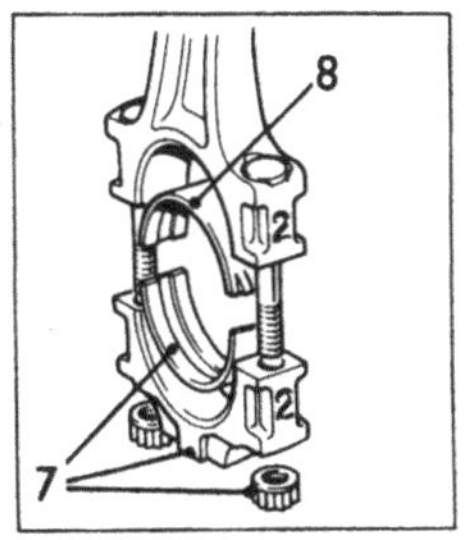

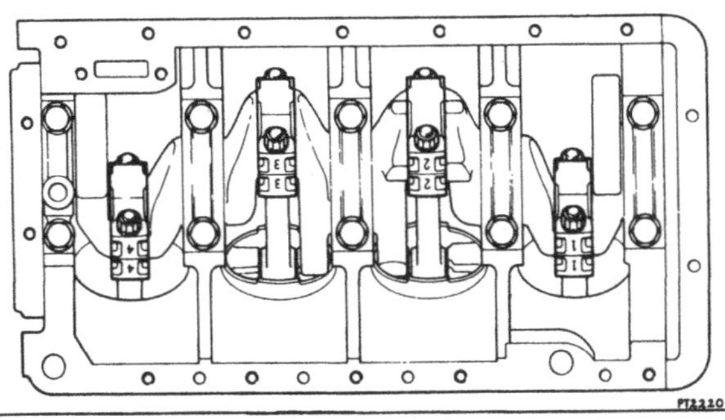

13 Fit the upper bearing shell to the connecting-rod ensuring that the keeper tag locates in the connecting-rod recess and pull the rod onto the crankpin.
14 Fit the lower bearing shell to the cap ensuring that the keeper tag locates correctly in the connecting-rod recess.
15 Remove the protective sleeves from the connecting-rod bolts and fit the bearing cap checking that:
   a The correct number cap is being fitted to the connecting-rod concerned.
   b The keeper tags are adjacent – i.e. on the same side of the bearing.
16 Fit and evenly tighten new nuts to the correct torque figure – see section 06.
17 Turn the crankshaft to bring numbers 1 and 4 big-ends to an accessible position.
18 Repeat instructions 12 to 18 on numbers 1 and 4 big-ends.
19 Fit the sump. 12.60.44.
20 Lower the ramp and fill the sump to high mark on the dipstick with oil of a recommended grade.
21 Reconnect the battery.
22 Drive the car from the ramp.

CRANKSHAFT PULLEY

– **Remove and refit** 12.21.01

Removing

1 Drive the car on to a ramp.
2 Disconnect the battery.
3 Slacken the alternator drive belt tension and slip it from the crankshaft pulley.
4 Place the car in gear and put the handbrake 'on'.
5 Raise the ramp.
6 Turn the special bolt – securing the pulley to the crankshaft – anti-clockwise and remove together with the washer.
7 Withdraw the crankshaft pulley.

Refitting

8 Ensure the key is in position in the crankshaft.
9 Lubricate the crankshaft and fit the pulley.
10 Secure the pulley with the special bolt and washer, and tighten to the correct torque figure – see Section 06.
11 Lower the ramp, fit and tension the drive belt.
12 Put the gear lever in the 'neutral' position and re-connect the battery and drive the car off the ramp.

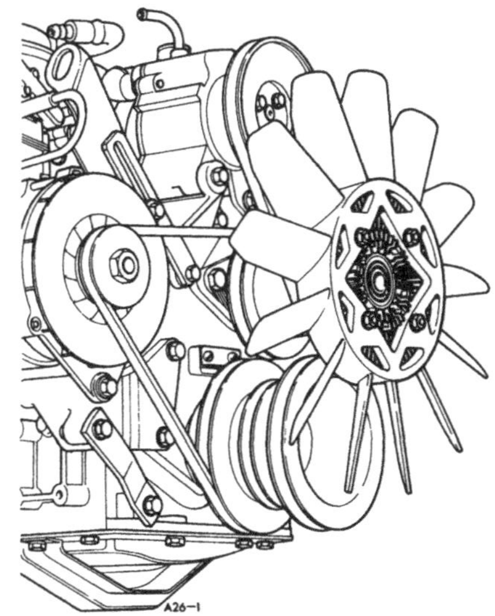

CRANKSHAFT PULLEY AND AUXILIARY DRIVES

USA SPECIFICATION

– Remove and refit 12.21.02

Removing

1 Drive the car onto a ramp, and disconnect the battery.
2 Remove the fan blades. 26.25.06.
3 Slacken the compressor adjustment bolts, and slip the belt from the pulley, see operation 82.10.02 instructions 1 to 3.
4 Slacken the alternator – fan drive belt adjustment bolts and release the belt from the pulley.
5 Place the car in gear and put the hand brake on.
6 Raise the ramp.
7 Remove the special bolt and washer securing the pulleys to the crankshaft.
8 Withdraw the auxiliary pulley cluster.
9 Remove the crankshaft pulley.

Refitting

10 Lubricate the crankshaft pulley boss and fit it to the crankshaft locating on the crankshaft key.
11 Fit the auxiliary pulley cluster locating on the crankshaft key.
12 Secure the assembly with the special bolt and washer and tighten to the correct torque figure see section 06.
13 Lower the ramp and put the gear lever in neutral.
14 Fit the alternator – fan drive belt and adjust the tension see operation 86.10.05 instructions 4 to 7.
15 Fit the compressor drive belt to the pulleys and adjust the tension see operation 82.10.01 instructions 7 to 11 and lower ramp.
16 Refit the fan blades 26.25.06 instructions 6 to 8.
17 Reconnect the battery and remove the car from ramp.

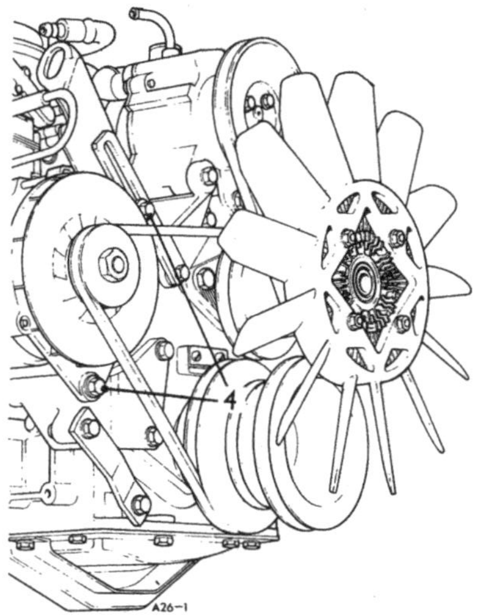

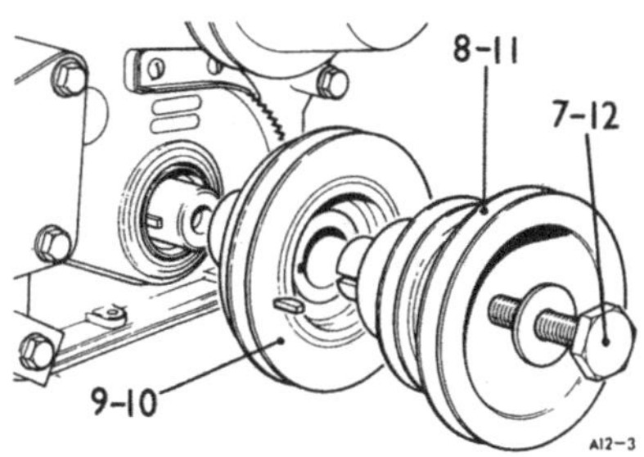

CRANKSHAFT REAR OIL SEAL

– Remove and refit 12.21.20

Removing

1 Drive the car onto a ramp.
2 Disconnect the battery.
3 † Remove the gearbox. 37.20.01.
4 † Remove the clutch. 33.10.01.
5 Remove the flywheel. 12.53.07.
6 Remove the two rear sump bolts.
7 Slacken the two rear-most R.H. side sump bolts.
8 Slacken the L.H. side sump rear-most nut and bolt.
9 Remove the six bolts securing the crankcase oil seal housing to the crankcase.
10 Press out the oil seal from the housing.

[NOTE: In some instances, oil leaks around the rear of the engine—which may seem to be caused by a faulty seal—are caused by rear sump bolts that have insufficient thread. That is, the unthreaded portions of the bolt shanks contact the engine block before the bolt heads have come into solid contact with the sump flange. Either select new bolts with fully-threaded shanks, or install an additional flat washer between each bolt head and the sump flange.]

Refitting

11 Lubricate the outer diameter of a new seal and press it squarely into the housing so that the lip faces the crankshaft.
12 Clean the crankcase and seal housing mating faces and using sealing compound fit a new gasket.
13 Lubricate the crankshaft and carefully ease the seal into position locating the housing on the two dowels.
14 Fit the six housing retaining bolts loosely noting that the two lower opposing bolts are longer.
15 Evenly tighten the bolts to the correct torque figure – see section 06.
16 Fit and tighten the two rear sump to seal bolts.
17 Tighten the R.H. side sump bolts.
18 Tighten the L.H. side nuts and bolts.
19 Fit the flywheel, see 12.53.07.
20† Fit the clutch assembly. 33.10.01.
21† Fit the gearbox. 37.20.01.
22 Reconnect the battery and remove the car from ramp.

† When automatic transmission is fitted refer to operations 44.20.01 and 44.17.01/07.

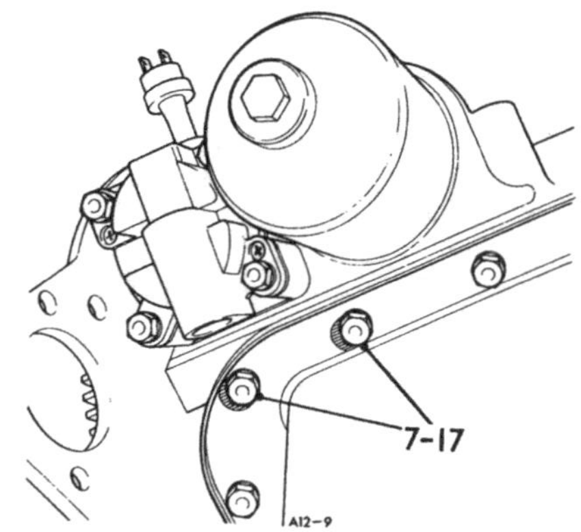

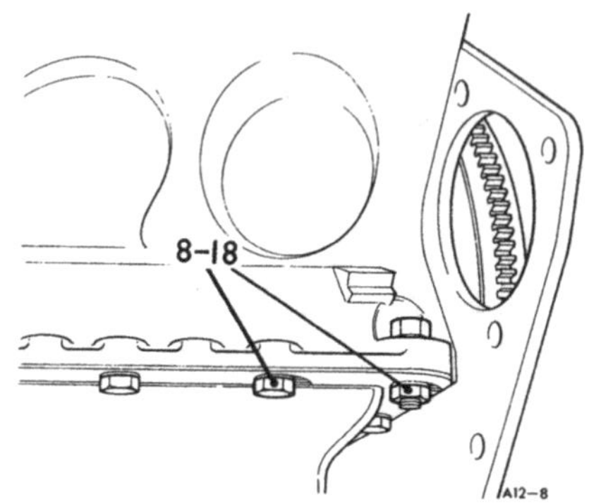

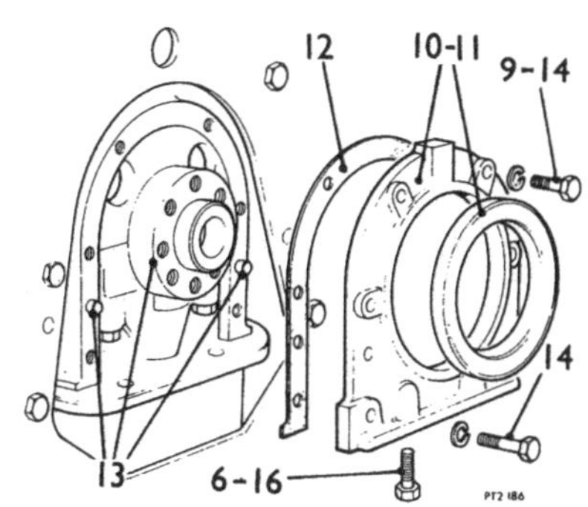

CRANKSHAFT END-FLOAT

– Check and adjust 12.21.26

Checking

1 Raise the car on a ramp.
2 Disconnect the battery.
3 Attach the magnetic base of a dial guage stand to the underside of the sump and arrange the dial gauge so that the stylus rests in a loaded condition on the front face of the crankshaft pulley.
4 Lever the crankshaft rearwards.
5 Zero the dial gauge and lever the crankshaft forward and note the reading.
6 Repeat instructions 4 and 5 several times until a constant reading is achieved – see data.
7 Remove the dial gauge and magnetic base.

[NOTE: Excessive crankshaft end-float may cause an irregular, rather sharp rapping noise while the engine is idling. The noise will usually stop when the clutch pedal is held down.]

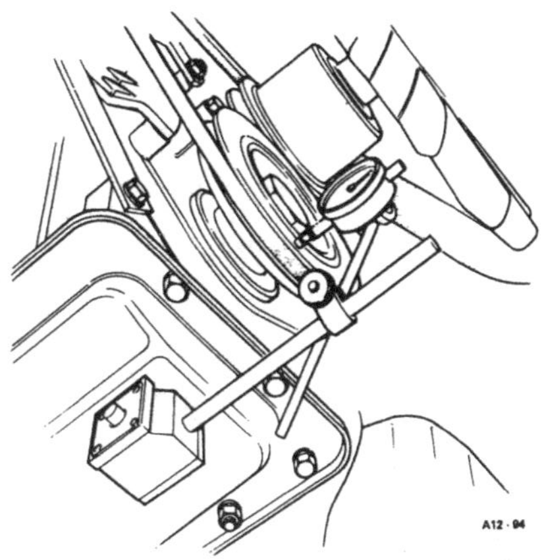

Adjusting

8 Drain the sump oil and remove the dipstick.
9 Remove the sump. 12.60.44.
10 Remove two bolts and withdraw number 3 main bearing cap and lower shell.
11 Using the blade of a thin screwdriver and taking care not to damage the crankshaft, remove the two crankshaft thrust bearings.

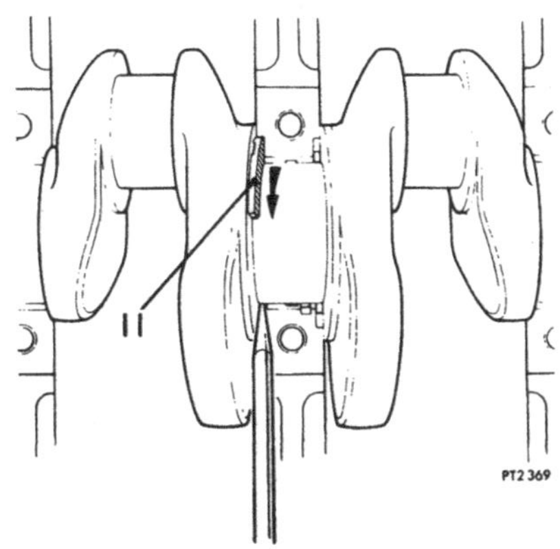

12 Lubricate and feed the thrust bearings of the appropriate size into the channel, reversing the method of removal. Ensure however, that the two grooves in the thrust bearing face outwards away from the bearing cap.
13 Fit the main bearing cap and lower shell noting that the keeper recesses in the cap and crankcase are adjacent. Temporarily tighten the two retaining bolts.
14 Using a feeler gauge or clock gauge check the crankshaft end-float by levering the crankshaft forwards or rearwards.
15 Repeat instructions 10 to 14 if necessary to achieve the correct end-float.
16 Finally tighten the number 3 main bearing cap bolts to the correct torque figure – see section 06.
17 Refit the sump. 12.60.44.
18 Refill the sump to the high mark on the dipstick with oil of a recommended grade.
19 Lower the car.
20 Connect the battery.

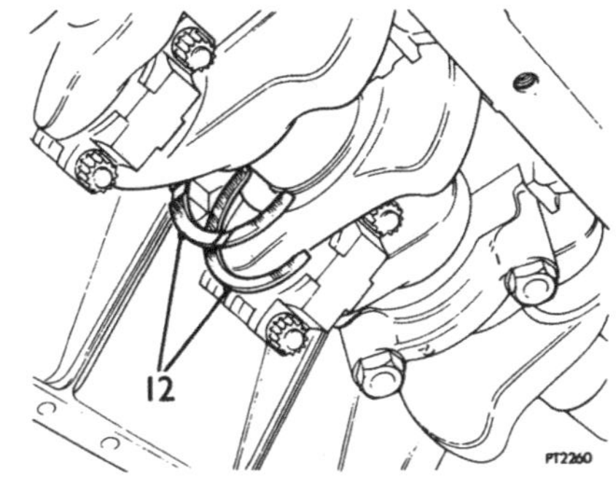

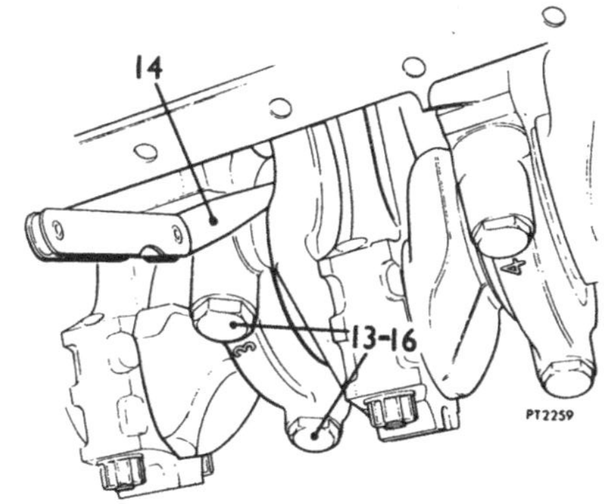

DATA

Crankshaft end-float	0.003 to 0.011 in (0,07 to 0,28 mm)
Adjustment	By selective thrust bearings 0.005 in (0,127 mm)

CRANKSHAFT

– Remove and refit 12.21.33

[**NOTE:** If the engine is to be stored for some time following repairs, it is advisable to use assembly lubricant (available from automotive supply stores) or a light coating or multipurpose grease on the crankshaft journals and crankpins rather than ordinary engine oil. Many experienced mechanics choose to use assembly lubricant whenever new bearing shells are installed.]

Removing

1 Remove the engine and gearbox assembly from the car. 12.37.01.
2 Drain the sump.
3 † Remove the gearbox, noting the position of the remaining bell housing bolts to assist refitting.
4 † Remove the clutch.
5 Remove the flywheel – 12.53.07 instructions 4 to 6.
6 Remove the engine rear adaptor plate.
7 Remove the sump, and dipstick.
8 Remove the crankshaft rear main oil seal – 12.21.20 instruction 9.
9 Remove the timing chain cover – 12.65.01
10 Remove the oil pick-up strainer.
11 Remove the oil thrower.
12 Remove the crankshaft sprocket.
13 Remove the drive key and shims.
14 Remove the connecting rod bearing caps and lower shells, and push the pistons up the bores just sufficiently to clear the crankshaft journals. 12.17.16.
15 Fit plastic or rubber sleeves over the connecting rod bolts to prevent damage to the crankshaft journals.
16 Remove the timing chain 12.65.14 instructions 2 to 15.
17 Remove the ten bolts retaining the main bearing caps.
18 Withdraw the five main bearing caps and lower shells.
19 Lift out the crankshaft.
20 Remove the spigot bush.
21 Remove the upper shells and thrust washers from the crankcase.

[**NOTE:** There is little use in fitting new bearing shells if the crankshaft journals and crankpins are worn undersize, worn out-of-round or worn to a taper. Measure each journal and crankpin with a micrometer. Check both ends of each journal or crankpin, making two measurements at 90° to one another at each point. If there is a difference of more than .002 in (0.05 mm) between any of these measurements, the crankshaft should either be replaced or reconditioned to accept one of the undersize bearing sets listed under Data.]

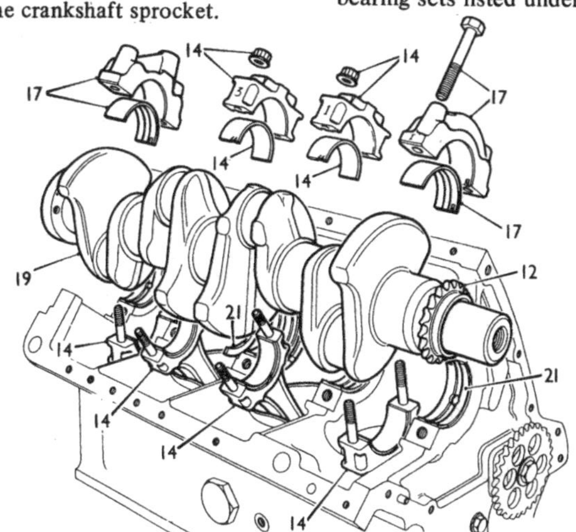

Refitting

22 Fit the upper shells to the crankcase ensuring that the keepertags locate in the recesses.
23 Fit the thrust washers to number three main bearing noting that the grooves face outwards.
24 Clean and lubricate the main bearing journals and lower the crankshaft into the crankcase.
25 Clean and fit the main bearing shells to the caps ensuring that the keeper tags locate in the cap recesses.
26 Fit the main bearing caps to the crankcase partially tightening the securing bolts and ensuring that the caps are fitted to their correct crankcase bearings.
27 Check the crankshaft end-float – see Data – by inserting a feeler gauge between the crankshaft and the thrust washers in number three bearing, see 12.21.26 instruction 14, or using a dial gauge.
28 Tighten the ten main bearing bolts to the correct torque.
29 Clean and lubricate the crankshaft journals, pull the connecting rods and upper shells onto the journals and remove the protective sleeves from the bolts.
30 Clean and fit the connecting rod caps ensuring:
  a The caps are fitted to their correct connecting rods.
  b The keeper tags in the rods and caps are adjacent.
  c The nuts are tightened to the correct torque.
31 Temporarily fit the crankshaft sprocket and check its alignment with the jackshaft sprocket using a straight-edge across the two sprockets. Adjust by fitting shims between the crankshaft and sprocket.
32 Check that the timing mark on the camshaft flange is in line with the groove on the camshaft front bearing cap.

CAUTION: While the timing chain is disconnected do not allow the pistons to reach T.D.C. otherwise damage to the valves and pistons may result.

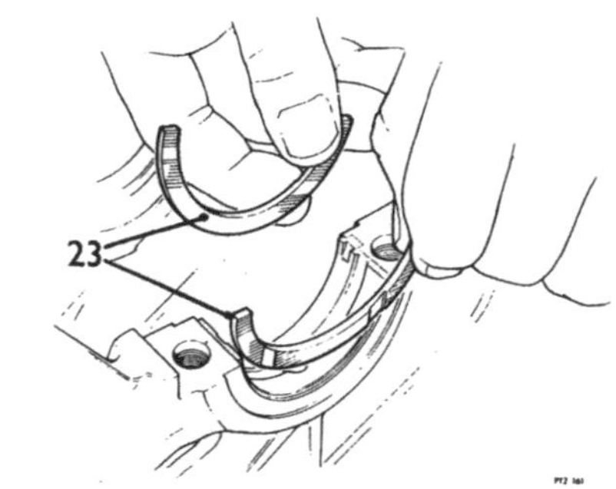

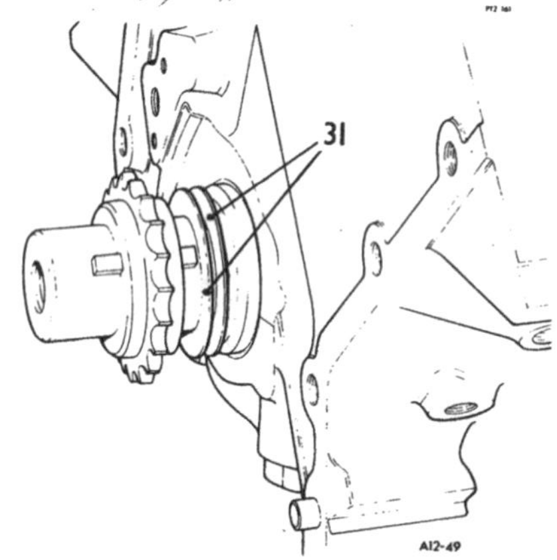

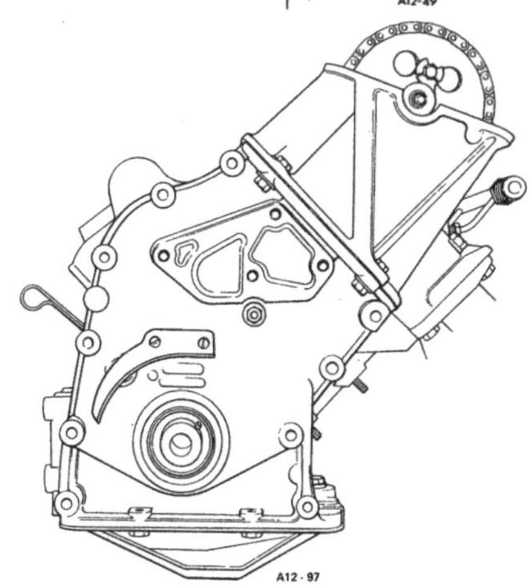

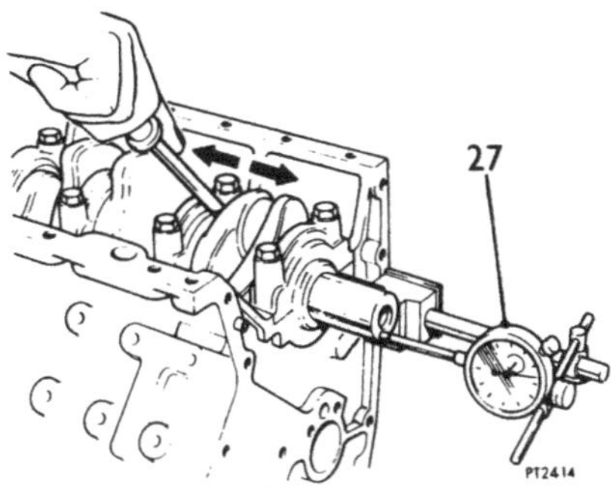

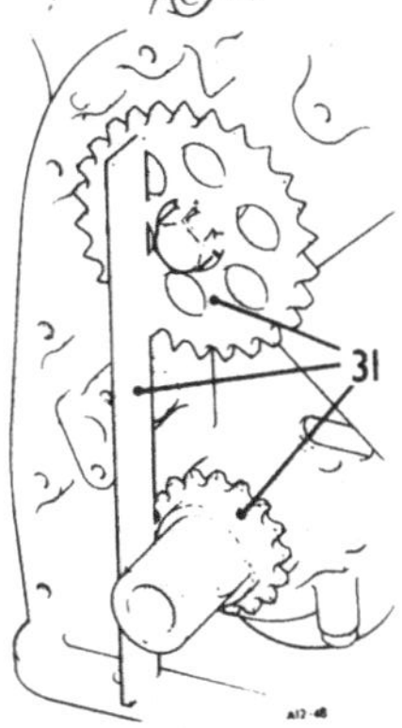

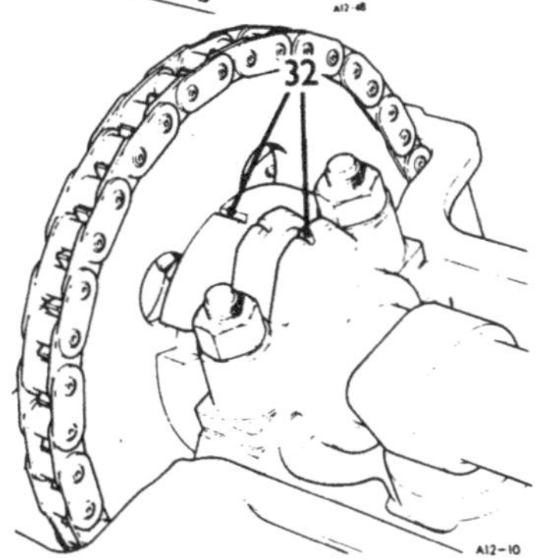

33 Fit the timing chain. 12.65.14.
34 Fit the oil thrower – dished face outwards.
35 Using a new gasket fit the timing chain cover locating it over the dowels. Secure with the centre retaining bolt, noting that it has a fibre washer under the head.
36 Fit the two nuts and bolts securing the timing cover to the cylinder head.
37 Fit the crankshaft pulley.
38 Fit the fan pulley assembly.
39 Fit the alternator mounting bracket. 12.25.21.
40 Fit the alternator and drive belt and adjust the tension. 86.10.02.
41 Fit a new crankshaft rear main oil seal 12.21.20 instructions 11 to 15.
42 Fit the engine rear adaptor plate 12.53.03 instructions 6 and 7.
43 Fit the oil pick-up strainer.
44 Fit the sump and dipstick.
45 Fit the spigot bush, (earlier models) or needle roller bearing (later models).
46 Fit the flywheel, see 12.53.07, instructions 7 to 11.
47† Fit the clutch. 33.10.01.
48† Fit the gearbox.
49 Fit the engine and gearbox to the car. 12.37.01.
50 Fill the sump with oil of a recommended grade to the high mark on the dipstick.
51 Fill the cooling system. 26.10.01.
52 Connect the battery.
53 Check and if necessary adjust the ignition timing. 86.35.15.

† Where automatic transmission is fitted refer to operations 44.20.01 and 44.17.01/07.

MAIN BEARINGS

Remove and refit – set	**12.21.39**
Remove and refit – each	**12.21.40**

Removing
1 Drive the car on to a ramp.
2 Disconnect the battery.
3 Drain the oil sump and remove the dipstick.
4 Remove the sump. 12.60.44.

CAUTION: It is important that during the following instructions the bearing caps, shells and bolts are not mixed but are kept identified with their respective bearings. It will be noted that the bearing caps are numbered 1 to 5 commencing at the front of the engine.

5 Remove the two bolts securing the bearing cap to the crankcase.
6 Withdraw the bearing cap complete with the lower shell.
7 With the tag end leading, carefully slide the upper shell bearing out from between the crankshaft journal and crankcase.
8 Remove the lower shell bearing from the bearing cap.
9 Repeat instructions 5 to 8 on the remaining bearings.

continued

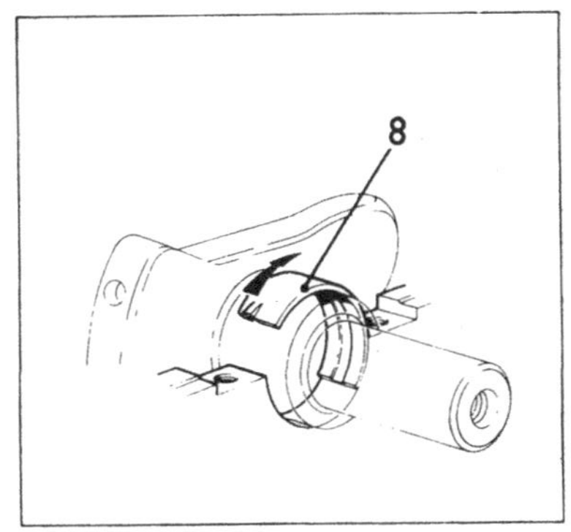

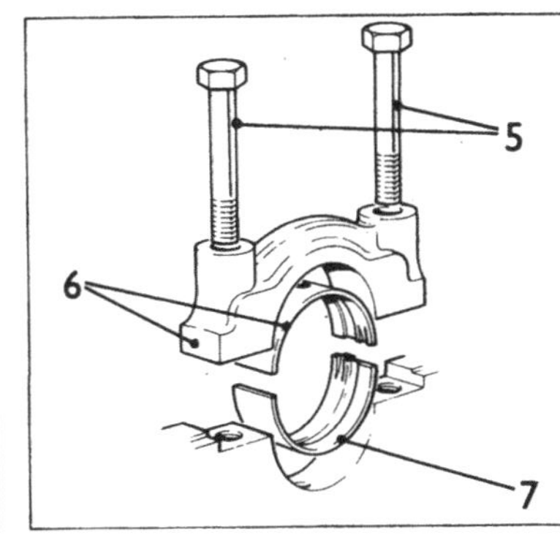

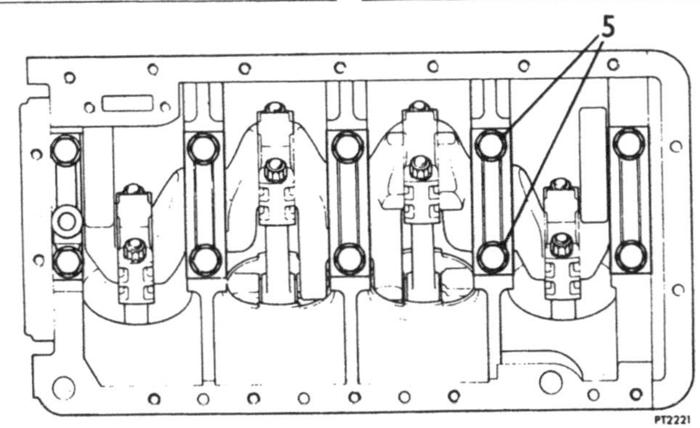

DATA

Main bearing journal diameter	2.1260 to 2.1265 in (54,000 to 54,013 mm)
Crankpin diameter	1.7500 to 1.7505 in (44,450 to 44,463 mm)
Minimum re-grind diameter	
– main bearings	2.0860 to 2.0865 in (52,984 to 52,997 mm)
– crankpins	1.7100 to 1.7105 in (43,434 to 43,447 mm)

Refitting

10 Lubricate with clean engine oil all bearing shells.

11 With the tag end trailing, feed the upper shell bearing between the crankshaft journal and the crankcase. Ensure that the keeper tag locates properly in the crankcase recess.

12 Fit the lower shell bearing into the bearing cap ensuring that the keeper tag locates correctly in the cap recess.

13 Fit the cap to the crankcase noting that the keeper tag is fitted adjacent to its counterpart in the crankcase.

14 Secure the cap with the bolts and tighten evenly to the correct torque figure – see **'TORQUE WRENCH SETTINGS'**.

15 Repeat instructions 11 to 14 on the remaining bearings.

16 Check and if necessary adjust the crankshaft end-float. 12.21.26.

17 Fit the sump. 12.60.44.

18 Lower the ramp and fill the sump with oil of a recommended grade to the high mark on the dipstick.

19 Reconnect the battery.

20 Drive the car from the ramp.

SPIGOT BUSH

Remove and refit 12.21.45

Note: From Engine No. CG8894, CL36771 and CV10 a needle roller bearing was fitted in place of the spigot bush shown. This bearing may be removed and refitted using the procedure below.

Removing

1 † Remove the gearbox. 37.20.01.

2 † Remove the clutch assembly. 33.10.01.

3 Remove the eight flywheel retaining bolts.

4 Remove the spigot bush retaining plate.

5 Withdraw the spigot bush.

Refitting

6 Insert the spigot bush into the crankshaft bore.

7 Fit the spigot bush retaining plate and the eight flywheel retaining bolts and tighten to the correct torque figure – see **'TORQUE WRENCH SETTINGS'**.

8 † Refit the clutch assembly. 33.10.01.

9 † Refit the gearbox. 37.20.01.

† Where automatic transmission is fitted, refer to operations 44.20.01 and 44.17.01/07.

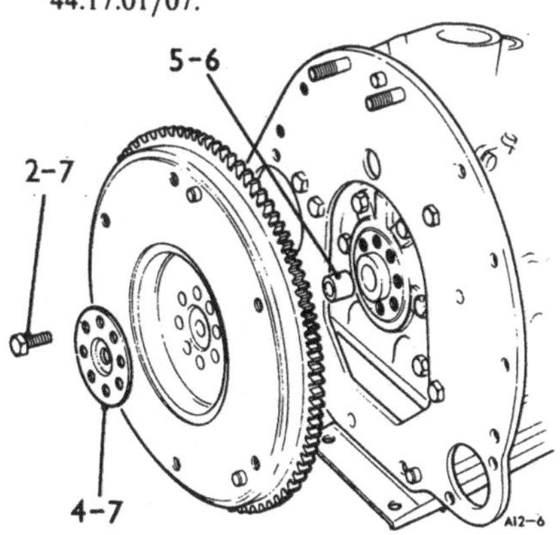

CYLINDER BLOCK DRAIN PLUG

Remove and refit 12.25.07

Removing

WARNING: This operation must only be carried out when the engine is cold.

1 Drive the car onto a ramp and disconnect the battery.

2 Raise the ramp and using a long extension with a universal joint and socket remove the drain plug slowly to allow the coolant to drain.

3 Remove the plug completely when the coolant flow lessens.

Refitting

4 Reverse instructions 1 to 3, using a new washer.

ALTERNATOR CARRIER BRACKET

Remove and refit 12.25.21

Removing

1 Disconnect the battery.
2 Remove the alternator. 86.10.02.
3 Remove the four bolts securing the carrier bracket to the engine.
4 Remove the support strap from the bracket.

Refitting

5 Secure the bracket to the engine with the two front bolts.
6 Fit the rear bolt ensuring that the correct thickness of washer is used between the bracket and engine.
7 Fit the support strap.
8 Refit the alternator and tension the belt. 86.10.02.
9 Reconnect the battery.

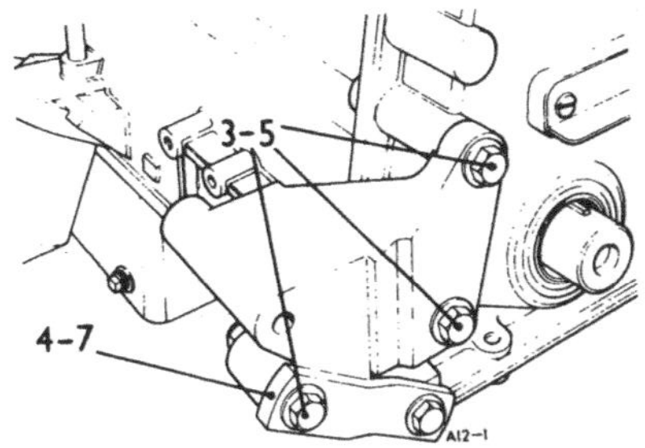

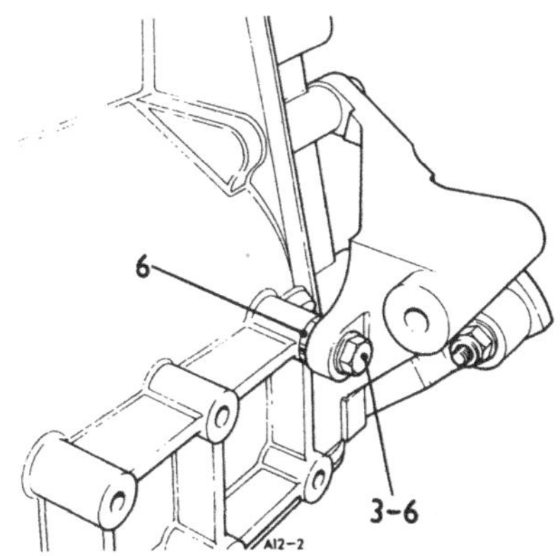

CYLINDER HEAD GASKET
U.K. AND EUROPE SPECIFICATION

– Remove and refit 12.29.02

Removing

1 Drive the car onto a ramp and disconnect the battery.
2 Drain the cooling system. 26.10.01.
3 Remove the air duct. 80.15.31.
4 Remove the air cleaner.
5 Remove the inlet manifold complete with carburetters. 19.15.15.
6 Remove the camshaft cover.
7 Remove the distributor cap.
8 Remove the semi-circular grommet.
9 Turn the engine over so that the camshaft sprocket bottom bolt is accessible.
10 Unlock and remove the bottom bolt.
11 Anchor the camshaft sprocket to the support bracket.
12 Turn the engine over so that the timing mark on the camshaft flange is in line with groove in the camshaft front bearing cap and the distributor rotor arm points to the manifold rear attachment hole in the cylinder head.
13 Unlock and remove the camshaft sprocket top retaining bolt.

continued

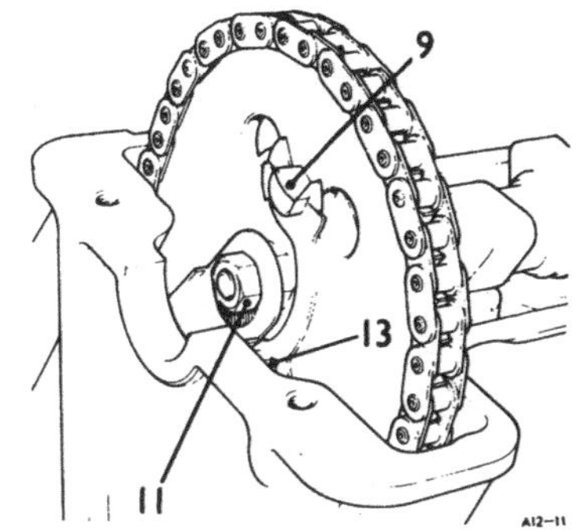

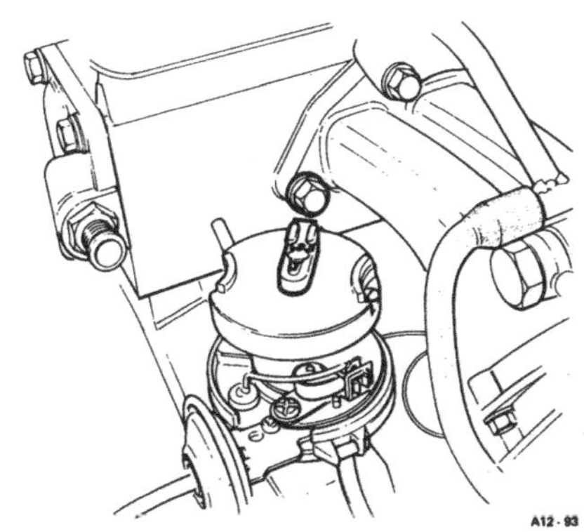

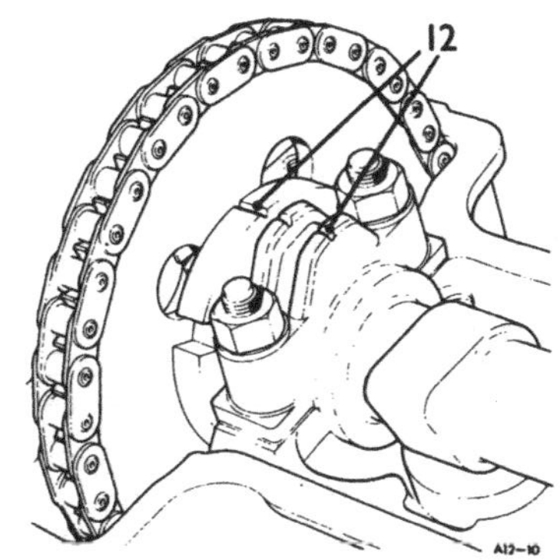

14 Disconnect the pipe from the water transfer housing.
15 Raise the ramp and disconnect the exhaust front pipe from the manifold flange.
16 Lower the ramp and remove the two cylinder head bolts to timing cover nuts and bolts.
17 Slacken the cylinder head retaining nuts and bolts in the reverse order as shown in operation 12.29.27.
18 Remove the five cylinder head studs and five bolts.
19 Withdraw the cylinder head complete with exhaust manifold.
20 Remove the cylinder head gasket.

Refitting

21 Clean the cylinder block and cylinder head mating faces.

[**NOTE:** If the old cylinder head gasket has been leaking, check the cylinder head and the cylinder block for warping before you reassemble the engine. To check, lay a straightedge diagonally across the mating surfaces of both the head and the block. Using feeler gauges of various thicknesses, measure the clearance, if any, between the surface and the straightedge. Then repeat the check with the straightedge at the opposite diagonal. If the clearance at either diagonal exceeds 0.004 in (0.10 mm), the part should be milled or ground by a competent machinist in order to obtain a true surface.]

22 Insert two guide studs in the cylinder block bolt holes as illustrated, to facilitate fitting the cylinder head and gasket.

NOTE: Suitable guide studs may be made to the dimensioned drawing and to enable the studs to be inserted and withdrawn a slot to accommodate a screwdriver blade should be made in one end.

23 Check that number 1 and 4 pistons are at T.D.C. and that the distributor rotor arm is in the same position as in instruction 12.
24 Fit the cylinder head gasket.
25 Fit the cylinder head locating it over the two guide studs.
26 Fit the five cylinder head studs.

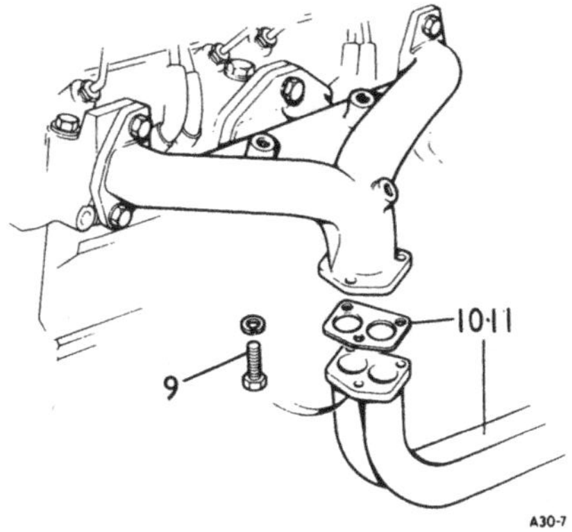

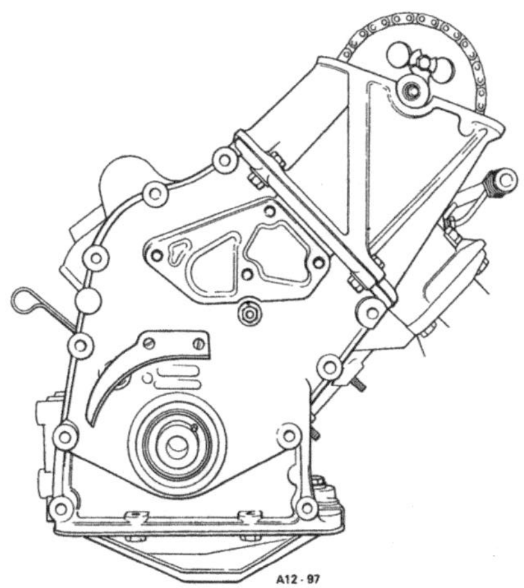

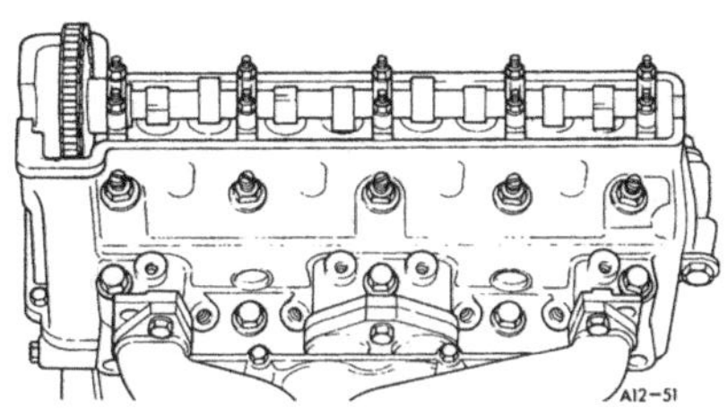

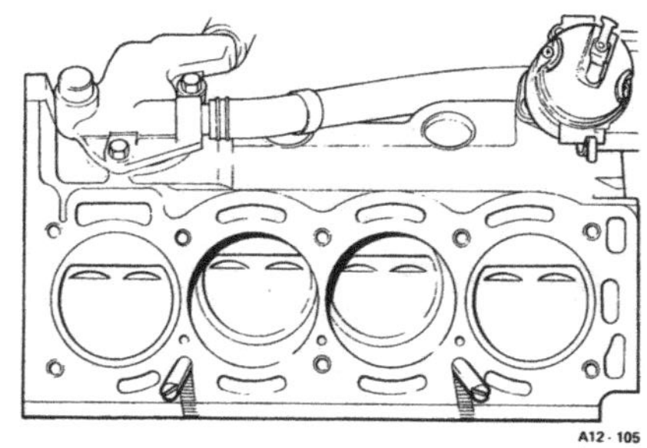

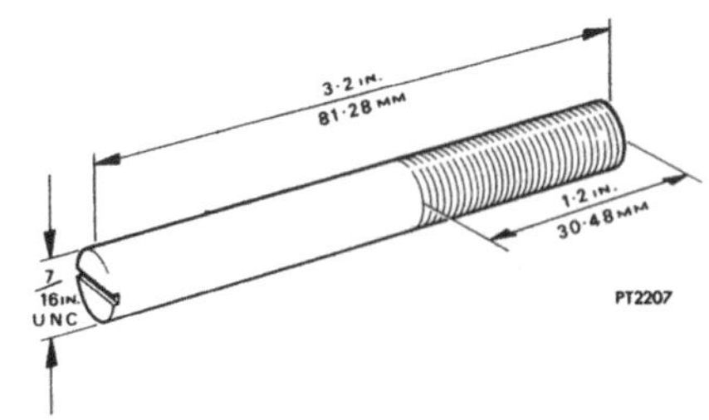

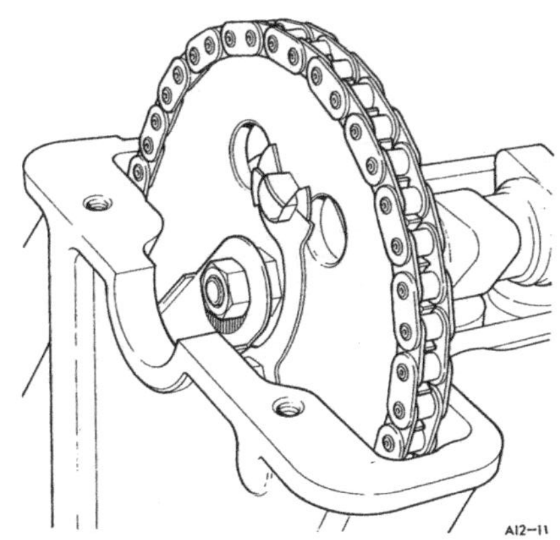

[**NOTE:** Before you install the cylinder head gasket, make certain that the cylinder head and cylinder block faces are clean. Install the new cylinder head gasket with its top side uppermost, which is usually marked. Under no circumstances should you ever reinstall any cylinder head gasket that has previously been compressed by torquing the head bolts and nuts. It is extremely important that the head bolts and nuts be torqued in the sequence given in 12.29.27. Go over the sequence three times, torquing the nuts and the bolts to 25 Nm (2.5 kgf m, or 20 lbf ft), then 50 Nm (5.0 kgf m, or 35 lbf ft), and finally 75 Nm (7.6 kgf m, or 55 lbf ft).]

27 Remove the two guide studs.
28 Fit and finger tighten only the five cylinder head bolts and washers.
29 Fit and finger tighten the five cylinder head nuts, washers and the two brackets for locating the spark plug leads.
30 Tighten the nuts and bolts to the correct torque figure – see section 06 – in the sequence shown in operation 12.29.27.
31 Fit the lock washer and camshaft sprocket top retaining bolt.
32 Turn the engine over to enable the bottom bolt to be fitted.
33 Tighten and lock the two bolts.
34 Remove the 'slave' nut from the camshaft sprocket.
35 Fit the semi-circular grommet to the cylinder head.
36 Fit the two timing covers to the cylinder head nuts and bolts.
37 Fit the camshaft cover.
38 Connect the heater hose to the water transfer housing.
39 Fit the plug leads and the distributor cap.
40 Raise the ramp and connect the exhaust down-pipe to the manifold flange.
41 Lower the ramp and fit the inlet manifold complete with carburetters. 30.15.02.
42 Fit the air cleaner. 19.10.02
43 Fit the air duct. 80.15.31.
44 Fill the cooling system. 26.10.01.
45 Connect the battery and drive the car off the ramp.

CYLINDER HEAD GASKET – USA Specification

Remove and refit 12.29.02

* *Applicable to carburetter engines only*
** *Applicable to fuel injection engines only*

Removing

1 Drive the car onto a ramp and disconnect the battery.
2 Drain the cooling system. 26.10.01.
3 Remove the air duct. 80.15.31.
4* Remove the air cleaner.
5 Remove the inlet manifold complete with carburetters or fuel rail.
6 Remove the camshaft cover.
7 Remove the distributor cap.
8 Remove the semi-circular grommet.
9 Turn the engine over so that the camshaft sprocket bottom bolt is accessible.
10 Unlock and remove the bottom bolt.
11 Anchor the camshaft sprocket to the support bracket.
12 Turn the engine over so that the timing mark on the camshaft flange is in line with groove in the camshaft front bearing cap and the distributor rotor arm points to the manifold rear attachment hole in the cylinder head.
13 Unlock and remove the camshaft sprocket top retaining bolt.

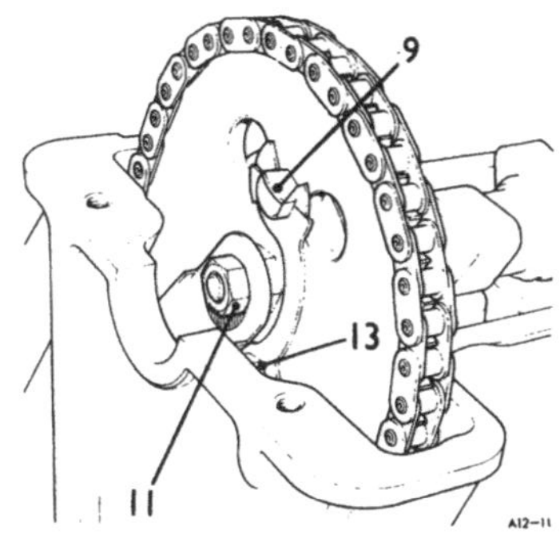

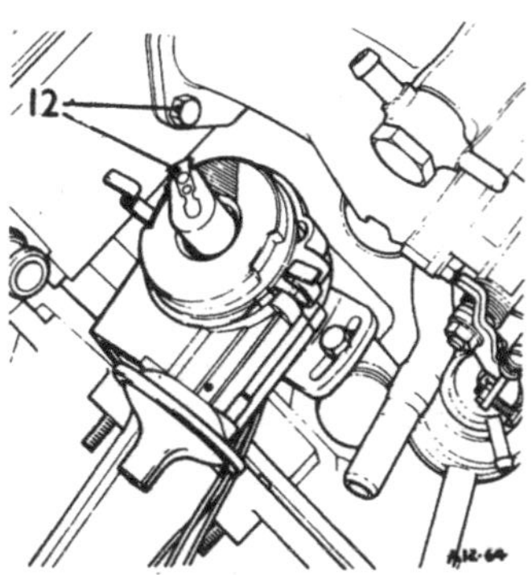

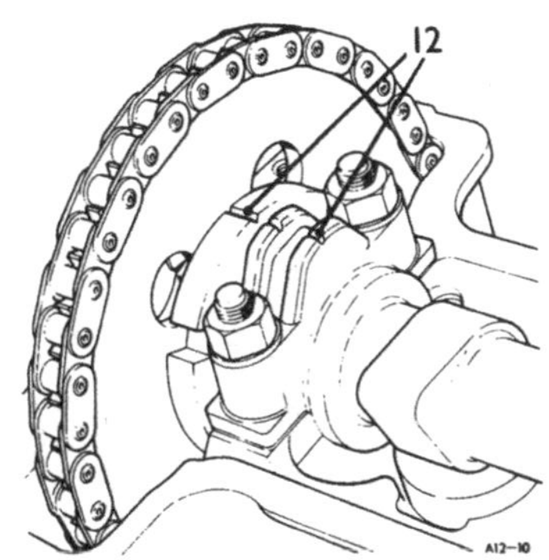

14* Disconnect the pipe to the check valve.
or
** Disconnect the electrical lead to the Lambda (oxygen) sensor.
15 Disconnect the pipe from the water transfer housing.
16 Raise the ramp and disconnect the exhaust front pipe from the manifold flange.
17 Lower the ramp and remove the two cylinder head bolts to timing cover nuts and bolts.
18 Slacken the cylinder head retaining nuts and bolts in the reverse order as shown in operation 12.29.27.

continued

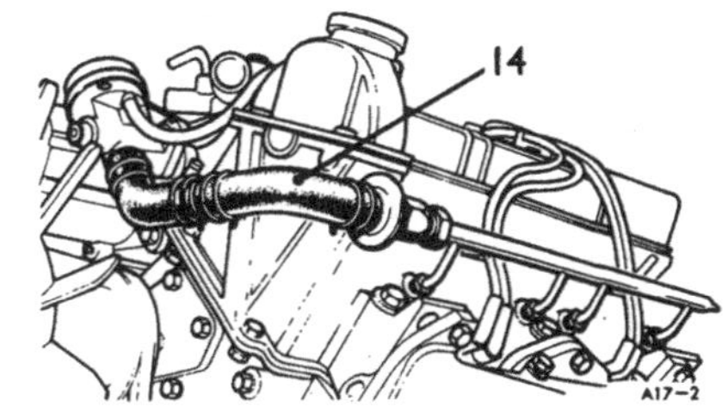

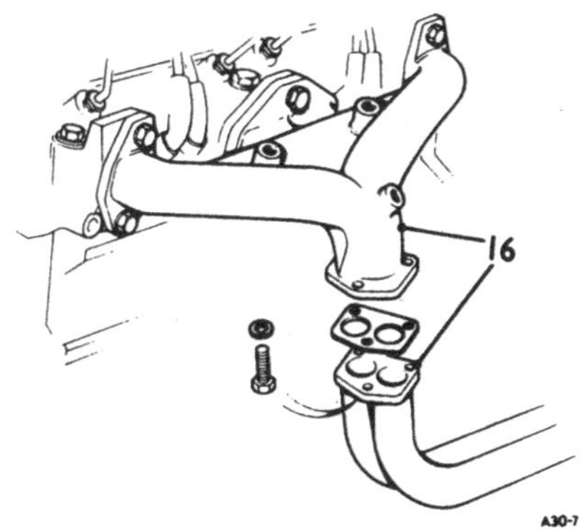

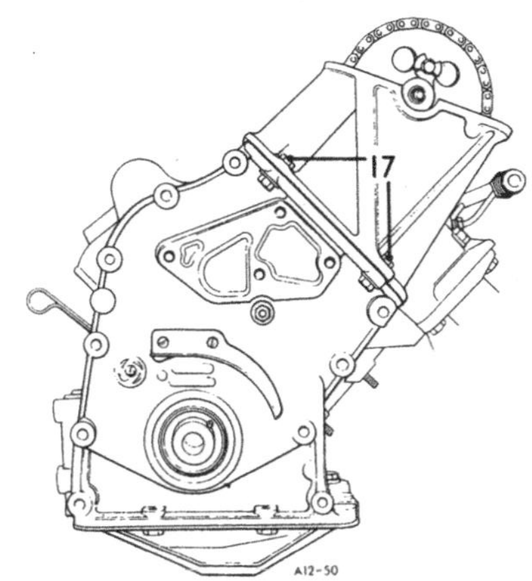

19 **Remove the five cylinder head studs and five bolts.**
20 **Withdraw the cylinder head complete with exhaust manifold.**
21 **Remove the cylinder head gasket.**

[**NOTE:** Before you install the cylinder head gasket, make certain that the cylinder head and cylinder block faces are clean. Install the new cylinder head gasket with its top side uppermost, which is usually marked. Under no circumstances should you ever reinstall any cylinder head gasket that has previously been compressed by torquing the head bolts and nuts. It is extremely important that the head bolts and nuts be torqued in the sequence given in 12.29.27. Go over the sequence three times, torquing the nuts and the bolts to 25 Nm (2.5 kgf m, or 20 lbf ft), then 50 Nm (5.0 kgf m, or 35 lbf ft), and finally 75 Nm (7.6 kgf m, or 55 lbf ft).]

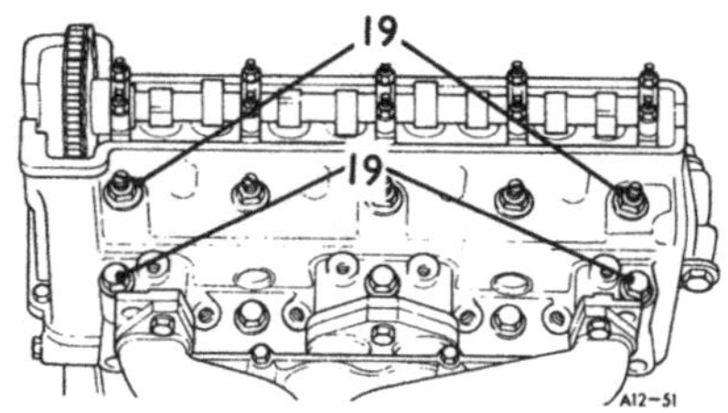

Refitting
22 Clean the cylinder block and cylinder head mating faces.

[**NOTE:** If the old cylinder head gasket has been leaking, check the cylinder head and the cylinder block for warping before you reassemble the engine. To check, lay a straightedge diagonally across the mating surfaces of both the head and the block. Using feeler gauges of various thicknesses, measure the clearance, if any, between the surface and the straightedge. Then repeat the check with the straightedge at the opposite diagonal. If the clearance at either diagonal exceeds 0.004 in (0.10 mm), the part should be milled or ground by a competent machinist in order to obtain a true surface.]

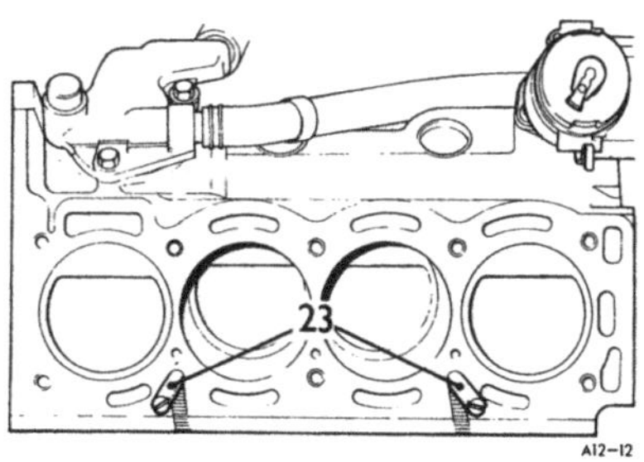

23 Insert two guide studs in the cylinder block bolt holes as illustrated, to facilitate fitting the cylinder head and gasket.
NOTE: Suitable guide studs may be made to the dimensioned drawing and to enable the studs to be inserted and withdrawn a slot to accommodate a screwdriver blade should be made in one end.
24 Check that number 1 and 4 pistons are at T.D.C. and that the distributor rotor arm is in the same position as in instruction 12.

25 Fit the cylinder head gasket.
26 Fit the cylinder head locating it over the two guide studs.
27 Fit the five cylinder head studs.
28 Remove the two guide studs.
29 Fit and finger tighten only the five cylinder head bolts and washers.
30 Fit and finger tighten the five cylinder head nuts, washers and the two brackets for locating the spark plug leads.
31 Tighten the nuts and bolts to the correct torque figure – see section 06 – in the sequence shown in operation 12.29.27.
32 **Fit the lock washer and camshaft sprocket top retaining bolt.**
33 **Turn the engine over to enable the bottom bolt to be fitted.**
34 **Tighten and lock the two bolts.**
35 **Remove the 'slave' nut from the camshaft sprocket.**
36 **Fit the semi-circular grommet to the cylinder head.**
37 **Fit the two timing covers to cylinder head nuts and bolts.**
38 **Fit the camshaft cover.**
39 **Connect the heater hose to the water transfer housing.**
40* **Connect the hose to the check valve.**
or
** **Connect the electrical lead to the Lambda (oxygen) sensor.**
41 **Fit the plug leads and the distributor cap.**
42 **Raise the ramp and connect the exhaust down-pipe to the manifold flange.**
43 **Lower the ramp and fit the inlet manifold complete with carburetters or fuel rail.**
44* **Fit the air cleaner. 19.10.02.**
45 **Fit the air duct. 80.15.31.**
46 **Fill the cooling system. 26.10.01.**
47 **Connect the battery and drive the car off the ramp.**

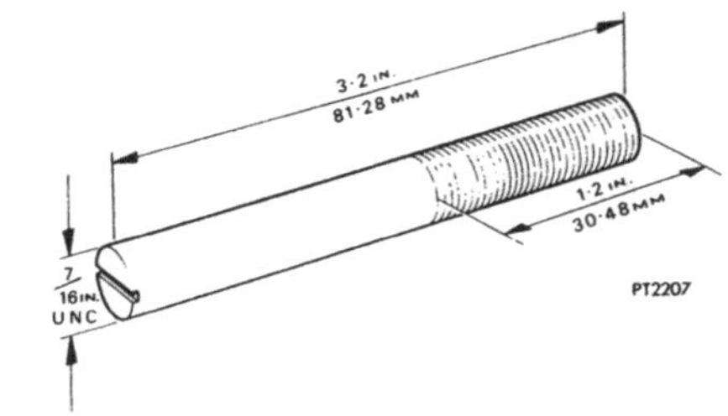

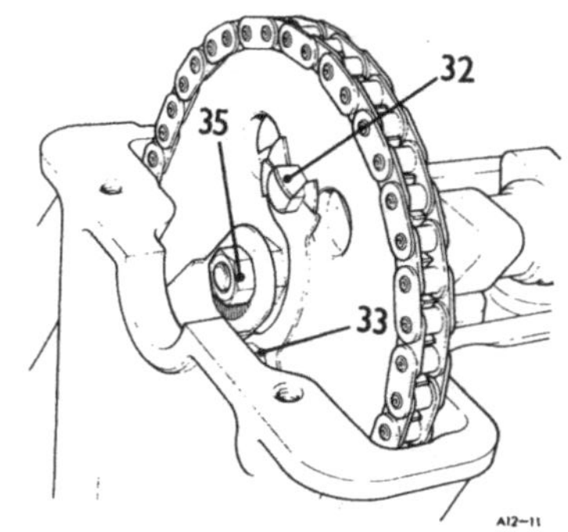

WATER TRANSFER HOUSING

– Remove and refit **12.29.04**

Removing
1 Disconnect the battery.
2 Drain the cooling system. 26.10.01.
3 Remove the fesh air duct. 80.15.31.
4 Disconnect the heater hose from the housing.
5 Remove the five bolts and plain washers securing the housing to the cylinder head.

Refitting
6 Clean the cylinder head and housing mating faces.
7 Using a new gasket fit the housing and secure with the five bolts and plain washers.
8 Connect the heater hose to the housing.
9 Fit the fresh air duct.
10 Fill the cooling system. 26.10.01.
11 Reconnect the battery.

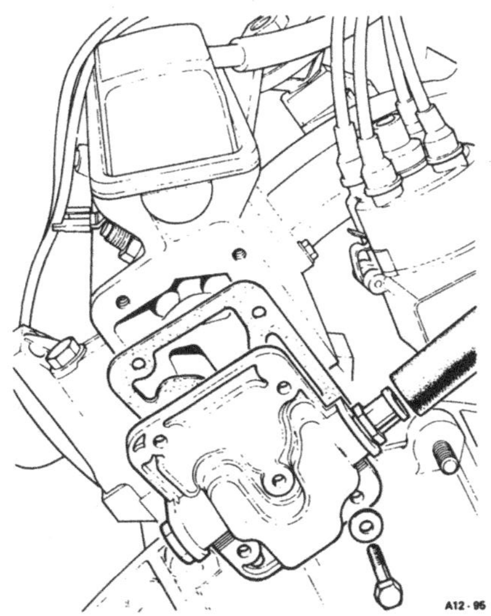

A12-95

CYLINDER HEAD

– Remove and refit **12.29.10**

This operation is covered under CYLINDER HEAD GASKET 12.29.02.

CYLINDER HEAD

– Overhaul **12.29.18**

Which includes:

Valves – exhaust – remove and refit	**12.29.60**
Valves – inlet and exhaust – remove and refit	**12.29.62**
Valves – inlet – remove and refit	**12.29.63**
Valve guide – inlet – remove and refit	**12.29.70**
Valve guide – exhaust – remove and refit	**12.29.71**
Inlet valve seat – remove and refit	**12.29.76**
Exhaust valve seat – remove and refit	**12.29.77**

Special tools: S352, S60A-8 with 60A and 560A-2

Dismantling

[**CAUTION:** Please read the instructions thoroughly before you attempt cylinder head repairs. If you lack the necessary skills or tools we suggest you leave this work to an Authorized Dealer or other qualified shop. Before taking a removed cylinder head in for repair, thoroughly clean its exterior but do not undertake any disassembly.]

1 Remove the cylinder head. 12.29.10.
2 Remove the exhaust manifold.
3 Remove the camshaft. 12.13.01 instructions 8 to 10.
4 Remove the tappets and pallets keeping them in their correct order.
5 Compress each valve spring in turn using special tool S352 and extract the collets, collars, springs, spring seats and valves.

NOTE: Keep all components in sets for reassembly in their original positions unless new items are to be fitted.

6 Remove all carbon deposits from the combustion chambers, exhaust ports and cylinder head face.

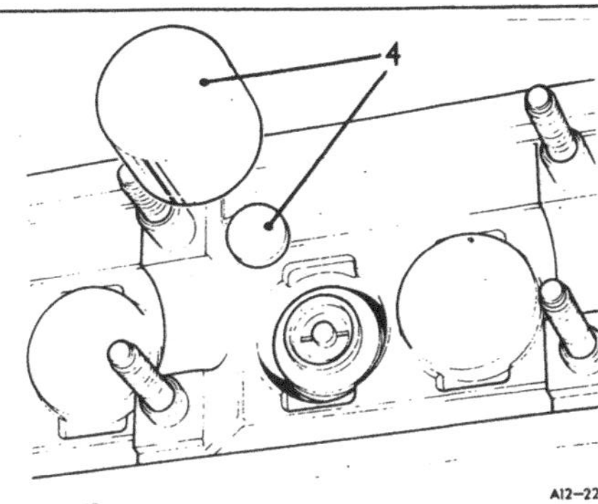

A12-22

[**NOTE:** A good way to keep the tappets and the pallets (shims) in order is to wipe each one dry as you remove it and then write a number on the part using waterproof ink or a waterproof felt-tip marker.]

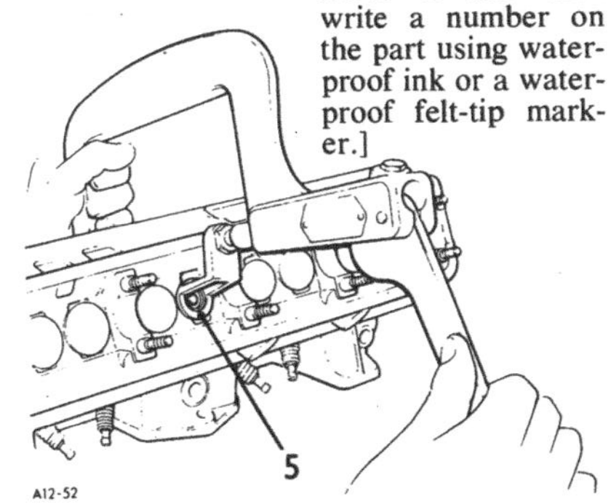

A12-52

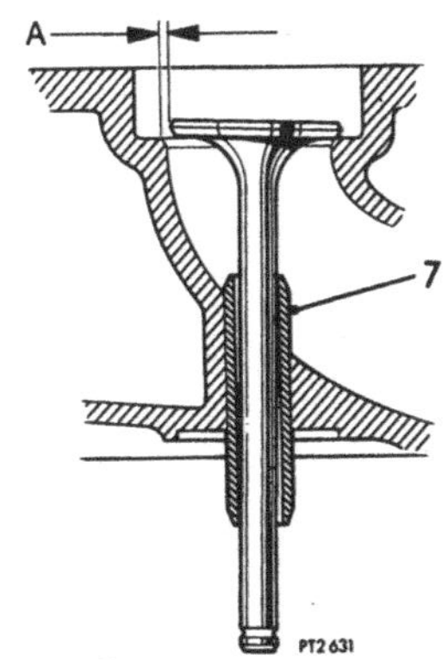

PT2631

Valve guides – checking
7 Check the inlet and exhaust valve guides for wear by inserting a new valve in each guide and tilting it. If movement across the valve seat – dimension A 0.20 in (0,508 mm) – is exceeded the valve guide should be renewed.

Valve guides – removing inlet and exhaust
8 Position main tool S60A on the top face of the cylinder head and using adaptor S60A-2 withdraw any worn guides.

Valve guides – fitting inlet and exhaust
9 Apply graphite grease to the new valve guides. Assemble main tool S60A and adaptors S60A-2 and S60A-8 on the combustion face of the cylinder head and draw-in the new guides.

continued

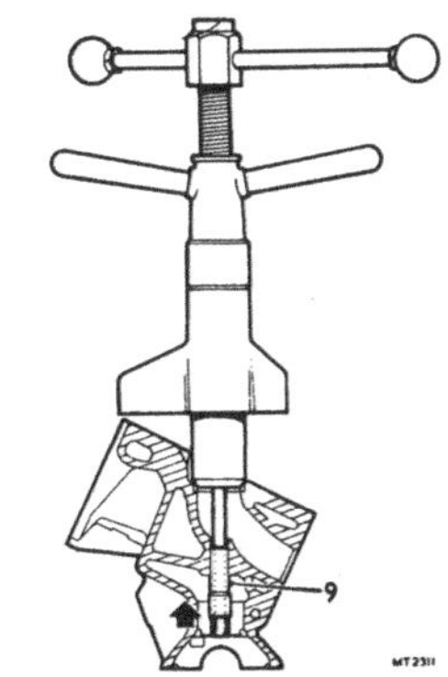

MT2311

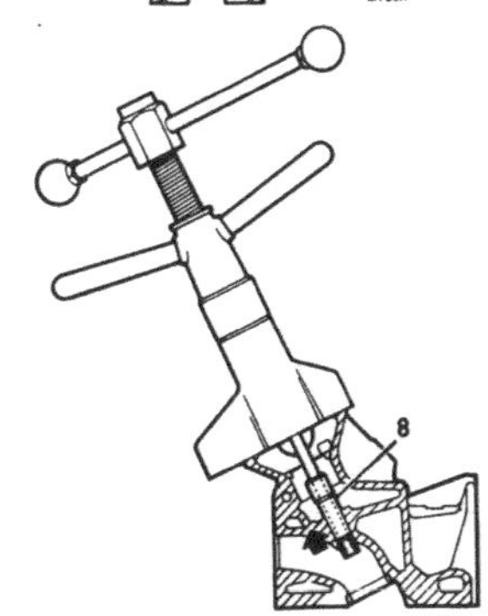

MT2312

Valve guides – reaming inlet and exhaust

10 Ream out the newly fitted guides using a 0.3125 in (7,95 mm) reamer.

11 Re-cut the valve seats and grind-in the valves where new guides have been fitted.

Valves

12 Examine the valves and discard any with worn or bent stems and badly pitted or burnt heads. Valves with the head thickness reduced below dimension 'A' should be renewed. Valves in an otherwise satisfactory condition may be refaced.

Valve springs

13 Examine the valve springs for cracks and distortion. Check the springs against the information in data and discard any that do not meet these requirements.

Valve seat inserts

14 Examine the valve seat inserts for wear pits, scores and pocketing. Reface where necessary, removing only the minimum of material to obtain a gas tight seal and a correctly seating valve.
A Correctly seating valve.
B Incorrectly seating valve.

15 Valve seat inserts that cannot be restored by machining to provide a correctly seating valve must be renewed as follows:
Machine-out the existing inserts taking care not to damage the insert bores in the cylinder head.

16 Machine the INLET valve seat bore in the cylinder head – dimension C to 1.665 to 1.666 in (42.29 to 42.32 mm).

17 Machine the EXHAUST valve seat bore – dimension B to 1.329 to 1.330 in (33.75 to 33.78 mm).

18 Heat the cylinder head uniformly to a temperature of 180°C and immediately fit the new inlet and exhaust valve seats squarely into the cylinder head.

19 Allow the cylinder head to cool and machine the inlet and exhaust seats to an inclusive angle of 89° – dimension D.

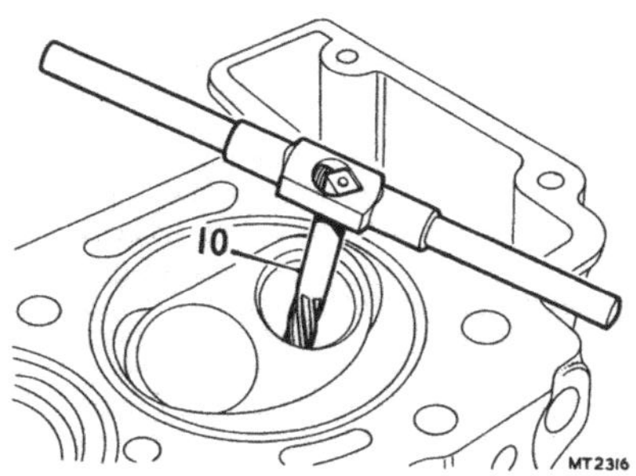

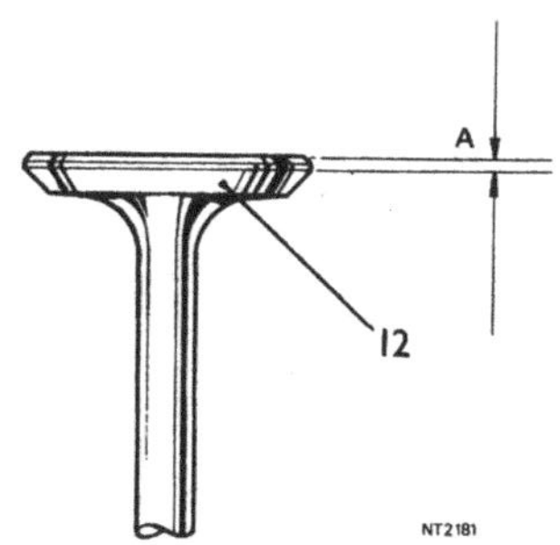

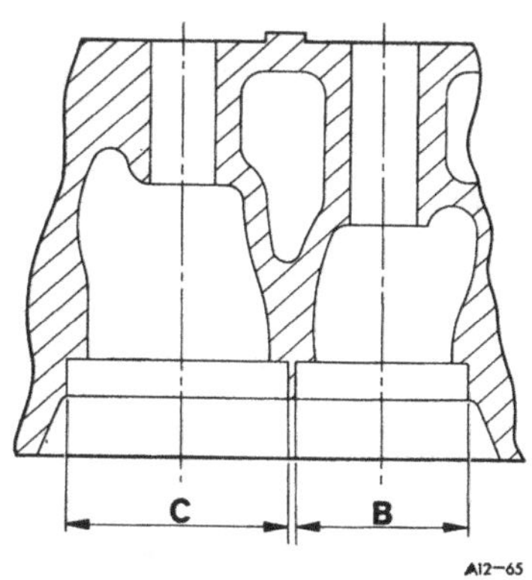

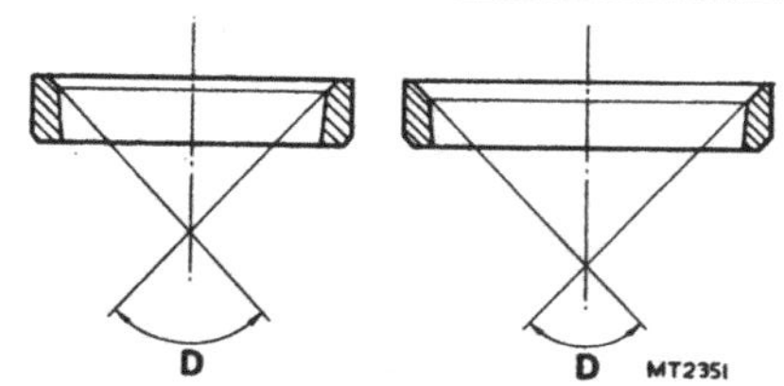

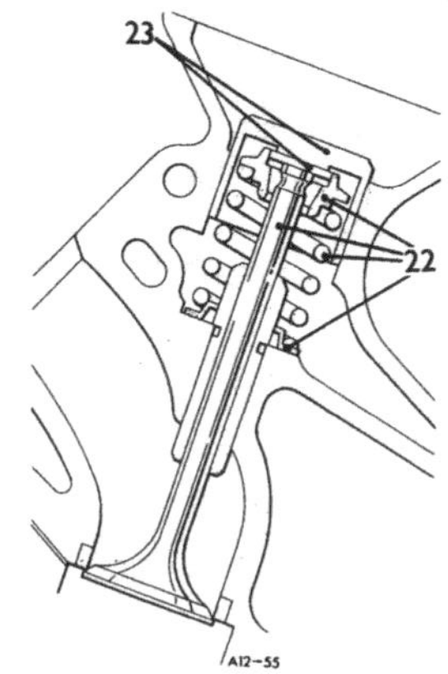

Lapping-in the valves

20 Insert each valve, in turn in its guide and lap-in using coarse followed by fine carborundum compound, until a continuous narrow band is obtained round the valve face and its seating.

21 Clean off all traces of compound from the valves and inserts.

Reassembling

22 Using the valve spring compressor S352, fit the valves, spring seats, spring collars, and secure with the split collets.

23 Fit the pallets and tappets.

24 Fit the camshaft and bearing caps ensuring that they are fitted in their correct positions and tighten the nuts to the correct torque figure – see section 06.

25 Check and if necessary adjust the valve clearances. 12.29.48.

26 Fit the exhaust manifold.

27 Fit the cylinder head. 12.29.10.

[**NOTE:** Lubricate the valve train components as you install them. If you are installing new tappets or a new camshaft, or if the engine will be stored for some time following repair, it is best to lubricate the parts with assembly lubricant (available from automotive supply stores) or a thin coating of multipurpose grease, rather than with ordinary engine oil.]

DATA

Valves	
Inlet	
Overall length	4.259 to 4.269 in (108,17 to 108,43 mm)
Head diameter	1.560 in (39,62 mm)
Angle of face	45°
Stem diameter	0.3107 to 0.3113 in (7,881 to 7,907 mm)
Stem to guide clearance	0.0017 to 0.0023 in (0,043 to 0,058 mm)
Exhaust	
Overall length	4.264 to 4.543 in (108,30 to 115,4 mm)
Head diameter	1.280 in (32,51 mm)
Angle of face	45°
Stem diameter	0.3100 to 0.3106 in (7,87 to 7,89 mm)
Stem to guide clearance	0.0014 to 0.0030 in (0,035 to 0,076 mm)
Valve springs	
Free length	1.600 in (40,40 mm)
Fitted length	1.440 in (36,57 mm)
Number of working coils	3¾
Valve seat inserts	
Outside diameter – Inlet	1.6695 to 1.6705 in (42,405 to 42,430 mm)
– Exhaust	1.3335 to 1.3345 in (33,870 to 33,896 mm)
Angle of seat	42½°

CYLINDER HEAD NUTS AND BOLTS

– Tighten **12.29.27**

1 To avoid distortion of the cylinder head it is important that the retaining nuts and bolts are tightened in the following alphabetical order, as illustrated, to the correct torque figure – see section 06.
A. B. C. D. E. F. G. H. I. J.
2 When releasing the nuts and bolts, prior to removing the cylinder head, the above sequence must be reversed i.e.
J. I. H. G. F. E. D. C. B.A.
3 When **checking** the torque loading of the nuts and bolts they should first be slackened off to overcome static friction and then re-tightened to the correct torque figure.

[**NOTE:** In reinstalling a removed cylinder head, always use a new cylinder head gasket. Never reinstall a head gasket that has been previously compressed by torquing the head bolts and nuts. Go over the tightening sequence three times, torquing the nuts and the bolts to 25 Nm (2.5 kgf m, or 20 lbf ft), then 50 Nm (5.0 kgf m, or 35 lbf ft), and finally 75 Nm (7.6 kgf m, or 55 lbf ft).]

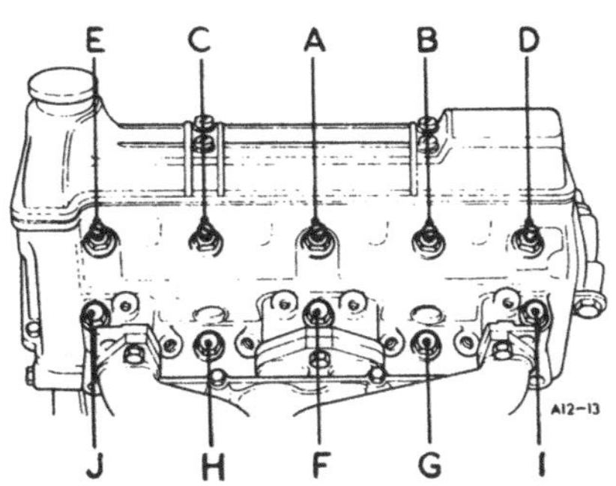

CAMSHAFT COVER

– Remove and refit **12.29.42**

Removing
1 Disconnect the H.T. leads from the spark plugs.
2 Disconnect the H.T. leads from the retainers on the camshaft cover and move the leads away.
3 Pull off the breather hose from the camshaft cover.
4 Remove the four slotted sleeve nuts and sealing washers.
5 Remove the two screws and washers.
6 Remove the camshaft cover.
7 Remove the gasket.

Refitting
8 Clean the camshaft cover and cylinder head mating faces.
9 Lightly smear both sides of a new gasket with grease.
10 Fit the gasket and camshaft cover, locating over the four studs.
11 Fit the four slotted sleeve nuts using, if necessary new:
a Washers
b Sealing rings.
12 Fit the two screws and spring washers.
13 Fit the breather hose.
14 Connect the H.T. leads to the spark plugs and attach the lead clips to the retainers.

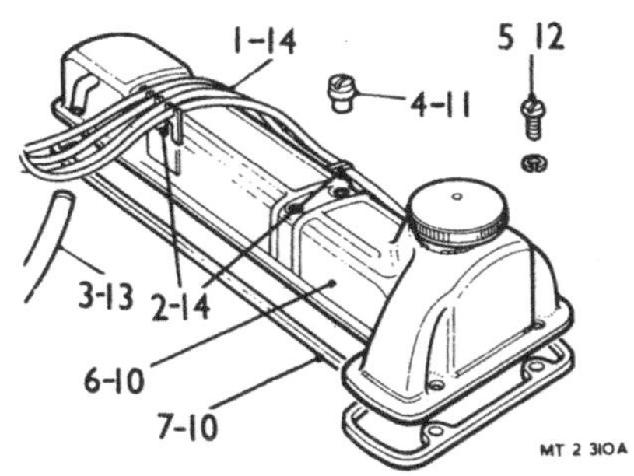

[**NOTE:** To prevent abrasive dirt from entering the engine, thoroughly clean the camshaft cover and the adjacent part of the cylinder head before you remove the cover.]

VALVE CLEARANCE

Check and adjust **12.29.48**

NOTE: This operation may be carried out with the cylinder head on the engine or on the bench. When on the bench turn the camshaft using a spanner on the hexagon at the rear of the shaft.

Checking
1 Disconnect the battery.
2 Remove the camshaft cover. 12.29.42.
3 Slacken off the camshaft bearing cap nuts and re-tighten to the correct torque figure – see section 06.
4 Rotate the engine and check and record the maximum clearance of each valve, in turn, using a feeler gauge between the cam heel and tappet. The maximum clearance exists when the cam is vertical to the cylinder head.

NOTE: If all the valve clearances are correct to that given in data, refit the camshaft cover and reconnect the battery. Should adjustments be necessary continue with the following instructions.

continued

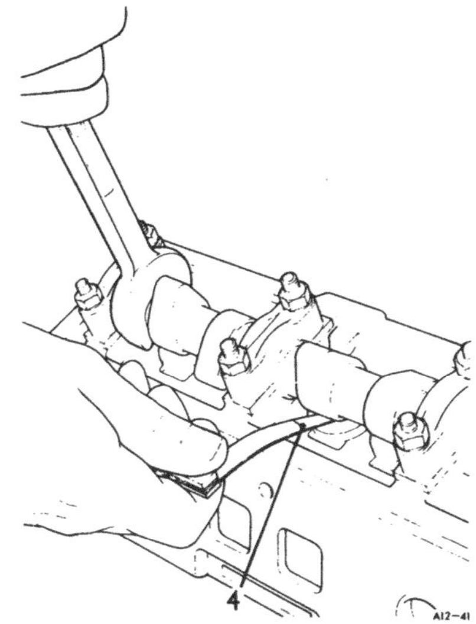

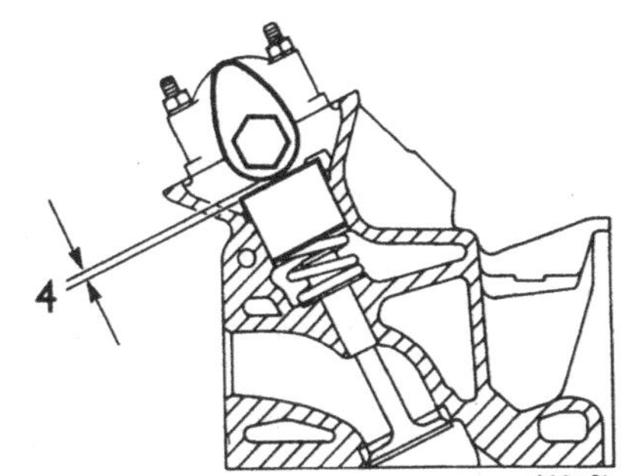

Adjusting
5 Remove the camshaft. 12.13.01.
6 Withdraw each tappet and pallet where the clearance requires adjustment and keep them in their numbered sequence.
7 Using a micrometer, measure and record the thickness of each pallet.
8 The following calculations should be followed, as an example, to select a new pallet of the appropriate thickness to give a correct valve clearance.

Excessive clearance (exhaust valve used for example)

Valve clearance recorded	0.023 in
Valve clearance required	0.018 in
Valve clearance excess	+ 0.005 in
Plus pallet thickness recorded	0.090 in
= Pallet thickness required	0.095 in

Insufficient clearance (inlet valve used for example)

Valve clearance recorded	0.005 in
Valve clearance required	0.008 in
Insufficient clearance	– 0.003 in
Pallet thickness recorded	0.100 in
Pallet thickness required	0.097 in

9 Fit the new pallets where necessary and refit the tappets 12.29.57 instructions 5 to 7.
10 Refit the camshaft. 12.13.01.
11 Fit the camshaft cover. 12.29.42.

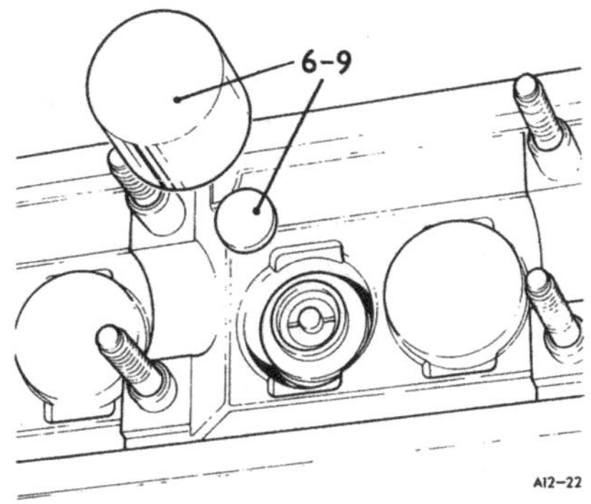

TAPPETS

– **Remove and refit** 12.29.57

Removing
1 Disconnect the battery.
2 Remove the camshaft cover. 12.29.42.
3 Remove the camshaft. 12.13.01.
4 Lift out the tappets and pallets, and identify them for reassembly.

Examination
5 Check the tappets for wear, scoring and pitting and discard as necessary.

Refitting
6 Fit the pallets into the valve collars.
7 Fit the tappets to the same valve assemblies renewing any faulty ones.
8 Refit the camshaft 12.13.01 ensuring that the valve clearances are checked and if necessary adjusted.
9 Fit the camshaft cover. 12.29.42.
10 Reconnect the battery.

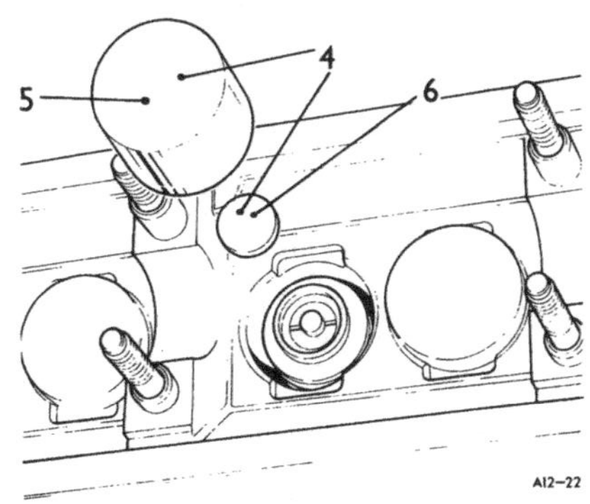

VALVES – EXHAUST –
remove and refit 12.29.60
VALVES – INLET AND EXHAUST –
remove and refit 12.29.62
VALVES – INLET –
remove and refit 12.29.63
VALVE GUIDE INLET –
remove and refit 12.29.70
VALVE GUIDE – EXHAUST –
remove and refit 12.29.71
INLET VALVE SEAT ––
remove and refit 12.29.76
EXHAUST VALVE SEAT –
remove and refit 12.29.77

The above operations are covered under CYLINDER HEAD OVERHAUL 12.29.18.

NOTE: It is always advisable to install new tappets whenever you install a new camshaft. If you are installing either, or if the engine will be stored for some time following repair, it is best to lubricate the parts with assembly lubricant (available from automotive supply stores) or a thin coating of multipurpose grease during installation, rather than with ordinary engine oil.]

DATA
Valve clearances

– Inlet	0.008 in (0.2 mm)
– Exhaust	0.018 in (0.5 mm)

ENGINE AND GEARBOX ASSEMBLY – UK and Europe Specification

Remove and refit 12.37.01

Removing
1 Disconnect the battery.
2 Remove the bonnet. 76.16.01.
3 Remove the fresh air duct.
4 Remove the radiator. 26.40.01.
5 Disconnect, from the engine, the heater hoses.
6 Disconnect the overflow hose from thermostat housing.
7 Pull off the air intake hoses from the air cleaner.
8 Disconnect the brake servo hose from the engine.
9 Disconnect the oil pressure lead from the switch.
10 Disconnect the H.T. and L.T. leads from the distributor.
11 Disconnect the water temperature transmitter lead from the thermostat housing switch.
12 Disconnect the harness plug from the alternator.
13 Remove the fuel pipe from the pump inlet, and blank-off the pipe and pump.
14 Disconnect the throttle cable.
15 Disconnect the mixture control cable.
16 Pull-off the float chamber spill pipe from the rear carburetter.
17 Remove the bonnet lock.76.16.21.
18 Remove the exhaust front pipe. 30.10.09.
19 Remove the gear selector lever, see 37.16.04 or 44.15.04.
20 Jack up the front of the car and lower onto axle stands.
21 Jack up the rear of the car and lower onto axle stands.
22 Mark the relationship of the propshaft to the gearbox drive flange and disconnect it from the gearbox.
23 Disconnect the two reverse light (and starter inhibitor) snap connectors.
24 Remove the speedometer cable from the gearbox.
25 Remove the clutch slave cylinder (where fitted) and move to one side.
26 Remove the engine L.H. mounting retaining nut.
27 Remove the stabiliser – where fitted. 12.45.16.
28 Remove starter motor shield.
29 Remove the battery lead from the starter motor solenoid.
30 Disconnect the Lucar connector from the starter motor solenoid.
31 Remove the bolt securing the battery earth lead to the bell housing.
32 Remove the propshaft guard.
33 Lower the propshaft.
34 Fit slings to engine lifting hooks and hoist to take the weight of the engine.
35 Support the gearbox with a trolley jack.
36 Remove the four nuts retaining the rear mounting to the body.
37 Remove the R.H. engine mounting nut.
38 Lower the jack supporting the gearbox.
39 Hoist the engine and gearbox assembly and manoeuvre from the engine bay.

Refitting
40 Fit slings to the engine lifting eyes, hoist engine and gearbox assembly and manoeuvre into the engine bay.
41 Place a trolley jack under the gearbox.
42 Lower the engine to line-up the L.H. mounting with its sub-frame location.
43 Continue lowering, so that the engine R.H. rubber mounting stud locates in the slot in the body bracket.
44 Raise the gearbox and connect up the cross member to the body locating studs.
45 Fit the nut and washer to the R.H. engine mounting stud.
46 Fit the nut and washer to the engine L.H. mounting stud.
47 Remove slings from engine lifting eyes.
48 Remove the jack from beneath gearbox.
49 Secure the propshaft to the gearbox lining up the identification marks. Tighten the nuts to the correct torque – see **'TORQUE WRENCH SETTINGS'**.
50 Fit the stabiliser – where fitted. 12.45.16.
51 Fit the propshaft guard.
52 Secure R.H. harness to its retaining clip.
53 Connect the reverse light (and starter inhibitor) snap connectors.
54 Fit the clutch slave cylinder (where fitted).
55 Secure the L.H. harness to its retaining clip.
56 Connect the negative earth lead from the battery to the bell housing.
57 Fit battery lead to starter motor solenoid.
58 Fit Lucar connector to the starter motor solenoid.
59 Fit the starter motor shield.
60 Fit the speedometer cable to the gearbox.
61 Fit the exhaust front pipe. 30.10.09.
62 Jack up rear of car and remove axle stands
63 Jack up front of car and remove axle stands.
64 Fit and align the bonnet lock. 76.16.21.
65 Fit spill pipe to the rear carburetter.
66 Fit the heater hoses to the engine connections.
67 Remove the blanking and fit the fuel pipe to the pump.
68 Fit the oil pressure warning light Lucar to the switch.
69 Connect the L.T. and H.T. leads to the distributor.
70 Connect the harness plug to the alternator.
71 Connect the coolant temperature transmitter lead.
72 Fit the vacuum hose to the engine.
73 Fit and adjust the throttle cable
74 Fit the mixture control cable.
75 Fit the air intake hoses to the air cleaner.
76 Connect the overflow hose to the thermostat housing.
77 Fit the radiator. 26.40.01.
78 Fit the fresh air duct.
79 Fit the bonnet. 76.16.01
80 Fit the gear selector lever, see 37.16.04 or 44.15.04.
81 Check the coolant level.
82 Check and if necessary top up the sump with oil of a recommended make and grade.
83 Check and if necessary top up the gearbox oil level – see **'MAINTENANCE'**.
84 Connect the battery.
85 Start the engine and run until normal operating temperature is reached whilst checking for coolant and oil leaks.
86 Check and adjust the ignition timing. 86.35.15.
87 Check and if necessary tune and adjust the carburetters. 19.15.02.
88 Road test the car.

ENGINE AND GEARBOX ASSEMBLY – USA and Canada Specification

Remove and refit 12.37.01

* *Applicable to carburetter engines only*
** *Applicable to fuel injection engines only*

Removing
1 Disconnect the battery.
2 Disconnect the bottom hose at the radiator and allow the collant to drain.
3 Remove the bonnet. 76.16.01.
4 Remove the radiator. 26.40.01.
5 Remove the fresh air duct. 80.15.31.
6 Remove the hot air hose.
7 Disconnect the heater hoses at the bulkhead connections.
8 Disconnect the brake servo hoses at the inlet manifold/plenum chamber connection.
9 Disconnect two hoses from the adsorption canister.
10 Disconnect the cooling system expansion hose from the thermostat housing.
11* Disconnect the vacuum hose from the inlet manifold to the anti-run-on valve.
12 Disconnect the electrical leads from the following as applicable:
a The oil pressure switch.
b Alternator.
c Temperature transmitter.
d Temperature sensor.
e Thermotime switch.
f Anti-run-on valve.
g L.T. leads from ignition coil.
h Lead from alternator to the throttle jack.
j H.T. lead from the coil to the distributor.
13 Disconnect the main fuel feed from the pump to the carburetter fuel line or fuel rail.

continued

14* Disconnect the choke cable from the rear carburetter.
15 Disconnect the throttle cable from the throttle linkage.
16 Remove the gear selector lever assembly. 37.16.04 or 44.15.04.
17 Jack-up the front and rear of the car and lower onto chassis stands.
18 Disconnect the propeller shaft from the gearbox.

[**NOTE:** Before you unbolt the propeller shaft from the flange on the gearbox, mark both the propeller shaft flange and the gearbox flange with a center punch. This will enable you to reinstall the propeller shaft in its original position on the gearbox flange, thus preserving the balance.]

19 Disconnect the reverse light (starter inhibitor) and seat belt interlock wires at the multiway connector.
20 Remove the exhaust down-pipe. 30.10.09. *and*
** Disconnect the electrical lead to the Lambda (oxygen) sensor.
21 Disconnect the speedometer cable from the gearbox.
22 Remove the clutch slave cylinder (where fitted) see 33.35.01.
23 Remove the engine stabilizer complete (where fitted). 12.45.16.
24 Remove the nut securing the L.H. engine mounting rubber to the sub-frame.
25 Disconnect the starter motor leads.
26 Release the harness from the bell housing clips.
27 Remove bell housing bolts necessary to remove the clips and release the clutch hydraulic pipe (as applicable).
28 Disconnect the main positive earth lead.
29 Remove the bonnet lock from the bulkhead. 76.16.21.
30 De-pressurize the air conditioning system – where fitted – and disconnect the hoses from the compressor. 82.30.05.
31 Using a sling with a 23 in leg to the rear lifting eye and an 18 in leg to the front eye or similar ratio raise the hoise to take the weight of the engine.
32. Disconnect the R.H. engine mounting rubber from the engine bracket.
33. Remove the five bolts and release the engine rear mounting cross-member from the body, complete with the tie bar and restraint cable – nut A.
34 Raise the rear end of the car.
35 Hoist the engine and remove the L.H. engine mounting.
36 Continue hoisting the engine and manoeuvre it from the car.
37 Lower the rear end of the car.

Refitting
38 Using a sling as in instruction 31 hoist the engine and gearbox assembly to a position over the engine bay.
39 Fit the L.H. engine mounting rubber to the sump cross-member.
40 Raise the rear end of the car and lower the engine unit into the car.
41 When the rear engine mounting is approximately 8 in from the floor lower the rear end of the car and jack-up the gearbox.

continued

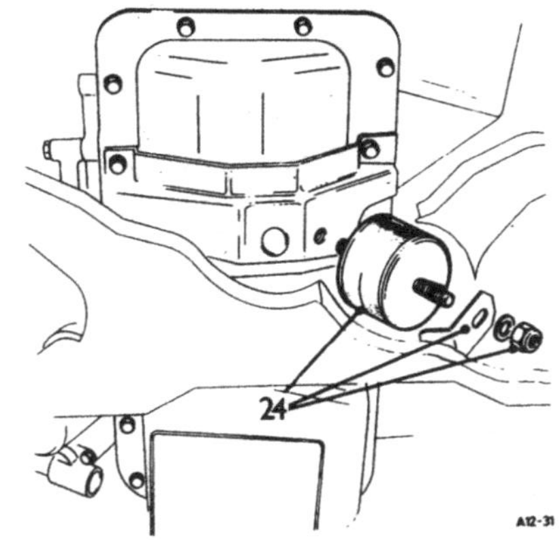

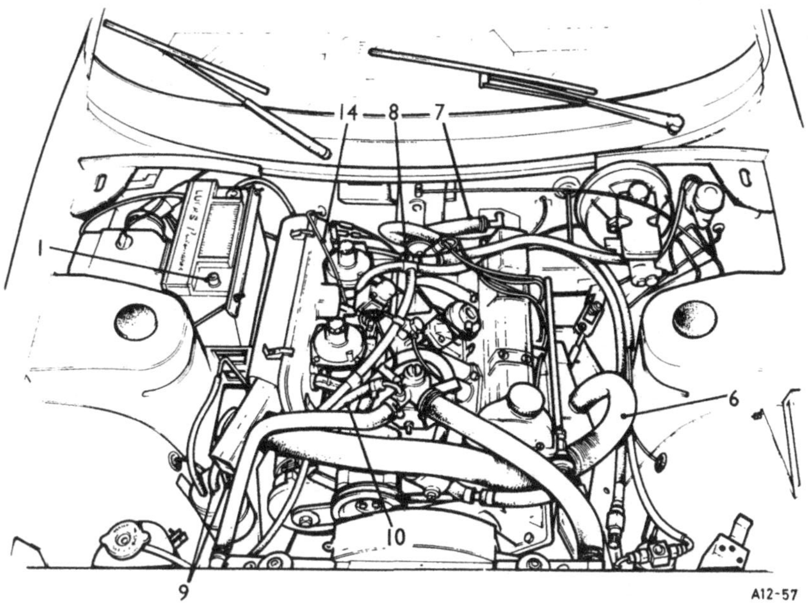

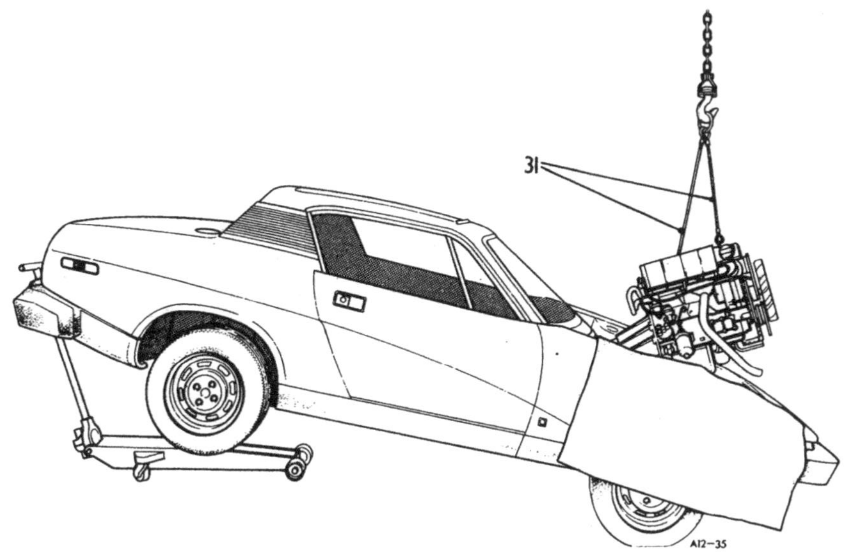

42 Continue lowering the engine whilst guiding the L.H. engine mounting stud into its sub-frame location.
43 Remove the slings from the engine.
44 Connect up the R.H. engine mounting to the engine bracket.
45 Bolt the engine and gearbox assembly rear mounting to the body with the four bolts.
46 Connect the propeller shaft to the gearbox.
47 Fit the clutch slave cylinder (as applicable).
48 Fit the bell housing bolts and secure the clutch hydraulic pipe over the bell housing with the clips, with an earth wire (as applicable).
49 Fit the L.H. engine mounting retaining nut.
50 Connect the gearbox harness to the multi-way connector.
51 Connect up the main earth lead to the bell housing.
52 Fit the engine stabilizer and adjust to the correct setting – where fitted. 12.45.16.
53 Fit the starter motor heat shield.
54 Fit the exhaust down-pipe. 30.10.09
and
** Connect the electrical lead to the Lambda (oxygen) sensor.

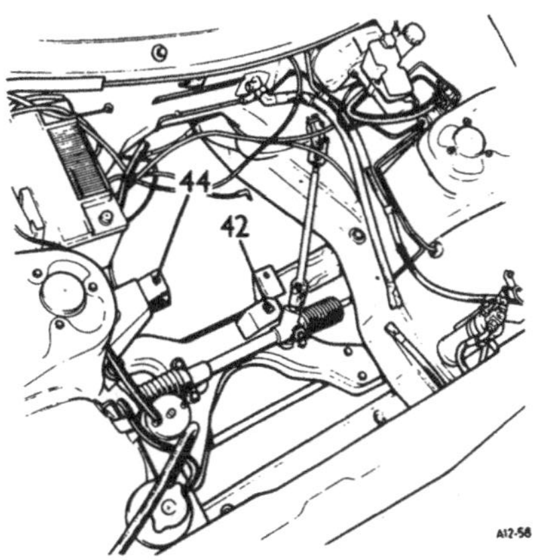

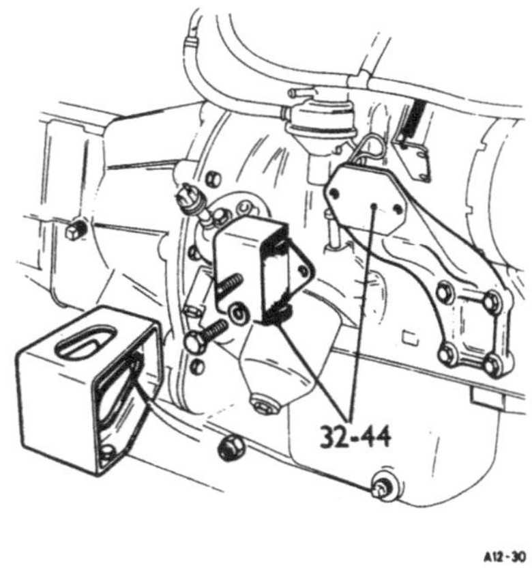

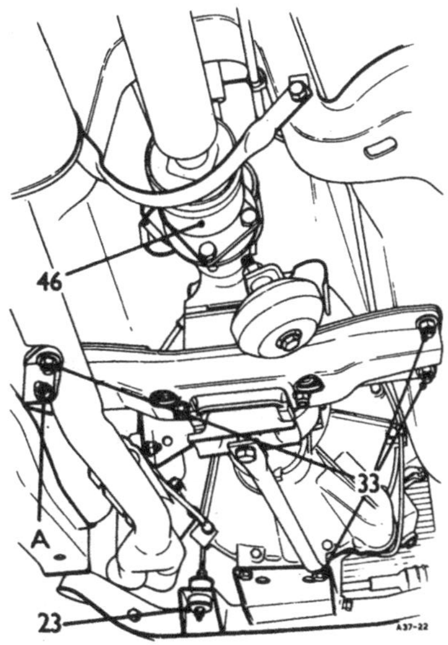

55 If fitted connect the air conditioning hoses to the compressor.
56 Connect up the compressor electro-magnet clutch lead.
57 Fit the heater pipes to the bulkhead connections. 80.25.07.
58 Fit the brake servo hose to the manifold/plenum chamber.
59 Fit the mixture control cable.
60 Fit the throttle cable to the carburetter/throttle linkage.
61* Fit the anti-run-on vacuum pipe to the manifold.
62 Fit the main-line fuel pipe to the inlet side of the pump or the fuel rail.
63 Connect up the electrical leads – reversing instruction 12.
64 Fit the cooling system expansion pipe to the thermostat housing.
65 Fit the pipes to the adsorption canister.
66 Connect the H.T. lead to the distributor.
67* Remove the blanking plug and fit the fuel pipe to the pump.
68 Fit the fresh air duct. 80.15.31.
69 Fit the radiator and connect the top and bottom hoses. 26.40.01.
70 Fit the hot air hose.
71 Fit the bonnet. 76.16.01.
72 Fit the gear selector lever assembly, see 37.16.04 or 44.15.04.
73 Fill the cooling system. 26.10.01.
74 Fill the sump with oil of the correct grade to the high mark on the dipstick.
75 Check the gearbox oil level.
76 Connect the battery.

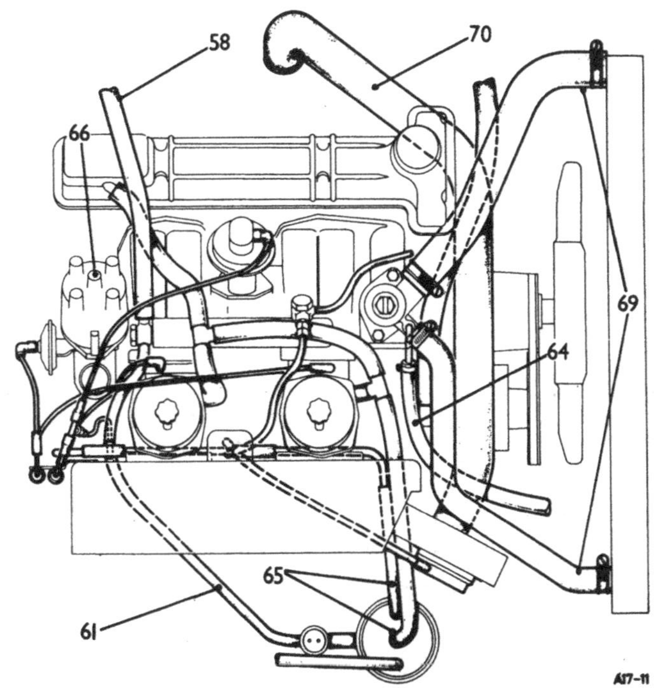

ENGINE ASSEMBLY – UK & Europe Specification

Strip and rebuild **12.41.05**

Special tool: 38 U3

Stripping

[**CAUTION:** There are many parts of the engine that must not be intermixed or which require special techniques to remove. If you are not an experienced Triumph mechanic, please read carefully the other parts of this Manual where each specific removal task is described.]

1 Remove tne engine and gearbox assembly from the car 12.37.01 and drain the sump.
2 Remove the gearbox – noting the position of the remaining bolts for refitting.
3 Drain the cylinder block.
4 Remove the starter motor.
5 Remove the clutch.
6 Remove the inlet manifold complete with carburetters, air cleaner and dipstick and tube.
7 Remove the water pump and remove the bottom hose
8 Remove the fuel pump.
9 Remove the distributor.
10 Remove the R.H. engine mounting bracket.
11 Remove the alternator complete with bracket.
12 Remove the alternator link.
13 Remove the exhaust manifold.
14 Remove the fan pulley and viscous coupling.
15 Remove the water pump.
16 Remove the spark plugs.
17 Remove the flywheel – 12.53.07 instructions 4 to 6.
18 Remove the oil pressure switch.
19 Remove the crankshaft pulley.
20 Remove the front lifting eye.
21 Remove the oil filter bowl and sealing ring.
22 Remove the oil pump and hexagon drive.
23 Remove the oil transfer housing.
24 Withdraw the six bolts and remove the rear adaptor plate.
25 Remove the camshaft cover.
26 Unlock and remove one of the camshaft sprocket retaining bolts.
27 Anchor the sprocket with a 'slave' nut to the support bracket and unlock and remove the remaining bolt.
28 Remove the two nuts and bolts securing the timing cover to the cylinder head.
29 Slacken the cylinder head retaining nuts and bolts in the reverse order as shown in operation 12.29.27.
30 Remove the five cylinder head studs and five bolts.
31 Withdraw the cylinder head complete with the camshaft and discard the gasket.
32 Remove the sump.
33 Remove the oil pick-up strainer.
34 Withdraw the remaining bolts – noting their positions and remove the timing cover.
35 Remove the oil thrower.
36 Remove the two retaining bolts and withdraw the timing chain tensioner.
37 Remove the three bolts securing the two chain guides and support bracket.
38 Remove the guides and support bracket complete with the chain and camshaft sprocket.

continued

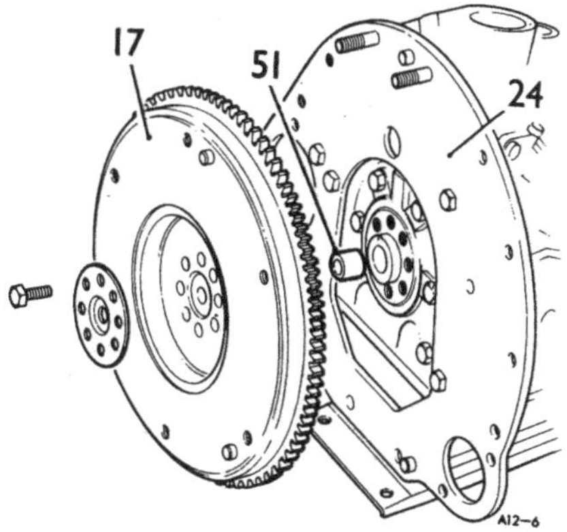

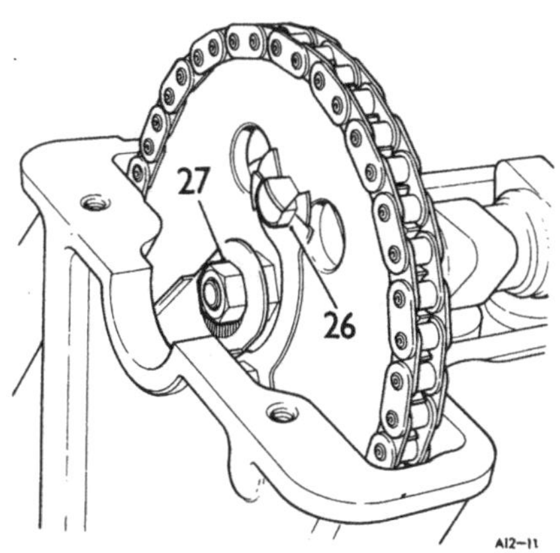

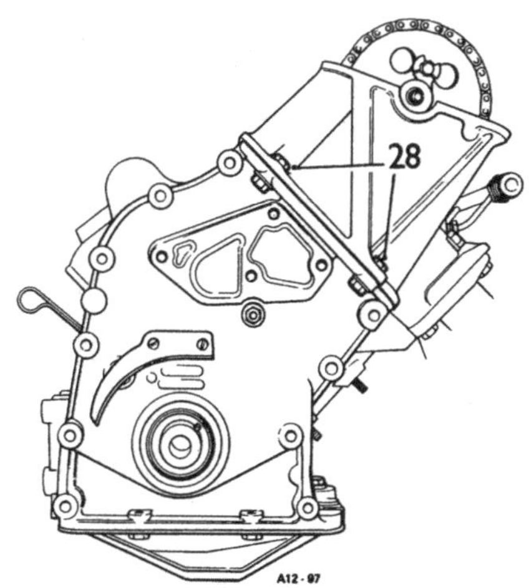

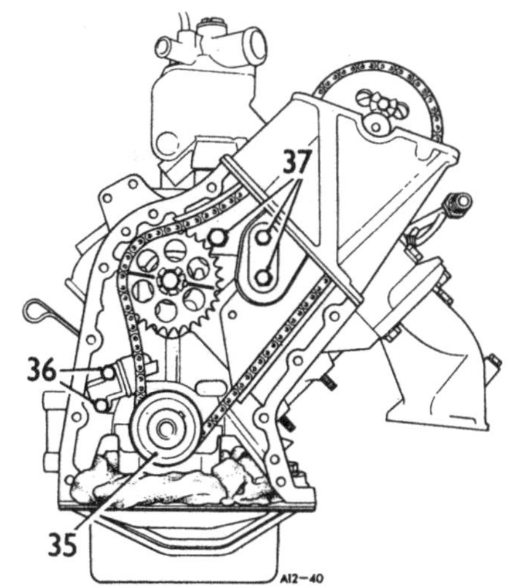

39 Turn the jackshaft sprocket to expose the two Allen screws retaining the jackshaft keeper plate. Remove the screws and withdraw the plate.
40 Withdraw the jackshaft and sprocket from the cylinder block.
41 Unlock the tab washer and remove the bolt and sprocket from the jackshaft.
42 Withdraw the crankshaft sprocket.
43 Remove the drive key and shims from the crankshaft.
44 Remove the six bolts and remove the rear main oil seal and housing.

45 Remove the nuts securing the connecting rod bearing caps and withdraw the caps and lower shells ensuring that if the shells are to be refitted they are kept identified with their respective caps.
46 Push the connecting rods upwards and withdraw the piston assemblies through the top of the bores, keeping the shells identified with their respective connecting-rods.
47 Remove the ten bolts securing the five main bearing caps and withdraw the caps and lower shells.

48 Lift out the crankshaft.
49 Remove the upper shells from the crankcase.
50 Remove the thrust washers from number three main bearing.
51 Withdraw the spigot bush from the crankshaft.

NOTE: The following rebuilding instructions assume that all the individual components and assemblies have been examined, worn parts renewed and assemblies overhauled. Moreover, all joint faces are clean and parts lubricated before assembly.

52 Fit the five main bearing upper shells ensuring that the keeper tags locate in the recess.
53 Fit the crankshaft thrust washers to number 3 main bearing noting that the oil grooves face outwards.
54 Lower the crankshaft into the crankcase.
55 Fit the lower shells to the main bearing caps and fit them to the crankcase. Partially tighten the securing bolts.

continued

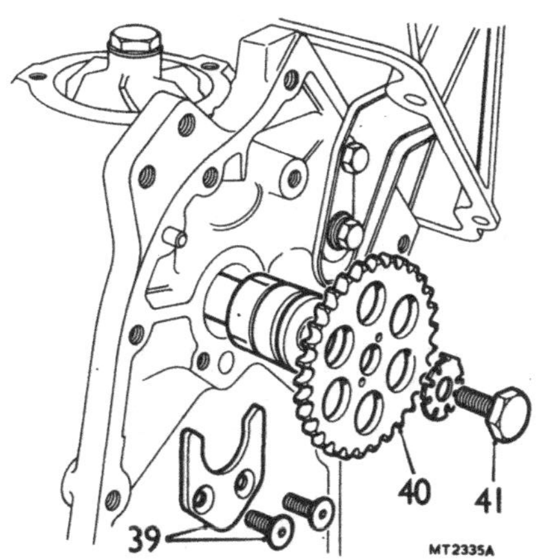

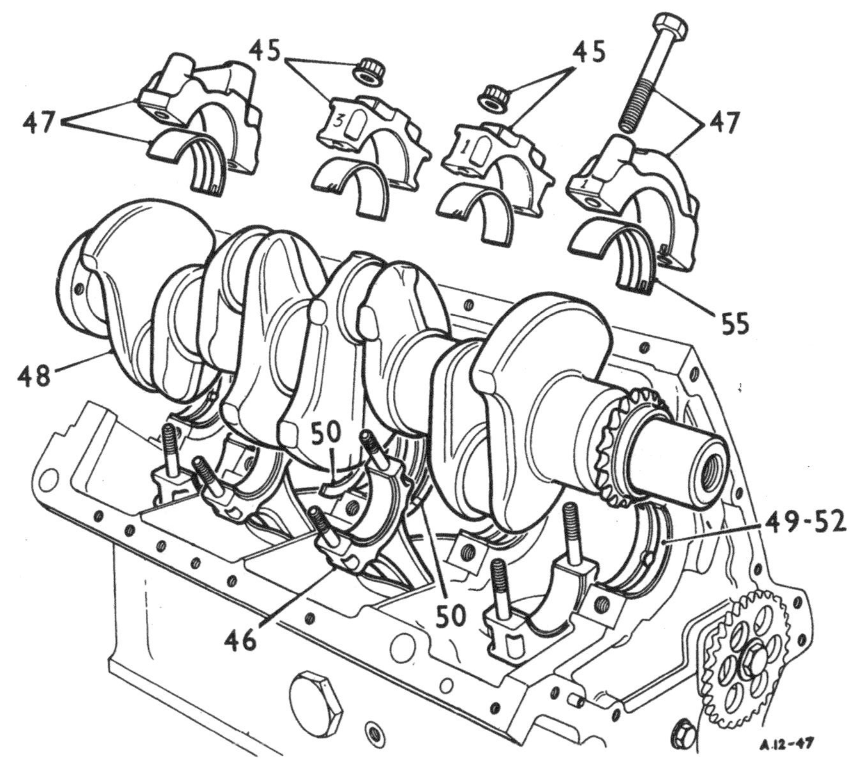

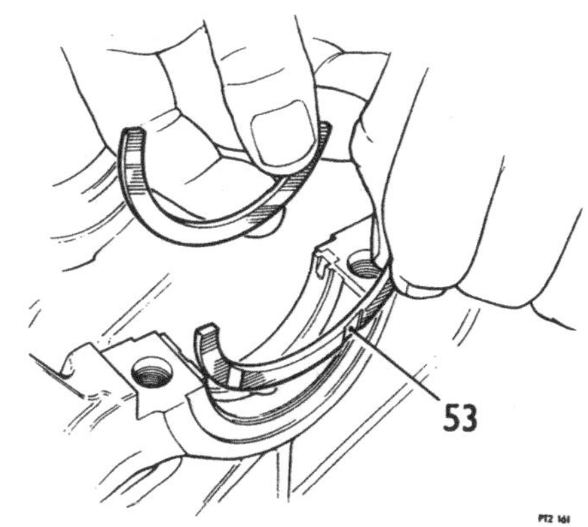

56 Using either a dial gauge or feelers check the crankshaft end-float – see data.

57 Tighten the main bearing cap bolts to the correct torque – see **'TORQUE WRENCH SETTINGS'**.

58 Stagger the piston ring gaps, avoiding a gap on the thrust side and compress the rings using special tool number 38 U3.

59 Turn the crankshaft to position numbers 2 and 3 crankpins at B.D.C. Insert the respective connecting rod and piston assemblies into the bores and tap the pistons home noting that the flat raised part and the valve head recesses on the piston crown are fitted to the L.H. side of the engine.

60 Fit the upper bearing shells to the connecting rods ensuring that the keeper tags locate in the recesses and pull the connecting rods onto the journals.

61 Fit the lower shells to the connecting rod caps ensuring that the keeper tags locate in the recesses and fit them to the connecting rods. Tighten new nuts to the correct torque – see **'TORQUE WRENCH SETTINGS'**.

62 Repeat instructions 58 to 61 on numbers 1 and 2 pistons and connecting rod assemblies.

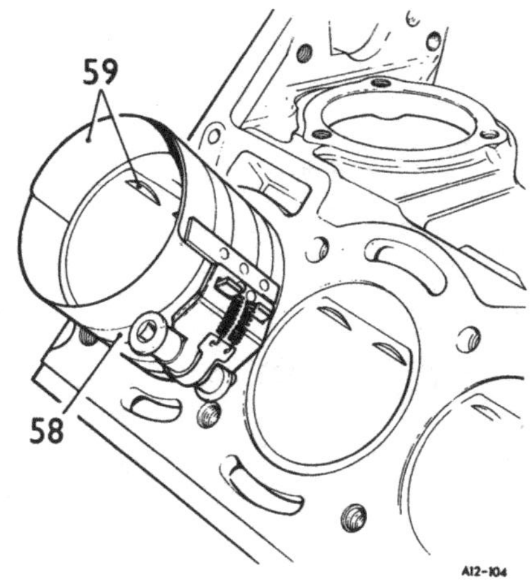

63 Fit a new seal to the rear main oil seal housing (lip towards the crankshaft) and using a new gasket and jointing compound fit the housing locating it over two dowels. Secure with the six bolts and spring washers noting that the two longer bolts are fitted at the bottom.

64 Fit the rear adaptor plate locating it on two dowels and secure with the six bolts.

65 Fit the spigot bush to the crankshaft.

66 Fit the flywheel – lining up the punch marks – if the original flywheel or crankshaft is being refitted.

67 Secure with the spigot bush retaining plate and eight bolts and tighten to the correct torque – see **'TORQUE WRENCH SETTINGS'**.

68 Check the flywheel for run-out using a dial gauge – see DATA.

continued

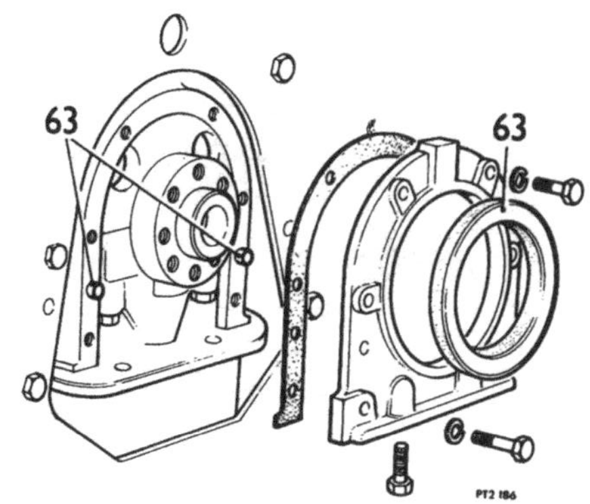

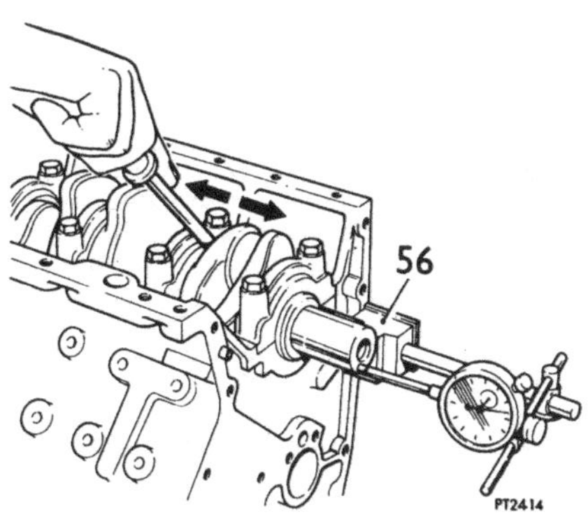

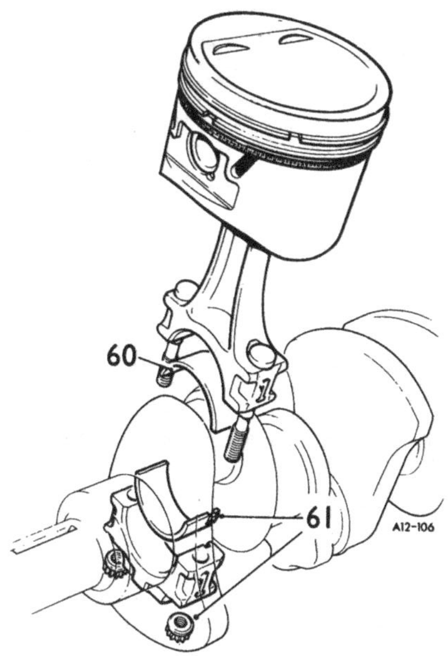

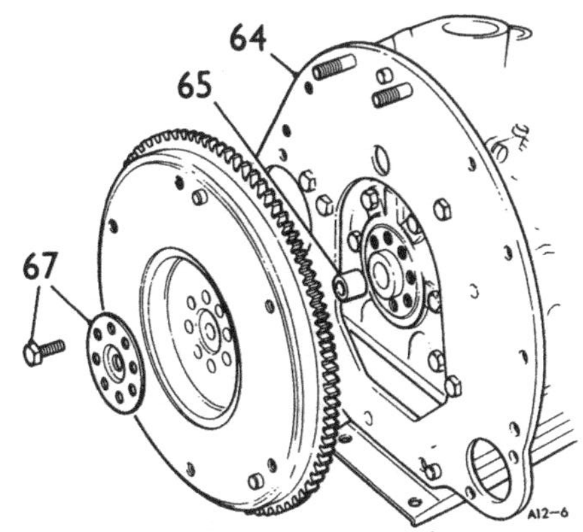

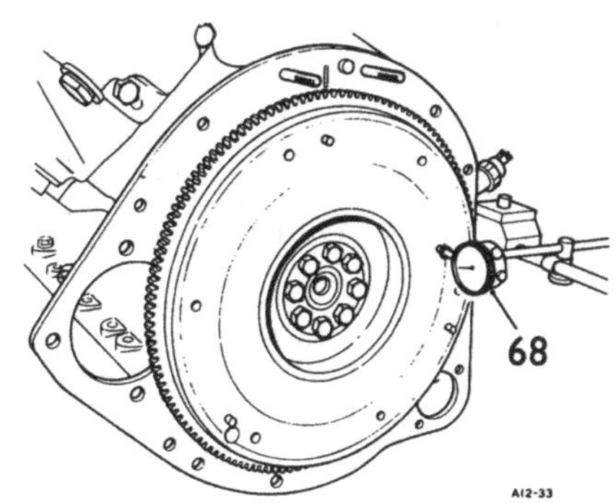

69 Fit the clutch driven and pressure plates locating the pressure plate assembly over the three dowels.
70 Centralize the driven plate using a dummy primary shaft and evenly tighten the six bolts and spring washers to the correct torque. See 33.10.01.
71 Fit the sprocket to the jackshaft locating it over the single dowel and secure with bolt and new lock washer.
72 Fit the jackshaft to the cylinder block, locate it with the keeper plate and secure the plate with the two Allen screws.
73 Fit the water pump impeller and cover assembly checking the clearance as described in the refitting instructions. 26.50.01.
74. Fit the oil transfer adaptor ensuring that new 'O' rings are fitted and correctly located. Secure with the single bolt.
75 Ensure that the pistons are **not** at T.D.C. see cautionary note below.
76 Fit two 'slave' guide studs to the bolt holes in the cylinder block to facilitate the fitting of the cylinder head and gasket.

NOTE: See 12.29.02 for dimensional drawing of suitable studs.

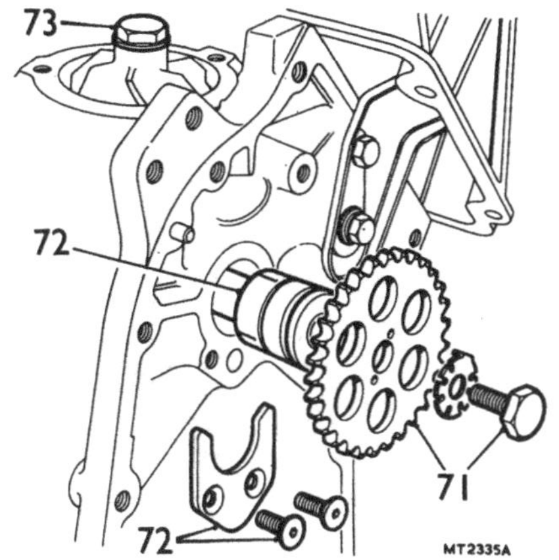

MT2335A

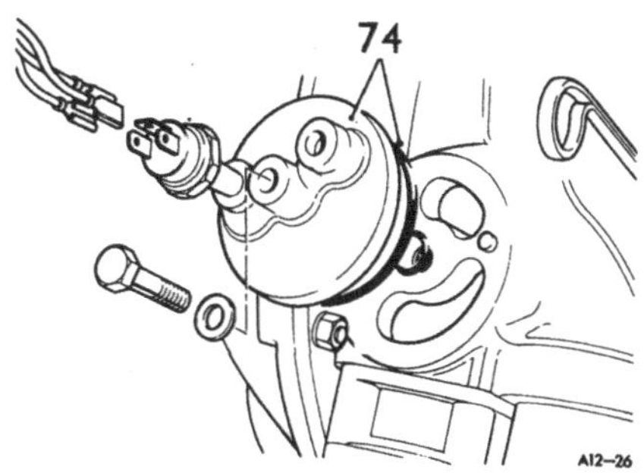

A12–26

77 Fit the cylinder head gasket locating it over the studs.
78 Fit the cylinder head easing it over the guide studs.
79 Fit the five cylinder head studs.
80 Remove the two 'slave' studs and fit the five cylinder head retaining bolts and plain washers.
81 Fit the nuts and plain washers to the cylinder head studs.
82 Tighten the nuts and bolts to the correct torque and in the correct sequence. 12.29.27.

CAUTION: Once the pistons are fitted care must be taken to ensure that the crankshaft is not allowed to turn a complete revolution until the valve timing has been completed, otherwise damage to the valves and pistons will occur.

83 Temporarily fit the crankshaft sprocket and check its alignment with the jackshaft sprocket by placing a straight edge across the two sprockets.
84 Adjust any misalignment by removing the crankshaft sprocket and fitting shims of suitable thickness between it and the crankshaft. Ensure that the sprocket is pushed fully home when checking the alignment.
85 Remove the sprocket, fit the crankshaft drive key and refit the sprocket.

continued

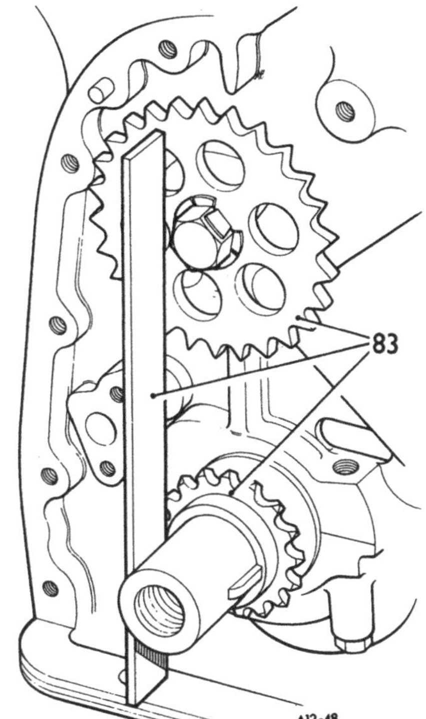

A12-48

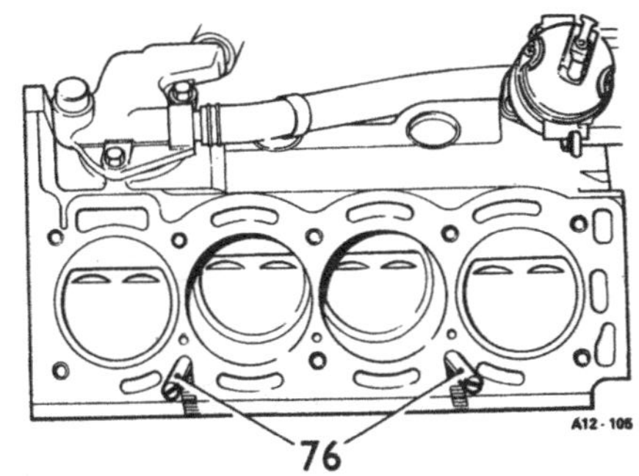

A12 - 105

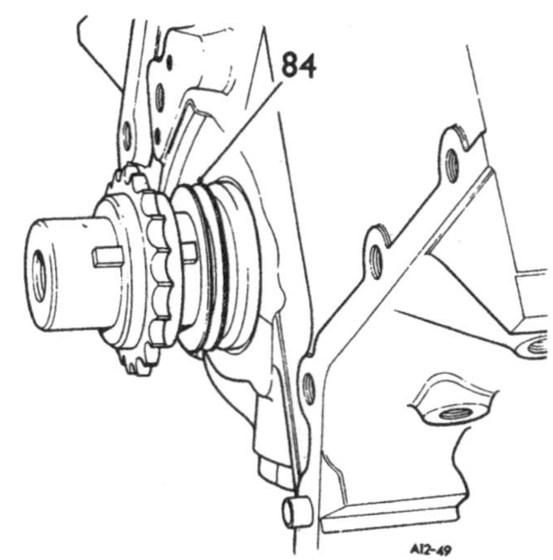

A12-49

86 Turn the camshaft until the timing mark on the flange is in line with the groove on the camshaft front bearing cap.

87 Temporarily locate the timing cover and crankshaft pulley and turn the crankshaft (having regard for the above CAUTION) until number 1 and 4 pistons are at T.D.C. Remove the pulley and timing cover.

88 Turn the jackshaft until the scribed line across the sprocket is equidistant between bolt A and hole B with the dowel to the left. Remove the pulley and timing cover.

89 Encircle the camshaft sprocket with the timing chain and insert through the cylinder head aperture. Locate the sprocket on the camshaft flange and secure with the lock plate and one bolt.

90 Keeping the chain taut on the drive side (i.e. the run between the camshaft and crankshaft sprockets) fit the chain to the crankshaft and jackshaft sprockets. Check and if necessary make any adjustment to the position of the jackshaft sprocket to maintain it in its correct position as in instruction 88.

91 Fit and loosely secure the guides as follows:
- a The adjustable guide.
- b The straight fixed guide.
- c The support bracket.

92 Fit a 'slave' bolt to the lower hole in the fixed guide to ensure alignment when fitting the timing cover centre bolt.

93 Fit the timing chain hydraulic tensioner and secure with the two bolts and spring washers. 12.65.28.

NOTE: To prevent the tensioner releasing while fitting, fit a spacer between the tensioner body and the back of the slipper.

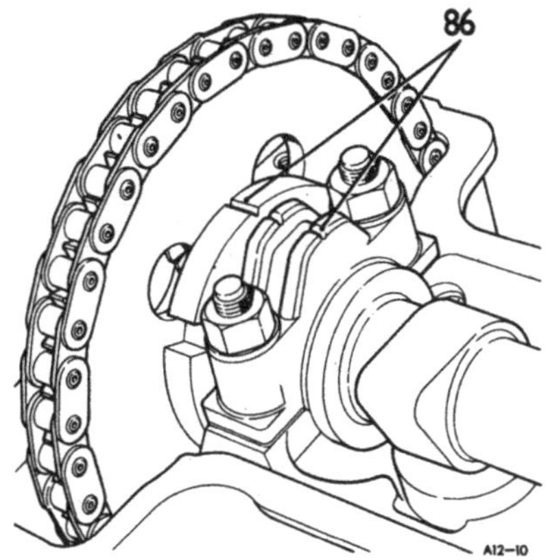

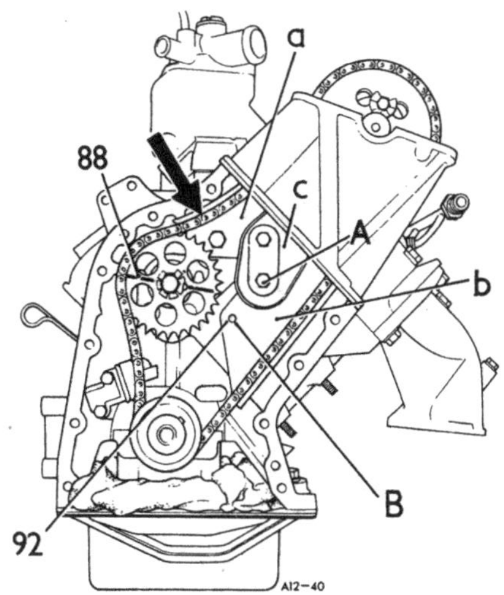

94 Remove the spacer and insert a 0.100 in (2,54 mm) feeler gauge or slip gauge in its place, dimension A.

95 Adjust the chain tension by applying pressure, in the direction of the arrow, to the adjustable guide while tightening the clamp bolt.

96 Remove the gauge and check that the scribed line on the jackshaft sprocket is still correctly positioned.

97 Tighten all the guide retaining bolts and remove the 'slave' bolt.

98 Turn the engine sufficiently to enable the remaining camshaft sprocket retaining bolt to be fitted and locked.

99 Check that the threaded spigot on the camshaft sprocket does not foul the support bracket.

100 Fit the oil thrower with the dished face outwards.

101 Using sealing compound fit a new timing cover gasket and position the timing cover over the two dowels.

102 Secure the cover with the centre bolt noting that it has a fibre washer under the head

103 Fit the two nuts and bolts retaining the timing cover to the cylinder head.

104 With the lip face leading press-in a new timing cover oil seal flush with the outer face of the cover.

continued

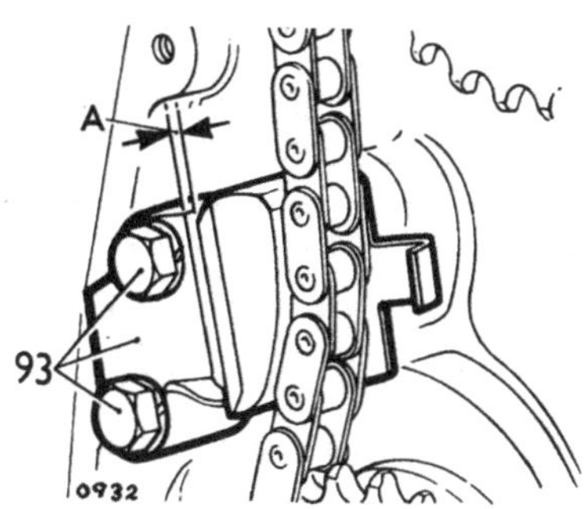

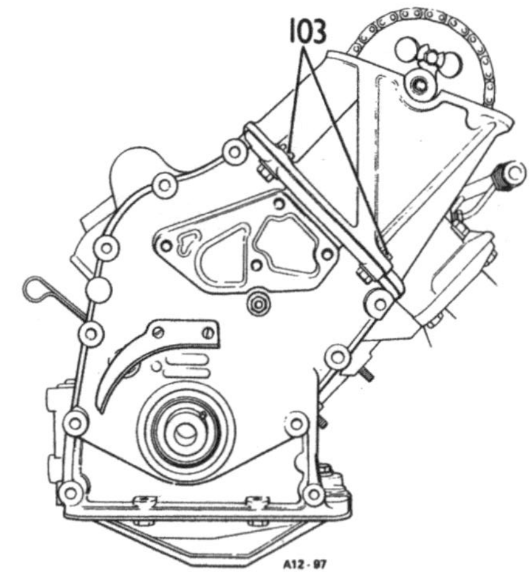

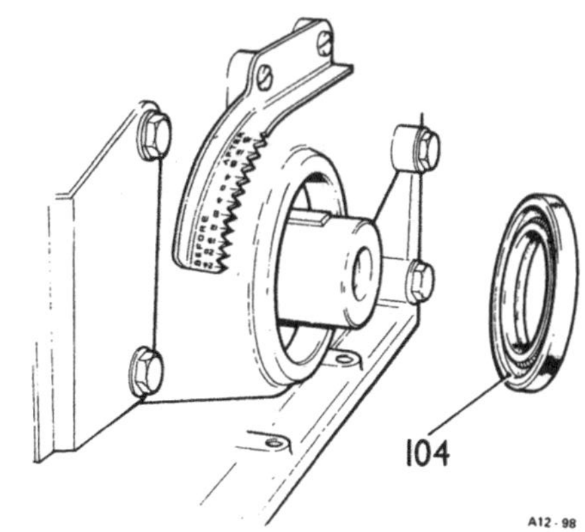

105 Fit the crankshaft pulley.
106 Turn the crankshaft to bring numbers 1 and 4 pistons to T.D.C. with the timing mark on the camshaft flange in line with the groove on the front bearing cap.
107 Fit the distributor so that the vacuum advance unit is pointing directly rearward and the rotor arm is in line with the manifold rear retaining bolt hole in the cylinder head.
108 Temporarily tighten the distributor retaining bolts pending final adjustment when the engine is installed.
109 Fit the camshaft cover.
110 Fit the oil pick-up strainer.
111 Fit the sump.
112 Fit the fuel pump using a new gasket and ensuring that the actuating lever rides on top of the jackshaft cam.
113 Fit the fan pulley assembly and the drive belt on the pulley.
114 Fit the alternator complete with bracket and adjusting link.
115 Fit the front lifting eye.
116 Fit and adjust the alternator belt tension. 86.10.05.
117 Fit the exhaust manifold.
118 Fit the hexagon drive and oil pump using a new sealing ring.
119 Fit the oil pressure switch.
120 Insert a new filter element in the filter bowl and fit the assembly with a new sealing rubber.
121 Fit the R.H. engine mounting bracket noting that the long bolts are fitted to the top two holes.
122 Fit the water pump to heater and inlet manifold pipe and hose.
123 Fit the bottom hose to the water pump cover.

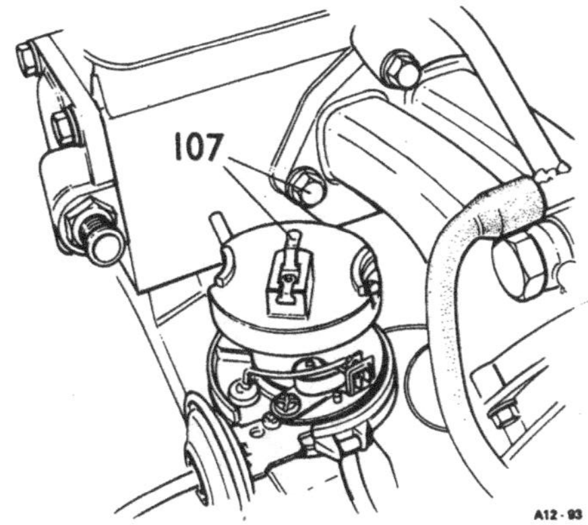

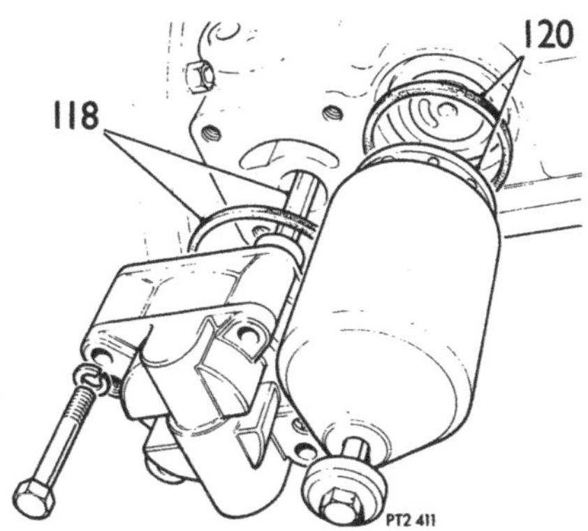

124 Fit the inlet manifold complete with carburetters 19.15.15 and the rear lifting eye.
125 Connect the fuel pump to carburetter hose.
126 Connect the engine breather pipes.
127 Fit the dipstick.
128 Fit the spark plugs.
129 Fit the distributor cap and connect the H.T. leads to the plugs (firing order 1, 3, 4, 2 distributor turns anti-clockwise).
130 Fit the gearbox.
131 Fit the starter motor.
132 Fit the engine and gearbox assembly to the car 12.37.01.
133 Check and if necessary adjust the ignition timing 86.35.15.

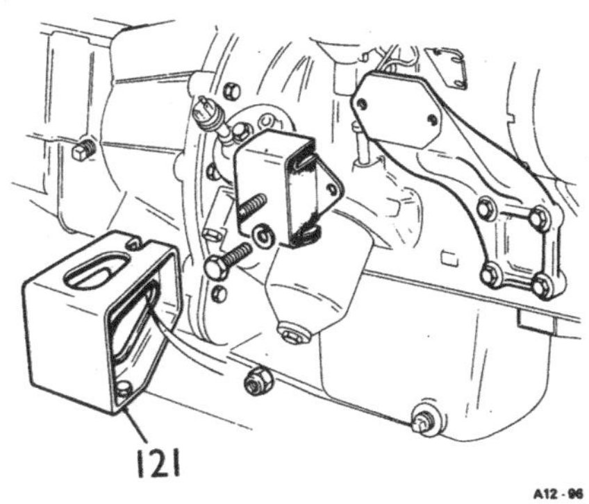

ENGINE ASSEMBLY – USA & Canada Specification

Strip and rebuild **12.41.05**

Special tool: 38 U3

* *Applicable to carburetter engines only*
** *Applicable to fuel injection engines only*

Stripping

1 Remove the engine and gearbox assembly from the car 12.37.01 and drain the sump.
2 Remove the gearbox – noting the position of the remaining bolts for refitting.
3 Drain the cylinder block.
4 Remove the starter motor.
5 Remove the clutch (as applicable).
6 Remove the inlet manifold complete with carburetters and air cleaner or fuel rail.
7 Remove the water pump and remove the bottom hose.
8* Remove the fuel pump.
9 Remove the distributor.
10 Remove the R.H. engine mounting bracket.
11 Remove the alternator complete with bracket.
12 Remove the alternator link.
13* Remove the air distribution manifold complete with diverter and relief valve, check valve and hoses.
14 Remove the exhaust manifold.
15 Remove the air conditioning compressor – where fitted – complete with brackets.
16* Remove the air pump.
17 Remove the fan pulley and torquatrol assembly.
18 Remove the water pump.
19 Remove the spark plugs.
20 Remove the flywheel – 12.53.07 instructions 4 to 6.
21 Remove the oil pressure switch.
22 Remove the dipstick.
23 Remove the auxiliary and crankshaft pulleys. 12.21.01.
24 Remove the front lifting eye.
25 Remove the oil filter bowl and sealing ring.
26 Remove the oil pump and hexagon drive.
27 Remove the oil transfer housing.
28 Withdraw the six bolts and remove the rear adaptor plate.
29 Remove the camshaft cover.
30 Unlock and remove one of the camshaft sprocket retaining bolts.
31 Anchor the sprocket with a 'slave' nut to the support bracket and unlock and remove the remaining bolt.
32 Remove the two nuts and bolts securing the timing cover to the cylinder head.

continued

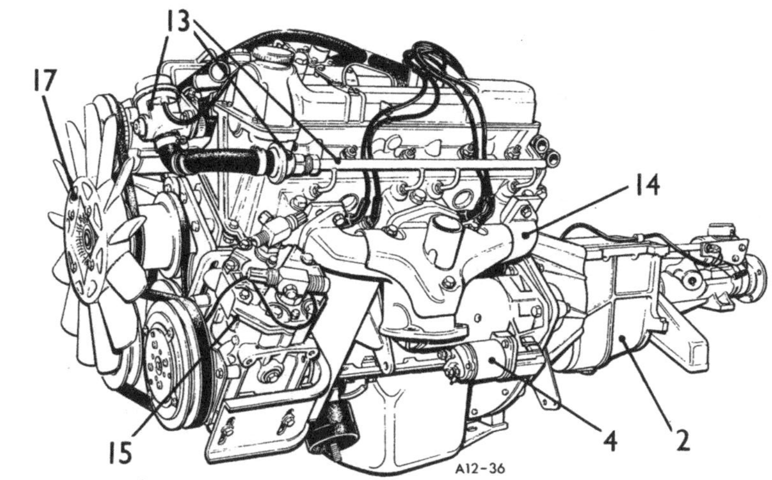

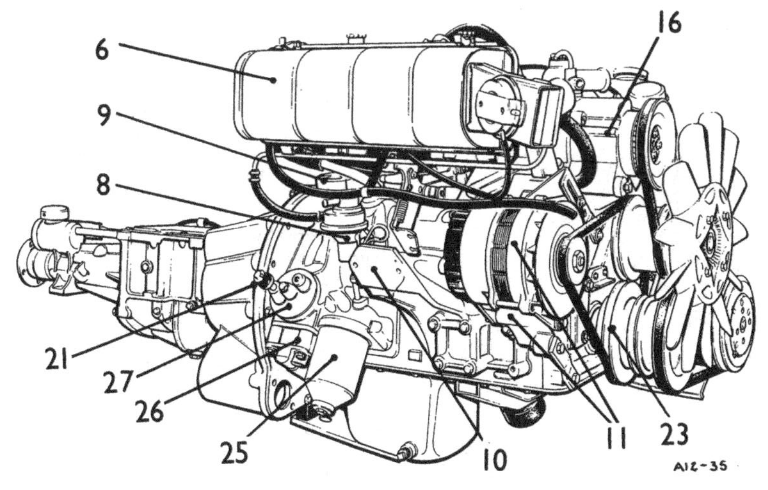

33 Slacken the cylinder head retaining nuts and bolts in the reverse order as shown in operation 12.29.27.
34 Remove the five cylinder head studs and five bolts.
35 Withdraw the cylinder head complete with the camshaft and discard the gasket.
36 Remove the sump.
37 Remove the oil pick-up strainer.
38 Withdraw the remaining bolts – noting their positions and remove the timing cover. Remove the oil seal.
39 Remove the oil thrower.
40 Remove the two retaining bolts and withdraw the timing chain tensioner.
41 Remove the three bolts securing the two chain guides and support bracket.
42 Remove the guides and support bracket complete with the chain and camshaft sprocket.
43 Turn the jackshaft sprocket to expose the two Allen screws retaining the jackshaft keeper plate. Remove the screws and withdraw the plate.
44 Withdraw the jackshaft and sprocket from the cylinder block.
45 Unlock the tab washer and remove the bolt and sprocket from the jackshaft.
46 Withdraw the crankshaft sprocket.
47 Remove the drive key and shims from the crankshaft.
48 Remove the six bolts and remove the rear main oil seal and housing.
49 Remove the nuts securing the connecting rod bearing caps and withdraw the caps and lower shells ensuring that if the shells are to be refitted they are kept identified with their respective caps.
50 Push the connecting rods upwards and withdraw the piston assemblies through the top of the bores, keeping the shells identified with their respective connecting-rods.
51 Remove the 10 bolts securing the five main bearing caps and withdraw the caps and lower shells.
52 Lift out the crankshaft.
53 Remove the upper shells from the crankcase.
54 Remove the thrust washers from number 3 main bearing.
55 Withdraw the spigot bush from the crankshaft.

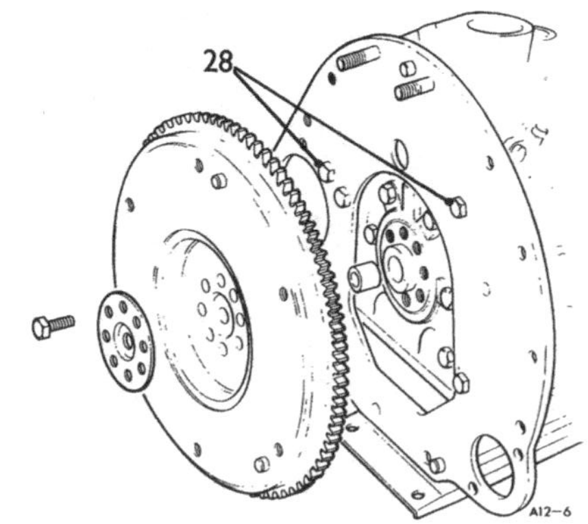

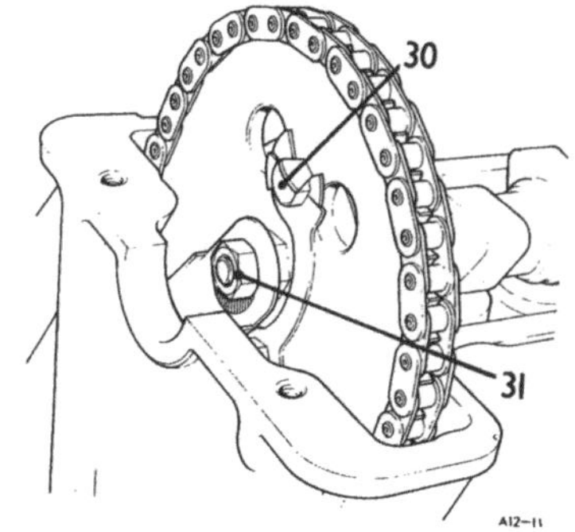

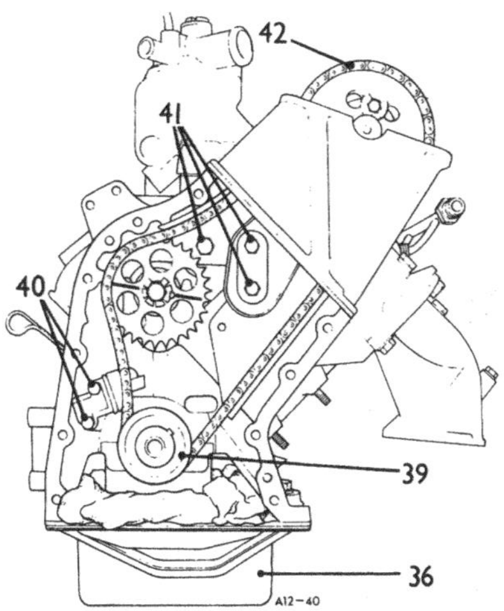

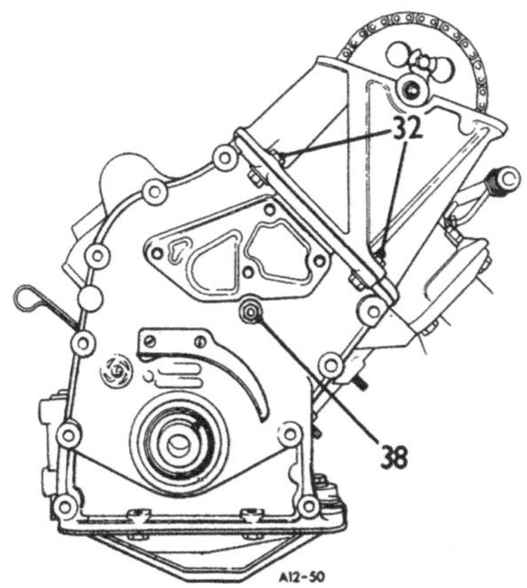

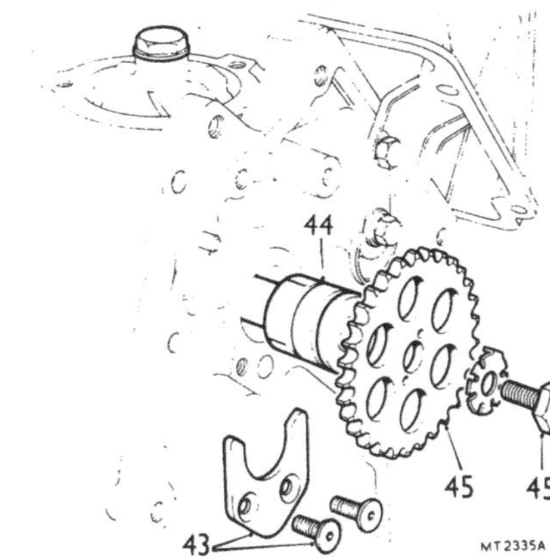

Rebuilding

NOTE: The following rebuilding instructions assume that all the individual components and assemblies have been examined, worn parts renewed and assemblies overhauled. Moreover, all joint faces are clean and parts lubricated before assembly.

56 Fit the five main bearing upper shells ensuring that the keeper tags locate in the recesses.
57 Fit the crankshaft thrust washers to number 3 main bearing noting that the oil grooves face outwards.
58 Lower the crankshaft into the crankcase.
59 Fit the lower shells to the main bearing caps and fit them to the crankcase. Partially tighten the securing bolts.

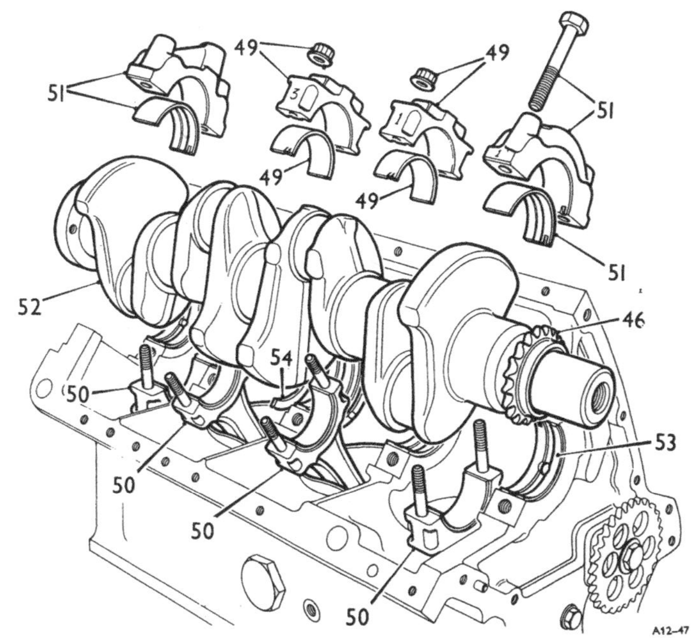

60 Using either a dial gauge or feelers check the crankshaft end-float – see data.
61 Tighten the main bearing cap bolts to the correct torque – see section 06.
62 Stagger the piston ring gaps, avoiding a gap on the thrust side and compress the rings using special tool number 38 U3.
63 Turn the crankshaft to position numbers 2 and 3 crankpins at B.D.C. Insert the respective connecting rod and piston assemblies into the bores and tap the pistons home noting that the flat raised part of the piston crown is fitted to the L.H. side of the engine.
64 Fit the upper bearing shells to the connecting rods ensuring that the keeper tags locate in the recesses and pull the connecting rods onto the journals.
65 Fit the lower shells to the connecting rod caps ensuring that the keeper tags locate in the recesses and fit them to the connecting rods. Tighten new nuts to the correct torque – see section 06.
66 Repeat instructions 62 to 65 on numbers 1 and 2 pistons and connecting rod assemblies.

continued

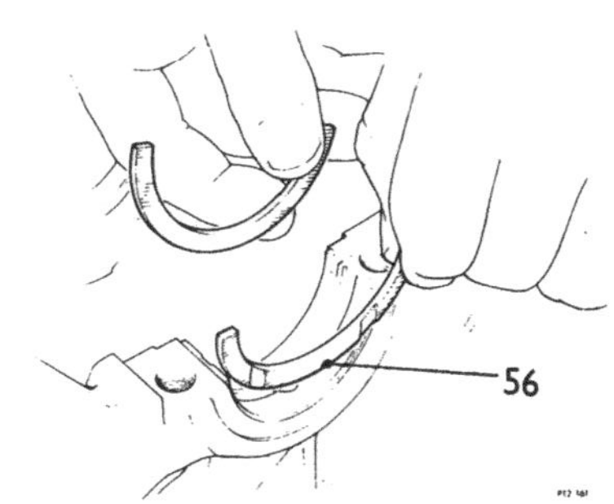

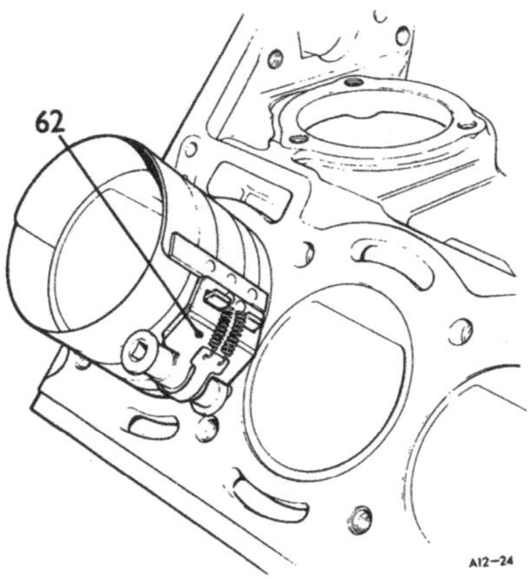

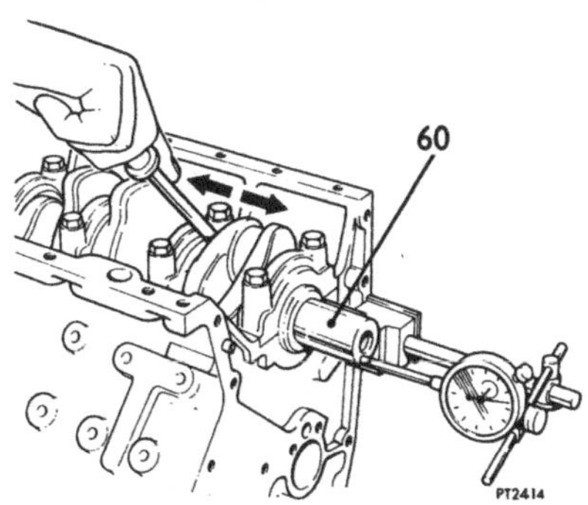

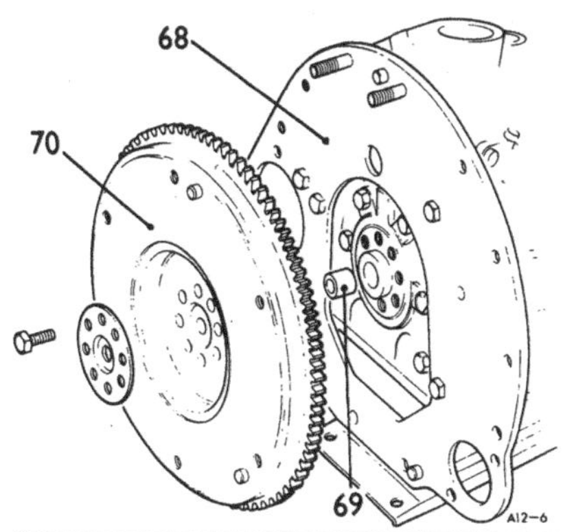

67 Fit a new seal to the rear main oil seal housing (lip towards the crankshaft) and using a new gasket and jointing compound fit the housing locating it over two dowels. Secure with the six bolts and spring washers noting that the two longer bolts are fitted at the bottom.
68 Fit the rear adaptor plate locating it on two dowels and secure with the six bolts.
69 Fit the spigot bush to the crankshaft.
70 Fit the flywheel – lining up the punch marks – if the original flywheel or crankshaft is being fitted. Secure with the spigot bush retaining plate and eight bolts and tighten to the correct torque – see section 06.
71 Check the flywheel for run-out using a dial gauge – see data.
72 Fit the clutch driven and pressure plates locating the pressure plate assembly over the three dowels.
73 Centralize the driven plate using a dummy primary shaft and evenly tighten the six bolts and spring washers to the correct torque. See 33.10.01.
74 Fit the sprocket to the jackshaft locating it over the single dowel and secure with bolt and new lock washer.
75 Fit the jackshaft to the cylinder block, locate it with the keeper plate and secure the plate with the two Allen screws.
76 Fit the water pump impeller and cover assembly checking the clearance as described in the refitting instructions 26.50.01.
77 Fit the distributor mounting plate.
78 Fit the oil transfer adaptor ensuring that new 'O' rings are fitted and correctly located. Secure with the single bolt.
79 Ensure that the pistons are **not** at T.D.C. see cautionary note.
80 Fit two 'slave' guide studs to the bolt holes in the cylinder block to facilitate the fitting of the cylinder head and gasket.

NOTE: See 12.29.02 for dimensional drawing of suitable studs.

81 Fit the cylinder head gasket locating it over the studs.
82 Fit the cylinder head easing it over the guide studs.
83 Fit the five cylinder head studs.
84 Remove the two 'slave' studs and fit the five cylinder head retaining bolts and plain washers.
85 Fit the nuts and plain washers to the cylinder head studs.
86 Tighten the nuts and bolts to the correct torque and in the correct sequence. See 12.29.27.

CAUTION: Once the pistons are fitted care must be taken to ensure that the crankshaft is not allowed to turn a complete revolution until the valve timing has been completed, otherwise damage to the valves and pistons will occur.

87 Temporarily fit the crankshaft sprocket and check its alignment with the jackshaft sprocket by placing a straight edge across the two sprockets.

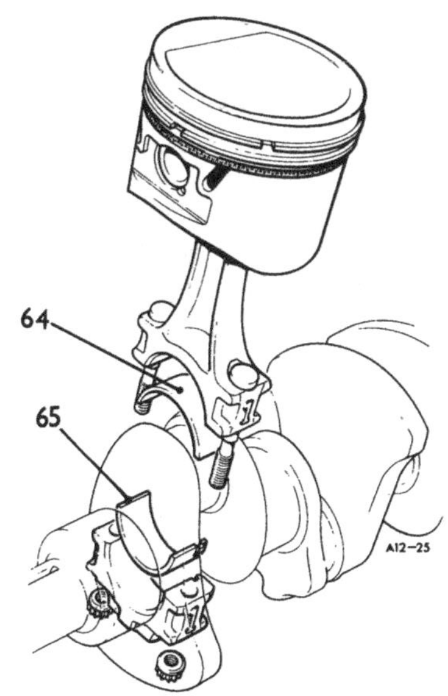

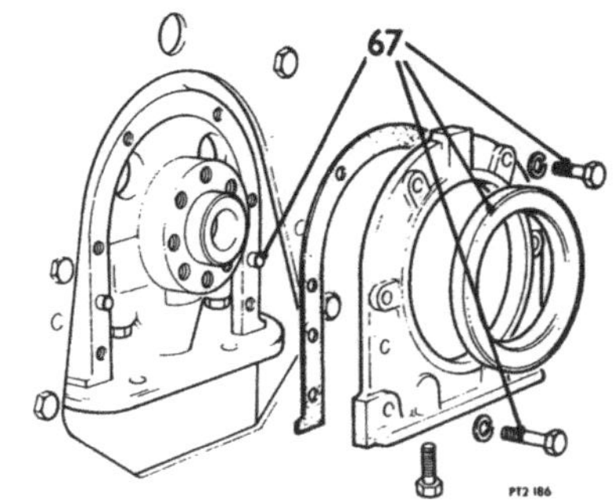

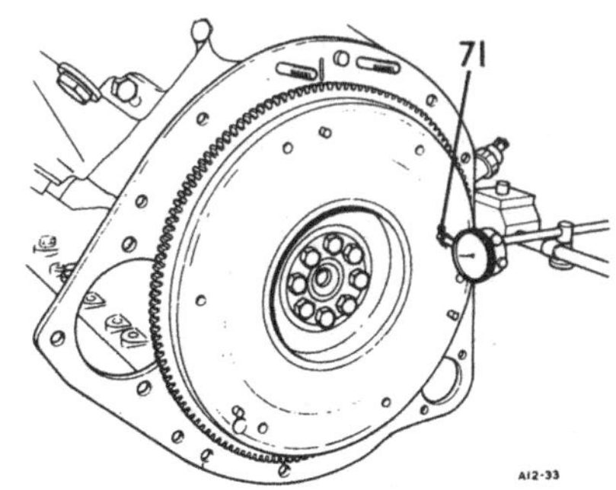

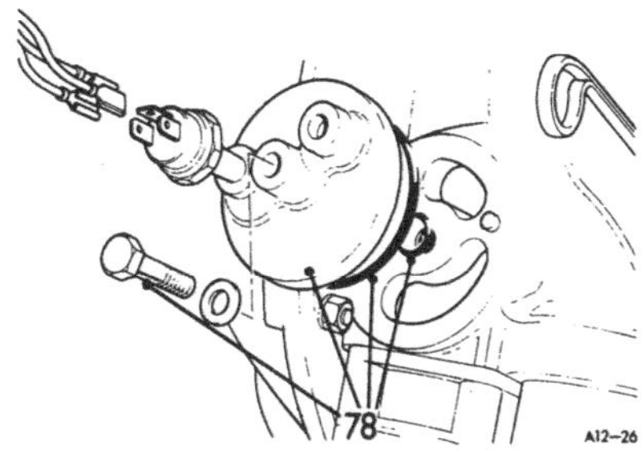

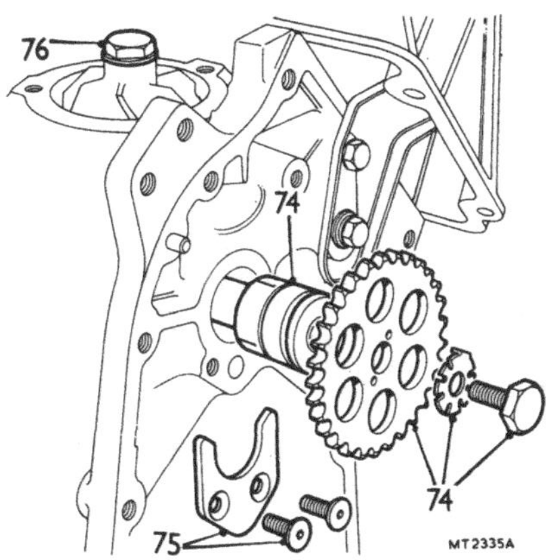

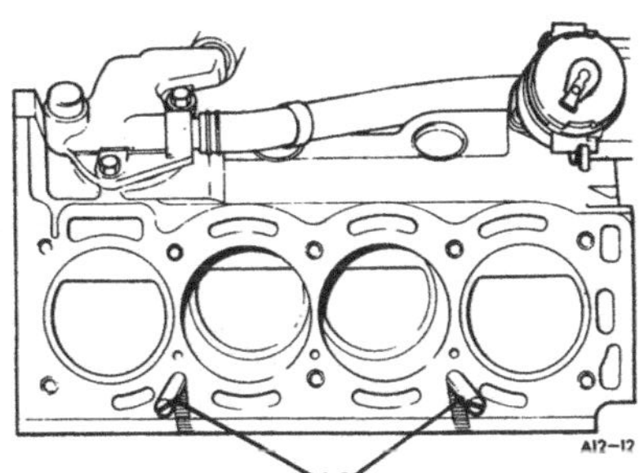

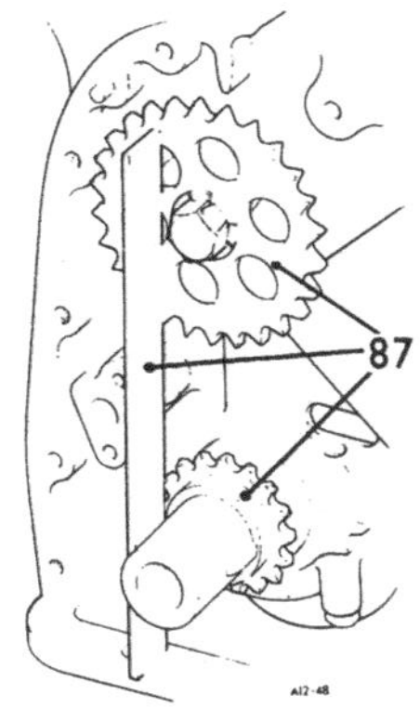

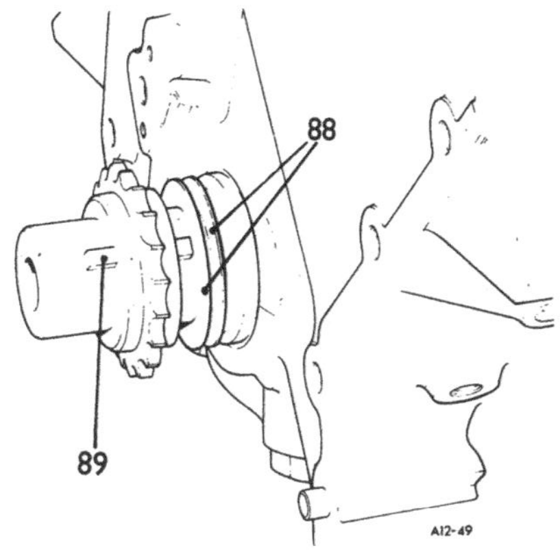

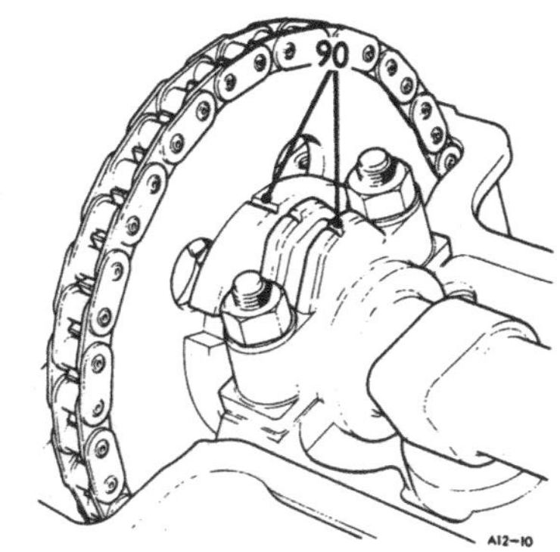

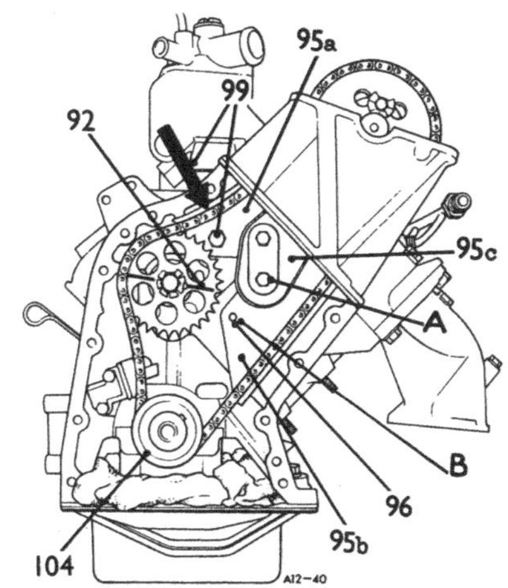

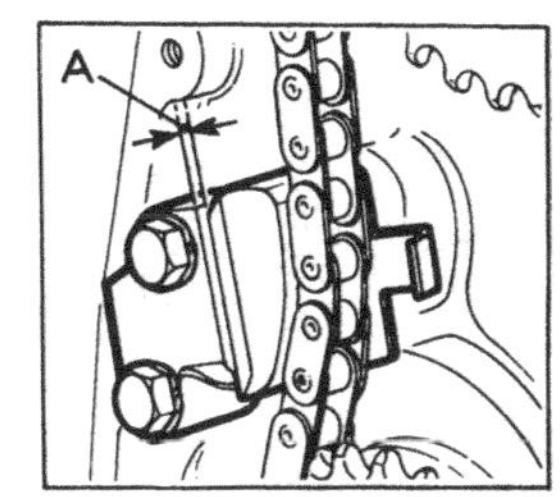

88 Adjust any misalignment by removing the crankshaft sprocket and fitting shims of suitable thickness between it and the crankshaft. Ensure that the sprocket is pushed fully home when checking the alignment.

89 Remove the sprocket, fit the crankshaft drive key and refit the sprocket.

90 Turn the camshaft until the timing mark on the flange is in line with the groove on the camshaft front bearing cap.

91 Temporarily locate the timing cover and crankshaft pulley and turn the crankshaft (having regard for the preceding CAUTION) until number 1 and 4 pistons are at T.D.C. Remove the pulley and timing cover.

92 Turn the jackshaft until the scribed line across the sprocket is equidistant between bolt A and hole B with the dowel to the left. Remove the pulley and timing cover.

93 Encircle the camshaft sprocket with the timing chain and insert through the cylinder head aperture. Locate the sprocket on the camshaft flange and secure with the lock plate and one bolt – instruction 30.

94 Keeping the chain taut on the drive side fit the chain to the crankshaft and jackshaft sprockets. Check and if necessary make any adjustment to the position of the jackshaft sprocket to maintain it in its correct position as in instruction 92.

95 Fit and loosely secure the guides as follows:
- a The adjustable guide.
- b The straight fixed guide.
- c The support bracket.

96 Fit a 'slave' bolt to the lower hole in the fixed guide to ensure alignment when fitting the timing cover centre bolt.

97 Fit the timing chain hydraulic tensioner and secure with the two bolts and spring washers. 12.65.28.

NOTE: To prevent the tensioner releasing while fitting, fit a spacer between the tensioner body and the back of the slipper.

98 Remove the spacer and insert a 0.100 in (2,54 mm) feeler gauge or slip gauge in its place, dimension A.

continued

99 Adjust the chain tension by applying pressure, in the direction of the arrow, to the adjustable guide while tightening the clamp bolt.
100 Remove the gauge and check that the scribed line on the jackshaft sprocket is still correctly positioned.
101 Tighten all the guide retaining bolts and remove the 'slave' bolt.
102 Turn the engine sufficiently to enable the remaining camshaft sprocket retaining bolt to be fitted and locked.
103 Check that the threaded spigot on the camshaft sprocket does not foul the support bracket.
104 Fit the oil thrower with the dished face outwards.
105 Using sealing compound fit a new timing cover gasket and position the timing cover over the two dowels.
106 Secure the cover with the centre bolt noting that it has a fibre washer under the head.
107 Fit the two nuts and bolts retaining the timing cover to the cylinder head.
108 Wit the lip face leading press-in a new timing cover oil seal flush with the outer face of the cover.
109 Fit the crankshaft and auxiliary pulleys.
110 Turn the crankshaft to bring numbers 1 and 4 pistons to T.D.C. with the timing mark on the camshaft flange in line with the groove on the front bearing cap.
111 Fit the distributor so that the vacuum retard unit is pointing directly rearward and the rotor arm is in line with the manifold rear retaining bolt hole in the cylinder head.
112 Temporarily tighten the distributor retaining bolts pending final adjustment when the engine is installed.
113 Fit the camshaft cover.
114 Fit the oil pick-up strainer.
115 Fit the sump.
116* Fit the fuel pump using a new gasket and ensuring that the actuating lever rides on top of the jackshaft cam.
117 Fit the fan pulley assembly and hang the drive belts on the pulleys.
118 Fit the alternator complete with bracket and adjusting link.
119 Fit the front lifting eye.
120* Fit the air pump.

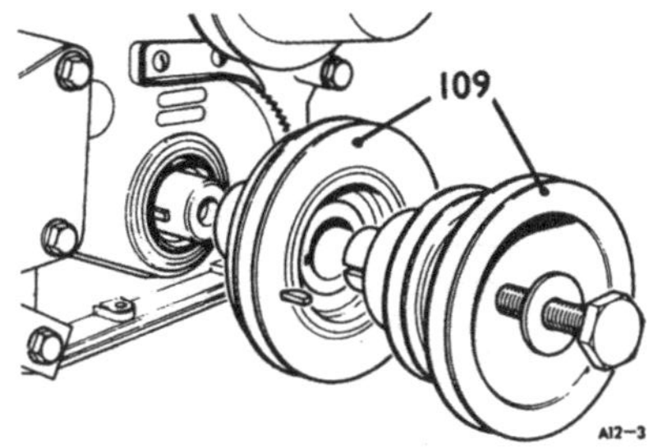

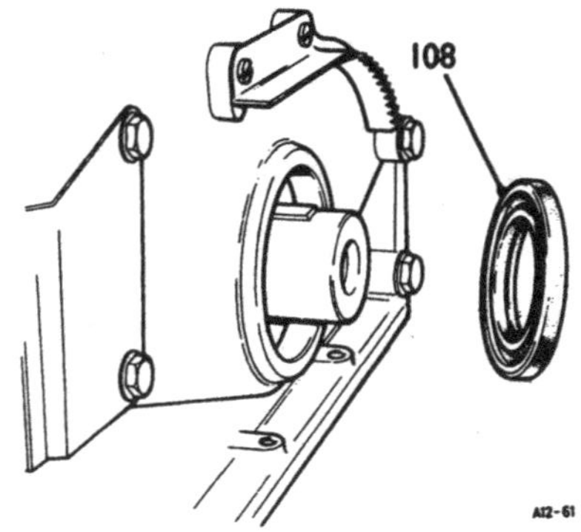

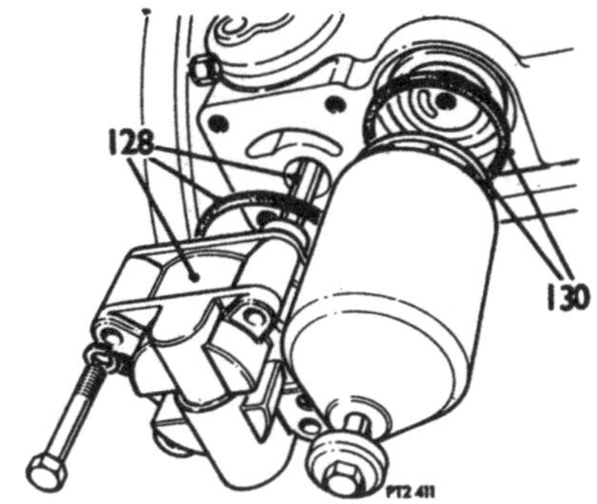

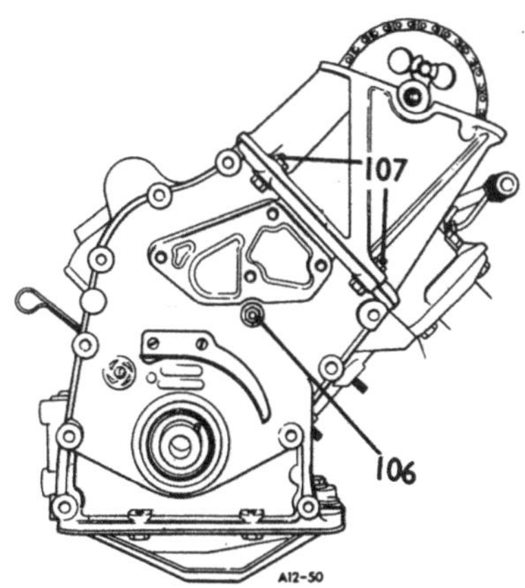

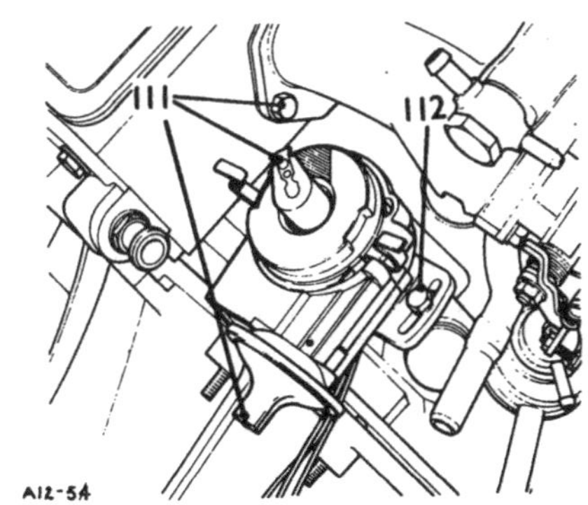

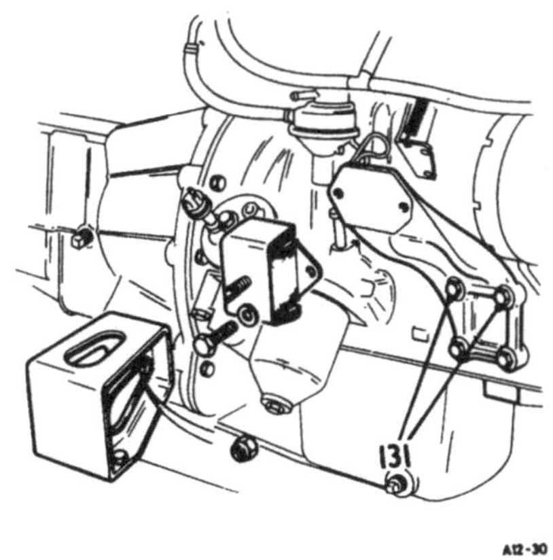

continued

121*Fit the diverter and relief valve complete with the air distribution manifold and hoses.
122*Fit the air pump to diverter and relief valve hose.
123*Fit and adjust the air pump drive belt tension, see 17.25.13, see instructions 4 to 6.
124 Fit and adjust the alternator belt tension, see 86.10.05.
125 Install the air conditioning compressor and brackets — if fitted.
126 Adjust air conditioning compressor drive belt tension, see 82.10.01.
127 Fit the exhaust manifold.
128 Fit the hexagon drive and oil pump using a new sealing ring.
129 Fit the oil pressure switch.
130 Insert a new filter element in the filter bowl and fit the assembly with a new sealing rubber.
131 Fit the R.H. engine mounting bracket noting that the long bolts are fitted to the top two holes.
132 Fit the water pump to heater and inlet manifold pipe and hose.
133 Fit the bottom hose to the water pump cover.
134 Fit the inlet manifold complete with carburetters, see 19.15.15., instructions 19 to 23 or the fuel rail and the rear lifting eye.
135*Connect the fuel pump to carburetter hose.
136 Connect the engine breather pipes.
137 Fit the dipstick.
138 Fit the spark plugs.
139 Fit the distributor cap and connect the H.T. leads to the plugs (firing order 1, 3, 4, 2 distributor turns anti-clockwise).
140 Connect the vacuum pipe to the diverter and relief valve.
141 Fit the gearbox.
142 Fit the starter motor.
143 Fit the engine and gearbox assembly to the car 12.37.01.
144 Check and if necessary adjust the ignition timing 86.35.15.

ENGINE MOUNTING – FRONT L.H.

Remove and refit **12.45.01**

Removing
1 Disconnect the battery.
2 Remove the fan guard.
3 Remove the fresh air duct. 80.15.31.
4 Raise the car on a ramp.
5 Remove the engine stabilizer. 12.45.16.
6 Remove the single nut securing the engine mounting to the sub-frame.
7 Place a jack to support the engine, under the sump coupling plate.
8 Remove the engine R.H. mounting retaining nut.
9 Jack up the engine sufficiently to enable the L.H. rubber mounting to be unscrewed from the sump cross-member threaded hole.
10 Remove the mounting.
11 Remove the packing piece – where fitted.

Refitting
12 Screw the mounting into the sump cross-member, inserting the packing piece – where fitted.
13 Lower the jack and tighten the engine R.H. mounting nut.
14 Remove the jack and fit and tighten the single nut securing the L.H. engine mounting to the sub-frame.
15 Fit and adjust the engine stabilizer. 12.45.16.
16 Lower the ramp.
17 Fit the fan guard.
18 Fit the fresh air duct. 80.15.31.
19 Connect the battery.

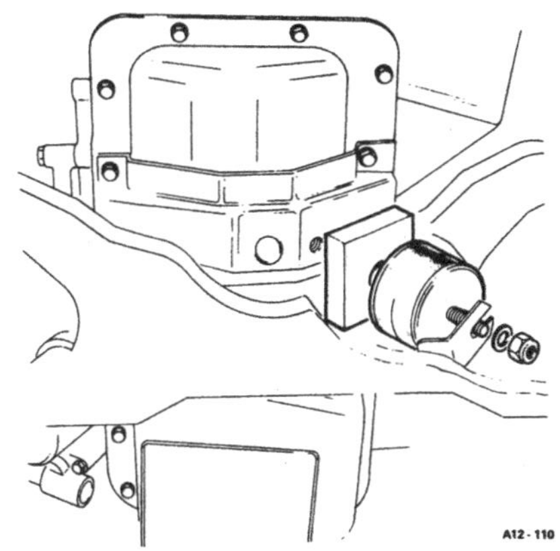

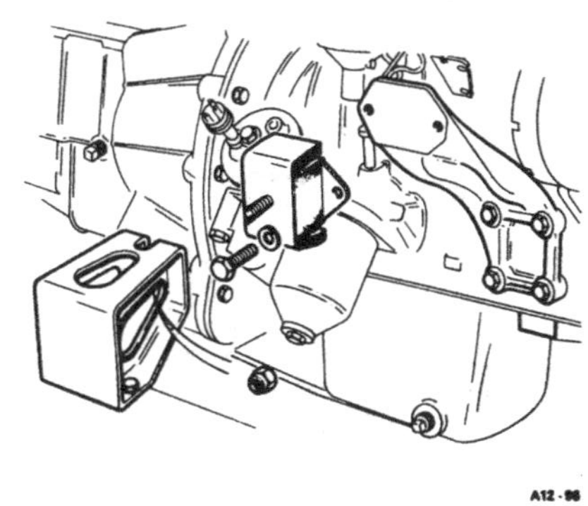

ENGINE MOUNTING – FRONT R.H.

Remove and refit **12.45.03**

Removing
1 Disconnect the battery.
2 Support the weight of the engine under the sump coupling plate.
3 Remove the single nut securing the mounting to the body bracket.
4 Remove the two bolts retaining the rubber mounting to the engine bracket.
5 Withdraw the mounting.

Refitting
6 Fit the mounting to the engine bracket with the two bolts and spring washers.
7 Secure the mounting to the body bracket with the single nut and plain washer.
8 Remove the engine support.
9 Connect the battery.

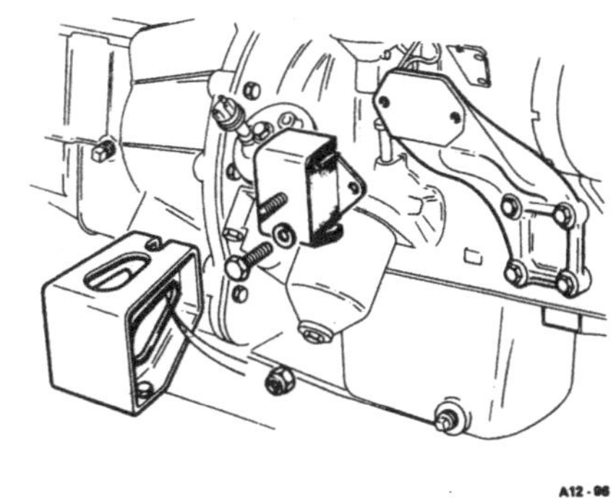

ENGINE MOUNTING – REAR CENTRE U.K. AND EUROPE SPECIFICATION

– Remove and refit 12.45.08

Removing
1 Drive the car onto a ramp and disconnect the battery.
2 Raise the ramp and place a jack in support under the gearbox.
3 Remove the two nuts and bolts retaining the rear mounting assembly to the gearbox extension.
4 Remove the two distance plates.
5 Remove the four bolts retaining the mounting cross-member to the body and remove the assembly from the car.
6 Remove the centre bolt and nut washer.
7 Remove the bracket.
8 Remove the distance piece.
9 Remove the restrictor plate.
10 Remove the two nuts and bolts securing the rubber mounting to the cross-member.
11 Remove the rubber mounting rubber from the cross-member.

Refitting
12 Fit the rubber mounting to the cross-member with the two nuts and bolts.
13 Reverse instructions 6 to 9 leaving the centre bolt and nut slack.
14 Fit the rear mounting assembly to the gearbox extension with the two nuts and bolts and four plain washers (one under the head of each bolt and one under the nut). Ensure that the two distance plates are fitted between the bracket and gearbox extension.
15 Fit the cross-member to the body with the four nuts.
16 Tighten the centre nut and bolt.
17 Remove the jack, lower the car and connect the battery.

ENGINE MOUNTING – REAR CENTRE USA SPECIFICATION

– Remove and refit 12.45.08

Removing
1 Drive the car onto a ramp and disconnect the battery.
2 Raise the ramp and place a jack in support under the gearbox.
3 Disconnect the restraint cable from the crossmember.
4 Remove the two nuts securing the mounting assembly to the gearbox extension.
5 Disconnect the tie bar from the sump coupling plate.
6 Remove the four bolts retaining the mounting crossmember to the body and remove the assembly from the car, complete with the tie bar.
7 Remove the centre bolt.
8 Remove the tie bar.
9 Remove the distance piece.
10 Remove the restrictor plate.
11 Remove the two nuts and bolts securing the rubber mounting to the cross-member.

Refitting
12 Fit the rubber mounting to the cross-member with the two nuts and bolts.
13 Reverse instructions 7 to 11 leaving the centre bolt slack.
14 Fit the cross-member to the body with the four nuts.
15 Fit the assembly to the gearbox extension with the two nuts.
16 Connect the tie bar to the sump coupling plate.
17 Tighten the centre bolt.
18 Connect and adjust the restraint cable 37.20.01 instruction 18.
19 Remove the jack, lower the car and connect the battery.

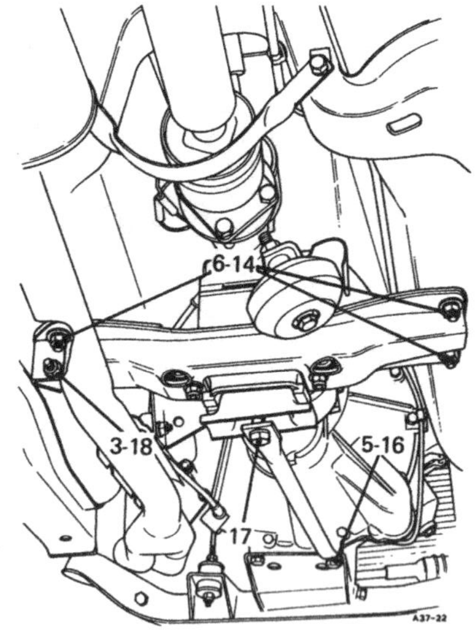

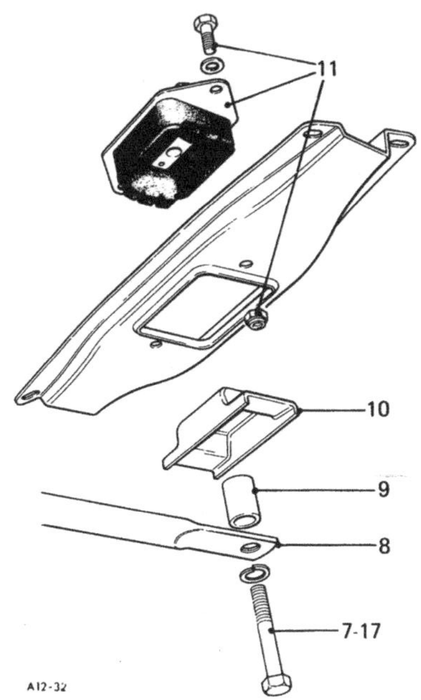

ENGINE STABILIZER (EARLY MODELS)

– Remove and refit 12.45.16

Removing
1 Raise the car on a ramp.
2 Remove the nyloc nut securing the engine stabilizer to the cylinder block bracket.
3 Remove the nyloc nut securing the stabilizer to the sub-frame and withdraw:
   a The lower washer.
   b The lower rubber.
   c The nylon distance piece.
4 Remove the stabilizer complete with:
   a The top rubber.
   b The top washer.
   c The top retaining nut.
   d The lock nut.

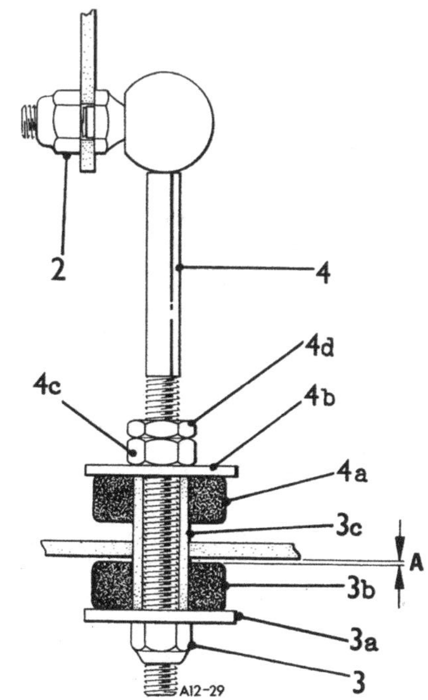

Refitting
5 Slacken off the lock nut and upper retaining nut.
6 Insert the stabilizer into the sub-frame location complete with the upper retaining nut and its lock-nut and upper washer and rubber.
7 Locate the stabilizer in the cylinder block bracket and secure with the nyloc nut.
8 Fit the nylon distance piece.
9 Fit the lower rubber washer and nyloc nut.
10 Tighten the lower nyloc nut until a gap of 0.040 in (1.0 mm) – dimension A – exists between the sub-frame and lower rubber.
11 Tighten the top retaining nut down until the upper washer just nips the nylon distance piece and tighten the lock nut.
12 Check the dimension A and lower the ramp.

ENGINE REAR ADAPTOR PLATE

Remove and refit 12.53.03

Removing
1 Remove the gearbox, see 37.20.01 or 44.20.01.
2 Remove the clutch assembly (as applicable).
3 Remove the flywheel, see 12.53.07, instructions 4 to 6.
4 Remove the six bolts securing the adaptor plate to the cylinder block.
5 Withdraw the rear adaptor plate.

Refitting
6 Clean the engine and adaptor plate mating faces and locate the plate on the engine over the dowels and studs.
7 Secure the plate with the six bolts.
8 Fit the flywheel 12.53.07 – instructions 7 to 11.
9 Fit the clutch (as applicable).
10 Fit the gearbox, see 37.20.01 or 44.20.01.

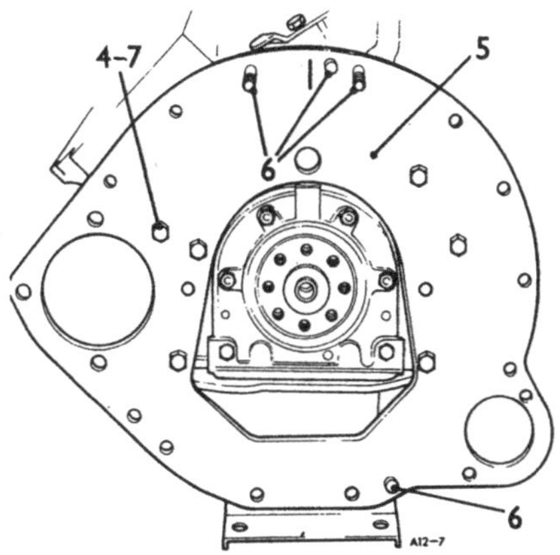

FLYWHEEL

Remove and refit 12.53.07

Removing
1 Isolate the battery.
2 Remove the gearbox, see 37.20.01 or 44.20.01
3 Remove the clutch assembly (as applicable).
4 Remove the eight bolts and spigot bush retaining plate.
5 Mark for reassembly the relationship between the flywheel hub and crankshaft flange.
6 Withdraw the flywheel and ring gear assembly.

Refitting
7 Locate the flywheel assembly on the crankshaft so that the identification marks line up.
8 Secure the flywheel to the crankshaft with the eight bolts and retaining plate. Tighten the bolts to the correct torque – see **'TORQUE WRENCH SETTINGS'**.
9 If a new flywheel has been fitted, turn the engine over so that numbers 1 and 4 pistons are at T.D.C. with number 1 cylinder firing.
10 Mark a chisel mark on the outside edge of the flywheel in line with the vertical mark on the engine rear adaptor plate.
11 Using a dial gauge check the flywheel for run-out – see DATA.
12 Refit the clutch (as applicable).
13 Refit the gearbox, see 37.20.01 or 44.20.01.
14 Reconnect the battery.

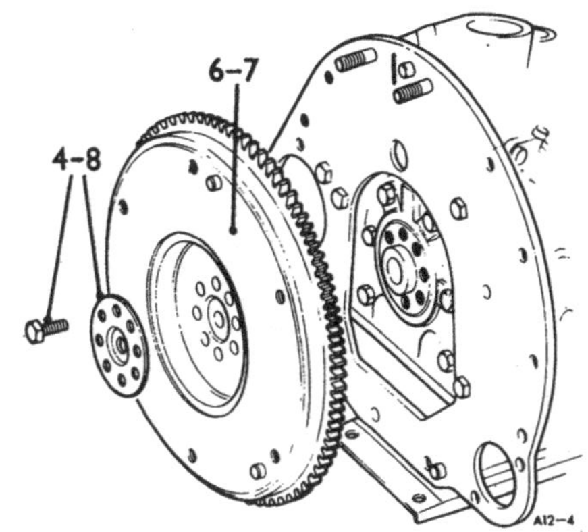

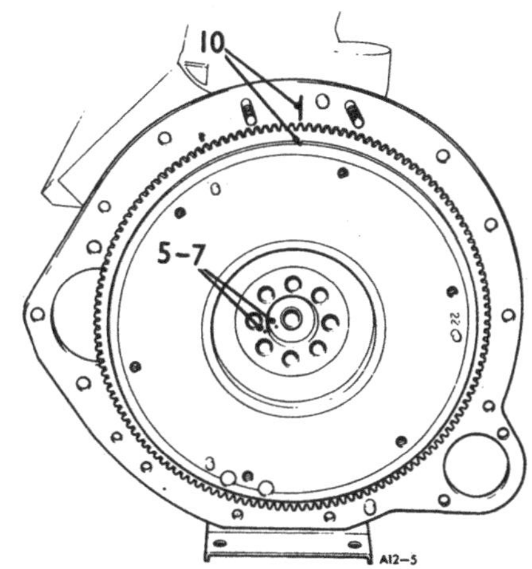

DATA

Flywheel run-out maximum . . . . 0.002 (0,050 mm) at a radius of 4 in (101 mm) on the friction face.

STARTER RING GEAR

– Remove and refit **12.53.19**

Removing
1 Remove the flywheel. 12.53.07.
2 Drill a hole approximately 0.375 in (10 mm) between the root of any tooth and inner rim of the ring gear. Drill sufficiently to weaken the ring gear without damaging the flywheel.
3 Secure the flywheel in a soft-jawed vice.
4 Place a heavy cloth over the ring gear for protection against flying fragments when splitting the ring gear.

WARNING: Take adequate precaution to avoid injury from flying fragments when splitting the ring gear.

5 Place a chisel immediately above the drilled hole and strike sharply to split the gear.

Refitting
6 Place the flywheel friction face downwards on a flat surface.
7 Heat the ring gear uniformly between 170° and 175°C (338° to 347°F) but do not exceed this temperature.
8 Locate the ring gear on the flywheel and retain it in position until it contracts sufficiently to grip the flywheel.
9 Allow the gear to cool gradually to avoid distortion. A maximum permissible gap of 0.025 in (0,6 mm) is allowed between the flywheel and ring gear in any one length of 6 in (15 cm) only of the circumference.
10 Fit the flywheel. 12.53.07

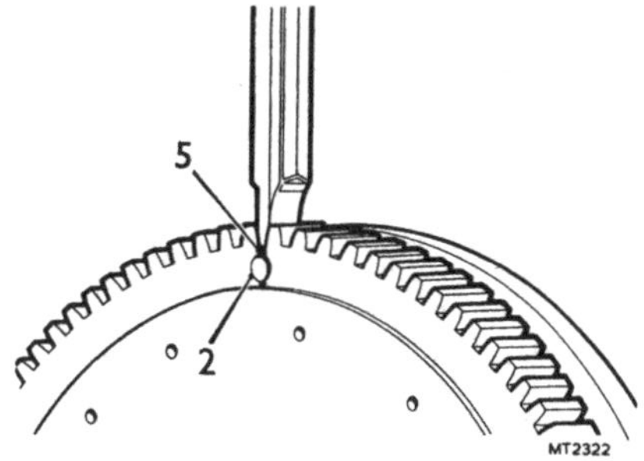

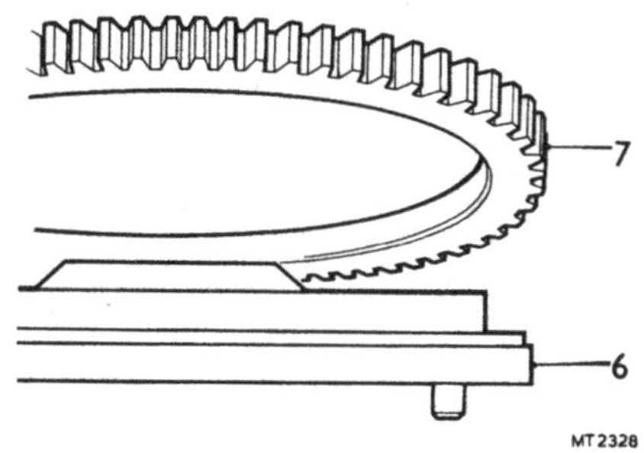

OIL FILTER ELEMENT

– Remove and refit **12.60.02**
– Overhauling filter assembly – instructions 6 to 9 **12.60.08**

Removing
1 Drive the car onto a ramp.
2 Disconnect the battery and raise the ramp.
3 Place a suitable receptacle under the filter bowl to catch surplus oil.
4 Remove the filter bowl centre retaining bolt, and withdraw the assembly.
5 Extract the seal from the annular groove in the crankcase.

Overhauling
6 Remove the element and discard.
7 Dismantle the centre bolt assembly and clean all components including the bowl.
8 Assemble the centre bolt assembly to the bowl as illustrated, using new seals and if necessary renew the spring.
9 Secure the assembly with the retaining clip ensuring that it locates properly in the annular groove in the bolt.

Refitting
10 Clean the crankcase mating face and annular groove and fit a new seal to the groove ensuring that it seats properly and is not twisted.
11 Insert a new element into the filter bowl and secure the assembly to the crankcase and whilst tightening the centre bolt rotate the bowl to ensure that it seats correctly in the groove. Tighten the bolt to the correct torque figure – see section 06.
12 Reconnect the battery, start the engine and check for oil leaks.
13 Stop the engine and lower the ramp.
14 Allow sufficient time for the oil to drain back into the sump, check the oil level and top-up to the high mark on the dipstick with oil of a recommended make and grade.

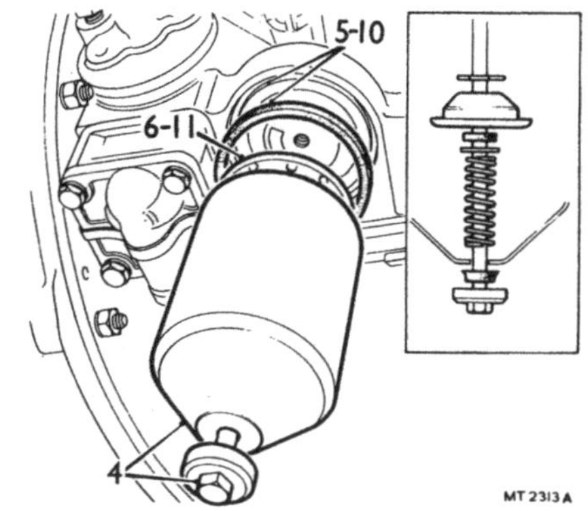

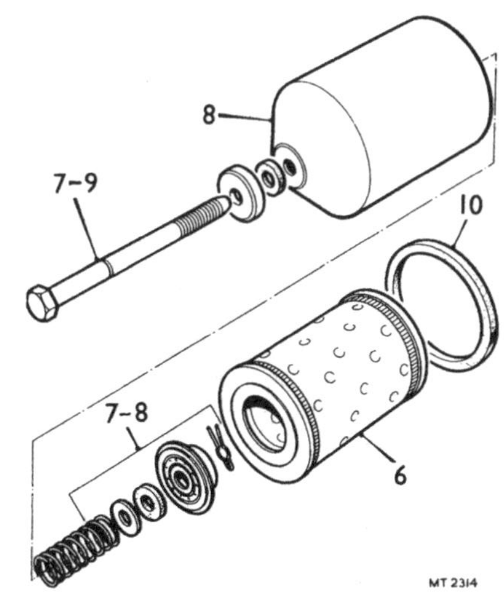

OIL TRANSFER HOUSING

– **Remove and refit** 12.60.14

Removing

1 Disconnect the battery.
2 Disconnect the three Lucar leads from the oil pressure switch.
3 Remove the centre retaining bolt and withdraw the transfer housing.
4 Remove the 'O' ring seals from the housing.

Refitting

5 Clean the transfer housing and crankcase mating faces.
6 Using new 'O' ring seals fit the transfer housing engaging the dowel in the crankcase in the hole in the housing.
7 Fit and tighten the centre retaining bolt and plain washer.
8 Reconnect the three electrical leads to the correct connectors. 86.65.30.
9 Reconnect the battery.

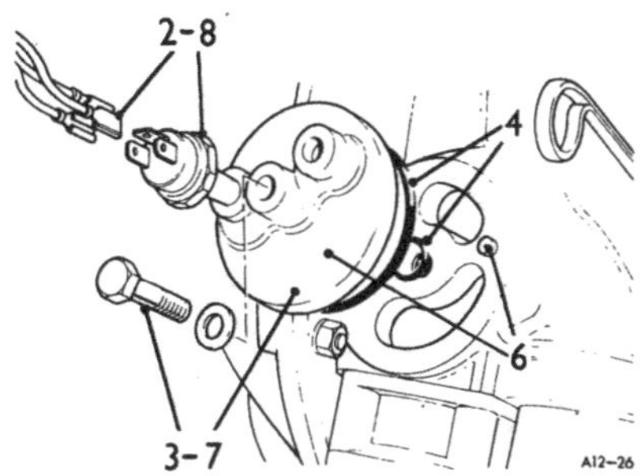

OIL PICK-UP STRAINER

– **Remove and refit** 12.60.20

Removing

1 Remove the sump. 12.60.44.
2 Remove the two bolts and plain washers.
3 Withdraw the strainer.
4 Remove the gasket.

Refitting

5 Clean the strainer gauge.
6 Clean the strainer and crankcase mating faces.
7 Fit a new gasket.
8 Fit the strainer and secure with the two bolts and plain washers.
9 Refit the sump. 12.60.20.

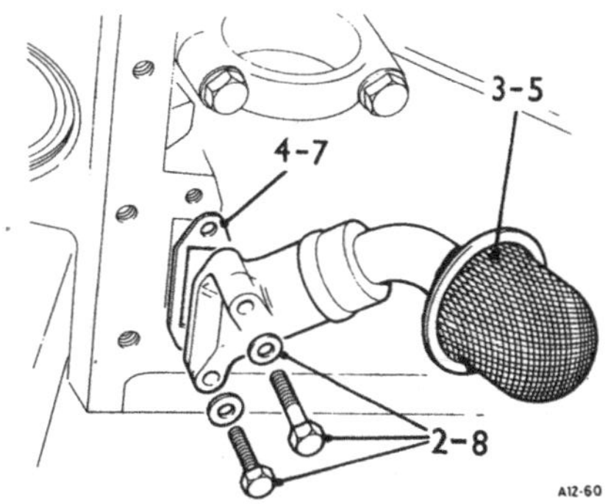

OIL PUMP

– **Remove and refit** 12.60.26

Removing

1 Drive the car onto a ramp and disconnect the battery.
2 Raise the ramp.
3 Remove the two bolts securing the clutch slave cylinder to the bell housing.
4 Carefully withdraw the clutch slave cylinder, complete with fluid pipe.
5 Withdraw the slave cylinder completely and secure it in a convenient position to gain access to the oil pump.
6 Remove the bell housing nut and bolt.
7 Remove the four oil pump retaining bolts and spring washers.
8 Withdraw the pump complete with the hexagonal drive shaft.
9 Remove the 'O' sealing ring.

Refitting

10 Clean the oil pump and crankcase mating faces.
11 Fit a new 'O' ring to the groove in the oil pump.
12 Engage the hexagonal drive shaft in the oil pump rotor.
13 Fit the oil pump ensuring that the drive shaft engages with the distributor.
14 Fit and evenly tighten the four oil pump retaining bolts and spring washers to the correct torque figure – see section 06.
15 Fit the bell housing nut and bolt.
16 Fit the clutch slave cylinder ensuring that the push rod enters the hole in the rubber boot.
17 Tighten the two retaining bolts and spring washers.
18 Lower the ramp and reconnect the battery.
19 Run the engine for three minutes, raise the ramp and check for oil leaks.

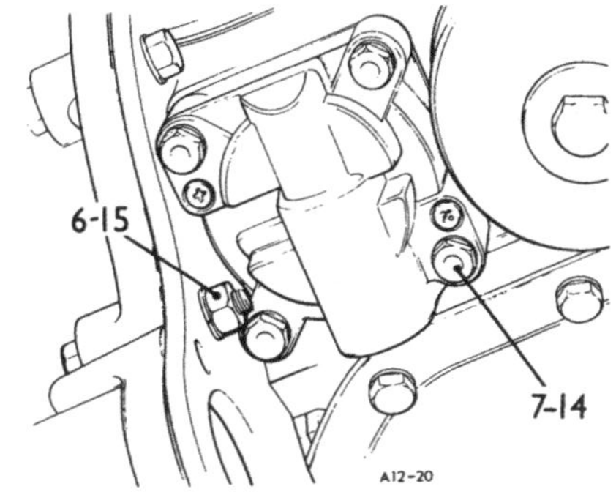

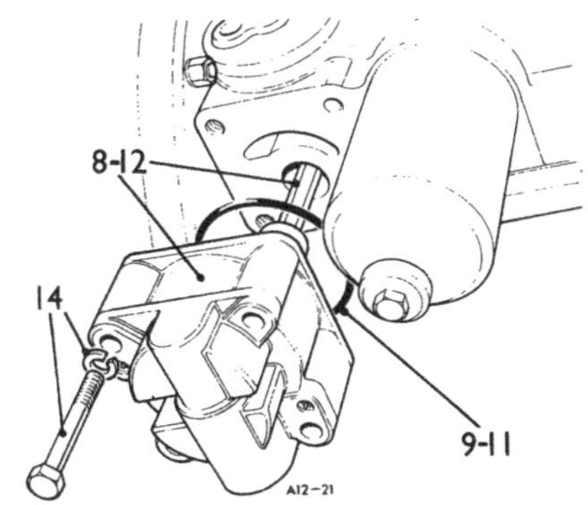

OIL PUMP

Overhaul **12.60.32**

Dismantling
1 Remove the oil pump. 12.60.26.
2 Withdraw the hexagonal drive shaft.
3 Remove the two screws and lift off the pump cover from the pump body.
4 Remove the rotors.
5 Remove the 'O' sealing ring from the pump body.
6 Remove the split pin from the oil pump cover.
7 Remove the locating plug, spring and relief valve.
8 Remove the 'O' sealing ring from the locating plug.

Inspecting
9 Clean all components
10 Install the rotors in the pump body, ensuring that the chamfered edge of the outer rotor is at the driving end of the rotor pocket.
11 Check the end-float of the inner and outer ring.
12 Check the outer ring to pump body diametrical clearance.
13 Check the rotor lobe clearances.
14 Check the length of the relief valve spring.
15 Check the relief valve and its bore for scoring or damage.
16 Renew the pump assembly if the clearances or end-floats measured in operations 11 to 13 exceed the figures given in DATA.
17 Check the bush in the pump cover, renew if scored or worn.

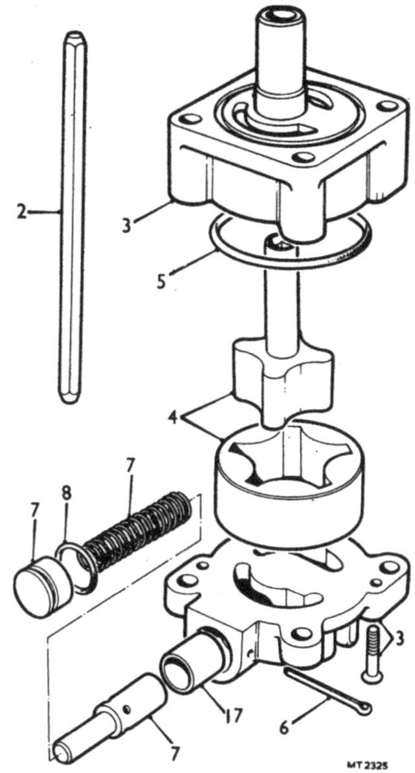

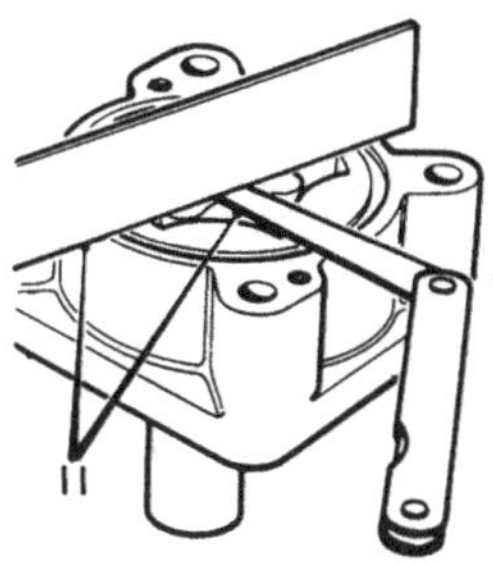

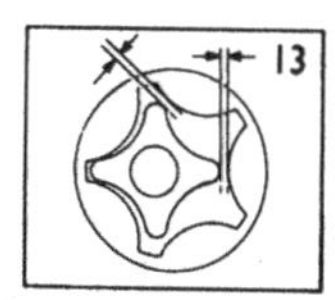

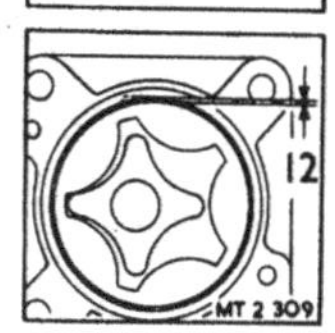

Reassembling
18 Lubricate all parts in clean engine oil before assembling.
19 Reverse the procedure in 1 to 8, noting:
   a Fit the relief valve with its large diameter first, so that its small spigot will engage with spring.
   b Fit a new 'O' sealing ring to the oil pressure relief valve locating plug.
   c Ensure that the outer rotor is installed in the pump body with its chamfered edge towards the driving end.
   d Fit a new 'O' sealing ring to the pump body.
20 Check the pump for freedom of action.
21 Fit the oil pump. 12.60.26.

DATA

Oil pump – Hobourn-Eaton	
Outer ring end-float	0.004 in (0.1 mm)
Inner ring end-float	0.004 in (0.1 mm)
Outer ring to pump body diametrical clearance	0.008 in (0.2 mm)
Rotor lobe clearance.	0.010 in (0.25 mm)
Relief valve spring free length . . .	1.70 in (43.18 mm)

OIL SUMP

Remove and refit **12.60.44**

Removing
1 Drive the car onto a ramp and disconnect the battery.
2 Remove the fresh-air duct.
3 Remove the fan guard.
4 Raise the ramp.
5 Drain the sump oil and refit the drain plug.
6 Remove the coupling plate bolts.
7 Remove the engine stabilizer. 12.45.16. (Where fitted).
8 Make up a support bracket from angle-iron in accordance with the dimensional drawing.

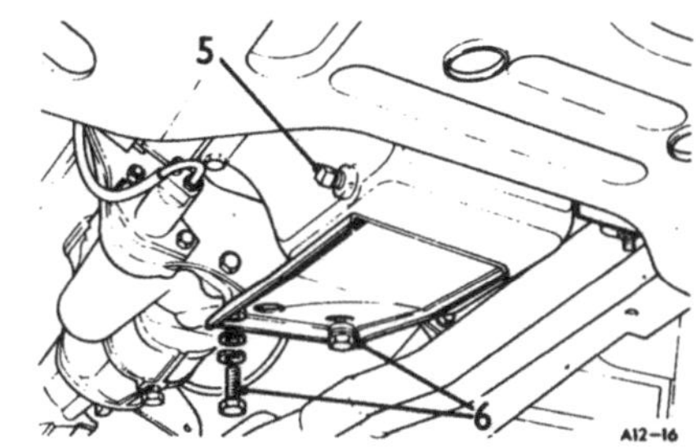

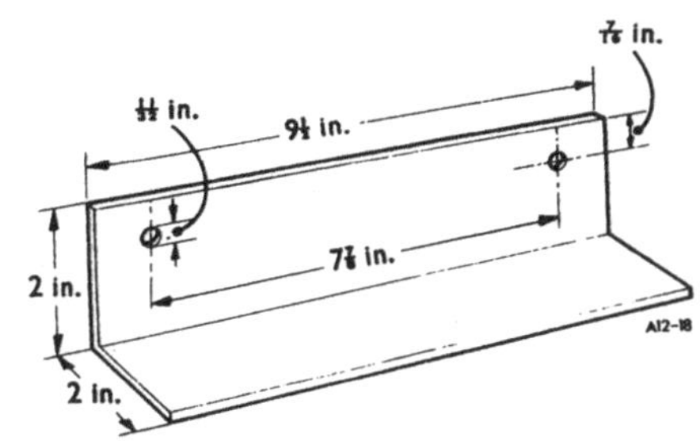

continued

9 Remove the alternator support strap bolt and move the link away.
10 Remove the opposite timing cover lower bolt.
11 Using these bolts attach the support bracket to the timing cover.
12 Support the engine under the bracket with a jack.
13 Remove the two engine R.H. mounting bolts.
14 Remove the engine L.H. mounting to sub-frame nut.
15 Remove the sump nuts and bolts.
16 Raise the engine sufficiently to enable the sump complete with the L.H. engine mounting and cross-member to be withdrawn.

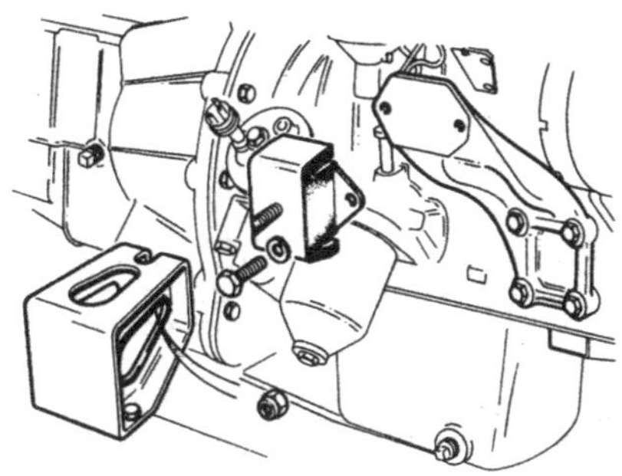

A12-96

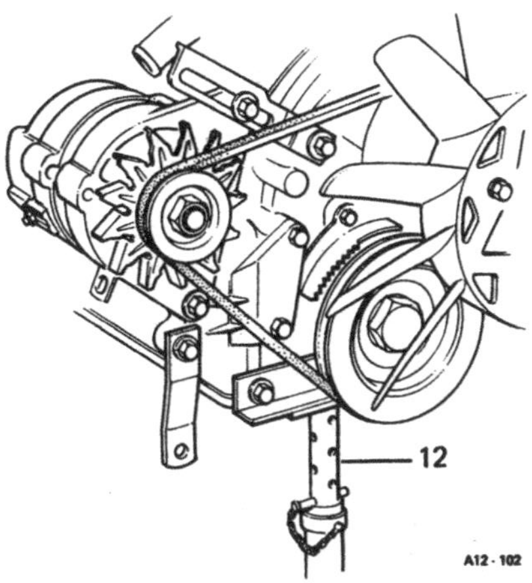

A12-102

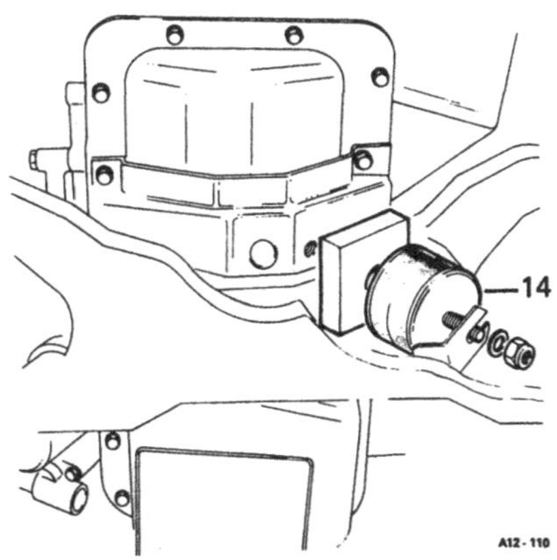

A12-110

Refitting

17 Using a new gasket, manoeuvre the sump into position on the crankcase and loosely secure in position with a few bolts.
18 Fit the cross-member with mounting to the sump and fit and tighten the remaining sump retaining nuts and bolts.
19 Lower the engine and secure the L.H. rubber mounting to the sub-frame.
20 Connect the R.H. side engine mounting.
21 Fit the engine stabilizer. 12.45.16. (Where fitted).
22 Remove the jack, and the support bracket and refit the timing cover bolts and alternator strap.
23 Fit the coupling plate bolts.
24 Lower the ramp.
25 Fit the fan guard.
26 Fit the fresh air duct.
27 Fill the sump with engine oil of recommended make and grade to the high mark on the dipstick.
28 Reconnect the battery, run the engine and check for leaks.

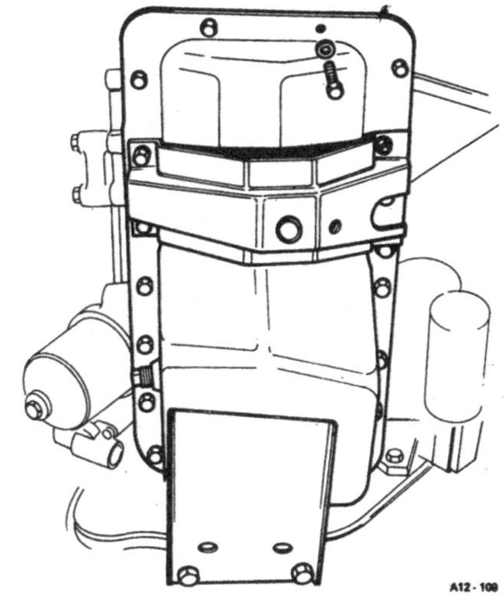

A12-109

OIL PRESSURE RELIEF VALVE

Remove and refit 12.60.56

Removing
1 Remove the oil pump. 12.60.26.
2 Withdraw the hexagonal drive shaft.
3 Remove the split pin from the oil pump casing.
4 Remove the locating plug, spring and plunger by tapping the pump cover.

NOTE: Since in some instances the 'O' ring on the locating plug may stick to the bore, in order to accomplish instruction 4 the pump cover will have to be removed and the plug tapped out from the underside of the cover. Follow instructions 5 and 6 if this is necessary.

5 Remove the two screws securing the pump cover to the main body and lift off the cover.
6 Using a suitable soft drip tap out of the plug from the underside of the pump cover.
7 Remove the 'O' ring from the locating plug.

Examination
8 Examine the plunger and its bore for scores or wear.
9 Check the plunger spring length and renew if not in accordance with data.

Refitting
10 Observing absolute cleanliness fit the plunger to the pump cover ensuring that it moves freely in its bore.
11 Fit the spring over the plunger spigot end noting that the close coiled end is inserted first.
12 Fit a new 'O' ring to the locating plug and insert the plug.
13 Fit a new split pin.
14 Fit the cover to the pump and secure with the two screws.
15 Fit the hexagonal shaft to the crankcase.
16 Refit the oil pump to the engine. 12.60.26.

DATA
Relief valve spring free length . . . . 1.70 in (43,18 mm)

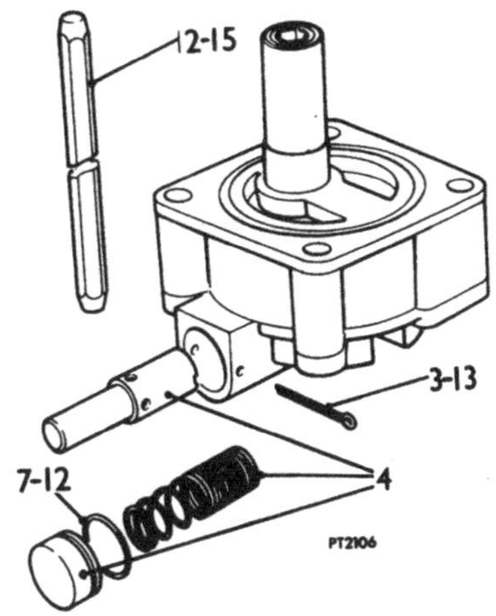

TIMING CHAIN COVER – UK & Europe Specification

Remove and refit 12.65.01

Removing
1 Remove the bonnet, drive the car on ramp and disconnect the battery.
2 Remove the crankshaft pulley. 12.21.02.
3 Remove the alternator. 86.10.02.
4 Remove the alternator mounting bracket. 12.25.21.
5 Remove the alternator adjusting link.
6 Remove the two front sump bolts and one each side.
7 Remove the fan and viscous coupling assembly. 26.25.21.
8 Remove the two bolts and nuts securing the timing cover to the cylinder head.
9 Remove the remaining timing cover retaining bolts, including the centre bolt.
10 Withdraw the timing cover and gaskets.
11 Place a piece of rag over the access to the sump to prevent items falling in by accident.

Refitting
12 Remove the protective rags.
13 Clean the engine and timing cover mating faces and place new gaskets in position on the timing cover.
14 Locate the timing cover on the engine over the dowels and fit and tighten the centre retaining bolt, noting that it has a fibre washer under the head.
15 Fit the fan and viscous coupling unit.
16 Fit, but leave loose, the cylinder head to timing cover nuts and bolts.
17 Fit, but leave slack, the alternator adjusting link bracket.
18 Fit the alternator mounting bracket.
19 Fit the alternator but leave slack.
20 Connect the alternator harness plug to the alternator.
21 Loosely place in position the alternator drive belt.
22 Finally tighten the cylinder head to timing cover nuts and bolts.
23 Raise the ramp and fit and tighten the sump bolts.
24 Fit the crankshaft pulley.
25 Adjust the alternator drive belt tension. 86.10.05.
26 Secure the radiator with the two top brackets and four nuts.
27 Fit the fan guard.
28 Fit the bonnet.
29 Reconnect the battery.

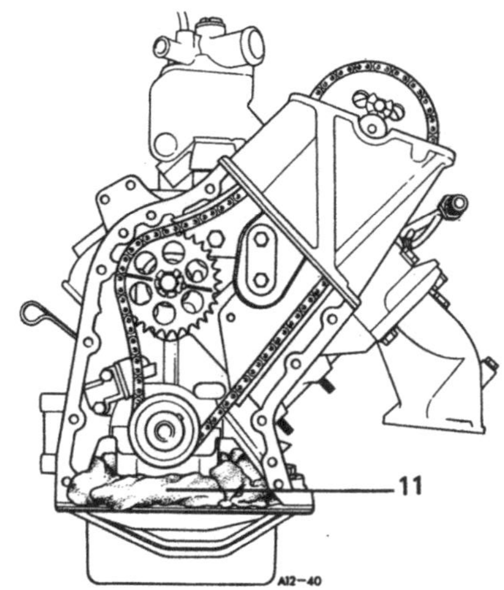

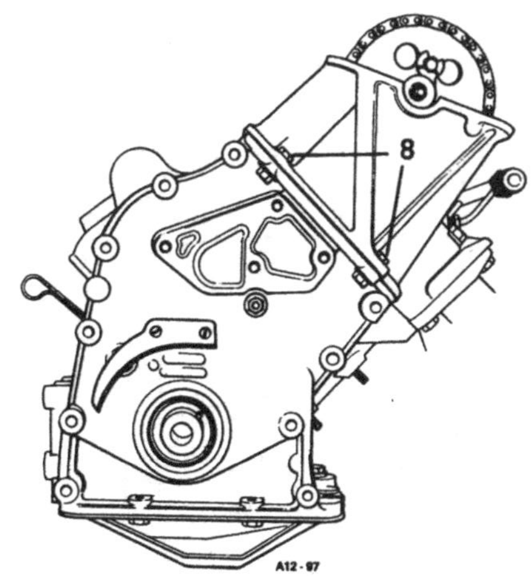

TIMING CHAIN COVER – USA & Canada Specification

Remove and refit 12.65.01

* *Applicable to carburetter engines only*

Removing

1 Remove the bonnet, drive car on ramp and disconnect the battery.
2 Remove the crankshaft pulleys. 12.21.02.
3 Remove the alternator. 86.10.02.
4 Remove the alternator mounting bracket. 12.25.21.
5 Remove the alternator adjusting link.
6 * Remove the air pump, see 17.25.07.
7 * Remove the 'diverter and relief valve' complete with attachment bracket.
8 * Remove the air pump bracket and lifting eye, see 17.25.07.
9 * Remove the air conditioning compressor steady bracket.
10 Remove the two bolts and nuts securing the timing cover to the cylinder head.
11 Slacken the four bolts securing the air conditioning compressor carrier bracket to the engine.

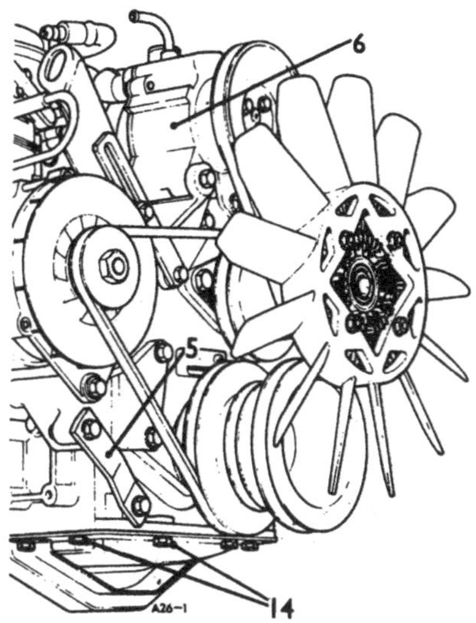

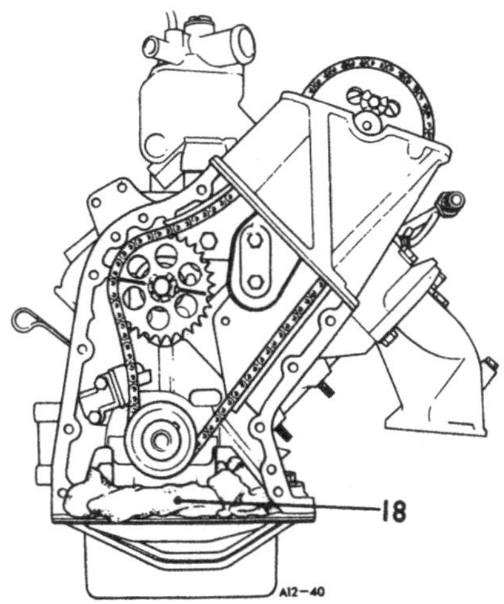

12 Remove the three air conditioning compressor adjusting bolts.
13 Remove the air conditioning compressor adjusting bracket.
14 Remove the two front sump bolts and one each side.
15 Remove the timing cover centre attachment bolt and bottom left hand bolt (looking at engine from front).
16 Remove the fan and Torquatrol assembly – instructions 6 to 9. 26.25.21.
17 Withdraw the timing cover and gaskets.
18 Place a piece of rag over the access to the sump to prevent items falling in by accident.

Refitting

19 Remove the protective rags.
20 Clean the engine and timing cover mating faces and place new gaskets in position on the timing cover.
21 Locate the timing cover on the engine over the dowels and fit and tighten the centre retaining bolt, noting that it has a fibre washer under the head.
22 Fit the fan and torquatrol assembly securing it to the timing cover with the two uncommitted bolts.
23 Fit the air conditioning compressor adjustment bracket securing it with the rear bolt but leaving it slack.
24 Fit, but leave loose, the cylinder head to timing cover nuts and bolts.
25 Fit the air conditioning compressor steady bracket and tighten the bolts.
26 Tighten the air conditioning compressor adjustment bracket rear bolt.
27 Fit the air conditioning compressor to the adjustment bracket but leave the bolts slack.
28* Fit the air pump bracket together with the lifting hook.
29* Fit the diverter and relief valve bracket complete with valve.
30 Fit, but leave slack, the alternator adjusting link bracket.
31 Fit the alternator mounting bracket.
32 Fit the alternator but leave slack.
33 Connect the alternator harness plug – instructions 16 and 17. 86.10.02.
34* Fit the air pump, but leave slack.

continued

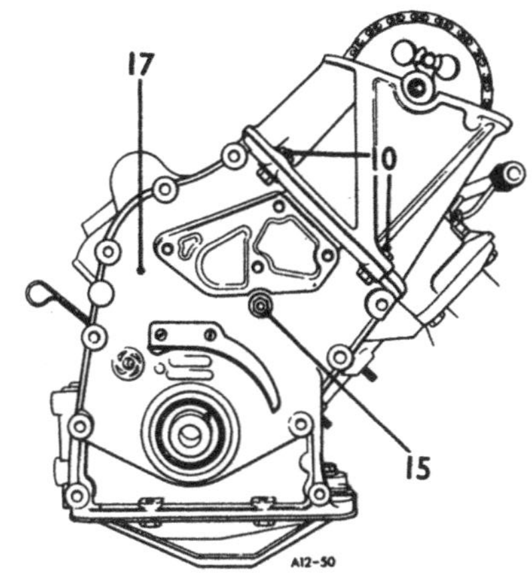

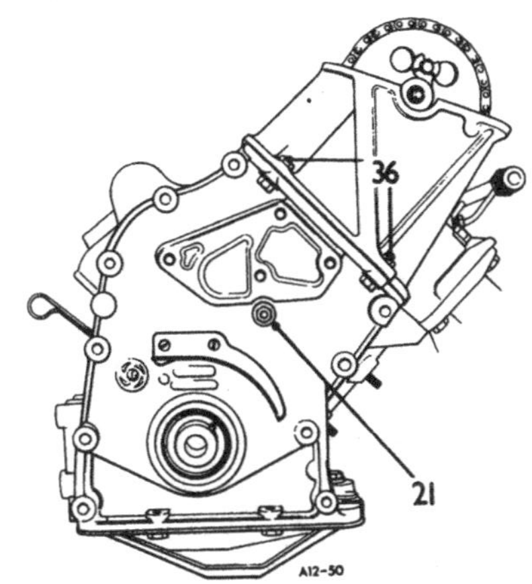

35 Loosely place in their correct relative positions the following:
  a The alternator drive belt.
  b* The air pump drive belt.
  c The air conditioning compressor drive belt.
36 Finally tighten the cylinder head to timing cover nuts and bolts.
37 Raise the ramp and fit and tighten the sump bolts.
38 Fit the crankshaft and auxiliary pulleys – instructions 9 to 11. 12.21.02.
39 Tighten the four air conditioning compressor to carrier engine bolts.
40 Fit air conditioning compressor drive belt to the pulleys and adjust tension. 82.10.01.
41 Adjust the alternator drive belt tension, see 86.10.05.
42* Adjust air pump drive belt tension, see 17.25.13.
43* Connect air pump hose to 'diverter and relief valve'.
44 Fit the fan blades to the Torquatrol unit.
45 Secure the radiator with the two top brackets and four nuts.
46 Fit the fan guard.
47 Fit the bonnet.
48 Reconnect the battery.

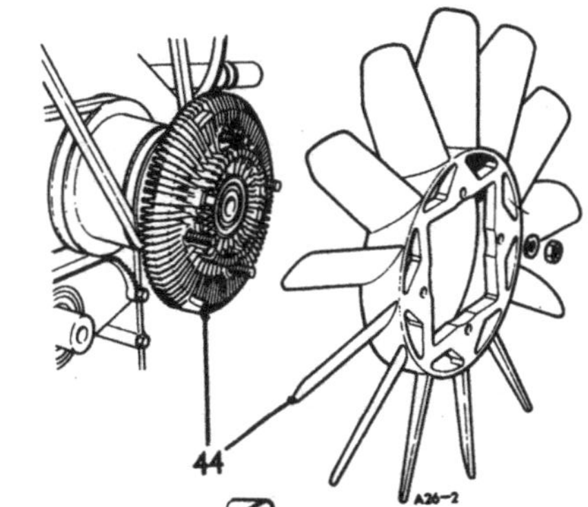

TIMING COVER OIL SEAL

Remove and refit 12.65.05

Removing
1 Disconnect the battery.
2 Remove the crankshaft pulleys. 12.21.02.
3 Lever out the old seal.

Refitting
4 Dip a new seal in engine oil and with the lip face leading tap it squarely into its housing until flush with the outside of the timing cover. Check by placing a straight edge across the face of the seal and timing cover.
5 Refit the crankshaft pulleys. 12.21.07.
6 Reconnect the battery.

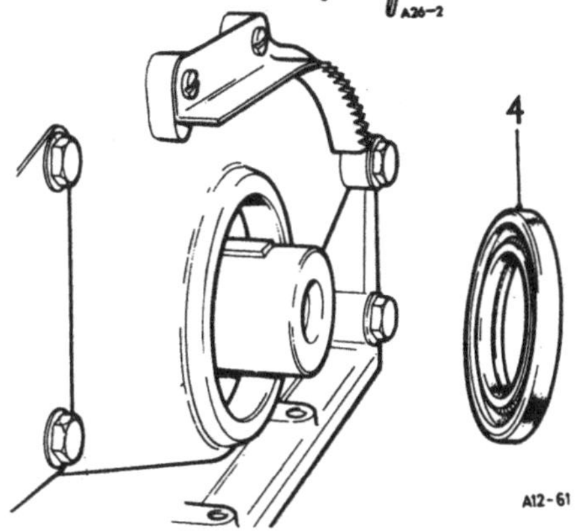

USA Specification

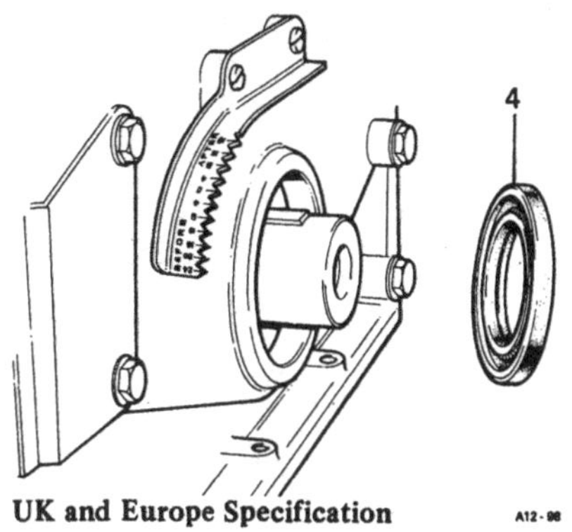

UK and Europe Specification

VALVE TIMING – UK and Europe Specification

Check and adjust 12.65.08

Checking
1 Disconnect the battery.
2 Turn the engine over until the timing mark on the crankshaft pulley coincides with the zero mark on the timing cover scale.
3 Remove the fresh air duct. 80.15.31.
4 Remove the distributor cap and check that the rotor arm points to the rear bolt securing the inlet manifold to the cylinder head.
5 Remove the camshaft cover. 12.29.42.
6 The valve timing is correct when the timing mark on the camshaft flange is in line with the corresponding groove in the camshaft front bearing cap.
7 Refit the distributor cap.
8 Refit the camshaft cover. 12.29.42.
9 Refit the fresh air duct. 80.15.31.
10 Reconnect the battery.

continued

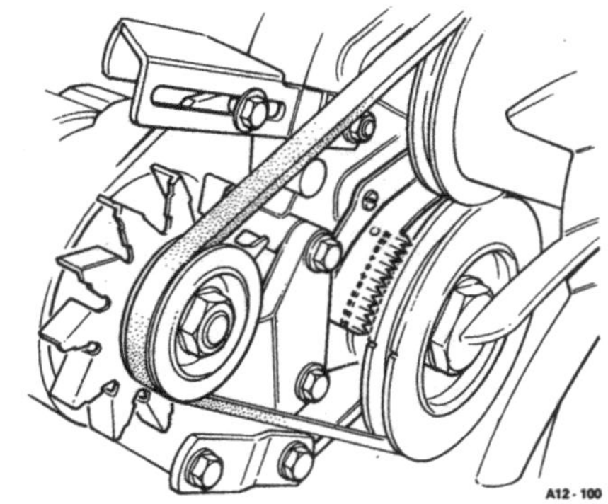

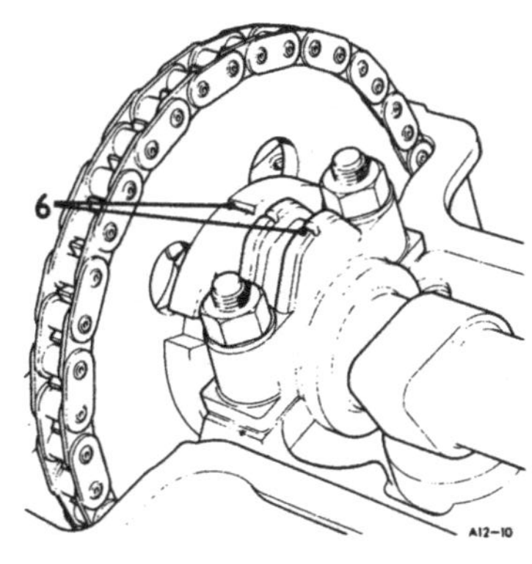

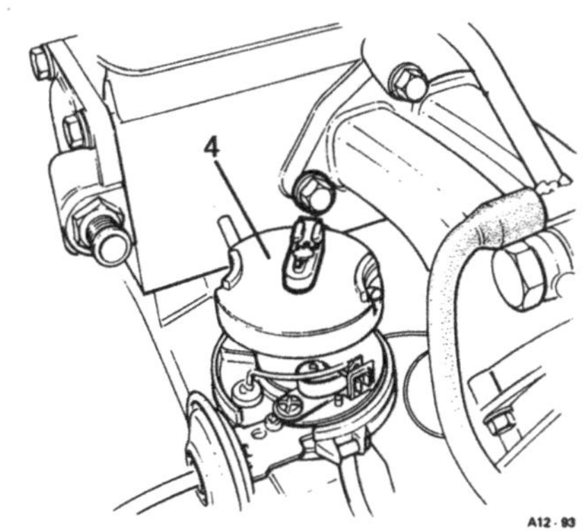

Adjusting
11 Disconnect the battery.
12 Remove the fresh air duct. 80.15.31.
13 Remove the timing chain cover. 12.65.01.
14 Remove the distributor cap.
15 Remove the camshaft cover.
16 Turn the engine over so that the timing mark on the camshaft flange is in line with the groove in the camshaft front bearing cap, i.e. number 1 cylinder T.D.C. firing – instruction 6.
17 Check that the distributor rotor arm points to the rear bolt securing the inlet manifold – instruction 4.
18 Remove the hydraulic timing chain tensioner. 12.65.28 instructions 2 to 4.
19 Fit a 'slave' bolt to the timing cover centre hole in the cylinder block to ensure alignment when refitting the cover.
20 Remove the two bolts securing the adjustable chain guide.
21 Slacken the remaining guide bolt.
22 Remove the adjustable guide.
23 Slip the timing chain from the jackshaft and crankshaft sprockets.
24 Temporarily fit the timing chain cover locating it on the two dowels.
25 Temporarily fit the crankshaft pulley.
26 Turn the crankshaft until the timing mark on the crankshaft pulley coincides with the zero figure on the timing cover scale – instruction 2.
27 Remove the timing cover and pulley.
28 Fit the timing chain to the sprockets keeping it taut on the drive side, i.e. the run between the camshaft and the crankshaft sprockets.
29 Since the jackshaft sprocket may have been moved, whilst disconnecting the chain, re-set to its correct position by turning the sprocket until the scribed line is equidistant between bolts A and B and the distributor rotor arm points to the rear bolt head securing the inlet manifold to the cylinder head.
30 Fit the adjustable guide locating it loosely with the two bolts – noting that the locking bolt has a plain and spring washer.
31 Fit the hydraulic timing chain tensioner. 12.65.28 instructions 12 to 20.
32 Remove the 'slave' bolt.
33 Refit the timing chain cover. 12.65.01.

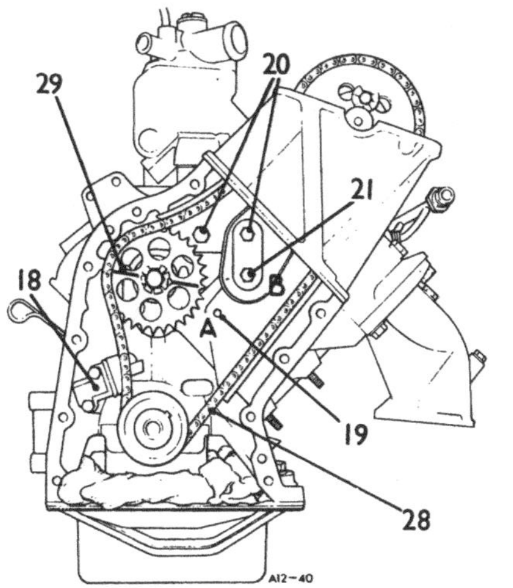

DATA

Valve timing

Inlet valves open	16° B.T.D.C.
Inlet valves close	56° A.B.D.C.
Exhaust valves open	56° B.B.D.C.
Exhaust valves close	16° A.T.D.C.

VALVE TIMING – USA & Canada Specification

Check and adjust 12.65.08

Checking
1 Disconnect the battery.
2 Turn the engine over until the timing mark on the crankshaft pulley coincides with the zero mark on the timing cover scale.
3 Remove the fresh air duct. 80.15.31.
4 Remove the distributor cap and check that the rotor arm points to the rear bolt securing the inlet manifold to the cylinder head.
5 Remove the camshaft cover. 12.29.42.
6 The valve timing is correct when the timing mark on the camshaft flange is in line with the corresponding groove in the camshaft front bearing cap.
7 Refit the distributor cap.
8 Refit the camshaft cover. 12.29.42.
9 Refit the fresh air duct. 80.15.31.
10 Reconnect the battery.

Adjusting
11 Disconnect the battery.
12 Remove the fresh air duct. 80.15.31.
13 Remove the timing chain cover. 12.65.01.
14 Remove the distributor cap.
15 Remove the camshaft cover.
16 Turn the engine over so that the timing mark on the camshaft flange is in line with the groove in the camshaft front bearing cap, i.e. number 1 cylinder T.D.C. firing – instruction 6.
17 Check that the distributor rotor arm points to the rear bolt securing the inlet manifold – instruction 4.
18 Remove the hydraulic timing chain tensioner. 12.65.28 instructions 2 to 4.
19 Fit a 'slave' bolt to the timing cover centre hole in the cylinder block to ensure alignment when refitting the cover.
20 Remove the two bolts securing the adjustable chain guide.
21 Slacken the remaining guide bolt.
22 Remove the adjustable guide.
23 Slip the timing chain from the jackshaft and crankshaft sprockets.

continued

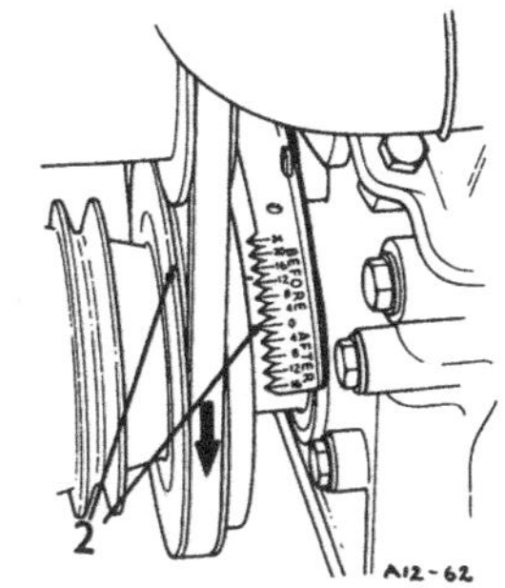

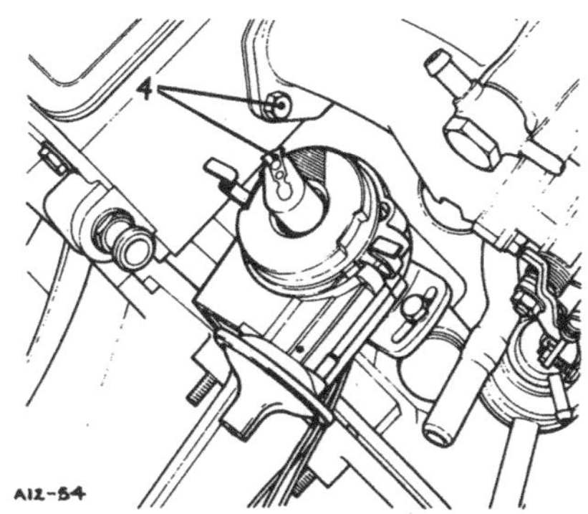

USA & Canada Specification

24 Temporarily fit the timing chain cover locating it on the two dowels.
25 Temporarily fit the crankshaft pulley.
26 Turn the crankshaft until the timing mark on the crankshaft pulley coincides with the zero figure on the timing cover scale – instruction 2.
27 Remove the timing cover and pulley.
28 Fit the timing chain to the sprockets keeping it taut on the drive side, i.e. the run between the camshaft and crankshaft sprockets.
29 Since the jackshaft sprocket may have been moved, whilst disconnecting the chain, re-set to its correct position by turning the sprocket until the scribed line is equidistant between the bolts A and B and the distributor rotor arm points to the rear bolt head securing the inlet manifold to the cylinder head.
30 Fit the adjustable guide locating it loosely with the two bolts – noting that the locking bolt has a plain and spring washer.
31 Fit the hydraulic timing chain tensioner 12.65.28 instructions 12 to 20.
32 Remove the 'slave' bolt.
33 Refit the timing chain cover 12.65.01 instructions 19 to 48.

DATA
Valve timing

Inlet valves open . . .	16° B.T.D.C.
Inlet valves close . . .	56° A.B.D.C.
Exhaust valves open . .	56° B.B.D.C.
Exhaust valves close . .	16° A.T.D.C.

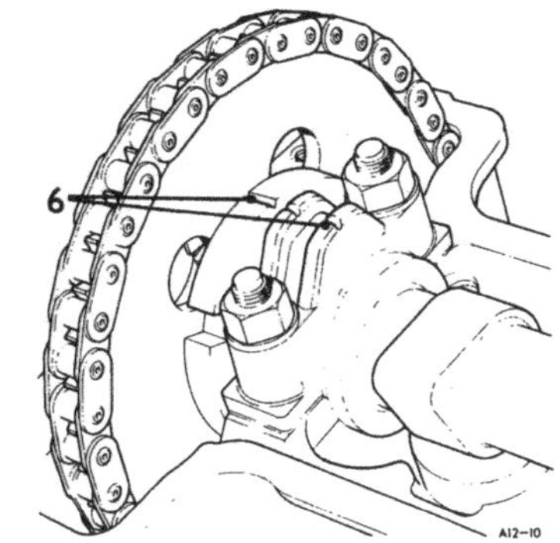

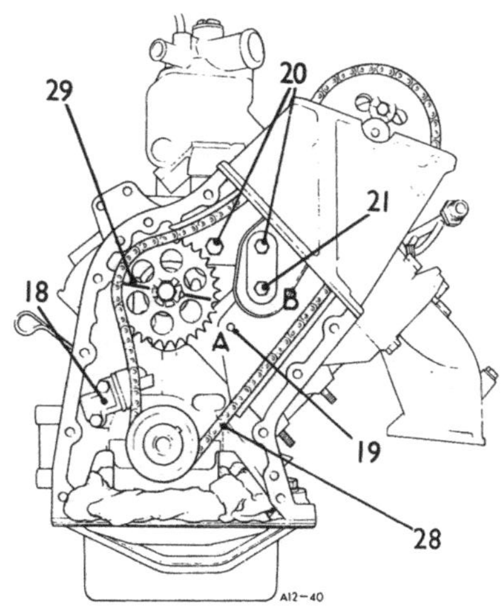

TIMING CHAIN AND SPROCKETS

– Remove and refit 12.65.12

Removing
1 Remove the timing chain. 12.65.14.
2 Unlock and remove the jackshaft retaining bolt.
3 Withdraw the jackshaft sprocket.
4 Remove the oil thrower.
5 Remove the crankshaft sprocket.
6 Remove the crankshaft key.
7 Remove the sprocket alignment shims.
8 Detach the camshaft sprocket from the support bracket.

[**NOTE:** To prevent abrasive dirt from entering the engine, thoroughly clean the camshaft cover and the timing cover before you remove them. Lubricate the chain during installation.]

Refitting
9 Using a new locking washer, fit the jackshaft sprocket, ensuring that it locates correctly over the dowel.
10 Fit and tighten the retaining bolt to the correct torque figure, see section 06, and lock it.
11 Temporarily fit the crankshaft sprocket and check its alignment with the jackshaft sprocket by using a straight edge across the two sprockets as illustrated.
12 Adjust any misalignment by the addition of shims behind the crankshaft sprocket.
13 Remove the crankshaft sprocket, fit the crankshaft key and refit the sprocket.
14 Fit the oil thrower noting that the dished side faces outwards away from the engine.
15 Attach the camshaft sprocket to the support bracket with a 'slave' nut.
16 Refit the timing chain. 12.65.14.

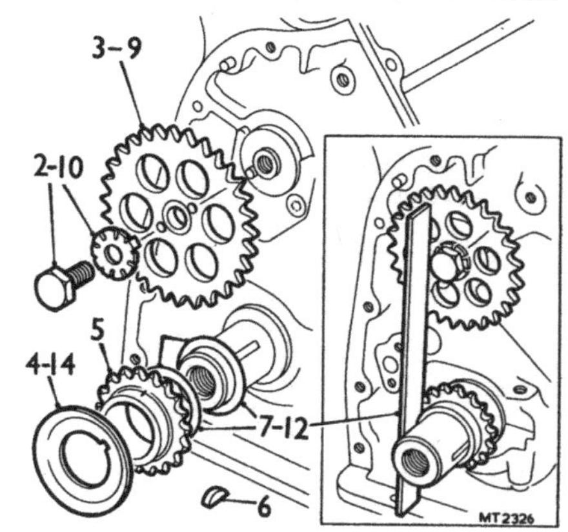

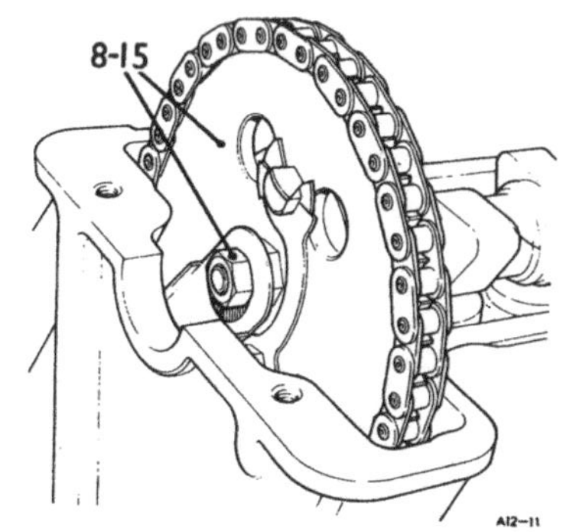

TIMING CHAIN

– Remove and refit **12.65.14**

Removing

1 Remove the timing chain cover. 12.65.01.
[**NOTE:** To prevent abrasive dirt from entering the engine, thoroughly clean the camshaft cover and the timing chain cover before you remove them. Lubricate the chain during installation.]
2 Remove the camshaft cover.
3 Remove the distributor cap.
4 Turn the engine over so that the timing mark on the camshaft is at the bottom i.e. 180° distant from the groove in the camshaft front bearing cap, to enable instruction 5 to be carried out.
5 Unlock and remove the exposed camshaft retaining bolt.
6 Turn the engine over until the mark on the camshaft flange is in line with the groove on the camshaft front bearing cap.
7 Check that the distributor rotor arm points to the head of the rear bolt securing the inlet manifold i.e. T.D.C. number 1 cylinder firing.

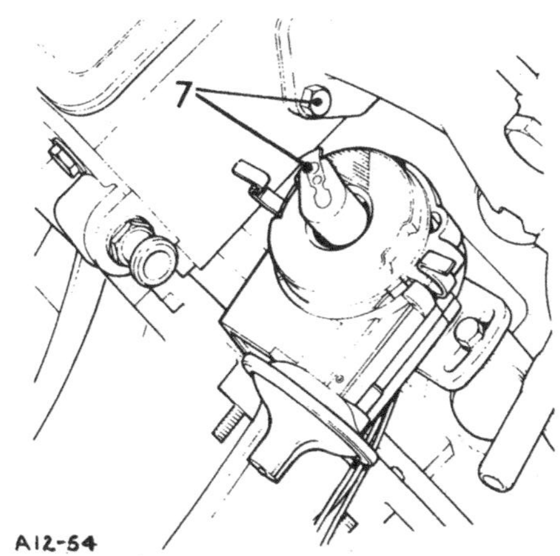

USA SPECIFICATION

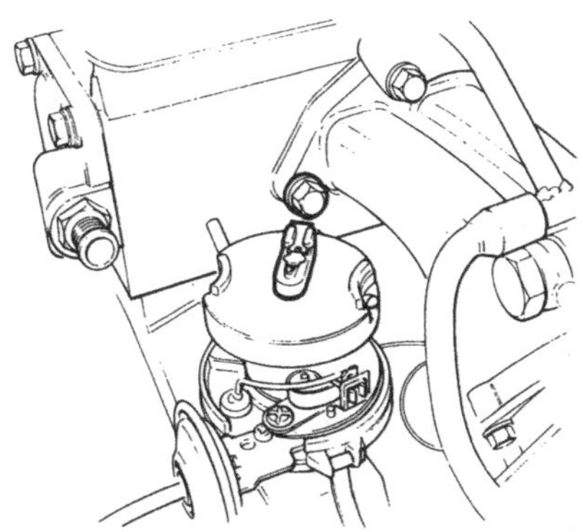

UK & EUROPE SPECIFICATION

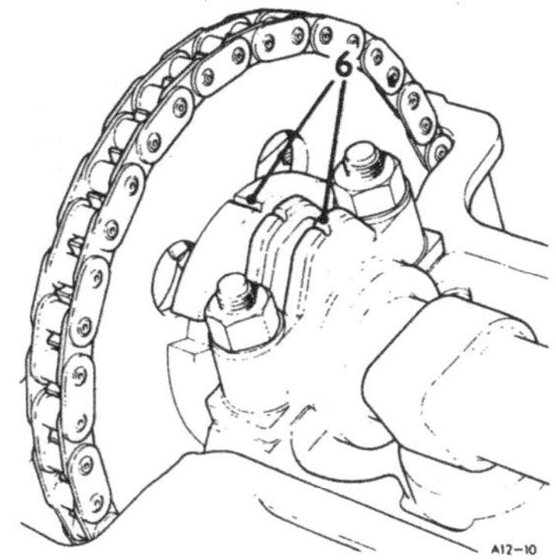

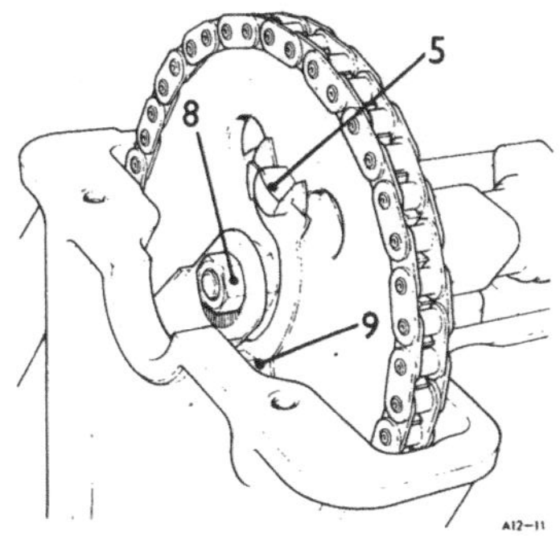

8 Secure the camshaft sprocket to the support bracket with a 'slave' nut.
9 Unlock and remove the remaining bolt retaining the camshaft sprocket.
10 Remove the hydraulic timing chain tensioner and guide plate.
11 Remove the locking bolt from the adjustable chain guide.
12 Remove the common bolt securing the adjustable guide and camshaft sprocket support bracket.
13 Remove the adjustable guide.
14 Remove the bolt securing the camshaft support bracket and fixed guide whilst holding the camshaft sprocket.
15 Remove the fixed guide and release the chain from the jackshaft and camshaft sprockets and withdraw upwards the camshaft sprocket and bracket together with the timing chain.

Refitting

16 Encircle the camshaft sprocket with the timing chain and insert it through the aperture with the support bracket and pass the chain over the jackshaft and crankshaft sprockets.
17 Fit the fixed guide and support bracket and loosely secure with the common bolt and spring washer.
18 Fit the adjustable guide and loosely secure with the two bolts noting that the locking bolt has a plain and spring washer.
19 Fit a slave bolt to the timing cover centre retaining bolt hole.
20 Secure the camshaft sprocket to the camshaft with one bolt and a new tab washer.
21 Turn the jackshaft sprocket so that the scribed line across it is equidistant between the 'slave' bolt A and bolt 'B'.
22 Check that the distributor arm is in the position as described in instruction 7.

NOTE: If it is suspected that the crankshaft may have been turned, temporarily fit the timing cover and crankshaft pulley and check that the timing mark on the pulley coincides with the zero figure on the timing cover scale.

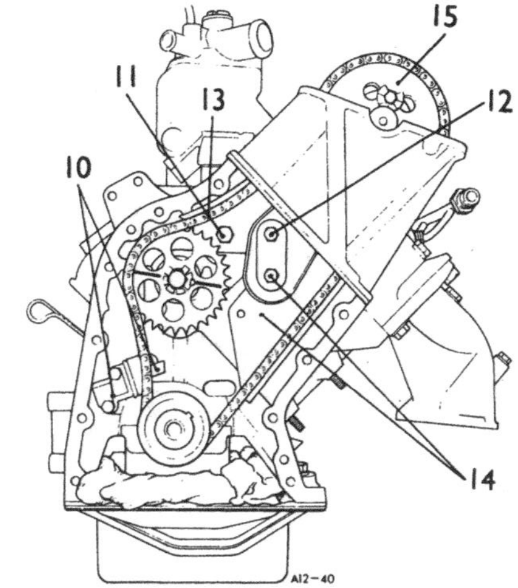

23 Fit the hydraulic timing chain tensioner 12.65.28 reversing instructions 2 to 4 ensuring that the three retaining bolts for the chain guides are tightened and that the position of the jackshaft sprocket has not altered.

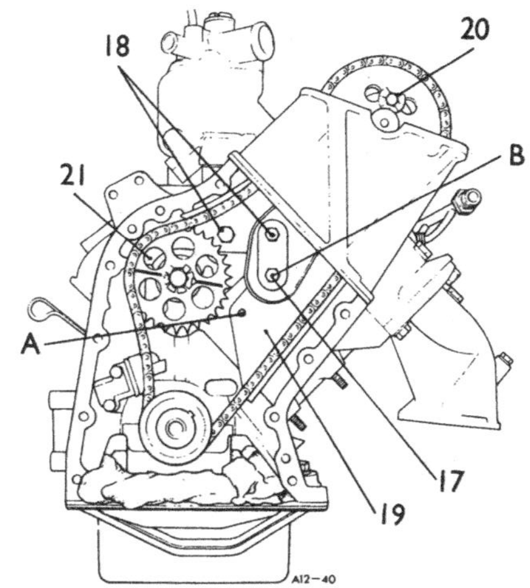

continued

24 Remove the 'slave' nut supporting the camshaft sprocket and ensure that the threaded spigot does not foul in the support bracket hole.
25 Turn the engine over sufficiently to enable the remaining camshaft sprocket retaining bolt to be fitted.
26 Tighten and lock the bolt and turn the engine back again and finally tighten and lock the first bolt.
27 Fit the camshaft cover.
28 Fit the distributor cap.
29 Fit the timing chain cover. 19.65.01.

TIMING CHAIN SPROCKETS

– Remove and refit 12.65.22

This operation is covered under operation 12.65.12.

TIMING CHAIN TENSIONER

– Remove and refit 12.65.28

Removing
1 Remove the timing chain cover. 12.65.01.
2 Remove the tensioner retaining bolts.
3 Remove the tensioner from the engine.
4 Remove the backplate between the tensioner body and cylinder block.

Dismantling
5 Press-in the slipper and remove it from the tensioner body.
6 Remove the ratchet and spring.

Examination
7 Check the slipper pad for wear and renew if necessary.

[**NOTE:** It is normal for the slipper pad to become grooved, due to contact with the chain. However, you should replace the pad if there is wear other than grooves or when the grooves have become so deep that the chain rollers are striking the pad. Under no circumstances should you dress the surface of the pad to restore a smooth surface, as doing this will accelerate pad wear.]

8 Renew the ratchet if worn.
9 Check the spring and renew if it is not in accordance with data.
10 Examine the tensioner body and check that the oil inlet hole is clear.
11 Ensure that the oil outlet hole is clear.

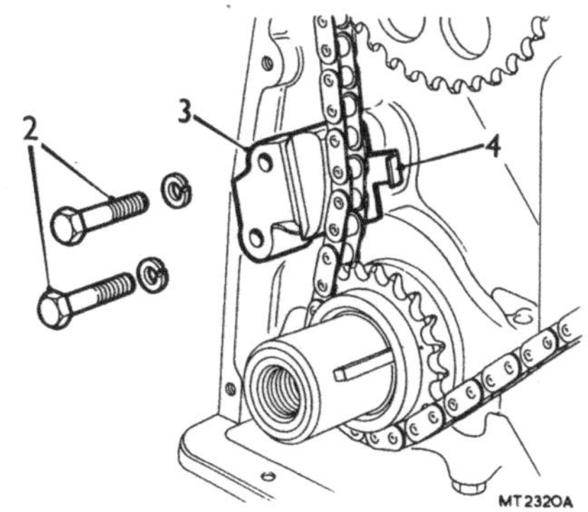

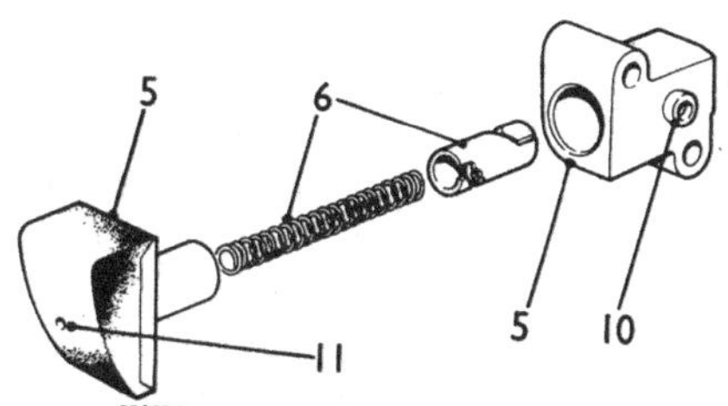

Reassembling
12 Insert the slipper into the bore.
13 Fit the ratchet into the bore and turn clockwise with an Allen key to lock it in a retracted position.

NOTE: To prevent the tensioner releasing while fitting, insert a spacer between the tensioner body and the back of the slipper.

Refitting
14 Fit the tensioner assembly and back plate to the cylinder block taking care not to allow the tensioner to release. Secure with the two bolts and spring washers. Remove the spacer.
15 Insert a 0.100 in (2.54 mm) feeler or slip gauge between the slipper and tensioner body – dimension A.
16 Slacken the three chain guide retaining bolts.
17 Press down, in the direction of the arrow, on top of the adjustable chain guide until the feeler or slip gauge is a sliding fit.
18 Whilst holding the guide in this position tighten the adjustable guide bolt first, and then the two remaining bolts.
19 Remove the feeler or slip gauge.
20 Remove the camshaft cover and check that the camshaft spigot is positioned centrally in the support bracket and does not foul when the camshaft is revolved.
21 Refit the timing chain cover. 12.65.01.

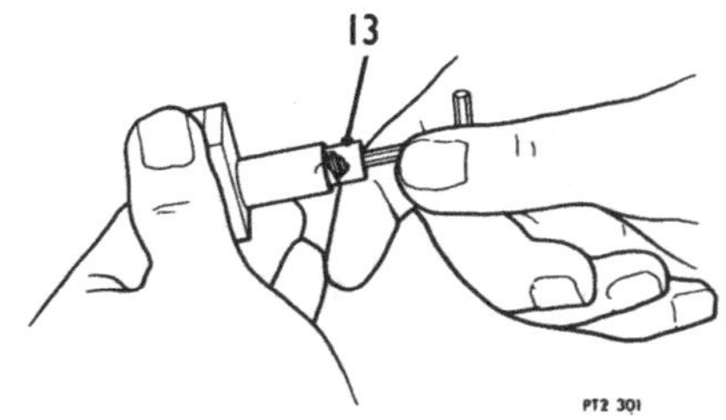

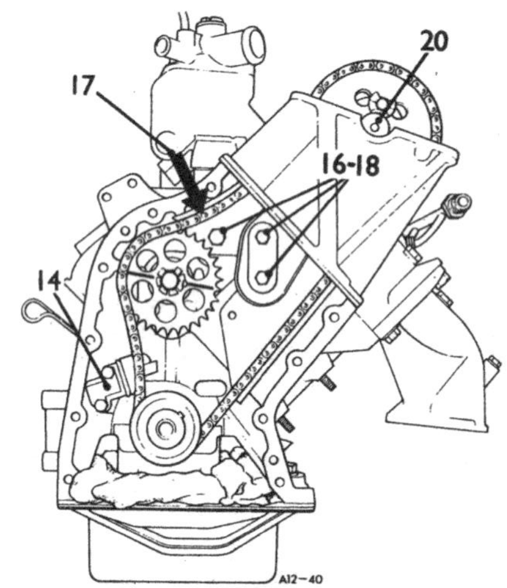

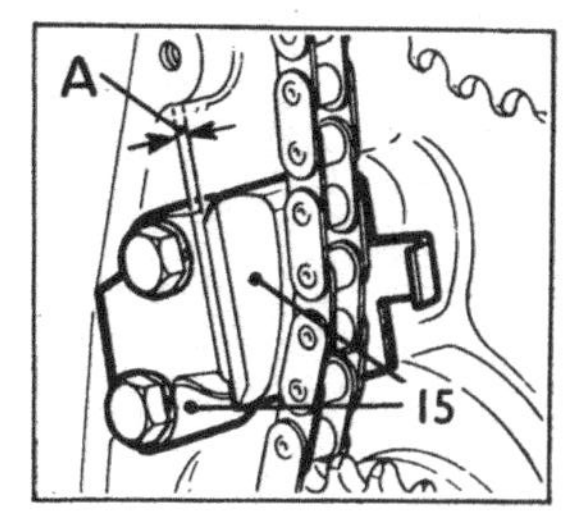

DATA

Tensioner spring free length	2.750 in (69,8 mm)

TIMING CHAIN GUIDES

Remove and refit **12.65.50**

Removing

1 Remove the timing chain cover. 12.65.01.
2 Remove the fresh air duct. 80.15.31.
3 Remove the camshaft cover.
4 Remove the distributor cap.
5 Turn the engine over until the following conditions are achieved:
  a The timing mark on the camshaft flange is in line with the groove in the camshaft front bearing cap.
  b The scribed line across the jackshaft sprocket is equal distance between bolt A and hole B.
  c The distributor rotor arm points to the head of the rear bolt securing the inlet manifold to the cylinder head.

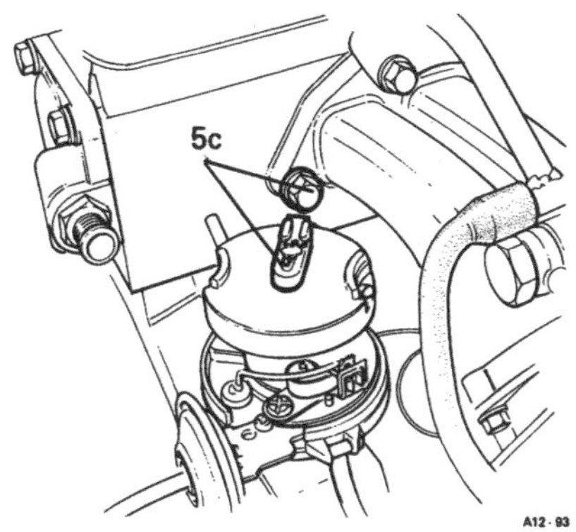

UK & Europe Specification

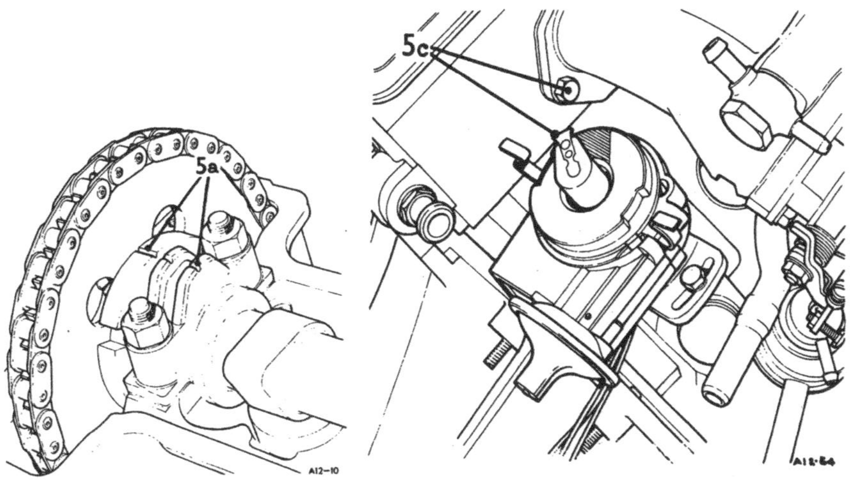

USA & Canada Specification

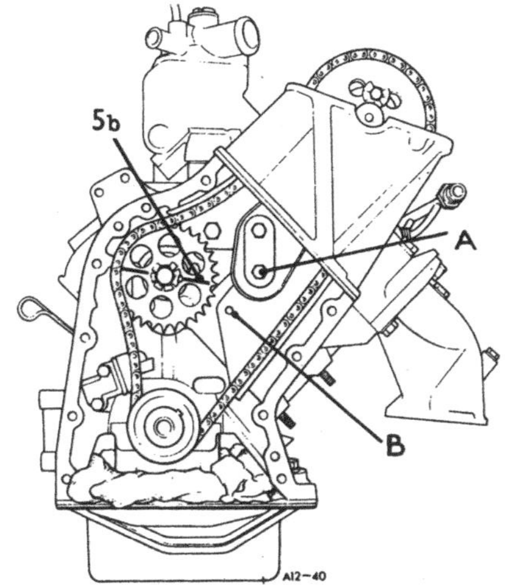

6 Secure the camshaft sprocket to the support bracket with a 'slave' nut.
7 Insert a 0.100 slip gauge between the body of the hydraulic tensioner and the back of the slipper to prevent the tensioner releasing when the chain pressure is released, and remove the tensioner.
8 Remove the three chain guide retaining bolts and withdraw:
  a The adjustable guide.
  b The fixed guide.

Refitting

9 Fit the fixed guide and support bracket securing it loosely with the common bolt.
10 Fit a 'slave' bolt to the timing cover centre retaining bolt to facilitate refitting the cover.
11 Fit the adjustable guide retaining it loosely with the two bolts, noting that the locking bolt has a plain and spring washer.
12 Fit the timing chain tensioner.
13 Insert a 0.100 in feeler gauge between the body of the tensioner and the back of the slipper.

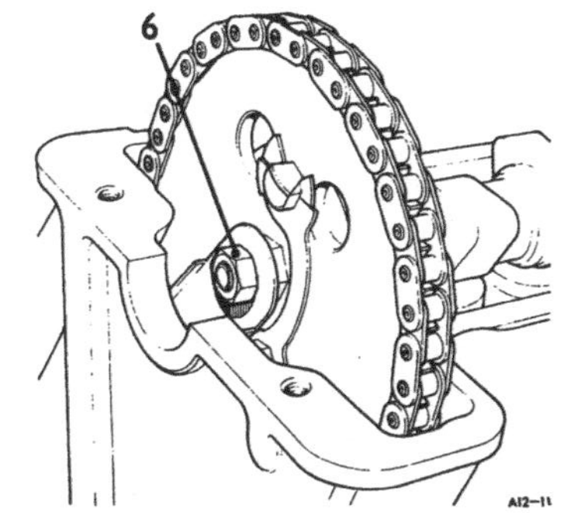

14 Press downwards on the adjustable guide whilst tightening the locking bolt.
15 Tighten the two remaining bolts.
16 Remove the 'slave' bolt.
17 Remove the 'slave' nut and ensure that the threaded spigot does not foul in the support bracket hole.
18 Check that the position of the jackshaft sprocket has not altered as described in instruction 5b.
19 Refit the timing chain cover – 12.65.01 – instructions 19 to 48.
20 Refit the camshaft cover.
21 Refit the distributor cap.

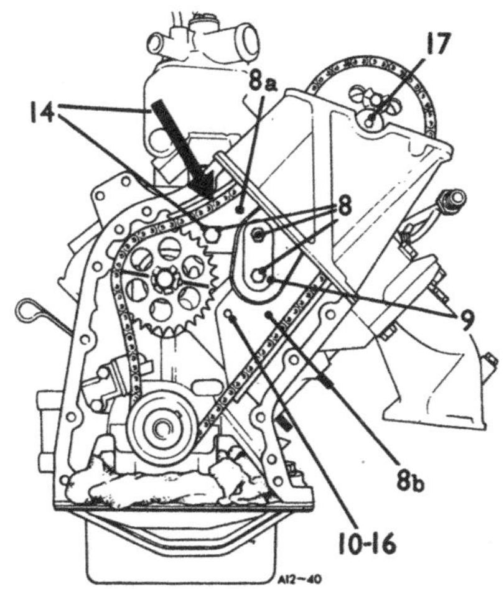

EMISSION AND EVAPORATIVE CONTROL OPERATIONS

[**NOTE:** The instructions given in this section of the Manual are mainly for removing and installing emission control components. Please see Section 10, MAINTENANCE, for emission control routine maintenance operations.]

CRANKCASE EMISSION CONTROL

Description **17.00.00**

Crankcase breathing and evacuation of 'blow by' gases is achieved by utilizing the characteristic partial vacuum in the constant depression carburetter. By this method crankcase emissions are burned in the engine combustion process. A wire gauze strainer in the engine top cover acts as an oil separator/flame trap.

On fuel injection engines a depression is maintained in the crankcase under all operating conditions by connecting the crankcase breathing housing to a point between the air meter flap and the throttle plate (i.e. a constant depression region).

[**TESTING:** Ensure restrictors are clear. Check hoses for signs of blockage, security, and deterioration.]

CRANKCASE BREATHING AND CANISTER PURGE SYSTEMS – USA MARKET 1975 AND 1976 MODEL YEARS AND LATER CANADIAN MARKET SPECIFICATIONS

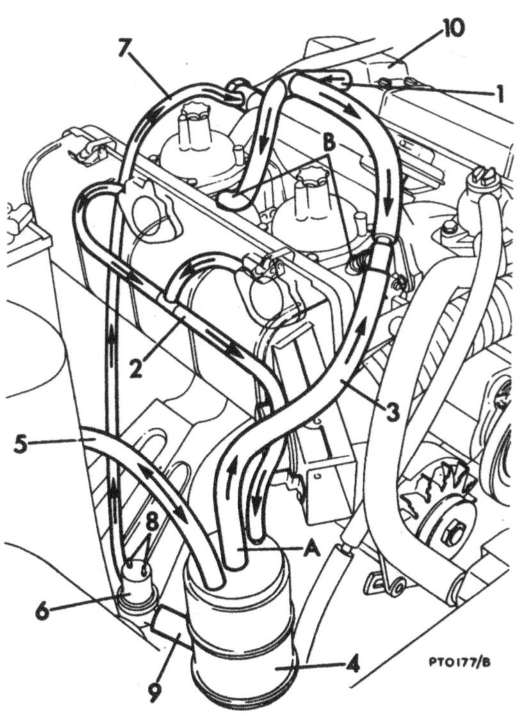

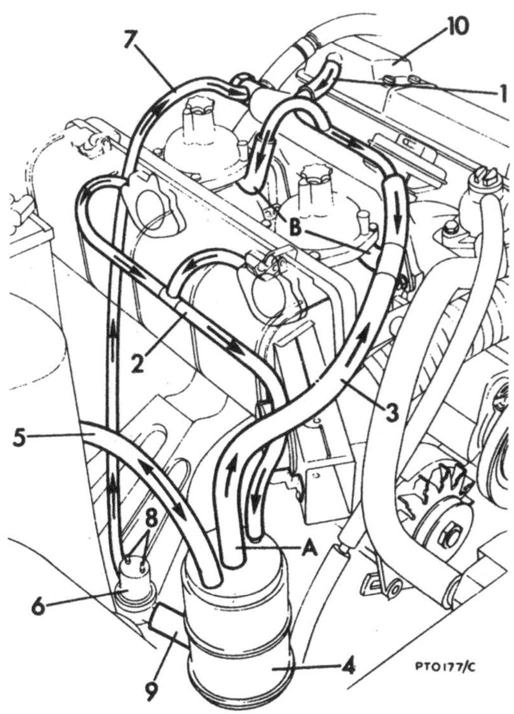

USA Federal and Canadian Market Specifications

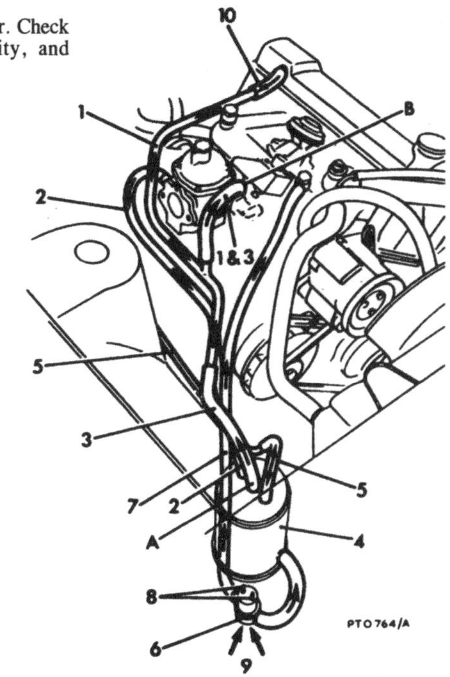

California Market Specification

1 Crankcase purge line
2 Carburetter float chamber vent pipe
3 Canister purge line
4 Adsorption canister
5 Fuel tank vent pipe
6 Anti-run-on valve
7 Manifold vacuum line
8 Electrical connections for anti-run-on valve
9 Purge air to canister
10 Flame arrestor
A. $\frac{3}{32}$ in restrictor
B. $\frac{5}{16}$ in restrictor

CRANKCASE BREATHING AND CANISTER PURGE SYSTEMS – USA MARKET 1977 MODEL YEAR

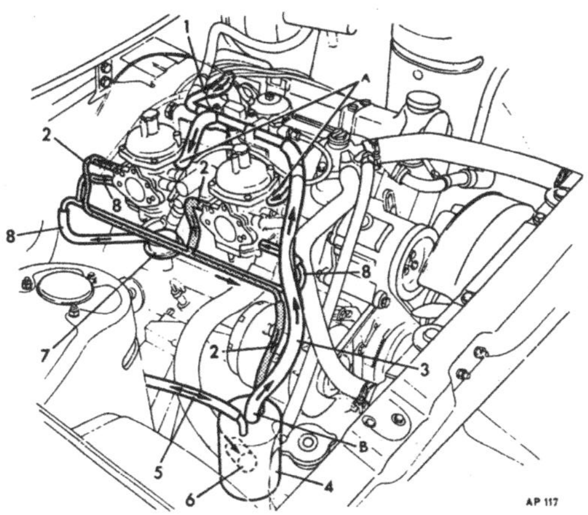

Federal Market Specification

1 Crankcase purge line
2 Carburetter float chamber vent pipe
3 Canister purge pipe
4 Adsorption canister
5 Fuel tank vent pipe
6 Purge air to canister
7 Fuel pump
8 Fuel pipe
9 Anti-run-on valve
10 Electrical connections for anti-run-on valve
11 Manifold vacuum line
   A. 0.280 to 0.300 in restrictors
   B. 3/32 in restrictor

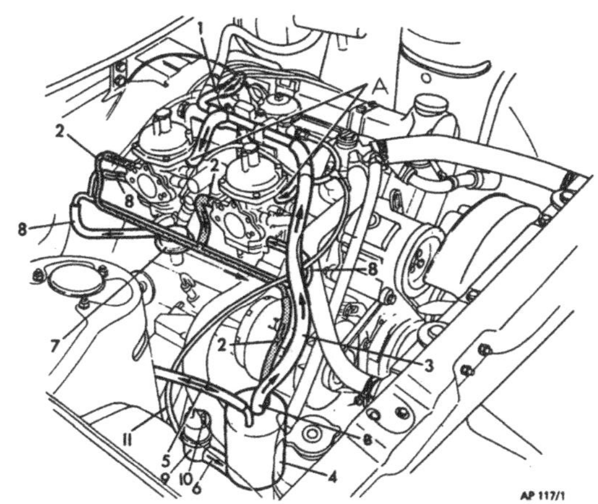

California Market Specification

CRANKCASE BREATHING AND CANISTER PURGE SYSTEMS – USA MARKET 1978, 1979 AND 1980 MODEL YEARS

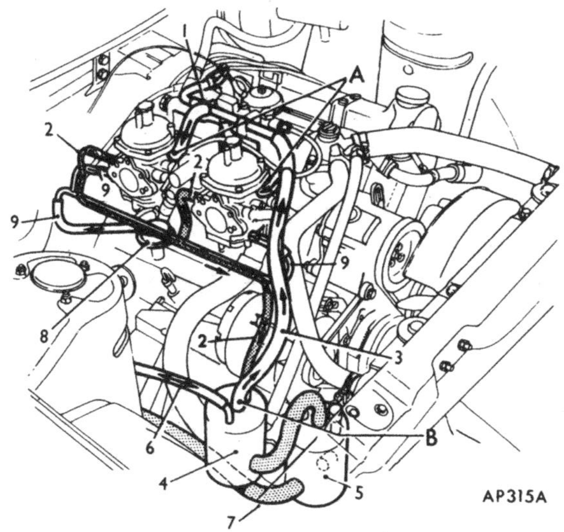

AP315A

Federal Market Specification—1978 and 1979

1 Crankcase purge line
2 Carburetter float chamber vent pipe
3 Canister purge pipe
4 Primary canister
5 Secondary canister
6 Fuel tank vent pipe
7 Purge air to canister
8 Fuel pump
9 Fuel pipe
10 Anti-run-on valve
11 Electrical connections for anti-run-on valve
12 Manifold vacuum line
A. 0,280 to 0.300 in restrictors
B. $\frac{3}{32}$ in restrictor

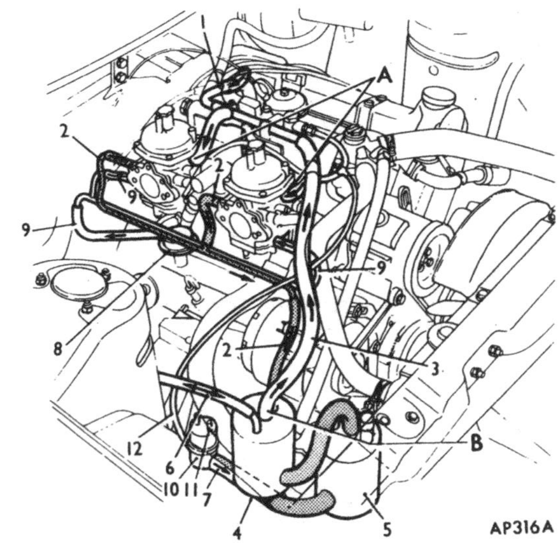

AP316A

California Market Specification—1978 and 1979

Federal Market Specification—1980

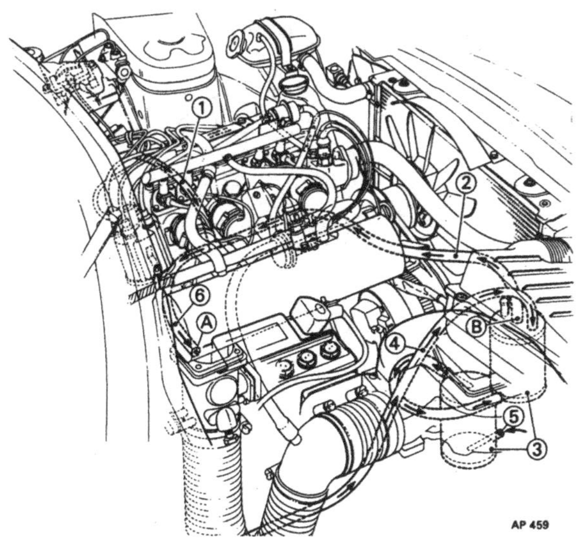

AP 459

California Market Specification — 1980

1. Crankcase purge line
2. Canister purge line
3. Charcoal canisters
4. Fuel tank vent pipe
5. Purge air to canister
6. Crankcase and canister purge line
   A 0.433 inch restrictor
   B $\frac{3}{32}$ inch restrictor

EVAPORATIVE LOSS CONTROL SYSTEM
USA MARKETS UP TO 1978 MODEL YEAR

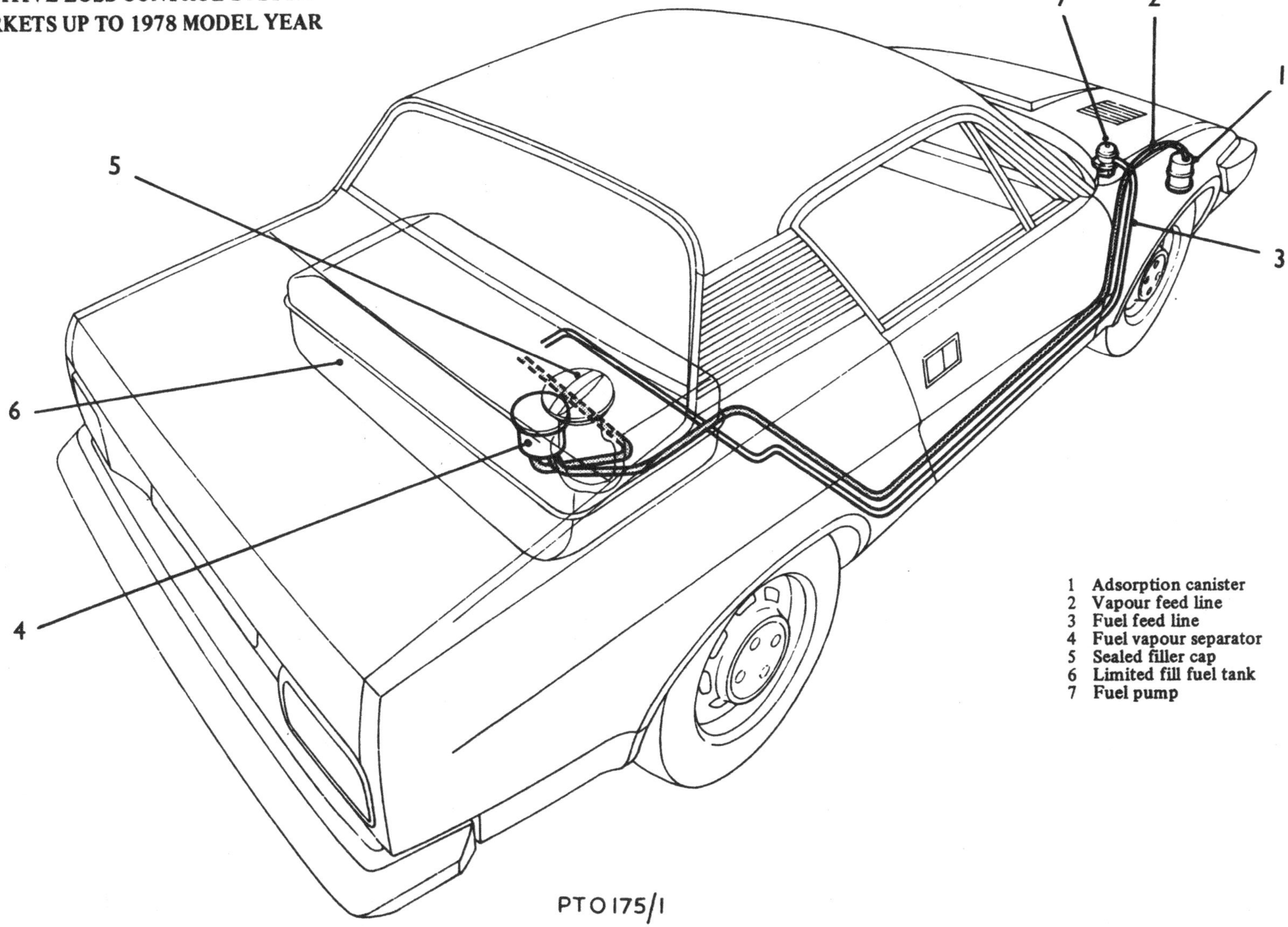

1 Adsorption canister
2 Vapour feed line
3 Fuel feed line
4 Fuel vapour separator
5 Sealed filler cap
6 Limited fill fuel tank
7 Fuel pump

EVAPORATIVE LOSS CONTROL SYSTEM
USA MARKETS — 1978, 1979 AND 1980 MODEL YEAR
(EXCLUDING CALIFORNIA MARKET 1980 MODEL YEAR)

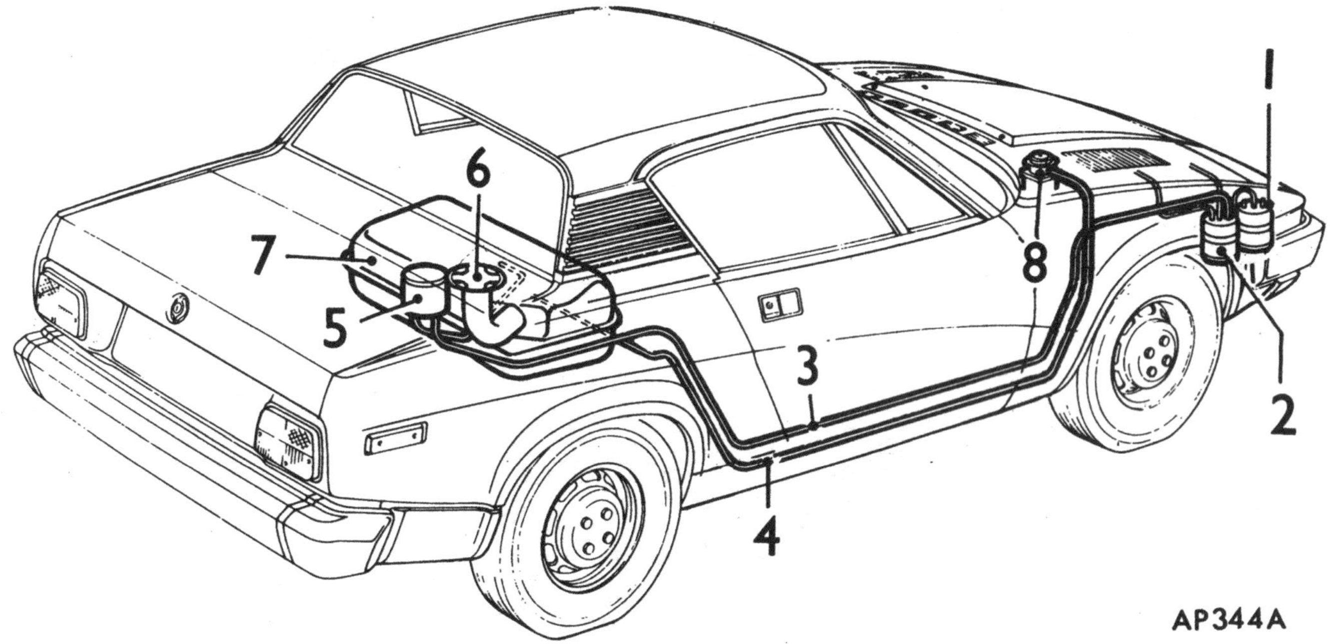

AP344A

1 Primary canister
2 Secondary canister
3 Vapour feed line
4 Fuel feed line
5 Fuel vapour separator
6 Sealed filler cap
7 Limited fill fuel tank
8 Fuel pump

EVAPORATIVE LOSS CONTROL SYSTEM
USA FEDERAL MARKETS—1981 MODEL YEAR
USA CALIFORNIA MARKET—1980 AND LATER MODEL YEARS

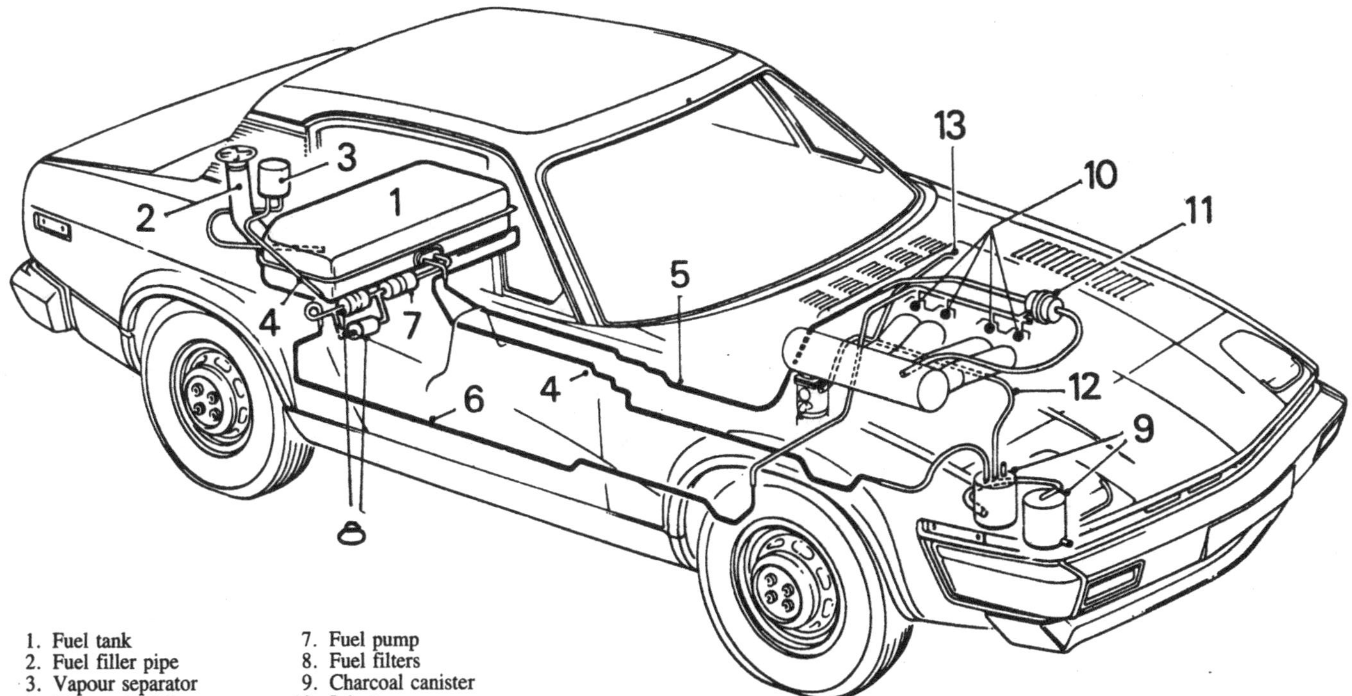

1. Fuel tank
2. Fuel filler pipe
3. Vapour separator
4. Vapour feed line
5. Fuel return line
6. Fuel feed line
7. Fuel pump
8. Fuel filters
9. Charcoal canister
10. Injectors
11. Fuel pressure regulator
12. Purge line
13. Crankcase purge line

CRANKCASE & EVAPORATIVE SYSTEM—SCHEMATIC
CARBURETTER MODELS

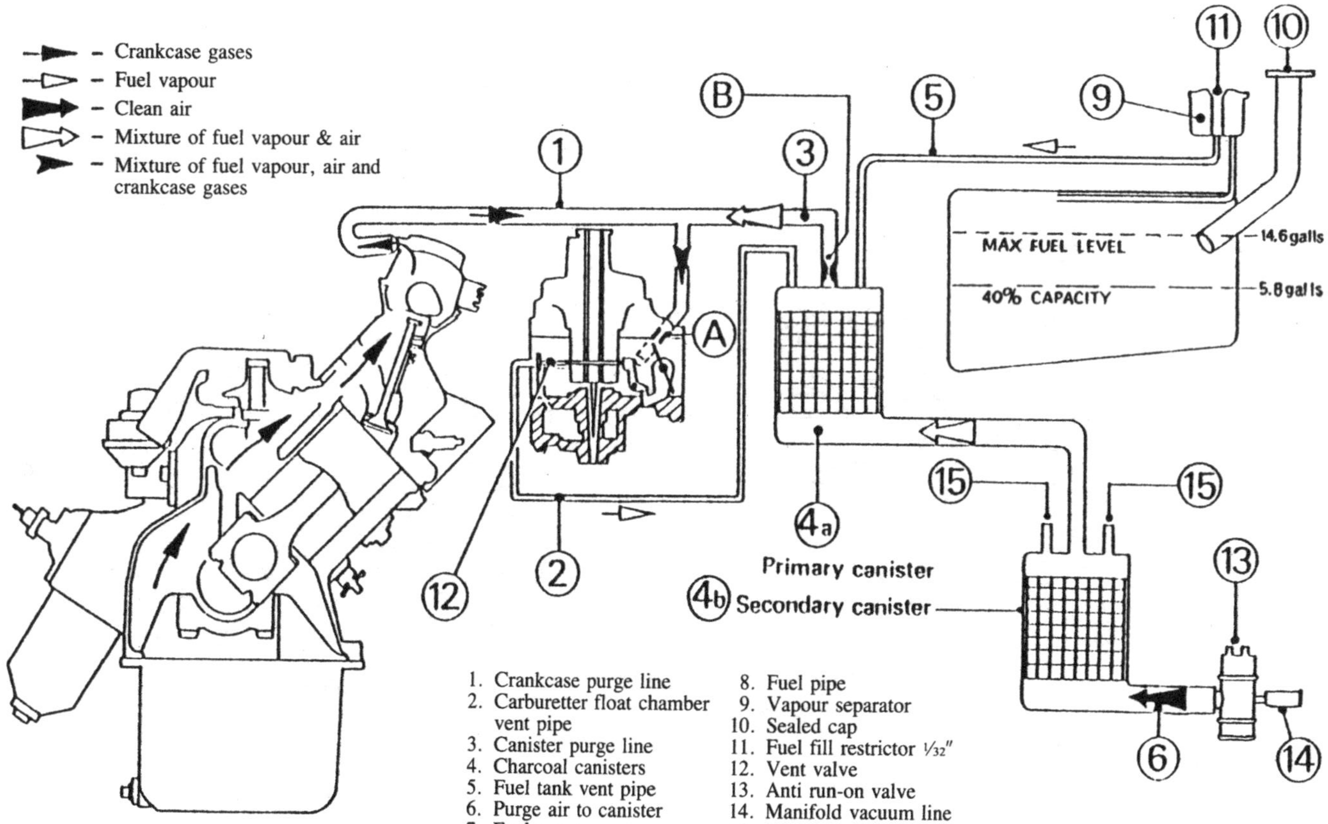

1. Crankcase purge line
2. Carburetter float chamber vent pipe
3. Canister purge line
4. Charcoal canisters
5. Fuel tank vent pipe
6. Purge air to canister
7. Fuel pump
8. Fuel pipe
9. Vapour separator
10. Sealed cap
11. Fuel fill restrictor 1/32″
12. Vent valve
13. Anti run-on valve
14. Manifold vacuum line
15. Pipes blocked off

A. .280–.300″ restrictor
B. 3/32″ restrictor

EVAPORATIVE LOSS CONTROL SYSTEM

Description **17.15.00**

The system is designed to prevent the emission to atmosphere of fuel vapour. The vapour collected from the fuel tank and carburetters is stored in an adsorption canister while the engine is at rest. Once the engine is running the vapour is purged from the canister and passed to the combustion chambers.
On some models twin canisters are fitted.

Fuel tank
When the fuel tank filler cap is correctly fitted the system is sealed and venting of the tank can only take place via a vapour separator to the adsorption canister(s).
To ensure that sufficient space is available to accommodate fuel displaced by expansion due to increased temperatures, a fuel filling restrictor is incorporated in the vapour separator which prevents the tank being completely filled with fuel.

Adsorption canister(s)
The adsorption canister provides a means of storing fuel vapour whilst the car is parked. The canister, which is not serviceable, contains active charcoal granules. Vapour tubes from the fuel tank, carburetter float chambers (where applicable) and the purge pipe from the engine breathing system are connected to ports on top of the canister. The manifold vacuum pipe can be connected to an anti-run on valve at the base of the canister.
Fuel vapour entering the canister through the vapour pipes is adsorbed by the charcoal granules. When the engine is started air is drawn by the engine breather system through the purge tube, at the bottom of the canister. The resulting purging action by the air passing over the granules carries the vapours through the engine breather system to the combustion chambers.

Vapour separator and restrictor
Vapour from the tank passes into the vapour separator and through two breather notches in a seat on which a taper plug rests. These small notches by virtue of their size allow only a limited flow of vapour as would be the case under normal running and stationary conditions. When the tank is being filled, however, the vapour air flow increases and the small notches create a restriction. Once the fuel has reached a level just above the filler tube this restriction and resulting pressure limits the quantity of fuel that the tank will accommodate to a predetermined maximum level.

continued

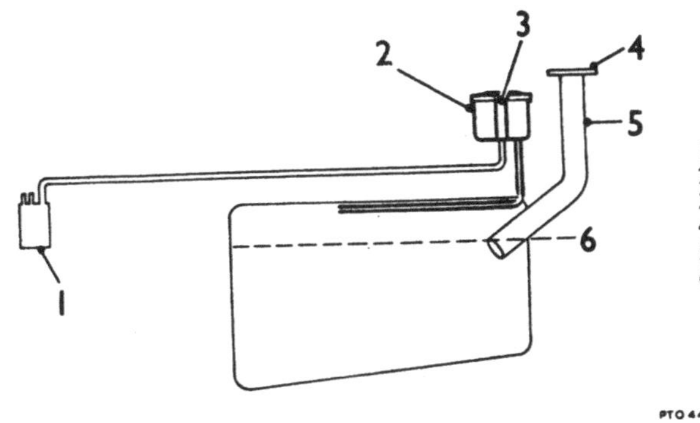

1 Adsorption canister.
2 Fuel vapour separator.
3 Fuel fill restrictor.
4 Sealed cap.
5 Fuel filler pipe.
6 Maximum fuel level.

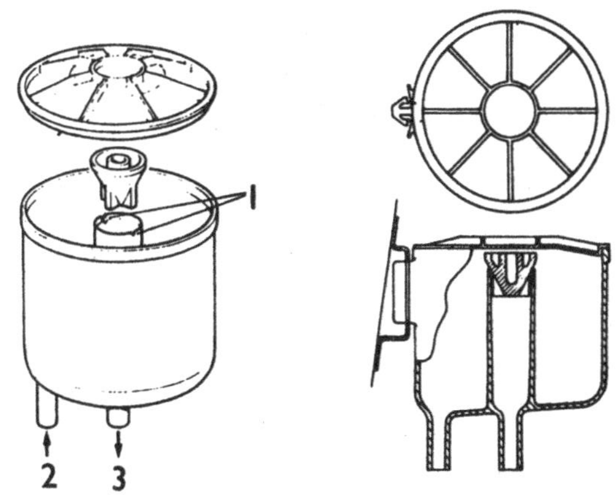

1 Breather notches.
2 Fuel tank pipe connection.
3 Adsorption pipe connection.

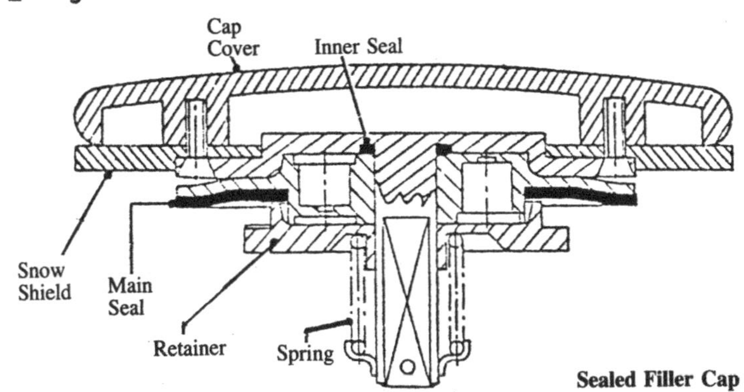

Sealed Filler Cap

Anti-run-on valve

The purpose of the valve is to prevent the engine 'running-on' once the ignition has been switched off.

The solenoid operated valve is connected by a hose to the base of the adsorption canister. The manifold vacuum hose is connected to the opposite side of the valve. The solenoid is connected electrically by two leads, one to the oil pressure switch and the other to the ignition switch.

When the ignition is turned-off a contact in the switch closes and energizes the solenoid thus opening the valve to manifold vacuum. At the same instant the purge vent to atmosphere closes.

Since the throttle is now closed the carburetter vent valve has opened the passage to the float chamber. Partial vacuum has now been transferred from the base of the adsorption canister, via the float chamber vent pipe, to the float chamber thus preventing fuel entering the carburetter jet.

When the engine stops revolving the drop in oil pressure activates the oil pressure switch, opens the circuit and de-energizes the solenoid valve.

DIAGRAMMATIC ARRANGEMENT OF FUEL AND VAPOUR CUT-OFF ANTI RUN-ON SYSTEM

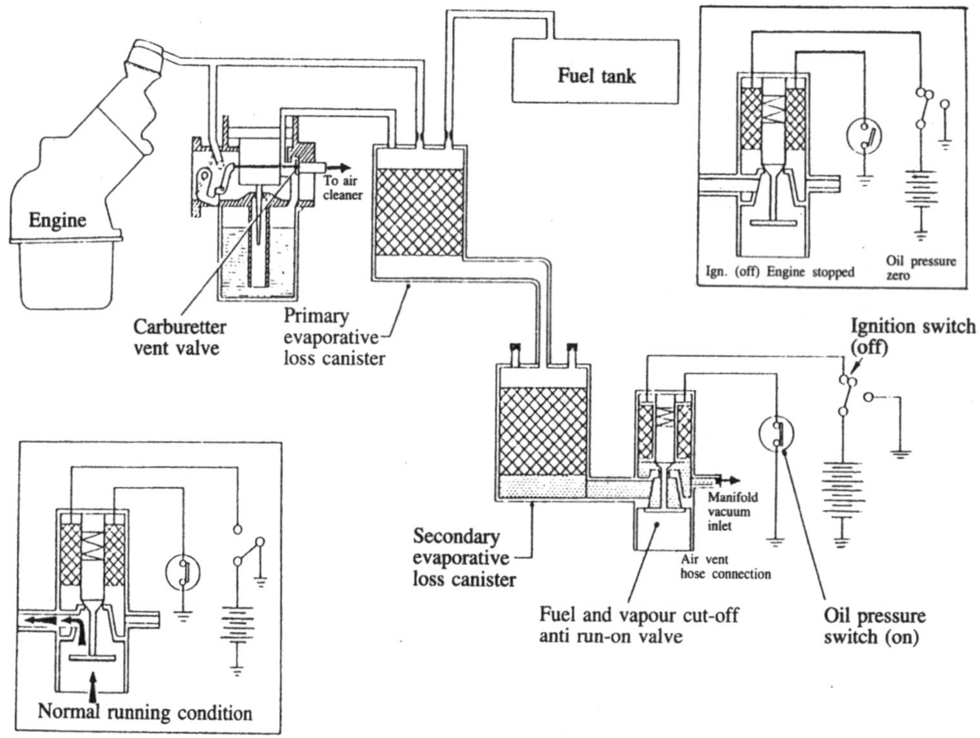

VAPOUR SEPARATOR

Remove and refit 17.15.02

1 Open the boot and remove the four screws retaining the access panel.
2 Remove the fuel filler cap and filler assembly. 19.55.08.
3 Pull-off the two hoses from the vapour separator.
4 Pull the vapour separator laterally from its retaining clip.

Refitting
5 Push the vapour separator into its retaining clip.
6 Fit the hose from the tank to the L.H. connection.
7 Fit the hose to the adsorption canister to the R.H. connection.
8 Fit the fuel filler assembly and cap. 19.55.08.
9 Refit the access panel.

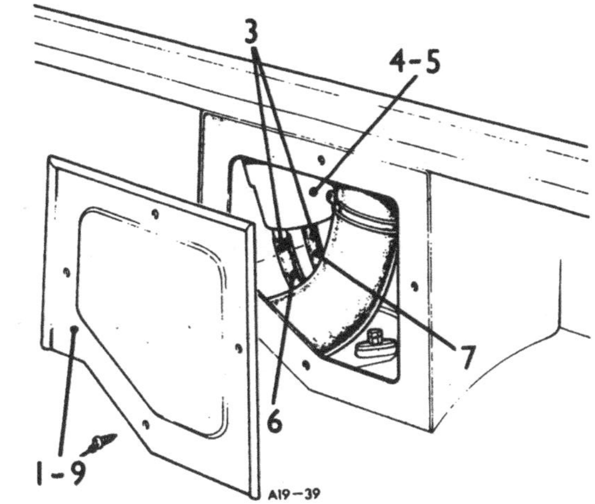

ADSORPTION CANISTERS – Twin Canister Specification

Remove and refit 17.15.13

Removing
1 Disconnect from the primary canister:
  a The canister purge pipe.
  b The carburetter float chamber vent pipe.
  c The fuel tank vent pipe.
  d The connecting pipe from the lower canister.
2 Slacken the clamp nut.
3 Remove the canister.
4 Remove the connecting pipe between the primary and secondary canisters.
5 Remove the purge air canister pipe from the secondary canister.
6 Slacken the clamp nut.
7 Remove the canister.

Refitting
8 Reverse instructions 1 to 7.

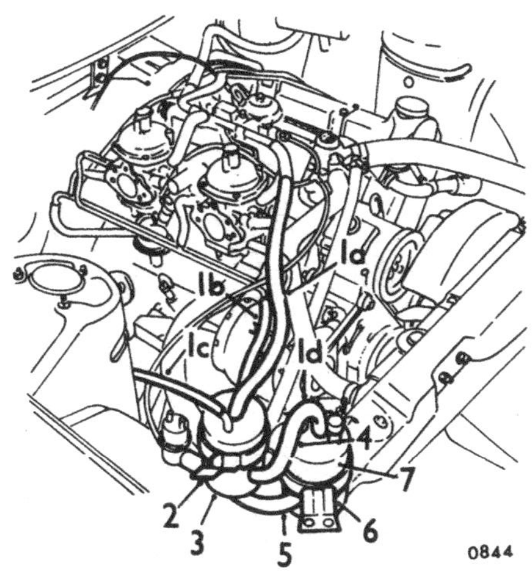

ADSORPTION CANISTER
Single Canister Specification

Remove and refit 17.15.13

Removing
1 Disconnect the canister to run-on valve pipe. (where fitted).
2 Disconnect from the canister:
  a The canister to fuel tank pipe.
  b The canister purge pipe.
  c Carburetter vent pipe.
3 Slacken the clamp nut.
4 Remove the canister.

Refitting
5 Secure the canister in the clamp.
6 Reverse instructions 1 to 3.

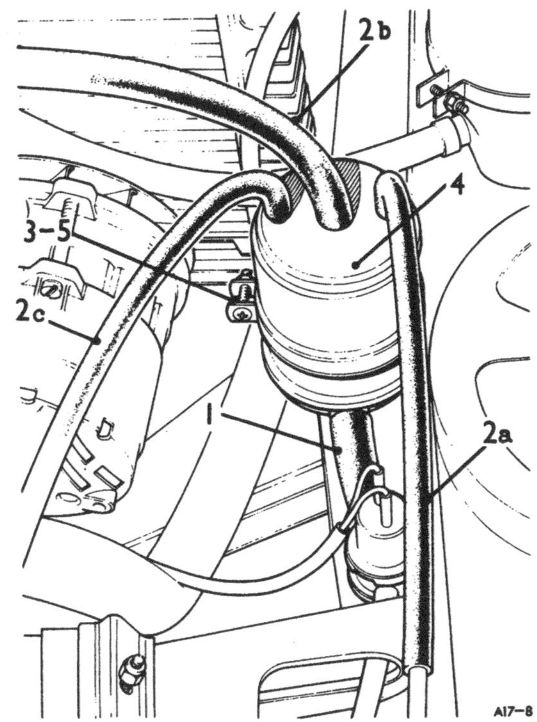

AIR INJECTION SYSTEM

Description 17.20.00

Air pump
The rotary vane type air pump is fitted at the front of the engine and driven by a belt from a crankshaft pulley.
The pump delivers air under pressure to each of the four exhaust ports via a diverter and relief valve, check valve and air inlet manifold.
The pressurized air combines with the exhaust gases to continue and assist in making more complete the oxidization process in the exhaust system.

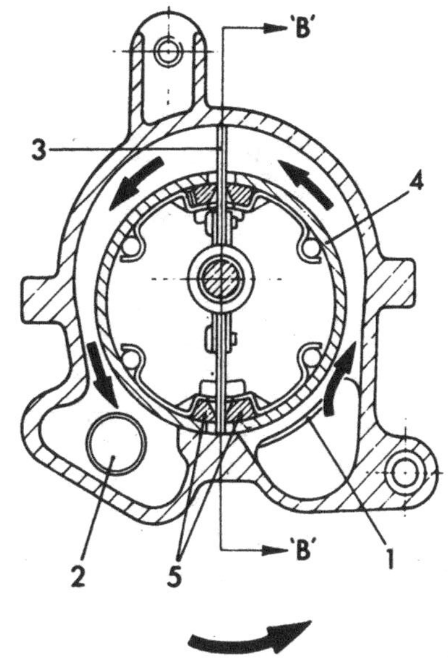

1 Inlet port.
2 Exhaust port.
3 Vanes.
4 Rotor.
5 Carbon sealing shoes.
6 Bearings.

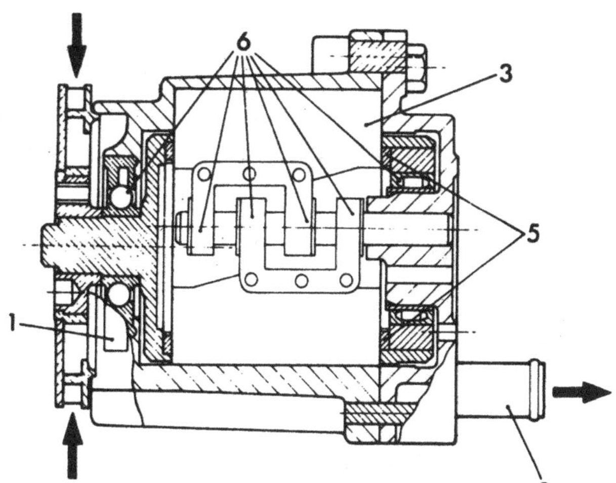

B–B

Diverter and Relief valve
Fitted to earlier models only.
Later models have a relief valve incorporated into the air pump.
This combined valve is incorporated to divert the air from the pump during deceleration to prevent backfire. The relief valve allows excessive air pressure at high engine speeds to discharge to the atmosphere.

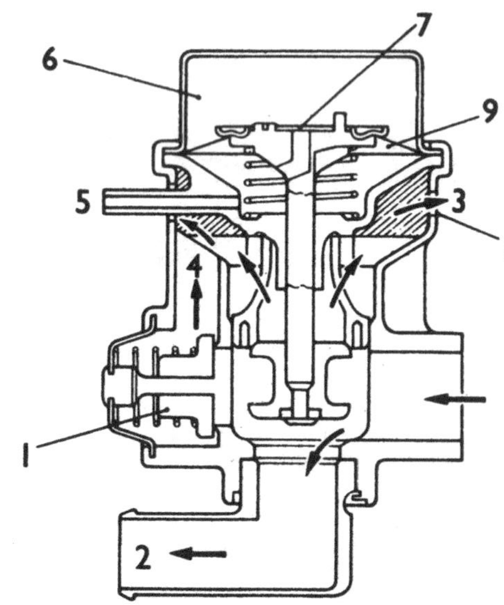

1 Relief valve.
2 Normal flow from air pump.
3 Diverted air.
4 Relief valve air.
5 Manifold vacuum.
6 Timed vacuum.
7 Timing feature.
8 Diverter and relief valve ports.
9 Diaphragm.

Check valve
The check valve is a one-way valve positioned between the diverter and relief valve and the air manifold. Its purpose is to protect the pump from back-flow of exhaust gases. The valve closes if the pump pressure falls while the engine is running, should, for example, the drive belt break.

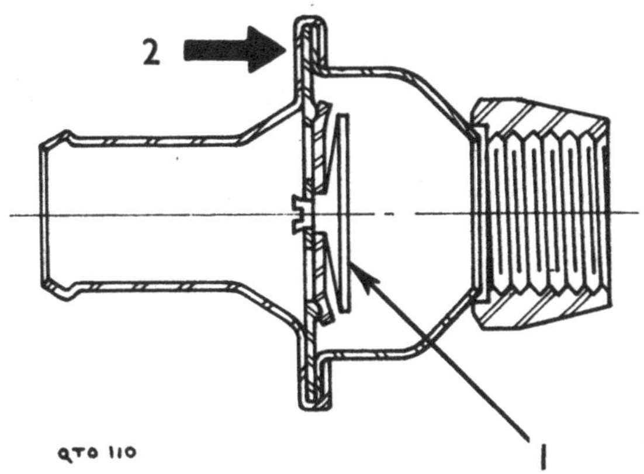

QTO 110

1 Valve diaphragm.
2 Direction of flow.

Air Manifold
The air manifold is finally responsible for delivering the pumped air directly into the exhaust ports through four small branch pipes.

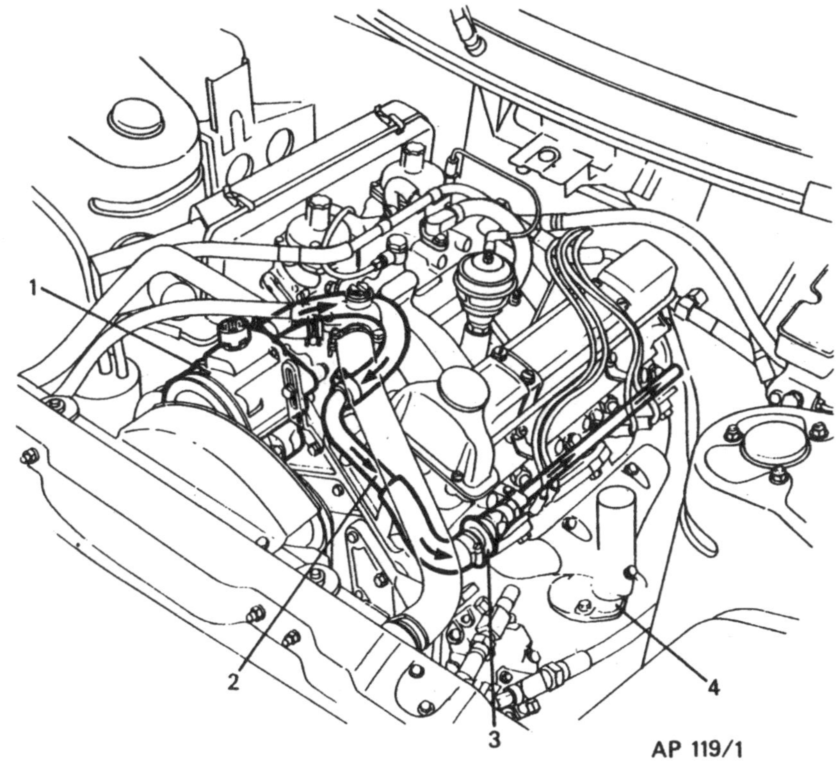

AP 119/1

TR7 AIR INJECTION SYSTEM

1 Air pump
2 Air rail
3 Check valve
4 Exhaust manifold

PUMP AND RELIEF VALVE

— Testing **17.25.07[A]**

Check that relief pressure operates at 8.2–10.5 lbs./in. by using a pressure gauge between the pump and check valve. If air is not relieved at the specified figures, renew valve. If pump pressure does not reach minimum figure, replace pump.

AIR PUMP

– Remove and refit **17.25.07[B]**

Removing

1 Disconnect the battery.
2 Disconnect the hose from the air pump.
3 Remove the adjusting nut and bolt.
4 Remove the pivot bolt and nut.
5 Remove the belt from the pump pulley.
6 Withdraw the pivot bolt until the head contacts the radiator.
7 Remove the pump.

Refitting

8 Fit the pump to its carrier bracket and loosely secure with the pivot nut and bolt.
9 Fit the adjusting nut and bolt.
10 Fit the belt onto the pulleys and correctly tension the belt. 17.25.13.
11 Reconnect the hose.
12 Reconnect the battery.

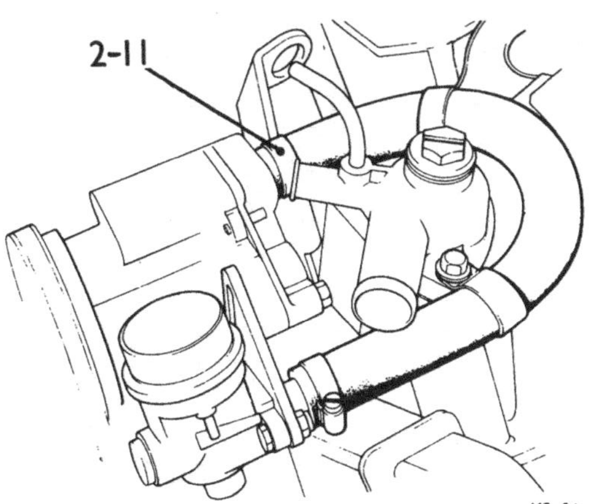

A19-64

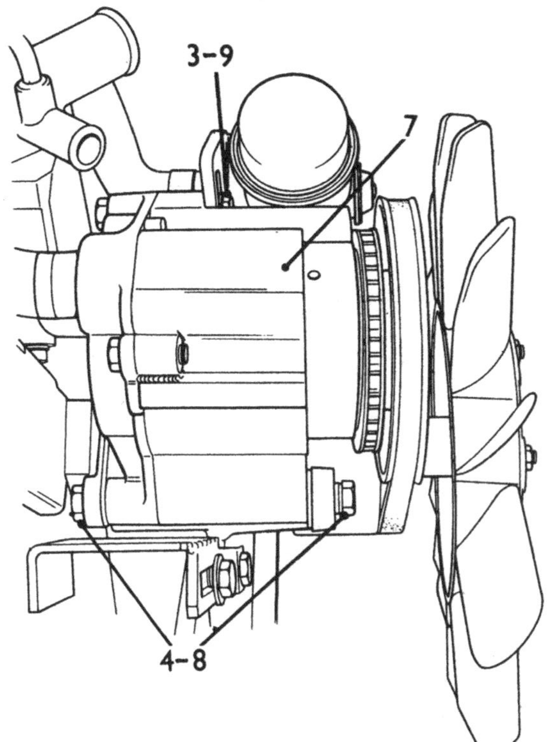

A17-1

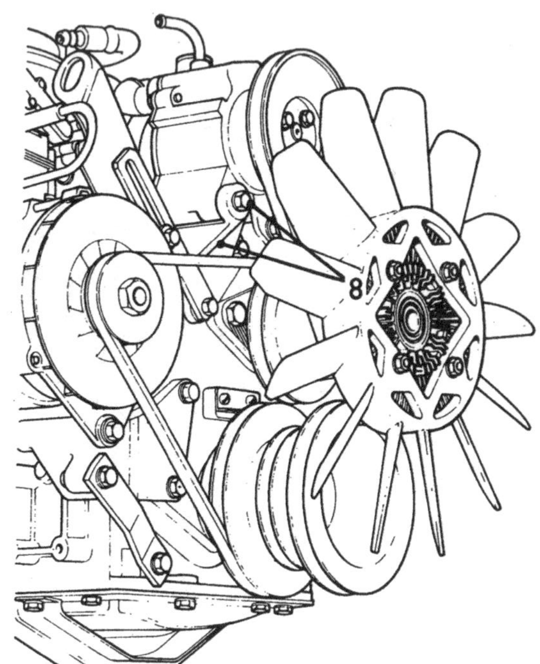

A26-1

AIR PUMP DRIVE BELT

– Tensioning 17.25.13

1 Disconnect the battery.
2 Slacken the adjusting nut and bolt.
3 Slacken the pivot bolt and nut.
4 With care use a lever on the pump to tension the belt and tighten the adjusting nut and bolt.
5 Tighten the pivot bolt.
6 Check the belt tension which is correct when 3/8 in. (9.5 mm) overall lateral movement is possible at the mid-point of the run.
7 Reconnect the battery.

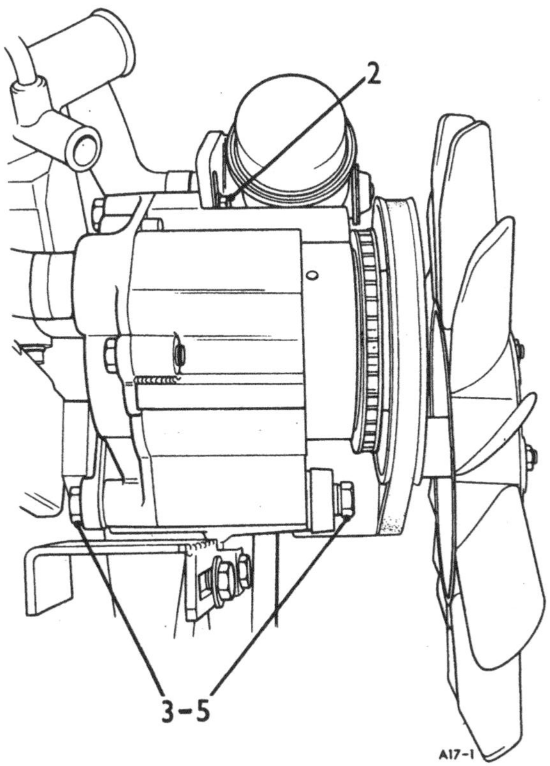

AIR PUMP DRIVE BELT

– Remove and refit 17.25.15

Removing

1 Disconnect the battery.
2 Remove the fan guard.
3 Remove the fan blades. 26.25.06.
4 Slacken off air pump belt tension.
5 Remove drive belt from the pulleys.

Refitting

6 Feed the belt onto the pulleys.
7 Tension the drive belt. 17.25.13.
8 Fit the fan blades. 26.25.06.
9 Reconnect the battery.

AIR DISTRIBUTION MANIFOLD

–Remove and refit 17.25.17

Removing

1 Disconnect the hose from the diverter and relief valve.
2 Unscrew the four union nuts securing the air injection tubes into the exhaust ports.
3 Withdraw the air distribution manifold complete with the check valve.
4 Hold the air distribution manifold in a vice by the hexagon and remove the check valve.

Refitting

5 Reverse instructions 1 to 4.

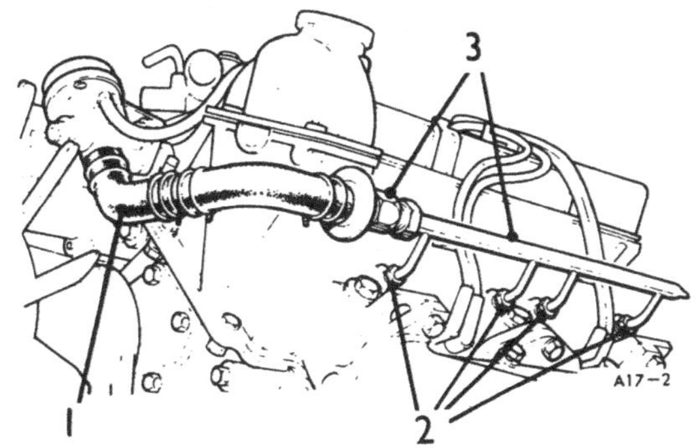

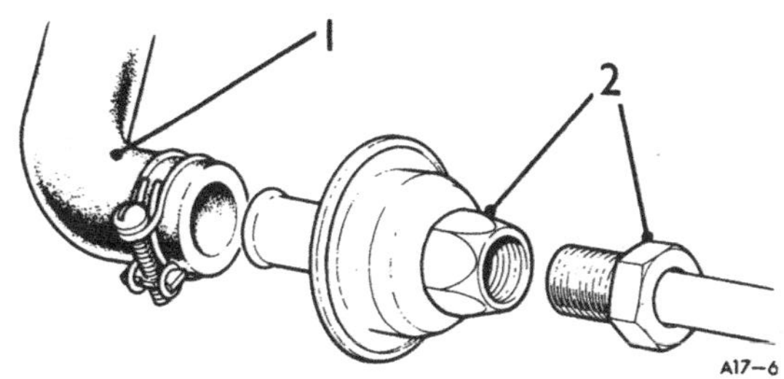

CHECK VALVE

– Remove and refit 17.25.21

Removing

1 Disconnect the air hose from the diverter and relief valve from the check valve.
2 Using two open-ended spanners – one on the air distribution manifold hexagon, to support the manifold, and the other to remove the check valve anti-clockwise.

CAUTION: Do not impose any strain on the air manifold.

Refitting

3 Reverse instructions 1 and 2.

CHECK VALVE
– Test 17.25.22

Testing

CAUTION: Do not use a pressure air supply for this test.

1 Remove the check valve. 17.25.21.
2 Blow through the valve orally in both directions in turn. Air should only pass through the valve when blown from the hose connection end. Should air pass through the valve when blown from the air manifold end, renew the valve.
3 Refit the check valve. 17.25.21.

DIVERTER AND RELIEF VALVE

– Remove and refit 17.25.25

Removing
1 Disconnect the battery.
2 Disconnect the pump to valve hoses.
3 Disconnect the air hose from the valve to the check valve.
4 Disconnect the diverter valve triggering vacuum pipe.
5 Remove the air pump adjusting nut and bolt.
6 Remove the bolt securing the common bracket holding the valve and air pump drive belt adjustment.
7 Remove the valve complete with bracket.
8 Remove the two nuts and bolts retaining the valve to the bracket.
9 Remove the valve and gasket.

Refitting
10 Fit the valve to the bracket using a new gasket.
11 Fit the bracket complete with valve to the engine.
12 Fit the drive belt tensioning nut and bolt.
13 Connect the check valve to diverter air hose.
14 Connect the air pump to diverter valve hose.
15 Connect the diverter air valve triggering vacuum pipe to the valve.
16 Check and if necessary adjust the air pump drive belt tension.
17 Reconnect the battery.

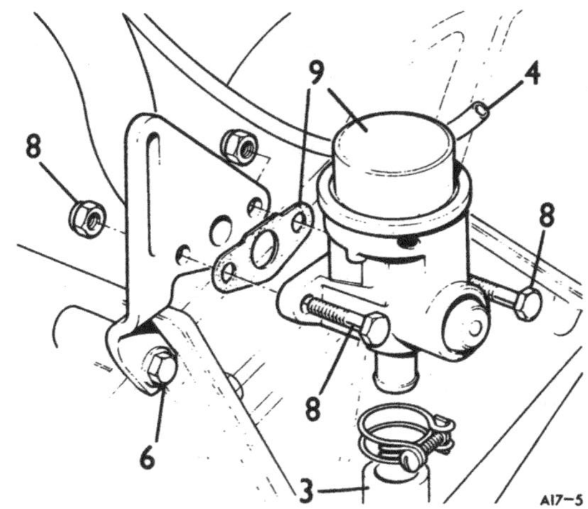

AIR INTAKE TEMPERATURE CONTROL SYSTEM

Description 17.30.00

The carburetters are tuned to function most efficiently at an air intake temperature of 100°F (38°C).

Earlier Models

The temperature is maintained by a sponge rubber flap valve which seals off the air cleaner cold air intake, permitting only warm air drawn from around the exhaust manifold to enter the engine after it is started up from cold. As the engine temperature rises, the flap valve moves under the influence of a bi-metal strip to first allow a mixture of warm air and air at ambient temperature to enter the air cleaner, and finally sealing off the warm air intake when the engine has reached operating temperature.

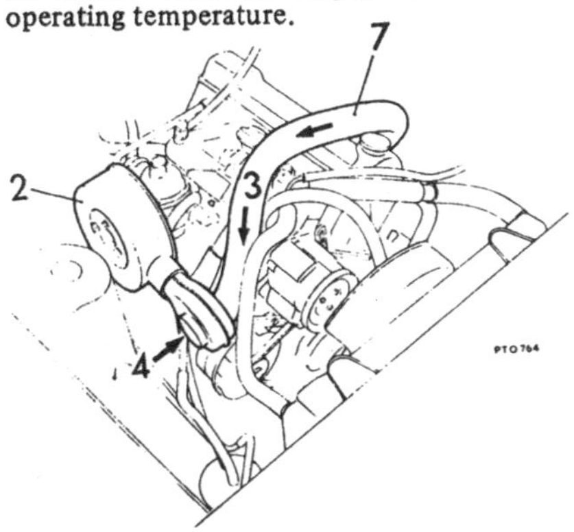

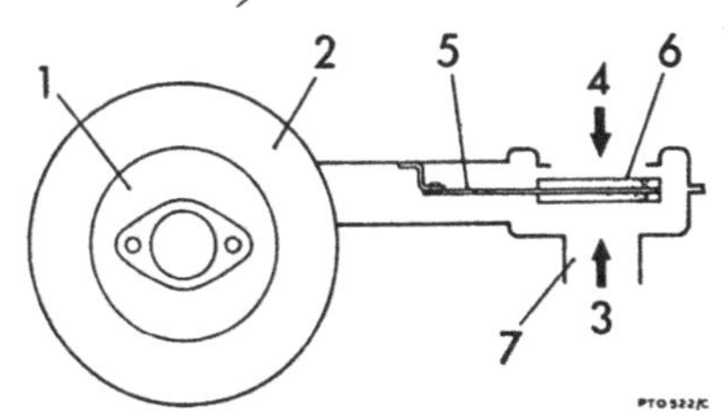

1 Air cleaner
2 Air cleaner cover
3 Hot air intake
4 Cold air intake
5 Bi-metal
6 Sponge rubber seal
7 Hot air pipe

Later models

The temperature is maintained by a sensing device incorporated in the air cleaner. The sensor allows inlet manifold vacuum to operate a flap valve in the air cleaner intake. The valve controls the entry of cold air at under bonnet temperature and hot air drawn from a duct on the exhaust manifold.

In order to maintain full vacuum influence on the flap valve when the manifold depression is temporarily destroyed during sudden throttle openings, a one-way valve is fitted in the vacuum line from the inlet manifold to the temperature sensor.

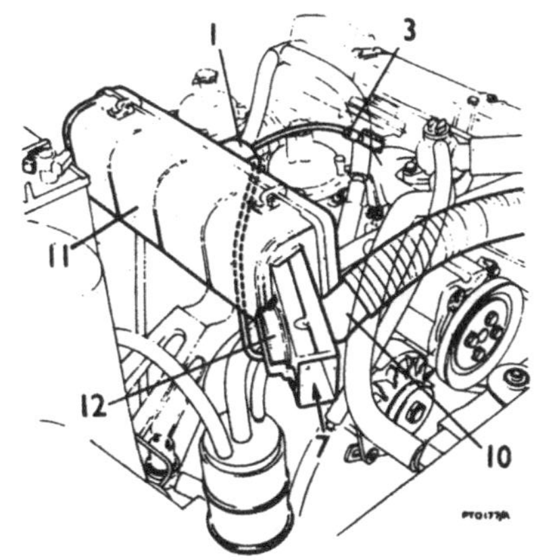

1 Temperature sensor
3 One-way valve
7 Cold air intake
10 Hot air hose
11 Air cleaner cover
12 Motor case

AIR INTAKE TEMPERATURE CONTROL SYSTEM

Function test 17.30.01

Earlier models

1. With the engine and air cleaner in a cold condition check that the sponge rubber valve effectively closes the aperture to atmosphere at the front of the intake.
2. Start the engine and check that the valve progressively closes the aperture to the hot air supply as the engine running temperature increases.

Later models

1 Inspect the condition and security of the following:
   a Hot air inlet hose.
   b The vacuum hose from the one-way valve to the temperature sensor unit.
   c The vacuum hose from the temperature sensor unit to the servo motor in the air cleaner outer cover.

2 With the engine and air cleaner cold i.e. below 100°F (38°C) check that the flap valve is in the cold air position i.e. parallel with the rectangular air intake – illustration A thus allowing cold air (ambient) only to enter the air cleaner.

3 Start the engine and allow it to idle for a few seconds. The flap valve should move immediately to the hot air position – illustration B thereby permitting only hot air via the hot air hose to enter the air cleaner.

4 With the flap valve in the hot air position – illustration B, attempt to lower the engine vacuum by increasing the engine speed suddenly and holding for a few seconds and check that the flap valve has remained in the hot air position.

5 Run the engine until normal operating temperature is reached and under bonnet temperature is above 100°F (38°C) and check that the flap valve has moved to the cold air position – illustration A.

continued

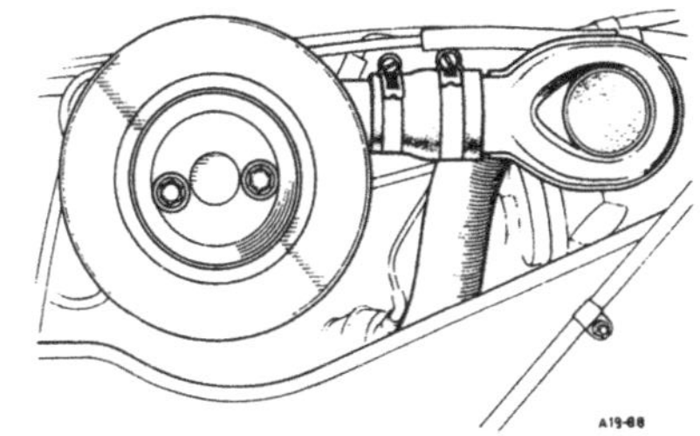

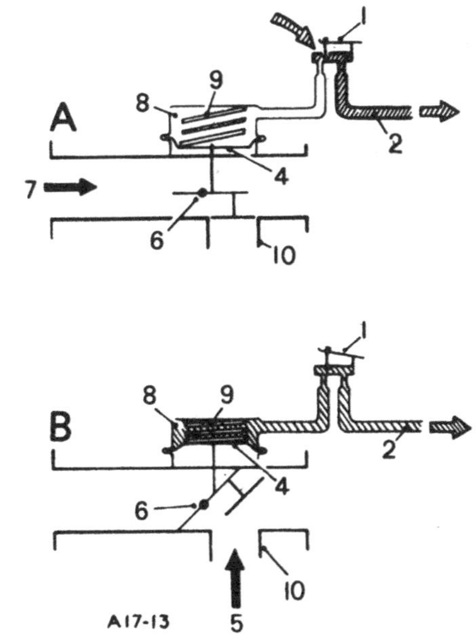

1 Temperature sensor
2 Inlet manifold
4 Valve diaphragm
5 Hot air intake
6 Flap valve
7 Cold air intake
8 Vacuum side of motor
9 Return spring
10 Hot air hose

AIR INTAKE TEMPERATURE CONTROL – FAULT FINDING

SYMPTOM	POSSIBLE CAUSE	ACTION
Poor or erratic idle Hesitation or flat spot (cold engine) Excessive fuel consumption Lack of power Engine overheating	Hot air inlet hose loose, adrift or blocked	Check hot air inlet hose for condition and security. Renew if necessary
Poor or erratic idle Hesitation or flat spot (cold engine) Excessive fuel consumption Lack of power Engine cuts-out or stalls (at idle) Engine 'runs-on' Engine 'knocks or pinks' Rich running (excess CO)	Flap valve jammed	Check operation of flap valve 17.30.01 If fault cannot be rectified renew air cleaner outer cover which includes flap valve. See 19.10.01
Poor or erratic idle Hesitation or flat spot (cold engine) Excessive fuel consumption Lack of engine power Engine cuts-out or stalls (at idle) Engine misfires Lean running (low CO)	Vacuum pipes disconnected or leaking*	Check the vacuum pipes for security and deterioration. Renew if necessary*

*These operations are not applicable to earlier type air cleaners

SYMPTOM	POSSIBLE CAUSE	ACTION
Hesitation or flat spot (cold engine) Excessive fuel consumption Lack of power Engine overheating Engine cuts-out or stalls (at idle) Engine 'runs-on' Engine knocks or pinks Rich running (excess CO)	One-way valve faulty*	Blow through valve to check 'one way' action. If the valve leaks fit a new valve. 17.30.05.*
Poor or erratic idle Hesitation or flat spot (cold engine) Excessive fuel consumption Lack of power Engine overheating Engine cuts-out or stalls (at idle) Engine 'runs-on' Engine knocks or pinks Rich running (excess CO)	Temperature sensor faulty, leaking or jammed*	Check and renew if necessary. 17.30.10.*
Poor or erratic idle Hesitation or flat spot (cold engine) Excessive fuel consumption Lack of power Engine cuts-out or stalls Rich running	Flap valve diaphragm leaking	Check with a distributor vacuum test unit. If leakage is apparent, renew air cleaner outer cover which includes the servo motor*

ONE-WAY VALVE

– Remove and refit 17.30.05

Removing
1 Disconnect the pipe from the one-way valve.
2 Remove the one-way valve, complete with union nut, from the brass banjo union.
3 Separate the union nut from the valve body taking care not to lose the ball and spring.

Refitting
4 Clean the valve internally and ensure the hole is clear.
5 Insert the spring and ball.
6 Fit and tighten the union nut.
7 Fit the valve assembly to the banjo union using a new fibre washer.
8 Reconnect the pipe to the one-way valve.

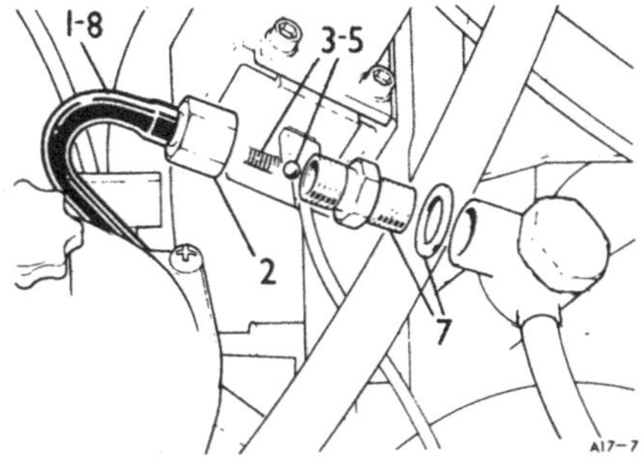

TEMPERATURE SENSOR UNIT

– Remove and refit 17.30.10

Removing
1 Remove the air cleaner. 19.10.08.
2 Prise off the clip securing the temperature sensor unit to the air cleaner back-plate.
3 Remove the temperature sensor unit.
4 Remove the felt washer.

Refitting
5 Fit the felt washer to the sensor unit.
6 Fit the sensor unit to the back-plate noting that it can only be fitted in one position.
7 Secure the unit with the spring clip.
8 Refit the air cleaner, 19.10.08, ensuring that the two vacuum pipes are correctly positioned on the sensor unit i.e.
   a The pipe from the one-way valve is fitted to the connection closest to the back-plate.
   b The pipe from the air inlet temperature control, unit furthest from the back-plate.

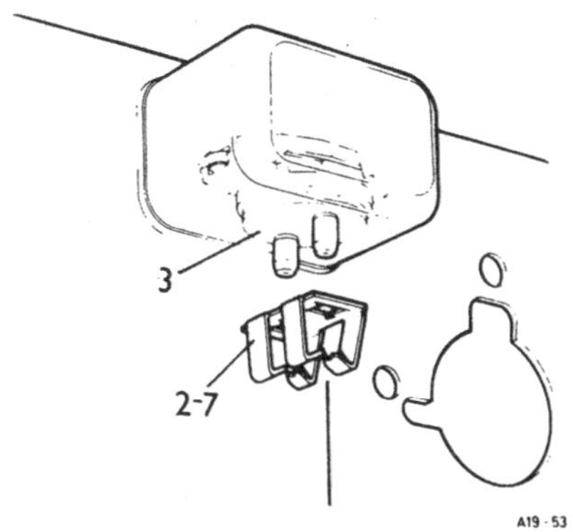

RUNNING-ON CONTROL VALVE
—Testing **17.40.01[A]**

With engine running, apply 12 volts to solenoid terminal. If engine does not stop, valve, wiring, ignition switch and oil pressure switch are suspect.
NOTE: If vent is not correctly adjusted, air will bleed into float chamber preventing vacuum — hence engine shutdown even if valve is operating correctly.

RUNNING-ON CONTROL VALVE – CANADIAN AND CALIFORNIA SPEC ONLY

– Remove and refit **17.40.01[B]**

Removing
1 Disconnect the solenoid electrical leads.
2 Disconnect the vacuum control pipe.
3 Disconnect the canister to run-on valve pipe.
4 Twist the valve and withdraw it from its retaining bracket.

Refitting
5 Reverse instructions 1 to 4.

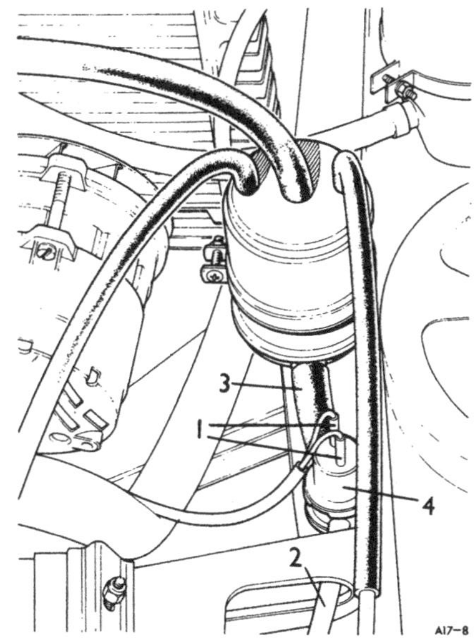

EXHAUST GAS RECIRCULATION SYSTEM

Description **17.45.00**

Exhaust gas recirculation system
Exhaust gas is taken from numbers 2 and 3 cylinders by two drillings to each side of the cylinder head. An exhaust gas recirculation valve (E.G.R. valve) mounted on a passage on the inlet manifold controls the flow to the balance pipe on the inlet manifold. The control signal is taken from a throttle-edged tapping which gives no recirculation at idle speed or full load, but gives an amount of recirculation dependent on the vacuum signal and metering profile of the valve. An E.G.R control valve cuts the signal to the E.G.R. valve when the choke is in operation by opening an air bleed into the vacuum line.

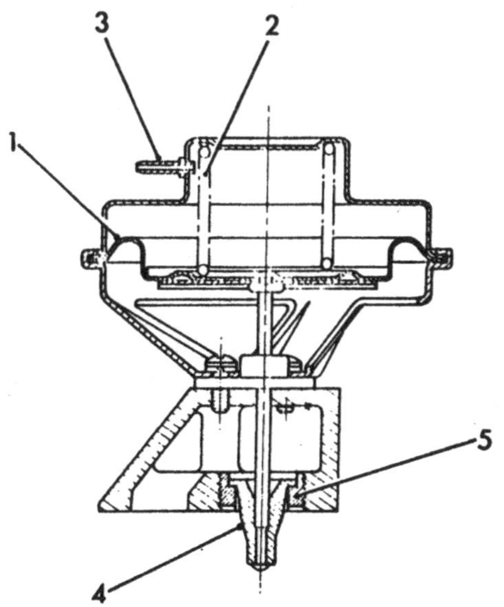

1 Diaphragm
2 Spring
3 To vacuum source
4 Metering pintle
5 Valve seat adjustable for production setting and sealed

E.G.R Control valve
This valve is attached to the rear carburetter and is actuated by a heel on the fast idle cam. The choke knob on the facia must be pulled out approximately ¼ in. before the control valve operates and must be pushed-in within 1/10 in. or fully home before the E.G.R. valve is operational.

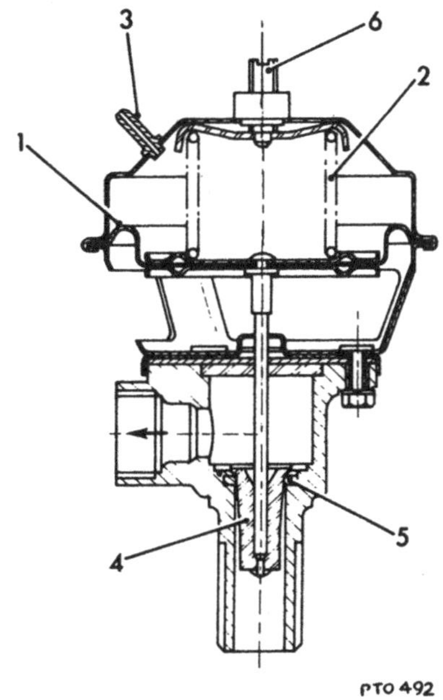

1 Diaphragm
2 Spring
3 To vacuum source
4 Metering pintle
5 Valve seat
6 Production adjustment – sealed after setting

EXHAUST GAS RECIRCULATION SYSTEMS – USA FEDERAL MARKET 1975 AND 1976 MODEL YEAR AND LATER. CANADIAN MARKET SPECIFICATIONS

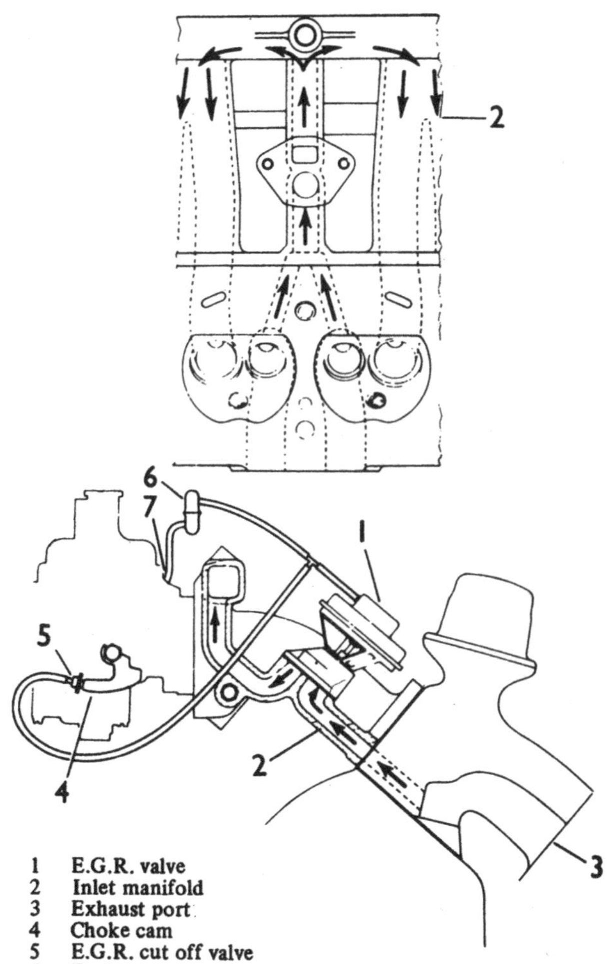

1 E.G.R. valve
2 Inlet manifold
3 Exhaust port
4 Choke cam
5 E.G.R. cut off valve
6 Fuel tap.
7 Vacuum source throttle edge tapping.

EXHAUST GAS RECIRCULATION SYSTEMS

USA CALIFORNIA MARKET
1975 AND 1976 MODEL YEARS

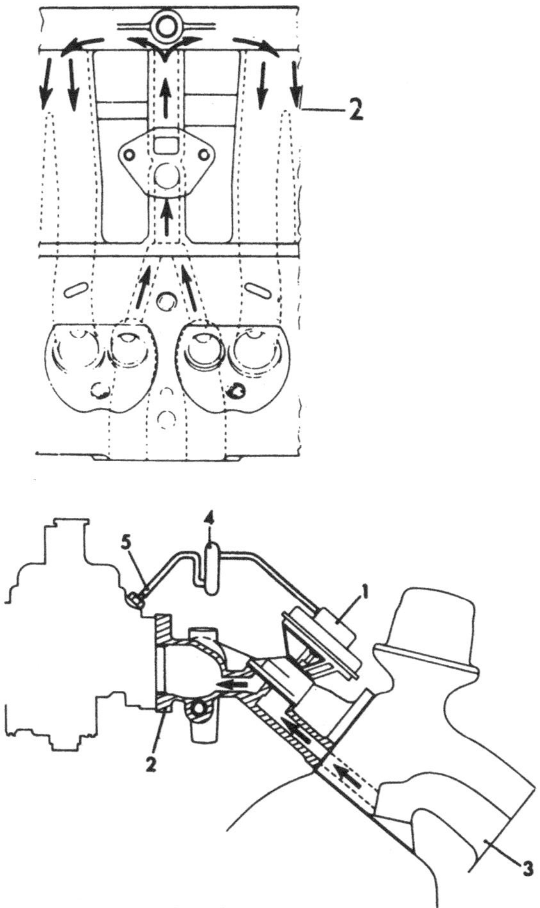

1 E.G.R. valve
2 Inlet manifold
3 Exhaust port
4 Petrol trap
5 Vacuum source – throttle edge tapping

USA FEDERAL AND CALIFORNIA MARKETS— 1977–1979 MODEL YEARS, and USA FEDERAL MARKET—1980 MODEL YEAR

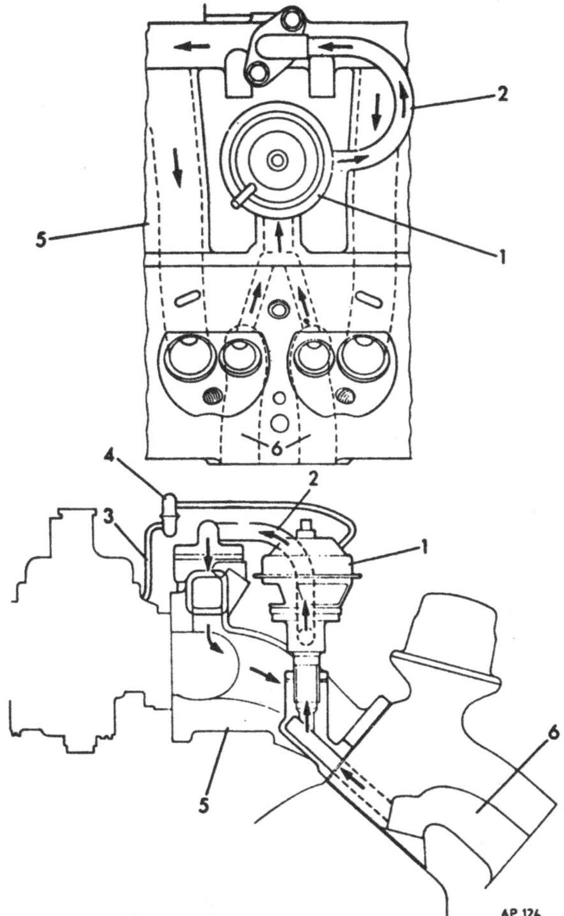

1 E.G.R. valve
2 E.G.R. pipe (asbestos lagged)
3 Throttle edge tapping
4 Fuel trap
5 Inlet manifold
6 Exhaust port

USA FEDERAL AND CALIFORNIA MARKET— 1978–1979 MODEL YEARS and USA FEDERAL MARKET—1980 MODEL YEAR

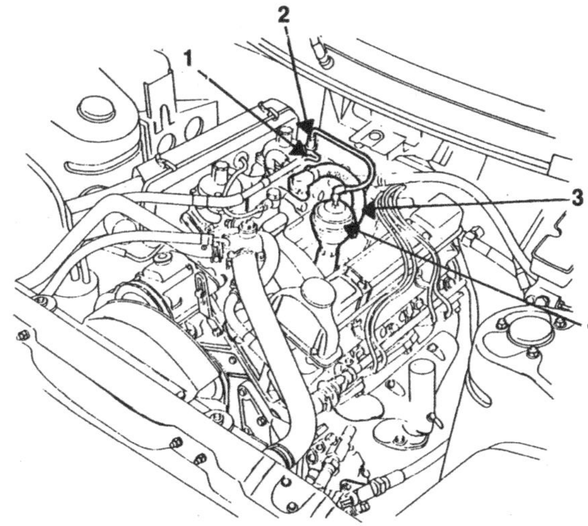

1. Throttle edge tapping
2. Fuel trap
3. E. G. R. pipe (asbestos lagged)
4. E. G. R. valve

EXHAUST GAS RECIRCULATION VALVE

Remove and refit 17.45.01

Earlier Models

Removing
1 Remove the vacuum pipe.
2 Remove the two bolts securing the valve to the manifold location.
3 Lift off the valve complete with gasket.

Refitting
4 Clean the valve port and the valve to manifold mating faces.
5 Place a new gasket in position ensuring that it is fitted with the correct face towards the manifold as stamped on the gasket i.e. 'MANIFOLD SIDE'.
6 Secure the valve with the two bolts and plain washers.
7 Connect the vacuum pipe.

Later models

Removing
1 Disconnect the vacuum pipe from the valve.
2 Disconnect the asbestos lagged pipe from the valve.
3 Unscrew the valve from the manifold.

Refitting
4 Reverse instructions 1 to 3, ensuring that the valve is securely sealed to the manifold.

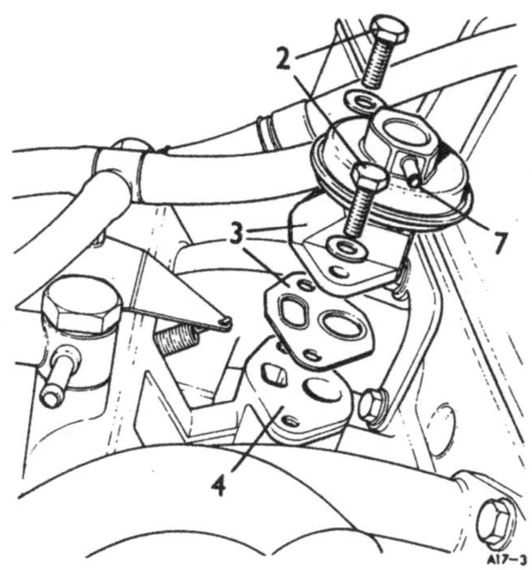

EXHAUST GAS RECIRCULATION CONTROL VALVE

Remove and refit 17.45.05

Removing
1 Disconnect the pipe from the valve.
2 Remove the screw and shake proof washer securing the valve bracket to the rear carburetter.
3 Remove the bracket and valve complete.

Refitting
4 Reverse instructions 1 to 3.

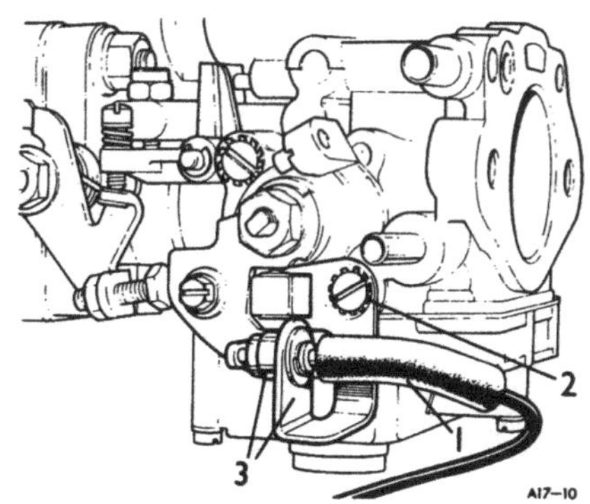

CATALYTIC CONVERTER

Remove and refit 17.50.01

Removing
1 Disconnect the battery and raise the car on a ramp, and allow converter to cool.
2 Remove the three nuts and bolts at both ends of the converter.
3 Ease converter and extract the olives from the exhaust pipe.
4 Withdraw the converter.

Refitting
5 Position the olives in the down-pipe and main pipe.
6 Ease the two pipes apart and insert the catalyst converter ensuring that the deflection plate is facing towards the front of the car.
7 Secure the converter with three nuts and bolts at both ends of the converter.
8 Lower the ramp and reconnect the battery.

FUEL SYSTEM UK & EUROPE

TAMPERPROOF CARBURETTERS (Non USA Markets)

Description 19.00.00

Later vehicles may be fitted with tamper-proof carburetters and the amount of attention that can be given to these units can be the subject of legislation in various territories.

They can be identified by a sealed mixture control nut (A), throttle return springs (B) acting directly on the throttle butterfly spindles and may have incorporated in them a capstat temperature compensated jet (C), a ball bearing suction chamber assembly (D), and a sealed slow-running adjustment screw (E).

The purpose of these carburetters is to control more stringently the air/fuel mixture entering the engine combustion chambers and, in consequence, the exhaust gas emissions leaving the engine.

For this reason the only readily accessible external adjustment on these carburetters is to the throttle settings for fast idle speed, using the screw (F).

Dismantling the carburetters
Should it be necessary, the float-chamber cover may be removed for access to the float or the float-chamber needle valve, refer to operation 19.15.24.

Otherwise, dismantling these carburetters is normally limited to removing the piston damper assembly, removing the suction chamber, and withdrawing the piston complete with needle assembly, for the purpose of cleaning these components or replacing the needle assembly.

CAUTION: Dismantling the carburetter unit beyond that described above, or unauthorized breaking of the mixture control or slow-running control seals and adjustment of the settings; or the fitting of an incorrect needle assembly, may render the vehicle user liable to legal penalties according to local territory legislation.

Should it be unavoidable and be permitted by local territory legislation, the mixture and slow-running adjustments may be altered. Remove the respective seals and adhere strictly to operation 19.15.02. Use an approved type CO meter as a final emissions level check and fit new sealing devices as required by local territory legislation.

Needle assembly
The needle is of a specific type, and with its related components is an assembly which may be peculiar to the model. Therefore it should always be replaced as an assembly.

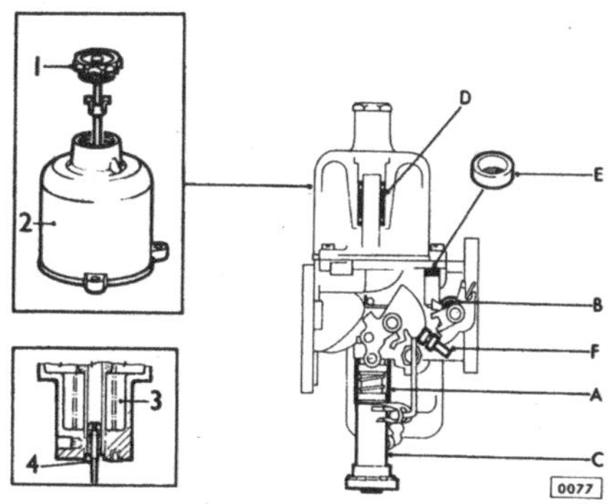

AIR CLEANER

Remove and refit 19.10.01

Later vehicles may be fitted with an air cleaner incorporating air intake temperature control similar to the USA and Canada later carburetter specification.

Removing
1 Disconnect the two air inlet hoses from the front of the air cleaner.
2 Disconnect the engine dipstick support from the air cleaner.
3 Slacken the nut securing the throttle linkage bracket to the air cleaner.
4 Release the two air cleaner cover clips and remove the cover.
5 Withdraw the filter element.
6 Remove the six bolts and spring washers (three each carburetter) securing the air cleaner body to the carburetters.
7 Remove the air cleaner body and gaskets.

Refitting
8 Offer up the air cleaner body to the carburetters ensuring that the rear support lug engages the throttle bracket bolt and that the carburetter flange gaskets are correctly positioned.
9 Fit the six securing bolts and spring washers to the carburetter flanges.
10 Tighten the nut securing the throttle bracket to the air cleaner body.
11 Install the filter element (metal frame to carburetters).
12 Fit the cover and secure with the two clips.
13 Fit the air inlet pipes.
14 Attach the dipstick guide to the cover.

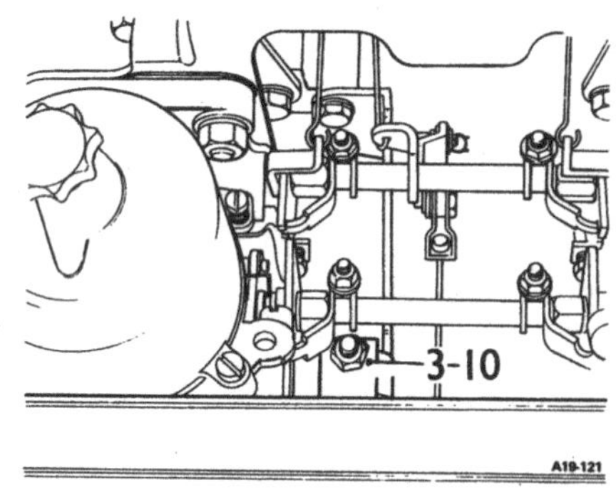

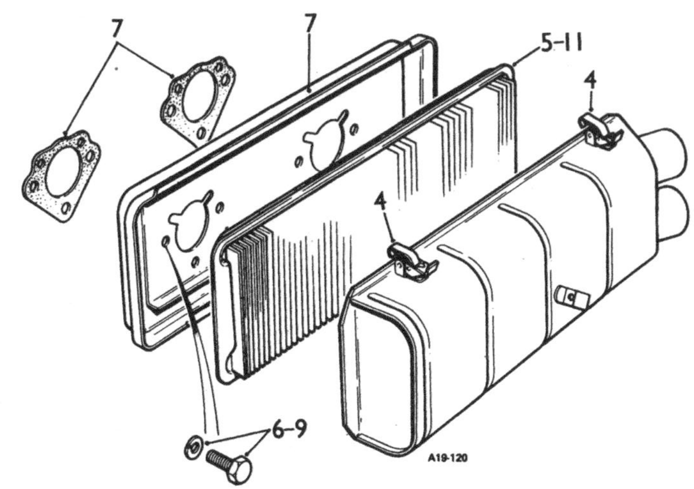

AIR CLEANER

Renew element **19.10.08**

Note: Later vehicles may be fitted with an air cleaner incorporating air intake temperature control similar to the USA and Canada later carburetter specification.

1. Disconnect the two air inlet pipes.
2. Remove the screw and nut securing the engine dipstick to the filter cover.
3. Release the two filter cover clips and remove the cover.
4. Withdraw the filter element.
5. Thoroughly clean interior of filter body and cover.
6. Install new filter element (metal frame to carburetters).
7. Fit the cover.
8. Fit the air inlet pipes.
9. Secure the dipstick clip to the cover.

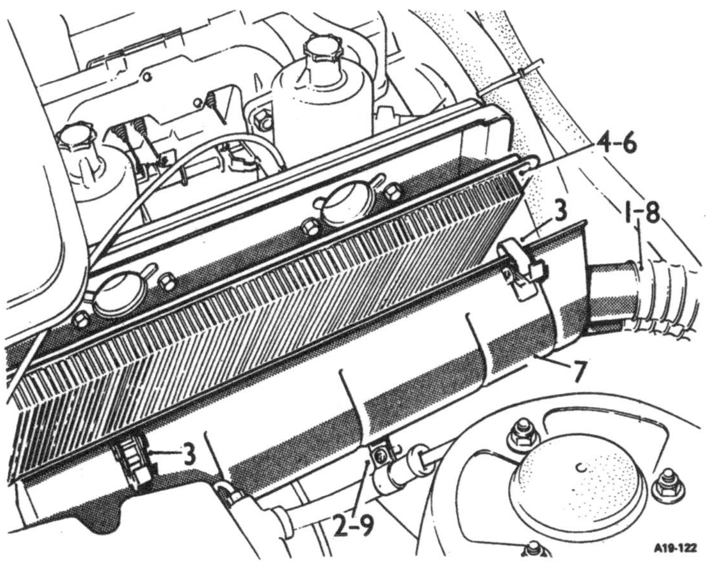

CARBURETTERS (TWIN SU INSTALLATION)

Tune and adjust 19.15.02

Where tamperproof carburetters are fitted, refer to 19.00.00.

1. Remove the air cleaner, see 19.10.02.
2. Check the throttle for correct operation ensuring that it does not stick.
3. Turn the fast idle adjusting screws anti-clockwise until well clear of the cams, and disconnect the mixture control cable at the trunnion.
4. Turn the throttle adjusting screws until just clear of the throttle levers with the throttles closed, then turn the screws 1½ turns clockwise.
5. Raise the piston of each carburetter in turn (use the lifting pin if fitted) and check that it falls freely onto the carburetter bridge. If either piston shows any tendency to stick, remove and clean the piston and suction chamber, see 19.15.30.
6. Lift and support the piston clear of the carburetter bridge so that the jet is visible.

continued

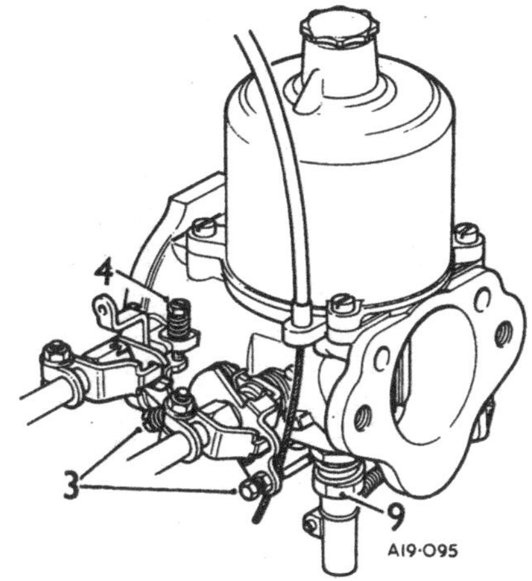

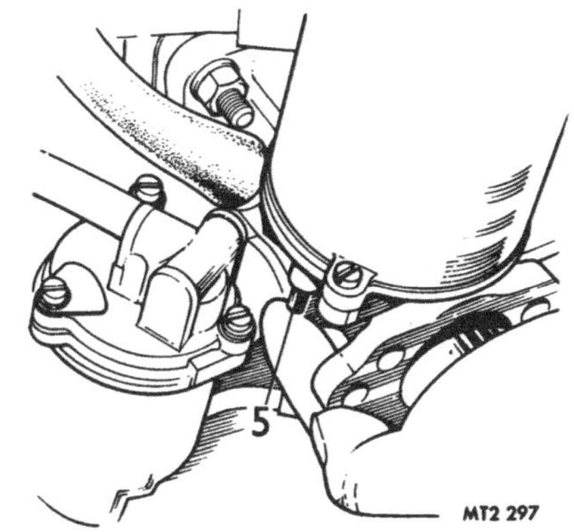

7 Turn the jet adjusting nuts to bring the jets flush with the carburetter bridge.
8 Check that the needle shank is flush with the underside of the piston.
9 Turn the jet adjusting nuts down two turns.
10 Check the piston damper oil level as follows:
a Unscrew the cap and withdraw the damper.
b Top-up the hollow piston rod with S.A.E. 20 oil until the level is ½ in (13 mm) above the top of the hollow piston rod – dimension A.

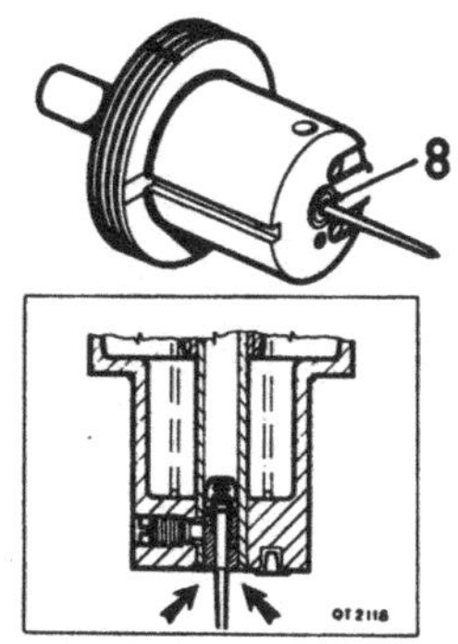

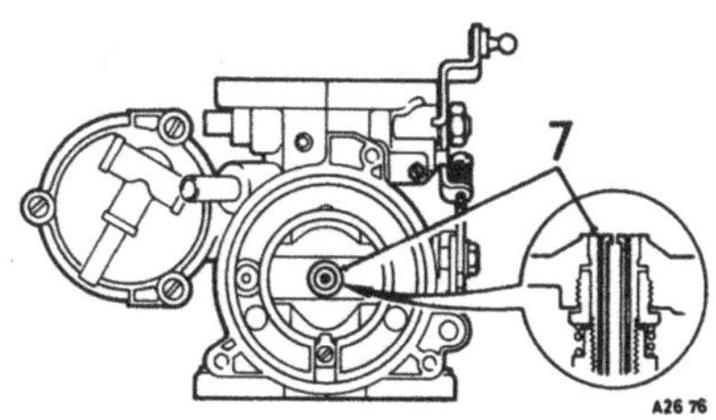

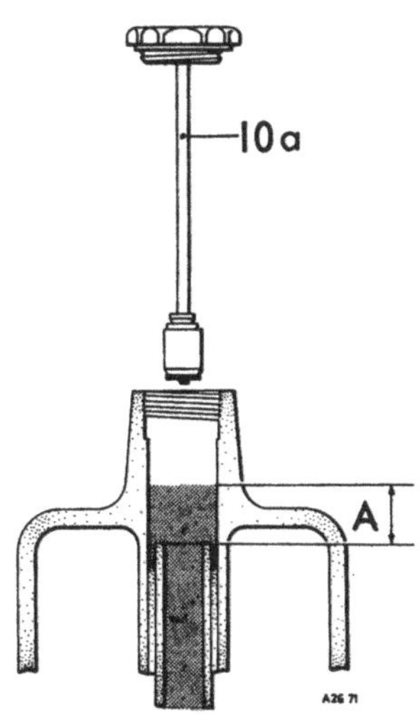

11 Start the engine and run at a fast idle speed until it attains normal running temperature and continue for a further five minutes.
12 Increase the engine speed to 2500 r.p.m. for thirty seconds.
13 Using an exhaust gas analyser insert the probe into the exhaust pipe in accordance with the manufacturers instructions.
NOTE: Tuning can now commence, but if adjustments cannot be achieved within three minutes increase the engine speed to 2500 r.p.m. and then continue. Repeat this clearing procedure at three minute intervals until tuning is complete.
14 Slacken both clamping nuts and bolts on the throttle spindle interconnections.
15 Slacken both nuts and bolts on the jet control interconnections.
16 Using a balancing meter check the carburetters for balance and adjust by turning the throttle adjusting screws whilst maintaining the correct idling speed – see DATA.

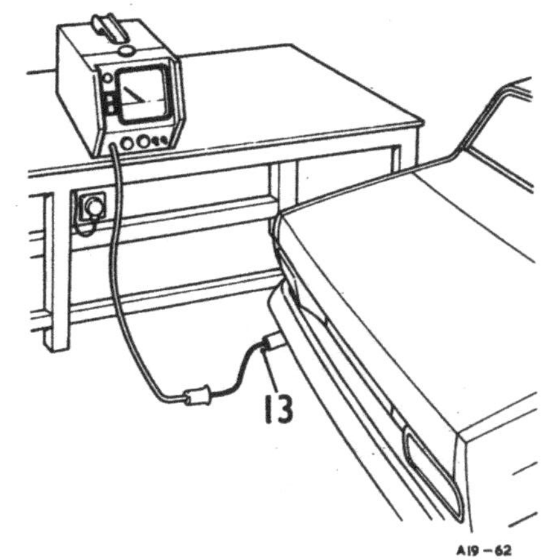

continued

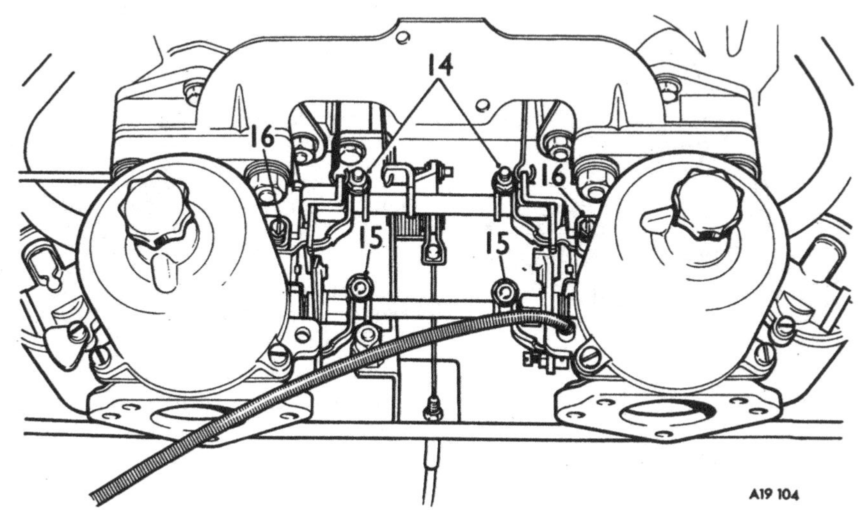

17 Turn the jet adjusting nut on each carburetter down to enrich or up to weaken the mixture by the same amount until the fastest speed is indicated on the tachometer.
18 Turn each adjusting nut up one flat at a time until the engine speed just commences to fall.
19 Turn the adjusting nuts down by the minimum amount until the fastest speed is regained.
20 Check the idle speed and adjust if necessary by turning the throttle adjusting screws by the same amount.
21 Check that the CO reading on the exhaust gas analyser is within the accepted limits at idling, see DATA.
22 If the CO reading falls outside the limits, reset both jet adjusting nuts by the minimum amount necessary.

NOTE: If an adjustment exceeding three flats or half a turn is required to achieve this the carburetters must be removed and overhauled.

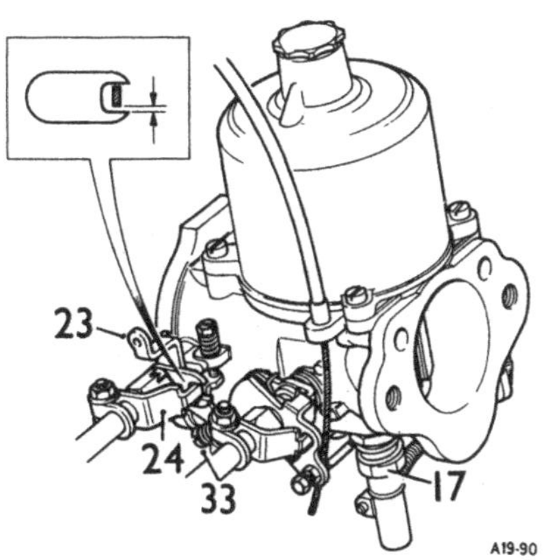

Throttle and mixture control linkage setting
23 Hold the throttle lever against its stop.
24 Rotate the lever pins on the throttle interconnection shaft until a gap of 0.010 in (0.25 mm) exists between the pins and the lower arm of the forks.
25 Ensure that the link rod is at the top of the elongated hole in throttle shaft actuating lever.
26 Whilst maintaining the positions in instructions 24 and 25 tighten the lever pin clamping nuts and bolts on both carburetters.
27 Adjust the throttle cable to take-up any slack.
28 Check and if necessary adjust the accelerator pedal stop to ensure full throttle opening.
29 Using the balancing meter re-check the carburetters for balance by running the engine at 1500 r.p.m.
30 With the fast idle cams against their respective stops tighten the jet control interconnection clamps, so that both cams begin to move simultaneously.
31 Connect the mixture control cable to the trunnion and ensure that there exists a $\frac{1}{16}$ in (1.5 mm) free movement of the cable before the cams move.
32 Pull out the mixture control knob until the linkage is about to move the jet.
33 Using the balancing meter to ensure equal adjustment turn the fast idle adjusting screws to give the correct fast idle speed – see DATA – and tighten the lock nuts.

Automatic kick-down
34 Where applicable, adjust the kick-down cable so that there is no slack, but that a gap of 0.020 to 0.025 in (0.050 to 0.63 mm) exists between the ferrule and the cable adjuster.
35 Fit the air cleaner, see 19.10.02.

CARBURETTERS

Remove and refit 19.15.11

Removing
1 Disconnect the battery.
2 Remove the air cleaner complete, see 19.10.01.
3 Remove the fresh air duct.
4 Disconnect the mixture control cable from the front carburetter.
5 Pull-off the fuel over-flow pipes from the front and rear float chambers.
6 Pull-off the vacuum pipe from the rear carburetter to the distributor advance capsule.
7 Disconnect the engine breather pipes from the carburetters.
8 Disconnect the fuel feed pipes to the carburetters.
9 Disconnect the vertical link from the throttle interconnection shaft.
10 Disconnect the three throttle shaft return springs.
11 Remove the eight (four per carburetter) nuts, plain and spring washers securing the carburetters to the inlet manifold.
12 Withdraw the carburetters complete.
13 Remove and discard the gaskets.

Refitting
14 Clean the carburetter and flexible mounting mating faces and fit new gaskets.
15 Fit the carburetters ensuring that the mixture control pins are located in their respective forks.
16 Fit the plain and spring washers and evenly tighten, to the correct torque, the eight retaining nuts.
17 Reverse instructions 1 to 10.
18 Tune and adjust the carburetters, see 19.15.02.

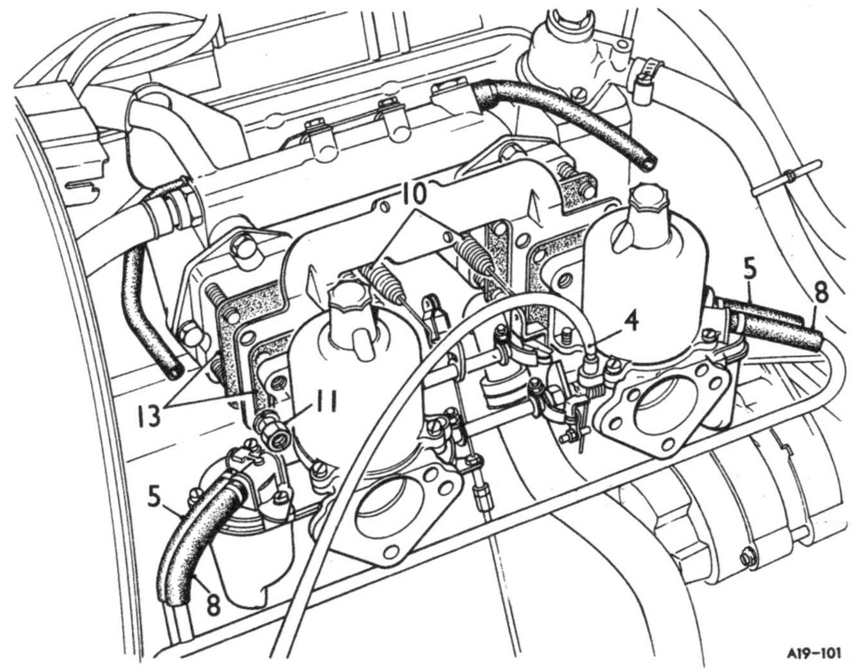

DATA

Idling speed	650 to 850 r.p.m.
Fast idling speed	1100 to 1300 r.p.m.
CO level at idling speed	2.0 to 4.5%

CARBURETTERS (TWIN SU INSTALLATION)

Overhaul and adjust 19.15.17

Dismantling

1 Remove the carburetter assembly, see 19.15.11.
2 Thoroughly clean the outside of the carburetter.
3 Mark the suction chamber to ensure it is refitted to the same body.
4 Remove the damper and its washer.
5 Unscrew the suction chamber securing screws and lift off the chamber.
6 Remove the piston spring.
7 Carefully lift out the piston assembly and empty the oil from the piston rod.
8 Remove the guide locking screw and withdraw the needle assembly, taking care not to bend the needle. Withdraw the needle from the guide and remove the spring from the needle.
9 Where fitted, push the piston lifting pin upwards, detach its securing circlip and washers and withdraw the pin and spring downwards.
10 Release the pick-up lever return spring from its retaining lug.
11 Support the plastic moulded base of the jet and remove the screw retaining the jet pick-up link and link bracket.
12 Unscrew the flexible jet tube sleeve nut from the float-chamber and withdraw the jet assembly. Note the gland, washer and ferrule at the end of the jet tube.
13 Remove the jet adjusting nut and spring.
14 Unscrew the jet locking nut and detach the nut and jet bearing; withdraw the bearing from the nut.
15 Unscrew and remove the lever pivot bolt and spacer.
16 Detach the lever assembly and return springs, noting the pivot bolt tubes and the location of the cam and pick-up lever springs.
17 Unscrew the securing bolt and remove the float chamber and spacer.
18 Mark the float chamber lid location.
19 Remove the lid securing screws and detach the lid with its joint washer and float. Retain the part number tag.

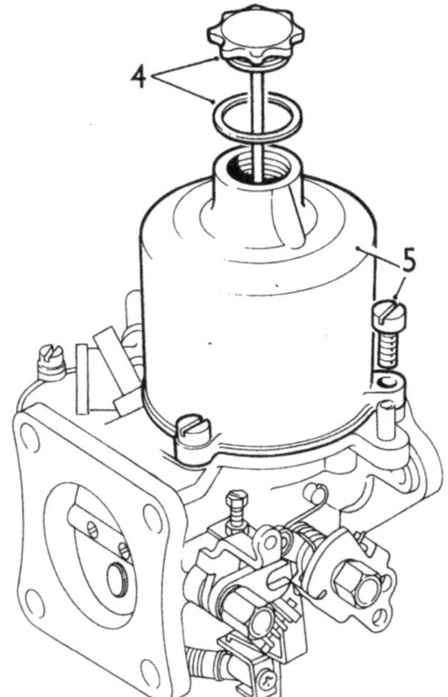

A19-105

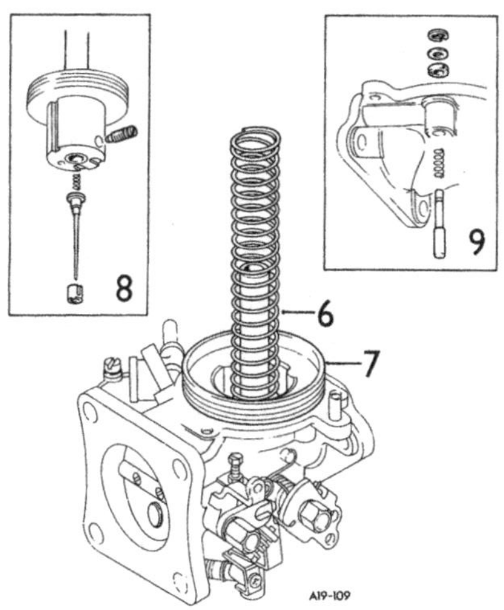

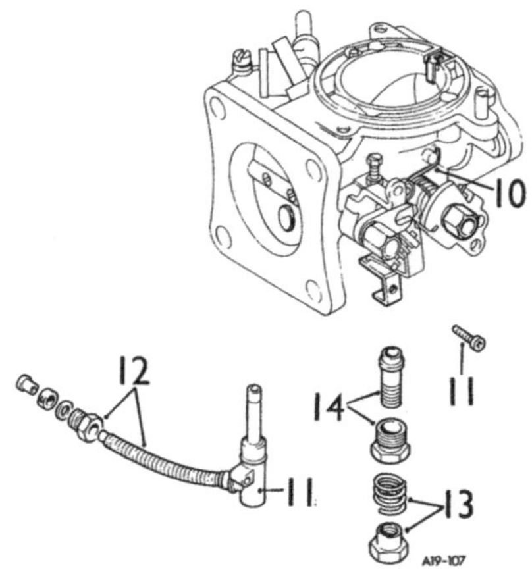

20 Hold the float hinge pin at its serrated end and withdraw the pin and float.
21 Extract the float needle from its seating and unscrew the seating from the lid.
22 Close the throttle and mark the relative position of the throttle disc and the carburetter flange.
23 Press the split ends of the disc retaining screws together and remove the screws. Open the throttle and remove the disc from its slot in the throttle spindle.

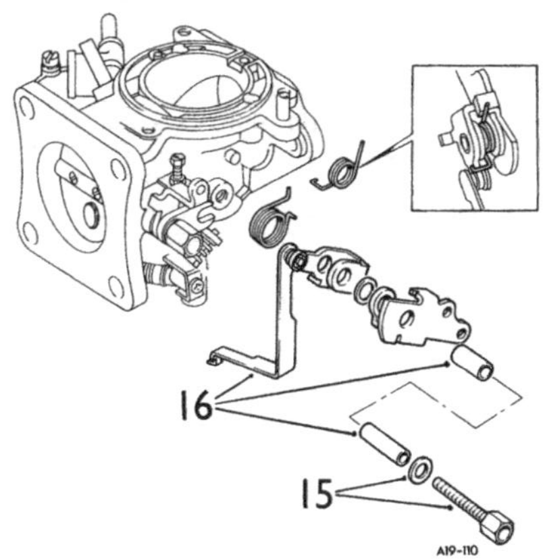

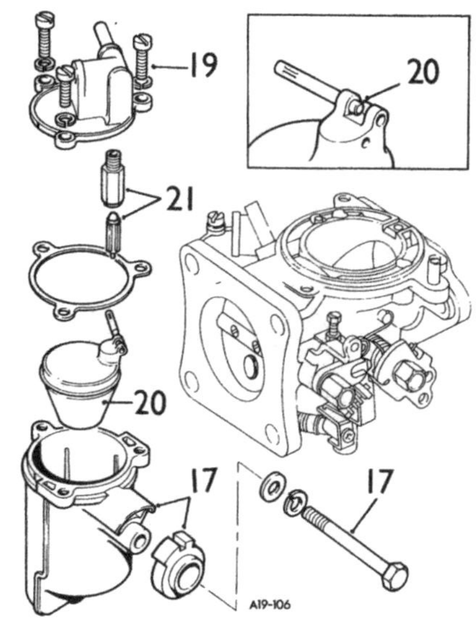

24 Release the lock washer tabs securing the spindle nut; remove the nut and detach the fork lever, lever arm, washer and throttle spindle; note the location of the lever arm in relation to the spindle and carburetter body.

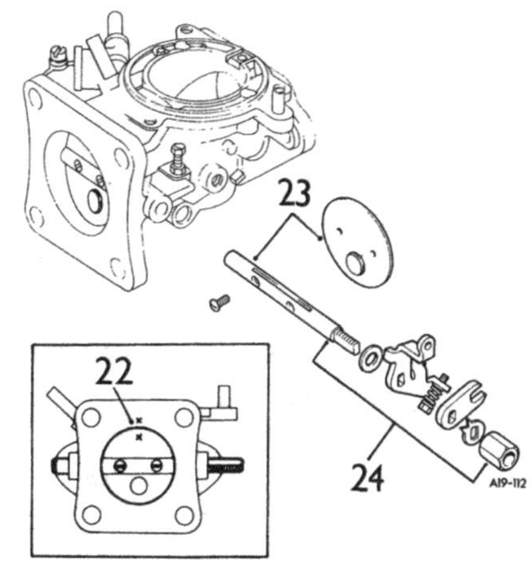

Inspecting

25 Examine the components as follows:

a Check the throttle spindle in the body for excessive play, and renew if necessary.

b Examine the float needle for wear; i.e. small ridges or grooves in the seat of the needle, and ensure that the spring-loaded plunger on the opposite end operates freely. Replace the needle and seating if necessary.

c Inspect all other components for wear and damage; renew unserviceable components.

Reassembling

26 Refit the spindle to the body, with the countersunk holes in the spindle facing outwards.

27 Assemble the spacing washer, lever, fork lever, lock washer and securing nut, ensure that the idling stop on the lever is against the idling screw abutment on the body in the closed throttle position. Tighten the spindle nut and lock with tab washer.

28 Insert the throttle disc into the spindle slot; note the markings for reassembling. Manoeuvre the disc in the spindle until the throttle can be closed, snap the throttle open and closed to centralize it in the bore of the carburetter.

29 Fit new disc retaining screws but do not fully tighten, check that the disc closes fully and adjust its position as necessary. Tighten the screws fully and spread their split ends just enough to prevent them turning.

30 Screw the seating into the float chamber lid; do not overtighten. Insert the needle coned-end first into the seating.

31 Refit the float to the chamber lid and insert the hinge pin.

32 Refit the float chamber lid with a new joint washer, noting the assembly markings, tighten the securing screws evenly.

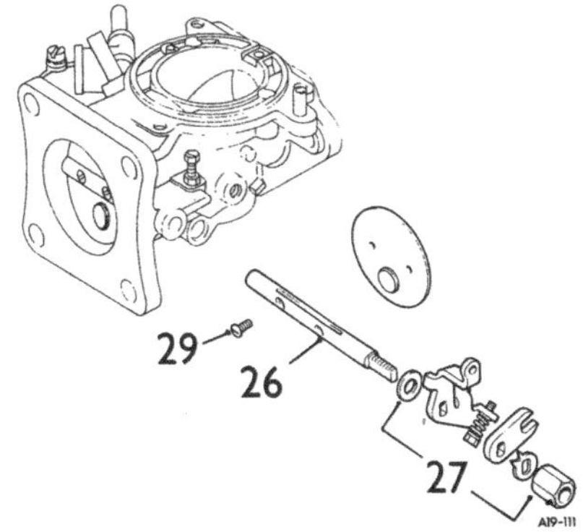

33 Refit the float chamber to the body and tighten the retaining bolt.

34 Where fitted, refit the piston lifting pin, spring and circlip.

35 Clean fuel deposits off the suction chamber and piston with fuel or methylated spirit and wipe dry.

CAUTION: Do not use an abrasive.

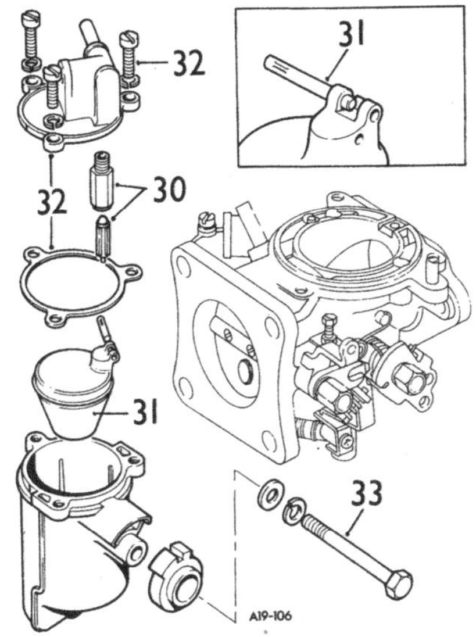

36 Check the operation of the suction chamber and piston (without the spring fitted) as follows:

a Refit the damper and washer to the suction chamber; temporarily plug the piston transfer holes with rubber plugs or Plasticine and insert the piston fully into the suction chamber.

b Secure a large flat washer to **one** of the fixing holes with a screw and nut so that it overlaps the bore.

c With the assembly upside-down, hold the piston and check the time taken for the suction chamber to fall the full extent of its travel. The time taken should be five to seven seconds; if this time is exceeded, check the piston and chamber for cleanliness and mechanical damage. Renew the assembly if the time taken is still not within these limits.

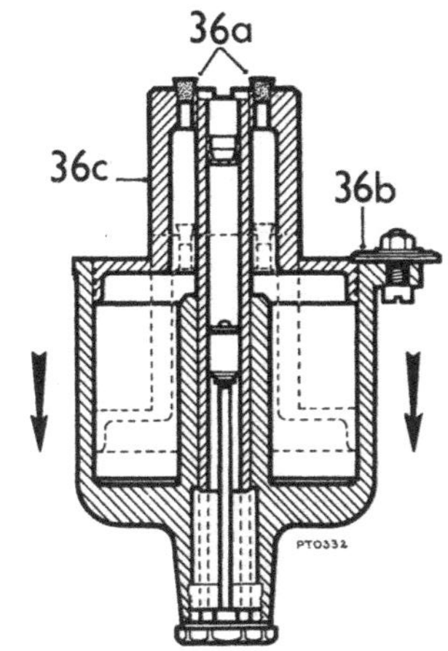

37 Fit the spring and guide to the needle and insert the assembly into the piston (with the shoulder of the needle flush with the underside face of the piston) and the slot in the guide positioned adjacent to the needle guide locking screw. Fit a new guide locking screw.

continued

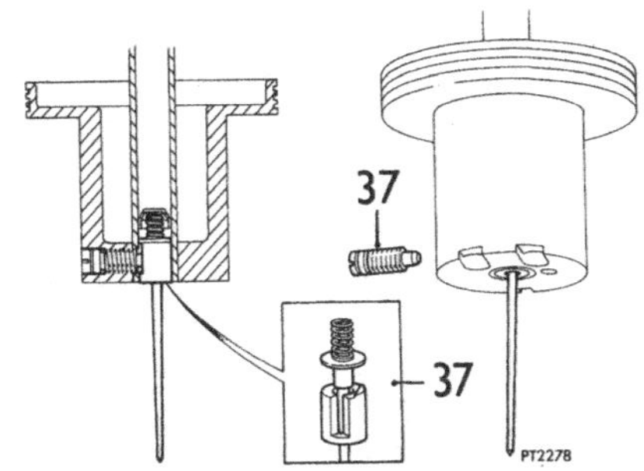

38 Check the piston key in the body for security.
39 Refit the jet bearing; fit and tighten the jet locking nut.
40 Refit the spring and jet adjustment nut; screw the nut up as far as possible.
41 Insert the jet into the bearing, fit the sleeve nut, washer and gland to the end of the flexible tube (if removed). The tube must project a minimum of $\frac{3}{16}$ in (4.8 mm) beyond the gland. Tighten the sleeve nut until the gland is compressed; over-tightening can cause leakage.
42 Refit the piston, spring and suction chamber to the body (noting the assembly marks) and tighten the securing screws evenly.
43 Reverse the procedure in 15 and 16.
44 Hold up the choke lever to relieve pressure on the jet pick-up link, refit the link bracket; support the end of the moulded jet and tighten the securing screw.

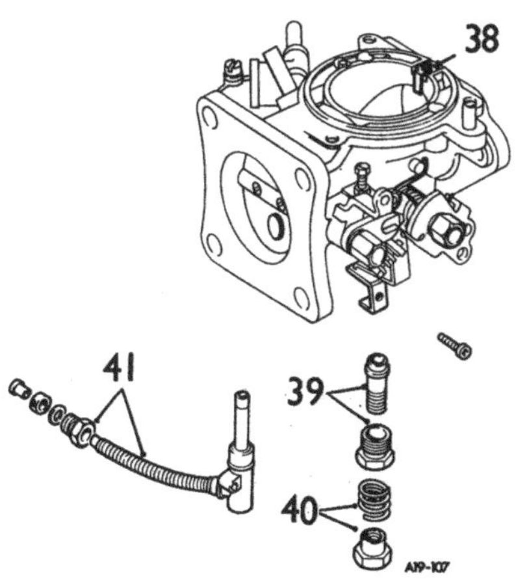

45 Screw the jet adjusting nut down two complete turns (12 flats) to provide the initial setting.
46 Refit the carburetters, see 19.15.11.
47 Tune and adjust the carburetters, see 19.15.02.

CARBURETTER FLEXIBLE MOUNTINGS

Remove and refit **19.15.19**

Removing
1 Remove the air cleaner, see 19.10.01.
2 Remove the carburetters, see 19.15.11.
3 Remove the six bolts (three per manifold) securing the mountings to the intakes.
4 Remove the mountings.

Refitting
5 Clean the manifold intake and flexible mounting faces and secure the mounting with the bolts and spring washers.
6 Fit the carburetters and tune, see 19.15.11.
7 Fit the air cleaner, see 19.10.01.

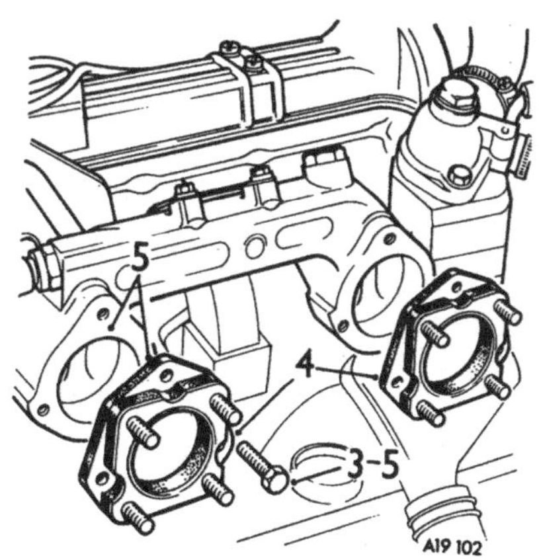

FLOAT CHAMBER NEEDLE AND SEAT

Remove and refit **19.15.24**

Removing
1 Remove the fresh air duct for rear carburetter, see 80.15.31.
2 Disconnect the fuel over-flow pipe.
3 Disconnect the fuel feed pipe to the float chamber.
4 Disconnect the breather hose from the carburetter.
5 Mark the lid and float-chamber for assembly.
6 Remove the lid securing screws and detach the lid.
7 Hold the float hinge pin at its serrated end and withdraw the pin and float.
8 Extract the float needle from its seating.

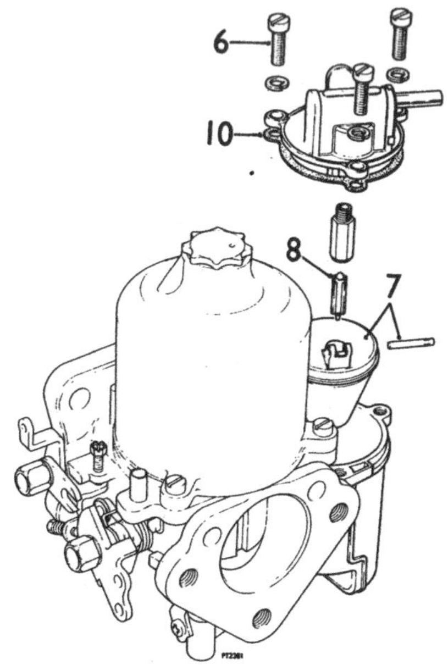

Inspecting
9 Examine the float needle for wear, i.e. small ridges or grooves in the seat of the needle, also check that the spring-loaded plunger on the opposite end operates freely. Renew the needle and seating if the needle is worn.

Refitting
10 Clean any sediment from the float-chamber, and fit a new joint washer if required.
11 Reverse instructions 1 to 9.

PISTON AND SUCTION CHAMBER (SU CARBURETTER)

Remove, clean and refit **19.15.30**

Where tamperproof carburetters are fitted refer to 19.00.00.

Removing
1 Remove the fresh air duct—for rear carburetter only, see 80.15.31.
2 Mark the relative position of the suction chamber and the carburetter body.
3 Remove the damper and its washer.
4 Unscrew the suction chamber securing screws and lift off the chamber.

continued

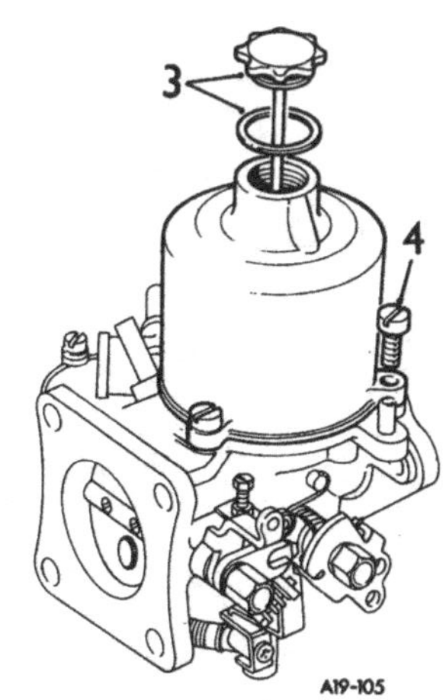

5 Remove the piston spring.
6 Carefully lift out the piston assembly and empty the oil from the piston rod.
7 Clean fuel deposits off the suction chamber and piston with fuel or methylated spirit and wipe dry.

CAUTION: Do not use abrasives.

8 Check the operation of the suction chamber and piston (without the spring fitted) as follows:
   a Refit the damper and washer to the suction chamber, temporarily plug the piston transfer holes with rubber plugs or Plasticine and insert the piston fully into the suction chamber.
   b Secure a large flat washer to one of the fixing holes with a screw and nut so that it overlaps the bore.
   c With the assembly upside-down, hold the piston and check the time taken for the suction chamber to fall the full extent of its travel. The time taken should be five to seven seconds, if this time is exceeded, check the piston and chamber for cleanliness and mechanical damage. Renew the assembly if the time taken is still not within these limits.

Note: This operation is not applicable to tamperproof carburetters.

Refitting
9 Refit the piston, spring and suction chamber to the carburetter (noting the assembly marks) and tighten the screw evenly.
10 Top up each piston damper with a recommended engine oil until the level is ½ in (13 mm) above the top of the hollow piston rod.
11 Refit each piston damper with its washer.

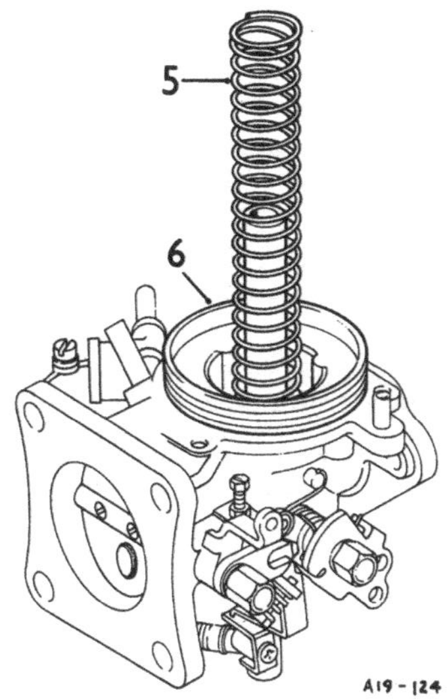

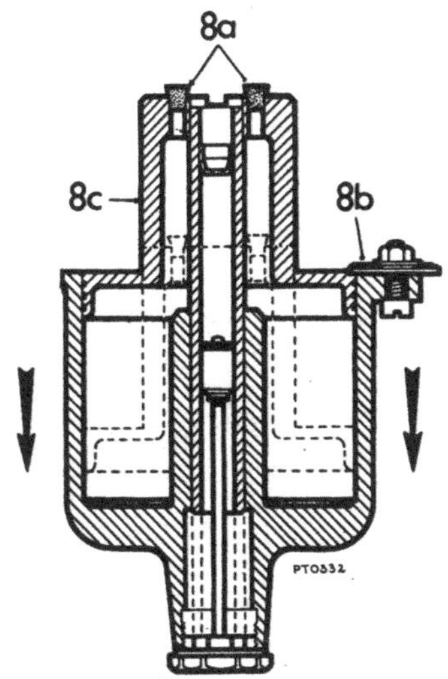

THROTTLE PEDAL ASSEMBLY

Remove and refit 19.20.01

Removing
1 Disconnect the battery.
2 Working from inside the car remove the spring clip securing the throttle cable.
3 Remove the cable from the pedal.
4 Open the bonnet and remove the two bolts securing the pedal assembly to the bulkhead.
5 Remove the pedal assembly from the car.
6 Strip the pedal assembly, and discard worn parts.

Refitting
7 Locate the spring in the bracket.
8 Fit the pedal to the bracket.
9 Insert the clevis pin.
10 Fit the anti-rattle washer.
11 Fit the plain washer.
12 Secure the assembly with a new split pin.
13 Fit the pedal to the bulkhead and secure with the two bolts.
14 Fit the cable to the pedal and retain with the spring clip.
15 Re-connect the battery.

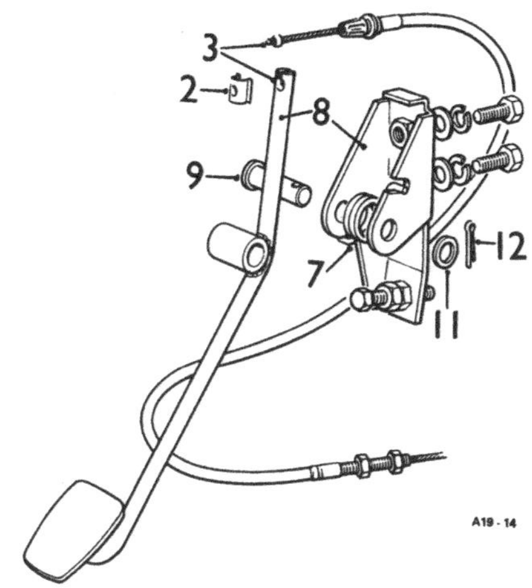

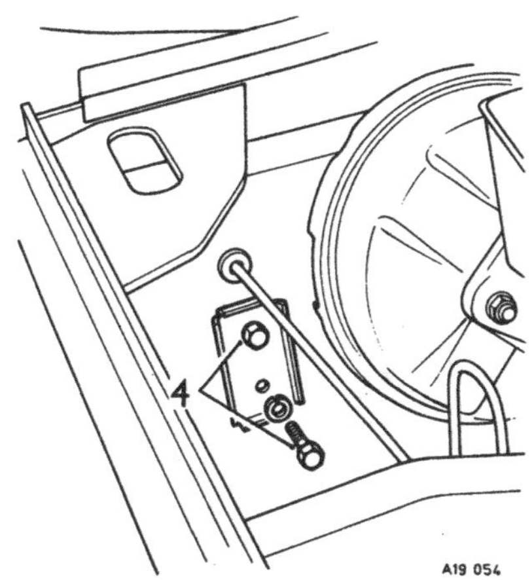

MIXTURE CONTROL CABLE ASSEMBLY

Remove and refit **19.20.13**
Inner cable – instructions 1 to 2 **19.20.14**

Removing
1 Disconnect the exposed inner cable from the carburetter trunnion.
2 Working inside the car, pull the mixture control knob and withdraw the inner cable complete.
3 Pull-off the choke indicator Lucar from the switch.
4 Slacken the outer cable retaining nut.
5 Unscrew the bezel from the outer cable.
6 Remove the outer cable complete from the bracket and car.
7 Slacken the grub screw and remove the choke indicator switch from the cable.

Refitting
8 Reverse instructions 1 to 7 ensuring that the choke indicator switch is fitted correctly, see 86.65.53.

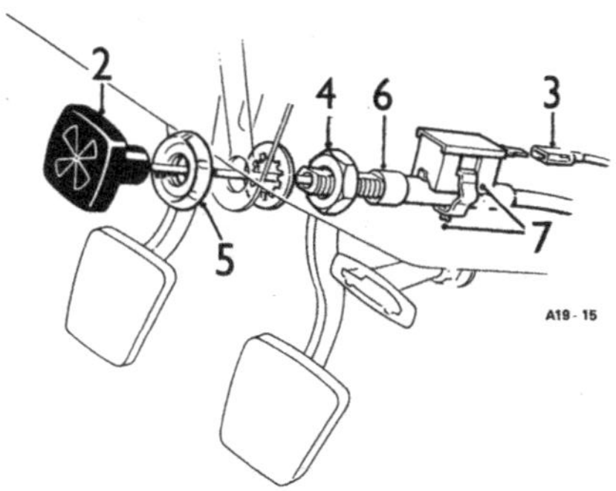

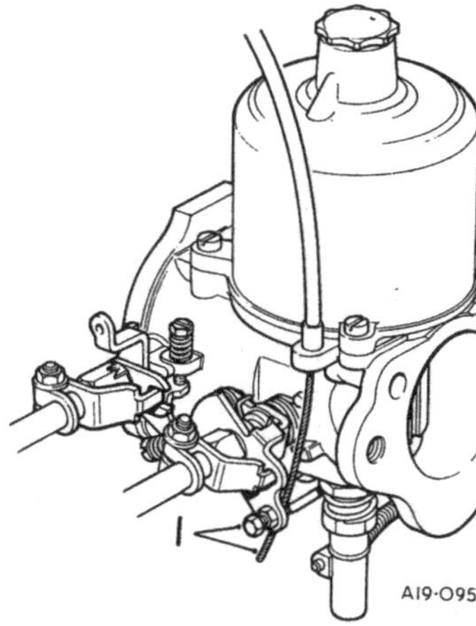

PETROL PIPE – MAIN LINE – TANK END SECTION

Remove and refit **19.40.02**

Removing
1 Drive the car on to a ramp and raise.
2 Drain the petrol tank.
3 Pull off the rubber connection from the tank.
4 Release the tank-end section from the engine-end section at the union (right-hand rear wing).
5 Release the pipe from the retaining clips and remove from the car.

Refitting
6 Clip the pipe into position.
7 Connect the pipe to the engine-end section at the union.
8 Connect the pipe to the tank with the rubber connector.
9 Lower the ramp and refill the tank.

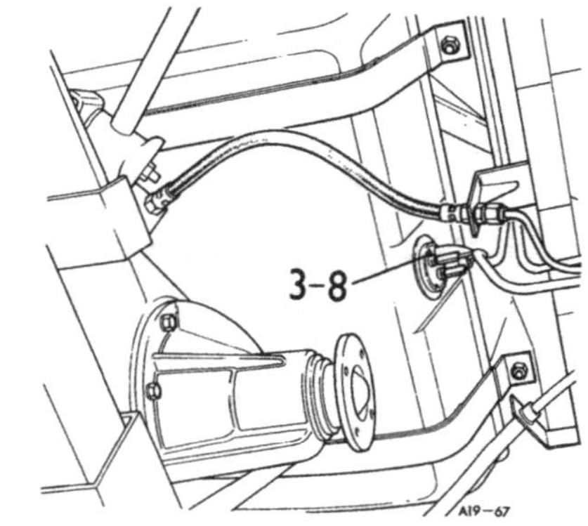

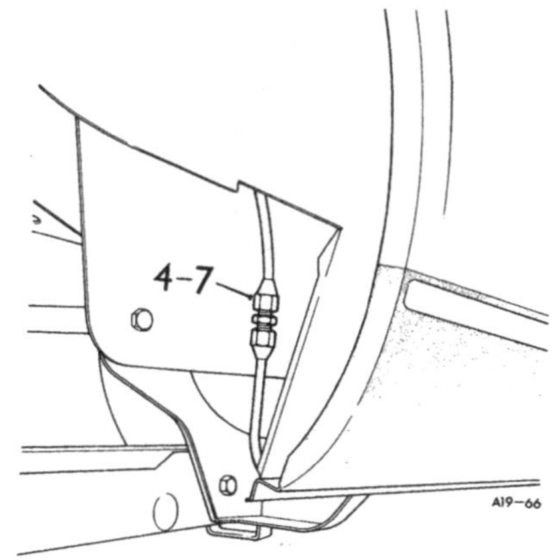

PETROL PIPE – MAIN LINE – ENGINE END SECTION

Remove and refit **19.40.04**

Removing
1 Drive the car onto a ramp.
2 Raise the ramp and drain the petrol tank.
3 Disconnect the engine-end section from the petrol tank-end section at the union (right-hand wing).
4 Pull off the rubber hose connecting engine-end section to the fuel pump.
5 Remove the pipe from the retaining clips.

Refitting
6 Reverse instructions 3 to 5.
7 Start the engine and checks for leaks.
8 Stop the engine and lower the ramp.

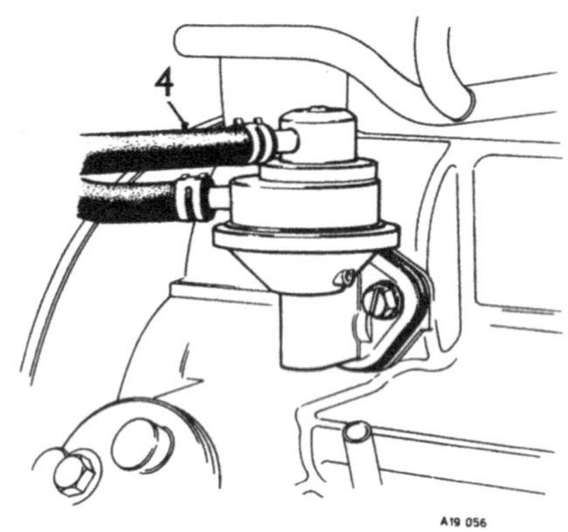

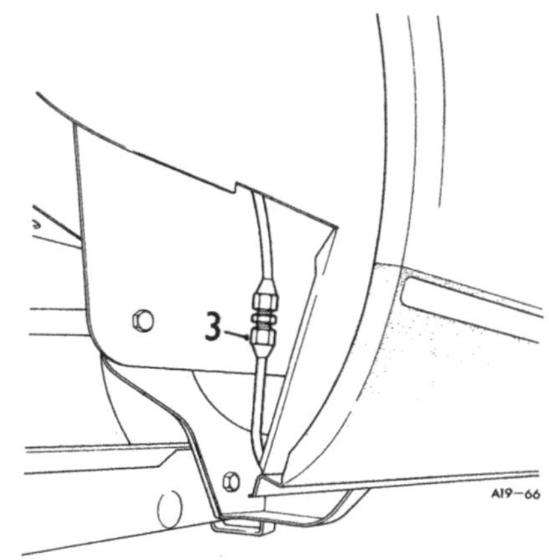

HOSE – FILLER TO TANK

Remove and refit 19.40.19

Removing
1 Open the boot and remove the access panel.
2 Remove the filler cap assembly, see 19.55.08.
3 Release the hose clip securing the filler hose to the fuel tank, visible through the access aperture.
4 Pull off the filler hose.

Refitting
5 Fit the filler hose to the tank and secure with the hose clip.
6 Fit the fuel filler cap assembly – instructions 7 to 10, see 19.55.08.
7 Fit the access panel.

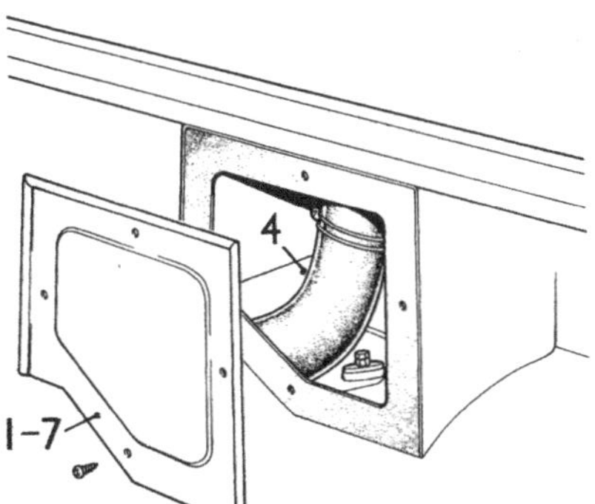

FUEL PUMP

Clean filter 19.45.05

1 Remove the fresh air duct, see 80.15.31.
2 Remove the pump cover retaining screw using a long shafted screw driver.
3 Lift-off the cover.
4 Remove the sealing ring.
5 Lift out the filter.
6 Clean sediment from the cover.
7 Clean and examine the filter and refit.
8 Fit a new sealing ring – if necessary.
9 Fit the cover and secure it with the screw and a new washer.

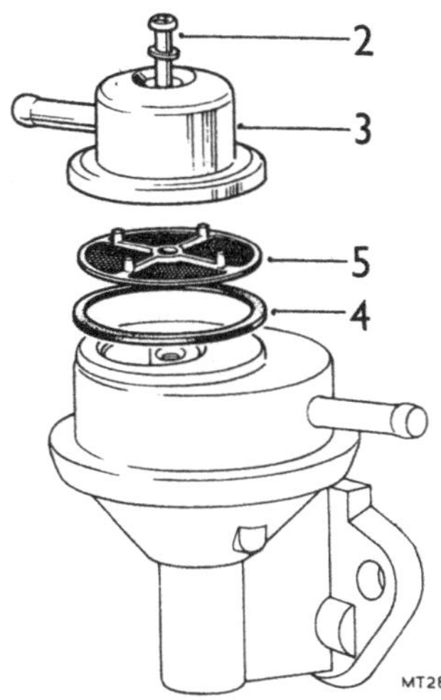

FUEL PUMP

Remove and refit 19.45.08

Removing
1 Remove the air cleaner outer cover and element.
2 Disconnect the pipe from the fuel tank to the pump inlet connection.
3 Disconnect the outlet pipe from the pump.
4 Remove the bolts securing the pump to the cylinder block.
5 Remove the pump.
6 Remove the gasket.

Refitting
7 Clean the pump and cylinder block mating faces.
8 Fit a new gasket.
9 Fit the pump ensuring that the actuating arm rides on top of the cam.
10 Fit and tighten the retaining nuts and spring washers.
11 Fit the fuel inlet and outlet pipes.

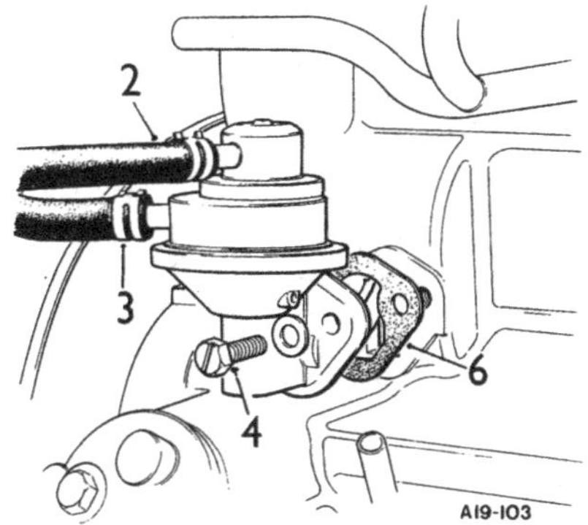

FUEL TANK

Remove and refit 19.55.01

Removing
1 Drive the car onto a ramp and disconnect the battery.
2 Remove the fuel filler cap assembly, see 19.55.08.
3 Raise the ramp and unclip the fuel tank breather hose from the LH side of the rear chassis member.
4 Drain the fuel tank.
5 Support the body with jacks both sides of the car forward of the trailing arm attachment points.
6 Disconnect the dampers at their lower attachment to the axle.
7 Disconnect the tank unit electrical leads.
8 Pull off the rubber connection to the main fuel line.
9 Disconnect the R.H. radius rod from the body bracket.
10 Remove the L.H. radius rod from the car.

continued

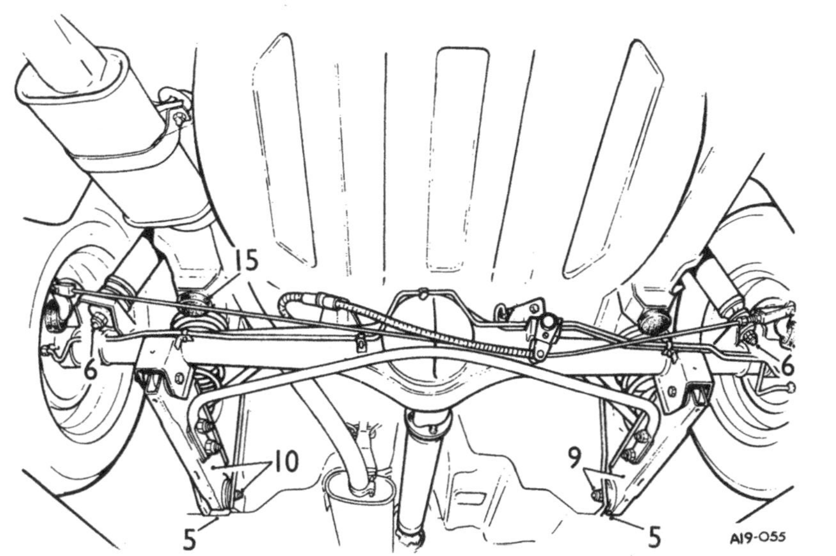

11 Jack up the body sufficiently to remove both road springs.

CAUTION: Care must be taken to ensure that whilst jacking up the body the brake hydraulic system jump hose is not stretched–shown arrowed.

12 Remove the L.H. rear road wheel and allow the axle to drop having regard for the above cautionary note.
13 Support the axle with a jack to relieve any tension in the hydraulic jump hose.
14 Remove the tail pipe and silencer assembly, see 30.10.22.
15 Remove the L.H. bump stop.
16 Remove the four nuts and tank retaining straps.
17 Withdraw the tank from the L.H. side of the car.
18 Remove the filler hose.
19 Remove the breather hose.
20 Remove the tank unit, see 88.25.32 instructions 6 to 8.

NOTE: Instructions 18 to 20 are only necessary if the tank is to be renewed.

Refitting

NOTE: If the tank is not being renewed ignore instructions 18 to 22. However, renew all hoses if any deterioration exists.

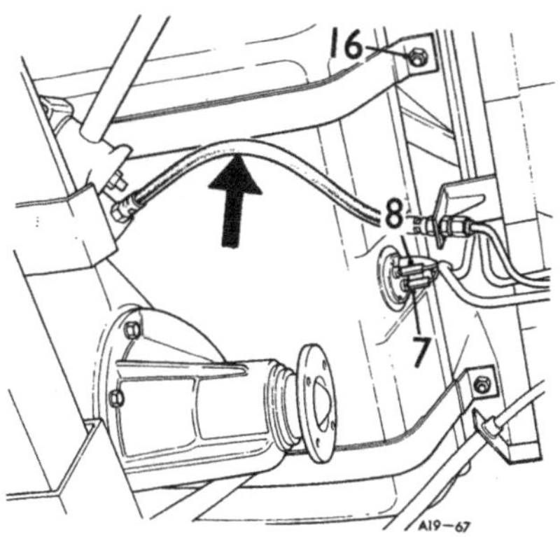

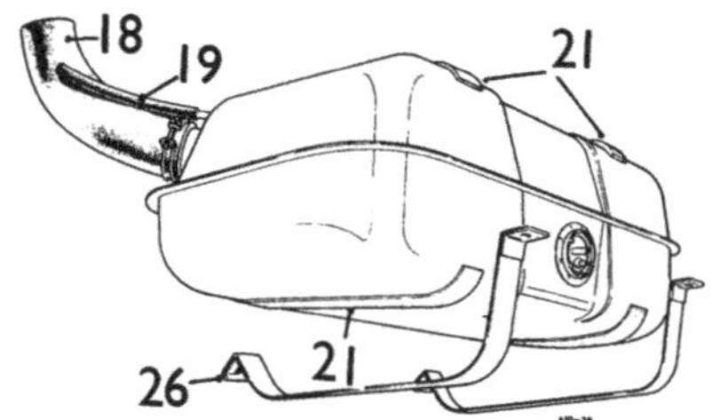

21 Using adhesive fit new cushion strips to the tank noting that thicker strips are fitted to the top of the tank.
22 Fit the filler hose.
23 Fit the breather hose.
24 Fit the tank unit – reversing instructions 6 to 8, see 88.25.32.
25 Manoeuvre the tank into position from the L.H. side of the car.
26 Fit the supporting straps noting that the elongated holes in the straps are fitted to the rear studs in the body.
27 Fit and tighten the four retaining nuts.
28 Fit the L.H. radius rod to the axle–including the handbrake cable bracket–leaving the body-end free for the time being.
29 Fit the springs and jack-up the axle both sides and connect the dampers to their axle locations.
30 Fit the L.H. bump stop.
31 Fit the L.H. and R.H. radius rods to their body locations.
32 Fit the tail pipe and silencer assembly, see 30.10.22.
33 Fit the L.H. rear road wheel.
34 Lower and remove the body and axle jacks.
35 Connect the fuel tank to the main line pipe with the rubber connector.
36 Fit the electrical leads to the tank unit–instruction 10, see 88.25.32.
37 Fit the breather pipe.
38 Lower the ramp and fit the fuel filler assembly, see 19.55.08 instructions 7 to 10.
39 Fit the access panel.
40 Refill the fuel tank, connect the battery, start the engine and check for leaks.

FUEL FILLER CAP AND FILLER ASSEMBLY

Remove and refit 19.55.08

Removing

1 Turn the fuel filler cap anti-clockwise and withdraw.
2 Remove the three screws securing the filler tube assembly to the body.
3 Open the boot and remove the access panel–four screws.
4 Slacken the top hose clip.
5 Withdraw the filler tube.
6 Remove the finisher.

Refitting

7 Place the finisher in position.
8 Insert the filler tube into the filler hose and secure with the three screws.
9 Refit the filler cap.
10 Tighten the filler hose top clip.
11 Refit the access panel and close the boot.

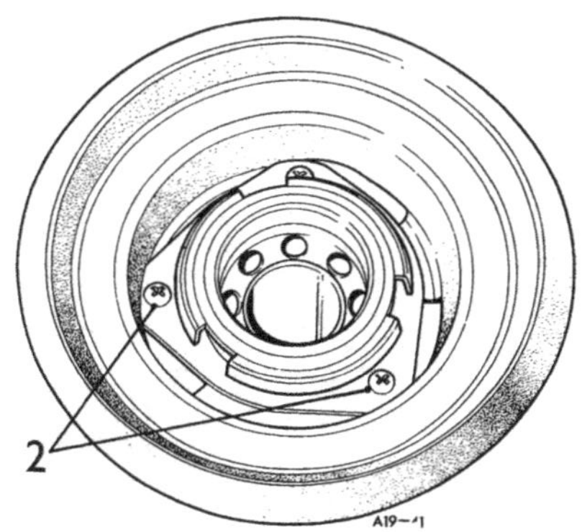

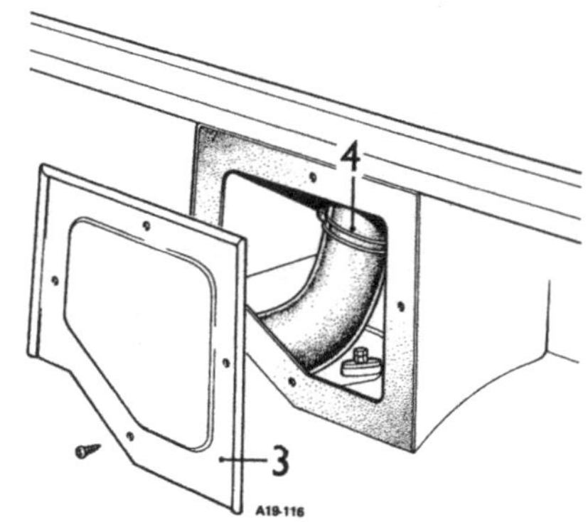

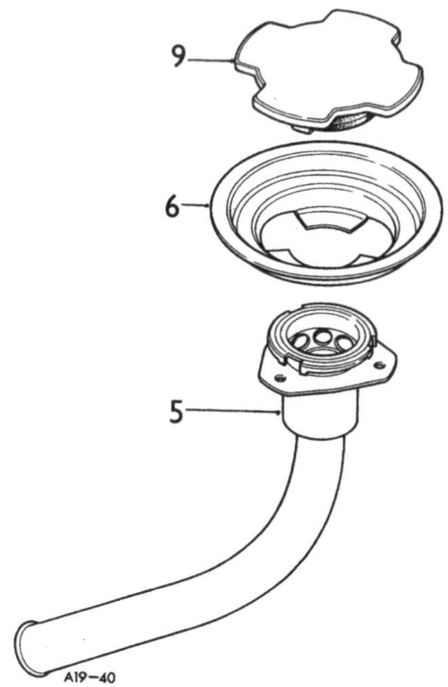

FUEL SYSTEM OPERATIONS—U. S. A. AND CANADA SPECIFICATION —CARBURETTER VEHICLES

Air cleaner	
—remove and refit	19.10.01
—renew element	19.10.08
Automatic choke (single carburetters)—remove and refit	19.15.38[A]
(F.A.S.D.)—remove and refit	19.15.38[B]
Carburetter	
—complete with inlet manifold—remove and refit	19.15.15
—remove and refit	19.15.11
—rubber mountings—set—remove and refit	19.15.19
—starter assembly—remove and refit	19.15.52
—tune and adjust (single)	19.15.01
—tune and adjust (twin)	19.15.02
Deceleration and by-pass valve (twin carburetter)	
– adjust	19.15.02
– remove and refit	19.15.64
Diaphragm – remove and refit	19.15.35
Float-chamber level – check and adjust	19.15.32
Float-chamber needle valve – remove and refit	19.15.24
Fuel filler cap and filler assembly – remove and refit	19.55.08
Fuel Pump	
– clean filter	19.45.05
– remove and refit	19.45.08
Fuel tank – remove and refit	19.55.01
Hose – filler to tank – remove and refit	19.40.19
Petrol pipe	
– engine end section – remove and refit	19.40.04
–main line – tank end section – remove and refit	19.40.02
Temperature compensator – remove and refit	19.15.59
Throttle pedal – remove and refit	19.20.01

AIR CLEANER

Remove and refit **19.10.01**

Earlier Models

Removing

1 Disconnect the emission control pipes from the clips attached to the air cleaner back plate.
2 Disconnect the hot air hose from the air cleaner intake.
3 Remove the two bolts, plain and rubber washers, securing the air cleaner to the carburetter.
4 Withdraw the air cleaner complete with the air temperature control unit.
5 Slacken the hose clip and remove the air temperature control unit.

Refitting

6 Reverse instructions 1 to 5, using a new gasket between the air cleaner and carburetter intake.

Later Models

Removing

1 Disconnect the hot air pipe from the air cleaner.
2 Remove the vacuum pipe from air inlet temperature control unit.
3 Release the two retaining clips.
4 Separate the air cleaner outer cover from the backplate and withdraw the element.
5 Remove the outer cover.
6 Remove the two fuel traps from the clips.
7 Disconnect the vacuum pipe from the temperature sensor unit.
8 Remove the single retaining nut behind the back-plate.
9 Remove the six bolts securing the back-plate to the carburetter intakes.
10 Withdraw the back-plate complete with gaskets.

Refitting

11 Using new gaskets reverse instructions 1 to 10.

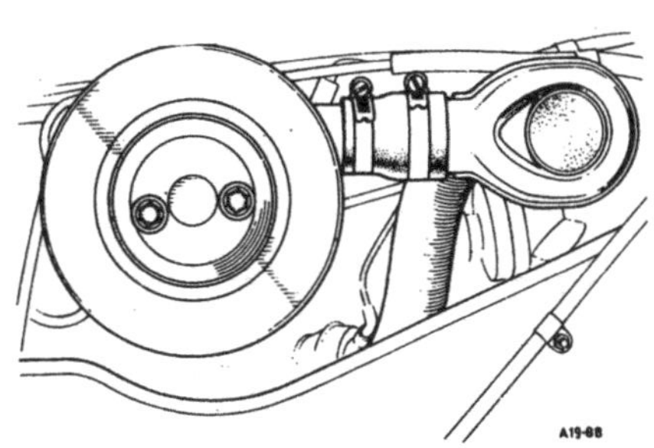

A19-88

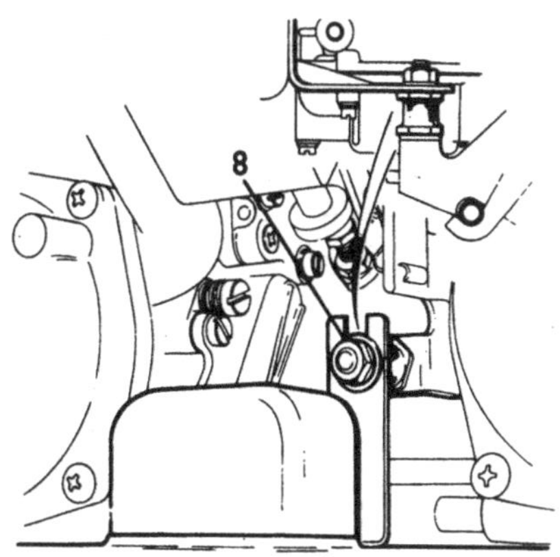

A19-52

AIR CLEANER

Renew element **19.10.08**

Earlier Models

1 Remove the air cleaner see 19.10.01, omitting instruction 5.
2 Separate the air cleaner back plate from the container.
3 Remove the used element and rubber sealing rings.
4 Clean the back plate and container.
5 Assemble the new element and sealing rings into the container.
6 Fit the back plate, ensuring that the peg on the container locates in the back plate slot.
7 Refit the air cleaner see 19.10.01, using a new gasket.

AHT 83

Later Models

1 Disconnect the hot air pipe from the air cleaner.
2 If necessary, disconnect engine dipstick support and slacken the nut, securing the throttle linkage bracket to the air cleaner.
3 Release the two retaining clips and hinge back the outer cover from the back-plate.
4 Withdraw the air cleaner element.
5 Clean the inside of the box and insert a new element noting that it is fitted with the rubber seal towards the engine.
6 Hinge forward the outer cover and secure with the two clips.
7 Connect the hot air pipe.
8 Check that the vacuum pipe is connected to the temperature sensor unit at one end and the air inlet temperature control unit at the other.
9 Ensure that the throttle linkage bracket is secured, and reconnect the dipstick support, if necessary.

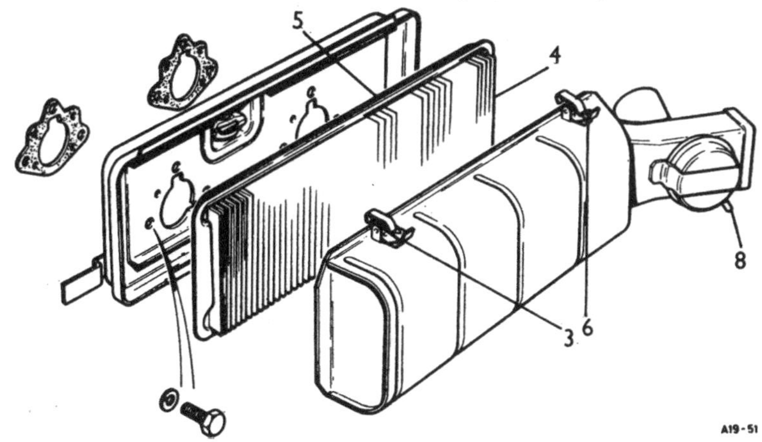

A19-51

CARBURETTER (SINGLE)

– Tune and adjust **19.15.01**

Special tools: Non dispersive infra-red exhaust gas analyser.
Mixture adjusting tool S353

CAUTION: To ensure compliance with the exhaust emission legislative requirements the following items MUST NOT be changed or modified in any way:

The fuel jet assembly.
The piston assembly.
The depression cover.

The following items must not be adjusted in service but should be replaced completely by factory-set units:

The piston assembly return spring.
The autochoke.

1. Run the engine until normal operating temperature is reached.

NOTE: During the course of the following instructions do not allow the engine to idle for longer than three minutes without a 'clear-out' burst of one minute duration at 2000 r.p.m.

Checking and adjusting idle speed

2. Start the engine and check that the idle speed is steady at 800 r.p.m.
3. If adjustment is required, turn the throttle adjusting screw clockwise until the correct idle speed is obtained. Stop the engine.

Checking and adjusting the C.O. at idle

4. Disconnect and plug the air pump outlet hose.
5. Start the engine and check that the idle speed is 800 r.p.m.
6. Check and if necessary adjust the ignition timing.
7. Re-check the idle speed.
8. Insert the gas analyser probe as far as possible into the exhaust pipe.
9. Check the CO reading against the figure quoted on the emission control decal on the car (under the bonnet).
10. Adjust the mixture if necessary – see mixture adjustment.
11. Check and if necessary adjust the idle speed.
12. Withdraw the analyser probe.
13. Stop the engine.
14. Un-plug air injection hose and connect to the pump.

continued

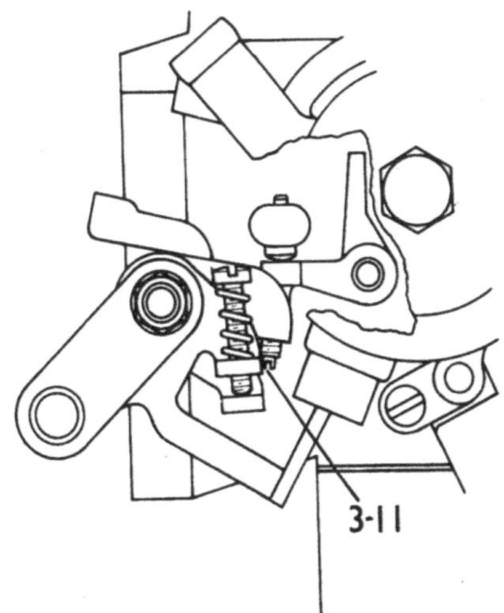

A19-63

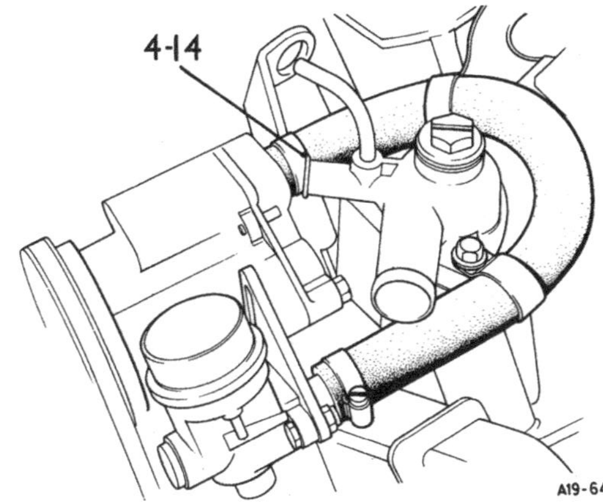

A19-64

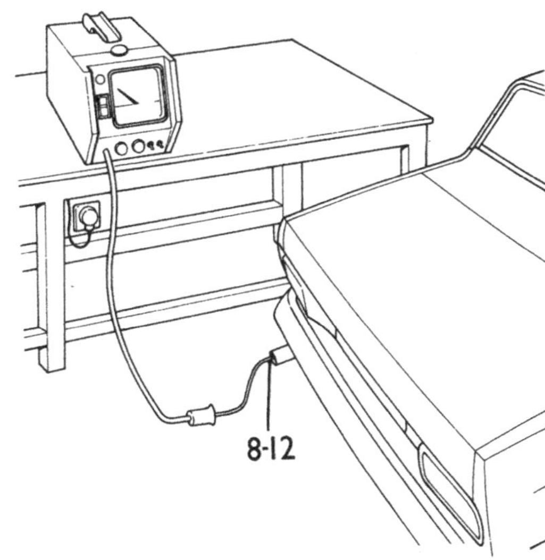

A19-62

Mixture adjustment

15. If the CO reading is only just outside the permitted limits adjust the fine idle CO screw until the CO reading is correct.

CAUTION: The setting must always be checked by means of a non-dispersive infra-red gas analyser.
Do not attempt to alter the coarse idle CO adjustment **A.** This is set at the factory and sealed with a plastic cap, **B.**
For significant deviation outside the specified limits the following instructions should be carried out:

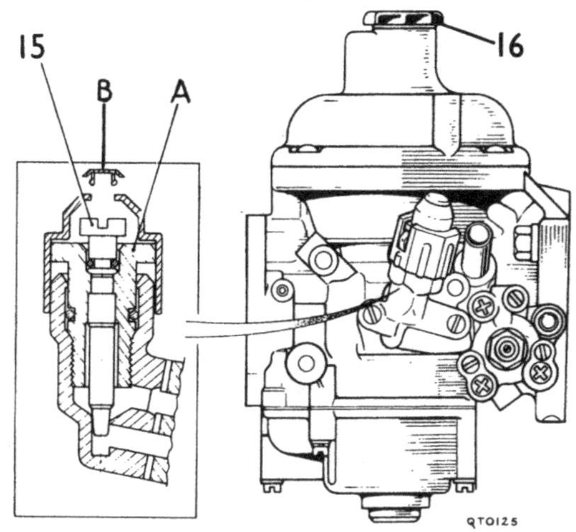

16. Maintain the engine at the normal running temperature and remove the piston damper.
17. Slowly insert tool S353 into the dashpot until the outer tool engages in the air valve and the inner tool engages the hexagon in the needle adjuster plug.

[**NOTE:** Tool BLT 2010, a tool similar to tool S 353, is also suitable for this job.]

CAUTION: The outer tool must be correctly engaged and held otherwise the diaphragm may be damaged.

18. Hold the tool firmly and turn the inner tool:
    a) Clockwise to enrich the mixture.
    b) Anti-clockwise to weaken.
19. Top-up the carburetter damper, ensuring that the special instructions in Section 10 are followed.
20. Recheck the CO reading and repeat instructions 16 to 19 until the CO reading is within the specified limits.

continued

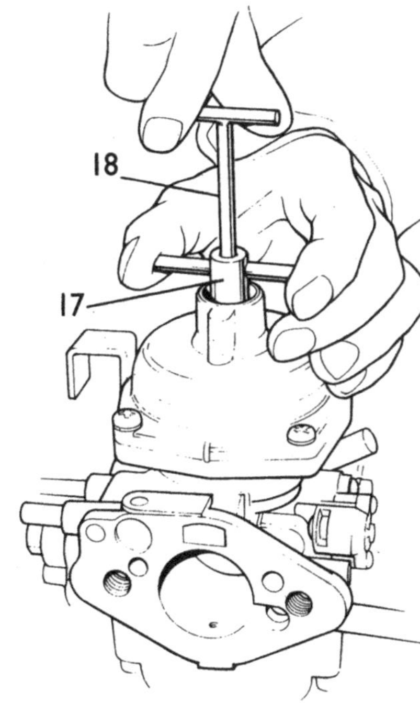

Checking and adjusting deceleration and by-pass valve

21. Maintain the engine at normal running temperature.
22. Remove the vacuum pipe from the distributor and place a finger over the end of the pipe.
23. The valve is operating correctly if the engine idle speed increases to approximately 1300 r.p.m.
24. Should the valve be 'floating' and therefore out of adjustment, the engine idle speed will increase rapidly to approximately 2000 to 2500 r.p.m.
25. If adjustment is necessary leave the vacuum pipe disconnected from the distributor and plug the end.
26. Start the engine and with the spring clip in position turn the by-pass valve screw clockwise until the engine increases abruptly to approximately 2000 to 2500 r.p.m. The valve is now 'floating' on its seat.

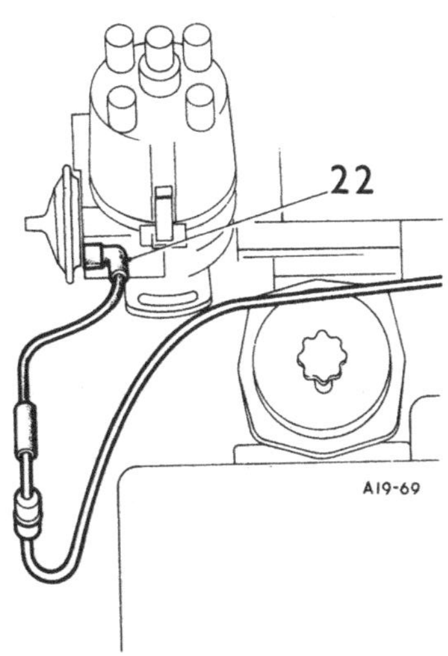

27. Turn the adjustment screw anti-clockwise until the engine speed falls to approximately 1300 r.p.m.
28. Using the throttle, suddenly increase the engine speed and immediately release the throttle. The engine speed should drop to approximately 1300 r.p.m. If not the valve is still 'floating'. Re-adjust repeating instructions 25 to 28.
29. When correct adjustment of the by-pass valve has been achieved, turn the valve adjustment screw anti-clockwise half a turn to finally seat the valve.
30. Unplug the vacuum pipe and connect it to the distributor.

CAUTION: The by-pass valve must be changed as a complete unit if renewal is necessary. 19.15.64.

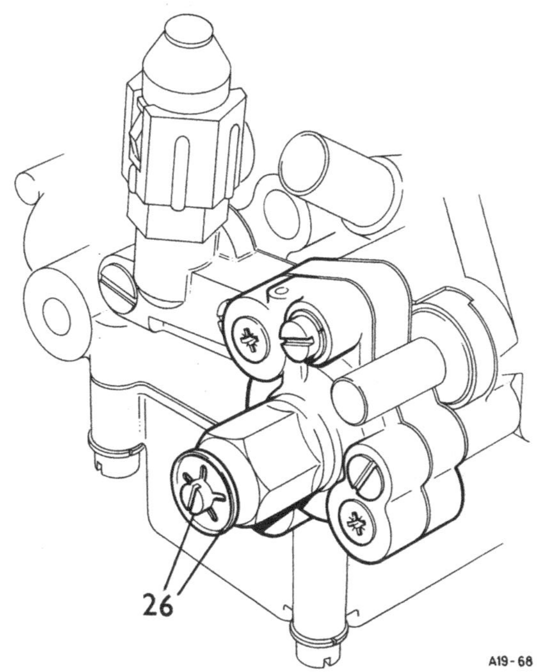

CARBURETTERS—[TWIN]

– Tune and adjust **19.15.02**
Special tools: Non dispersive infra-red exhaust gas analyser and mixture adjusting tool S353.

CAUTION: To ensure compliance with exhaust emission legislative requirements the following items must not be changed or modified in any way:
The fuel jet assembly.
The piston assembly.
The depression cover.

The following items must not be adjusted in service but should be replaced completely by factory-set units:

The temperature compensator.
The piston assembly return spring.
The starter assembly.

NOTE: During the course of the following instructions do not allow the engine to idle for longer than three minutes without a 'clear-out' burst of one minute duration at 2000 r.p.m.

1 Remove the fresh air duct – 80.15.31.
2 Run the engine until normal operating temperature is reached.
3 Remove the air cleaner. 19.10.01.

Adjusting idle speed and air flow for balance
4 Ensure that the fast idle screw is clear of the fast idle cam.
5 Using an air flow meter check that the air flow through both carburetters is the same. If not, adjust as follows:
6 Hold the roller into the corner of the progression lever. – (Canadian Specification only).
7 Adjust the spring loaded screws equally until the adjusting bar is in the centre of both forks.
8 Unscrew the throttle adjusting screws on both carburetters to permit the throttles to close completely.
9 Rotate the throttle adjusting screws so that they just touch the throttle levers and give a further half-turn to provide a datum setting.

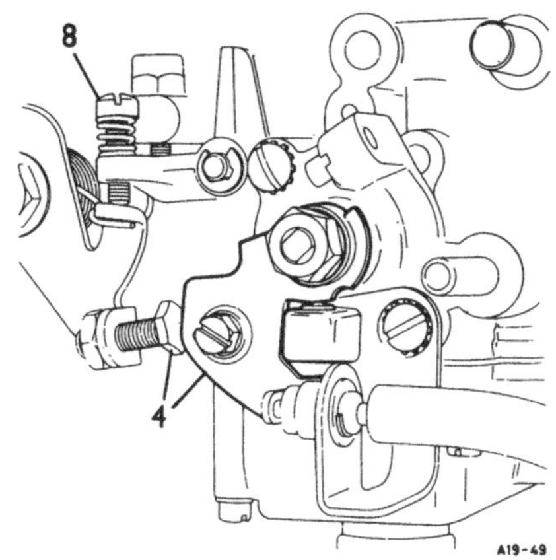

A19-49

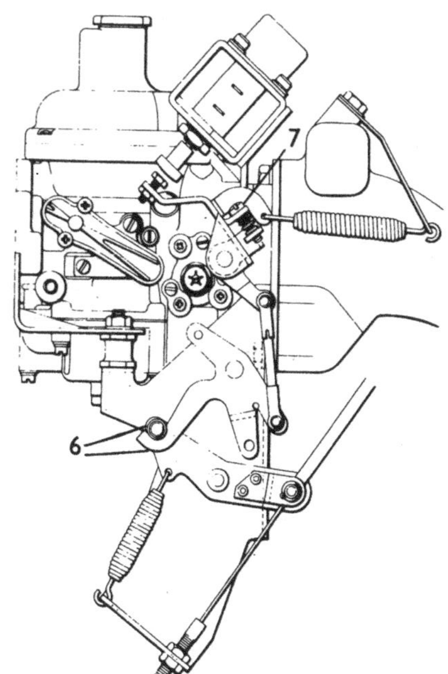

A19-47

10 Continue rotating the screws by equal amounts while holding an air flow meter to each carburetter intake in turn, until a balanced air flow is achieved at an engine speed of 800 r.p.m.
11 Hold the roller into the corner of the progression lever. – (Canadian Specification only).
Canadian Specification Vehicles only
12 Adjust the spring loaded screws to give a gap of 0.010 in. (0,25 mm) dimension A between the forks and the adjusting bar.
13 If necessary turn the throttle adjusting screws on both carburetters an equal amount to maintain the idle speed of 800 r.p.m.
14 Increase the engine speed to 1600 r.p.m. and check the balance with the air flow meter. If necessary turn the throttle adjusting screws by equal amounts to achieve a balance.
15 Re-check the air flow balance at idle speed.

Checking and adjusting CO level at idle – all vehicles
16 Disconnect and plug the outlet hose from the air pump.
17 Maintain the engine at normal operating temperature, and check that the idle speed is 800 r.p.m.
18 Check and if necessary adjust the ignition timing. 86.35.15.
19 Re-check the idle speed.
20 Insert the gas analyser probe as far as possible into the exhaust pipe.
21 Check the CO reading against the emission control decal on the car (under the bonnet lid).
22 Adjust the mixture if necessary – see mixture adjustment.
23 Check and if necessary adjust the idle speed.
24 Withdraw the analyser probe.
25 Switch off the ignition.
26 Unplug air injection hose and reconnect to the pump.

NOTE: Do not allow the engine to idle for longer than 3 minutes without a 'clear out burst' of 1 minute at 2000 rev/min.

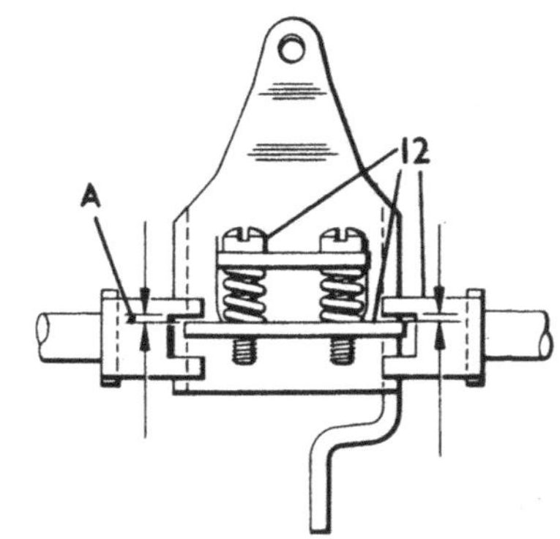

A19-48

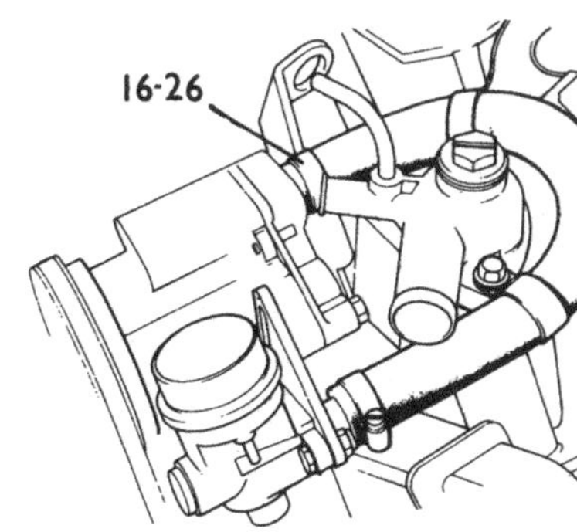

A19-64

Mixture adjustment
NOTE: If the CO reading is only slightly outside the permitted limits, adjust the idle trim screw A on both carburetters by equal amounts until the CO reading is to within the limits.

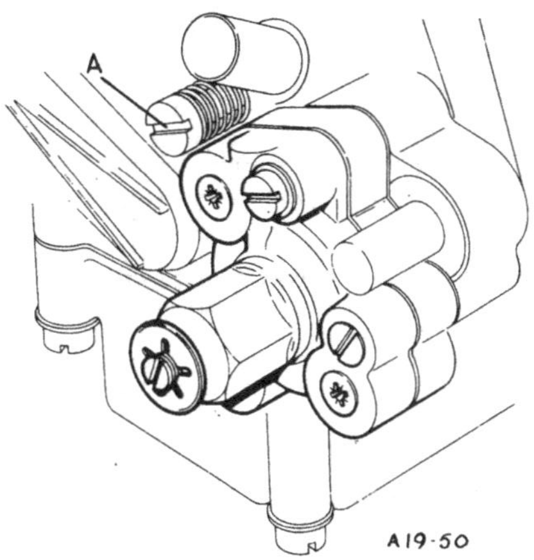

A19-50

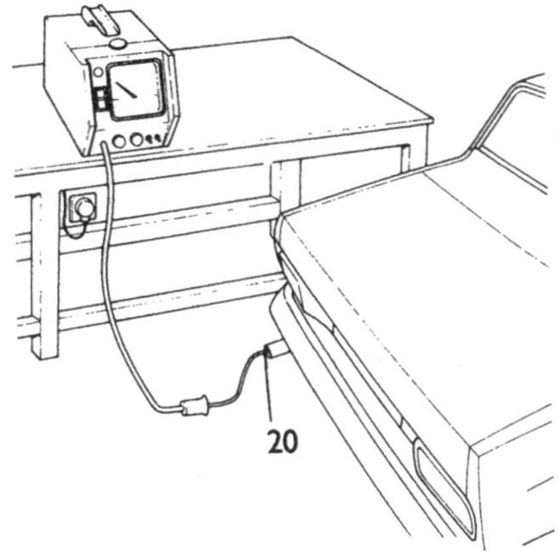

A19-62

CAUTION: The setting MUST ALWAYS be checked by means of a non-dispersive infra-red exhaust gas analyser.
For significant deviation outside the specified CO limits the mixture should be adjusted as follows:

27 Remove the piston damper from both carburetters.
28 Carefully insert special tool S353 into the dashpot until the outer tool engages in the air valve and the inner tool engages the hexagon in the needle adjuster plug.

[**NOTE:** Tool BLT 2010, a tool similar to tool S 353, is also suitable for this job.]

CAUTION: The outer tool must be correctly engaged and held in position otherwise damage to the diaphragm may result.

29 Holding the outer tool turn the inner tool:
   a Clockwise to enrich the mixture.
   b Anti-clockwise to weaken.
30 Repeat instruction 28 and 29 on the remaining carburetter ensuring that the adjustment made is by the same amount.
31 Top-up the carburetter dampers, ensuring that the special instructions are followed in Section 10.
32 Re-check the CO reading -- instructions 16 to 26 and if necessary repeat instructions 27 to 30 until the CO reading is within the specified limits.

Checking and adjusting fast idle speed –
Canadian Specification Vehicles only
33 Maintain the engine at normal running temperature.
34 Check that the mixture control cam lever on both carburetters returns to its stop.
35 Check and if necessary adjust the mixture control cable so that there is free movement.
36 Pull-out the mixture control cable knob on the control cowl approximately ¼ in. until the fast idle cams are correctly engaged with the ball locators.

continued

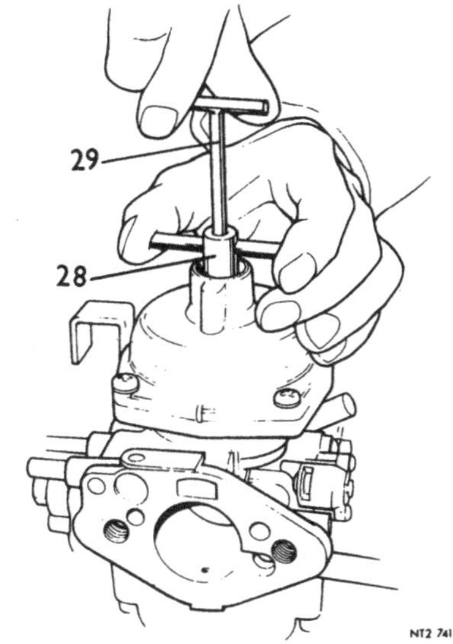

NT2 741

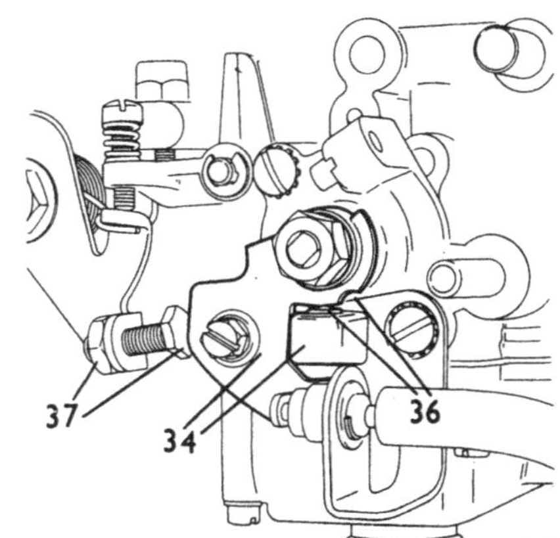

A19-49

37 Slacken the fast idle screw lock-nut on both carburetters, start the engine and rotate the screw head (on both carburetters) against the cam until the engine revolutions reach a steady 1600 r.p.m.
38 Use the air flow meter to check that the carburetters are in balance at 1600 r.p.m. if necessary adjust the fast idle screws as required.
39 Tighten the fast idle screw lock-nut on both carburetters and push the mixture control knob fully home.

NOTE: An engine set to 1600 r.p.m. while hot is equivalent to a fast idle speed of approximately 1300 r.p.m. when cold.

Checking and adjusting deceleration by-pass valves –
California and Canadian Specification Vehicles only

40 Maintain the engine at normal running temperature.
41 With the engine idling, disconnect the vacuum pipe from the distributor and place a finger over the end of the pipe.
42 If the valves are correctly adjusted the engine idle speed should increase to approximately 1300 r.p.m.
43 Should one or both of the valves be 'floating' and therefore out of adjustment the engine revolutions will increase quickly to approximately 2000–2500 r.p.m. Furthermore if the throttle is momentarily opened the resulting increase in r.p.m. will be slow to fall.
If adjustment is required continue with the following instructions:
44 Without removing the circular spring clip, turn the by-pass valve adjusting screw anti-clockwise fully on to its seat on the carburetter NOT being adjusted.

NOTE: This procedure prevents the one by-pass valve from working while the valve on the other carburetter is being adjusted. It is not however, necessary to repeat this procedure when adjusting the valve on the second carburetter.

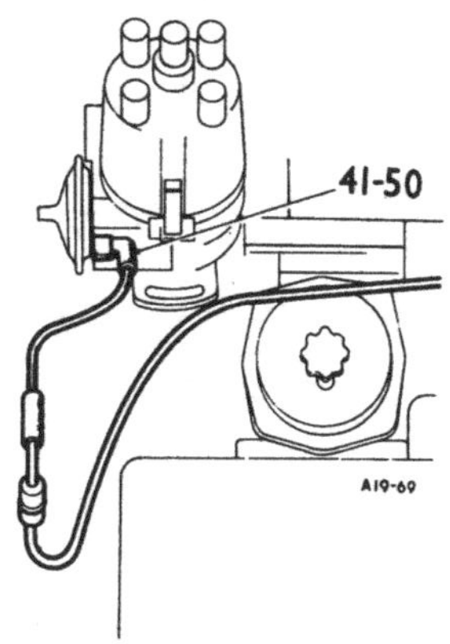

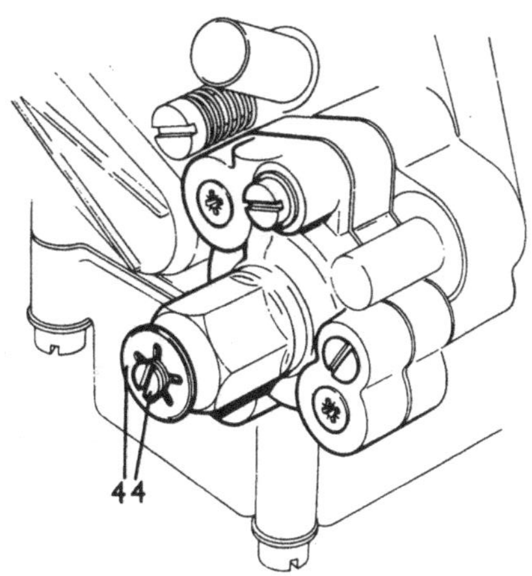

45 Remove and plug the vacuum pipe from the distributor.
46 With the engine running turn the by-pass valve clock-wise until the engine speed increases abruptly to approximately 2000 to 2500 r.p.m. thereby causing the valve to 'float' on its seat.
47 Turn the adjustment screw anti-clockwise until the engine speed falls to approximately 1300 r.p.m.
48 Using the throttle, momentarily increase the engine speed and immediately release the throttle. If the valve is correctly adjusted the engine speed should fall to approximately 1300 r.p.m. If not, the valve is still floating and the above adjusting sequence must be repeated.
49 When a correct adjustment has been achieved on both carburetters, turn the by-pass valve adjustment screw, on both carburetters, anti-clockwise half a turn to finally seat the valves.
50 Unplug the vacuum pipe and connect it to the distributor.

Checking and adjusting deceleration by-pass
Federal Specification Vehicles only – excluding California

51 Maintain the engine at normal running temperature and idling speed.
52 Slowly turn the rear carburetter by-pass valve adjusting crew clockwise from the fully loaded position until the engine revolutions begin to increase.
53 Turn the by-pass valve adjusting screw anti-clockwise 3 full revolutions.
54 Repeat operation No. 1 on the front carburetter by-pass valve adjusting screw.
55 Turn the front carburetter adjusting screw anti-clockwise 2 full revolutions.
56 Turn the rear carburetter adjusting screw anti-clockwise 1 full revolution.

CAUTION: Renewal of the by-pass valves, should it be necessary, must only be done as complete units.

57 Refit the air cleaner.
58 Refit the fresh air duct. 80.15.31.

CARBURETTERS – CAR SET

Remove and refit **19.15.11**

Which includes:
Carburetter rubber mountings–remove and refit omitting instructions 13, 17 and 25

Removing
1 Disconnect the battery.
2 Remove the fresh-air duct. 80.15.31.
3 Remove the hot air hose from the air cleaner.
4 Remove the air cleaner. 19.10.02.
5 Disconnect all the piping from the carburetters.
6 Disconnect the fuel feed pipe.
7 Disconnect the throttle return spring.
8 Disconnect the throttle link.
9 Remove the eight nuts, plain and spring washers (four per carburetter) securing the carburetters to the manifold.
10 Withdraw the carburetters complete with gaskets.
11 Vehicles with air conditioning and automatic transmission only: Remove the electrical leads from the throttle jack.

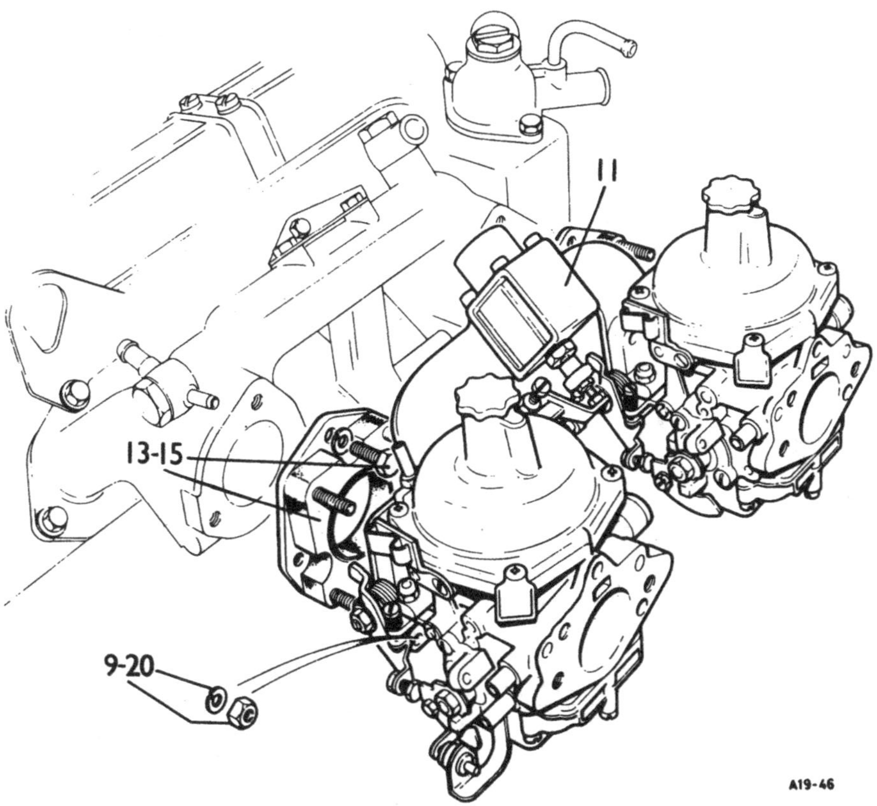

12 Vehicles with air conditioning and automatic transmission only: Remove the throttle jack mounting bracket.
13 Remove the six bolts and spring washers (three per mounting) securing the carburetter rubber mountings to the manifold.
14 Remove the throttle interconnection linkage and separate the carburetters.

Refitting
15 Fit the carburetter rubber mountings securing with the six (three each side) bolts and spring washers.
16 Vehicles with air conditioning and automatic transmission only: Fit the throttle jack mounting bracket.
17 Vehicles with air conditioning and automatic transmission only: Connect the electrical leads to the throttle jack.
18 Join the carburetters together keeping the interconnection clamps slack.
19 Clean the carburetter mating faces and fit new gaskets.
20 Fit the carburetters and secure with the eight nuts, plain and spring washers (four per carburetter) tightening the nuts evenly to the correct torque figure.
21 Connect the throttle link.
22 Connect the fuel feed pipe.
23 Connect the mixture control cable.
24 Connect the throttle return spring.
25 Connect all the piping to the carburetters.
26 Tune and adjust the carburetters. 19.15.02.
27 Refit the air cleaner. 19.10.02.
28 Refit the fresh air duct.

INLET MANIFOLD COMPLETE WITH CARBURETTERS

– Remove and refit **19.15.15**

Removing

1 Disconnect the battery.
2 Drain the cooling system including the cylinder block. 26.10.01.
3 Remove the fresh air duct. 80.15.31.
4 Remove the air cleaner. 19.10.02.
5 Disconnect the top water hoses from the engine.
6 Disconnect all emission control hoses.
7 Remove water temperature transmitter lead. 88.25.20. Instruction 3.
8 Disconnect petrol main feed pipe to carburetters.
9 Disconnect brake servo pipe from manifold connection.
10 Disconnect the mixture control cable from the carburetters.

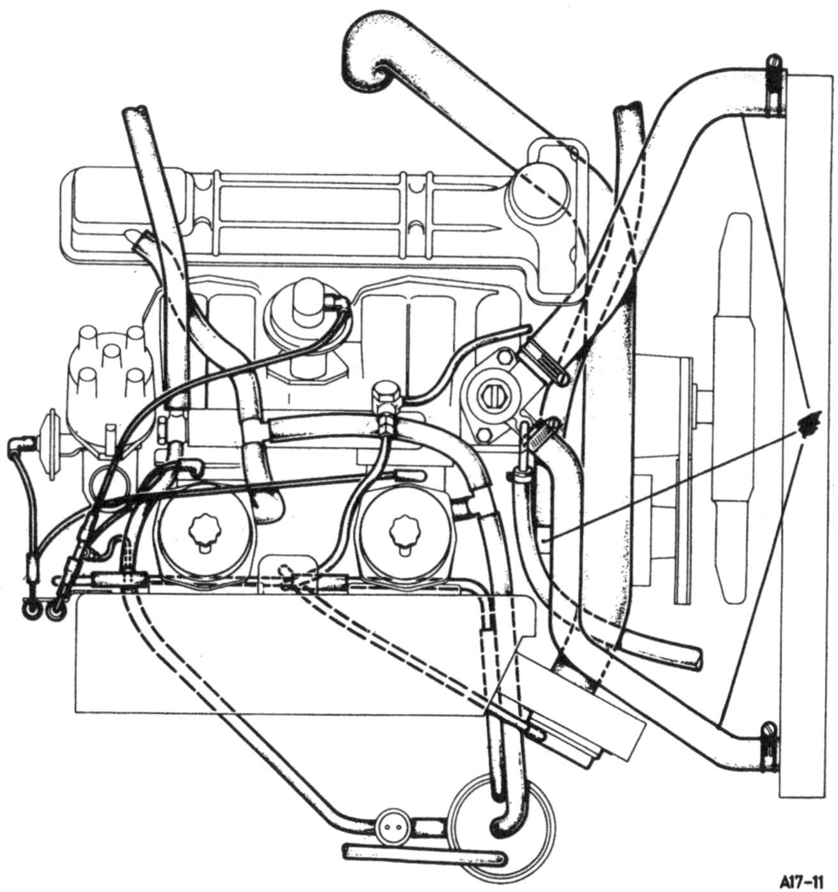

A17-11

11 Disconnect the throttle cable from the carburetters.
12 Remove the distributor cap.
13 Remove the six manifold attachment bolts.
14 Disconnect the heater pipe at the union.
15 Lift the manifold complete with carburetters from the cylinder head.
16 Remove the 'O' ring.
17 Remove the water pump connecting tube, and 'O' rings.
18 Remove the gasket.

Refitting

19 Fit a new 'O' ring to the manifold.
20 Fit two new 'O' rings to the connecting tube and insert into the pump cover.

NOTE: Early engines may be fitted with a moulded type tube, in which case ignore instruction 20 and follow instructions 32 and 33.

21 Clean the cylinder head and manifold mating faces and using a new gasket fit the manifold complete with carburetters ensuring that the connecting tube is properly located. Fit the manifold retaining bolts but leave slack.
22 Connect the heater pipe to the union but leave the nut slack.
23 Tighten the manifold attachment bolts evenly to the correct torque figure and finally tighten the heater union nut.
24 Fit the distributor cap.
25 Fit the petrol feed pipe to the carburetters.
26 Connect the throttle cable and adjust.
27 Connect the mixture control cable.
28 Connect the lead to the water temperature sensor.
29 Reconnect the emission and evaporative control pipe work.
30 Fit the air cleaner. 19.10.02.
31 Fit the brake servo pipe to the manifold connection.

continued

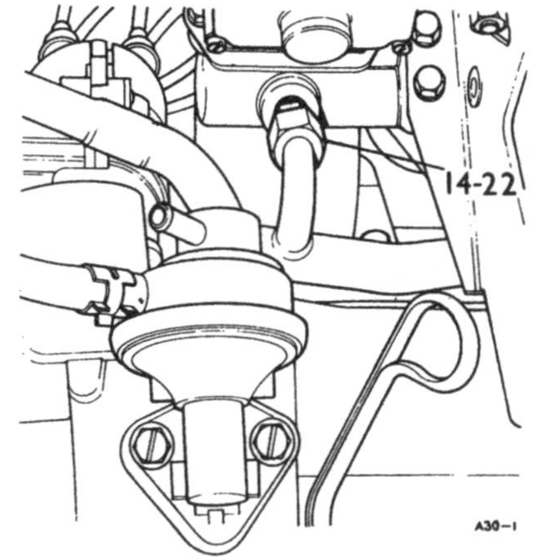

A30-1

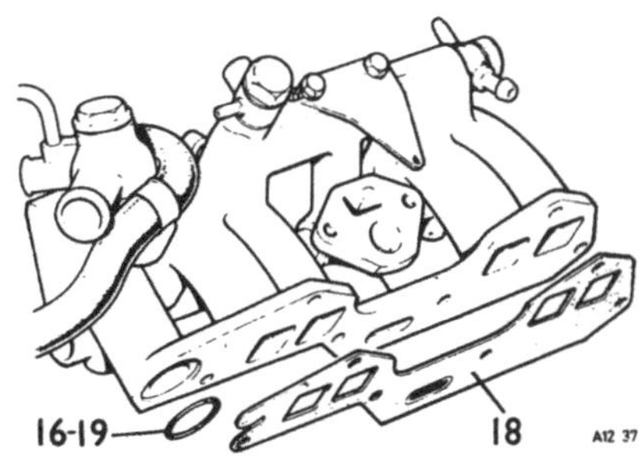

A12 37

NOTE: Instructions 32 and 33 should only be followed if a moulded type connecting tube is fitted.

32 Remove the thermostat housing and thermostat and push the connecting tube fully home.
33 Refit the thermostat and housing using a new gasket.
34 Reconnect the top water hoses.
35 Refit the fresh air duct. 80.15.31.
36 Refill the cooling system. 26.10.01.
37 Reconnect the battery.
38 Run the engine and check for coolant leaks.

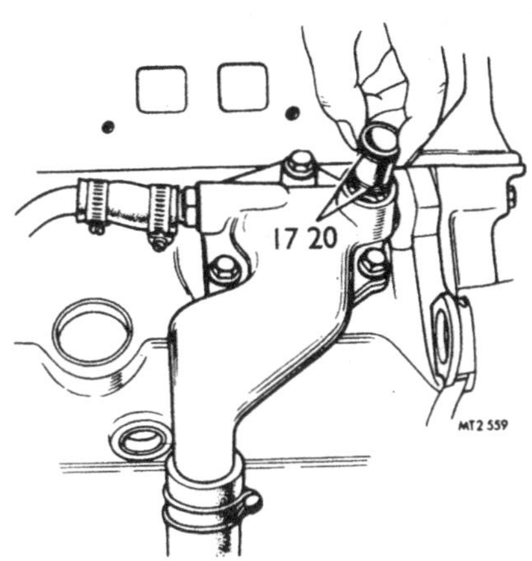

CARBURETTER RUBBER MOUNTINGS

– Remove and refit 19.15.19

This operation is included in Operation 19.15.11

FLOAT CHAMBER NEEDLE VALVE

– Remove and refit 19.15.24

Removing
1 Remove the carburetters. 19.15.11.
or
2 Remove the six screws securing the float chamber to the body.
3 Remove the float chamber.
4 Remove the gasket.
5 Remove the float assembly by gently prising the spindle from the locating clips.
6 Remove the needle valve and washer.

Refitting
7 Fit the needle valve and renew the washer.
8 Fit the float assembly.
9 Check and if necessary, adjust the height of both floats. Instruction 5. 19.15.32.
10 Renew the gasket and refit the float chamber.
11 Refit the carburetters.

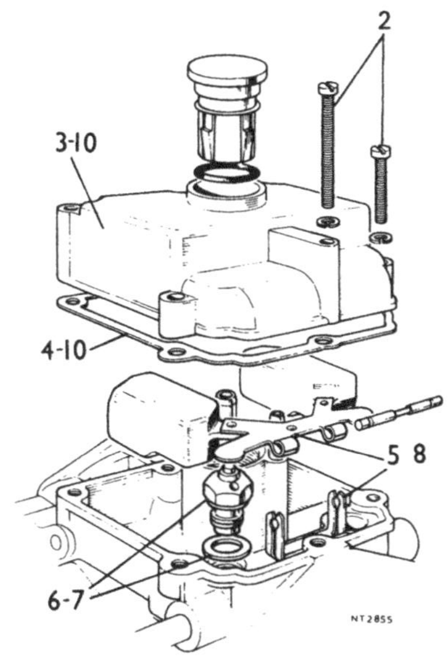

FLOAT CHAMBER LEVELS

– Check and adjust 19.15.32

Check
1 Remove the carburetters. 19.15.11.
2 Remove the six screws securing the float chamber to the body.
3 Remove the float chamber.
4 Remove the gasket.
5 With the carburetter in the inverted position check the distance between the gasket face on the carburetter body to the highest point of each float A.

NOTE: The height of both floats must be the same i.e. 0.625 to 0.627 in. (16 to 17 mm).

Adjust
6 Bend the tab that contacts the needle valve but ensure that it sits at right angles to the valve to prevent the possibility of sticking.
7 Fit a new gasket and reverse instructions 1 to 3.

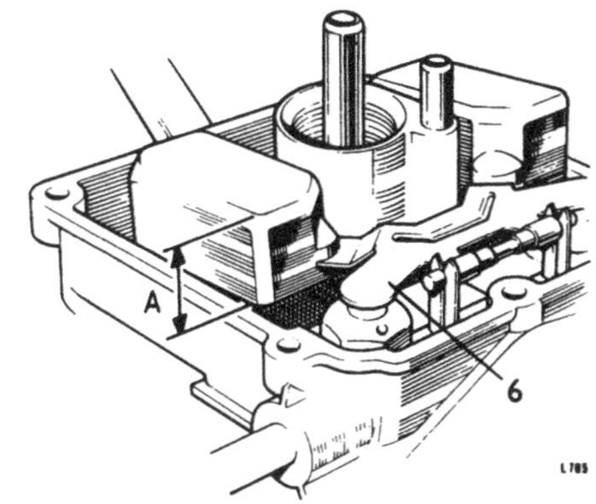

DIAPHRAGM

– Remove and refit 19.15.35

Removing
1 Remove the four screws securing the top cover to the carburetter body.
2 Lift off the top cover.
3 Remove the diaphragm spring.
4 Remove the diaphragm retaining plate.
5 Remove the diaphragm.

Refitting
6 Fit the diaphragm, locating the inner tag in the air valve recess.
7 Fit the retaining plate and ensure the correct diaphragm seating and tighten the screws.
8 Locate the diaphragm outer tag in the recess in the carburetter body.
9 Fit the top cover and evenly tighten the screws.
10 Check and if necessary top up damper – see special instructions, Section 10.

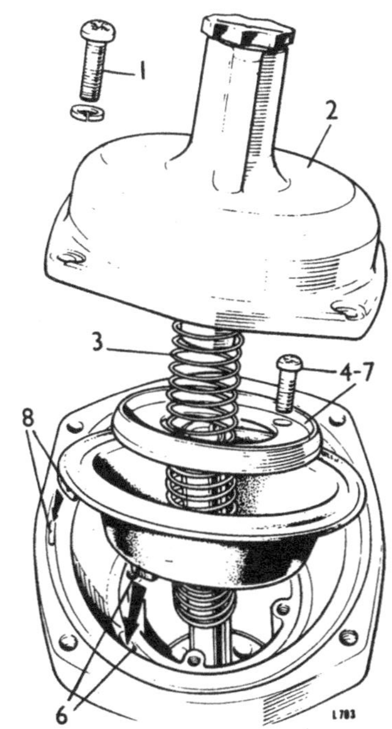

AUTOMATIC CHOKE (SINGLE CARBURETTER)

– Remove and refit **19.15.38[A]**

CAUTION: The automatic choke (auto choke) must only be renewed as a complete unit.

Removing

1. Remove the air cleaner.
2. Remove the carburetter from the engine.
3. Open the throttle and prevent closure by inserting a suitable stop (plastic, rubber or soft wood) between the throttle bore and butterfly.
4. Remove the three retaining screws securing the auto choke to the carburetter, noting that the lower screw is shorter.
5. Lift off the auto choke and gasket.

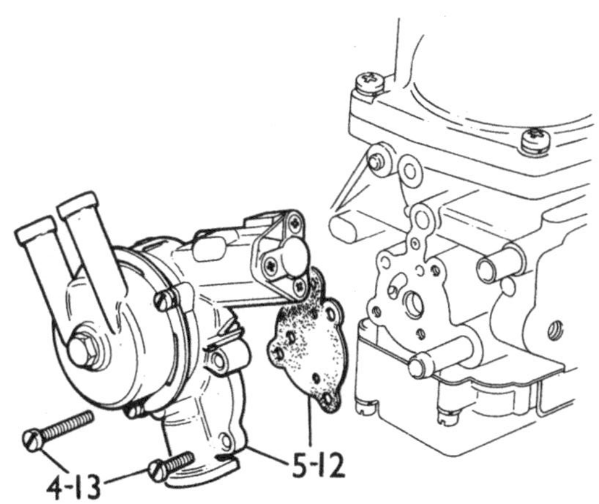

A19-54

Refitting

6. Clean the carburetter and auto choke mating faces.
7. Remove the central bolt and washer on the water jacket and remove:
   a) The water jacket.
   b) The rubber sealing ring.
8. Remove the three screws and spring washers retaining the clamp ring.
9. Remove the clamp ring.
10. Carefully remove the finned aluminium heat mass, ensuring that the attached temperature sensitive bi-metal coil does not become strained.
11. Remove the heat insulator.
12. Fit a new gasket to the carburetter.

13. Fit the auto choke body to the carburetter tightening the retaining screws progressively and evenly to 40–45 in lb (46–52 cm kg). Note the lower screw is shorter.

continued

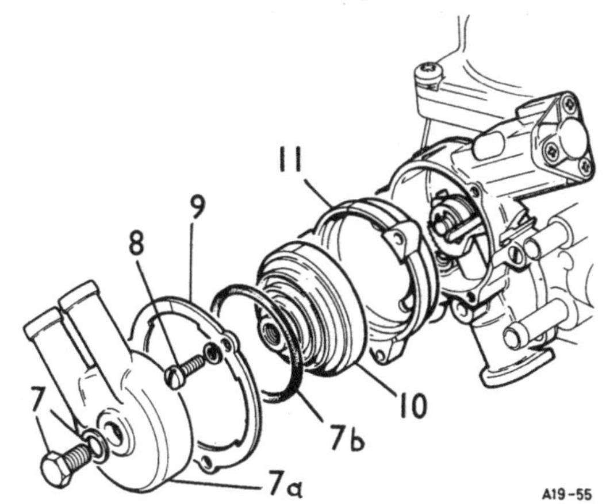

A19-55

14. Adjust the fast idle screw until the gap between the base circle of the cam and the fast idle pin is within 0.035 in (0,889 mm) dimension A.
15. Position the heat insulator ensuring that the bi-metal lever protrudes through the slot in the insulator.

NOTE: Provided the auto choke is in the fully 'ON' position, this insulator can only be fitted in one position; the back of it locating in the auto-choke body and the three holes aligned with the threaded holes in the starter body.

16. Carefully position the aluminium heat mass with the fins facing outwards and in such a way that the rectangular loop on the outer end of the temperature sensitive bi-metal coil fits over the bi-metal lever.
17. Without lifting the heat mass from its location (this is necessary to prevent the rectangular loop disengaging from the bi-metal lever) rotate it in both directions 30° to 40° ONLY and in both cases it should spring back to its static position. If not repeat instruction 16.

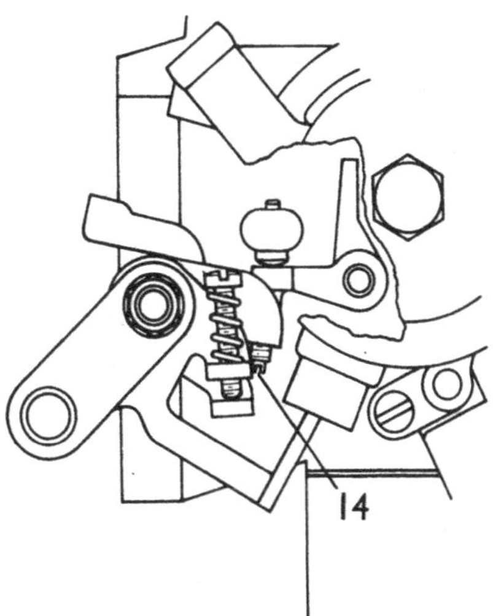

A19-63

18. Fit the clamp ring locating it with the three screws and spring washers, but leaving the screws slack.
19. Rotate the heat-mass in an anti-clockwise direction to align the scribed line on its outermost edge with the datum mark on the insulator and auto-choke body. Whilst holding it in this position evenly tighten the clamp ring retaining screws to lock the heat mass in this position.

CAUTION: It should only be necessary to turn the heat mass through some 30° to 40°. Excessive rotation must be avoided to prevent permanent damage being caused to the temperature sensitive coil.

20. Fit the sealing ring.
21. Fit the water-jacket leaving the centre retaining bolt slack until the carburetter is fitted to the engine so that the water jacket may be rotated to line-up with the water pipes.
22. Refit the carburetter to the engine.
23. Connect up the water pipes to the water jacket and tighten the centre retaining bolt.
24. Check and if necessary top-up damper – see special instructions in Section 10.
25. Operate the throttle before attempting to start the engine to enable the auto-choke to reset itself.
26. Fit the air cleaner.
27. Start the engine and run until normal operating temperature is reached.
28. Adjust the engine idle speed setting screw to give an idling speed of 800 r.p.m.
29. When the engine is cold check and if necessary top-up the cooling system.

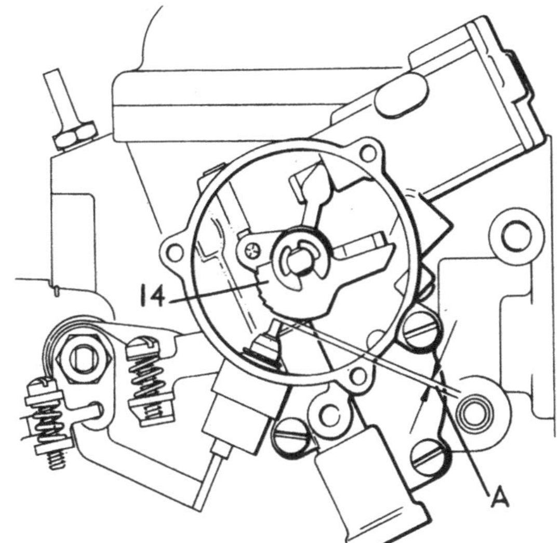

A19-56

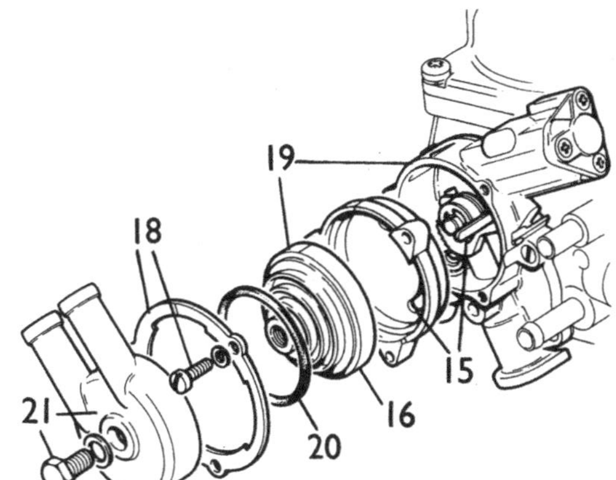

A19-55

AUTOMATIC CHOKE (Fully Automatic Starter Device)

– Remove and refit **19.15.38[B]**

CAUTION: The automatic choke must be renewed as a complete unit.

Removing

1 Remove the air cleaner. 19.10.01.
2 Slacken the clip securing the auto-choke fuel mixture outlet pipe hose, to the inlet manifold.
3 Disconnect the lead from the auto-choke heater.
4 Remove the carburetters. 19.15.11.

NOTE: The automatic choke is fitted to the front carburetter only.

5 Remove the 3 setscrews securing the auto-choke to the carburetter.
6 Remove the gasket and auto-choke assembly.

Refitting

7 Reverse instructions 1 to 6.

STARTER ASSEMBLY

[Canadian Specification and U.S.A. Specification 1975 & 1976 models]
– Remove and refit 19.15.52

CAUTION: This component must only be renewed as a complete unit.

Removing

1 Remove the fresh air duct. 80.15.31.
2 Remove the air cleaner. 19.15.01.
3 Disconnect the mixture control cable (depending upon whether it is the front or rear carburetter).
4 Remove the two screws securing the starter assembly to the carburetter body.
5 Remove the exhaust gas recirculation control valve (rear carburetter only).
6 Withdraw the starter assembly.

Refitting

7 Clean the carburetter and starter mating faces.
8 Reverse instructions 1 to 6.

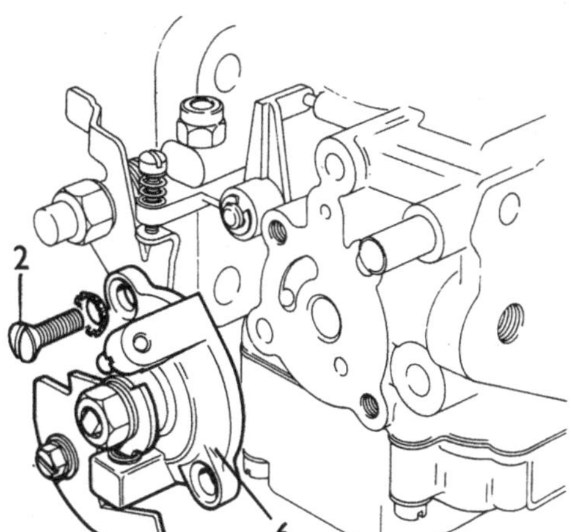

A19-58

TEMPERATURE COMPENSATOR

– Remove and refit **19.15.59**

CAUTION: This component must only be renewed as a complete new unit.

Removing

1 Remove the air cleaner assembly. 19.10.01.
2 Remove the two screws and shake proof washers securing the temperature compensator to the carburetter.
3 Withdraw the compensator complete.
4 Remove and discard the outer rubber washer.
5 Remove the inner rubber washer from the carburetter body and discard.

Refitting

6 Clean the carburetter and temperature compensator mating faces.
7 Insert a new inner rubber washer into the bore in the carburetter body.
8 Fit a new outer rubber washer.
9 Fit the compensator to the carburetter and secure with the two screws and shakeproof washers.
10 Refit the air cleaner.

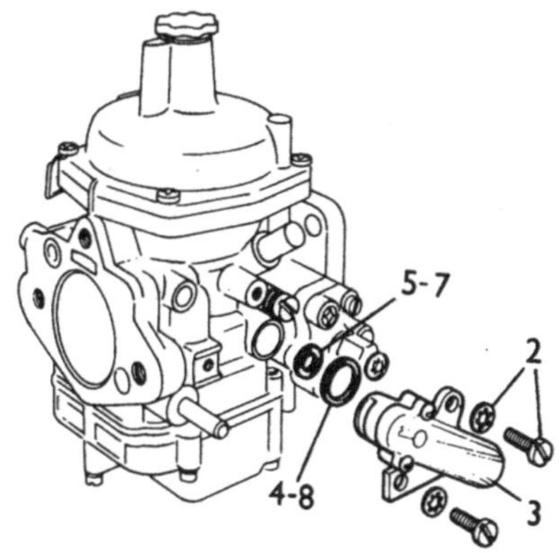

A19-65

DECELERATION AND BY-PASS VALVE

– Remove and refit **19.15.64**

CAUTION: This component must only be renewed as a complete unit.

Removing

1 Remove the carburetters. 19.15.11.
2 Remove the two cheese headed screws and the single countersunk slotted screw (not cross slotted) securing the by-pass valve assembly to the carburetter.
3 Withdraw the valve assembly complete.
4 Remove the gasket.

Refitting

5 Clean the carburetter and valve assembly mating faces.
6 Using a new gasket fit the assembly to the carburetter with the three screws and washers.
7 Refit the carburetters to the engine – less the air cleaner and fresh air duct.
8 Check and if necessary adjust the deceleration and by-pass valves – 19.15.02 instructions 40 to 52.

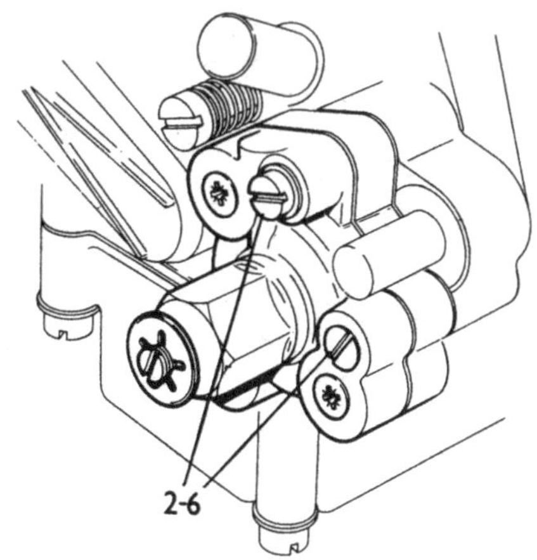

A19-50

THROTTLE PEDAL ASSEMBLY

– Remove and refit 19.20.01

Removing

1 Disconnect the battery.
2 Working from inside the car remove the spring clip securing the throttle cable.
3 Remove the cable from the pedal.
4 Open the bonnet and remove the two bolts securing the pedal assembly to the bulkhead.
5 Remove the pedal assembly from the car.
6 Strip the pedal assembly, and discard worn parts.

Refitting

7 Locate the spring in the bracket.
8 Fit the pedal to the bracket.
9 Insert the clevis pin.
10 Fit the anti-rattle washer.
11 Fit the washers (plain).
12 Secure the assembly with a new split pin.
13 Fit the pedal to the bulkhead and secure with the two bolts.
14 Fit the cable to the pedal and retain with the spring clip.
15 Re-connect the battery.

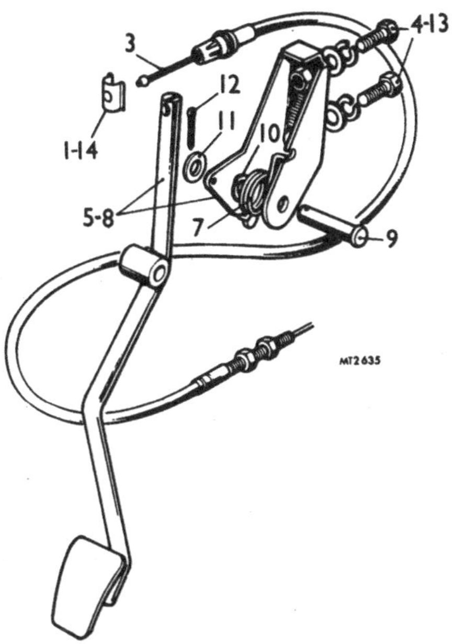

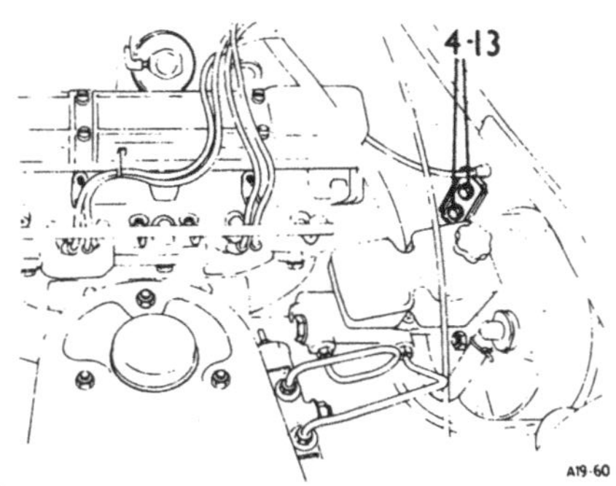

PETROL PIPE – MAIN LINE – TANK END SECTION

– Remove and refit 19.40.02

Removing

1 Drive the car on to a ramp and raise.
2 Drain the petrol tank.
3 Pull off the rubber connection between the tank and main line pipe.
4 Release the tank-end section from the engine-end section at the union (right-hand rear wing).
5 Release the pipe from the retaining clips and remove from the car.

Refitting

6 Clip the pipe into position.
7 Connect the pipe to the engine-end section at the union.
8 Connect the pipe to the tank with the rubber connector.
9 Lower the ramp and refill the tank.

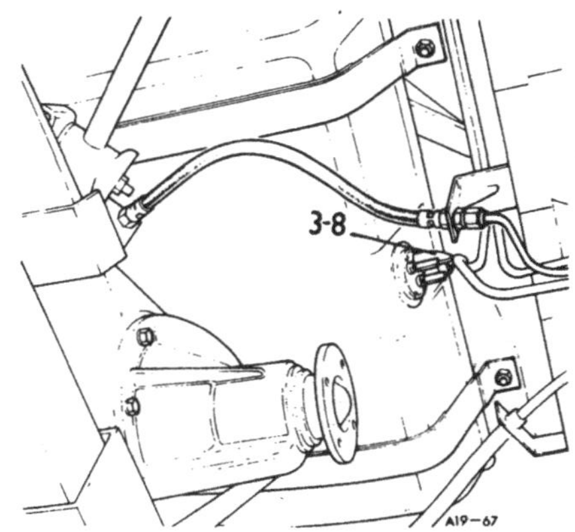

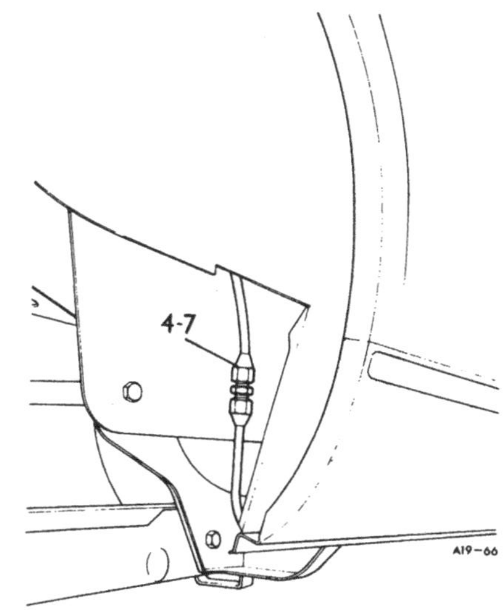

PETROL PIPE – MAIN LINE – ENGINE END SECTION

– Remove and refit **19.40.04**

Removing

1. Drive the car onto a ramp.
2. Raise the ramp and drain the petrol tank.
3. Disconnect the engine-end section from the petrol tank-end section at the union (right-hand wing).
4. Pull off the rubber hose connecting the engine-end section to the fuel pump.
5. Remove the pipe from the retaining clips.

Refitting

6. Reverse instructions 3 to 5.
7. Start the engine and check for leaks.
8. Stop the engine and lower the ramp.

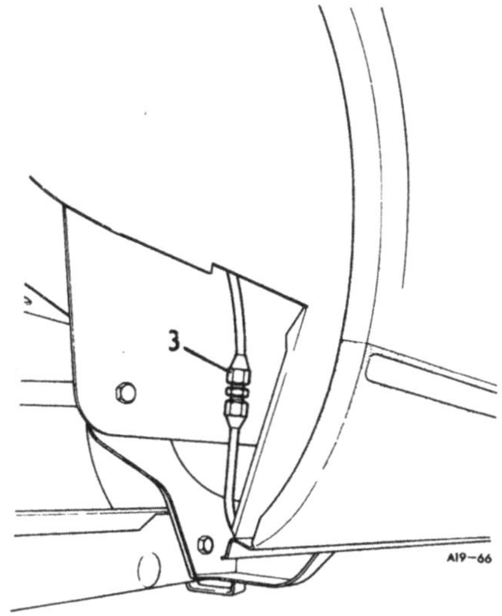

HOSE – FILLER TO TANK

– Remove and refit **19.40.19**

Removing

1. Open the boot and remove the access panel.
2. Remove the filler cap assembly. 19.55.08.
3. Remove the hoses from the vapour separator.
4. Remove the vapour separator by pulling it laterally from its retaining clip.
5. Release the hose clip securing the filler hose to the fuel tank, visible through the access aperture.
6. Pull off the filler hose.

Refitting

7. Fit the filler hose to the tank and secure with the hose clip.
8. Fit the vapour separator by pushing it laterally into its retaining clip.
9. Fit the tank breather hose to the vapour separator L.H. connection.
10. Fit the hose to the adsorption canister pipe run to the R.H. connection on the vapour separator.
11. Fit the fuel filler cap assembly – instructions 7 to 10. 19.55.08.
12. Fit the access panel.

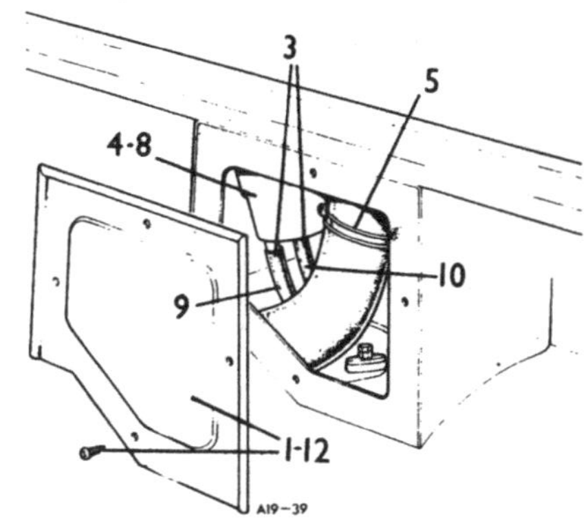

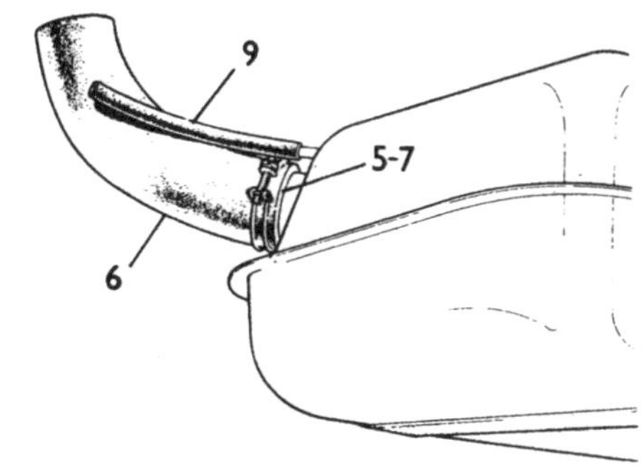

FUEL PUMP

– Clean filter 19.45.05

1 Remove the fresh air duct. 80.15.31.
2 Remove pump cover retaining screw.
3 Lift off the cover.
4 Lift out the filter.
5 Clean sediment from the filter and cover.
6 Refit the cover renewing, if necessary the rubber seal.
7 Secure the cover with the retaining screw and fit, if necessary, a new fibre washer under the screw head.
8 Refit the fresh air duct.

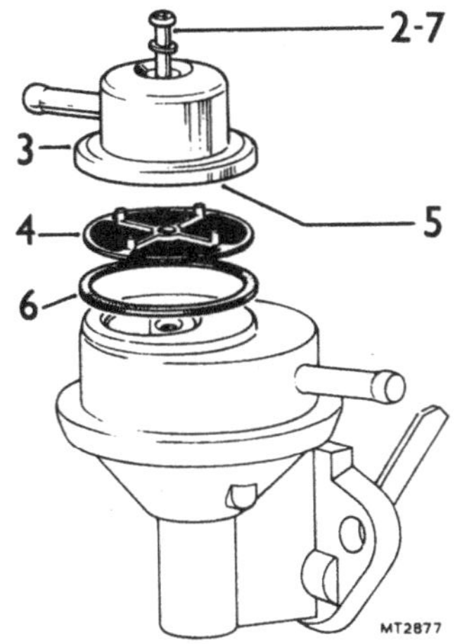

FUEL PUMP

– Remove and refit 19.45.08

Removing
1 Remove the fresh air duct. 80.15.31.
2 Remove the inlet hose to the pump.
3 Remove the outlet hose from the pump.
4 Remove the two fuel pump retaining bolts and plain washers.
5 Withdraw the pump complete with gasket.

Refitting
6 Reverse instructions 1 to 5 ensuring that when fitting, the pump actuating lever rides on top of the jackshaft cam.

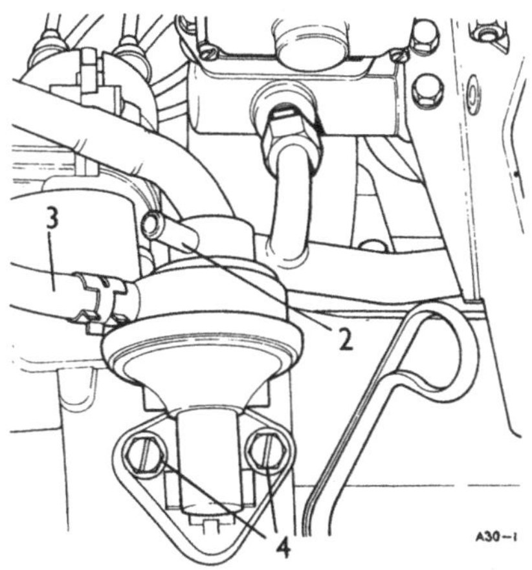

FUEL TANK

– Remove and refit 19.55.01

Removing
1 Drive the car onto a ramp and disconnect the battery.
2 Remove the fuel filler cap assembly. 19.55.08.
3 Disconnect the pipe from the tank to the vapour separator.
4 Raise the ramp and drain the fuel tank.
5 Support the body with jacks both sides of the car forward of the trailing arm attachment points.
6 Disconnect the dampers at their lower attachment to the axle.
7 Disconnect the tank unit electrical leads.
8 Pull off the rubber connection to the main fuel line.
9 Disconnect the R.H. radius rod from the body bracket.
10 Remove the L.H. radius rod from the car.
11 Jack up the body sufficiently to remove both road springs.

CAUTION: Care must be taken to ensure that whilst jacking up the body the brake hydraulic system jump hose is not stretched–shown arrowed.

12 Remove the L.H. rear road wheel and allow the axle to drop having regard for the above cautionary note.
13 Support the axle with a jack to relieve any tension in the hydraulic jump hose.
14 Remove the tail pipe and silencer assembly. 30.10.22.
15 Remove the L.H. bump stop.
16 Remove the four nuts and tank retaining straps.
17 Withdraw the tank from the L.H. side of the car.
18 Remove the filler hose.
19 Remove the breather hose.
20 Remove the tank unit – 88.25.32 instructions 6 to 8.

NOTE: Instructions 18 to 20 are only necessary if the tank is to be renewed.

continued

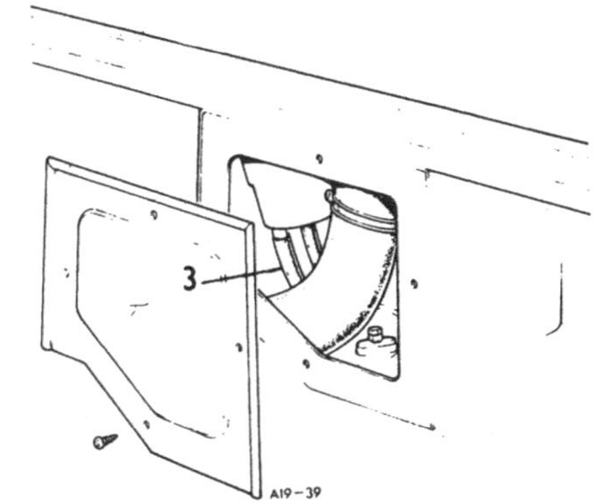

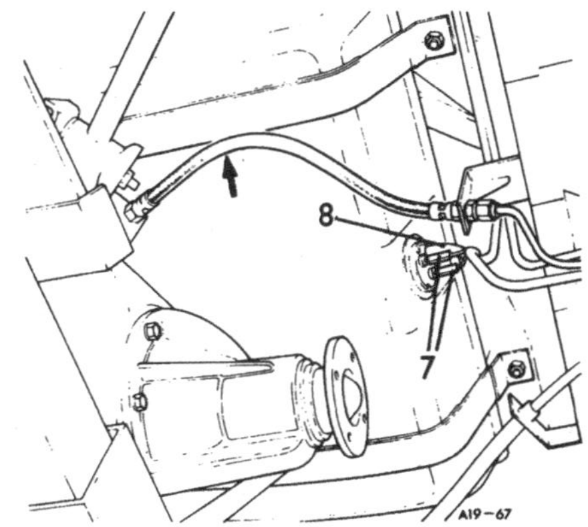

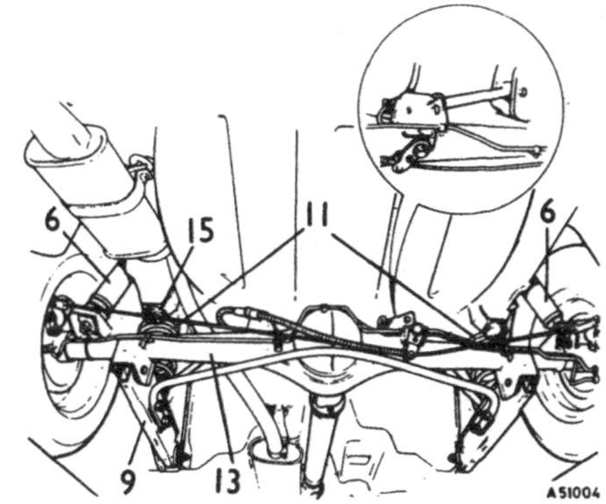

Refitting

NOTE: If the tank is not being renewed ignore instructions 19 to 22. However, renew all hoses if any deterioration exists.

21 Using adhesive fit new cushion strips to the tank noting that thicker strips are fitted to the top of the tank.
22 Fit the filler hose.
23 Fit the breather hose.
24 Fit the tank unit – reversing instructions 6 to 8. 88.25.32.
25 Manoeuvre the tank into position from the L.H. side of the car.
26 Fit the supporting straps noting that the elongated holes in the straps are fitted to the rear studs in the body.
27 Fit and tighten the four retaining nuts.
28 Fit the L.H. radius rod to the axle – including the handbrake cable bracket – leaving the body-end free for the time being.
29 Fit the springs and jack-up the axle both sides and connect the dampers to their axle locations.
30 Fit the L.H. bump stop.
31 Fit the L.H. and R.H. radius rods to their body locations.
32 Fit the tail pipe and silencer assembly. 30.10.22.
33 Fit the L.H. rear road wheel.
34 Lower and remove the body and axle jacks.
35 Connect the fuel tank to the main line pipe with the rubber connector.
36 Fit the electrical leads to the tank unit – instruction 10. 88.25.32.
37 Fit the breather pipe from the tank to the L.H. connection on the vapour separator.
38 Lower the ramp and fit the fuel filler assembly. 19.55.08 instructions 7 to 10.
39 Fit the access panel.
40 Refill the fuel tank, connect the battery, start the engine and check for leaks.

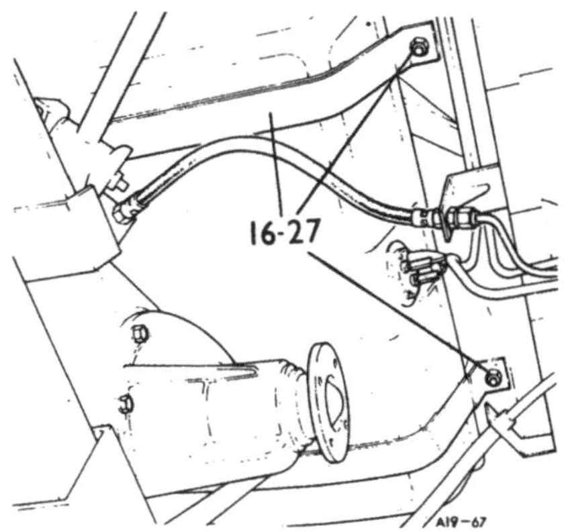

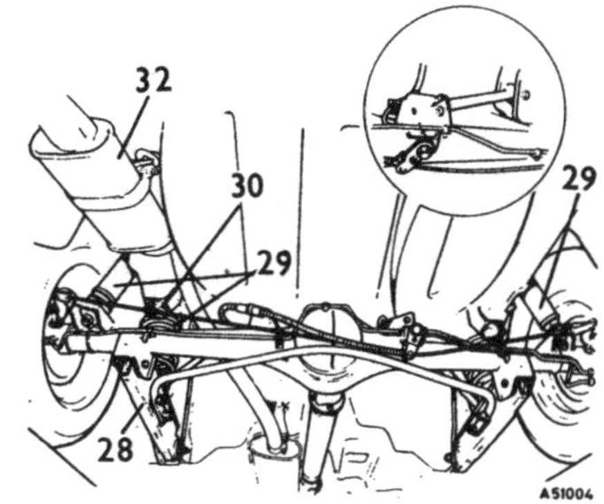

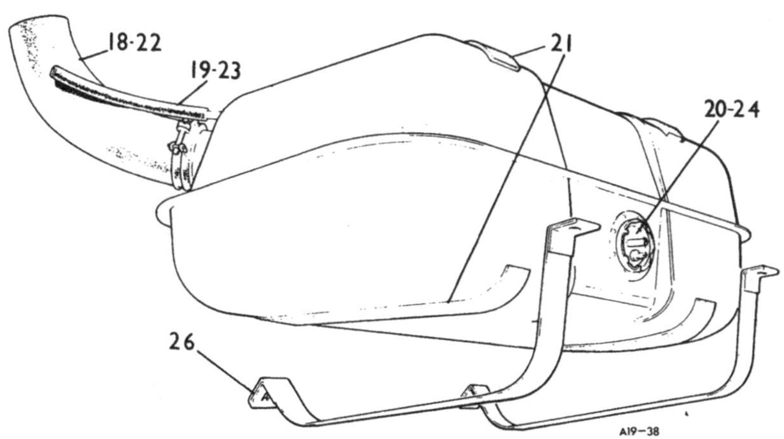

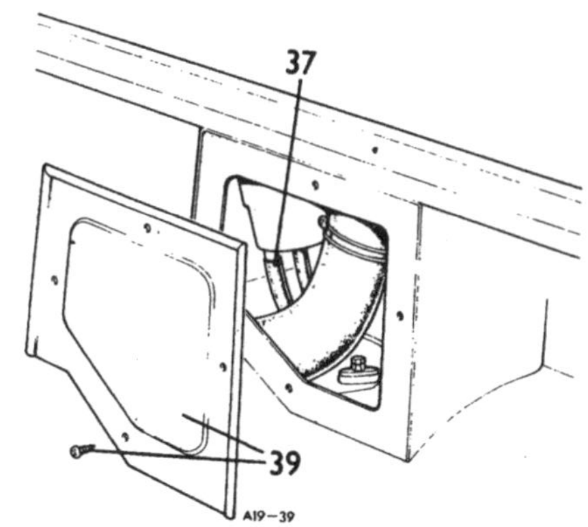

FUEL FILLER CAP AND FILLER ASSEMBLY

– Remove and refit 19.55.08

Removing

1 Turn the fuel filler cap anti-clockwise and withdraw.
2 Remove the three screws securing the filler tube assembly to the body.
3 Open the boot and remove the access panel – four screws.
4 Slacken the top hose clip.
5 Withdraw the filler tube.
6 Remove the finisher.

Refitting

7 Place the finisher in position.
8 Insert the filler tube into the filler hose and secure with the three screws.
9 Refit the filler cap.
10 Tighten the filler hose top clip.
11 Refit the access panel and close the boot.

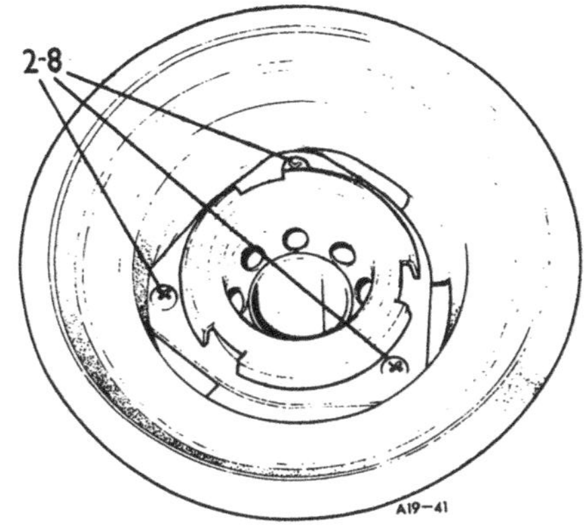

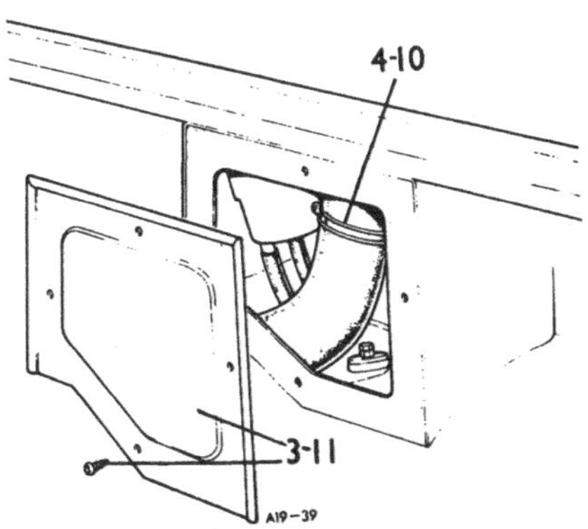

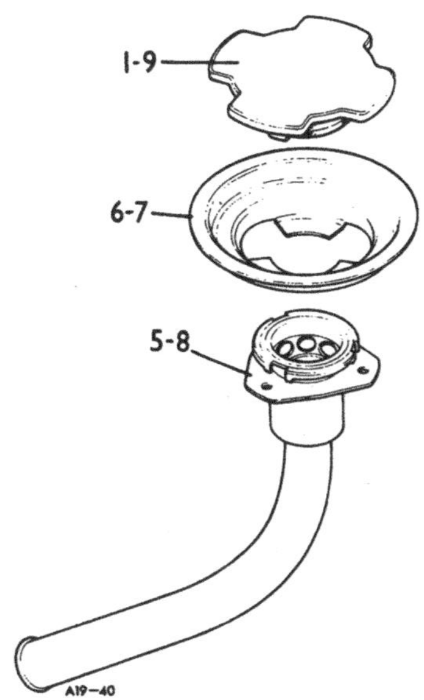

FUEL SYSTEM OPERATIONS—U.S.A. & CANADA SPECIFICATION —FUEL INJECTION VEHICLES

ELECTRONIC FUEL INJECTION SYSTEM

Description **19.00.00**

The electronic fuel injection system, fitted as an alternative to carburetters, comprises of two parts: a fuel injection system and an electronic control for the fuel injection system.

Fuel injection system components
Fuel is drawn from a tank at the rear of the vehicle and pressurized to approximately 36 lbf/in² (2.5 kgf/cm²) by an electric fuel pump located beneath the car floor. The fuel pump will only operate when the ignition and/or the starter motor circuits are energized. From this pump fuel passes through fuel filters located beneath the car floor to a pressure regulator, the spring chamber of which is connected to the engine intake manifold. As a result, the difference between the intake manifold pressure and the fuel pressure is held constant, excess fuel being returned to the fuel tank via an anti-surge pot.

A fuel rail links the pressure regulator with the fuel injectors, one injector being fitted to each inlet manifold spur. The injectors may be either 'open' or 'closed' and are solenoid operated. The injector solenoids are energized through a relay actuated by the ignition circuit and are pulsed to 'open' by the electronic control unit (E.C.U.) completing a circuit to 'earth'. When 'open', the injectors spray fuel into the inlet manifold to be drawn into the engine cylinders at the next induction stroke of the working cycle.

The E.C.U. computes the open time from data received from various sensors fitted to the engine. These sensors are connected electrically to the E.C.U.

The main sensor is the airflow meter. Data from this sensor is used by the E.C.U. to compute the basic fuel requirement.

Other sensors, coolant and air temperature, throttle position, engine speed and exhaust gas provide additional information to the E.C.U., which automatically adjusts the injector open time to suit engine conditions and requirements.

Therefore there needs to be no fixed relationship between the injector timing and the engine ignition or valve timing.

The injectors are programmed to 'open' as a bank of four, in unison, twice per engine operating cycle (two revolutions). The time that the injectors are 'open' governs the amount of fuel supplied to the engine and this 'open' time is computed by the electronic control unit from the input it receives from various sensors.

To assist cold starting, a separate cold start injector sprays a fine jet of fuel against the air stream entering the plenum chamber before fuel is added to it by the main injectors. The cold start injector is energized from the engine starter motor circuit and has in series with it a thermotime switch. This switch is dual activated by engine coolant temperature (heat) and a heater coil around a bi-metal strip (time), the coil being again energized from the starter motor circuit. The purpose of the thermotime switch is to ensure that the cold start injector will not be energized when the engine is at normal operating temperature or should the starter motor be used for prolonged periods when the engine is below normal operating temperature. Thus the switch prevents extra fuel being supplied to the engine when it is not required. The switch will isolate the cold start injector after approximately 8 to 12 seconds at −20°C (−4°F) decreasing this time as the engine approaches its normal operating temperature.

continued

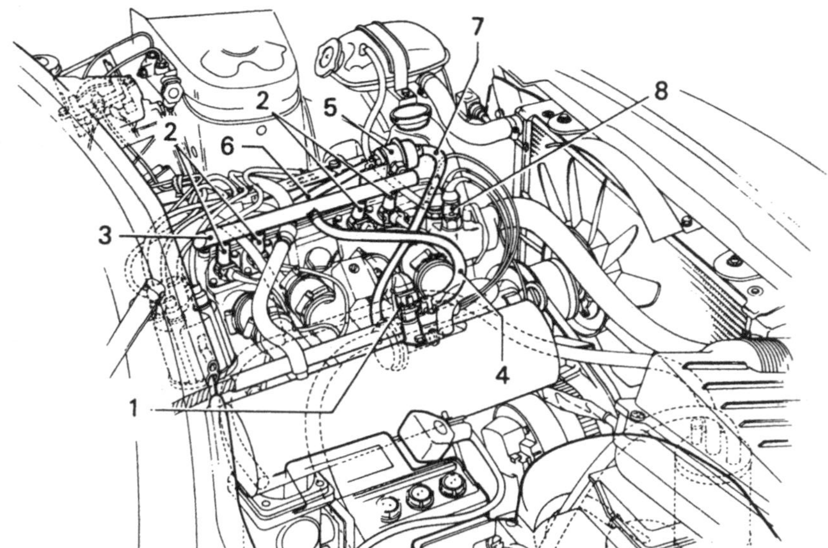

Fuel injection system components

1 Cold start injector
2 Injectors
3 Fuel rail
4 Cold start injector—fuel feed pipe
5 Fuel pressure regulator
6 Return line to fuel tank
7 Pipe to plenum chamber
8 Thermotime switch

Injection Components and Fuel System

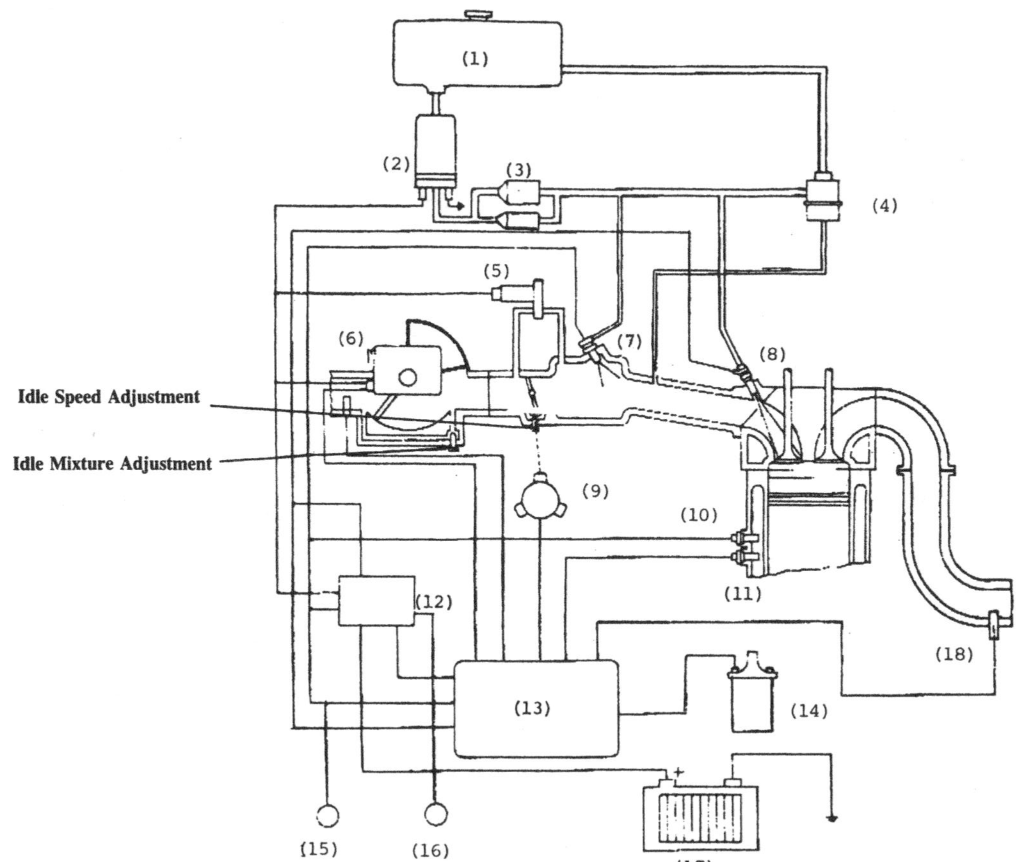

1. Fuel tank
2. Fuel pump
3. Twin fuel filters
4. Fuel pressure regulator—influenced by vacuum
5. Auxiliary air valve
6. Air flow meter—incorporates air temperature sensor
7. Cold Start Injector
8. Injector
9. Throttle switch
10. Thermotime switch
11. Water temperature sensor
12. System relay
13. Electronic control unit (E.C.U.)
14. Ignition coil
15. Starter switch
16. Ignition switch
17. Battery
18. Oxygen sensor

Electric fuel pump operation

The fuel pump is energized independently of the electronic control unit, from an output terminal on the combined relay. The combined relay is the component that provides an interface between the main vehicle electrical harness and those items that are specifically related to the electronic fuel injection system. An inertia switch is included in the circuit to isolate the fuel pump and prevent it from operating in the event of an impact-type accident. The circuit is also routed through the electronic control system air-flow meter where a simple contact switch ensures that the fuel pump cannot operate when no air is flowing into the engine, i.e. the engine is not running. This contact switch is by-passed when the starter motor circuit is energized.

Once the engine is running a circuit from the ignition switch passes through a relay to 'earth', via the electronic control unit. When energized, this relay permits a circuit to be made to the air-flow meter contact switch. Providing the contact switch is closed, a circuit is completed through a second relay to 'earth'. When energized this second relay completes the circuit to operate the fuel pump.

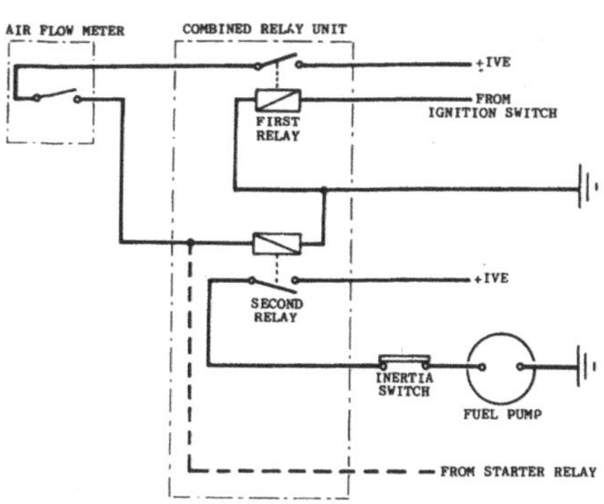

Circuit principles of fuel pump operation

Under engine starting conditions the air-flow meter contact switch would normally isolate the fuel pump as no air is flowing into the engine. To overcome this an input is taken direct from the starter motor circuit to energize the second relay and thus permit the fuel pump to operate during the engine starting operation.

The pump is a wet design unit, i.e., fuel flows through the complete pump, there is no risk of combustion occurring as it does not contain ignitable mixture (no air). *The pump is designed as a sealed unit and is not repairable.*

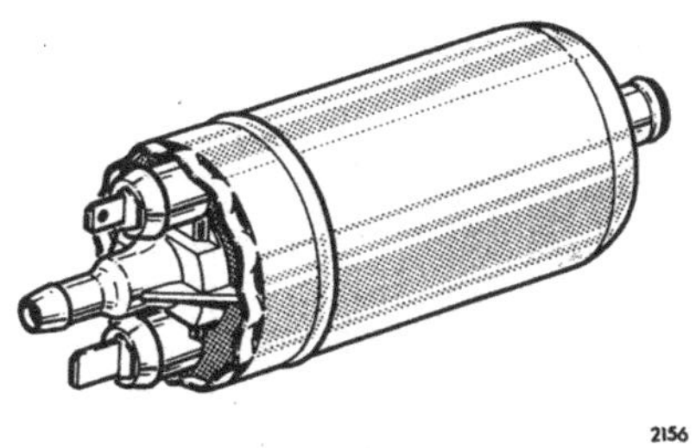

Electric fuel pump

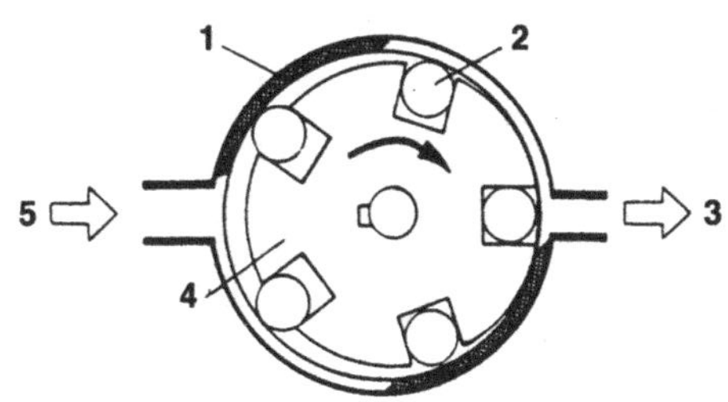

1. Housing
2. Roller
3. Outlet (pressure side)
4. Rotor
5. Inlet

Roller-cell pump element

Electronic control of the fuel injection system

At the heart of the electronic control system is the electronic control units (E.C.U.) which is a box approximately 9 × 7 × 2 inches (23 × 18 × 5 cm) located under the glove box. [The E.C.U. consist of the electronic components, i.e., integrated circuits, diodes, resistors and capacitors, etc. in the form of a small analogue computer (reads values, ohms, volts, etc.).] The E.C.U. receives input signals from various sensors and computes from these an output signal to the fuel injector solenoid circuits. When activated, the solenoids 'open' the injectors to spray fuel into the engine inlet manifold, the injectors remaining open for between 1.5 and 10 milliseconds depending upon engine running conditions.

An automatic function of the E.C.U. is to provide fuel enrichment whenever the engine is cranked regardless of engine temperature. This achieves easier starting and is sensed by the E.C.U. via an electrical connection direct to the E.C.U. from the starter operating circuit.

NOTE: Service for the E.C.U. is by replacement only.

The electronic control unit is sealed; it requires no maintenance and should not be opened or tampered with.

Engine speed

This input is very simply obtained by taking a tapping from the ignition coil low tension circuit output (−ive). Thus the ignition low tension circuit pulses are passed to the E.C.U. to be computed into an engine speed input.

The speed sense is used by the control unit for two purposes:

(1) Engine speed—the control unit ascertains speed, compares it with the air flow rate (data from the airflow meter) and controls fueling accordingly.

(2) Injection points at the start of each injection cycle are computed by the E.C.U.

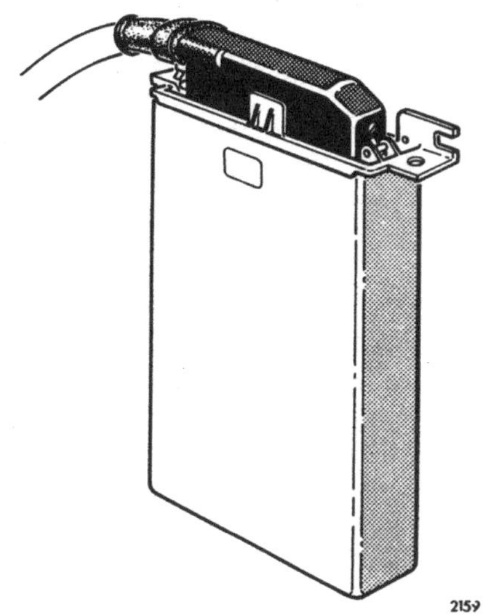

Electronic control unit

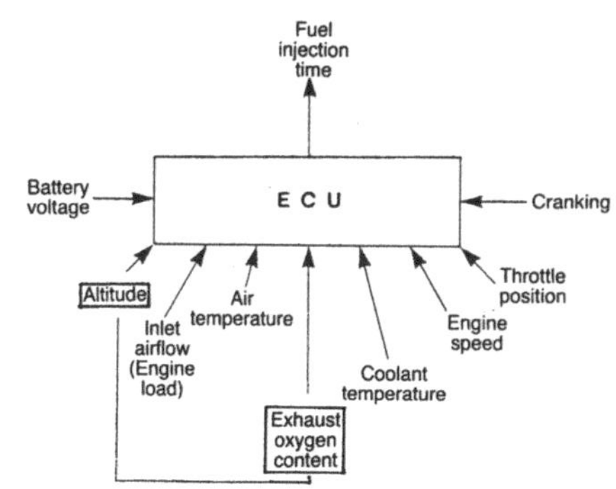

Fuel pressure regulator

The pressure regulator consists of a sealed metal housing incorporating fuel inlet and outlet apertures. Internally, a spring-loaded diaphragm exposes the aperture to the outlet allowing fuel to return to the tank, whenever the pressure setting is exceeded.

The fuel pressure regulator has a pipe connection to the plenum chamber. As a result, the difference between intake pressure and fuel pressure is held constant. The pressure drop across the injector nozzles is therefore equalized under all load conditions.
LOCATION: Forward end of the fuel rail.

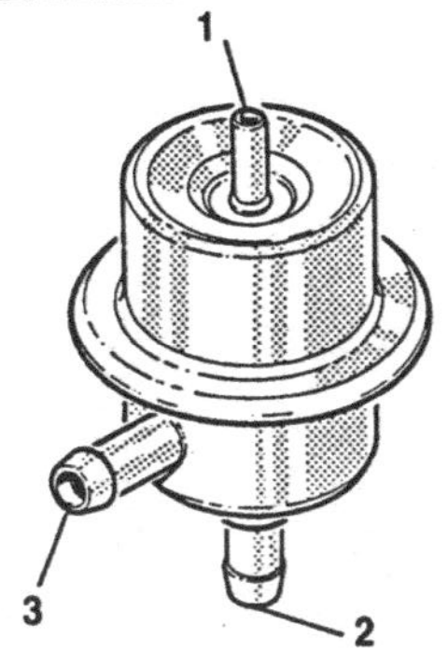

1. Pipe connection to inlet manifold
2. Fuel return to tank
3. Fuel inlet

Fuel pressure regulator

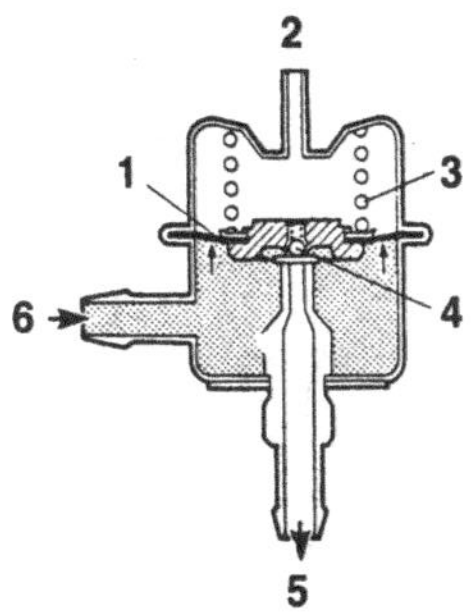

1. Diaphragm
2. Pipe connection to inlet manifold
3. Pressure spring
4. Valve
5. Fuel return line to tank
6. Fuel inlet

Fuel pressure regulator, sectioned

Air-flow meter

Apart from the E.C.U., the airflow meter is the most important item of the control system. All the air drawn into the engine passes through the meter. Accurate data regarding engine load is passed on the E.C.U. The airflow meter consists of a rectangular tunnel in which a similarly shaped flap swings on an axis in the air stream against the reverse pressure of a spiral spring. Therefore, as air is drawn into the engine, the flap opens and closes in relation to the amount of air variable through the tunnel. The flap is linked to a variable potentiometer which regulates a voltage signal fed to the E.C.U. depending on the angular displacement of the flap.

To counteract pressure oscillations that may occur in the meter, a compensating flap, similar to the main flap is used. The flaps are part of the same casting, approximately 90° apart, and swing on the same axis.

Since air density is dependent on temperature, an air temperature sensor is incorporated in the air flow meter feeding information to the E.C.U.

A plenum chamber or air box must be used to provide a sufficient quantity of air to accurately measure the air and to compensate for the pulsating nature of the air stream entering the engine.

Two other functions of the airflow meter are:

(1) *Fuel Pump Switch* - The first movement of the flap (approx. 5°) is controlled by a weak spring before the main spiral spring takes effect. The relatively free play of the flap, which is responsive to airflow induced by idling, is used to close the switch and apply battery voltage to the fuel pump via the relay. This method of switching the fuel pump minimizes the fire risk in the event of a collision.

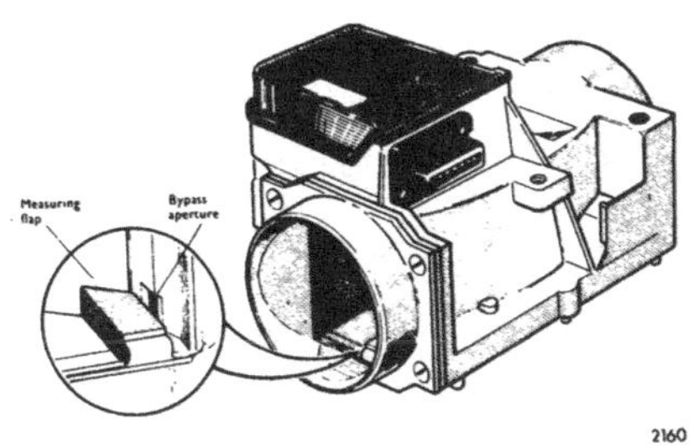

Air-flow meter

External circuitry arranges for this switch to be overridden when the engine is being cranked to ensure rapid build up of fuel pressure.

(2) *Idling* - When the engine is idling, some of the air entering the airflow meter is drawn off the main flow through an adjustable bypass. Air enters the bypass through a small hole made in the tunnel wall, but sufficient air will still be available through the main tunnel to operate the flap and switch on the fuel pump.

The adjustable bypass influences the CO setting at low engine speed.

LOCATION: On R/H inner fender well.

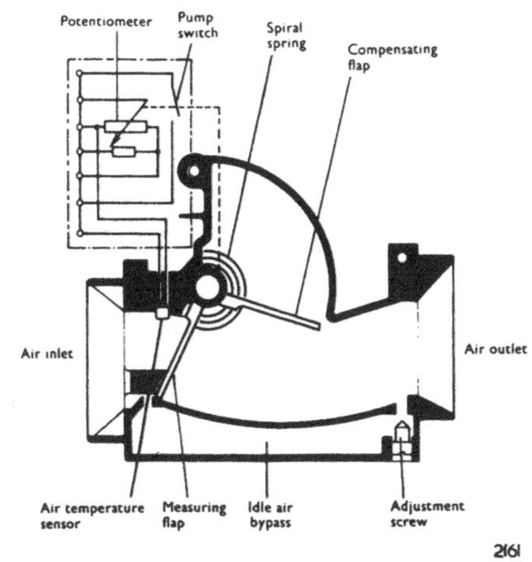

Schematic view of air-flow meter

Throttle switch

A throttle switch forms part of the electronic control for the fuel injection system and provides the E.C.U. with information on throttle operating conditions.

This switch is of the contact type and is located on the throttle body in the engine compartment. The switch contacts close when the accelerator pedal is fully depressed, signalling to the E.C.U. to lengthen the time that the main injectors are 'open', thus supplying extra fuel for the acceleration required.

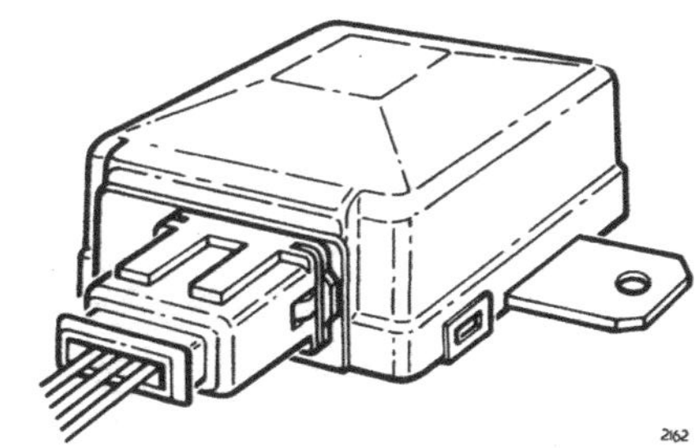

Throttle switch

Coolant temperature sensor

This sensor is located at the side of the engine and provides coolant temperature information to the E.C.U. The signal from the sensor serves two purposes. First, it causes the E.C.U. to slightly lengthen the time that the main injectors are 'open', reducing this time as the engine warms up and cutting it off when normal engine operating temperature is reached. Second, it completes an 'earth' return circuit in the E.C.U. for the heater element in the extra air valve when the engine is running below normal operating temperature. In practice the sensor functions by modifying an output voltage from the E.C.U. through an 'earth' return circuit.

continued

Coolant temperature sensor

Extra air valve

This valve is mounted above a water passage in the inlet manifold and registers the same temperature as the engine coolant. Its purpose is to provide the additional air required to maintain satisfactory engine idle speed until the engine reaches normal operating temperature. This air is taken from a point before the throttle butterfly (but after the air-flow meter, so that the air is registered by the E.C.U.) and returned to the plenum chamber after the throttle butterfly.

To allow air to pass through the valve, and thus by-pass the throttle butterfly, an opening in a rotatable metal disc is aligned with the inlet and outlet tubes on the valve. The position of this disc is controlled by a bi-metal strip which deflects according to the temperature it experiences. As the bi-metal strip heats up it rotates the metal until its opening no longer lines up with the air valve tubes and the extra air source is reduced and finally terminated as normal engine operating temperature is reached. The bi-metal strip is heated from two sources, the coolant temperature and a heater coil around the strip. The heater coil is energized from the ignition circuit and comes into operation whenever the coolant temperature sensor causes the E.C.U. to complete the heater coil circuit to 'earth'.

LOCATION: Under forward end of intake manifold.

Over-run valve

This second air-flow control device bleeds air into the engine inlet manifold, via the plenum chamber, when the manifold depression is high and thus maintains combustion during engine over-run.

The valve operates independently of the electronic control system and is incorporated in the throttle butterfly connecting the constant depression region between the throttle and the air-flow meter measuring flap.

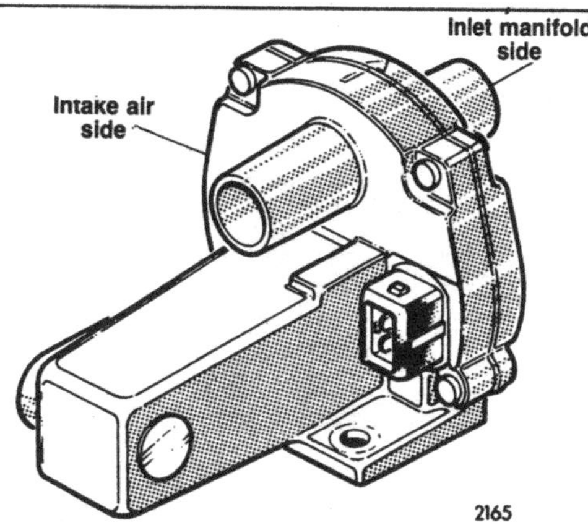

Extra air valve

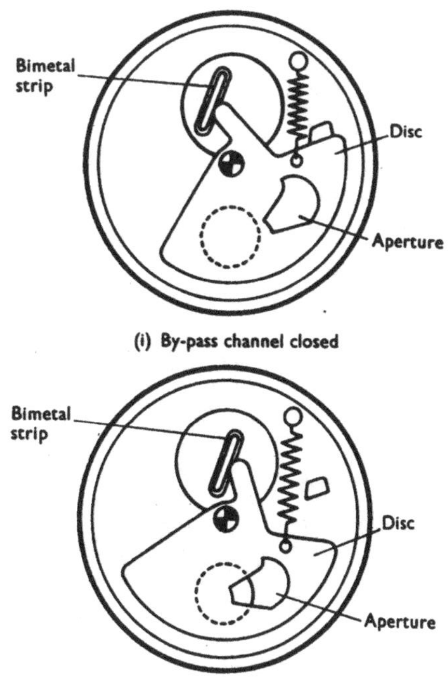

Extra air valve operation

Lambda sensor

A Lambda sensor is fitted to the exhaust system in addition to a 3 way catalytic convertor. The Lambda sensor is a device which measures the oxygen content of exhaust gases, relaying this information to the ECU via an electrical connection. The device can only be used with unleaded fuels.

The internal working surfaces of the sensor are coated with a thin platinum layer which is permeable to gas. A special ceramic layer protects the electrodes against corrosion and this becomes conductive to oxygen atoms at about 600°C (1112°F). If the concentration of oxygen inside the sensor (from the exhaust gases) differs from that outside the sensor (atmosphere) a voltage is developed between the two surfaces that changes when the outer electrode has catalytic activity. This voltage is passed to the E.C.U. which compares it against a reference voltage for ideal combustion and adjusts the main injector 'open' time accordingly to permit more or less fuel to be used by the engine. More fuel will use up the excess oxygen in the exhaust gas, less fuel will allow more oxygen, and so the ideal oxygen content is supplied to permit the catalyst to operate at its best efficiency.

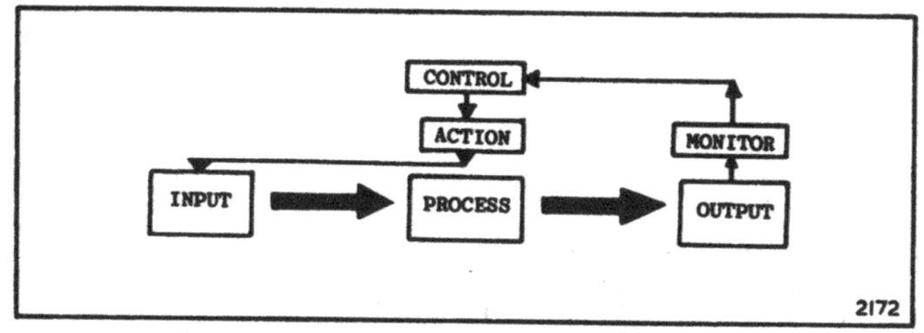

The general 'closed loop' feed back control system

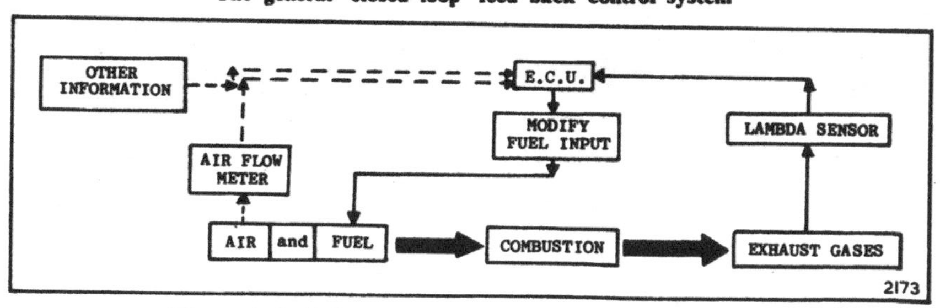

Electronic fuel injection 'closed loop' feed back control system

By using Lambda Sensors to monitor the exhaust gases in this way a feed back 'closed loop' type of control system can be introduced for the fuel injection operation. Should for any reason the Lambda Sensors become inoperative or be disconnected the electronic control system will continue to function as an 'open loop' system without the fine tuning effect of the sensors.

LOCATION: Forward end of exhaust manifold.

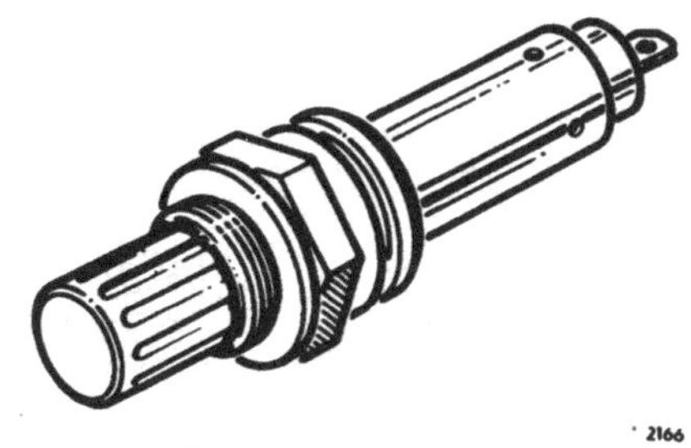

Lambda (oxygen) sensor

Three-way catalyst

The catalyst will oxidize CO and HC as does the two-way, but will also reduce NO_x to its parts of nitrogen and oxygen. This catalyst will give the lowest overall emission figures, but for correct operation it must be fed with exhaust gases from an engine working very closely to stoichimetric (perfect) air/fuel ratio. For this reason, it is normally used in conjunction with a feed-back control system, having an oxygen sensor in the exhaust manifold. The amount of oxygen in the exhaust gas is dependent on the air/fuel ratio and therefore the oxygen sensor monitors the level of oxygen in the gases. The signal is sent back to the electronic unit, which continuously adjusts the fueling to maintain the air/fuel ratio.

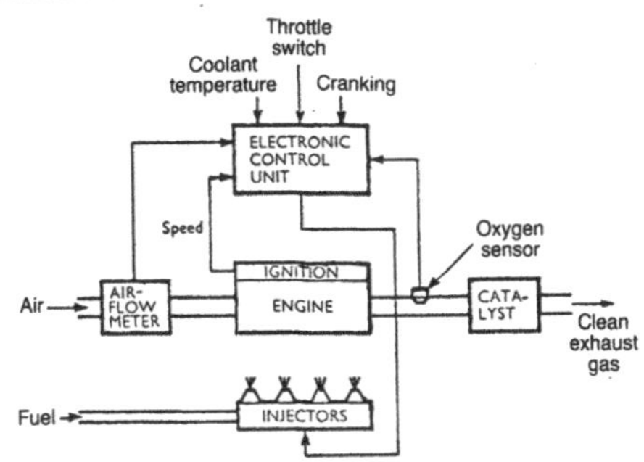

Closed loop EFI control system

Injectors

Injectors, (one per cylinder) are solenoid-operated valves. The solenoid plunger is linked to a nozzle valve which is opened when the winding is energized, with 3 volts maximum controlled by the E.C.U. The injectors are secured in rubber mountings in the inlet manifold adjacent to each cylinder so that the fuel spray is directed towards the back of each inlet valve.

The injectors are of the half-flow type, injecting half the amount of fuel required for one operating cycle twice during each rotation of the crankshaft.

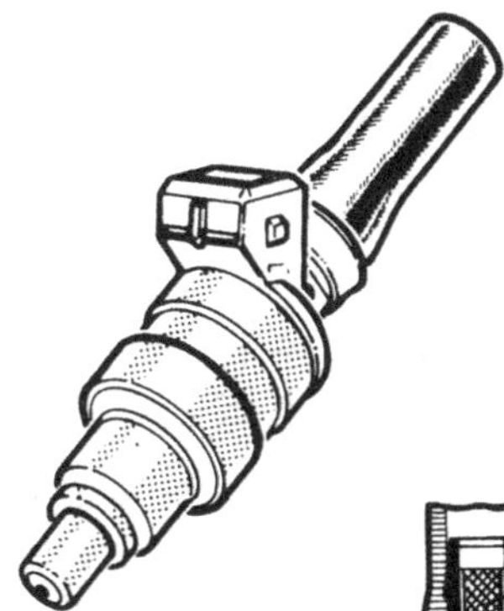

Fuel Injector

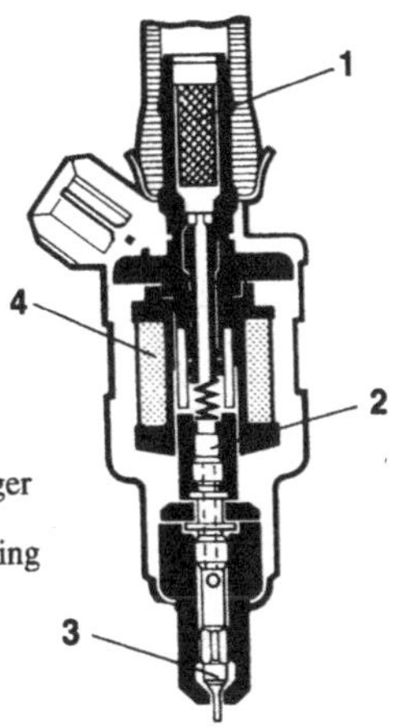

1. Filter
2. Solenoid plunger
3. Nozzle valve
4. Solenoid winding

Fuel Injector, sectioned

Cold start system

During cold starting, some of the fuel injected by the main injectors will condense on the cylinder/manifold walls. As a result, less fuel is mixed with the air drawn in, resulting in a weak mixture.

To overcome these conditions, a cold-start injector, thermotime swicth, extra air valve and coolant temperature sensor are used to supply the additional fuel and air necessary to ensure a rapid start. The cold start injector is only operative when cranking and when the thermotime switch sensitive to engine coolant temperature completes the electrical circuit. On engine cranking, an electrical signal to the E.C.U. increases the open time of the injectors to provide cranking enrichment.

Cold start injector

The cold start injector is fitted below the intake manifold plenum. It is a solenoid-operated valve and has a turbulence nozzle to inject atomized fuel into the air stream. Its operation is controlled through the action of the thermotime switch.
LOCATION: On bottom side of plenum chamber.

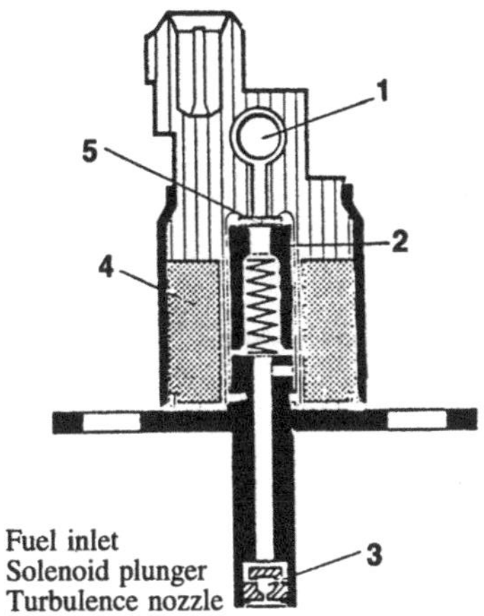

1. Fuel inlet
2. Solenoid plunger
3. Turbulence nozzle
4. Solenoid winding
5. Seal

Cold start Injector, Sectioned

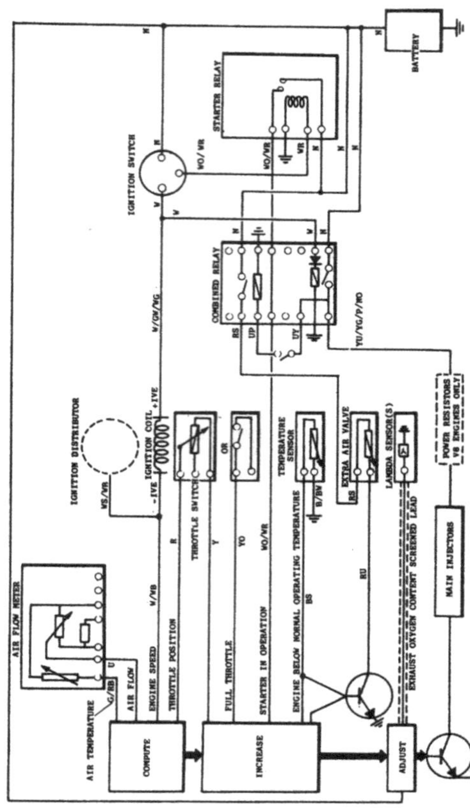

2175

Schematic wiring diagram of electronic fuel injection system components

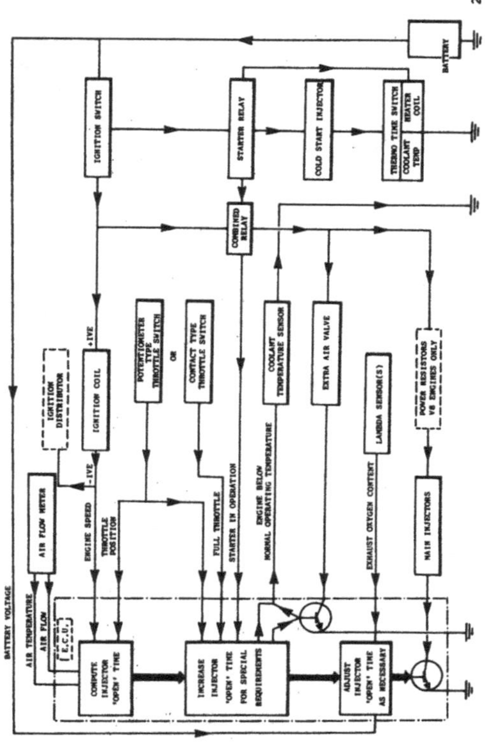

2141

Function diagram of electronic fuel injection system components

Test equipment

Special test equipment is required to carry out a full and accurate check of the electronic fuel injection system.

Lucas has developed the EPITEST so that fault diagnosis on the vehicle can be quickly carried out. The EPITEST meter, part number 60073009, is supplied complete with fuel pressure gauge and operating instructions. An adaptor, part number 60973063, is required to check the TR7 fuel injection system.

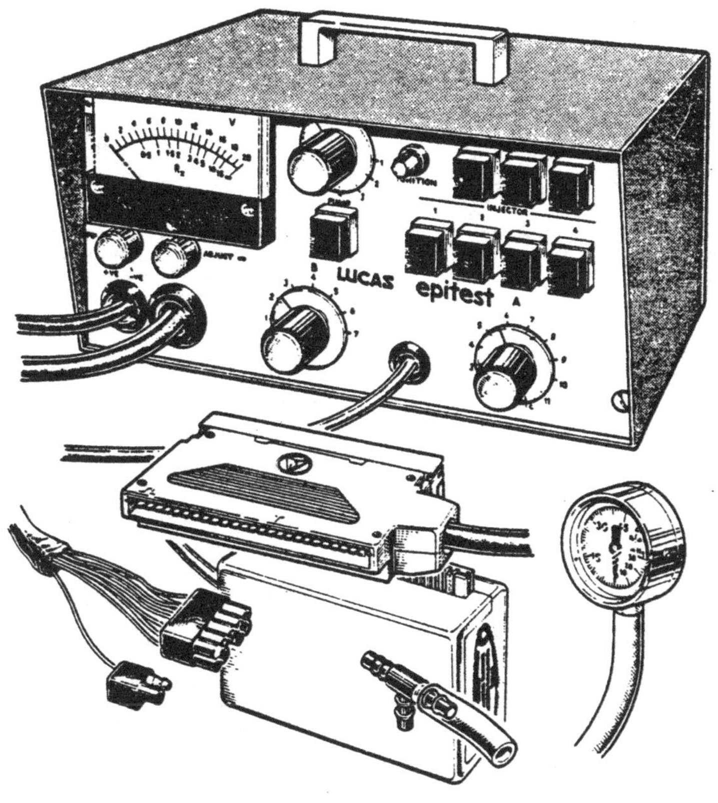

Lucas EPITEST—Part Number 60073009

Feedback monitor—Part Number 60973066

The feedback monitor is used to set the engine fuelling at idle speed. The following procedure is used in place of CO/HC checking with infrared equipment.

IMPORTANT: Checks must be made with engine at normal operating temperature.

NOTE: If engine is already hot, run for two minutes before commencing test.

1. Connect feedback monitor to diagnostic connector below E.C.U. in glove locker.
2. Position switch to the high position.
3. Fuelling is correct when No. 2 light in either Row A and B is lit.
   NOTE: No. 1 light in row A may also light, but should be ignored.
4. Fuelling adjustments are made by removing the blanking plug from the mixture screw. Clockwise turn of the screw enriches the mixture, counter-clockwise weakens the mixture.

Checking the Lambda sensor operation

Remove fuel pressure regulator vacuum pipe. Feedback monitor light should move towards "RICH" (i.e., light 2 to light 1). If indicator lights do not change, the Lambda Sensor or wiring circuit is suspect.

Replace fuel pressure vacuum pipe after completion of check.

AIR CLEANER

Remove and refit **19.10.01**

Removing

1 Locate the air cleaner at the front right-hand side of the engine compartment forward of the front suspension turret.
2 Remove the air cleaner element.
3 Disconnect the air cleaner case from the air-flow meter.
4. Remove four bolts connecting the case to the body and lift off the case.

Refitting

5 Reverse instructions 1 to 4.

AIR CLEANER

Renew element **19.10.08**

1 Remove the air intake pipe from the air cleaner case.
2 Remove eight screws to release the two halves of the air cleaner case.
3 Carefully lift off the top half of the case to expose the element.
4 Remove and discard the element, noting its position for refitting.
5 Clean the interior of the air cleaner case and fit a new element.
6 Reverse instructions 1 to 3.

THROTTLE PEDAL ASSEMBLY

Remove and refit **19.20.01**

Removing

1 Disconnect the battery.
2 Working from inside the car remove the spring clip securing the throttle cable.
3 Remove the cable from the pedal.
4 Open the bonnet and remove the two bolts securing the pedal assembly to the bulkhead.
5 Remove the pedal assembly from the car.
6 Strip the pedal assembly, and discard worn parts.

Refitting

7 Locate the spring in the bracket.
8 Fit the pedal to the bracket.
9 Insert the clevis pin.
10 Fit the anti-rattle washer.
11 Fit the washers (plain).
12 Secure the assembly with a new split pin.
13 Fit the pedal to the bulkhead and secure with the two bolts.
14 Fit the cable to the pedal and retain with the spring clip.
15 Reconnect the battery.

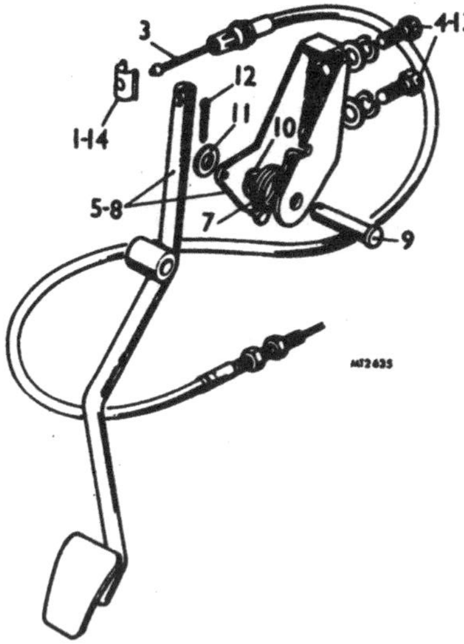

THROTTLE CABLE

Remove and refit **19.20.06**

Removing

1 Remove the plenum chamber for access.
2 Release the outer cable from the throttle body bracket.
3 Release the inner cable from the throttle body bracket.
4 From inside the car remove the clip and disconnect the inner cable from the accelerator pedal.
5 Remove the cable into the engine compartment.

Refitting

6 Reverse instructions 1 to 5.

AKM3079B

EXTRA AIR VALVE

Remove and refit **19.20.16**

CAUTION: This operation should only be completed on a cold or cool engine.

Removing
1 Remove the plenum chamber for access.
2 Disconnect the electrical multi-pin plug from the valve.
3 Disconnect two air hoses from the valve.
4 Remove two mounting bolts securing the valve to the inlet manifold.
5 Lift off the valve.

Refitting
6 Clean the mating faces on the valve and manifold.
7 Reverse instructions 1 to 5.

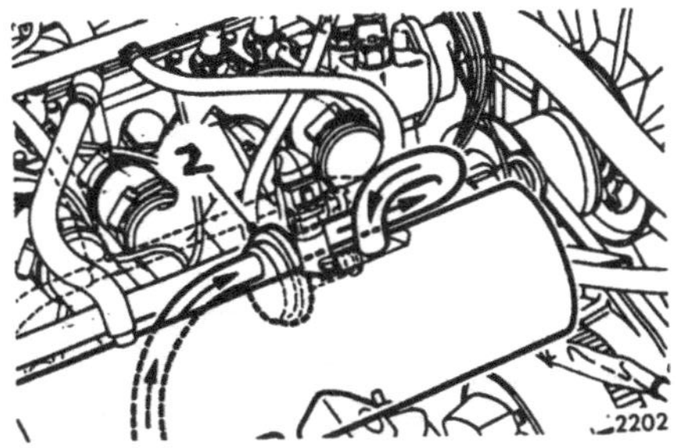

EXTRA AIR VALVE

Test **19.20.17**

1 Remove the electrical connector from the valve.
2 Connect a voltmeter across the terminals of the connector.
3 Crank the engine over: battery voltage should be obtained. If there is no voltage there is a fault in the electrical system and the wires should be checked for leakage and deterioration. When it has been established that power is reaching the valve, the heating coils resistance should be checked.
4 Connect an ohmmeter between the terminals of the air valve. A resistance of 33 ohms should be obtained. If there is no resistance, the valve should be replaced.

IDLE SPEED

Adjust **19.20.18**

1 Run the engine until it has been working at normal operating temperature for at least two minutes.
NOTE: Before, and every three minutes during, the fuel setting procedure below the engine should be given a clear outburst by running it at approximately 2000 rev/min on light load for a minimum of 30 seconds to maintain normal operating temperature.
2 Ensure that the engine ignition timing is correct (refer to Section 05, Engine Tuning Data) and that the throttle linkage is correctly set.

Idle speed
3 Using a separate proprietary tachometer, connected following the manufacturer's instructions, or if necessary the vehicle fitted tachometer, check the engine idle speed; refer to Section 05, Engine Tuning Data.
4 If adjustment is necessary, slacken the locknut and turn the idle adjustment bolt (clockwise to decrease speed, anti-clockwise to increase speed). Retighten the locknut.

Idle fuel setting
Two alternative methods may be used depending on the equipment available.

Approved type exhaust gas analyser
5 Insert analyser probe into exhaust pipe.
6 Check analyser CO reading which must not exceed that given in Section 05, Engine Tuning Data.
7 If adjustment is necessary follow instructions 13 to 15 below.
8 Remove analyser probe.

Approved type fuel setting indicator
9 Connect the instrument to the fuel setting diagnostic connector, located behind the right-hand glove box.
10 Check the instrument reading which, if correct, will show lamp No. 2 in rows A or B alight.
11 If adjustment is necessary follow instructions 13 to 15 below.
12 Disconnect the instrument.

Idle fuel setting adjustment
13 Remove the blanking plug from the air flow meter to expose the recessed adjustment screw.
14 Turn the adjustment screw until the required reading is obtained.
15 Reseal the adjustment screw as required by local territory legislation.

LAMBDA (OXYGEN) SENSOR

Remove and refit **19.22.16**

Removing
1 Disconnect the electrical lead from the sensor.
2 Unscrew the sensor from the exhaust pipe, taking care not to strain the exhaust system.

Refitting
3 Lubricate the threads of the sensor and fit it to the exhaust pipe.
4 Reconnect the electrical leads to the sensor.
5 Reset the service interval counter, using the special service tool key.

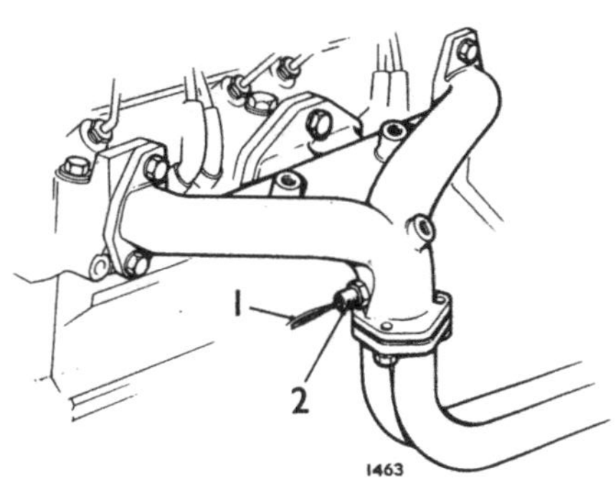

COOLANT TEMPERATURE SENSOR

Remove and refit **19.22.18**

Removing
CAUTION: Before commencing this procedure ensure that the cooling system is depressurized. Remove the cooling system header tank filler cap taking care to avoid scalding; refer to Section 26, Cooling System.

1 Locate the sensor, referring to the illustration and the vehicle wiring diagram.
2 Disconnect the electrical leads from the sensor.
3 Unscrew the sensor from the engine.

Refitting
4 Reverse instructions 2 and 3, ensuring that the sensor is sufficiently tightened to prevent water leaks without over straining the threads.
5 Check/top-up the cooling system water level.

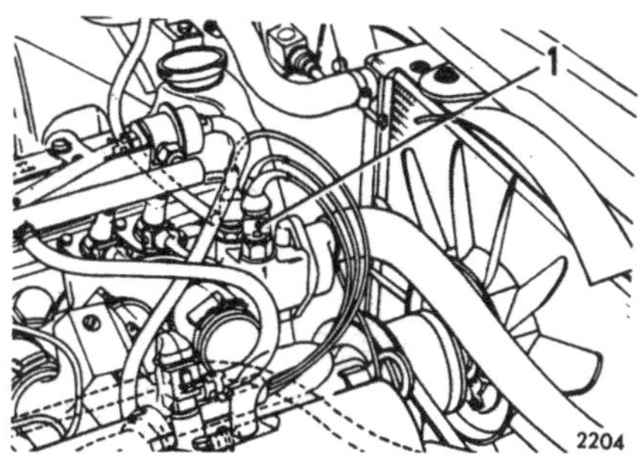

COOLANT TEMPERATURE SENSOR

Test **19.22.19**

1 Disconnect battery.
2 Pull connector from temperature sensor.
3 Connect suitable ohmmeter between terminals; note resistance reading. The reading is subject to change according to temperature, and should closely approximate to the relevant resistance value given in the table.
4 Disconnect ohmmeter.
5 Check resistance between each terminal in turn and body of sensor. A very high resistance reading (open circuit) must be obtained.
6 Re-connect cable.
7 Re-connect battery.

Coolant Temperature (°C)	Resistance (kilohms)
−10	9.2
0	5.9
+20	2.5
+40	1.18
+60	0.60
+80	0.325

THERMOTIME SWITCH

Remove and refit **19.22.20**

Removing
CAUTION: Before commencing this procedure ensure that the cooling system is depressurized. Remove the cooling system header tank filler cap taking care to avoid scalding; refer to Section 26, Cooling System.

1 Locate the switch, referring to the illustration and the vehicle wiring diagram.
2 Disconnect the electrical leads from the switch.
3 Unscrew the switch from the engine.

Refitting
4 Reverse instructions 2 and 3, ensuring that the switch is sufficiently tightened to prevent water leaks without over-straining the threads.
5 Check/top-up the cooling system water level.

THERMOTIME SWITCH

Test **19.22.21**

Equipment required: Stop watch, ohmmeter, single-pole switch, jump lead for connecting switch to battery and Thermotime switch, and a thermometer.
NOTE: Check coolant temperature with thermometer and note reading before carrying out procedures detailed below. Check rated value of Thermotime switch (stamped on body flat).
1 Disconnect battery earth lead.
2 Pull electrical connector from Thermotime switch.

'A' Coolant temperature higher than switch rated value
3 Connect ohmmeter between terminal 'W' and earth. A very high resistance reading (open circuit) should be obtained.
4 Renew switch if a very low resistance reading (short circuit) is obtained.

'B' Coolant temperature lower than switch rated value
5 Connect ohmmeter between terminal 'W' and earth. A very low resistance reading (closed circuit) should be obtained.
6 Connect 12V supply via isolating switch to terminal 'G' of Thermotime switch.
7 Using stop watch, check time delay between making isolating switch and indication on ohmmeter changing from low to high resistance. Delay period must closely approximate to time indicated in table, see table for specific coolant temperature (noted above).
8 Renew Thermotime switch if necessary.
9 Re-connect Thermotime switch.
10 Re-connect battery earth lead.

Coolant Temperature	Delay
−20°C	8 sec.
0°C	4½ sec.
+10°C	3½ sec.
+35°C	0

AIR TEMPERATURE SENSOR

Remove and refit **19.22.22**

The air temperature sensor is integral with the air-flow meter and as such cannot be replaced as a separate item.

AIR TEMPERATURE SENSOR

Test **19.22.23**

1 Disconnect the battery.
2 Remove the multi-pin electrical connector from air-flow meter.
3 Connect a suitable ohmmeter between terminals 6 and 27 of the air-flow meter.

Ambient Air Temperature (°C)	Resistance (kilohms)
−10	9.2
0	5.9
+20	2.5
+40	1.18
+60	0.60

4 Note the resistance reading. The reading is subject to change according to the temperature and should closely approximate to the relevant resistance value given in the table above.
5 Disconnect the ohmmeter.
6 Re-connect the multi-pin connector.
7 Re-connect the battery.

AIR-FLOW METER

Remove and refit **19.22.25**

Removing
1 Remove the air cleaner for access.
2 Disconnect the air outlet pipe from the air-flow meter.
3 Disconnect the electrical multi-pin plug from the meter.
4 Remove four bolts securing the meter to the body.
5 Lift out the meter.

Refitting
6 Reverse instructions 1 to 5.

ELECTRONIC CONTROL UNIT (E.C.U.)

Remove and refit **19.22.34**

Removing
1 Locate the E.C.U. below the passenger glove box inside the car and remove the cover (if fitted).
2 Peel back the passenger footwell carpet for access.
3 Disconnect the electrical multi-pin plug from the E.C.U.
4 Remove three bolts securing the E.C.U. and lower it into the footwell.

Refitting
5 Reverse instructions 1 to 4.

THROTTLE SWITCH

Remove and refit **19.22.36**

Removing
1 Locate the throttle switch on the throttle bracket at the left-hand end of the plenum chamber.
2 Pull the electrical connector from the throttle switch.
3 Remove the two screws, plain and shakeproof washers securing the throttle switch and lift the switch from the spindle.

Refitting
4 Locate the switch on the spindle.
5 Secure the switch with the two screws, plain and shakeproof washers.

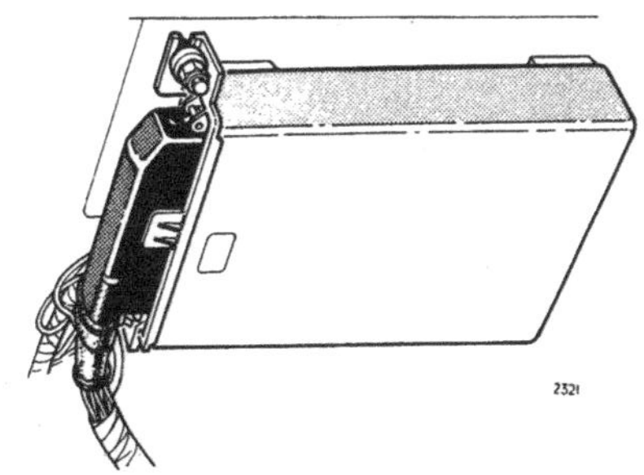

THROTTLE SWITCH

Test **19.22.37**

NOTE: Before commencing the following tests ensure that the throttle linkage is correctly adjusted.

1 Disconnect the battery.
2 Remove the electrical connector from the throttle switch.
3 Connect a powered test lamp between terminals 3 and 18 of the throttle switch.
4 Open the throttle; the bulb should light up when the throttle nears the wide open position.
6 If the bulb does not light, replace the throttle switch.
7 Refit the electrical connector to the switch.
8 Re-connect the battery.

PLENUM CHAMBER

Remove and refit **19.22.46**

Removing

NOTE: If a water-heated plenum chamber is NOT fitted, instructions 1 and 2 should be omitted.

1 Partially drain the cooling system, see 26.10.01.
2 Disconnect the water pipes from the plenum chamber.
3 Remove the battery for access.
4 Release the throttle bracket from the side of the chamber.
5 Release the engine oil dipstick tube from the plenum chamber bracket.
6 Release the interconnecting pipes from the plenum to the inlet manifold.
7 Disconnect from the plenum chamber the brake servo vacuum pipe, the extra air valve pipe, the pressure regulator vacuum pipe and the cold start injector.
8 Remove five bolts to release the chamber from the throttle housing and lift off the chamber.

Refitting

9 Reverse instructions 1 to 8.

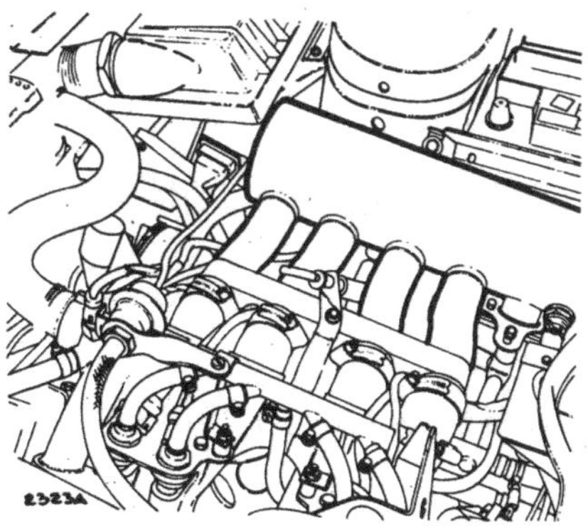

FUEL SYSTEM

WARNING: The fuel in the system is pressurized by the fuel pump to approximately 36 lbf/in² (2.5 kgf/cm²) whilst the engine is running and experience has shown that this pressure has a very slow decay rate. Therefore it is essential to depressurize the fuel system, following the procedure under 19.50.02 before disconnecting any component in the system.

WARNING: During all operations connected with the fuel system make provision to minimize any risk of fire or explosion.

FUEL LINE FILTER(S)

Remove and refit **19.25.03**

NOTE: A single filter or twin filters may be fitted, and later vehicles have a protective shield over the filter(s).

Removing

1 Depressurize the fuel system, see 19.50.02.
2 Locate the filter(s) on the right-hand underside of the body, forward of the rear axle, just below the fuel pump.
3 Where fitted, remove four screws and lift off the filter cover.
4 Disconnect the input fuel pipe(s) and plug them to prevent fuel loss.
5 Disconnect the output fuel pipe(s) and plug them to prevent fuel loss.
6 Remove the filter retaining clamp bolt and nut and manoeuvre the filter(s) off the car.

Refitting

7 Reverse instructions 2 to 6.

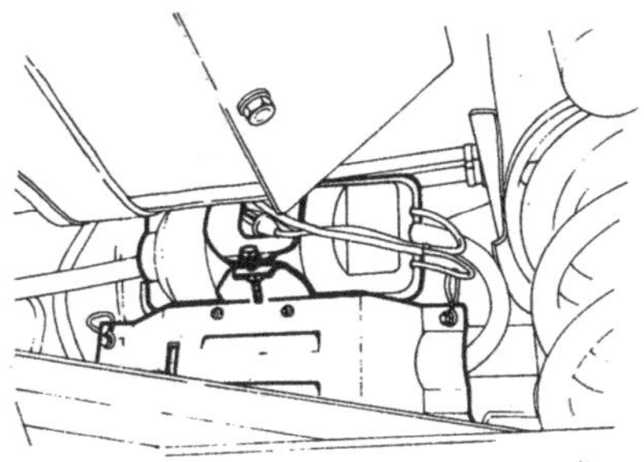

FUEL PIPES

Petrol pipe — main line — engine end section — remove and refit **19.40.02**

Petrol pipe — main line — tank end section — remove and refit **19.40.04**

Petrol pipe — fuel return valve to petrol tank — remove and refit **19.40.54**

FUEL PUMP

Remove and refit **19.45.08**

Removing

1 Depressurize the fuel system, see 19.50.02.
2 Locate the fuel pump on the right-hand underside of the body, forward of the rear axle.
3 Disconnect the electrical leads to the fuel pump.
4 Clamp the input fuel pipe to the pump to prevent the fuel tank draining when it is disconnected. Disconnect the pipe.
5 Disconnect the output fuel pipe from the pump and plug it to prevent fuel loss.
6 Remove the single bolt and nut securing the clamping bracket.
7 Manoeuvre the pump from the car.

Refitting

8 Reverse instructions 2 to 7.

FUEL PRESSURE REGULATOR

Remove and refit 19.45.11

Removing
1 Depressurize the fuel system, see 19.50.02.
2 Disconnect the vacuum pipe from the regulator.
3 Disconnect and plug the fuel inlet pipe.
4 Remove the screw and nut securing the regulator bracket.
5 Using two spanners release the regulator from the fuel rail, taking care not to strain the rail.

Refitting
6 Reverse instructions 2 to 5.

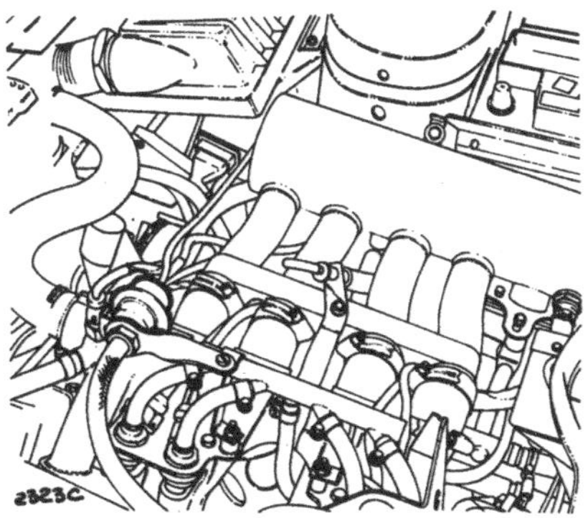

FUEL PRESSURE REGULATOR

Test 19.45.12

1 Depressurize the fuel system, see 19.50.02.
2 Slacken pipe clip securing cold start injector supply pipe to fuel rail and pull pipe from rail.
3 Connect pressure gauge pipe to fuel rail and tighten pipe clip.
CAUTION: Pressure gauge must be checked against an approved standard at regular intervals.
4 Pull '−ve' L.T. lead from ignition coil and switch ignition on.
Check reading on pressure gauge; reading must be 36.25 ± 0.725 lbf/in^2 (2,55 ± 0,05 kgf/cm^2).
NOTE: The pressure reading may slowly drop through either the regulator valve seating or the pump non-return valve. A slow, steady drop is permissible; a rapid fall **MUST** be investigated.
If satisfactory results have been obtained, depressurize the fuel system and continue with operations 5 to 9. If satisfactory results have not been obtained, replace the regulator with a new unit.
5 Slacken pipe clip and remove pressure gauge from fuel rail.
6 Re-connect cold start injector supply pipe and secure pipe clip.
7 Re-connect the '−ve' L.T. lead to ignition coil.
8 Switch ignition on and check for leaks.
9 Switch ignition off.

FUEL SYSTEM

Depressurize 19.50.02

CAUTION: The fuel system MUST always be depressurized before disconnecting any fuel system components.

1 Remove the fuel pump earth lead.
2 Switch on and crank engine for a few seconds.
3 Switch the ignition off and re-connect the pump to earth.

Pressure test, see operation 19.45.12.

ALTERNATIVELY: The fuel cut-off inertia switch may be used to disconnect the fuel pump before cranking the engine.

FUEL TANK

Remove and refit 19.55.01

Follow the instructions for carburetter fitted vehicles, operation 19.55.01, noting that the fuel return pipe from the pressure regulator must additionally be disconnected and reconnected to the fuel tank.

INJECTORS

Injector winding check **19.60.02**

1 Use ohmmeter to measure resistance value of each injector winding, which should be 2.4 ohms at 20°C (68°F).
2 Check for short-circuit to earth on winding by connecting ohmmeter probes between either injector terminal and injector body. Meter should read infinity. If any injector winding is open-circuited or short-circuited, replace the injector.

FUEL RAIL

Remove and refit **19.60.05**

Removing
1 Depressurize the fuel system, see 19.50.02.
2 Remove the pressure regulator, see 19.45.11.
3 Release the fuel pipe connections to the rail.
4 Remove two bolts securing the rail to the inlet manifold and lift off the rail.

Refitting
5 Reverse instructions 2 to 4.

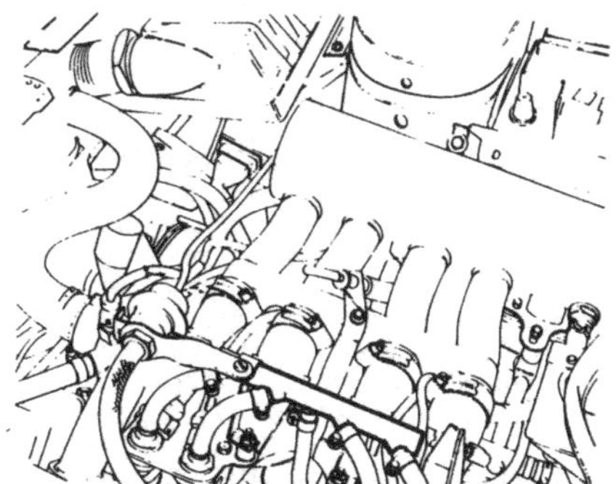

23238

COLD START INJECTOR

Remove and refit **19.60.06**

Removing
1 Depressurize the fuel system, see 19.50.02.
2 Locate the injector beneath the plenum chamber.
3 Disconnect the electrical leads from the injector.
4 Remove the injector clamp bracket screws and withdraw the injector and bracket.

Refitting
5 Renew the sealing ring and reverse instructions 2 to 4.

INJECTORS

Remove and refit **19.60.10**

Removing
1 Depressurize the fuel system, see 19.50.02.
2 Disconnect the electrical lead from the injector.
3 Disconnect the fuel feed pipe to the injector.
4 Remove the injector clamp bracket securing screws and lift off the bracket.
5 Withdraw the injector.

Refitting
6 Renew the sealing ring and reverse instructions 2 to 5.

FAULT FINDING 19.70.00

Within the following pages are suggested procedures for tracing faults associated with poor engine running and failure to start.

In these circumstances and before suspecting, checking or changing fuel injection system components, it is recommended that common engine malfunction/failure to start causes are checked out, including, for example:

a. the presence and quality of fuel in the fuel tank;

b. the adequate and correct functioning of the ignition system up to the point of a good spark at *each* spark plug;

c. correct ignition and valve timing and adequate compression in *each* cylinder;

and, in the case of an engine failing to start:

d. that the battery has sufficient power reserve to operate the starter motor, supply the ignition system and power the fuel injection components.

SECTION 1. ENGINE FAILS TO START

	PROCEDURE	RESULT	ACTION
	If the engine fails to crank	Engine does not turn over	Check battery leads for security and cleanliness; battery, for state of charge
B	Ignition ON. Attempt to start engine	Engine starts	Go to N
		Engine fails to start	Go to C
C	Check that coolant temperature sensor is correctly connected	Connected	Go to D
		Disconnected	Re-connect sensor. Start engine and drive away
D	Check for spark at No. 1 spark plug lead. Disconnect lead from plug and hold end approx. 1/8 in from bare metal of engine block and crank engine	Good spark	Go to E
		No spark	Check ignition system
E	Ignition ON. Check fuel tank contents	Fuel in tank	Go to F
F	Ensure fuel cut-off inertia switch is closed	Switch closed	Go to G
		Switch open	Close switch by pressing button on top of switch. Start engine and drive away
G	Switch ignition OFF. Switch ignition ON. Crank engine and listen for fuel pump running	Pump does not run	Go to M
		Pump runs	Go to K
H	Switch ignition OFF. Remove cover from inertia switch and ensure cables are connected. If cables secure, disconnect and short together. Switch ignition ON. Pump should run when engine cranked	Pump does not run	Go to J
		Pump runs	Secure cables together and make safe with insulating tape etc. Start engine and drive away. Contact Dealer or Distributor as soon as possible

	PROCEDURE	RESULT	ACTION
J	Re-connect inertia switch. Ensure fuel injection system 12 volt cable is connected. Check connections	Loose or detached connections	Secure connections. Start engine and drive away
		All connections secure	Contact Dealer or Distributor
K	Crank engine in neutral gear position. Listen carefully to ensure all injectors click	Injectors click	Go to M
		Injectors do not click	Go to L
L	Check electrical connections to relay and ignition coil check multi-pin connectors to E.C.U. and air meter	Loose or disconnected connectors or wires	Re-connect, secure connectors and drive away
		Connectors secure	Go to M
M	Check for possible air leakage into manifold	Loose or disconnected	Secure. Start engine and drive away
		Connections secure	Go to O Contact Dealer or Distributor
N	Engine starts and runs	Normally	Drive away
		Unevenly	Check following connections for condition and security. Coolant temperature sensor. Auxiliary air valve. Throttle switch and all injector plugs. If all in order go to Section 2
O	Check for free movement of air meter control flap after removing hose at rear of air meter	Operates normally	Contact Dealer or Distributor
		Sticking	Free flap. Drive away. Contact Dealer or Distributor

SECTION 2. ENGINE CUTS OUT, RUNS UNEVENLY OR LOSES POWER

	PROCEDURE	RESULT	ACTION
A	Vehicle operating normally	Engine cuts out	Pull to side of road, switch OFF ignition. Go to C of Section 1
		Engine runs unevenly or loses power	Pull to side of road, leave engine running. Go to B
B	Engine running, car at standstill. Ensure manual transmission cars are in neutral, automatic cars have either 'N' or 'P' selected and handbrake applied	Engine idles normally	Go to D
		Engine idles unevenly	Go to C
C	Ignition OFF Check—Spark plug leads Injector connectors Start engine	Engine runs normally	Drive away
		Engine runs unevenly	Go to F and return if necessary to C. Stop engine. Check for excessive oil fumes at breather or exhaust. Contact Distributor or Dealer as soon as possible
D	Apply pressure to throttle pedal and increase engine revolutions slowly	Engine runs normally	Go to E
		Engine runs unevenly	Go to F
E	Engage forward gear and drive at moderate speed	Engine runs unevenly	Go to F
		Engine runs normally	Fuel filter or supply pipe probably obstructed. Drive away, contact Distributor or Dealer as soon as possible
F	Check for possible air leakage into manifold. 1. Convolute hose 2. Induction elbow 3. Auxiliary air valve, top No. 13 and bottom hose	Loose or disconnected	Secure. Start engine and drive away
		Connections secure	Return to C

POSSIBLE CAUSES IN ORDER OF CHECKING

SYMPTOMS

Will not start*	Difficult cold start	Difficult hot start
Starts but will not run	Misfires and cuts out	Runs rough
Idle speed to fast	Hunting at idle	Low power and top speed
High fuel consumption		

* Before proceeding with checks, hold throttle fully open and attempt a start. If the engine then starts and continues to run, no further action is necessary.

PROCEDURES FOR RECTIFICATION OF CAUSES SHOWN IN TABLE

Battery	Battery depleted, giving insufficient crank speed or inadequate spark. Check battery condition with hydrometer. Re-charge, clean and secure terminals, or renew as necessary.
Connections	Ensure all connector plugs are securely attached. Pull back rubber boot and ensure plug is fully home. While replacing boot press cable towards socket. Ensure electronic control unit (E.C.U.) multi-pin connector is fully made. Ensure all ground connections are clean and tight.
Ignition System	Check ignition system as detailed in electrical section of Repair Operations Manual.
Fuel System	Open filler cap of fuel tank. Check for fuel pipe failure (strong smell of fuel) and retention of in-line fuel pressure. Check inertia switch closed. If necessary, clear fuel tank vents or supply pipe.
Cold Start System	Fault conditions could cause cold start system to be inoperative on a cold engine, or operative on a hot engine. If engine is either very hot, or cold, these particular faults will cause the engine to run very rich. Check cold start system.
E.C.U./ Amplifier	If the E.C.U. is faulty it is possible that injectors will be inoperative. The E.C.U. may also be responsible for any degree of incorrect fuelling. Before suspecting the E.C.U. for fuelling problems, however, all other likely components should be proved good.
Air Leaks	Ensure all hose and pipe connections are secure. Engine is, however, likely to start more easily with air leaks if cold, as air leaking augments that through the auxiliary air valve. A leak, or failed air valve is shown up, however, by a very high idle speed when the engine is warm and air valve main passage should be closed.
Temperature Sensors	If either sensor is short-circuited, starting improves with higher engine temperature. Engine will run very weak, improving as temperature rises, but still significantly weak when fully hot. If a sensor is open-circuit, or disconnected, engine will run very rich, becoming worse as temperature rises. Engine may not run when fully hot, and will almost certainly not restart if stalled. Effort of air temperature sensor will be less marked than coolant temperature sensor.
Auxiliary Air Valve	Check opening throttle. If engine immediately starts, unscrew idle speed adjustment, and re-check start with closed throttle. Re-set idle speed when engine hot. Check cold start. Check throttle return springs and linkage for sticking or maladjustment as a sticking throttle may have enforced incorrect idle speed adjustment on a previous occasion.
Throttle Switch	Check operation of throttle switch. Incorrect function or sequence of switching will give this fault.
Throttle Butterfly	Check adjustment of the throttle butterfly valve, ensure return springs correctly fitted, and throttle not sticking open.
Compression	Low compressions; a general lack of engine tune could cause this fault. Check engine timing, ignition timing, and function of ignition system complete. If necessary, check valve condition.
Idle Fuel Control Setting	Check exhaust gas CO level. If necessary adjust. **CAUTION:** This MUST NOT be moved unless correct test equipment and skilled personnel are in attendance to monitor changes made.
Air Filters	Remove air cleaner element and check for choked filter element.
Throttle Linkage	Check throttle linkage adjustment and ensure that throttle butterfly valves can be fully operated.

COOLING SYSTEM

COOLING SYSTEM

Description **26.00.00**

Two types of cooling system have been fitted to TR7 models, the change taking place during the 1979 Model Year manufacture.

Earlier vehicles are fitted with an expansion tank, later models have a header tank. Both are 'no loss' systems allowing for coolant expansion and contraction.

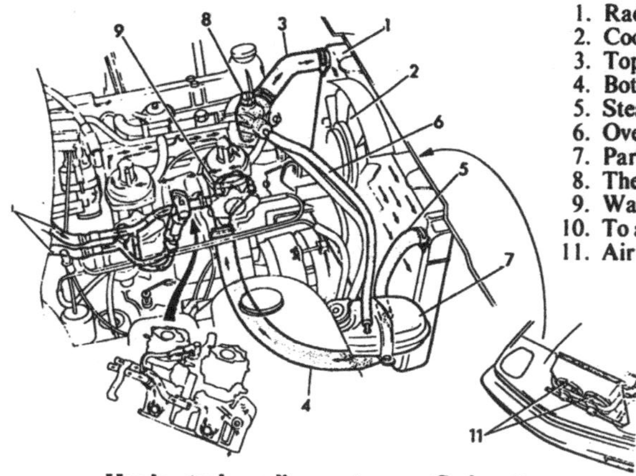

1. Radiator
2. Cooling fan
3. Top hose
4. Bottom hose
5. Steam bleed
6. Overflow pipe
7. Partial flow tank
8. Thermostat
9. Water pump
10. To and from heater
11. Air conditioning fans

Header tank cooling system — Carburetter engines

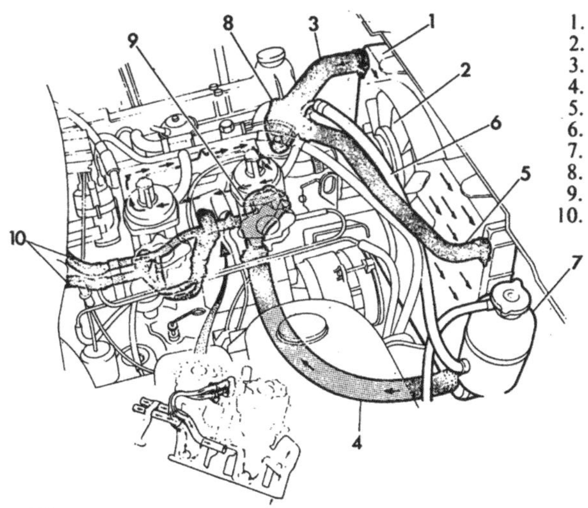

1. Radiator
2. Cooling fan
3. Top hose
4. Bottom hose
5. Steam bleed
6. Overflow pipe
7. Expansion bottle
8. Thermostat
9. Water pump
10. To and from heater

Expansion tank cooling system — Carburetter engines

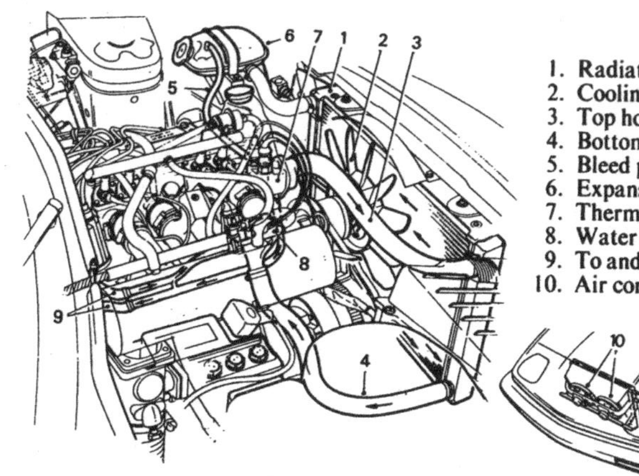

1. Radiator
2. Cooling fan
3. Top hose
4. Bottom hose
5. Bleed pipe
6. Expansion tank
7. Thermostat
8. Water pump
9. To and from heater
10. Air conditioning fans

Header tank cooling system — Fuel Injection engines AP 458

COOLANT

Drain and refill 26.10.01

Draining

WARNING: This operation must only be carried out when the engine is cold.

1 Disconnect the battery.
2 Place a suitable container beneath the radiator bottom hose connection.
3 Remove the thermostat housing filler plug.
4 Disconnect the bottom hose at the radiator and allow the coolant to drain.
5 Remove the cylinder block drain plug. 12.25.07.

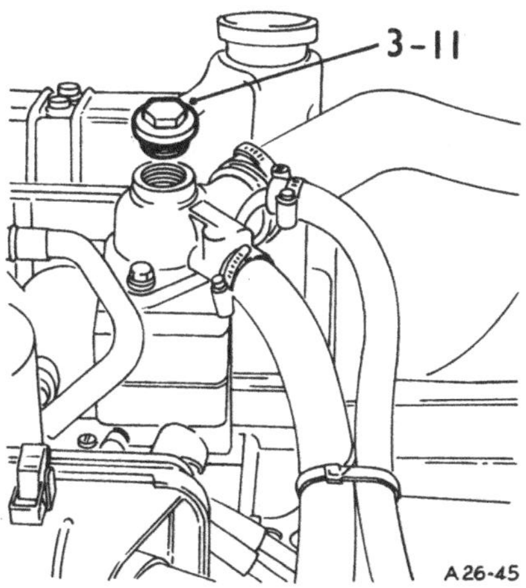

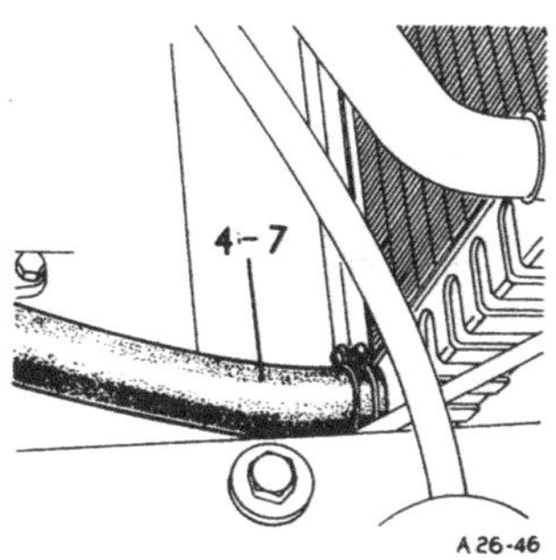

Refilling (earlier models with expansion tank)

6 Fit and tighten the cylinder block drain plug using a new washer. 12.25.07.
7 Connect the bottom hose to the radiator.
8 Check that the expansion tank is half-full.
9 Set the heater control to 'HOT'.
10 Fill the system through the thermostat housing to the bottom of the threads, with the correct mixture of a recommended anti-freeze solution see **'LUBRICANTS & ANTI-FREEZE SOLUTIONS'**.
11 Refit the thermostat housing filler plug carefully; do not overtighten.
12 Remove the pressure cap from the expansion tank.
13 Run the engine for three minutes at approximately 1200 rev/min.
14 Stop the engine.
15 Refit the expansion tank pressure cap.
16 Remove the filler plug from the thermostat housing.
17 Gently squeeze the large top hose between the thermostat housing and radiator to expel any trapped air in the hose.
18 Top up the system to the bottom of the filler plug threads in the thermostat housing.
19 Refit the filler plug using a new sealing washer if necessary. Do not overtighten.
20 Recheck the expansion tank coolant level, and top up to half full if necessary.

NOTE: Subsequent topping-up of the cooling system should be carried out, when the engine is cool, via the expansion tank which must be maintained at half full.

Refilling (later models with header tank)

1 Fit the bottom hose, cylinder block drain plug and thermostat housing filler plug.
2 Remove the header tank filler cap.
3 Set the interior heater controls to the maximum heat position.
4 Fill the system until the coolant level is 1 inch (25 mm) below the neck of the header tank.
5 Refit the header tank filler cap and run the engine at approximately 1500 rev/min until the coolant temperature rises sufficiently to open the thermostat.
6 Stop the engine and, taking care to avoid scalding, remove the header tank filler cap.
7 Top up the coolant level as necessary until it is 1 inch (25 mm) below the header tank filler neck.
8 Refit the filler cap.

NOTE: The coolant level should be maintained at 1 inch (25 mm) below the neck of the header tank (1).

HEADER TANK–LATER MODELS

Remove and refit **26.15.01[A]**

Removing

WARNING: This operation must only be carried out when the engine is cold.

1 Disconnect the lower hose from the tank and drain the coolant into a suitable container.
2 Disconnect the upper hose and overflow hose from the tank.
3 Remove two bolts securing the tank to the mounting bracket.
4 Lift off the header tank.

Refitting

5 Reverse instructions 1 to 4 and refill the cooling system, see 26.10.01.

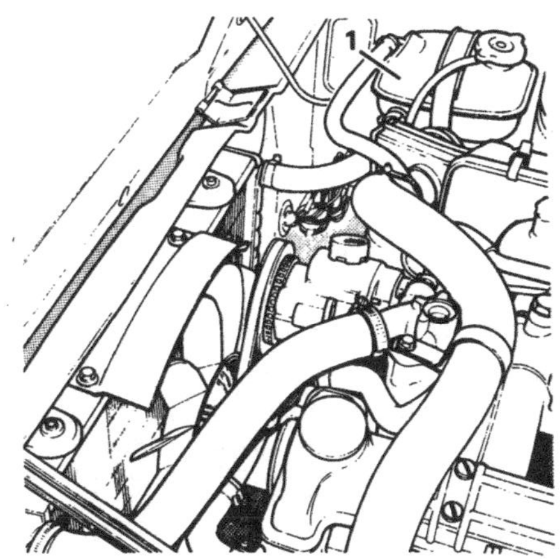

EXPANSION TANK – EARLIER MODELS

Remove and refit **26.15.01[B]**

Removing

WARNING: This operation must only be carried out when the engine is cold or cool.

1. Disconnect the expansion pipe from the thermostat housing.
2. Remove the expansion tank pressure cap and allow the coolant to drain into a suitable clean container.
3. Remove the expansion pipe from the expansion tank.
4. Slacken the expansion tank retaining clamp nut and bolt.
5. Withdraw the expansion tank.
6. Remove the overflow pipe from the tank.

Refitting

7. Fit the tank into the clamp and secure.
8. Fit the overflow pipe.
9. Fit the expansion pipe to the tank and thermostat housing.
10. Half-fill the expansion tank with the correct mixture of anti-freeze and water.
11. Refit the pressure cap.
12. Refill the cooling system, see 26.10.01.

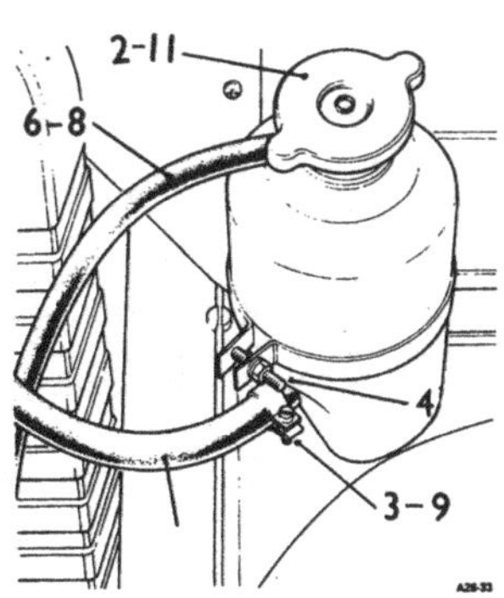

FAN AND ALTERNATOR DRIVE BELT – UK and Europe

Remove and refit **26.20.07**

Removing

1. Disconnect the battery.
2. Slacken the alternator adjustment bolts and move the alternator towards the engine.
3. Slip the belt from the pulleys and feed it over the fan blades to remove it from the engine.

Refitting

4. Reverse instructions 1 to 3 ensuring that the drive belt is correctly tensioned. 86.10.05.

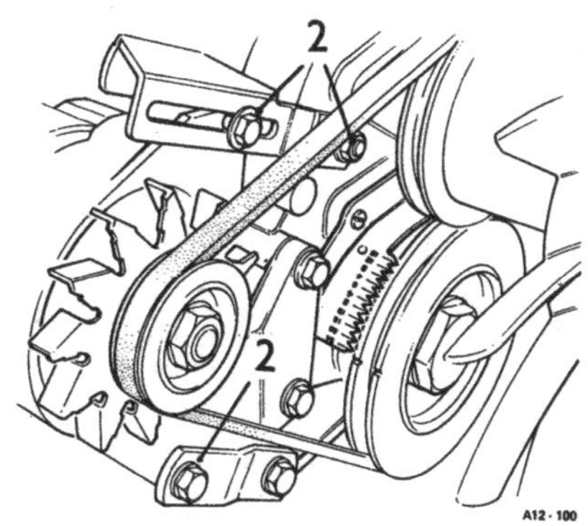

FAN AND ALTERNATOR DRIVE BELT – USA Specification Vehicles

Remove and refit **26.20.07**

Removing

1. Disconnect the battery.
2. Raise the car and slacken off the air conditioning compressor belt adjustment bolts–see operation 82.10.02 instructions 1 to 3.
3. Remove the fan guard–two bolts.
4. Slacken off the four fan blade attachment nuts and bolts to provide a clearance for removing the belt.
5. *Carburetter vehicles only:* Slacken off the air pump adjustment (see operation 17.25.13) and remove the belt from the pump pulley only.
6. Slacken off the alternator adjustment bolts and remove the drive belt, feeding it over the air pump and compressor belts.

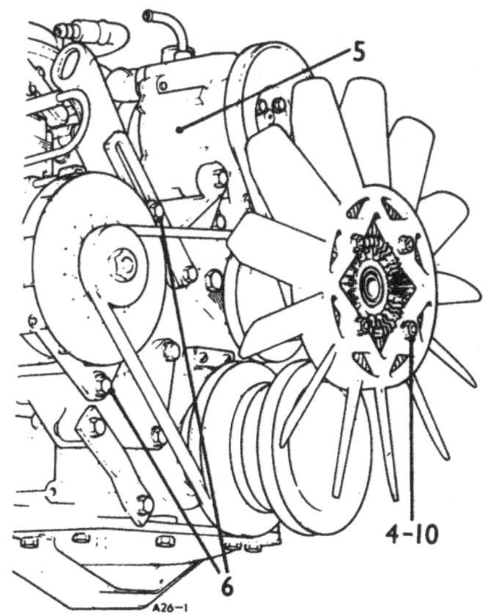

Refitting

7. Refit the alternator–fan drive belt–reversing instruction 6 and correctly tension the belt. 86.10.05.
8. *Carburetter vehicles only:* Fit the air pump belt and correctly tension, see 17.25.13.
9. Fit and adjust the compressor belt, see operation 82.10.01. instructions 7 to 11.
10. Tighten the fan blade attachment bolts.
11. Refit the fan guard.
12. Reconnect the battery.

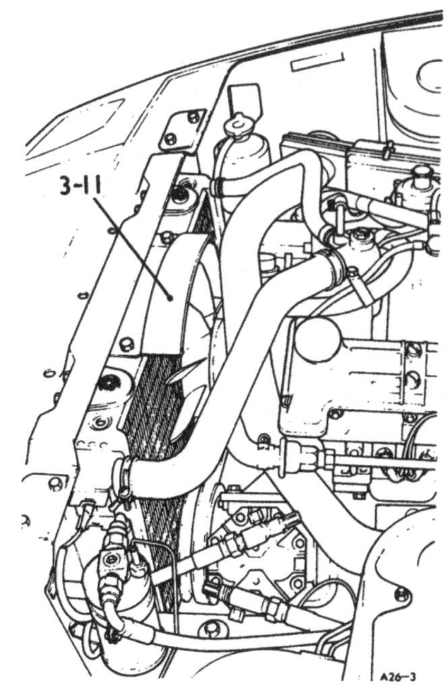

FAN BLADES

Remove and refit 26.25.06

Removing
1 Disconnect the battery.
2 Remove the fan blade shield.
3 Slacken off the nuts securing the radiator top support brackets sufficiently to move the radiator forward–away from the engine.
4 Remove the four nuts and bolts securing the blades to the viscous coupling.
5 Move the radiator forward whilst withdrawing the fan blades vertically.

Refitting
6 Fit the fan blades to the Torquatrol unit and tighten the retaining nuts and bolts evenly.
7 Tighten the four radiator support bracket nuts.
8 Refit the fan blade shield.
9 Reconnect the battery.

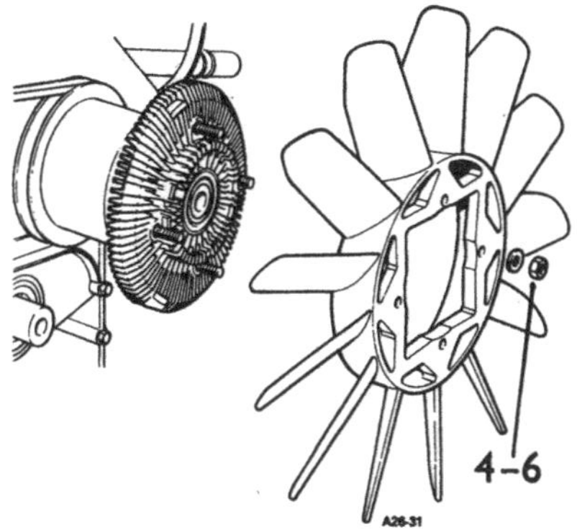

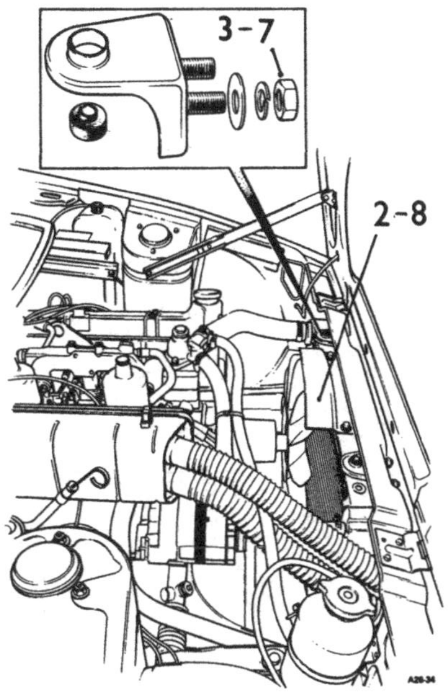

VISCOUS COUPLING

Remove and refit 26.25.19

Removing
1 Remove the fan blades and pulley assembly. 26.25.21.
2 Remove the fan blades. 26.25.06.
3 Place a support at the back of the torquatrol unit as close to the centre as possible and using a suitable press tool, press the pulley assembly from the viscous coupling.
4 Remove the tolerance ring.

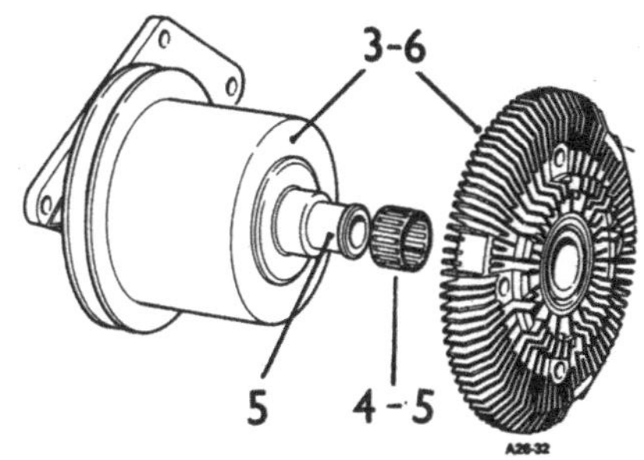

Refitting
5 Fit a new tolerance ring in position on the pulley bearing shaft.
6 Compress the tolerance ring with the fingers whilst locating the viscous coupling on the bearing shaft. Ensure that the coupling is fitted the correct way round – see illustration.
7 Press the unit squarely onto the shaft keeping the tolerance ring compressed.
8 Refit the fan blades.
9 Fit the fan blades and pulley assembly to the engine. 26.25.21.

FAN PULLEY AND BLADE ASSEMBLY – U.K. AND EUROPE

– Remove and refit **26.25.21**

Removing
1 Disconnect the battery.
2 Slacken the alternator adjustment and slip the belt from the pulleys.
3 Remove the four bolts retaining the pulley assembly to the timing cover.
4 Withdraw the pulley assembly from the left-hand side of the engine.

Refitting
5 Reverse instructions 1 to 4.

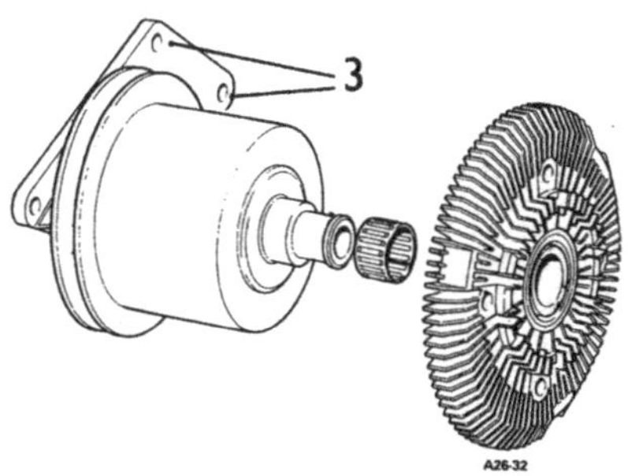

FAN PULLEY AND BLADE ASSEMBLY – USA Specification

Remove and refit **26.25.21**

Removing
1 Disconnect the battery.
2 Remove the bonnet. 76.16.01.
3 *Carburetter vehicles only:* Remove the air pump, see 17.25.07.
4 Slacken the alternator adjustment and slip the belt from the pulleys.
5 Remove the five bolts securing the air compressor steady bracket.
6 Remove the remaining two bolts securing the pulley assembly to the timing chain cover, i.e. one on L.H. side and one on R.H. side of pulley assembly.
7 Remove the fan blade guard (two bolts).

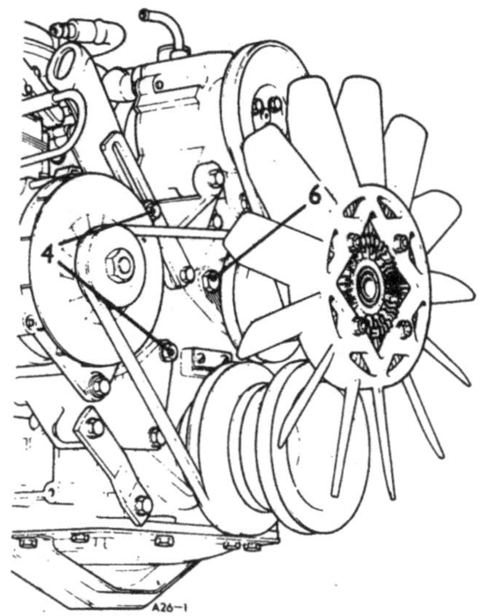

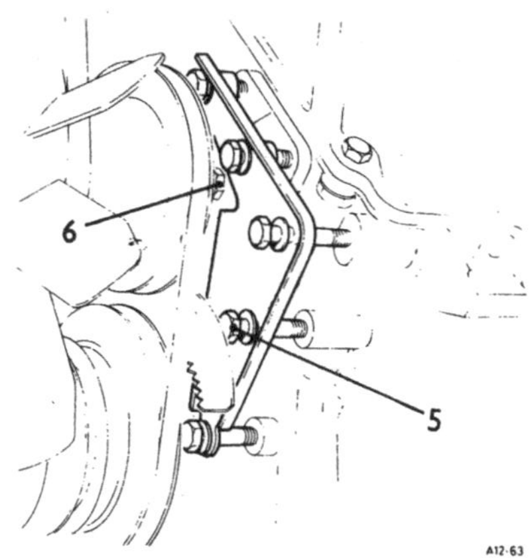

8 Slacken the four bolts securing the radiator top location brackets sufficiently to enable the radiator to be moved forward.
9 Pull the radiator forward–away from the engine–and withdraw the fan pulley assembly complete with fan blades and torquatrol unit.

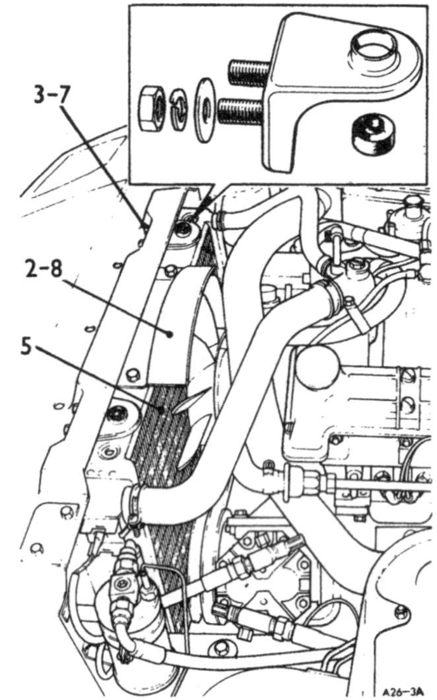

Refitting

NOTE: Before refitting the assembly ensure that the alternator and air pump drive belts are in their correct relative positions.

10 Place the drive belt on the fan pulley and offer the assembly to the engine.
11 Locate the assembly with the two short bolts, i.e. those not common to securing the compressor bracket and fan pulley assembly.
12 Fit the steady bracket and tighten the remaining five bolts.
13 *Carburetter vehicles only:* Fit the air pump and correctly tension the drive belt.
14 Tighten the radiator top location brackets.
15 Fit the fan guard.
16 Fit the bonnet. 76.16.01.
17 Reconnect the battery.

RADIATOR TOP HOSES

– Remove and refit **26.30.01**

Removing

WARNING: This operation must only be carried out when the engine is cold or cool.

1 Disconnect the battery.
2 Partially drain the radiator.
3 Disconnect and remove the two top hoses.

Refitting
4 Fit the two hoses and tighten the clips.
5 Top-up the cooling system–26.10.01–with the correct mixture of anti-freeze.
6 Reconnect the battery.
7 Check for leaks after the engine has been run and attained normal operating temperature.

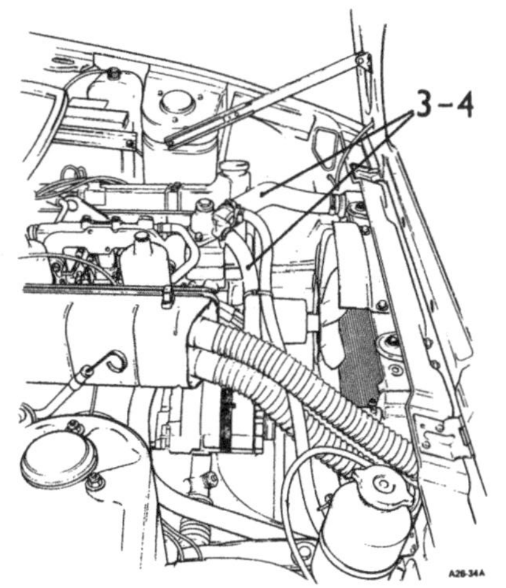

RADIATOR BOTTOM HOSE

– Remove and refit **26.30.07**

Removing

WARNING: This operation must only be carried out when the engine is cold or cool.

1 Disconnect the battery.
2 Place a suitable container below the hose connection to the radiator, disconnect the hose from the radiator and allow the coolant to drain.
3 Disconnect and remove the hose from the water pump cover connection.

Refitting
4 Place the hose clips over the hose and fit the hose to the pump and radiator. Tighten the clips.
5 Connect the battery.
6 Refill the cooling system. 26.10.01.

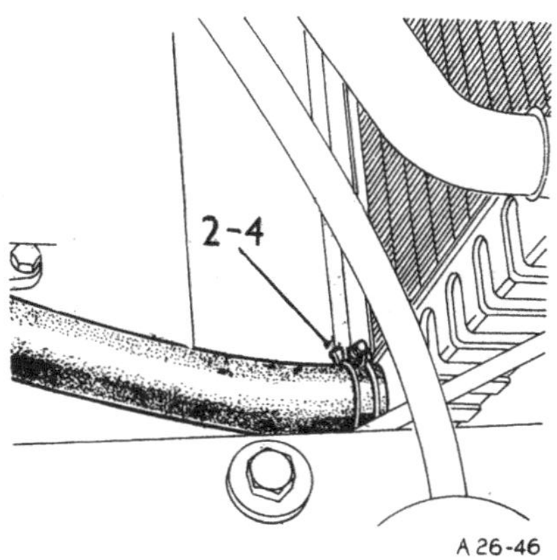

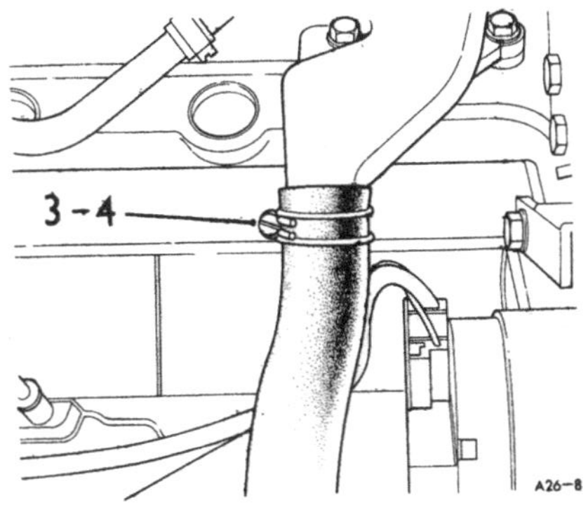

CONNECTING TUBE

– Remove and refit **26.30.25**

See Operation 30.15.02

RADIATOR

– Remove and refit **26.40.01**

Removing
1 Disconnect the battery.
2 Drain the cooling system. 26.10.01.
3 Disconnect from the radiator the two top hoses.
4 Remove the temperature sensor leads from the L.H. side of the radiator. – Air conditioning specification vehicles.
5 Remove the four nuts and washers securing the two radiator attachment brackets.
6 Remove the brackets.
7 Withdraw the radiator vertically.

Refitting
8 Ensure that the four radiator mounting rubbers are in position on the radiator–two top, two bottom.
9 Lower the radiator into position ensuring that the bottom mounting rubbers locate properly in the holes in the radiator support crossmember.
10 Refit the radiator top support brackets.
11 Refit the top hoses.
12 Reconnect the bottom hose.
13 Refill the cooling system. 26.10.01.
14 Reconnect the temperature sensor leads.
15 Reconnect the battery.

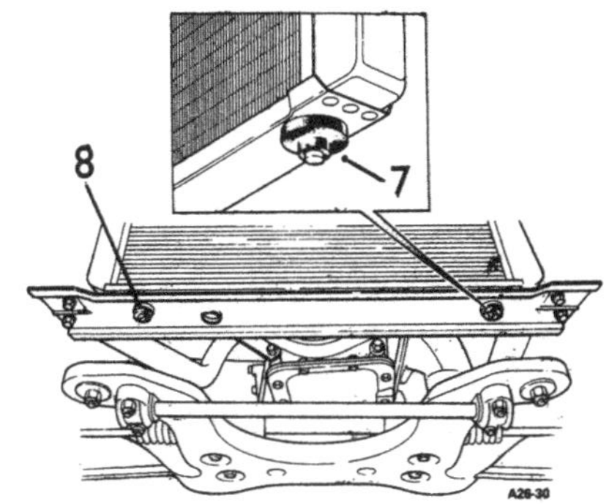

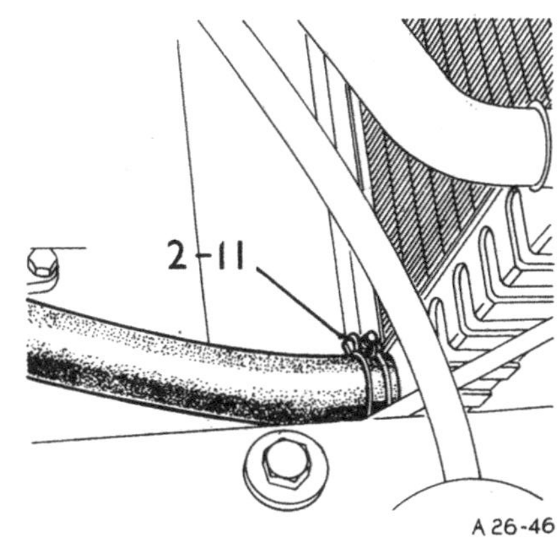

THERMOSTAT

Remove and refit 26.45.01

WARNING: This operation must only be carried out when the engine is cold.

Carburetter engines

Removing

1 Disconnect the battery.
2 Partially drain the cooling system.
3 Remove the two bolts securing the thermostat housing dome.
4 Lift off the dome and withdraw the thermostat.

Refitting

5 Locate the thermostat in the housing.
6 Using a new gasket fit the domed cover and evenly tighten the retaining bolts.
7 Reconnect the battery.
8 Top-up and check the cooling system.

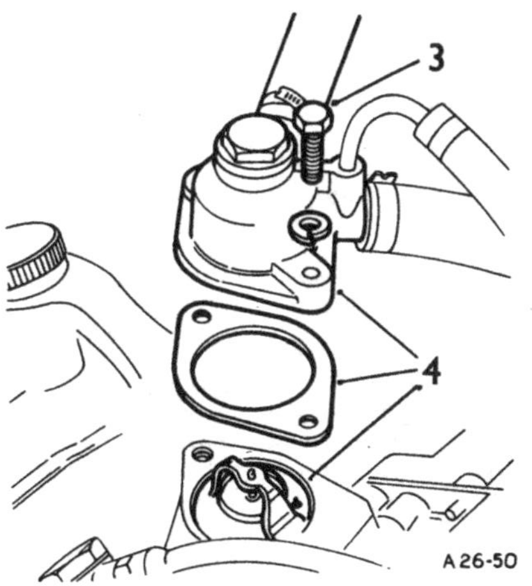

Fuel injection engines
The thermostat is located at the radiator top hose connection on the inlet manifold. The thermostat cover is secured by two bolts. The thermostat is removed and refitted following the procedure above.

THERMOSTAT

– Test 26.45.10

1 Remove the thermostat. 26.45.01.
2 Note the temperature stamped on the thermostat at which it should be fully open.
3 Place the thermostat and a centigrade thermometer in a laboratory beaker and heat the water, observe the temperature at which the thermostat opens.
4 Refit or renew the thermostat as necessary. 26.45.01–instructions 5 to 8.

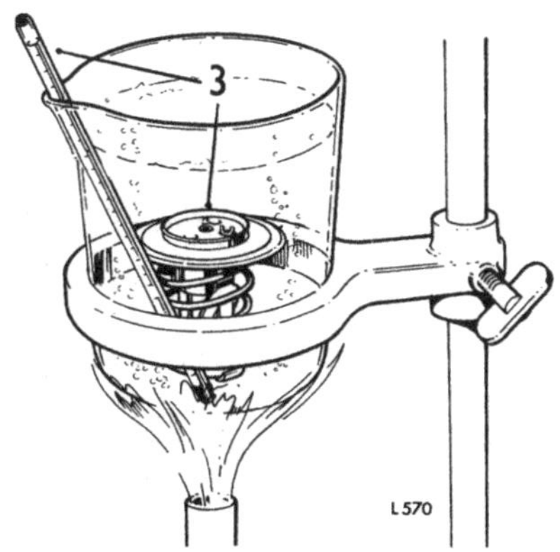

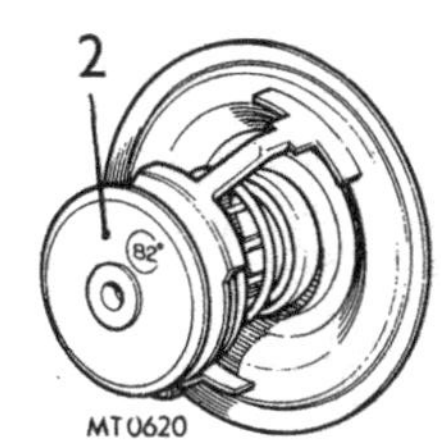

WATER PUMP

– Remove and refit 26.50.01

Service tools S4235A/10, 4235A.

WARNING: This operation must only be carried out when the engine is cool.

1 Disconnect the battery.
2 Remove the inlet manifold complete with carburetters. 30.15.02.
3 Remove the connecting tube from the water-pump cover.
4 Disconnect the bottom hose from the water pump cover.
5 Remove the three bolts securing the pump cover to the cylinder block.
6 Lift off the pump cover complete with gaskets.
7 Using a spanner on the impeller centre bolt turn clockwise until either:
   a the water pump is released from the jackshaft gear and can be withdrawn.
   or
   b the centre bolt is removed.
8 If (b) applies fit special tool S4235A/10 and 4235A impact tool and adaptor to remove the pump.

continued

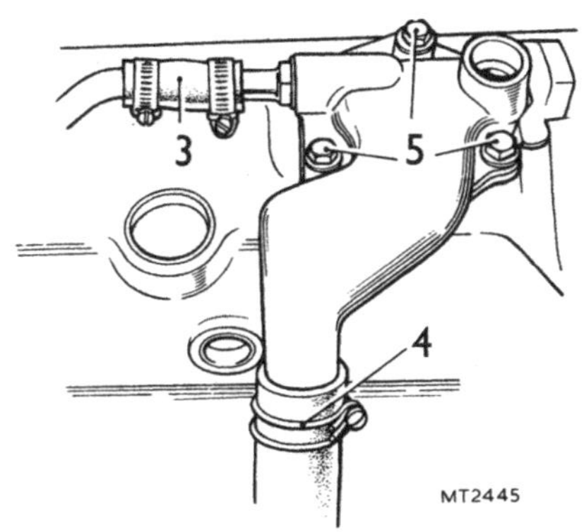

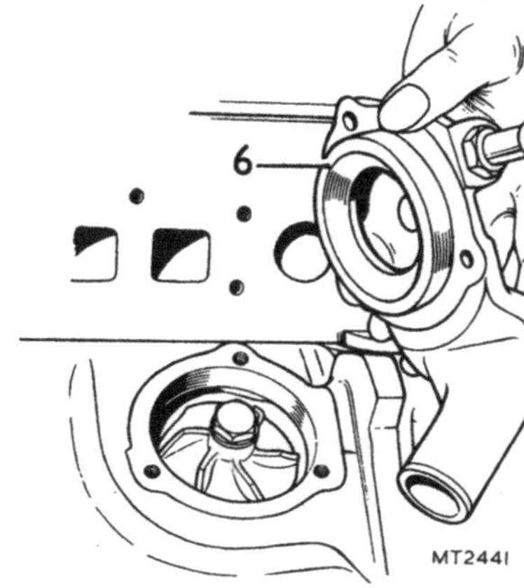

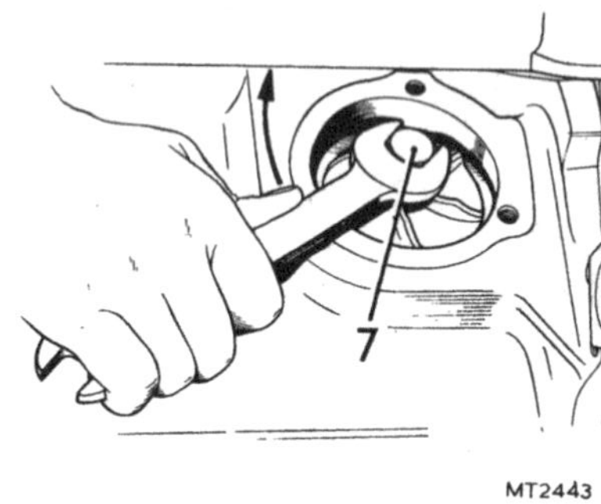

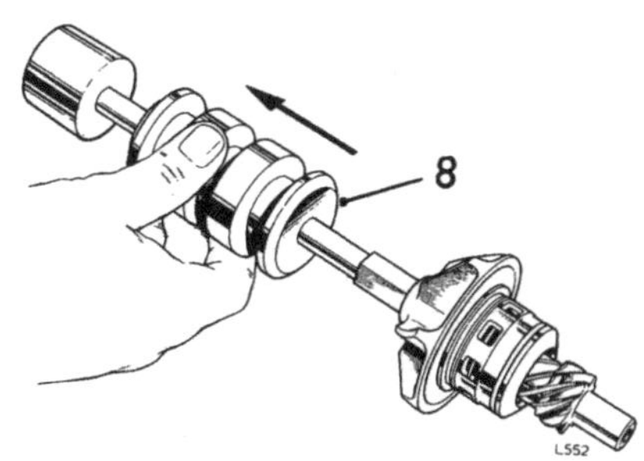

Refitting

9 Fit the pump into the cylinder block housing, ensuring that the pump and jackshaft gears mesh correctly and that the pump is properly seated. Check by turning the impeller centre bolt anti-clockwise.
CAUTION: The use of force or impact to seat the pump will damage the pump and graphite seal.

10 Ensure that the cylinder block housing and mating faces are clean.

11 Temporarily fit the pump cover leaving the three bolts finger tight.

12 Using feeler gauges check and note that the gap between the pump cover and cylinder block is equal. Equalize the gap by adjusting the bolts.

13 Select water pump gaskets to equal the gap noted in instruction 12 plus 0.010 in to 0.025 in (0.25 to 0.5 mm) to obtain the correct running clearance.
NOTE: Water pump gaskets are available in the following thicknesses, 0.010, 0.020, 0.030 in.

14 Remove the pump cover, fit the selected gaskets, fit the cover and tighten the bolts evenly to the correct torque figure—see section 06.

15 Refit the inlet manifold complete with carburetters. 30.15.02.

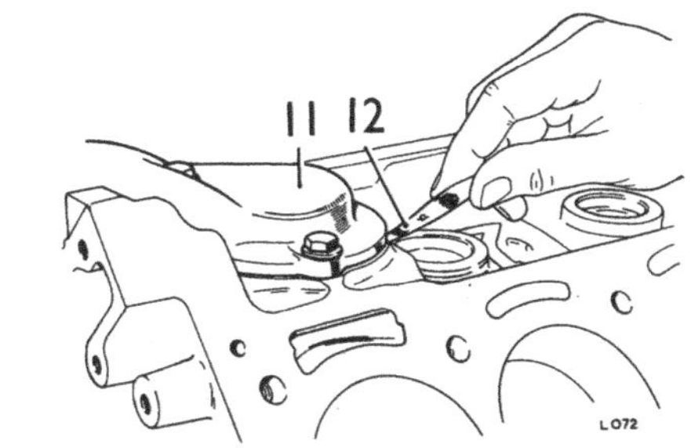

L072

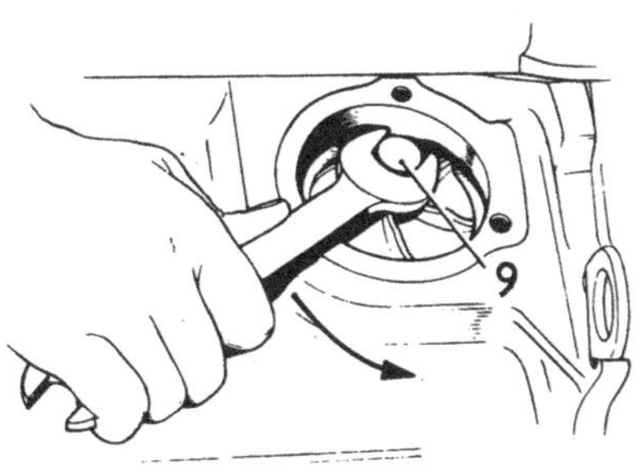

MT2443A

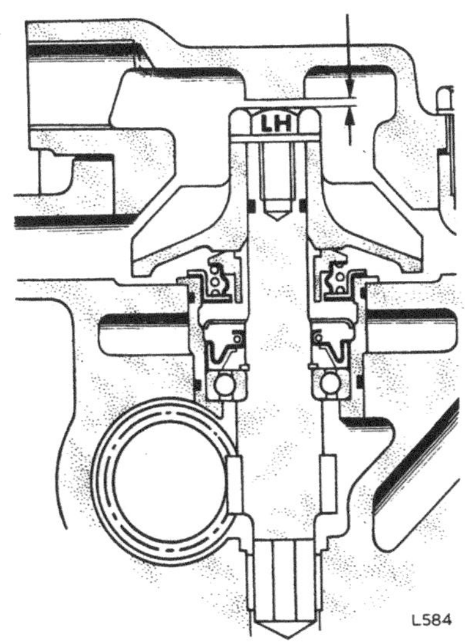

L584

WATER PUMP

– Overhaul **26.50.06**

Service Tool Kit: S348

Dismantling

1 Remove the water pump. 26.50.01.
2 Remove the centre bolt—noting that it has a left-hand thread.
3 Insert the assembly into the large hole of special tool S348/1.
4 Using a special tool S348/6 drift the unit from the impeller.
5 Insert the assembly, gear uppermost, into the small hole of tool S348/1.
6 Drift the unit from the housing.

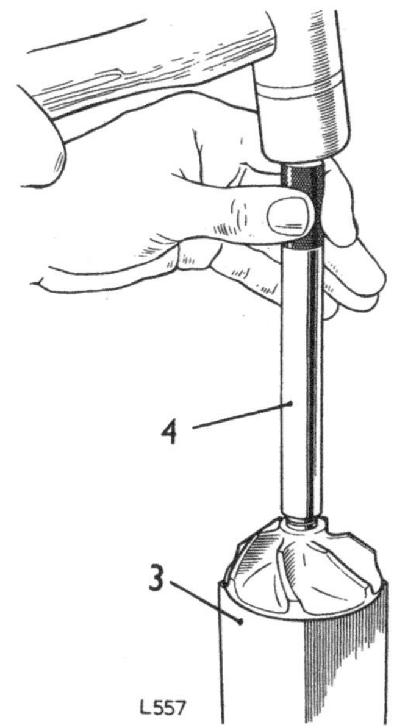

L557

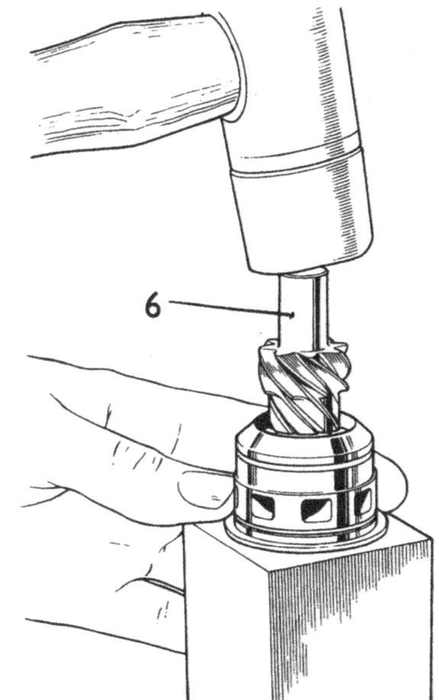

L558

7 Remove from the shaft in the following order:

a 'O' ring.
b graphite seal.
c water flinger.
d oil seal.
e the circlip.

continued

8 Insert the shaft, gear downwards, into tool S348/7.
9 Drift the shaft from the bearing.
10 Remove the oil flinger.
11 Remove the 'O' rings from the housing.
12 Discard the 'O' rings, seals and bearing.

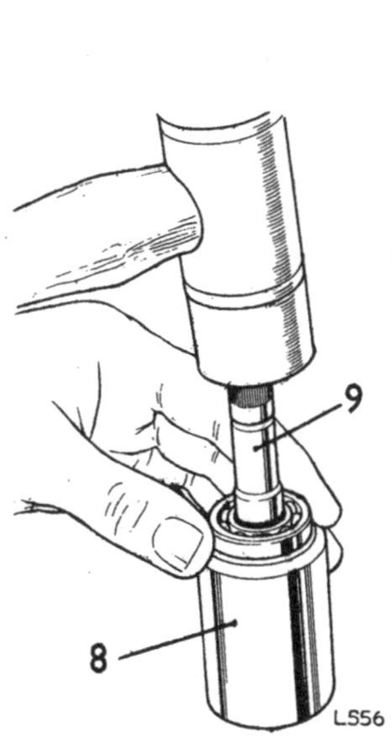

Examination

13 Inspect the shaft, housing and impeller for serviceability and renew if necessary.

Assembling

14 Fit the oil flinger to the shaft dished face towards the gear.
15 Fit the bearing to the shaft using tool S348/7.
16 Fit the circlip ensuring that it seats correctly.

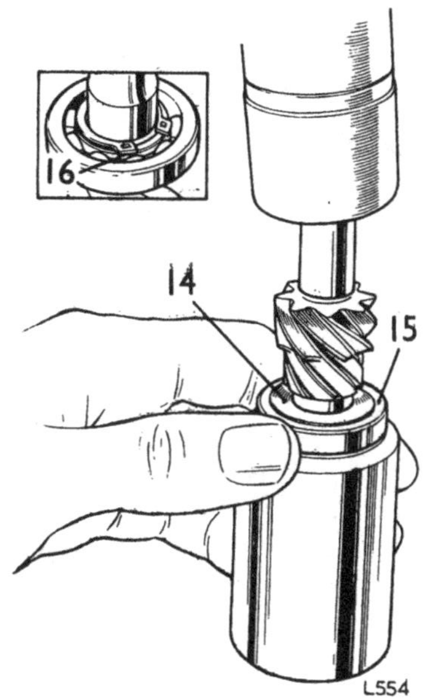

17 Fit the water pump body into the small hole in tool S348/1.
18 Fit the shaft unit, gear downwards, into the housing and using tool S348/2 gently drift into position.
19 Fit the oil seal, flat face towards the bearing.

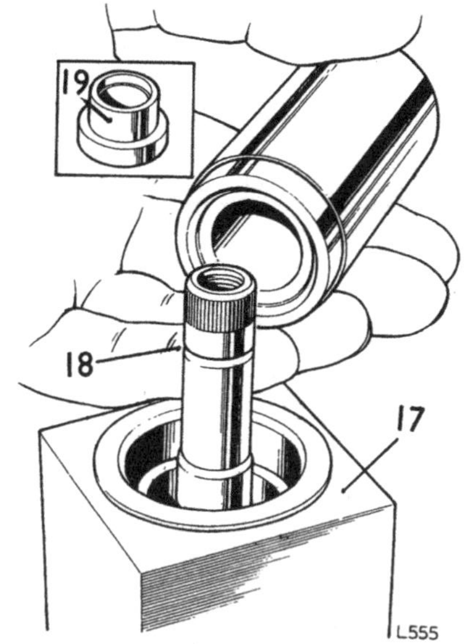

20 Fit the water flinger, dished face towards the bearing using tool S348/2.

CAUTION: The water flinger will seat on the shoulder of the shaft. Excessive force applied when fitting, will cause distortion and fouling of the water-pump body.

21 Fit the graphite seal, flat face downwards, over the shaft and seating in the housing.
22 Fit the 'O' ring to the shaft.
23 Press the impeller on to the shaft.
24 Fit the centre bolt and washer (left-hand thread) and tighten to the correct torque figure – see section 06.
25 Fit the two 'O' rings to the water pump housing, smaller one nearest to the gear.
26 Refit the water pump. 26.50.01.

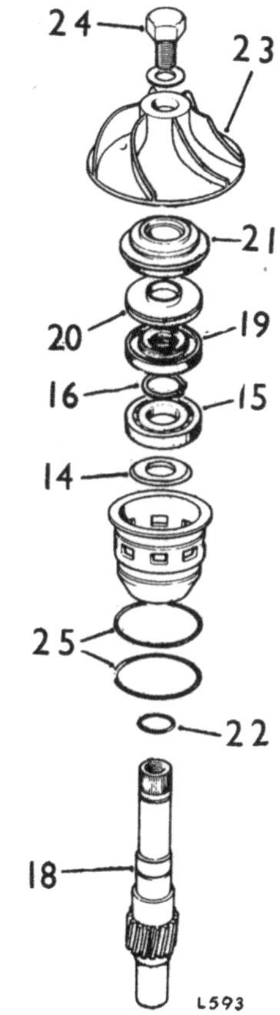

MANIFOLD AND EXHAUST SYSTEM

FRONT PIPE

– Remove and refit 30.10.09

Removing
1 Drive the car on to a ramp.
2 Raise the ramp.
3 Slacken the silencer to down-pipe clamp bolt.
4 Un-hook the silencer from its body hanger.

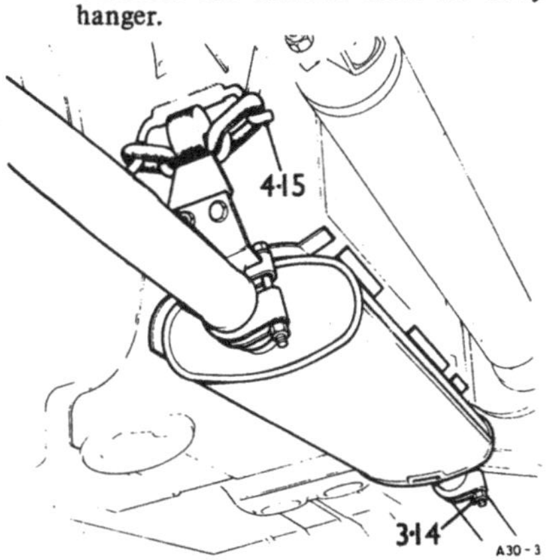

5 Un-hook the tail-pipe from its hanger.
6 Drive the silencer from the down-pipe.
7 Remove the nut and bolt securing the rear-end of the down pipe.
8 Remove the two nuts and bolts retaining the front end of the pipe.
9 Remove the three bolts securing the down-pipe to the manifold flange.

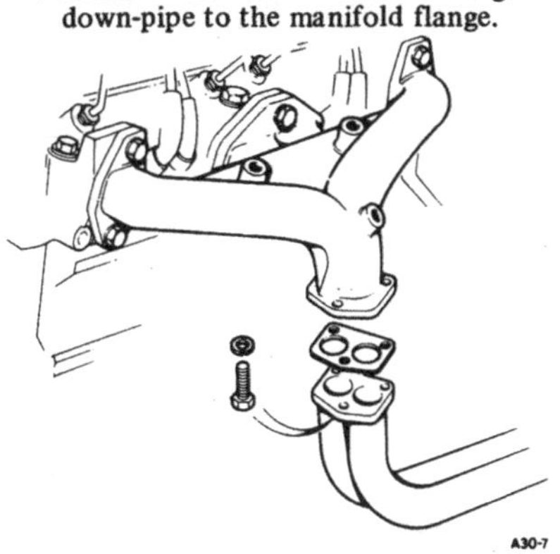

10 Remove the pipe complete with gasket.

Refitting
11 Fit the down pipe to the manifold flange using a new gasket.
12 Secure the pipe to the front bracket using the two nuts and bolts.
13 Connect the pipe to its rear mounting with the single nut and bolt.
14 Fit the silencer to the down-pipe leaving the clamp slack.
15 Fit the silencer to its body hanger.
16 Secure the tail pipe to its hanger.
17 Tighten the silencer front clamp bolt.
18 Lower the ramp.

SILENCER – FRONT

– Remove and refit 30.10.14

Removing
1 Raise the car on a ramp.
2 Release the clips clamping the silencer to the tail and down-pipes.

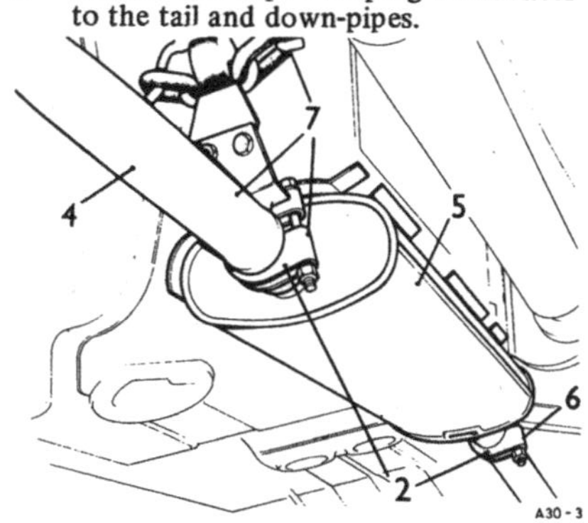

3 Disconnect the tail pipe from the body.
4 Withdraw the tail pipe assembly from the silencer.
5 Withdraw the silencer from the front-pipe.

Refitting
6 Fit the silencer to the front-pipe and tighten the retaining clip.
7 Fit the tail pipe assembly to the silencer and attach it to the body. Tighten the tail-pipe to silencer clip.
8 Lower the ramp.

TAIL-PIPE AND REAR SILENCER

– Remove and refit 30.10.22

Removing
1 Raise the car on a ramp.
2 Release the clip clamping the tail-pipe to the silencer.
3 Unhook the tail-pipe from its forward hanger.
4 Unhook the tail-pipe from its rear hanger.
5 Withdraw the tail-pipe from the silencer.
6 Remove the two nuts and bolts securing the hanger bracket assembly to the tail-pipe bracket, and separate the two assemblies.

Refitting
7 Fit the hanger bracket assembly to the tail-pipe bracket with the two nuts and bolts.
8 Fit the tail-pipe to the silencer.
9 Hang the tail-pipe to its forward and rear hanger brackets using, if necessary, new rubbers.
10 Tighten the silencer-to-tail-pipe clamp bolt.
11 Lower the ramp.

FRONT PIPE FLANGE GASKET

– Remove and refit 30.10.26

Removing
1 Raise the car on a ramp.
2 Remove the three front-pipe to manifold flange retaining bolts.

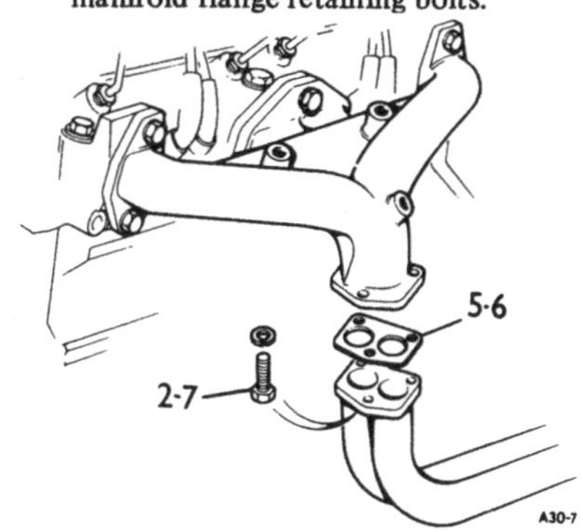

3 Remove the two nuts and bolts securing the down-pipe to its suspension bracket.
4 Remove the nut and bolt securing the intermediate pipe to its bracket.
5 The down-pipe will now drop sufficiently to remove the gasket.

Refitting
6 Insert the gasket into position.
7 Fit the flange manifold bolts but leave slack.
8 Fit the two nuts and bolts securing the down-pipe to its bracket, but leave slack.
9 Fit and tighten the nut and bolt retaining the intermediate pipe to its bracket.
10 Tighten the flange bolts to the correct torque figure – see Section 06.
11 Tighten the front-pipe bracket bolts.
12 Lower the ramp.

EXHAUST MANIFOLD – U.K. AND EUROPE

– Remove and refit 30.15.01

Removing
1 Drive the car on to a ramp and disconnect the battery.
2 Raise the ramp and remove the three bolts securing the exhaust front-pipe to the manifold.
3 Remove the four manifold lower retaining bolts.
4 Lower the ramp.
5 Remove the three manifold top retaining bolts and lift-off the manifold.

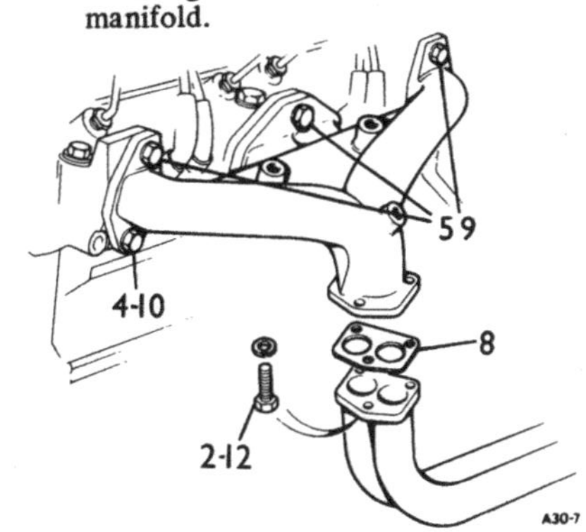

Refitting

6 Place a new front pipe to manifold gasket in position.
7 Fit the manifold securing it loosely with the three top bolts.
8 Raise the ramp and fit the four lower retaining bolts.
9 Finally, evenly tighten the manifold bolts to the correct torque figure – see Section 06.
10 Fit and tighten the three front-pipe to manifold bolts.
11 Lower the ramp and reconnect the battery.

EXHAUST MANIFOLD – U.S.A. Specification Vehicles

– Remove and refit 30.15.01

Removing

1 Drive the car on to a ramp and disconnect the battery.
2 Raise the ramp and remove the three bolts securing the exhaust front-pipe to the manifold.
3 Remove the manifold four lower retaining bolts.
4 Lower the ramp and remove the hot air hose from the manifold.
5 Remove the three manifold top retaining bolts and lift-off the manifold.
6 Withdraw the three bolts and remove the hot air collection pressing.

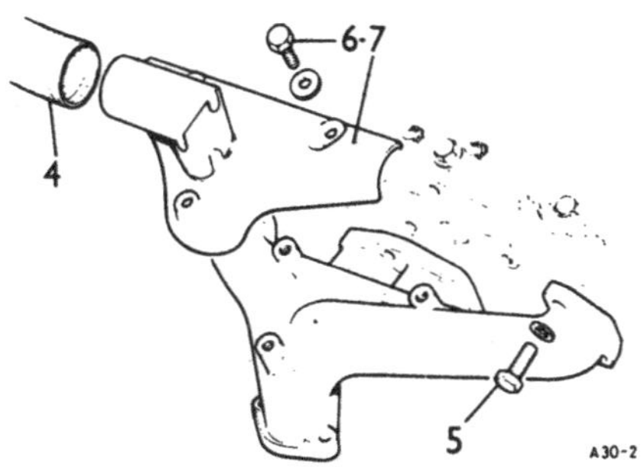

Refitting

7 Fit the hot air collection pressing to the manifold.
8 Place the new front pipe to manifold gasket in position.
9 Fit the manifold securing it loosely with the three top bolts.
10 Raise the ramp and fit the four lower retaining bolts.
11 Finally, evenly tighten the manifold bolts to the correct torque figure – see Section 06.
12 Fit and tighten the three front-pipe to manifold bolts.
13 Lower the ramp and fit the hot air hose.
14 Reconnect the battery.

INDUCTION MANIFOLD – CARBURETTER ENGINES

Remove and refit 30.15.02

Removing

1 Disconnect the battery.
2 Drain the cooling system. 26.10.01.
3 Remove the fresh air duct. 80.15.31.
4 Remove the air cleaner. 19.10.02.
5 Disconnect the top water hoses from the engine.
6 Disconnect the water temperature transmitter lead.
7 Disconnect the main fuel feed to the carburetters.
8 Disconnect the brake servo hose from the manifold.
9 Disconnect the mixture control cable from the front carburetter. – Not U.S.A.
10 Pull-off the spill pipe from the rear carburetter float chamber.
11 Disconnect the throttle cable from the carburetter linkage.
12 Pull-off the breather hose from the camshaft cover.
13 Remove the distributor cap.
14 Pull-off the distributor vacuum pipe from the rear carburetter.
15 Remove the six manifold attachment bolts and remove the engine lifting bracket.
16 Disconnect the heater pipe at union with the induction manifold.
17 Lift the manifold complete with carburetters from the engine.
18 Remove the 'O' ring between manifold and cylinder head.
19 Remove the connecting tube.
20 Remove the gasket.
21 Disconnect the engine breather pipes from the carburetters.
22 Disconnect the throttle linkage springs from the manifold bracket.
23 Remove the bolts retaining the throttle linkage bracket to the manifold.
24 Remove the eight nuts securing the carburetters to the manifold (four per carburetter).
25 Withdraw the carburetters and gaskets.
26 Remove the throttle jack bracket.
27 Remove the rubber mountings and gaskets.

NOTE: If the manifold is to be renewed remove the following items:–

28 Remove the two bolts and lift-off the thermostat housing cover.
29 Withdraw the thermostat.
30 Remove the two bolts and remove the common throttle spring and breather pipe bracket.
31 Remove the water temperature transmitter.
32 Remove the brake servo pipe banjo.
33 Remove the banking plug.

Refitting

34 Reverse instructions 28 to 33.
35 Using new gaskets fit the rubber mountings.
36 Fit the throttle jack bracket.
37 Fit the carburetters using new gaskets, and secure with eight nuts and spring washers.
38 Fit the throttle linkage bracket to the manifold.
39 Connect the throttle linkage springs.
40 Connect the engine breather pipes to the carburetters.
41 Place a new gasket in position on the manifold face.
42 Fit a new 'O' ring to the manifold.
43 Fit the manifold complete with carburetters and engine lifting bracket and secure with the six bolts, tightening evenly to the correct torque.
44 Connect the heater pipe at the union with the induction manifold.
45 If fitted, remove the thermostat housing cover and withdraw the thermostat.
46 Insert the connecting tube squarely through the thermostat housing until fully home.
47 Refit the thermostat and cover using a new gasket.
48 Reverse instructions 1 to 14.

INDUCTION MANIFOLD — FUEL INJECTION ENGINES

Remove and Refit **30.15.02**

Removing

1 Disconnect the battery.
2 Drain the cooling system, see 26.10.01.
3 Remove the plenum chamber, see 19.22.46, disconnecting the air pipes at the manifold.
4 Disconnect the temperature sensor, temperature switch and Thermotime switch from the inlet manifold, noting their connections for refitting. If required, unscrew the units from the manifold.
5 Depressurize the fuel system, see 19.50.02.
6 Remove the fuel rail, see 19.60.05.
7 Remove the fuel injectors, see 19.60.10.
8 Disconnect the radiator top hose and outlet water pipes from the manifold.
9 Remove the thermostat, see 26.45.01.
10 Remove four bolts and lift off the manifold.

Refitting

11 Clean all mating faces on the manifold and cylinder head. Replace any gasket that is suspect or damaged.
12 Reverse instructions 1 to 4 and 5 to 9.

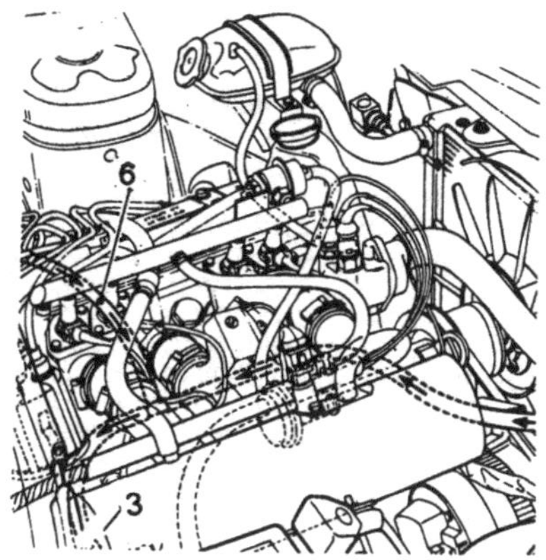

CLUTCH OPERATIONS

CLUTCH ASSEMBLY

– Remove and refit **33.10.01**

Removing
1 Remove the gearbox. 37.20.01
2 Evenly slacken and remove the six bolts and spring washers securing the clutch to the engine flywheel.
NOTE: If the clutch is to be refitted, scribe or mark clutch and flywheel to identify original relationship.
3 Remove the clutch assembly and drive plate.

Refitting
4 Using a substitute first motion shaft offer up the drive plate to the flywheel. Note that the longer boss of the drive plate must be adjacent to the gearbox.
5 Engage the clutch assembly on the flywheel dowels. Fit the six securing bolts and spring washers and tighten evenly to the correct torque figure – see Section 06.
6 Remove the substitute first motion shaft.
7 Fit the gearbox. 37.20.01.

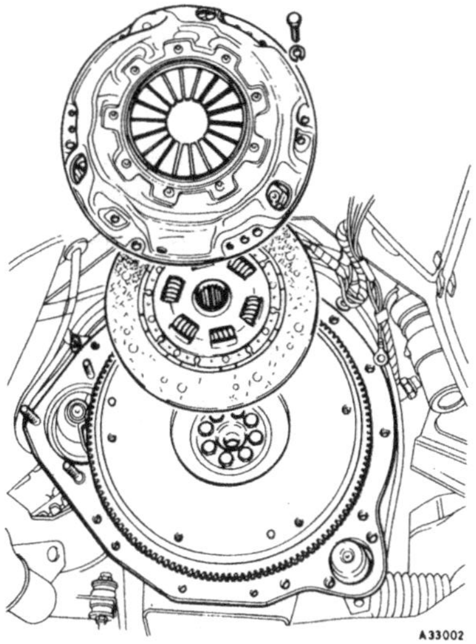
A33002

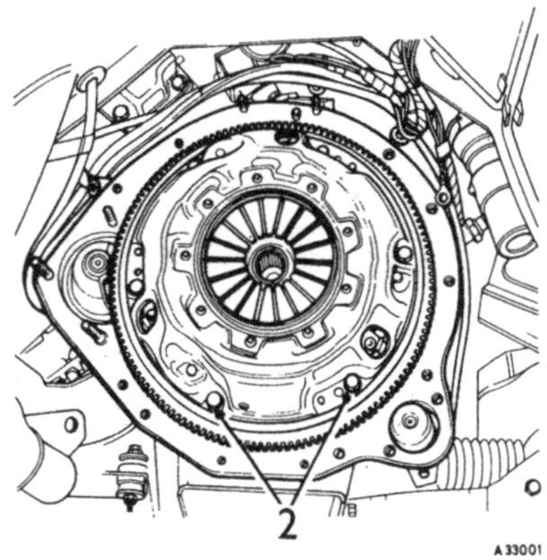

A33001

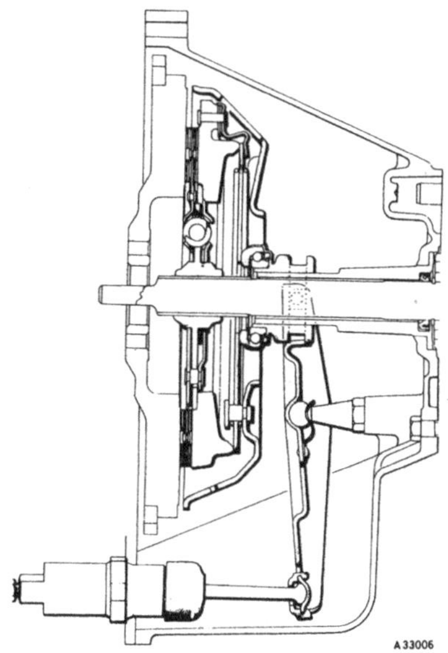
A33006

DRIVE PLATE

– Remove and refit **33.10.02**

As operation 33.10.01.

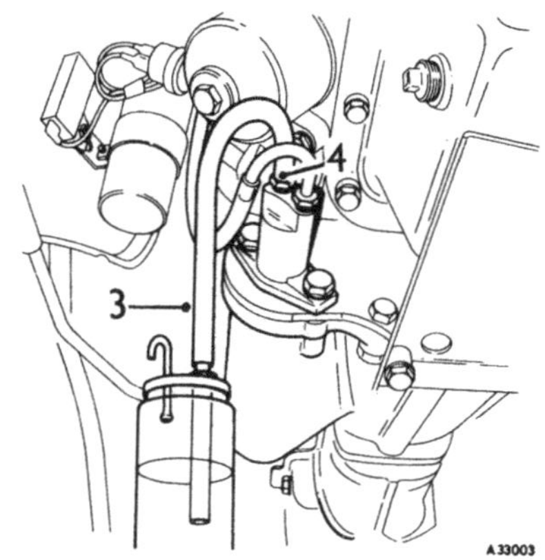

A33003

HYDRAULIC SYSTEM

– Bleeding **33.15.01**
1 Thoroughly clean the areas in the vicinity of the master cylinder reservoir filler cap and the slave cylinder bleed nipple.
2 Ensure that the master cylinder reservoir is topped up.
3 Attach one end of a bleed tube to the slave cylinder bleed nipple and immerse the other end in a transparent vessel containing brake fluid.
4 Open the bleed nipple approximately three-quarters of a turn. Depress and release the clutch pedal, pausing momentarily at each down stroke until fluid free of air issues from the slave cylinder. Close the bleed nipple with pedal depressed and release the pedal.
Care must be taken to ensure that the level of fluid in the master cylinder reservoir is never permitted to fall to less than half capacity.
5 Remove the bleed tube and top up the reservoir.

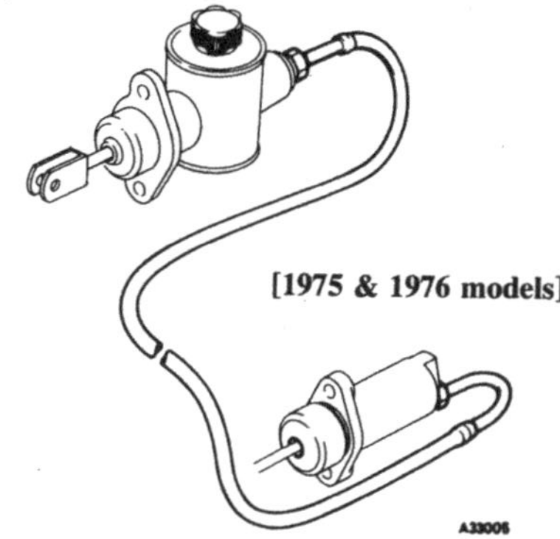
[1975 & 1976 models]
A33005

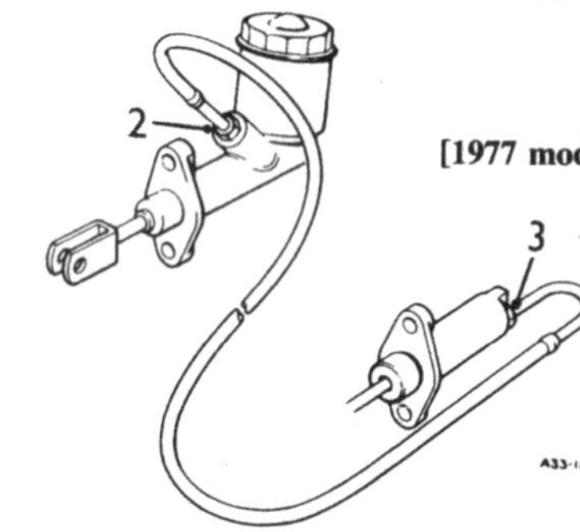

[1977 models]
A33-11

FLUID PIPE

– Remove and refit **33.15.09**

Removing
1 Drain the hydraulic system.
2 Unscrew the pipe from the master cylinder.
3 Unscrew the pipe from the slave cylinder.
4 Remove the pipe from the car.

Refitting
5 Align the new pipe in the approximate position in the car.
6 Screw the pipe into the master cylinder.
7 Screw the pipe into the slave cylinder.
8 Top up the master cylinder with new brake fluid.
9 Bleed the system. 33.15.01.

MASTER CYLINDER

Remove and refit 33.20.01

Removing

1 Unscrew the pipe from the master cylinder.
CAUTION: Plug the master cylinder outlet and the end of the pipe to prevent spillage of fluid.
2 Remove the split pin, washer and clevis pin securing the push rod to the clutch pedal.
3 Remove the two nuts, spring washers and bolts securing the master cylinder to the body.
4 Withdraw the master cylinder.

Refitting

5 Reverse instructions 1 to 4.
6 Bleed the system. 33.15.01.

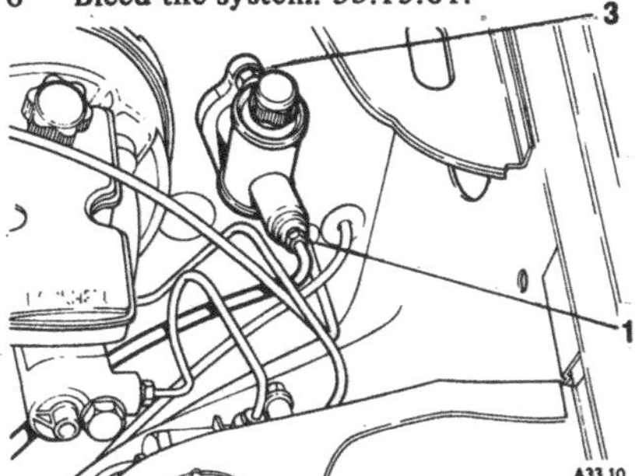

[1975 and 1976 models]

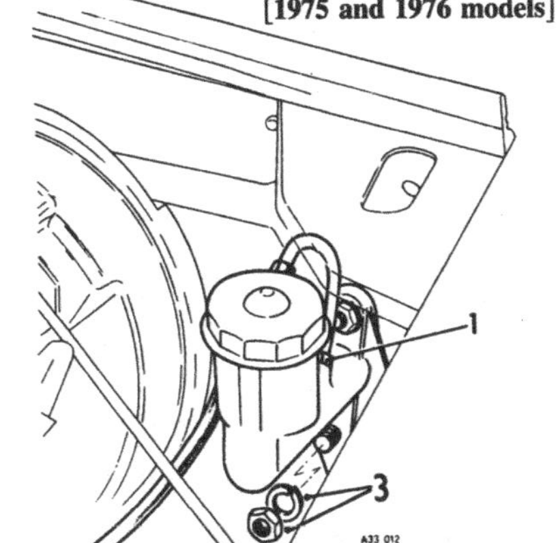

[1977 and later models]

MASTER CYLINDER (Earlier Models)

Overhaul 33.20.07

Dismantling

1 Remove the master cylinder, see 33.20.01.
2 Slide the rubber boot clear of the mounting flange to expose the push-rod retaining circlip.
3 Remove the circlip and withdraw the push-rod assembly. Remove the rubber boot.
4 Extract the piston and rear cup seal, front cup seal and washer, spring and spring retainer button.

Inspection

5 Discard the rubber boot, front and rear cup seals and thoroughly clean remaining components in clean brake fluid or methylated spirit.
6 Carefully examine the piston and cylinder bore. Renew either or both components if there is evidence of corrosion or scoring.

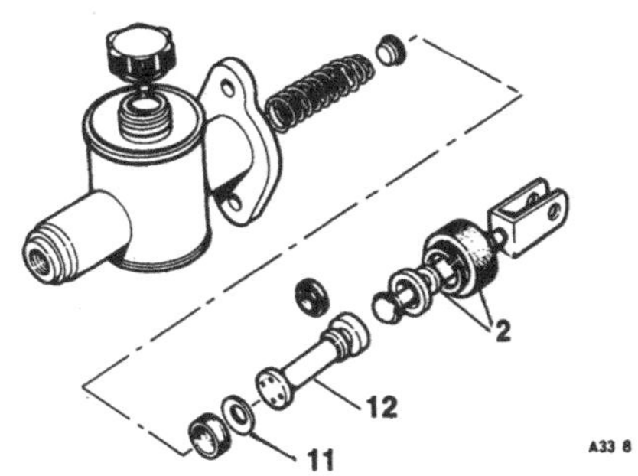

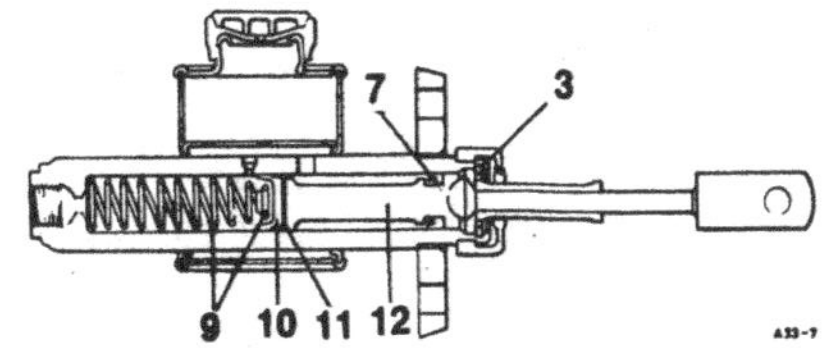

Reassembling

7 Fit a new rear cup to the piston.
8 Lubricate the cylinder bore with clean brake fluid.
9 Insert the spring (larger diameter leading) and spring retainer button into the cylinder.
10 Insert the front cup (cup lips leading) into the cylinder.
11 Insert the dished spring washer.
12 Insert the piston complete with rear cup.
13 Fit a new rubber boot to the push-rod assembly and smear the ball-end and shaft with disc brake lubricant or rubber grease.
14 Fit the push-rod assembly and retaining circlip.
15 Slide the rubber boot into position on the master cylinder.
16 Install the master cylinder on the car, see 33.20.01.
17 Bleed the circuit.

MASTER CYLINDER (Later Models)

Overhaul 33.20.07

1 Remove the master cylinder. 33.20.01.
2 Slide the rubber along the push-rod.
3 Remove the circlip from the end of the master cylinder and withdraw the push-rod and washer.
4 Withdraw the piston spring and seal assembly from the master cylinder. Withdrawal may be facilitated by applying a compressed air line to the fluid outlet union.
5 Straighten the prong of the spring thimble and remove the thimble and spring from the piston.
6 Release the valve stem from the keyhole slot in the thimble.
7 Slide the valve seal spacer along the valve stem.
8 Remove the valve seal from the valve stem and fit a new seal.

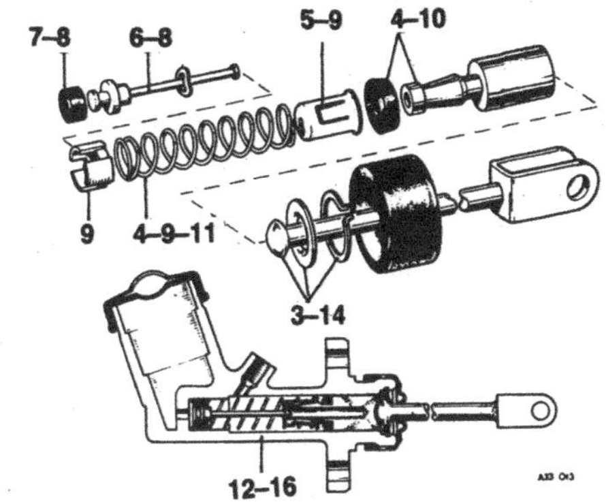

9 Assemble the spacer, spring and thimble to the valve stem.
10 Remove the seal from the piston and fit a new seal (seal lip towards the spring).
11 Engage the spring thimble on the piston and carefully depress the thimble prong.
12 Lubricate the bore of the master cylinder with clean brake fluid and insert the seal assembly spring and piston.
13 Fit a new rubber to the push-rod.
14 Fit the push-rod and washer to the master cylinder and secure with the clip.
15 Slide the rubber into position on the master cylinder.
16 Refit the master cylinder. 33.20.01.
17 Bleed the system. 33.15.01.

RELEASE BEARING ASSEMBLY

Remove and refit 33.25.12

Service tools: ST 1136

Removing

1 Remove the gearbox. 37.20.01.
2 Using Tool ST 1136 unscrew the clutch release lever pivot bolt from the clutch housing.
3 Withdraw the release lever complete with pivot bolt and the release bearing.
4 Detach the release bearing from the release lever.

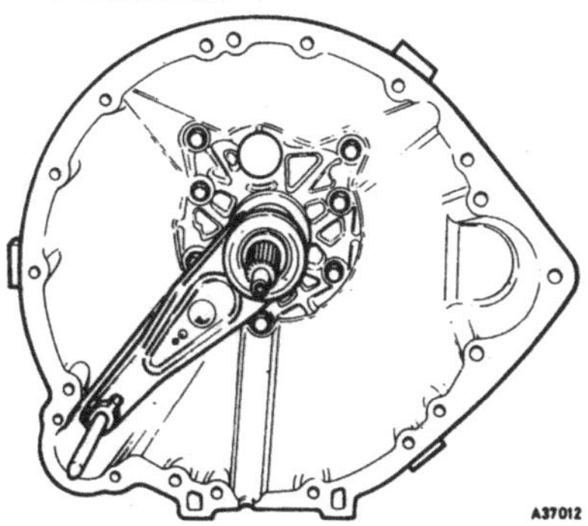

A37012

Refitting

5 Offer up the release bearing to the release lever ensuring that the rectangular slippers engage the bearing collar.
6 Slide the bearing complete with release lever and pivot bolt into position in the clutch housing.
7 Engage the pivot bolt in the clutch housing.
8 Using Tool ST 1136, tighten the pivot bolt.
9 Fit the gearbox to the car. 37.20.01.

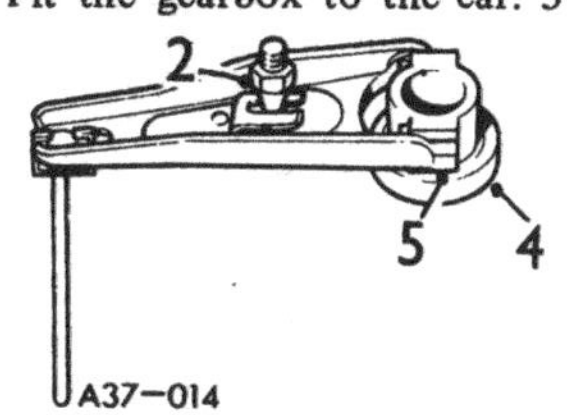

A37–014

RELEASE BEARING

Overhaul 33.25.17

Removing

1 Remove the release bearing assembly. 33.25.12.
2 Mount the assembly in a press and extract the sleeve from the bearing.

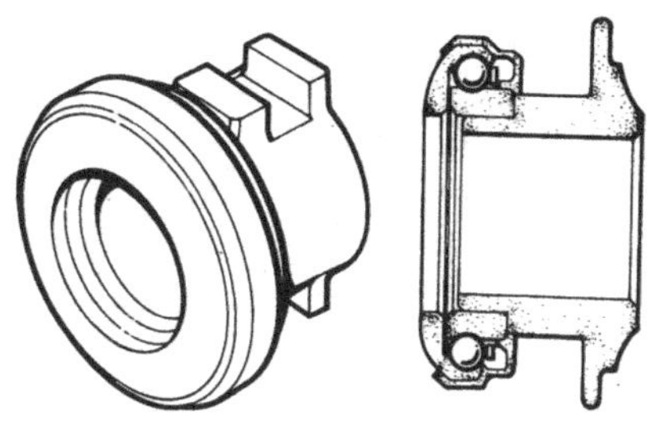

A33 9

Refitting

3 Stand the sleeve on its collar end and evenly engage the new release bearing on the sleeve. (Release face of bearing uppermost.)
4 Press the bearing onto the sleeve until it abuts against the sleeve shoulder.
CAUTION: The bearing *must not* be assembled to the sleeve by applying a load to the outer race.
5 Fit the release bearing assembly to the gearbox. 33.25.12.

CLUTCH AND BRAKE PEDAL ASSEMBLY

Remove and refit 33.30.01

As operation 70.35.03.

CLUTCH PEDAL RETURN SPRING

Remove and refit 33.30.03

Removing

1 Release the spring ends from the clutch pedal and pedal bracket.

Refitting

2 The longer leg of the spring attaches to the clutch pedal. Engage the spring hook in the pedal bracket then clip the other hook through the drilling in the clutch pedal.

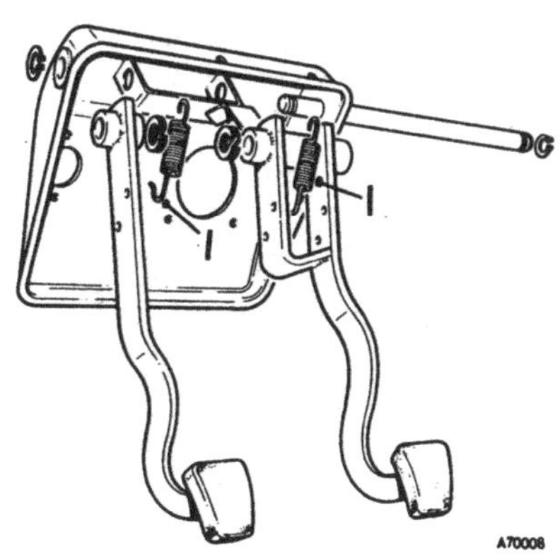

A70008

CLUTCH AND BRAKE PEDAL ASSEMBLY

Overhaul 33.30.06

As operation 70.35.04.

SLAVE CYLINDER

Remove and refit 33.35.01

Removing

1 Raise or jack up the car and support securely.
2 Clean the slave cylinder removing all loose mud etc. in the vicinity of the fluid pipe and nipple.
3 Disconnect the fluid pipe and plug the cylinder and pipe union.
4 Remove the two bolts and nuts securing the slave cylinder to the engine.
5 Gently withdraw the slave cylinder.
CAUTION: When the slave cylinder is removed do not attempt to move the operating push rod in a forward direction as this may cause the clutch release lever to be dislodged, necessitating the removal of the gearbox to permit refitting of the release lever.

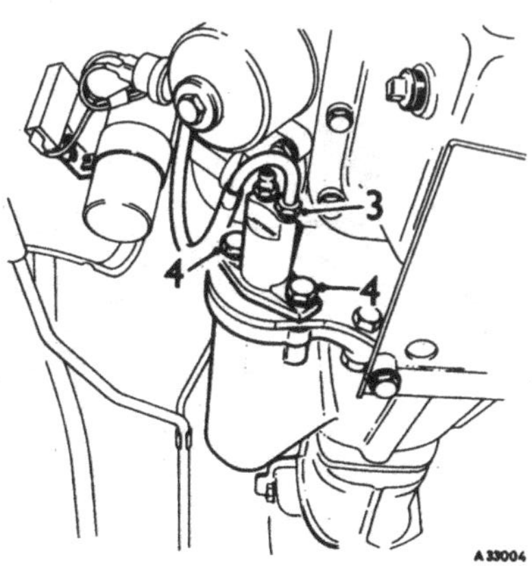

A33004

Refitting

6 Reverse instructions 3 to 5. The slave cylinder must be mounted with the bleed nipple *above* the fluid pipe.
7 Bleed the system. 33.15.01.
8 Lower the car.

SLAVE CYLINDER

Overhaul **33.35.07**

Dismantling

1 Remove the slave cylinder. 33.35.01.
2 Remove the rubber boot.
3 Extract the circlip piston, cup and spring.

Inspection

4 Discard the rubber boot and cup and clean remaining components in clear brake fluid or methylated spirit.
5 Carefully examine the piston and cylinder bore. Renew either or both components if there is evidence of corrosion or scoring.

Assembling

6 Fit a new cup to the piston.
7 Lubricate the cylinder bore with clean brake fluid.
8 Fit the smaller diameter of the spring to the piston and fit the spring (larger diameter leading) piston and cup into the cylinder. Fit the circlip.
9 Smear the piston and bore with disc brake lubricant or rubber grease.
10 Fit the rubber boot.
11 Refit the slave cylinder. 33.35.01.
12 Bleed the system. 33.15.01.

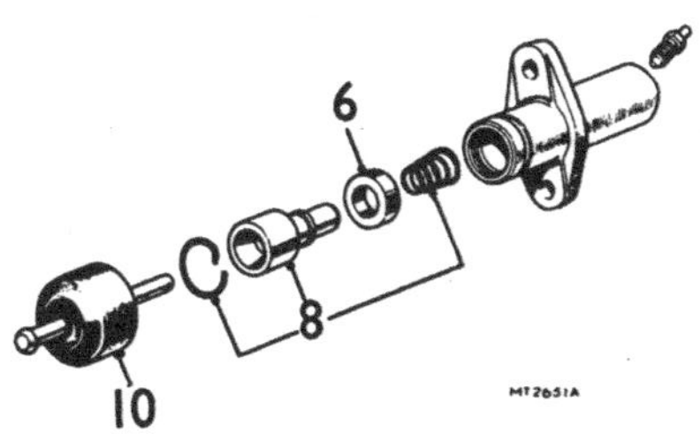

MANUAL GEARBOX – 4 SPEED

[**CAUTION:** Please read the instructions thoroughly before you attempt transmission repairs. If you lack the necessary skills or tools we suggest you leave this work to an Authorized Dealer or other qualified shop. Before taking a removed transmission in for repair, thoroughly clean its exterior but do not undertake any disassembly.]

DRIVE FLANGE

Remove and refit **37.10.01**

Service tools: S337 or 18G 1205

Removing

1 Raise the car and support securely.
2 Slacken one bolt securing the propeller shaft safety strap and remove the remaining bolt. Swing the strap clear of the propeller shaft.
3 Scribe the relationship of the propeller shaft and gearbox flanges and remove the four securing nuts and bolts.
4 Release the propeller shaft from the driving flange.
5 Using tool S337 to prevent rotation of the driving flange remove the securing nut and washer.
6 Withdraw the driving flange.

Refitting

7 Reverse instructions 1 to 6.
The drive flange nut should be torqued to: 90-120 lbf ft (12.44-16.60 kgf m).
Propeller shaft bolts to: 26-34 lbf ft (3.60-4.70 kgf m).

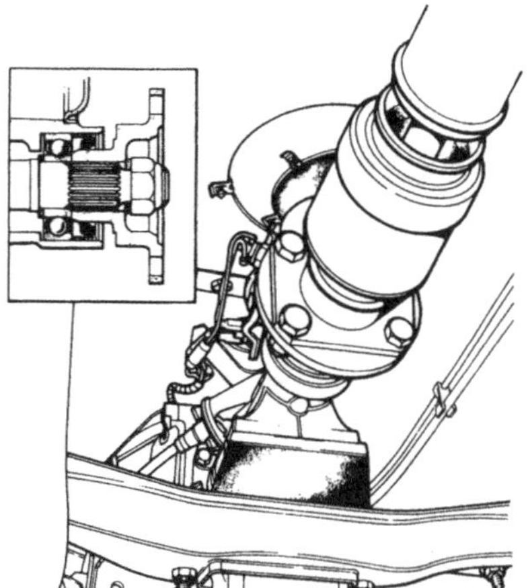
A37002

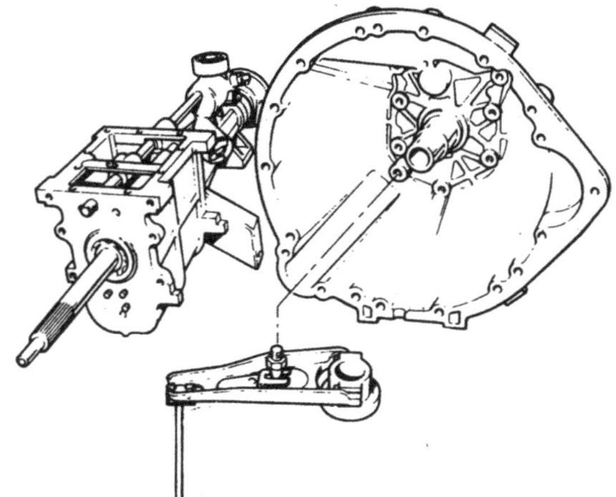
A37–014

REAR EXTENSION

Remove and refit **37.12.01**

Service tools : S337 or 18G 1205

Removing

1 Raise the car, drain the gearbox and refit the drain plug.
2 Remove the gearbox from the car, see 37.20.01.
3 Support the gearbox in a vice by means of the drain plug and remove the seat belt interlock switch and reverse switch.
4 Remove the clutch housing and gasket, see 37.12.07.
5 Withdraw the three springs from the layshaft front thrust washer.
6 Carefully drift out the roll pin from the selector rod.

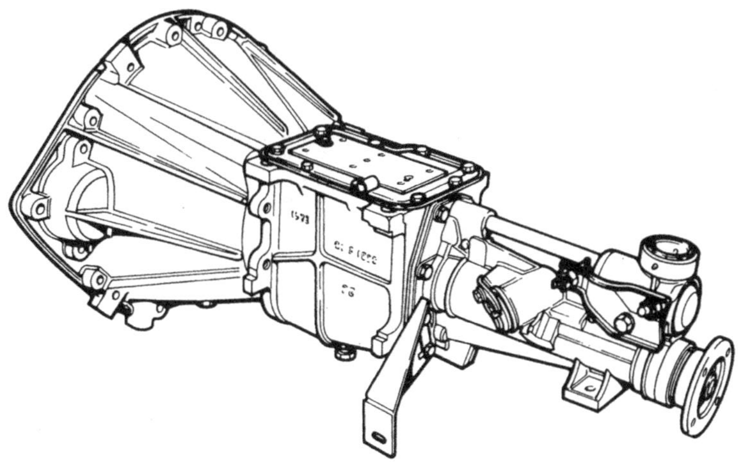
A37003

7 Remove the speedometer drive pinion, see 37.25.05.
8 Using tool S337 to prevent rotation of the driving flange remove the securing nut and washer.
9 Withdraw the driving flange.
10 Remove the gearbox top cover, gasket and spool plate.
11 Locate the gear selector in the reverse gear position and ensure that the selector shaft pins clear the interlock spool and the gear selector forks.

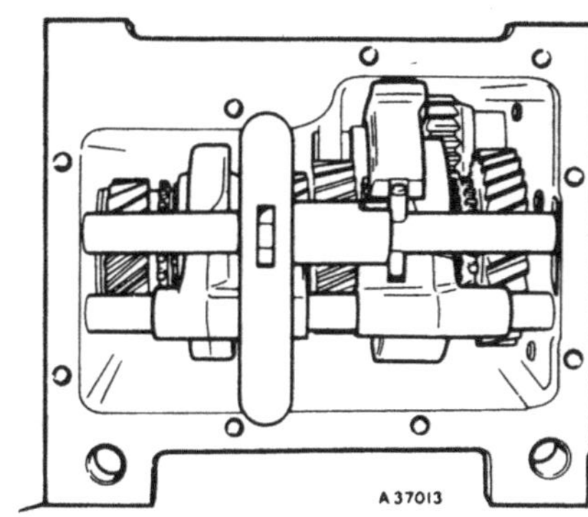
A37013

12 Remove the bolts and spring washers securing the rear extension to the gearbox and remove the rear extension exhaust bracket.
13 Remove the rear extension from the gearbox. Ensure that the selector pins do not foul and that the layshaft is not displaced. Lift out the interlock spool as the rear extension and selector shaft are withdrawn.
14 Remove the gasket.
15 Remove the distance washer from the mainshaft.
16 Remove the oil seal and bearing from the rear extension case.

Refitting
17 Ensure that the mating faces of the gear case and rear extension housing are clean.
18 Locate a new gasket in position on the rear of the gear case.
19 Fit the distance washer to the mainshaft.
20 Offer up the rear extension to the gear case and guide the selector rail into position remembering to fit the selector spool.
21 Fit the rear extension securing bolts and washers and exhaust bracket and lock plate.
22 Fit the rear bearing and a new oil seal to the extension casing.
The lip of the seal must be fitted towards the gearbox. When installed the seal should be flush with the casing.
23 Lubricate the seal lip and running surface of the drive flange. Fit the drive flange, washer and nut. Using tool S337 to prevent flange rotation, torque the nut to 900-120 lbf ft (12.44-16.60 kgf m).
24 Fit the roll pin to the front end of the selector rail. This pin must be positioned centrally.
25 Fit the speedometer drive pinion, see 37.25.05.
26 Fit the seat belt interlock switch and reverse switch.
27 Fit the clutch housing, gasket, clutch release fork and bearing, see 37.12.07.
28 Refill the gearbox with fresh lubricant. Ensure drain plug is tight.
29 Fit the interlock spool plate, gasket and top cover.
30 Fit the gearbox to the car, see 37.20.01.

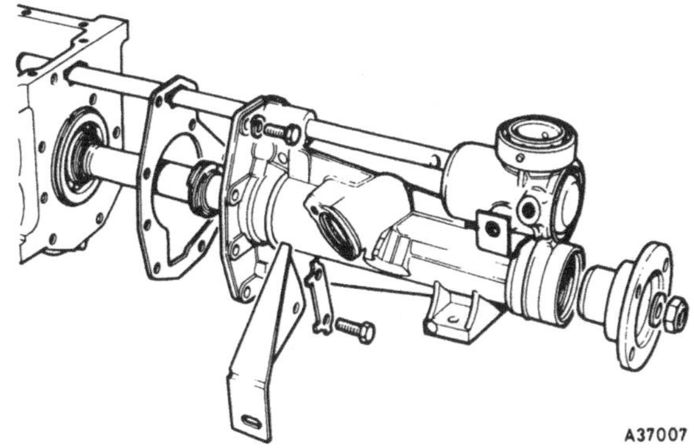
A37007

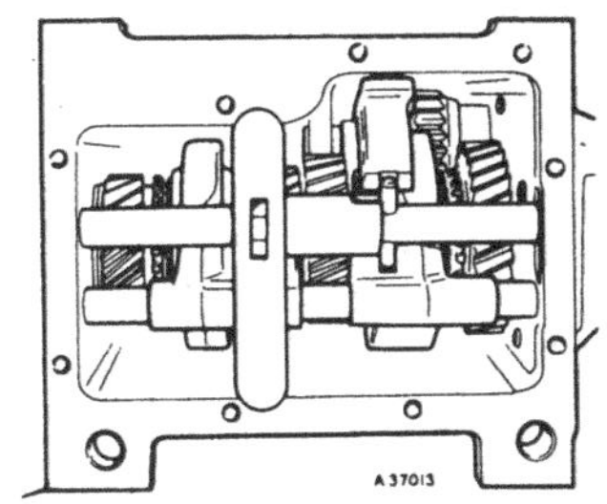
A37013

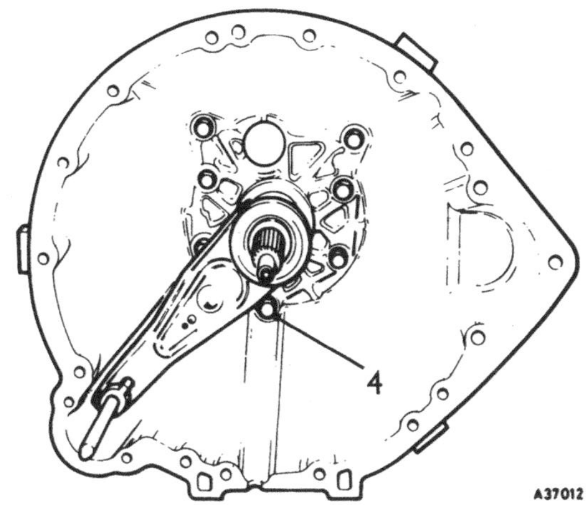

A37012

CLUTCH/BELL HOUSING

Remove and refit **37.12.07**

Removing
1 Remove the gearbox from the car, see 37.20.10.
2 Withdraw the clutch release lever and release bearing, see 33.25.12.
3 Tape the splines of the first motion shaft to prevent damage to the oil seal when the clutch housing is withdrawn.
4 Remove the seven bolts and washers securing the clutch housing to the gearbox. Note that the lower bolt is fitted with a copper washer; the remaining bolts are fitted with spring washers.
5 Withdraw the clutch housing and gasket. Observe that three compression springs are inserted in the front of the gearbox casing to provide a thrust load for the layshaft front bearing.

Refitting
6 Reverse instructions 2 to 5.
7 Remove the tape from the splines of the first motion shaft.
8 Fit the gearbox to the car, see 37.20.01.

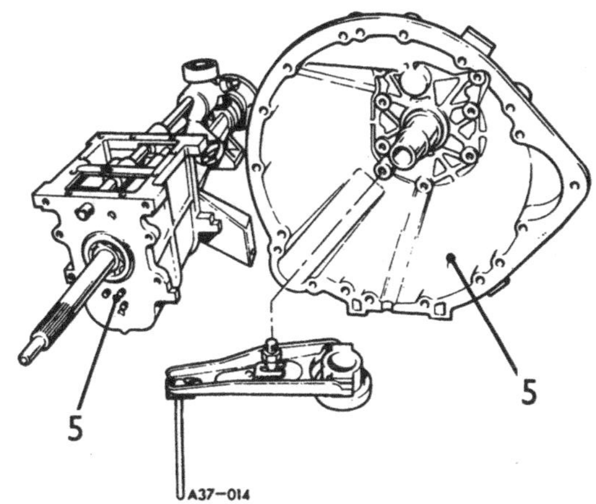

A37–014

MAIN CASE

Remove and refit **37.12.40**

As operation 37.20.04 but excluding instructions 21 to 25, 27 to 35, and 38.

GEAR CHANGE LEVER

Remove and refit **37.16.04**

Removing

1 Select neutral and unscrew and remove the gear lever knob.
2 Remove the two screws securing the gear lever gaiter and top panel assembly to the rear of the console.
3 Remove the gear lever gaiter and panel assembly.
4 Remove the four screws retaining the gear lever draught excluder and flange assembly.
5 Slide the draught excluder flange to the left to clear the console and lift off.
6 Release the bayonet cap securing the gear lever to the gearbox extension and lift out the gear lever. Care must be taken not to lose the nylon plunger and spring (anti-rattle) from the gear lever ball swivel.

Refitting

7 Reverse instructions 1 to 6.

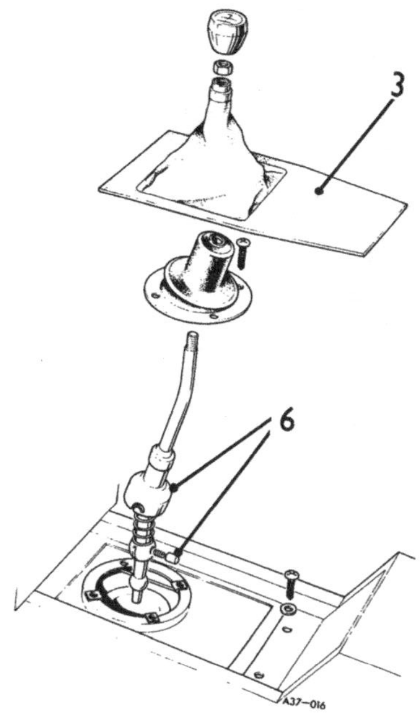

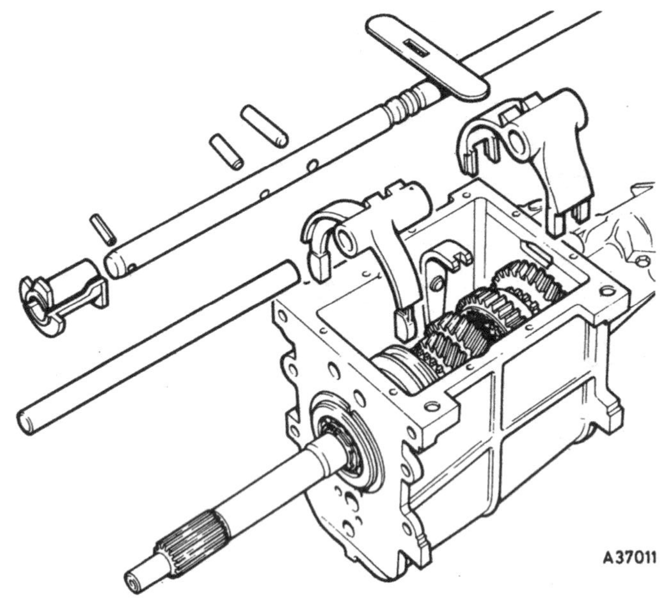

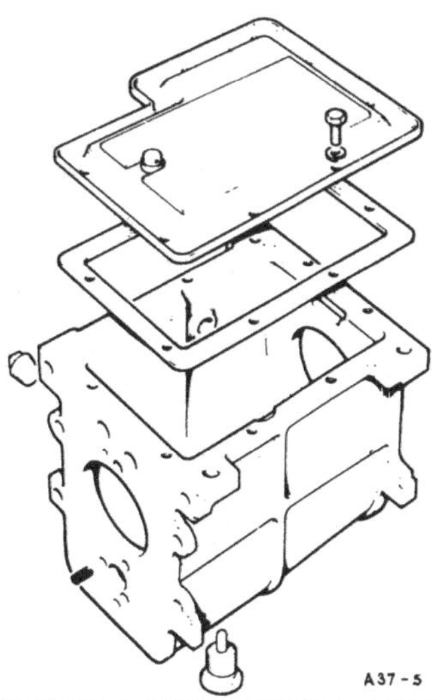

GEAR CHANGE LEVER DRAUGHT EXCLUDER

Remove and refit **37.16.05**

As operation 37.16.04 instructions 1 to 5.

GEAR CHANGE SELECTORS

Remove and refit **37.16.31**

Service tools: S337 or 18G 1205

Removing

1 Remove the gearbox rear extension, see 37.12.01.
2 Withdraw the selector fork shaft.
3 Lift out the selectors.

Refitting

4 Fit the selectors to the gears.
5 Fit the selector shaft.
6 Fit the gearbox rear extension, see 37.12.01.

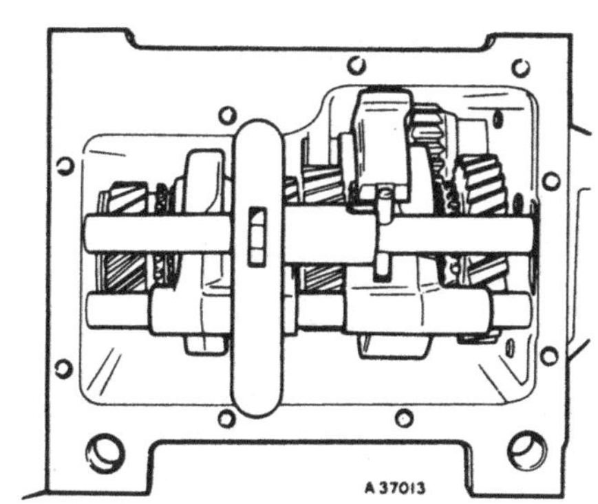

SELECTOR SHAFT YOKE PLASTIC BUSH

Reinstalling **[37.16.32]**

The selector shaft yoke plastic bush (see illustration, 37.16.37) can work loose. If it does, reinstall it as described in this procedure.

1 Remove the gear change lever. 37.16.04.
2 Fully remove the plastic bush. Then thoroughly clean and dry the hole in the yoke.
3 Apply Loctite® 636 or Loctite® 307 to the hole in the yoke.
4 Firmly support the yoke. Then drive in the plastic bush. 37.16.37.

CAUTION: The yoke is made of cast metal and will break if it is not supported while the plastic bush is being driven in.

5 Reinstall the gear change lever. 37.16.04.

GEAR SELECTOR MECHANISM AND HOUSING

Remove and refit **37.16.37**

Service tools: S337 or 18G 1205

Removing

1 Remove the gearbox rear extension, see 37.12.01.
2 Slide the selector shaft rearward until it contacts the rear blanking plug.
3 Gently tap the selector shaft to remove the blanking plug.
4 Slide the selector shaft rearward until the selector shaft yoke is exposed.
5 Remove the roll pin securing the yoke to the shaft and withdraw the yoke.
6 Withdraw the selector shaft, moving it towards the front of the rear extension. Ensure that the roll pin hole at the rear of the selector shaft is maintained in an horizontal position to prevent the selector plunger trapping the shaft as it is withdrawn.
7 Remove the nylon plug, plunger, spring and 'O' ring from the rear extension casing.

Refitting

8 Reverse instructions 1 to 7.

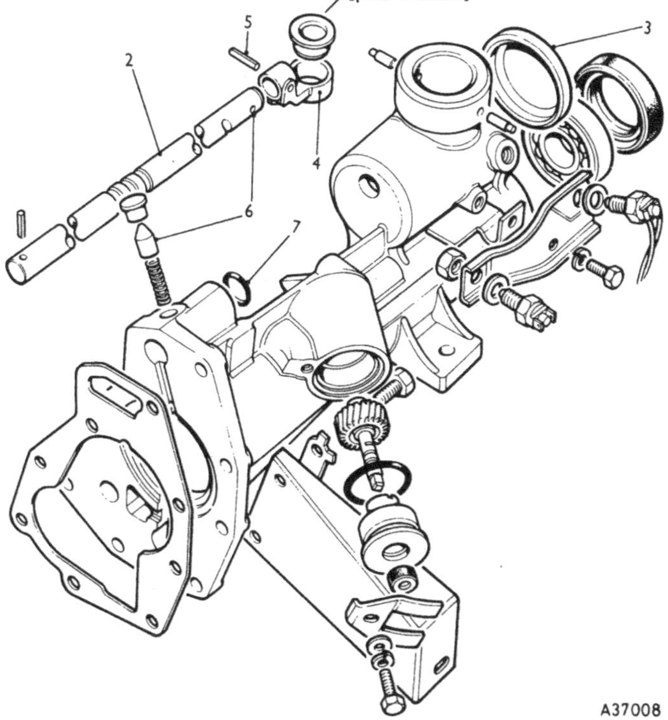

A37008

GEARBOX ASSEMBLY

Remove and refit **37.20.01**

Removing

1 Disconnect the battery.
2 Remove the gear lever, see 37.16.04.
3 Remove the radiator fan guard.
4 Raise the car on a ramp or jack up and support securely.
5 Disconnect the propeller shaft at the gearbox flange and tie the propeller shaft so that it will not obstruct gearbox withdrawal.
6 Remove the exhaust down pipe, see 30.10.09.
7 Disconnect the speedometer cable at the gearbox.
8 Disconnect the electrical harness plug to the gearbox.
9 Position a jack under the engine sump and support the engine. To avoid damage to the sump a suitable piece of wood should be interposed between the jack pad and the sump.
10 Remove the starter motor.
11 Remove the clutch slave cylinder. Do not allow the cylinder to hang suspended by the fluid pipe.
12 Remove the two bolts securing the sump stiffening plate to the clutch housing.

[**NOTE:** On U.S.A. cars, disconnect the restraint cable from the crossmember.]

13 Remove the four bolts and nuts securing the gearbox rear mounting bracket to the body.
14 Carefully lower the jack to permit access to the clutch housing bolts. Ensure that the water hoses and the engine stabiliser are not strained.
15 Remove the bolts and nuts securing the clutch housing to the engine. Note the location of the fitted bolt adjacent to the clutch slave cylinder.
16 Carefully withdraw the gearbox ensuring that weight and stress are not imparted to the clutch driving plate.
17 Remove the gearbox mounting bracket.

continued

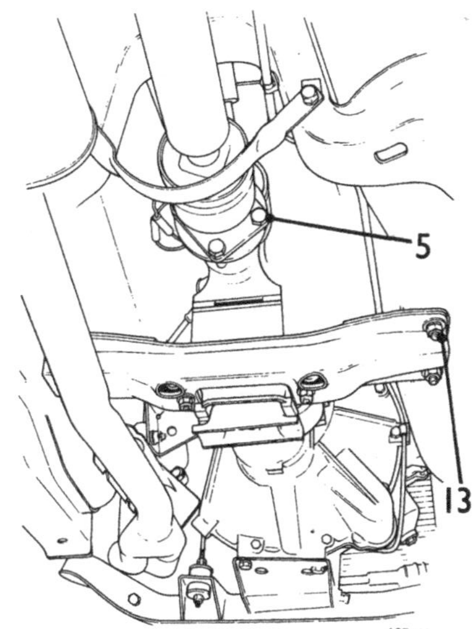

A37-45

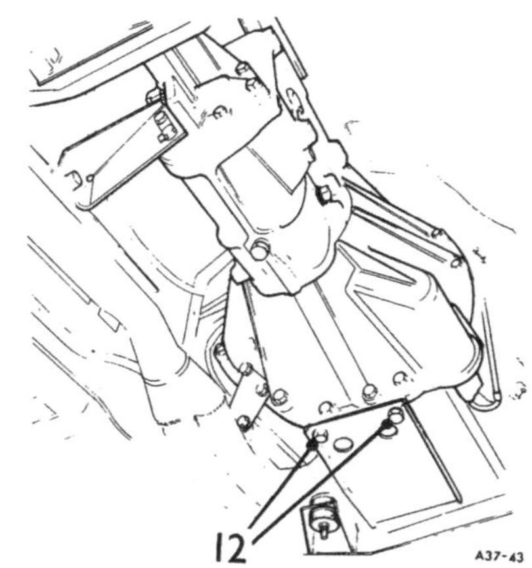

A37-43

Refitting

18. Reverse instruction 1 to 17. Ensure that the engine stabiliser is adjusted as described in Operation 12.45.16. Adjust the restraint cable on U.S.A. cars as follows:
    a. Ensure that the front nut at the rear end of the cable is fully slackened off.
    b. Tighten the rear nut to a torque of 5–8 lbf ft (0.69–1.11 Kgf m) to settle the cable in position.
    c. Slacken the rear nut and holding the cable taut by hand, position the rear nut so that a clearance of 1/32 - 1/16 in (0.5–1.5 mm) exists between the nut and the cable bracket.
    d. Tighten the front nut.

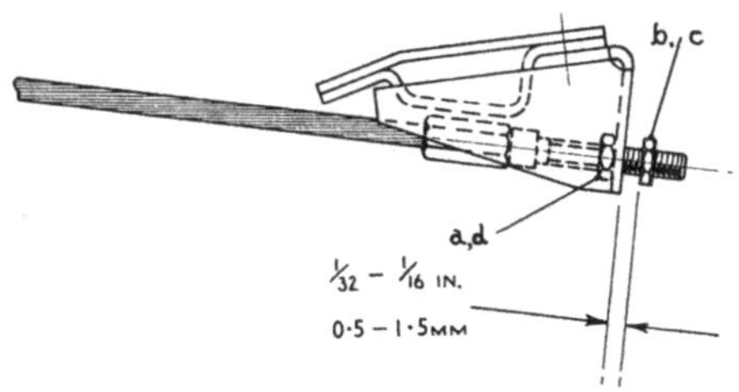

A37-28

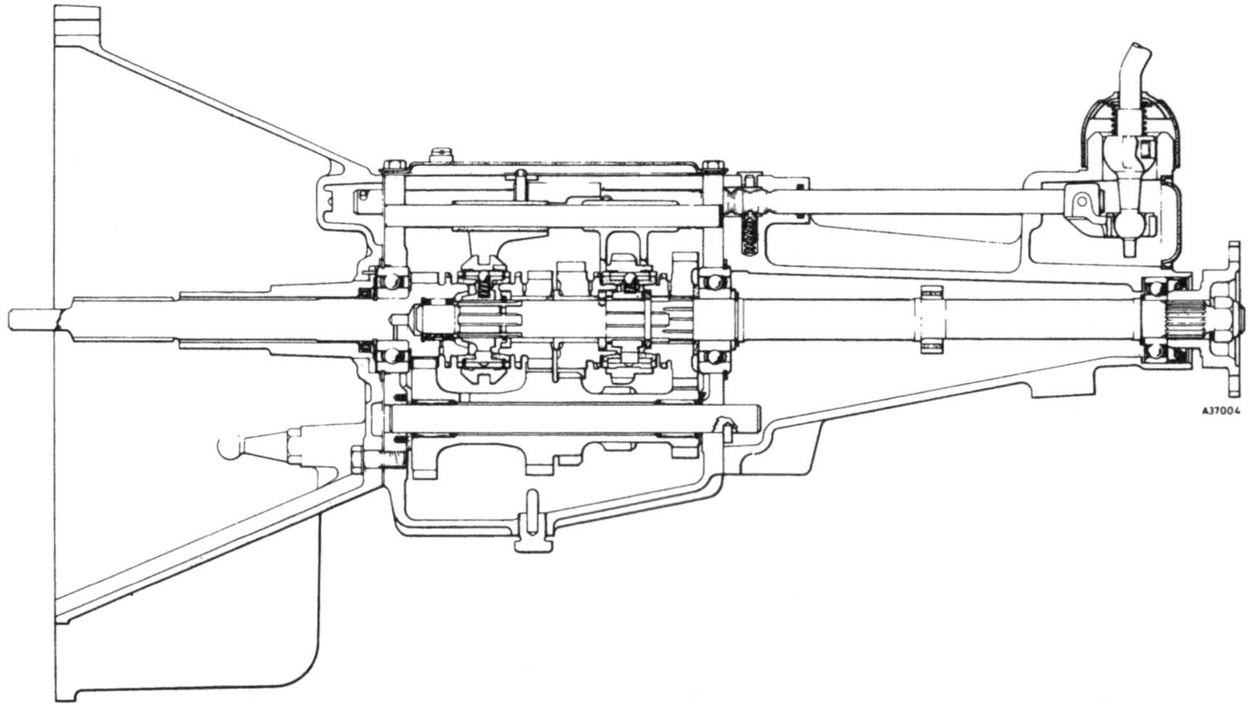
A37004

GEARBOX ASSEMBLY

Overhaul **37.20.04**

Service tools: ST1136 47, S337 or 18G 1205, 18G 1208, 18G 1199, S145C, 18G 1197, 3072 or S4235 A, S4235 A-2, 18G 47 BP, 18G 1208, 18G 1198 A

Dismantling

1 Remove the gearbox from the car, see 37.20.01.

2 Remove the clutch housing, see 37.12.07.

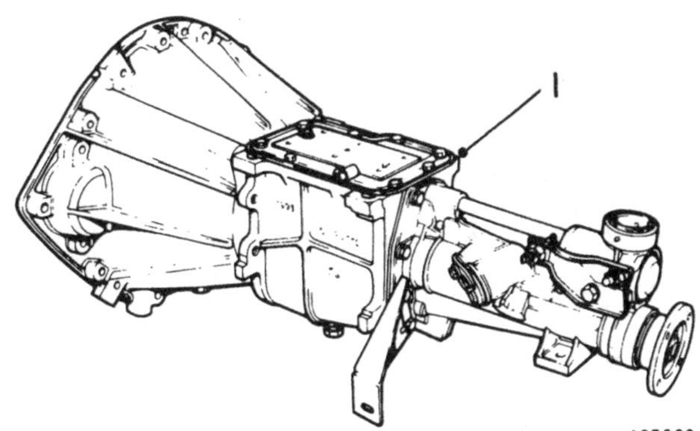

A37003

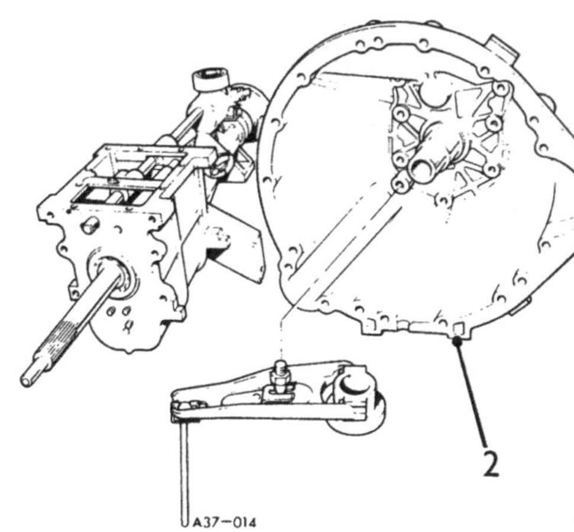

A37-014

3 Remove the top cover and spool interlock plate.
4 Remove the rear extension, see 37.12.01.

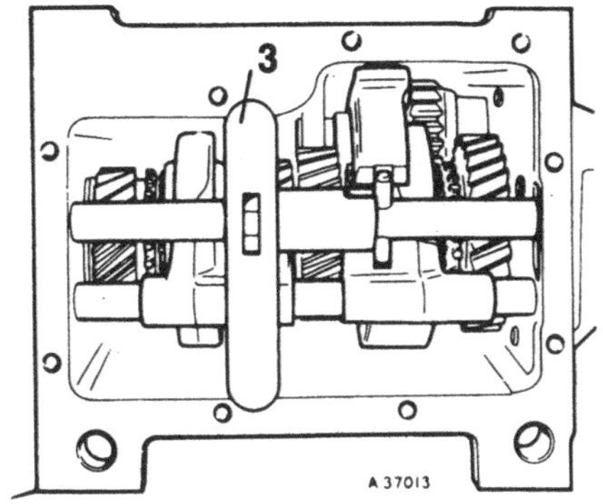

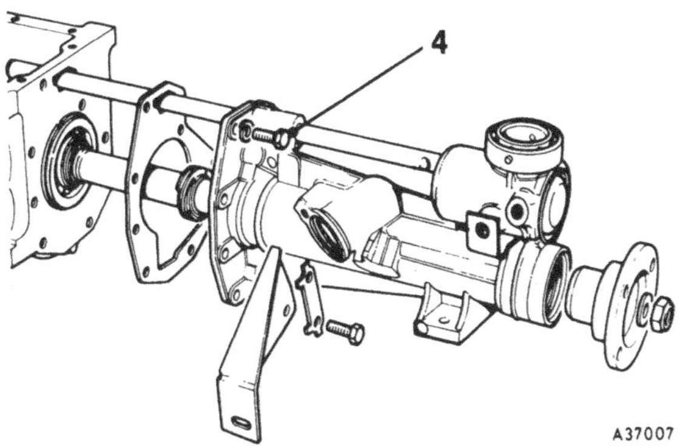

5 Remove the gear selector mechanism, see 37.16.37.
6 Remove the selector shaft and selector forks.

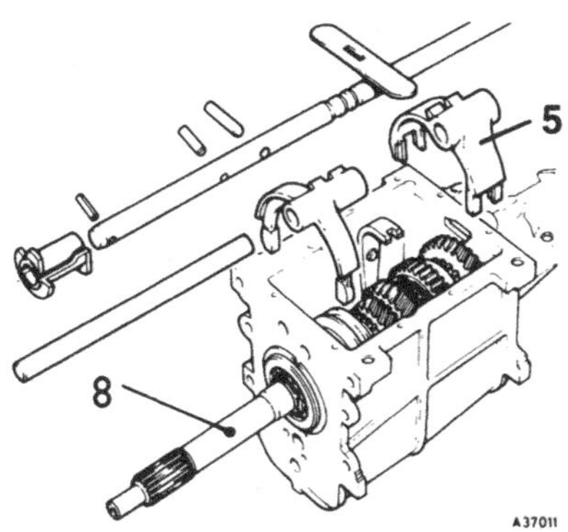

7 Using tool 18G 1208 displace and remove the layshaft, allowing the laygear to drop to the bottom of the gear case.
8 Remove the first motion shaft, see 37.20.16.

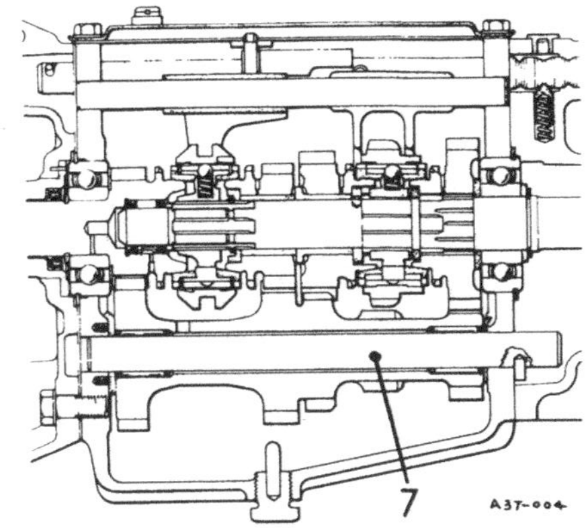

9 Fit the steady bracket of 18G 47 BP to the front of the gear case using two clutch housing bolts. Ensure that the centre adjusting bolt and locknut are fully released before tightening the tool securing bolts.
10 Adjust the centre bolt to locate and support the mainshaft spigot. Tighten the locknut.

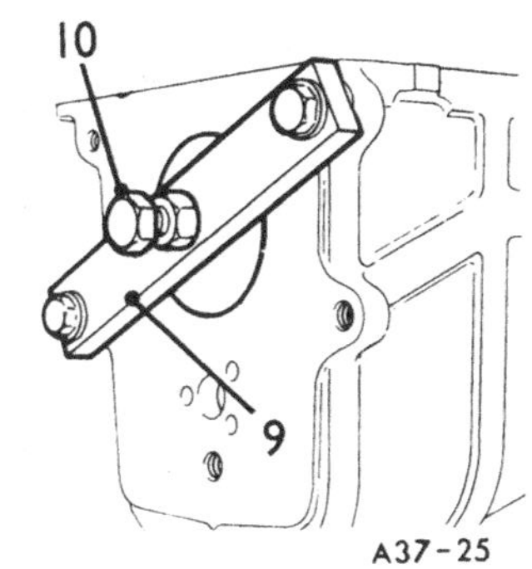

11 Remove the bolt and spring washer securing the reverse/idler gear spindle.
12 Withdraw the reverse/idler gear spindle, and spacer and remove the reverse/idler gear.

continued

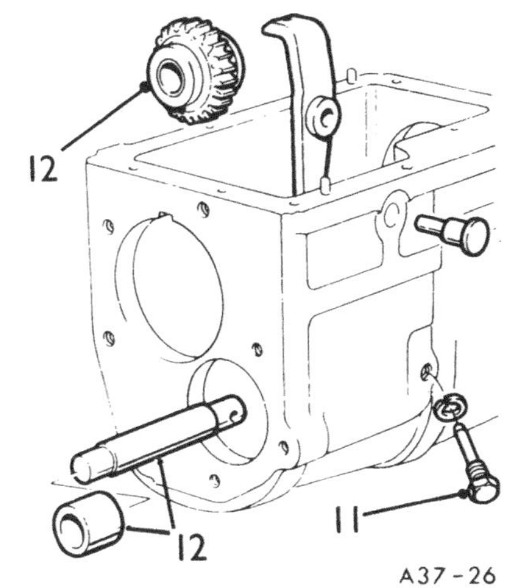

13 Release the circlip securing the mainshaft bearing to the mainshaft.
14 Remove the snap ring fitted to the mainshaft bearing.
15 Using tool 47 and bearing remover of 18G 47 BP, remove the mainshaft bearing, selective washer, circlip and speedometer gear from the mainshaft.
16 Remove the steady bracket of tool 18G 47 BP from the front of the gear case. Do not alter the setting of the centre bolt and locknut prior to reassembly.
17 Tilt the mainshaft and withdraw complete with gears through the top cover aperture.
18 Remove the laygear and thrust washers.
19 Remove the reverse gear operating lever.
20 Remove from the mainshaft.
  a Thrust washer and 1st speed gear.
  b 1st speed synchro – cup.
  c Two, split collars.
  d 3rd/4th speed synchro-hub and sleeve assembly.
  e 3rd speed synchro-cup.
  f Using tool 18G 1199 expand the circlip retaining 3rd speed gear. Carefully lever off 3rd speed gear, and bush, thrust washer, circlip, and tool.
  g 2nd speed gear and bush.
  h 2nd speed synchro-cup.
  i Selective washer.
  j Using a magnet extract the ball locating the selective washer from its recess in the mainshaft.
  k 1st/2nd speed synchro-hub and sleeve assembly.

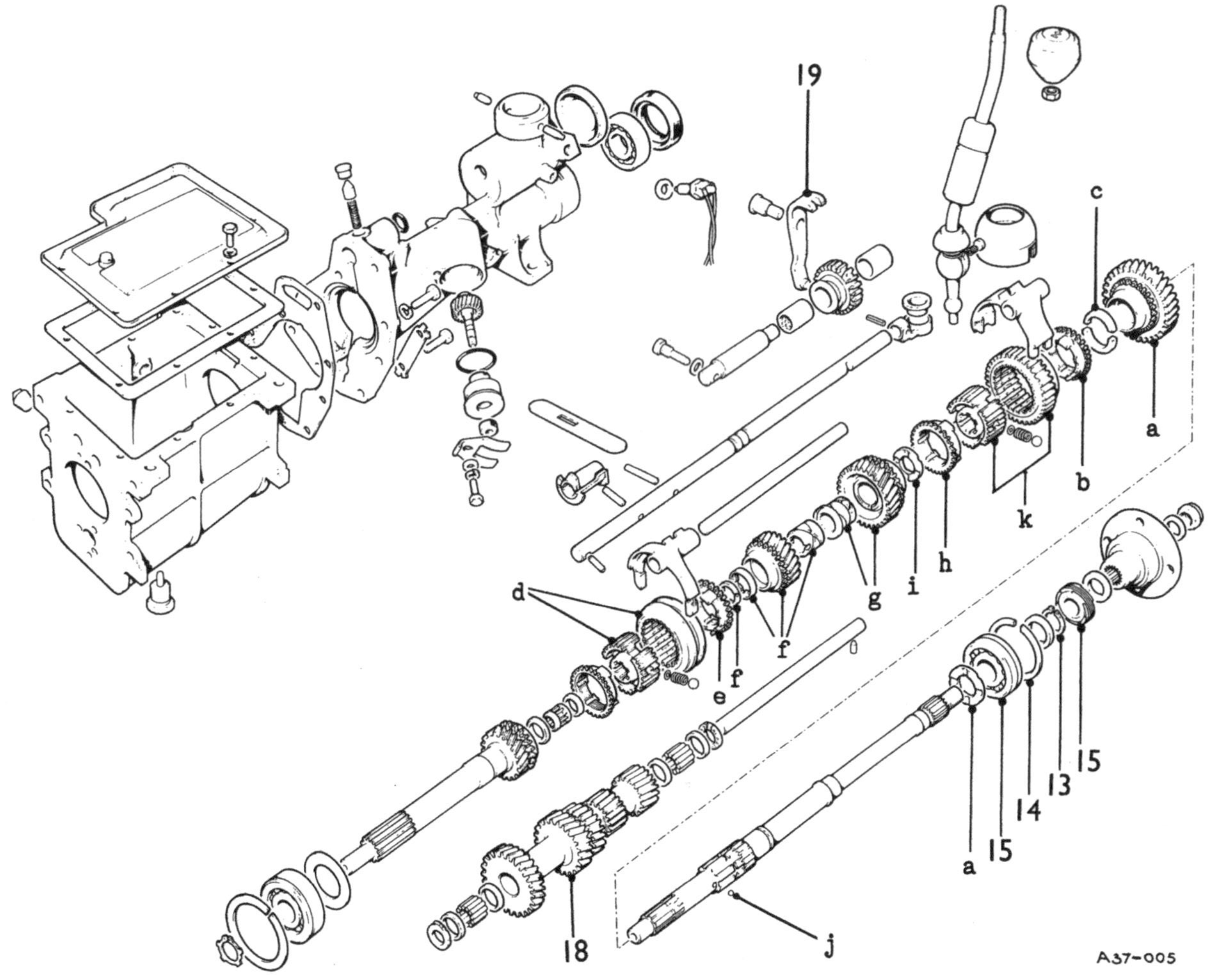

21 Scribe the sleeve and synchro-hub assemblies of 1st/2nd and 3rd/4th speeds to ensure reassembly in the original locations. Separate the sleeves from the hubs ensuring that care is taken to capture the three balls and springs in each assembly.
Note: In some instances shims may be fitted below the springs.
22 Remove the 25 needle rollers from each end of the laygear cluster.
23 Remove the needle roller retaining rings from the laygear (only if renewal is intended).
24 Press out the bush from the reverse/idler gear.
25 Remove the reverse operating lever pivot pin from the gear case (only if renewal is intended).

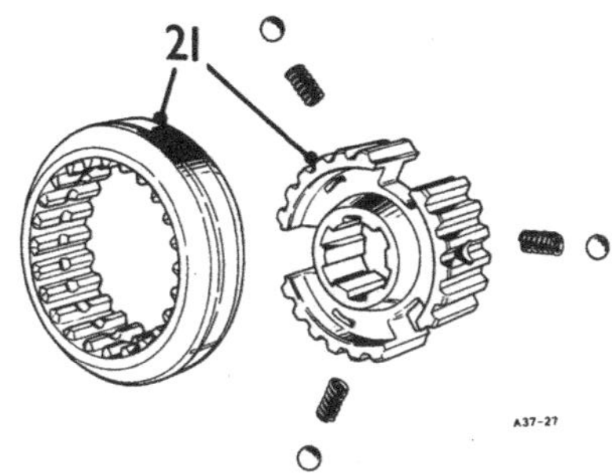

Preparation for reassembly
26 Thoroughly clean and examine all components. Obtain new parts as necessary. Renew all gaskets and seals.
27 If required fit new needle roller retaining rings to the laygear to the dimensions shown. Care must be taken not to damage the laygear bore.
28 Using grease install the needle rollers (25 each side) in the laygear. Fit dummy layshaft, tool 18G 1208.
29 Fit the shims (if removed), springs and balls to the 1st/2nd speed synchro-hub and slide the sleeve into position observing the scribe marks made prior to dismantling. The teeth of the outer member must be fitted adjacent to the larger boss of the synchro-hub.
30 Check the load required to shift the sleeve in either direction. This should be within 19-27 lb (8.7-12.2 kg).
Add or remove shims to obtain required effort.
31 Fit the shims (if removed), springs and balls to the 3rd/4th speed synchro-hub and slide the sleeve into position observing the scribe marks made prior to dismantling.
32 Check the load required to shift the sleeve in either direction. This should be within 19-21 lb (8.7-9.5 kg). Add or remove shims to obtain required effort.

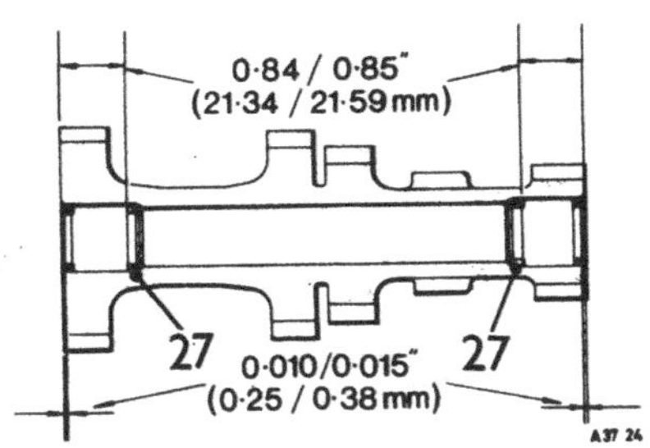

33 Fit a new brush to the reverse/idler gear. The bush should be flush with the boss opposite the collar of the operating lever. Ream the bush within 0.6585-0.6592 in diameter (16.7279-16.8011 mm).

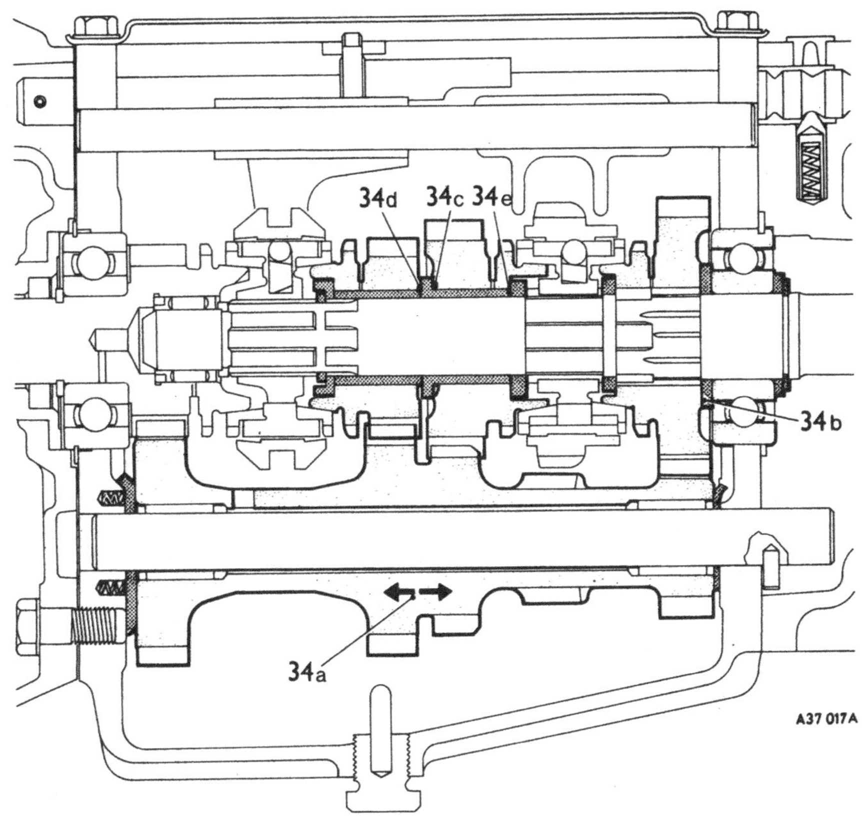

34 Check that the following requirements obtain:
a Laycluster end-float – 0.007-0.015 in (0.178-0.381 mm).
b 1st gear end-float (between split collars and thrust washer) – 0.004-0.013 in (0.102-0.33 mm).
c 2nd gear end-float (on bush) 0.002 in (0.051 mm).
d 3rd gear end-float (on bush) 0.002-0.006 in (0.051-0.152 mm).
e Mainshaft bushes (2nd and 3rd gears) – 0-0.006 in (0-0.051 mm). Adjust clearances by means of selective washer. Four alternative washers are available in 0.003 in (0.076 mm) steps.
35 Renew the first motion shaft bearing, see 37.20.17.

continued

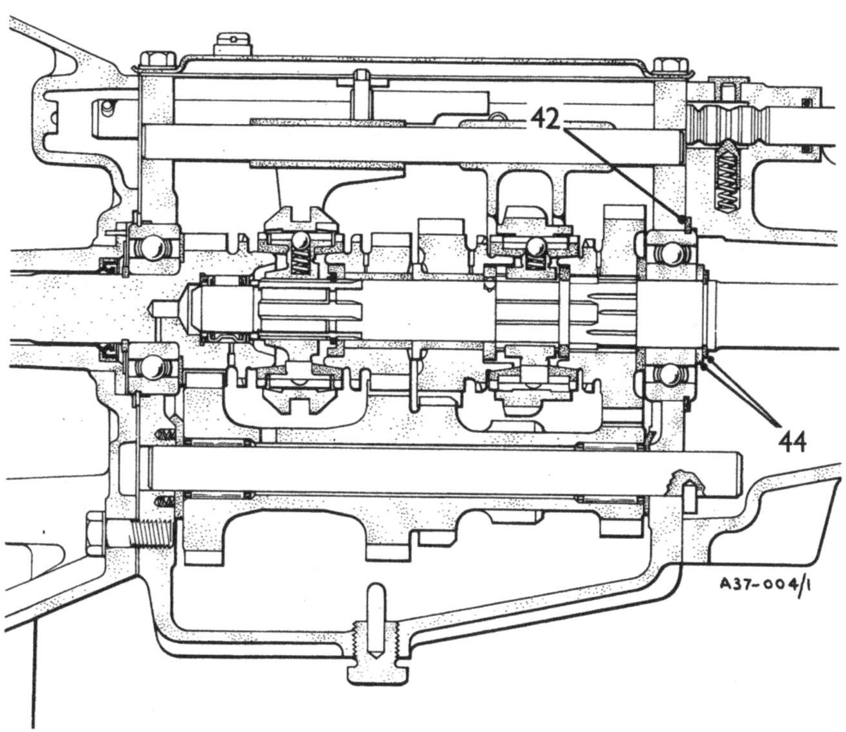

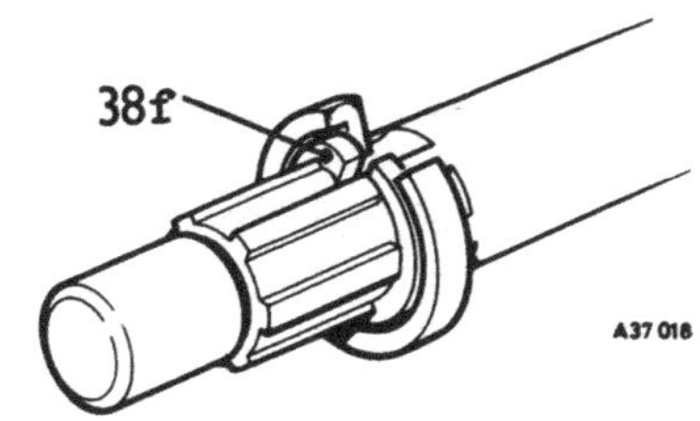

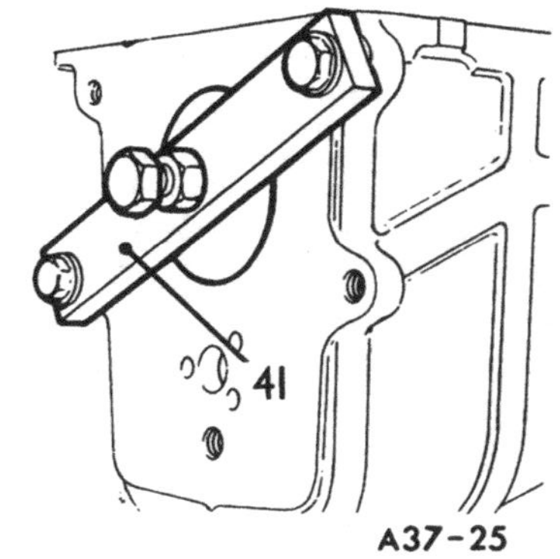

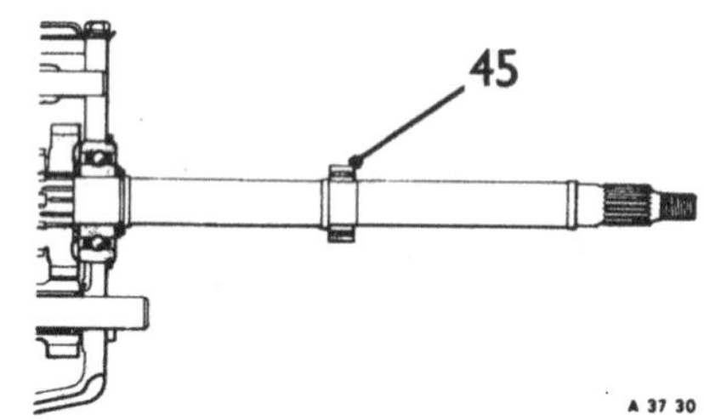

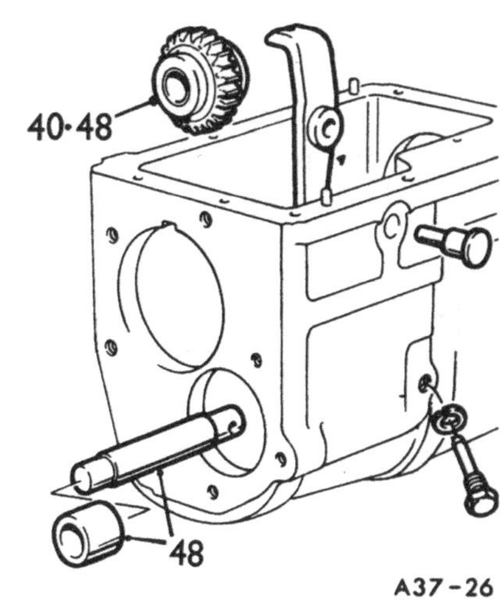

Reassembling

36 Locate the laygear front and rear thrust washers in position ensuring that their respective tabs engage in the casing slots.

37 Place the laygear cluster complete with dummy shaft in the bottom of the gear case. (Large gear to front of case.)

38 Fit the following components to the mainshaft.

a 1st/2nd speed synchro assembly (larger hub boss to front of gearcase).

b 2nd speed synchro-cup.

c Locating ball for selective washer.

d Selective washer, ensuring groove on washer inner diameter aligns with ball.

e 2nd gear and bush (bush collar to front of gearbox), 3rd gear and bush, and thrust washer (rim of washer to front of gearbox).

f Using tool 18G 1198 fit the retaining circlip ensuring that the inclined end of the clip faces forward and the clip end aligns with the edge of the mainshaft spline.

g 3rd gear synchro-cup, 3rd gear synchro assembly (larger boss to hub of front of gearbox).

h 1st speed synchro-cup, split collars, 1st gear and thrust washer.

39 Tilt the mainshaft (rear end leading) into position through the top cover aperture.

40 Place the reverse gear in the bottom of the gear case and fit the reverse operating lever.

41 Fit the steady bracket of 18G 47 BP to the front of the gear case engaging the mainshaft spigot.

42 Fit the snap ring to the mainshaft centre bearing and slide the bearing (snap ring trailing) on to the mainshaft.

43 Drive the bearing into position using tool 18G 1197 and the bearing replacer of 18G 47 BP.

44 Fit the selective washer and circlip. Four selective washers are available in 0.003 in (0.076 mm) steps. Mainshaft end-float should not exceed 0.002 in (0.051 mm).

45 Fit the speedometer gear.

46 Remove the steady bracket of 18G 47 BP from the front of the gear case and fit spigot bearing and rings, 4th speed synchro-cup, and the 1st motion shaft.

47 Fit the layshaft, removing the dummy shaft in the process.

48 Fit the reverse gear, reverse gear shaft and spacer.

49 Fit the selector forks and shaft.
50 Fit the gear selector mechanism, see 37.16.37.
51 Fit the rear extension and drive flange, see 37.12.01.
52 Fit the three thrust springs to the laygear front bearing.
53 Fit the clutch housing gasket, clutch housing, clutch release fork and bearing, see 37.12.07. Ensure splines of 1st motion shaft do not damage seal.
54 Fit the gear case drain plug and fill to level plug with fresh oil.
55 Fit the spool interlock plate, gasket and top cover.

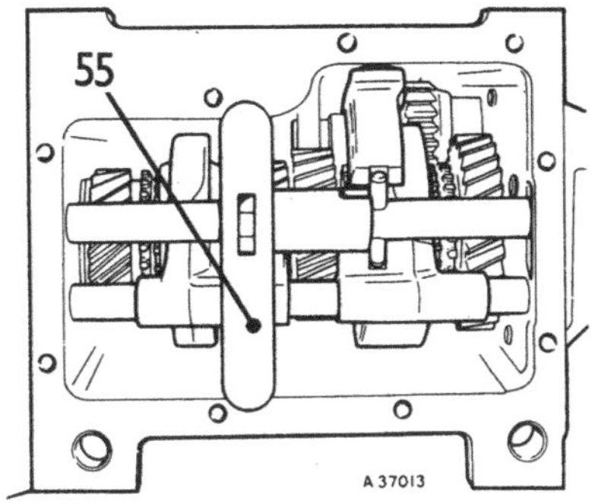

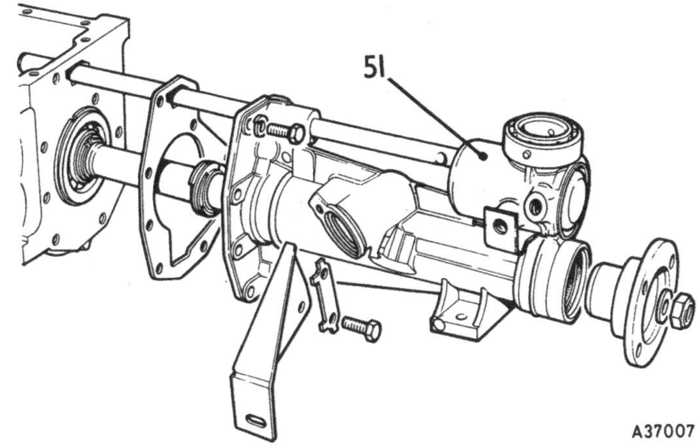

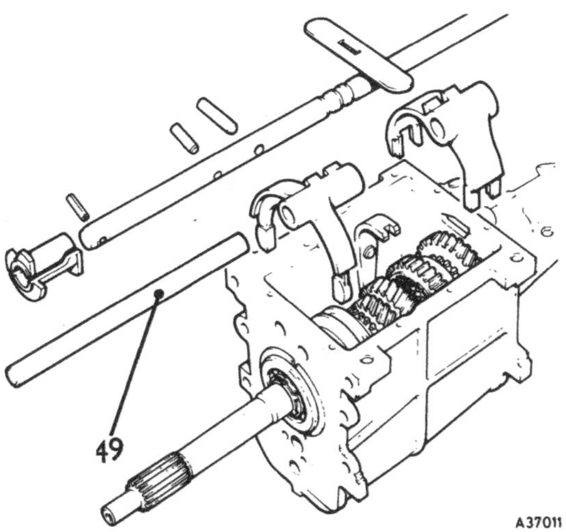

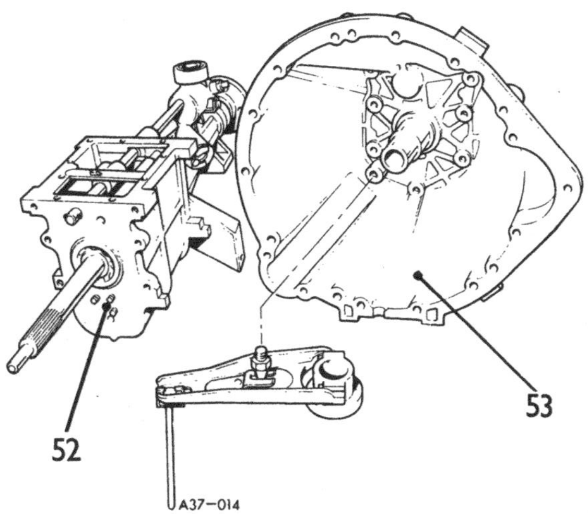

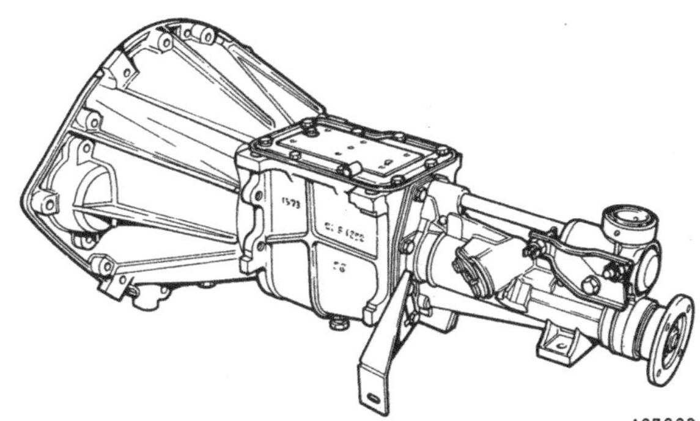

SYNCHRONISER ASSEMBLIES

Remove and refit **37.20.07**

Service tools: Refer operation 37.20.04

Removing
Instructions 1 to 17 and 20a to 20h.

Refitting
Instructions 38 to 40, 42 to 50 and 52 to 55.

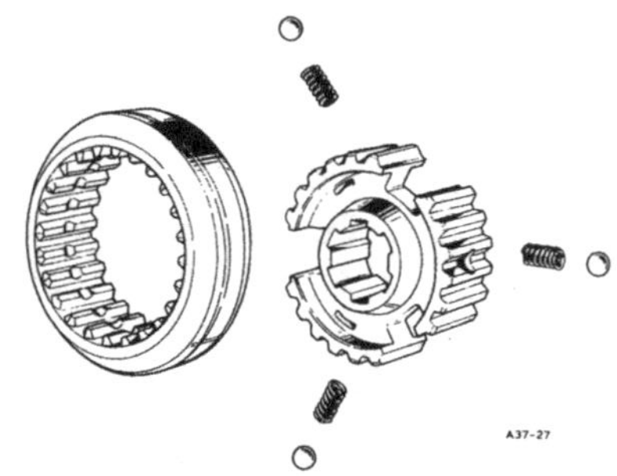

A37-27

FIRST MOTION SHAFT

Remove and refit **37.20.16**

Service tools: 18G 1208, 3072 or S4235 A, S4235 A - 2

Removing
1 Remove the gearbox from the car, see 37.20.01.
2 Remove the rear extension, see 37.12.01.
3 Remove the clutch housing, see 37.12.07.
4 Insert tool 18G 1208 and push out the layshaft.
5 Allow the laygear with tool 18G 1208 to drop to the bottom of the gearbox case.
6 Fit tool 3072 and adaptor S4235 A - 2 to the first motion shaft.
7 Extract the first motion shaft from the gearbox.
8 Remove the spigot bearing and spacers (2).
9 Remove the 4th gear synchromesh cup.

Refitting
10 Reverse instructions 1 to 9.

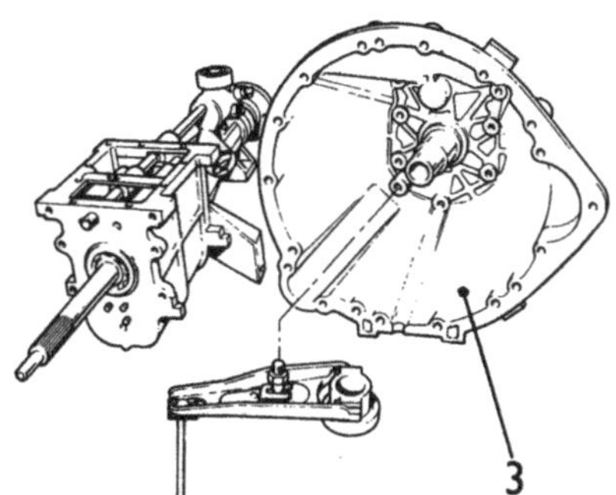

A37-014

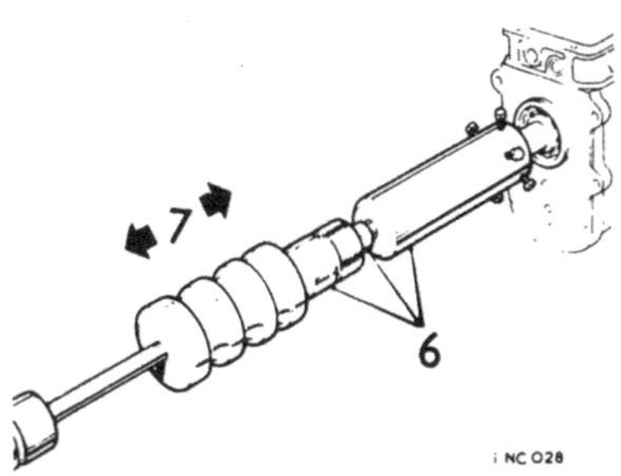

NC028

FIRST MOTION SHAFT BEARING

Remove and refit **37.20.17**

Service tools 18G 1208, 3072 or S4235 A, S4235-2

Removing
1 Remove the first motion shaft, see 37.20.16.
2 Remove the bearing circlip.
3 Remove the bearing snap ring.
4 Using tool S4235 A and adaptor S4235 A-2 remove the bearing from the shaft.
5 Remove the oil flinger.

Refitting
6 Reverse instructions 1 to 5. Use grease to retain the oil flinger when fitting the bearing to the shaft.

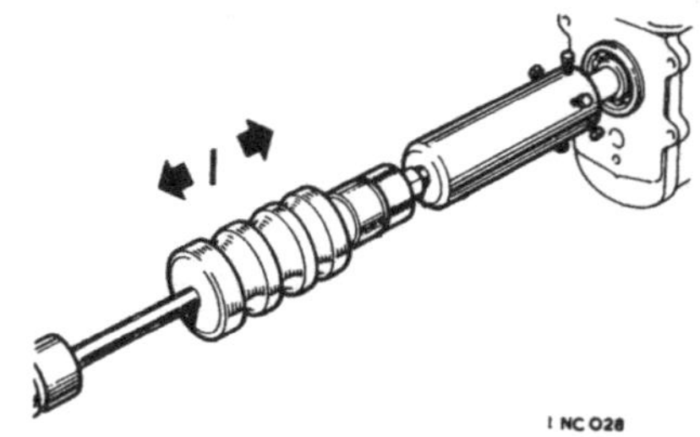

NC028

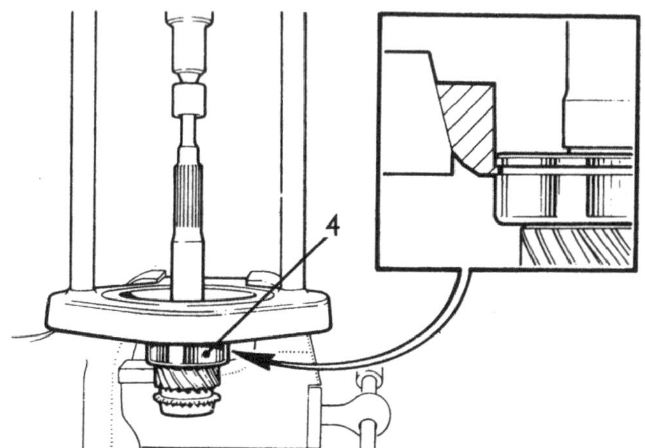

A37-23

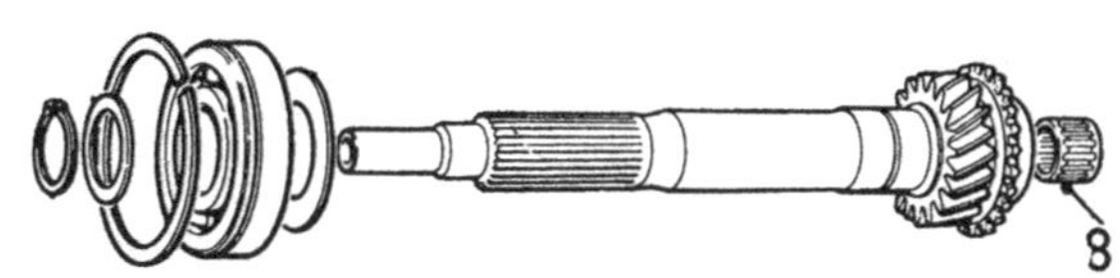

A37-061

LAYSHAFT

Remove and refit **37.20.19**

Service tools: S337 or 18G 1205, 18G 1208

Removing

1 Remove the gearbox from the car, see 37.20.01.
2 Remove the clutch housing, see 37.12.07.
3 Remove the top cover and spool interlock plate.
4 Remove the rear extension, see 37.12.01.
5 Using tool 18G 1208 displace and remove the layshaft.

Refitting

6 Reverse instructions 1 to 5.

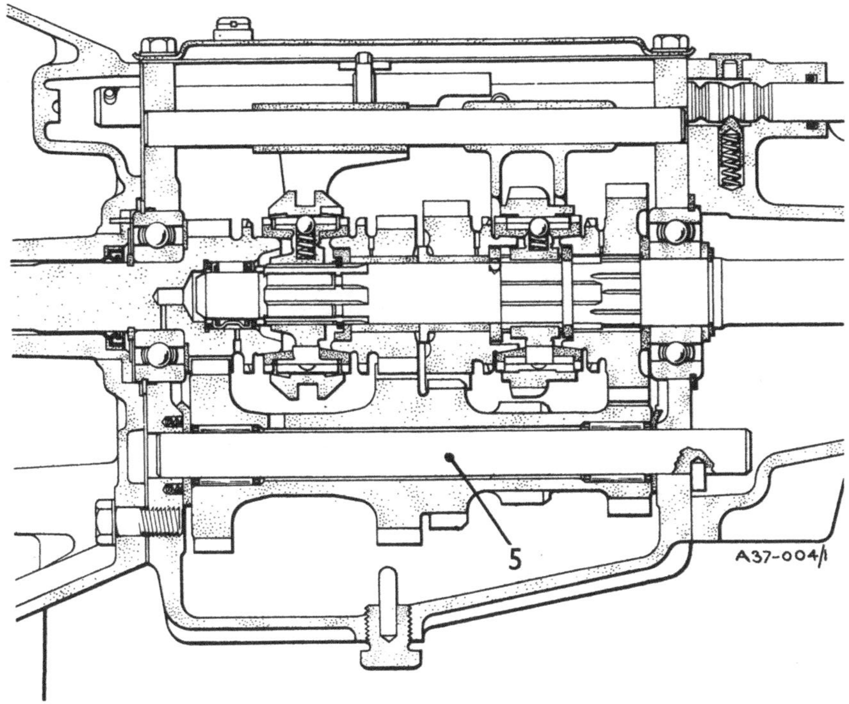

LAYSHAFT BEARING

Remove and refit **37.20.22**

Service tools: S337 or 18G 1205, 18G 1208, 18G 1199, S145 C, 18G 1197, 3072 or S4235 A, S4235 A - 2

As operation 37.20.04 instructions 1 to 10, 13 to 18, 22, 23, 27, 28, 36, 37 and 41 to 55.

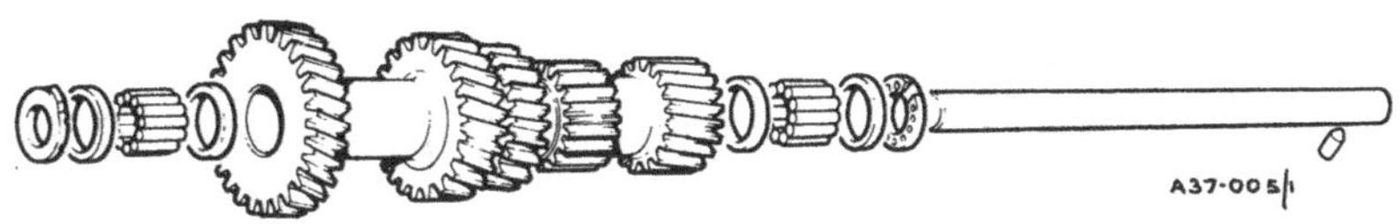

MAINSHAFT BEARING

Remove and refit **37.20.26**

Service tools: S337 or 18G 1205, 18G 1208, 18G 1197.

Removing

As operation 37.20.04 Instructions 1 to 4 and 6 to 16.

Refitting

As operation 37.20.04 Instructions 42 to 48, 50, and 52 to 55.

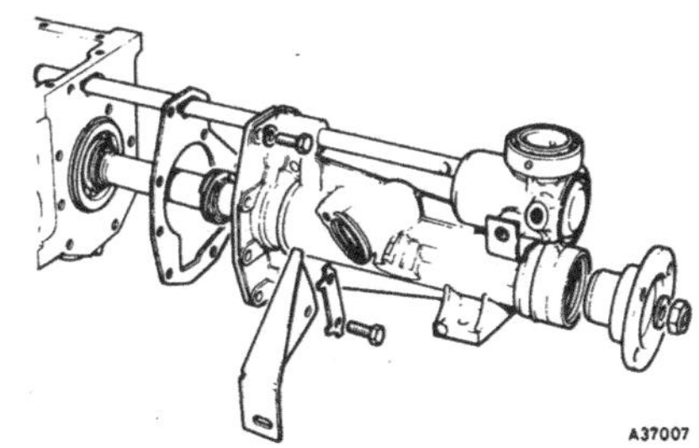

REAR OIL SEAL

Remove and refit **37.23.01**

Service tools: S337 or 18G 1205

Removing

1 Raise the car.
2 Remove the drive flange, see 37.10.01.
3 Prise out the oil seal.

Refitting

4 Evenly install the new seal in the gearbox rear extension ensuring that it is flush with the rear face of the extension housing. Note that the seal lips must be towards the gearbox.
5 Lubricate the seal lips and driving flange seal contact area.
6 Fit the drive flange, see 37.10.01.
7 Lower the car.

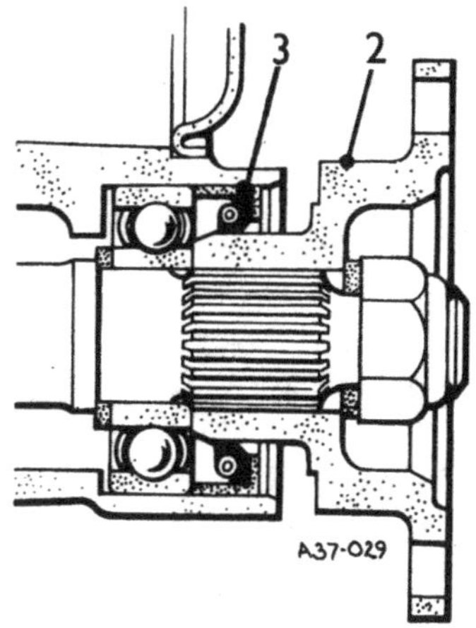

FIRST MOTION SHAFT OIL SEAL

Remove and refit **37.23.04**

Removing

1 Remove the gearbox from the car, see 37.20.01.
2 Remove the clutch housing, see 37.12.07.
3 Remove the oil seal.

Refitting

4 Evenly install a new oil seal ensuring that the seal lips are towards the clutch.
5 Lubricate the seal lips and ensure that the splines of the first motion shaft are protected by tape to prevent damage to the seal lip when the clutch housing is fitted.
6 Fit the clutch housing, see 37.12.07.
7 Remove the tape from the splines of the first motion shaft.
8 Fit the gearbox to the car, see 37.20.01.

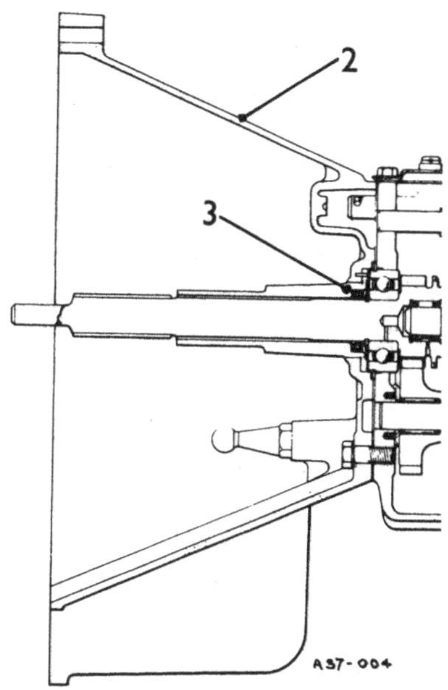

GEAR CHANGE ROD SEAL

Remove and refit **37.23.10**

Service tools: S337 or 18G 1205

As operation 37.16.37.

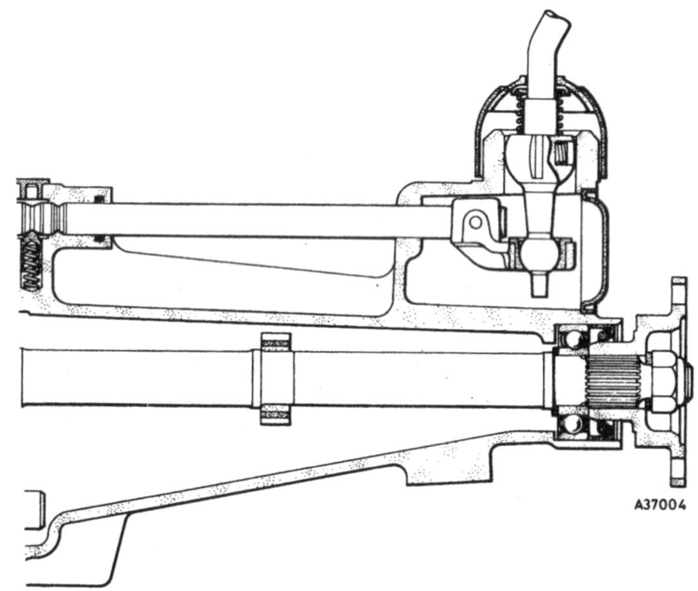

SPEEDOMETER DRIVE GEAR

Remove and refit **37.25.01**

Service tools: S337 or 18G 1205, 18G 1197

Removing
1 Remove the gearbox from the car, see 37.20.01.
2 Remove the rear extension, see 37.12.01.
3 Remove the speedometer drive gear.

Refitting
4 Using tool 18G 1197 drive the speedometer gear into position on the mainshaft ensuring that it registers against the step on the mainshaft.
5 Reverse instructions 1 and 2.

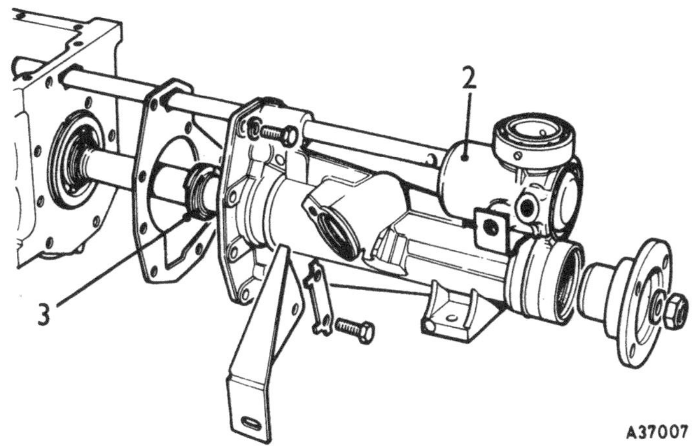

SPEEDOMETER DRIVE PINION

Remove and refit **37.25.05**

Removing
1 Remove the bolt and spring washer retaining the speedometer cable clamp plate.
2 Release the speedometer cable from the drive pinion.
3 Remove the pinion housing and pinion from the gearbox rear extension.

Refitting
4 Reverse instructions 1 to 3. Renew the pinion housing 'O' ring if necessary.

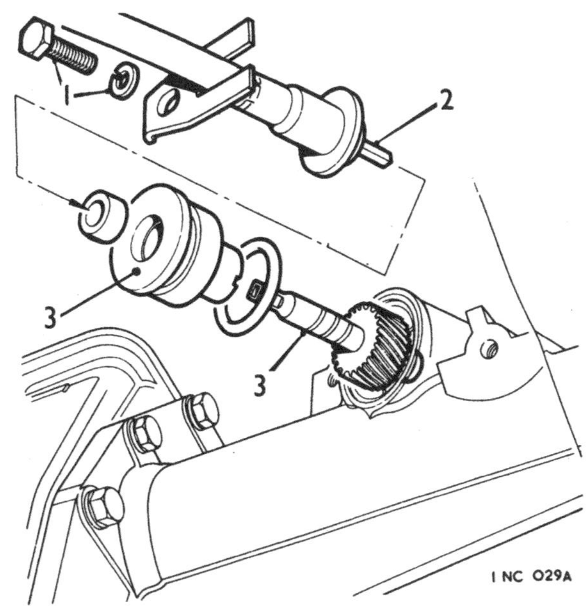

GEARBOX – 5-SPEED

[**CAUTION:** Please read the instructions thoroughly before you attempt transmission repairs. If you lack the necessary skills or tools we suggest you leave this work to an Authorized Dealer or other qualified shop. Before taking a removed transmission in for repair, thoroughly clean its exterior but do not undertake any disassembly.]

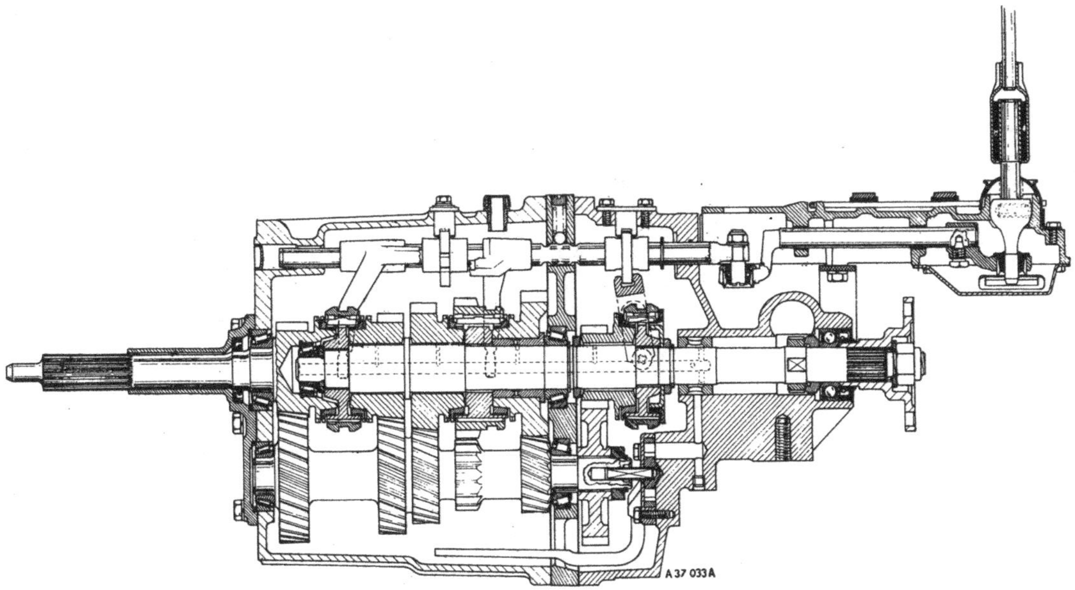

DRIVE FLANGE

Remove and refit **37.10.01**

Service tools: RG 421 or 18G 1205

Removing

1 Raise the car and support securely.
2 Slacken the bolt securing the propeller shaft strap and remove the remaining bolt. Swing the strap clear of the propeller shaft.
3 Scribe the relationship of the propeller shaft and gearbox flanges and remove the four securing nuts and bolts.
4 Release the propeller shaft from the gearbox flange.

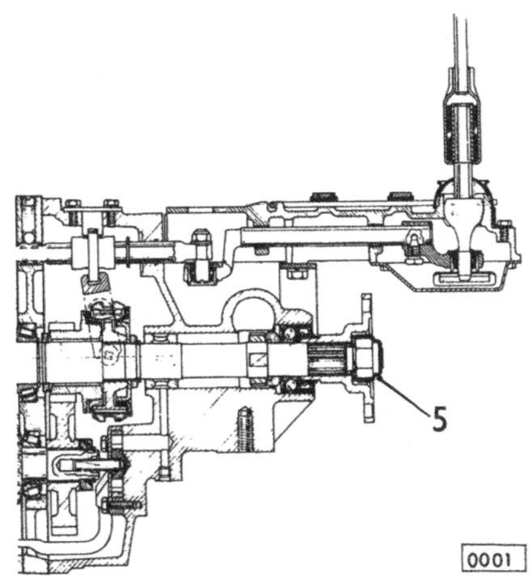

5 Using tool RG 421 or 18G 1205 to prevent rotation of the gearbox driving flange remove the securing nut and washer.
6 Withdraw the driving flange.

Refitting

7 Reverse instructions 1 to 6. The gearbox flange nut should be torqued to: 90-120 lbf ft (12.44-16.60 kgf m). Propeller shaft bolts to: 26-34 lbf ft (3.60-4.70 kgf m).

CLUTCH/BELL HOUSING

Remove and refit **37.12.07**

Service tools: ST 1136

Removing

1 Remove the gearbox from the car, see 37.20.10/55.
2 Withdraw the clutch release lever and release bearing, see 33.25.12.
3 Remove the six bolts, plain and spring washers securing the clutch housing to the gearcase.
4 Remove the clutch housing.

Refitting

5 Reverse instructions 1 to 4.

REAR COVER

Remove and refit **37.12.42**

Service tools: RG 421 or 18G 1205

Removing

1 Disconnect the battery.
2 Remove the gear lever.
3 Raise the car on a ramp or jack and support securely.
4 Remove one bolt from the propeller shaft safety strap, slacken the remaining bolt and swing the strap aside.
5 Disconnect the propeller shaft from the gearbox.
6 Tie the propeller shaft to the vehicle in a position where it allows access to the gearbox.
7 Release the rubber rings securing the exhaust system.
8 Detach the exhaust system from the manifold down-pipe.
9 Remove the gearbox drain plug and drain the oil.
10 Disconnect the speedometer cable and remove the speedometer drive pinion and housing, see 37.25.05.
11 Using tool RG 421 or 18G 1205 to prevent shaft rotation remove the nut and washer securing the gearbox drive flange.
12 Withdraw the drive flange.

continued

13 Locate the jack under the engine flywheel and support the weight of engine and gearbox.
14 Remove the four bolts, spring washers and plate washers securing the gearbox rear crossmember to the body.
15 Carefully lower the jack slightly to facilitate access to the top of the gearbox.
16 Remove the nut, washer and pin securing the remote control linkage to the gearbox selector rod and detach the reverse switch lead.
17 Remove the two bolts and spring washers securing the flange of the fifth gear spool locating boss.
18 Withdraw the fifth gear spool locating boss.

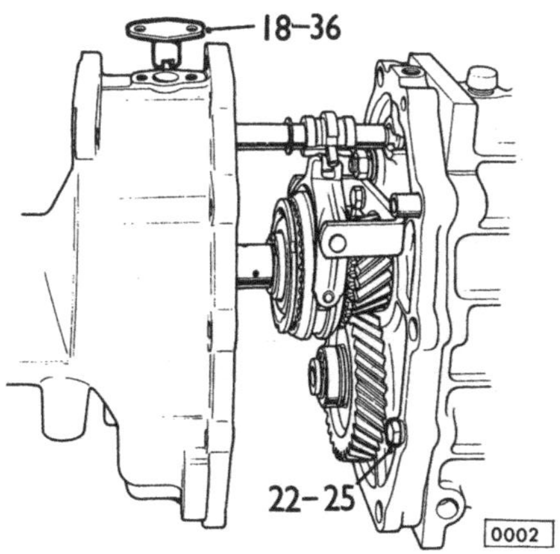

19 Place a container under the gearbox centre plate/rear cover to catch residual oil when the rear cover is disturbed.
20 Remove the bolts and spring washers securing the rear cover to the gearbox.
21 Withdraw the rear cover and gasket ensuring that the centre plate is not disturbed.
22 Fit temporary slave bolts to retain the centre plate in position.
23 Remove the oil pump drive shaft.

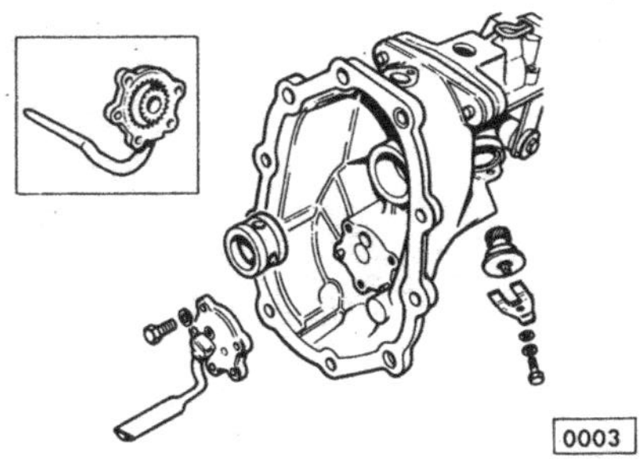

24 Remove the rear oil seal, bearing, spacer, ring and speedometer driving gear from the rear cover.

Refitting
25 Remove the slave bolts from the gearbox centre plate.
26 Ensure that the centre plate and rear cover mating faces are clean and fit a new gasket to the centre plate.

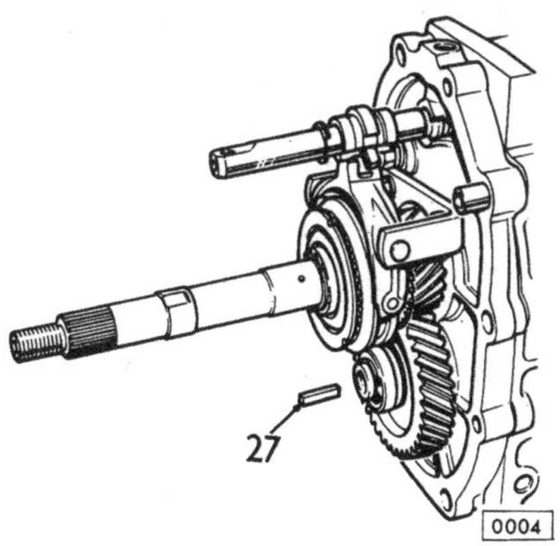

27 Engage the oil pump drive shaft in the layshaft.

28 Note the radial relationship of the square oil pump drive and align the oil pump gear centre.
29 Offer up the rear cover to the gearbox mainshaft and slide carefully into position. Ensure that the oil pump shaft engages the oil pump.
30 Fit and tighten the rear cover securing bolts.

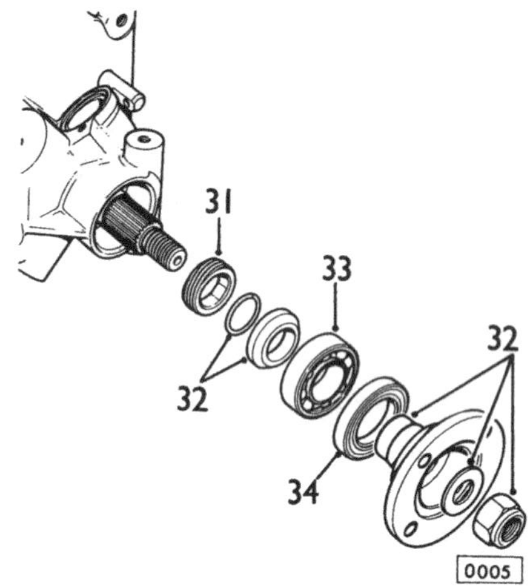

31 Fit the speedometer driving gear ensuring that it properly engages the flats on the mainshaft and that the tapered head of the gears is to the rear.
32 Fit the circlip and the spacer, (machined recess in spacer towards the circlip).
33 Fit the rear bearing.
34 Fit a new oilseal. Lubricate the seal lip.
35 Lubricate the seal contact area on the driving flange and fit the flange, washer and nut.
36 Fit the 5th gear spool locating boss.
37 Fit the pin, washer and nut securing the remote control linkage to the gearbox selector rod. Connect the reverse switch lead.
38 Raise the jack supporting the engine and bolt the gearbox crossmember to the body.
39 Remove the engine jack.
40 Connect the propeller shaft to the gearbox.
41 Fit the propeller shaft safety strap.
42 Connect up the exhaust system.
43 Fit and tighten the gearbox drain plug.
44 Refill the gearbox with fresh oil.
45 Lower the car.
46 Fit the gear lever.
47 Connect the battery.

OIL PUMP

Remove and refit 37.12.47

Removing
1 Remove the rear cover, see 37.12.42.
2 Remove the four bolts and spring washers securing the oil pump cover to the gearbox rear cover.
3 Withdraw the oil pump cover and oil intake pipe.
4 Remove the internal and external toothed gears.

Refitting
5 Reverse instructions 1 to 4.

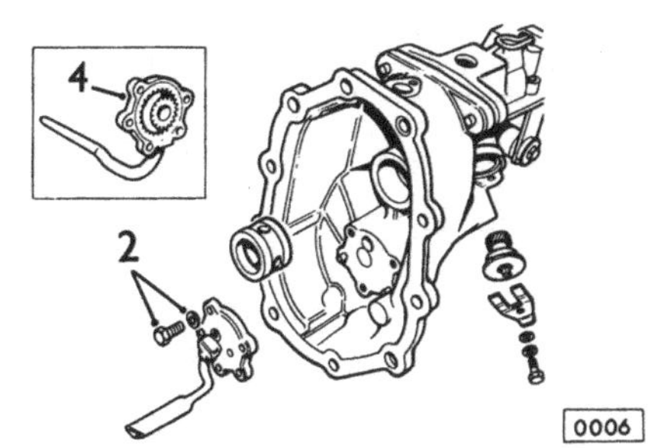

OIL PUMP

Test 37.12.50

1 Ensure that the gearbox oil level is correct.
2 Start the engine and allow to idle.
3 Remove the threaded plug in the rear cover, a steady flow of oil should be expelled.
4 Switch off the engine.
5 Prime the plug with 'Locquic Primer Grade T' then applying 'Loctite 270' immediately prior to assembly, refit the threaded plug.
6 Check and top up the gearbox level.

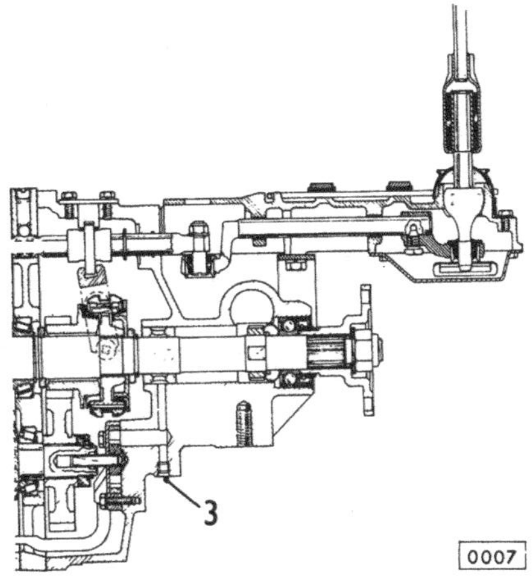

GEAR CHANGE LEVER

Remove and refit 37.16.04

Removing
1 Select neutral and unscrew and remove the gear lever knob.
2 Slacken the two screws securing the gear lever gaiter and top panel assembly to the rear of the console.
3 Remove the gear lever gaiter and panel assembly.
4 Remove the four screws retaining the gear lever draught excluder and flange assembly.
5 Withdraw the draught excluder and flange assembly.
6 Remove the dome cover securing the gear lever to the correct extension housing.
7 Remove the countersunk screw and bolt securing the bias spring rear bridge, and withdraw the bridge and liner.
8 Carefully prise the bias spring legs clear of the gear lever pins.
9 Gently lift out the gear lever taking care not to lose the nylon plunger and spring (anti-rattle) from the gear lever pivot ball.

Refitting
10 Reverse instructions 1 to 9.

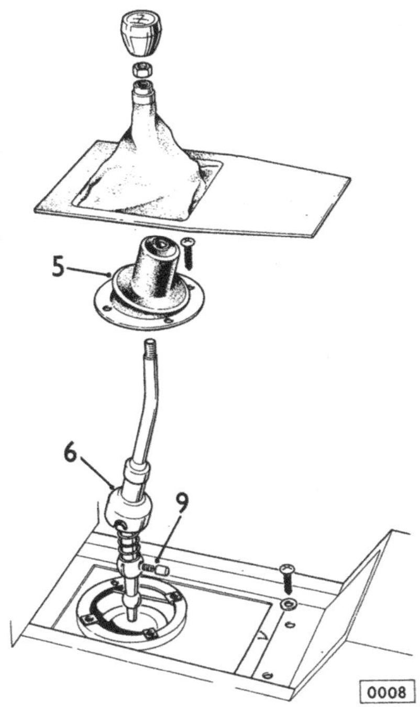

GEAR CHANGE REMOTE CONTROL ASSEMBLY

Remove and refit 37.16.19

Removing
1 Disconnect the battery.
2 Remove the gear lever, see 37.16.04/55.
3 Raise the car on a ramp, or jack up the rear of the vehicle and support securely.
4 Remove one bolt from the propeller shaft safety strap and slacken the remaining bolt and swing the strap aside.
5 Disconnect the propeller shaft from the gearbox.
6 Tie the propeller shaft to the vehicle in a position where it allows access to the gearbox.
7 Release the rubber rings securing the exhaust system.
8 Detach the exhaust system from the manifold down pipe.
9 Locate a jack under the gearbox and support weight.
10 Disconnect the two reverse light switch leads.
11 Disconnect the speedometer cable at the gearbox.
12 Remove the four bolts securing the gearbox rear mounting bracket.
13 Carefully lower the jack and gearbox sufficient to obtain access to the remote control assembly.
14 Disconnect the nut, washer and pin securing the gearbox selector shaft to the remote control shaft.
15 Remove the four bolts, spring and plain washers securing the remote control assembly to the gearbox rear cover.
16 Withdraw the remote control assembly.

Refitting
17 Reverse instructions 1 to 16.

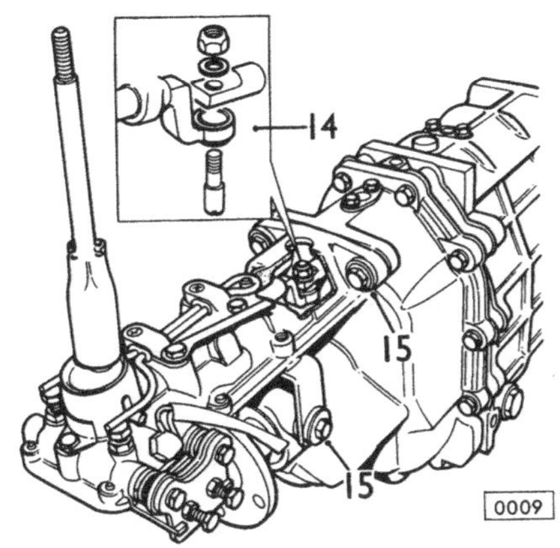

GEAR CHANGE REMOTE CONTROL ASSEMBLY

Overhaul **37.16.20**

Dismantling

1 Remove the remote control assembly from the gearbox.

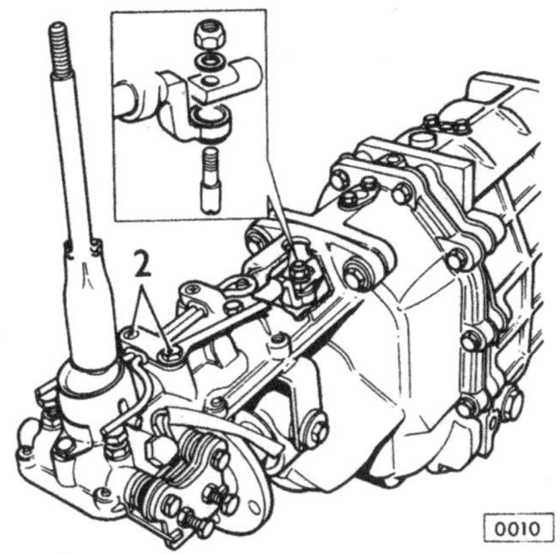

2 Remove the two bolts and two countersunk screws securing the bias spring bridge plates.
3 Remove the two bridge plates, bridge plate liners and the bias spring.
4 Remove the bias spring adjusting bolts and locknuts.
5 Remove the two bolts and washers securing the reverse baulk plate assembly and withdraw the reverse baulk plate, springs and spacers.
6 Remove the four bolts and washers securing the bottom cover plate.
7 Remove the bottom cover plate.
8 Remove the reverse light switch and locknut.
9 Remove the square-headed pinchbolt securing the selector shaft elbow.
10 Remove the selector shaft elbow.
11 Withdraw the selector shaft.
12 Press out the two selector shaft bushes in the remote control casing.
13 Remove the circlips securing the pivot ball and bushes in the selector shaft elbows and press out the bushes and pivot balls.

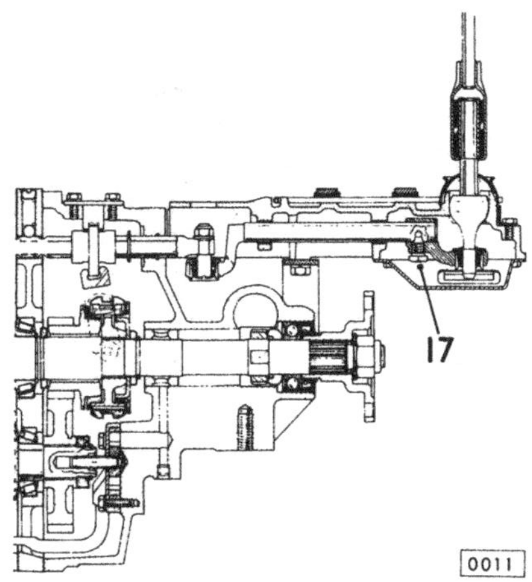

Assembling

14 Press in new selector shaft bushes in the remote control casing.
15 Fit new bushes, pivot balls and circlips to the selector shaft elbow.
16 Fit the selector shaft to the casing.
17 Fit the rear elbow and secure with the square-headed pinchbolt.
18 Fit the baulk plate assembly.
19 Fit the reverse switch and locknut.
20 Fit the bottom cover plate.
21 Fit the bias spring adjusting bolts and locknuts.
22 Fit the bias spring, bridge plate liners and bridge plates.
23 Fit the remote control assembly to the gearbox.
Operations to be carried out following the fitting of the remote control assembly to the gearbox.
24 Fit the gear lever.

Adjusting the reverse baulk plate

Adjustment of the reverse baulk plate must be carried out on a complete gearbox assembly.

25 Remove the bottom cover plate of the gear lever remote control assembly.
26 Locate the gear lever in neutral in a vertical position.
27 Slacken the baulk plate adjusting bolts and locknuts until the baulk plate is in contact with the backing plate.
28 Tighten the adjusting bolts *equally* until they just start to move the baulk plate out of contact with the backing plate.
29 Adjustment should be such that an effort of 30 to 35 lbf (13.6 to 15.9 kgf) is required to overcome the resistance of the baulk plate. This may be checked using a spring balance attached to the threaded end of the gear lever (gear knob end). Adjust by tightening or slackening the two adjacent bolts and locknuts located on the right-hand side of the reverse baulk plate.
NOTE: A minimum clearance of 0.010 in (0.254 mm) must exist between the upper face of the baulk plate and the lower edge of the gear lever bush (b).

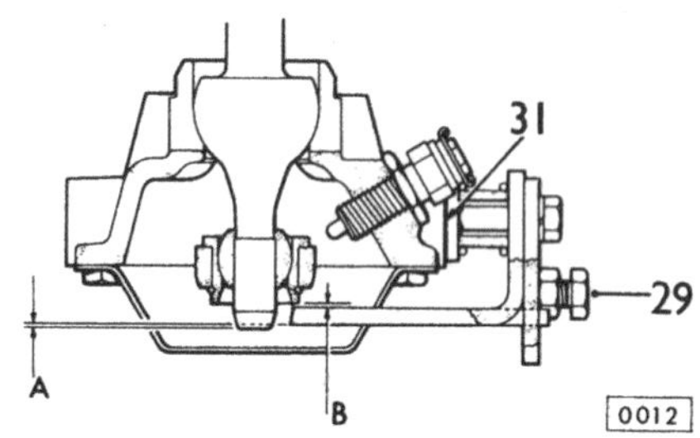

Adjustment of 1st/2nd gate stop

This operation must be carried out following the adjustment of the reverse baulk plate.

30 Engage 1st gear.
31 Check the clearance between the side of the gear lever and the edge of the baulk plate. This should be 0.010 to 0.049 in (0.25 to 1.25 mm). Adjust by adding or removing shims as necessary.
32 Engage 2nd gear and repeat the above operation.
33 Fit bottom cover plate.

Adjustment of gear lever bias spring

34 Unit completely assembled, engage 3rd gear.
35 Adjust the screws to position both legs of spring 0.5 mm clear of lever crosspin.
36 Apply a light load to gear lever in LH direction taking up play. Adjust RH screw downward until RH spring leg just makes contact with crosspin.
37 Repeat instruction 36 on the other side. Play will still be present but at extremes of gear lever travel the crosspin should make contact with the spring legs.
38 Return lever to neutral and rock across gate several times. Lever should return to 3rd/4th gate.
39 Tighten the locknuts.
40 Adjust the reverse switch, see 86.65.20.

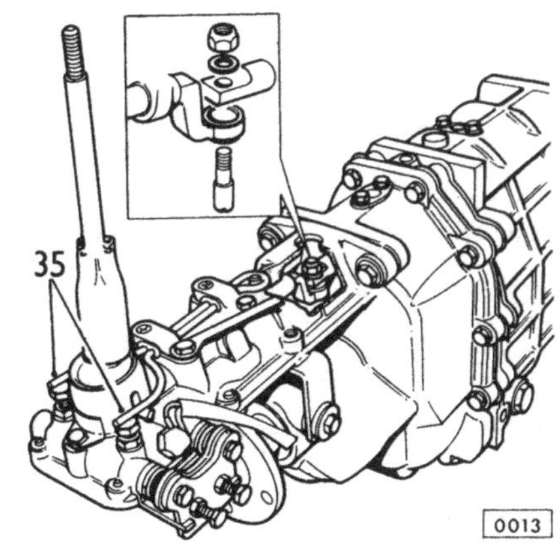

GEARBOX

Remove and refit 37.20.01

Removing

1 Disconnect the battery.
2 Remove the gear lever, see 37.16.04/55.
3 Raise the car on a ramp, or jack up and support securely.
4 Remove the propeller shaft safety strap.
5 Disconnect the propeller shaft at the gearbox flange and tie the propeller shaft so that it will not obstruct gearbox withdrawal.
6 Disconnect the electrical connections at the reverse switch.
7 Release the speedometer cable at the gearbox.
8 Remove the exhaust down pipe.
9 Withdraw the starter motor heat shield. (Spring clips to solenoid.)
10 Position a jack under the engine sump and support the engine. To avoid damage to the sump a suitable piece of wood should be interposed between the jack pad and the sump.
11 Remove the two bolts securing the clutch slave cylinder to the clutch housing and withdraw the slave cylinder. Do not allow the weight of the cylinder to hang suspended on the fluid hose.
12 Remove the two bolts securing the engine sump stiffening plate to the clutch housing.
13 Remove the four nuts securing the gearbox rear crossmember to the body.
14 Lower the jack located under the engine sump ensuring that radiator hoses are not strained. Ensure also that the gear lever remote control housing does not foul the brake pipe.
15 Disconnect the wires to the starter motor and remove the starter motor.
16 Remove the bolts and nuts securing the clutch housing to the engine. Note the location of the dowel bolt adjacent to the lower bolt of the clutch slave cylinder mounting flange and the positions of the clips for the electrical harness.

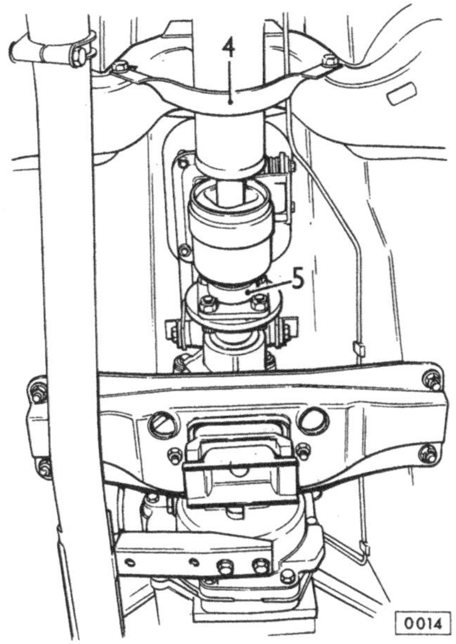

0014

0015

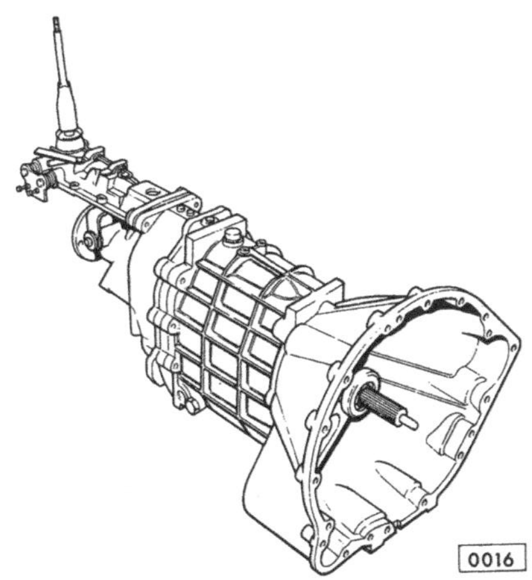
0016

17 Carefully withdraw the gearbox and clutch housing assembly ensuring that weight and stress are not imparted to the clutch driving plate. Mention of the fact that the dry weight of the gearbox and clutch housing assembly is approximately 60.5 lb (27.5 kg) provides some idea of the effort required to handle this unit.

Refitting

18 Reverse instructions 1 to 17.

GEARBOX ASSEMBLY

Overhaul **37.20.04**

Service tools: 47, 284, 18G 705-1, RG 421 or 18G 1205, RTR 47-23, LC 370-2, 18G 705 or S323, 18G 284 AAH, ST 1136

Dismantling

1 Place the gearbox on a bench or a gearbox stand ensuring that the oil is first drained.

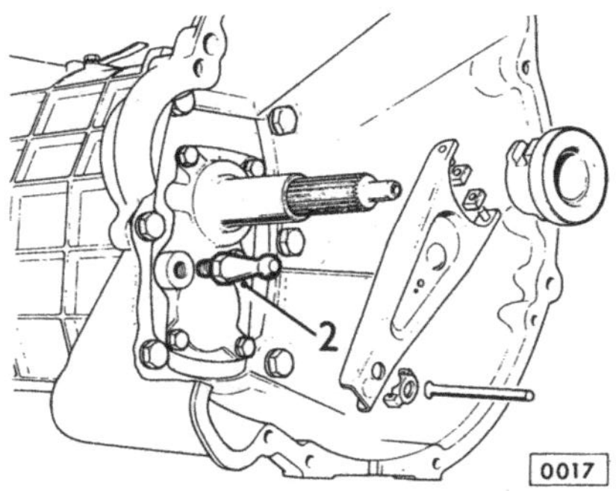

2 Using tool ST 1136 unscrew the clutch release lever pivot bolt and withdraw the clutch release lever complete with pivot bolt and release bearing slippers.
3 Detach the release bearing and slippers.
4 Remove the six bolts, plain and spring washers securing the bell housing to the gearcase and remove the bell housing.
5 Remove the nut and connecting pin linking the selector shaft to the remote control shaft.
6 Remove the four bolts, spring and plain washers (2 top, 1 either side) securing the remote control housing to the gearcase rear cover.
7 Remove the nut and plain washer securing the drive flange to the mainshaft. Use tool RG 421 or 18G 1205 to prevent shaft rotation.
8 Withdraw the driving flange.

9 Remove the speedometer driven gear and housing.
10 Remove the two bolts and spring washers securing the locating boss for the selector rear spool and withdraw the locating boss.
11 Remove the ten bolts, spring and plain washers securing the rear cover to the gearcase and withdraw the rear cover and gasket.

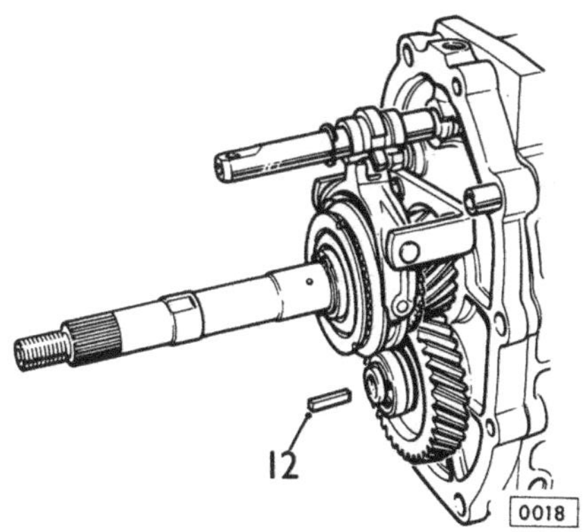

12 Withdraw the oil pump drive shaft.
13 Remove the fifth gear selector fork and bracket (two bolts and spring washers).

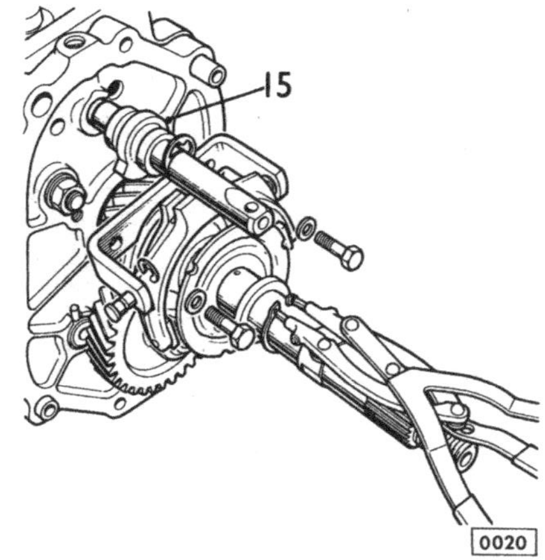

14 Remove the circlip from the selector shaft.
15 Withdraw the fifth gear selector spool. Note that the longer cam of the spool is fitted towards the bottom of the gearbox.
16 Remove the circlip retaining the fifth gear synchro assembly to the mainshaft.

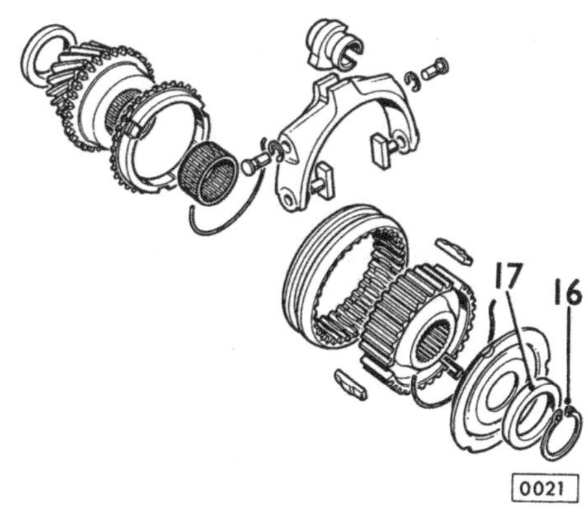

17 Withdraw the synchro assembly, fifth gear (driven) and spacer from the mainshaft.

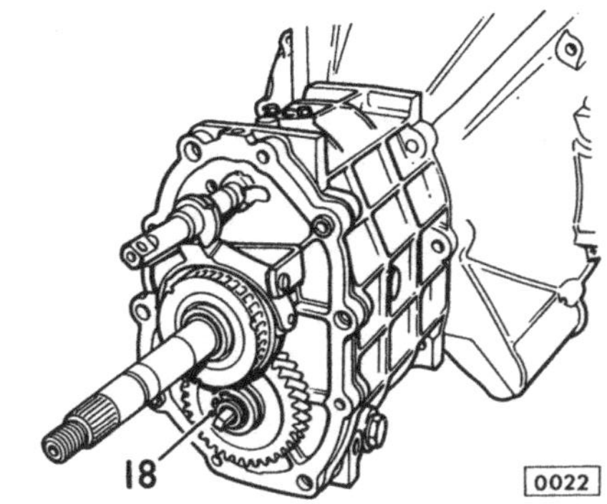

18 Remove the circlip retaining the fifth gear (driving) from the layshaft.
19 Using tool 18G 705 and adaptors 18G 705-1 remove the fifth gear and spacer from the layshaft.

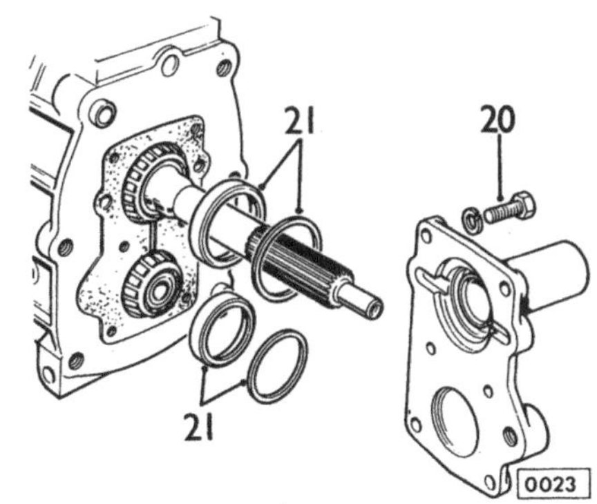

20 Remove the front cover and gasket (six bolts and spring washers).
21 Remove the input shaft selective washer, bearing track, layshaft selective washer and bearing track from the gearcase.
22 Remove the two bolts and spring washers securing the locating boss for the selector shaft front spool and withdraw the locating boss.
23 Remove the selector plug, spring and ball from centre plate.

24 Supporting the gearbox on the centre plate withdraw the gearcase.
25 Remove the input shaft and 1st gear synchro cone.
26 Withdraw the layshaft cluster.

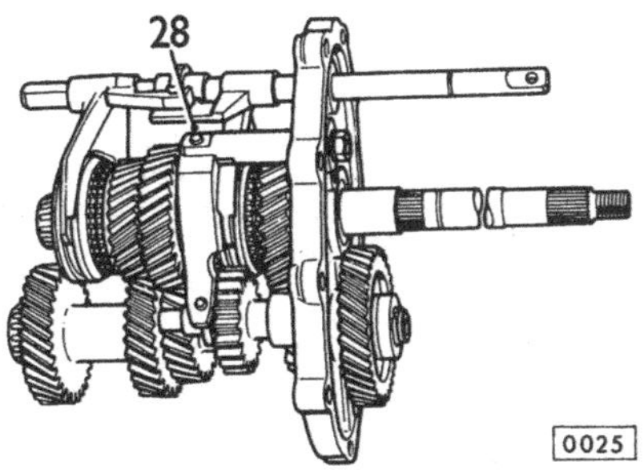

27 Support the centre plate complete with gears in protected vice jaws.
28 Remove the reverse lever pivot pin circlip and pivot pin.

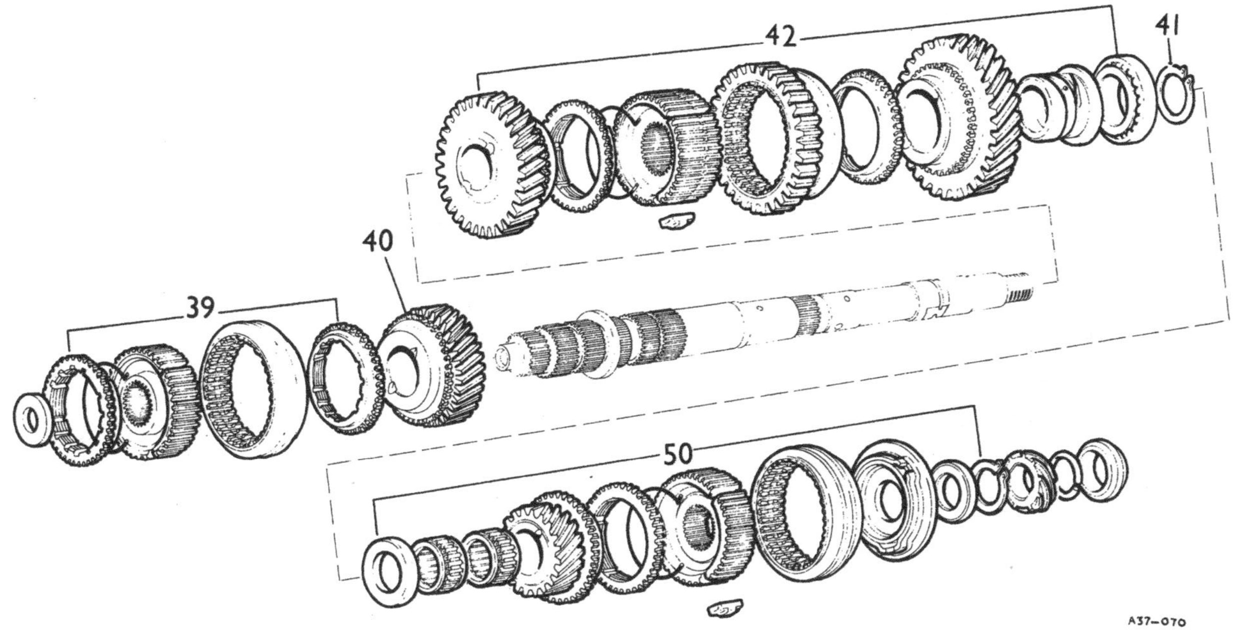

29 Remove the reverse lever and slipper pad.
30 Slide the reverse shaft rearwards and withdraw the reverse gear spacer, mainshaft, selector shaft, selector shaft fork and spool in a forward direction clear of the centre plate.
31 Withdraw the selector fork and spool. Note that the shorter cam of the spool is fitted towards the bottom of the gearbox.
32 Remove the nut and spring washers securing the reverse gear pivot shaft and remove the pivot shaft (only if renewal of the pivot shaft and/or the centre plate is intended).
33 Remove the centre plate from the vice and extract the two dowels (only if dowels and/or centre plate renewal is intended).

Input Shaft and Front Cover
34 Using tool RTR 47-23 remove the external bearing.
35 Using tools 18G 284 AAH and 284 withdraw the internal bearing track.
36 Remove the oil seal from the front cover.

Layshaft
37 Using tools LC 370-2 remove the layshaft bearings.

Mainshaft
38 Remove the pilot bearing and spacer.
39 Remove the 3rd and 4th speed synchroniser hub and sleeve.
40 Remove the 3rd speed gear.
41 Remove the circlip securing the mainshaft bearing.
42 Remove the bearing, 1st gear and bush, 1st and 2nd speed hub, sleeve and synchromesh cones, and 2nd gear.

Rear Cover
43 Remove the oil seal, bearing, speedo gear, circlip and sleeve and oil sleeve. Remove the oil pump drive, pump cover and gears.
44 Thoroughly clean and examine all components. Obtain new parts as necessary.

continued

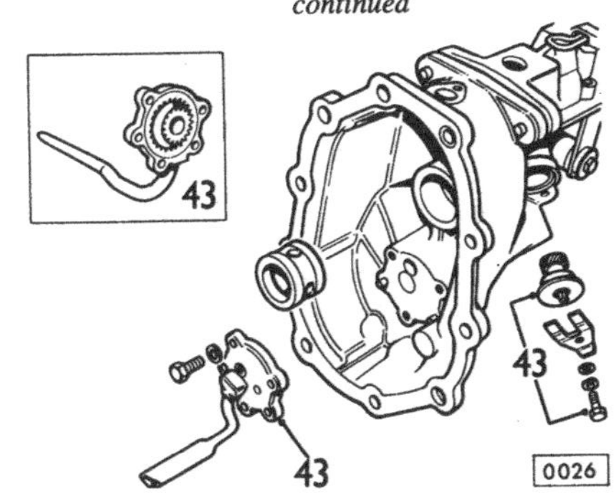

Layshaft

45 Fit the bearings to the layshaft.

Mainshaft

46 *Synchro Assemblies.* With the outer sleeve held, a push-through load applied to the outer face of the synchro hub should register 18 – 22 lb (18.2 – 10 kgm) to overcome spring detent in either direction.

47 *Checking 1st speed bush end-float.* Fit 2nd gear, 1st/2nd speed synchro hub and 1st gear bush to the mainshaft.

Manufacture a spacer to the dimensions illustrated and slide the spacer on the mainshaft.

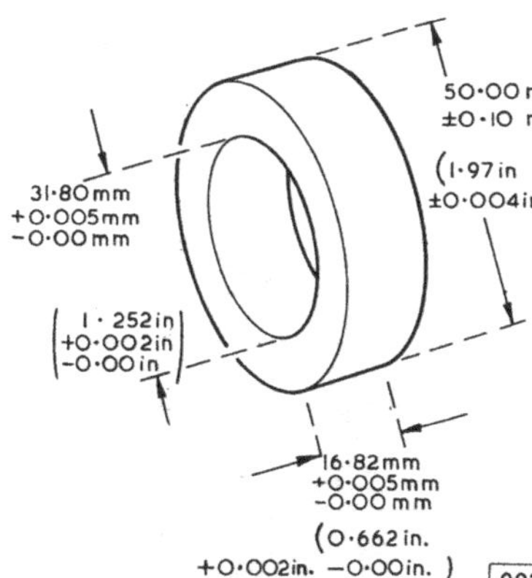

0027

Using an oil circlip and feeler gauges check the clearance existing between the spacer and the circlip, which should be within 0.005 to 0.055 mm (0.0002 to 0.002 in). The first speed bush is available with collars of different thickness. Select a 1st speed bush with a collar which will give the required end-float.

48 Remove the circlip, spacer, bush, synchro hub and 2nd gear from the mainshaft.

49 *Checking 5th gear end-float.* Fit the 5th gear assembly to the mainshaft, i.e. front spacer, 5th gear, synchro hub, rear plate and spacer. Fit an old circlip and using feeler gauges check the end-float which should be within 0.005 to 0.055 mm (0.0002 to 0.002 in). The rear spacer is available in a range of sizes. Select a rear spacer which will ensure the required clearance.

50 Remove the circlip spacer and 5th gear assembly.

Assembly

51 It is important that 1st/2nd synchro is assembled correctly (short splines on inner member) towards 2nd gear. Fit 2nd gear, baulk ring, synchro hub and sleeve (selector fork annulus to rear of gearbox), baulk ring, 1st gear and selective bush, bearing and a *new* circlip. When fitting the circlip care must be taken to ensure that it is not stretched (opened) beyond the *minimum* necessary to obtain entry over the shaft. The internal diameter of an expanded circlip *must not* exceed 32.30 mm.

52 Fit 3rd gear, baulk ring, and synchro hub and sleeve (longer boss of synchro hub to front of gearbox) to the mainshaft.

53 Fit the spacer and bearing to front of mainshaft.

54 Fit the layshaft bearing track to the centre plate.

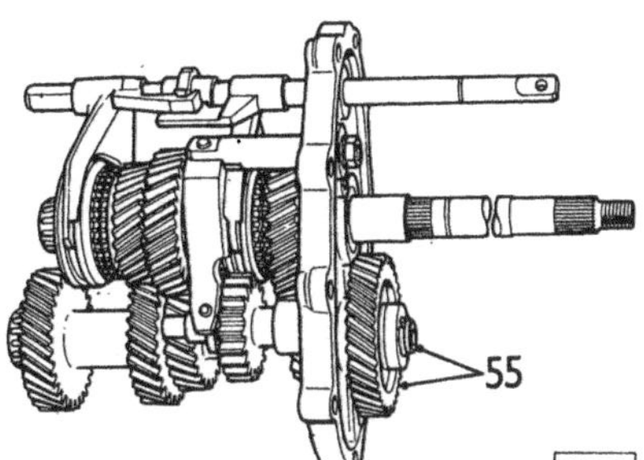

0028

55 Fit the layshaft to the centre plate and fit the fifth gear, spacer and a new circlip. When fitting the circlip care must be taken to ensure that it is not stretched (opened) beyond the *minimum* necessary to obtain entry. The internal diameter of an expanded circlip *must not* exceed 22.5 mm.

56 Fit the mainshaft bearing track to the centre plate.

57 Locate the centre plate in protected vice jaws.

58 Take the selector shaft complete with 1st and 2nd selector fork, front spool and 3rd and 4th selector fork and engage both forks in their respective synchro sleeves on the mainshaft. Simultaneously engage the selector shaft and mainshaft assemblies in the centre plate.

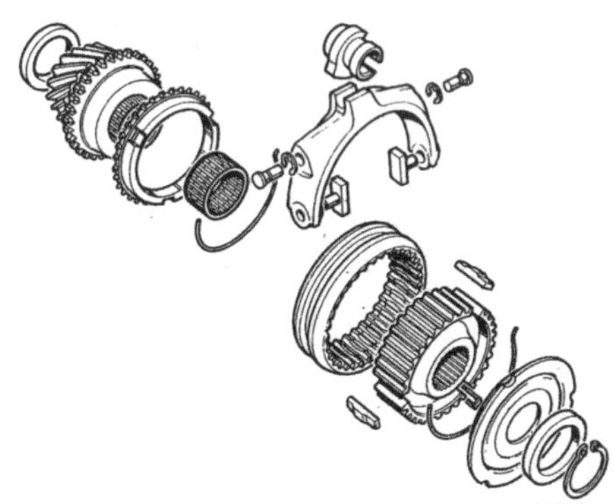

0029

59 Fit the spacer, 5th gear, baulk ring, synchro hub and sleeve end-plate, selective spacer (instruction 48), and a new circlip. When fitting the circlip care must be taken to ensure that it is not stretched (opened) beyond the *minimum* necessary to obtain entry. The internal diameter of an expanded circlip *must not* exceed 27.63 mm.

60 Fit the reverse gear (lip for slipper pad to front of box) front and rear spacers and the reverse shaft.

61 Fit the reverse lever, slipper pad, pivot pin and circlip. If a new reverse gear pivot shaft is to be fitted it is necessary to ensure that its radial location is consistent with reverse pad slipper engagement/clearance. Radial location is determined on assembly. Secure with spring washer and nuts, subsequently checking movement of reverse lever and ensuring slipper pad is properly engaged.

62 Remove the centre plate and gear assembly from the vice and locate on a suitable stand with the front of the mainshaft vertically uppermost. Ensure that the reverse shaft does not slide out of position.

63 Fit the centre plate front gasket.

64 Fit the external bearing and internal bearing track to the input shaft.

65 Fit the input shaft to the gearcase.

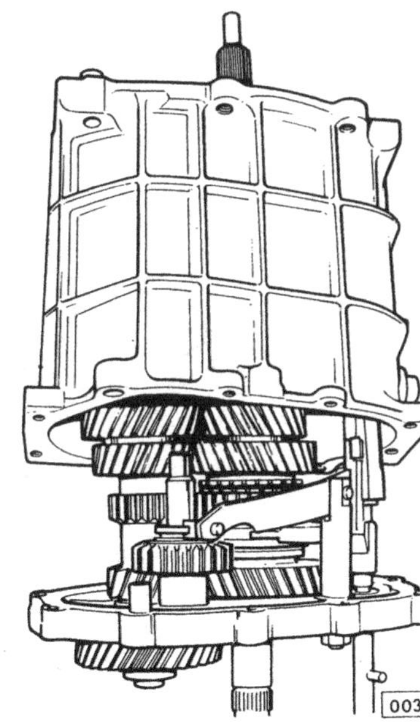

0030

66 Carefully slide the gearcase and input shaft into position over the gear assemblies. Do not use force. Ensure that the centre plate dowels and selector shaft are engaged in their respective locations.

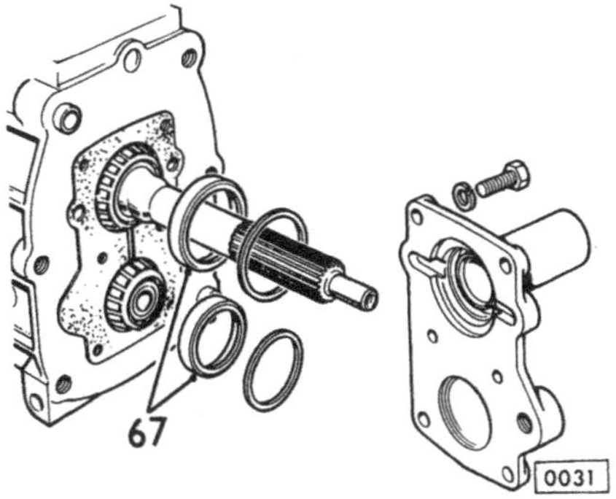

67 Fit the layshaft and input shaft bearing outer tracks.
68 Using seven slave bolts and plain washers to prevent damaging the rear face of the centre plate evenly draw the gearcase into position on the centre plate.

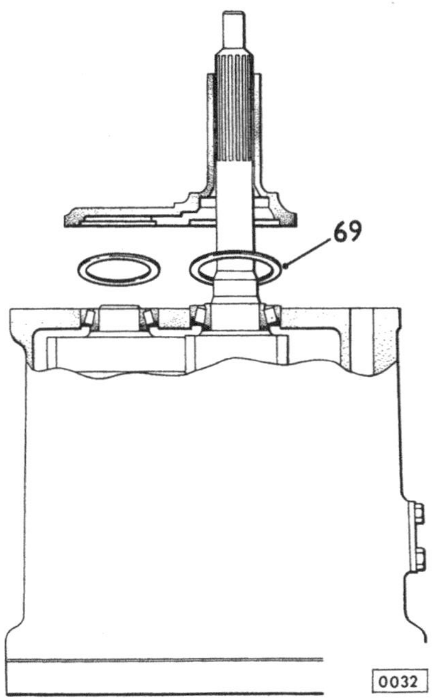

69 Place a layshaft spacer of nominal thickness (0.040 in, 1.02 mm) on the layshaft bearing track, and fit the front cover and gasket, securing with six bolts.
70 Using a dial gauge check layshaft end-float.
71 Remove the front cover and provisional spacer. The required layshaft end-float is 0.005 to 0.055 mm (0.0002 to 0.002 in). Check the thickness of the provisional spacer. Spacer thickness required is: provisional spacer thickness, plus end-float obtained, minus 0.055 mm (0.002 in).
72 Again fit the front cover and gasket, this time with the correct spacer arrived at in instruction 71.
73 Check layshaft end-float to ensure it is within the limits specified in instruction 71.
74 Place a ball bearing in the centre of the input shaft. This facilitates checking mainshaft end-float using a dial gauge.
75 Mount the dial gauge on the gearcase with the stylus resting on the ball. Zero the gauge.
76 Check the mainshaft and input shaft combined end-float. Care must be taken when checking dial gauge readings to ensure that end-float only – as distinct from side movement of the input shaft – is recorded. If difficulty is encountered in differentiating between end-float and side movement remove the front cover and wrap the plain portion of the input shaft below the splines with six turns of masking tape.
Refit the front cover and again check end-float ensuring that rise and fall of the input shaft is not restricted by the tape.
77 Having ascertained end-float select the spacer required as follows:
End-float minus 0.055 mm (0.002 in) – spacer thickness required. Fit the spacer thus determined and again check end-float which must be within 0.005 to 0.055 mm (0.0002 to 0.002 in).
78 Remove the front cover and tape (if employed).
79 Fit the oil seal to the front cover and lubricate the seal lips.
80 Mask the splines and fit the front cover. Remove the splines masking.
81 Place the gearbox on a bench or stand and remove the slave bolts and washers from the centre plate.
82 Fit the 5th gear spool and circlip to the selector shaft.
NOTE: The longer cam of the spool is fitted towards the bottom of the gearbox.
83 Fit the 5th gear selector fork and bracket.
84 Renew the selector shaft 'O' ring in the rear cover and fit the oil ring bush.
85 Fit the rear gasket to the centre plate and engage the oil pump shaft in the layshaft.

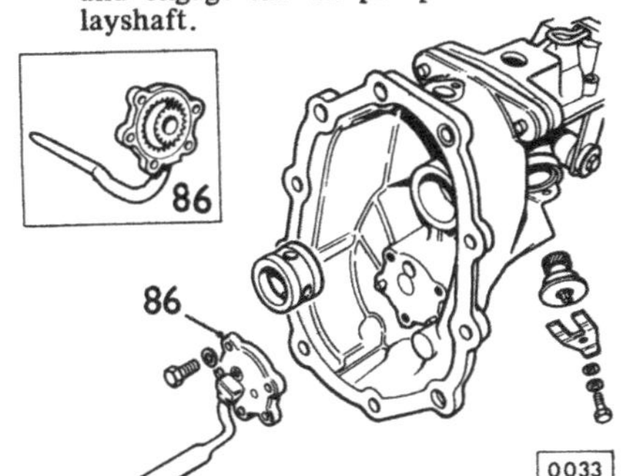

86 Fit the oil pump gears and cover to the gearbox rear cover.
87 Fit the rear cover ensuring that the oil pump shaft engages the oil pump.
88 Fit the selector shaft ball, spring and plug to the centre plate.
89 Fit the spool locating bosses (2) to the 1st/2nd spool and 5th gear spool.
90 Fit the speedometer driving gear to the mainshaft ensuring that it properly engages the mainshaft flats.
91 Fit the circlip and sleeve and ball race to the mainshaft.
92 Fit the rear oil seal. Lubricate seal lip.
93 Fit the driving flange, washer and nut.
94 Fit the speedometer driven gear and housing.
95 Fit the bell housing.
96 Fit the clutch release bearing and withdrawal lever.
97 Fit the remote control housing.

FIRST MOTION SHAFT OIL SEAL

Remove and refit **37.23.04**

Service tools: ST 1136

Removing

1 Remove the gearbox, see 37.20.01.

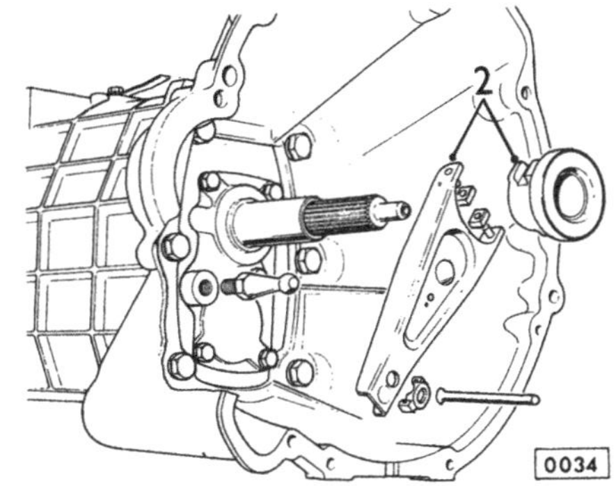

2 Using tool ST 1136 remove the clutch release fork and bearing.
3 Remove the bolts and washers securing the front cover to the gearbox.
4 Remove the front cover and gasket.

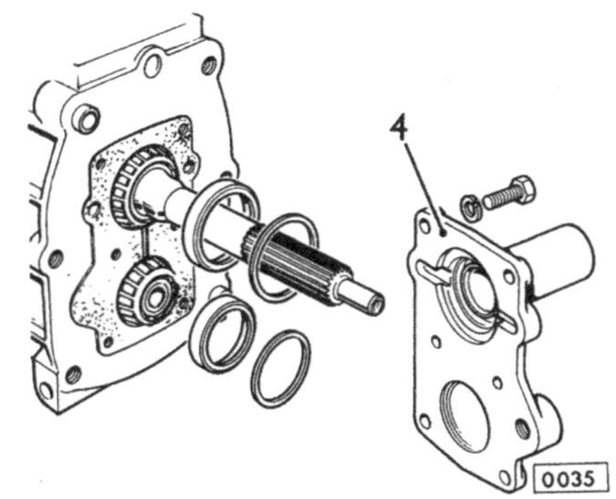

5 Remove the oil seal from the front cover. Ensure that the spacers for the first motion shaft and layshaft bearings are not intermixed.

continued

Refitting

6 Fit a new oil seal to the front cover (seal lip towards gearbox).
7 Lubricate the oil seal and fit the front cover and gasket.
8 Fit the clutch release fork and bearing.
9 Fit the gearbox to the car, see 37.20.01.

SPEEDOMETER DRIVE GEAR

Remove and refit **37.25.01**

As Operation 37.12.42.

SPEEDOMETER DRIVE PINION

Remove and refit **37.25.05**

Removing

1 Remove the bolt and washer securing the speedometer clamp plate.
2 Release the speedometer cable from the drive pinion.
3 Withdraw the pinion housing and pinion from the gearbox.
4 Remove the pinion from the housing.

Refitting

5 Reverse instructions 1 to 4. Renew the pinion housing 'O' ring if necessary.

AUTOMATIC GEARBOX OPERATIONS

Automatic Transmission

[**CAUTION:** Please read the instructions thoroughly before you attempt transmission repairs. If you lack the necessary skills or tools we suggest you leave this work to an Authorized Dealer or other qualified shop. Before taking a removed transmission in for repair, thoroughly clean its exterior but do not undertake any disassembly.]

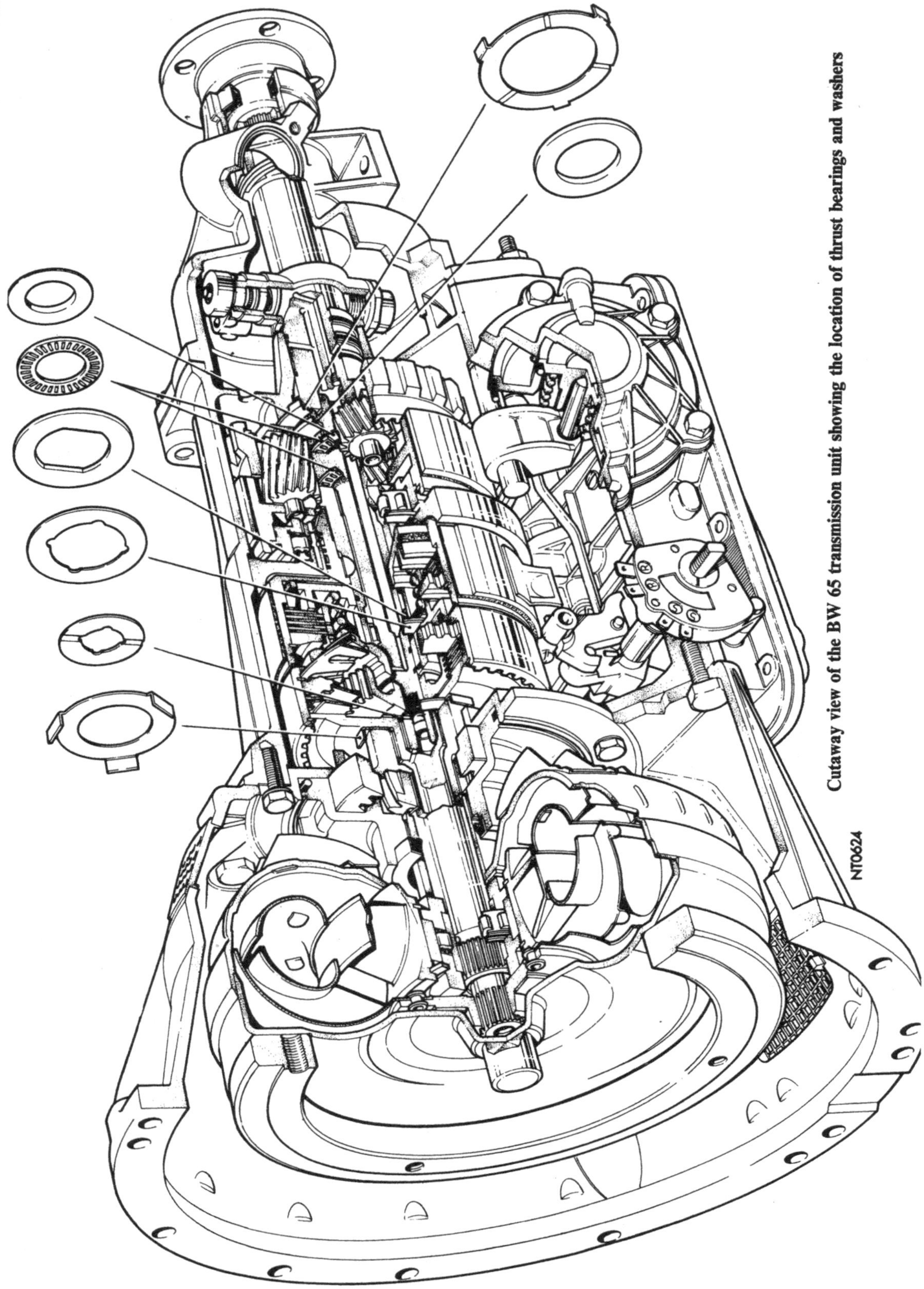

NT0624

Cutaway view of the BW 65 transmission unit showing the location of thrust bearings and washers

IMPORTANT

Under agreements existing between Borg-Warner Limited and the car manufacturers, the former does NOT undertake the servicing of automatic transmission units, nor do they supply spare parts or service tools. All matters appertaining to service or spares must therefore be dealt with by Triumph Distributors or Dealers within the organisation.

TYPE – Borg-Warner 65 Unit – P.R.N.D.2.1. System

UNIT IDENTIFICATION

A serial number prefix 027 appears on a dark admiralty grey nameplate on the left hand side of the transmission case.

AUTOMATIC TRANSMISSION – SHIFT SPEEDS

Throttle Position	Zero Throttle	Light Throttle		Part Throttle	KICK-DOWN						
Selector	–1	2	D	D	D	D	D	1	D	2	2
Shift	2–1	1–2	2–3	3–2	1–2	2–3	3–2	2–1	3–1	1–2	2–1
Road Speed M P H	30–38	8–12	12–16	30–40	38–44	65–71	58–66	42–50	30–38	38–44	30–38
KM/H	48–61	13–19	13–26	48–64	61–71	105–114	93–106	68–81	48–61	61–71	48–61

Capacity 5.3 litres (9½ Imperial pints) [11¼ U.S. pints] including cooler of 0.3 litres (½ Imperial pint).

TRANSMISSION DATA

	TOP (3rd)	Intermediate (2nd)	Low (1st)	Reverse
Gearbox ratios	1:1	1.45:1	2.39:1	2.09:1
Convertor reduction (1.91)	Infinitely variable between operating in all gears		1–1.91:1	
OVERALL RATIOS	3.27:1	4.74:1	7.83:1	6.85:1

217 K Convertor – dia 242 mm (9½") – RATIO 1.91:1

EXAMINATION OF COMPONENTS

Transmission Case and Servo Covers
Check for cracks and obstructions in passages.

Front and Rear Pump
Check for scoring and excessive wear.

Shafts
Check bearing and thrust faces for scoring.

Clutch Plates
Check for warping, scoring, overheating and excessive wear.

Bands
Check for scoring, overheating and excessive wear.

Drums
Check for overheating and scoring.

Gears
Check teeth for chipping, scoring, wear and condition of thrust faces.

Uni-directional Clutch and Races
Check for scoring, overheating and wear.

Valve Block and Governor
Check for burrs, crossed or stripped threads, and scored sealing faces.

Impeller Hub and Front Pump Drive Gear
Check for pitting and wear. Ensure good contact.

Thrust Washers
Check for burrs, scoring and wear.

White Metal Bushes
Check for scoring and loss of white metal.

Lip Seals
Check for cuts, hardening of rubber, leakage past outer diameter.

Rubber 'O' Rings and Seals
Check for hardening, cracking, cuts or damage.

Cast Iron Sealing Rings
Check fit in groove and wear (evident by lip overhanging the groove).

Teflon Sealing Rings
Check for cracking, cuts or damage.

SERVICING REQUIREMENTS

1 For all operations high standards of cleanliness are essential.
2 Rags and cloths must be clean and free from lint; nylon cloths are preferable.
3 Prior to assembly all components must be cleaned thoroughly with petrol, paraffin or an industrial solvent.
4 All defective items must be renewed.
5 Components should be lubricated with transmission fluid before assembly.
6 New joint washers should be fitted where applicable.
7 Where jointing compound is required, the use of Hylomar SQ32M, Hermetite or Wellseal is approved.
8 All screws, bolts and nuts must be tightened to the recommended torque figure.
9 Thrust washers and bearings should be coated with petroleum jelly to facilitate retaining them in position during assembly operations. Grease should not be used as it may be insoluble in the transmission fluid and could subsequently cause blockage of fluid passages and contamination of brake band and clutch facings.
IMPORTANT: Metric threads are used throughout most of the transmission unit and it is therefore essential that fastenings, and especially lock washers, are segregated into sets and not intermixed with those from other parts of the vehicle.

FRONT BRAKE BAND

Remove and refit 44.10.01

Service tools: CBW 60, CBW 547A-50

Removing

1 Remove the transmission unit. 44.20.01.
2 Wash the exterior of the unit in clean petrol or paraffin, invert it and place on a bench cradle CBW 60. Remove the switch. 44.15.15.
3 Unscrew the bolts securing the torque converter housing.
4 Remove the torque converter housing.
5 Unscrew 12 bolts.
6 Remove the oil pan, joint washer and magnet.

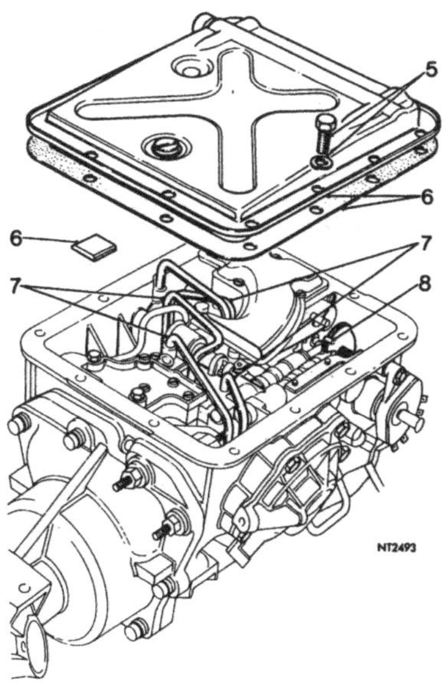

7 Pull out the five oil tubes.
8 Release the downshift inner cable from the downshift cam.
9 Take out three bolts and washers.
10 Lift off the valve block.
11 Unscrew two bolts.
12 Remove the oil tube locating plate.
13 Pull out the oil tubes. (Note the 'O' ring on the pump section tube.)
14 Take out five bolts.

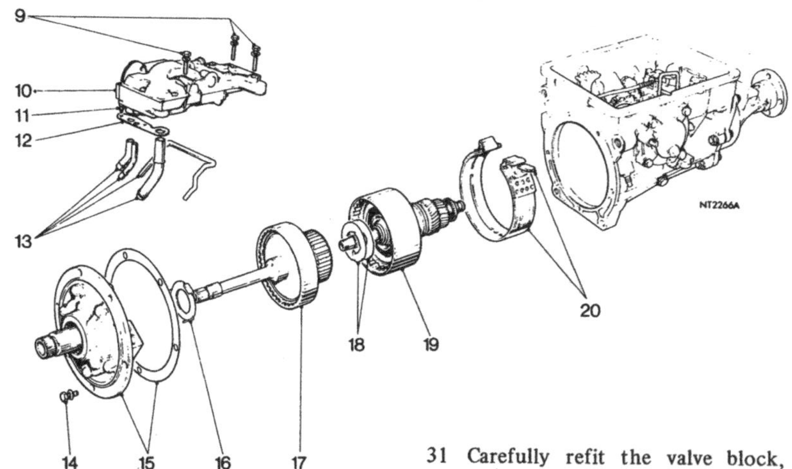

15 Remove the pump and joint washer.
16 Remove the thrust washer.
17 Withdraw the front clutch.
18 Remove the thrust washers.
19 Withdraw the rear clutch and forward sun gear.
20 Squeeze together the ends of the front brake band and remove it together with the strut.

Refitting

21 Squeeze together the ends of the front brake band and fit it in position together with the strut.
22 Refit the rear clutch and forward sun gear assembly.
23 Using petroleum jelly, stick the thrust washers to the rear clutch assembly (phosphor bronze towards the front clutch).
24 Refit the front clutch assembly.
25 Using petroleum jelly, stick the thrust washer to the pump assembly.
26 Refit the pump assembly and joint washer.
27 Fit and tighten the bolts.
28 Refit the oil tubes. (Note the 'O' ring on the pump suction tube.)
29 Refit the oil tube locating plate.
30 Fit and tighten the two bolts.
31 Carefully refit the valve block, ensuring that the oil tubes are not distorted.
32 Fit and tighten the three bolts and washers.
33 Connect the downshift inner cable to the downshift cam.
34 Refit the five oil tubes.
35 Replace the magnet and refit the oil pan and joint washer.
36 Fit and tighten 12 bolts.

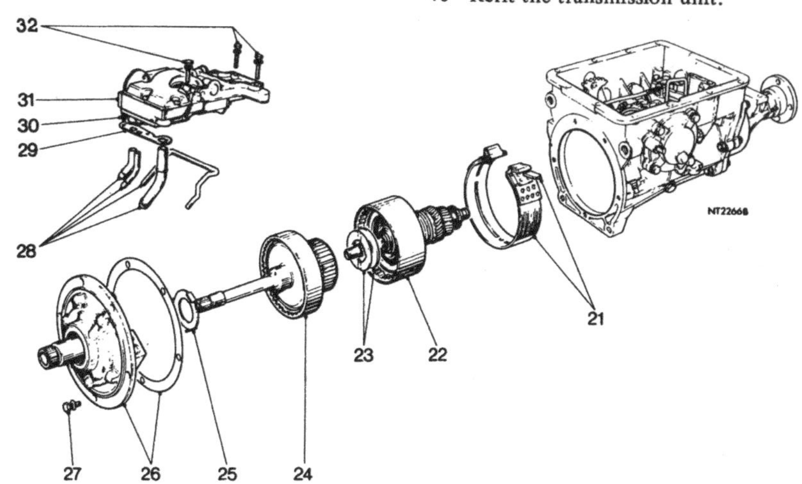

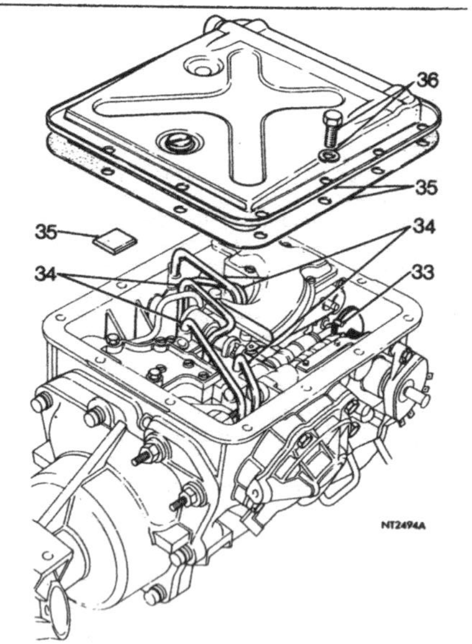

37 Locate the torque converter housing in place.
38 Fit and tighten four bolts securing the torque converter housing.
39 Refit the switch 44.15.15.
40 Refit the transmission unit.

REAR BRAKE BAND

Remove and refit 44.10.09

Service tools: CBW 60, CBW 547A-50

Removing

1 Remove the transmission unit. 44.20.01.
2 Wash the exterior of the unit in clean petrol or paraffin, invert it and place on a bench cradle CBW 60. Remove the switch. 44.15.15.
3 Unscrew the bolts securing the torque converter housing.
4 Remove the torque converter housing.
5 Unscrew 12 bolts.
6 Remove the oil pan, joint washer and magnet.
7 Pull out the oil tubes.
8 Release the downshift inner cable from the downshift cam.
9 Take out three bolts and washers.
10 Lift off the valve block.
11 Unscrew two bolts.
12 Remove the oil tube locating plate.
13 Pull out the oil tubes. (Note the 'O' ring on the pump suction tube.)
14 Take out five bolts.
15 Remove the pump and joint washer.
16 Remove the thrust washer.
17 Withdraw the front clutch.
18 Remove the thrust washers.
19 Withdraw the rear clutch and forward sun gear.
20 Squeeze together the ends of the front brake band and remove it together with the strut.
21 Unscrew the bolts.
22 Withdraw the centre support/planet gear assembly and thrust race.
23 Squeeze together the ends of the rear brake band, tilt and withdraw it from the casing together with the strut.

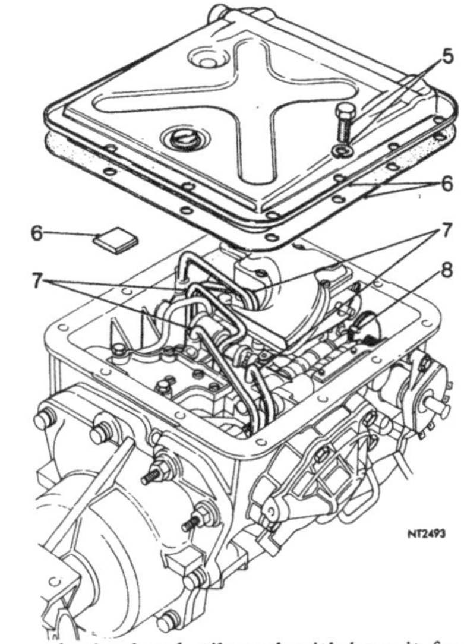

Refitting

24 Refit the rear brake band and strut.

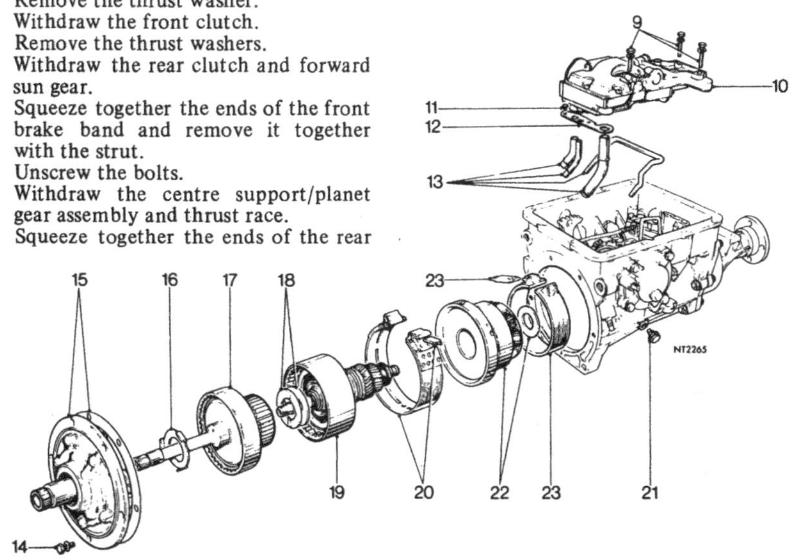

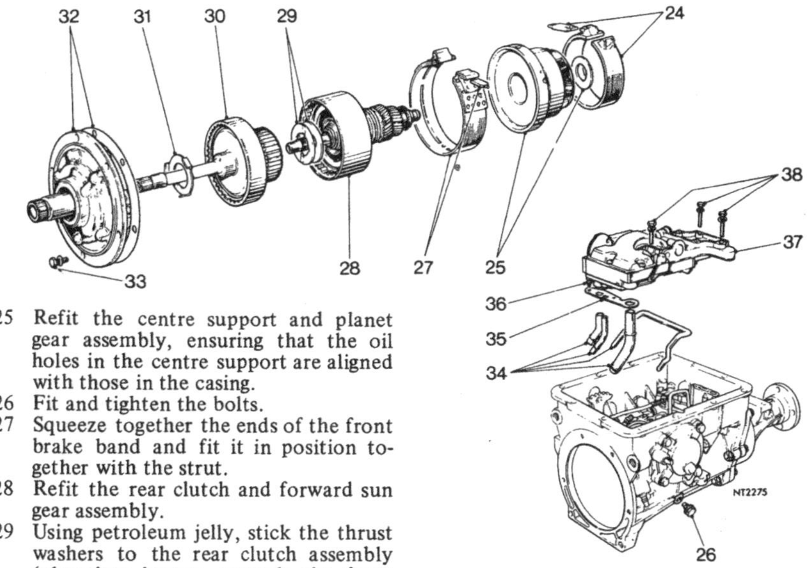

25 Refit the centre support and planet gear assembly, ensuring that the oil holes in the centre support are aligned with those in the casing.
26 Fit and tighten the bolts.
27 Squeeze together the ends of the front brake band and fit it in position together with the strut.
28 Refit the rear clutch and forward sun gear assembly.
29 Using petroleum jelly, stick the thrust washers to the rear clutch assembly (phosphor bronze towards the front clutch).
30 Refit the front clutch assembly.
31 Using petroleum jelly, stick the thrust washer to the pump assembly.
32 Refit the pump assembly and joint washer.
33 Fit and tighten the bolts.
34 Refit the oil tubes. (Note the 'O' ring on the pump suction tube.)
35 Refit the oil tube locating plate.
36 Fit and tighten the two bolts.
37 Carefully refit the valve block, ensuring that the oil tubes are not distorted.
38 Fit and tighten the three bolts and washers.
39 Connect the downshift inner cable to the downshift cam.
40 Refit the oil tubes.
41 Replace the magnet and refit the oil pan and joint washer.
42 Fit and tighten 12 bolts.
43 Locate the torque converter housing in place.
44 Fit and tighten four bolts securing the torque converter housing.
45 Refit the switch. 44.15.15.
46 Refit the transmission unit. 44.20.01.

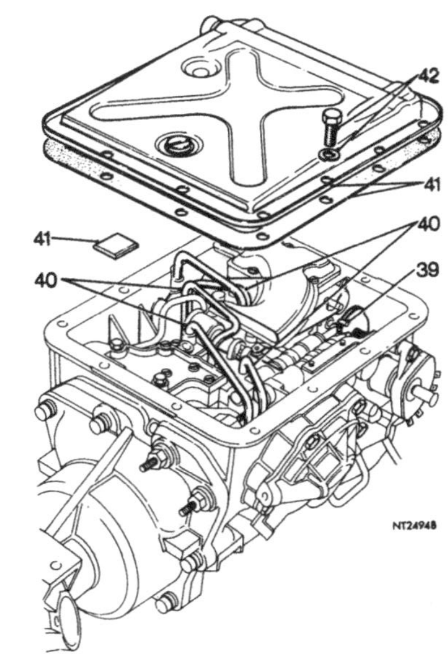

FRONT CLUTCH

Remove and refit 44.12.04

Service tools: CBW 60, CBW 547A-50

Removing

1 Remove the transmission unit. 44.20.01.
2 Wash the exterior of the unit in clean petrol or paraffin, invert it and place on a bench cradle CBW 60. Remove the switch. 44.15.15.
3 Unscrew the bolts securing the torque converter housing.
4 Remove the torque converter housing.
5 Unscrew 12 bolts.
6 Remove the oil pan, joint washer and magnet.
7 Pull out the oil tubes.
8 Release the downshift inner cable from the downshift cam.
9 Take out three bolts and washers.
10 Lift off the valve block.
11 Unscrew two bolts.
12 Remove the oil tube locating plate.
13 Pull out the oil tubes. (Note the 'O' ring on the pump suction tube.)
14 Take out five bolts.
15 Remove the pump joint washer.
16 Remove the thrust washer.
17 Withdraw the front clutch.
18 Remove the thrust washers.

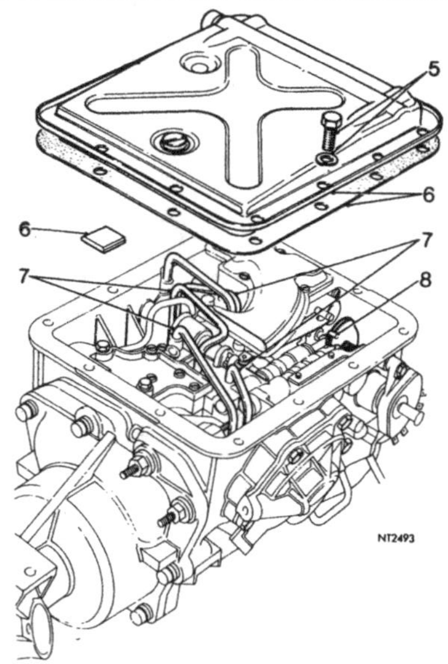

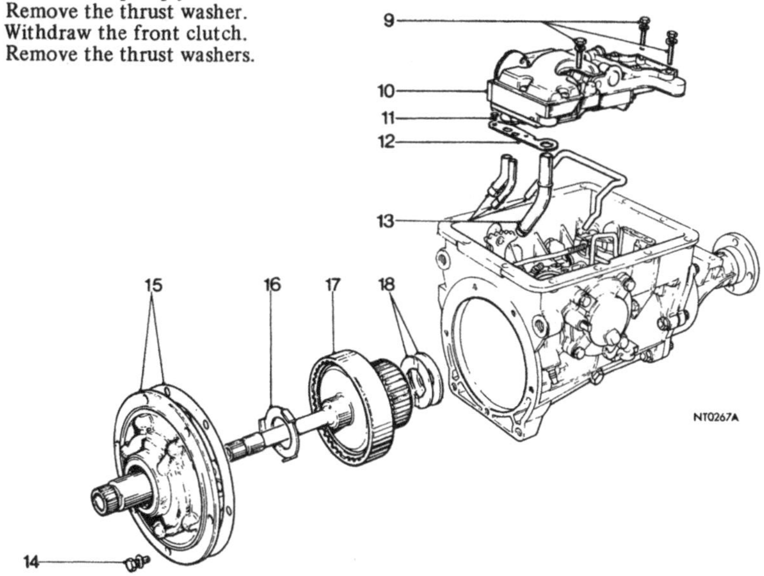

Refitting

19 Using petroleum jelly, stick the thrust washers to the rear clutch assembly (phosphor bronze towards the front clutch).
20 Refit the front clutch assembly.
21 Using petroleum jelly, stick the thrust washer to the pump assembly.
22 Refit the front assembly and joint washer.

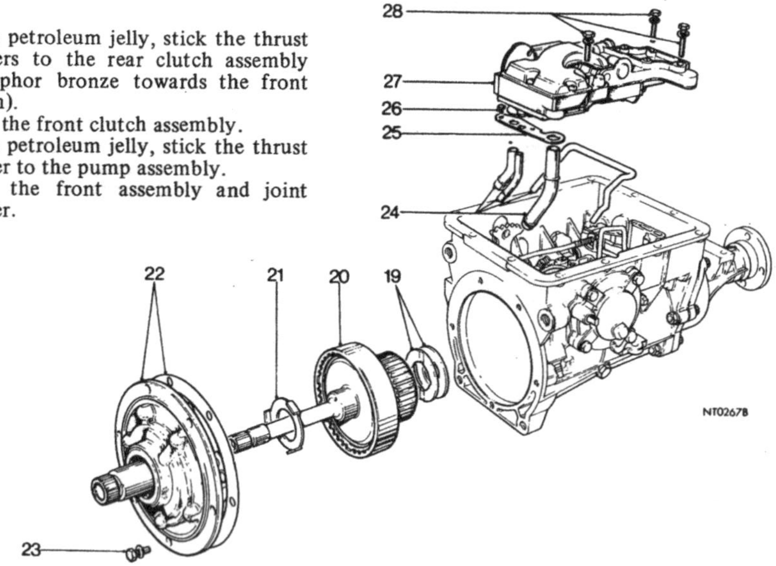

23 Fit and tighten the bolts.
24 Refit the oil tubes. (Note the 'O' ring on the pump suction tube.)
25 Refit the oil tube locating plate.
26 Fit and tighten the two bolts.
27 Carefully refit the valve block, ensuring that the oil tubes are not distorted.
28 Fit and tighten the three bolts and washers.
29 Connect the downshift inner cable to the downshift cam.
30 Refit the oil tubes.
31 Replace the magnet and refit the oil pan and joint washer.
32 Fit and tighten the 12 bolts.
33 Locate the torque converter housing in place.
34 Fit and tighten four bolts securing the torque converter housing.
35 Refit the switch. 44.15.15.
36 Refit the transmission unit. 44.20.01.

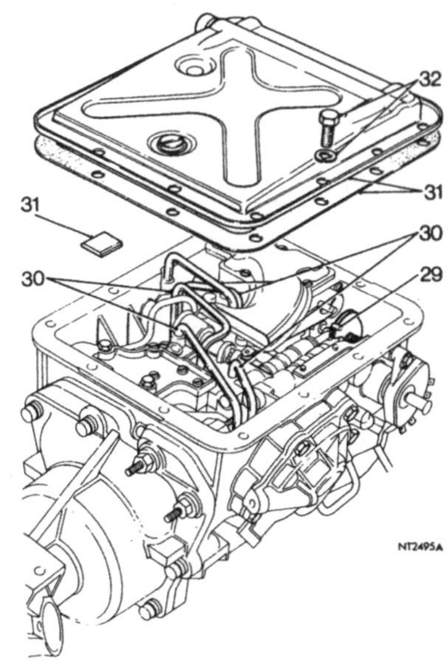

REAR CLUTCH

Remove and refit 44.12.07

Service tools: CBW 60, CBW 547A-50

Removing

1 Remove the transmission unit. 44.20.01.
2 Wash the exterior of the unit in clean petrol or paraffin, invert it and place on a bench cradle CBW 60. Remove the switch. 44.15.15.
3 Unscrew the bolts securing the torque converter housing.
4 Remove the torque converter housing.
5 Unscrew 12 bolts.
6 Remove the oil pan, joint washer and magnet.
7 Pull out the oil tubes.
8 Release the downshift inner cable from the downshift cam.
9 Take out three bolts and washers.
10 Lift off the valve block.
11 Unscrew two bolts.
12 Remove the oil tube locating plate.
13 Pull out the oil tubes. (Note the 'O' ring on the pump suction tube.)
14 Take out five bolts.
15 Remove the pump and joint washer.
16 Remove the thrust washer.
17 Withdraw the front clutch.
18 Remove the thrust washers.
19 Withdraw the rear clutch and forward sun gear.
20 Separate the forward sun gear assembly from the rear clutch.

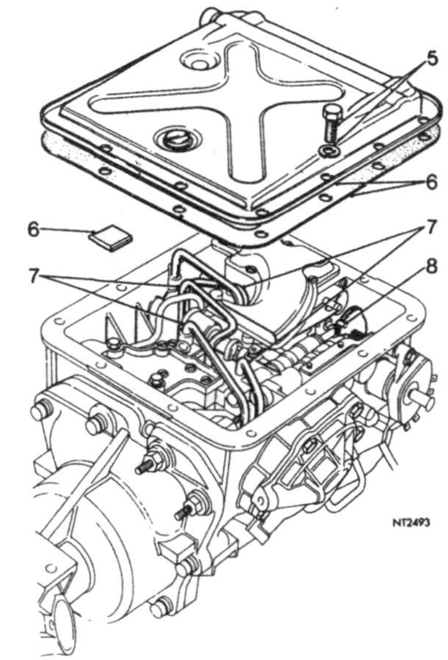

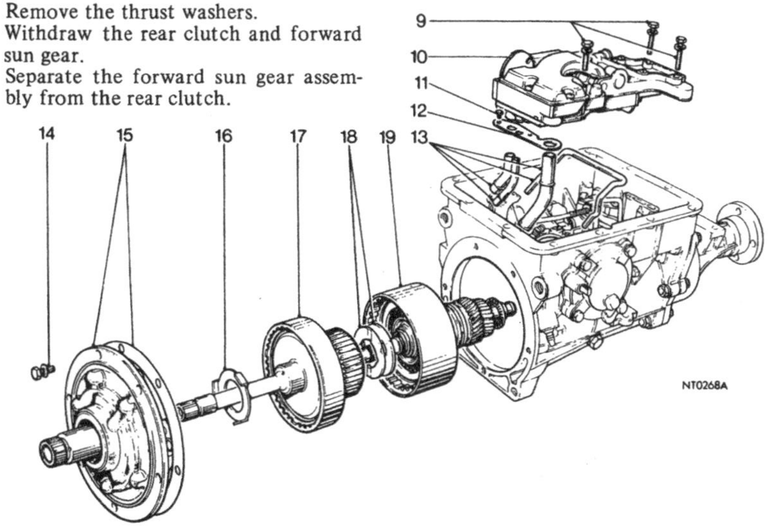

Refitting

21 Assemble the forward sun gear to the rear clutch.
22 Refit the rear clutch and forward sun gear assembly.

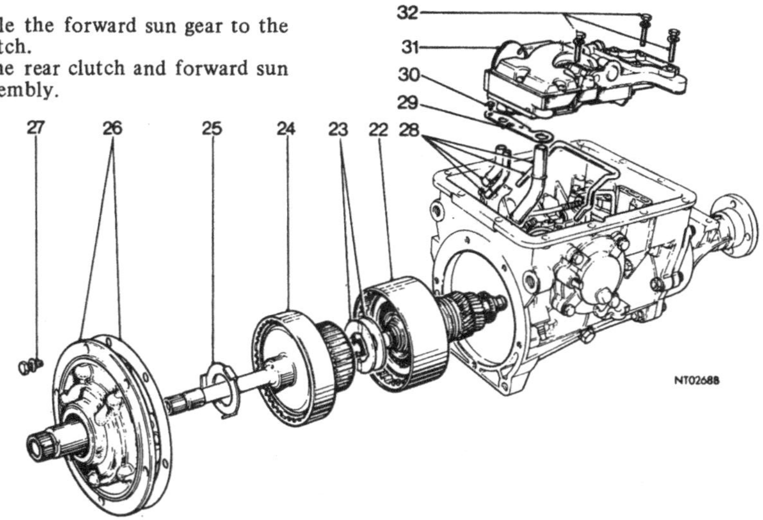

23 Using petroleum jelly, stick the thrust washers to the rear clutch assembly (phosphor bronze towards the front clutch).
24 Refit the front clutch assembly.
25 Using petroleum jelly, stick the thrust washer to the pump assembly.
26 Refit the pump assembly and joint washer.
27 Fit and tighten the bolts.
28 Refit the oil tubes. (Note the 'O' ring on the pump suction tube.)
29 Refit the oil tube locating plate.
30 Fit and tighten the two bolts.
31 Carefully refit the valve block, ensuring that the oil tubes are not distorted.
32 Fit and tighten the three bolts and washers.
33 Connect the downshift inner cable to the downshift cam.
34 Refit the oil tubes.
35 Replace the magnet and refit the oil pan and joint washer.
36 Fit and tighten 12 bolts.
37 Locate the torque converter housing in place.
38 Fit and tighten four bolts securing the torque converter housing.
39 Refit the switch. 44.15.15.
40 Refit the transmission unit. 44.20.01.

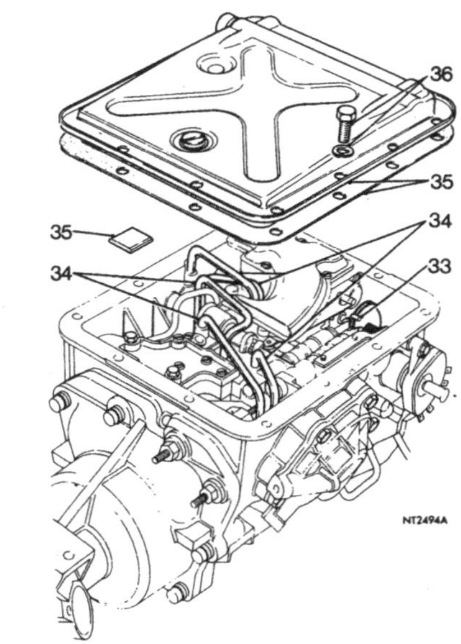

FRONT CLUTCH

Overhaul 44.12.10

Service tool: BW 42

1 Remove the front clutch. 44.12.04.

Dismantling

2 Remove the circlip.
3 Withdraw the input shaft.
4 Remove the thrust washer.
5 Remove the hub.
6 Take out the inner and outer friction plates.
7 Remove the pressure plate.
8 Remove the circlip.
9 Take out the spring.
10 Remove the spring bearing.
11 Withdraw the piston. (If necessary, blank off the bores of the clutch drum and apply a compressed air line to the piston valve hole.)
12 Remove the seal from the piston.
13 Remove the 'O' ring from the drum.

Reassembling

14 Refit the 'O' ring to the drum.
15 Refit the seal to the piston.
16 Fit the piston into tool no. BW 42 and place the tool in the drum. Push the piston into the drum and remove the tool.
17 Locate the spring bearing in position.
18 Refit the spring.

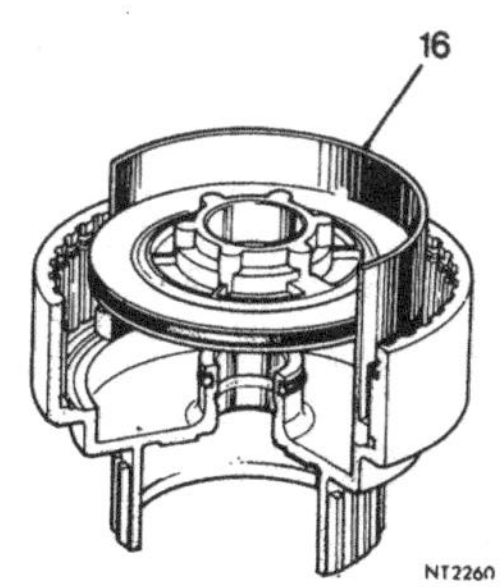

19 Fit the circlip.
20 Refit the pressure plate.
21 Refit the hub.
22 Fit the inner and outer friction plates in alternate sequence.
23 Using petroleum jelly, stick the thrust washer to the hub.
24 Locate the input shaft in position.
25 Refit the circlip.
26 Refit the front clutch. 44.12.04.

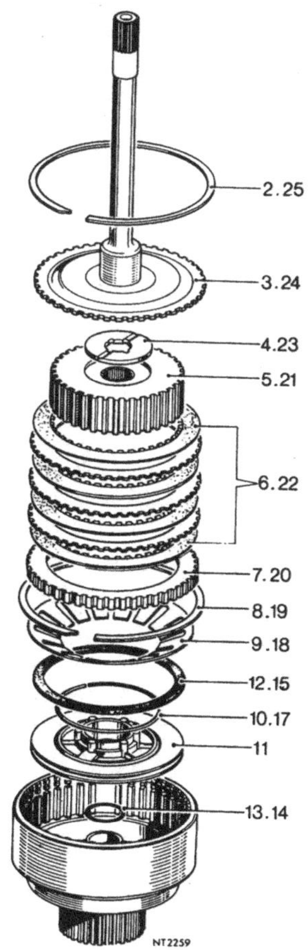

REAR CLUTCH

Overhaul 44.12.13

Service tools: BW 37A, BW 41.

1 Remove the rear clutch. 44.12.07.

Dismantling

2 Remove the circlip.
3 Take out the pressure plate.
4 Remove the inner and outer friction plates.
5 Using tool BW 37A as shown, compress the spring and remove the spring seat circlip. Remove the tool.
6 Take out the spring seat.
7 Remove the spring.
8 Withdraw the piston.
9 Remove the rubber sealing ring from the piston.
10 Remove the rubber 'O' ring from the drum.

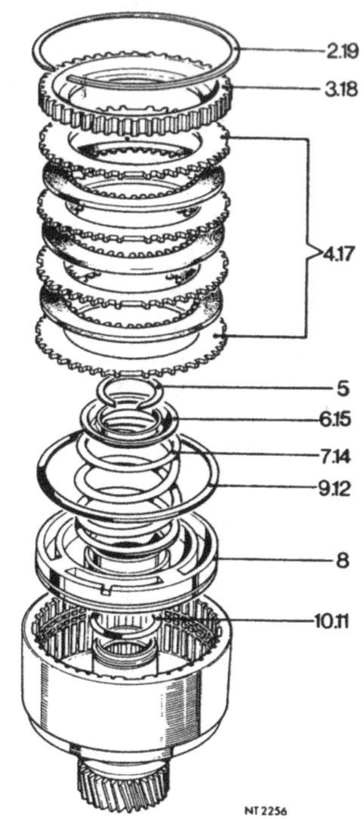

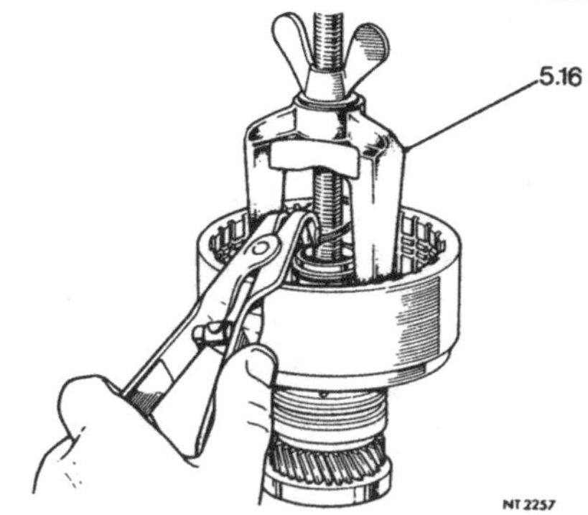

Reassembling

11 Fit the 'O' ring to the drum.
12 Fit the sealing ring to the piston drum.
13 Fit the piston assembly into tool BW 41 and locate the tool in the drum. Push the piston into the drum. Remove the tool.
14 Refit the spring.
15 Refit the spring seat.
16 Using tool BW 37A, compress the spring and fit the circlip. Remove the tool.
17 Refit the inner and outer clutch plates in alternate sequence.
18 Fit the pressure plate.
19 Refit the circlip.
20 Refit the rear clutch. 44.12.07.

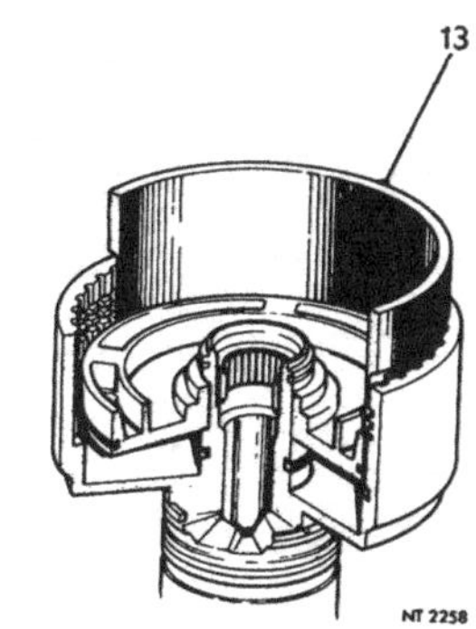

UNI-DIRECTIONAL CLUTCH

Remove and refit 44.12.16

Service tools: CBW 60, CBW 547A-50

Removing

1 Remove the transmission unit. 44.20.01.
2 Wash the exterior of the unit in clean petrol or paraffin, invert it and place on a bench cradle CBW 60. Remove the switch. 44.15.15.
3 Unscrew the bolts securing the torque converter housing.
4 Remove the torque converter housing.
5 Unscrew 12 bolts.
6 Remove the oil pan, joint washer and magnet.
7 Pull out the oil tubes.
8 Release the downshift inner cable from the downshift cam.
9 Take out three bolts and washers.
10 Lift off the valve block.
11 Unscrew two bolts.
12 Remove the oil tube locating plate.
13 Pull out the oil tubes. (Note the 'O' ring on the pump suction tube.)
14 Take out five bolts.
15 Remove the pump and joint washer.
16 Remove the thrust washer.
17 Withdraw the front clutch.
18 Remove the thrust washer.
19 Withdraw the rear clutch and forward sun gear.

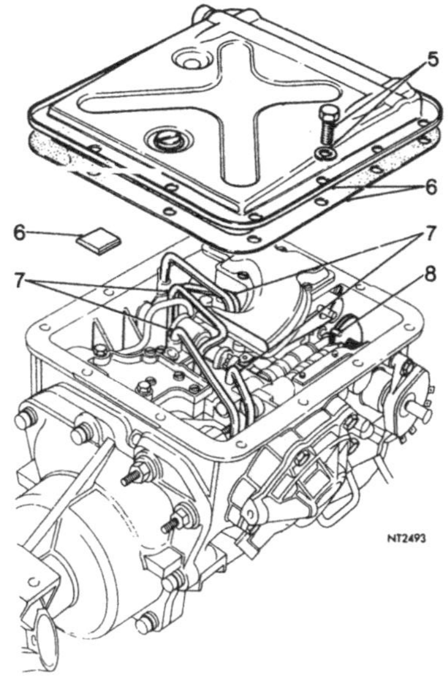

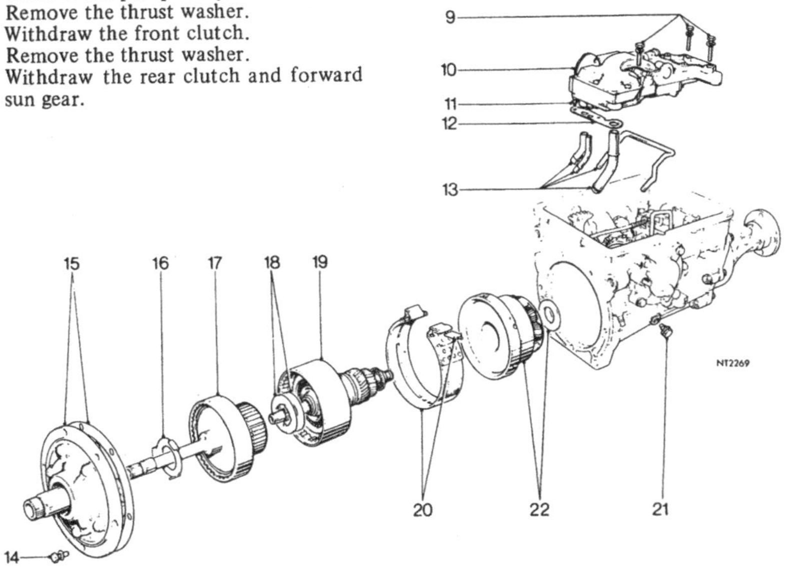

20 Squeeze together the ends of the front brake band and remove it together with the strut.
21 Unscrew the bolts.
22 Withdraw the centre support/planet gear assembly.
23 Separate the centre support from the planet gear assembly.
24 Withdraw the uni-directional clutch.
25 Remove the circlip.
26 Remove the uni-directional clutch outer race.

Refitting

27 Refit the uni-directional clutch outer race to the rear drum.
28 Refit the circlip.
29 Refit the uni-directional clutch.
30 Assemble the centre support and planet gear assembly.
31 Refit the assembly, ensuring that the oil and locating holes in the centre support align with those in the casing.
32 Fit and tighten three bolts.
33 Squeeze together the ends of the front brake band and fit it in position together with the strut.
34 Refit the rear clutch and forward sun gear assembly.
35 Using petroleum jelly, stick the thrust washer to the rear clutch assembly (phosphor bronze towards the front clutch).

continued

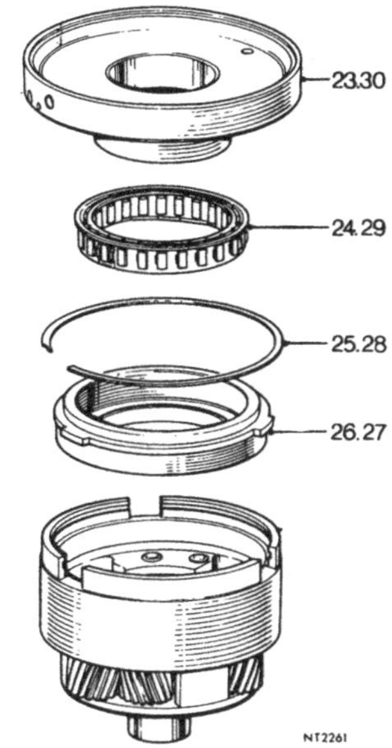

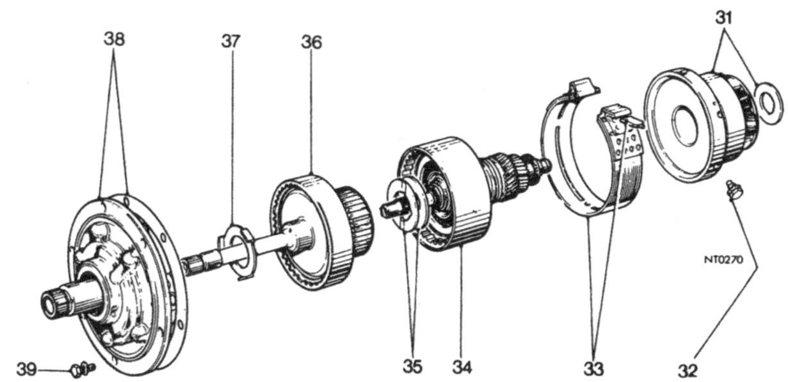

36 Refit the front clutch assembly.
37 Using petroleum jelly, stick the thrust washer to the pump assembly.
38 Refit the pump assembly and joint washer.
39 Fit and tighten the bolts.
40 Refit the oil tubes. (Note the 'O' ring on the pump suction tube.)
41 Refit the oil tube locating plate.
42 Fit and tighten the two bolts.
43 Carefully refit the valve block, ensuring that the oil tubes are not distorted.
44 Fit and tighten three bolts and washers.
45 Connect the downshift inner cable to the downshift cam.
46 Refit the oil tubes.
47 Replace the magnet and refit the oil pan and joint washer.
48 Fit and tighten 12 bolts.
49 Locate the torque converter housing in place.
50 Fit and tighten four bolts securing the torque converter housing.
51 Refit the switch. 44.15.15.
52 Refit the transmission unit. 44.20.01.

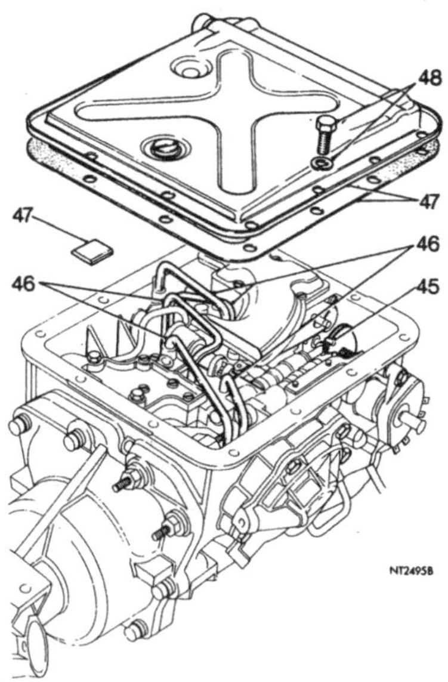

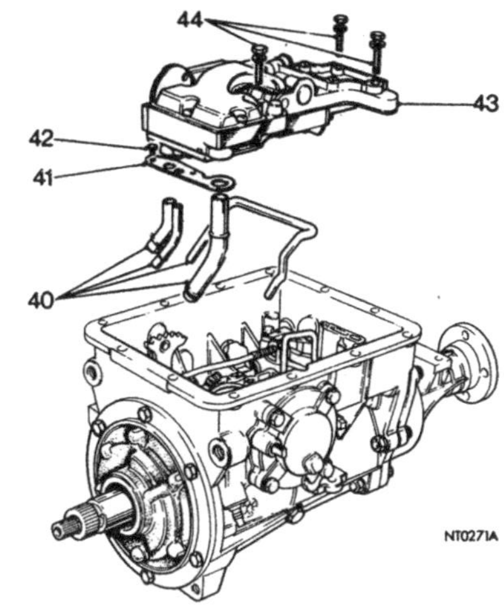

DOWNSHIFT CABLE

Remove and refit 44.15.01

Service tool: CBW 62

Removing

1 Drive the vehicle onto a ramp, select 'N', chock the wheels and open the bonnet.
2 Unscrew the locknut and remove the cable from the bracket.
3 Remove the split pin, washer and clevis pin.
4 Raise the ramp and remove the transmission sump pan. 44.24.04.
5 Disconnect the downshift inner cable from the cam.
6 Using tool no. CBW 62, remove the downshift outer cable from the gearbox casing.
7 Remove the downshift cable assembly.

Refitting

8 Clip the downshift outer cable into the gearbox casing.
9 Connect the inner cable to the downshift cam.
10 Refit the sump pan and lower the ramp.

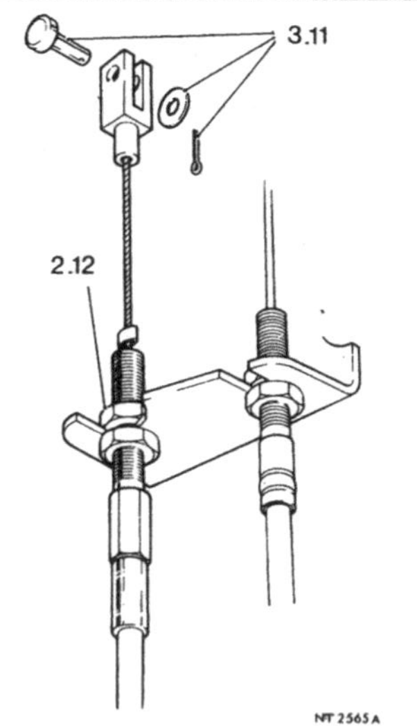

11 Refit the clevis pin, washer and split pin.
12 Refit the locknut and adjust the cable. 44.30.01.
13 Refill the unit with transmission fluid.

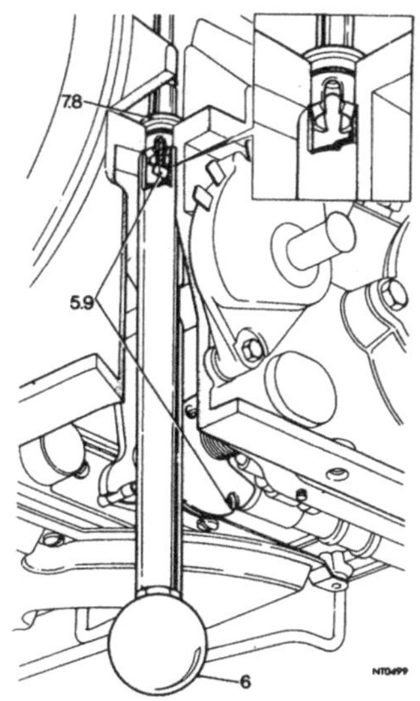

HAND SELECTOR LEVER

Remove and refit 44.15.04

Removing

1 Remove the push button cap.
2 Remove the push button securing screw.
3 Remove the push button and spring.
4 Unscrew and remove the gear knob.
5 Remove the two screws securing the gear lever surround.

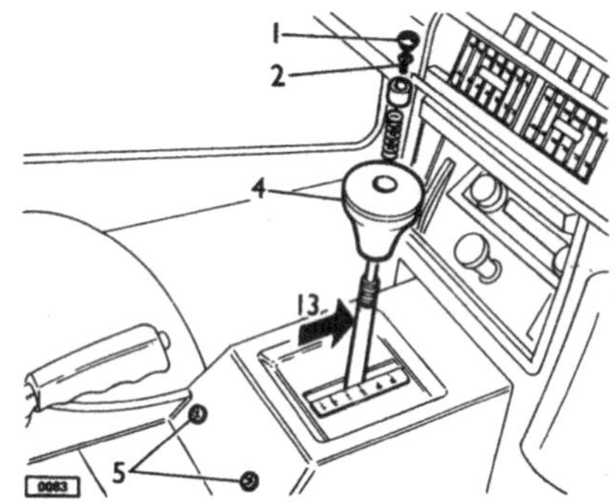

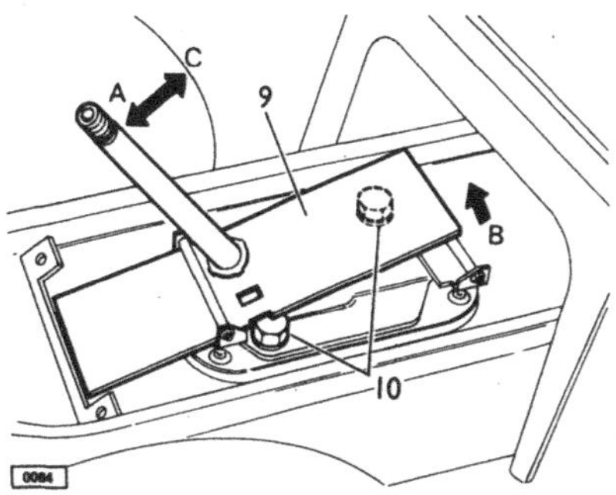

6 Disconnect the quadrant illumination light from the surround.
7 Remove the gear lever surround.
8 Disconnect the gear indication light from the socket.
9 Remove the steel cover by operating the gear lever to each extreme.
10 Remove the two bolts securing the roller assembly.
11 Remove the roller assembly.
12 Remove the two bolts securing the gear lever to the turret shaft and remove the lever.

Refitting
13 Reverse 1–12
(To facilitate the refitting of the push button securing screw, pull out the inner lever and hold in position against the quadrant stop.)

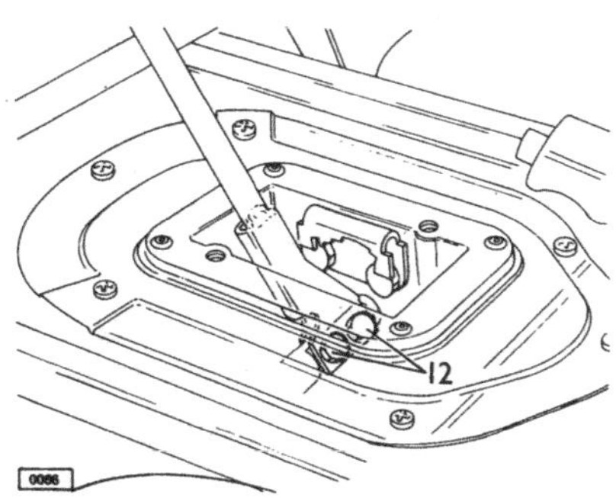

HAND SELECTOR LEVER

Overhaul 44.15.05

1 Remove the hand selector lever. 44.15.04.
2 Pull out the inner lever.
3 Clean, inspect and regrease the inner lever.
4 Refit the inner lever.
5 Refit the hand selector lever.

HAND LEVER TURRET ASSEMBLY

Overhaul 44.15.06

1 Jack up front of the vehicle and place on two axle stands.
2 Disconnect the clip securing the selector rod to the turret lever.
3 Disconnect the selector rod from the turret lever.
4 Remove the hand selector lever assembly. 44.15.04.
5 Remove the two screws securing the hand brake lever surround.
6 Carefully remove the hand brake surround over the hand brake lever.
7 With the centre glove box open, remove the two screws securing the glove box catch to the centre console
8 Remove the glove box catch.
9 Remove the four screws and rear plate securing the centre console.

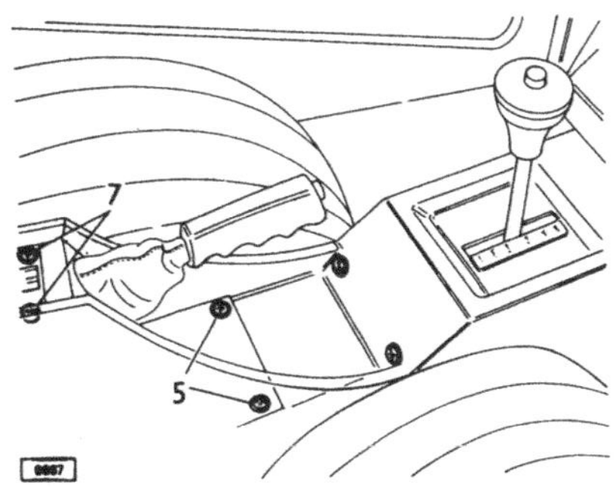

10 Unclip the illumination light harness wires from the console.
11 Remove the centre console.
12 Remove the screws securing the turret assembly to the transmission tunnel.
13 Lift out the turret assembly.
14 Remove the two nuts and bolts securing the quadrant to the turret.

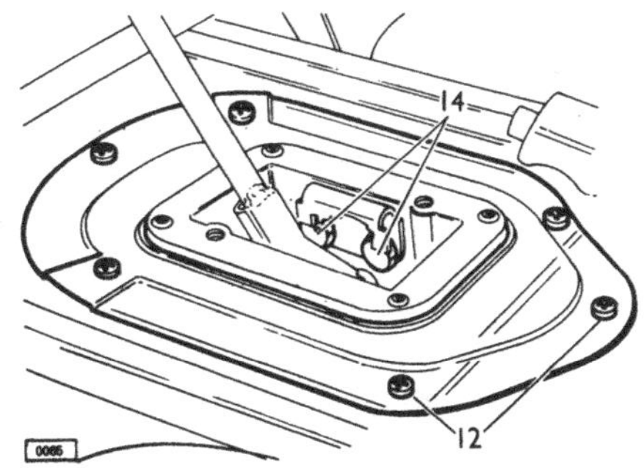

15 Remove the quadrant.
16 Remove the circlip and washer retaining the turret shaft in the tunnel.
17 Remove the turret shaft.
18 Clean and inspect all parts including the hand selector lever.
19 Grease all bearing surfaces.
20 Reverse 4–17.
21 Select 'P–Park' position on the lever.
22 Select 'P–Park' position on the gearbox selector lever.
23 Check/adjust the selector rod as necessary. 44.30.04.
24 Reconnect the selector rod to the turret lever.
25 Engage the securing clip.
26 Lower the vehicle.

SELECTOR ROD

Remove and refit 44.15.08

Removing
1 Drive the vehicle onto a ramp, lock the selector lever in 'N' and apply the hand brake.
2 Raise the ramp.
3 Push the clips clear of the levers.
4 Remove the selector rod from the gearbox selector and hand lever.

Refitting
5 Slacken the selector rod locknut.
6 Ensure that the gearbox selector lever and the hand lever are both in position 'N'.
7 Fit the selector rod to the gearbox selector lever.
8 Fit the clip onto the selector lever.
9 Alter the length of the rod by adjusting the turn-buckle until the end of the rod can be located in the hand lever.
10 Tighten the locknut.
11 Push the clip onto the lever and secure the rod.
12 Lower the ramp.

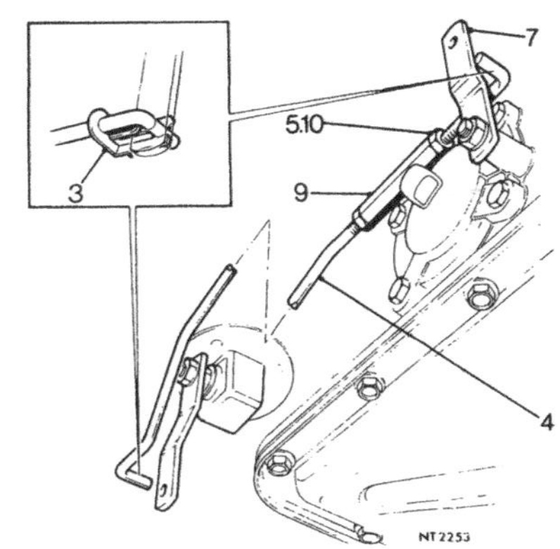

GEARBOX SELECTOR LEVER

Remove and refit 44.15.09

Removing

1 Drive the vehicle onto a ramp, select 'N' and apply the hand brake.
2 Raise the ramp.
3 Push the clip rearward.
4 Disconnect the selector rod from the lever.
5 Unscrew the nut and washer.
6 Remove the lever.

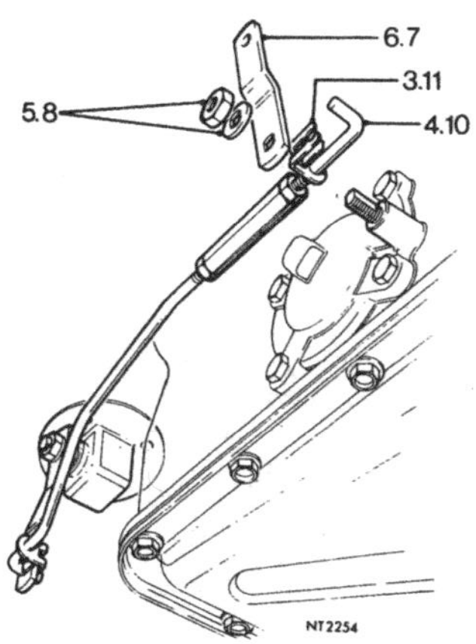

Refitting

7 Fit the lever to the shaft.
8 Fit and tighten the nut and washer.
9 Move the selector into the neutral position.
10 Connect the selector rod to the lever.
11 Push the clip onto the lever and secure the rod.
12 Lower the ramp.

STARTER INHIBITOR/REVERSE LAMP SWITCH

Remove and refit 44.15.15

Removing

1 Drive the vehicle onto a ramp, chock the wheels and raise the ramp.
2 Remove the three bolts securing the heat shield to the transmission sump.
3 Remove the heat shield and spacers.
4 Disconnect the switch from the block connector on the wiring harness.
5 Remove the thread protector (if fitted).
6 Remove the bolt securing the switch to the transmission unit.
7 Remove the switch.

Refitting

8 Reverse 1–7.

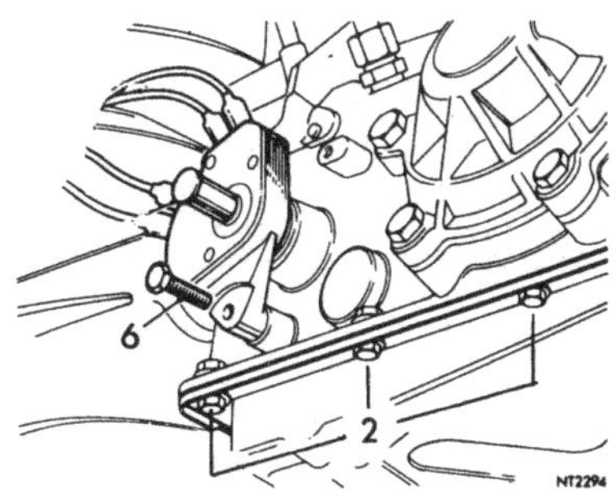

TORQUE CONVERTER HOUSING

Remove and refit 44.17.01

Service tool: CBW 547A-50 – Torque Wrench.

Removing

1 Remove the gearbox. 44.20.01
2 Unscrew the four bolts securing the torque converter housing to the transmission.
3 Remove the housing.

Refitting

4 Place the torque converter housing in position.
5 Fit and tighten the four bolts using Service tool CBW 547A-50.
6 Refit the gearbox. 44.20.01.

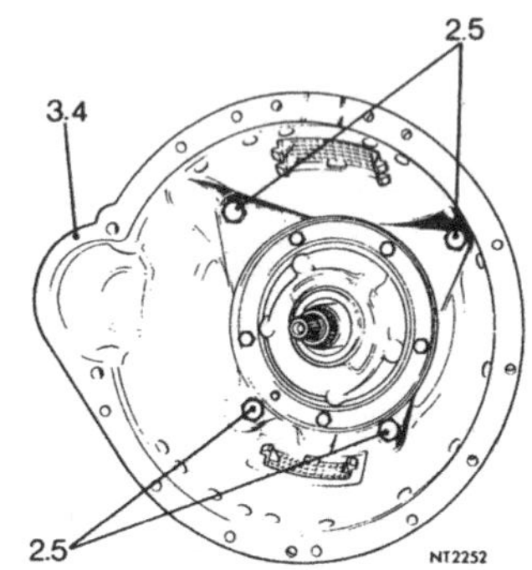

TORQUE CONVERTER

Remove and refit 44.17.07

Service tool: CBW 547A-50

Removing

1 Remove the transmission unit. 44.20.01.
2 Remove the torque converter from the transmission unit.

Refitting

3 Relocate the torque converter in the transmission unit.
4 Refit the gearbox. 44.20.01.

GEARBOX

Remove and refit 44.20.01

Removing

1 Drive the vehicle onto a ramp, select 'N' and chock the wheels.
2 Remove the split pin, washer and clevis pin and undo the cable lock nut.
3 Release the downshift cable from the throttle linkage.
4 Raise the ramp.
5 Remove the exhaust front pipe. 30.10.09.
6 Disconnect the dipstick tube from the transmission sump and drain the fluid into a suitable receptacle.
7 Disconnect the selector lever from the selector shaft.
8 Disconnect the breather hose from the gearbox.
9 Disconnect the starter inhibitor/reverse light switch at the block connector on the wiring harness.
10 Disconnect the cooler pipes.
11 Remove the clamp and disconnect the speedometer cable.
12 Remove the propshaft bolts and disconnect the propshaft.
13 Remove the four bolts securing the torque converter to the engine drive plate.
14 Using a suitable jack under the engine sump, support the engine and gearbox assembly.
15 Remove the centre bolt and plate from the rear mounting.

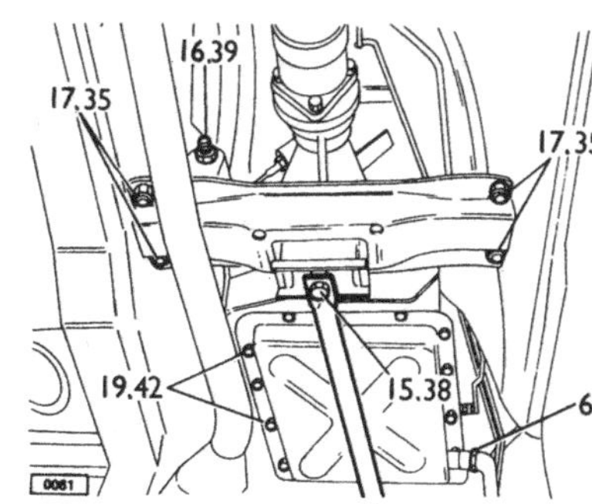

16 Remove the steady bar and disconnect the cable (if fitted).
17 Remove the four nuts securing the rear cross member.
18 Remove the cross member.
19 Remove the three bolts, heat shield and spacers.
20 Position the unit lift and support the gearbox.
21 Remove the four bolts securing the radiator lower mounting.
22 Detach the cooler pipes from the securing clips.
23 Detach the radiator from the upper mountings and withdraw forwards.
24 Remove the exhaust support bracket.

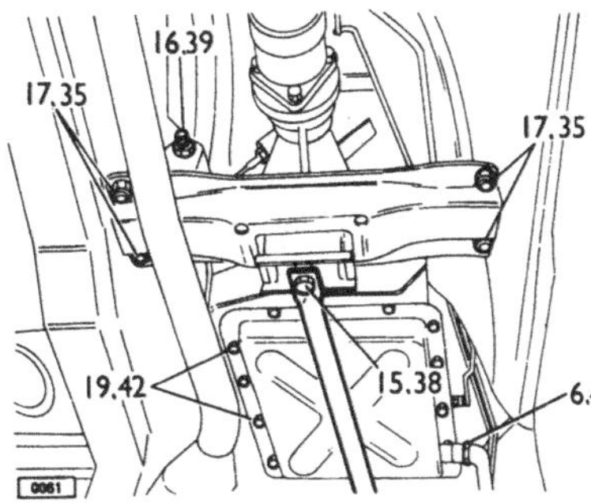

25 Lower the unit and remove all the bolts and nuts securing the bell housing and starter motor to the engine unit.
26 Remove the dipstick tube.
27 Remove the earth leads and all the harness securing clips.
28 Lower the gearbox rearwards on the unit lift.

Refitting

29 Raise the gearbox into position using the unit lift.
30 Fit and tighten the dowel bolt and nuts securing the bell housing to the engine.
31 Reposition the starter motor, earth leads, dipstick tube and harness clips to the converter housing.
32 Fit and tighten the remaining securing bolts.
33 Refit the exhaust bracket and tighten the fixing bolts.
34 Raise the engine and gearbox assembly into the mounting position.
35 Position the rear cross member and tighten the four securing bolts.
36 Remove the engine support jack and unit lift.
37 Position the rear mounting centre plate and steady bar (if fitted).
38 Tighten the rear mounting centre bolt.
39 Reconnect the steady cable (if fitted).
40 Fit and tighten the four bolts securing the torque converter to the engine drive plate.
41 Reconnect the propshaft and tighten the four bolts.
42 Fit the spacers, heat shield and securing bolts.
43 Reconnect and secure the speedo cable.
44 Reconnect the cooler pipes to the gearbox and tighten the union nuts.
45 Reconnect the starter inhibitor/reverse light switch at the block connector on the wiring harness.
46 Reconnect the breather pipe.
47 Reconnect the selector lever.
48 Reconnect and tighten the dipstick tube to the transmission sump.
49 Reposition the radiator and tighten the four bolts securing the lower mounting.
50 Clip the cooler pipes to the radiator.
51 Refit the exhaust front pipe.
52 Lower the ramp.
53 Reconnect the downshift cable to the throttle linkage.
54 Fill the gearbox with fluid. 44.24.02

GEARBOX

Overhaul 44.20.06

Service tools: CBW 60, CBW 33, CBW 34, CBW 62, CBW 547A-50, CBW 547A-50-2A, RG 421 or S 337, CBW 33.

1 Remove the transmission unit. 44.20.01.
2 Wash the exterior of the unit in clean petrol or paraffin, invert it and place on a bench cradle CBW 60.

Dismantling

3 Unscrew the bolts securing the torque converter housing.
4 Remove the torque converter housing.
5 Unscrew 12 bolts.
6 Remove the oil pan, joint washer and magnet.
7 Pull out the oil tubes.
8 Release the downshift inner cable from the downshift cam.
9 Take out three bolts and washers.
10 Lift off the valve block.
11 Unscrew two bolts.
12 Remove the oil tube locating plate.
13 Pull out the oil tubes. (Note the 'O' ring on the pump suction tube.)

continued

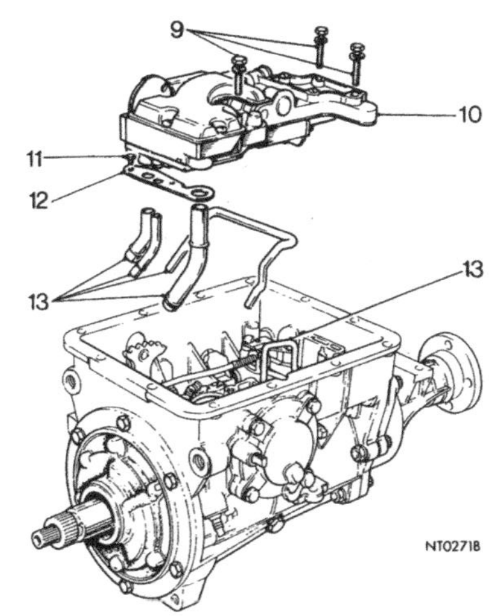

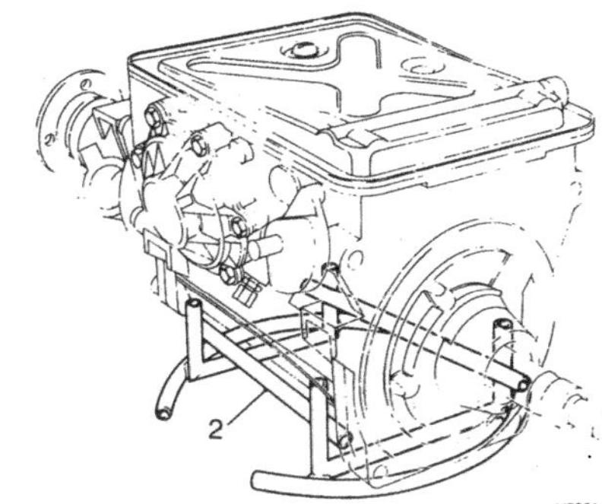

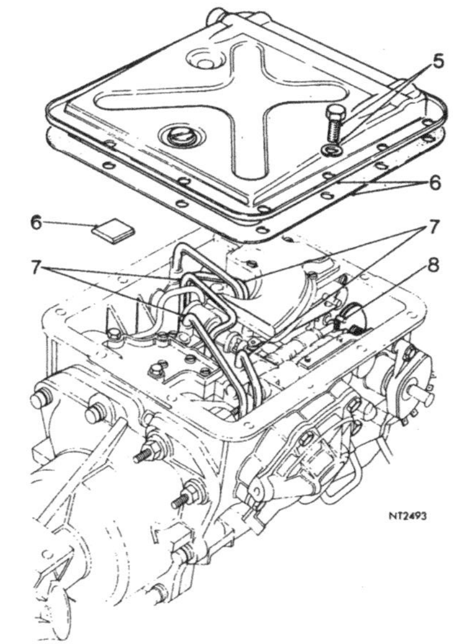

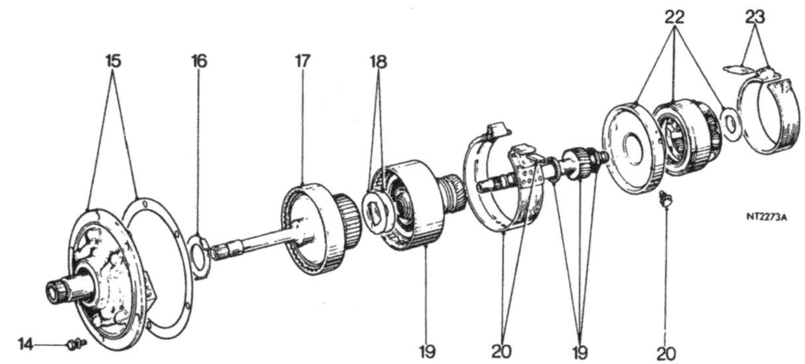

14 Take out five bolts.
15 Remove the pump and joint washer.
16 Remove the thrust washer.
17 Withdraw the front clutch.
18 Remove the thrust washers.
19 Withdraw the rear clutch and forward sun gear.
20 Squeeze together the ends of the front brake band and remove it together with the strut.
21 Unscrew the three bolts.
22 Withdraw the centre support/planet gear assembly.
23 Squeeze together the ends of the rear brake band, tilt and withdraw it together with the strut.

24 Using tool no. RG 421 or S 337 to hold the flange, unscrew the nut.
25 Withdraw the flange.
26 Unscrew the bolts.
27 Remove the rear extension and joint washer.
28 Remove the clamp tube.
29 Withdraw the speedometer drive gear.
30 Unscrew the counterweight.
31 Withdraw the governor assembly.
32 Withdraw the output shaft assembly and thrust washer.
33 Remove the oil tubes.
34 Unscrew the bolts.

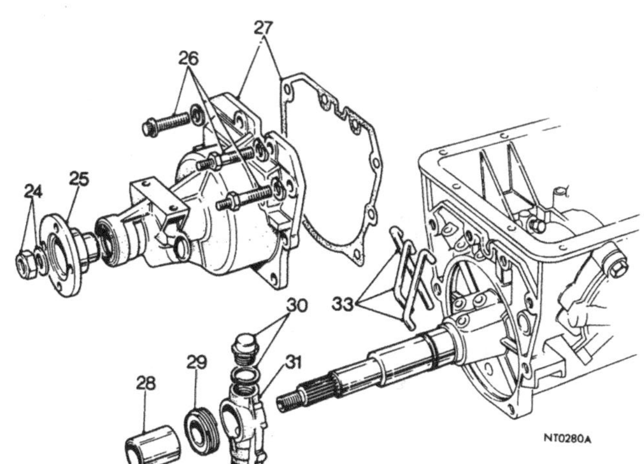

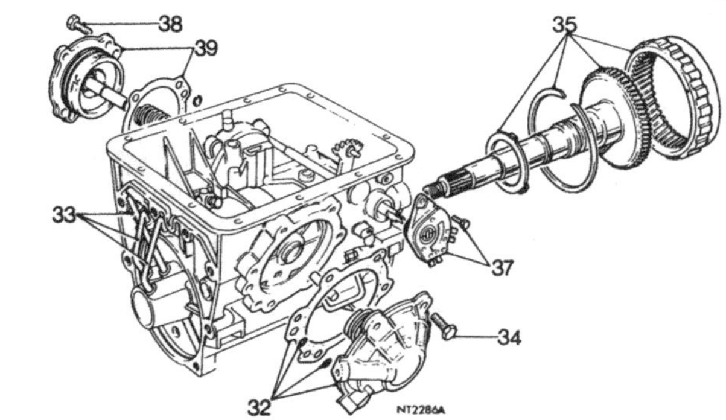

35 Remove the rear servo assembly, joint washer and 'O' rings.
36 Unscrew the nut and remove the selector lever.
37 Unscrew the bolt and remove the switch.
38 Unscrew the bolts.
39 Remove the front servo and joint washer.
40 Remove the spring clip.
41 Withdraw the pin.
42 Withdraw the cross-shaft and remove the 'O' ring.
43 Remove the detent lever, collar, washers and 'O' ring.
44 Remove the oil seal.
45 Unscrew two screws and remove the cam plate.
46 Remove the parking brake rod assembly.
47 Withdraw the parking brake pawl pivot pin.
48 Remove the parking brake pawl.
49 Remove the spring.
50 Remove the relay lever pivot pin.
51 Remove the relay lever.
52 Remove the torsion spring.
53 Using tool no. CBW 62, remove the downshift cable assembly.
54 Using tool no. CBW 62, remove the breather adaptor.

55 Unscrew the unions and remove the bridge pipe.
56 Unscrew the adaptor.
57 Unscrew the return valve.
58 Withdraw the rear servo lever pivot pin.
59 Remove the rear servo lever.
60 Unscrew the locknuts.
61 Unscrew the adjusting screws.
62 Unscrew the pressure take-off plug.

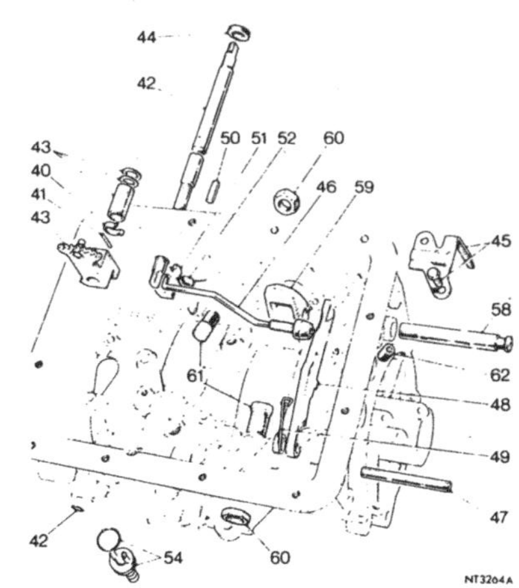

Reassembling

63 Fit the pressure take-off plug.
64 Refit the adjusting screws.
65 Loosely refit the locknuts.
66 Replace the rear servo lever.
67 Refit the rear servo lever pivot pin.
68 Fit the oil return valve.
69 Fit the adaptor.
70 Refit the bridge pipe and tighten the unions.
71 Refit the breather adaptor.
72 Refit the downshift cable.
73 Replace the relay lever and torsion spring.
74 Refit the relay lever pivot pin.
75 Replace the parking brake pawl and spring.
76 Refit the parking brake pawl pivot pin.
77 Refit the parking brake rod.
78 Refit the cam plate, ensuring that the tag end locates in the groove in the rear servo lever pivot pin.
79 Fit and tighten the bolts.
80 Fit a new cross-shaft oil seal.
81 Locate the cross-shaft through the oil seal and fit the washers.
82 Fit the collar and detent lever and push the cross-shaft fully home.
83 Refit the pin.
84 Refit the clip.
85 Refit the 'O' ring.
86 Refit the front servo and joint washer.
87 Fit and tighten the bolts.
88 Refit the switch and secure with the bolt.
89 Refit the selector lever and secure with the nut.

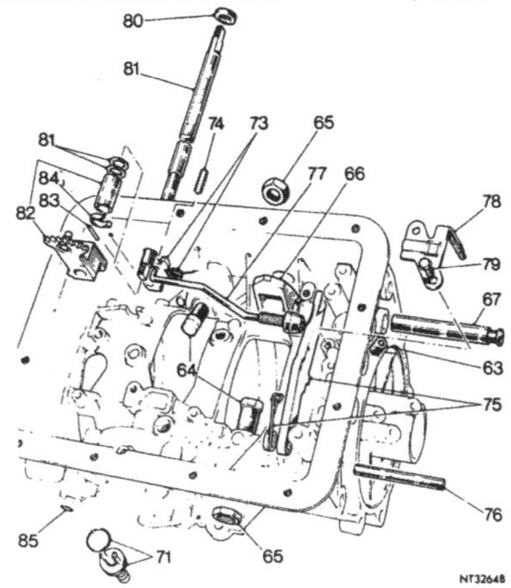

90 Refit the rear servo assembly, joint washer and 'O' rings, retaining them in position using petroleum jelly.
91 Fit and tighten the bolts.
92 Refit the oil tubes, ensuring that they are correctly located.
93 Locate the thrust washer on the end wall of the casing, using petroleum jelly.
94 Carefully refit the output shaft assembly.
95 Refit the governor assembly.
96 Fit and tighten the counterweight.
97 Refit the speedometer drive gear.
98 Refit the clamp tube.
99 Refit the rear extension housing and joint washer.

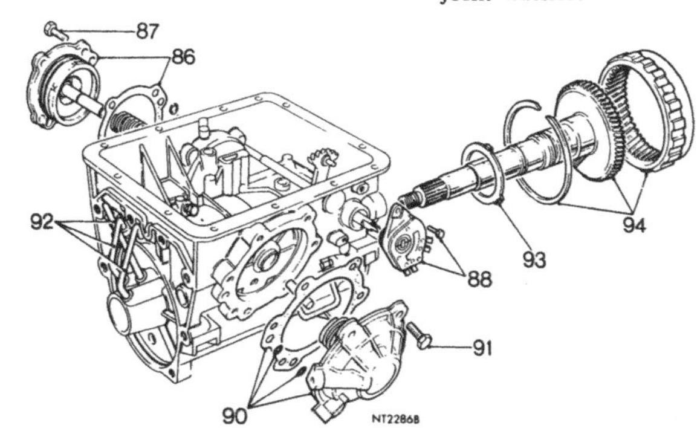

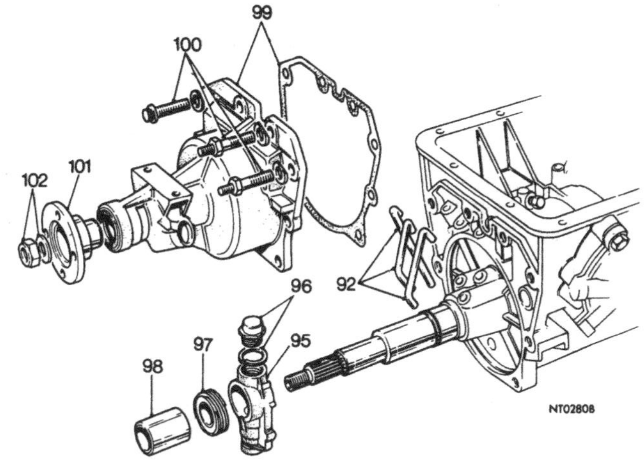

100 Fit and tighten the bolts.
101 Tap the drive flange into position.
102 Fit the washer and nut; using tool no. RG 421 or S 337 to hold the flange, tighten the nut to the correct torque.
103 Squeeze together the ends of the rear brake band, tilt and locate it in position.
104 Refit the rear brake band strut.
105 Using petroleum jelly, locate the thrust race on the rear drum spigot.
106 Refit the centre support/planet gear assembly, ensuring the oil and locating holes align with those in the casing.
107 Fit and tighten the three bolts.
108 Squeeze together the ends of the front brake band and fit it in position together with the strut.
109 Refit the rear clutch and forward sun gear assembly.
110 Using petroleum jelly, stick the thrust washers to the rear clutch assembly (phosphor bronze towards the front clutch).
111 Refit the front clutch assembly.
112 Using petroleum jelly, stick the thrust washer to the pump assembly.
113 Refit the pump assembly and joint washer.
114 Fit and tighten the bolts.

continued

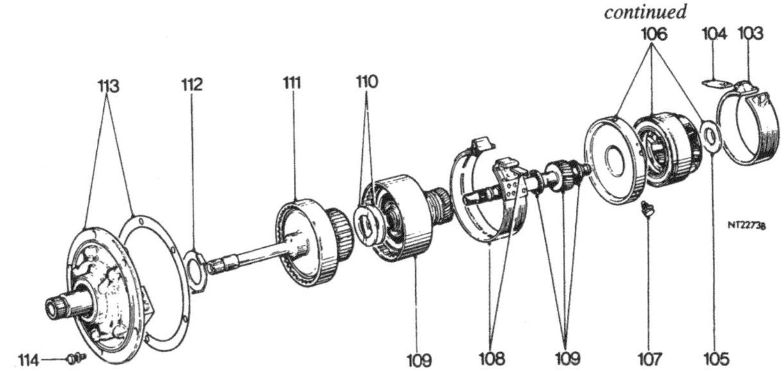

115 Using tool no. CBW 33, check the gear train end-float, and if necessary, adjust the selective use of the thrust washer fitted between the pump and the front clutch.
Recommended end-float 0.25 mm to 0.75 mm (0.010 to 0.030 in).

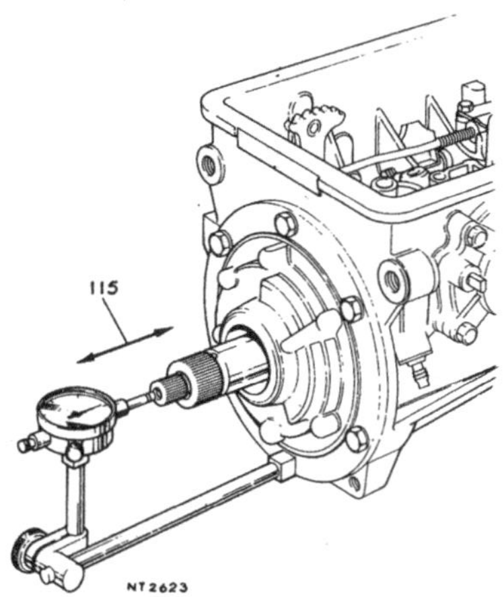

116 Adjust the front band as follows:
a, Slacken the adjusting screw and locknut.
b. Tighten the adjusting screw to 0.7 kgf m (5 lbf ft) and back off three-quarters of a turn.
c. Tighten the locknut.

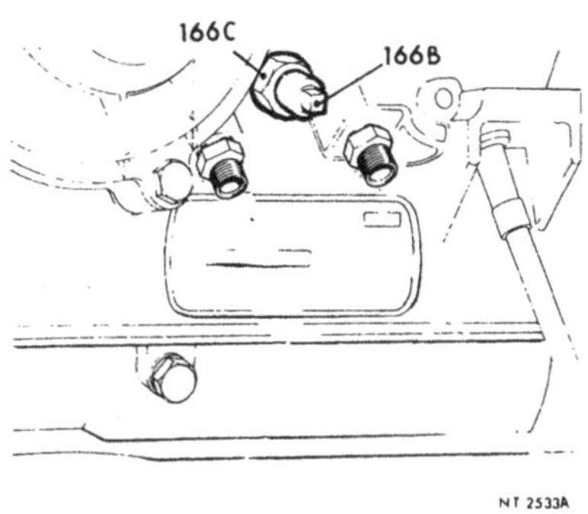

117 Adjust the rear band as follows:
a. Slacken the adjusting screw and locknut.
b. Tighten the adjusting screw to 0.7 kgf m (5 lbf ft) and back off three-quarters of a turn.
c. Tighten the locknut.

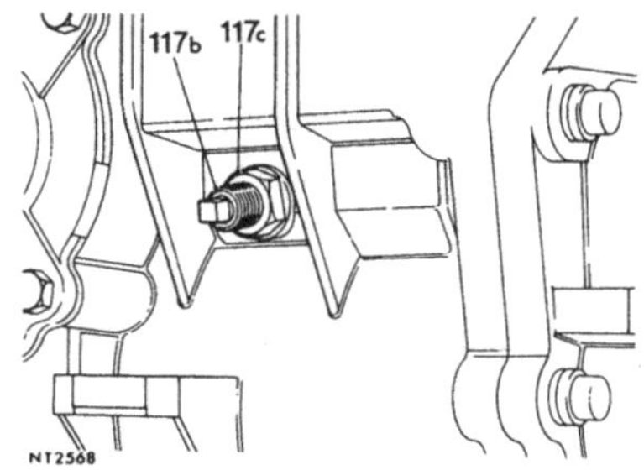

118 Refit the oil tubes. (Note the 'O' ring on the pump suction tube).
119 Refit the oil tube locating plate.
120 Fit and tighten the two bolts.
121 Carefully refit the valve block, ensuring that the oil tubes are not distorted.
122 Fit and tighten the three bolts and washers.
123 Connect the downshift inner cable to the downshift cam.
124 Refit the oil tubes.
125 Replace the magnet and refit the oil pan and joint washer.
126 Fit and tighten 12 bolts.
127 Locate the torque converter housing in place.
128 Fit and tighten four bolts securing the torque converter housing.
129 Refit the transmission unit.

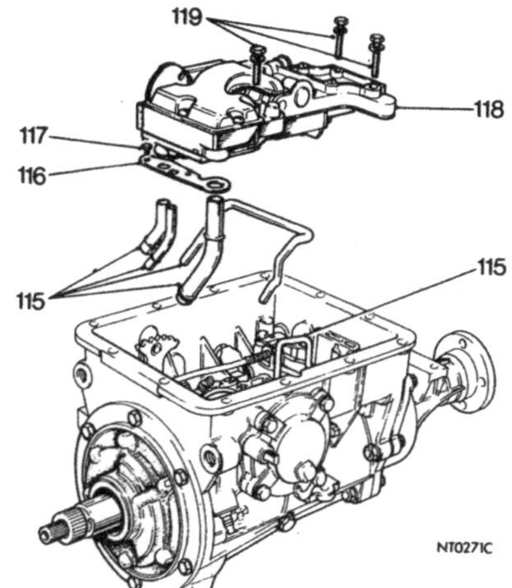

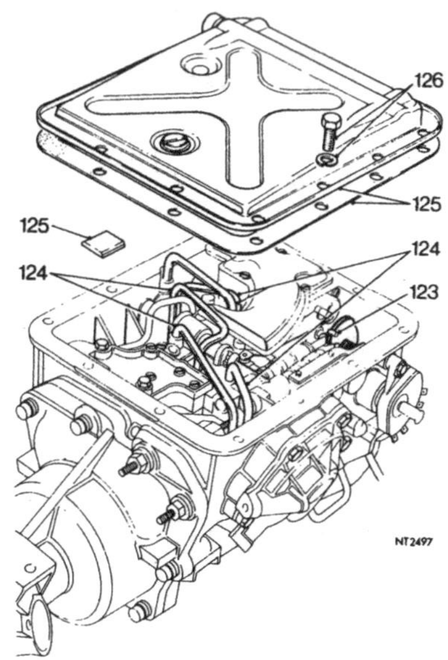

REAR EXTENSION HOUSING

Remove and refit 44.20.15

Service tool: RG 421 or 5337

Removing

1 Drive the vehicle onto a ramp, select 'N', chock the road wheels and raise the ramp.
2 By removing one of the bolts and slackening the other, swing the prop guard clear of the propshaft.
3 Remove the four nuts and bolts securing the propshaft to the drive flange and disconnect the propshaft.

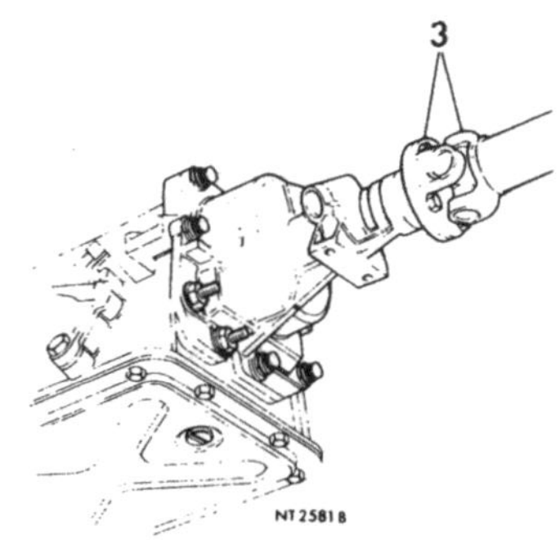

4 Using Service tool RG 421 or 5337 remove the nut securing the drive flange.
5 Remove the drive flange.
6 Using a ramp jack and suitable wooden block, support the transmission unit under the sump pan.
7 Remove the centre gearbox mounting bolt and plate.
8 Remove the steady bar and disconnect the cable (if fitted).
9 Remove the four nuts securing the cross member.
10 By raising the LH captive bolts, release the cross member and remove.

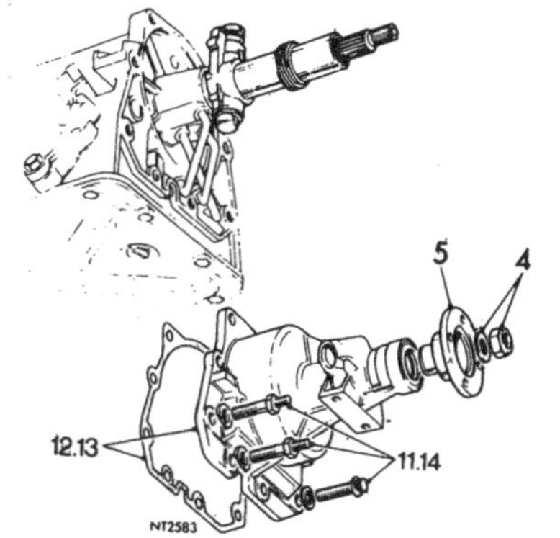

11 Remove the clamp bolt and clamp and disconnect the speedo cable.
12 Remove the nut and bolt securing the exhaust pipe to the mounting bracket.
13 Remove the two nuts securing the bracket to the rear extension housing and remove the bracket and spacers.
14 Remove the eight bolts securing the rear extension to the transmission case.
15 Withdraw the rear extension housing.

Refitting
16 Reverse 1–15.

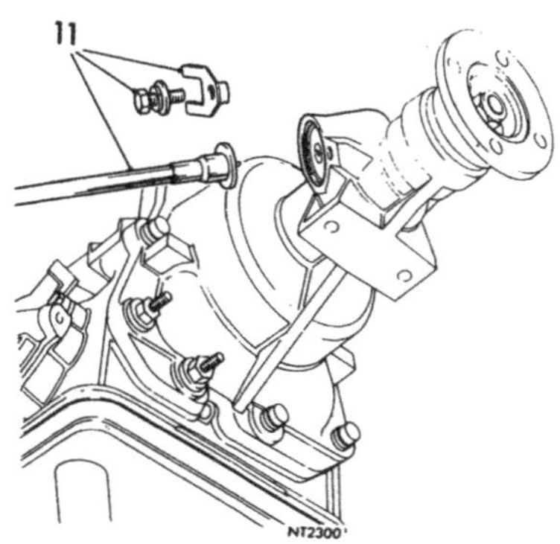

REAR OIL SEAL

Remove and renew 44.20.18

Removing
1 Drive vehicle onto a ramp, select 'P', chock the road wheels and raise the ramp.
2 Remove the four nuts and bolts securing the propshaft to the drive flange.
3 Disconnect the propshaft.
4 Remove the nut securing the drive flange.
5 Remove the drive flange.
6 Prise out the oil seal.
7 Clean the area surrounding the oil seal.

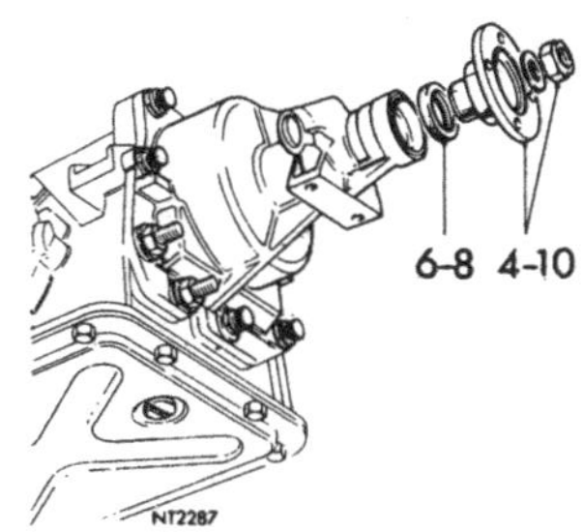

Renewing
8 Carefully fit the new oil seal into the rear extension.
9 Lubricate the lip of the oil seal.
10 Refit the drive flange and securing nut.
11 Reconnect the propshaft.
12 Refit and tighten the prop bolts and nuts.
13 Lower the ramp.

GOVERNOR

Remove and refit 44.22.01

Removing
1 Remove the rear flange, rear extension and speedometer drive gear. 44.38.07.
2 Unscrew the counterweight.
3 Note the position and withdraw the governor assembly.

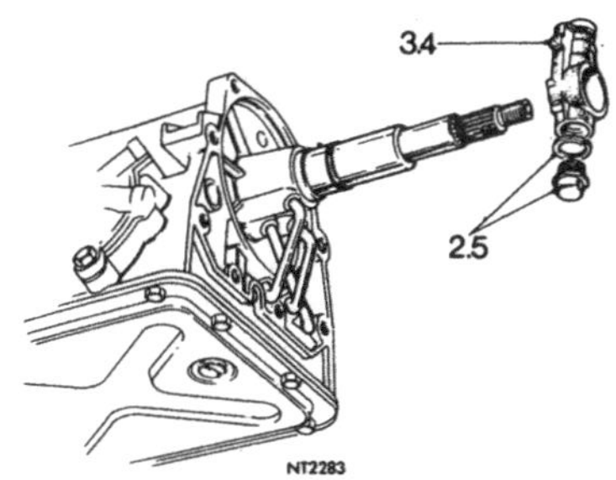

Refitting
4 Slide the governor assembly into the noted position.
5 Locate and secure the counterweight.
6 Refit the speedometer drive gear, rear extension and drive flange. 44.38.07.

GOVERNOR

Overhaul 44.22.04

Dismantling
1 Pull off the retainer.
2 Withdraw the weight.
3 Withdraw the stem.
4 Remove the spring.
5 Withdraw the valve.

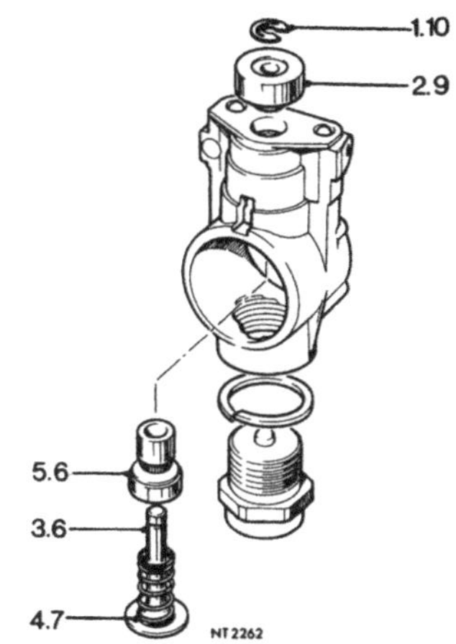

Reassembling
6 Insert the valve.
7 Refit the spring onto the stem.
8 Refit the stem and spring.
9 Refit the weight.
10 Refit the retainer.

DIPSTICK/FILLER TUBE

Remove and refit 44.24.01

Removing

1 Drive the vehicle onto a ramp, select 'P'-'Park', apply the hand brake and open the bonnet.
2 Withdraw the dipstick.
3 Raise the ramp.
4 Release the filler tube from the engine/transmission flange.
5 Unscrew the union nut from the sump pan, and release the filler pipe from the sump.
6 Withdraw the filler tube from below the vehicle.

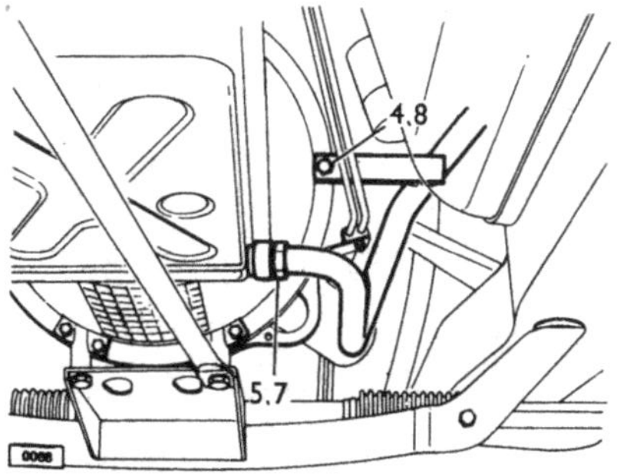

Refitting

7 Manoeuvre the filler tube into position and secure it to the sump pan with the union nut.
8 Secure the filler tube to the engine/transmission flange.
9 Lower the ramp.
10 Refill the unit with fluid. 44.24.02.

TRANSMISSION FLUID

Drain and refill 44.24.02

Draining

1 Drive the vehicle onto a ramp, select 'P'-'PARK' and apply the hand brake.
2 Raise the ramp.
3 Place a tray under the drain plug.
4 Unscrew the plug.
5 Drain the fluid into the tray.
NOTE: It is not possible to drain the torque converter.

Filling

If the sump has been drained it will be necessary to replenish the transmission unit until the fluid level is no higher than the 'cold high' mark 'C' (third mark down from top).

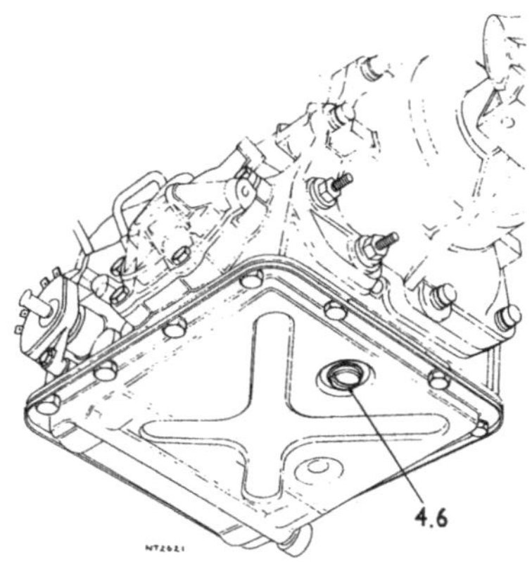

Check the level (hot) as follows:

7 Drive the vehicle for approximately 30 km (20 miles) until the transmission unit has reached its normal operating temperature. Park the vehicle on level ground, apply the hand brake and select 'P'-'PARK'.
8 Raise the bonnet and wipe clean around the dipstick/filler orifice.
9 Switch off the engine, withdraw the dipstick and wipe it clean, using clean paper or a non-fluffy cloth.
10 Push the dipstick home and again withdraw it for reading. The fluid level should be at the top mark 'A' ('Hot high').
DO NOT OVERFILL THE TRANSMISSION.

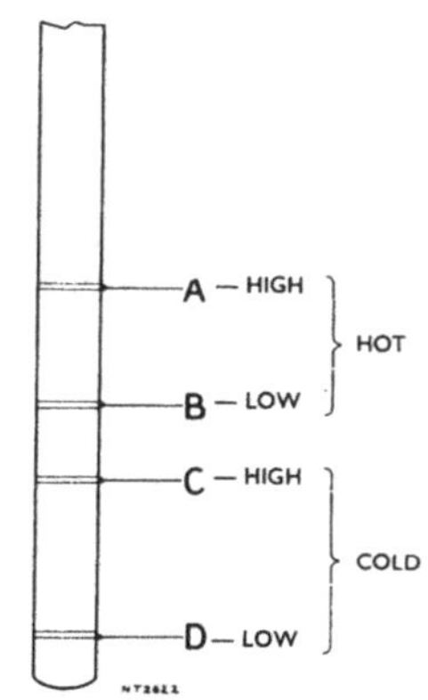

TRANSMISSION SUMP

Remove and refit 44.24.04

Removing

1 Unscrew the filler pipe union and drain the transmission unit.
2 Remove the bolts securing the heat shield.
3 Remove the heat shield spacers.
4 Remove the steady bar (if fitted).
5 Unscrew the bolts securing the sump.
6 Remove the sump and joint washer.

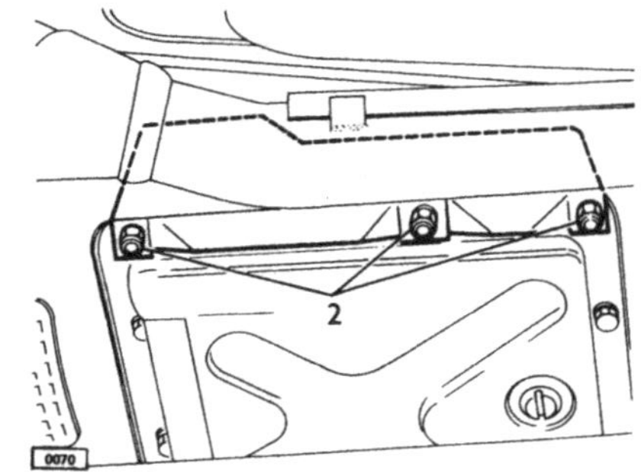

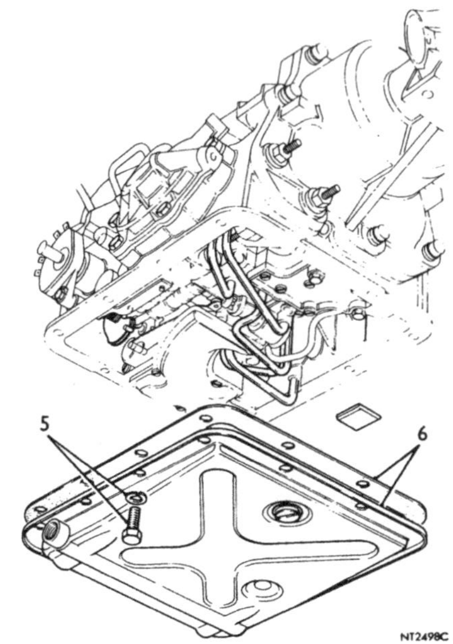

Refitting

7 Reverse 1–6.
8 Fill the transmission unit with fluid. 44.24.02.

OIL/FLUID FILTER

Remove and refit 44.24.07

Removing
1 Remove the transmission sump. 44.24.04.
2 Remove the four screws securing the filter in position.
3 Remove the filter.

Refitting
4 Reverse 1–3.

OIL/FLUID COOLER

Remove and refit 44.24.10

Removing
1 Drive the vehicle onto a ramp, select 'P'-'Park', chock the road wheels and raise the ramp.
2 Detach the cooler pipes from the securing clip beneath the radiator.
3 Undo the cooler pipe union nuts.
4 Disconnect the cooler pipes.
5 Remove the two bolts securing the cooler to the body.
6 Remove the cooler.

Refitting
7 Reverse 1–6
8 Check/top up the transmission fluid.

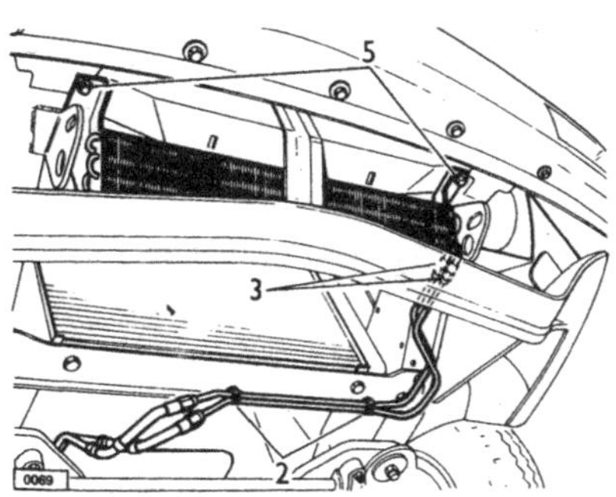

DOWNSHIFT CABLE

Initial setting 44.30.01

1 Check that the carburetter slow running and fast idle settings are satisfactory.
2 Slacken the locknut.
3 Adjust the outer cable in the bracket until the crimped stop is 1.5 mm (1/16 in) from the end of the outer cable ferrule.
4 Tighten the locknut.
5 Road test the vehicle and check the gear shift speeds.

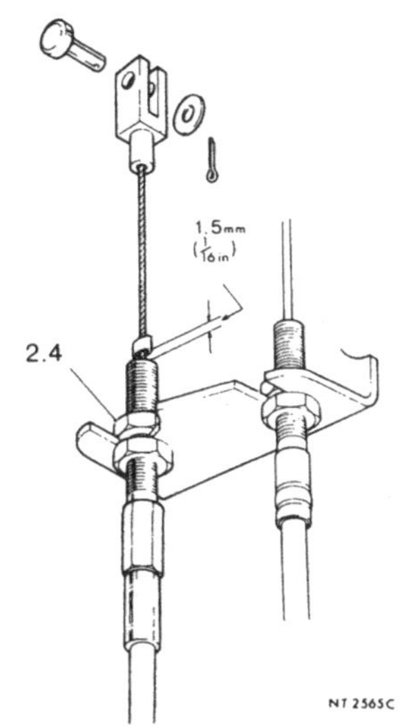

DOWNSHIFT CABLE

Adjust 44.30.02

1 Drive the vehicle onto a ramp, apply the hand brake and chock the wheels.
2 Start the engine, select 'D' and adjust the idling speed to 750 rev/min. Stop the engine.
3 Slacken the locknut.
4 Adjust the outer cable to 1.5 mm (1/16 in) from the stop. 44.20.01 (para. 3).
5 Remove the sump pan. 44.24.04.
6 Check that the downshift cam is in the idling position.
7 With the aid of an assistant in the driving seat, fully open the throttle and check that the downshift cam is in the kick-down position.
8 If necessary, adjust the outer cable until the idling and kick-down positions can be correctly obtained on the downshift cam. Tighten the locknut.
9 Refit the sump. 44.24.04.

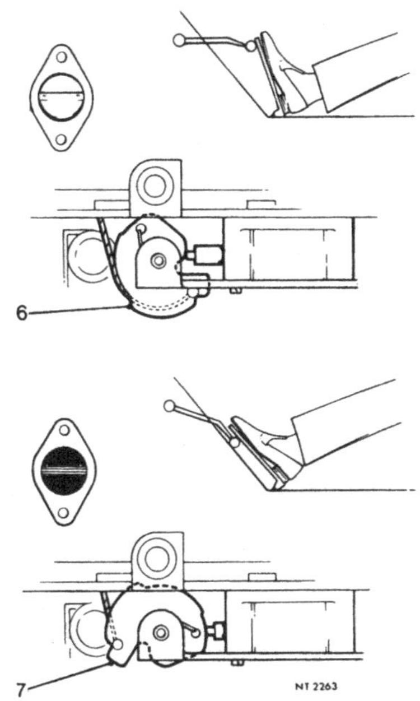

DOWNSHIFT CABLE

Pressure check 44.30.03

Service tools: CBW 1A, B, or C, and CBW 1B-2 or CBW 1C-2.
1 Start and run the engine until the transmission reaches its normal operating temperature.
2 Drive the vehicle onto a ramp and check that the engine idling speed is approximately 750 rev/min. Stop the engine.
3 Raise the ramp.
4 Remove the plug.
5 Connect the pressure gauge to the transmission unit.
6 Lower the ramp, chock the wheels and apply the hand brake and foot brake.
7 Start the engine and select 'D'

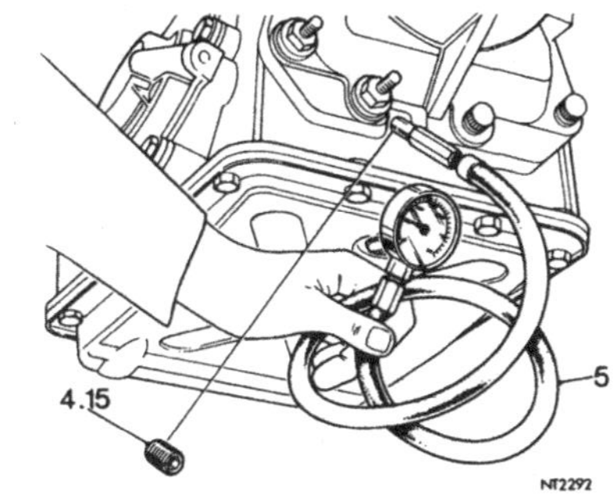

8 With the engine idling at 750 rev/min, note the pressure gauge reading which should be 4.2 to 5.3 kgf/cm^2 (60 to 75 lbf/in^2).
9 Increase the engine speed to 1,000 rev/min and note the pressure increase which should be 1.0 to 1.4 kgf/cm^2 (15 to 20 lbf/in^2).
10 Stop the engine.
11 If the pressure increase is less than 1.0 kgf/cm^2 (15 lbf/in^2), increase the effective length of the outer cable. If the pressure increase is more than 1.4 kgf/cm^2 (20 lbf/in^2), decrease the effective length of the outer cable.
12 Repeat operations 7 to 11 until the pressure increase is correct.
13 Raise the ramp.
14 Disconnect the pressure gauge.
15 Refit the plug.
16 Lower the ramp.

SELECTOR ROD

Adjust 44.30.04

1 Drive the vehicle onto a ramp, lock the selector lever in 'N' and apply the hand brake.
2 Raise the ramp.
3 Slacken the locknut.
4 Push the clip off the hand lever.
5 Disconnect the selector rod and check that the gearbox selector lever is in the neutral position.
6 Alter the length of the selector rod by adjusting the turnbuckle until the end of the rod can be located in the hand lever.
7 Tighten the locknut.
8 Push the clip onto the lever and secure the rod.
9 Lower the ramp.

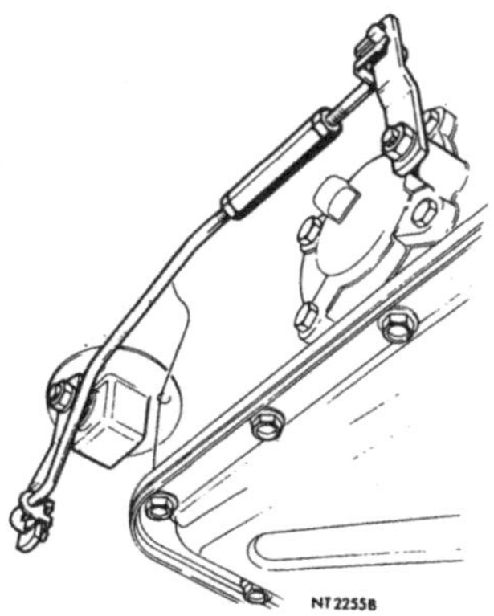

FRONT BRAKE BAND

Adjust 44.30.07

Service tools: 18G 307

1 Drive the vehicle onto a ramp, select P – 'Park', apply the hand brake and raise the ramp.
2 Slacken the locknut.
3 Tighten the adjusting screw to 0.41 kgf m (36 lbf in) and back off three-quarters of a turn.
4 Tighten the locknut to 3.1 kgf m (22 lbf ft).
5 Lower the ramp.

REAR BRAKE BAND

Adjust 44.30.10

Service tools: 18G 307

1 Drive the vehicle onto a ramp, select P – 'Park', apply the handbrake and raise the ramp.
2 Slacken the locknut.
3 Tighten the adjusting screw to 0.41 kgf m (36 lbf in) and back off three-quarters of a turn.
4 Tighten the locknut to 3.1 kgf m (22 lbf ft).
5 Lower the ramp.

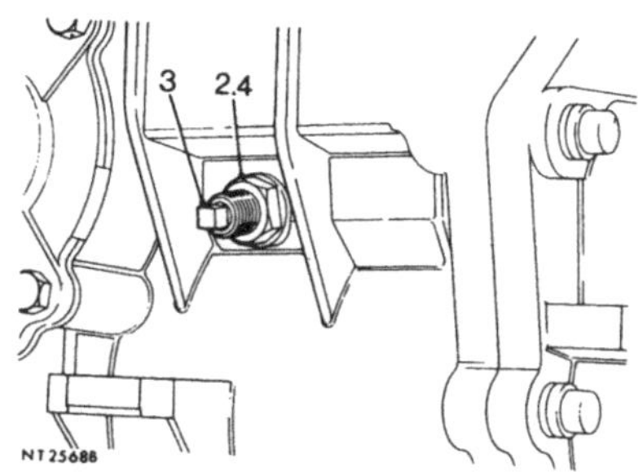

STALL TEST 44.30.13

The function of a stall test is to determine that the torque converter and gearbox are operating satisfactorily.

1 Check the condition of the engine. An engine which is not developing full power will affect the stall test readings.
2 Allow the engine and transmission to reach correct working temperatures.

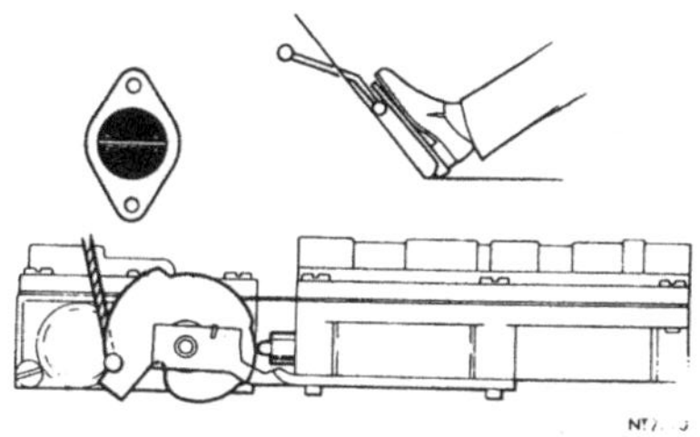

3 Chock the wheels and apply the hand brake and foot brake.
4 Select '1' or 'R' and depress the throttle to the 'kick-down' position. Note the reading on the tachometer which should be 2,200 rev/min. If the reading is below 1,400 rev/min, suspect the converter for stator slip. If the reading is down to 1,600 rev/min, the engine is not developing full power. If the reading is in excess of 2,400 rev/min, suspect the gearbox for brake band or clutch slip.

NOTE: Do not carry out a stall test for a longer period than 10 seconds, otherwise the transmission will become overheated.

ROAD TEST 44.30.17

Throughout the road test procedure the term 'full throttle' is equivalent to approximately seven-eighths of the available pedal movement and 'kick-down' is equivalent to the full movement.

Procedure

1 Check that the starter motor will operate only with the selector lever in 'PARK' or 'N' and that the reverse lights operate only in 'R'.
2 Apply the hand brake and with the engine idling select 'N–D', 'N–2', 'N–R'. Engagement should be positive. A cushioned 'thump' under fast idling conditions is normal.
3 With the transmission at normal running temperatures, select 'D', release the brakes and accelerate with minimum throttle. Check the 1–2 and 2–3 shift speeds and the quality of change.
4 Stop the vehicle, select 'D' and re-start using 'full throttle'. Check 1–2 and 2–3 shift speeds and the quality of change.
5 At 40 m.p.h. (65 km/h) apply 'full throttle'. The vehicle should accelerate in third gear and should not downshift to second.
6 At a maximum speed of 56 m.p.h. (90 km/h) 'kick-down' fully. The transmission should downshift to second gear.
7 At a maximum speed of 35 m.p.h. (56 km/h) 'kick-down' fully. The transmission should downshift to first gear.
8 Stop the vehicle, select 'D' and re-start using 'kick-down'. Check the 1–2 and 2–3 shift speeds.
9 At 35 m.p.h. (56 km/h) select 2 and release the throttle. Check the 3–2 downshift.
10 At 30 m.p.h. (50 km/h) select 1 and release the throttle. Check the 2–1 downshift.

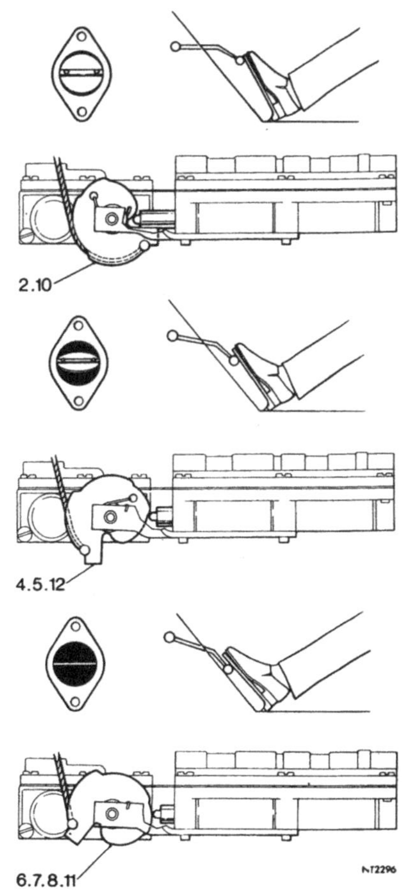

11 With 1 still engaged, stop the vehicle and using 'kick-down' accelerate to over 40 m.p.h. (65 km/h). Check for 'slip', 'squawk', and the absence of upshifts.

12 Stop tne vehicle and select 'R'. Reverse using 'full throttle' if possible. Check for 'slip' and 'squawk'.

13 Stop the vehicle on a gradient. Apply the hand brake and select 'P'-'PARK'. Release the hand brake and check the parking pawl hold. Check that the selector lever is held firmly in the gate in 'P'.

CONVERTER DIAGNOSIS

Inability to start on steep gradients, combined with poor acceleration from the rest and low stall speed (1,400 rev/min) indicates that the converter stator uni-directional clutch is slipping. This condition permits the stator to rotate in an opposite direction to the impeller and turbine, and torque multiplication cannot occur.

Poor acceleration in third gear above 30 m.p.h. (50 km/h) and reduced maximum speed, indicates that the stator uni-directional clutch has seized. The stator will not rotate with the turbine and impeller and the 'fluid flywheel' phase cannot occur. This condition will also be indicated by excessive overheating of the transmission although the stall speed will be correct.

ROAD TEST–FAULT DIAGNOSIS CHART

NOTE: The numbers indicate the recommended sequence of investigation

Diagnosis		Engagement of 1, 2, D or R			Take off						Upshifts			
	SYMPTOM	Bumpy	Delayed	None	None forward	None reverse	Seizure reverse	No neutral	Squawk	Slip	No 1–2	No 2–3	Above normal speed	Below normal speed
ADJUSTMENT FAULTS	Fluid level insufficient		1	1					1					
	Downshift cable incorrectly assembled or adjusted	2							2				1	1
	Manual linkage incorrectly assembled or adjusted		2	2	1	1		1	3	1	1	1		
	Incorrect engine idling speed	1	3											
	Incorrect front band adjustment						1				2			
	Incorrect rear band adjustment					2								
HYDRAULIC CONTROL FAULTS	Oil tubes incorrectly installed, missing or leaking		4	3		7			7		8	8		
	Sealing ring missing or broken		7	4	3	6			5		9	9	8	5
	Valve block screws missing or loose	5	6	5	2	5		3			10	10	9	6
	Primary regulator valve sticking	3	5	6									10	
	Throttle valve sticking	4									6	6	2	2
	Modulator valve sticking										7	7	7	
	Governor valve sticking, leaking or incorrectly assembled										3	2	3	3
	Orifice control valve sticking													
	1–2 shift valve sticking					3					4		4	
	2–3 shift valve sticking					4						3	5	
	2–3 shift valve plunger sticking											4	6	4
	Converter 'out' check valve sticking or missing		13											
	Check valve sticking or missing		8											
MECHANICAL FAULTS	Front clutch slipping		9		4				4					
	Front clutch seized or plates distorted	6					2	2						
	Rear clutch slipping		10			9			6			5		
	Rear clutch seized or plates distorted	7												
	Front band slipping due to faulty servo or worn band										5			
	Rear band slipping due to faulty servo or worn band		11			8								
	Uni-directional clutch slipping or incorrectly installed				5									
	Uni-directional clutch seized													
	Input shaft broken			7										
	Front pump drive tangs on converter hub broken			8										
	Front pump worn		12	9										
	Converter blading and/or uni-directional clutch failed			10										

Diagnosis		Upshift Quality						Downshifts					Downshift Quality			
	SYMPTOM	Slip 1–2	Slip 2–3	Rough 1–2	Rough 2–3	Siezure 1–2	Siezure 2–3	No 2–1	No 3–2	Involuntary 3–2	Above normal speed	Below normal speed	Slip 2–1	Slip 3–2	Rough 2–1	Rough 3–2
ADJUSTMENT FAULTS	Fluid level insufficient	1	1							1						
	Downshift cable incorrectly assembled or adjusted	2	2	1	1			1	1		1	1				
	Manual linkage incorrectly assembled or adjusted	3	3													
	Incorrect engine idling speed													1		1
	Incorrect front band adjustment	4	4	2	2		1									
	Incorrect rear band adjustment					1										
HYDRAULIC CONTROL FAULTS	Oil tubes incorrectly installed, missing or leaking	8	9				2			2				6		
	Sealing ring missing or broken	9	10			5	3				5	5		7	3	
	Valve block screws missing or loose	10	11	10	6	6	4				6	6		8		5
	Primary regulator valve sticking	6	7	3	3									4		3
	Throttle valve sticking	7	8	4	4						4	4		5		4
	Modulator valve sticking			5												
	Governor valve sticking, leaking or incorrectly assembled			6				3	3		2	2				
	Orifice control valve sticking													3		2
	1–2 shift valve sticking							2	2			7				
	2–3 shift valve sticking											8				
	2–3 shift valve plunger sticking										3	3				
	Converter 'out' check valve sticking or missing															
	Check valve sticking or missing															
MECHANICAL FAULTS	Front clutch slipping			9											2	
	Front clutch seized or plates distorted															
	Rear clutch slipping		5							3				9		6
	Rear clutch seized or plates distorted				5	2			4							7
	Front band slipping due to faulty servo or worn band	5	6						5					2		8
	Rear band slipping due to faulty servo or worn band							4								
	Uni-directional clutch slipping or incorrectly installed			7		3							1		1	
	Uni-directional clutch seized			8		4										
	Input shaft broken															
	Front pump drive tangs on converter hub broken															
	Front pump worn															
	Converter blading and/or uni-directional clutch failed															

PUMP

Remove and refit 44.32.01

Service tools: CBW 60, CBW 547A-50

Removing

1 Remove the transmission unit. 44.20.01.
2 Wash the exterior of the unit in clean petrol or paraffin, invert it and place on a bench cradle CBW 60. Remove the switch. 44.15.15.
3 Unscrew the bolts.
4 Remove the torque converter housing.
5 Unscrew 12 bolts.
6 Remove the oil pan, joint washer and magnet.
7 Pull out the oil tubes.
8 Release the inner downshift cable from downshift cam.
9 Take out three bolts and washers.
10 Lift off the valve block.
11 Unscrew two bolts.
12 Remove the oil tube locating plate.
13 Pull out the oil tubes. (Note the 'O' ring on the pump suction tube.)
14 Take out five bolts.
15 Remove the pump and joint washer.
16 Remove the thrust washer.

Refitting

17 Using petroleum jelly, stick the thrust washer to the pump assembly.
18 Refit the pump assembly and joint washer.
19 Fit and tighten the bolts.
20 Refit the oil tubes. (Note the 'O' ring on the pump suction tube.)
21 Refit the oil tube locating plate.
22 Fit and tighten the two bolts.
23 Carefully refit the valve block, ensuring that the oil tubes are not distorted.

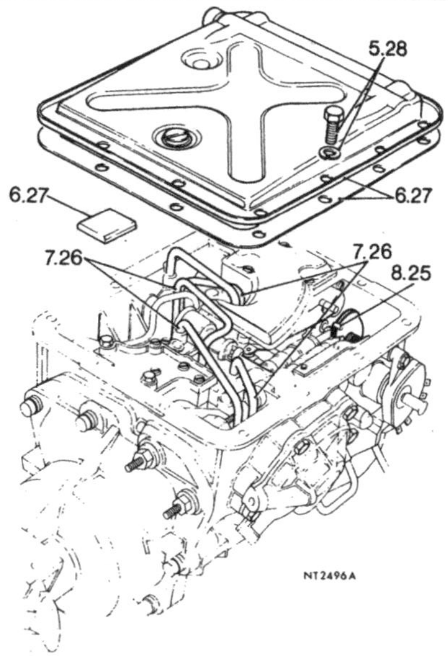

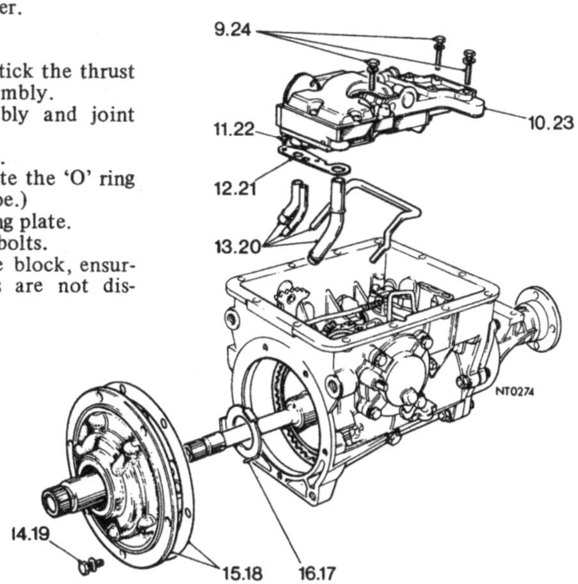

24 Fit and tighten the three bolts and washers.
25 Connect the downshift inner cable to the downshift cam.
26 Refit the oil tubes.
27 Replace the magnet and refit the oil pan and joint washer.
28 Fit and tighten 12 bolts.
29 Locate the torque converter housing in place.
30 Fit and tighten four bolts.
31 Refit the switch. 44.15.15.
32 Refit the transmission unit.

PUMP

Overhaul 44.32.04

Service tool: CBW 547A-50

1 Remove the pump. 44.32.01.

Dismantling

2 Unscrew the bolts.
3 Take out the locating screw.
4 Separate the stator support from the pump body assembly.
5 Mark the outside faces of the gears to facilitate correct assembly.
6 Remove the gears.
7 Remove the 'O' ring.
8 Extract the seal.

Reassembling

9 Renew the seal.
10 Refit the 'O' ring.
11 Fit the gears into the pump body.
12 Lightly lubricate the gears and the 'O' ring.
13 Refit the stator support.
14 Fit and tighten the locating screw and lock washer.
15 Fit and tighten the bolts and lock washers.
16 Refit the front pump. 44.32.01.

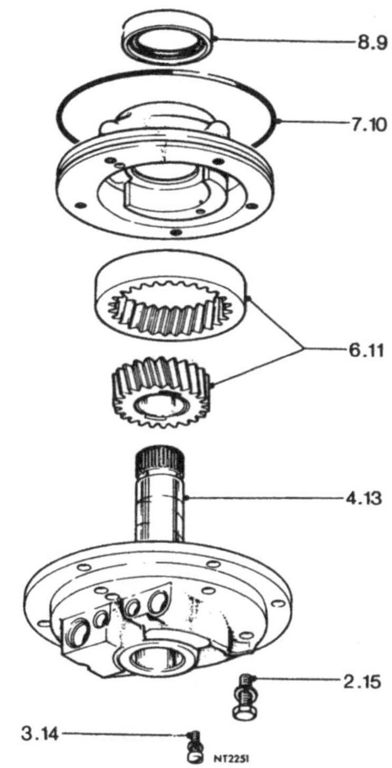

FRONT SERVO

Remove and refit 44.34.07

Service tool: CBW 547A-50

Removing

1 Drive the vehicle onto a ramp, select 'N' and apply the hand brake.
2 Disconnect the gearbox selector lever.
3 Take out the four bolts.
4 Withdraw the front servo assembly, spring and joint washer.

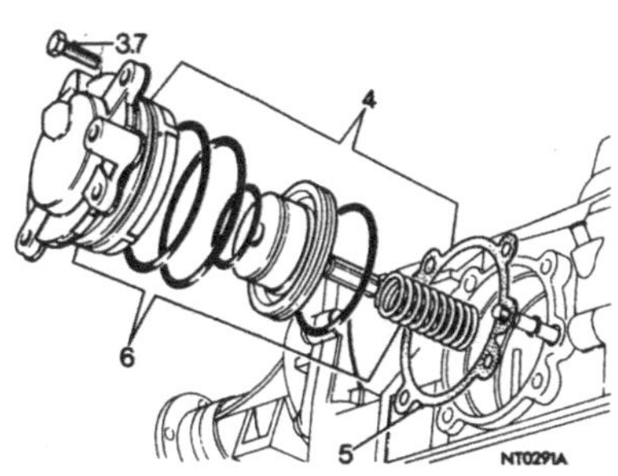

Refitting

5 Locate the joint washer onto the servo body flange.
6 Refit the servo and spring.
7 Fit the tighten the bolts using CBW 547A-50.

FRONT SERVO

Overhaul 44.34.10

1 Remove the front servo. 44.34.07

Dismantling

2 Remove the spring.
3 Withdraw the piston using air pressure.
4 Remove the 'O' rings from the body.
5 Remove the 'O' rings from the piston.

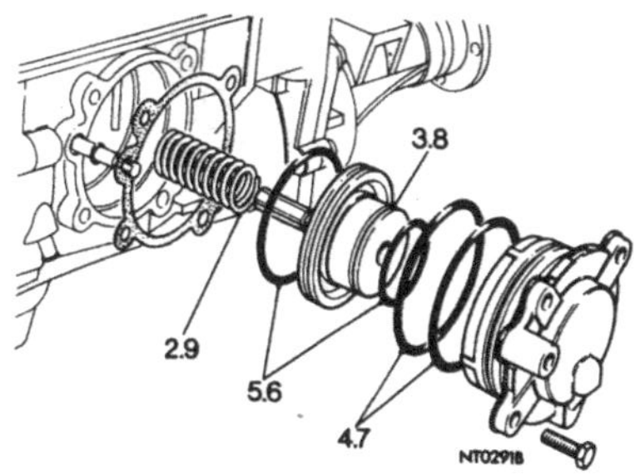

Reassembling

6 Fit the 'O' rings to the piston.
7 Fit the 'O' rings to the body.
8 Refit the piston.
9 Fit the spring.
10 Refit the servo assembly.

REAR SERVO

Remove and refit 44.34.13

Service tool: CBW 547A-50

1 Remove the exhaust front pipe. 30.10.09.

Removing

2 Unscrew the six bolts.
3 Withdraw the servo and joint washer together with 'O' rings, spring and push-rod.

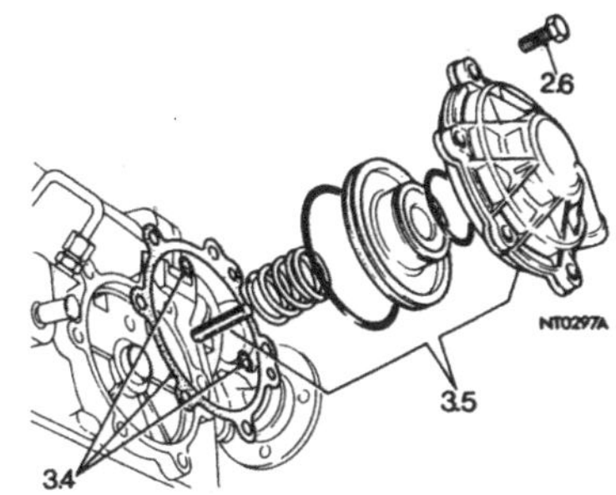

Refitting

4 Locate the 'O' rings and joint washer onto the gearbox casing.
5 Fit the servo assembly, spring and push-rod.
6 Fit and tighten the six bolts using CBW 547A-50.
7 Refit the exhaust front pipe. 30.10.09.

REAR SERVO

Overhaul 44.34.16

1 Remove the rear servo. 44.34.13.

Dismantling

2 Remove the push-rod.
3 Remove the spring.
4 Withdraw the piston using air pressure.
5 Remove the 'O' rings.

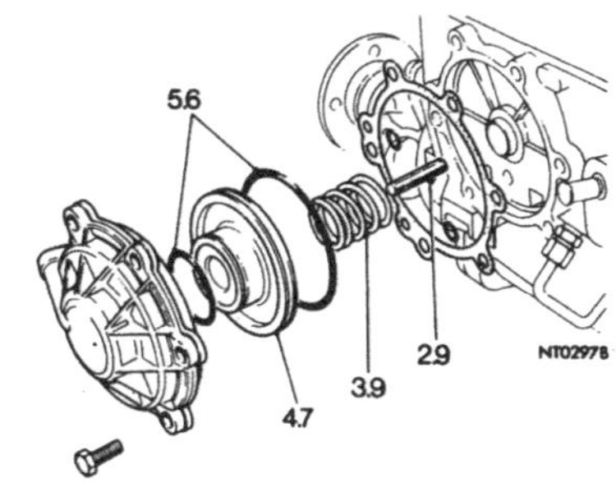

Reassembling

6 Fit the 'O' rings to the piston.
7 Refit the piston.
8 Refit the spring.
9 Refit the push-rod.
10 Refit the rear servo.

OUTPUT SHAFT

Remove and refit **44.36.01**

Service tools: CBW 60, RG 421 or S 337, CBW 547A-50

Removing

1 Remove the transmission unit. 44.20.01.
2 Wash the exterior of the unit in clean petrol or paraffin, invert it and place on a bench cradle CBW 60. Remove the switch. 44.15.15.
3 Unscrew the bolts securing the torque converter housing.
4 Remove the torque converter housing.
5 Unscrew 12 bolts.
6 Remove the oil pan, joint washer and magnet.
7 Pull out the oil tubes.
8 Release the downshift inner cable from the downshift cam.
9 Take out three bolts and washers.
10 Lift off the valve block.
11 Unscrew two bolts.
12 Remove the oil tube locating plate.
13 Pull out the oil tubes. (Note the 'O' ring on the pump suction tube.)
14 Take out five bolts.
15 Remove the pump and joint washer.
16 Remove the thrust washer.
17 Withdraw the front clutch.
18 Remove the thrust washers.
19 Withdraw the rear clutch and forward sun gear.
20 Squeeze together the ends of the front brake band and remove it together with the strut.

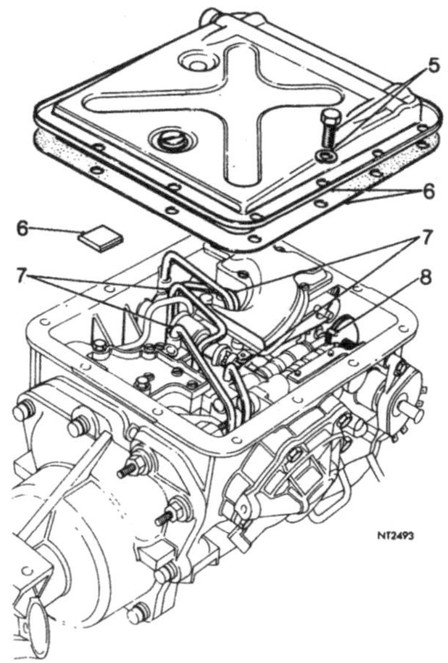

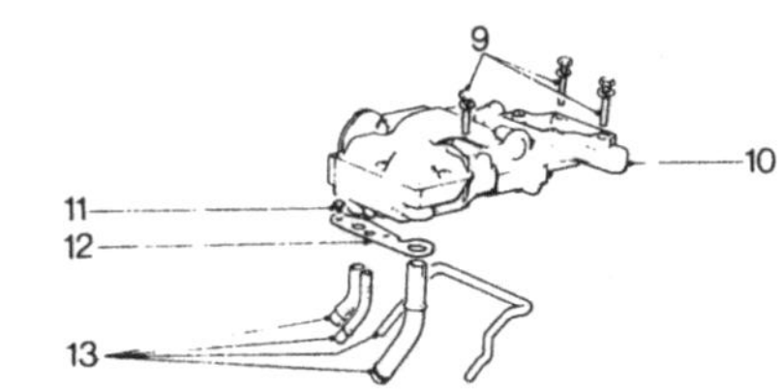

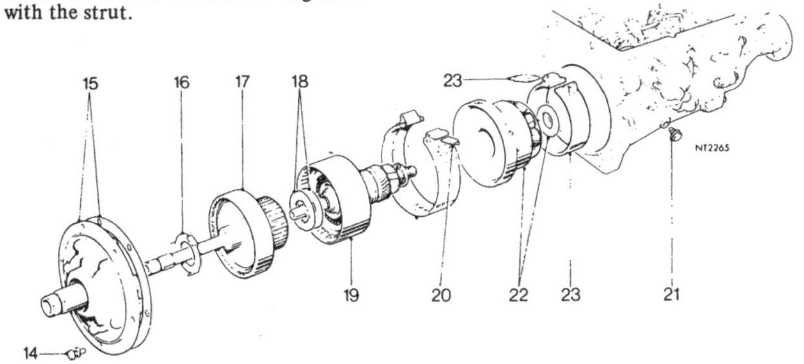

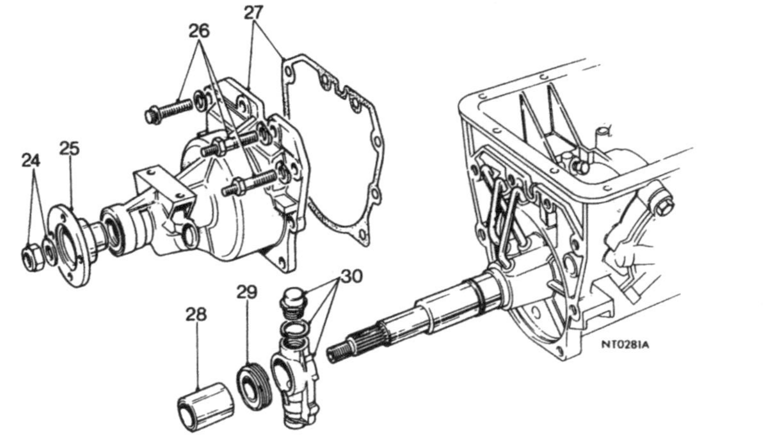

21 Unscrew the three bolts.
22 Withdraw the centre support/planet gear assembly and needle thrust assembly.
23 Squeeze together the ends of the rear brake band, tilt and withdraw together with the strut.
24 Using tool no. RG 421 or S 337 to retain the flange, unscrew the nut.
25 Withdraw the flange.
26 Unscrew the bolts.
27 Withdraw the rear extension and joint washer.
28 Remove the clamp tube.
29 Withdraw the speedometer drive gear.
30 Unscrew the counterweight and remove the governor.
31 Withdraw the output shaft assembly.
32 Remove the thrust washer.
33 Remove the circlip.
34 Detach the outer annulus from the output shaft.

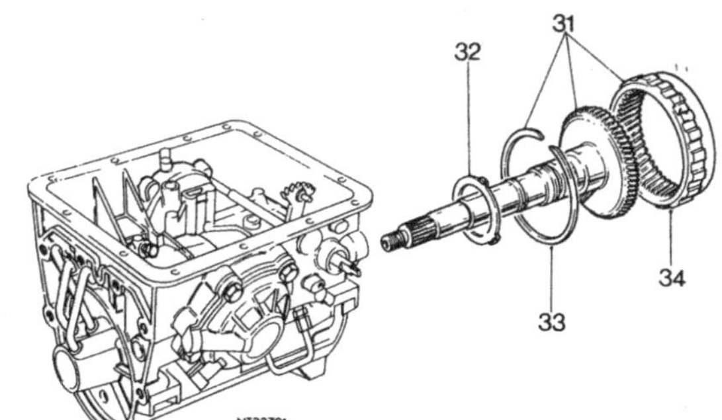

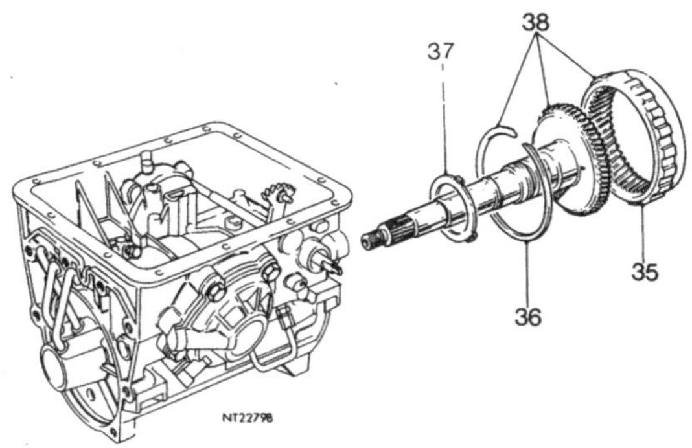

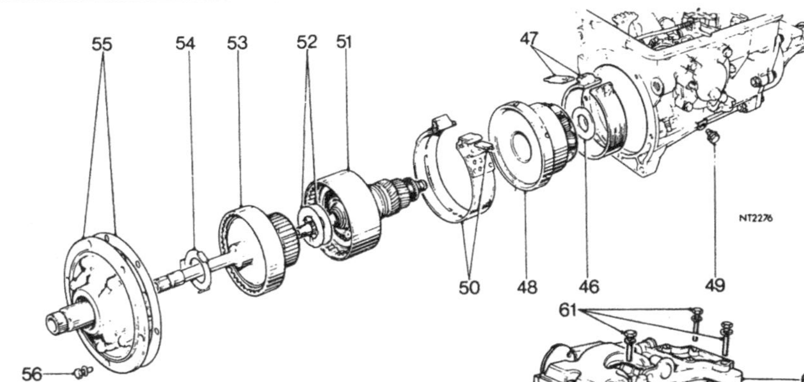

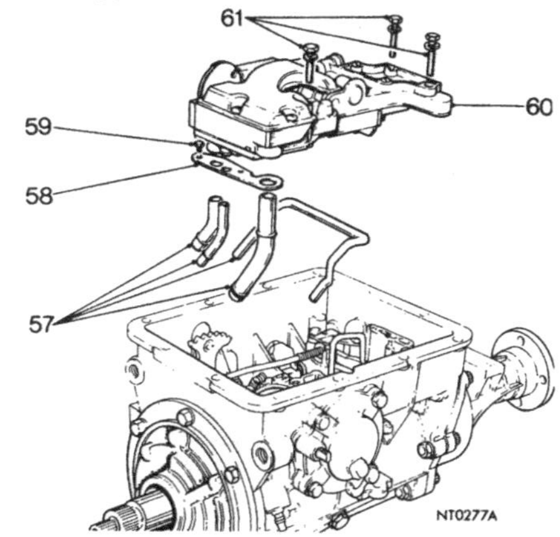

Refitting

35 Assemble the outer annulus and the output shaft.
36 Fit the circlip.
37 Using petroleum jelly, stick the thrust washer to the casing.
38 Refit the output shaft assembly.
39 Refit the governor and secure it with the counterweight.
40 Refit the speedometer drive gear.
41 Refit the clamp tube.
42 Refit the rear extension, using a new joint washer if necessary.
43 Fit and tighten the bolts.
44 Refit the flange.
45 Holding the flange with tool no. RG 421 or S 337, fit and tighten the nut.
46 Using petroleum jelly, stick the needle thrust bearing onto the planet gear case (rear drum).
47 Refit the rear brake band and strut.
48 Refit the centre support/planet gear assembly, ensuring that the oil and locating holes align with those in the casing.

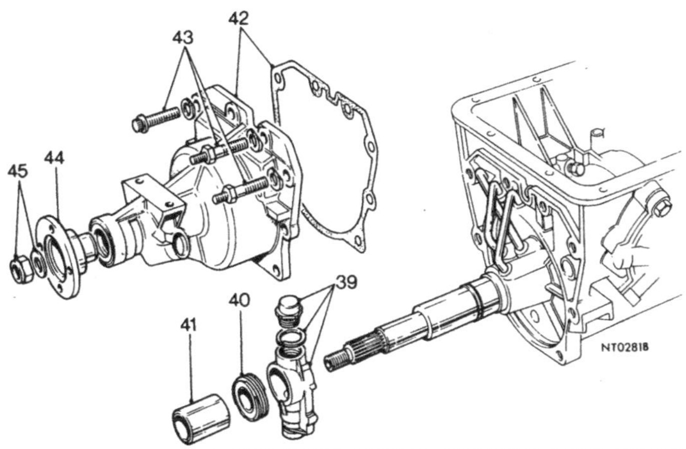

49 Fit and tighten the bolts.
50 Squeeze together the ends of the front brake band and fit it in position together with the strut.
51 Refit the rear clutch and forward sun gear assembly.
52 Using petroleum jelly, stick the thrust washers to the rear clutch assembly (phosphor bronze towards the front clutch).
53 Refit the front clutch assembly.
54 Using petroleum jelly, stick the thrust washer to the pump assembly.
55 Refit the pump assembly and joint washer.
56 Fit and tighten the bolts.
57 Refit the oil tubes. (Note the 'O' ring on the pump suction tube.)
58 Refit the oil tube locating plate.
59 Fit and tighten the two bolts.
60 Carefully refit the valve block, ensuring that the oil tubes are not distorted.
61 Fit and tighten the three bolts and washers.
62 Connect the downshift inner cable to the downshift cam.
63 Refit the oil tubes.
64 Replace the magnet and refit the oil pan and joint washer.
65 Fit and tighten 12 bolts.
66 Locate the torque converter housing in place.
67 Fit and tighten four bolts securing the torque converter housing.
68 Refit the switch. 44.15.15.
69 Refit the transmission unit. 44.15.15.

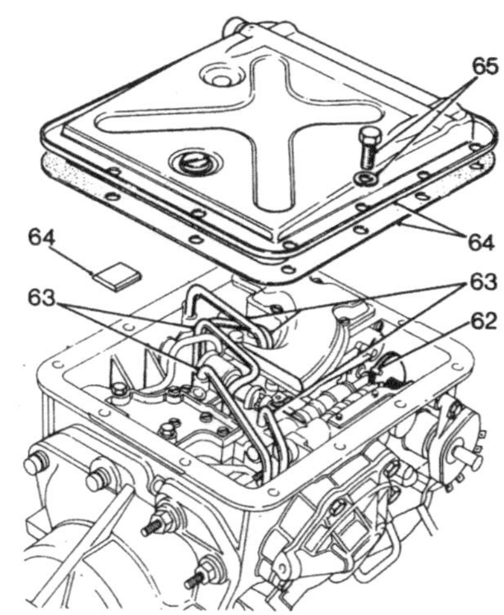

PLANET GEARS AND REAR DRUM ASSEMBLY

Remove and refit 44.36.04

Service tools: CBW 60, CBW 547A-50

Removing

1 Remove the transmission unit. 44.20.01.
2 Wash the exterior of the unit in clean petrol or paraffin, invert it and place on a bench cradle CBW 60. Remove the switch. 44.15.15.
3 Unscrew the bolts securing the torque converter housing.
4 Remove the torque converter housing.
5 Unscrew 12 bolts.
6 Remove the oil pan, joint washer and magnet.
7 Pull out the oil tubes.
8 Release the downshift inner cable from the downshift cam.
9 Take out three bolts and washers.
10 Lift off the valve block.
11 Unscrew two bolts.
12 Remove the oil tube locating plate.
13 Pull out the oil tubes. (Note the 'O' ring on the pump suction tube.)
14 Take out five bolts.
15 Remove the pump and joint washer.
16 Remove the thrust washer.
17 Withdraw the front clutch.
18 Remove the thrust washers.
19 Withdraw the rear clutch and forward sun gear.
20 Squeeze together the ends of the front brake band and remove it together with the strut.
21 Take out three bolts.

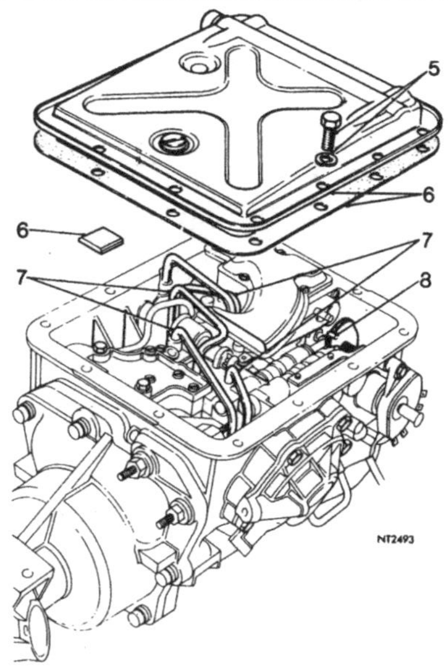

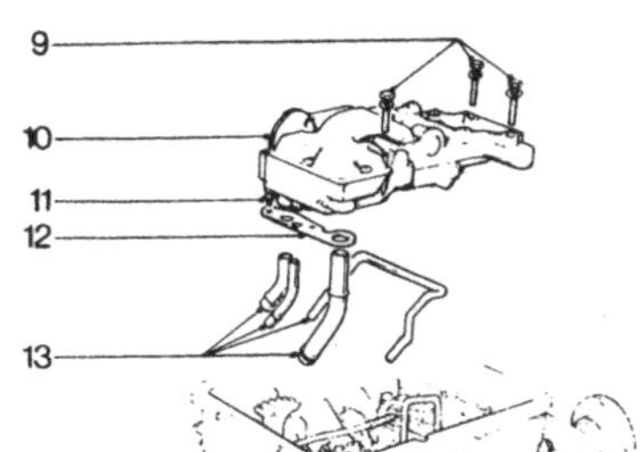

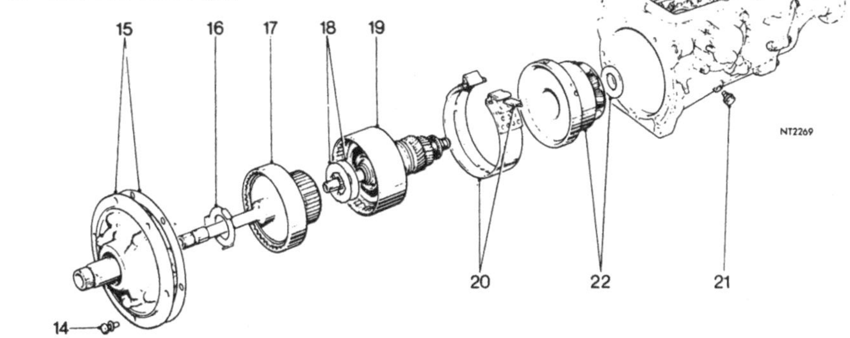

22 Withdraw the centre support/planet gear assembly.
23 Separate the centre support from the planet gear assembly.
24 Withdraw the uni-directional clutch.
25 Remove the circlip.
26 Detach the uni-directional clutch outer race.

Refitting

27 Fit the uni-directional clutch outer race to the rear drum assembly.
28 Fit the circlip.
29 Refit the uni-directional clutch.
30 Assemble the centre support and planet gear assembly.
31 Refit the centre support/planet gear assembly, ensuring that the oil and locating holes align with those in the casing.
32 Fit and tighten the bolts.
33 Squeeze together the ends of the front brake band and fit it in position together with the strut.
34 Refit the rear clutch and forward sun gear assembly.
35 Using petroleum jelly, stick the thrust washers to the rear clutch assembly (phosphor bronze towards the front clutch).

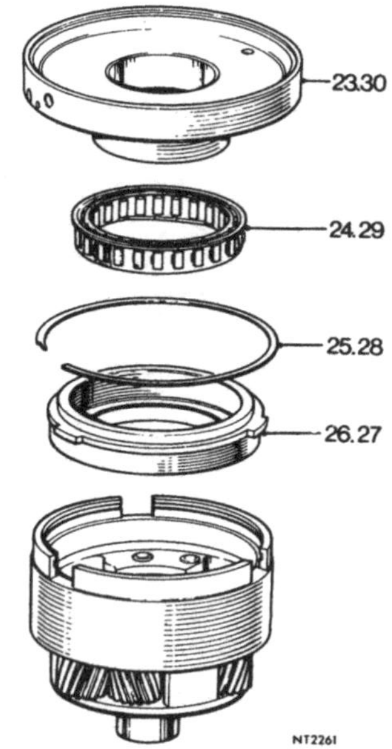

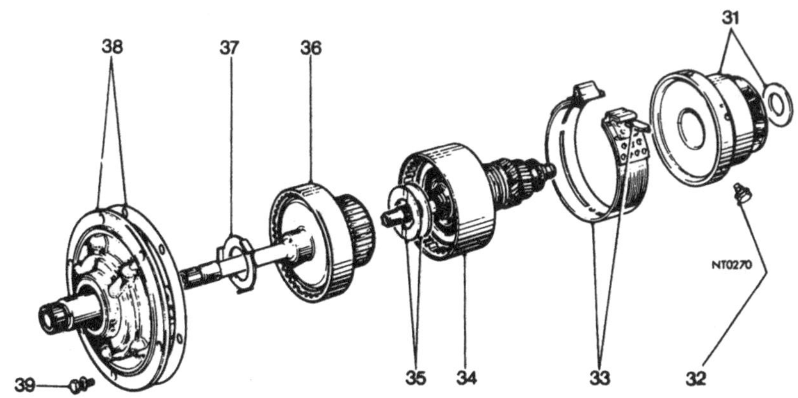

36 Refit the front clutch assembly.
37 Using petroleum jelly, stick the thrust washer to the pump assembly.
38 Refit the pump assembly and joint washer.
39 Fit and tighten the bolts.
40 Refit the oil tubes. (Note the 'O' ring on the pump suction tube.)
41 Refit the oil tube locating plate.
42 Fit and tighten the two bolts.
43 Carefully refit the valve block, ensuring that the oil tubes are not distorted.
44 Fit and tighten the three bolts and washers.
45 Connect the downshift inner cable to the downshift cam.
46 Refit the oil tubes.
47 Replace the magnet and refit the oil pan and joint washer.
48 Fit and tighten 12 bolts.
49 Locate the torque converter housing in place.
50 Fit and tighten four bolts securing the torque converter housing.
51 Refit the switch. 44.15.15.
52 Refit the transmission unit.

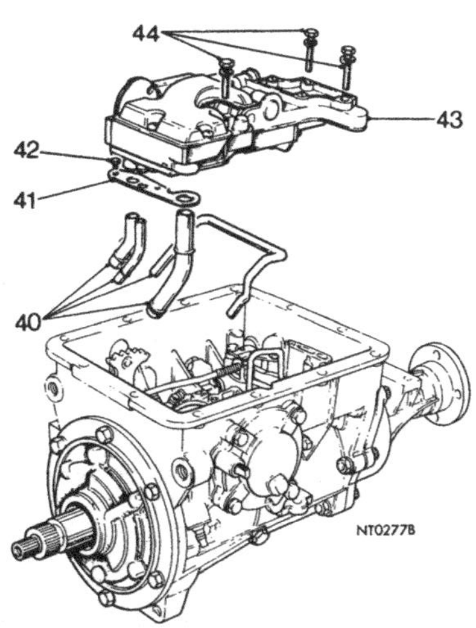

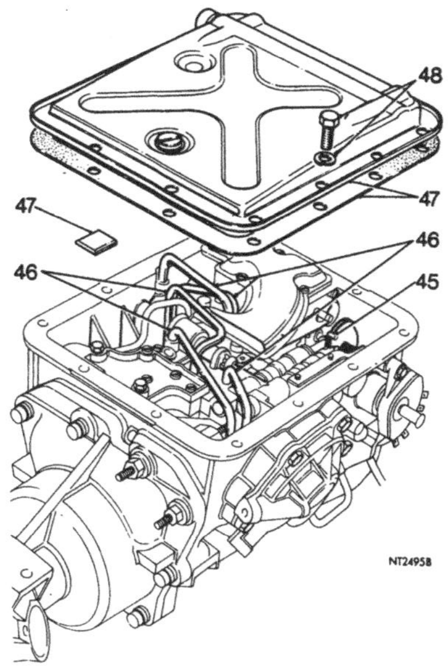

SPEEDOMETER DRIVE PINION

Remove and refit 44.38.04

Removing
1 Drive the vehicle onto a ramp, apply the hand brake and raise the ramp.
2 Disconnect the speedometer cable from the gearbox.
3 Carefully prise the speedometer pinion housing out of the extension.
4 Withdraw the pinion from the housing.
5 Remove the 'O' ring.
6 Extract the seal.

Refitting
7 Press a new seal into the housing.
8 Fit a new 'O' ring to the housing.
9 Fit the drive pinion into the housing.
10 Press the housing into the rear extension.
11 Refit the speedometer cable.
12 Lower the ramp.

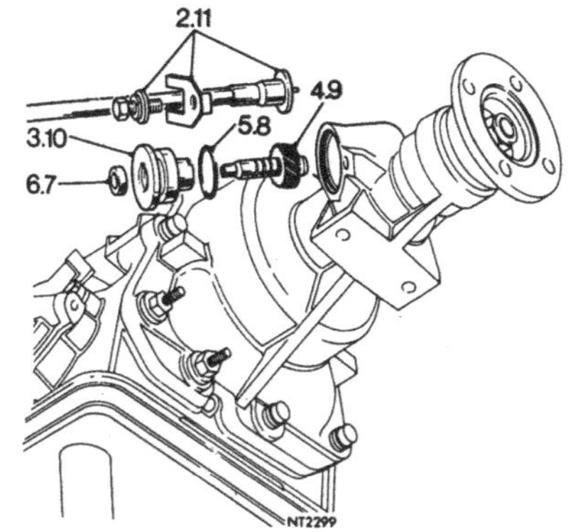

SPEEDOMETER DRIVE GEAR

Remove and refit 44.38.07

Removing
1 Remove the rear extension. 44.20.15.
2 Remove the clamp tube.
3 Withdraw the speedometer drive gear.

Refitting
4 Fit the speedometer drive gear.
5 Refit the clamp tube.
6 Refit the rear extension. 44.20.15.

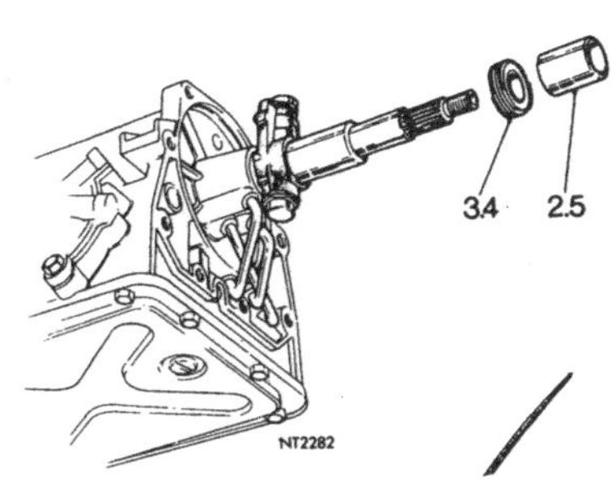

VALVE BLOCK

Remove and refit 44.40.01

Removing
1 Remove the sump pan, 44.24.04, and remove the switch. 44.15.15.
2 Remove the magnet.
3 Pull out the oil connector pipes.
4 Disconnect the downshift cable from the cam.
5 Take out three bolts.
6 Release the valve block.

Refitting
7 Ensure that the pipes are correctly located.
8 Fit the valve block to the unit.
9 Secure with three bolts.
10 Attach the downshift cable to the cam, ensuring that the cam is correctly located on the manual valve.
11 Refit the oil connector pipes.
12 Attach the magnet to one end of the bolt heads.
13 Refit the switch. 44.15.15.
14 Replace the sump pan. 44.24.04.

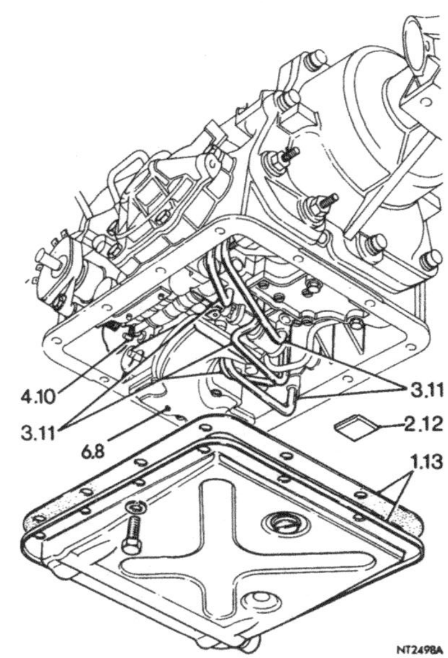

VALVE BLOCK

Overhaul **44.40.04**

Service tool: CBW 548

1 Remove the valve block. 44.40.01.

Dismantling

2 Take out two screws.
3 Remove the downshift cam assembly.
4 Take out four screws.
5 Remove the oil strainer and gasket.
6 Take out screw.
7 Remove detent spring and spacer.
8 Take out eight screws.
9 Remove upper valve body.
10 Take out eight screws.
11 Remove the oil tube collector.
12 Take out four screws.
13 Remove the governor line plate.
14 Remove the separating plate.
15 Remove the check valve.
16 Remove the check valve ball and spring.
17 Remove the servo orifice control valve spring and stop.
18 Remove the throttle valve stop and return spring.
19 Remove the throttle valve plate.
20 Withdraw the manual control valve.
21 Withdraw the downshift valve.
22 Remove the throttle valve spring.
23 Withdraw the throttle valve.
24 Tap out the dowel pin, applying light pressure to the plug.
25 Withdraw the modulator plug.
26 Withdraw the modulator valve.
27 Withdraw the modulator valve spacer.
28 Withdraw the modulator valve spring.
29 Withdraw the servo orifice control valve.
30 Slacken progressively the three screws.
31 Carefully remove the end plate.
32 Remove the spring.
33 Withdraw the sleeve.
34 Take out the primary regulator valve.
35 Remove the spring.
36 Withdraw the secondary regulator valve.
37 Remove the screws from the upper valve body.
38 Remove the front end plate.
39 Take out three screws.
40 Remove the rear end plate.
41 Withdraw the 2–3 shift valve from the rear.
42 Remove the spring.
43 Withdraw the plunger.
44 Withdraw the 1–2 shift valve from the rear.
45 Remove the spring.
46 Withdraw the plunger.

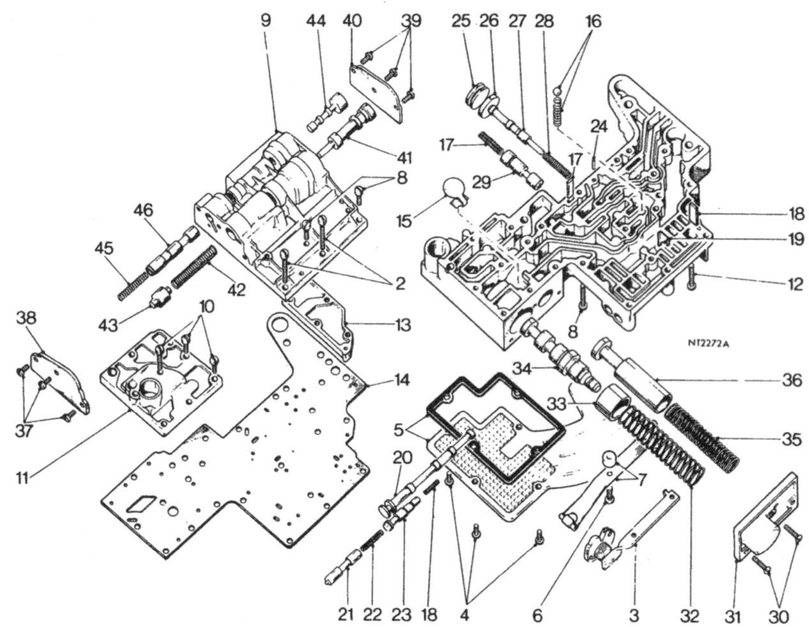

Reassembling

47 Insert the 1–2 shift valve.
48 Insert the 2–3 shift valve.
49 Replace the rear end plate.
50 Fit and tighten the three screws.
51 Insert the 1–2 shift valve plunger.
52 Insert the 2–3 shift valve spring.
53 Insert the 1–2 shift valve spring.
54 Insert the 2–3 shift valve plunger.
55 Locate the front end plate in position.
56 Fit and tighten three screws.
57 Insert the secondary regulator valve into the lower valve body.
58 Refit the spring.
59 Insert the primary regulator valve.
60 Insert the sleeve.
61 Insert the spring.
62 Hold the end plate in position.
63 Fit and tighten the three screws.
64 Insert the servo orifice control valve.
65 Insert the spring.
66 Depress the spring and fit the stop.
67 Insert the modulator control valve spring.
68 Insert the spacer.
69 Insert the modulator control valve.
70 Insert the plug.
71 Fit the dowel pin.
72 Insert the throttle valve.
73 Insert the spring.
74 Insert the downshift valve.
75 Insert the manual control valve.
76 Insert the throttle valve return spring and stop.
77 Refit the throttle valve plate.
78 Refit the check valve ball and spring.
79 Refit the check valve.
80 Place the separating plate in position.
81 Hold the governor line plate in position.
82 Fit and loosely tighten the four screws.
83 Replace the oil tube collector.
84 Fit and loosely tighten eight screws.
85 Replace the upper valve body.
86 Fit and tighten the eight screws.
87 Refit the oil strainer and gasket.
88 Fit and tighten four screws.
89 Refit the detent spring and spacer.
90 Fit and tighten the screw.
91 Tension the downshift cam and refit the assembly.
92 Fit and tighten two screws.
93 Refit the valve block. 44.40.01.

PROPELLER SHAFT

PROPELLER SHAFT

Remove and refit **47.15.01**

Removing

1 Scribe the gearbox flange, rear axle flange, and propeller shaft flanges to enable reassembly in original locations.
2 Remove the guard strap, as applicable.
3 Remove the four bolts and nyloc nuts securing the propeller shaft and gearbox flanges.
4 Remove the four bolts and nyloc nuts securing the propeller shaft and rear axle pinion flanges.
5 Remove the propeller shaft.

Refitting

6 Reverse instructions 1 to 5.

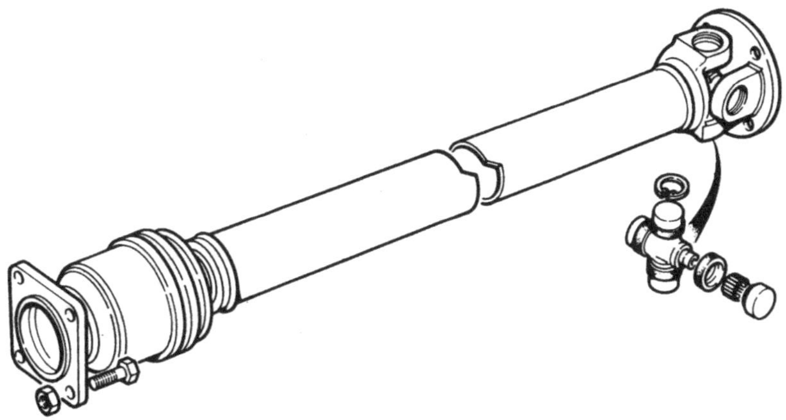

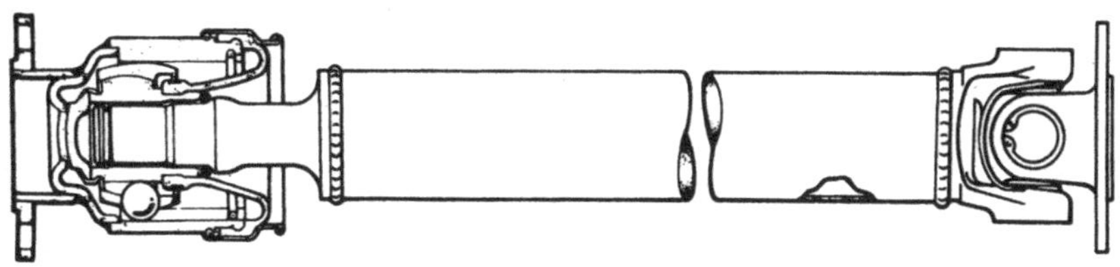

A47-4

Early type shown –

UNIVERSAL JOINT (Early types)

Remove and refit **47.15.18**

Removing

1 Remove the propeller shaft, see 47.15.01.
2 Remove paint, rust, etc., from the vicinity of the bearing cups and circlips.
3 Remove the circlips.
4 Tap the bearing yokes to eject the bearing cups.
5 Withdraw the bearing cups and spider.

Refitting

6 Remove the bearing cups from the new spider.
7 Check that all needles are in place.
8 Fit the spider to the yoke.
9 Engage the spider trunnion in the bearing cup and insert the cup in the yoke ensuring that the needle bearings are not displaced.
10 Fit the opposite bearing cup to the yoke and carefully press both cups into position ensuring that the spider trunnions engage both cups and that the needle bearings are not displaced.
11 Using two flat-faced adaptors of slightly smaller diameter than the bearing cups press the cups into the yoke until they reach the lower land of the circlip grooves. Do not press the cups below this point or damage may be caused to the cups and seals.
12 Fit the circlips.

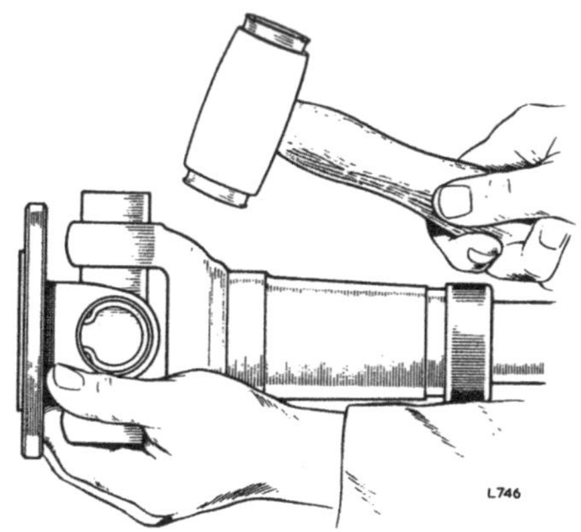

L746

L790

REAR AXLE OPERATIONS – 4 SPEED

[**CAUTION:** Please read the instructions thoroughly before you attempt rear axle repairs. If you lack the necessary skills or tools we suggest you leave this work to an Authorized Dealer or other qualified shop. Before taking a removed rear axle in for repair, thoroughly clean its exterior but do not undertake any disassembly.]

HALF SHAFT

– Remove and refit **51.10.01**

Service tools: S4235A, S4235A–1

Removing

1 Raise and support the rear of the car.
2 Remove the road wheel.
3 Remove the split pin, nut and washer securing the hub to the half shaft.
4 Remove the brake drum.
5 Remove the clevis pin linking the handbrake cable to the wheel cylinder operating lever.
6 Disconnect the brake pipe(s) from the wheel cylinder. Seal or blank off the pipe(s) and wheel cylinder to prevent the ingress of dirt.
7 Remove the four nuts, spring washers and bolts securing the backplate to the axle flange.
8 Using Tool S4235A and adaptor S4235A–1 withdraw the half shaft complete with hub, backplate, oil catcher plate, oil seal and housing and bearing retaining plate.

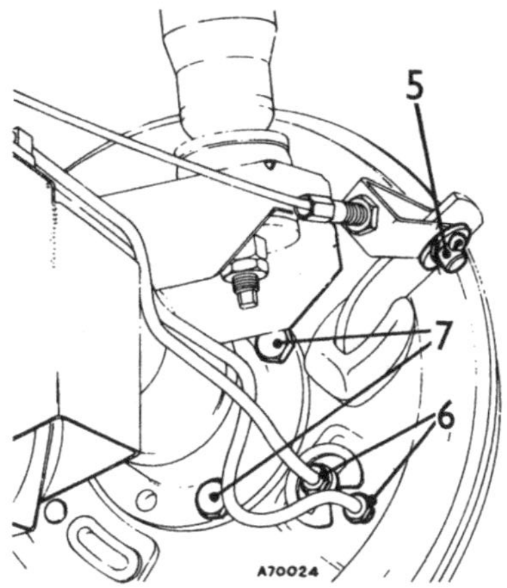

Refitting

9 Pack the half shaft bearing with a lithium base grease.
10 Fit the half shaft complete with hub, backplate, oil catcher plate, and oil seal and housing to the axle casing.
11 Align the holes in the oil seal housing plate, bearing retainer plate, backplate and oil catcher. Ensuring that the oil drain hole in the oil seal housing corresponds with the drain hole in the axle flange. Note that the trough of the oil catcher plate must be fitted below the half shaft.
12 Fit the four bolts, spring washers and nuts securing the backplate to the axle casing.
13 Fit the hub securing washer and nut and torque to 100–110 lbf ft (13,83–15,21 kgf m).
14 Connect the brake pipe(s).
15 Connect the hand brake cable.
16 Fit the brake drum.
17 Fit the road wheel.
18 Bleed the brakes.
19 Lower the car.

HALF SHAFT BEARING AND OIL SEAL

– Remove and refit **51.10.02**

Service tools: S356C, S4235A, S4235A-1

Removing

1 Raise and support the rear of the car.
2 Remove the road wheel.
3 Remove the nut and washer securing the hub to the half shaft.
4 Remove the brake drum. using Tool S356C.
5 Withdraw the hub from the half shaft
6 Remove the clevis pin and release the handbrake cable from the wheel cylinder operating lever.
7 Disconnect the brake pipe(s) from the wheel cylinder. Seal or blank off the pipe(s) and wheel cylinder connections to prevent the ingress of dirt.
8 Remove the four-nuts, spring washers and bolts securing the backplate to the axle flange.
9 Remove the oil catcher plate. Note that the elongated area of the plate aligns with drain holes in the backplate, oil seal housing and axle casing flange.
10 Remove the backplate.
11 Remove the hub oil seal housing and gasket. Note that the drain hole aligns with a corresponding hole in the axle casing flange.
12 Remove the oil seal from the housing.
13 Remove the key from the half shaft.
14 Withdraw the half shaft and bearing using Tool S4235A, S4235A-1.
15 Remove the inner oil seal from the axle casing.
16 Press the bearing from the half shaft.

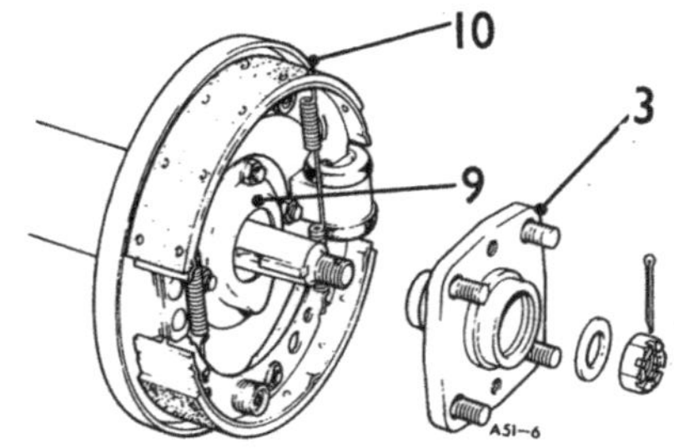

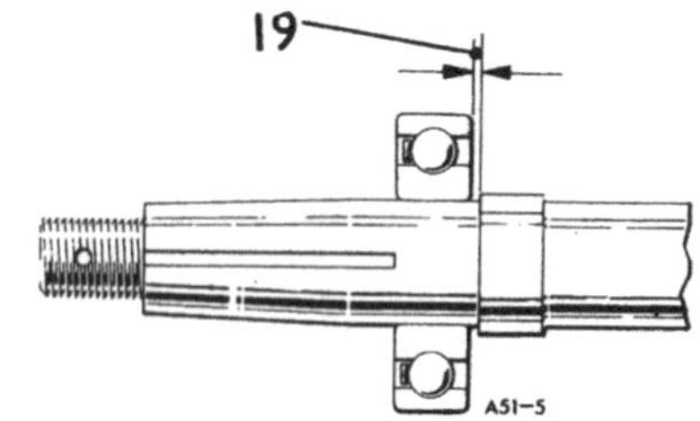

Refitting

17 Install a new inner seal in the axle casing. The lips of the seal must be towards the differential. Lubricate the seal lips.
18 Fit a new hub seal to the housing and lubricate the seal lips. The seal lips must be towards the differential.
19 Pack the half shaft bearing with a lithium-base grease and fit the bearing to the half shaft (shielded side of bearing to hub).
CAUTION: The bearing must not abut against the collar on the half shaft. A minimum clearance of 0.16 in (4,1 mm) between the bearing inner track and the half shaft collar is required. (The final, installed position of the bearing is determined by the hub when it is fully tightened).
20 Fit the half shaft, hub seal housing and gasket, backplate and oil catcher. Ensure oil drain holes align with the axle casing flange.
21 Fit the half shaft key.
22 Fit the hub washer and nut and torque to 100–110 lbf ft (13,83–15,21 kgf m).
23 Fit the brake drum.
24 Connect the brake pipe(s).
25 Connect the handbrake cable.
26 Fit the road wheel and hub cap.
27 Bleed the brakes.
28 Lower the car.

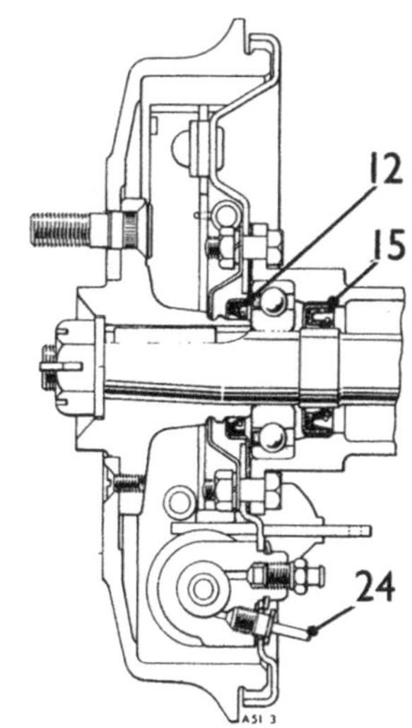

REAR HUB

– Remove and refit **51.10.18**

Service tools: S356C

Removing
1 Raise and support the rear of the car.
2 Remove the road wheel.
3 Remove the nut and washer securing the hub to the half shaft.
4 Remove the brake drum.
5 Withdraw the hub from the half shaft using Tool S356C and the nuts supplied. The standard wheel nuts must not be used for this purpose.

Refitting
6 Reverse instructions 1 to 5. The nut securing the hub to the half shaft must be torqued to 100–110 lbf ft (13,83–15,21 kgf m).

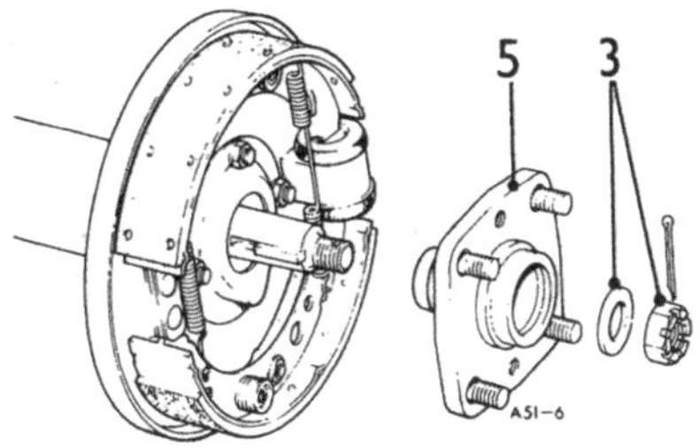

REAR HUB WHEEL STUDS

– Remove and refit **51.10.19**

Service tool: S356C

Removing
1 Raise the rear of the car and support securely.
2 Remove the rear hub. 51.10.18.
3 Drive out the wheel studs.

Refitting
4 Ensure that mating, tapered faces of studs and wheel hub are clean.
5 Enter the stud in the hub ensuring splines engage.
6 Drive or press the stud into position.
7 Fit the rear hub. 51.10.18.
8 Lower the car.

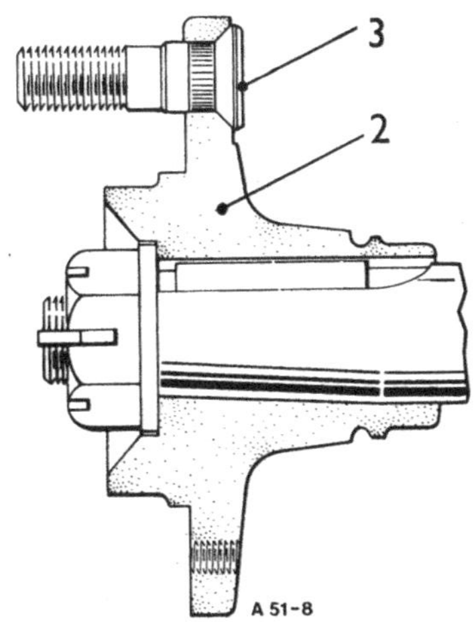

DIFFERENTIAL ASSEMBLY

– Remove and refit **51.15.01**

Service tools: S4235A, S4235A-1

Removing
1 Jack up or raise the rear of the car and support securely.
2 Remove both road wheels and hub caps.
3 Remove the split pins, nuts and washers securing the hubs to the half shafts.
4 Release the handbrake and remove both brake drums.
5 Release the handbrake cables from the rear brake cylinder operating levers.
6 Disconnect the three brake pipes from the rear wheel cylinders and pipe ends to prevent ingress of grit.
7 Remove the four bolts, nuts and spring washers securing each backplate to the axle flanges.
8 Fit Tool S4235A and adaptor S4235A-1 to the half shaft and withdraw the half shaft complete with hub, backplate and bearing. Repeat on other half shaft.
9 Remove the four bolts and nuts securing the propeller shaft flange to the differential pinion flange. Release the propeller shaft from the differential pinion.
10 Place a container under the differential to collect the oil when the differential is disturbed.
11 Remove the nuts and spring washers securing the differential assembly to the axle casing.
12 Ease the differential assembly flange from the axle casing and allow the oil to drain into the container.
13 Withdraw the differential assembly.

Refitting
14 Reverse instructions 1 to 13. Note: The hub securing nuts must be torqued to 100–110 lbf ft (13,83–15,21 kgf m).
15 Refill the axle with fresh lubricant.
16 Bleed the brakes.

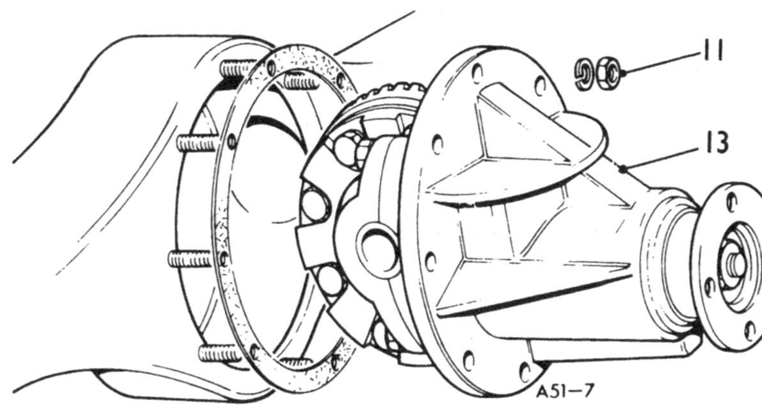

DIFFERENTIAL ASSEMBLY

– Overhaul **51.15.07**

Service tools: 47, 18G 47BD, S337 or 18G 1205, S4221A-17, 550, 18G 134DH, 18G 191, 18G 191M, S101

Dismantling

1 Remove the differential assembly. 51.15.01.

Crown wheel and differential unit

2 Place the differential assembly in a vice (crown wheel uppermost).
3 Mark the bearing caps to identify original locations.
4 Remove the bearing cap bolts and spring washers. (4).
5 Remove the bearing caps.
6 Carefully lever the crown wheel and differential unit clear of the assembly housing.
7 Lift out the crown wheel and differential unit complete with carrier bearings and shims.
8 Using press 47 and adaptor 18G 47BD withdraw the carrier bearings.
9 Remove the eight bolts and spring washers securing the crown wheel to the differential unit and withdraw the crown wheel.
10 Drive out the pin retaining the differential pinion pin.
11 Remove the differential pinion pin.
12 Rotate the differential sun wheels to bring the planet wheels and their respective selective thrust washers clear of the differential unit.
13 Remove the two planet wheels and thrust washers.
14 Remove the two sun wheels and thrust washers.

Pinion and bearings

15 Locate the differential assembly housing horizontally in the vice.
16 Remove the cap from the differential pinion flange.
17 Using Tool S337 to hold the pinion flange remove the pinion shaft nut and washer.
18 Withdraw the pinion flange.

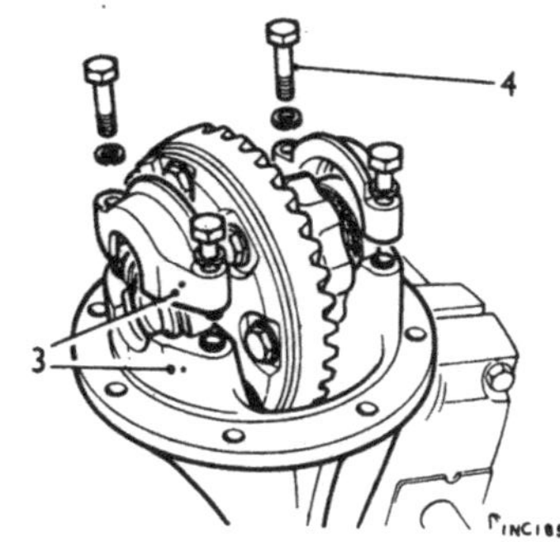

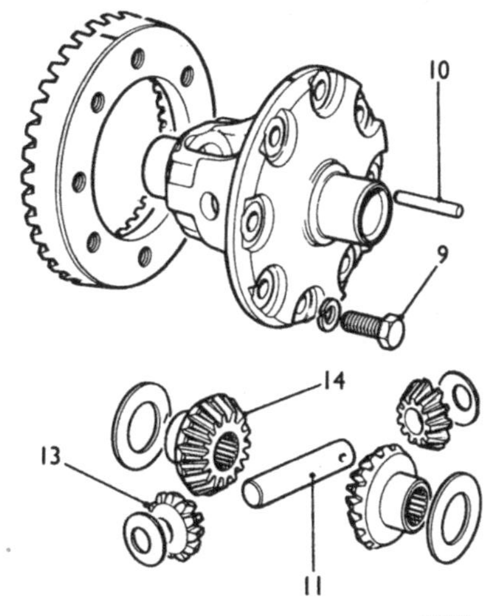

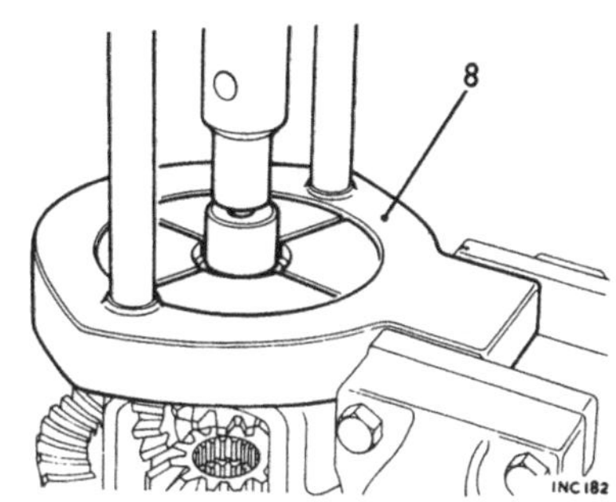

19 Using a hardwood block carefully tap out the pinion complete with pinion head bearing, selective washer and collapsible spacer.
20 Carefully drive out the pinion outer bearing and oil seal.
21 Evenly drift out both bearing outer tracks.
22 Using press 47 and adaptor S4221A-17 remove the pinion head bearing and selective washer.
23 Thoroughly clean all components.

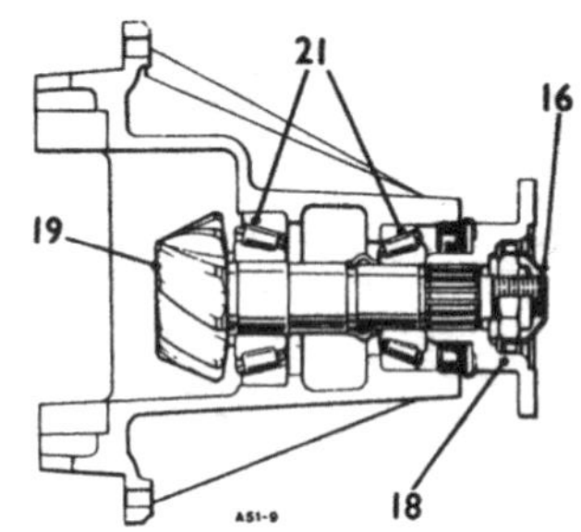

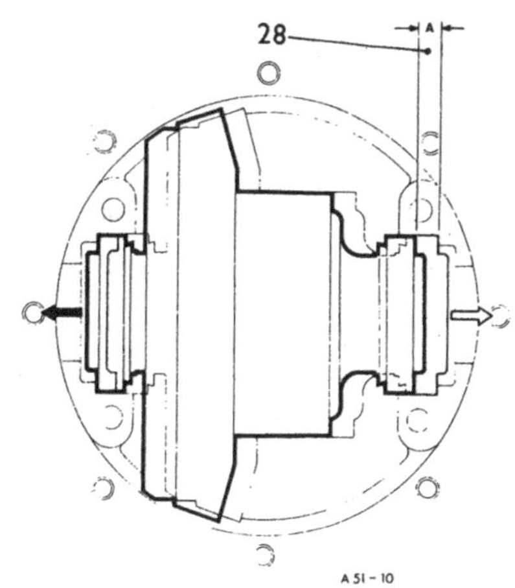

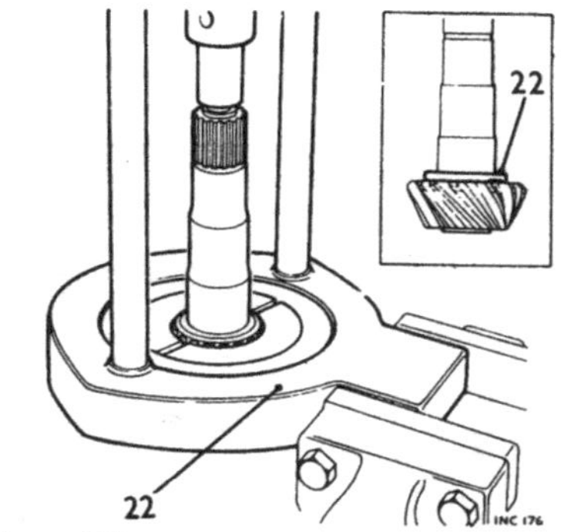

Assembling

Crown wheel and differential unit

24 Using Tool 550 and 18G 134DH fit the carrier bearings to the differential unit. (Tapered face of bearings towards half shafts.)
25 *Lightly* lubricate the carrier bearings and locate the differential unit in the differential casing.
26 Rotate the differential unit to allow the bearings to settle and push the differential unit complete with bearings to one side of the differential casing.
27 Using a dial gauge mounted against the crown wheel mounting flange check for 'run-out'. If 'run-out' is found to exceed 0.003 in (0,08 mm) it will be necessary to remove the carrier bearings and renew the differential unit housing. Ensure before discarding the differential unit housing that the excessive 'run-out' reading obtained is not due to improperly seated bearing cups. If 'run-out' is within 0.003 in (0,08 mm) proceed as follows:
28 Ensuring that the bearing cups are properly in contact with the bearings, float the differential unit laterally until it butts against the bearing seats in either direction. Using the dial gauge measure this lateral free movement. Call the measurement *Dimension 'A'*. Record Dimension 'A' as it will be used later to determine the required shim pack for the carrier bearings and to obtain correct pinion backlash.

continued

29 Remove the differential unit complete with bearings.
30 Lightly lubricate the sun wheel thrust faces and thrust washers and install them in the differential unit.
31 Mesh the planet wheels with the sun wheels ensuring the planets are diametrically opposite and rotate both sun wheels to bring the planets into their position in the differential unit casing.
32 Fit the differential pinion pin.
33 Check and assess the planet end-float.
34 Remove the pinion pin and rotate both sun wheels to bring the planets clear of the differential unit casing.
35 Lubricate the planet thrust washers selected and slide the planets and washers into position. Fit the pinion pin and again check the planets for end-float and backlash. Zero backlash is required. Choose thrust washers as appropriate from the nine thicknesses available.
36 Fit the retaining pin to the pinion pin and stave the casing.
37 Fit the crown wheel.
38 Smear the threads of the crown wheel securing bolts with 'Locquic' grade 'T' primer and 'Locquic 75' compound. Fit the bolts and spring washers and evenly tighten to the recommended torque.

Pinion and bearings
39 Fit the pinion inner bearing to the dummy pinion 18G 191M. *The standard pinion head spacer 0.077 in (1,95 mm) is incorporated in the dimensions of the dummy pinion.*
40 Lightly lubricate the bearings and fit the dummy pinion, bearings spacer, washer and nut.
41 Gradually tighten the nut until a bearing pre-load of 15 to 18 lbf in (0,17 to 0,21 kgf m) is obtained. This can be measured using a lbf in (kgf m) scale torque wrench and a suitable size socket spanner.
42 Ensure the face of the dummy pinion is clean. Position the dial gauge foot of 18G 191 on the dummy head and zero the gauge on to the head.

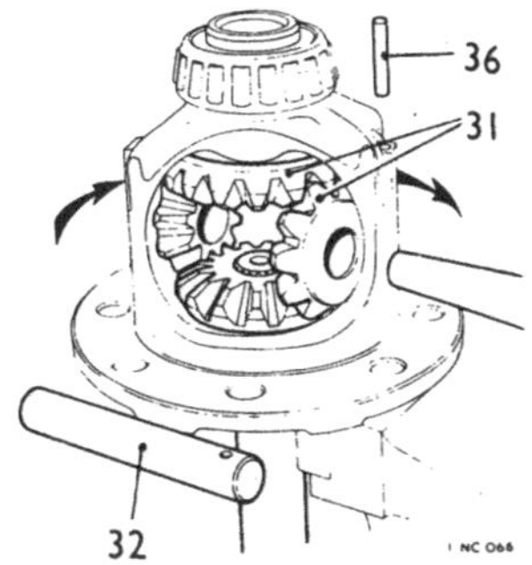

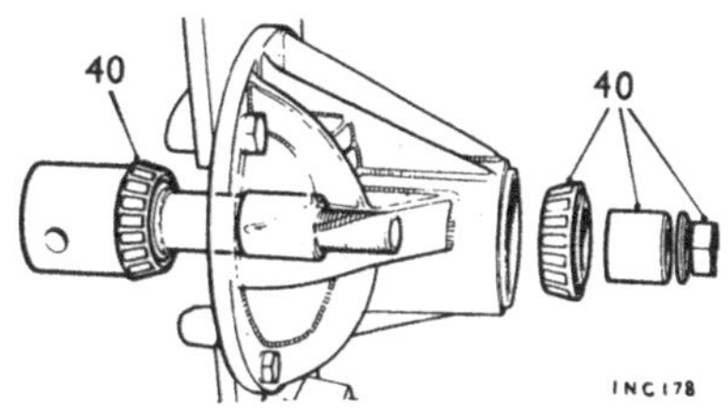

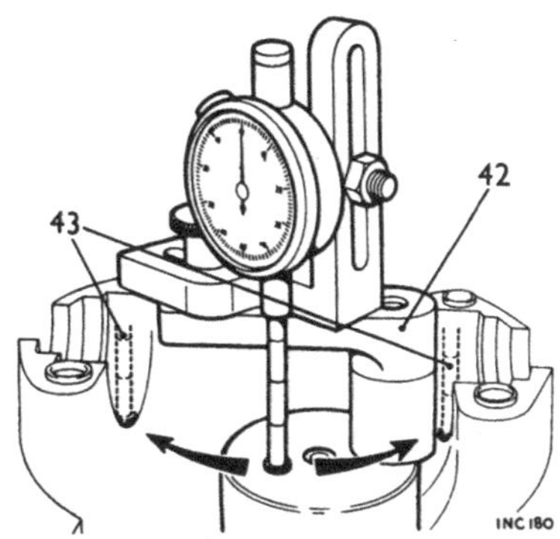

43 Move the gauge foot over the centre of one differential bearing bore. Note the indicated measurement. Repeat for the opposite bearing bore.
44 *Add the two measurements and divide by two.*
45 Twenty-two pinion head washers are available ranging from 0.075 to 0.096 in (1,91 to 2,44 mm).
46 Remove the dummy pinion 18G 191M and dismantle.

Calculating pinion head spacer size
Example
Sum of each bore measurement
divided by two . . . 0.002 in (0,051 mm)
Plus dummy pinion spacer
allowance . . . . . 0.077 in (1,956 mm)
Required size of pinion
head spacer . . . . 0.079 in (2,007 mm)

NOTE: Etched + or – markings will be found on the pinion face.
These markings should be ignored since they are allowed for in the design and method of using the dummy pinion.

47 Fit the correct pinion head spacer to the pinion.
48 Fit the inner bearing to the pinion using S4221A-17.
49 Insert the pinion, spacer and bearing in the differential casing.
50 Fit a new collapsible spacer and the outer bearing to the pinion shaft.
51 Fit a new oil seal (lips towards pinion). The seal should be soaked in clean oil for one hour before fitting.
52 Fit the flange, washer and nut to the pinion shaft.
53 Gradually tighten the nut checking the bearing pre-load.
54 Rotate the flange to settle the bearings. Using an in/lb (kg/m) torque wrench or S98A check the torque required to rotate the flange.
55 Tighten the nut as required to obtain a torque (flange rotation) of 13 to 20 lbf in (1,80 to 2,77 kgf m).
CAUTION: If the pinion nut is overtightened and the above torque figure exceeded it will be necessary to *renew* the collapsible spacer and repeat instructions 53 to 55.

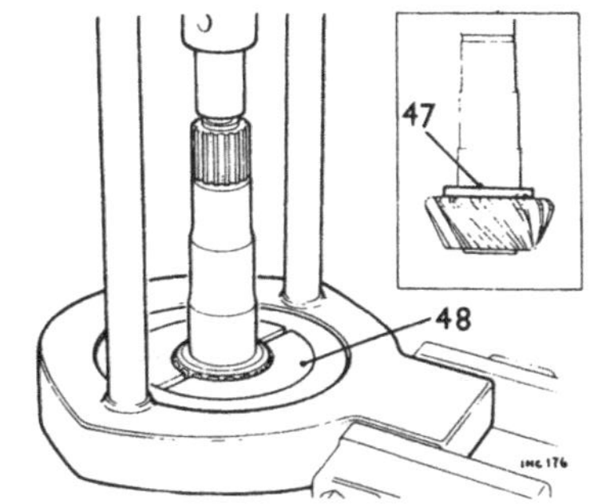

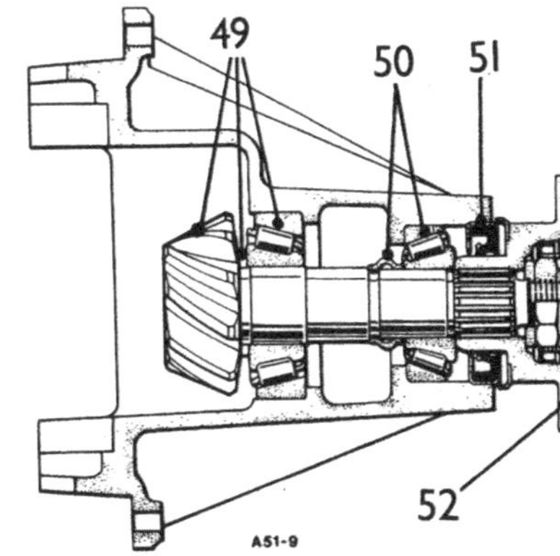

56 Place the differential unit/crown wheel assembly complete with bearing cups into position in the differential casing.
57 Move the crown wheel fully into mesh with the pinion and zero the dial gauge on the rear of the crown wheel.
58 Move the crown wheel and differential unit fully in the opposite direction. Note the gauge reading. This is the 'IN-OUT' of mesh clearance. Call this dimension 'B'.

continued

Setting Crown Wheel Backlash

Example
'IN-OUT' of mesh' clearance (Dimension B)
Instruction 58 . . . . . 0.025 in (0,63 mm)

Minus required backlash
0.005 in . . . . 0.005 in (0,127 mm)
0.020 in (0,503 mm)

Plus required carrier bearing pre-load divided by two 0.004 in / 2 . . 0.002 in (0,051 mm)
0.022 in (0,554 mm)

This equals shims required for crown wheel side carrier bearing.

Total side clearance (Dimension A)
Instruction 28 . . . . . 0.060 in (1,52 mm)
Minus above calculation for crown wheel side shims . . . 0.022 in (0,554 mm)
0.038 in (0,976 mm)

Plus carrier bearing pre-load divided by two 0.004 in / 2 . . . . . 0.002 in (0,051 mm)
0.040 in (2,027 mm)

This equals shims required for carrier bearing opposite crown wheel.

59 Allocate the shims calculated for each carrier bearing. Shims are available in sizes 0.003 in (0,076 mm), 0.005 in (0,127 mm), 0.010 in (0,25 mm), and 0.020 in (0,50 mm).
60 Fit the spreading Tool S101 to the differential casing.
61 Carefully stretch the differential casing to permit the differential unit complete with bearings and their respective shim packs to be placed in position.
IMPORTANT: Do not stretch more than is necessary or damage will be caused to the differential casing. Each flat on the turnbuckle is numbered. Do not exceed a stretch beyond three to four flats. The maximum permissible stretch is 0.008 in (0,20 mm). Do not lever against the stretcher.
62 Remove the stretcher.
63 Fit the bearing caps and the four securing bolts and spring washers. Evenly tighten the four bolts.
64 Fit the differential assembly to the axle casing. 51.15.01.

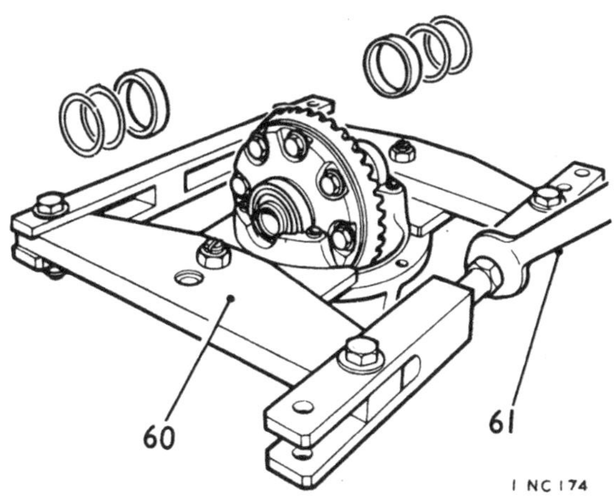

PINION OIL SEAL

– Remove and refit **51.20.01**

Service tools: S337 or 18G 1205

Removing
1 Raise the rear of the car and support securely.
2 Scribe the propeller shaft rear flange and differential pinion flange to identify original relationship.
3 Remove the four bolts and nuts securing the propeller shaft rear flange.
4 Release the propeller shaft from the differential pinion.
5 Remove the nut shield from the differential pinion flange.
6 Using a centre punch carefully mark the flange, pinion shaft and nut to identify original position.
7 Using Tool S337 to hold the differential pinion flange, unscrew and remove the nut and washer. Count the number of turns required for nut removal.
8 Withdraw the differential pinion flange.
9 Extract the pinion oil seal.

Refitting
10 Fit a new pinion oil seal (lip of seal towards axle).
Note: This seal should be soaked in engine oil for one hour before installation.
11 Fit the differential pinion flange, aligning the flange marking (instruction 6) to the mark in the pinion shaft.
12 Fit the washer and nut. Count the number of turns and using Tool S337 to hold the pinion flange, tighten the nut until the pop marks are aligned.
13 Fit the nut shield to the flange.
14 Fit the propeller shaft noting the flange alignment marks.
15 Fit and tighten the shaft four securing bolts and nuts.
16 Lower the car.

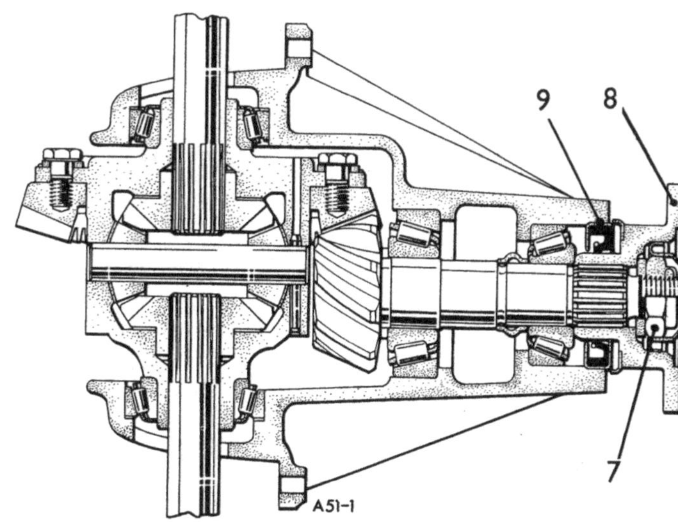

HALF SHAFT INNER OIL SEAL

– Remove and refit **51.20.14**

Removing
1 Remove the half shaft assembly. 51.20.14.
2 Remove the half shaft inner oil seal.

Refitting
3 Evenly install a new half shaft inner oil seal in the axle casing to lips of seal towards differential.
4 Lubricate the lips of the seal.
5 Refit the half shaft assembly. 51.20.14.

HUB BEARING OIL SEAL

– Remove and refit **51.20.17**

Service tool: S356C

Removing
1 Raise the car and support securely.
2 Remove the rear hub. 51.10.18.
3 Disconnect the handbrake cable from the rear wheel cylinder operating lever.
4 Disconnect the brake pipe(s) from the rear wheel cylinder, seal the pipe end(s) and cylinder to prevent ingress of dirt.
5 Remove the four bolts and nuts securing the backplate to the axle flange.
6 Withdraw the oil catcher plate, backplate and the hub oil seal housing and gasket.
7 Extract the hub oil seal from its housing. Renew the housing gasket.

Refitting
8 Fit a new oil seal to the seal housing (seal lips towards differential).
9 Lubricate the seal lips and reverse instructions 1 to 6.
10 Bleed the brakes.
11 Lower the car.

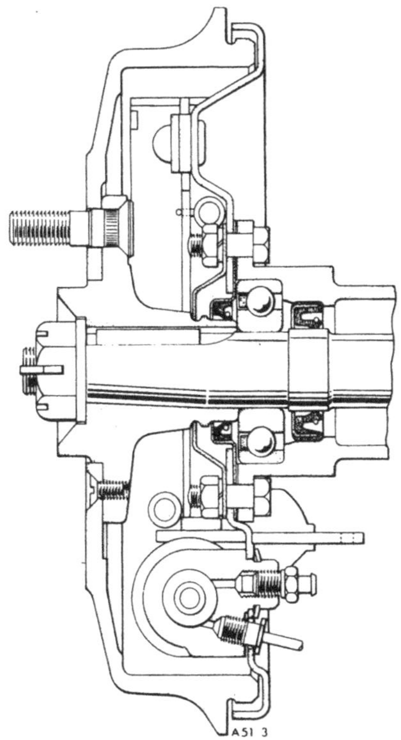

REAR AXLE ASSEMBLY

– Remove and refit **51.25.01**

Removing
1 Locate a trolley jack under the rear axle, raise the car and support the body securely on stands.
2 Remove the rear wheels.
3 Disconnect the propeller shaft at the rear axle.
4 Disconnect the forward end of the flexible rear brake hose.
5 Disconnect the handbrake cable forks at the backplate.
6 Remove the nut and bolt clamping the handbrake compensator and release the handbrake cables and trunnion. Withdraw the cables clear of the differential bracket.
7 Disconnect the rear dampers at the axle casing bracket.
8 Lower the jack and remove the rear road springs.
9 Remove the two bolts and nuts securing the radius rods to the axle and detach the handbrake cable bracket from the left hand side.
10 Remove the two bolts and nuts securing the axle to the rear suspension arms.
11 Lower the jack and release the radius rods from the axle brackets.
12 Lift the axle clear of the suspension arms and anti-roll bar and remove from the car.

Refitting
13 Reverse instructions 1 to 12.
14 Bleed the brakes.

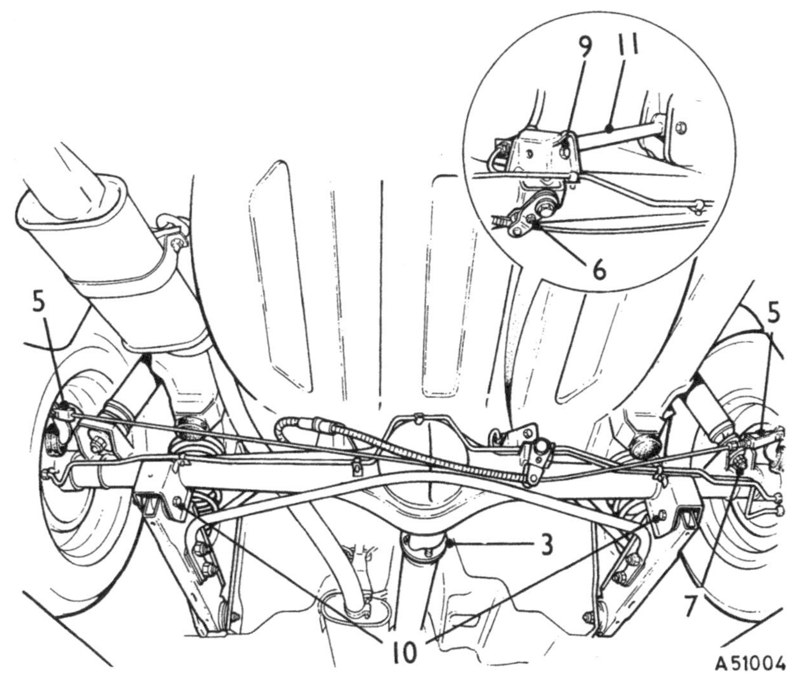

REAR AXLE CASING

– Remove and refit **51.25.04**

Service tools: S356C, S4235A, S4235A-1

Removing

1 Raise the rear of the car and support the body securely, on stands.
2 Remove the rear axle assembly. 51.25.01.
3 Remove the differential assembly. 51.15.01 (exclude instructions 1 and 2).
4 Detach the brake pipes from the five nylon clips.
5 Unscrew and remove the breather.
6 Remove the filler/level plug.

Refitting

7 Remove the brake pipe nylon clips (3 single, 2 double).
8 Renew the half shaft inner seals.
9 Reverse instructions 1 to 6.
10 Refill the casing with fresh lubricant.
11 Bleed the brakes.

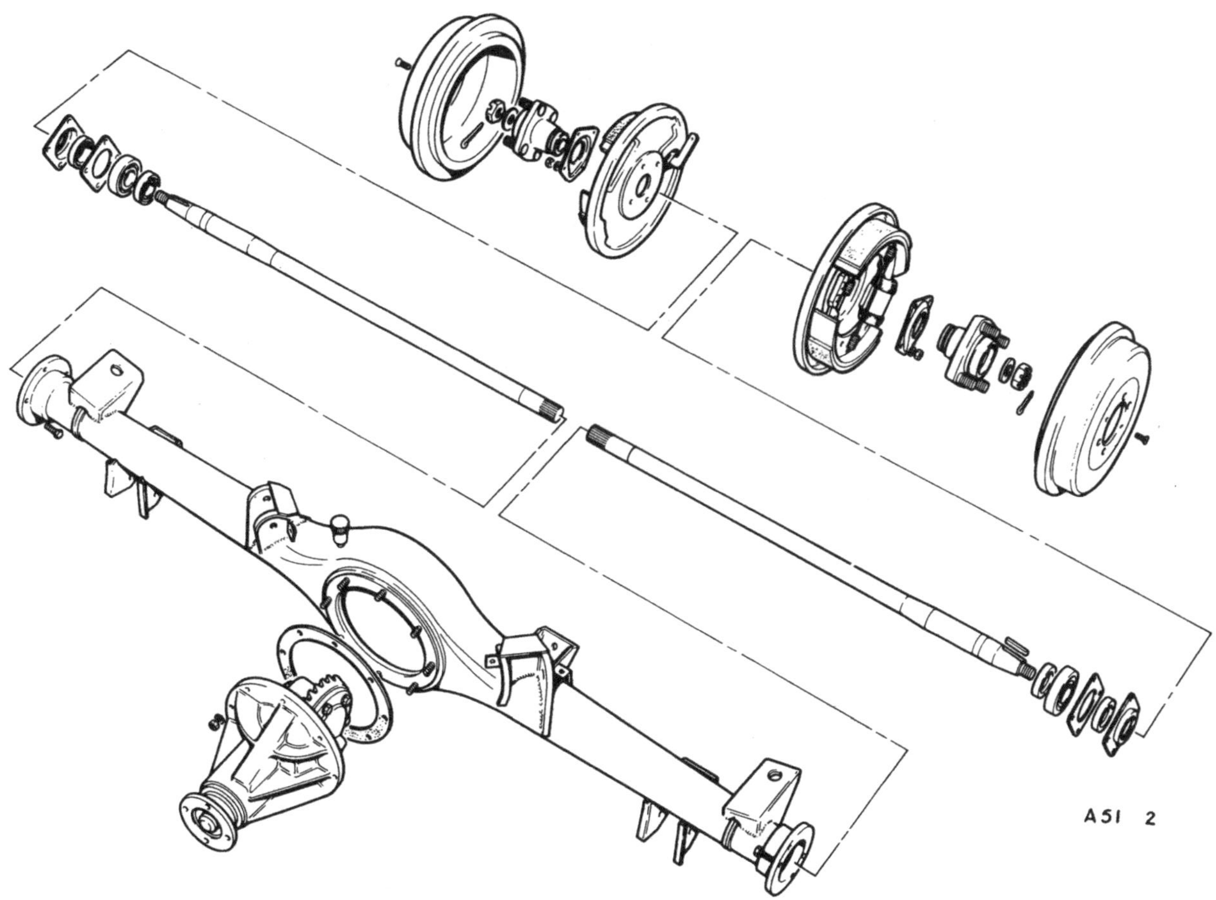

A51 2

REAR AXLE – 5 SPEED

[**CAUTION:** Please read the instructions thoroughly before you attempt rear axle repairs. If you lack the necessary skills or tools we suggest you leave this work to an Authorized Dealer or other qualified shop. Before taking a removed rear axle in for repair, thoroughly clean its exterior but do not undertake any disassembly.]

HALF SHAFT BEARING AND OUTER OIL SEAL

Remove and Refit 51.10.02

Service Tools: 18G 284 AR and 284

Removing

1 Jack up the rear of the car and support securely.
2 Remove the half shaft. 51.10.12
3 Using a drill, bore the retaining collar to weaken it.
CAUTION: Do not allow the drill to penetrate the collar as damage will be caused to the half shaft.

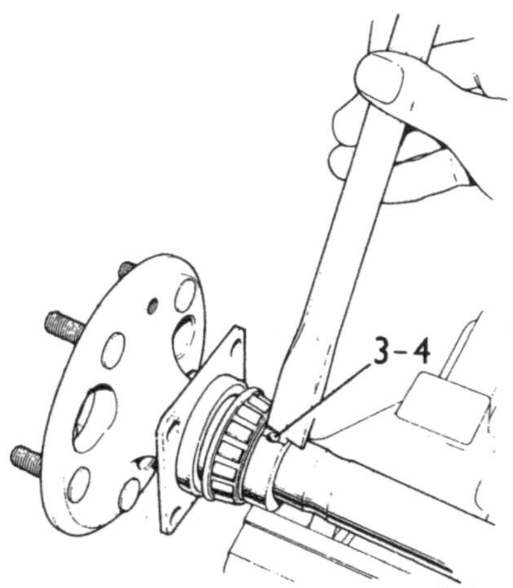

4 Using a hammer and chisel carefully burst the retaining collar and remove it from the half shaft.
5 Using a press remove the bearing, oil seal and retaining plate.
6 Using tools 18G 284 AR and 284 remove the half shaft bearing outer track from the axle casing.

Refitting

7 Fit the half shaft outer bearing track to the axle casing.
8 Fit the retaining plate to the half shaft (welded member adjacent to the shaft flange).
9 Lubricate the lip of the oil seal.
10 Slide the seal into position on the shaft.
11 Fit the bearing (tapered face of bearing towards the half shaft splines).
12 Wipe the shaft in front of the bearing clean of grease.
13 Smear the shaft in front of the bearing with Loctite 602 compound and also the bore of the new retaining collar.
14 Fit the new retaining collar and press the collar home until it butts against the bearing.
CAUTION: A force of not less than three tons should be required to slide the retaining collar into position over the last 0.125 in (3.175 mm) of its travel. If it is found that the interference fit of the collar is such that it can be fitted using a force of less than three tons the collar must be removed and another fitted.
15 Smear the bearing oil seal and the rear axle tube with a lithium base grease. See **NOTE** Operation No. 7 – 51.10.12.
16 Fit the half shaft. 51.10.12.
17 Lower the car.

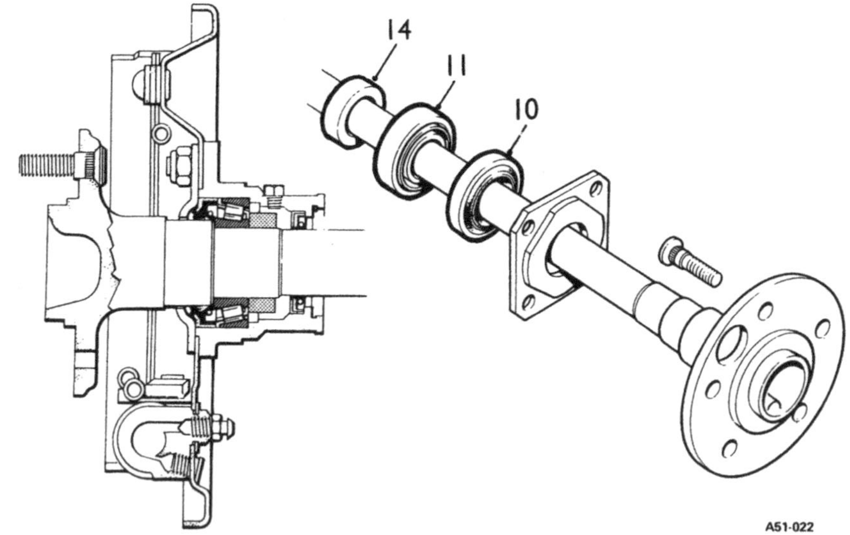

A51-022

HALF SHAFT ASSEMBLY

Remove and refit 51.10.12

Service Tools: 18G 284-1 and 284. (4235 or 3072 with S4235A-1 may be used in lieu of 284).

Removing

1 Jack up the rear of the car and support securely.
2 Remove the rear road wheel.
3 Release the handbrake and remove the brake drum.
4 Remove the four bolts and nuts securing the half shaft assembly and the back plate to the axle tube flange.
5 Using tools 18G 284-1 and 284 withdraw the half shaft.

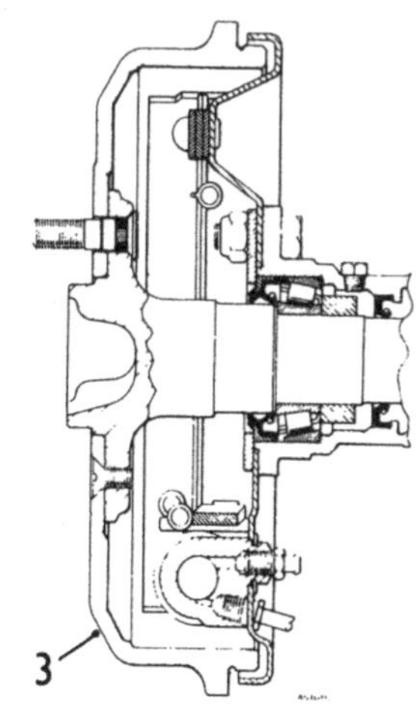

Refitting

6 Smear the interior of the axle tube/half shaft bearing area with a lithium base grease.
7 Similarly grease the half shaft bearing and oil seal.

NOTE: On initial build, when lubricating the half shaft bearing using a grease-gun, the shaft should be rotated during the operation to ensure that the bearing is properly 'primed'.
On initial assembly, a total of 40 gms of grease must be used.

8 Enter the half shaft in the axle tube and engage the differential splines.
9 Carefully slide the half shaft into position. Ensure that the bearing and the oil seal enter the axle tube squarely.
10 Fit the four securing bolts and nuts. Tighten them evenly.
11 Carefully wipe off surplus grease to prevent contamination of the brake linings.
12 Fit the brake drum.
13 Fit the road wheel.
14 Lower the car.

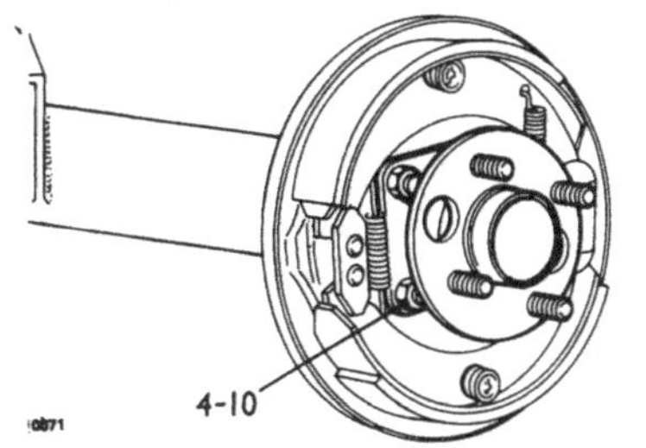

HALF SHAFT INNER OIL SEAL

Remove and refit 51.10.14

Service Tool: 18G 1271

Removing
1 Jack up the rear of the car and support securely.
2 Remove the half shaft assembly. 51.20.12.

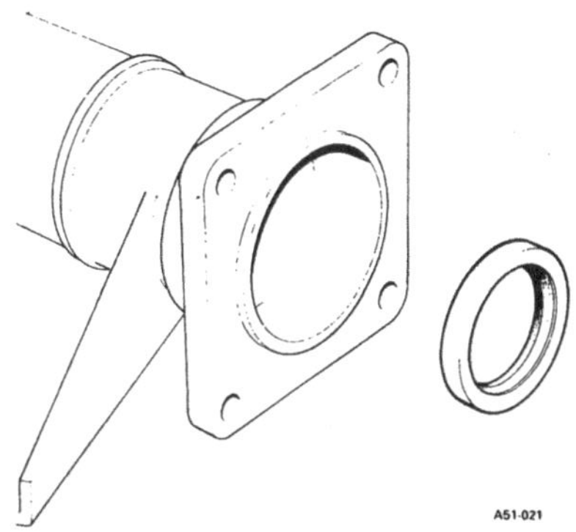

3 Remove the half shaft inner oil seal using tool 18G 1271

Refitting
4 Evenly install a new half shaft inner oil seal in the axle casing (lip of seal towards differential).
5 Lubricate the seal lip.
6 Fit the half shaft assembly.
7 Lower the car.

DIFFERENTIAL ASSEMBLY

Overhaul 51.15.07

Service Tools: S101, S101-1, 47, S4221A-16, 18G 191, 18G 191M, S98A, 18G 47AJ, 18G 1272, 6312, 18G 1273

1 Remove the rear axle assembly from the car. 51.25.01.
2 Remove the brake pipes.
3 Support the axle on stands.

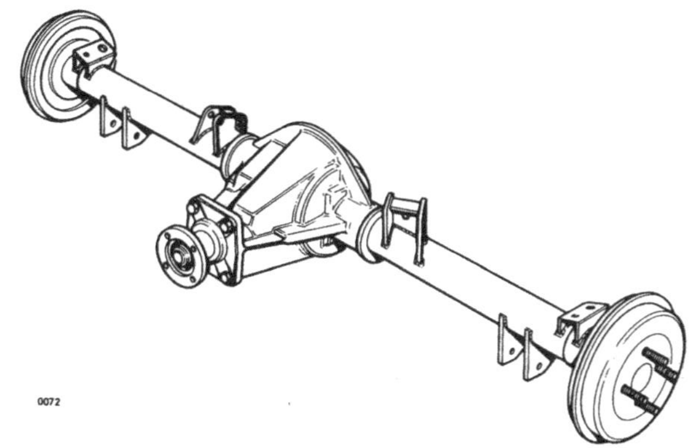

Dismantling
4 Remove the differential rear cover and drain the oil.
5 Remove the half shafts and brake backplates. 51.10.12.
6 Remove the nyloc nut securing the drive flange and remove the flange.
7 Remove the four bolts securing the pinion oil seal housing to the axle case.

Crown wheel differential unit
8 Check or mark the carrier bearing caps to establish original locations. Bearing caps must not be interchanged.
9 Remove the four bolts securing the bearing caps.
10 Remove the bearing caps.
11 Carefully lever the crown wheel and differential unit clear of the axle casing. If difficulty is experienced a spreader, tool S101 and S101-1 should be employed.
12 Lift out the crown wheel and differential unit complete with carrier bearings and shims.

continued

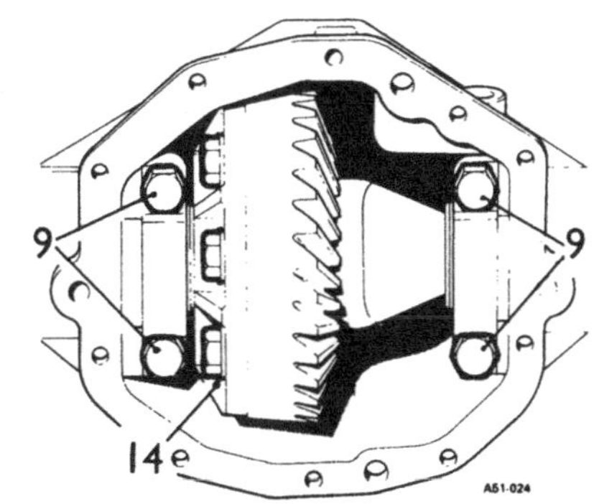

13 Using press 47 and adaptors S4221A-16 withdraw the carrier bearings.
14 Remove the eight bolts securing the crown wheel to the carrier flange and withdraw the crown wheel.
15 Withdraw the ball locating the differential pinion pin.
16 Remove the differential pinion pin.
17 Rotate the differential sun wheels to bring the two planet wheels and their respective thrust washers clear of the casing.
18 Remove the planet wheels and thrust washers.
19 Remove the two sun wheels and thrust washers.

Pinion and Bearings

20 Using tools 18G 1272 and S98A remove the pinion shaft nut.
21 Using a hardwood block carefully tap out the pinion complete with selective spacer, pinion head bearing and collapsible spacer.
22 Remove the pinion outer bearing from the axle casing.
23 Evenly drift out the pinion inner and outer bearing tracks from the axle casing. Care must be taken not to damage the axle casing.
24 Remove the collapsible spacer from the pinion shaft.
25 Using tool 47 and adaptor 18G 47AJ remove the pinion head bearing and spacer.

Assembling

Crown Wheel and Differential Unit

26 Fit the carrier bearings to the differential unit.
27 Lightly lubricate the carrier bearings, fit the bearing outer tracks and place the differential unit in the axle casing.
28 Slide the differential unit to one side of the axle casing and rotate the unit to allow the bearings to centralise.
29 Using a dial gauge check the crown wheel mounting flange for 'run-out'. 'Run-out' should not exceed 0.003 in (0.08 mm). Ensure that the bearing tracks are properly seated. When satisfied that 'run-out' is correct proceed as follows:
30 With the differential unit and bearings pressed to one side of axle casing note the reading on the dial gauge.
31 Slide the differential unit and bearings to the other side of the casing. Note this lateral travel registered on the dial gauge. Record this lateral travel (dimension 'A') as it is subsequently used to determine the shim pack required for the carrier bearings.
CAUTION: The lateral travel in some cases may be found to be restricted due to a foul condition between the axle case and the short side of the differential unit. In the event of this, a nominal shim should be used between the case and the bearing on the short side of the differential unit. The value of this shim should be added to the gauge reading to give the lateral travel.
32 Remove the differential complete with bearings.
33 Lightly lubricate the sun wheel thrust washers and install the sun wheel and thrust washers in the differential unit.
34 Mesh the planet wheels with the sun wheels ensuring that the planet wheels are diametrically opposite. Rotate the sun wheels to bring the planet wheels into position in the differential unit casing.

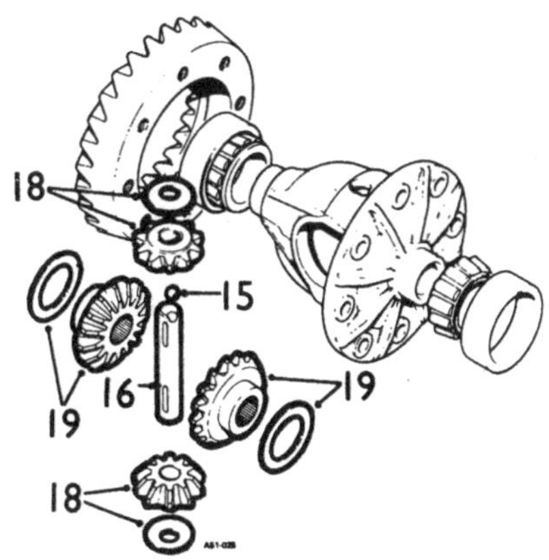

35 Fit the differential pinion pin.
36 Check and assess planet wheel end-float.
37 Remove the pinion pin and rotate both sun wheels to bring the planet wheels clear of the differential unit casing.
38 Lubricate the planet wheel thrust washers selected and slide the planet wheels and thrust washers into position. Fit the pinion pin ensuring that the groove aligns with the location for the retaining ball and again check the planet gears for backlash and end-float. Zero backlash is required. Choose thrust washers as necessary from the selection available.
39 Fit the crown wheel and the ball retaining the pinion pin.
40 Smear the threads of the crown wheel securing bolts with 'Locquic T' grade primer and 'Loctite 601' compound. Fit the bolts and evenly tighten. (See Torque Wrench Settings.)

Pinion and Bearings

41 Fit the pinion shaft inner and outer bearing tracks to the axle casing.
42 Fit the pinion inner bearing to the dummy pinion 18G 191-1.
NOTE: The dimension of the dummy pinion incorporates a built-in allowance for the maximum, pinion head bearing spacer available, 0.0492 in (1.25 mm).
43 Lightly lubricate the bearings and fit the dummy pinion, bearings, spacer, washer and nut.
44 Tighten the dummy pinion nut until a bearing pre-load of 15 to 18 lbf in (0.17 to 0.21 kgf m) is obtained. This can be checked using a lbf in (kgf m) scale torque wrench and a socket.
45 Mount a dial gauge (tool 18G 191) on the dummy pinion and zero the gauge using the dummy pinion head as a base.
46 Move the gauge stylus over the centre of one carrier bearing bore. Note the indicated measurement. Repeat for the opposite carrier bearing bore.
47 Add the two measurements and divide by two.

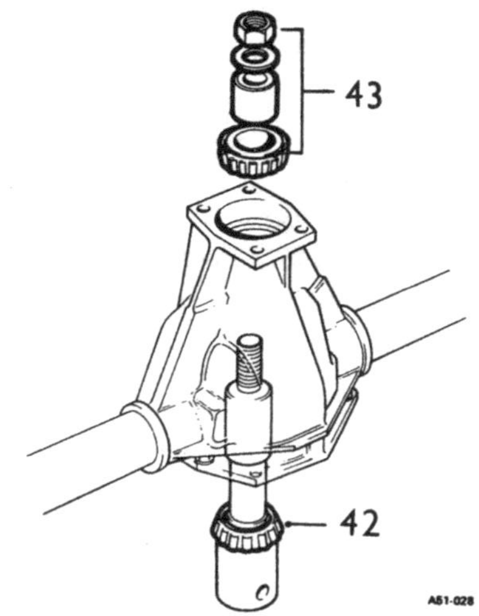

48 Remove the gauge and dummy pinion.
49 Calculating the spacer required for the pinion head bearing.
Example:
Sum of carrier bearing bore readings
= 0.002 in (0.051 mm) + 0.002 in (0.051 mm)
= 0.004 in (0.102 mm)
Divide by two
$= \frac{0.004 \text{ in } (0.102 \text{ mm})}{2} = 0.002 \text{ in } (0.051 \text{ mm})$

Add dummy pinion spacer allowance. 0.049 in (1.25 mm) (Inbuilt in dummy pinion dimensions)
= 0.002 in (0.051 mm) + 0.049 in (1.25 mm)
= 0.051 in (1.301 mm)
Examine the markings on the pinion head. Three markings will be found. These are:
X Identification for matched crown wheel/pinion set.

Y Variation in pinion head thickness from nominal.

Z This is a 'boxed' figure which indicates the variation from nominal setting to obtain the best running position. This dimension must be used when calculating the thickness of the pinion head spacer. If the 'boxed' figure Z is plus, subtract this value (+0.003 in) from the previous calculation. In the example 0.051 in (1.301 mm) – 0.003 in (0.076 mm) = 0.048 in (1.225 mm) = pinion head spacer required. If the 'boxed' figure Z is minus, add this value to the previous calculation. In the example 0.051 in (1.301 mm) 0.003 in (0.076 mm) = 0.054 in (1.337 mm) = pinion head spacer required.

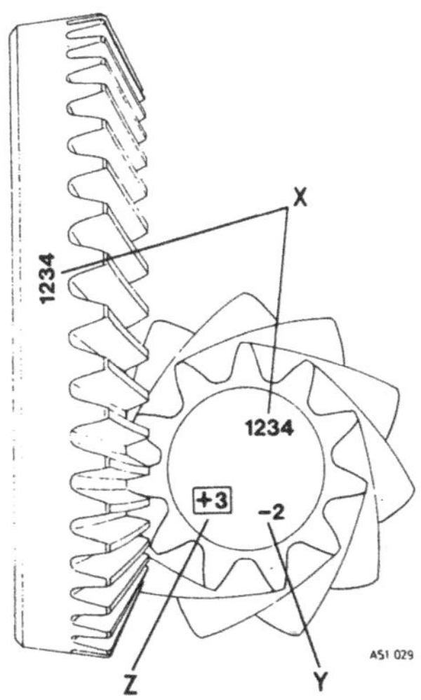

50 Fit a spacer of the required thickness to the pinion.

51 Fit the pinion head (inner) bearing to the pinion.

52 Fit a new collapsible spacer to the pinion and enter the pinion in the axle casing.

53 Fit the outer bearing, washer and nut.

54 Using tools 18G 1272 and S98 A carefully tighten the nut, periodically checking the torque required to rotate the pinion. Rotation torque should be within 13 to 20 lbf in (0.91 to 1.41 kgf m).
IMPORTANT: Should the nut be overtightened so that a torque reading exceeding 20 lbf in (1.41 kgf m) is obtained it will be necessary to renew the collapsible spacer and repeat instructions 51 to 53. Under no circumstances must the nut be slackened to obtain the required torque.

Crown wheel and differential unit

55 Place the crown wheel and differential unit complete with outer bearing tracks into position in the axle casing.

56 Slide the unit towards the pinion until the crown wheel and pinion are fully meshed.

57 Zero a dial gauge on the rear (plain face) of the crown wheel ensuring that the crown wheel remains in contact with the pinion.

58 Slide the crown wheel and differential unit complete with bearings fully in the opposite direction. Note the reading indicated on the dial gauge. Call this dimension 'B'.
CAUTION: Refer to Caution in Operation Number 31.

59 To calculate the shim pack thickness required for each carrier bearing the following information is necessary:
Required crown wheel/pinion backlash:
0.004 to 0.006 in (0.102 to 0.15 mm)
Required carrier bearing pre-load
0.002 to 0.004 in (0.051 to 0.102 mm)
Carrier bearing shims are available in 0.0016 in (0.04 mm) steps from 0.112 in (2.85 mm) to 0.136 in (3.45 mm).

60 Shim thickness required for carrier bearing on toothed side of crown wheel = Dimension 'A' (Instruction 31) minus Dimension 'B' (Instruction 58) plus

$$\frac{\text{carrier bearing preload}}{2}$$

Example:
Using the mean value of 0.003 in for carrier bearing pre-load and assumed values of 0.240 in for dimension 'A' the calculation is:
0.240 (dimension 'A') – 0.115 (dimension 'B')

$$\frac{+.003 \text{ (bearing pre-load)}}{2}$$

= 0.240 – 0.115 + 0.0015
= 0.1265 in (3.21 mm) = thickness of shims required.

61 Shim thickness required for carrier bearing on plain side of the crown wheel = Dimension 'B' (Instruction 58) minus crown wheel backlash

$$\frac{\text{plus carrier bearing pre-load}}{2}$$

Example:
Using the mean values of 0.005 in for the crown wheel backlash, 0.003 in for carrier bearing pre-load and an assumed valve of 0.115 in for dimension 'B' the calculation is:
0.115 (dimension 'B') – 0.005 (backlash)

$$+\frac{0.003 \text{ (preload)}}{2}$$

= 0.115 – 0.005 + 0.0015
= 0.1115 in (2.8321 mm) = thickness of shim required.
In the example quoted the required shim thickness falls outside the shim sizes available. The nearest shim thickness is 0.112 in (2.85 mm). It is thus thicker by 0.0005 in (0.028 mm) than the calculated requirement. Since the calculation was made using the mean values for backlash and bearing preload and the increased thickness of the shim available falls within the tolerance specified (Instruction 60) the 0.112 in (2.85 mm) shim is acceptable.

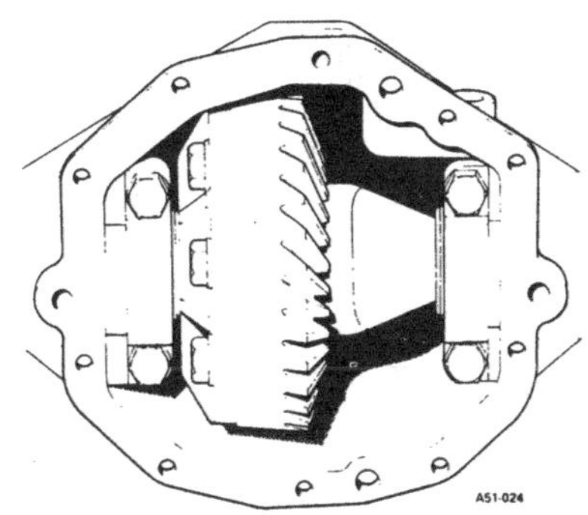

62 Allocate the selected shims to their respective carrier bearings. Ensure the shims are not interchanged.

63 Fit the spreading tool S101 and S101-1 to the axle casing.

64 Carefully expand the spreading tool to allow the crown wheel/differential unit to be placed in position complete with carrier bearing shims. Do not expand more than is necessary or irreparable damage will be caused to the axle casing. Do not exceed a stretch exceeding three or four flats of a finger tight turnbuckle. Do not lever against the spreader.

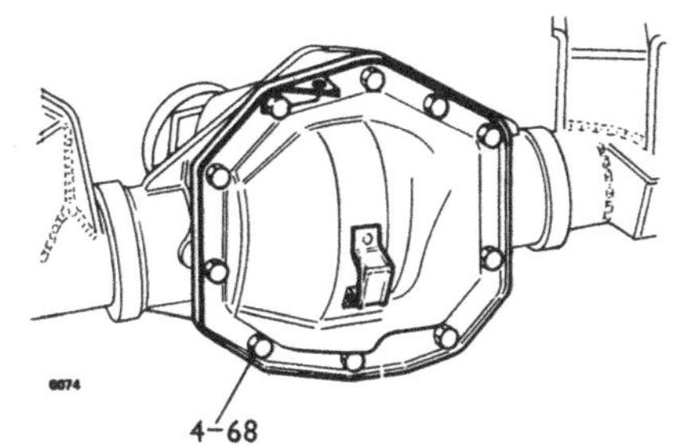

65 Remove the spreading tool.

66 Fit the carrier bearing caps to their original marked position and the four securing bolts. Evenly tighten the bolts. (See Torque Wrench Settings.)

continued

67 Check crown wheel/pinion backlash.
68 Fit the rear cover and gasket to the axle casing.
69 Fit the halfshafts and backplate assemblies, 51.10.12.

Pinion oil seal housing
70 Remove the old oil seal from the pinion seal housing and fit a new seal (tool 6312, seal lip facing away from the front face of the seal housing).
71 Using a strip of masking tape place it over the machined step on the pinion shaft as illustrated. This protects the seal when the housing is eased into position. Apply the tape marginally to facilitate tape removal.
72 Lubricate the seal lip and masking tape.
73 Enter the seal housing in the axle casing and tap gently into position. Remove the masking tape.
74 Fit the brake pipes.
75 Fill the axle to the level of the filler plug with fresh oil.
76 Fit the axle to the car. 51.25.01
77 Bleed the brakes.
78 Lower the car.

DIFFERENTIAL PINION OIL SEAL

Remove and refit 51.20.01

Service tools: 6321 and 18G 1273 18G 1205

Removing
1 Jack up the rear of the vehicle and support securely on stands.
2 Remove the four bolts securing the rear propshaft coupling.
3 Remove the nut securing the drive flange and remove the flange.
4 Remove the four bolts and spring washers securing the pinion oil seal housing to the axle case.
5 Remove the pinion oil seal (Tool 6312).

Refitting
6 Fit a new oil seal to the housing ensuring that the lip of the seal faces away from the front face of the housing (Tool 18G 1273).

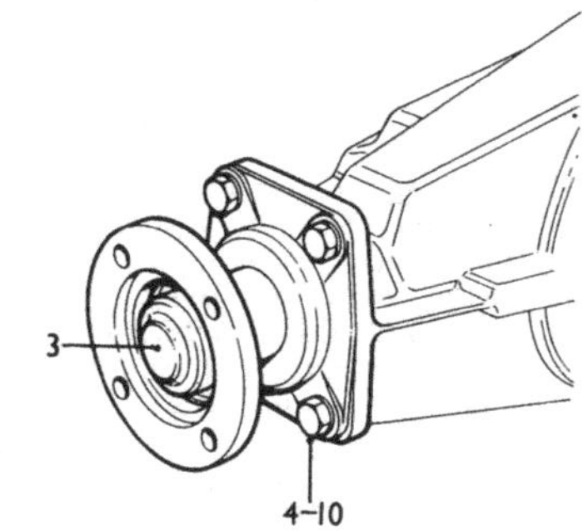

7 Wrap a narrow strip of masking tape over the machined step on the pinion shaft. This prevents damage to the seal when fitting. Apply the tape marginally to facilitate tape removal.
8 Lubricate the seal lip and the masking tape.

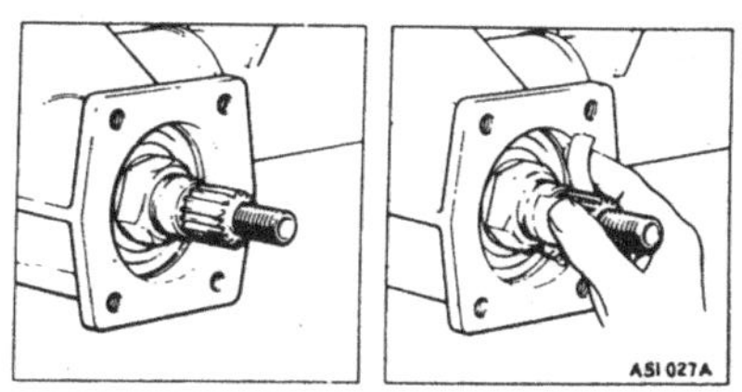

9 Fit the seal and housing, evenly entering the housing in the axle casing. Gently and evenly tap into position.
10 Fit and tighten the four bolts and spring washers securing the pinion oil seal housing to the axle case.
11 Refit the flange and nyloc nut and washer.
12 Tighten the flange nut.
13 Refit and tighten the rear propshaft coupling bolts.
14 Lower the vehicle.
15 Check the oil level.

REAR AXLE ASSEMBLY

Remove and refit 51.25.01

Removing
1 Jack up the rear of the vehicle and support the body securely on two stands.
2 Remove the rear wheels and release the handbrake.
3 Co-relate the drive flanges and remove the four rear propshaft securing nuts and bolts.
4 Fit a brake pipe clamp to the flexible brake hose.
5 Undo the fixed brake pipe and union nut and displace the flexible hose from the axle.

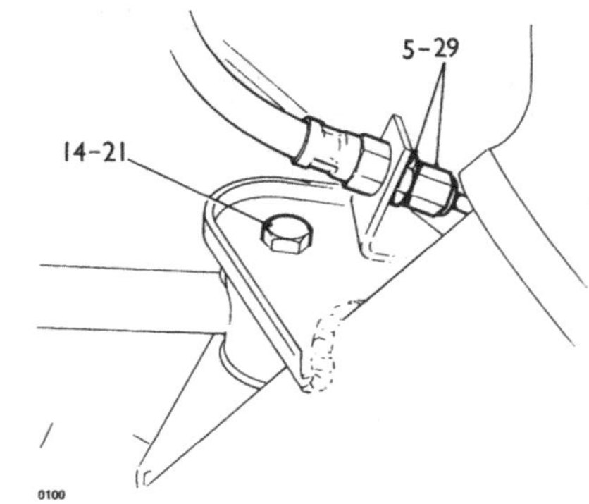

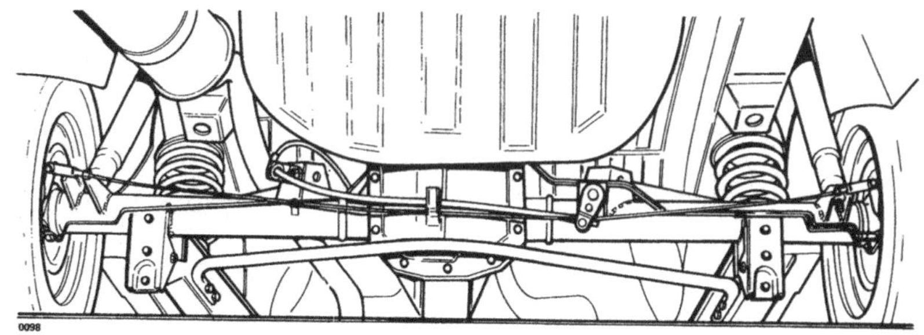

6 Remove the split pins and clevis pin from the handbrake cable clevis forks.
7 Slacken the compensator pinch bolt.
8 Remove the compensator trunion.
9 Feed the handbrake cables through the rear bracket on the axle.
10 Support the axle on a jack.
11 Disconnect the rear dampers from their lower fixing brackets.

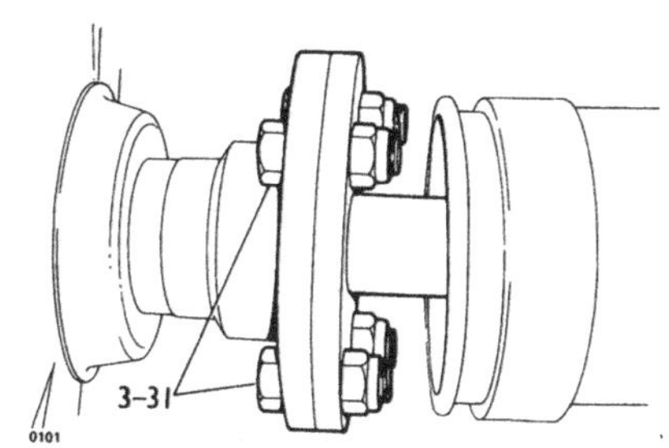

12 Lower the axle on the jack.
13 Remove the rear road springs.
14 Remove the nuts and bolts securing the radius arms to the axle.
15 Displace the handbrake cables and bracket to one side.
16 Remove the nuts and bolts securing the trailing arms to the axle.
17 Manoeuvre the axle over the anti-roll bar and clear of the vehicle.

Refitting
18 Position the jack and manoeuvre the axle into position over the anti-roll bar.

19 Grease all bushes with rubber grease.
20 Refit and tighten the trailing arm securing bolts and nuts.
21 Reposition the handbrake cable bracket and refit and tighten the radius arm securing bolts and nuts.
22 Refit the rear road springs.
23 Jack up the axle and locate the dampers in their lower fixing brackets.
24 Fit the damper rubbers and washers and tighten the securing nuts and lock nuts.
25 Lower the jack.

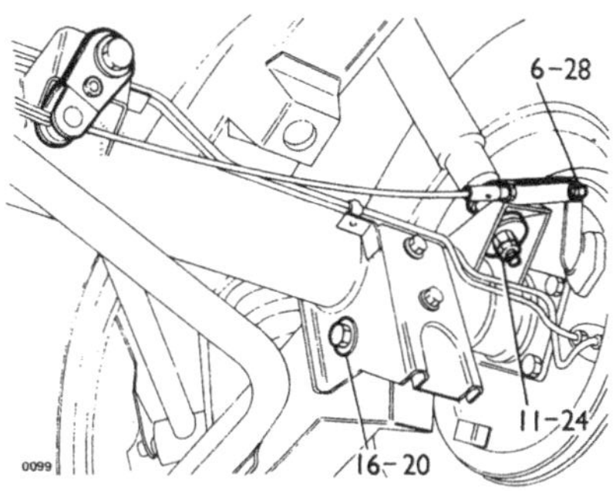

26 Feed the handbrake cables through the rear bracket and refit the compensator trunion.
27 Tighten the pinch bolt.
28 Connect the handbrake cable clevis forks, pins and split pins.
29 Refit the flexible brake hose to the axle and connect the fixed brake pipe.
30 Remove the brake pipe clamp.
31 Refit the propshaft and tighten the four nuts and bolts.
32 Bleed the brakes.
33 Refit the rear wheels.
34 Remove the stands and lower the rear of the vehicle.
35 Check the axle oil level.

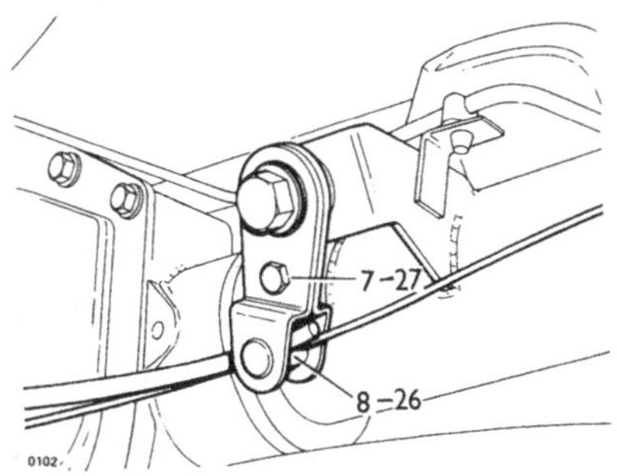

TORQUE FIGURES FOR REAR AXLE

	Nm	*lbf ft*	
Rear cover to axle case	22–28	16–21	
Pinion oil seal housing to axle case	40–50	30–37	
Differential unit bearing caps to axle case	80–100	60–75	
Crown wheel to differential unit (with loctite)	108–122	80–90	
Pinion Flange to pinion	120–160	90–120	
Axle shafts/backplates to axle casing	40–60	35–40	
Oil capacity	1.60 Imp. pts.	0.91 litres	1.9 U.S. pts.

Loctite specification – axle shaft collars and crown wheel bolts

Hub grease – as specification SD 1

STEERING

STEERING RACK AND PINION

Remove and refit 57.25.01

Removing

1 Drive the vehicle onto a ramp.
2 Set the road wheels to the straight-ahead position.
3 Raise the ramp.
4 Scribe the pinion shaft and the lower steering coupling to ensure original spline location on refitting.
NOTE: This instruction applies only if the pinion and rack are not to be dismantled.
5 Disconnect the rack tie-rod outer ball joints from the steering arms.
6 Remove the pinch bolt securing the steering coupling to the rack pinion.
7 Remove the two bolts, spring washers and plain washers securing the pinion end of the rack to the sub-frame.
8 Remove the two nyloc nuts and plain washers, and withdraw the bolts securing the rack to the sub-frame.
9 Disconnect the lower coupling from the pinion shaft.
10 Withdraw the rack from the driver's side.

Refitting

11 Locate the rack on the sub-frame from the driver's side. With the steering-wheel in the straight-ahead position, align the previously scribed markings and engage the pinion shaft into the lower coupling;
or
If no scribe lines were made, or the rack and pinion were dismantled, it is advised that the rack shaft is centralized before beginning installation on the car. Remove the centre plug from the thrust pad and using a length of stiff wire (e.g. welding rod), locate the dimple in the rack shaft. When the dimple is aligned with the wire, the rack shaft is centred. Fit a bolt to the pinion plug and gently tighten until the pinion shaft is pinched. This will hold the rack shaft in its centred position. The rack assembly can now be offered to the sub-frame.

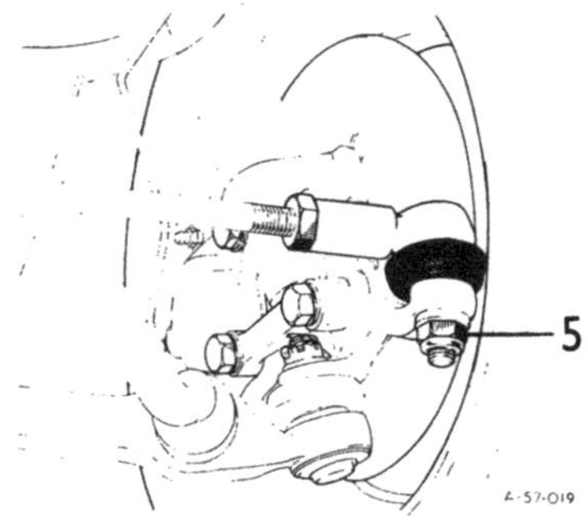

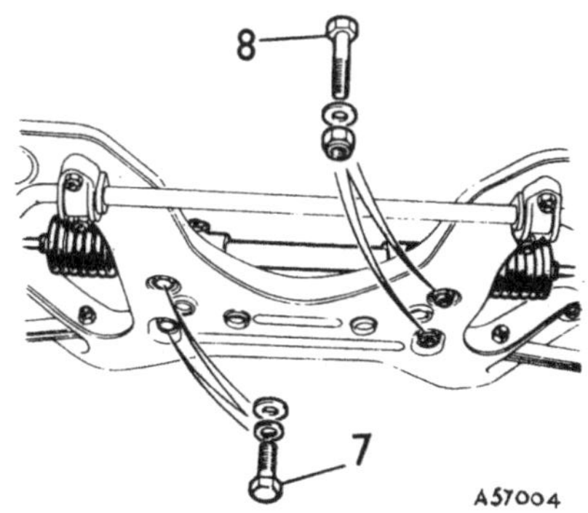

12 Secure the rack to the sub-frame with the bolts, washers and nyloc nuts.
13 Fit the tie-rod outer ball joints to the steering-arms.
14 Fit and tighten the pinion coupling pinch bolt.
15 Remove the locating bolt used to temporarily lock the pinion shaft, and fit the centre plug.
16 Lower the ramp.
17 Check and adjust the front wheel track as necessary, see 57.65.01.

STEERING RACK GAITERS

Remove and refit 57.25.02

Removing

1 Slacken the locknut securing both tie-rod outer ball joints.
2 Remove the nut and washer securing the tie-rod outer ball joint to the steering-arm.
3 Release the ball joint from the steering-arm.
4 Unscrew the ball joint from the tie-rod and remove the retaining locknut.
5 Remove the inner and outer clips retaining the gaiter to the rack and tie-rod respectively.
6 Withdraw the gaiter.
7 Repeat instructions 2 to 6 on the opposite tie-rod.

Refitting

8 Lubricate the tie-rod inner ball joint with fresh grease.
9 Slide the new gaiter along the tie-rod into position on the rack.
10 With the rack centralized, fit the inner clip to the gaiter and rack housing.
11 Position the outer end of the gaiter on the tie-rod so that it is capable of accommodating movement of the tie-rod from lock to lock.
12 Secure the outer end of the gaiter to the tie-rod end with the clip.
13 Fit the locknut to the tie-rod, locating it as near as possible to its original location.
14 Fit the outer ball joint to the tie-rod.
15 Connect the tie-rod outer ball joint to the steering-arm and secure it with the plain washer and nut.
16 Repeat instructions 8 to 15 on the opposite tie-rod.
17 Check, and re-set the front wheel track as necessary.
18 Tighten the locknut securing the tie-rod outer ball joint.

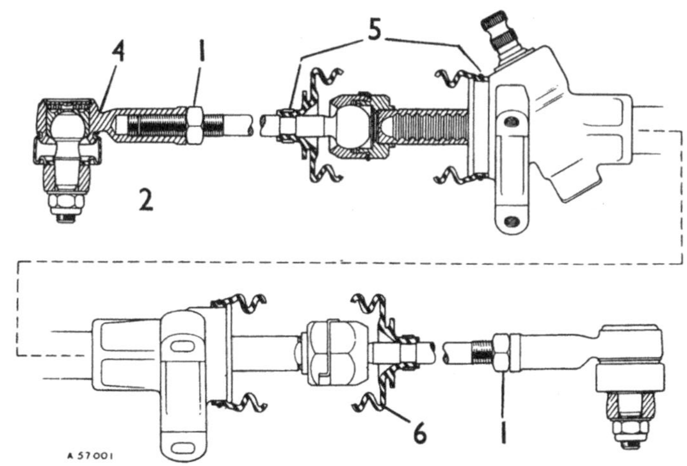

STEERING RACK AND PINION

Overhaul 57.25.07

Special tool: 18G 1261

Dismantling
1 Remove the rack, see 57.25.01.

Rack damper
2 Remove the plug securing the damper assembly to the rack housing.
3 Withdraw the spring, shim(s) and damper.

Pinion
4 Remove the rubber seal.
5 Remove the pinion retaining plug using the special tool 18G 1261.
6 Withdraw the pinion and bearing from the rack housing.
7 Remove the circlip securing the ball race to the pinion shaft and withdraw the ball race.

Tie-rods and rack shaft
8 Release the clips and tie-wires securing the gaiters to the rack housing and the tie-rods, and slide the gaiter clear of the rack.
9 Slide the pinion end of the rack housing towards its adjacent tie-rod inner ball joint.
10 Grip the exposed rack shaft in protected vice jaws.
11 Unscrew the tie-rod inner ball joint assemblies from both ends of the rack shaft.
12 Withdraw the rack shaft.

Rack housing bush
13 Using a suitable drift, remove the bush from the rack housing.

Reassembling
Rack housing bush
14 Fit a new bush to the rack housing.

Rack shaft and tie-rods
15 Hold the plain (toothless) portion of the rack shaft in protected vice jaws.
16 Fit the tie-rod inner ball joint assembly to the rack shaft, tighten to 35 to 45 lbf ft (4.84 to 6.22 kgf m), and deform the collar to secure the ball joint housing to the rack shaft.

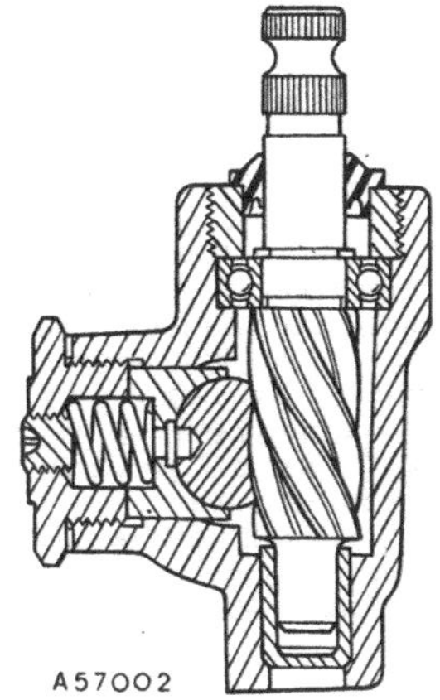

A57002

17 Fit the rack shaft into the housing, ensuring that the toothed end of the shaft is located at the pinion housing.
18 Fit the tie-rod assembly to the pinion end of the rack shaft – refer to operation 15.

Pinion
19 Fit the ball race to the pinion and secure it with the circlip. Fit the rubber seal.
20 Position the rack shaft teeth to permit pinion entry and engage the rack with the pinion.
21 Lubricate the pinion shaft, refit and tighten the end plug.
22 Pack the ends of the rack shafts and tie-rod inner ball joints with clean grease.
23 Fit and secure the gaiters.
24 Fit the rack damper, see 57.35.10.
25 Fit the rack to the car, see 57.25.01.

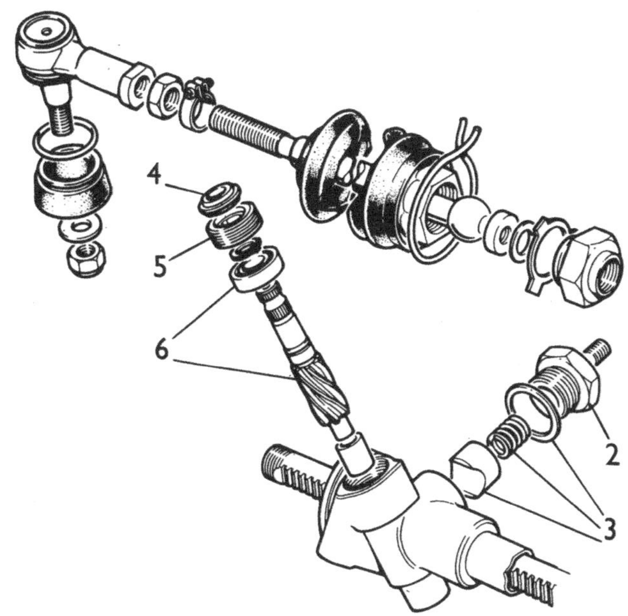

A57005

STEERING RACK DAMPER

Adjust 57.35.09
1 Remove the rack, see 57.25.01.
2 Locate the rack shaft in mid-position.
3 Remove the centre plug from the damper plug and insert a stylus or dial gauge and check the rack shaft for side movement (90° to shaft axis). Side movement should be within 0.001 to 0.007 in (0.03 to 0.18 mm).
4 Adjust if necessary by removing the damper plug and adding or subtracting shim(s) as required;
or
4a In the absence of a dial gauge, remove the damper plug and shims.
4b Remove the shims and spring and replace the damper plug.
4c Gently tighten the damper plug until the plunger grips the rack eliminating all side-play.
4d With feeler gauges inserted between the rack pinion housing and the underside of the damper plug flange, check the clearance existing.
4e To the thickness of the feeler gauge pack, add the rack side movement required, 0.001 to 0.007 in (0.03 to 0.18 mm). This gives the thickness of the shims to be fitted under the damper plug flange.
4f Remove the damper plug and fit the required shim pack. Tighten the damper plug to 45 to 60 lbf ft (6.22 to 8.30 kgf m).
5 Refit the rack, see 57.25.01.

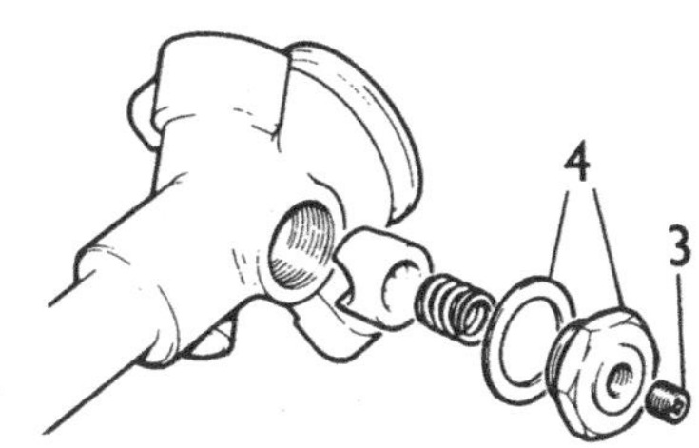

A57006

STEERING RACK DAMPER

Remove and refit **57.35.10**

Removing

1 Remove the rack, see 57.25.01.
2 Remove the clip and tie-wire securing the gaiter at the pinion end of the rack and slide the gaiter along the tie-rod to expose the rack shaft.
3 Unscrew and remove the damper plug and shims.
4 Withdraw the spring and plunger.

Refitting

5 Locate the rack shaft in mid-position.
6 Fit the plunger and spring.
7 Check, and adjust the rack shaft side movement as required (instructions 4 to 6, operation 57.35.09).
8 Refit the rack, see 57.25.01.

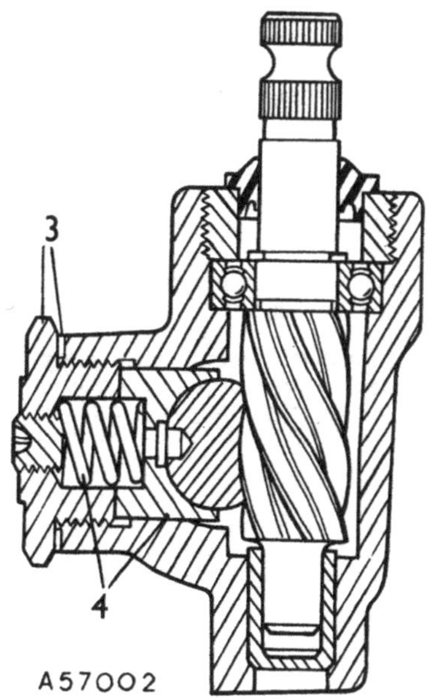

A57002

STEERING-COLUMN ASSEMBLY

Remove and refit **57.40.01**

Removing

1 Open the bonnet and disconnect the battery.
2 Remove the pinch bolt securing the upper universal coupling to the steering mast.
3 Remove the cleat securing the electrical harness to the steering-column.
4 Disconnect the plug-in connectors (4) for the ignition/starter, horn/trafficator/lights and windscreen wiper/washer switches.
5 Remove the two screws securing the nacelle to the steering-column. Remove the nacelle.
6 Using a centre-punch, mark the centre of the two shear-head bolts securing the column housing to the body.
7 Using a small chisel, unscrew the shear-head bolts;
   or
   a If instruction 7 proves to be unsuccessful drill into the shear-head bolts where previously marked by the centre punch and unscrew using an Easiout extractor.
8 With the road wheels in the straight-ahead position, withdraw the steering-column assembly, noting the positions of the flat and wavy washers.

Refitting

9 Enter the steering-column in the lower bush. Ensure that the flat and wavy washers are in position.
10 With the road wheels in the straight-ahead position and the steering-wheel centralized, engage the steering-column splines in the upper universal coupling.
11 Fit and tighten the pinch bolt.
12 Locate the steering-column on the body and align the mounting holes.
13 Fit two new shear-head bolts and tighten evenly until both heads shear.
14 Re-connect the plug in connectors and secure the electrical harness to the steering-column housing with the plastic cleat.
15 Fit the nacelle and secure with the two screws.
16 Connect the battery.

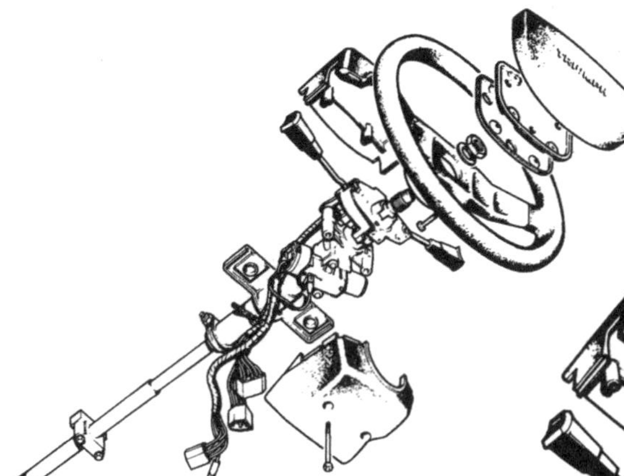
A57007

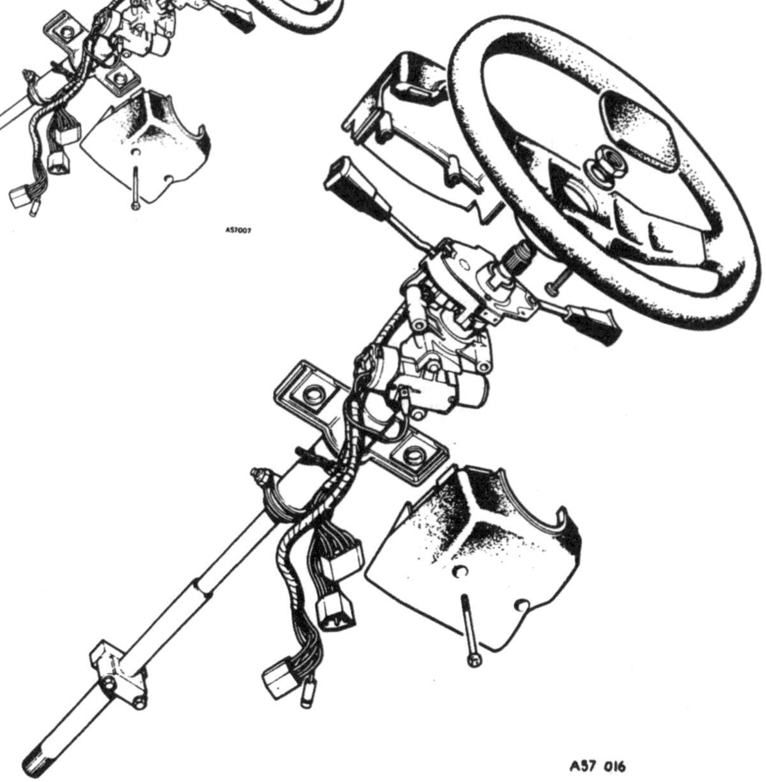
A57 016

STEERING COLUMN ASSEMBLY

Overhaul **57.40.10**

Dismantling

1 Remove the steering-column assembly from the car, see 57.40.01.
2 Remove the steering-wheel centre cover.
3 Slacken and remove the nut and washer securing the steering-wheel hub to the steering-column.
4 Using a suitable extractor, remove the steering-wheel.
5 Slacken the clamping screw and withdraw the steering-column multi-purpose switch assembly from the column.
6 Using a centre-punch, mark the centre of the two shear-head bolts securing the steering lock to the column housing.
7 Using a small chisel, unscrew the shear head bolts;
or
a If instruction 7 proves to be unsuccessful, drill into the shear-head bolts where previously marked by the centre-punch and unscrew using an Easiout extractor.
8 Withdraw the column housing off the steering mast.
9 Remove the nut and bolt securing the clamp to the steering mast. Remove the clamp.
10 Using a suitable drift, remove the top and bottom bushes from the steering-column housing.

Reassembling

11 Align the slots in the bushes with the lugs in the column housing. Press in the bushes.
12 Fit the clamp to the steering mast. Fit and tighten the pinch bolt.
13 Fit the steering mast into the housing.
14 Fit the steering lock to the column housing and secure with two new shear-head bolts. Evenly tighten the bolts until the heads shear off.
15 Fit the multi-purpose switch assembly to the column housing and secure by tightening the clamp screw.
16 Align the arrow on the trafficator cancelling collar with the centre of the trafficator stalk.
17 Align the lugs of the steering-wheel with the cut-outs in the cancelling cam. Fit the steering-wheel.
18 Fit and tighten the plain washer and nut securing the steering-wheel hub to the steering-column.
19 Fit the steering-wheel centre cover.
20 Fit the steering-column assembly to the car, see 57.40.01.

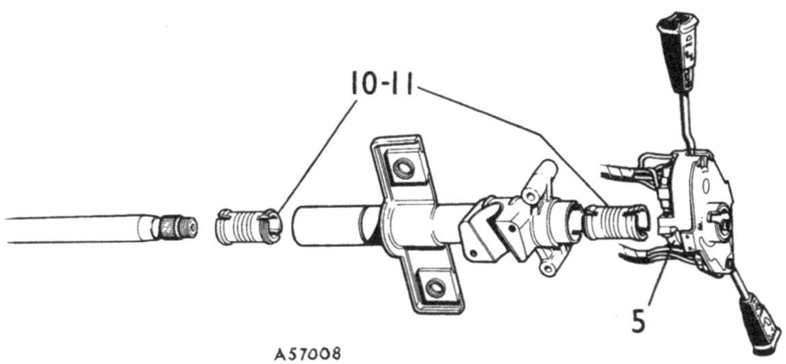

A57008

INTERMEDIATE SHAFT

Remove and refit **57.40.22**

Removing

1 Remove the pinch bolt securing the intermediate shaft to the upper universal coupling.
2 Remove the pinch bolt securing the intermediate shaft universal joint to the rack pinion.
3 Set the road wheels to the straight-ahead position.
4 Slide the intermediate shaft upwards to disengage the universal joint from the pinion shaft.
5 Withdraw the intermediate shaft downwards and disengage from the upper universal coupling.

Refitting

6 Engage the splines of the intermediate shaft in the upper universal coupling.
7 Ensure that the steering-wheel is in the straight-ahead position and engage the intermediate shaft universal joint in the splines of the rack pinion.
8 Fit and tighten the two pinch bolts.

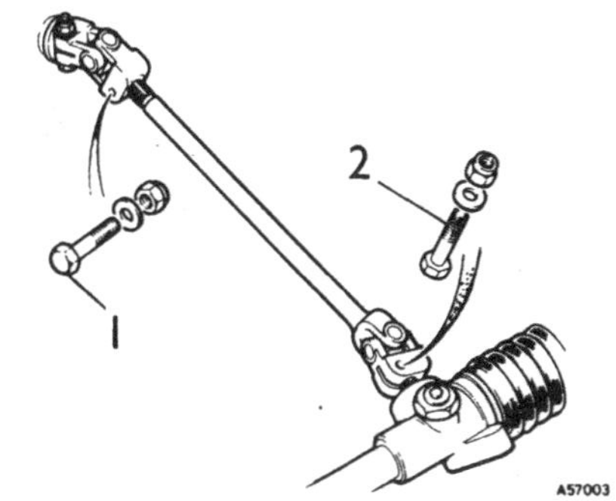

A57003

STEERING-COLUMN UPPER UNIVERSAL COUPLING

Remove and refit **57.40.26**

Removing

1 Slacken the pinch bolt securing the upper universal coupling to the steering mast.
2 Remove the pinch bolt securing the upper universal coupling to the intermediate shaft.
3 Turn the steering-wheel to facilitate the removal of the top pinch bolt.
4 Set the road wheels to the straight-ahead position.
5 Slide the upper universal coupling down the intermediate shaft and remove the two washers from the steering mast (noting position for reassembly).
6 Remove the upper universal coupling from the intermediate shaft.

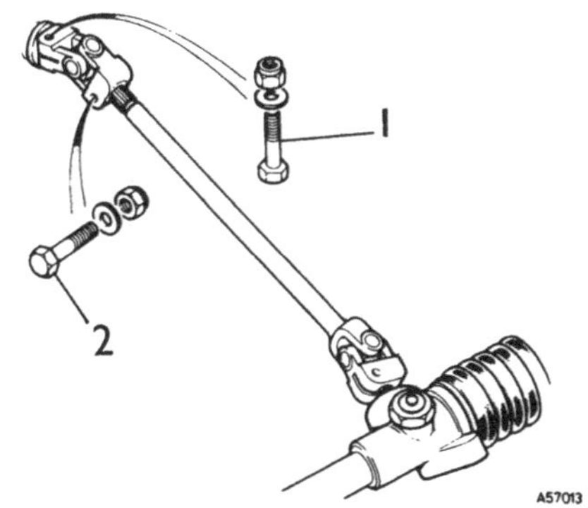

A57013

Refitting

7 Engage the upper universal coupling in the splines of the intermediate shaft.
8 Fit the two washers to the steering mast.
9 Ensuring that the steering-wheel and the road wheels are in the straight-ahead position, engage the coupling in the splines of the steering mast.
10 Fit and tighten the two pinch bolts.

STEERING-COLUMN NACELLE

Remove and refit 57.40.29

Removing

1 Withdraw the key from the steering lock/ignition switch.
2 Remove the two screws clamping the nacelle halves.
3 Remove the nacelle halves.

Refitting

4 Reverse instructions 1 to 3.

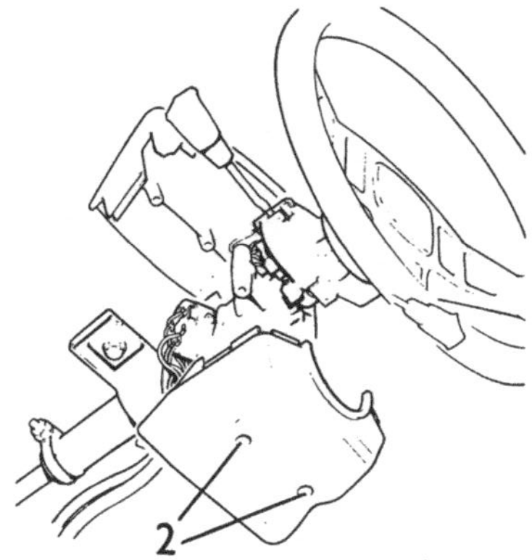

A57 017

STEERING LOCK/IGNITION SWITCH

Remove and refit 57.40.31

Removing

1 Remove the nacelle, see 57.40.29.
2 Using a centre-punch, mark the centre of the two shear-head bolts securing the steering lock to the column.
3 Using a small chisel, unscrew the shear-head bolts;
or
a If instruction 3 proves to be unsuccessful, drill into the shear-head bolts where previously marked by the centre-punch, and unscrew using an Easiout extractor.
4 Disconnect the plug-in connector to the ignition switch.
5 Remove the steering lock.

Refitting

6 Locate the steering lock on the column and align the mounting holes.
7 Fit two new shear-head bolts. Evenly tighten until both heads shear.
8 Connect the plug-in connector for the ignition switch.
9 Fit the nacelle, see 57.40.29.

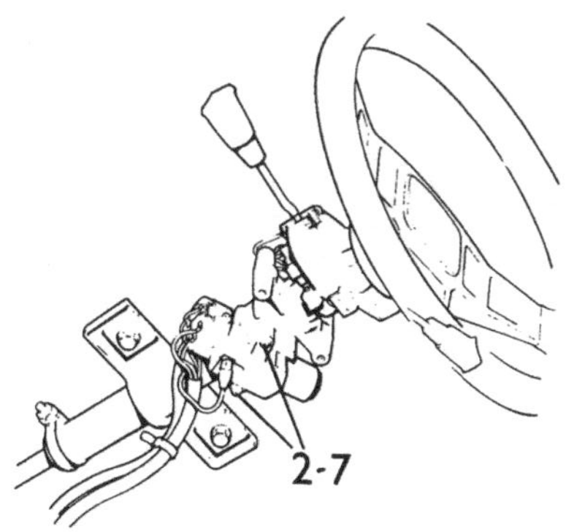

A57 018

TIE-ROD BALL JOINT – OUTER

Remove and refit 57.55.02

Removing

1 Slacken the locknut securing the tie-rod to the outer ball joint.
2 Remove the nut and washer securing the ball joint to the steering-arm.
3 Release the ball joint from the steering-arm.
4 Unscrew the ball joint from the tie-rod.

Refitting

5 Screw the ball joint onto the tie-rod. (The distance between tie-rod ball joint centres [inner to outer] is $13\frac{1}{3}$ in (338 mm.)
6 Connect the ball joint to the steering-arm and secure with the washer and nut.
7 Check and adjust the front wheel track as necessary.
8 Tighten the tie-rod locknut.

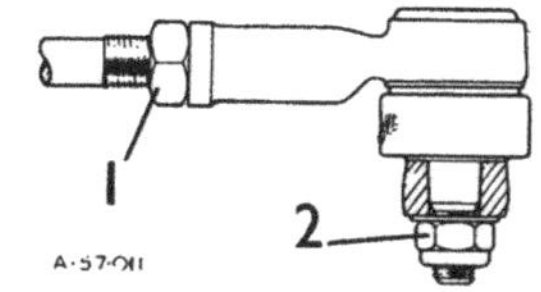

A57 011

TIE-ROD BALL JOINT – INNER

Remove and refit 57.55.03

Removing

1 Remove the tie-rod outer ball joint, see 57.55.02.
2 Release the wire and clip, and remove the gaiter from the inner ball joint to be renewed.
3 Release the tie-wire and clip, and slide the gaiter along the tie-rod at the opposite end of the rack.
4 Wipe the inner ball joints clean of grease.
5 Unscrew the tie-rod inner ball joint from the rack shaft. To prevent stress being applied to the rack pinion the opposite inner ball joint assembly should be held with a spanner.

Refitting

6 Fit the new inner tie-rod ball joint assembly to the rack shaft and tighten to 35 to 45 lbf ft (4.84 to 8.22 kgf m). To prevent stress being applied to the rack pinion the opposite inner ball joint assembly should be held with a spanner.

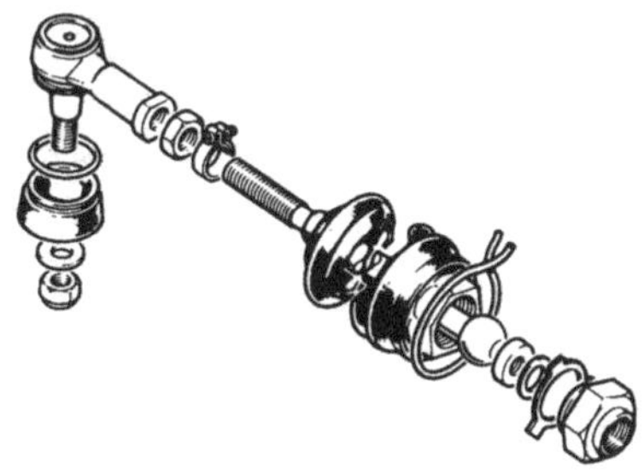

A57009

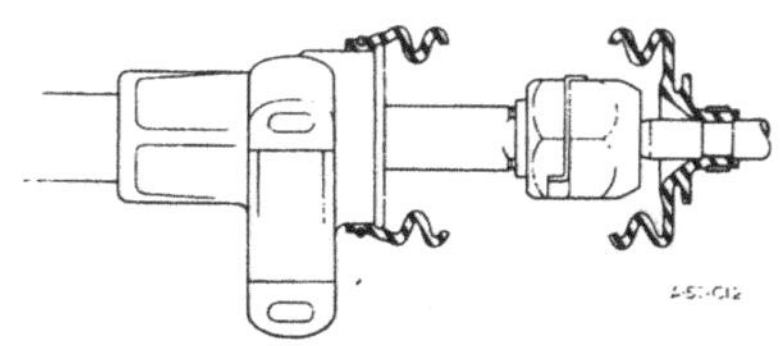

A57 012

7 Using a suitable drift, deform the lock collar to secure the inner ball joint assembly to the rack shaft.
8 Fit the gaiters, tie-wires and clips.
9 Fit the outer ball joint to the tie-rod, see 57.55.02.
10 Check and adjust the front wheel track as necessary.

STEERING-WHEEL

Remove and refit **57.60.01**

Removing

1. Remove the steering-wheel centre cover.
2. Locate the road wheels in the straight-ahead position.
3. Slacken and remove the nut and washer securing the steering-wheel hub to the steering-column.
4. To ensure that the steering-wheel (hub) will be replaced in its original spline location, scribe both the hub centre and the top of the steering mast.
5. Using a suitable extractor, withdraw the steering-wheel. DO NOT attempt to drive or tap the steering-wheel from the mast.

Refitting

6. Ensure that the arrow on the trafficator cancelling collar aligns with the centre of the trafficator stalk. Reverse instructions 1, 2, 3 and 5. If the steering-wheel was withdrawn without the spline location being marked, set the road wheels to the straight-ahead position and centralize the steering-wheel.

A57 016

STEERING WHEEL PAD [1975 & 1976 USA models and all UK/European models]

Remove and refit **[57.60.03]**

Removing

1. Remove the three screws (underside of the steering wheel spokes) securing the steering-wheel pad.
2. Withdraw the steering-wheel pad.
3. Remove the four screws at the back of the pad and part the two plates securing the pad.

Refitting

4. Reverse instructions 1 to 3.

[**NOTE:** For other USA models—the steering wheel pad may be carefully prised out of the wheel assembly.]

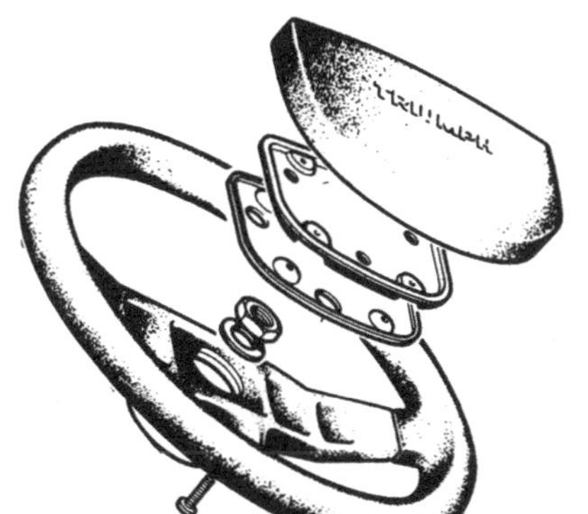

A57007

STEERING GEOMETRY

DATA **57.65.00**

	Kerb condition	
Camber	¼° negative ± 1°	
Castor	3½° positive ± 1°	
King pin inclination	11¼°	± 1°
Wheel alignment . .	0-$\frac{1}{16}$"	Toe-in

FRONT WHEEL ALIGNMENT

Check and adjust **57.65.01**

Checking

1. Locate the car on level ground and position the front wheels in the straight-ahead position.
2. Using wheel alignment equipment, check the front wheels for toe-in. Four requirements should be met:
   - a Centralized steering-wheel.
   - b Centralized steering-rack.
   - c Front wheels parallel to $\frac{1}{16}$ in (1.59 mm) toe-in.
   - d Ball centres of both tie-rods equal.

Adjusting

3. Slacken the outer clips on the rack gaiters.
4. Slacken the locknut at the tie-rod outer ball joints.
5. Shorten or extend both tie-rods by an equal amount to obtain the required setting (0 to $\frac{1}{16}$ in, 0 to 1.59 mm toe-in).
6. Tighten the locknuts at the tie-rod outer ball joints.
7. Tighten the gaiter clips.

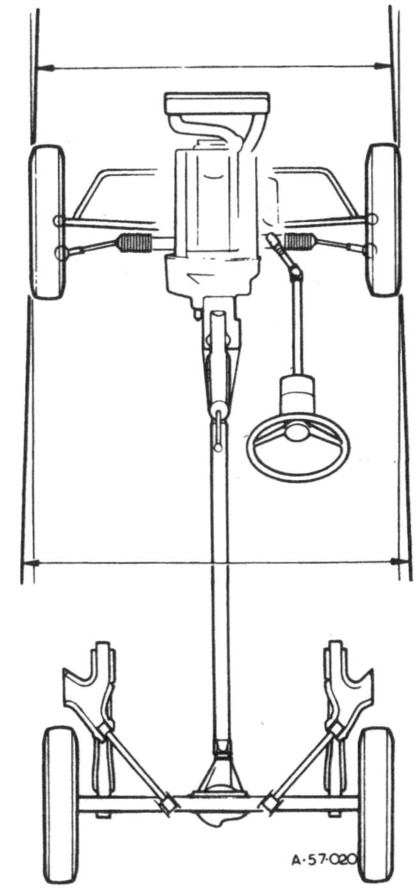

A-57-020

FRONT SUSPENSION

FRONT SUSPENSION

Anti-roll bar

– Remove and refit **60.10.01**
Bottom Link Rubbers
– remove and refit **60.10.06**

Removing

1. Raise the car and support securely.
2. Remove the bolts and nyloc nuts (two each side) securing the anti-roll bar brackets to the sub frame.
3. Remove the spring pin, nyloc nut, flat washer, dished washer and outer rubber bush, securing each end of the anti-roll bar to the bottom link.
4. Withdraw the anti-roll bar, adjusting the jacks as necessary to facilitate removal.
5. Remove the inner bush and dished washer from each end of the anti-roll bar.

Refitting

6. Fit the inner dished washer (dish towards the bush) and the inner rubber bush to each end of the anti-roll bar.
7. Offer up the anti-roll bar and align the ends with the mounting holes in each of the bottom links.
8. Fit the outer rubber bush, dish washer, flat washer and nyloc nut to each end of the anti-roll bar.
9. Tighten to stop and fit the spring pin.
10. Position the mounting brackets and secure the anti-roll bar to the sub frame with the bolts and nyloc nuts.
11. Lower the car.

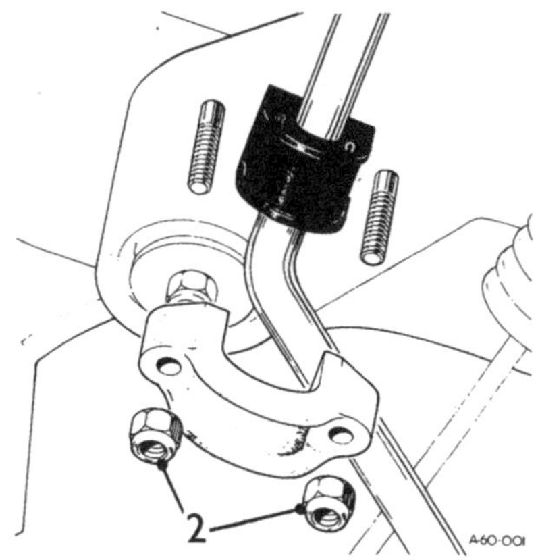

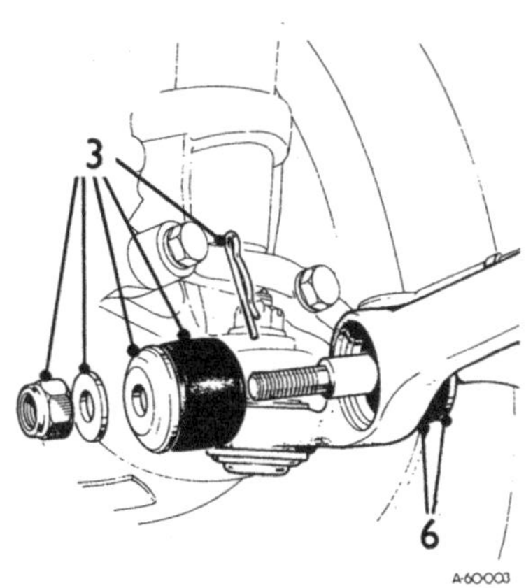

ANTI-ROLL BAR MOUNTING RUBBERS

– Remove and refit **60.10.05**

Removing

1. Raise the car and support securely.
2. Remove the anti-roll bar. 60.10.01.
3. Cut the old mounting bushes and remove them from the anti-roll bar.

Refitting

4. Ensure that the anti-roll bar is clean throughout its length.
5. Smear the anti-roll bar with the approved rubber grease.
6. Slide the new mounting bushes into position along the anti-roll bar.
7. Fit the anti-roll bar to the car. 60.10.01.
8. Lower the car.

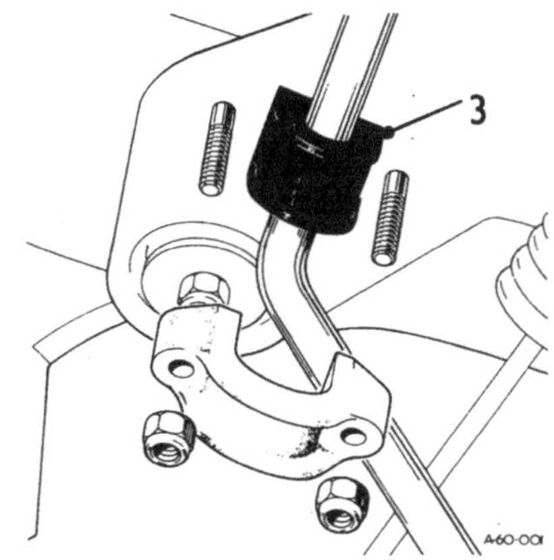

BALL JOINT

– Remove and refit **60.15.03**

Removing

1. Remove the bottom link. 60.40.02.
2. Remove the plastic boot from the ball joint.
3. Remove the circlip.
4. Press out the ball joint housing.

Refitting

5. Using a short length of suitable bore steel tubing, press a new ball joint and housing squarely into the bottom link. Do not apply pressure to the centre of the housing end cap.
6. Fit the circlip and plastic boot.
7. Fit the bottom link. 60.40.02.

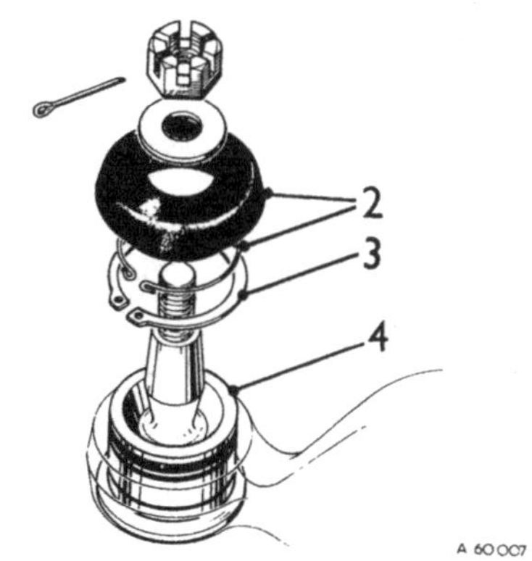

FRONT ROAD SPRING

– Remove and refit 60.20.01
Bump stop – remove and refit 60.30.10

Service tools: P.5045, RTR 360

Removing

1 Jack up the car and support the body on stands.
2 Remove the road wheel.
3 Detach the steering arm from the stub axle assembly (two bolts).
4 Slacken the locknut securing the brake hose to the bracket on the damper tube.
5 Remove the remaining bolt securing the brake caliper to the stub axle assembly, and support the brake caliper.
6 Remove the split pin, slotted nut and flat washer and release the ball joint from the stub axle assembly.
7 Remove the three nyloc nuts securing the damper and spring to the wing valance.
8 Pull the strut clear of the car.
9 Fit the two clamps P.5045 to the spring and compress coils evenly.
10 Remove the nut from the damper piston rod.
11 Lift off the spring pan complete with the top mounting and swivel assembly.
12 Withdraw the road spring from the damper strut.
13 Progressively slacken the spring clamps.

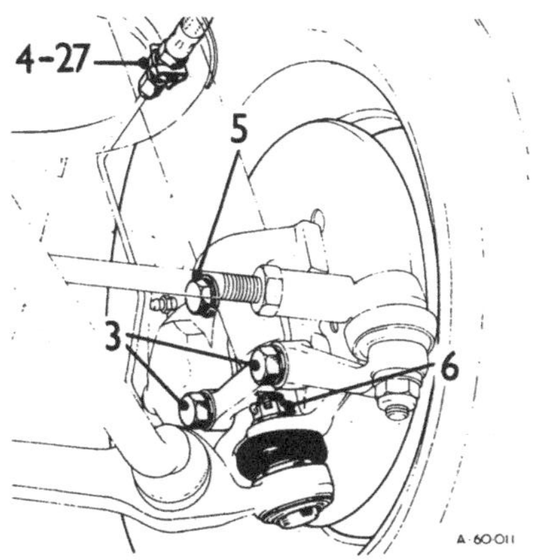

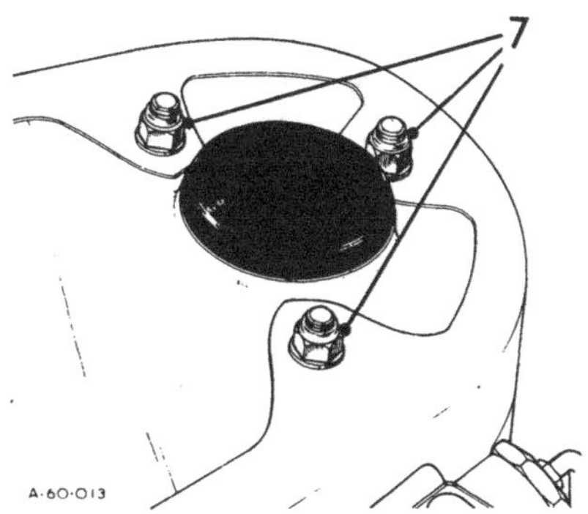

Refitting

14 Using the clamps, P.5045 compress the road spring.
15 Ensure that the bump stop rubber is in position.
16 Extend fully the damper piston rod and fit the lower insulating ring, rubber gaiter, road spring, upper insulating ring and spring pan.
17 Fit the seal to the thrust collar and position on the upper spring pan.

[**NOTE:** To keep the upper strut bearing from binding and causing hard steering, owing to water ingress, pack the thrust collar and its adjacent parts with grease.]

18 Fit the large plain washer. (Ground surface facing spring pan.)
19 Fit the rubber mounting to the damper piston rod and secure with the dished washer and nut using the special tool RTR 360. Tighten to the correct torque.

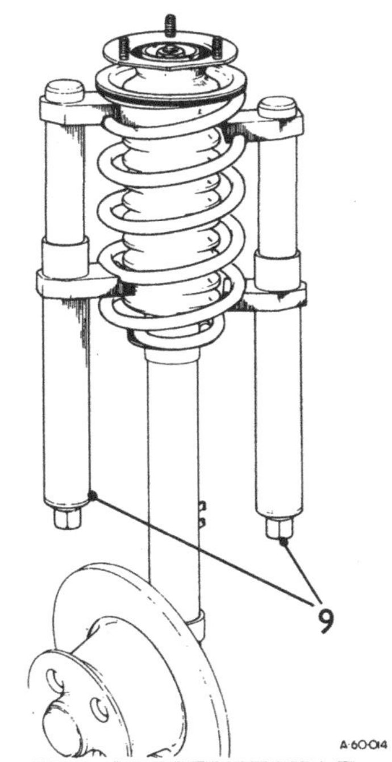

20 Slacken progressively the spring clamps, ensuring correct seating of the spring.
21 Ensure that the rubber gaiter is correctly fitted.
22 Thoroughly clean the spring turret and apply plasti-seal to the damper upper mounting flange.
23 Offer up the damper to the spring turret with the cut out facing outboard.
24 Engage the three studs, fit the plain washers, nyloc nuts and tighten.
25 Fit the ball joint into the stub axle assembly and secure with the flat washer, slotted nut and split pin.
26 Position the brake caliper onto the stub axle assembly and insert the upper bolt – do not tighten at this stage.
27 Position the brake pipe to the bracket on the damper tube.
28 Fit the steering arm to the stub axle assembly (the rear bolt also secures the brake caliper) and tighten to the correct torque.
29 Tighten the brake hose locknut to the fixing bracket.
30 Fit the road wheel and lower the car.

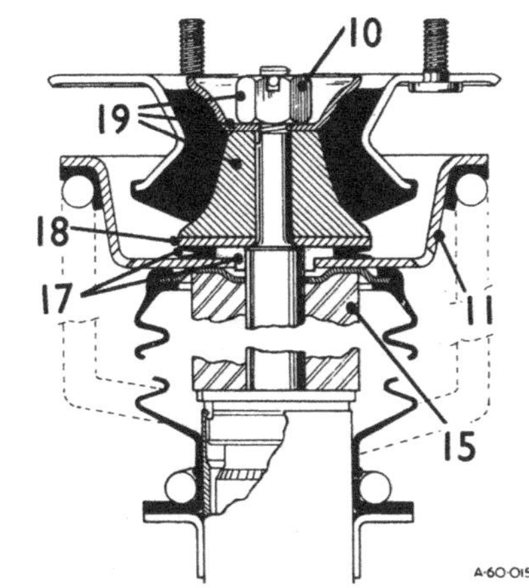

FRONT HUB

– **Remove and refit** **60.25.01**

Removing
1 Jack up the car and support the body on stands.
2 Remove the front road wheel.
3 Slacken the locknut securing the brake hose to the bracket on the damper tube.
4 Remove the two bolts securing the steering arm to the stub axle assembly.
5 Remove the bolt securing the brake caliper to the stub axle assembly.
6 Withdraw the caliper clear of the brake disc, ensuring that the strain is not imposed on the brake hose and support the caliper.
7 Prise off the hub cap and wipe grease from the end of the stub axle.
8 Remove the split pin, nut retaining cap, nut and washer from the stub axle.
9 Withdraw the hub complete with disc, bearings and oil seal.

Refitting
10 Partially pack the hub with fresh grease.
11 Locate the oil seal in the hub and enter the hub and bearings on the stub axle.
12 Fit the washer and slotted nut to the stub axle.
13 Tighten the slotted nut to a torque of 5 lbf ft (0.691 kgf m) back off one flat, fit the nut retaining cap and secure with a split pin.
14 Position the brake pipe to the fixing bracket on the damper tube.
15 Fit the caliper to the stub axle assembly with the upper bolt, do not tighten.
16 Fit the steering arm to the stub axle assembly.
17 Tighten the three bolts.
18 Tighten the locknut securing the brake hose to the fixing bracket.
19 Fit the road wheel and lower the car.

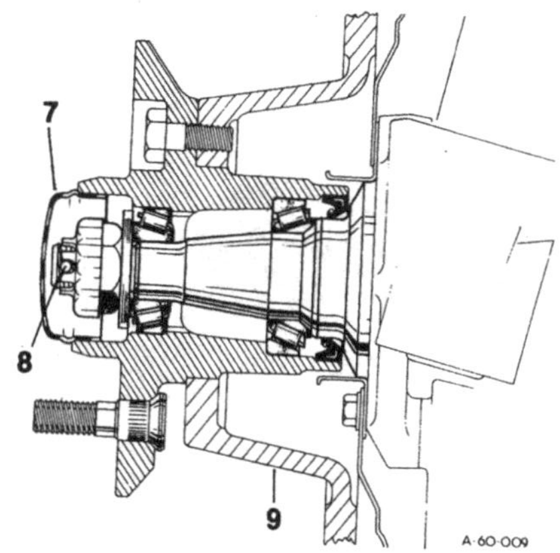

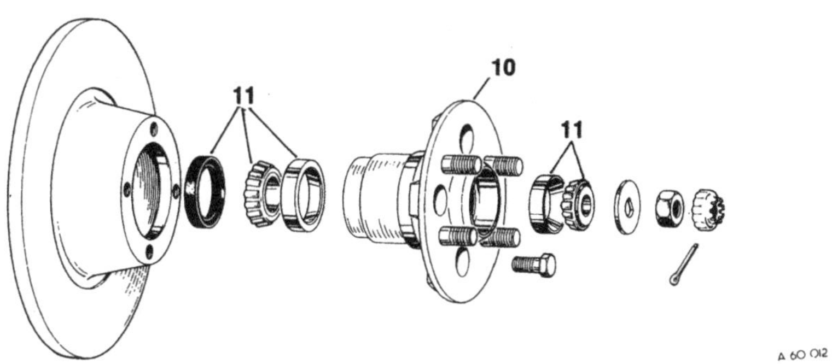

FRONT HUB BEARING END-FLOAT

– **Check and adjust** **60.25.13**

1 Remove the road wheel.
2 Check the hub for end-float.
3 If adjustment is required remove hub cap and split pin.
4 Tighten the slotted nut as required to eliminate end-float. A torque of 5 lbf ft (0.691 kgf m) must not be exceeded or damage may be caused to the bearings and bearing tracks. Back off one flat and fit the nut retaining cap.
5 Insert and lock the split pin.
6 Clean the hub cap and refit.
7 Fit the road wheels.

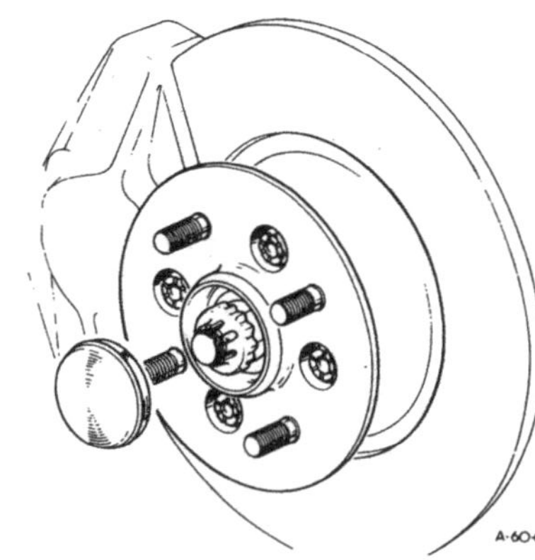

FRONT HUB BEARINGS

– **Remove and refit** **60.25.14**

Removing
1 Remove the front hub. 60.25.01.
2 Remove the outer bearing, inner oil seal and inner bearing.
3 Thoroughly clean the hub.
4 Drift the outer and inner bearing tracks from the hub.

Refitting
5 Clean the bearing track recesses in the hub.
6 Install the new tracks in the hub, ensuring that they abut against the machined lip.
7 Fit a new oil seal. 60.25.15.
8 Fit the hub assembly to the car. 60.25.01.

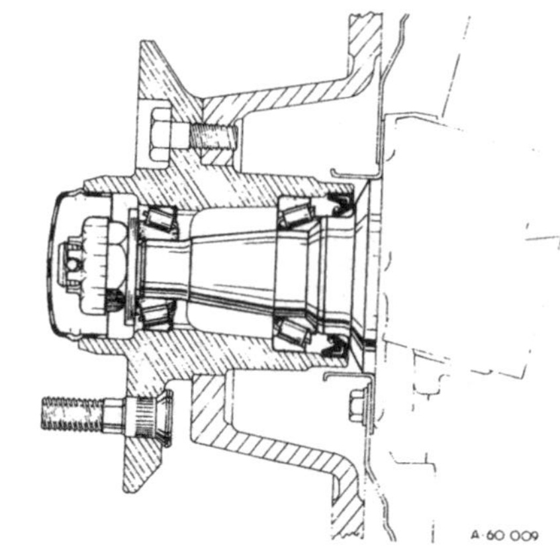

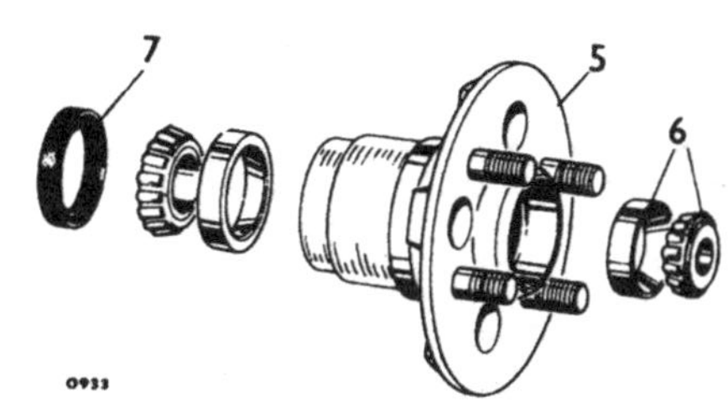

FRONT HUB OIL SEAL

– Remove and refit **60.25.15**

Removing
1 Remove the front hub. 60.25.01.
2 Extract the oil seal from the hub.

Refitting
3 Insert the new oil seal (lip towards bearing) into hub and press or drift evenly into position.
4 Lubricate the seal lip.
5 Ensure that the seal deflector ring, incorporated in the disc shield is not damaged and does not foul the hub.
6 Fit the hub to the car. 60.25.01.

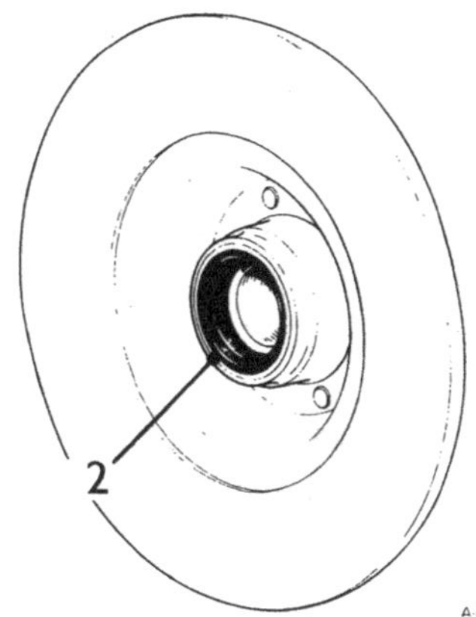

STUB AXLE ASSEMBLY

– Remove and refit **60.25.22**

Removing
1 Remove the road spring. 60.20.01.
2 Remove the front damper cartridge. 60.30.02.
3 Remove the front hub. 60.25.01, instructions 7 to 9.
4 Remove the three bolts securing the disc shield.

Refitting
5 Secure the disc shield with the three bolts and spring washers.
6 Refit the front hub. 60.25.01, instructions 10 to 15.
7 Fit the front damper cartridge. 60.30.02.
8 Fit the road spring to the car. 60.20.01.

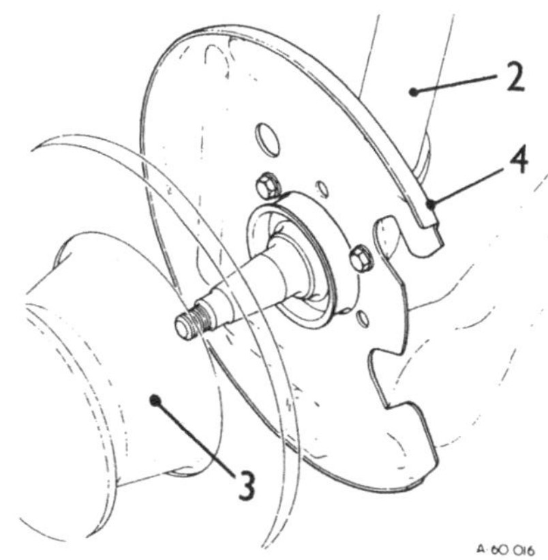

WHEEL STUD

– Remove and refit **60.25.29**

Removing
1 Remove the front hub. 60.25.01.
2 Remove the four bolts retaining the hub to the brake disc and remove the hub from the disc.
3 Extract the stud from the hub.

Refitting
4 Ensure that the mating countersunk faces of the stud and the flange are clean.
5 Enter the stud from the rear of the hub flange, align the splines and press into position.
6 Fit the front hub to the disc and evenly tighten the four bolts 25-32 lbf ft (3.46-4.42 kgf m).
7 Fit the front hub to the stub axle. 60.25.01.

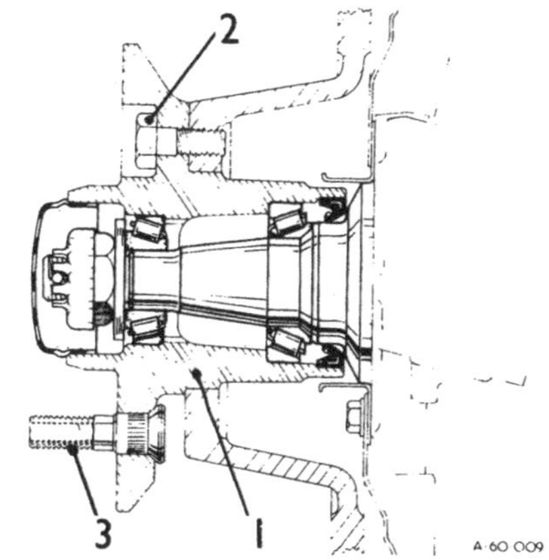

FRONT DAMPER

– Remove and refit **60.30.02**

Service tool: RTR 359

Removing
1 Remove the front road spring. 60.20.01.
2 Using the special tool undo the closure nut.
3 Remove the damper cartridge.

Refitting

[**NOTE:** To keep the upper strut bearing from binding and causing hard steering, owing to water ingress, pack the thrust collar and its adjacent parts with grease.]

4 Fit the damper cartridge.
5 Fit the closure nut, and tighten to the correct torque.
6 Fit the road spring. 60.20.01.

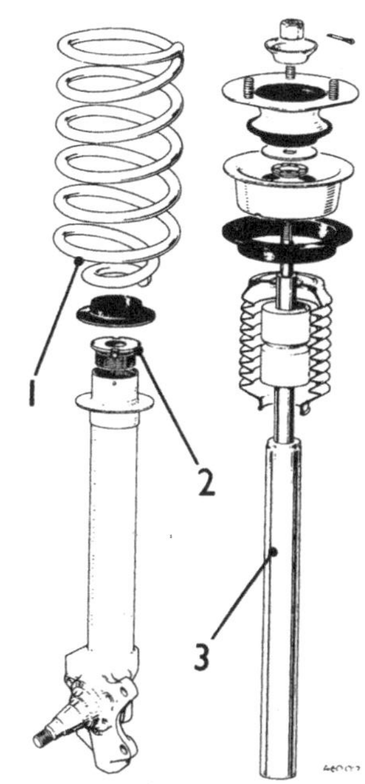

FRONT STRUT UPPER SWIVEL ASSEMBLY

– Remove and refit **60.30.04**

As operation 60.20.01.

BOTTOM LINK

Remove and refit **60.40.02**

Removing

1 Jack up the car and support the body on stands.
2 Remove the road wheel.
3 Remove the spring pin, nyloc nut, flat washer and outer rubber bush from the end of the anti-roll bar.
4 Remove the two bolts securing the steering arm to the stub axle assembly and push clear.
5 Remove the split pin, slotted nut and plain washer and release the ball joint from the stub axle assembly.
6 Remove the bolt and nyloc nut securing the bottom link to the sub frame. Withdraw the bottom link.

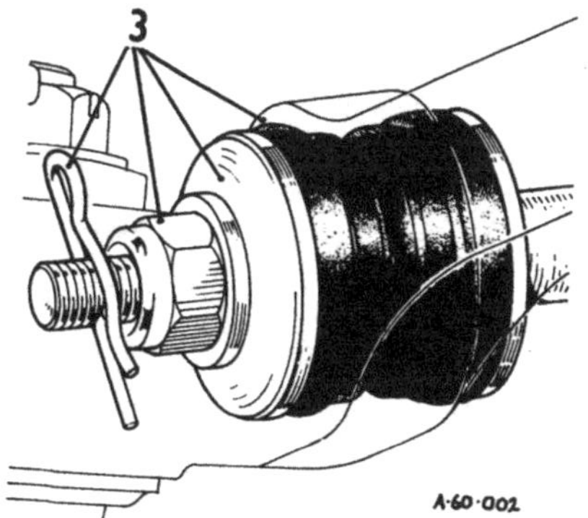

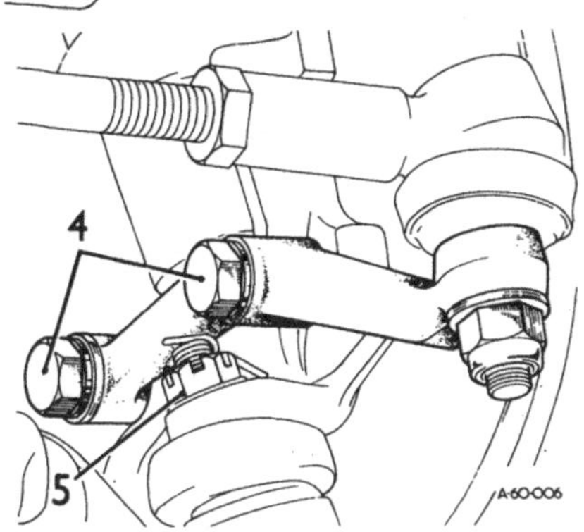

Refitting

7 Locate the anti-roll bar into the mounting hole in the bottom link.
8 Position the bottom link and secure to the sub frame with the bolt and nyloc nut. DO NOT tighten fully until the car is resting on its wheels.
9 Fit the ball joint into the stub axle assembly and secure with the flat washer, slotted nut and split pin.
10 Place the jack under the bottom link and carefully raise the link to locate the outer rubber bush, dished washer, flat washer and nyloc nut onto the end of the anti-roll bar.
11 Tighten the nyloc nut to the stop and fit the spring pin.
12 Lower the jack.
13 Fit the steering arm and tighten the two bolts.
14 Fit the road wheel and lower the car.

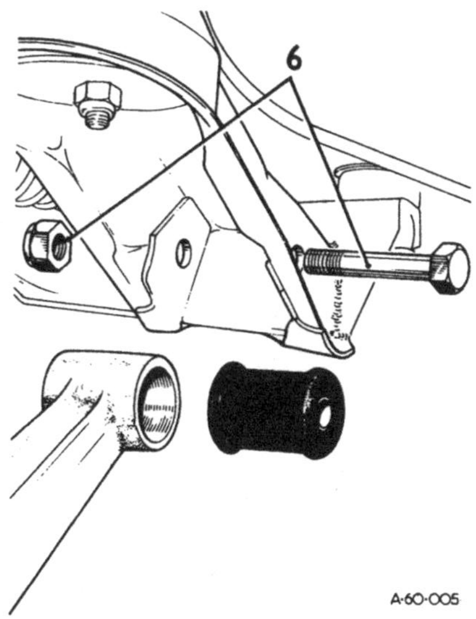

BOTTOM LINK

Overhaul **60.40.06**

1 Remove the bottom link. 60.40.02.
2 Remove the plastic gaiter and rubber ring from the ball joint.
3 Remove the circlip retaining the ball joint housing to the bottom link.
4 Press or drive out the ball joint and housing.
5 Enter the new ball joint and housing from the underside of the bottom link ensuring that the housing is squarely located.
6 Press the housing into the bottom link taking care not to damage the bottom of the housing. (A short length of suitable diameter tube is recommended.)
7 Fit the circlip, new plastic gaiter and gaiter retaining ring.
8 Press out the rubber bush and sleeve from the fulcrum end of the bottom link.
9 Press the new bush and sleeve into position.

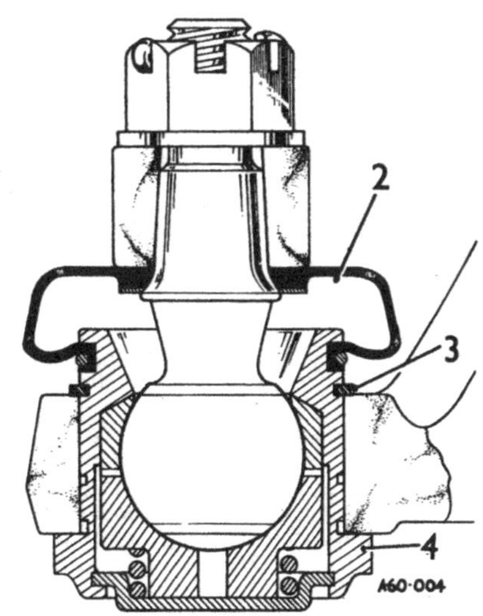

REAR SUSPENSION

ROAD SPRING

– **Remove and refit** **64.20.01**
– **Insulating rings** **64.20.17**

Removing
1 Jack up the car and support the body on stands.
2 Remove the road wheel.
3 Transfer the jack to support the suspension arm and partially compress the road spring, taking care not to relieve the weight on the stands.
4 Remove the two nuts and bolts securing one side of the anti-roll bar to the suspension arm.
5 Remove the nut and bolt securing the rear end of the suspension arm to the axle bracket.
6 Carefully lower the jack.
7 Remove the spring and its upper and lower insulating rubbers.

Refitting
8 Ensure that the spring insulating rubbers are correctly positioned and fit the spring.
9 Position the jack under the suspension arm.
10 Carefully raise the jack and engage the rear end of the suspension arm in the axle bracket.
11 Fit the bolt and nut.
12 Connect the anti-roll bar to the suspension arm.
13. Fit the road wheel.
14 Remove the stands and lower the car.
15 Tighten the nut and bolt securing the rear end of the suspension arm to the axle bracket.

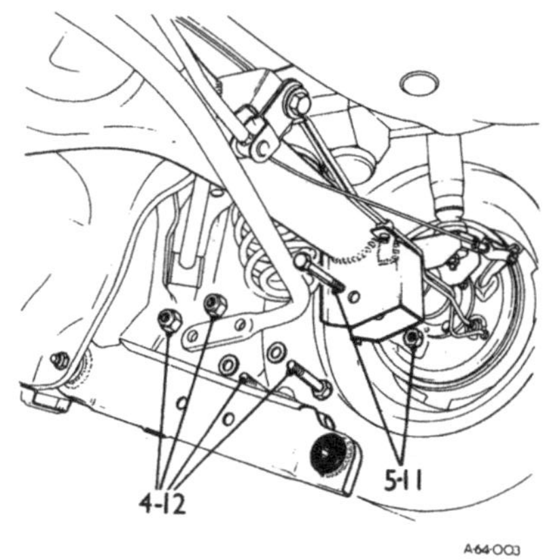

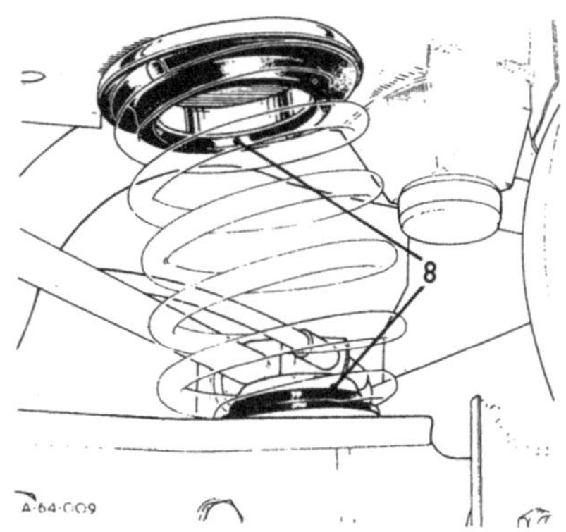

REAR DAMPER LEFT HAND

– **Remove and refit** **64.30.02**

Removing
1 Jack up the car and support the body on stands.
2 Remove the rear wheel.
3 Remove the three screws securing the damper access plate to the body in the boot. Remove the plate.
4 Remove the locknut, nut, plain washer and rubber bush securing the upper end of the damper to the body.
5 Remove the locknut, nut, plain washer and rubber bush securing the lower end of the damper to the axle bracket.
6 Withdraw the damper.
7 Remove the rubber and plain washer from each end of the damper.

Refitting
8 Fit the plain washer and rubber bush to the upper end of the damper.
9 Position the damper on the car. Fit the rubber, plain washer, nut and locknut securing the upper end of the damper to the body.
10 Apply a plasti-seal, to the damper cover plate.
Fit the plate and secure with the three screws.
11 Fit the plain washer and rubber to the lower end of the dampers. Position the damper in the axle bracket and secure with the rubber, bush, plain washer, nut and locknut.
12 Refit the road wheel.
13 Remove the stands and lower the car.

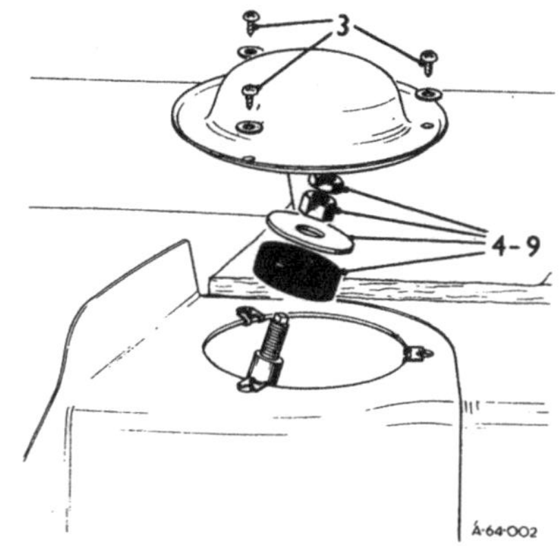

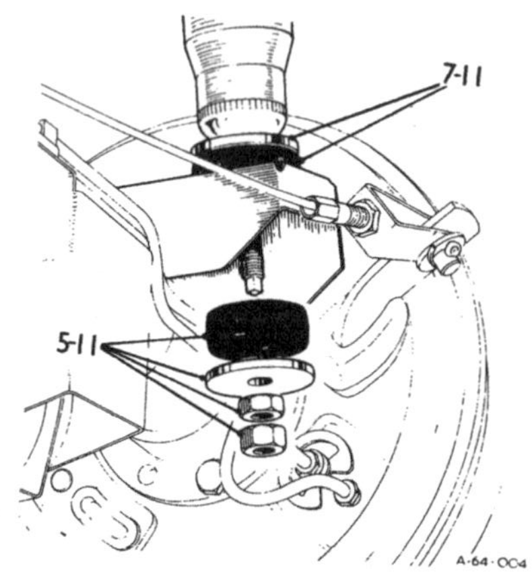

REAR DAMPER RIGHT HAND

– Remove and refit **64.30.03**

Removing
1 Jack up the car and support the body on stands.
2 Remove the rear road wheel.
3 Remove the fuel filler cap and filler assembly. 19.55.08.
4 Remove the locknut, nut, plain washer and rubber bush securing the upper end of the damper to the body.
5 Remove the locknut, nut, plain washer and rubber bush securing the lower end of the damper to the axle bracket.
6 Withdraw the damper.
7 Remove the rubber bush and plain washer from each end of the damper.

Refitting
8 Fit the plain washer and rubber bush to the upper end of the damper.
9 Position the damper on the car. Fit the rubber bush, plain washer, nut and locknut, securing the upper end of the damper to the body.
10 Fit the fuel filler cap and filler assembly. 19.55.08.
11 Fit the plain washer and rubber bush to the lower end of the damper. Position the damper in the axle bracket and secure with the rubber bush, plain washer, nut and locknut.
12 Refit the road wheel.
13 Remove the stands and lower the car.

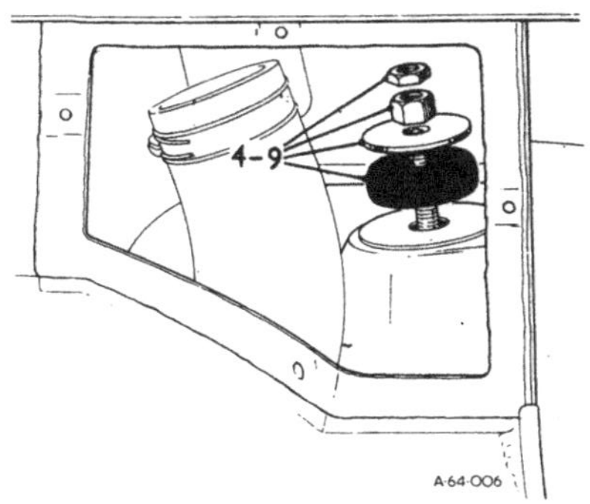

BUMP STOP

– Remove and refit **64.30.15**

Removing
1 Remove the bump stop from its mounting.

Refitting
2 Press new bump stop into position.

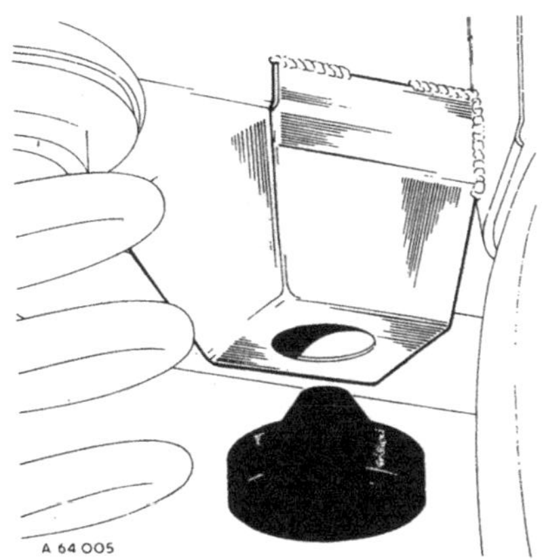

SUSPENSION ARM

– Remove and refit **64.35.02**

Removing
1 Jack up the car and support the body on stands.
2 Remove the road wheel.
3 Transfer the jack to support the suspension arm and partially compress the road spring, taking care not to relieve the weight on the stands.
4 Remove two nuts and bolts securing the anti-roll bar to the suspension arm.
5 Remove the nut and bolt securing the rear end of the suspension arm to the axle bracket.
6 Carefully lower the jack.
7 Remove the spring.
8 Remove the nut and bolt securing the forward end of the suspension arm to the body bracket.
9 Detach the suspension arm from the bracket.

Refitting
10 Engaging the forward end of the suspension arm in the body bracket, fit the bolt and nut. Do not tighten at this stage.
11 Place the jack under the suspension arm.
12 Ensure that the spring insulating rubbers are correctly positioned and fit the spring.
13 Raise the jack and engaging the rear end of the suspension arm in the axle bracket, fit the bolt and nut. Do not tighten at this stage.
14 Connect the anti-roll bar to the suspension arm.
15 Fit the road wheel.
16 Remove the stands and lower the car.
17 Tighten the front and rear suspension arm, nuts and bolts.

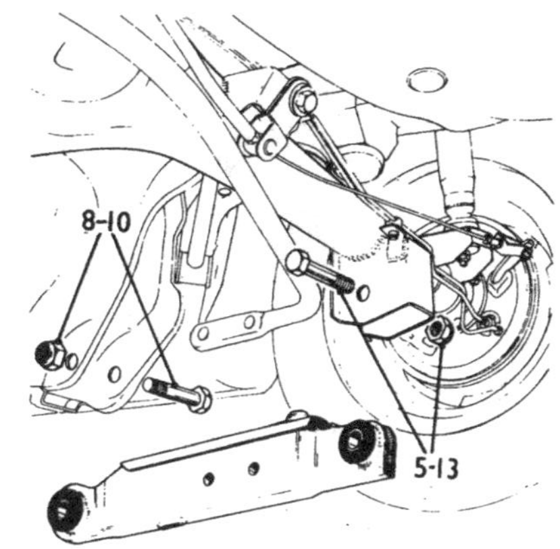

A-64-010

SUSPENSION ARM BUSHES

– Remove and refit **64.35.05**

Removing
1 Remove the suspension arm. 64.35.02.
2 Press out the old bushes.

Refitting
3 Press in the new bushes ensuring that they are centralized in the suspension arm. Note that the front bush must be installed in the position illustrated.
4 Fit the suspension arm. 64.35.02.

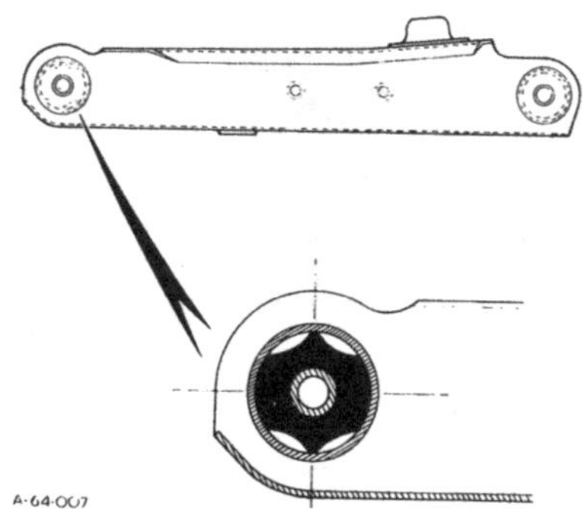
A-64-007

ANTI-ROLL BAR

– Remove and refit **64.35.08**

Removing
1 Raise the car and support it safely.
2 Remove the four bolts and nuts (two either side) securing the anti-roll bar to the rear suspension arms.
3 Withdraw the anti-roll bar and shim(s), if fitted.

Refitting
4 Locate the anti-roll bar (and shim(s), if removed) in position on the rear suspension arms.
5 Align the mounting holes and fit and tighten the four securing bolts and nuts.
6 Lower the car.

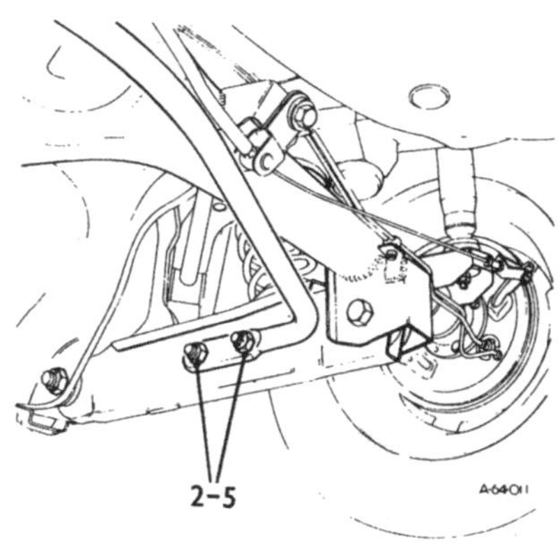

RADIUS ROD

– Remove and refit **64.35.28**

Removing
1 Jack up the car and support the body on stands.
2 Remove the nut and bolt securing the rear end of the radius rod to the rear axle bracket.
3 Remove the nut and bolt securing the forward end of the radius rod to the body bracket.
4 Withdraw the radius rod.

Refitting
5 Refit the radius rod.
6 Engage the forward end of the radius rod in the body bracket and fit the bolt and nut.
7 Engage the rear end of the radius rod in the axle tube bracket. Fit the bolt and nut.
8 Tighten both nuts.
9 Remove the stands and lower the car.

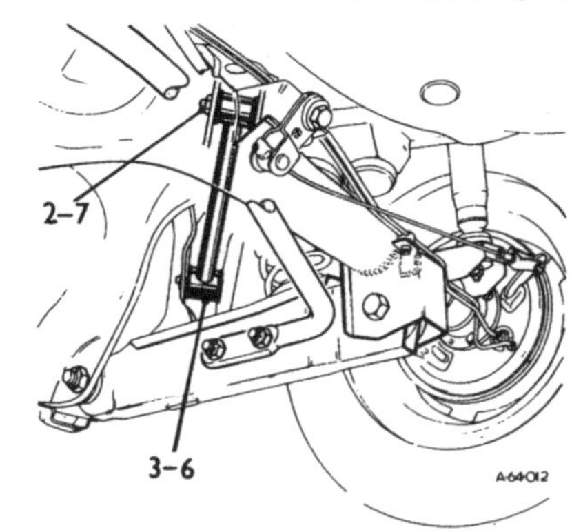

RADIUS ROD BUSHES

– Remove and refit **64.35.29**

Removing
1 Remove the radius rod from the car. 64.35.28.
2 Press out the bush from the radius rod.

Refitting
3 Fit a new bush to the radius rod.
4 Install the radius rod in the car. 64.35.28.

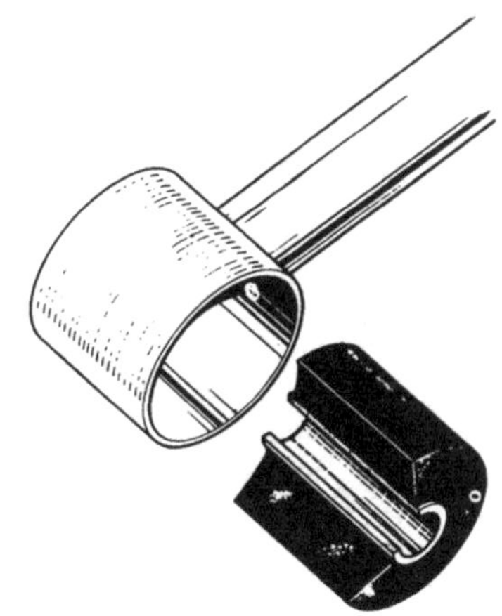
A-64-008

BRAKES

Brake hose	
– front L.H. – remove and refit	70.15.02
– front R.H. – remove and refit	70.15.03
– rear – remove and refit	70.15.17
Brakes	
– adjust	70.25.03
– bleed	70.25.02
Front brake – disc	
– caliper – remove and refit	70.55.02
– caliper seals – remove and refit	70.55.13
– disc – remove and refit	70.10.10
– disc shield – remove and refit	70.10.18
– pads – remove and refit	70.40.02
Handbrake	
– cable – remove and refit	70.35.16
– cables – adjust	70.35.10
– lever assembly – remove and refit	70.35.08
Hydraulic pipes	70.20.00
Pedal box	
– overhaul	70.35.04
– remove and refit	70.35.03
Pressure reducing valve	70.01.01
– remove and refit	70.25.21
Master cylinder – tandem	
– overhaul	70.30.09
– remove and refit	70.30.08
Rear brake	
– backplate – remove and refit	70.10.26
– drum – remove and refit	70.10.03
– shoes – remove and refit	70.40.03
– wheel cylinder – overhaul	70.60.26
– wheel cylinder – remove and refit	70.60.18
Servo	
– non-return – remove and refit	70.50.15
– remove and refit	70.50.01
– renewing filter	70.50.25
– vacuum hose – remove and refit	70.50.14

PRESSURE REDUCING VALVE

70.01.01

The pressure reducing valve is installed in the brake circuit between the master cylinder and the front and rear brakes. Its function is to limit the pressure applied to the rear brakes relative to the pressure applied to the front brakes, thus minimizing the possibility of rear wheel locking. In the event of a failure in the front brake circuit the cut-off pressure is increased and the pressure reduction ratio changes.

Operation

Fluid from the primary chamber of the master cylinder is fed into the pressure reducing valve at port A and out to the front brakes via ports C and D. The master cylinder secondary chamber feeds into port B, through the internal passages in the valve plunger, past the metering valve and out to the rear brakes via port E. The large spring S is pre-loaded to bias the valve plunger to the left. Hydraulic pressure therefore acts on the annular area (a1 – a2) forcing the plunger to the left while the force acting on area a1 and annular area (a4 – a3) tends to move the plunger to the right where it is opposed by spring S. When the net force acting to the right overcomes the pre-load provided by spring S the plunger assembly shifts to the right thereby closing the metering valve F. Pressure at the rear outlet port E therefore falls relative to the input pressure. As pressure is increased at ports A and B the plunger is forced to the left, opening the metering valve F and permitting a small quantity of fluid to be fed to the rear brakes. The resultant increase in pressure acting on area a1 causes the plunger to again shift to the right closing the metering valve. This procedure continues until there is no further increase in pressure from the master cylinder.

The pressure at outlet E is reduced after cut-off in proportion to the areas a2 and the difference between the two annular areas (a1 – a2) and (a4 – a3). The cut-off pressure is equal to the pre-load in the spring S divided by the combined areas a2 and (a4 – a3). Should the front brake circuit fail there will be no pressure acting on annular area (a4 – a3) so that the net force tending to move the plunger to the right will be equivalent to the product of the input pressure and area a2. Thus, as the value of the pre-load spring S is unchanged, the cut-off pressure will increase considerably (approximately threefold).

As the annular area (a4 – a3) is now redundant, the reduction ratio after cut-off changes to a value which is proportional to the areas a1 and (a1 – a2). Should the rear brake circuit fail the pressure reducing valve is completely inoperative and pressure is fed to the front brakes in the normal manner.

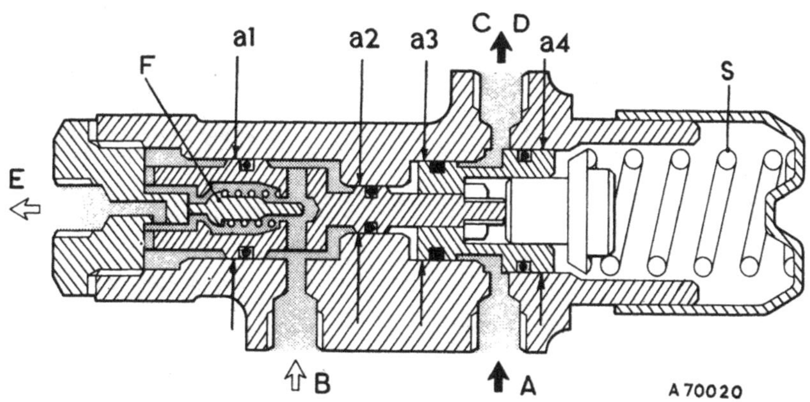

A70020

REAR BRAKE-DRUM

Remove and refit **70.10.03**

Removing

1 Jack up the car and support the body on stands.
2 Remove the road wheel.
3 Release the hand brake.
4 Remove the countersunk screw(s) securing the brake-drum to the hub and withdraw the brake-drum.

Refitting

5 Align the countersunk hole(s) in the drum with the tapped hole(s) in the hub.
6 Engage the wheel studs in the drum.
7 Slide the drum into position. If the brake-shoes were disturbed, they may require to be centralized on the backplate to allow drum entry.
8 Fit and tighten the countersunk screw(s).
9 Fit the road wheel and lower the jack.
10 Apply the foot brake several times to adjust the rear brakes.

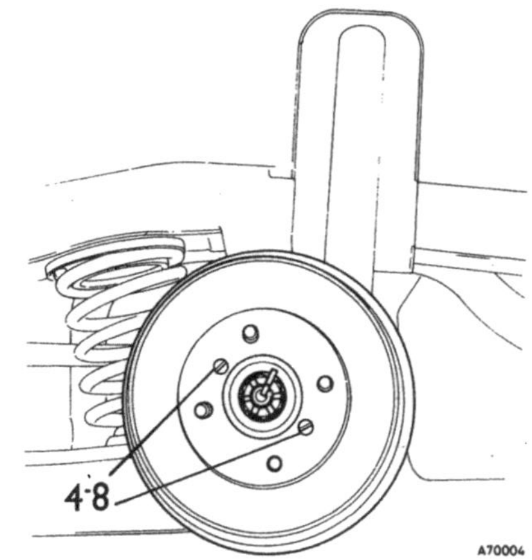

A70004

BRAKE DISC

Remove and refit **70.10.10**

Removing

1 Remove the front hub, see 60.25.01.
2 Remove the four bolts securing the disc to the hub.
3 Withdraw the disc.

Refitting

4 Offer up the disc to the hub.
5 Fit and evenly tighten to 25 to 32 lbf ft (3.5 to 4.4 kgf m) the four bolts securing the disc to the hub.
6 Fit the front hub to the stub axle and adjust, see 60.25.01.

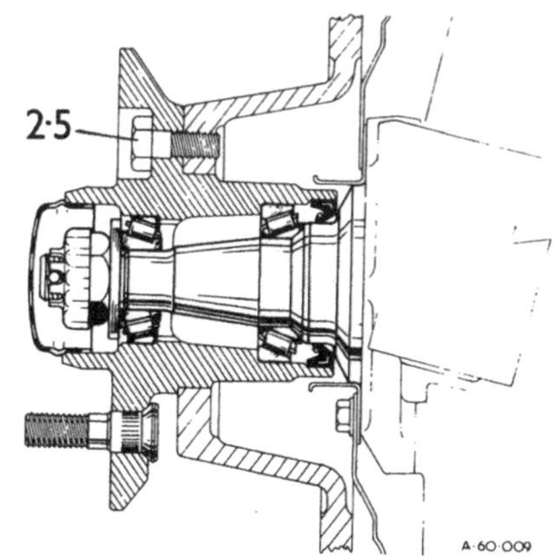

A-60 009

DISC SHIELD

Remove and refit 70.10.18

Removing
1 Remove the front hub, see 60.25.01.
2 Remove the three bolts and spring washers securing the disc shield to the vertical link assembly.
3 Remove the disc shield.

Refitting
4 Position the disc shield over the stub axle and secure to the vertical link with the three bolts and spring washers.
5 Fit the front hub to the stub axle and adjust, see 60.25.01.

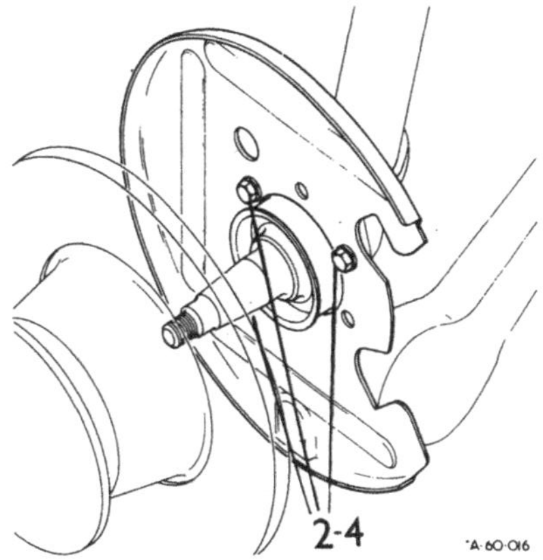

REAR BRAKE BACKPLATE

Remove and refit 70.10.26

Removing
1 Jack up the car and support the body on stands.
2 Remove the rear road wheel and release the hand brake.
3 Remove the rear hub, see 51.10.18.
4 Remove the clevis pin securing the hand brake cable fork to the back plate lever.
5 Disconnect the fluid feed pipe union at the wheel cylinder (left-hand side only);
or
Disconnect the fluid feed and transfer pipe unions at the wheel cylinder (right-hand side only).
6 Remove four nuts, spring washers, and bolts securing the backplate to the axle casing flange.
7 Withdraw the deflector plate and the backplate.

Refitting
7 Reverse instructions 1 to 7.
8 Bleed the brakes.

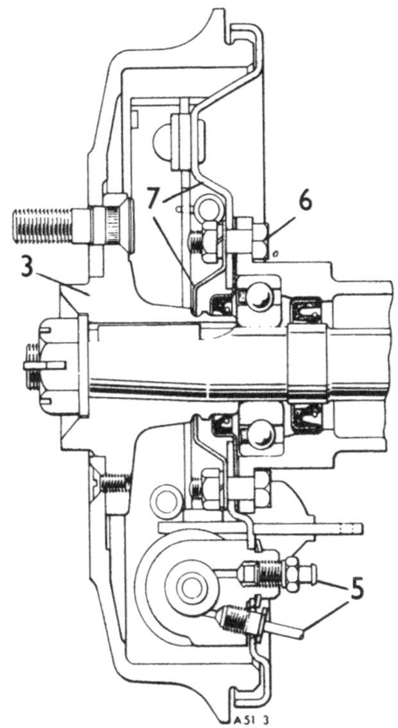

BRAKE HOSE – FRONT

Remove and refit
Left-hand **70.15.02**
Right-hand **70.15.03**

Removing
1 Disconnect the brake pipe and union from the inboard end of the flexible hose.
2 Disconnect the brake pipe and union from the outboard end of the flexible hose.
3 Using two spanners, remove the locknuts and washers securing the hose to the support brackets, one on the wheel arch the other on the damper tube, and remove the hose.

Refitting
4 Reverse instructions 1 to 3. Ensure that the hose is neither kinked nor twisted when installed.
5 Bleed the brakes.

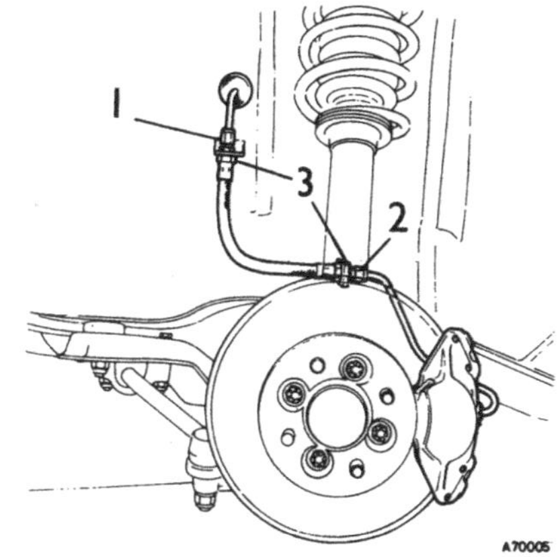

BRAKE HOSE – REAR

Remove and refit 70.15.17

Removing
1 Disconnect the brake pipe and union at the front end of the brake hose.
2 Disconnect the brake pipe and union at the rear end of the brake hose.
3 Using two spanners, remove the locknuts and washers securing the hose to the brackets, one on the body, the other on the axle tube. Remove the brake hose.

Refitting
4 Reverse instructions 1 to 3. Ensure that the hose is neither kinked nor twisted when installed.
5 Bleed the brakes.

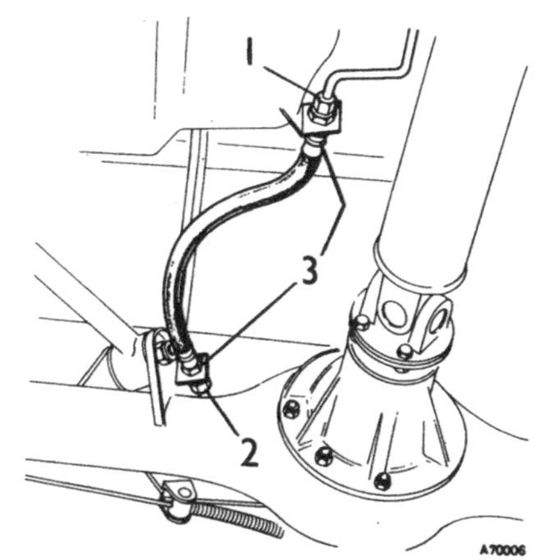

HYDRAULIC PIPES

Remove and refit 70.20.00

To aid identification of individual pipes, operation numbers are included in the illustration showing the general arrangement of the brake system.

Pipe – master cylinder to pressure reducing valve – front brakes	70.20.44
Pipe – master cylinder to pressure reducing valve – rear brakes	70.20.45
Pipe – pressure reducing valve to L.H. front hose	70.20.02
Pipe – pressure reducing valve to R.H. front hose	70.20.03
Pipe – L.H. front hose to caliper	70.20.04
Pipe – R.H. front hose to caliper	70.20.05
Pipe – Pressure reducing valve to rear hose	70.20.13
Pipe – R.H. rear wheel cylinder to L.H. rear wheel cylinder	70.20.17
Pipe – rear hose to R.H. rear wheel cylinder	70.20.18

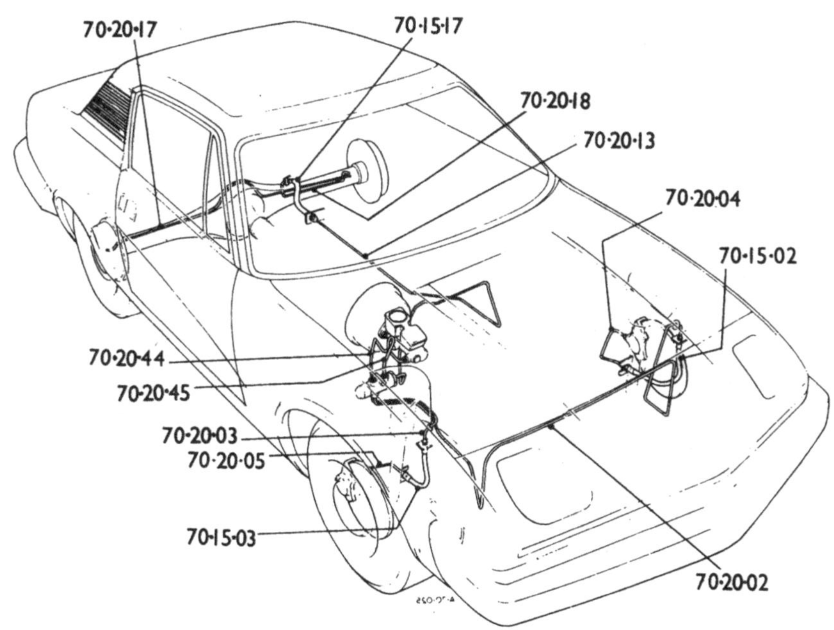

WARNING: Metric brake fittings are used and it is imperative that only metric replacements are fitted.

The illustrations show metric and the equivalent Unified (U.N.F.) parts for comparison.

UNIFIED	METRIC	COMMENTS
		Metric pipe nuts, hose ends, unions and bleed screws are coloured **BLACK** and most are also identified with the letter **'M'**.
		Metric pipe flares must be used.
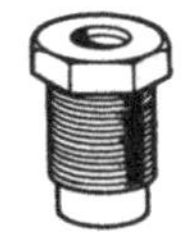		

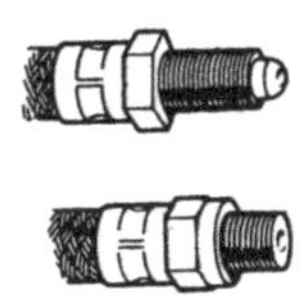	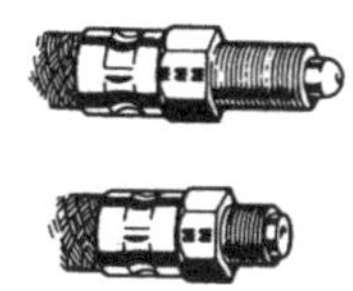	End of a metric hose is coloured **BLACK**.
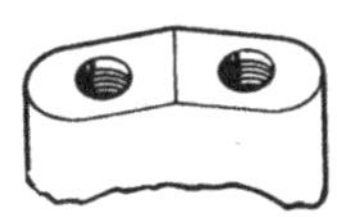		Metric parts are not counterbored; some U.N.F. threaded parts also had no counterbore; always check.
		If the thread of a component is in doubt, screw the hydraulic connections and bleed screw fully in with the fingers. If they cannot be screwed right in or if they are unduly slack, the threads may not be compatible. Metric hose seals against the bottom of the port, gap between hose hexagon and face of cylinder or calliper.

BRAKES

Bleed 70.25.02

Do not allow the fuel level in the reservoir to fall below half capacity. When topping-up during the bleeding process, DO NOT USE aerated fluid exhausted from the system. DO NOT bleed the system with the servo in operation (engine running).

1. Disconnect the wires to the pressure failure switch and remove the pressure failure switch from the underside of the master cylinder.
2. Release the hand brake.
3. Attach the bleed tube to the bleed nipple of the front caliper farthest from the master cylinder, allowing the free end of the bleed tube to hang submerged in brake fluid in a transparent container.
4. Open the bleed nipple (90 to 180 degrees).
5. Fully depress the brake pedal and follow with three rapid successive strokes. Allow the pedal to return. Repeat this procedure until fluid free from air bubbles issues from the wheel cylinder.
6. Depress the brake pedal, close the nipple and release the pedal.
7. Remove the bleed tube.
8. Attach the bleed tube to the opposite front caliper and repeat instructions 4 to 7.
9. Attach the bleed tube to the single nipple on the rear backplate (R.H. Stg. – left-hand backplate; L.H. Stg. – right-hand backplate) and repeat instructions 4 to 7.
10. In order to ensure the efficiency of the brake bleeding operation, the following check should be carried out:
    a. Start the engine and apply the handbrake.
    b. With a load of 25 lbf (11.3 kgf) applied to the brake pedal, check that the pedal travel does not exceed 1.90 in (48.26 mm).
    c. If the pedal travel exceeds this amount, repeat instructions 2 to 9.
11. Fit the pressure failure switch to the master cylinder and connect the wires. The P.D.W.A. shuttle fitted to this vehicle is self-centering.

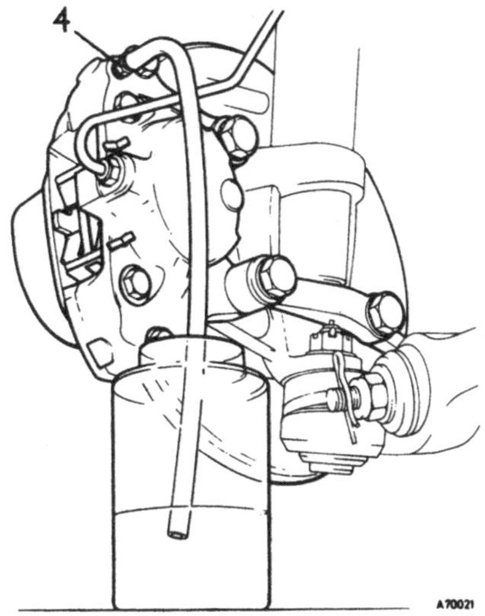

BRAKES

Adjust 70.25.03

Self-adjusting brakes are fitted to the front and rear. Front adjustment is hydraulically self-compensating to provide for brake pad wear. In the rear brakes a self-adjusting mechanism incorporated in the brake-shoe hand brake linkage maintains a fixed brake liner/drum running clearance; self-adjustment occurs on the application of the footbrake.

PRESSURE REDUCING VALVE

Remove and refit 70.25.21

Removing

1. Slacken the brake pipe unions at the master cylinder.
2. Remove the two inlet pipes from the top of the pressure reducing valve.
3. Remove the rear brake outlet pipe from the end plug of the pressure reducing valve.
4. Remove the two front brake outlet pipes from the underside of the pressure reducing valve. To facilitate removal of the L.H. front brake pipe, first remove the R.H. front brake pipe from the pressure reducing valve and detach the pipe from the clip on the inner wheel arch.
5. Remove the nut, plain washer, spring washer and bolt, and remove the pressure reducing valve with the bracket from the suspension turret.

Refitting

6. Align the lugs on the bracket with the holes in the suspension turret.
7. Fit the pressure reducing valve and secure to the suspension turret with the bolt, plain washer, spring washer and nut.
8. Fit the brake pipes, tighten the pressure reducing valve unions, and attach the R.H. front brake pipe to the clip on the inner wheel arch.
9. Tighten the brake pipe unions at the master cylinder.
10. Bleed the brakes.

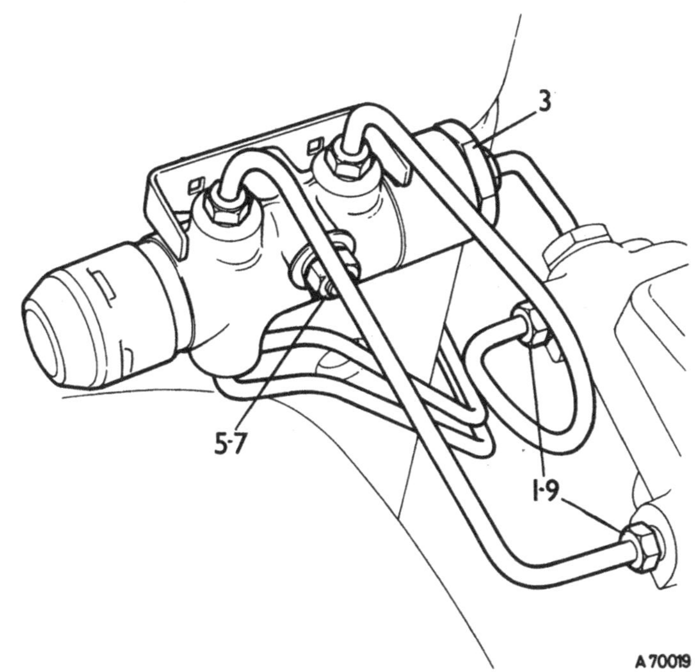

MASTER CYLINDER – TANDEM

Remove and refit **70.30.08**

Removing

1 Noting their positions, disconnect the brake pipes at the master cylinder. Plug the master cylinder ports to prevent fluid discharge from the reservoir. Seal the brake pipes to prevent ingress of foreign matter.
2 Disconnect the wires to the pressure failure switch.
3 Remove the two nuts and spring washers securing the master cylinder to the servo and withdraw the master cylinder.

Refitting

4 Reverse instructions 1 to 3.
5 Bleed the brakes.

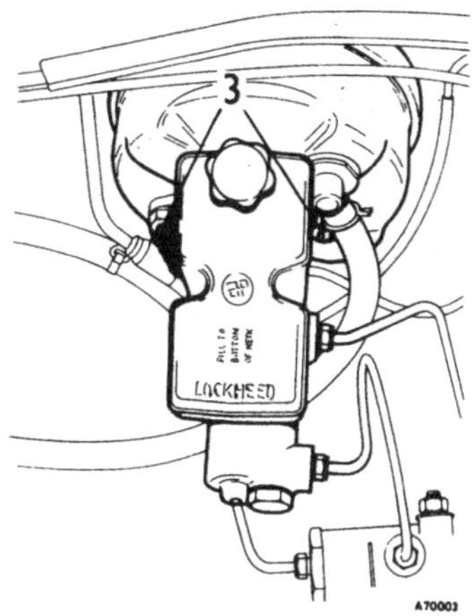

MASTER CYLINDER – TANDEM

Overhaul **70.30.09**

Dismantling

1 Remove the tandem master cylinder from the car, see 70.30.08.
2 Drain the fluid from the reservoir.
3 Remove the two screws securing the fluid reservoir to the master cylinder and remove the reservoir.
4 Noting their original positions, remove the two rubber seals from the cylinder body recesses.
5 Remove the metal cap and rubber boot from the end of the master cylinder.
6 Remove the circlip, withdraw the primary piston and return spring.
7 Insert a soft metal rod into the cylinder and depress the secondary piston. This will release the stop pin which is seated adjacent to the secondary piston fluid feed port.
8 Remove the pin and withdraw the secondary piston, the spring retainer and the return spring.
9 Carefully note the sizes and positions of the rubber seals, the piston washers and the spring retainers. Remove the seals using only the fingers.

[CAUTION: Wash hydraulic cylinder components in brake fluid only. Never use gasoline or other petroleum-based solvents since they are damaging to rubber parts in the system.]

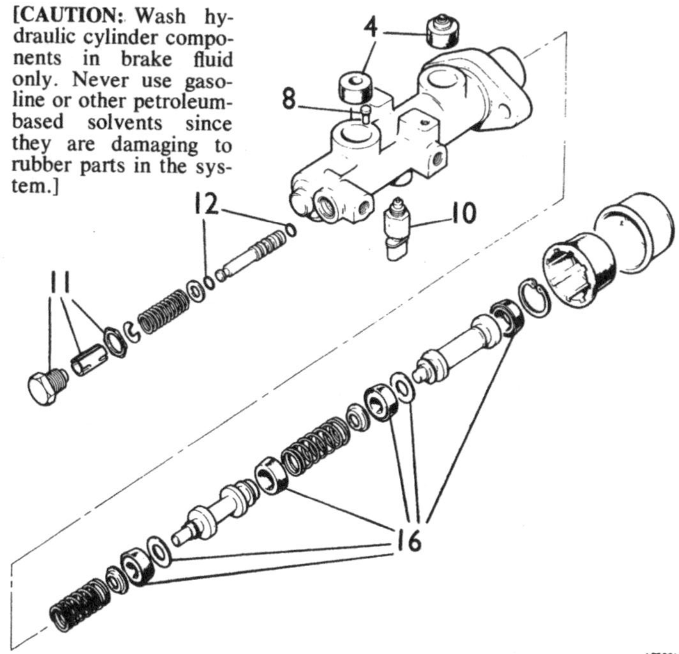

10 Unscrew the pressure failure switch from the cylinder body.
11 Remove the end plug and copper washer, withdraw the distance piece, and the piston and spring sub assembly.
12 Carefully remove the two rubber seals.

Inspection

13 Clean all the components thoroughly in a recommended brake fluid and dry using a lint-free cloth.
14 Carefully inspect the metal components for faults and wear. A replacement assembly must be fitted if the cylinder bores show the slightest signs of corrosion, ridging or scoring.
15 Ensure that all the ports and drillings in the cylinder body, piston heads and the vent hole in the filler cap are clear of any obstructions.

Reassembly

CAUTION: Scrupulous cleanliness is essential. Immerse all components in a recommended brake fluid and assemble when wet.

16 Fit new seals and washers to the primary and secondary pistons, using only the fingers.
17 Fit the secondary return spring, spring retainer and piston into the cylinder bore, taking care not to bend back the lip of the seal.
18 With a soft metal rod, depress the secondary piston. When the head of the piston passes the secondary fluid feed port, fit the piston stop pin.
19 Fit the primary return spring, spring retainer and piston into the cylinder bore, ensuring that the lip of the seal is not bent back.
20 Fit the circlip at the mouth of the cylinder bore and check that it is correctly seated in the groove.
21 Fit the two rubber seals into the cylinder body recesses.
22 Ensure that the reservoir is clean and fit to the master cylinder body with the two screws. Tighten to a torque of 5 lbf ft (0.69 kgf m). DO NOT overtighten.
23 Fit two new 'O' rings into the grooves on the P.D.W.A. piston and spring sub-assembly.
24 Insert the piston and spring sub-assembly into its respective bore, taking care not to damage the 'O' rings.
25 Fit the metal distance piece.
26 Fit a new copper washer to the end plug and screw into the bore. Tighten to a torque of 33 lbf ft (4.56 kgf m).
27 Fit the master cylinder to the car, see 70.30.08.
28 Bleed the brakes.

PEDAL BOX

Remove and refit 70.35.03

Removing

1 Disconnect the two spade terminals from the brake stop light switch.
2 Remove the speedo cable from its retaining clip.
3 Remove the clevis pin securing the brake pedal to the brake master cylinder rod.
4 Remove the clevis pin securing the clutch pedal to the clutch master cylinder rod.
5 Remove the four nuts and spring washers securing the servo to the pedal box.
6 Remove the two nuts, bolts and spring washers securing the clutch master cylinder to the pedal box.
7 Remove the cleat securing the harness to the stabilizer bar.
8 Remove the nut and bolt securing the stabilizer bar to the fascia rail.
9 Remove the three bolts and spring washers securing the top of the pedal box to the body.
10 Withdraw the pedal box assembly.

Refitting

11 Reverse instructions 1 to 10.

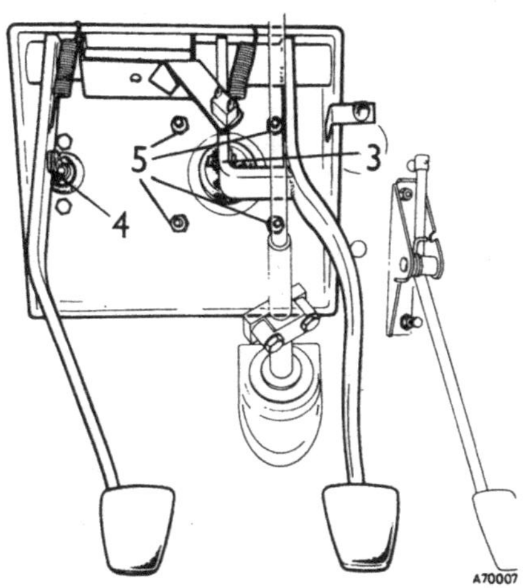

PEDAL BOX

Overhaul 70.35.04

1 Remove the pedal box from the car, see 70.35.03.
2 Remove the brake stop light switch.
3 Remove the circlip from one end of the pedal pivot rod.
4 Push the rod through the bracket.
5 Remove the brake and clutch pedals and the two anti-rattle springs.
6 Remove and renew the two Teflon coated bushes in each pedal.
7 Remove and renew the pedal pad rubbers.
8 Locate both the clutch and brake pedals in the pedal box, ensuring that the anti-rattle springs are fitted.
9 Push through the pivot rod and secure using the circlip.
10 Fit the brake stop light switch.
11 Fit the pedal box to the car, see 70.35.03.

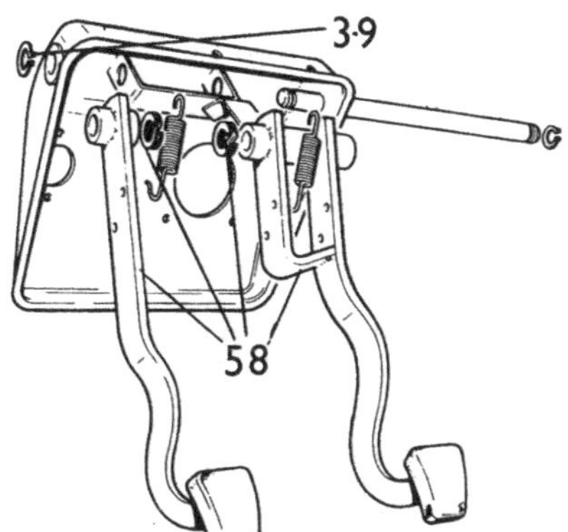

HANDBRAKE LEVER ASSEMBLY

Remove and refit 70.35.08

Removing

1 Drive the vehicle onto a ramp. Release the hand brake.
2 Raise the ramp.
3 Using a screw jack, raise the body to allow access to the underside of the transmission tunnel.
4 Pull back the rubber gaiter and release the hand brake cable locknut; note original position.
5 Unscrew the hand brake cable from the operating rod.
6 Lower the ramp.
7 Remove the centre console, see 76.25.01.
8 Remove the Lucar connector from the handbrake warning light switch.
9 Remove the four bolts and spring washers securing the hand brake lever assembly to the transmission tunnel.
10 Withdraw the hand brake lever assembly, the lower plate and the rubber gaiter.

Refitting

11 Reverse instructions 1 to 10.
12 Check the adjustment of the hand brake and rectify as necessary, see 70.35.10.

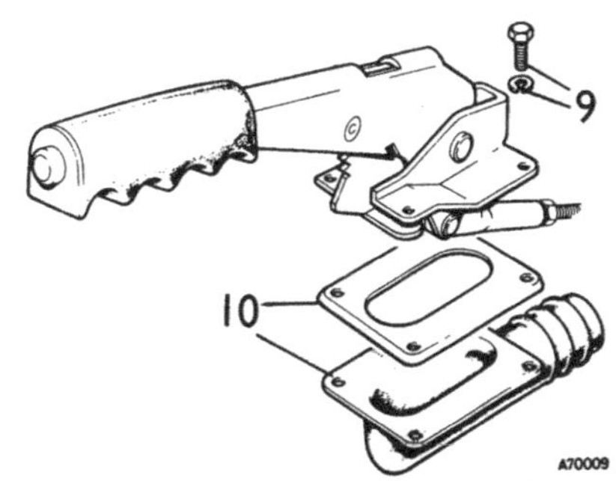

HAND BRAKE CABLES

Adjust 70.35.10

1 Jack up the rear of the vehicle and support the axle on stands.
2 Release the hand brake.
3 Disconnect the hand brake cables from the rear brake backplate levers.
4 Applying finger pressure, push the brake operating levers inboard to ensure that the operating levers are in contact with the brake-shoe webs.
5 Maintaining the compensator in the vertical position, adjust the cable forks so that the clevis pin can be entered through them and the operating levers without straining the cables. Ensure that the rear wheels do not drag.
6 Insert and fix the clevis pins.
7 Screw in each fork and adjuster 3½ complete revolutions and tighten the locknuts.
8 Apply alternately the hand and foot brakes several times.
9 With 25 lbf effort applied to the hand brake, the travel of the lever should be between four and seven notches.

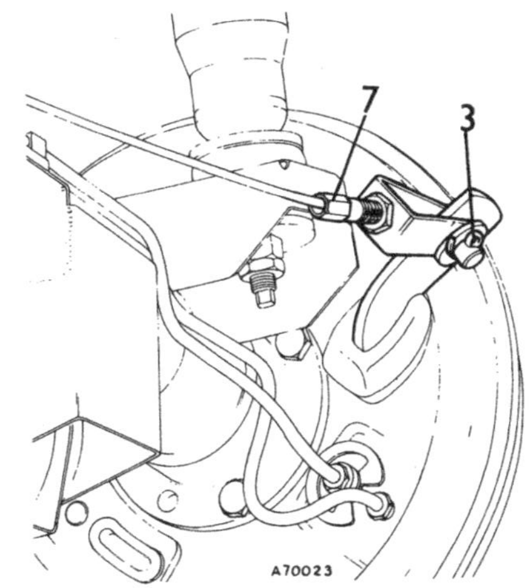

HANDBRAKE CABLE ASSEMBLY

Remove and refit **70.35.16**

Removing

1 Drive the vehicle onto a ramp. Release the handbrake.
2 Raise the ramp.
3 Using a screw jack, raise the body to allow access to the underside of the transmission tunnel.
4 Pull back the rubber gaiter and release the handbrake cable locknut.
5 Unscrew the handbrake cable from the operating rod.
6 Slacken the nut to release the cable from the abutment bracket.
7 Remove the split pin, washer and clevis pin retaining the cable to each rear brake operating lever.
8 Remove the trunnion locating nut, spring washer and bolt.
9 Slacken the nut securing the compensating levers.
10 Remove the cable assembly from the compensating levers, and the retaining brackets.

Refitting

11 If a new handbrake cable has been fitted the following setting procedure should be carried out:
 a Feed the cable forward through the guide bracket at the left-hand upper link, then through the guide bracket on the heelboard.
 b Locate the cable abutment into the rear of the handbrake bracket but do not tighten the nut.
 c Fit the cable adjuster (complete with jam nut) to the handbrake lever and screw fully into the connection. Lock the jam nut and pull the rubber boot over the joint.
 d Secure the abutment to the bracket.
 e Fit rubber sleeves on the cable into the guide brackets.
 f Pivot the compensator lever on the axle towards the left-hand wheel so that the lever is offset approximately ½ in (12 mm) from vertical.
 g Pull the left-hand operating lever away from the backplate to eliminate free movement and screw the adjuster into the fork-ends with fingers until all slack is removed from the cable.
 h Repeat (g) for the right-hand brake.
 i Apply the handbrake lever five times with a good double-handed pull (equal to approximately 100 lbf or 45 kgf) for a full stroke of the lever on each occasion to compress the handbrake outer cable.
 j Repeat (g) and (h) to finally adjust the fork-ends.
 k Secure the locknuts to the fork-ends.
 l Apply Retinax grease to the fork-ends.
12 Reverse instructions 1 to 10.
13 Adjust the handbrake cable, see 70.35.10.

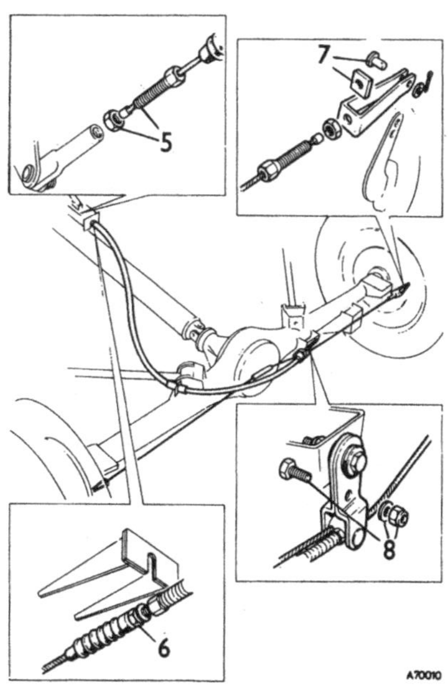

FRONT BRAKE PADS

Remove and refit **70.40.02**

Removing

1 Jack up the car and remove the front road wheel.
2 Depress the pad-retaining spring and withdraw the split pins.
3 Taking note of their positions, lift the pads and shims out of the caliper recesses.
The shims need not be renewed provided they are undamaged and are not corroded.

Refitting

4 Ease the caliper pistons into the bores to provide the extra clearance to accommodate the new unworn brake pads. During this operation brake fluid will be displaced and to prevent the reservoir overflowing, open the caliper bleed screw as pressure is applied to the piston. Close the bleed screw when the piston has moved the required amount. Repeat on the opposite piston in the caliper.
5 Remove dust and clean the brake pad locations in the caliper.
6 Insert the new pads and shims (smaller cut-out uppermost) into the caliper recesses.
7 Fit the new pad-retaining spring and split pins.
8 Firmly depress the footbrake pedal several times to correctly locate the friction pads.
9 Fit the road wheel and lower the car.
10 Check the fluid level in the reservoir and top up as necessary.

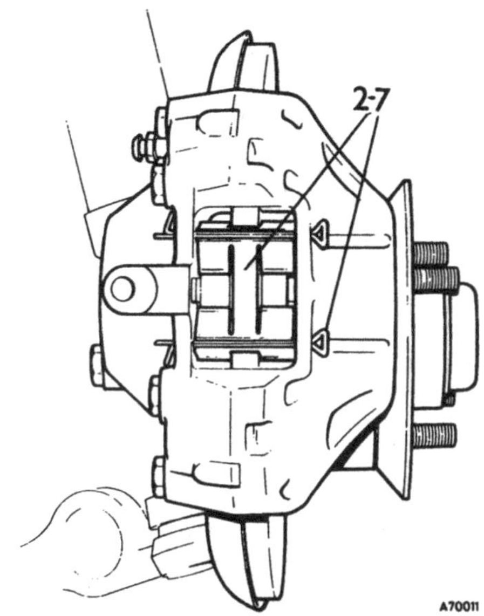

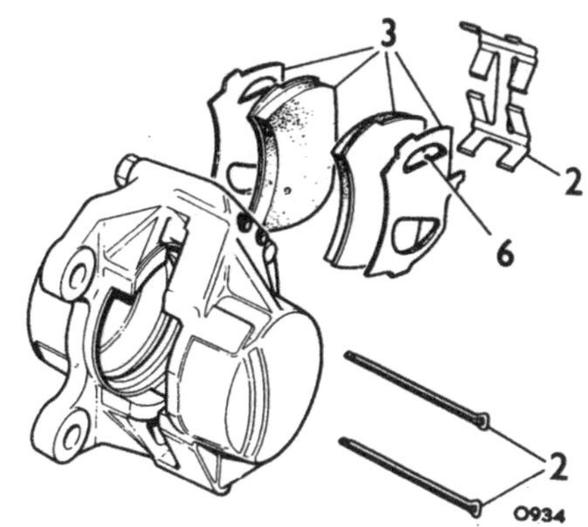

REAR BRAKE-SHOES

Remove and refit 70.40.03

Removing

1 Jack up the car and support the body on stands.
2 Remove the road wheel.
3 Release the hand brake.
4 Remove the two countersunk screws securing the brake-drum and withdraw the brake-drum.
When brake-drum removal is found to be difficult due to wear or ridging, remove the rubber plug on the inboard side of the backplate. Insert a small screwdriver, and engage it in the slotted hole in the small adjusting lever. Press down to release the mechanism.
5 Carefully note the position of the shoes and springs.
6 Remove the shoe steady pin cups and springs and extract the shoe steady pin from the rear of the backplate.
7 Ease the toe of the leading shoe followed by the heel of the trailing shoe out of the slotted piston heads.
8 Unlock the pull-off springs, and the cross lever tension spring and remove the brake-shoes.
To avoid possible ejection of the wheel cylinder pistons restrain them in position with a twist of wire or a suitable clamp. Take care not to damage the rubber boots.

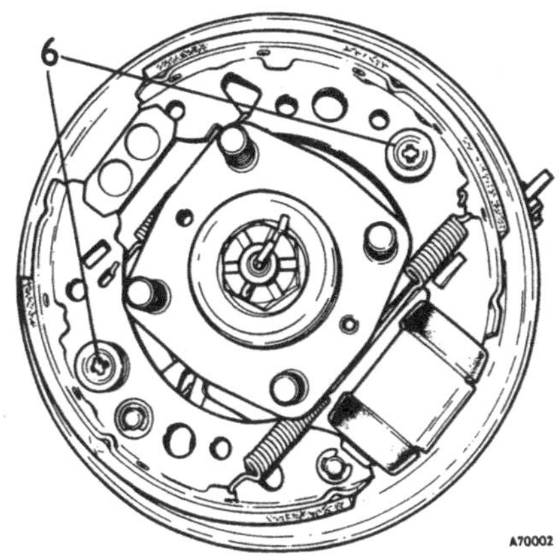

A70002

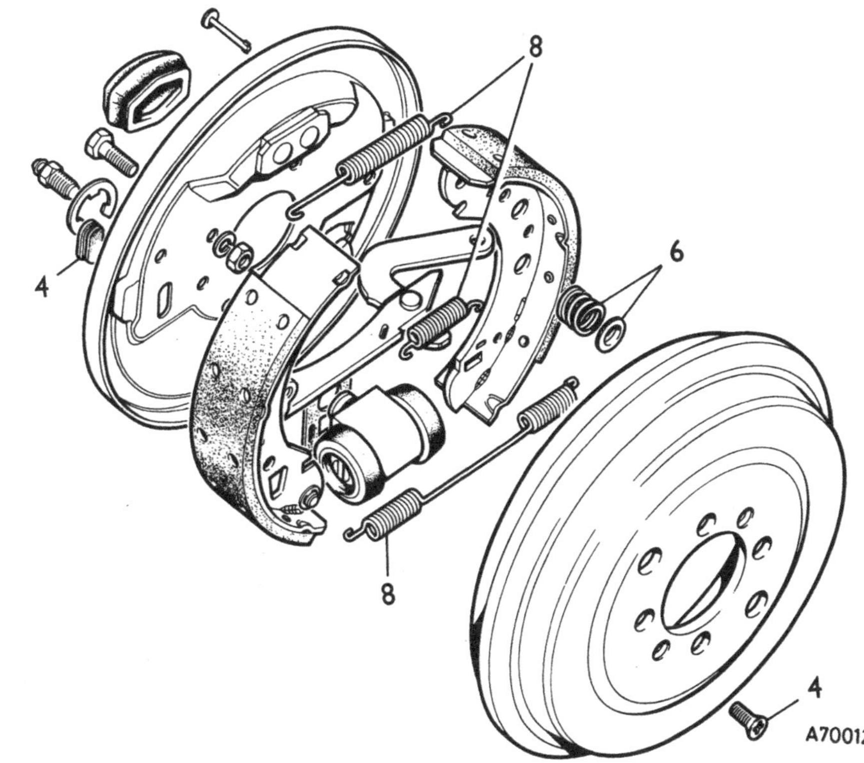

A70012

Refitting

9 Insert the cross lever tension spring hook in the cross lever, engage the other end in the leading shoe web in the previously noted hole position.
This spring is not interchangeable with the spring on the opposite brake.
10 Ease the brake-shoe and the cross lever towards the backplate, engage the toe of the shoe into the slot in the piston and the heel of the shoe into the abutment.
11 Hold the cross lever and shoe against the backplate, fit the steady pin, spring and cup.
12 Hook the pull-off springs into the holes in the shoe webs. The spring nearest the abutment is fitted on the backplate side of the shoes.
13 Pull the trailing shoe against the resistance of the springs. Position the heel into the slot in the piston and the toe into the abutment. Ensure that the cut-out in the cross lever engages with the slot in the adjuster plate.
14 Fit the remaining steady pin, spring and cup.
15 The functioning of the adjuster can be checked by gently operating the foot brake with the drum removed. Following expansion of the brake-shoes the ratchet will be seen to operate. Brake-shoes expansion can be cancelled by raising the ratchet plate to separate the ratchet teeth and allowing the pull-off springs to retract the shoes.
16 Replace the brake-drum and road wheel. Lower the car.
17 Apply the foot brake heavily several times to centralize and adjust the brake-shoes.
18 Road-test the car. If the operation of the brakes, including the hand brake, is poor, make four brake applications applying moderately high pedal efforts to decelerate the vehicle from 20 m.p.h. to rest. This will ensure correct adjustment of the rear brakes.

SERVO

Remove and refit **70.50.01**

Removing

1 Remove the master cylinder from the servo, see 70.30.01.
2 Disconnect the vacuum hose from the non-return valve.
3 Remove the clevis pin securing the servo push-rod to the brake pedal.
4 Remove the four nuts and spring washers securing the servo to the pedal bracket.
5 Withdraw the servo.

Refitting

6 Reverse instructions 1 to 5.
7 Bleed the brakes.

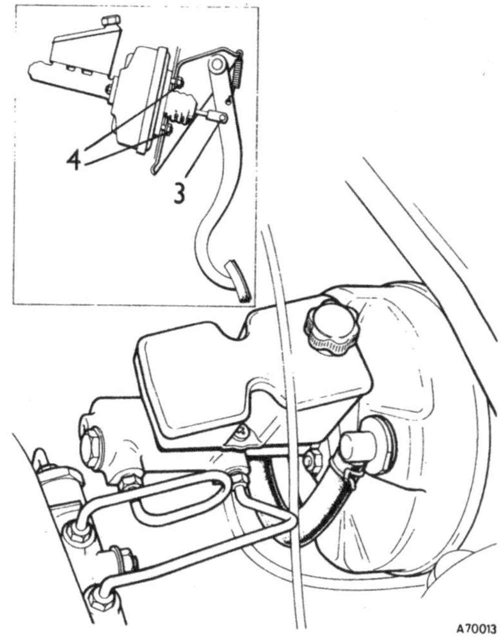

NON-RETURN VALVE

Remove and refit **70.50.15**

Removing

1 With the engine stopped, depress the brake pedal to destroy the vacuum in the servo.
2 Release the hose clip securing the vacuum hose to the non-return valve and disconnect the hose.
3 Withdraw the non-return valve from the servo.

Refitting

4 Renew the sealing rubber as necessary, and press the non-return valve into position in the servo.
5 Connect the vacuum hose to the non-return valve and secure with the hose clip.

VACUUM HOSE

Remove and refit **70.50.14**

Removing

1 Release the vacuum hose clips at the manifold and the servo non-return valve.
2 Remove the hose from the manifold and the non-return valve.

Refitting

3 Reverse instructions 1 and 2.

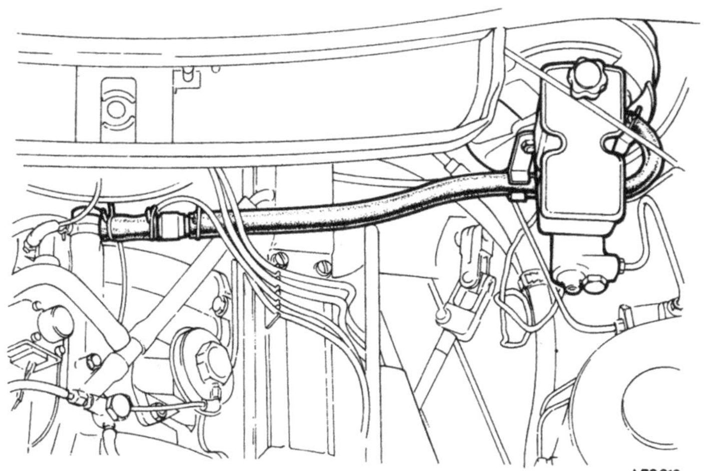

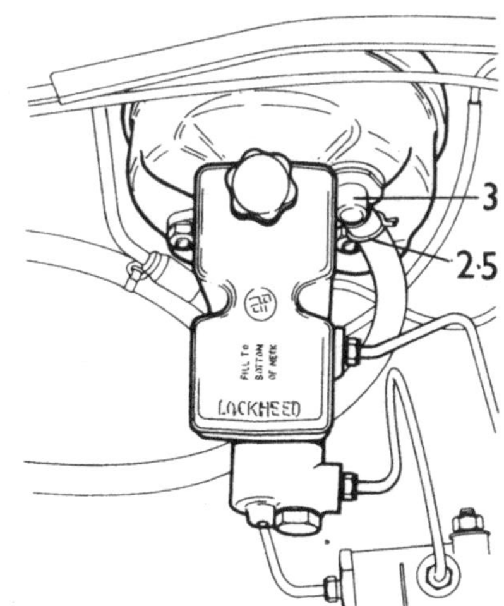

SERVO FILTER

Remove and refit 70.50.25

Removing

1 Remove the brake stop light switch.
2 Remove the split pin, plain washer and clevis pin securing the servo rod to the brake pedal.
3 Remove the rubber boot from the push-rod.
4 Withdraw the filter.

Refitting

5 Reverse instructions 1 to 4.

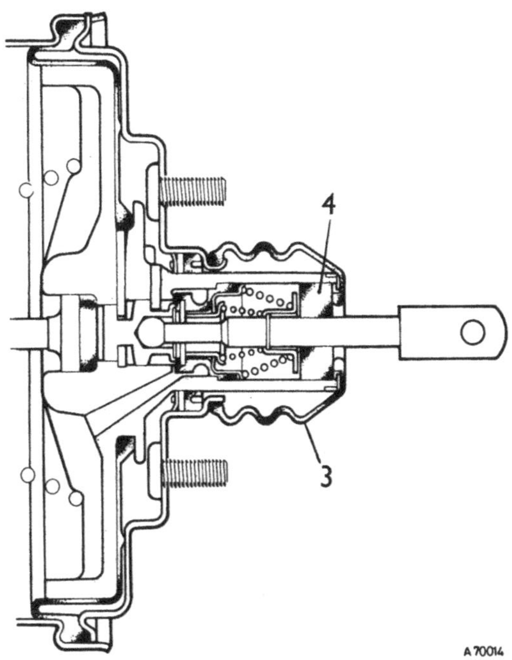

A 70014

BRAKE CALIPER – FRONT

Remove and refit 70.55.02

Removing

1 Jack up the car and remove the front wheel.
2 Disconnect the brake union at the caliper and seal the fluid connections to prevent entry of grit.
3 Remove the two bolts and spring washers retaining the steering-arm and the caliper lower mounting lug to the stub axle assembly.
4 Push the steering-arm clear.
5 Remove the bolt and spring washer securing the caliper upper mounting lug to the stub axle assembly.
6 Withdraw the caliper.

Refitting

7 Engage the caliper on the disc and align the locating lugs.
8 Position the upper bolt and spring washer; do not tighten.
9 Fit the steering-arm to the stub axle assembly.
10 Tighten the three bolts.
11 Bleed the brakes.
12 Fit the road wheel and remove the jack.

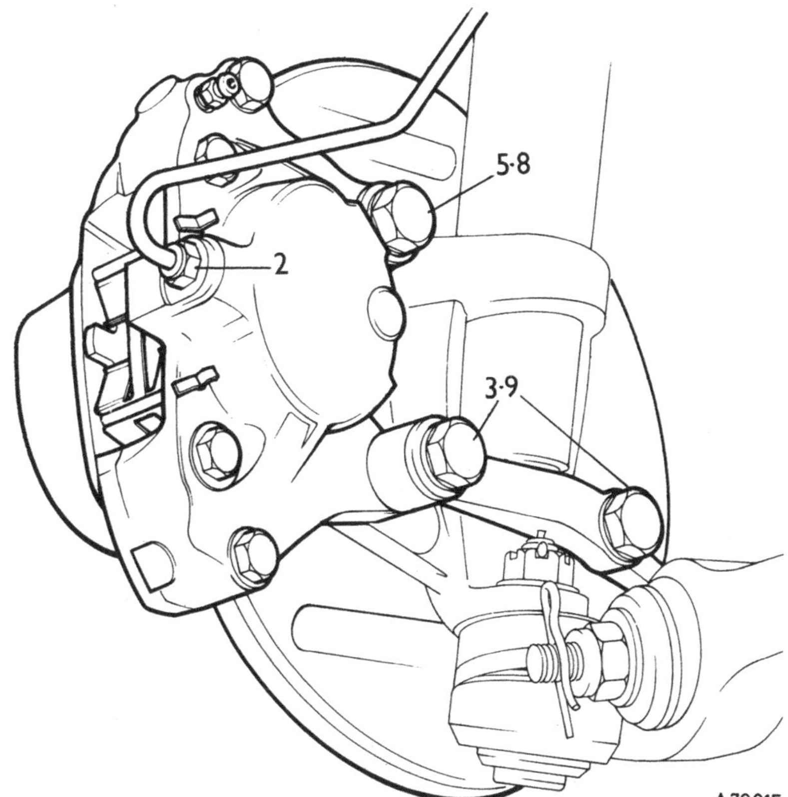

A 70015

FRONT CALIPER

Renew seals **70.55.13**

Dismantling

1 Remove the caliper from the car, see 70.55.02.
2 Remove the brake pads and shims, see 70.40.02.
3 Extract the caliper pistons. Piston removal may be affected by using a low pressure air line. DO NOT interchange the pistons. If either piston is seized, the whole caliper assembly must be renewed.
4 Using a blunt screwdriver carefully prise out the wiper seal retainers, taking care not to damage the seal grooves in the caliper bores.
5 Extract the wiper dust seal and the fluid seal from each caliper bore.
6 Thoroughly clean the caliper pistons and caliper bores with new clean brake fluid or methylated spirit.
7 Carefully inspect the caliper bores and pistons. If they show any signs of corrosion, fault or wear, the parts affected must be renewed.

Reassembling

8 Fit the new fluid seals, using only the fingers, into the grooves in the caliper bores, ensuring that they are properly located. The fluid seal grooves and the seals are not the same in section, therefore even when located correctly the seal feels proud at the edge farthest away from the mouth of the caliper bore.
9 Lubricate the bores with new, clean brake fluid.
10 Squarely insert the pistons into the caliper bores. Leave approx. $\frac{5}{16}$ in (7.94 mm) of each piston projecting from the mouth of each bore.
11 Fit a new wiper seal into each of the seal retainers and slide the assemblies, seal first, carefully into the mouth of each bore using the pistons as a guide.
12 Carefully press home the seals, taking care not to distort the retainers.
13 Push home the pistons.
14 Fit the caliper to the car, see 70.55.02.
15 Fit the brake pads and shim(s), see 70.40.02.
16 Bleed the brakes.

[CAUTION: Wash hydraulic cylinder components in brake fluid only. Never use gasoline or other petroleum-based solvents since they are damaging to rubber parts in the system.]

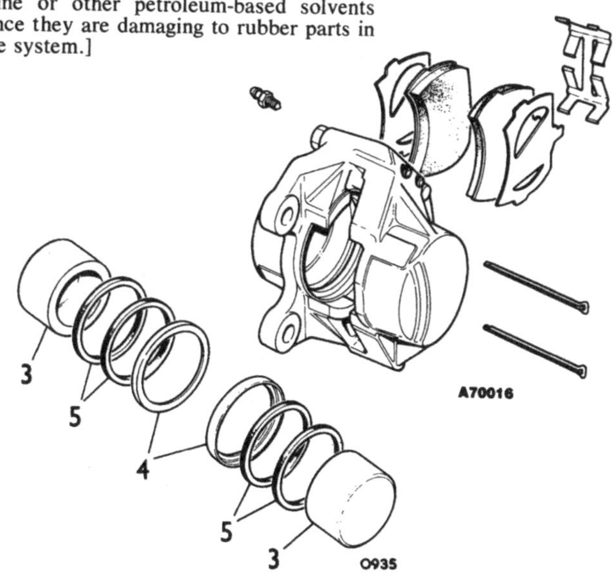

A70016

0935

REAR WHEEL CYLINDER

Remove and refit **70.60.18**

Removing

1 Jack up the car and support the body on stands.
2 Remove the rear wheel.
3 Remove the brake-drum, see 70.10.03.
4 Remove the brake-shoes, see 70.40.03.
5 Disconnect the hand brake cable at rear of the backplate.
6 Disconnect the fluid feed pipe union at the wheel cylinder and remove the bleed screw (left-hand side only);
or
Disconnect the fluid feed and transfer pipe unions at the wheel cylinder (right-hand side only).
7 Remove the spring clip securing the wheel cylinder to the rear of the backplate.
8 Remove the wheel cylinder and gasket.

Refitting

9 Reverse instructions 1 to 8, fitting a new gasket and circlip.
10 Apply the foot brake several times to adjust the rear brakes.
11 Bleed the brakes.
12 Road-test the car. If the operation of the brakes, including the hand brake, is poor, make four brake applications applying moderately high pedal efforts to decelerate the vehicle from 20 m.p.h. to rest. This will ensure correct adjustment of the rear brakes.

A70012

REAR WHEEL CYLINDER

Overhaul **70.60.26**

1 Remove the wheel cylinder from the backplate, see 70.60.18.
2 Remove the rubber boots from the cylinder body and remove the pistons.
3 Withdraw the pistons from the bore and retrieve the spring located between the two pistons.
4 Carefully inspect all components for faults and wear. A replacement wheel cylinder assembly must be fitted if, after cleaning, the bore of the old unit shows the slightest signs of corrosion or scoring.
5 Smear the cylinder bore with clean brake fluid.
6 Renew the seal on each piston, carefully fit the seal into the larger groove on each piston, with the lip of the seal facing away from the slotted head.
7 Locate the rubber boots into the smaller groove on each of the pistons.
8 Insert the pistons into the cylinder bore ensuring that the spring locates between them in the counter-bored ends.
9 Refit the wheel cylinder, see 70.60.18.

[**CAUTION:** Wash hydraulic cylinder components in brake fluid only. Never use gasoline or other petroleum-based solvents since they are damaging to rubber parts in the system.]

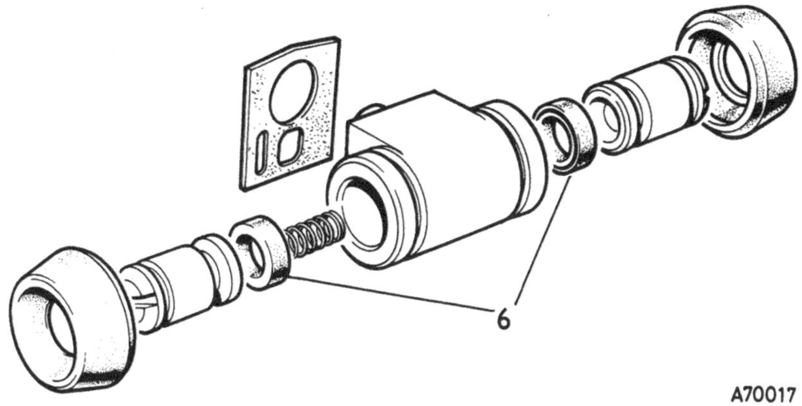

A70017

BODY OPERATIONS

BODY UNIT

Alignment check **76.10.01**

1 A preliminary check of the alignment should be carried out by dropping a plumb-bob from the centre of points A, B, C, D on each side of the car.
2 Establish a centre line by means of a large pair of compasses at points B and D.
3 Check measurements against those given in DATA.
4 Construct diagonals.

Incorrect alignment will be evident by the failure of the diagonals to intersect on the centre line or by considerable deviation from the dimensions given.

DATA

A	B	C	D
552 mm	594 mm	964 mm	1042 mm
(21.73 in)	(23.38 in)	(37.95 in)	(41.02 in)

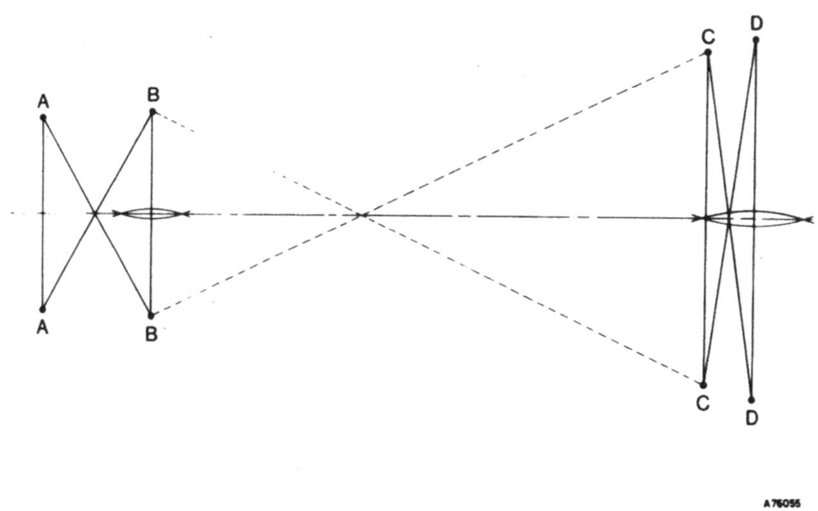

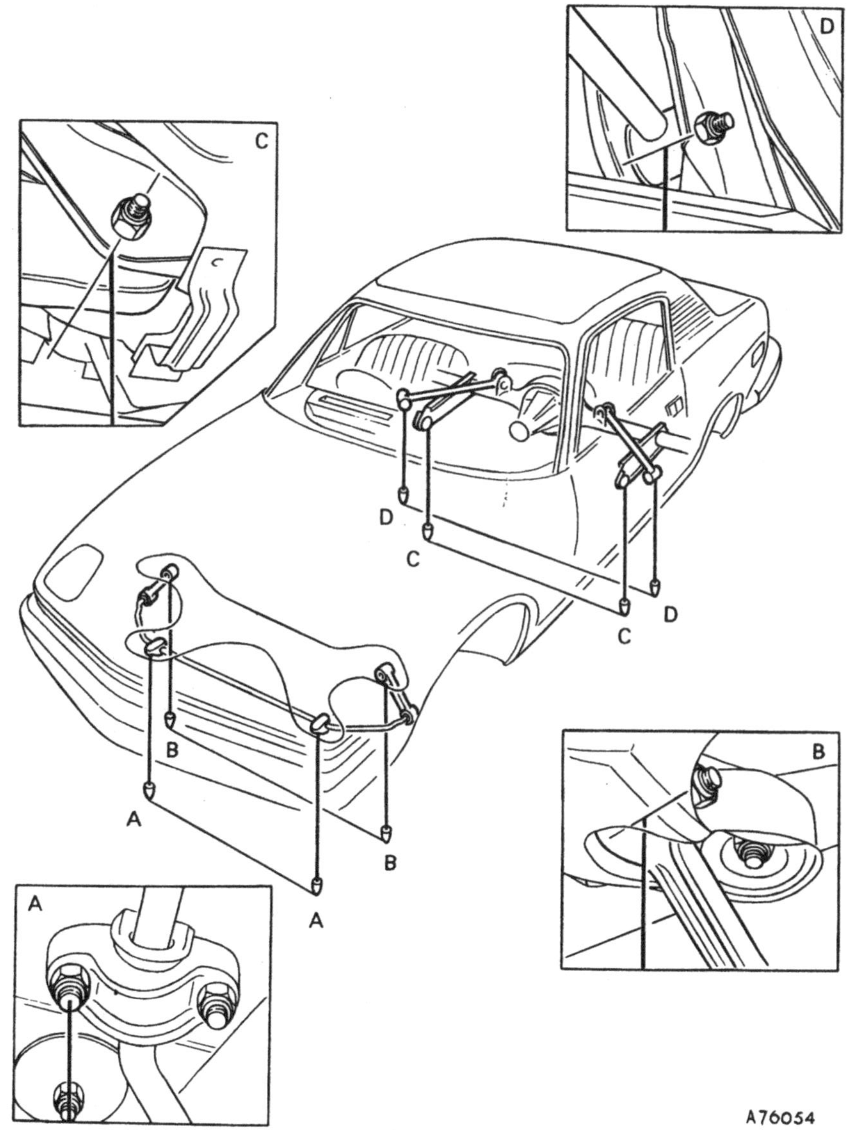

continued

Vertical alignment

The dimensions given below are for cars in showroom condition, unladen and without fuel.
It should be noted that the important point is the relative positions of the vertical datum points to each other and not their actual height from the wheel hub centres.

Code	Dimension	Location
A	346 mm (13.62 in)	Anti-roll bar front mounting to suspension pod cap
B	141,5 mm (5.57 in)	Rear suspension arm mounting to radius rod mounting
a	33 mm (1.29 in)	Anti-roll bar clamp front fixing bolt to lower link
b	361 mm (14.20 in)	Top of front wheel arch to wheel hub centre – Europe – 4 speed gearbox
	356 mm (14.02 in)	Top of front wheel arch to wheel hub centre – Europe – 5 speed gearbox
	353 mm (13.90 in)	Top of front wheel arch to wheel hub centre – Europe – Automatic transmission
	354 mm (13.96 in)	Top of front wheel arch to wheel hub centre – USA – 4 speed gearbox
	350 mm (13.78 in)	Top of front wheel arch to wheel hub centre – USA – 5 speed gearbox
	347 mm (13.67 in)	Top of front wheel arch to wheel hub centre – USA – Automatic transmission
	356 mm (14.02 in)	Top of front wheel arch to wheel hub centre – USA – 4 speed gearbox – air conditioning
	352 mm (13.88 in)	Top of front wheel arch to wheel hub centre – USA – 5 speed gearbox – air conditioning
	349 mm (13.75 in)	Top of front wheel arch to wheel hub centre – USA – Automatic transmission – air conditioning

Code	Dimension	Location
c	571 mm (22.48 in)	Lower link to front suspension pod cap
d	476.5 mm (18.75 in)	Front suspension pod cap to datum line
e	338 mm (13.3 in)	Top of wheel arch to datum line
f	136 mm (5.35 in)	Rear suspension arm mounting to radius rod mounting
g	95 mm (3.74 in)	Rear suspension arm mounting to datum line
h	319 mm (12.56 in)	Top of rear wheel arch to datum line
j	348 mm (13.71 in)	Top of rear wheel arch to wheel hub centre – Europe – 4 speed gearbox
	350 mm (13.80 in)	Top of rear wheel arch to wheel hub centre – Europe – 5 speed gearbox
	345 mm (13.60 in)	Top of rear wheel arch to wheel hub centre – Europe – Automatic transmission
	346 mm (13.63 in)	Top of rear wheel arch to wheel hub centre – USA – 4 speed gearbox
	348 mm (13.72 in)	Top of rear wheel arch to wheel hub centre – USA – 5 speed gearbox
	343 mm (13.52 in)	Top of rear wheel arch to wheel hub centre – USA – Automatic transmission
Z-Z		Datum line

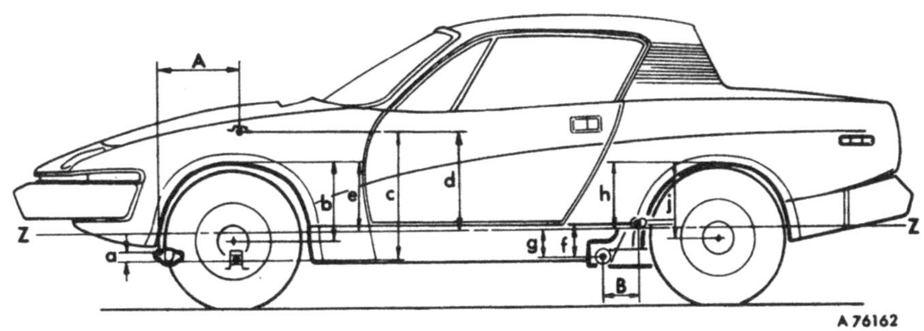

BODY UNIT

Alignment check **76.10.01**

(Using Churchill 700 or 707 system)

Whilst severe underframe damage is readily detected, less serious damage may cause distortion that is not visually apparent.
If steering or suspension checks indicate a fault which cannot be attributed to anything other than underframe distortion, initial checking should be carried out to determine the area and extent of distortion.

Initial check

1 Clip the location tape to the right-hand side of the jig and make a chalk mark on the floor at each required location for initial checking.
2 Remove the tape to avoid damage.
3 Position the car centrally over the jig with the front wheel centres approximately 76 cm (30 in) from the front of the jig.

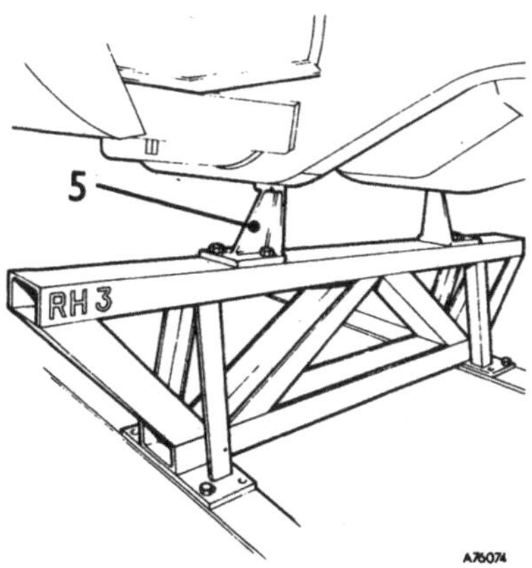

4 Raise the front of the car and fit transverse member number 3 to the jig with the rear mounting holes at tape position 12/7.
5 Fit brackets S700–39/1 (LH) and S700–39/2 (RH) to the transverse member, locating the inner bolt in hole J.

6 Lower the car to locate the bracket pegs in the front tooling holes in the floor side-member.
7 Raise the rear of the car by jacking under the differential.

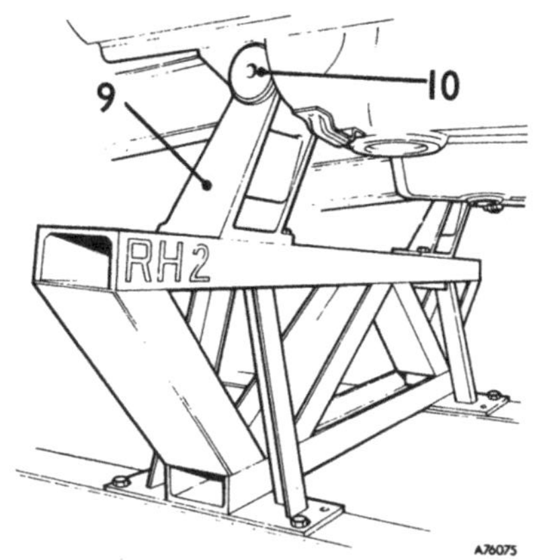

8 Fit transverse member number 2.
9 Fit brackets S700-29/1 (LH) and S700-29/2 (RH) to the transverse member, locating the inner bracket bolts in holes 'L'.
NOTE: These holes are not marked on Churchill 700 systems. Distance across bolt centres should be 819 mm (32.25 in).
10 Lower the car and fit the two bushes S700-29/4 over the trailing arm hanger bolt heads.
11 Fit transverse member number 4 with the rear mounting holes at tape position 18.

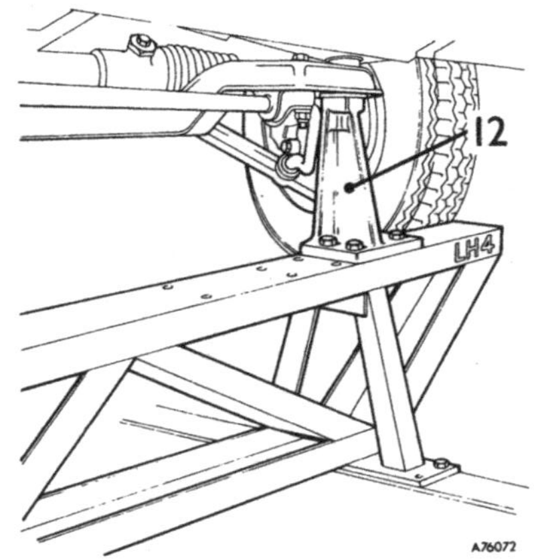

12 Fit bracket S700-401/1 (LH) and S700-401/2 (RH) locating the inner bracket bolts in holes 'G'.

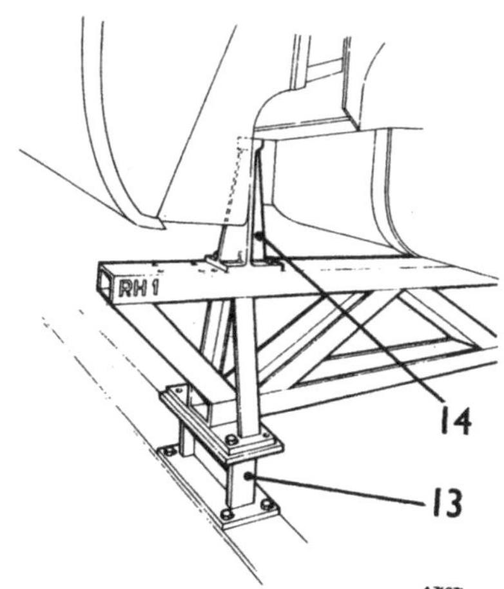

13 Fit risers S700-2B and transverse member number 1 with the rear mounting holes at tape position 4/28.
14 Fit brackets S700-19 (LH) and S700-19/2 (RH) locating the outer bracket bolts in holes 'B'.
NOTE: These holes are not marked on Churchill 700 systems. Distance across bolt centres should be 952 mm (37-50 in).

The following operations are only necessary if repairs are required.

Repair Stage

It may not be necessary to fit the full set of repair brackets. If damage is confined to the front end of the car, repair brackets can be fitted at the front and the initial check brackets retained at the rear or vice-versa in the case of rear end damage. Where it is necessary to remove sub-assemblies before fitting repair brackets, reference should be made to the appropriate workshop manual section.
For front end repairs, the transverse members and brackets used for initial checking are used again in their original positions with the following additions:-

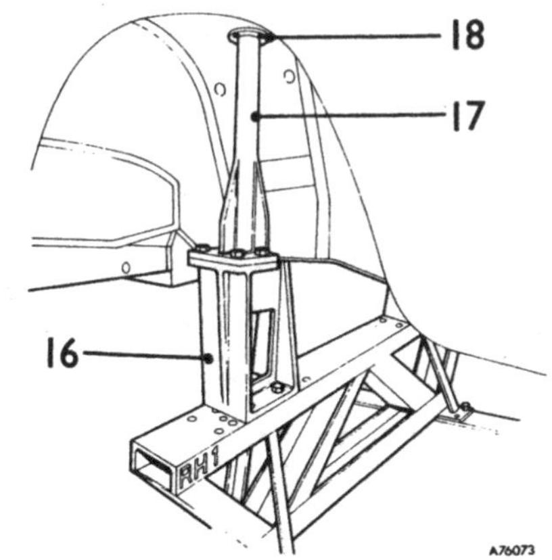

15 Fit transverse member number 1 with the rear mounting holes at tape position 15.

16 Fit risers S700-17/1 (LH) and S700-17/2 (RH).

17 Fit brackets S700-18/1 (LH) and S700-18/2 (RH) to the risers, using the eight screws supplied.

18 Secure the brackets to the spring turrets, using the damper attachment nuts.

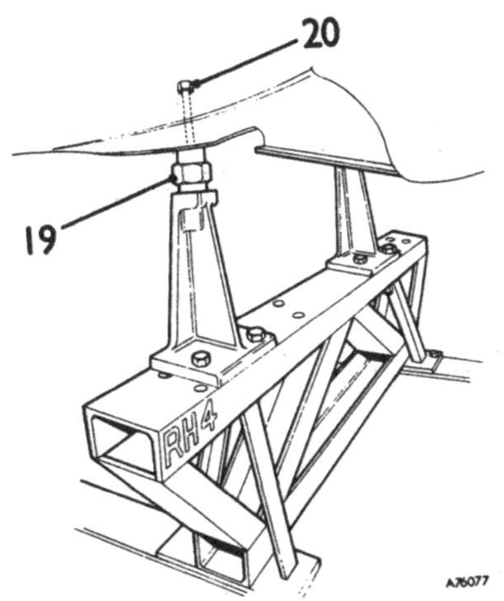

19 Fit adaptors S700-401/3 into brackets S700-401/1 (LH) and S700-401/2 (RH).

20 Install the two original bolts through the sub-frame front mounting holes in the longitudinal members and screw them into the adaptors.

For rear end repairs:-

21 Remove the two bushes S700-29/4 and the two bolts and nuts from the trailing arm hangers. Refit the bushes and fit the two pins S700-29/5.

continued

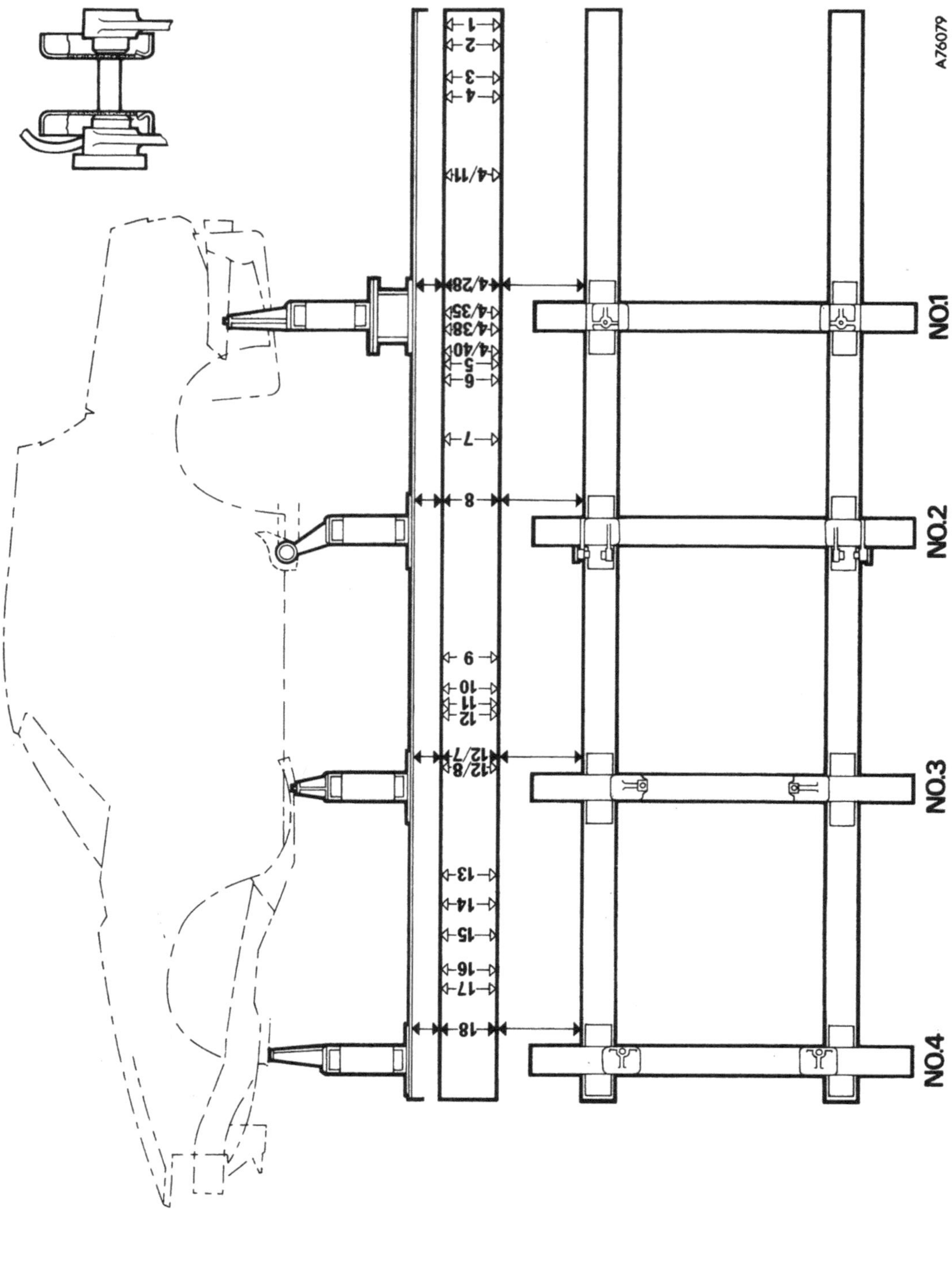

Transverse member locations for initial check

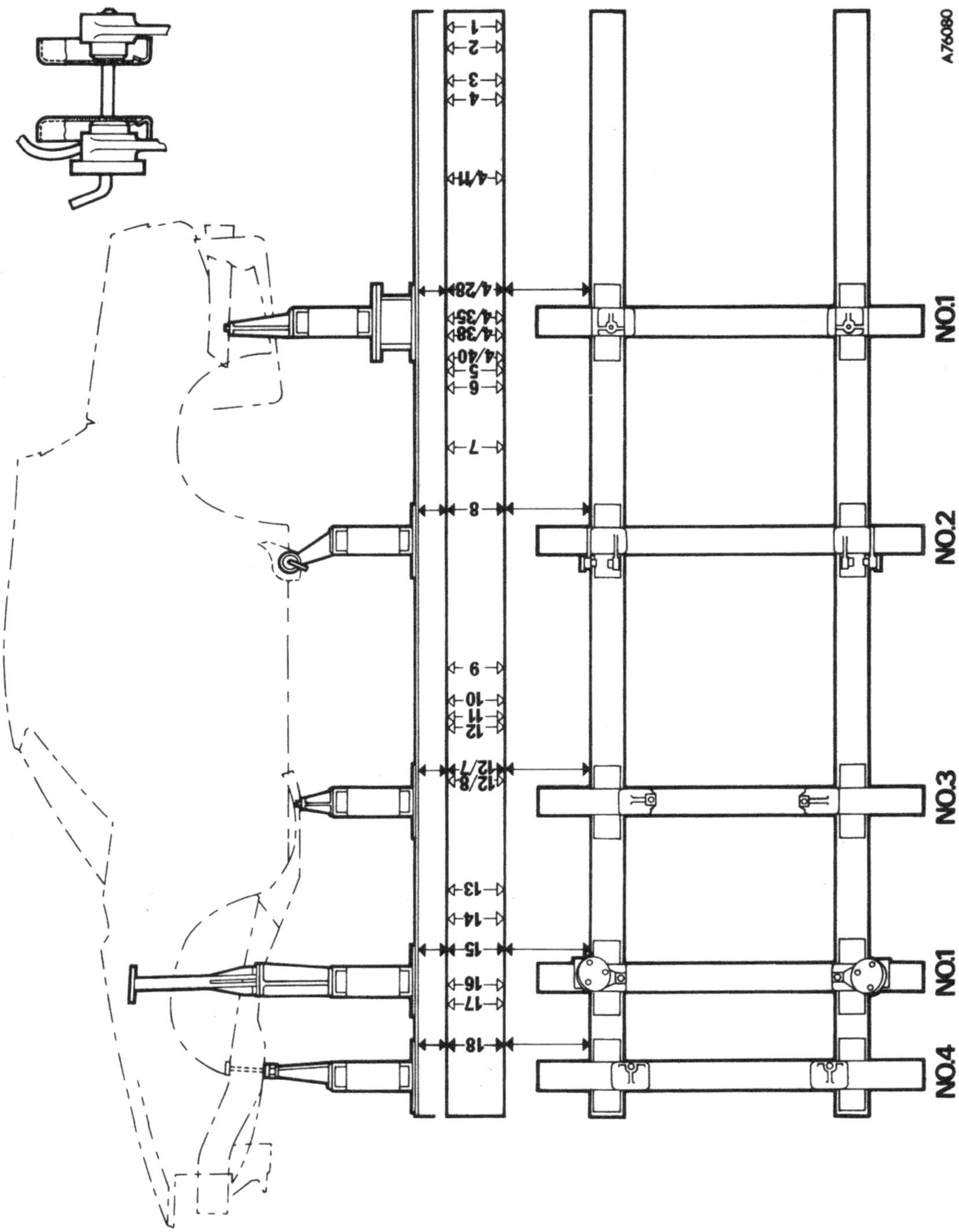

Transverse member locations for repair

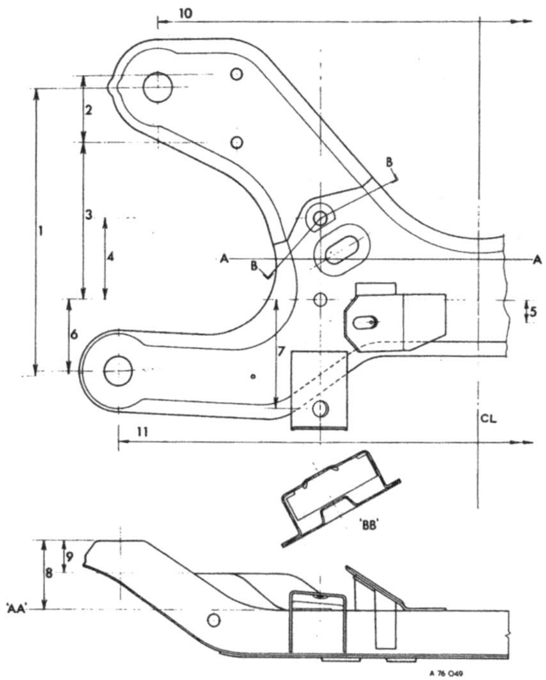

SUB FRAME

Alignment check **76.10.04**

Diagram Number	Millimetres	Inches
AA Datum line		
BB Datum face		
CL Centre line		
1	310 ± 0.50	12.20 ± 0.020
2	74 ± 0.25	2.91 ± 0.010
3	171 ± 1	6.73 ± 0.040
4	88 ± 0.25	3.46 ± 0.010
5	25 ± 0.50	0.98 ± 0.020
6	78 ± 1	3.07 ± 0.040
7	118.5	4.64
8	75 ± 1	2.95 ± 0.040
9	36 ± 1	1.41 ± 0.040
10	730 ± 0.50	28.74 ± 0.020
11	812 ± 0.50	31.96 ± 0.020

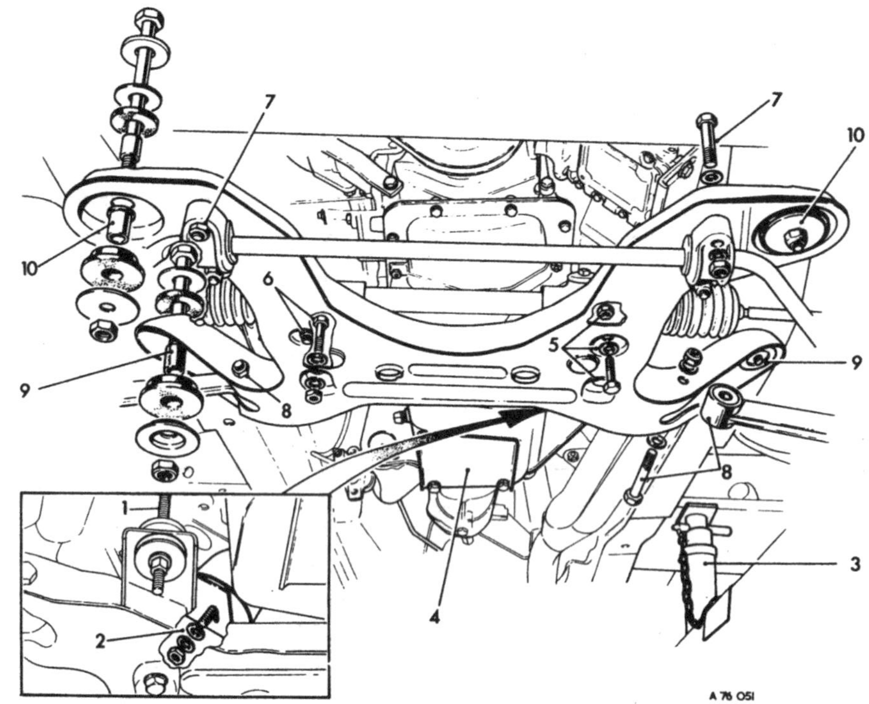

SUB-FRAME

Remove and refit **76.10.29**

Removing

1 Remove the engine stabilizer, see 12.45.16.

2 Remove the nut, spring washer and plain washer securing the engine mounting bracket to the sub-frame.

3 Jack up the front of the vehicle sufficiently to take the weight of the body off the front suspension.

4 Support the engine by placing a jack under the sump coupling plate.

5 Remove the two bolts, spring washers and plain washers securing the pinion end of the steering rack to the sub-frame.

6 Remove the two nuts, bolts and plain washers securing the steering rack to the sub-frame.

7 Remove the four nuts, bolts, spring washers, plain washers and mounting brackets securing the anti-roll bar to the sub-frame.

8 Remove the two nuts, spring washers and plain washers securing the lower links to the sub-frame. Adjust the jacks if necessary to ensure that the bolts are not under load. Withdraw the two bolts and release the lower links from the sub-frame.

9 Remove the two nuts, retainers, mounting rubbers and sleeves securing the rear of the sub-frame to the body.

10 Support the sub-frame and remove the two nuts, retainers, mounting rubbers and sleeves securing the front of the sub-frame to the body.

11 Lower the sub-frame and remove it from the vehicle.

continued

Refitting

12 Position the sub-frame on the mounting bolts ensuring that the four mounting rubbers and washers are correctly positioned between sub-frame and body.

13 Tighten the nuts 10 and 9 to torque figures given in '**TORQUE WRENCH SETTINGS**'.

14 Reverse instructions 2 to 8.

15 Refit and adjust the engine tie-rod, see 12.45.16.

REAR QUARTER TRIM PAD

Remove and refit 76.13.12

Removing

1 Pull the weatherstrip away from the door aperture in the area of the trim pad.

2 Carefully pull the trim pad edging away from the body flange.

3 Remove the cap, screw and retainer.

4 Prise off the trim pad–5 clips.

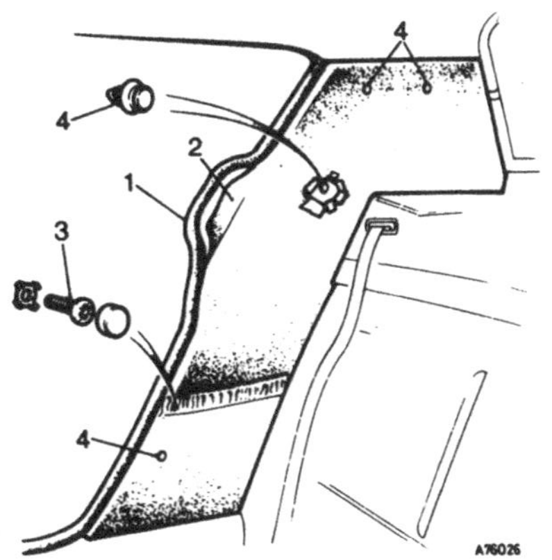

Refitting

5 Reverse instructions 1 to 4, using Dunlop SP758 adhesive on the mating surfaces of trim pad edging and body flange.

REAR COMPARTMENT TRIM PAD

Remove and refit 76.13.20

Removing

1 Move both seats and seat squabs to the fully forward position.

2 Open the rear console lid and remove the two screws and plain washers securing the console and trim pad to the body.

3 Remove the two screws and cup washers securing the trim pad to the body.

4 Remove the four screws and cup washers securing the parcel tray and the trim pad to the body.

5 Carefully raise the forward edge of the parcel tray, raise the trim pad clear of the console and remove it from the car.

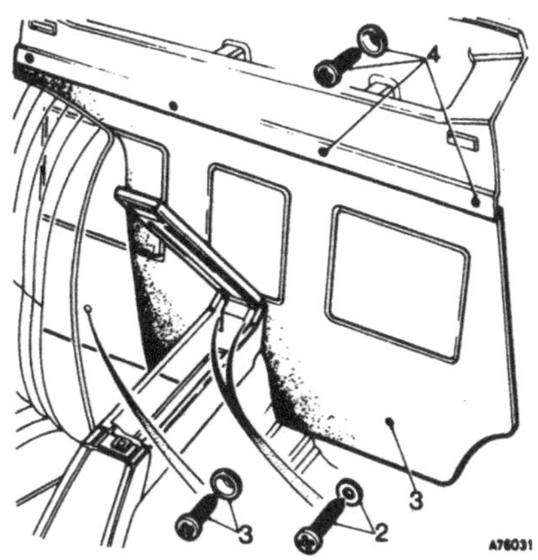

Refitting

6 Reverse instructions 1 to 5.

BONNET

Remove and refit 76.16.01

Removing

1 Pull the tubing from the screenwasher pump; attach a suitable length of string to the end of the tubing to facilitate refitting, and pull the tubing through the holes in the inner wheel arch.

2 Mark the hinge positions on the body.

3 Support the bonnet and remove the single bolt and nut (earlier models) or two screws and spring washers (later models) securing the stay to the bonnet.

4 Remove the four bolts securing the hinges to the adjuster plates and lift off the bonnet.

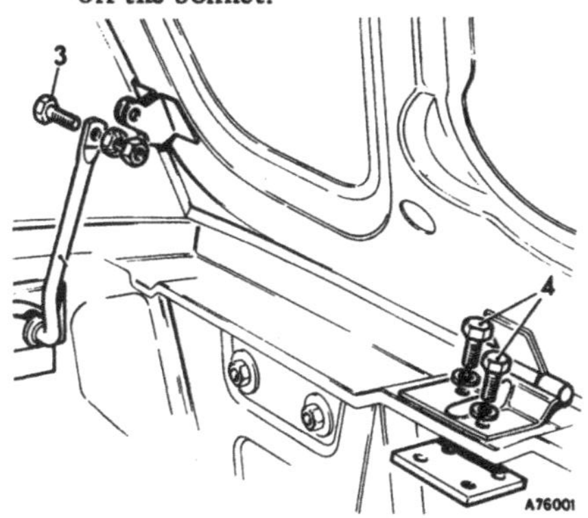

Refitting

5 Reverse instructions 1 to 4, adjusting if necessary to ensure correct alignment before fully tightening bolts 4.

BONNET STAY

Remove and refit 76.16.14

Removing

1 Support the bonnet and remove the single bolt and nut (earlier models) or two screws and spring washers (later models) securing the stay to the bonnet.

2 Disengage the lower end of the stay from the body (earlier models) or remove the bolt and spring washer securing the lower end of the stay to the body (later models).

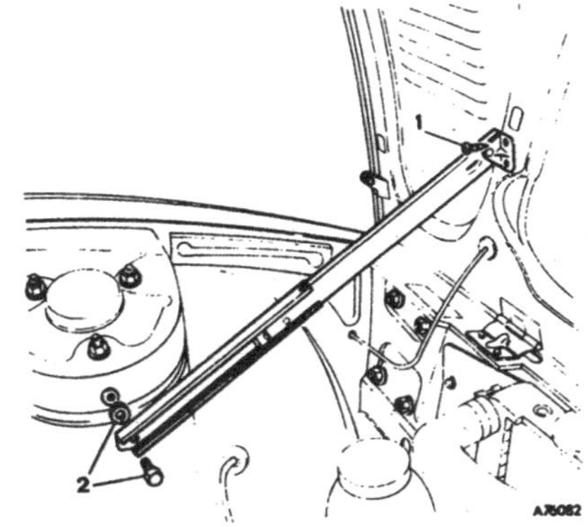

Refitting

3 Reverse instructions 1 and 2.

BONNET CATCH

Adjust 76.16.20

To ensure positive locking and eliminate free movement at the closing face, adjust the bonnet catch as follows:

1 Pull back the spring and slacken the locknut at the base of the shaft.

2 Using a screwdriver, screw the shaft in or out as required.

3 Retighten the locknut.

4 Check the bonnet closing action and repeat instructions 1 to 3 if necessary.

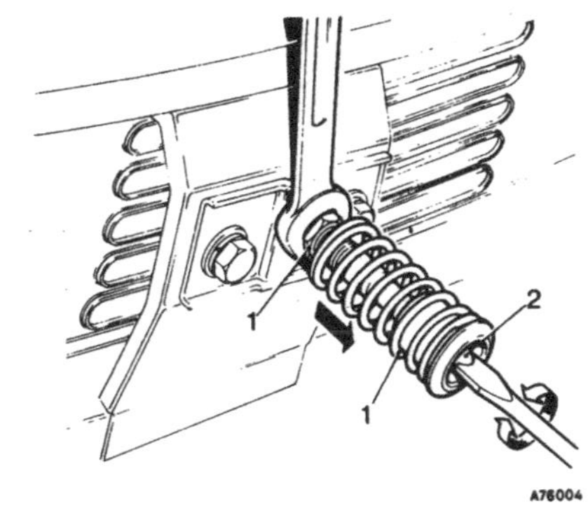

BONNET LOCK

Remove and refit 76.16.21

Removing

1 Remove the trunnion from the cable.
2 Slacken the pinch bolt and detach the cable from the lock.
3 Remove the four bolts, spring washers and plain washers securing the lock to the body.

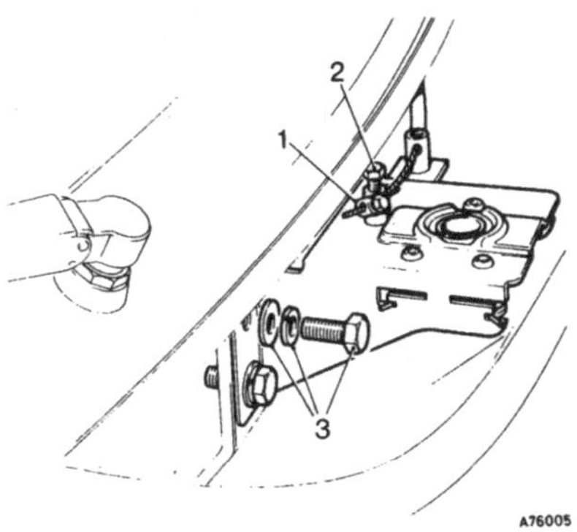

Refitting

4 Reverse instructions 1 to 3.

BONNET RELEASE CABLE

Remove and refit 76.16.29

Removing

1 Remove the trunnion from the cable.
2 Slacken the pinch bolt and detach the cable from the lock.
3 Unscrew the nut securing the outer cable to the bracket beneath the facia.
4 Withdraw the cable through the grommet on the bulkhead and collect the nut and shakeproof washer.
CAUTION: Do not close the bonnet with the cable removed or loose.

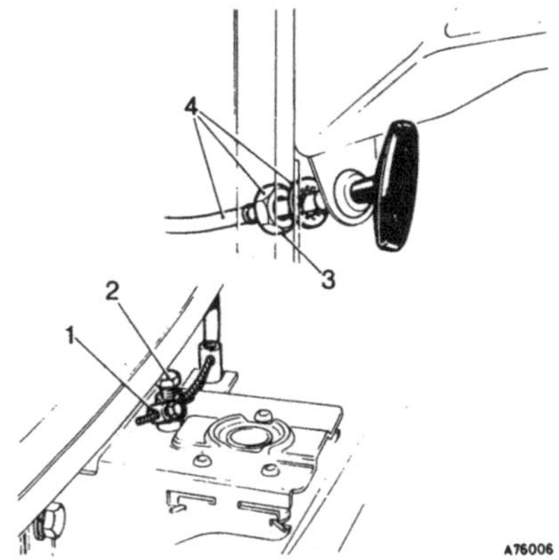

Refitting

5 Reverse instructions 1 to 4, ensuring that the lock release lever is not pre-loaded by the cable.

BONNET CATCH

Remove and refit 76.16.34

Removing

1 Remove the two bolts, spring washers and plain washers. Remove the catch from the bonnet.

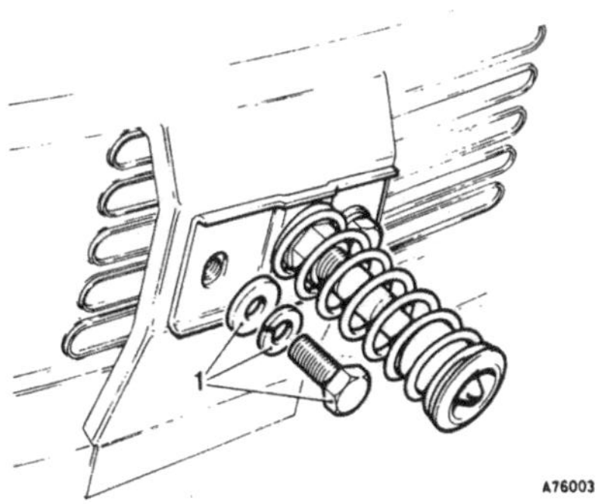

Refitting

2 Refit in reverse order and adjust if necessary, see 76.16.20.

LUGGAGE COMPARTMENT LID

Remove and refit 76.19.01

Removing

1 Mark the hinge positions on the lid.
2 Disconnect the two rear number plate lamp leads from the connector at the R.H. side of the luggage compartment.
3 *Earlier Models:* Support the lid and remove the two bolts securing the stay to the lid.
4 Remove the four bolts, spring washers and plain washers securing the lid to the hinges and lift off the lid.

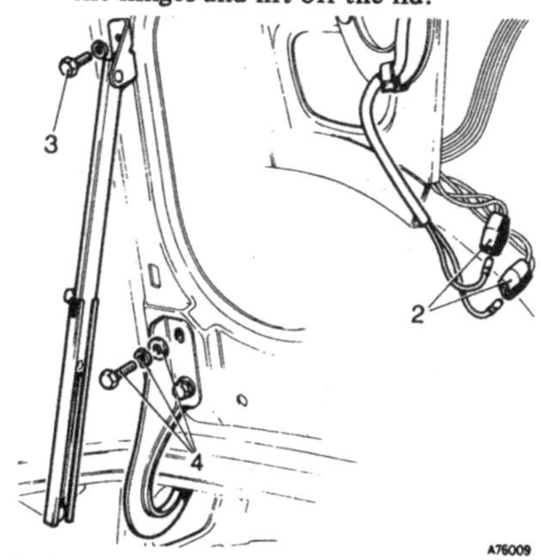

Refitting

5 Reverse instructions 1 to 3, ensuring correct alignment of the lid in the horizontal plane before fully tightening bolts 4.

LUGGAGE COMPARTMENT LID HINGES

Remove and refit 76.19.07

Removing

1 a Remove the luggage compartment lid, see 76.19.01.
b *Later models only:* Remove one bolt, nut and washers to release the gas strut from each hinge.
2 Remove (two each side) the four bolts, spring washers and plain washers securing the hinges to the body and lift off the hinges.

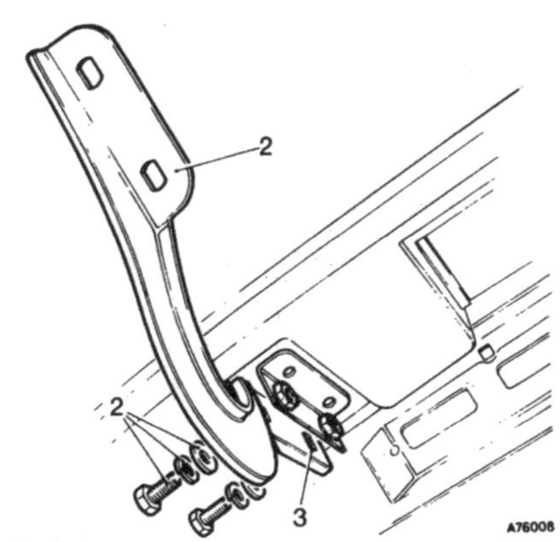

Refitting

3 Reverse instructions 1 and 2 ensuring correct alignment in the vertical plane before fully tightening bolts 2.

LUGGAGE COMPARTMENT LOCK

Remove and refit 76.19.11

Removing

1 Remove the three bolts, spring washers and plain washers and lift off the latch.
2 Pull the spring clip to one side to disengage it from the lock.
3 Carefully withdraw the lock and gasket.

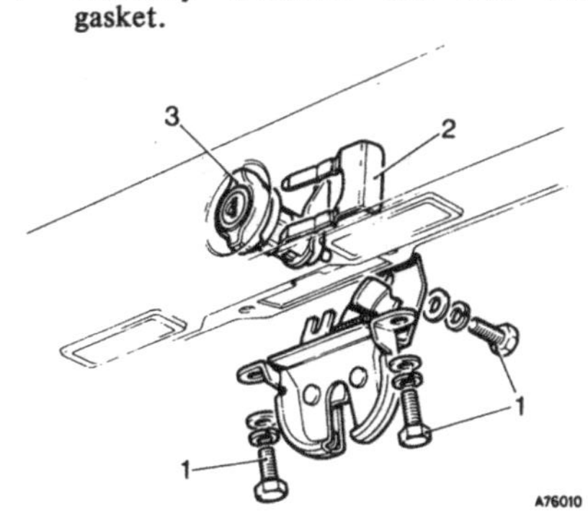

Refitting

4 Reverse instructions 1 to 3.

LUGGAGE COMPARTMENT LOCK STRIKER

Remove and refit 76.19.12

Removing
1 Pull the weatherstrip away from the body panel in the area of the striker.
2 Mark the position of the striker bolts on the body panel.
3 Remove the three bolts, spring washers and plain washers.
4 Withdraw the striker from the body panel aperture.

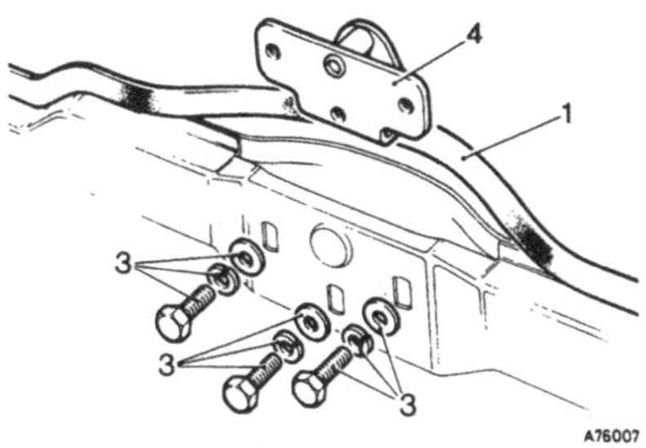

Refitting
5 Reverse instructions 1 to 4, adjusting the striker in the vertical plane if necessary, before fully tightening bolts 3.

BUMPER–FRONT

Remove and refit 76.22.08

Removing
1 Disconnect the six snap connectors from the parking and flasher lamp leads.
2 Remove the outer four nuts, spring washers and plain washers.
3 Support the bumper and remove the remaining two nuts, spring washers and plain washers. Lift off the bumper.

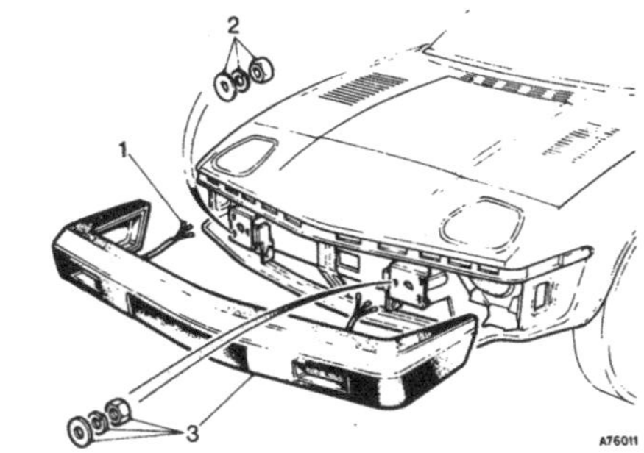

Refitting
4 Reverse instructions 1 to 3.

BUMPER–REAR

Remove and refit 76.22.15

Removing
1 Remove the four nuts, spring washers and plain washers securing the side support brackets to the bumpers.
2 Remove the three nuts, spring washers and plain washers and lift off the handling brackets. – U.S.A. models.
3 Support the bumper and remove the remaining two nuts, spring washers and plain washers. Lift off the bumper.

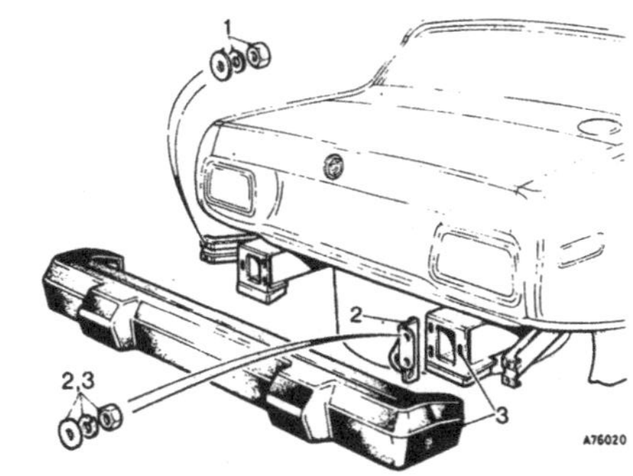

Refitting
4 Reverse instructions 1 to 3.

CONSOLE ASSEMBLY

Remove and refit 76.25.01

Removing
1 Raise the handgrip lever and pull off the grip.
2 Unscrew the gear lever knob.
3 Remove the two screws and lift off the handbrake lever surround trim panel.
4 Prise off the gear lever gaiter–4 fasteners.
5 Remove the two screws and plain washers securing the front console to the transmission tunnel.

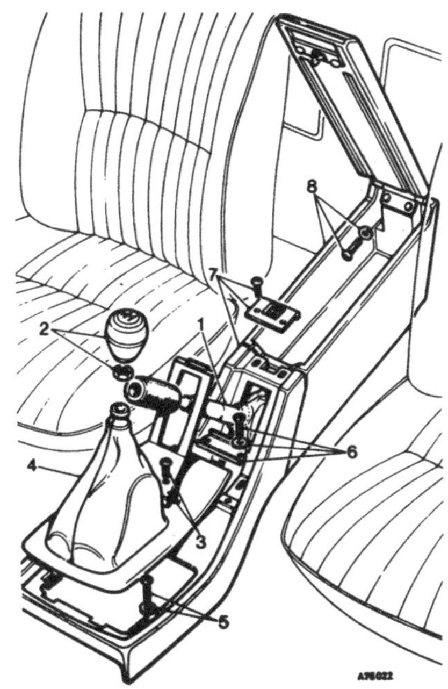

6 Remove the two screws and plain washers securing the consoles and bridge plate to the floor and lift off the bridge plate.
7 Raise the lid/armrest and remove the two screws securing the catch to the consoles. Lift off the catch and the front console.
8 Remove the two screws and plain washers securing the rear console to the rear parcel shelf and body–lift off the rear console.

Refitting
9 Reverse instructions 1 to 8.

CONTROL COWL

Remove and refit 76.25.03

Removing
1 Remove the console assembly, see 76.25.01.
2 Pull the four knobs from the heating and ventilation control levers.
3 Remove the two screws securing the cowl to the control levers.
4 Remove the two screws securing the control illumination panel to the cowl.
5 Remove the screw securing the cowl to the bracket beneath the fascia.
6 Remove the two screws securing the cowl to the bracket on the transmission tunnel.

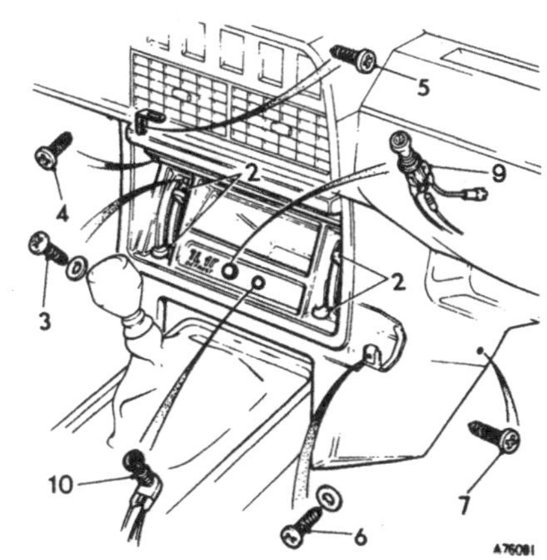

continued

7 Remove the two screws securing the cowl to the body.
8 Engage top gear and pull the cowl clear of the heater.
9 Disconnect the bulbholder and two leads from the cigar lighter.
10 Disconnect the two lucar connectors from the rheostat.
11 Remove the cowl from the vehicle.

Refitting
12 Reverse instructions 1 to 11.

DOOR

Remove and refit 76.28.01

Removing
1 Isolate the battery.
2 Remove the door trim pad, see 76.34.01.
3 Support the door and remove the six nuts, spring washers, plain washers and two adjuster plates securing it to the hinges. Lift off the door.

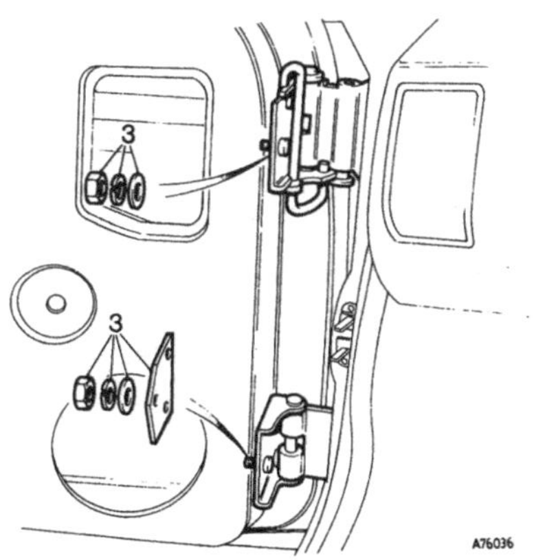

Refitting
4 Reverse instructions 1 to 3. Check the door closing action and alignment and adjust if necessary before fully tightening nuts 3.

DOOR HINGE – FRONT UPPER

Remove and refit 76.28.42

The body half of the hinge is welded to the 'A' post and will only be renewed during extensive body repair.
This operation renews the door half of the hinge and the two hinge pins.
To ensure satisfactory retention of the serrated pins, the door half of the hinge and the two hinge pins should always be fitted as an all new set.

Removing
1 Using a chisel under the pin head, drive the upper pin upwards until the serrations are clear of the assembly.
2 Drive the lower pin downwards until the serrations are clear of the assembly.
3 Remove the door, see 76.28.01.
4 Remove the two pins.
5 Remove the door half of the hinge and collect up the spring.

Refitting
6 Apply grease to the bore of the body half of the hinge.
7 Position the spring and the door half of the hinge such that the door adopts its half open position.
8 Compress the spring by hand and fit the two pins finger-tight.
9 Engage the pin serrations either by using a hammer and drift or by pressing both pins in together using a suitable clamp.
10 Fit the door, see 76.28.01.

DOOR GLASS

Remove and refit 76.31.01

Removing
1 Remove the door trim pad, see 76.31.45.
2 Remove the door lock remote control, see 76.37.31
3 Remove the door glass regulator, see 76.31.45.

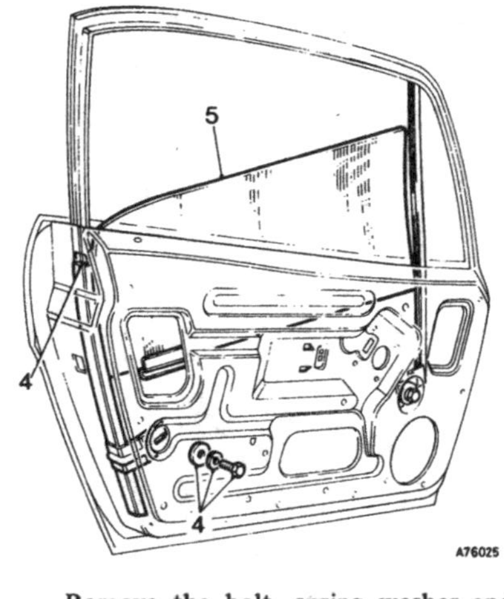

4 Remove the bolt, spring washer and plain washer securing the rear glass channel to the door. Detach the insert from the channel and carefully withdraw the channel and weather curtain.
5 Turn the glass anti-clockwise until it can be withdrawn through the door aperture. Lift out the glass taking care to avoid scratching it on the seal clips.

Refitting
6 Reverse instructions 1 to 5, ensuring that the glass channel is positioned to enable the glass to be moved freely.

DOOR QUARTER LIGHT

Remove and refit 76.31.29

Removing
1 Fully lower the door glass.
2 Remove the door trim pad, see 76.34.01.
3 Drill out the rivet securing the top of the glass channel to the door.
4 Remove the two bolts and plain washers securing the front glass channel to the door.
5 Pull the top of the glass channel away from the quarter light weatherstrip.

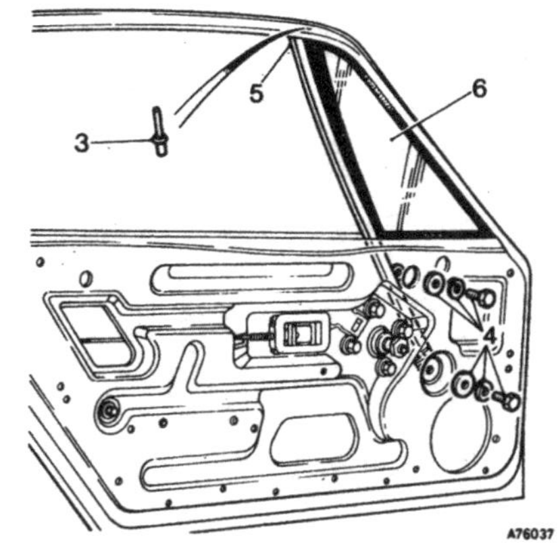

6 Carefully ease the quarter light and weatherstrip rearwards and upwards and remove them from the door.

Refitting
7 Reverse instructions 1 to 6, ensuring that the weatherstrips are correctly positioned.

DOOR GLASS REGULATOR

Remove and refit 76.31.45

Removing
1 Remove the door trim pad, see 76.34.01.
2 Remove the door lock remote control, see 76.37.31.
3 Refit the regulator handle, fully raise the glass and then turn the handle back (anti-clockwise) slightly to ensure that the regulator mounting bolts are not under load. Remove the handle.
4 Support the glass.
5 Remove the four bolts, spring washers and plain washers securing the regulator to the door.
6 Carefully push the regulator handle shaft inside the door aperture and slide the regulator assembly towards the rear of the door, lowering the glass by hand sufficiently to enable the rollers to be disengaged from the glass channels and the channel on the door interior.

continued

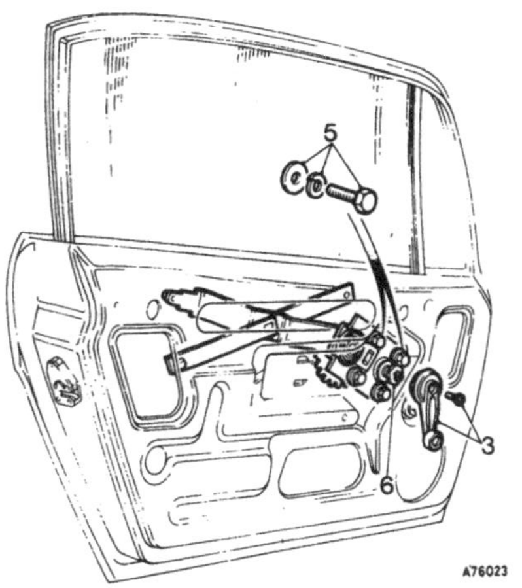

7 Manoeuvre the regulator out of the door through the lower aperture, taking care to avoid scratching the glass.

Refitting
8 Reverse instructions 1 to 7.

DOOR TRIM PAD

Remove and refit 76.34.01

Removing
1 Unscrew and remove the plunger knob.

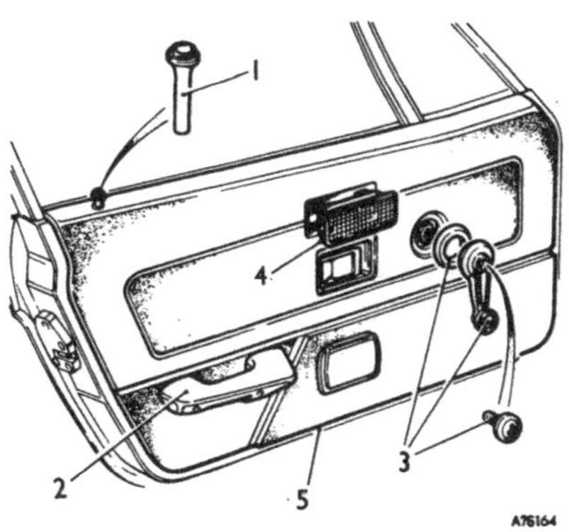

2 Remove the arm-rest, see 76.34.23.
3 Remove the screw securing the regulator handle. Lift off the handle and bezel.
4 Remove the map/courtesy lamp, see 86.45.10.
5 Using a large screwdriver, carefully prise out the nine clips securing the trim pad to the door.
6 Carefully lift the trim pad over the plunger rod.
7 Later models: Disconnect the two speaker leads.

Refitting
8 Reverse instructions 1 to 7.

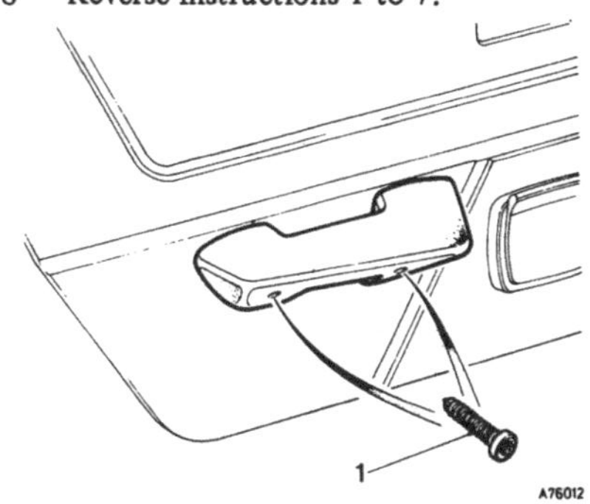

DOOR ARM-REST

Remove and refit 76.34.23

The door arm-rest is secured to the door shell by two screws.

DOOR LOCK

Remove and refit 76.37.12

Removing
1 Remove the door trim pad, see 76.34.01.
2 Remove the bolt, spring washer and plain washer securing the rear glass channel to the door. Detach the insert from the channel and carefully withdraw the channel and weather curtain.
3 Remove the four screws securing the lock assembly to the door and lift off the disc latch.

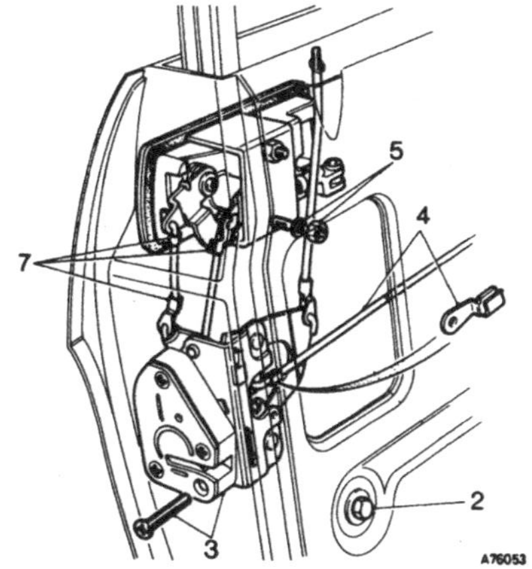

4 Remove the retaining clip and detach the remote control rod from the lock.
5 Remove the two nuts and spring washers securing the clamp bracket to the outside handle.
6 Manoeuvre the outside handle and seal, together with the door lock, through the handle aperture and out of the door.
CAUTION: Care must be taken to avoid straining locks or linkages.
7 Remove the retaining clips from the door handle and private lock control rods and separate the outside handle and door lock.

Refitting
8 Reverse instructions 1 to 7 ensuring that the plunger rod is correctly located before securing the lock to the door.

DOOR LOCK STRIKER

Remove and refit 76.37.23

Removing
1 Remove the rear quarter trim pad, see 76.13.12.
2 Remove the two screws securing the striker assembly to the 'B' post.

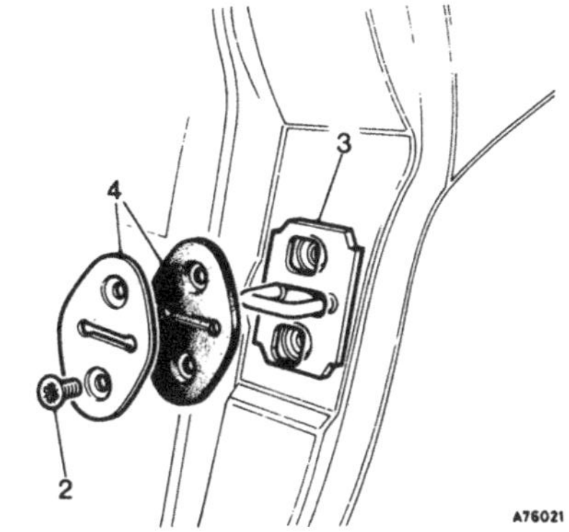

3 Withdraw the striker from inside the 'B' post.
4 Lift off the outer striker plate and the seal.

Refitting
5 Reverse instructions 1 to 4.

DOOR LOCK STRIKER

Adjust 76.37.27

1 Slacken the two screws.
2 Adjust the striker position as necessary to ensure correct door locking action and alignment.
3 Re-tighten the screws.

DOOR LOCK REMOTE CONTROL

Remove and refit 76.37.31

Removing
1 Remove the door trim pad, see 76.34.01.
2 Remove the retaining clip and detach the control rod from the lock.
3 Remove the screw and lift off the escutcheon.
4 Remove the screw and push the control assembly rearwards to disengage it from the door.

continued

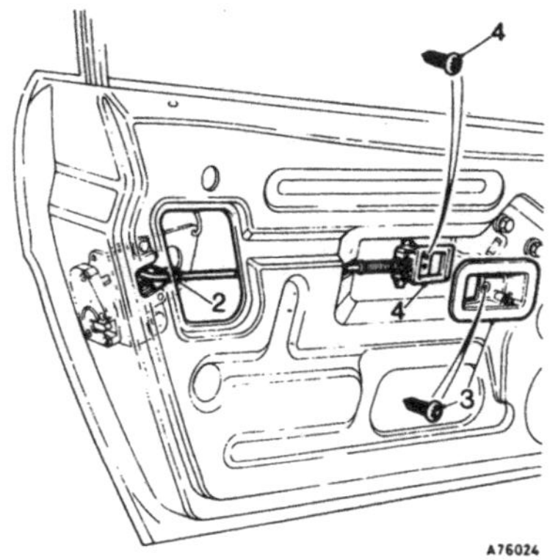

5 Withdraw the control assembly from the door.

Refitting
6 Reverse instructions 1 to 5.

DOOR PRIVATE LOCK

Remove and refit 76.37.39

Removing
1 Remove the door outside handle, see 76.58.01.
2 Remove the circlip.

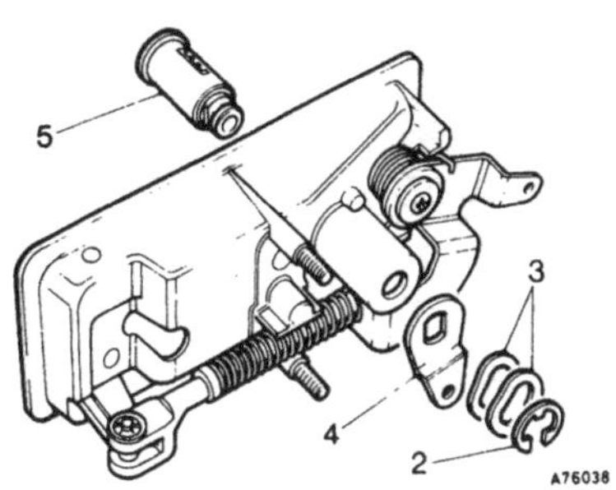

3 Remove the two spring washers.
4 Remove the locking lever.
5 Push out the lock.

Refitting
6 Reverse instructions 1 to 5.

WINDSCREEN FINISHER–LOWER

Remove and refit 76.43.41

Removing
1 Remove the drivers wiper arm, see 84.15.02
2 Remove the passengers wiper arm, see 84.15.03.
3 Remove the nut, distance piece and rubber washer from the passenger's side wheelbox spindle.
4 Pull off the two bulkhead weatherstrips.
5 Remove the two screws and plain washers and detach the finisher from the body taking care to avoid scratching the body and windscreen.

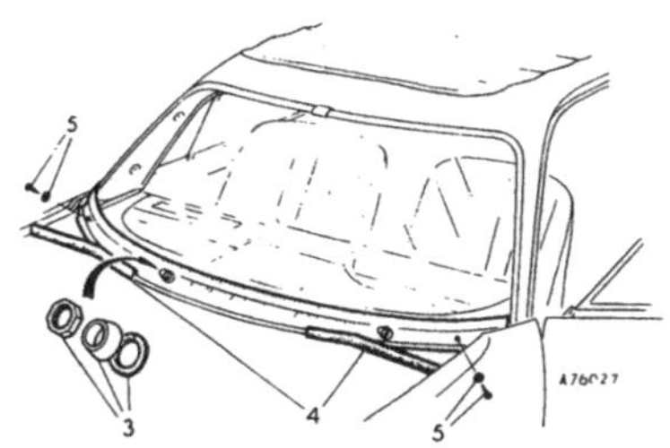

Refitting
6 Reverse instructions 1 to 5.

FASCIA

Remove and refit 76.46.01

Removing
1 Isolate the battery.
2 Remove the instrument cowl, see 76.46.17.
3 Remove the fascia switch panel, see 86.65.66, noting the positions of the three harness plugs and two switch identification bulb holders for refitting.
4 Remove the two 'A' post trim pads–two screws–together with the fascia corner finishers.
5 Remove the four screws securing the demister vents to the fascia.
6 Remove the two screws securing the fascia to the brackets on the bulkhead
7 Remove the two bolts and plain washers securing the tongues on the fascia to the bulkhead.

continued

8 Remove the steering nacelles–two screws.
9 Remove the two shear-head bolts securing the steering column housing to the body. Operation 57.40.01, instructions 6 and 7.
10 Remove the five screws (two each side of the steering column–one below bonnet release) securing the fascia to the support rail beneath the instrument panel.
11 Remove the two screws securing the fascia to the support rail beneath the glovebox.
12 Remove the two screws securing the control illumination panel to the cowl.
13 Remove the screw securing the fascia to the bracket on the control cowl.
14 Remove the lid from the component mounting panel inside the glovebox–two screws.
15 Remove the three screws and two brackets securing the component mounting panel to the fascia.
16 Remove the two screws securing the wiring harness clips to the top of the fascia.
17 Disconnect the cable from the speedometer by depressing the lever to release the catch from the annular groove in the boss.
18 Unscrew the knurled nut and release the speedometer trip reset from the bracket slot.
19 Unscrew the knurled nut and release the clock reset from the bracket slot.
20 Disconnect the two multi-contact harness plugs
21 Ease the fascia rearwards and pull the air hoses from the outer swivelling vents.
22 Carefully manoeuvre the fascia out of the car, simultaneously feeding the three harness plugs and two switch identification bulb holders through the four apertures above the switch panel aperture

Refitting
23 Ensure that the demister vents are correctly positioned.
24 Reverse instructions 1 to 22.

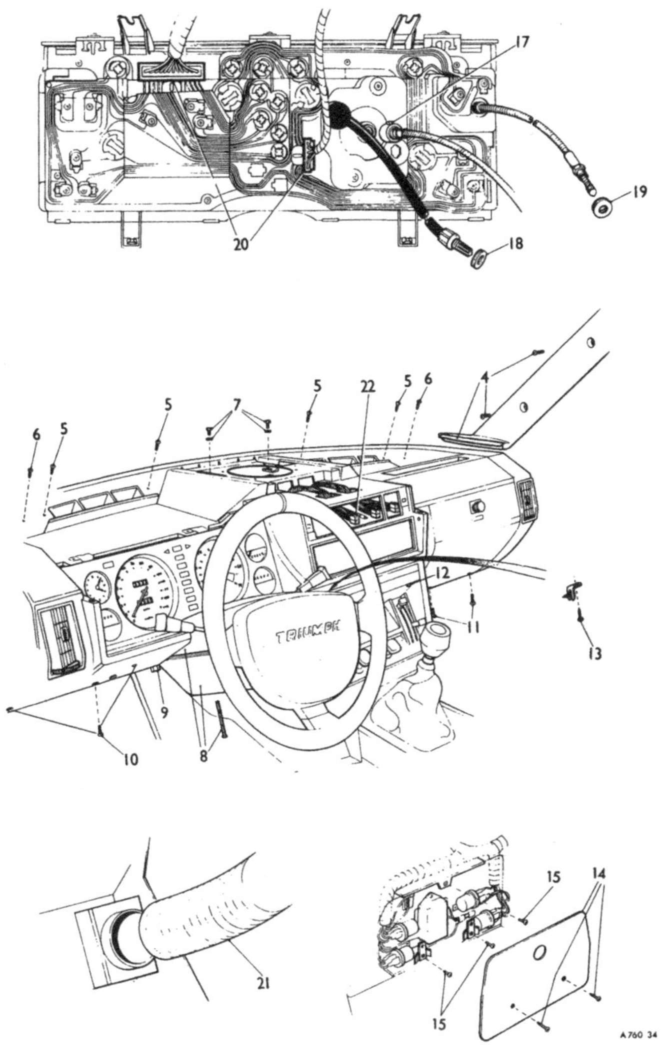

FASCIA INSTRUMENT COWL

Remove and refit **76.46.17**

Removing
1 Remove the fascia centre grille, see 76.55.14.
2 Remove the two screws securing the cowl to the fascia above the switch panel.
3 Remove the two screws securing the cowl to the fascia above the instrument panel.
4 Remove the screw securing the underside of the cowl to the bracket above the switch panel.
5 Remove the two screws securing the outer tongues of the cowl to the fascia.
6 Swing the cowl rearwards to disengage the side tongues from the fascia.
7 Remove the cowl.

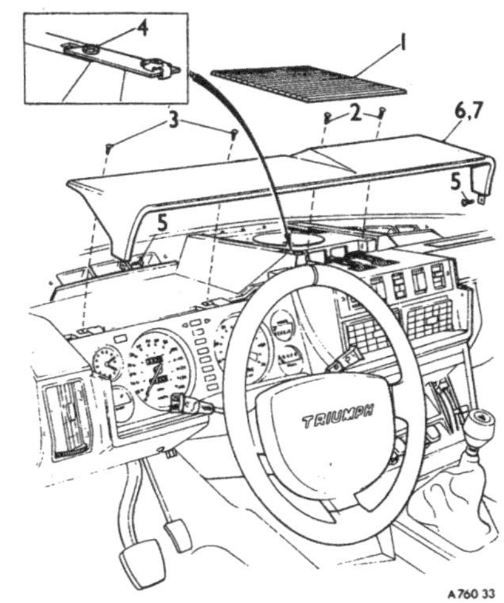

Refitting
8 Reverse instructions 1 to 7.

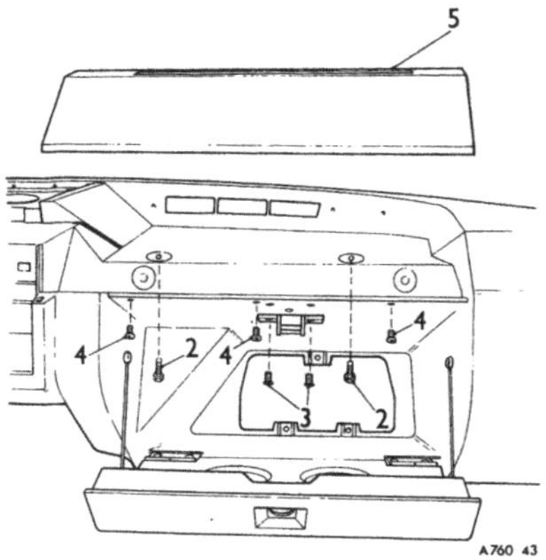

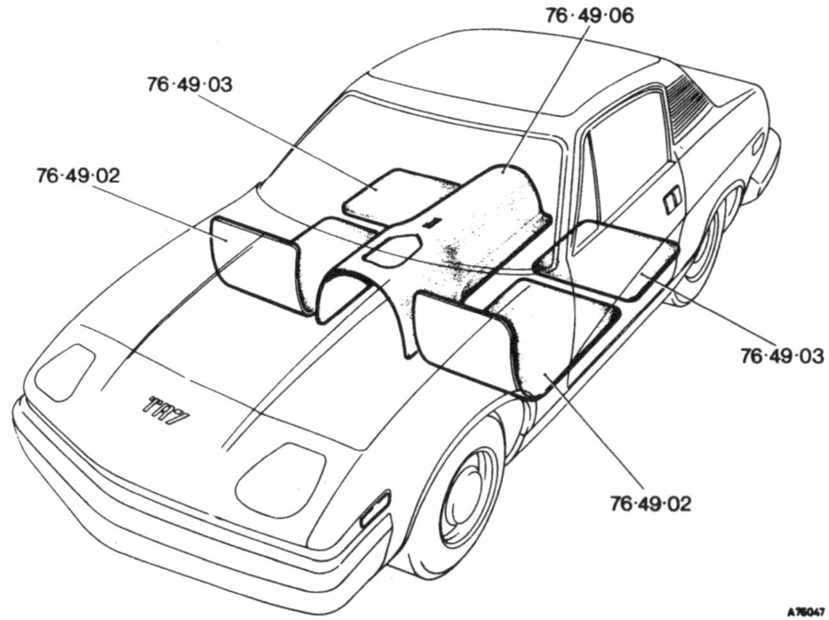

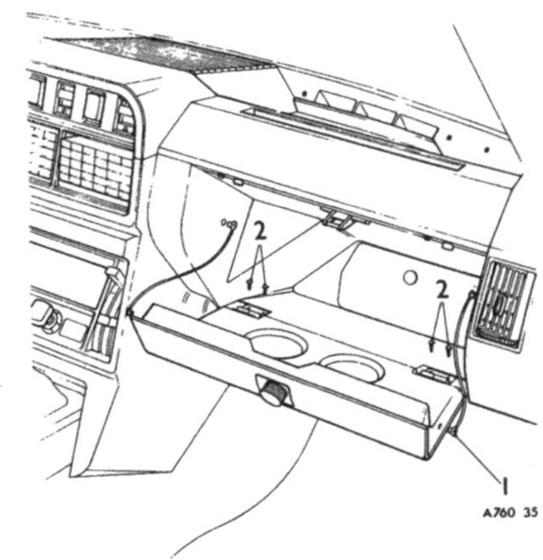

FASCIA GLOVEBOX COWL

Remove and refit 76.46.18

Removing

1 Remove the fascia, see 76.46.01.
2 Remove the two bolts securing the cowl to the inside of the fascia.
3 Remove the three screws securing the glovebox lid latch.
4 Remove the three screws from inside the glovebox.
5 Lift off the cowl.

Refitting

6 Reverse instructions 1 to 5.

CARPET–REAR FLOOR

Remove and refit 76.49.03

Removing

1 Remove the seat, see 76.70.04/05.
2 Lift out the carpet.

Refitting

3 Reverse instructions 1 and 2.

CARPET–TRANSMISSION TUNNEL

Remove and refit 76.49.06

Removing

1 Remove the console assembly, see 76.25.01.
2 Lift out the carpet.

Refitting

3 Reverse instructions 1 and 2.

GLOVEBOX LID ASSEMBLY

Remove and refit 76.52.02

Removing

1 Press out the centres of the rokut rivets and detach the support straps from the lid and the fascia.
2 Remove the four screws securing the hinges to the fascia and lift off the lid.

Refitting

3 Position the lid and refit the four hinge screws.
4 Fit new support straps using a suitable tool to install the rokut rivets.

GLOVEBOX LOCK

Remove and refit 76.52.08

Removing
1 Pull off the knob.
2 Remove the two bolts securing the lock to the lid interior and withdraw the lock.
3 Remove the latch if necessary by removing the two screws securing it to the fascia.

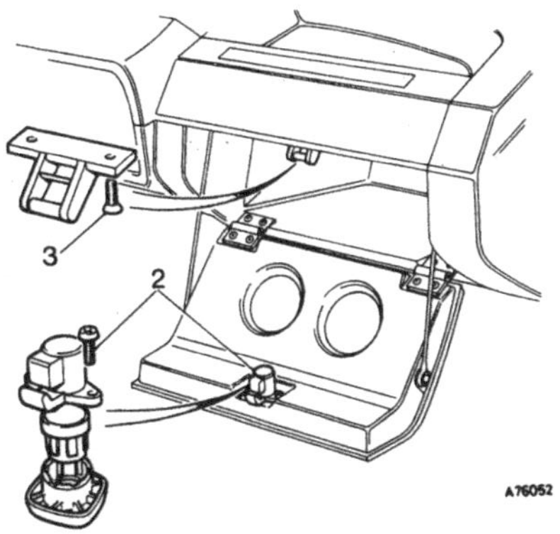

Refitting
4 Reverse instructions 1 to 3.

FRONT GRILLE

Remove and refit 76.55.03

Removing
1 Remove the blanking plate–two screws.
2 Press out the centres of the three rokut rivets securing the grille to the body.
3 Remove the grille.

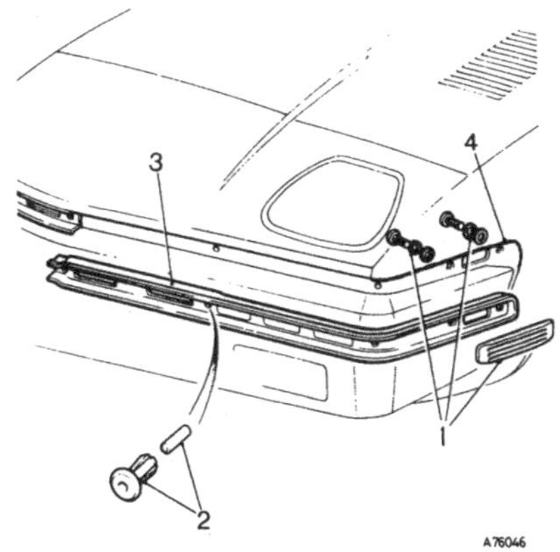

Refitting
4 Apply Seelastik to the body area shown.
5 Reverse instructions 1 to 3, using a suitable tool to install the rokut rivets.

FASCIA CENTRE GRILLE

Remove and refit 76.55.14

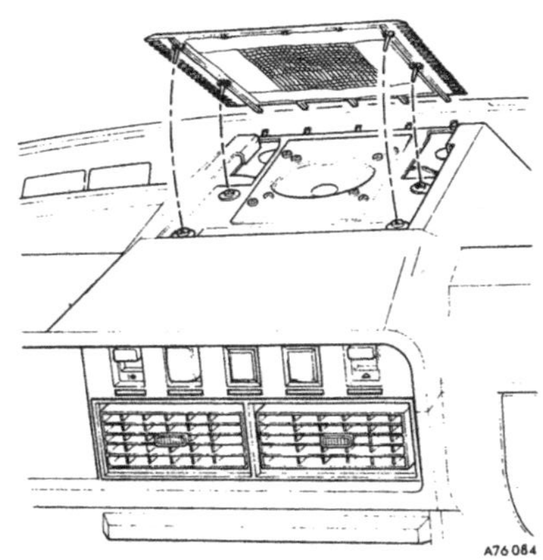

Removing
1 The grille is retained by five forward projections which locate into slots in the fascia and by four plastic spigots which locate into 'snap backs' secured to the fascia.
2 Using a wide bladed screwdriver, carefully prise up the grille adjacent to the spigots.
3 Move the grille rearwards to disengage the forward projections.

Refitting
4 Locate the forward projections into the fascia slots.
5 Locate the spigots into the 'snap backs' and press down into position.

AIR VENT GRILLE

Remove and refit 76.55.17

Removing
1 Remove the rear quarter trim pad, see 76.13.12.
2 Remove the four nuts, spring washers, plain washers and two spacers. Note spacer positions for refitting–long spacer on front lower fixing–short spacer on rear lower fixing.
3 Lift off the moulding together with the four 'T' shaped bolts.

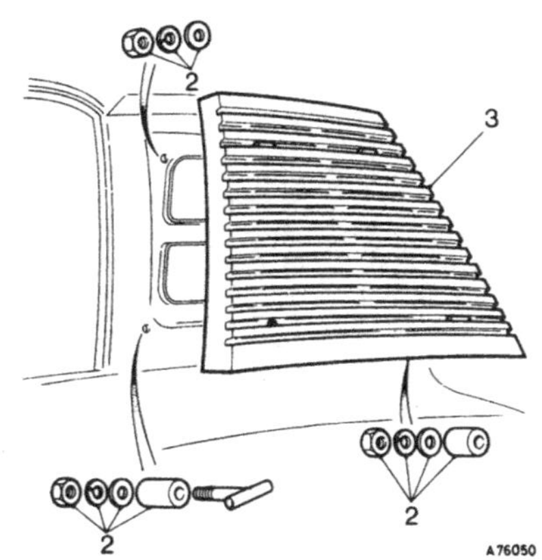

Refitting
4 Reverse instructions 1 to 3.

DOOR OUTSIDE HANDLE

Remove and refit 76.58.01

Removing
1 Remove the door trim pad, see 76.34.01.
2 Remove the bolt, spring washer and plain washer securing the rear glass channel to the door. Detach the insert from the channel and carefully withdraw the channel and weather curtain.
3 Remove the four screws securing the lock assembly to the door and lift off the disc latch.
4 Remove the retaining clip and detach the remote control rod from the lock.
5 Remove the two nuts and spring washers securing the clamp bracket to the outside handle.
6 Manoeuvre the outside handle and seal, together with the door lock, through the handle aperture and out of the door.
CAUTION: Care must be taken to avoid straining locks or linkages.
7 Remove the retaining clips from the door handle and private lock control rods and separate the outside handle and door lock.

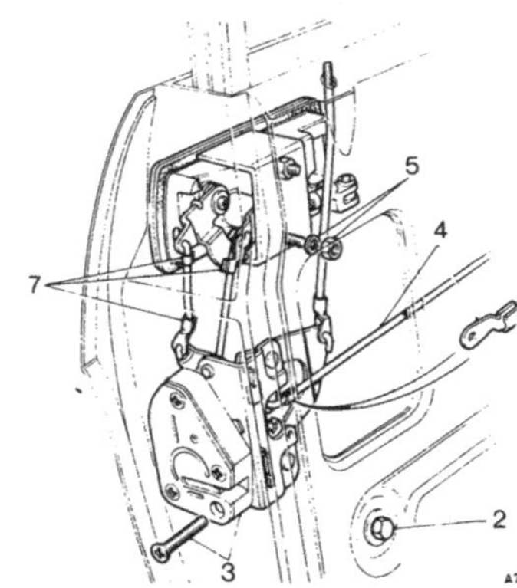

Refitting
8 Reverse instructions 1 to 7 ensuring that the plunger rod is correctly located before securing the lock to the door.

HEADLINING

Remove and refit
(Cars without sliding roof) 76.64.01

Removing

1. Isolate the battery.
2. Remove the roof lamp, see 86.45.02.
3. Remove the heated backlight, see 76.81.11.
4. Remove the sun visors and retainers.
5. Remove the interior mirror – two screws.
6. Pull off the windscreen header rail finisher.
7. Remove the two 'A' post trim pads–two screws each.
8. Pull off the door weatherstrips.
9. Remove the rear quarter trim pads, see 76.13.12.
10. Pull the lining edges away from the body flanges.
11. Detach the listing rails and remove the headlining.

Refitting

12. Apply a 2 in border of Dunlop SP 758 adhesive to the headlining, around the roof light aperture, and on the body flanges. Allow ten minutes for the adhesive to become tacky.
13. Reverse instructions 1 to 11, cutting off any excess material, to leave approximately 13 mm (½ in) overlap on all flanges and apertures.

HEADLINING

Remove and refit
(Cars with sliding roof) 76.64.01

Removing

1. Isolate the battery.
2. Remove the sliding roof, see 76.82.01.
3. Remove the roof lamp, see 86.45.02.
4. Remove the heated backlight, see 76.81.11.
5. Remove the sun visors and retainers.
6. Remove the interior mirror – two screws.
7. Pull off the windscreen header rail finisher.
8. Remove the two 'A' post trim pads – two screws each.
9. Pull off the door weatherstrips.
10. Remove the rear quarter trim pads, see 76.13.12.
11. Pull the lining edges away from the body and remove the headlining from the vehicle.

Refitting

12. Apply a 2 in border of Dunlop SP 758 adhesive to the headlining edges and around the roof light and sliding roof apertures. Allow ten minutes for the adhesive to become tacky.
13. Reverse instructions 1 to 11, cutting off any excess material to leave approximately 13 mm (½ in) overlap on all flanges and apertures.

HOOD (USA only)

Remove and refit 76.61.08

Removing

1. **Release the front of the hood by turning the levers.**
2. **Disconnect the eight fasteners securing the sides of the hood to the body.**
3. **Remove the eight screws securing the rear trimboard to the body and pull the trimboard forwards.**
4. **Remove the seven nuts securing the hood retaining strip to the rear deck.**
5. **Remove the two hood linkage covers – three screws each.**
6. **Lower the hood into the rearmost position ensuring that the hood material does not become damaged.**
7. **With the aid of an assistant, remove the four bolts, spring washers and plain washers and lift off the hood.**

Refitting

8. **Install the hood and loosely fit bolts 7.**
   Ensure that the hood moves freely, and fully tighten the bolts.
9. **Reverse instructions 1 to 6.**

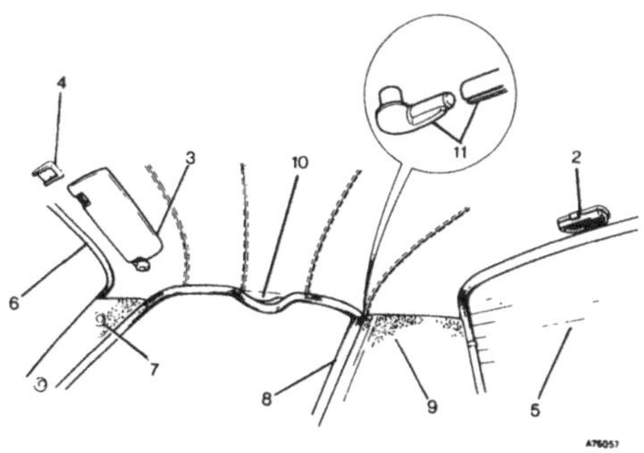

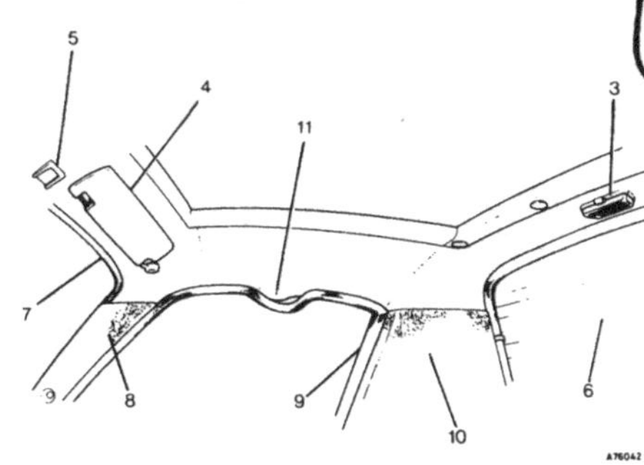

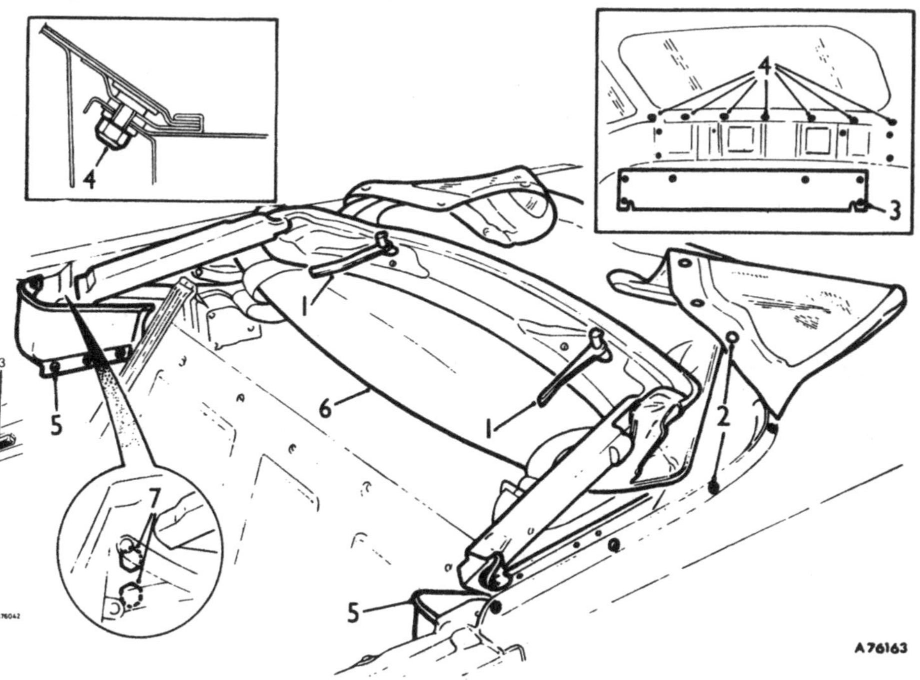

AKM3079B

PARCEL TRAY–REAR

Remove and refit 76.67.06

Removing

1 Move both seats and seat squabs to the fully forward position.
2 Remove the two bolts and spring washers securing the seat belt swivel brackets to the seats.
3 Feed the seat belts through the apertures in the parcel tray.
4 Remove the four screws and cup washers securing the parcel tray and rear compartment trim pad to the body.
5 Remove the four screws and cup washers securing the rear of the parcel tray to the body.
6 Carefully manoeuvre the parcel tray upwards and forwards and remove it from the car.

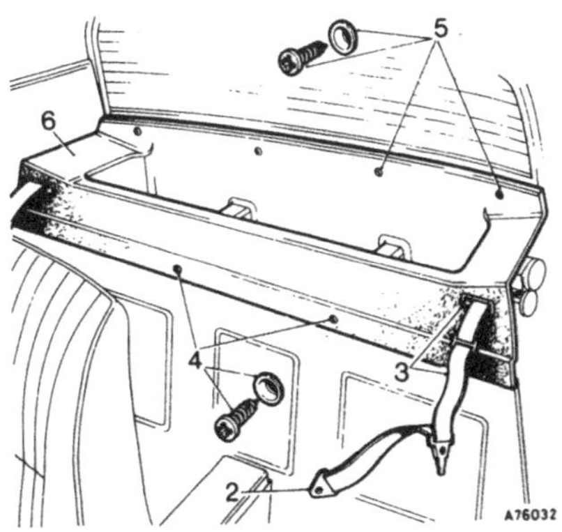

Refitting

7 Install the parcel tray in the car, feeding both seat belts through the apertures before finally positioning it.
8 Reverse instructions 1, 2, and 4 to 6.

ASHTRAY

Remove and refit 76.67.13

Removing

1 Depress the stubber to release the top edge of the bowl and carefully disengage the retainers on the bottom edge from the surround.
Using a small screwdriver carefully prise up the retainers on the surround sufficiently to enable it to be pulled clear of the trim pad.

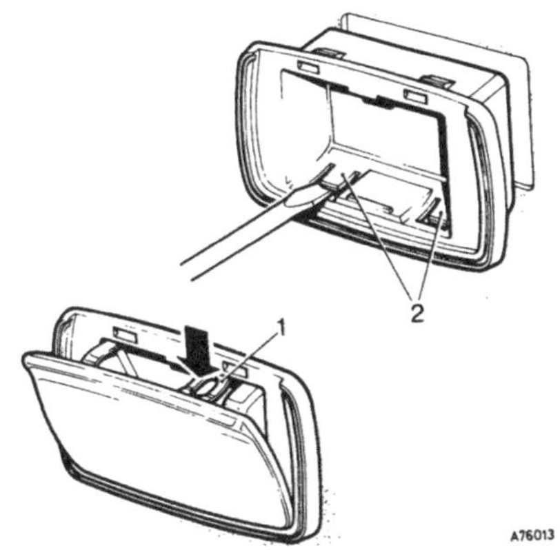

Refitting

3 Press the surround into the trim pad sufficiently to engage the retainers.
4 Refit the bowl.

SEATS

Remove and refit

Driver's seat 76.70.04
Passenger's seat 76.70.05

Removing

1 Move the seat fully rearwards.
2 Remove the two capscrews (long), and rectangular washers securing the runners to the floor.

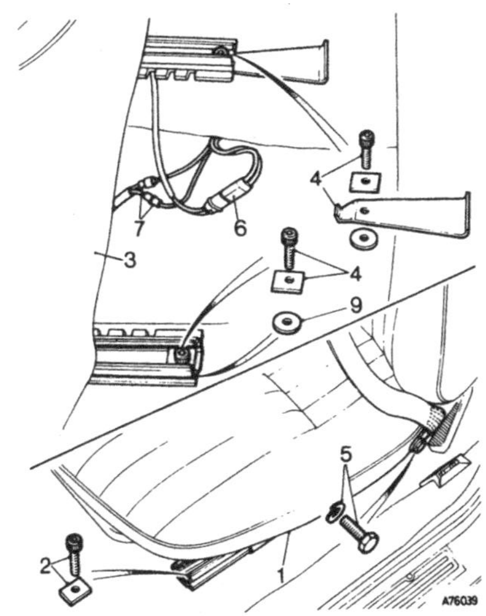

3 Move the seat fully forwards.
4 Remove the two capscrews (short), one rectangular washer and one stop plate securing the seat to the floor.
5 Remove the bolt and spring washer securing the seat belt bracket to the seat.
6 Disconnect the seat belt buckle unit harness plug (if fitted).
7 Disconnect the two seat belt warning light switch harness plugs (if fitted).
8 Lift out the seat complete with runners.

Refitting

9 Ensure that the packing washers are correctly positioned.
10 Reverse instructions 1 to 8.

SEAT RUNNERS

Remove and refit 76.70.21

Removing

1 Remove the seat, see 76.70.04/ 76.70.05.
2 Remove the six capscrews and spring washers securing the runners to the seat, moving the slides as necessary to obtain access.

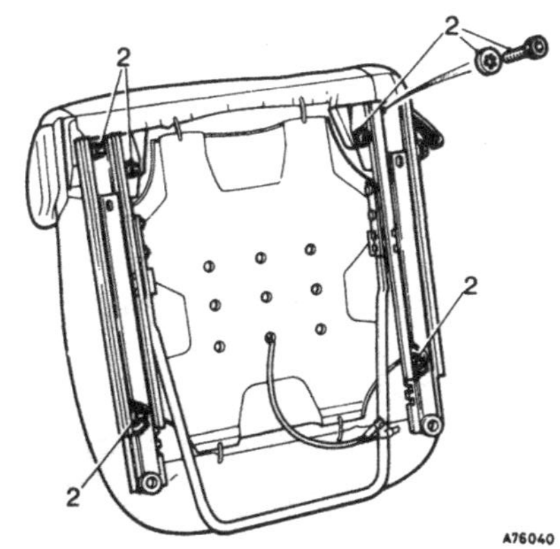

Refitting

3 Reverse instructions 1 and 2.

SEAT BELTS

Remove and refit — 76.73.10

Removing

1 Remove the bolt and spring washer securing the swivel bracket to the seat.
2 *Earlier Models:* Feed the seat belt through the aperture in the parcel shelf.
3 *Earlier models:* Working inside the luggage compartment remove the trim panel above the wheel arch—two fasteners.
4 Remove the bolt and spring washer securing the reel unit mounting bracket to the wheel arch.
NOTE: On later models a cover over the reel must be prised up and lifted off to gain acess to the reel securing bolt.

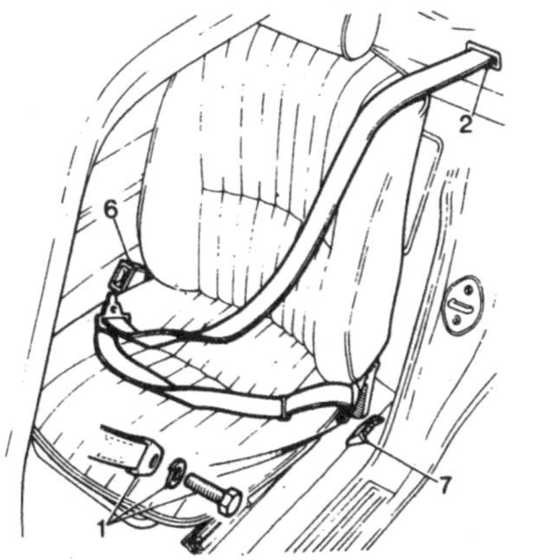

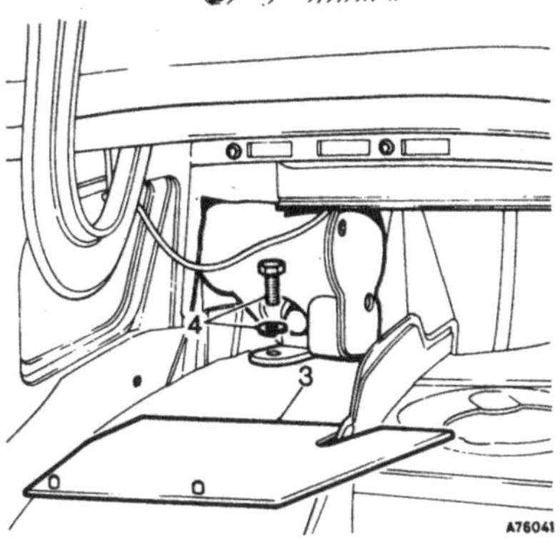

5 Withdraw the seat belt and reel from the vehicle.
6 Remove the seat belt switch if necessary, see 86.65.31/86.65.32.
7 Remove the tongue retainer, if necessary—two screws.

Refitting

8 Reverse instructions 1 to 7.

WINDSCREEN

Introduction — 76.81.00

The Thermo Electric Windscreen Sealer is an uncured 'Neoprene' based material supplied in round strip form, which has a thin insulated resistance wire running through its centre core. The compound surrounding the wire also has a heat activated accelerator incorporated in it and has sufficient initial tackiness to adhere it to the painted metal aperture.
The service kit contains a preformed Solbit sealing strip of sufficient length to lap completely around the windscreen aperture flange. Also included are a bottle of primer, wire for cutting out laminated windscreens and two rubber spacers for supporting the new windscreen. The internal resistance wire is exposed at each end of the strip and connected to a low voltage electrical supply. This is obtained by connecting two fully charged 12 volt batteries in series (+ to –) or alternatively using a variable output transformer capable of giving 11 amps at 24 volts. A lower current would be insufficient to effect a satisfactory cure of the Solbit strip. The resistance wire becomes warm and softens the strip, enabling the windscreen and finishers to be bedded into position. Further heat helps to cure the sealer and after 1½ to 2 hours the adhesion is strong enough for the car to be used.

NOTE:

A The aid of an assistant is required for this operation.
B Glass lifters should be used for handling the screen.
C Cleanliness is essential.

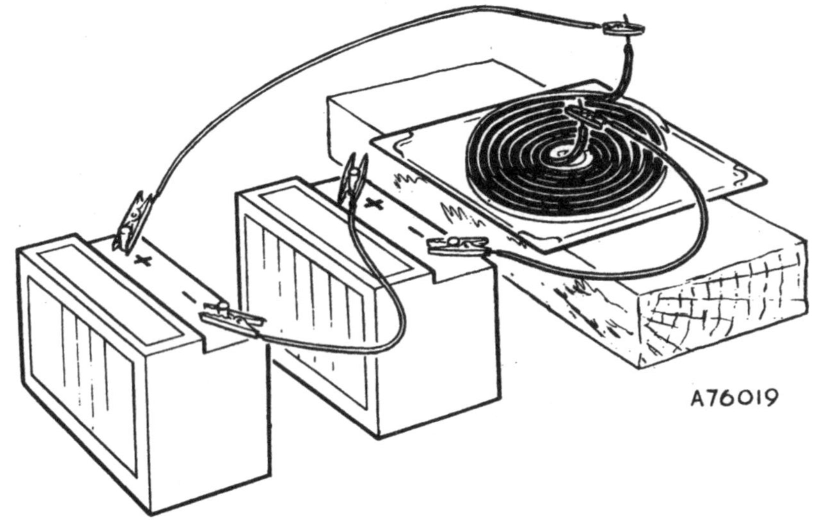

A76019

WINDSCREEN GLASS

Remove and refit — 76.81.01

Removing

1 Cover the fascia and interior trim to protect them from broken glass.
2 Remove the windscreen wiper arms, see 84.15.02/03.
3 Remove the lower windscreen finisher, see 76.43.41.
4 Remove the 'A' post trim pads – two screws, together with the fascia corner finishers.
5 Pull off the header rail finisher.
6 Push the finisher moulding cover to one side to expose the Solbit.

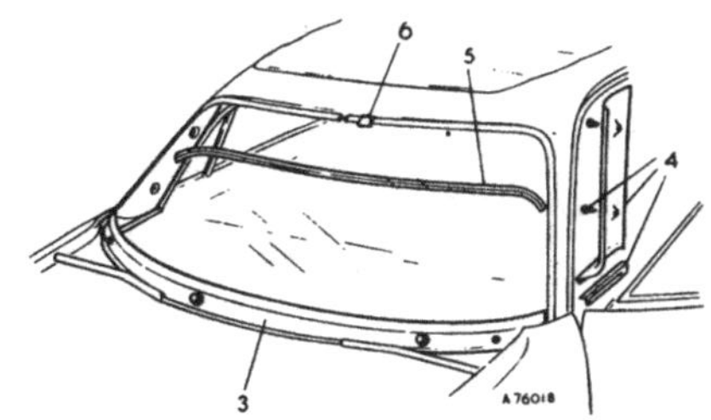

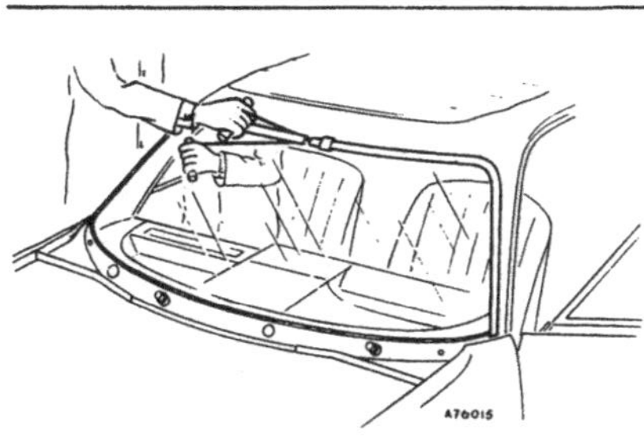

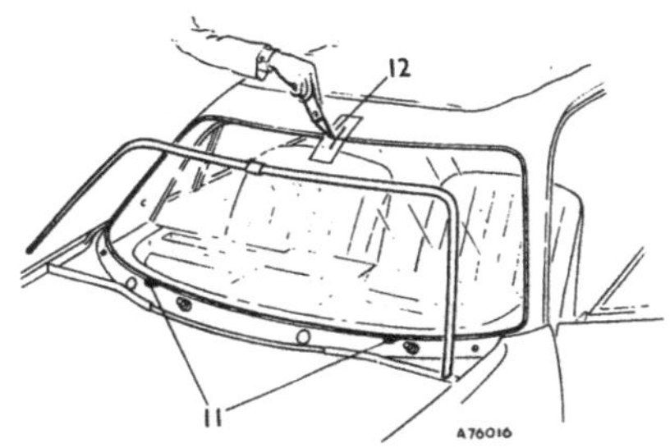

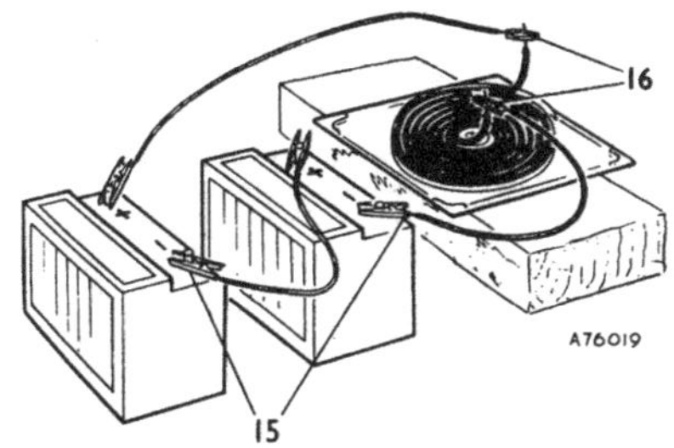

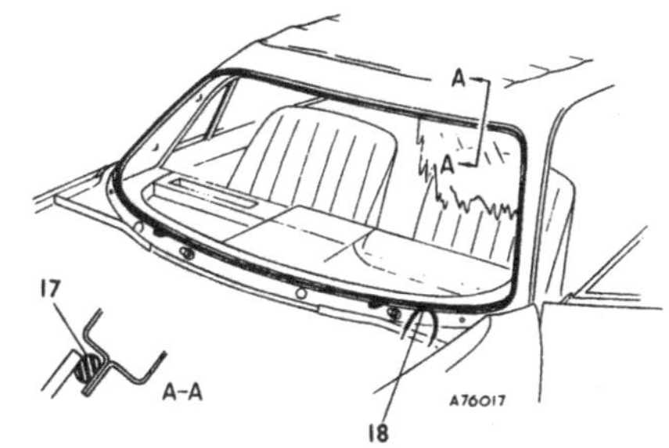

7 From both outside and inside, cut or pierce a hole through the existing Solbit and thread one end of the wire through the hole to an assistant inside the car. At each end of the wire twist on a small piece of wood to act as a handle or form a handle by threading the wire through a small hole drilled in the end of an old screwdriver. By pulling the wire in a saw like motion, cut through the Solbit. Use long steady pulls rather than short quick ones otherwise the wire will overheat and break. Narrow the angle of the cut by keeping the wire ends as close as possible to the glass. Ensure that the wire outside the car is pulled along the moulding rather than the painted body otherwise the latter will be damaged.

8 With a sharp knife, cut through the old Solbit remaining on the flange and remove it by a combination of cutting and pulling.

9 Cut away any sharp edges or remaining lumps of old Solbit to ensure an even surface.

10 Cut away and remove Solbit from the finisher mouldings. Avoid distorting the mouldings during the cleaning operation. It may be necessary to obtain replacements. Remove any dirt or loose material from the flange using a clean lint-free cloth moistened with methylated spirit.

Refitting

11 Place the two rubber spacers from the kit on to the bottom flange about 150 mm (6 in) to 225 mm (9 in) from the pillars. Using glass lifters, offer the windscreen up to rest on the spacers; positioning it to give equal gaps all the way round. Position finisher mouldings and join them together at the top with the cover to ensure that they fit well and have not been distorted during removal. Remove the finisher mouldings.
NOTE: If any straightening of the mouldings is necessary, it should be carried out at this stage.

12 Place a piece of masking tape approximately on the top centre of the glass, sticking this across the gap between the glass and the body. Mark the masking tape so as to facilitate location when finally placing the windscreen for fitting. Then cut through the tape between the glass and body and remove the windscreen on to a bench.
NOTE: Take careful note of the size of the gap because when applying the Solbit strip to the flange, there must be sufficient Solbit along the whole length of the gap into which the reveal mouldings are embedded and sufficient Solbit under the glass for a watertight bond line to be made.

13 Apply a thin coat of primer to the remaining Solbit or painted flange using a clean lint-free cloth. Allow to dry for approximately one minute.

14 Clean the new windscreen with a solvent (e.g. methylated spirit) if necessary, and apply a thin coating of primer, no more than 13 mm (½ in) wide, to the periphery of the inner face of the windscreen. Allow to dry for approximately one minute.

15 Connect two *fully charged* 12 volt batteries in series (+ to –).
NOTE: Alternatively a variable output transformer set to give 11 amps at 24 volts may be used.

16 Expose the bared ends of the wire in the Solbit strip whilst in the container pack and connect to the batteries or transformer until the Solbit is just sufficiently tacky to adhere to the flange. This time can vary between 15 and 90 seconds depending on the temperature and age of the Solbit. Disconnect the current.
NOTE: Do not overheat the Solbit as this will make it both difficult to remove from the pack and to handle.

17 Starting about 75 mm (3 in) from one of the bottom corners of the screen aperture, place the Solbit strip on the flange so that when the windscreen is fitted, there is sufficient exudation of Solbit to the right and the left of the edges of the screen aperture for bonding the glass and for embedding the chrome finishers.

18 Cross over the ends of the Solbit, approximately 50 mm (2 in) in length at the lower right corner of the windscreen aperture, having first placed a small strip of masking tape on the body to prevent marking by the ends of the Solbit.
NOTE: Care must be taken with this operation otherwise a hairline crack could be left on the surface of the Solbit after compression.

19 Using glass lifters install the windscreen in the aperture resting the bottom edge on the spacers and taking care to position it exactly where previously marked with the masking tape. Avoid fingering the primed areas. Remove approximately 25 mm (1 in) of the exposed excess Solbit from the wire.

20 Connect the two bared wire ends of the Solbit strip to the 24 volt supply. This will begin to soften the Solbit and after 1½ to 2 minutes, press the glass into the Solbit, starting with pressure in the centre of the glass and out to the pillars. Do not press the glass too far into the Solbit at this stage because, if somewhat proud, the depth can be rectified when the chrome finishers are embedded into the Solbit.

continued

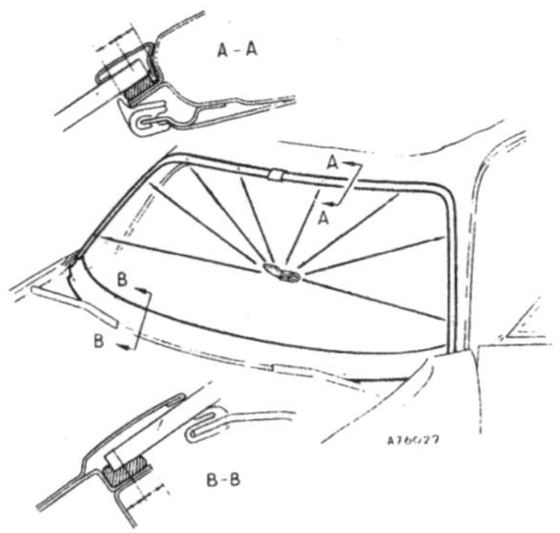

21 Inspect the bond line (which should be no less than 7 mm (¼ in) wide) through the glass during early stages of the heating cycle. Potential leak sources can easily be seen and rectified by a little extra pressure in the area while the heat is on or by pressing some of the previously removed excess Solbit into the particular leak source.
22 After 10 minutes position and embed the chrome finishers into the Solbit; very little pressure is required. If at this stage mouldings and windscreen are proud, additional pressure can be applied. Disconnect electrical current after 1½ to 2 hours.
23 Allow the glass to cool down.
24 Test for water leaks.
25 Trim off exposed ends of Solbit and rubber spacers with snips or press out of sight into the rebate.
26 Refit the header rail.
27 Refit the 'A' post trim pads–2 screws, together with the fascia corner finishers.
28 Refit the lower windscreen finisher, see 76.43.41.
29 Refit the windscreen wiper arms, see 84.15.02/03.

HEATED BACKLIGHT

Remove and refit 76.81.11

Removing
1 Break the seal, using a suitably blunt tool.
2 Carefully pull the rear edges of the rear quarter trim pads away from the body to expose the Lucar connectors. Disconnect the two Lucar connectors from the backlight.
3 Push the glass outwards.
CAUTION: Take care to avoid scratching the glass, which must be steadied by an assistant.
4 Remove the weatherstrip from the glass.

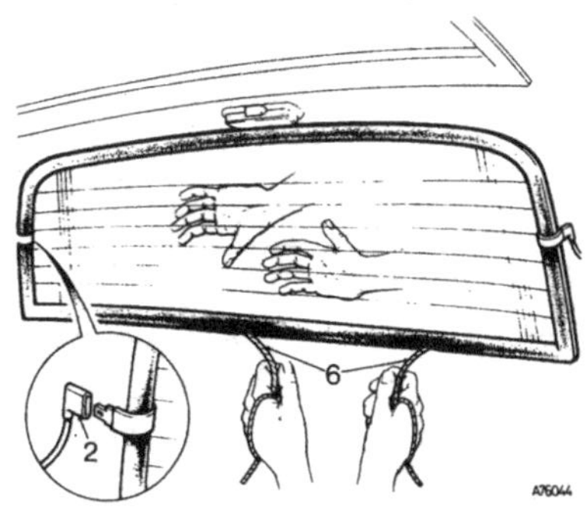

Refitting
5 Reverse instruction 4 using a new weatherstrip if necessary, and apply Seelastik to the glass channel before fitting.
6 Insert a strong cord into the weatherstrip inner channel, allowing the ends to protrude from the lower edge.
7 Have an assistant position the glass centrally in the aperture, and maintain a steady pressure whilst the cord ends are pulled to locate the weatherstrip on the body flange.
8 Seal the outer channel of the weatherstrip to the body using Seelastik.
9 Reconnect the two Lucar connectors and ensure that the rear quarter trim pads are correctly positioned.

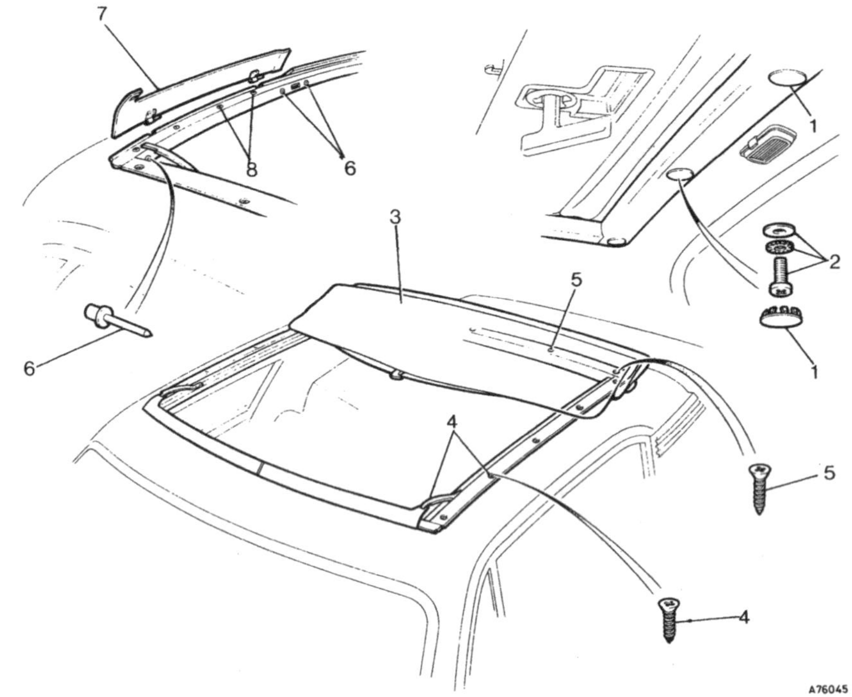

SLIDING ROOF

Remove and refit 76.82.01

Removing
1 Prise out the four chrome caps.
2 Remove the four screws, spring washers and plain washers.
3 Open the roof and carefully pull it rearwards on one side at a time until the four slides are disengaged from the runners. Lift off the roof.
4 Remove the five screws securing one of the side runners to the roof. Remove the runner taking care to avoid breaking the spring when disengaging it. Remove the opposite side runner in the same manner.
5 Remove the five screws securing the rear frame to the roof. Remove the frame.
6 Drill out the four rivets on the vertical face of the front frame.
7 Carefully pull the two hinge brackets outwards to disengage them.
8 Drill out the eight rivets on the top face of the frame. Remove the frame.

Refitting
9 Ensure that the headlining is correctly positioned with an overlap of approximately 13 mm (½ in).
10 Apply Kelseal to the forward top face of the roof aperture.
11 Reverse instructions 3 to 8.
12 Close the roof and adjust if necessary to obtain correct tension before fully tightening screws 2.
13 Replace the four chrome caps.

HEATING AND VENTILATION

DEMISTER DUCT

– Remove and refit
– passenger's side 80.15.03

Removing
1 Remove the fascia. 76.46.01.
2 Remove the screw securing the control cowl to the body.
3 Detach the duct from the heater unit.

Refitting
4 Reverse instructions 1 to 3.

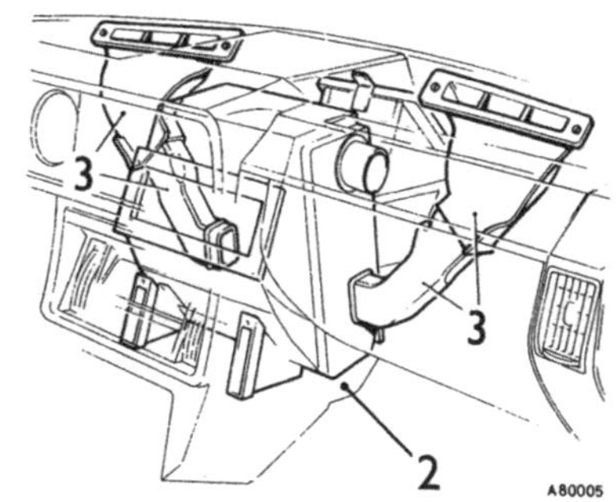

DEMISTER DUCT

– Remove and refit – driver's side 80.15.08
(L.H.D. omit instructions 1 and 2)

Removing
1 Slacken the knurled nut and detach the speedo trip control from the bracket.
2 Depress the lever to release the catch from the annular groove in the boss and pull the speedometer cable from the instrument.
3 Remove the screw securing the side of the control cowl to the body.
4 Remove the two screws securing the duct to the fascia.
5 Detach the duct from the heater and fascia and remove it from the vehicle. **NOTE**: It may be necessary to separate the upper and lower halves of the duct to facilitate removal.

Removal
6 Reverse instructions 1 to 5 ensuring that the speedo inner cable is engaged in the instrument and that the catch is engaged on the annular groove.

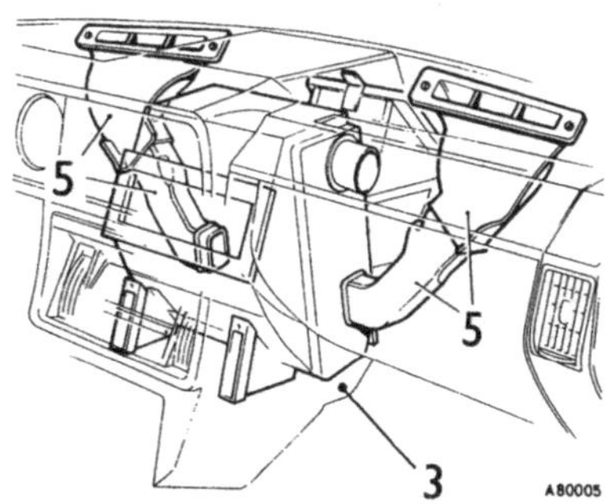

AIR HOSES

– Remove and refit – R.H. 80.15.11
– Remove and refit – L.H. 80.15.12

Removing
1 *For access to passenger side air hose only:* remove the fascia. 76.46.01.
2 Pull the hose ends from the vent duct and heater.

Refitting
3 Reverse instructions 1 and 2 ensuring that components behind the fascia are not damaged when the hoses are installed.

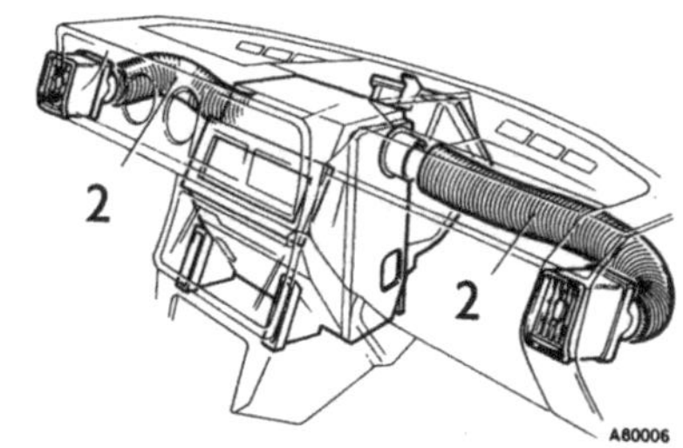

SWIVELLING VENTS – OUTER

– Remove and refit–L.H. 80.15.22
– Remove and refit–R.H. 80.15.23

Removing
1 Pull the air hose from the vent duct.
2 Push the vent assembly outwards to release the four tongues securing the vent to the fascia.
3 Depress the sides of the vent sufficiently to release the four tongues securing it to the duct.
4 Remove the duct and pull out the vent.

Refitting
5 Reverse instructions 1 to 4.

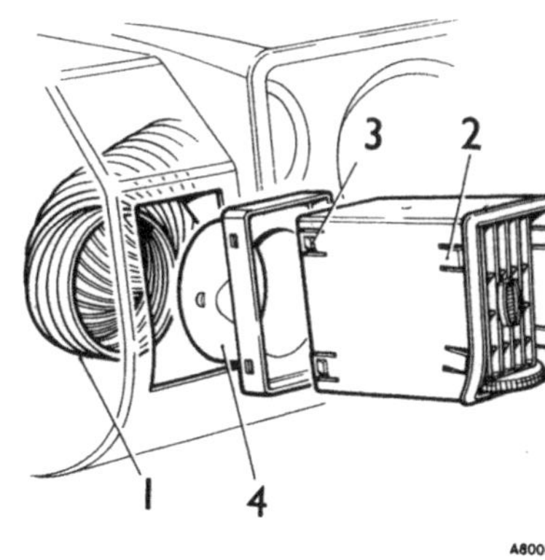

SWIVELLING VENT – CENTRE

– Remove and refit 80.15.24

Removing
1 Depress one side of the vent frame to detach the retaining boss from the surround,
2 Lift out the vent.

Refitting
3 Reverse instructions 1 and 2.

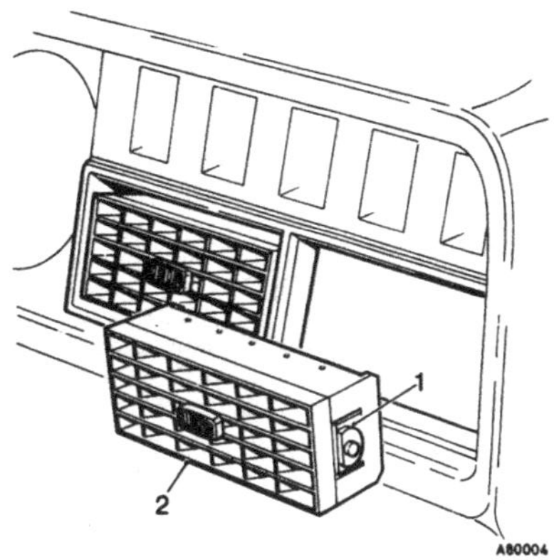

FRESH AIR DUCT

– Remove and refit 80.15.31

Earlier models—

Removing
1 Pull the bonnet lock cable clear of the duct.
2 Remove the two screws, spring washers and plain washers and carefully manoeuvre the duct clear of the engine compartment.

Refitting
3 Reverse instruction 1 and 2 ensuring that the duct is located in the retaining clips on the bulkhead.

Later models—

Removing
1 Release the two clips.
2 Disengage the bonnet lock cable from the duct and carefully manoeuvre the duct clear of the engine compartment.

Refitting
3 Reverse instruction 1 and 2 ensuring that the duct is located in the retaining clips on the bulkhead.

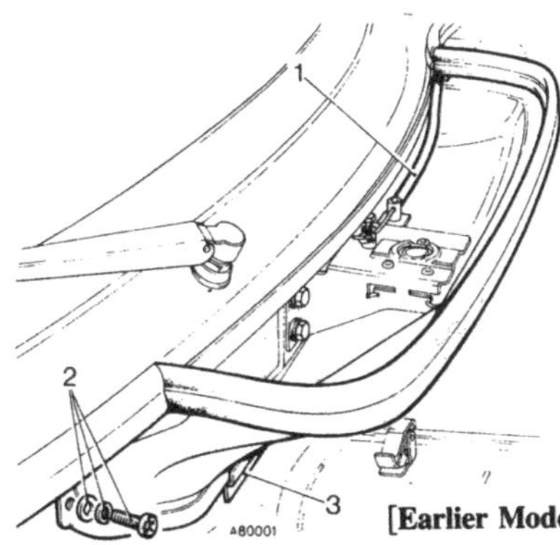

[Earlier Models]

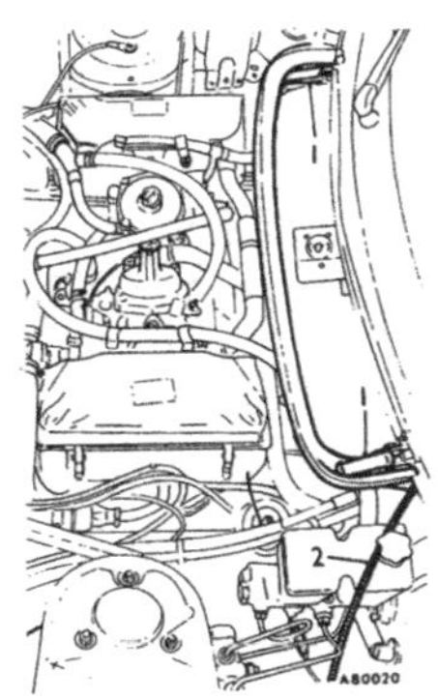

[Later Models]

HEATER UNIT

– Remove and refit 80.20.01

Removing
1 Isolate the battery.
2 Drain the coolant.
3 Remove the fresh air duct. 80.15.31.
4 Slacken the two clips and disconnect both water hoses from the heater.
5 Remove the fascia. 76.46.01.
6 Remove the two air hoses from the heater.
7 Remove the console assembly. 76.25.01.
8 Remove the control cowl. 76.25.03.
9 Remove the two demister ducts from the heater unit.
10 Remove the one bolt securing the heater air intake to the bulkhead.
11 Remove the control cowl support bracket – two bolts, spring washers and plain washers.
12 Slacken the two bolts securing the heater to the front of the heater support bracket on the transmission tunnel.
13 Remove the two heater support brackets – two bolts, spring washers, and plain washers.
14 Remove the two nuts, bolts, spring washers and plain washers securing the fascia support rails to the support bracket on the transmission tunnel.
15 Disconnect the two leads from the fan motor. (1 Black 1 Green).
16 Remove the nut, spring washer and plain washer securing the water pipe bracket to the bulkhead.
17 Remove the nut, spring washer and plain washer securing the rear of the heater to the bulkhead.
18 Remove the heater from the vehicle taking care to avoid spillage of coolant remaining in the matrix.

Refitting
19 Reverse instructions 1 to 18.

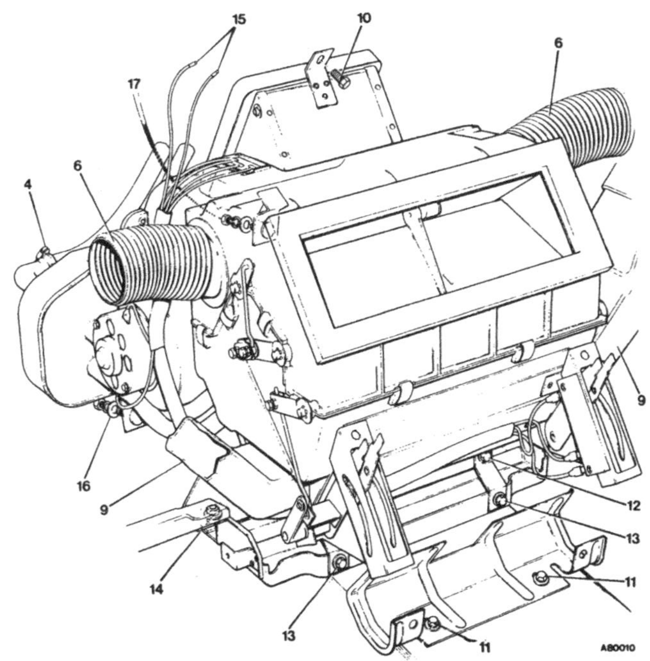

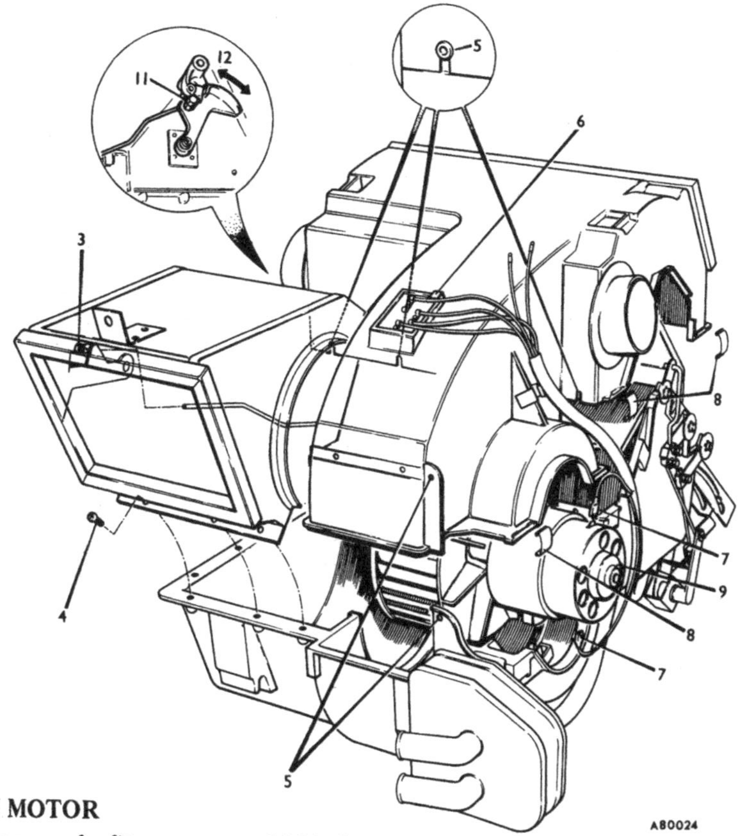

FAN MOTOR

Remove and refit 80.20.15

Removing

1 Isolate the battery.
2 Remove the heater, see 80.20.01.
3 Slacken the trunnion and disconnect the air intake control rod.
4 Remove the six screws and detach the air intake from the heater.
5 Drill out the four rivets.
6 Release the clip and detach the resistor from the heater.
7 Disconnect the two Lucar connectors from the fan motor.
8 Detach the fourteen clips and lift off the upper half of the casing.
9 Lift out the fan motor.

Refitting

10 Reverse instructions 9 to 4.
11 With the cam set in the position shown, set the air inlet flap against the aperture with the control lever in the 'OFF' position and tighten the trunnion screw.
12 Operate the control lever to check engagement of detents and full movement of the flap, then return the lever to the 'OFF' position and check sealing of the flap.
13 Reverse instructions 1 and 2.

HEATER MATRIX

Remove and refit 80.20.29

Removing

1 Isolate the battery.
2 Remove the heater unit, see 80.20.01.
3 Remove the fan motor, see 80.20.15.
4 Remove the seal from the pipe flange plate.
5 Drill out the rivet securing the pipe flange plate to the heater.
6 Remove the two screws securing the pipe mounting bracket to the heater. Lift off the bracket and packing piece.
7 Carefully ease the pipe assembly out of the matrix bushes.
8 Slacken the three trunnions and disconnect the flap operating rods from the levers.
9 Lift off the face level flap.
10 Lift off the heater inlet flap.
11 Drill out the four rivets securing the control level mounting plate to the heater.
12 Remove the two screws securing the lower flange of the control lever mounting plate to the heater. Lift off the mounting plate.
13 Remove the spire nut securing the lower side flap operating rod to the lever and disconnect the rod from the lever.
14 Remove the spire nut securing the lower flap spindle to the matrix housing.
15 Carefully pull the operating lever and spring clip off the lower flap spindle.
16 Detach the lower lap from the matrix housing and collect the spacer.

continued

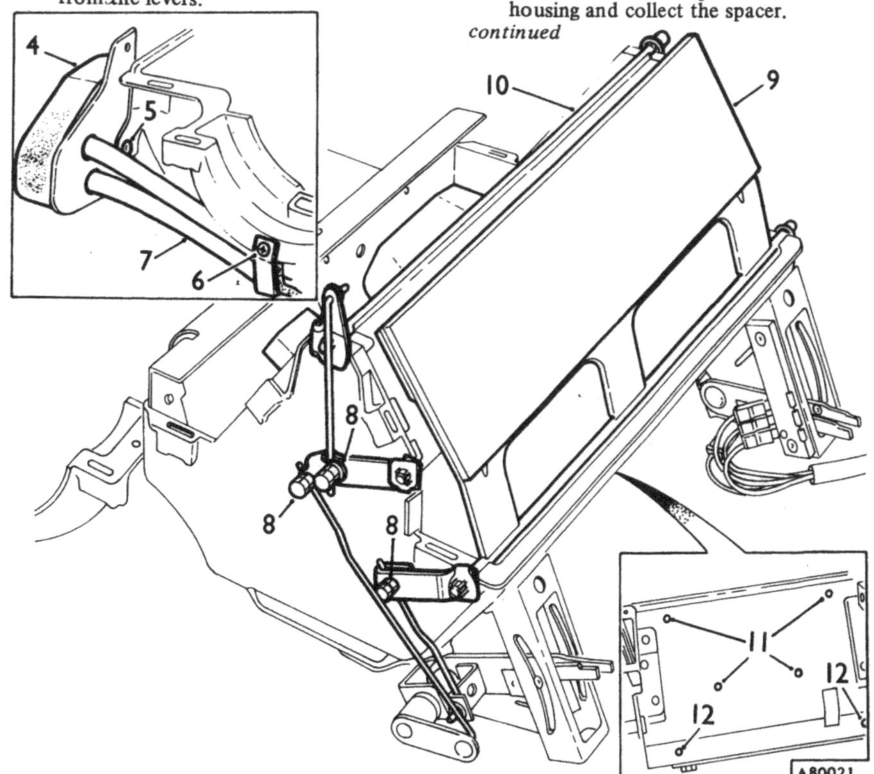

17 Drill out the two rivets securing the lower side flap assembly to the heater box.
18 Remove the lower side flap assembly.
19 Drill out the seven rivets securing the matrix housing to the heater box.
20 Remove the matrix complete with housing from the heater box.
21 Remove the matrix from the housing.

Refitting
22 Fit two new pipe seals into the matrix and replace the foam packing piece.
23 Reverse instructions 8 to 21.
24 Lubricate the pipes and seals with an anti-freeze solution.
25 Reverse instructions 7 to 3.
26 With the face level ventilation flap closed against the aperture as shown and the control lever in the 'OFF' position, tighten the trunnion screw.
27 Operate the control lever to check engagement of detents and full movement of the flap, then return the lever to the 'OFF' position and check sealing of the flap.
28 With the lower flap closed against the aperture as shown and the control lever in the 'SCREEN' position tighten the trunnion screw.
29 Operate the control lever to check engagement of detents and full movement of the flap, then return the lever to the 'SCREEN' position, and check sealing of the flap.
30 With the cam in the position shown set the air inlet flap against the aperture with the control lever in the 'OFF' position and tighten the trunnion screw.

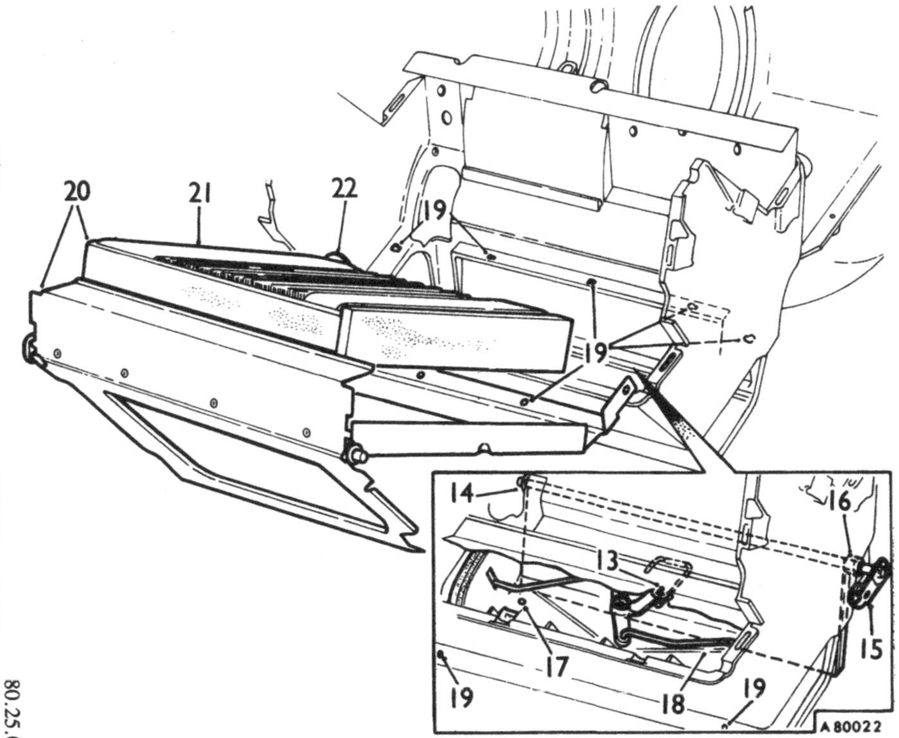

31 Operate the control lever to check engagement of detents and full movement of the flap, then return the lever to the 'OFF' position and check sealing of the flap.
32 With the heater inlet flap set as shown and the outlet flap closed as shown, tighten the trunnion screw.
33 Set the control level in the 'COLD' position and tighten the trunnion screw.
34 Operate the control lever to check engagement of detents and full movement of both flaps, then return the lever to the 'COLD' position and check sealing of the outlet flap.
35 Reverse instructions 1 and 2.

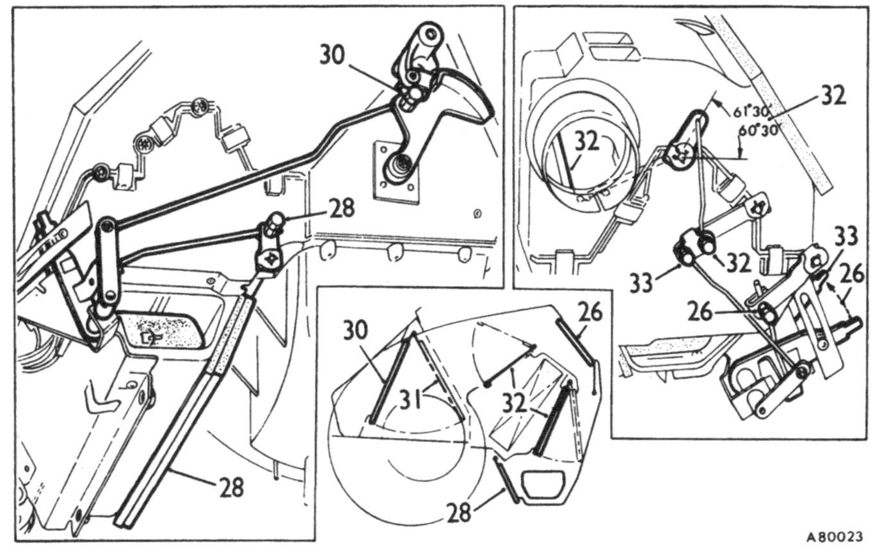

WATER HOSES

– Remove and refit
– hose – feed
– engine to heater **80.25.07**
– Hose – return – heater to engine return pipe **80.25.12**

Removing
1 Drain the cooling system. 26.10.01.
2 Remove the fresh air duct. 80.15.31.
3 Slacken the clips and remove the hoses.

Refitting
4 Reverse instructions 1 to 3.

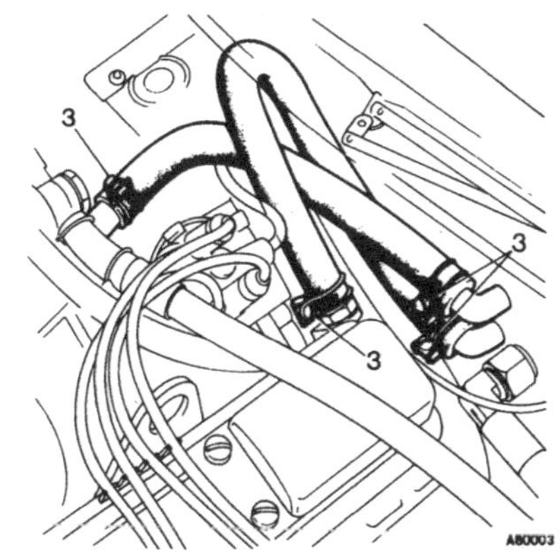

AIR CONDITIONING OPERATIONS

Introduction

The air conditioning system is designed to provide a two level output. The upper level provides cool air at face level for increased comfort in hot climates. The lower level provides either cool or warm air at foot level and an air supply available to the screen.

The principal component of the system is the air conditioner unit. It governs all airflows and contains one blower motor and two matrixes. A cold matrix is cooled by the cold refrigerant circuit and a hot matrix heated by the hot water circuit.

All incoming air is accelerated by the blower motor running at one of three speeds. When the blower motor is selected OFF a flap prevents outside air entering the system. All incoming air is first cooled, dehumidified and cleaned by passing through the cold matrix. The air is then passed to the distribution and hot matrix area.

Cold air is delivered from the central fascia vent and the two end fascia vents at a temperature controlled by the cold temperature control system.

Cold air or hot air is delivered from the footwell outlets and screen outlets at a temperature controlled by a combination of both the cold temperature control system and the hot temperature control system.

Air extraction is from two air vent grilles in the vehicle body rear quarter panels.

The refrigerant circuit consists of a compressor at the front of the engine driven from a drive belt and electromagnetic clutch, a condenser with two fans in the nose of the vehicle, a receiver drier cylinder at the front of the engine bay and the cold matrix in the air conditioner unit. Hoses join the components.

The system is filled with refrigerant which must be subject to special precautions. It exists in the circuit both as a liquid and a vapour.

Service personnel who are not familiar with air conditioned vehicles must study Servicing 82.30.00. A full understanding of this section must be obtained before breaking into the system. Failure to observe this instruction may result in severe personal injury.

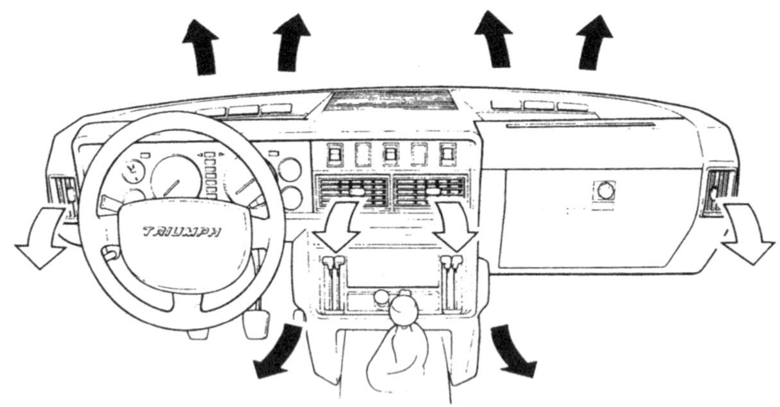

A82 012

Hot water circuit

The function of the hot water circuit is to heat the hot matrix. The hot water flow is induced by the engine water pump. Hot water is drawn from the water transfer housing at the rear of the cylinder head and passes through the water flow valve to the hot matrix. From the hot matrix outlet the flow is forward below the carburetters to enter the water pump cover. This entry is on the suction side of the water pump.

Cold refrigeration circuit

Introduction

The function of the refrigeration circuit is to cool the cold matrix. The circuit comprises the following main components:

Compressor
Condenser
Receiver drier
Expansion valve and cold matrix

Hoses are employed to transport the refrigerant between components.

Compressor

The compressor draws vaporized refrigerant from the cold matrix. It is compressed, and thus heated, and passed on to the condenser as a hot, high pressure vapour.

Condenser

The condenser is mounted at the front of the car. Its function is to remove heat from the refrigerant and disperse it into the atmosphere. It is delivered with hot, high pressure vapour. Air flow across the tubes, induced by vehicle movement and assisted by two electric fans, cools the vapour, causing it to condense into a high pressure liquid. As this change of state occurs a large amount of latent heat is released.

Receiver drier

This unit filters, removes moisture, and acts as a reservoir for the liquid. To prevent icing inside the system, extreme precautions are taken during servicing to exclude moisture. The receiver drier should be considered as a second stage insurance to prevent the serious consequences of ice obstructing the flow. A sight glass provided in the unit top enables a visual check to be made of the high pressure liquid flow.

Expansion valve and cold matrix

High pressure liquid refrigerant is delivered to the expansion valve. A severe pressure drop occurs across the valve and as the refrigerant enters the cold matrix space at a temperature of approximately -6° C it boils and vaporizes. As this change of state occurs, a large amount of latent heat is absorbed. The cold matrix is therefore cooled and as a result heat is extracted from the air flowing across the matrix.

Second cycle

Vaporized refrigerant is then drawn from the the cold matrix by the compressor and a second cycle commences.

1 Compressor
2 Condenser
3 Receiver drier
4 High pressure cut out
5 Air conditioner unit

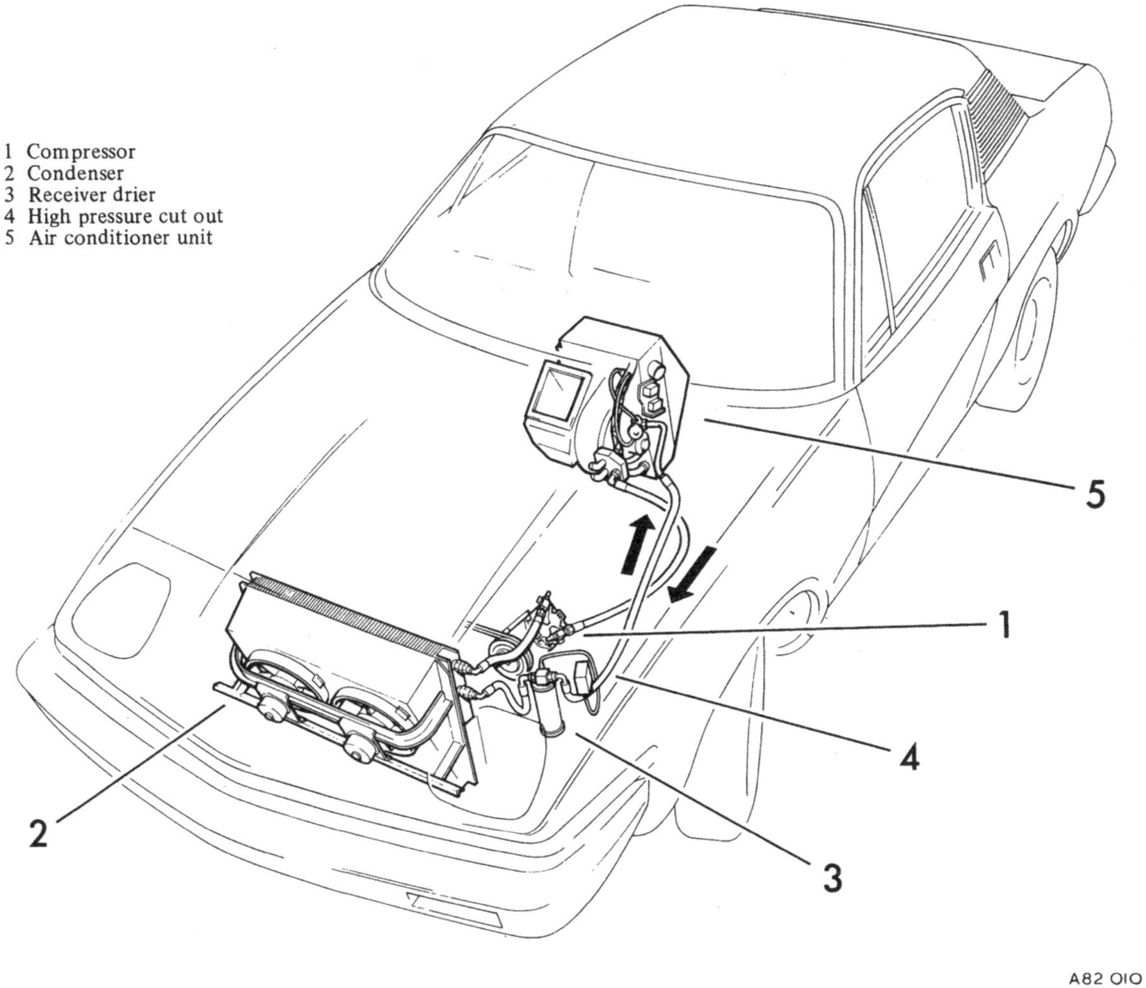

COMPRESSOR 82.10.00

Data

Compressor	
Manufacturer	York Division of Borg Warner, York, U.S.A.
Model	210
Triumph part no.	RKC 0456
Cylinders	2 in line
Capacity	10.3 cu in/rev
Bore	1.875 in
Stroke	1.866 in
Crankcase and liners	Die-cast light alloy with case in cast iron liners
Head	Die-cast light alloy
Valves and valve plate	Two-way non-return valve unit above each cylinder. Assembled onto a common steel valve plate
Front plate, rear plate and base plate	Die-cast light alloy
Main bearings	Two ball bearings
Crankshaft nose tapping	$\frac{5}{16}$ –24 UNF 1.250 deep
Lubrication	Refrigerant compressor oil. Distributed by splash and pressure differential
Lubrication plugs	Two located on crankcase sides.
Mounting tappings	$\frac{3}{8}$–16 UNC 0.620 in deep
Maximum permissible speed	6,000 rev/min
Weight	14.6 lb
Clutch	
Manufacturer	Warner Electric
Triumph part no.	TKC 2019
Current	4.5 amp.
Pulley nominal diameter	6.0 in

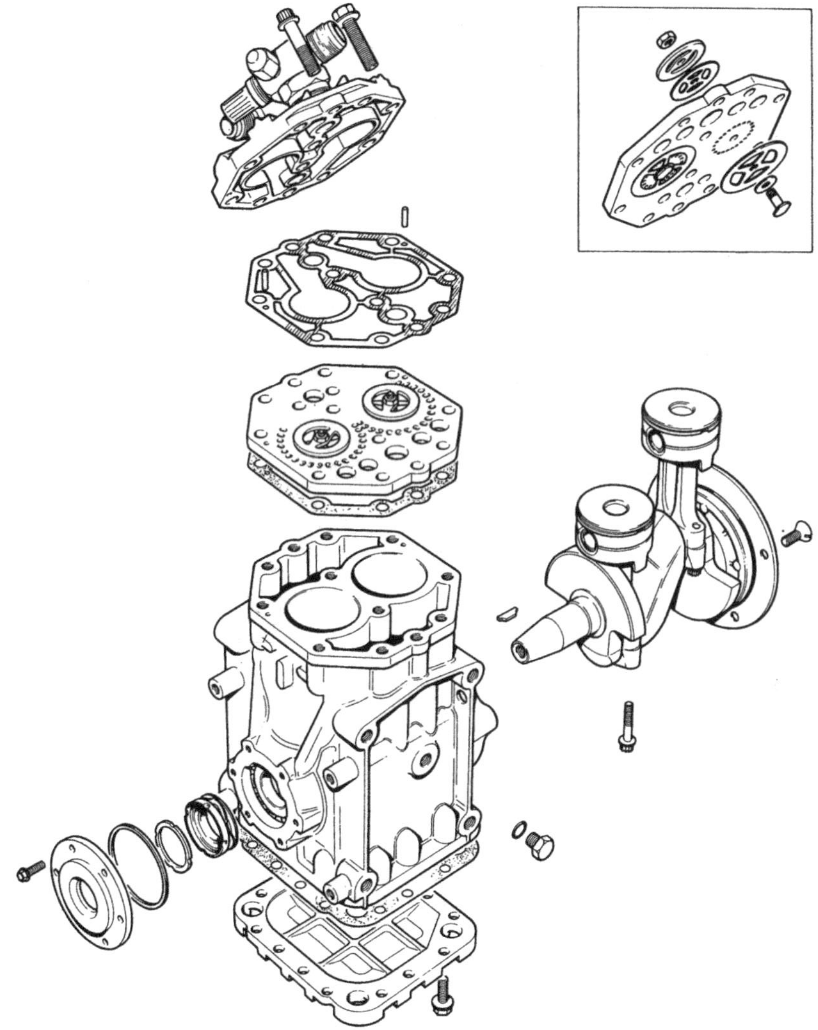

COMPRESSOR

Drive belt – adjust **82.10.01**

1 Drive the vehicle onto a ramp.
2 Raise the ramp.
3 At the transverse support bracket slacken two bolt assemblies.
4 At the longitudinal support bracket slacken two bolt assemblies.
5 Lower the ramp.
6 At the exhaust bracket slacken one bolt assembly.
7 Slacken the locknut.
8 Slacken three main mounting nuts.
9 Rotate the adjustment nut – clockwise to tighten the belt or anti-clockwise to slacken the belt.
10 Tighten three main mounting nuts.
11 Check the belt tension. This should be 0.75 to 1.00 in (19 to 25 mm) at the mid-point of the belt run.
12 If a correction is required repeat operations 8 to 11.
13 Tighten the locknut.
14 At the exhaust bracket tighten one bolt assembly.
15 Raise the ramp.
16 At the longitudinal support bracket tighten two bolt assemblies.
17 At the transverse support bracket tighten two bolt assemblies.

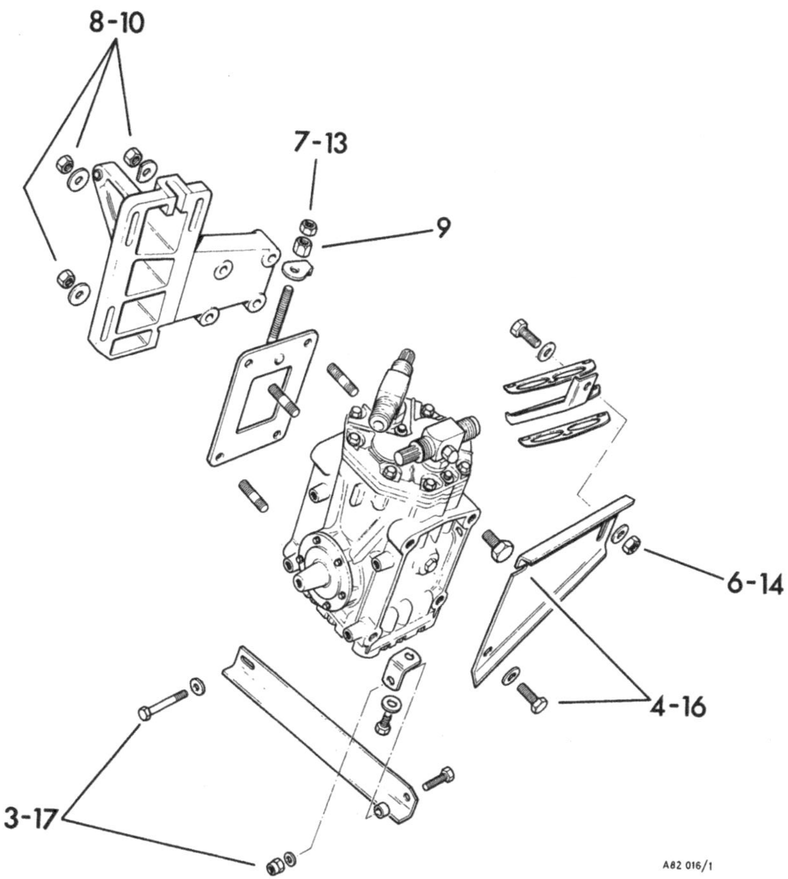

COMPRESSOR

Drive belt – remove and refit **82.10.02**

Remove
1 Perform 82.10.01, operations 1 to 8.
2 Rotate the adjustment nut anti-clockwise to lower the compressor.
3 Slacken the fan blade bolt assemblies to enable the fan blades to be slid forward to provide the required clearance.
4 Remove the belt. Do not stretch the belt over the wide section of the clutch pulley. Run the belt off the crankshaft pulley.

Refit
5 Fit the belt.
6 Tighten the fan blade bolt assemblies.
7 Perform 82.10.01, operations 9 to 17.

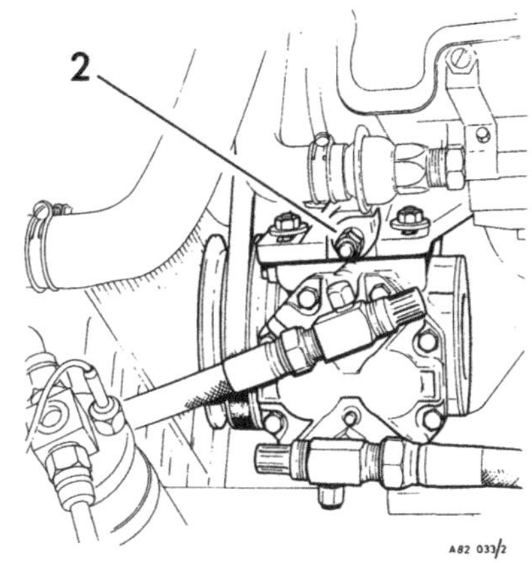

COMPRESSOR

Clutch – remove and refit 82.10.08

Remove

1 Remove the radiator, see 26.40.01.
2 Disconnect the clutch electrical connection.
3 Use a slave wire between the battery positive terminal and the clutch electrical connection to energize the clutch. This action will lock the compressor crankshaft to the pulley.

WARNING: Do not energize the clutch through the vehicle harness as this would require the ignition circuit to be energised with the danger of an engine start during the operation.

4 Retain the pulley and remove the centre bolt.
5 Retain the pulley and screw a 0.625 in UNC slave bolt into the thread provided to withdraw the pulley from the tapered shaft.

CAUTION: Take care to screw the slave bolt in square to prevent a 'cross thread' condition.

6 Remove four bolts and lift away the clutch coil.

Refit

7 Position the clutch coil with the wire emerging upwards. Secure with four bolts.
8 Ensure that the key is fitted to the tapered shaft.
9 Fit the pulley to the tapered shaft.
10 Electrically energise the clutch.
11 Retain the pulley and fit the centre bolt. Torque load to 16 lbf ft (2.2 kgf m).
12 Connect the clutch electrical connection.
13 Adjust the drive belt, see 82.10.01.
14 Fit the radiator, see 26.40.01.

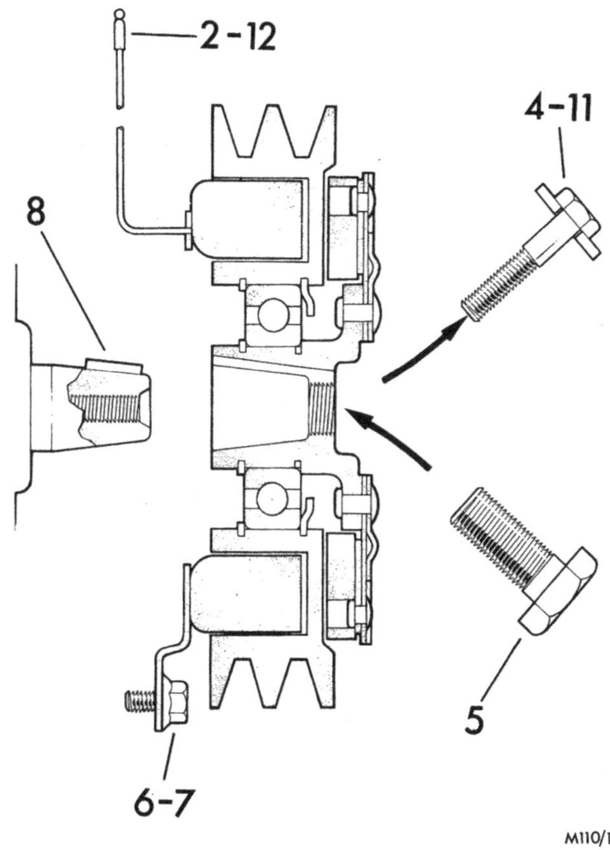

COMPRESSOR

Remove and refit **82.10.20**

Remove

1 Drive the vehicle onto a ramp.
2 Connect gauge set, see 82.30.01.
3 Discharge – compressor only, see 82.30.20.
4 Protect the eyes with safety goggles and wear gloves during operation 5.
5 Unscrew two service valve union nuts. Blank exposed compressor connections immediately.
6 Remove the fresh air duct, see 80.15.31.
7 Remove two bolts, spring washers and washers. Lift away the fan guard.
8 Raise the ramp.
9 At the transverse support bracket slacken the right-hand bolt assembly and remove the left-hand bolt assembly.
10 At the longitudinal support bracket slacken the upper bolt assembly and remove the lower bolt assembly
11 *Early vehicles only:* Remove the engine stabilizer, see 12.45.16.
12 At the front left-hand engine mounting remove one nut and washer.
13 Provide a jack under the engine with a large block of wood to distribute the load under the engine sump.
14 Jack up the engine to the position where the front left-hand engine mounting stud remains just retained in the sub frame bracket.

NOTE: Do not allow the stud to come out of the bracket as insertion may prove difficult.

15 Lower the ramp.
16 At the exhaust bracket remove one bolt assembly.
17 Withdraw the longitudinal support bracket.
18 Slacken the locknut.
19 Slacken three main mounting nuts.
20 Rotate the adjustment nut anti-clockwise to lower the compressor.
21 Remove the belt from the compressor pulley. Do not stretch the belt over the wide section of the clutch pulley. Run the belt rearwards off the crankshaft pulley.
22 Remove three main mounting nuts and washers. Lift the compressor from the vehicle.
23 If necessary detach the following items from the compressor:
a Adjustment plate assembly.
b One bolt.
c One bolt, washer and right angle bracket.
d Three studs.
e Clutch, see 82.10.08.

continued

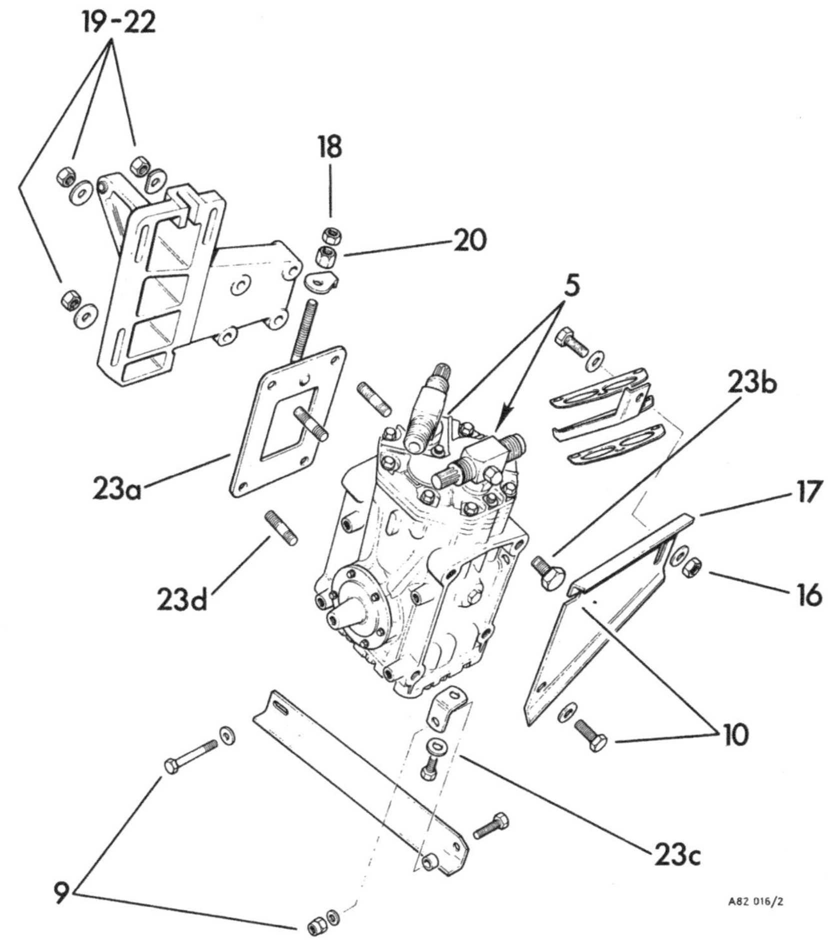

Refit

24 If necessary attach the items listed at operation 23 above to the compressor.

NOTE: Item d, the shorter stud threads screw into the compressor tappings. The lower rear tapping is not fitted with a stud.

25 Position the compressor to the vehicle. Secure with three washers and nuts finger tight.
26 Fit the belt.
27 Adjust the belt, see 82.10.01, operations 9 to 12.
28 Tighten the locknut.
29 Position the longitudinal support bracket.
30 At the exhaust bracket fit one bolt assembly finger tight.
31 Raise the ramp.
32 Lower the jack.
33 At the front left-hand engine mounting fit one washer and nut.
34 *Early vehicles only:*
Fit the engine stabilizer, see 12.45.16.
35 At the longitudinal support bracket fit and tighten the lower bolt assembly and tighten the upper bolt assembly.
36 At the transverse support bracket fit and tighten the left-hand bolt assembly and tighten the right-hand bolt assembly. Ensure that the right angle bracket is correctly aligned.
37 Lower the ramp.
38 At the exhaust bracket tighten one bolt assembly.
39 Position the fan guard. Secure with two bolts, spring washers and washers.
40 Fit the fresh air duct, see 80.15.31.
41 Connect two service valve union nuts. Use refrigerant compressor oil on all mating surfaces to assist leakage prevention. Position the service valves for the best hose alignment.
42 Evacuate – compressor only, see 82.30.21.
43 Charge – compressor only, see 82.30.23.
44 Perform a leak test on any disturbed joints, see 82.30.09.
45 Functional check, see 82.30.16.
46. Disconnect gauge set, see 82.30.01.

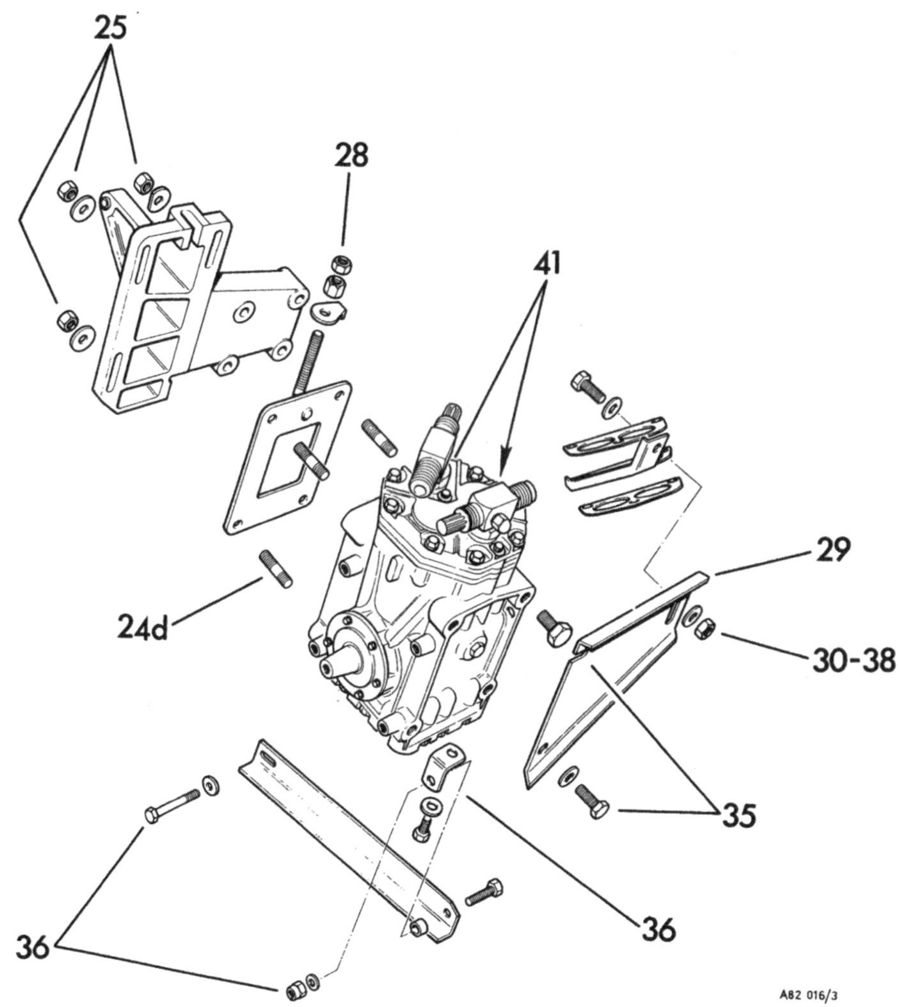

CONDENSER FAN MOTOR

Remove and refit 82.15.01

Remove

1 Drive the vehicle onto a ramp.
2 Raise the ramp.
3 Using a small screwdriver slacken the grub screw to release the fan.
4 Remove the fan from the motor shaft.
5 Disconnect the fan motor harness plug.
6 Remove three nuts, anti-vibration washers, washers, washers and bolts.
7 Withdraw the fan motor from the mounting assembly. Take care not to damage the wires. To allow the grommet to pass through the aperture tilt the fan motor. If the grommet remains obstructive pull it from the aperture.
8 Remove the fan motor from the vehicle. It may be necessary to slightly bend the vehicle body front cross-member to achieve this operation.

Refit

9 Ensure that the fan is positioned in the cowling.
10 Reverse 1 to 8. Insert the fan motor to the mounting assembly with the wires emerging downwards for improved water sealing.
Fit the fan to the motor shaft with the grub screw aligned to a motor shaft recess.

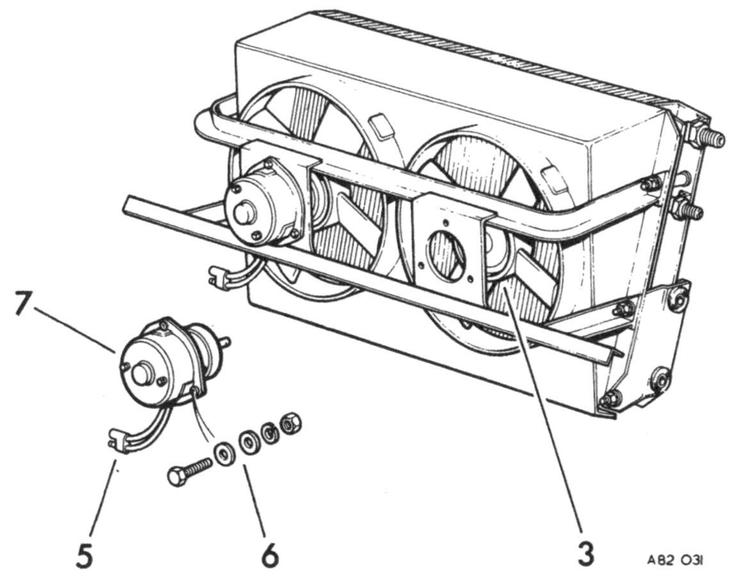

A82 031

CONDENSER FAN MOTOR

Overhaul 82.15.02

[**CAUTION:** In refitting the electric cooling fan, be sure to install the motor so that the wires are pointed down. Otherwise, water may run down the wires, enter the housing, and ruin the motor.]

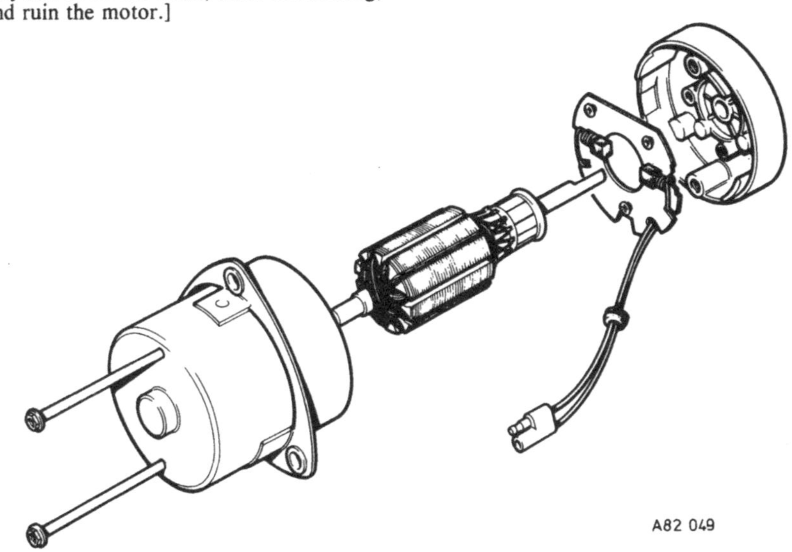

A82 049

CONDENSER

Remove and refit **82.15.07**

Remove

1 Remove the condenser and fan assembly, see 82.15.10.
2 Remove two top nuts, anti-vibration washers, washers, washers and shorter bolts.
3 Remove two middle nuts, anti-vibration washers, washers, washers and longer bolts.
4 Remove the fan motor mounting assembly.
5 Remove one nut, anti-vibration washer, washer, washer and bolt. Remove the left-hand bracket which includes two mounting rubbers.
6 Remove one nut, anti-vibration washer, washer, washer and bolt. Remove the right-hand bracket which has no mounting rubbers.

Refit

7 Reverse 1 to 6.

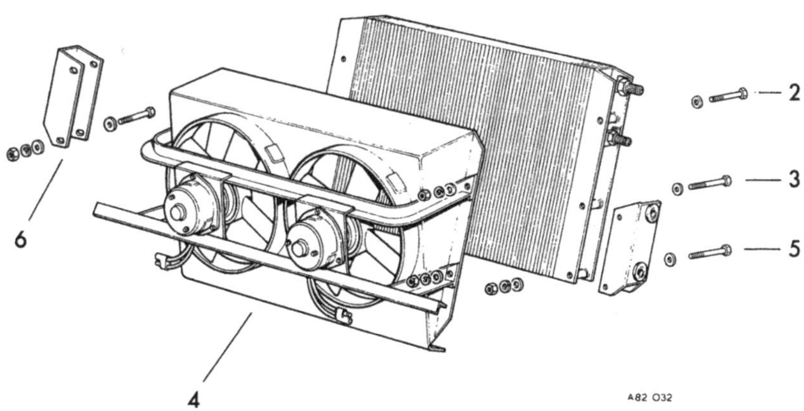

CONDENSER AND FAN ASSEMBLY

Remove and refit **82.15.10**

Remove

1 Drive the vehicle onto a ramp.
2 Select the master light switch to raise the headlamps.
3 Isolate the battery to extinguish the headlamps.
4 Connect gauge set, see 82.30.01.
5 Discharge – complete system, see 82.30.05.
6 Remove the radiator, see 26.40.01.
7 Protect the eyes with safety goggles and wear gloves during operation 8.
8 Carefully disconnect two hose connections. Use two spanners at each joint to protect the delicate condenser pipe joints. Blank exposed connections immediately.
9 Raise the ramp.
10 Remove four bolts, lock washers and washers. Remove the radiator mounting cross-member.
11 Disconnect two fan motor harness plugs.
12 Slacken two right-hand bracket bolt assemblies.

NOTE: The right-hand bracket is provided with slots as shown to facilitate withdrawal and positioning of the assembly and to ensure that the brackets are drawn up firmly against the body longitudinal members. Note that the left-hand bracket has no slots.

13 Remove two nuts, lock washers and washers securing the fan motor mounting assembly rail.
14 Remove two left-hand longer side bolts, anti-vibration washers and washers.
15 Collect up the nut plate.
16 Remove two right-hand shorter side bolts, anti-vibration washers and washers.
17 Carefully withdraw the condenser and fan assembly downwards from the vehicle. Handle with care as the fins are easily damaged.

Refit

18 Ensure that the two right-hand bracket bolt assemblies are slackened.
19 Reverse 9 to 17.
20 Connect two hose connections. Use refrigerant compressor oil on all mating surfaces to assist leakage prevention.
21 Fit the radiator, see 26.40.01.
22 Evacuate – complete system, see 82.30.06.
23 Connect the battery.
24 Charge – complete system, see 82.30.08.
25 Perform a leak test on any disturbed joints, see 82.30.09.
26 Select the master light switch to lower the headlamps.
27 Functional check, see 82.30.16.
28 Disconnect gauge set, see 82.30.01.

Illustration on following page

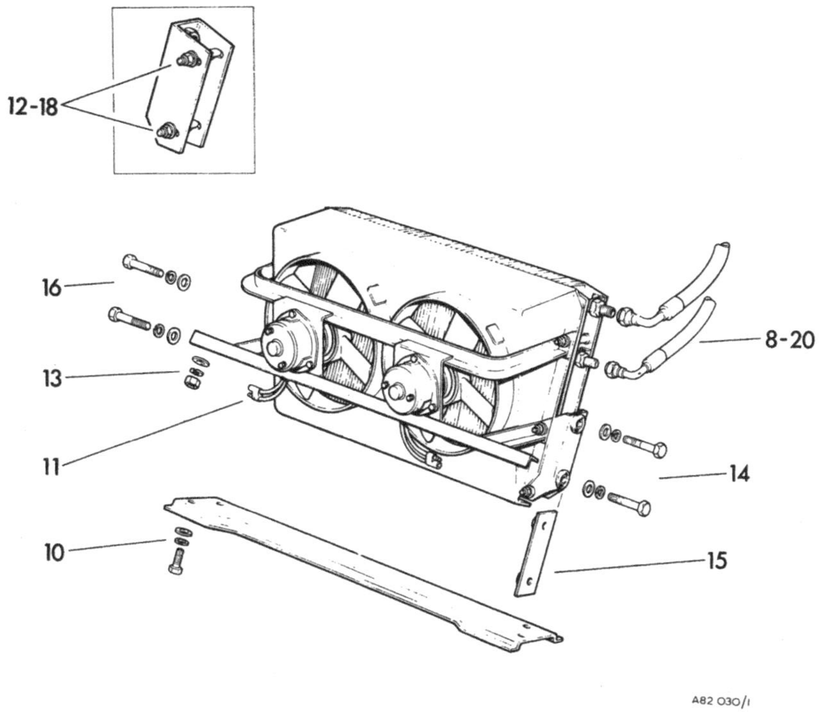

RECEIVER DRIER

Remove and refit 82.17.01

CAUTION: Immediate blanking of the receiver drier is important. Exposed life of the unit is only 15 minutes.

Remove

1 Connect gauge set, see 82.30.01.
2 Discharge – complete system, see 82.30.05.
3 Protect the eyes with safety goggles and wear gloves during operations 4 and 5.
4 Carefully disconnect the capillary tube from the receiver drier. Blank exposed connections immediately.
5 Carefully disconnect two hose connections. Use a second spanner to support the squared hose adaptor. Blank exposed connections immediately.
6 Remove the clamp screw, washer, washer and nut.
7 Withdraw the receiver drier from the mounting bracket.

Refit

8 Insert the receiver drier into the mounting bracket with the inlet and outlet connections correct to the refrigerant circuit flow as shown.
9 Connect two hose connections finger tight. Use refrigerant compressor oil on all mating surfaces to assist leakage prevention.
10 Fit the clamp screw, washer, washer and nut.
11 Tighten two hose connections. Use a second spanner to support the squared hose adaptor.
12 Carefully connect the capillary tube to the receiver drier. Use refrigerant compressor oil on all mating surfaces to assist leakage prevention.
13 Evacuate – complete system, see 82.30.06.
14. Charge – complete system, see 82.30.08.
15 Perform a leak test on any disturbed joints, see 82.30.09.
16 Functional check, see 82.30.16.
17 Disconnect gauge set, see 82.30.01.

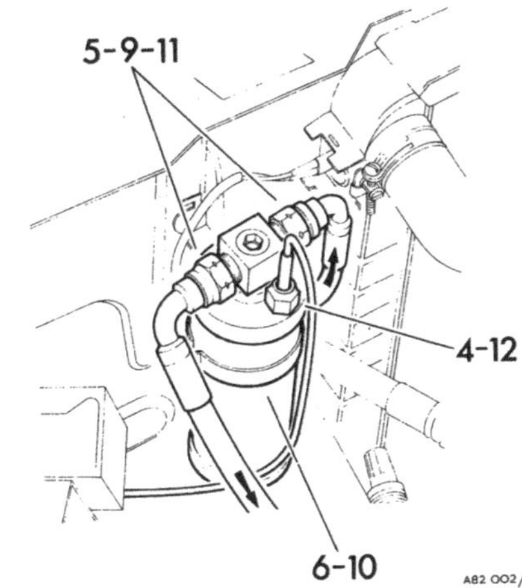

CONTROLS

Description **82.20.00**

Control levers

Lever A – controls the flow of cold air delivered from the central fascia vent. Moving lever A up to the OFF position will terminate the airflow. Moving lever A down to the MAX position will provide the maximum airflow. Any intermediate position may be used. This control operates through the direct linkage A to position flap A.

Lever B – directs the temperature required. Moving lever B up to the COLD position will provide maximum cooling. Moving lever B down to the HOT position will provide maximum heating. Any intermediate position may be used. This control operates through the non-direct linkage B to provide an input to both the cold thermostat and the hot thermostat. Lever B therefore influences the airflow from both the cold outlets and the hot outlets. After setting the position of lever B the cold temperature control system and/or the hot temperature control system will maintain an approximately constant interior temperature.

Lever C – is the master control to bring into action the cold refrigerant circuit and the blower motor. It also selects the input air – either fresh air at ambient temperature through the under bonnet fresh air duct or recirculated air from the vehicle interior.

Moving lever C up to the OFF position will electrically switch off the compressor electro-magnetic clutch, throttle jack (automatic vehicles only) and the two condenser fan motors. Flap C will be in the recirculate position so vehicle movement will not induce an airflow through the system.

Moving lever C slightly down to position MAX will electrically switch on the cold refrigerant circuit and the blower motor at its fastest speed. Flap C will remain in the recirculate position. This selection of lever C will provide the maximum cooling or the maximum heating condition.

Moving lever C further down to position 3 will maintain the cold refrigerant circuit electrically on and the blower motor at its fastest speed. Flap C will be traversed by the non-direct 'quick changeover' linkage C to the fresh position. Any intermediate position between MAX and 3 may be used.

Moving lever C slightly further down to position 2 will maintain the cold refrigerant circuit electrically on and the flap C at the fresh position. The blower motor will drop to its middle speed.

Moving lever C slightly further down to position 1 will maintain the cold refrigerant circuit electrically on and flap C at the fresh position. The blower motor will drop to its slowest speed.

Lever D – controls the flow of cold air or hot air delivered from the screen vents and footwell. Moving lever D up to the SCREEN position will provide the maximum airflow to the screen vents. Moving lever D down to the CAR position will provide the maximum airflow to the footwell. Any intermediate position may be used. This control operates through the direct linkage D to position flap D.

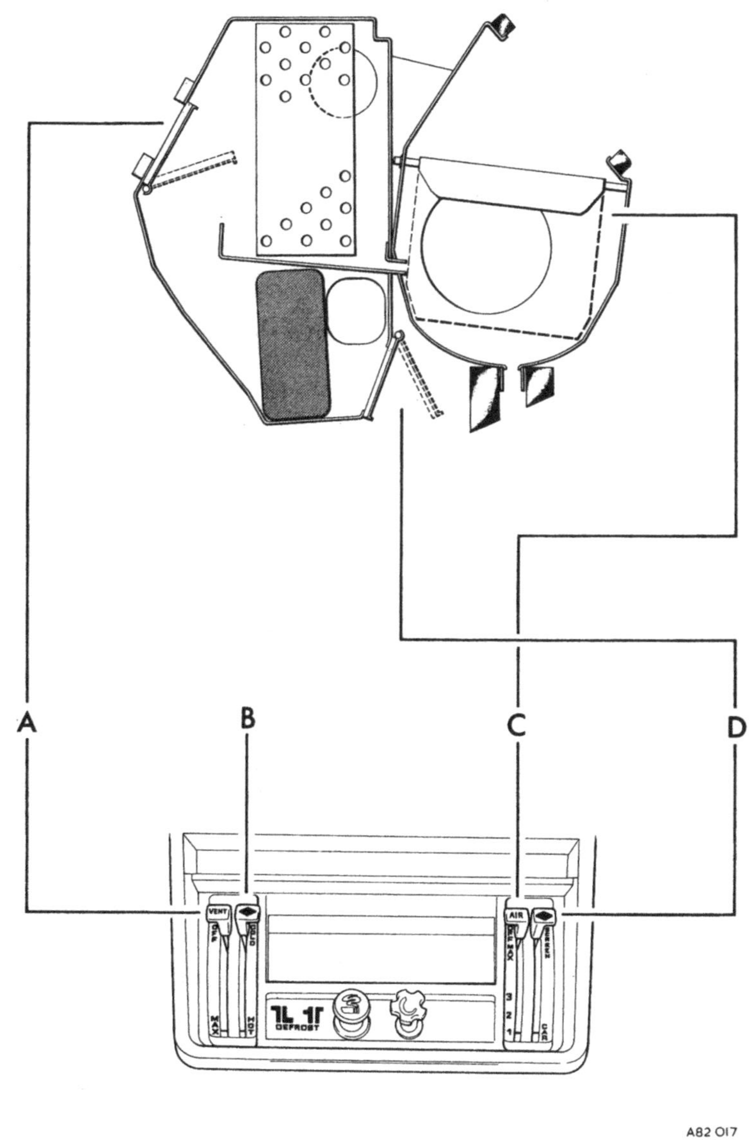

Cold temperature control system

The principal unit of the system is the cold thermostat mounted on the left-hand side of the air conditioner unit. The thermostat receives two inputs. A capillary tube inserted into the air space of the cold matrix senses the cold matrix temperature. The driver's direction of the temperature required is transmitted by lever B and linkage B to position the thermostat lever. The output of the thermostat is an electric switch in the air conditioning electrical circuit. This controls indirectly the on-off switching of the compressor electro-magnetic clutch, throttle jack (automatic vehicles only) and the two condenser fan motors.

Hot temperature control system

The principal unit of the system is the hot thermostat mounted on the left-hand side of the air conditioner unit. The thermostat receives two inputs. A capillary tube mounted against the downstream face of the hot matrix senses the hot matrix temperature. The driver's selection of the temperature required is transmitted by lever B and linkage B to position the thermostat lever. The output of the thermostat is a water valve in the hot water circuit. This controls the flow of hot water from the engine into the hot matrix.

continued

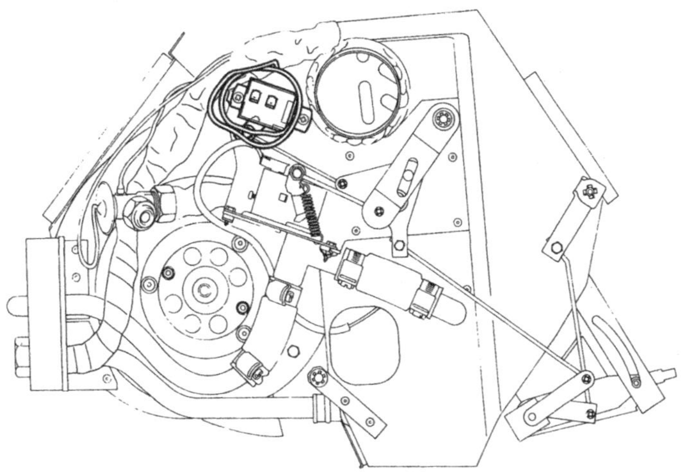

Cold thermostat

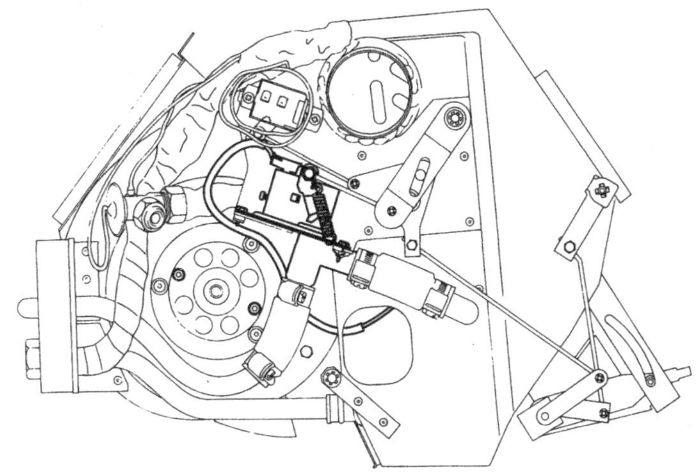

Hot thermostat

Electrical circuit

The function of the circuit is to control the on-off switching of the compressor electro-magnetic clutch, throttle jack (automatic vehicles only) and the two condenser fan motors. The circuit may be considered in two sections. The relay control circuit and the power circuit.

Relay control circuit–supply to the relay winding is from an ignition controlled fuse. The earth path may be interrupted by the cold thermostat and by the driver's selection of lever C. With lever C up to the OFF position the circuit is broken. With lever C selected to any other position, MAX, 3, 2 or 1 an earth path exists across the blower motor switch. The relay control circuit current is small so the introduction of a blower motor speed control resistor in the circuit at positions MAX, 3 and 1 will have no consequence.

Power circuit–actuation of the relay connects supply from the air conditioning fuse to the compressor clutch, throttle jack and the two condenser fan motors. The circuit to the compressor clutch and throttle jack may be interrupted by the delay circuit as detailed below and by the high pressure cut-out which is a safety feature not subject to continuous cycling. The two condenser fan motors are controlled directly by the relay but these units may also be selected independently by the radiator switch as detailed below.

Delay circuit–this circuit is included to prevent a severe momentary voltage drop in the vehicle electrical system that would occur if the compressor clutch, throttle jack and the two condenser fan motors were allowed to 'cut in' simultaneously.

The circuit allows the two condenser fan motors to 'cut in' when the relay energises. An acceptable first voltage drop occurs and the recovery is quick as the alternator output is rebalanced to the additional load.

The delay circuit flasher unit also first receives current when the relay energises. This current heats a metal ribbon causing expansion which finally allows a metal vane to relax and the contacts to close.

The delay circuit relay is energised to hold the condition and the compressor clutch and throttle jack 'cut in'. A second acceptable voltage drop occurs and again recovery is quick as the alternator output is rebalanced to the additional load.

This action is repeated each time the air conditioning circuit is manually selected by lever C, or automatically switched on-off by the cold temperature control system.

High pressure cut out

The unit is a safety feature to protect the refrigeration circuit from excessive pressure. It is not subject to continuous cycling during normal system operation.

A capillary tube senses the refrigerant pressure at the receiver drier. This pressure governs an electrical switch within the unit.

The switch is included in the electrical circuit. If the refrigerant pressure rises to the 'cut out' pressure the electrical circuit to the compressor clutch and throttle jack (automatic vehicles only) is interrupted. The compressor electro magnetic clutch reverts to the off condition and the refrigerant pressure is prevented from rising further.

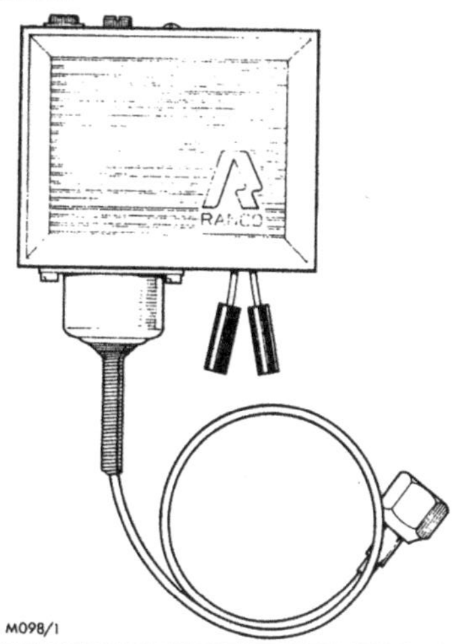

M098/1

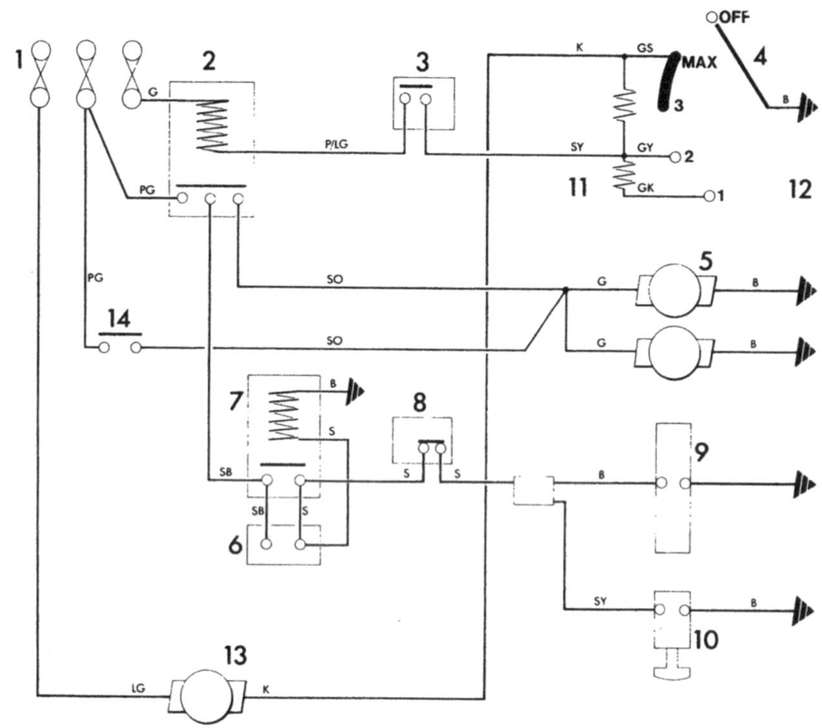

Earlier Models

A82 028

1 Fuses

Relay control circuit

2 Relay
3 Cold thermostat
4 Lever 'C'

Power circuit

5 Condenser fan motors
6 Delay circuit flasher unit
7 Delay circuit relay
8 High pressure cut out
9 Compressor clutch
10 Throttle jack (automatic vehicles only)

Other components

11 Blower motor resistors
12 Blower motor switch
13 Blower motor
14 Radiator switch

Colour code

B Black
G Green
K Pink
LG Light green
O Orange
P Purple
S Slate
Y Yellow

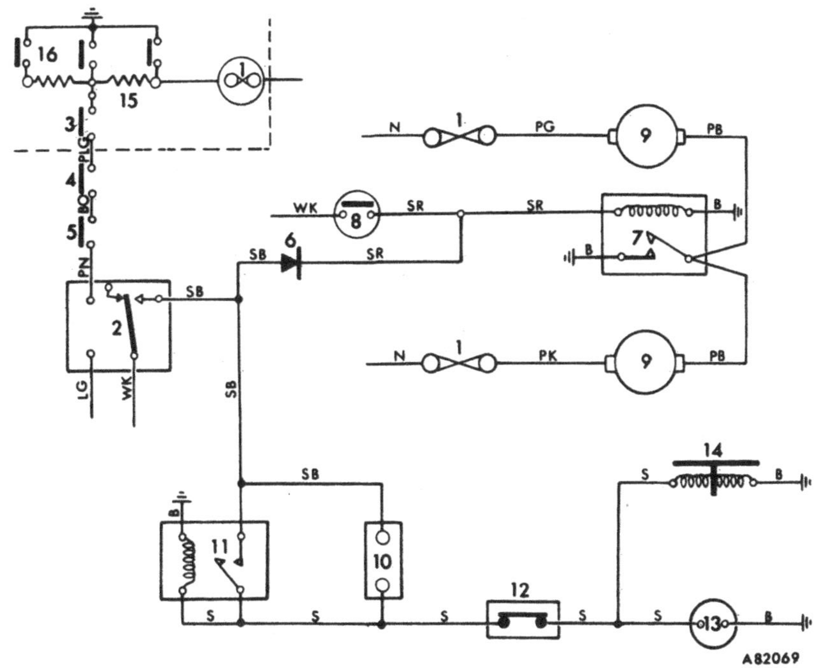

Later Models

1 Fuses

Relay control circuit
2 Relay
3 Cold thermostat
4 Full throttle cut-off switch
5 Air conditioning cut-out switch
6 Diode
7 Fan relay
8 Radiator thermostat

Power circuit
9 Condenser fan motors
10 Delay circuit flasher unit
11 Delay circuit relay
12 High pressure cut-out
13 Compressor clutch
14 Throttle jack (automatic vehicles only)

Other components
15 Blower motor resistors
16 Blower motor switch

Colour code
B Black
G Green
K Pink
LG Light Green
N Brown
O Orange
P Purple
R Red
S Slate
W White
Y Yellow

Throttle jack

The throttle jack is fitted to vehicles equipped with both air conditioning and automatic transmission.

This unit consists of a solenoid coil and plunger. It is included to maintain an approximately constant engine idle speed both with the compressor clutch free and with the compressor clutch engaged to apply the compressor drive load to the engine.

With the air conditioning system switched out the plunger is retracted to an unemployed position under the action of a spring.

With the air conditioning system switched in the solenoid coil is energised to extend the plunger to provide a new stop position for the carburetter throttle linkage.

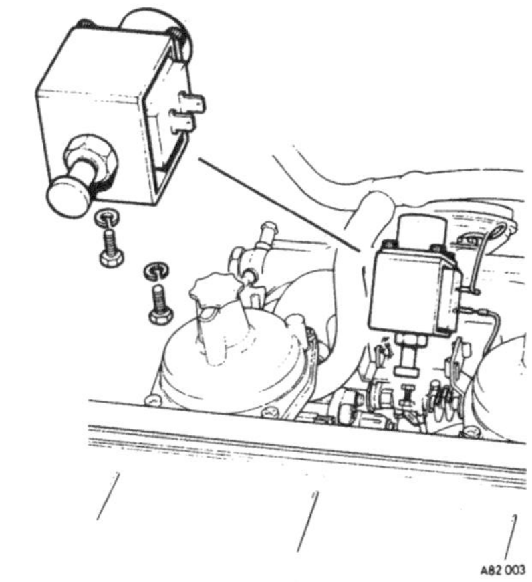

Radiator switch

This item is not strictly a component of the air conditioning system. Its function is to actuate if the engine coolant temperature should reach a high level. Advantage of the two condenser fan motors is then taken to assist in cooling the radiator.

NOTE: As the temperature of the coolant in the upper section of the radiator may reach a high level immediately after the vehicle is parked, and the electrical supply to the radiator switch is direct from a battery control fuse, it is possible for the two fans to start running while the vehicle is parked and unattended. The run should only continue for a short period until the radiator has cooled. Under certain vehicle operating conditions this action may be considered as normal.

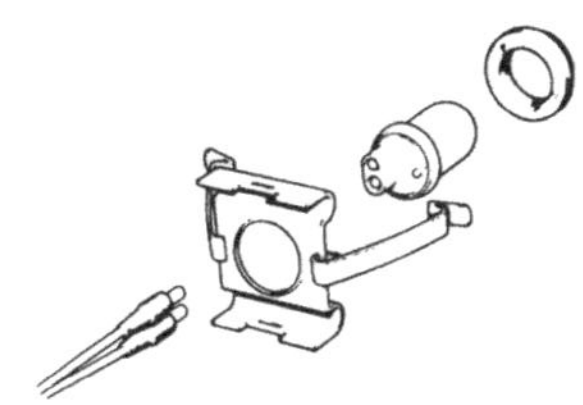

CONTROL SUMMARY

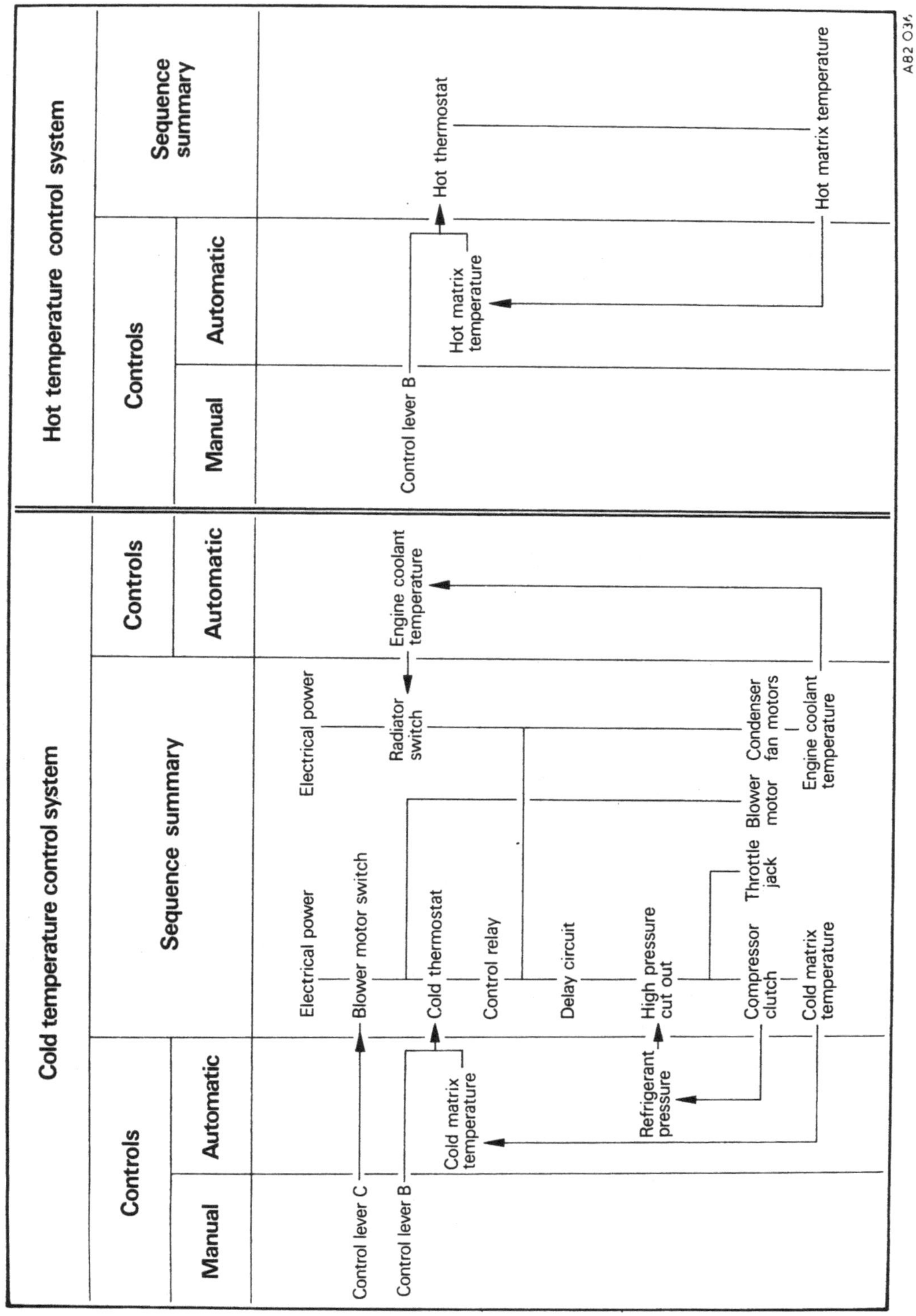

CONTROLS

Radiator switch – remove and refit **82.20.09**

Remove

1 Isolate the battery.
2 Pull off two electrical connections.
3 Release the rear spring clip from the radiator flange. Swing the bracket forward. Release the front spring clip from the radiator flange.
4 Using a wide blade screwdriver, carefully prise the switch outwards from the rubber seal.
5 Remove the rubber seal.

Refit

6 Fit the rubber seal. Fit a new rubber seal if available.
7 Insert the switch. Position the electrical connection pedestal forward as shown to facilitate operation 8 below.
8 Engage the front spring clip to the radiator flange. Swing the bracket rearward to align correctly to the switch. Engage the rear spring clip to the radiator flange.
9 Push on two electrical connections. The connections may be fitted either way round.
10 Connect the battery.

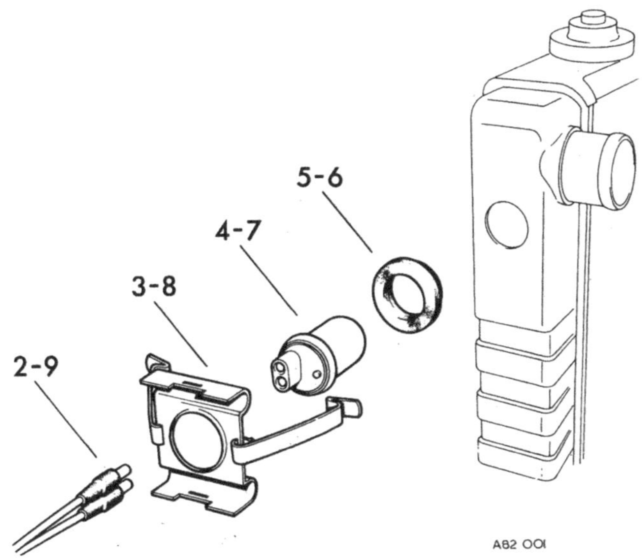

CONTROLS

Cold thermostat – remove and refit 82.20.18

Remove

1 Isolate the battery.
2 Remove the fascia, see 76.46.01.
3 Disconnect two Lucar connectors.
4 Hold the cold thermostat trunnion hexagon and slacken the screw. Detach the control rod.
5 Free the capillary tube from the sealer below the refrigerant outlet pipe.
6 Centre punch and drill out two rivets. Use a ⅛ in drill.
7 Carefully withdraw the capillary tube from the air space of the cold matrix.
8 Collect up the split rubber grommet.
9 Remove the cold thermostat from the vehicle.

Refit

10 Bend the capillary tube of the new cold thermostat to the correct shape. Use the old unit as a guide and bend the end as shown.
11 Fit two spire nuts FJ 2544 to the mounting bracket.
12 Thread the capillary tube between the refrigerant outlet pipe and the case. Insert the capillary tube end into the air space of the cold matrix to achieve the position shown.
13 Fit the split rubber grommet AAP 0165.
14 Coil the surplus length of the capillary tube as shown. Ensure that the coil does not extend outboard of the cold thermostat. Position the cold thermostat against the mounting bracket. Secure with two screws YZ 3404.
15 If necessary restore the sealer below the refrigerant outlet pipe.
16 Attach the control rod. Adjust linkage B, see 82.25.08.
17 Connect two Lucar connectors. The connectors may be fitted either way round.
18 Fit the fascia, see 76.46.01.
19 Connect the battery.
20 Functional check, see 82.30.16.
21 If the operation of the cold thermostat is suspect an adjustment may be made by amending the position of the capillary tube in the air space of the cold matrix.
Access to achieve this operation may be obtained by removing the fascia centre grille, see 76.55.14.

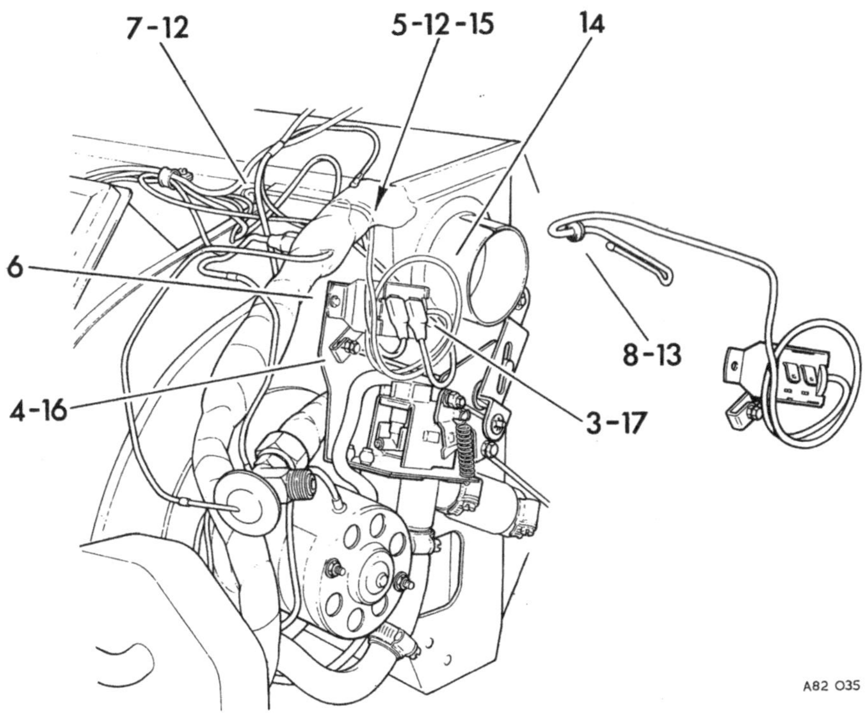

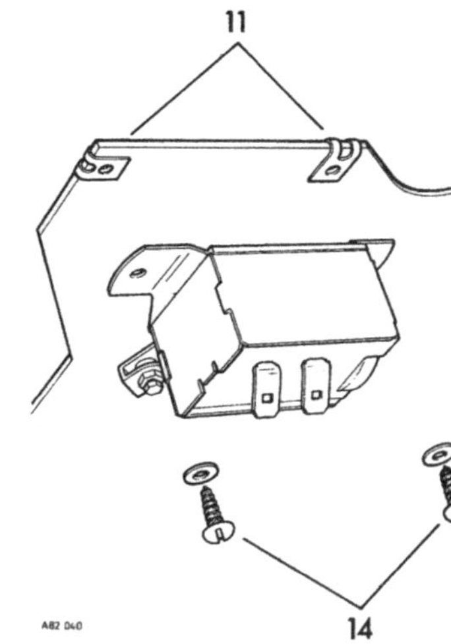

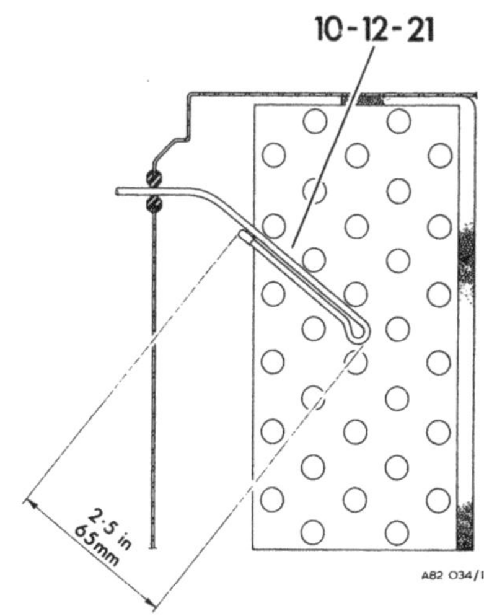

CONTROLS

High pressure cut out – remove and refit **82.20.20**

CAUTION: Immediate blanking of the receiver drier is important. Exposed life of the unit is only 15 minutes.

Remove

1 Connect gauge set, see 82.30.01.
2 Discharge – complete system, see 82.30.05.
3 Protect the eyes with safety goggles and wear gloves during operation 4.
4 Carefully disconnect the capillary tube from the receiver drier. Blank exposed connections immediately.
5 Disconnect two electrical snap connectors.
6 Remove two bolts and spring washers. Lift the high pressure cut out from the vehicle.

Refit

7 Position the high pressure cut out to the bracket. Secure with two bolts and spring washers.
8 Connect two electrical snap connectors as follows:
Slate/black wire to slate/black wire. Green/orange wire to four way snap connector. Other wires to this snap connector are slate/yellow and black.
9 Carefully connect the capillary tube to the receiver drier. Use refrigerant compressor oil on all mating surfaces to assist leakage prevention.
10 Evacuate – complete system, see 82.30.06.
11 Charge – complete system, see 82.30.08.
12 Perform a leak test on any disturbed joints, see 82.30.09.
13 Functional check, see 82.30.16.
14 Disconnect gauge set, see 82.30.01.

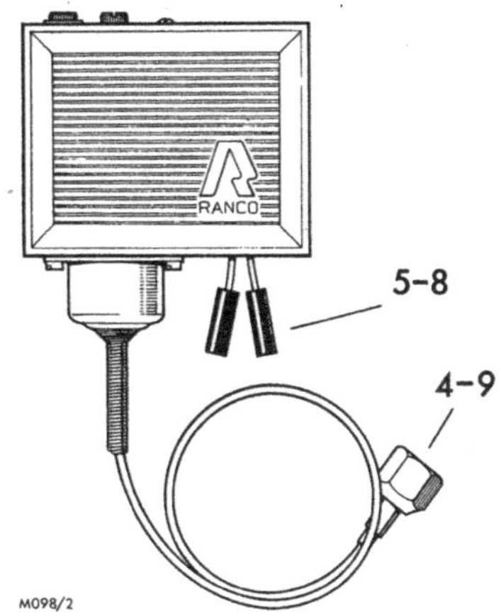

M098/2

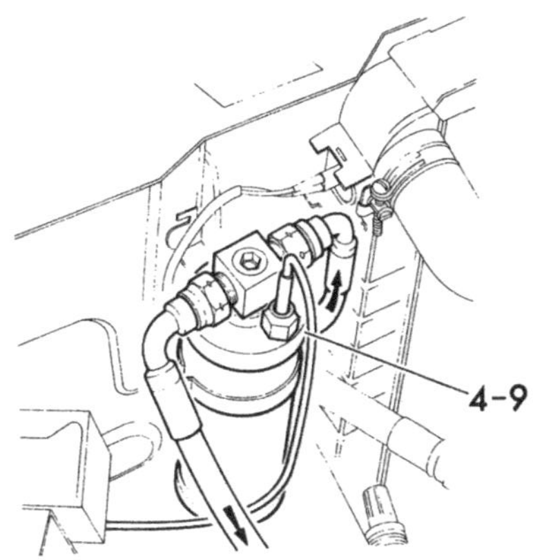

A82 011/1

CONTROLS

Hot thermostat – remove and refit 82.20.31

Remove

1 Isolate the battery.
2 Remove the fascia, see 76.46.01.
3 Drain the coolant, see 26.10.01.
4 Note the run of the capillary tube.
5 Cut the metal case between the screen outlet and grommet aperture. This will allow the capillary tube to be withdrawn sideways.
6 Hold the hot thermostat trunnion hexagon and slacken the screw. Detach the control rod.
7 Centre punch and drill out three rivets. Use a short $\frac{3}{16}$ in drill and right angle drive to remove the forward two rivets. Use a $\frac{3}{16}$ in drill from below to remove the rear rivet.
8 Slacken four hose clips.
9 Using a wide bladed screwdriver carefully prise the capillary tube loop from the four plastic clips securing it against the face of the hot matrix.
10 Carefully remove the hot thermostat from the vehicle. Maintain the shape of the capillary tube loop as a guide for the new unit.
11 Collect up the split rubber grommet.

Refit

12 Bend the capillary tube of the new hot thermostat to the correct shape. Use the old unit as a guide.
13 Fit three spire nuts FJ 2544 to the mounting bracket.
14 Insert the capillary tube loop into position against the face of the hot matrix. Secure to the four plastic clips.
15 Manoeuvre the hot thermostat to position the two hose connections and achieve the capillary tube run noted at operation 4 above. Position the hot thermostat on the mounting bracket. Secure with three screws YZ 3404 and washers WP 0005.
16 Tighten four hose clips.
17 Attach the control rod. Adjust linkage B. see 82.25.08.
18 Fit the split rubber grommet.
19 Apply suitable sealer to the cut in the metal case.
20 Refill the coolant, see 26.10.01.
21 Fit the fascia, see 76.46.01.
22 Connect the battery.

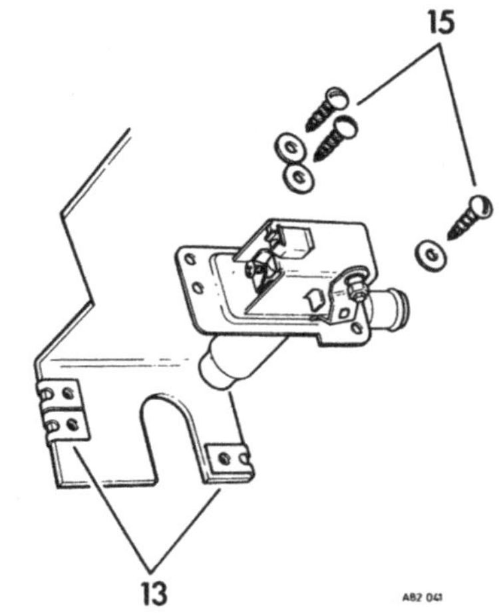

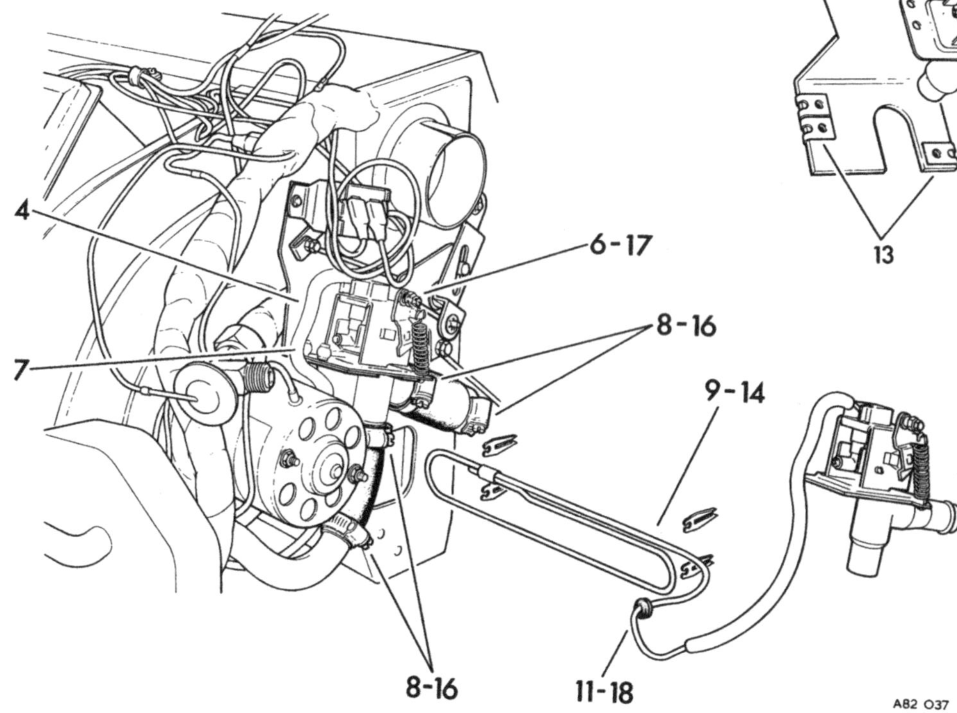

CONTROLS

Throttle jack – remove and refit 82.20.36

The throttle jack is fitted to vehicles equipped with both air conditioning and automatic transmission.

Remove

1 Disconnect two Lucar connectors.
2 Remove two bolts and lock washers.

Refit

3 Position the throttle jack. Secure with two bolts and lock washers.
4 Connect two Lucar connectors. The connectors may be fitted either way round.
5 Adjust the throttle jack, see 82.20.37.

Fuel injection engines

The throttle jack is located on the throttle linkage bracket at the rear of the plenum chamber.

To remove, refit and adjust the throttle jack, follow the procedure in 82.30.36 and 82.30.37.

CONTROLS

Throttle jack – adjust 82.20.37

The throttle jack is fitted to vehicles equipped with both air conditioning and automatic transmission.

1 Connect a tachometer to the engine as instructed by the manufacturer.

NOTE: The vehicle instrument panel tachometer may be used if no other instrument is available.

2 Run the engine.
3 Select lever C to 'OFF'.

NOTE: This action will switch out the air conditioning system. The compressor clutch will disengage removing the compressor drive load from the engine. The throttle jack plunger will be retracted.

4 The idle speed should now be 650 to 850 r.p.m. Nominal 750 r.p.m.
5 If a correction is required refer to Carburetters – tune and adjust, see 19.15.02.
6 Select lever C to any other position, 1, 2 or 3.

NOTE: This action will switch in the air conditioning system. The compressor clutch will engage applying the compressor drive load to the engine. The throttle jack plunger will be extended.

7 The idle speed should be maintained at 650 to 850 r.p.m. Nominal 750 r.p.m.
8 If a correction is required perform the following: Slacken the lock nut. Rotate the screw to adjust the idle speed. Tighten the lock nut.
9 Check that the correct idle speed has been maintained.

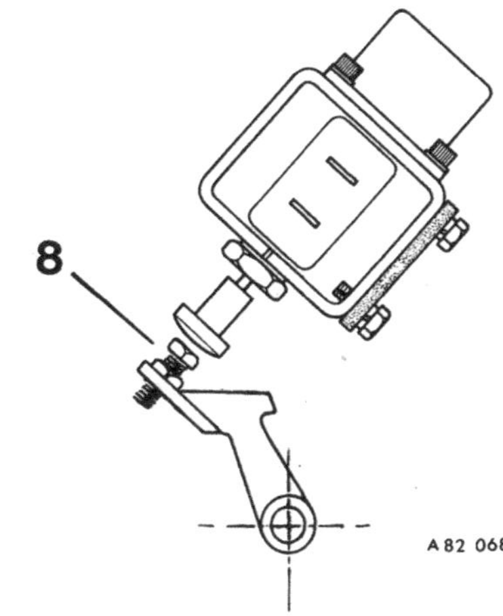

AIR CONDITIONER UNIT

Description **82.25.00**

The air conditioner unit is positioned on the centre line of the vehicle between the bulkhead and the fascia/centre console. The function of the unit is to receive air, process and deliver it to the outlets as directed by the control positions.

Controls – to comprehend the system it should be appreciated that control lever 'A' positions flap 'A' via linkage 'A'. Similarly control lever 'B' operates linkage 'B'. Control lever 'C' positions flap 'C' via linkage 'C'. Finally control lever 'D' positions flap 'D' via linkage 'D'.

Intakes – the system draws fresh air at ambient temperature through the fresh air duct, or recirculated air from the vehicle interior, into the blower motor intake. The choice is directed by control lever and flap C.

Blower motor – one blower motor which may be considered as an integral component of the air conditioner unit transfers air into the cold matrix.

Cold matrix – this unit is cooled by the cold refrigeration circuit. All airflow passes across the cold matrix to be conditioned. Cooling is achieved by heat being absorbed by the cold surfaces. Dehumidifying is achieved by moisture carried in the air condensing on the pipes of the cold matrix. Cleaning is achieved by the dust suspended in the air tending to be retained by the moisture. Water drains into a tray below the cold matrix. From the tray it escapes into the air inlet chamber via a short drain pipe. A second drain pipe runs vertically through the mounting rubber to the underside of the vehicle.

Hot matrix – this unit is heated by the hot water circuit. All airflow which is not permitted to escape cold through the three fascia vents is passed through the hot matrix to be heated.

1 Fresh air inlet
2 Recirculated air inlet
3 Screen outlet
4 Refrigerant inlet
5 Water inlet
6 Refrigerant outlet
7 Water outlet
8 Drain pipe
9 Drain pipe
10 Footwell outlet
11 Linkage D
12 Linkage C
13 Control levers
14 Central facia vent outlet
15 End facia vent outlet

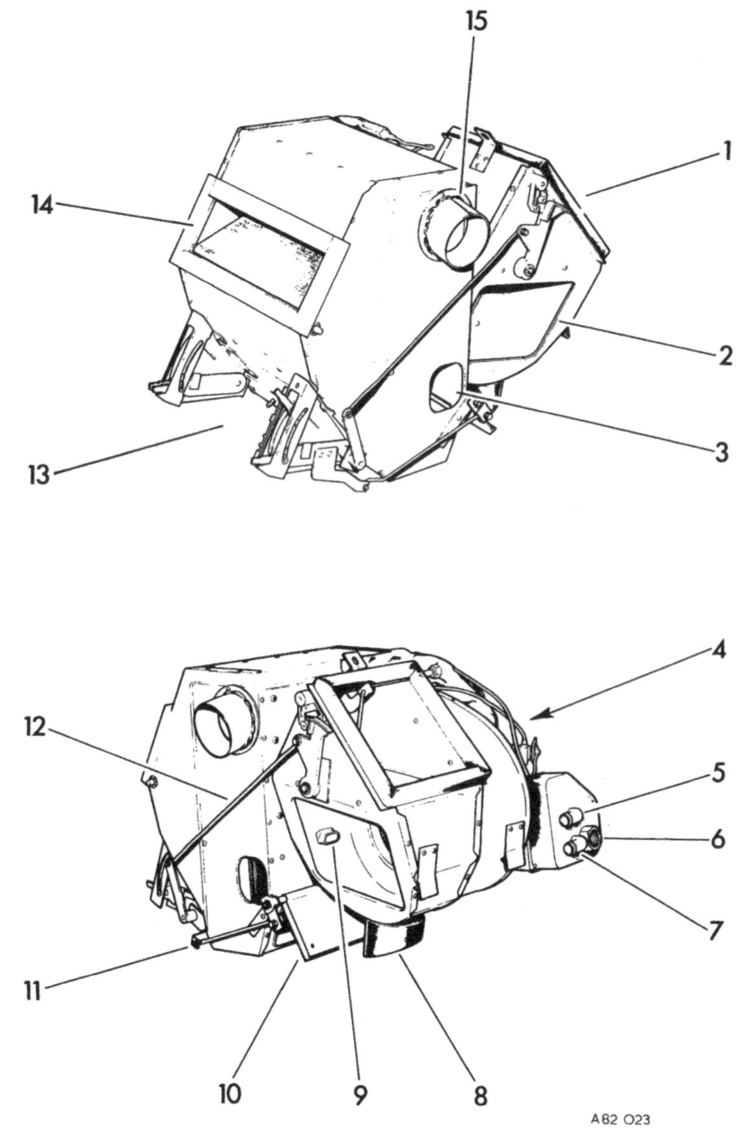

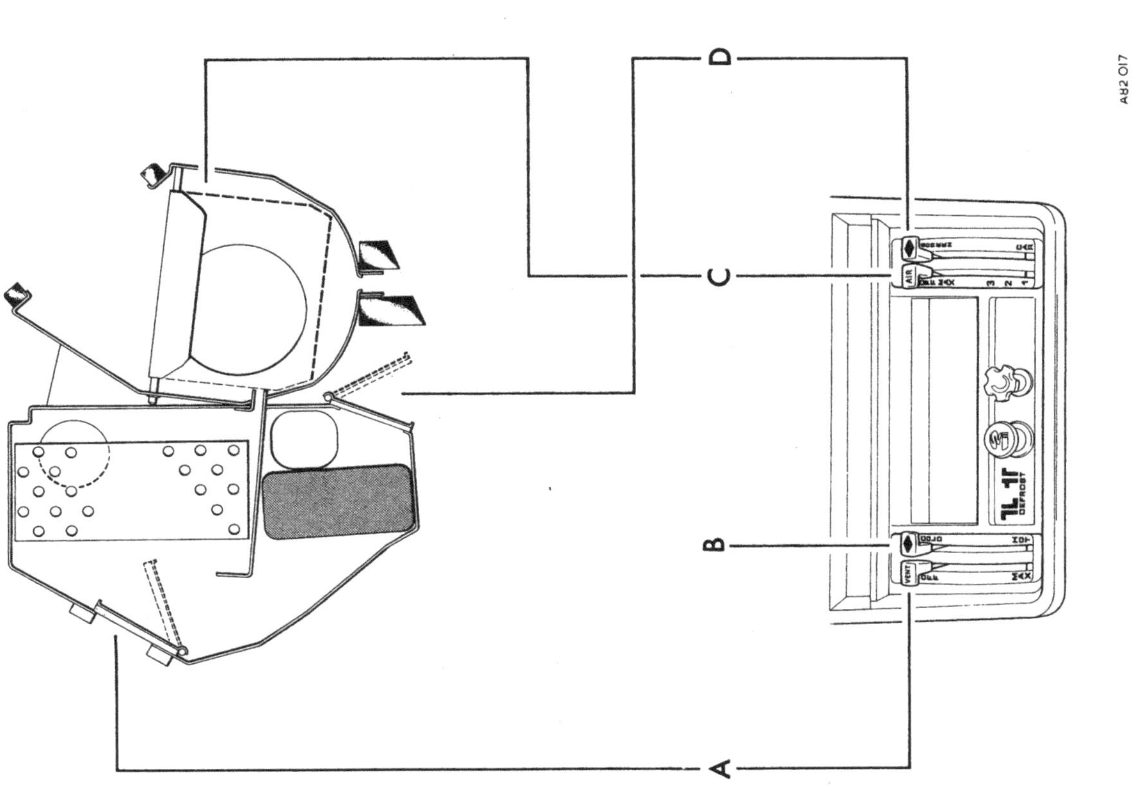

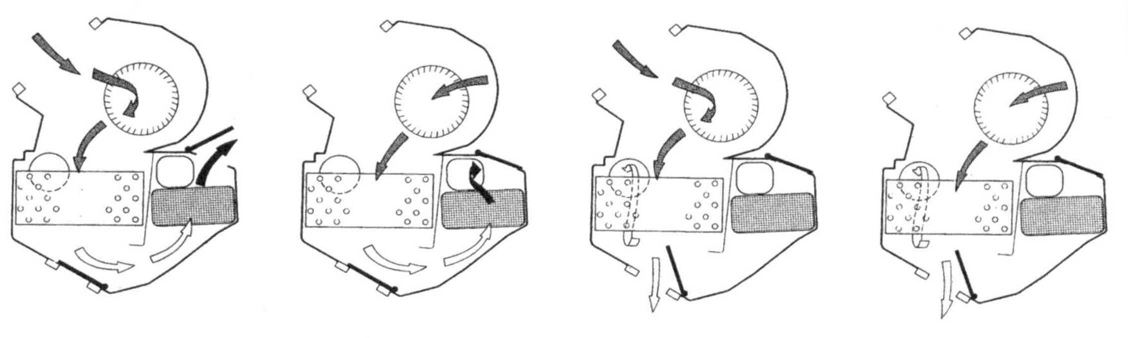

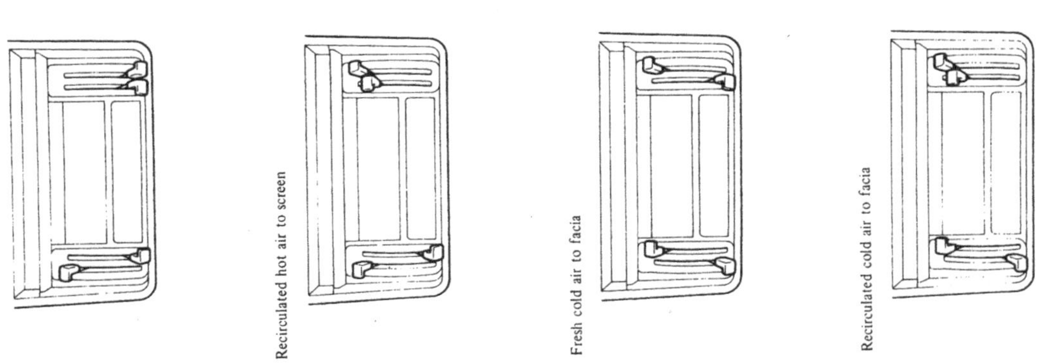

Fresh hot air to footwell

Recirculated hot air to screen

Fresh cold air to facia

Recirculated cold air to facia

AIR CONDITIONER UNIT

Expansion valve – remove and refit **82.25.01**

Remove

1 Connect gauge set, see 82.30.01.
2 Discharge – complete system, see 82.30.05.
3 Isolate the battery.
4 Remove the fascia, see 76.46.01.
5 Note the run of the two capillary tubes.
6 Carefully cut and pull back the insulating material to expose the temperature capillary tube clamp and the pressure capillary tube union.
7 Slacken the clamp screw. Carefully withdraw the temperature capillary tube coil.
8 Carefully disconnect the pressure capillary tube union. Blank the exposed connections immediately.
9 Carefully disconnect the hose connection. Blank the exposed connections immediately.
10 Carefully disconnect the expansion valve mounting connection. Use two spanners at the joint to protect the delicate air conditioner unit pipe joint. Note that the larger rearward hexagon is the union to be rotated. Blank the exposed connections immediately.

Refit

11 Apply refrigerant compressor oil to the mating surfaces of the three connections to assist leakage prevention.
12 Position the expansion valve with the run of the two capillary tubes as noted at operation 5 above. Carefully insert the temperature capillary tube coil in the clamp. Assemble the three connections finger tight.
13 Tighten the expansion valve mounting connection.
14 Tighten the hose connection.
15 Tighten the pressure capillary tube union.
16 Ensure that the temperature capillary tube coil is clean and in good contact with the refrigerant outlet pipe. Tighten the clamp screw.
17 Evacuate – complete system, see 82.30.06.
18 Connect the battery.
19 Charge – complete system, see 82.30.08.
20 Isolate the battery.
21 Perform a leak test on any disturbed joints, see 82.30.09.
22 Restore the insulating material. Ensure that the refrigerant outlet pipe is fully covered.
23 Fit the fascia, see 76.46.01.
24 Connect the battery.
25 Functional check, see 82.30.16.
26 Disconnect gauge set, see 82.30.01.

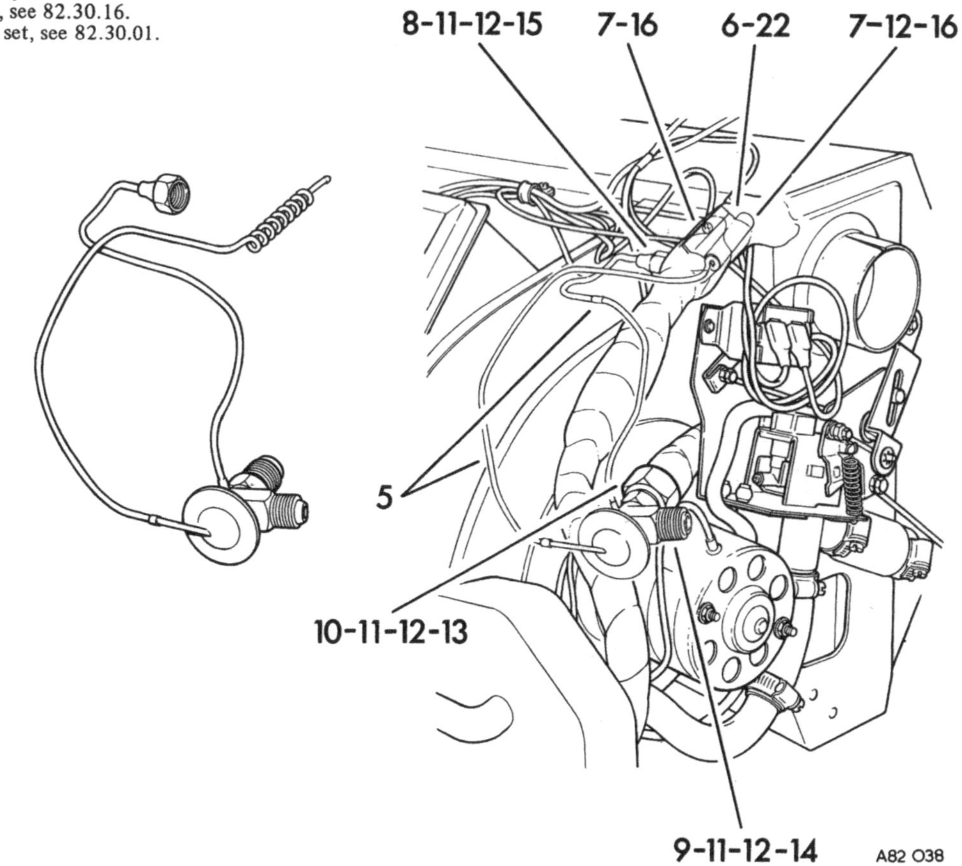

AIR CONDITIONER UNIT

Linkages – adjust 82.25.08

Linkage A

This is the linkage between control lever A and flap A.

1 Isolate the battery.
2 Remove the front console, see 76.25.01. This is necessary to obtain access to two screws.
3 Remove the control cowl, see 76.25.03.
4 Hold the trunnion hexagon and slacken the screw.
5 Position control lever A up to the OFF position.
6 Hold the flap arm in the close position. Tighten the screw.
7 Fit the control cowl, see 76.25.03.
8 Fit the front console, see 76.25.01.
9 Connect the battery.

Linkage B

This is the linkage between control lever B to the cold thermostat and the hot thermostat.

10 Isolate the battery.
11 Remove the instrument panel, see 88.20.01.
12 Pull the facia end vent hose from the air conditioner unit outlet.
13 Hold the hot thermostat trunnion hexagon and slacken the screw.
14 Hold the cold thermostat trunnion hexagon and slacken the screw.
15 Hold the bell crank trunnion hexagon and slacken the screw.
16 Position control lever B up to the COLD position. Maintain in this position during operations 17 to 19.
17 Hold the bell crank so that the bell crank pin is positioned at the lower end of the swinging link slot. Tighten the screw.
18 Hold the cold thermostat input lever at its rearmost limit of travel. Tighten the screw.
19 Hold the hot thermostat input lever at its rearmost limit of travel with the forward projection against the metal stop. Tighten the screw.
20 Push the facia end vent hose onto the air conditioner unit outlet.
21 Fit the instrument panel, see 88.20.01.
22 Connect the battery.

continued

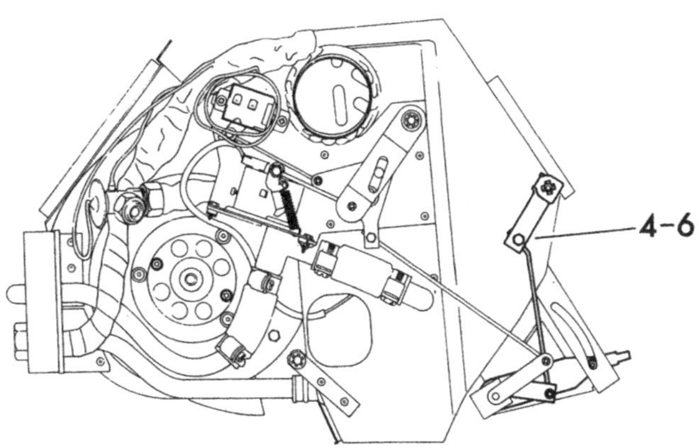

A82 025/1

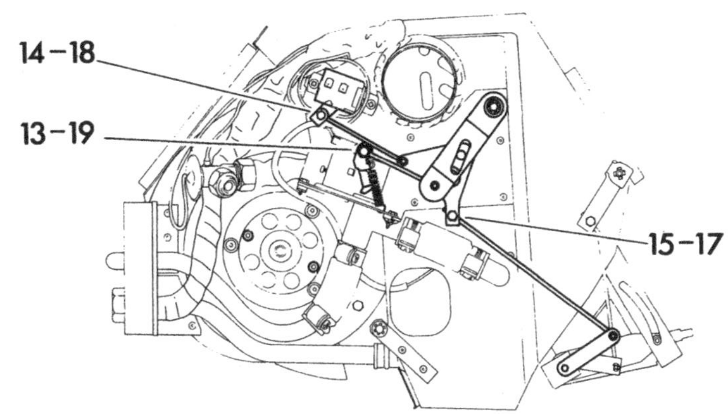

A82 026/1

Linkage C

This is the linkage between control lever C and the recirculate/fresh air flap C.

23 Isolate the battery.
24 Remove the facia, see 76.46.01.
25 Hold the trunnion hexagon and slacken the screw.
26 Position control lever C to the MAX position. This is one stop down from the full up position.
27 Hold the position of the 'quick changeover' lever to achieve the dimension shown. This is a nominal dimension of 2.75 mm from the lever reference face to the follower hole centre line. Tighten the screw.
28 Fit the facia, see 76.46.01.
29 Connect the battery.

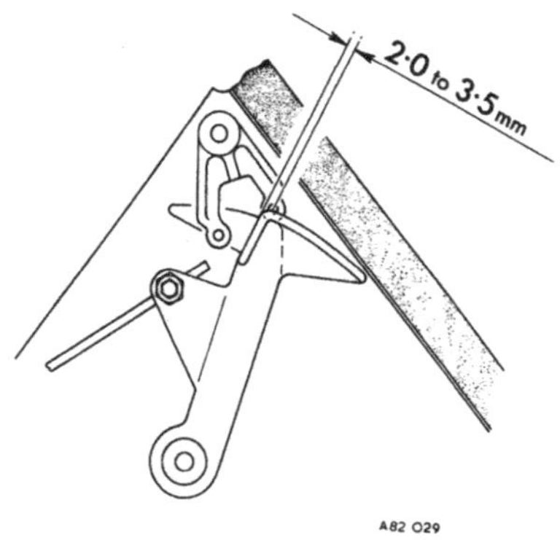

Linkage D

This is the linkage between control lever D and flap D.

30 Isolate the battery.
31 Remove the front console, see 76.25.01. This is necessary to obtain access to two screws.
32 Remove the control cowl, see 76.25.03.
33 Hold the trunnion hexagon and slacken the screw.
34 Position control lever D up to the SCREEN position.
35 Hold the flap arm in the close position. Tighten the screw.
36 Fit the control cowl, see 76.25.03.
37 Fit the front console, see 76.25.01.
38 Connect the battery.

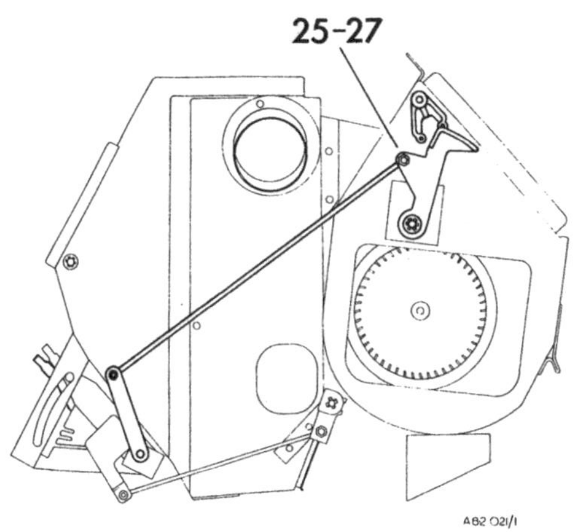

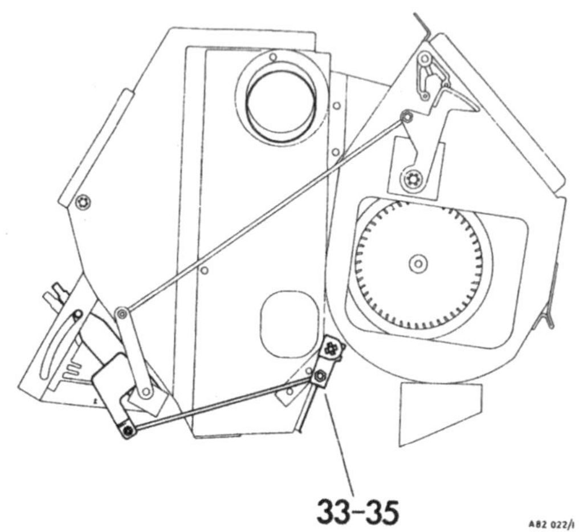

SERVICING 82.30.00

WARNING:

The refrigeration circuit must only be disturbed by a qualified refrigeration engineer possessing the required special servicing equipment. Failure to observe this instruction may result in severe personal injury.

When discharging or subsequently breaking open any pipe connections, protect the eyes with safety goggles and wear gloves.

If any liquid refrigerant should contact the eyes, splash with cold water to slowly raise the temperature. Mineral or cod liver oil on the area will reduce the chance of infection. Consult an eye specialist as soon as possible.

Ensure that no refrigerant vapour comes into contact with an open flame. Should this occur a poisonous, corrosive gas may be produced. This vapour may attack metal.

Refrigerant in containers must be protected from heat. Do not expose to radiant heat from the sun. Do not place in water above 50° C (122° F). Do not heat with any flame. Do not carry a container in the vehicle interior.

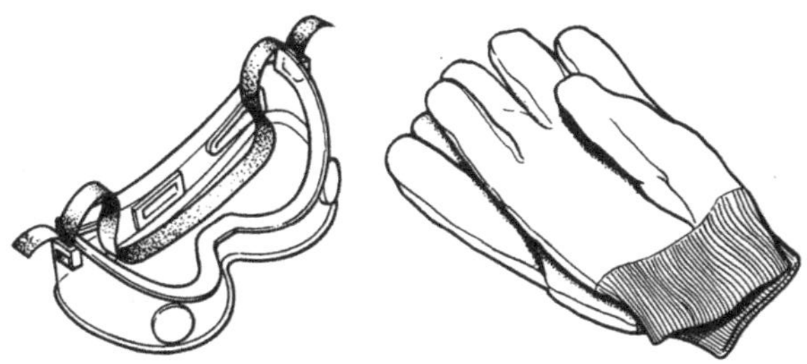

A82 045

Data

Refrigerant

Type	Refrigerant 12
Approved trade names	Arcton 12 Freon 12 Isceon 12
Properties at normal atmospheric pressure and temperature	Vapour Odourless Colourless Heavier than air Non-corrosive Non-explosive Non-inflammable Non-poisonous
Dangerous at normal atmospheric pressure and temperature	
– contact with the skin	Liquid refrigerant will freeze anything it contacts. Severe burns may result. Especially dangerous to the eyes. Always protect with safety goggles.
– contact with an open flame	A poisonous, corrosive gas may be produced. This vapour may attack metal.

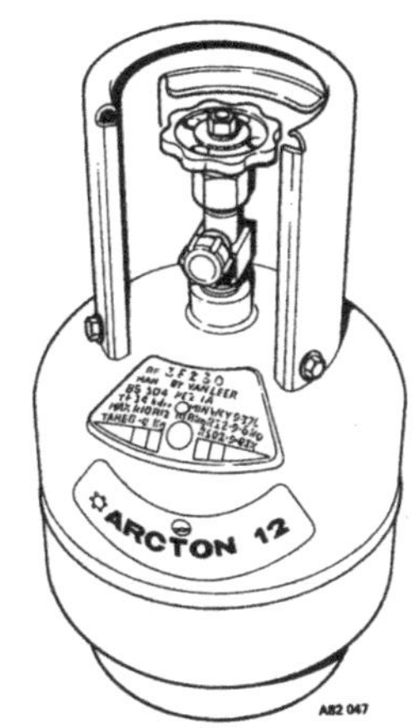

A82 047

Description 82.30.00

Servicing equipment must include a suitable gauge set or service trolley. Other tools are shown below.

When it is necessary to 'break into' the refrigeration circuit the system must first be discharged.

Either the complete system or the compressor only may be discharged. The choice will depend on the operation to be undertaken. When the compressor only is discharged the remainder of the system stays charged and undisturbed. A saving of refrigerant is also achieved.

To prevent icing or corrosion inside the refrigeration system extreme precautions must be observed during servicing to exclude moisture. Component connections and hose ends must only be open to atmosphere for a brief period. Blanking caps must be fitted immediately to any exposed connections. Replacement components will be supplied sealed and must only be opened immediately prior to making the connections.

After assembly the system must be evacuated. This should remove air, moisture and old refrigerant from the system.

The system should then be immediately charged with fresh refrigerant.

Refrigerant may be provided from single cans, from a multi-can manifold or from a service trolley container. The container should be replenished from a heavy bottle.

A method of calculating the weight of refrigerant introduced into the system is required. This may be by a spring balance or a service trolley container with a graduated scale.

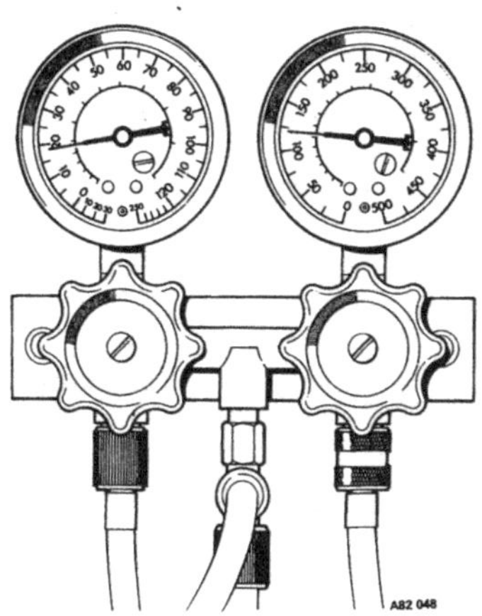

Gauge set

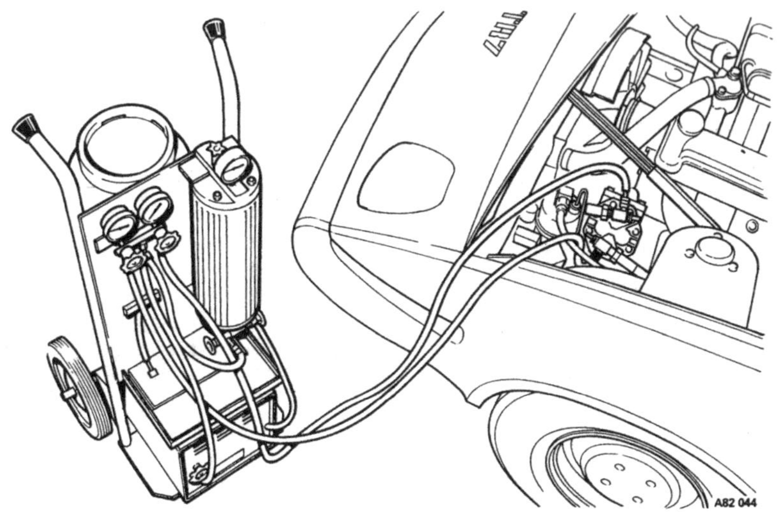

Service trolley

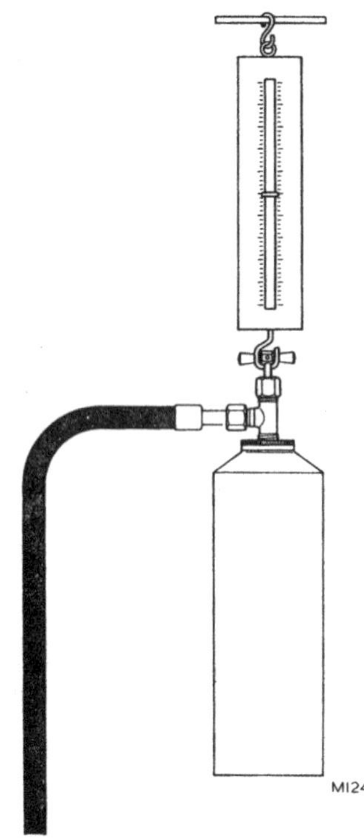

Single can on spring balance

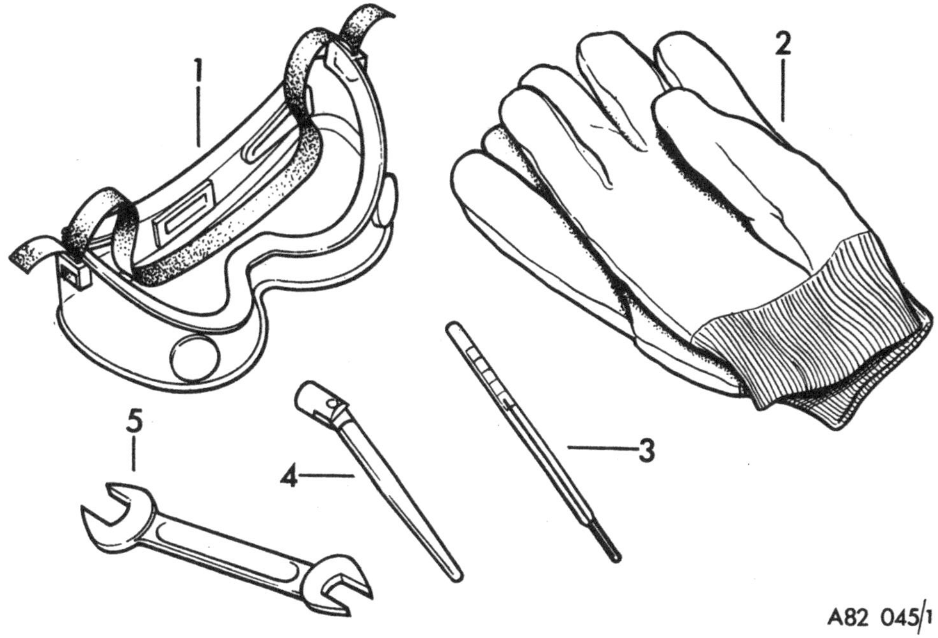

A82 045/1

1 Goggles
2 Gloves
3 Thermometer
4 Square ¼ AF tool to fit service valve stem.

NOTE: Girling brake adjuster spanner – Girling Part No. 64947051 – is ideal for this operation.

5 Spanner 5/16 AF to fit service valve hexagon cap.

Service valves

Refrigerant is introduced and removed from the refrigeration circuit by way of two service valves located on the compressor top plate.

One service valve communicates with the input side of the compressor and is designated the 'SUCTION' or low side service valve. The other service valve is associated with the compressor output side and is designated the 'DISCHARGE' or high side service valve.

The service valves are controlled by square ended shafts normally covered by protective covers.

Each service valve has three basic positions summarized as follows:

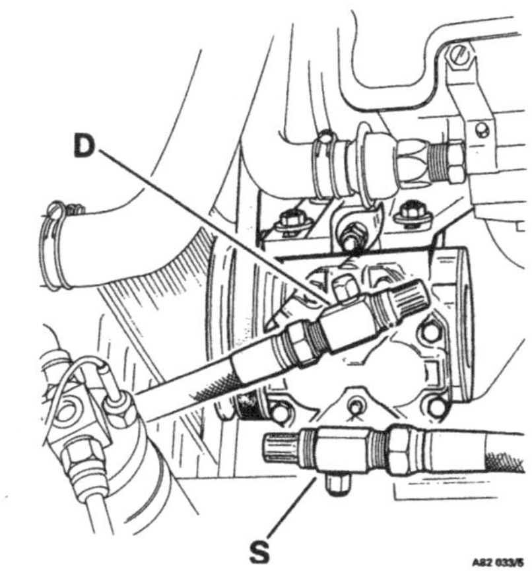

D Discharge valve
S Suction valve

SERVICE VALVE POSITIONS

Back seat position	Turn full anti-clockwise	Normal system operating position Compressor communicating with refrigeration circuit Service port sealed	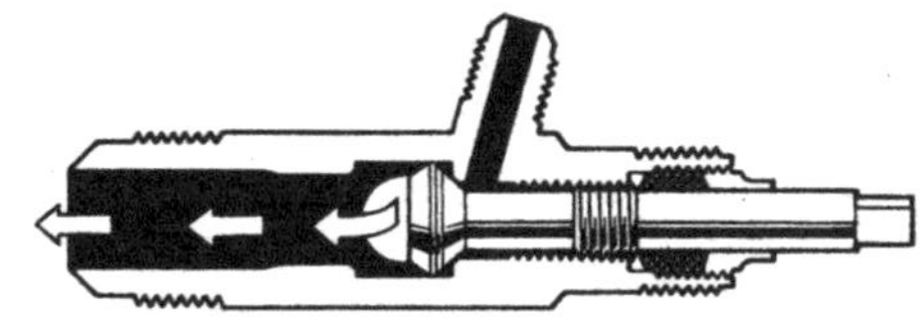
Mid position	From back seat position turn clockwise 3 turns	Compressor and refrigeration circuit communicating with service port	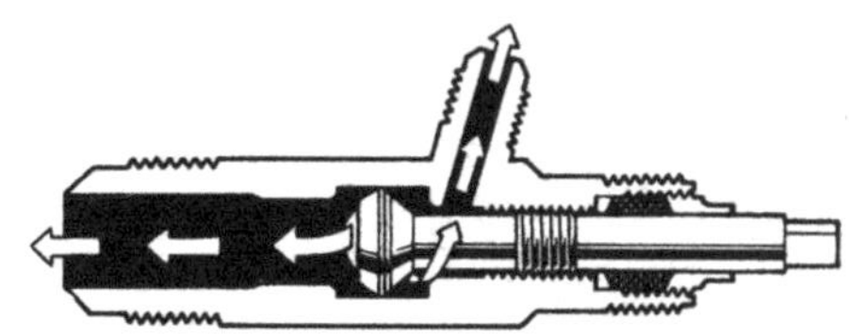
Front seat position	Turn full clockwise	Compressor communicating with service port Refrigeration circuit sealed	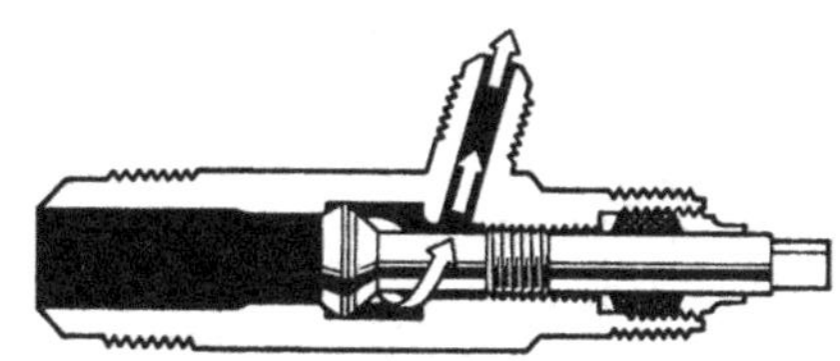

MO84

SERVICING

Gauge set – connect 82.30.01

1 Position the vehicle in a suitable area. This should be:

- a Well ventilated for the discharge of refrigerant.
- b Away from any naked flame.
- c Suitable for an engine run.

2 Provide a suitable gauge set or service trolley.

3 Close all gauge set or service trolley valves.

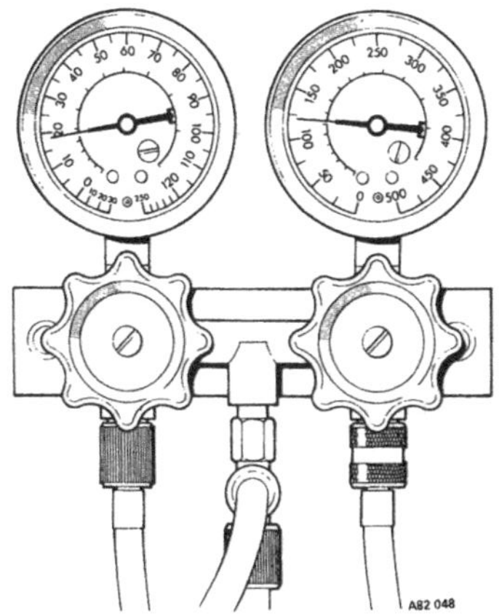

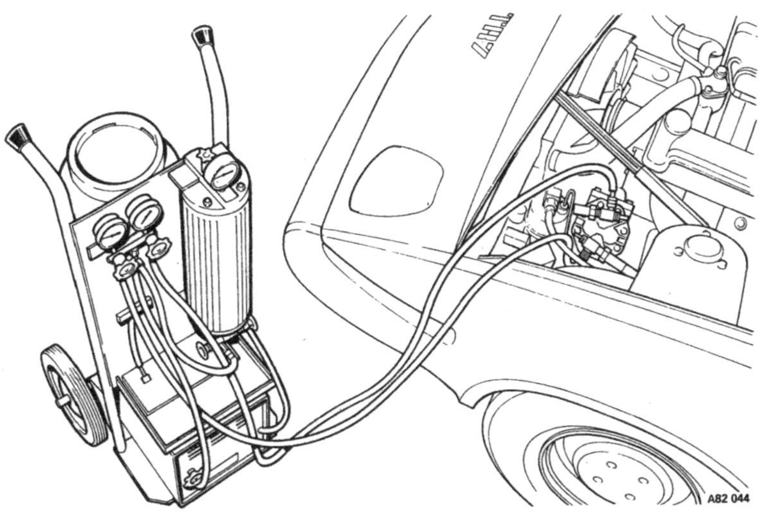

4 Remove the two service valve protective covers.

5 Ensure that both service valves are in the back seat position – turn anti-clockwise.

6 Remove the two hexagon caps.

7 Connect the gauge set discharge hose to the discharge service valve.

8 Connect the gauge set suction hose to the suction service valve.

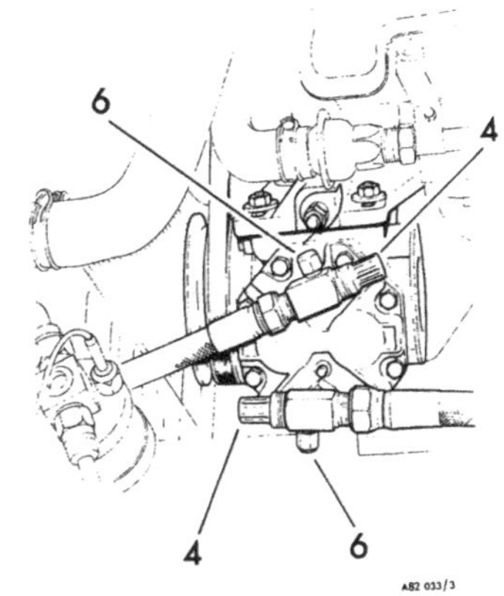

9 Purge the air from the two gauge set hoses as follows:

- a Provide a supply of refrigerant to the gauge set centre manifold hose.

  NOTE: This may be from a single can, from a multi-can manifold or from a service trolley container.

- b Protect the eyes with safety goggles and wear gloves during operations c and d.
- c Loosen the gauge set hose connection to the discharge service valve. Carefully open the gauge set discharge valve. When refrigerant is seen to be escaping tighten the hose connection. Close the gauge set discharge valve.
- d Loosen the gauge set hose connection to the suction service valve. Carefully open the gauge set suction valve. When refrigerant is seen to be escaping tighten the hose connection. Close the gauge set suction valve.

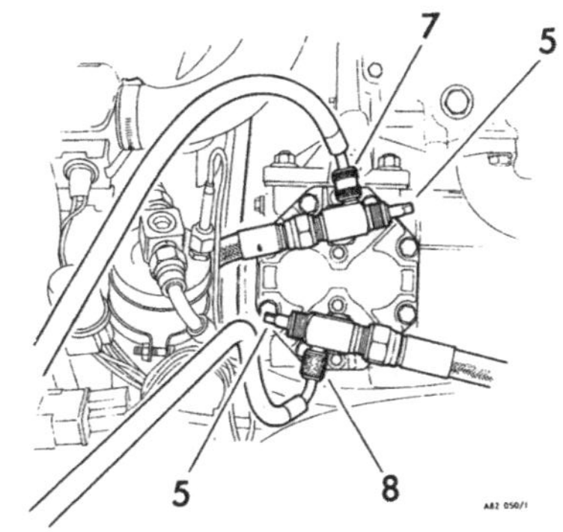

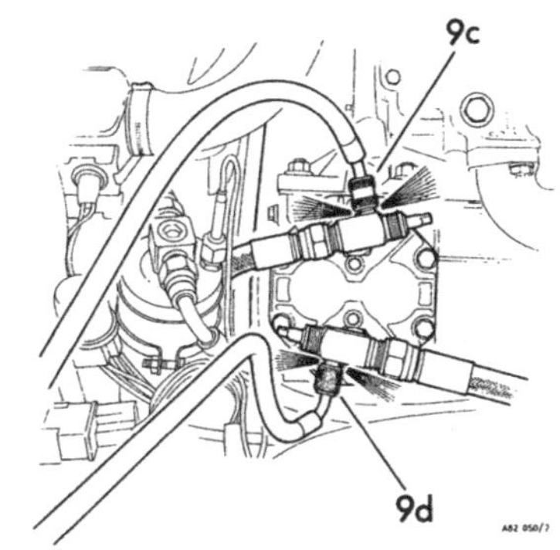

SERVICING

Gauge set – disconnect 82.30.01

10 Close all gauge set or service trolley valves.
11 Ensure that both service valves are in the back seat position – turn anti-clockwise.
12 Protect the eyes with safety goggles and wear gloves during operation 13.
13 Disconnect both gauge set hoses from the service valves.
14 Fit the two hexagon caps.
15 Fit the two service valve protective covers.

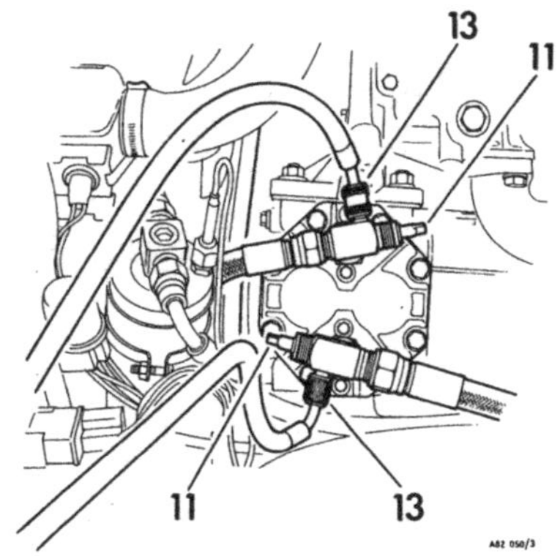

A82 050/3

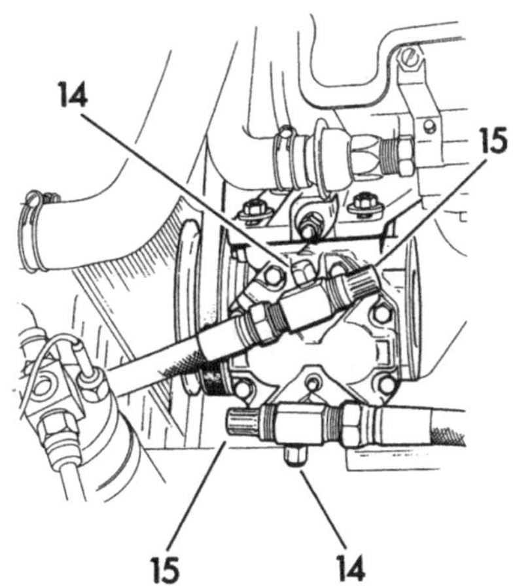

A82 033/4

SERVICING

Discharge – complete system 82.30.05

1 Connect gauge set, see 82.30.01.
2 Immobilize the refrigerant circuit by disconnecting the electrical connection to the compressor clutch.
3 Open the discharge service valve to mid-position – turn clockwise three turns.
4 Open the suction service valve to mid-position – turn clockwise three turns.
5 Protect the eyes with safety goggles and wear gloves during operations 6 to 8.
6 Hold the gauge set centre manifold hose end in a suitable rag.
7 Slightly open the gauge set discharge valve to allow the refrigerant vapour to slowly discharge to atmosphere via the hose end. If oil from the compressor is discharged, reduce the gauge set discharge valve opening.
8 When the discharge has nearly stopped open the gauge set suction valve.
9 When the discharge has stopped ensure that both gauges read zero.
10 Close both gauge set valves.
11 Close the discharge service valve to the back seat position – turn anti-clockwise.
12 Close the suction service valve to the back seat position – turn anti-clockwise.

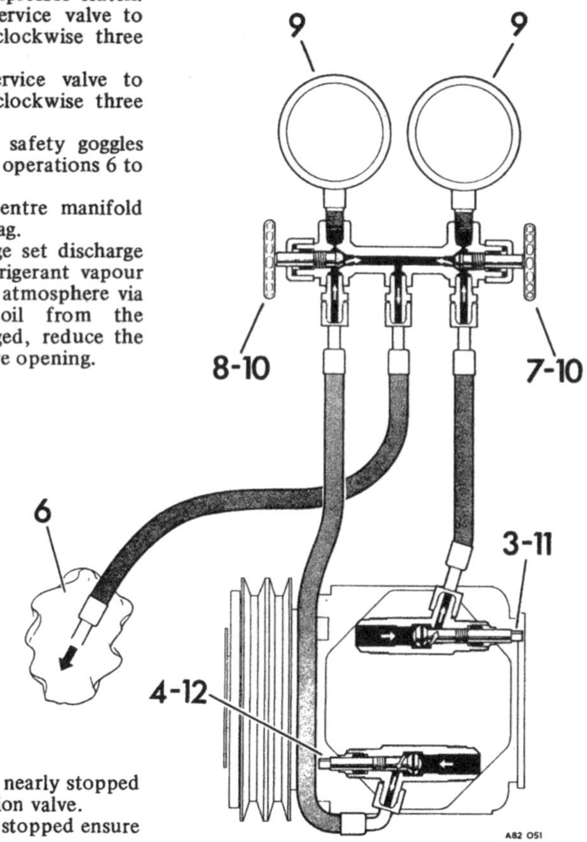

A82 051

SERVICING

Evacuate – complete system 82.30.06

1 Connect gauge set, 82.30.01.
2 Discharge – complete system, see 82.30.05.
3 Connect a vacuum pump to the gauge set centre manifold hose.
4 Open the discharge service valve to mid-position – turn clockwise three turns.
5 Open the suction service valve to mid-position – turn clockwise three turns.
6 Open both gauge set valves fully.

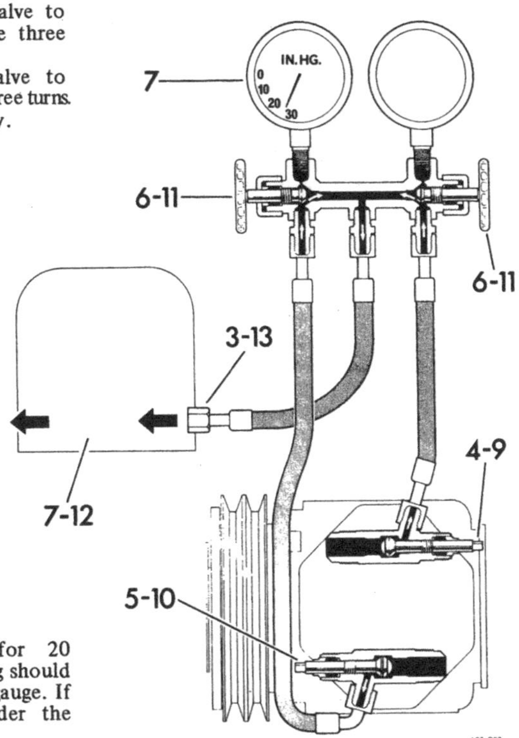

7 Run the vacuum pump for 20 minutes. A vacuum of 28 in Hg should be indicated on the suction gauge. If this is not achieved, consider the possibility of a system leak.
8 Perform operations 9 to 11 with the vacuum pump running.
9 Close the discharge service valve to the back seat position – turn anti-clockwise.
10 Close the suction service valve to the back seat position – turn anti-clockwise.
11 Close both gauge set valves.
12 Stop the vacuum pump.
13 Disconnect the vacuum pump from the gauge set centre manifold hose.

SERVICING

Charge – complete system 82.30.08

CAUTION: Charge with refrigerant 12 of approved trade names:

Arcton 12
Freon 12
Isceon 12

Do not charge with methylchloride refrigerant. This would react undesirably with aluminium parts used in the system.

1 Connect gauge set, see 82.30.01.
2 Discharge – complete system, see 82.30.05.
3 Evacuate – complete system, see 82.30.06
4 Provide a supply of refrigerant to the gauge set centre manifold hose.

NOTE: *This may be from single cans, from a multi-can manifold or from a service trolley container.*

5 If the centre manifold hose contains air purge with refrigerant as follows:

a Protect the eyes with safety goggles and wear gloves during operation b.
b Carefully loosen the gauge set centre manifold hose connection. When refrigerant is seen to be escaping tighten the hose connection.

continued

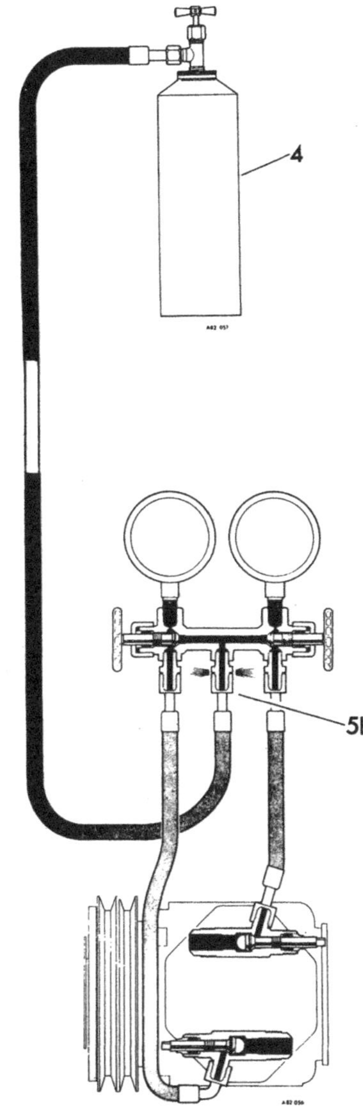

6 Provide a method of calculating at any time during the operation, the weight of refrigerant put into the system.

NOTE: *A single can or a multi-can manifold may be weighed by hanging from a spring balance. A service trolley container should have a graduated scale.*

7 Note the datum refrigerant weight.

8 Open the suction service valve to mid-position – turn clockwise three turns.

9 Slightly open the gauge set suction valve. Allow approximately ½ lb (230 g) of refrigerant to enter the system. Close the gauge set suction valve.

NOTE: *High pressure liquid refrigerant from the container will vaporize on entering the evacuated low pressure system. Flow will continue until container pressure equals system pressure or until the gauge set suction valve is closed.*

10 Perform a first leak test on any disturbed joints, see 82.30.09.

11 Slightly open the gauge set suction valve. If the pressure differential between the container and the system permits allow the flow to continue until a total of 2½ lb (1130 g) of refrigerant – from the datum weight operation 7 – has entered the system. Close the gauge set suction valve.

NOTE: *High pressure liquid refrigerant from the container will vaporize on entering the evacuated low pressure system. Flow will continue until container pressure equals system pressure or until the gauge set suction valve is closed.*

12 If a total of 2½ lb (1130 g) of refrigerant is in the system perform the following:

a Close the suction service valve to the back seat position – turn anti-clockwise.

b Ignore operations 13 to 19.

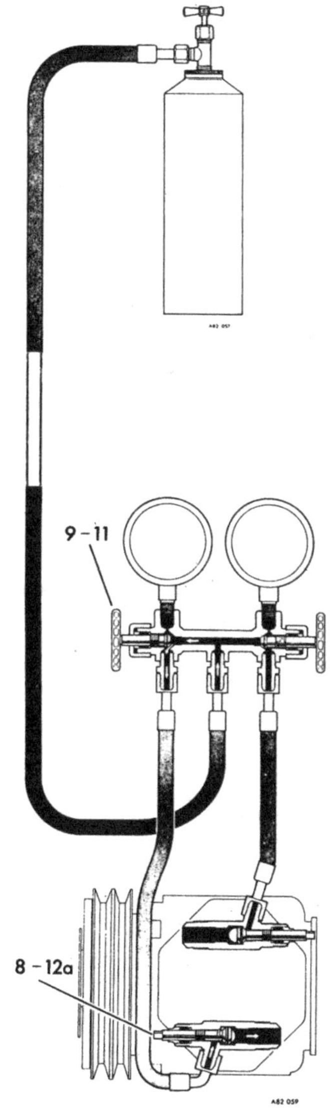

13 If the container pressure and the system pressure equalize before a total of 2½ lb (1130 g) of refrigerant is in the system perform operations 14 to 19.

14 Ensure that the gauge set suction valve is closed.

15 Connect the electrical connection to the compressor clutch.

16 Run the engine at 1000 to 1500 rev/min for 5 minutes with the control levers set as follows:

Lever A to MAX.
Lever B to COLD
Lever C to 3
Lever D to CAR

This is to warm the engine and stabilize the system.

NOTE: *Warm air from above the engine will enter the fresh air duct to be presented to the cold matrix. This condition will cause the system to operate hard and prevent frequent cutting in and out of the compressor clutch.*

17 Slightly open the gauge set suction valve. Allow the flow to continue until a total of 2½ lb (1130 g) of refrigerant – from the datum weight operation 7 – has entered the system. Close the gauge set suction valve.

NOTE: *Compressor suction will draw further vaporized refrigerant into the system. The refrigeration circuit will commence to function and vapour passed into the condenser will accumulate as liquid refrigerant in the receiver drier.*

18 Close the suction service valve to the back seat position – turn anti-clockwise.

19 Stop the engine.

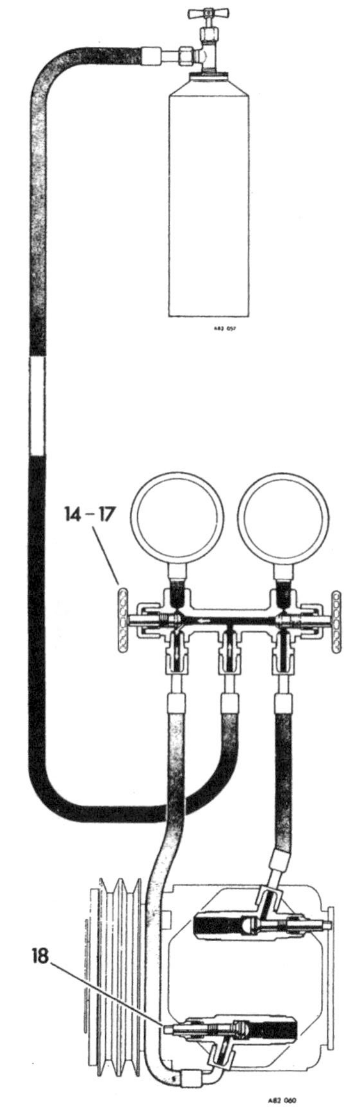

SERVICING

Leak test 82.30.09

A major leak in the system should be shown up during the evacuate operation prior to charging with fresh refrigerant.

Minor leaks should be searched for as instructed in the charge operation using one of the two basic types of leak testing equipment in common use.

The burner type has a hand held burner connected by hose to a cylinder of gas. A second hose attached to the burner is the search hose which draws in air or refrigerant vapour. This hose is of some length so its end may be positioned close to the unions while the burner is held and observed by a second operator a safe distance from the vehicle. A leak is indicated by the flame changing colour to green or purple. The product of burning gas and refrigerant is a poisonous, corrosive gas which should not be inhaled.

The electronic type may be semi portable with mains electric power or fully portable with batteries. The sensors are sensitive to refrigerant vapour. The sensor may be positioned at the end of a search cable or the air sample may be drawn through a search hose by a small electric air pump to be passed across a sensor in the unit. The sensor signal is amplified. A leak is indicated by audible warning, a light signal or meter reading.

Whether a burner type or electronic type unit is used the equipment should be employed as detailed by the manufacturer. The following instructions are provided to assist leak testing.

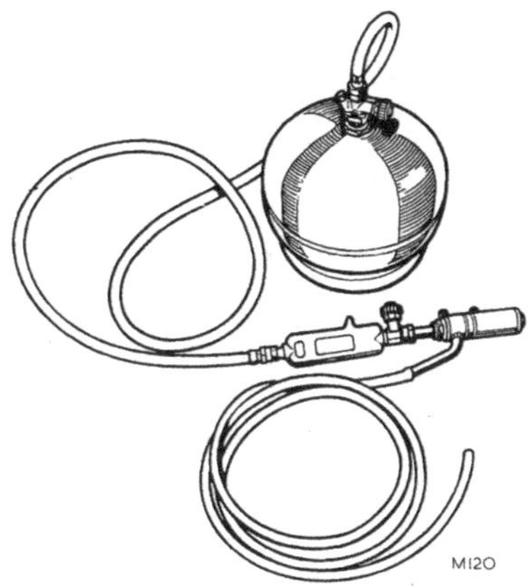

Typical burner leak tester.

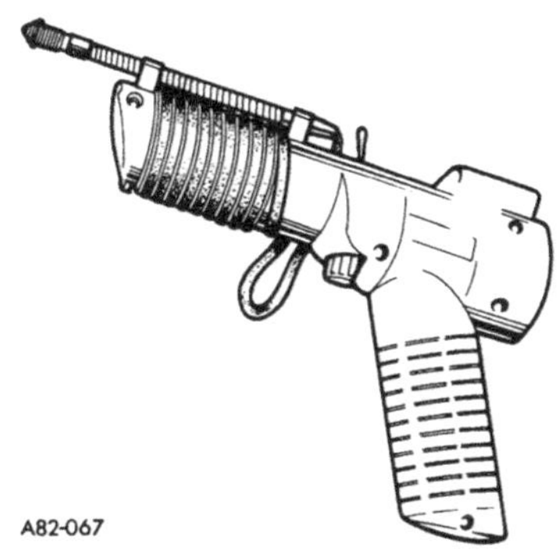

Typical electronic leak tester.

General

Place the vehicle in a well ventilated area or refrigerant may persist in the vicinity and give misleading results.

Strong draughts should be avoided as a leak may be dissipated without detection.

Refrigerant is heavier than air. When checking each union pass the detector slowly round each joint with special attention to the underside.

Compressor

Check two hose connections, service valves, head joint, rear bearing plate joint and base joint.

Check the shaft seal by positioning the detector between the clutch and seal plate. As access is poor, leave the detector in the vicinity for one minute.

Condenser

Check two hose connections. Check all soldered joints and pass detector across underside of unit.

Receiver drier

Check two hose connections and high pressure cut-out capillary.

Expansion valve and air conditioner unit pipes

Check all accessible pipes and joints. If a known leak can not be located it may finally be necessary to perform the following:

Remove the facia. Carefully cut and pull back the insulating material to expose the pipe runs. After the leak test restore the insulating material. Ensure that the refrigerant outlet pipe is fully covered.

Cold matrix

Select lever 'A' to MAX. Insert the probe through the central facia vent.

Select lever 'D' to CAR. Insert the probe into the footwell outlet.

Check below the matrix by positioning the probe into the recirculated air inlet and checking near the drain pipe. (See 82.25.00, Sheet 1, Item 9).

SERVICING

Functional check 82.30.16

1 Ensure that the compressor drive belt is correctly adjusted, see 82.10.01.
2 Connect gauge set, see 82.30.01.
3 Open the discharge service valve towards mid-position – turn clockwise half a turn.
4 Open the suction service valve forwards mid-position – turn clockwise half a turn.
5 Note the ambient air temperature.
6 Run the engine at 1000 to 1500 rev/min for 5 minutes with the control levers set as follows:

Lever A to MAX.
Lever B to COLD

NOTE: *In humid conditions it may be necessary to reduce lever 'B' setting from full 'COLD' towards 'HOT' to prevent icing of the cold matrix. Slightly higher central fascia vent air temperatures should then be expected.*

Lever C to MAX.
Lever D to CAR

This is to warm the engine and stabilize the system.

7 Check that the two condenser fan motors run when the compressor clutch pulls in and stop when the compressor clutch drops out.
8 Check that the receiver drier sight glass is clear of bubbles or foam.
9 Note the maximum discharge gauge reading when the compressor clutch is pulled in.
10 Note the minimum suction gauge reading when the compressor clutch is pulled in.
11 Insert a thermometer into the central fascia vent and note the minimum temperature.
12 Stop the engine.
13 Compare all readings with the values given in the table.
14 Close the discharge service valve to the back seat position – turn anti-clockwise.
15 Close the suction service valve to the back seat position – turn anti-clockwise.
16 Disconnect gauge set, see 82.30.01.

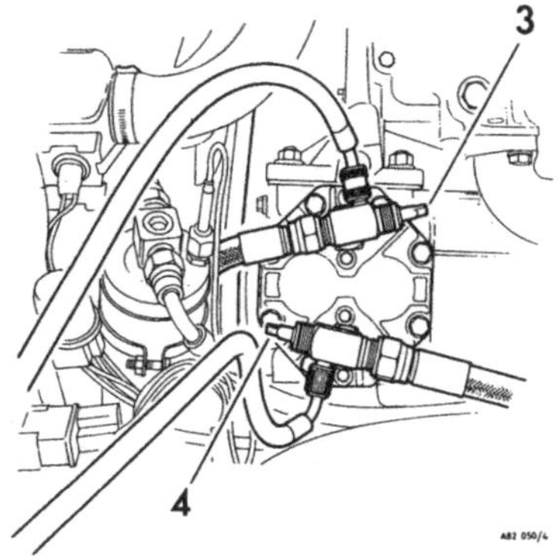

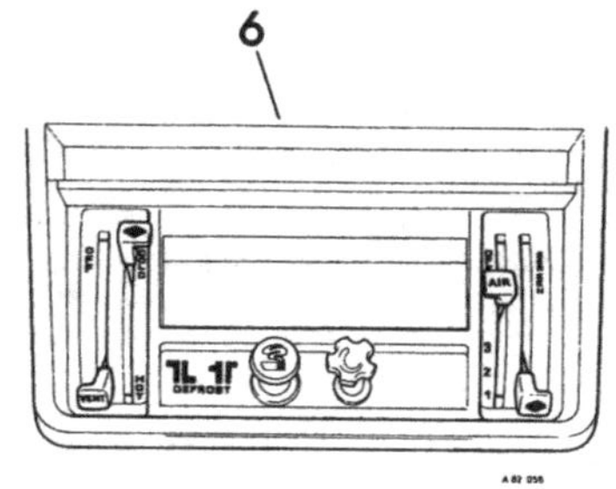

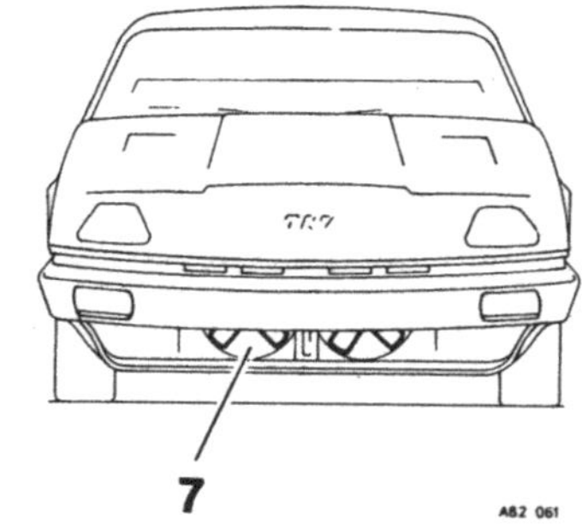

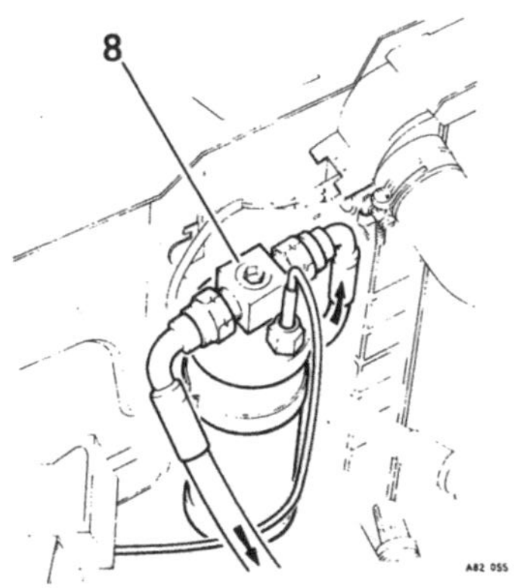

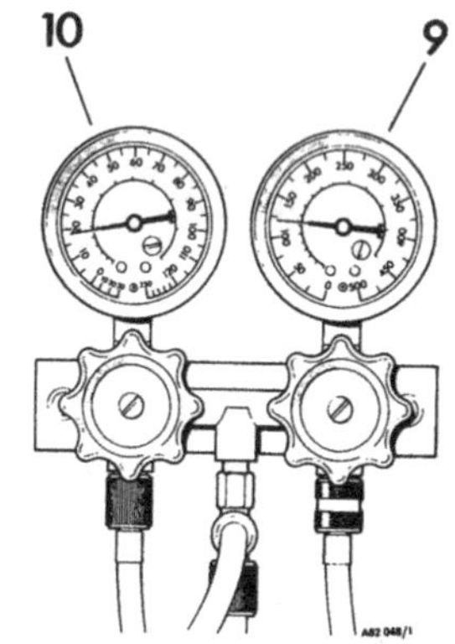

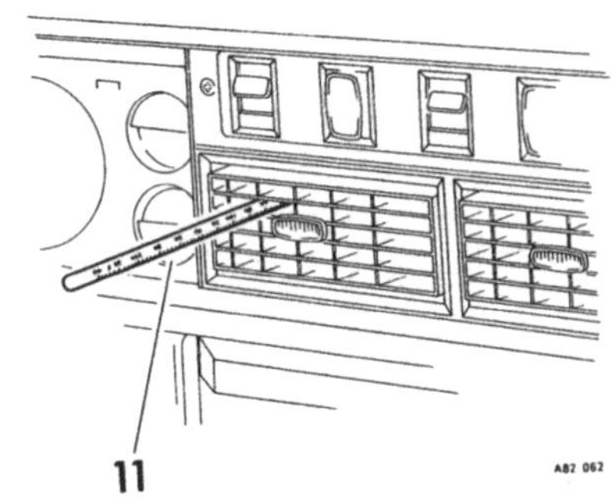

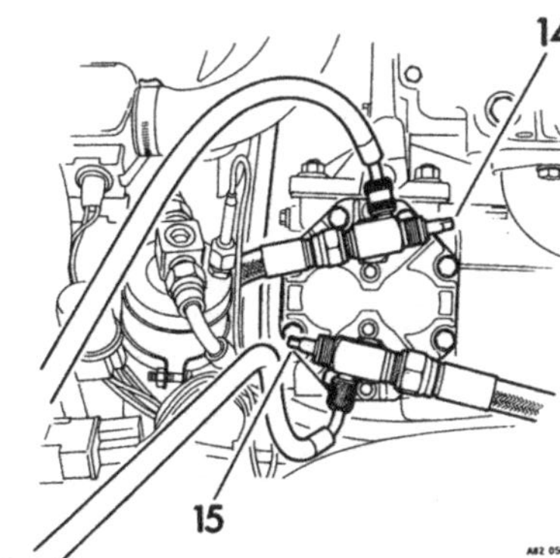

Ambient air temperature		Discharge gauge – maximum		Suction gauge – minimum		Central facia vent temperature – minimum	
°F	°C	psi	kg/cm²	psi	kg/cm²	°F	°C
60	16	12 to 20	0.8 to 1.4	80 to 130	5.6 to 9.1	32 to 40	0 to 4
80	27	16 to 25	1.1 to 1.8	130 to 180	9.1 to 12.7	35 to 44	2 to 7
100	38	18 to 28	1.3 to 2.0	200 to 240	14.1 to 16.9	38 to 47	3 to 8
110	43	19 to 29	1.3 to 2.0	240 to 275	16.9 to 19.3	39 to 48	4 to 9

The above figures are not specific to the TR7 installation. All readings obtained should be approximately equal to the values given in the table.

SERVICING

Discharge – compressor only 82.30.20

1 Connect gauge set, see 82.30.01.
2 Immobilize the refrigerant circuit by disconnecting the electrical connection to the compressor clutch.
3 Close the discharge service valve to the front seat position – turn clockwise.
4 Close the suction service valve to the front seat position – turn clockwise.
5 Protect the eyes with safety goggles and wear gloves during operations 6 to 8.
6 Hold the gauge set centre manifold hose end in a suitable rag.
7 Slightly open the gauge set discharge valve to allow the refrigerant vapour to slowly discharge to atmosphere via the hose end. If oil from the compressor is discharged, reduce the gauge set discharge valve opening.
8 When the discharge has nearly stopped open the gauge set suction valve.
9 When the discharge has stopped ensure that both gauges read zero.
10 Close both gauge set valves.

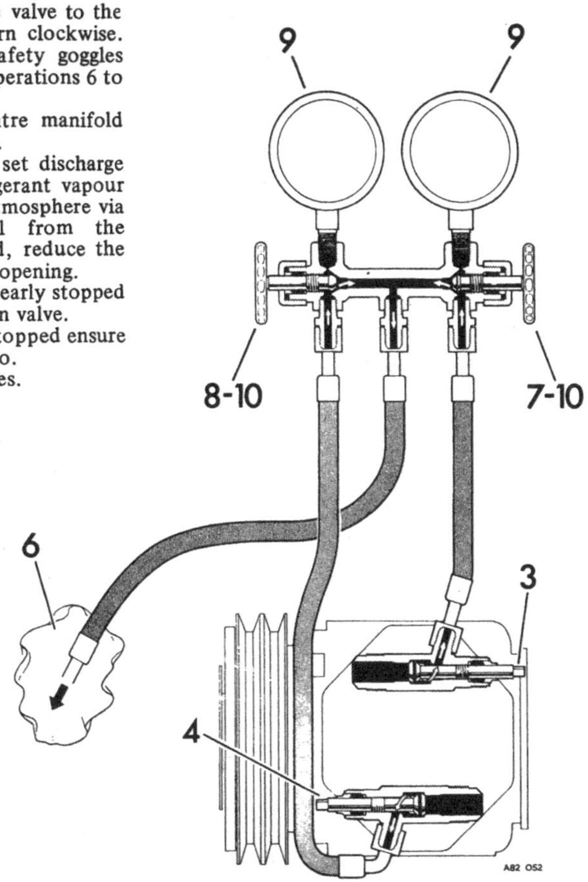

SERVICING

Evacuate – compressor only 82.30.21

1 Connect gauge set, see 82.30.01.
2 Discharge – compressor only, see 82.30.20.
3 Connect a vacuum pump to the gauge set centre manifold hose.
4 Open both gauge set valves fully.

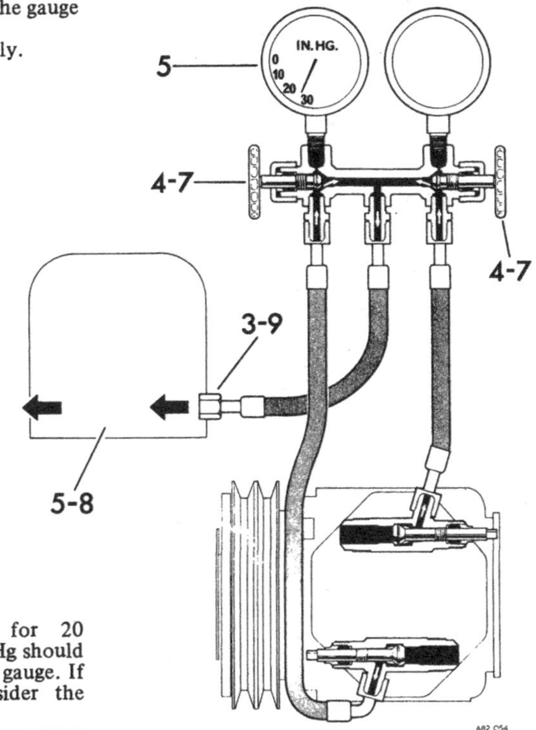

5 Run the vacuum pump for 20 minutes. A vacuum of 28 in Hg should be indicated on the suction gauge. If this is not achieved, consider the possibility of a system leak.
6 Perform operation 7 with the vacuum pump running.
7 Close both gauge set valves.
8 Stop the vacuum pump.
9 Disconnect the vacuum pump from the gauge set centre manifold hose.

SERVICING

Charge – compressor only 82.30.23

CAUTION: Charge with refrigerant 12 of approved trade names:

Arcton 12
Freon 12
Isceon 12

Do not charge with methylchloride refrigerant. This would react undesirably with aluminium parts used in the system.

1 Connect gauge set, see 82.30.01.
2 Discharge – compressor only, see 82.30.20.
3 Evacuate – compressor only, see 82.30.21.
4 Provide a supply of refrigerant to the gauge set centre manifold hose.

NOTE: *This may be from single cans, from a multi-can manifold or from a service trolley container.*

5 If the centre manifold hose contains air purge with refrigerant as follows:

a Protect the eyes with safety goggles and wear gloves during operation b.
b Carefully loosen the gauge set centre manifold hose connection. When refrigerant is seen to be escaping tighten the hose connection.

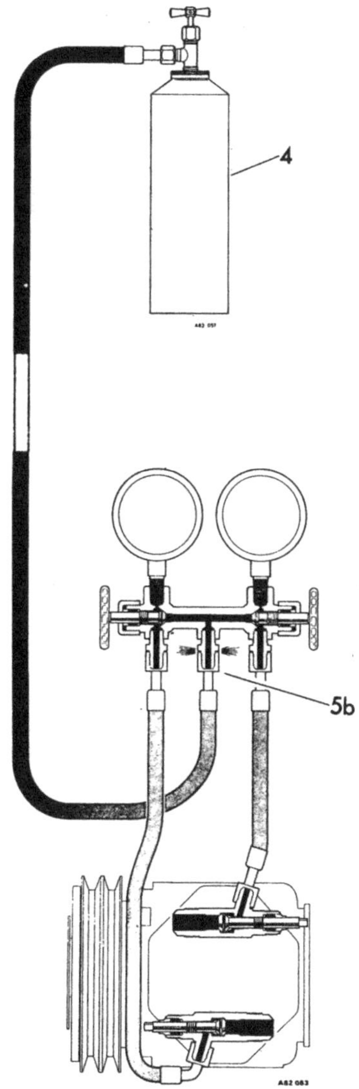

6 Provide a method of calculating at any time during the operation the weight of refrigerant put into the system.

NOTE: *A single can or a multi-can manifold may be weighed by hanging from a spring balance. A service trolley container should have a graduated scale.*

7 Note the datum refrigerant weight.
8 Open the discharge service valve to mid-position – turn anti-clockwise five turns.
9 Open the suction service valve to mid-position – turn anti-clockwise five turns.

NOTE: *Operation 8 and 9 will allow high pressure refrigerant in the remainder of the sytem to enter the evacuated low pressure compressor.*

10 Perform a first leak test on any disturbed joints, see 82.30.09.

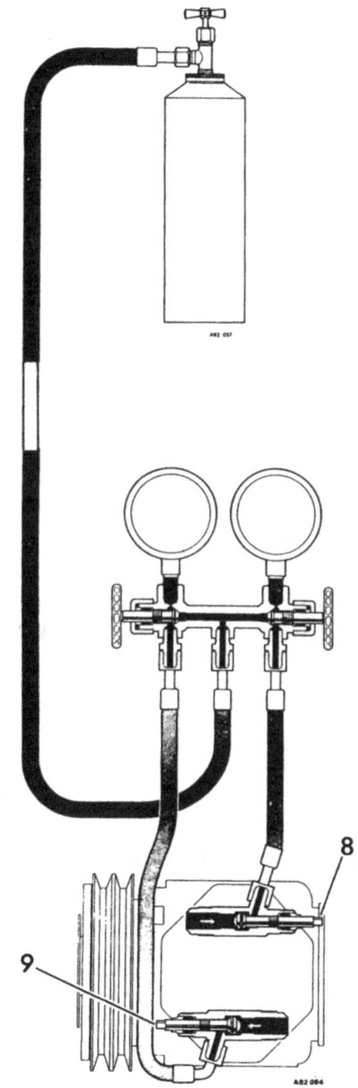

11 Close the discharge service valve to the back seat position – turn anti-clockwise.

12 Slightly open the gauge set suction valve. If the pressure differential between the container and the system permits, allow the flow to continue until 6 oz (170 g) of refrigerant has entered the system. Close the gauge set suction valve.

NOTE: *High pressure liquid refrigerant from the container will vaporize on entering the system. Flow will continue until container pressure equals system pressure or until the gauge set suction valve is closed.*

13 If 6 oz (170 g) of refrigerant has entered the system perform the following:

a Close the suction service valve to the back seat position – turn anti-clockwise

b Ignore operations 14 to 20.

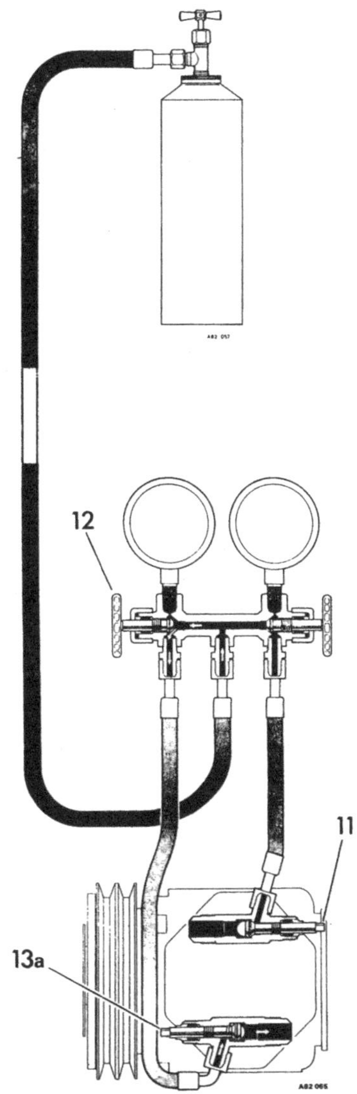

14 If the container pressure and the system pressure equalize before 6 oz (170 g) of refrigerant has entered the system perform operations 15 to 20.

15 Ensure that the gauge set suction valve is closed.

16 Connect the electrical connection to the compressor clutch.

17 Run the engine at 1000 to 1500 rev/min for 5 minutes with the control levers set as follows:

Lever A to MAX.
Lever B to COLD
Lever C to 3
Lever D to CAR

This is to warm the engine and stabilize the system.

NOTE: *Warm air from above the engine will enter the fresh air duct to be presented to the cold matrix. This condition will cause the system to operate hard and prevent frequent cutting in and out of the compressor clutch.*

18 Slightly open the gauge set suction valve. Allow the flow to continue until 6 oz (170 g) of refrigerant – from the datum weight operation 7 – has entered the system. Close the gauge set suction valve.

NOTE: *Compressor suction will draw further vaporized refrigerant into the system. The refrigeration circuit will commence to function and vapour passed into the condenser will accumulate as liquid refrigerant in the receiver drier.*

19 Close the suction service valve to the back seat position – turn anti-clockwise.

20 Stop the engine.

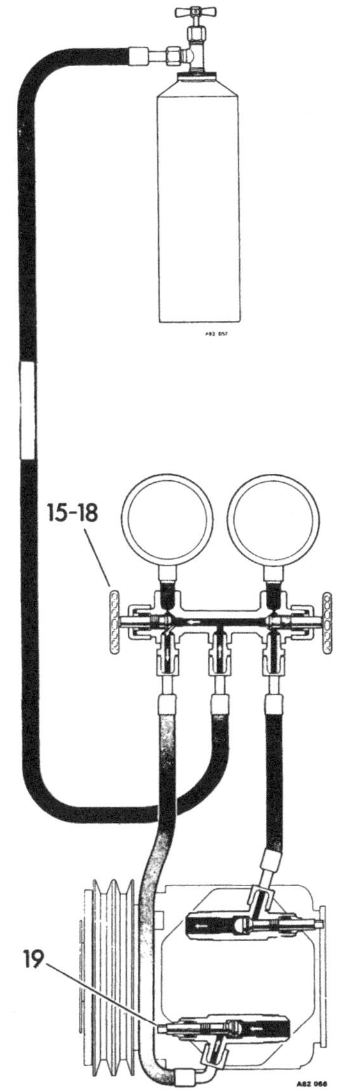

WINDSCREEN WIPERS AND WASHERS

WINDSCREEN WASHER RESERVOIR

Remove and refit **84.10.01**

Remove
1 Pull the windscreen washer reservoir pipe from the windscreen washer pump.
2 Lift the windscreen washer reservoir from the retaining bracket.

Refit
3 Reverse 1 to 2.

WINDSCREEN WASHER JET

Remove and refit **84.10.09**

Remove
1 Pull the pipe from the jet.
2 Remove the nut and anti-vibration washer.
NOTE: Take care not to drop the nut and anti-vibration washer as retrieval may prove difficult.
3 Remove the jet and rubber sealing washer.

Refit
4 Reverse 1 to 3. Fit the jet with the jet outlet in the correct location to the bonnet.
5 Close the bonnet and operate the pump to check the jet aim.
6 If necessary adjust the jet aim by using a screwdriver in the slot to slightly rotate the jet outlet.

WINDSCREEN WASHER PUMP

Remove and refit **84.10.21**

Remove
1 Remove the windscreen washer reservoir, see 84.10.01 to obtain improved access.
2 Disconnect two Lucar connectors.
3 Note the positions of the inlet and outlet pipes. IN and OUT are stamped on the mounting bracket.
4 Pull the outlet pipe from the windscreen washer pump.
5 Remove two Pozidriv screws and washers. Remove the windscreen washer pump.
NOTE: Take care not to drop the two Pozidriv screws and washers as retrieval may prove difficult.

Refit
6 Position the windscreen washer pump. Secure with two Pozidriv screws and washers.
7 Push the outlet pipe onto the windscreen washer pump in the position noted at operation 3 above.
8 Connect two Lucar connectors as follows:
Light green/black wire to the positive terminal.
Black wire to the negative terminal.
9 Refit windscreen washer reservoir, see 84.10.01.

WINDSCREEN WASHER PUMP

Overhaul **84.10.24**

The motor is a sealed unit and cannot be serviced. It is possible to dismantle and clean the interior of the pump as detailed below.

Disassemble
1 Remove three screws.
2 Lift off the bracket.
3 Remove the pump cover.
4 Remove the rubber disc.
5 Remove the metal disc.
6 Remove the plastic disc.
7 Carefully withdraw the rotor.
8 Withdraw the rotor housing.

Assemble
9 Reverse 1 to 8. Ensure that all the components are assembled the correct way round as shown.

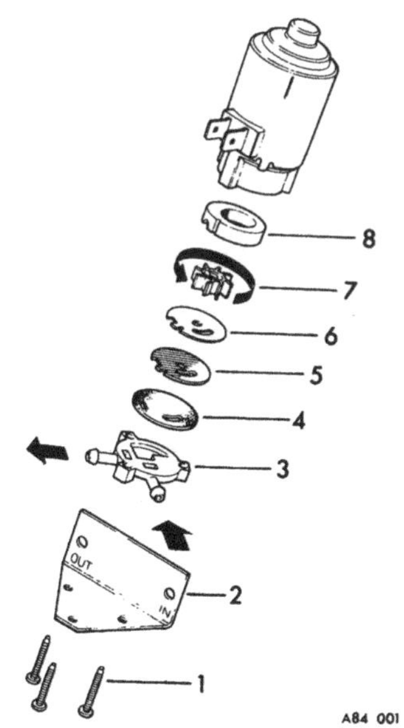

Description 84.15.00

The unit consists of a two speed permanent magnet motor and a gearbox unit which drives a cable rack mechanism. Rotation of the motor armature is converted to a reciprocating motion of the cable rack by a single stage worm and gear, a connecting rod and a cross head contained in a guide channel.

Two speed operation is provided by a third brush. When high speed is selected the positive supply is transferred from the normal speed brush to the high speed brush.

A switching feature stops the blades in the park position irrespective of their position when the steering column switch is selected OFF. This is effected by a two stage limit switch unit attached to the gearbox. The contacts are actuated by a straight cam slope on a slider block which is traversed by a projection from the cross head.

When the steering column switch is selected **OFF**, the motor will continue to run until the limit switch first stage contacts open. A momentary period follows during which no contact is made. The second stage contacts then close causing regenerative braking of the armature which maintains consistent parking of the blades.

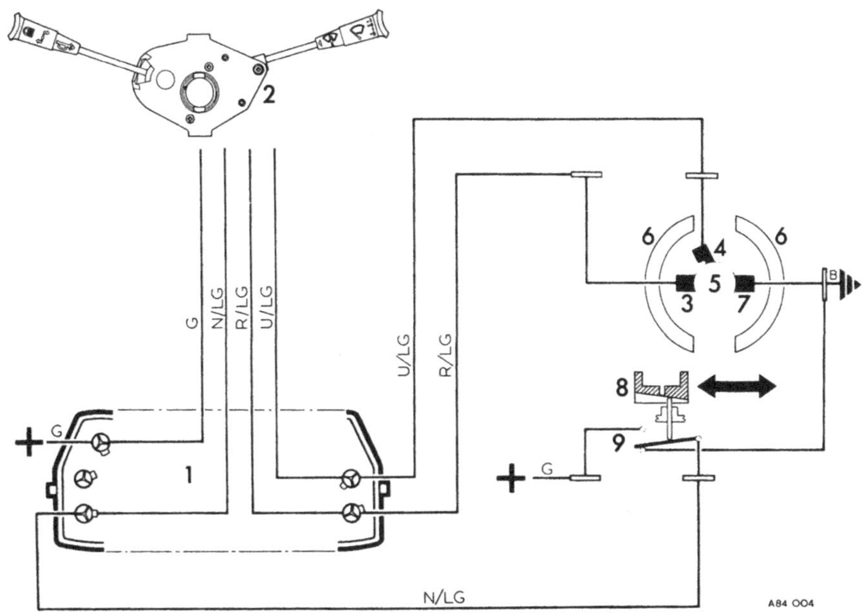

+ Supply
1. Steering column harness plug
2. Steering column multi-purpose switch–washer/wiper unit:

2	HIGH SPEED	G to U/LG
1	NORMAL SPEED	G to R/LG
0	OFF	N/LG to R/LG
1	SWEEP WIPE FACILITY	G to R/LG

3. Normal speed brush
4. High speed brush
5. Commutator
6. Permanent magnet
7. Earth brush
8. Slider block
9. Limit switch unit

DATA

Motor

Manufacturer	Lucas
Type	16W
Lucas Part No.: motor minus gear assembly	
gear assembly	
Triumph Part No.: motor minus gear assembly	
gear assembly	
Running current–after 60 seconds from cold with connecting rod removed:	
Normal speed	1.5 amp
High speed	2.0 amp
Running speed–final gear after 60 seconds from cold with connecting rod removed:	
Normal speed	46 to 52 rev/min.
High speed	60 to 70 rev/min.
Armature end-float	0.002 to 0.008 in. (0.05 to 0.20 mm)
Brush length–normal speed: new	0.380 in. (9.65 mm)
renew if less than	0.180 in. (4.76 mm)
high speed: new	0.380 in. (9.65 mm)
renew if less than	0.280 in. (7.11 mm) (i.e. when narrow section is worn to step into full width section)
earth: new	0.380 in. (9.65 mm)
renew if less than	0.180 in. (4.76 mm)
Brush spring pressure–when compressed so brush bottom is aligned with brushbox slot end	5 to 7 ozf (140 to 200 gf)
Maximum permissible force to move cable rack in tubing–arms and blades removed	6 lbf (3 kgf)

WINDSCREEN WIPER ARM

Driver's – remove and refit 84.15.02

Remove

1 Position a screwdriver as shown and impart a twisting action to lift the clip from the spindle groove.
2 The arm may now be removed from the spindle by hand.
3 Remove the spindle nut.
4 Remove the distance piece.
5 Remove the wiper arm pivot plate.

Refit

6 Position the wiper arm pivot plate.
7 Position the distance piece with the tongue correctly located through the pivot plate, rubber gasket and body slot.
8 Fit the spindle nut.
9 Ensure that the spindle is in the 'park' position.
10 Locate the splines for a suitable 'park' position. Push on to engage the clip to the spindle groove.

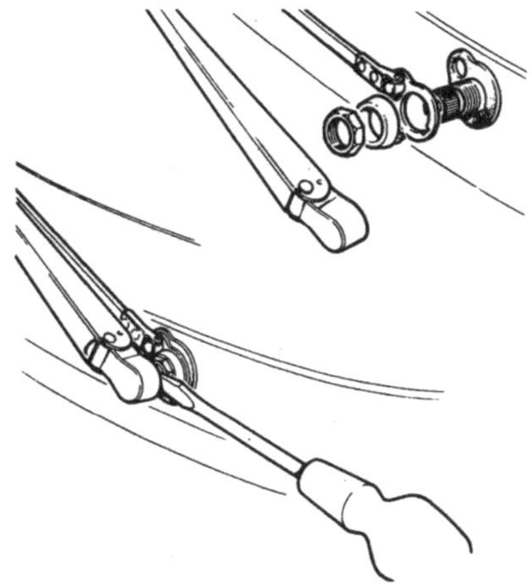
A86 014

WINDSCREEN WIPER BLADE

Driver's – remove and refit 84.15.06

Remove

1 Lift the wiper arm and blade away from the screen.
2 Simultaneously depress the clip and withdraw the blade pin from the pivot block.
CAUTION: If the wiper blade refit is not to take place immediately protect the windscreen glass as follows: Bind the arm end with suitable tape. Isolate the battery to ensure that the wiper motor is not energised.

Refit

3 Reverse 1 to 2.

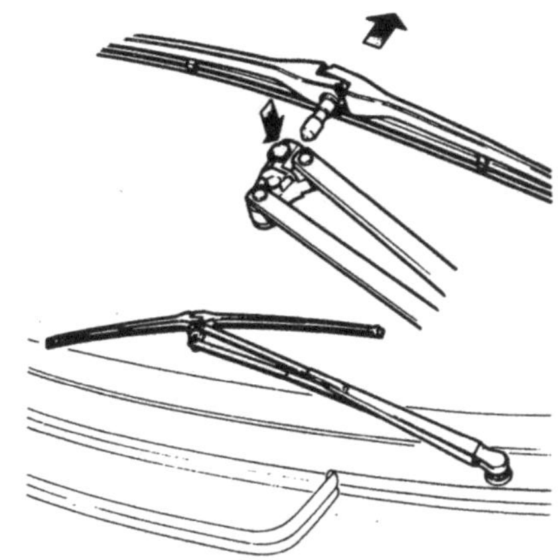
A86 016

WINDSCREEN WIPER ARM

Passenger's – remove and refit 84.15.03

Remove

1 Lift the wiper arm and blade from the screen so that it falls into its service position.
2 Position a screwdriver as shown and impart a twisting action to lift the clip from the spindle groove.
3 The assembly may now be removed by hand.

Refit

4 Ensure that the spindles are in the 'park' position.
5 Hinge the wiper arm against the spring to adopt its service position.
6 Locate the splines for a suitable 'park' position. Push on to engage the clip to the spindle groove.
7 Lower the wiper arm to the screen.

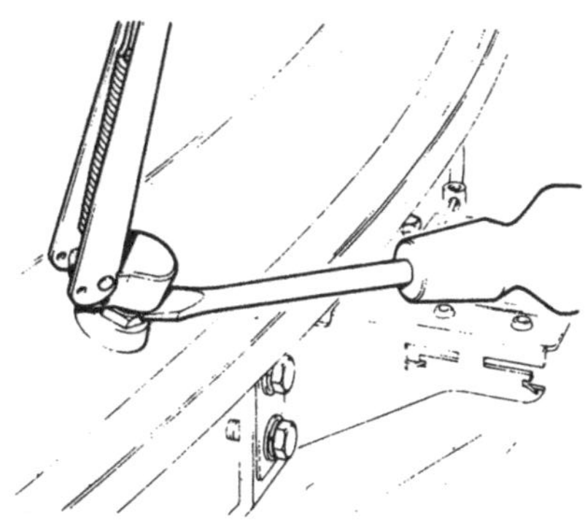
A86 015

WINDSCREEN WIPER BLADE

Passenger's–remove and refit 84.15.07

Remove

1 Lift the wiper arm and blade from the screen so that it falls into its service position.
2 Simultaneously lift the clip and withdraw the blade pin from the arm.

Refit

3 Reverse 1 to 2.

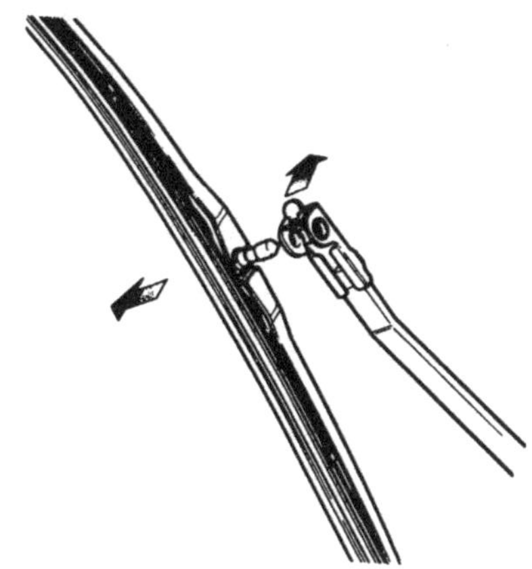
A86 017

WINDSCREEN WIPER SYSTEM

Motor and drive assembly – remove and refit **84.15.10**

Remove

1 Remove the passenger's wiper arm, see 84.15.03.
2 Remove the driver's wiper arm from the wheelbox spindle, see 84.15.02 operations 1 to 2.
3 Remove the fresh air duct, see 80.15.31.
4 Disconnect two 35 amp Lucar connectors from the battery lead connector.
5 Remove the windscreen washer reservoir tank.
6 Remove the battery, see 86.15.01.
7 Disconnect the harness plug.
8 Remove single screw. Disengage the clamp strap from the vehicle body slot.
9 Remove two Pozidriv screws and washers. Withdraw the plate adjacent to the brake master cylinder servo.
10 At the driver's wiper arm remove the spindle nut, distance piece, pivot plate and rubber gasket.
11 At the passenger's wiper arm remove the spindle nut, distance piece and rubber gasket.
12 Manoeuvre the motor and drive assembly from the vehicle.

Refit

13 Position the motor and drive assembly to the vehicle.
14 At the passenger's wiper arm fit the rubber gasket, distance piece and spindle nut.
15 At the driver's wiper arm fit the rubber gasket, pivot plate, distance piece and spindle nut.
16 Insert the plate adjacent to the brake master cylinder servo. Secure with two Pozidriv screws and washers.
17 Slacken the olive nut.
18 Ensure that the rubber pad is correctly positioned.
19 Position the motor and tubes for the best alignment.
20 Engage the clamp strap to the vehicle body slot. Secure with single screw.
21 Tighten the olive nut.
22 Connect the harness plug.
23 Fit the battery, see 86.15.01.
24 Fit the windscreen washer reservoir tank.
25 Connect two 35 amp Lucar connectors to the battery lead connector.
26 Fit the fresh air duct, see 80.15.31.
27 Run the motor and 'switch off' using the windscreen wiper switch so that the wheelbox spindles assume the park position.
28 Fit the driver's wiper arm to the wheelbox spindle, see 84.15.02 operations 9 to 10.
29 Fit the passenger's wiper arm, see 84.15.03.

WINDSCREEN WIPER SYSTEM

Motor – remove and refit 84.15.12

Remove

1 Remove the motor and drive assembly, see 84.15.10.
2 Remove five screws. Lift off the gearbox cover.
3 Remove the crankpin spring clip by withdrawing sideways. Remove the washer.
4 Carefully withdraw the connecting rod. Remove the washer.
5 Lift out the cross-head, rack and tube assembly.

Refit

6 Lubricate all moving parts of the motor during assembly as instructed on the illustration.
7 Position the slider block with the direction of cam slope as shown.
8 Position the cross-head, rack and tube assembly locating the projection in the slider block slot.
9 Fit the washer. Carefully insert the connecting rod.
10 Fit the washer. Fit the crankpin spring clip by inserting sideways.
11 Position the gearbox cover. Secure with five screws.
12 Fit the motor and drive assembly, see 84.15.10.

RLG Ragosine Listate grease
ST410 Shell turbo 41 oil

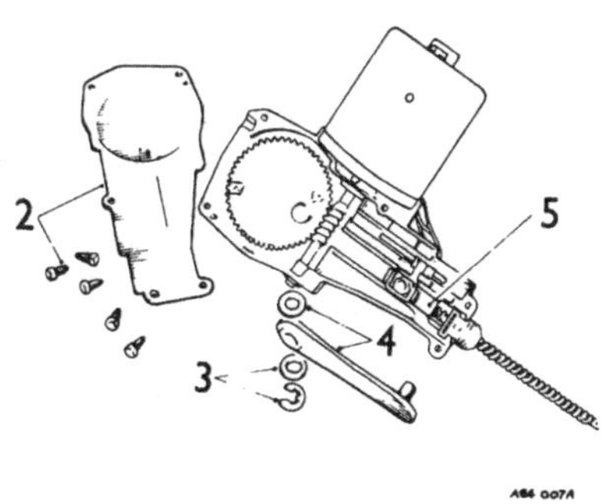

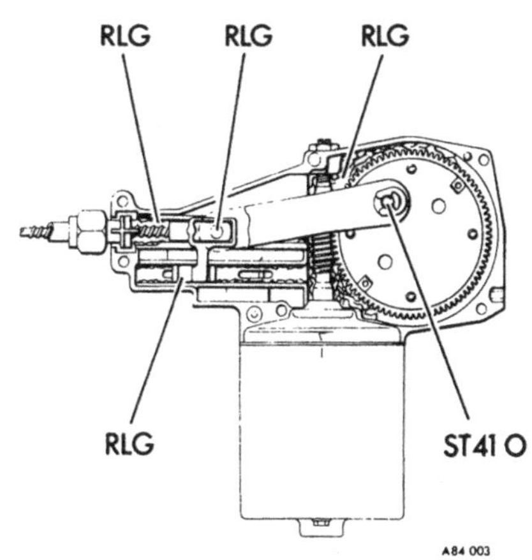

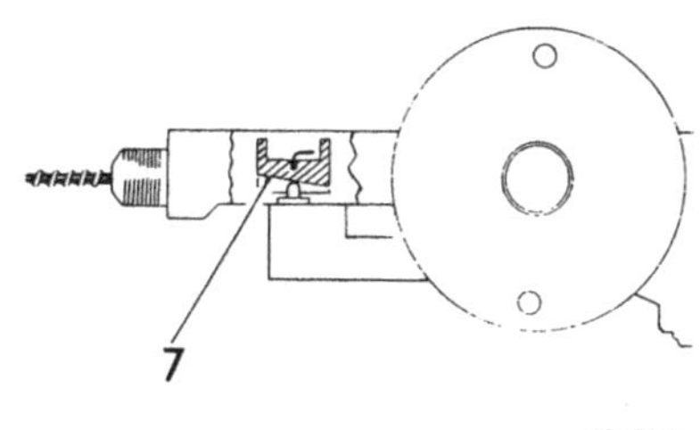

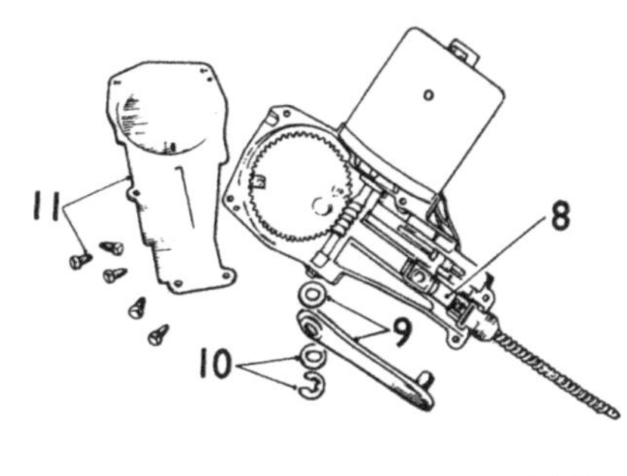

WINDSCREEN WIPER MOTOR

Overhaul 84.15.18

Disassemble

1 Remove five screws. Lift off the gearbox cover.
2 Lift out the slider block.
3 Remove the crankpin spring clip by withdrawing sideways. Remove the washer.
4 Carefully withdraw the connecting rod. Remove the washer.
5 Remove the final gear shaft spring clip by withdrawing sideways. Remove the washer.
6 Ensure that the shaft is burr-free and withdraw it. Remove the dished washer.
7 Remove the thrust screw and locknut.
8 Remove the through bolts.
9 Carefully withdraw the cover and armature about 0.2 in (5 mm). Continue withdrawal allowing the brushes to drop clear of the commutator. Ensure that the three bushes are not contaminated with grease.
10 Pull the armature from the cover against the action of the permanent magnet.
11 Scribe a line round the limit switch to note its position on the gearbox.
12 Remove three screws to release the brush assembly.
13 Remove two screws and washers to release the limit switch.
14 Remove both units joined together by the wires.
15 Remove the plate.

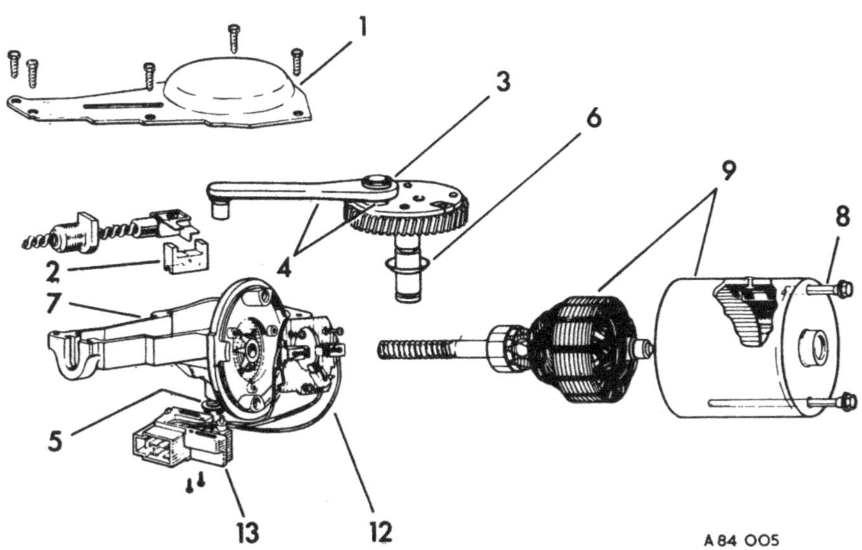

A84 005

Assemble

16 Lubricate all moving parts of the motor during assembly as instructed in the text and on the illustration.
17 Position the plate so that the round hole will accommodate the limit switch plunger.
18 Position the limit switch to the scribe lines made at operation 11 above. Secure with two screws and washers.
19 Secure the brush assembly with three screws.
20 Lubricate the cover bearing and saturate the cover bearing felt washer with Shell Turbo 41 oil.
21 Position the armature to the cover against the action of the permanent magnet.
22 Lubricate the self-aligning bearing with Shell Turbo 41 oil.
23 Ensure that the three brush springs and brushes are correctly positioned. Retain in position using slave clips locally made from paper clips or similar wire as shown.
24 Carefully insert the armature shaft through the bearing. Ensure that the brushes are not contaminated with lubricant. Ensure that the commutator clears the brushes.
25 With the brushes over the commutator remove the slave clips.

continued

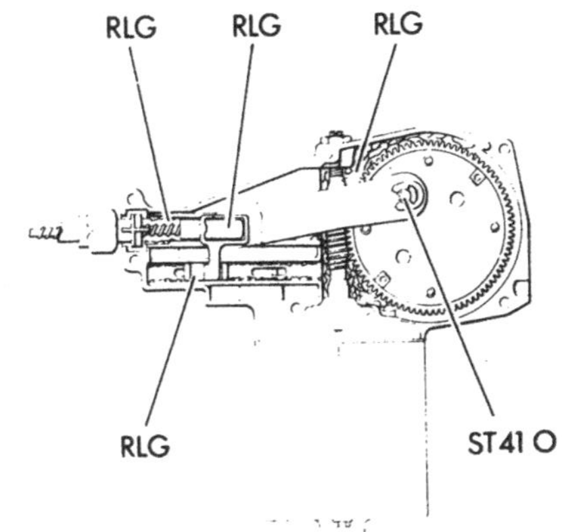

A84 003

RLG Ragosine Listate grease
ST410 Shell turbo 41 oil

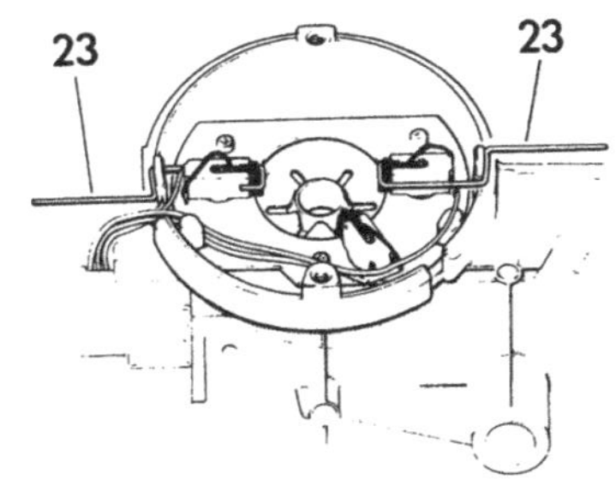

A84 002

26 Seat the cover against the gearbox. Turn the cover to align the marks shown. Fit the through bolts.
27 Fit the thrust screw and locknut.
28 Adjust the armature end-float as follows:
Slacken the locknut. Screw the thrust screw in until resistance is felt. Screw the thrust screw out a quarter of a turn – maintain in this position and tighten the locknut.
29 Lubricate the final gear bushes with Shell Turbo 41 oil.
30 Fit the dished washer with its concave surface facing the final gear. Insert the shaft.
31 Fit the washer. Fit the spring clip by inserting sideways.
NOTE: If the motor is to be immediately fitted to the motor and drive assembly, ignore operations 32 to 35. Refit the motor, see 84.15.12. If the motor is to be stored perform the remaining operations.
32 Position the slider block with the direction of cam slope as shown.
33 Fit the washer. Carefully insert the connecting rod.
34 Fit the washer. Fit the crankpin spring clip by inserting sideways.
35 Position the gearbox cover. Secure with five screws.

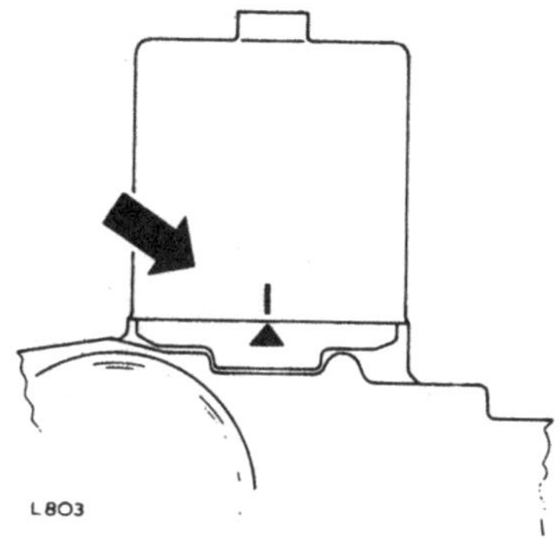

L803

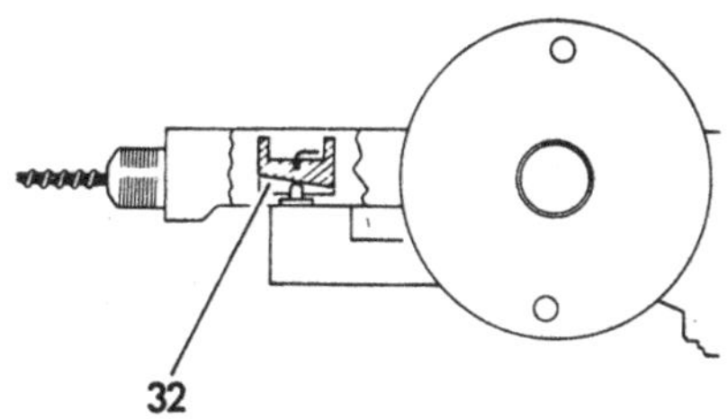

A84 010

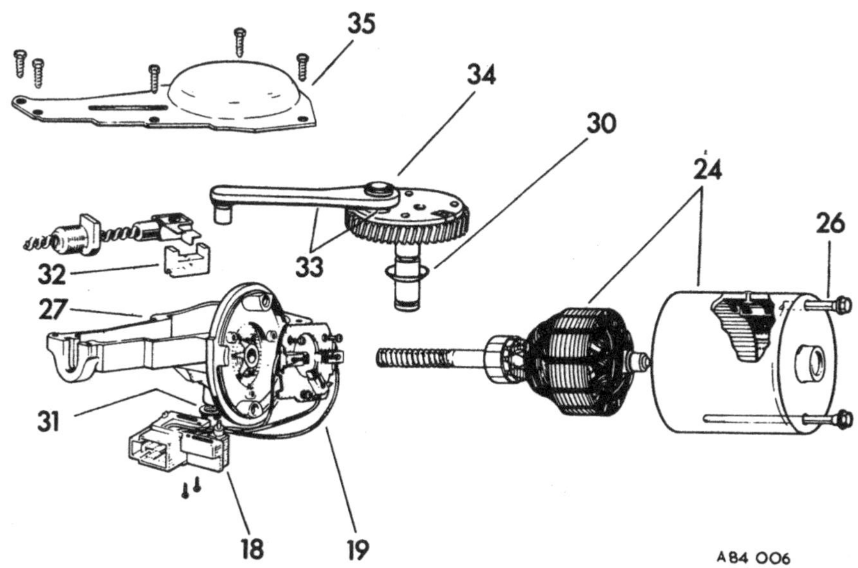

A84 006

WINDSCREEN WIPER SYSTEM

Rack – remove and refit **84.15.24**

Remove
1 Remove the motor and drive assembly, see 84.15.10.
2 Remove the motor, see 84.15.12.
3 Withdraw the rack from the tube assembly.
4 Remove the ferrule.

Refit
5 Fit the ferrule.
6 Lubricate the rack with Ragosine Listate grease.
7 Insert the rack into the tube assembly. If necessary slightly rotate the wheelbox spindles by hand to facilitate rack engagement.
8 Fit the motor, see 84.15.12.
9 Fit the motor and drive assembly, see 84.15.10.

WINDSCREEN WIPER SYSTEM

Wheelbox – driver's – remove and refit **84.15.28**

Earlier models

Removing

1 Remove the motor and rack assembly complete, see 84.15.10.
2 Remove two Pozidriv screws and washers. Withdraw the plate adjacent to the brake master cylinder servo.
3 Disengage the tube from the passenger's wheelbox.
4 Remove the spindle nut.
5 Remove the distance piece.
6 Remove the wiper arm pivot plate.
7 Remove the rubber gasket.
8 Withdraw the wheelbox spindle from the vehicle body aperture.
9 Remove the driver's wheelbox and two tubes as an assembly from the vehicle.
10 Note the position of the long tube and the short tube.
11 Scribe a line to note the radial position of the long tube.
12 Remove two nuts and remove the wheelbox plate.
13 Disengage two tubes from the wheelbox.

Refitting

14 Reverse 3 to 13.
15 Ensure that the position of the tube and wheelbox achieves the best alignment.
16 Insert the plate adjacent to the brake master cylinder servo. Secure with two Pozidriv screws and washers.
17 Fit the motor and rack assembly complete, see 84.15.10.

WINDSCREEN WIPER SYSTEM

Wheelbox – driver's – remove and refit **84.15.28**

Later models

Removing

1 Remove the motor and drive assembly, see 84.15.10.
2 Scribe a line to note the radial position of the tube.
3 Remove the two nuts. Remove the wheelbox plate.
4 Disengage and remove the short straight tube.
5 Disengage and remove the wheelbox.

Refitting

6 If a new wheelbox is to be fitted reproduce the scribe line on the new unit.
7 Fit the wheelbox the correct way round.
8 Lubricate the wheelbox and exposed section of the rack with Ragosine Listate grease.
9 Reverse 1 to 4.

WINDSCREEN WIPER SYSTEM

Wheelbox – passenger's – remove and refit **84.15.29**

Earlier models

Removing

1 Remove the motor rack assembly complete, see 84.15.10.
2 Disengage the tube from the wheelbox.
3 Remove the spindle nut.
4 Remove the distance piece.
5 Remove the rubber washer.
6 Withdraw the wheelbox spindle from the vehicle body aperture.

Refitting

7 Reverse 1 to 6.

WINDSCREEN WIPER SYSTEM

Wheelbox – passenger's – remove and refit **84.15.29**

Later models

Removing

1 Remove the motor and drive assembly, see 84.15.10.
2 Scribe two lines to note the radial position of the tube.
3 Remove two nuts. Remove the wheelbox plate.
4 Disengage and remove the tube and far wheelbox assembly.
5 Disengage and remove the wheelbox.

Refittting

6 If a new wheelbox is to be fitted reproduce the scribe lines on the new unit.
7 Fit the wheelbox the correct way round.
8 Lubricate the wheelbox and exposed section of the track with Ragosine Listate grease.
9 Reverse 1 to 4.

WINDSCREEN WIPER DELAY UNIT

—Remove and refit **84.25.31**

Removing

1 Disconnect both multi-pin plugs from the delay unit.
2 Remove the mounting screws and nuts, and withdraw the delay unit and insulating plate.

Refitting

3 Reverse instructions 1 and 2, ensuring that the earth lead is fitted to one of the mounting screws.

AKM3079B
AKM3079/1

ELECTRICAL OPERATIONS

continued

BULB CHART

	Watts	Lucas Part No.	Unipart No.	Triumph Part No.	
Headlamps:					
L.H. dip.	75/60		UKC 2423		†
R.H. dip – Normal	60/50	54523079	GLU 114	215735	†
France	45/40	411	GLB 411	510219	
U.S.A.	50/40			514578	†
Front parking lamp	4				
Front flasher lamp	21	382	GLB 382	502379	
Front marker lamp	3	504	GLB 504	518414	
Rear marker lamp	3	504	GLB 504	518414	
Tail lamp	5	207	GLB 207	57591	
Stop lamp	21	382	GLB 382	502379	
Rear flasher lamp	21	382	GLB 382	502379	
Reverse lamp	21	382	GLB 382	502379	
Plate illumination lamp	5	207			
Roof lamp	6			586438	
Instrument panel illumination	2.2				
Instrument panel warning light	1.2	286		UKC 2412	
Hazard warning light	1.2	286		UKC 2412	

† Sealed beam light unit

WIRING DIAGRAM—TR7 RIGHT- AND LEFT-HAND STEER
UK AND EUROPE—1976 AND 1977 MODEL YEARS

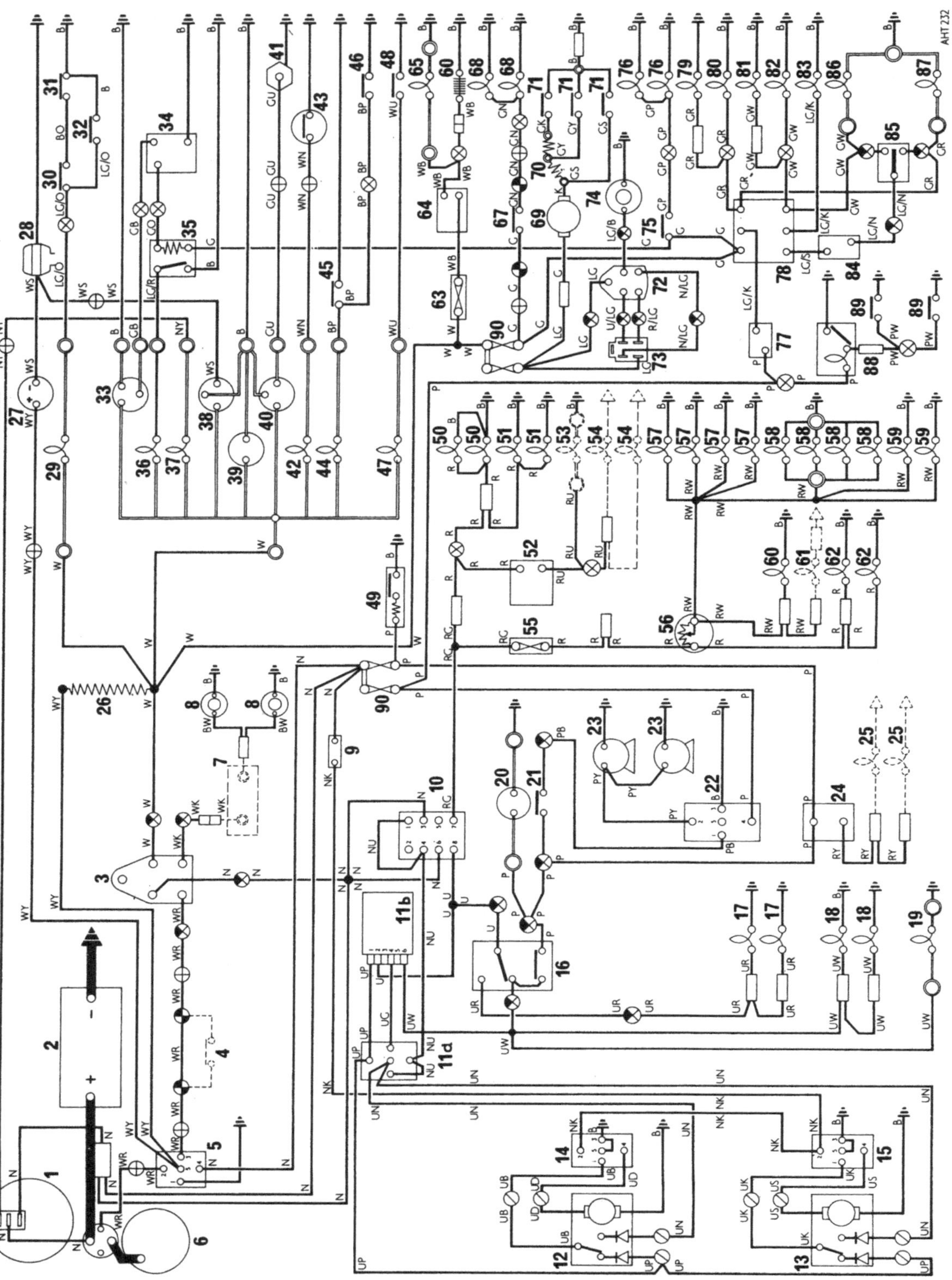

WIRING DIAGRAM KEY—TR7 RIGHT- AND LEFT-HAND STEER
UK AND EUROPE—1976 AND 1977 MODEL YEARS

1 Alternator
2 Battery
3 Ignition/starter switch
4 Starter inhibitor switch (automatic transmission only)
5 Starter motor relay
6 Starter motor
7 Radio
8 Speaker
9 Headlamp – circuit breaker
10 Master light switch
11a Headlamp – flash relay
11b Headlamp – flash control unit
12 Headlamp – actuator passenger's side
13 Headlamp – actuator driver's side
14 Headlamp – run/stop relay passenger's side
15 Headlamp – run/stop relay driver's side
16 Main/dip/flash switch
17 Dip beam
18 Main beam
19 Main beam warning light
20 Clock
21 Horn push
22 Horn relay
23 Horn
24 Fog lamp switch
25 Fog lamp
26 Ballast resistor wire
27 Ignition cowl
28 Ignition distributor
29 Fasten belts warning light
30 Passenger's seat switch
31 Passenger's belt switch
32 Driver's belt switch
33 Fuel indicator
34 Fuel tank unit
35 Fuel warning light delay unit
36 Fuel warning light
37 Ignition warning light
38 Tachometer
39 Battery condition indicator
40 Temperature indicator
41 Temperature transmitter
42 Oil pressure warning light
43 Oil pressure switch
44 Brake warning light
45 Brake line failure switch
46 Handbrake switch
47 Choke warning light
48 Choke switch
49 Cigarette lighter
50 Plate illumination lamp
51 Tail lamp
52 Rear fog lamp switch
53 Rear fog lamp warning light
54 Rear fog lamp
55 In line fuse
56 Panel rheostat
57 Heater control illumination
58 Instrument illumination
59 Facia switch panel illumination
60 Cigarette lighter illumination
61 Selector panel illumination (automatic transmission only)
62 Front parking lamp
63 In line fuse
64 Heated backlight switch
65 Heated backlight warning light
66 Heated backlight
67 Reverse lamp switch
68 Reverse lamp
69 Heater motor
70 Heater resistor
71 Heater switch
72 Washer/wiper switch
73 Wiper motor
74 Washer pump
75 Stop lamp switch
76 Stop lamp
77 Hazard flasher unit
78 Hazard switch
79 LH front flasher lamp
80 LH rear flasher lamp
81 RH front flasher lamp
82 RH rear flasher lamp
83 Hazard warning light
84 Turn signal flasher unit
85 Turn signal switch
86 LH flasher warning light
87 RH flasher warning light
88 Roof lamp
89 Door switch
90 Fuse

COLOUR CODE

B	Black	N	Brown	S	Slate
G	Green	O	Orange	U	Blue
K	Pink	P	Purple	W	White
LG	Light green	R	Red	Y	Yellow

WIRING DIAGRAM — TR7 — 1978 AND 1979 UK AND EUROPEAN MARKET
RIGHT-HAND AND LEFT-HAND STEERING

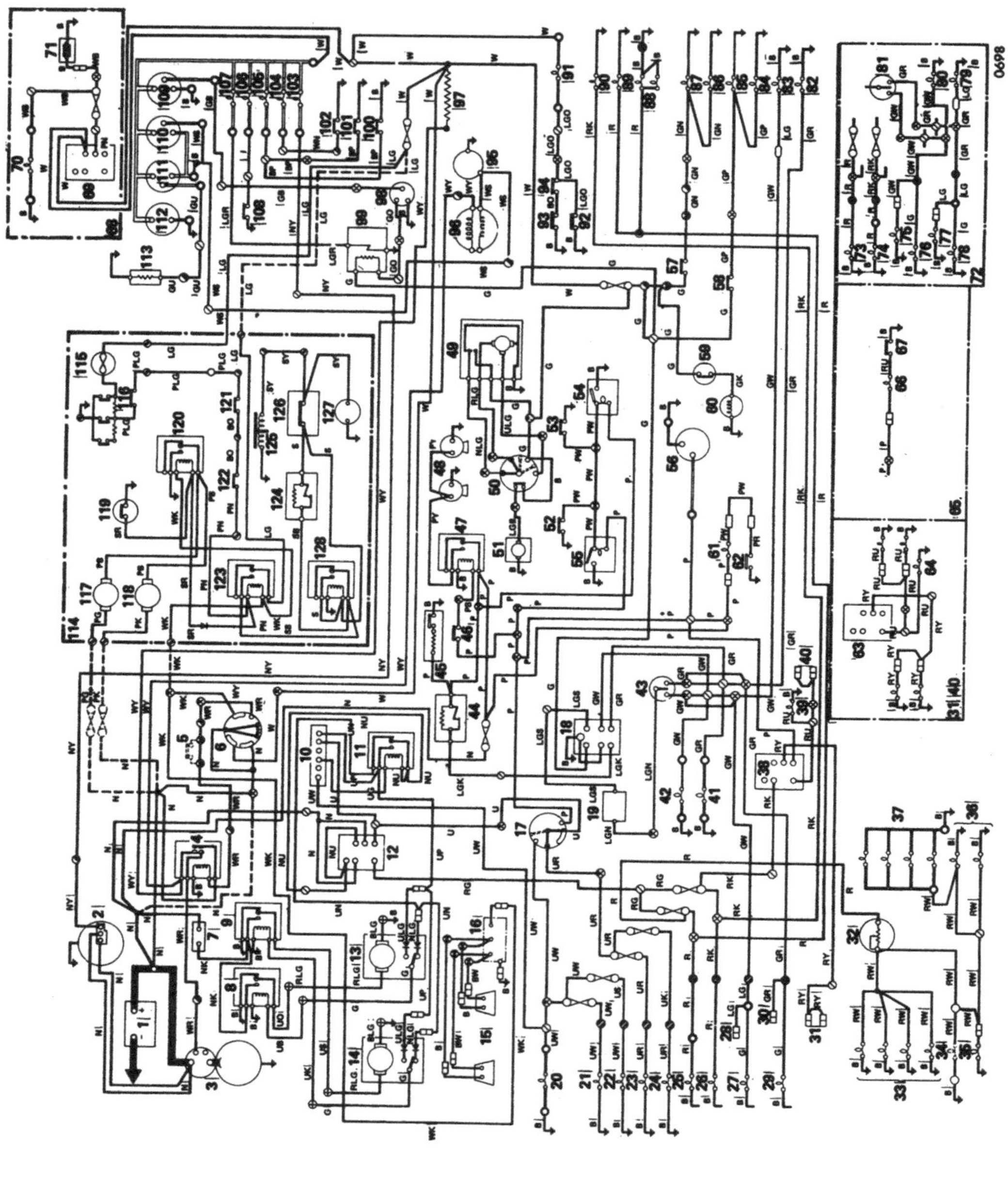

WIRING DIAGRAM KEY—TR7—1978 AND 1979—UK AND EUROPE—RIGHT- AND LEFT-HAND STEER

1 Battery
2 Alternator
3 Starter motor
4 Starter motor relay
5 Start inhibit switch (Automatic)
6 Ignition switch
7 Headlamp motor circuit breaker
8 Headlamp relay
9 Headlamp relay
10 Headlamp flash control
11 Headlamp flash relay
12 Master light switch
13 Headlamp motor
14 Headlamp motor
15 Loudspeakers
16 Radio
17 Main dip flash switch
18 Hazard switch
19 Flasher unit
20 Main beam warning light
21 Main beam
22 Main beam
23 Dip beam
24 Dip beam
25 Sidelight
26 Sidelight
27 Side indicator
28 Side repeater connection
29 Side indicator
30 Side repeater connection
31 Front fog lamps junction
32 Panel rheostat
33 Heater control illumination
34 Borg-Warner illumination
35 Cigar lighter illumination
36 Switch panel illumination
37 Panel illumination
38 Fog lamps switch
39 Rear fog lamps warning light
40 Rear fog lamps junction
41 L.H. indicator warning light
42 R.H. indicator warning light
43 Direction indicator switch
44 Hazard unit
45 Cigar lighter
46 Horn-push
47 Horn relay
48 Horns
49 Windscreen wiper motor
50 Windscreen wipe/wash switch
51 Windscreen washer motor
52 Door switch
53 Door switch
54 Courtesy light
55 Courtesy light
56 Clock
57 Reverse light switch
58 Stop light switch
59 Engine thermostat
60 F.A.S.D.
61 Boot lamp
62 Boot lamp switch
63 Fog switch
64 Rear fog lamp warning light
65 Boot lamp and switch wiring assembly – L.H. steering
66 Boot lamp
67 Boot lamp switch
68 Heated rear screen assembly – TR7 coupé
69 Heated rear screen switch
70 Heated rear screen warning light
71 Heated rear screen
72 Wiring condition for front, side and front and rear indicator lights – L.H. steering
73 Sidelight
74 Sidelight
75 Side repeater
76 Side indicator
77 Side repeater
78 Side indicator
79 Rear side indicator
80 Rear side indicator
81 Direction indicator switch
82 Side indicator
83 Side indicator
84 Stop light
85 Stop light
86 Reverse light
87 Reverse light
88 Number-plate lights
89 Tail light
90 Tail light
91 Seat belt warning light
92 Driver's buckle switch
93 Passenger belt switch
93 Passenger seat switch
94 Passenger seat switch
95 Distributor
96 6 volt coil
97 Eureka wire
98 Tank unit
99 Low fuel delay unit
100 Brake pressure differential switch
101 Hand brake switch
102 Oil pressure switch
103 Ignition warning light
104 Oil warning light
105 Brake warning light
106 Choke warning light
107 Low fuel warning light
108 Choke warning light switch
109 Fuel gauge
110 Tachometer
111 Temperature gauge
112 Battery condition indicator
113 Temperature transmitter
114 Air conditioning blower unit
115 Heater/air conditioning blower unit
116 Thermostat
117 R.H. condenser fan
118 L.H. condenser fan
119 Radiator thermostat
120 Fan relay
121 Full throttle cut-off switch
122 Air conditioning cut-out switch
123 Air conditioning relay
124 Delay unit
125 Throttle jack
126 Ranco valve high pressure cut-out
127 Clutch
128 Delay circuit relay

COLOUR CODE

B	Black	N	Brown	S	Slate
G	Green	O	Orange	U	Blue
K	Pink	P	Purple	W	White
LG	Light green	R	Red	Y	Yellow

WIRING DIAGRAM — TR7 — 1980 UK AND EUROPEAN MARKET

RIGHT-HAND AND LEFT-HAND STEERING

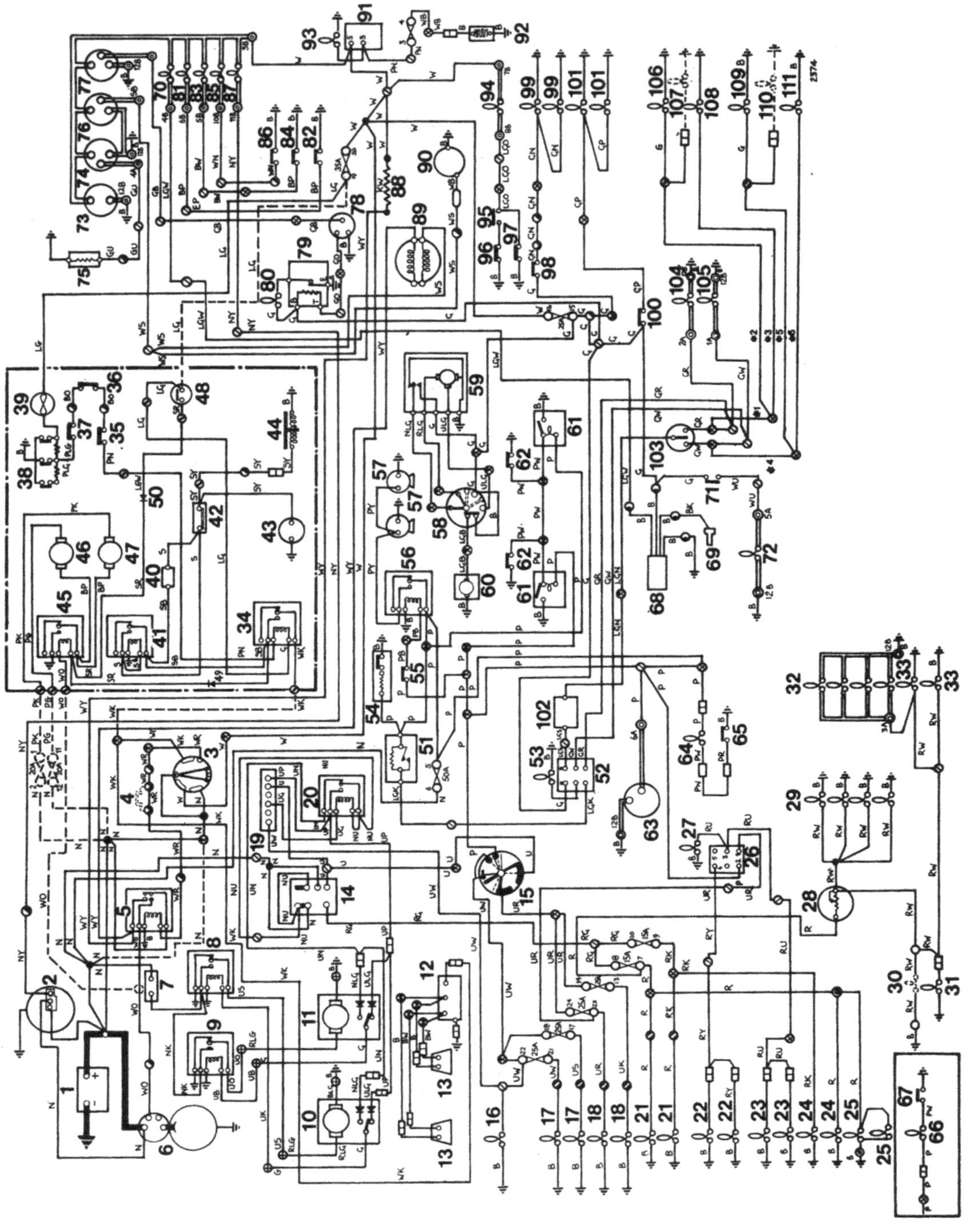

AKM3079B

WIRING DIAGRAM KEY—TR7—1980 UK AND EUROPE—RIGHT- AND LEFT-HAND STEER

1 Battery
2 Alternator
3 Ignition/starter switch
4 Start inhibit switch (automatic transmission only)
5 Starter relay
6 Starter motor
7 Headlamp motor circuit breaker
8 Headlamp run/stop relay — L.H.
9 Headlamp run/stop relay — R.H.
10 Headlamp actuator — L.H.
11 Headlamp actuator — R.H.
12 Radio
13 Speaker
14 Master light switch
15 Main/dip/flash switch
16 Main beam warning light
17 Main beam
18 Dip beam
19 Headlamp delay unit
20 Headlamp flash relay
21 Front parking lamp
22 Front fog lamp
23 Rear guard lamp
24 Tail lamp
25 Plate illumination lamp
26 Fog switch
27 Rear guard lamp warning light
28 Panel rheostat
29 Heater control illumination
30 Selector panel illumination (automatic transmission only)
31 Cigarette lighter illumination
32 Instrument illumination
33 Switch panel illumination

Air conditioning:
34 Master relay
35 Manual cut-out swich
36 Full throttle cut-out switch
37 Cold thermostat
38 Blower motor switch
39 Blower motor
40 Delay unit
41 Clutch delay relay
42 Ranco valve high pressure cut-out
43 Clutch
44 Throttle jack
45 Condenser fan relay
46 Condenser fan L.H.
47 Condenser fan R.H.
48 Radiator thermostat
49 Diode
50 Diode

51 Hazard flasher unit
52 Hazard switch
53 Hazard warning light
54 Cigarette lighter
55 Horn-push
56 Horn relay
57 Horn
58 Windscreen wash/wipe switch
59 Windscreen wiper motor
60 Windscreen washer pump
61 Map/courtesy lamp
62 Door switch
63 Clock

R.H. steering only:
64 Luggage boot lamp
65 Luggage boot lamp switch

L.H. steering only:
66 Luggage boot lamp
67 Luggage boot lamp switch

68 Coolant low indicator unit
69 Coolant low sensor
70 Coolant low warning light
71 Choke switch
72 Choke warning light
73 Battery condition indicator
74 Temperature indicator
75 Temperature transmitter
76 Tachometer
77 Fuel indicator
78 Tank unit
79 Low fuel delay unit
80 Low fuel warning light
81 Brake warning light
82 Brake pressure differential switch
83 Handbrake warning light
84 Handbrake switch
85 Oil pressure warning light
86 Oil pressure switch
87 Ignition warning light
88 Ballast resistor wire
89 Ignition coil — 6 volt
90 Ignition distributor
91 Heated back-light switch
92 Heated back-light
93 Heated back-light warning light
94 Seat belt warning light
95 Passenger seat switch
96 Passenger belt switch
97 Driver belt switch
98 Reverse lamp switch
99 Reverse lamp
100 Stop lamp switch
101 Stop lamp
101 Stop lamp
102 Turn signal flasher unit
103 Turn signal switch
104 L.H. turn signal warning light
105 R.H. turn signal warning light
106 L.H. front flasher lamp
107 L.H. front flasher repeater lamp
108 L.H. rear flasher lamp
109 R.H. front flasher lamp
110 R.H. front flasher repeater lamp
111 R.H. rear flasher lamp

WIRE COLOUR CODES OF FLASHER LAMP CIRCUIT

Wire identity	Item	Right-hand steering		Left-hand steering	
		Heater vehicles	Air conditioning vehicles	Heater vehicles	Air conditioning vehicles
*1	L.H. supply	G/LG	GR	LG/G	GR
*2	L.H. front lamp	G/LG	GR	LG/G	GR
*3	L.H. rear lamp	GR	GR	GR and LG	GR and LG
*4	R.H. supply	LG/G	GW	G/LG	GW
*5	R.H. front lamp	LG/G	GW	G/LG	GW
*6	R.H. rear lamp	GW and LG	GW and LG	GW	GW

COLOUR CODE

B	Black	N	Brown	S	Slate
G	Green	O	Orange	U	Blue
K	Pink	P	Purple	W	White
LG	Light green	R	Red	Y	Yellow

WIRING DIAGRAM — 1981 UK AND EUROPEAN MARKET
RIGHT-HAND AND LEFT-HAND STEERING

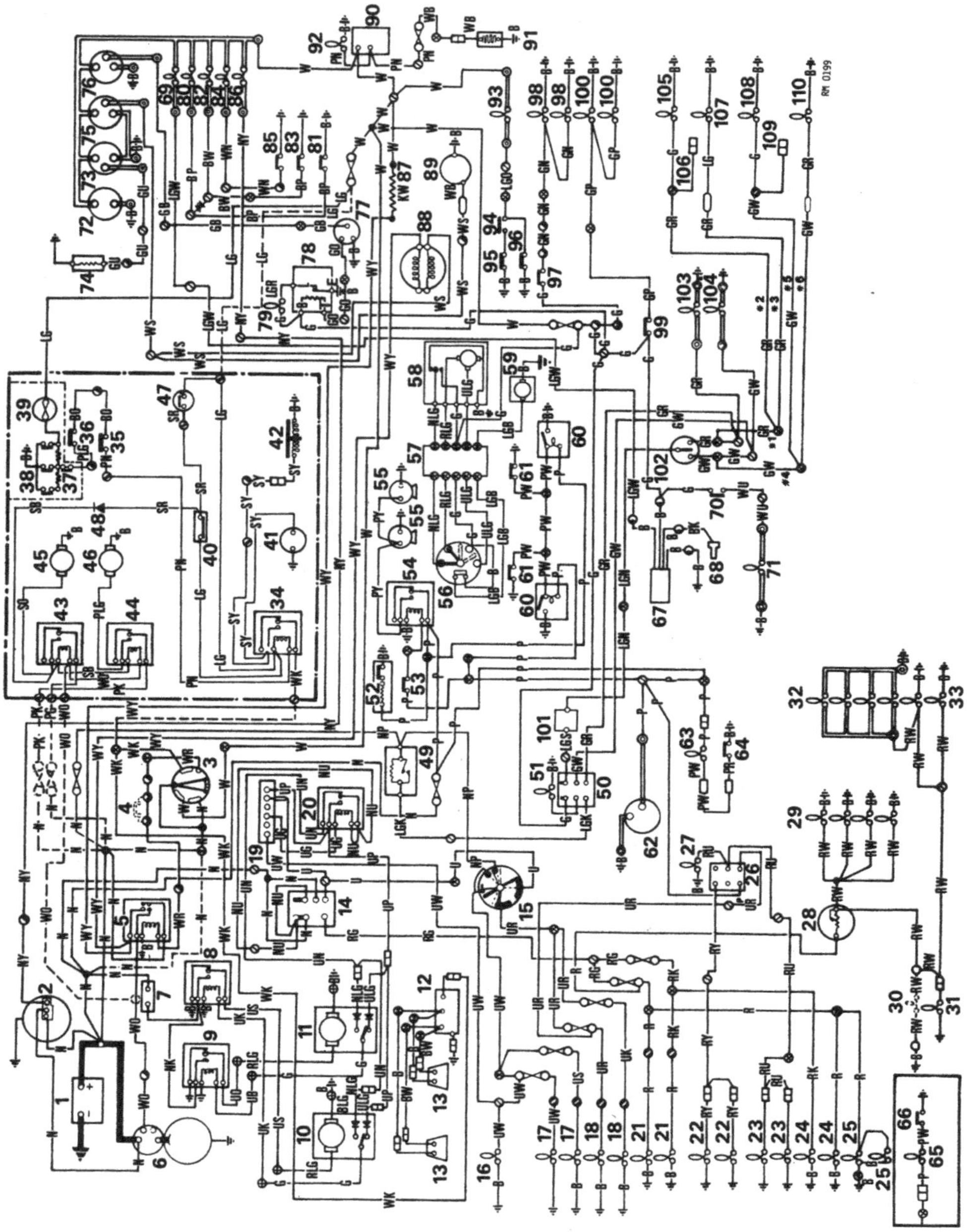

RM 0199

WIRING DIAGRAM KEY—TR7—1981 UK AND EUROPE—RIGHT- AND LEFT-HAND STEER

1 Battery
2 Alternator
3 Ignition/starter switch
4 Start inhibit switch (automatic transmission only)
5 Starter relay
6 Starter motor
7 Headlamp motor circuit breaker
8 Headlamp relay — L.H.
9 Headlamp relay — R.H.
10 Headlamp motor — L.H.
11 Headlamp motor — R.H.
12 Radio
13 Speaker
14 Master light switch
15 Main/dip/flash switch
16 Main beam warning light
17 Main beam
18 Dip beam
19 Headlamp flash unit
20 Headlamp flash relay
21 Front parking lamp
22 Front fog lamp
23 Rear fog guard lamp
24 Tail lamp
25 Number plate illumination lamp
26 Fog switch
27 Rear fog guard lamp warning light
28 Panel rheostat
29 Heater control illumination
30 Selector panel illumination (automatic transmission only)
31 Cigarette lighter illumination
32 Panel illumination
33 Switch panel illumination
34 Clutch relay
35 Air conditioning cut-out switch
36 Full throttle cut-out switch
37 Thermostat
38 Blower motor switch
39 Blower motor
40 Pressure cut in switch
41 Clutch
42 Throttle jack
43 Condenser fan relay — L.H.
44 Condenser fan relay — R.H.
45 Condenser fan — L.H.
46 Condenser fan — R.H.
47 Radiator thermostat
48 Diode
49 Hazard flasher unit
50 Hazard switch
51 Hazard warning light
52 Cigarette lighter
53 Horn-push
54 Horn relay
55 Horn
56 Windscreen wash/wipe switch
57 Wiper delay unit
58 Windscreen wiper motor
59 Windscreen washer motor
60 Map/courtesy lamp
61 Door switch
62 Clock
63 Luggage boot lamp } R.H. steering
64 Luggage boot lamp switch } R.H. steering
65 Luggage boot lamp } L.H. steering
66 Luggage boot lamp switch } L.H. steering
67 Coolant low indicator unit
68 Coolant low sensor
69 Coolant low warning light
70 Choke switch
71 Choke warning light
72 Voltmeter
73 Temperature gauge
74 Temperature transmitter
75 Tachometer
76 Fuel gauge
77 Tank unit
78 Low fuel delay unit
79 Low fuel warning light
80 Brake warning light
81 Brake pressure differential switch
82 Handbrake warning light
83 Handbrake switch
84 Oil pressure warning light
85 Oil pressure switch
86 Ignition warning light
87 Ballast resistor wire
88 Ignition coil — 6 volt
89 Ignition distributor
90 Heated rear screen switch
91 Heated rear screen
92 Heated rear screen warning light
93 Seat belt warning light
94 Passenger seat switch
95 Passenger belt switch
96 Driver belt switch
97 Reverse lamp switch
98 Reverse lamp
99 Stop lamp switch
100 Stop lamp
101 Direction indicator flasher unit
102 Direction indicator switch
103 Indicator warning light — L.H.
104 Indicator warning light — R.H.
105 Front flasher lamp — L.H.
106 Front flasher repeater lamp connection — L.H.
107 Rear flasher lamp — L.H.
108 Front flasher lamp — R.H.
109 Front flasher repeater lamp connection — R.H.
110 Rear flasher lamp — R.H.

WIRE COLOUR CODES OF FLASHER LAMP CIRCUIT

Wire identity	Item	Right-hand steering		Left-hand steering	
		Heater vehicles	Air conditioning vehicles	Heater vehicles	Air conditioning vehicles
*1	L.H. supply	G/LG	GR	LG/G	GR
*2	L.H. front lamp	G/LG	GR	LG/G	GR
*3	L.H. rear lamp	GW	GR	GR and LG	GR and LG
*4	R.H. supply	LG/G	GW	G/LG	GW
*5	R.H. front lamp	LG/G	GW	G/LG	GW
*6	R.H. rear lamp	GR and LG	GW and LG	GW	GW

COLOUR CODE

B	Black	N	Brown	S	Slate
G	Green	O	Orange	U	Blue
K	Pink	P	Purple	W	White
LG	Light green	R	Red	Y	Yellow

WIRING DIAGRAM, 1975—TR7 —U.S.A.—AIR CONDITIONING

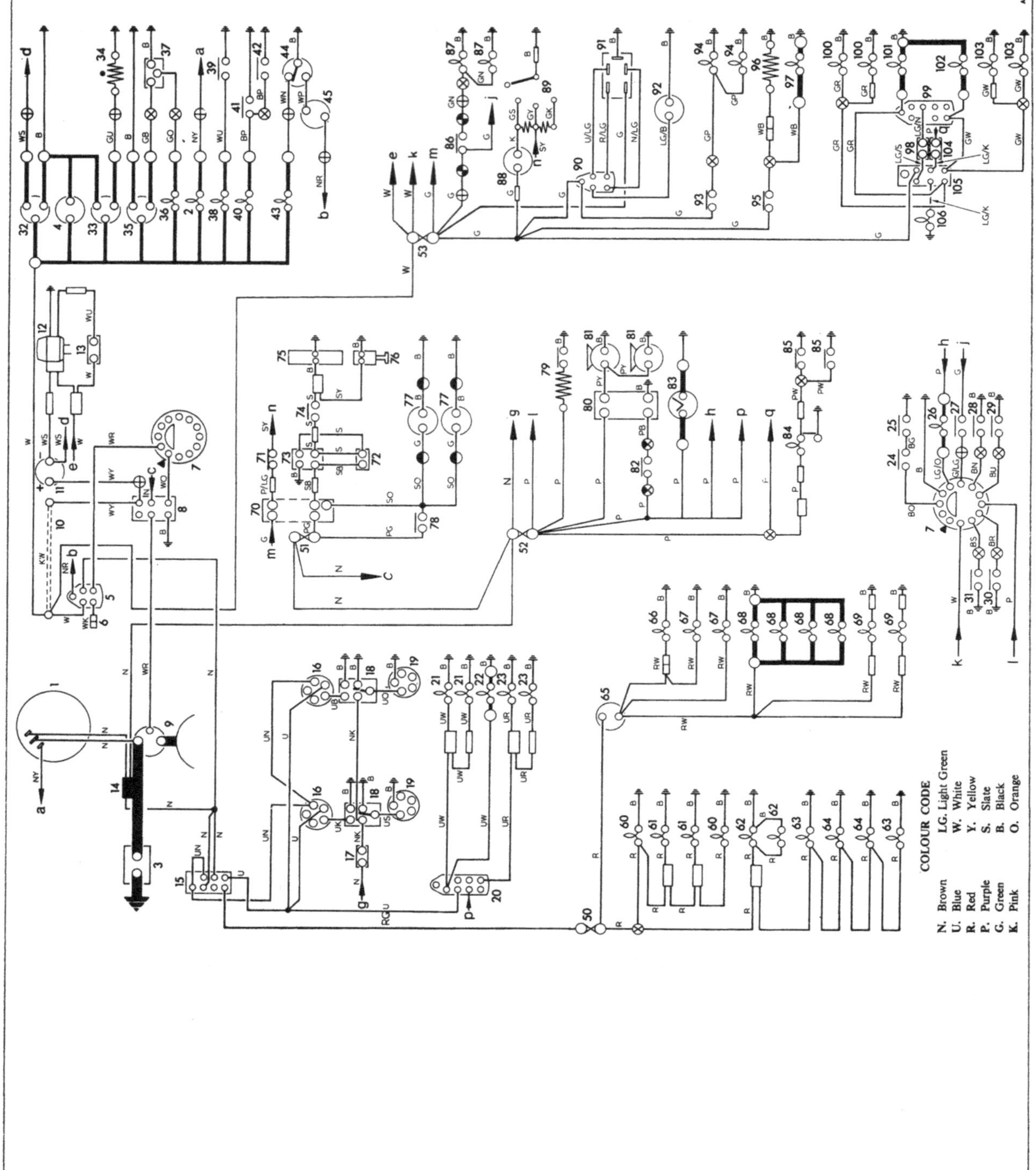

COLOUR CODE

N. Brown	LG. Light Green
U. Blue	W. White
R. Red	Y. Yellow
P. Purple	S. Slate
G. Green	B. Black
K. Pink	O. Orange

WIRING DIAGRAM KEY, 1975—TR7—U.S.A.—AIR CONDITIONING

1. Alternator
2. Ignition warning light
3. Battery
4. Battery condition indicator
5. Ignition/starter switch
6. Radio supply
7. Interlock module
8. Interlock starter motor relay
9. Starter motor
10. Ballast resistor wire
11. Ignition coil
12. Ignition distributor
13. Drive resistor
14. Battery lead connector
15. Master light switch
16. Actuator—limit switch
17. Circuit breaker
18. Headlamp—run/stop relay
19. Actuator—motor
20. Main/dip/flash switch
21. Main beam
22. Main beam warning light
23. Dip beam
24. L.H. door switch
25. Key switch
26. Fasten belts warning light
27. Interlock gearbox switch
28. Driver's belt switch
29. Driver's seat switch
30. Passenger's belt switch
31. Passenger's seat switch
32. Tachometer
33. Temperature indicator
34. Temperature transmitter
35. Fuel indicator
36. Fuel warning light
37. Fuel tank unit

Non-Catalytic Converter vehicles only:

38. Choke warning light
39. Choke switch

Catalytic Converter vehicles only:

38. Catalyst service warning light
39. Catalyst service interval indicator

40. Brake warning light
41. Brake line failure switch
42. Handbrake switch
43. Oil pressure warning light
44. Oil pressure switch
45. Anti-run-on valve
50. Fuse—35 amp
51. Fuse—50 amp
52. Fuse—50 amp
53. Fuse—50 amp
60. Front marker lamp
61. Front parking lamp
62. Plate illumination lamp
63. Rear marker lamp
64. Tail lamp
65. Panel rheostat
66. Cigarette lighter illumination
67. Heater control illumination
68. Instrument illumination
69. Facia switch panel illumination
70. Air conditioning—control relay
71. Cold thermostat
72. Air conditioning—delay circuit flasher unit
73. Air conditioning—delay circuit relay
74. High pressure cut-out
75. Compressor clutch
76. Throttle jack
77. Condenser fan motor
78. Radiator switch
79. Cigarette lighter
80. Horn relay
81. Horn
82. Horn push
83. Clock
84. Roof lamp
85. Door switch
86. Reverse lamp switch
87. Reverse lamp
88. Blower motor
89. Blower motor switch
90. Windscreen washer/wiper switch
91. Windscreen wiper motor
92. Windscreen washer pump
93. Stop lamp switch
94. Stop lamp
95. Heated backlight switch
96. Heated backlight
97. Heated backlight warning light
98. Turn signal flasher unit
99. Turn signal switch
100. L.H. flasher lamp
101. L.H. turn signal warning light
102. R.H. turn signal warning light
103. R.H. flasher lamp
104. Hazard flasher unit
105. Hazard switch
106. Hazard warning light

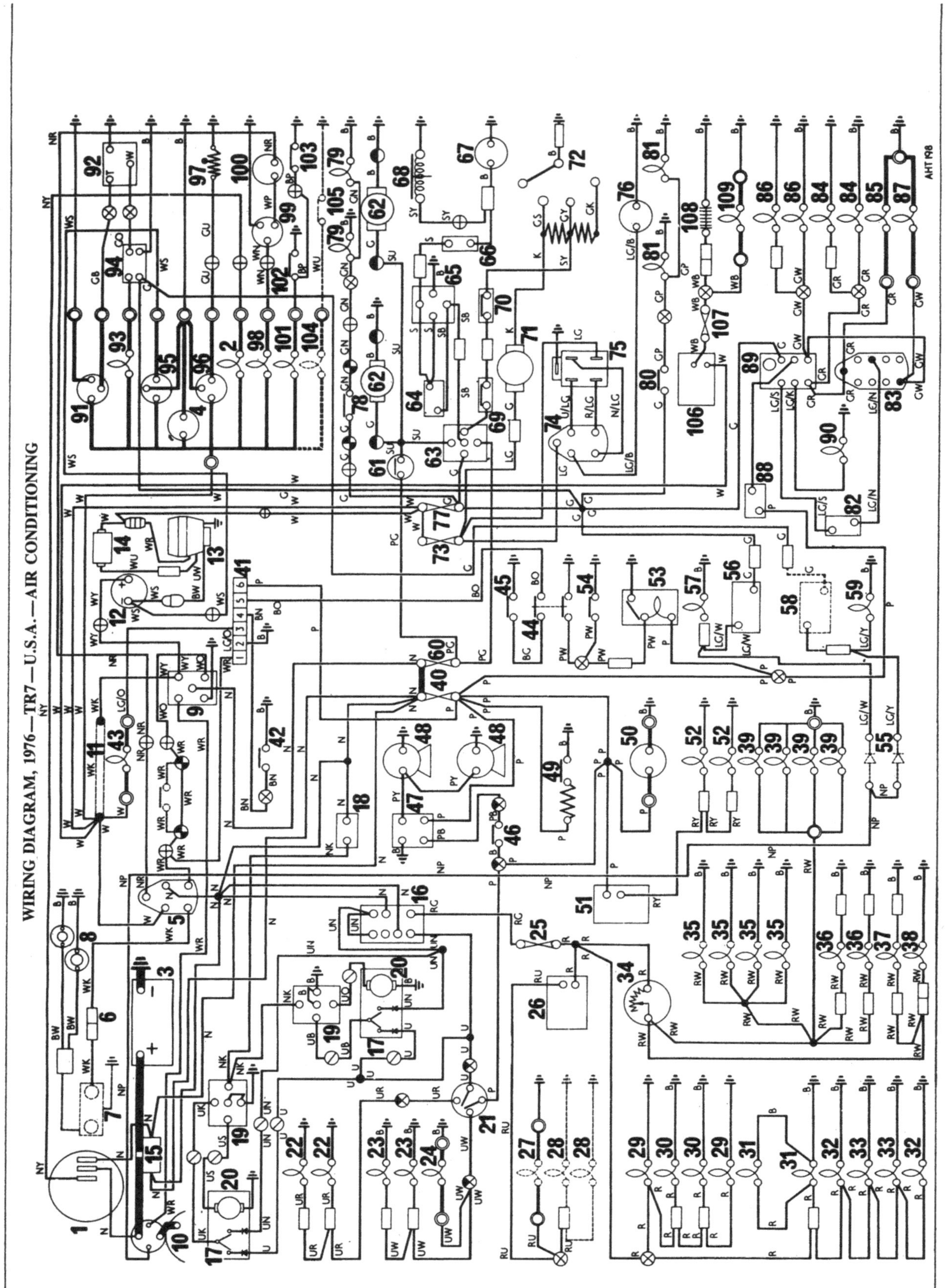

WIRING DIAGRAM, 1976—TR7—U.S.A.—AIR CONDITIONING

WIRING DIAGRAM KEY, 1976—TR7—U.S.A.—AIR CONDITIONING

COLOUR CODE

N.	Brown	LG.	Light Green
U.	Blue	W.	White
R.	Red	Y.	Yellow
P.	Purple	S.	Slate
G.	Green	B.	Black
K.	Pink	O.	Orange

1. Alternator
2. Ignition warning light
3. Battery
4. Battery condition indicator
5. Ignition/starter switch
6. Radio supply
7. Radio
8. Speaker
9. Starter motor relay
10. Starter motor
11. Ballast resistor wire
12. Ignition coil
13. Ignition distributor
14. Drive resistor
15. Battery lead connector
16. Master light switch
17. Actuator – limit switch
18. Circuit breaker
19. Headlamp – run/stop relay
20. Actuator motor
21. Main/dip/flash switch
22. Dip beam
23. Main beam
24. Main beam warning light
25. Fuse – 35 amp
26. Rear fog lamp switch
27. Rear fog lamp warning light
28. Rear fog lamp
29. Front marker lamp
30. Front parking lamp
31. Plate illumination lamp
32. Rear marker lamp
33. Tail lamp
34. Panel rheostat
35. Heater control illumination
36. Facia switch panel illumination
37. Selector panel illumination
38. Cigarette lighter illumination
39. Instrument illumination
40. Fuse – 50 amp
41. Buzzer/timer module
42. Driver's belt switch
43. Fasten belt warning light
44. L.H. door switch
45. Key switch
46. Horn push
47. Horn relay
48. Horn
49. Cigarette lighter
50. Clock
51. Front fog lamp switch
52. Front fog lamp
53. Roof lamp
54. R.H. door switch
55. Service indicator diodes – catalyst vehicles only
56. E G.R. service indicator
57. E.G.R. service warning light
58. Catalyst service indicator
59. Catalyst service warning light
60. Fuse – 50 amp
61. Radiator switch
62. Condenser fan motor
63. Air conditioning – control relay
64. Air conditioning – delay circuit flasher unit
65. Air conditioning – delay circuit relay
66. High pressure cut out
67. Compressor clutch
68. Throttle jack
69. Air conditioning cut out switch
70. Cold thermostat
71. Blower motor
72. Blower motor switch
73. Fuse – 35 amp
74. Windscreen washer/wiper switch
75. Windscreen wiper motor
76. Windscreen washer pump
77. Fuse – 50 amp
78. Reverse lamp switch
79. Reverse lamp
80. Stop lamp switch
81. Stop lamp
82. Turn signal flasher unit
83. Turn signal switch
84. L.H. flasher lamp
85. L.H. turn signal warning light
86. R.H. flasher lamp
87. R.H. turn signal warning light
88. Hazard flasher unit
89. Hazard switch
90. Hazard warning light
91. Fuel indicator
92. Fuel tank unit
93. Fuel warning light
94. Fuel delay unit
95. Tachometer
96. Temperature indicator
97. Temperature transmitter
98. Oil pressure warning light
99. Oil pressure switch
100. Anti-run-on valve
101. Brake warning light
102. Brake line failure switch
103. Handbrake switch
104. Choke warning light non-catalyst vehicles only
105. Choke switch
106. Heated backlight switch
107. Fuse – 15 amp
108. Heated backlight
109. Heated backlight warning light

WIRING DIAGRAM 1977 CARS WITH FOUR FUSES — AIR CONDITIONING

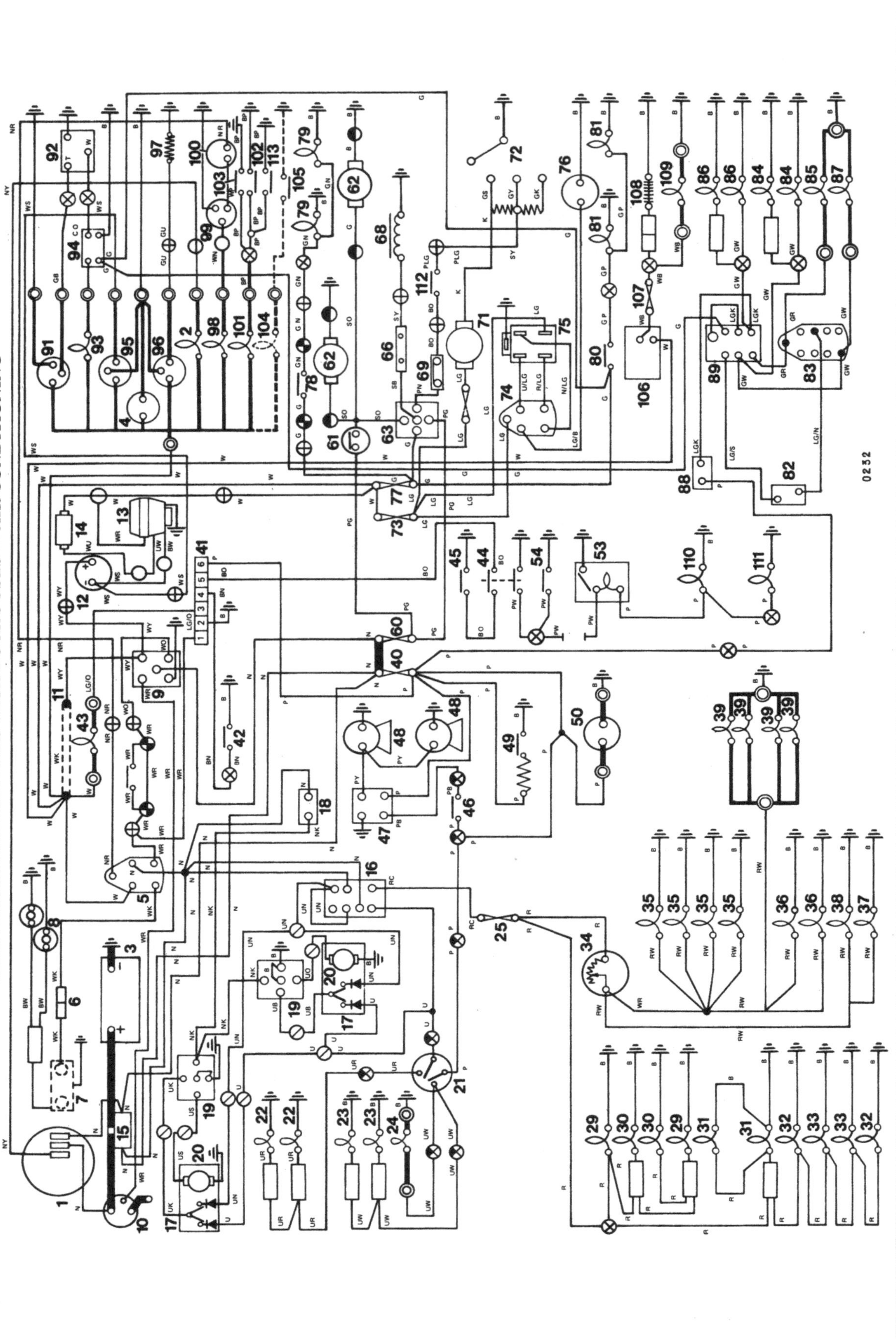

WIRING DIAGRAM KEY 1977 CARS WITH FOUR FUSES — AIR CONDITIONING

1 Alternator
2 Ignition warning light
3 Battery
4 Battery condition indicator
5 Ignition/starter switch
6 Radio supply
7 Radio
8 Speaker
9 Starter motor relay
10 Starter motor
11 Ballast resistor wire
12 Ignition coil
13 Ignition distributor
14 Drive resistor
15 Battery lead connector
16 Master light switch
17 Actuator – limit switch
18 Circuit breaker
19 Headlamp – run/stop relay
20 Actuator motor
21 Main/dip/flash switch
22 Dip beam
23 Main beam
24 Main beam warning light
25 Fuse – 35 amp
29 Front marker lamp
30 Front parking lamp
31 Plate illumination lamp
32 Rear marker lamp
33 Tail lamp
34 Panel rheostat
35 Heater control illumination
36 Facia switch panel illumination
37 Selector panel illumination
38 Cigarette lighter illumination
39 Instrument illumination
40 Fuse – 50 amp
41 Buzzer/timer module
42 Driver's belt switch
43 Fasten belt warning light
44 L.H. door switch
45 Key switch
46 Horn push
47 Horn relay
48 Horn
49 Cigarette lighter
50 Clock
53 Roof lamp
54 R.H. door switch
60 Fuse – 50 amp
61 Radiator switch
62 Condenser fan motor
63 Air conditioning – control relay
66 High pressure cut out
68 Throttle jack
69 Air conditioning cut out switch
71 Blower motor
72 Blower motor switch
73 Fuse – 35 amp
74 Windscreen washer/wiper switch
75 Windscreen wiper motor
76 Windscreen washer pump
77 Fuse – 50 amp
78 Reverse lamp switch
79 Reverse lamp
80 Stop lamp switch
81 STop lamp
82 Turn signal flasher unit
83 Turn signal switch
84 L.H. flasher lamp
85 L.H. turn signal warning light
86 R.H. flasher lamp
87 R.H. turn signal warning light
88 Hazard flasher unit
89 Hazard switch and warning light
91 Fuel indicator
92 Fuel tank unit
93 Fuel warning light
94 Fuel delay unit
95 Tachometer
96 Temperature indicator
97 Temperature transmitter
98 Oil pressure warning light
99 Oil pressure switch
100 Anti-run-on valve
101 Brake warning light
102 Brake line failure switch
103 Handbrake switch
104 Choke warning light – non-catalyst vehicles only
105 Choke switch
106 Heated backlight switch
107sFuse – 15 amp
108 Heatedsbacklight
109 Heated backlight warning light
110 Map light
111 Boot light
112 Full throttle cut-out switch
113 Ignition switch

COLOUR CODE

N	Brown	G	Green	Y	Yellow
U	Blue	K	Pink	S	Slate
R	Red	LG	Light green	B	Black
P	Purple	W	White	O	Orange

WIRING DIAGRAM—1977, 1978, 1979 CARS WITH 12 FUSES AIR CONDITIONING

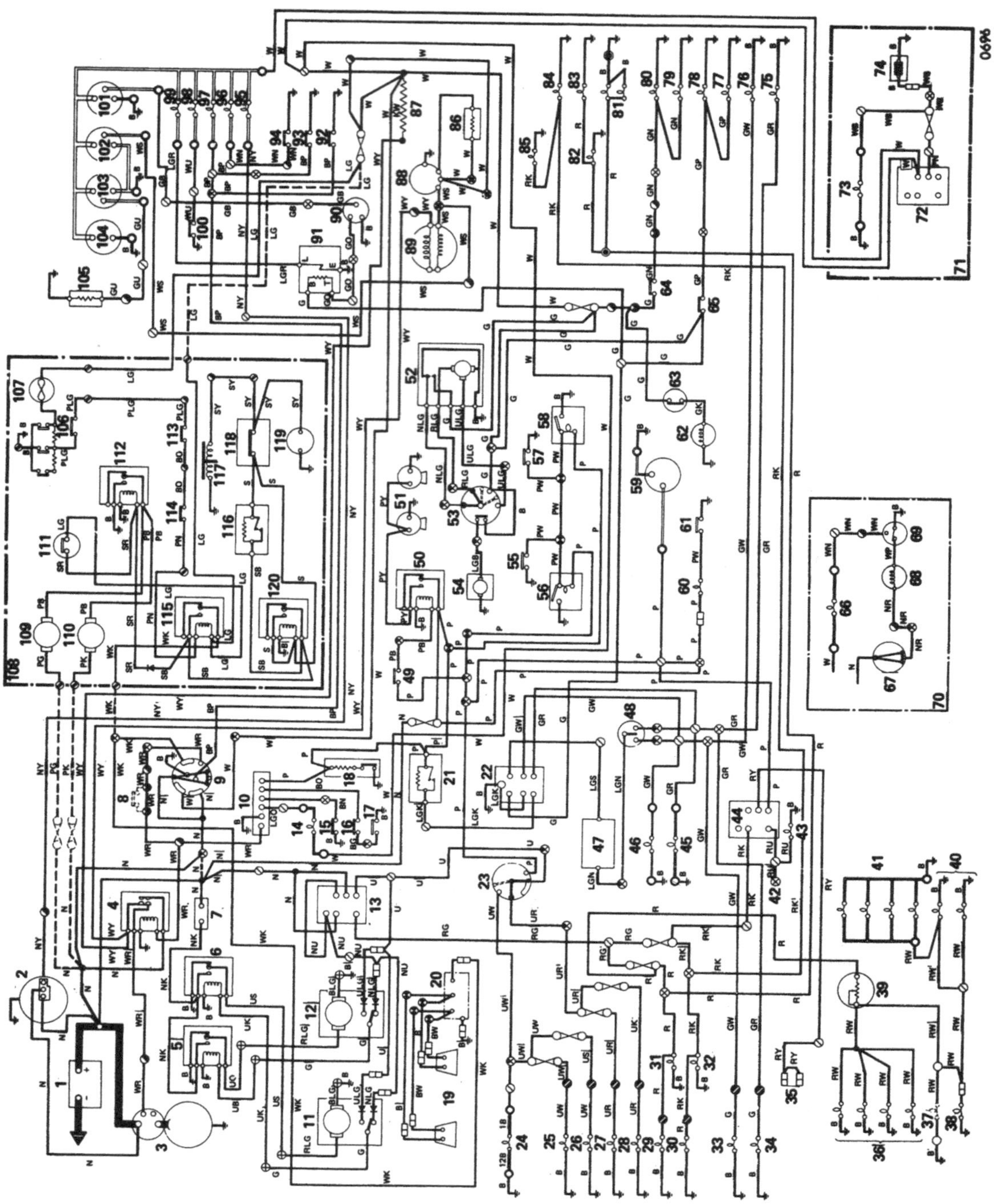

WIRING DIAGRAM KEY—1977, 1978, 1979 CARS WITH 12 FUSES
AIR CONDITIONING

1. Battery
2. Alternator
3. Starter motor
4. Starter relay
5. R.H. head lamp relay
6. L.H. head lamp relay
7. Head lamp motor circuit breaker
8. Start inhibit switch (Auto)
9. Ignition switch
10. Buzzer timer unit
11. L.H. Head lamp motor
12. R.H. Head lamp motor
13. Master light switch
14. Seat belt warning light
15. Buckle switch
16. Door switch
17. Audible warning
18. Cigar lighter
19. Loudspeakers
20. Radio
21. Hazard unit
22. Hazard switch
23. Main/dip/flash switch
24. Main beam warning light
25. R.H. main beam
26. L.H. main beam
27. R.H. dip beam
28. L.H. dip beam
29. R.H. side light
30. L.H. side light
31. R.H. side marker
32. L.H. side marker
33. R.H. side indicator
34. L.H. side indicator
35. Front fog lamps junction
36. Heater control illumination
37. Borg Warner illumination
38. Cigar lighter illumination
39. Panel rheostat
40. Switch panel illumination
41. Panel illumination
42. Rear fog lamps illumination
43. Rear fog lamps warning light
44. Fog Switch
45. L.H. indicator warning light
46. R.H. indicator warning light
47. Flasher unit
48. Direction indicator switch
49. Horn push
50. Horn relay
51. Horns
52. Windscreen wiper motor
53. Windscreen wipe/wash switch
54. Windscreen washer motor
55. Door switch
56. Courtesy light
57. Door switch
58. Courtesy light
59. Clock
60. Boot lamp
61. Boot lamp switch
62. F.A.S.D.
63. Engine thermostat
64. Reverse light switch
65. Stop light switch
66. Oil warning light
67. Ignition switch
68. Anti run on valve
69. Oil pressure switch
70. Anti run on valve—California market only
71. Heated rear screen—TR7 coupe
72. Heated rear screen switch
73. Heated rear screen warning light
74. Heated rear screen
75. L.H. side indicator
76. R.H. side indicator
77. L.H. stop light
78. R.H. stop light
79. L.H. reverse light
80. R.H. reverse light
81. Number plate lights
82. R.H. sidemarker
83. R.H. tail lights
84. L.H. tail lights
85. L.H. sidemarker
86. Drive resistor
87. Eureka wire (1.3-1.5Ω)
88. Distributor
89. 6v. coil
90. Tank unit
91. Low fuel delay unit
92. Brake pressure differential switch
93. Handbrake switch
94. Oil pressure switch
95. Ignition warning light
96. Oil warning light
97. Brake warning light
98. Choke warning light
99. Low fuel warning light
100. Choke warning light switch
101. Fuel gauge
102. Tachometer
103. Temperature gauge
104. Battery condition indicator
105. Temperature transmitter
106. Thermostat
107. Heater/air conditioning blower unit
108. Air conditioning circuit
109. R.H. condenser fan
110. L.H. condenser fan
111. Radiator thermostat
112. Fan relay
113. Full throttle cut-off switch
114. Air conditioning cut-out
115. Air conditioning relay
116. Delay unit
117. Throttle jack
118. Ranco valve high pressure cut-off
119. Clutch
120. Clutch relay

FUSE BLOCK

1	2	3	4	5	6
Left Low Beam	1) Turn Signals 2) Windshield washers 3) Wipers 4) Choke heater 5) Reverse lights 6) Brake lights	Left High Beam	1) Main air cond. relay 2) Heater/AC blower	Right High Beam	Right Low Beam
10A	25A	10A	35A	10A	10A
7	**8**	**9**	**10**	**11**	**12**
Right Condensor Fan	Rear Demister	1) Hazzard 2) Horn relay 3) Seat belt warning unit 4) Fog lamps 5) Boot lamp 6) Cigar 7) Courtesy lights	Right Side Marker Lamps and Tail Lights	Left Side Marker Lamps and Tail Lights	Left Condensor Fan
35A	15A	50A	10A	3A	35A

COLOUR CODE

N.	Brown	LG.	Light Green
U.	Blue	W.	White
R.	Red	Y.	Yellow
P.	Purple	S.	Slate
G.	Green	B.	Black
K.	Pink	O.	Orange

WIRING DIAGRAM — TR7 — 1980 USA MARKET CARBURETTER VEHICLES

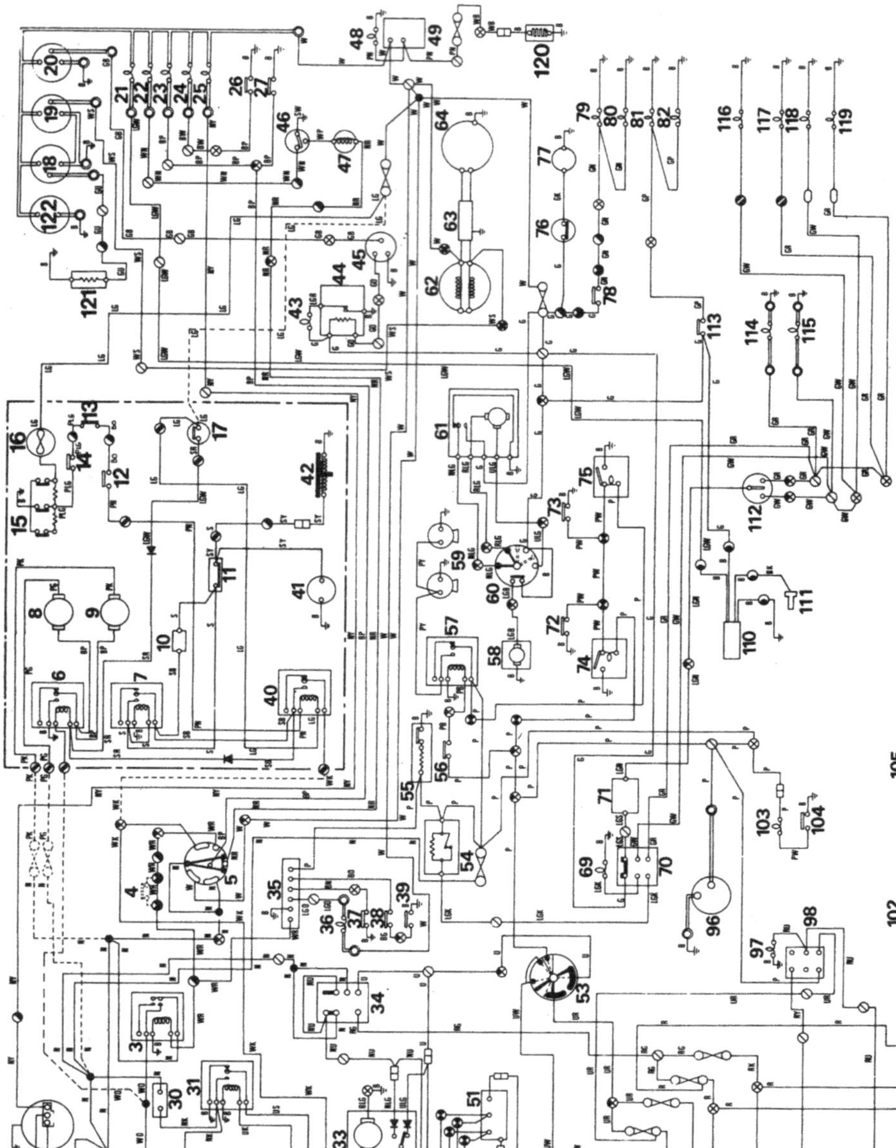

WIRING DIAGRAM KEY—TR7—1980 USA MARKET—CARBURETTER VEHICLES

1 Battery
2 Alternator
3 Starter relay
4 Starter inhibitor switch (automatic transmission only)
5 Ignition/starter switch

Air conditioning:
6 Condenser fan relay
7 Clutch delay relay
8 Condenser fan — L.H.
9 Condenser fan — R.H.
10 Delay unit
11 Ranco valve high pressure cut-out
12 Manual cut-out switch
13 Full throttle cut-out switch
14 Cold thermostat
15 Blower motor switch
16 Blower motor
17 Radiator thermostat

18 Temperature indicator
19 Tachometer
20 Fuel indicator
21 Coolant low warning light
22 Oil pressure warning light
23 Brake warning light
24 Handbrake warning light
25 Ignition warning light
26 Handbrake switch
27 Brake pressure differential switch
28 Starter motor
29 Headlamp run/stop relay — R.H.
30 Headlamp motor circuit breaker
31 Headlamp run/stop relay — L.H.
32 Headlamp actuator — L.H.
33 Headlamp actuator — R.H.
34 Master light switch
35 Seat belt timer/buzzer unit
36 Seat belt warning light
37 Belt switch
38 Door switch
39 Key switch

Air conditioning:
40 Master relay
41 Clutch
42 Throttle jack

43 Low fuel warning light
44 Low fuel delay unit
45 Tank unit
46 Oil pressure switch
47 Anti run-on valve
48 Heated back-light warning light
49 Heated back-light switch
50 Speakers
51 Radio
52 Main beam warning light
53 Column light switch
54 Hazard flasher unit
55 Cigarette lighter
56 Horn-push
57 Horn relay
58 Windscreen washer pump
59 Horn
60 Windscreen wash/wipe switch
61 Windscreen wiper motor
62 and 63 Ignition coil and electronic module assembly
64 Ignition distributor
65 Main beam — R.H.
66 Main beam — L.H.
67 Dip beam — R.H.
68 Dip beam — L.H.
69 Hazard warning light
70 Hazard switch
71 Turn signal flasher unit
72 Door switch
73 Door switch
74 Map/courtesy lamp
75 Map/courtesy lamp
76 Engine thermostat
77 Carburetter fuel heater
78 Reverse lamp switch
79 Reverse lamp
80 Reverse lamp
81 Stop lamp
82 Stop lamp
83 Front marker lamp — R.H.
84 Front parking lamp — R.H.
85 Front marker lamp — L.H.
86 Front parking lamp — L.H.
87 Front fog lamp
88 Front fog lamp
89 Rear guard lamp
90 Rear guard lamp
91 Tail lamp — L.H.
92 Rear marker lamp — L.H.
93 Plate illumination lamp
94 Rear marker lamp — R.H.
95 Tail lamp — R.H.
96 Clock
97 Rear guard lamp warning light
98 Fog switch
99 Panel rheostat
100 Selector panel illumination (automatic transmission only)
101 Cigarette lighter illumination
102 Heater control illumination
103 Luggage boot lamp
104 Luggage boot lamp switch
105 Instrument illumination
106 Instrument illumination
107 Instrument illumination
108 Instrument illumination
109 Switch panel illumination
110 Coolant low indicator unit
111 Coolant low sensor
112 Turn signal switch
113 Stop lamp switch
114 Turn signal warning light — L.H.
115 Turn signal warning light — R.H.
116 Front flasher lamp — R.H.
117 Front flasher lamp — L.H.
118 Rear flasher lamp — R.H.
119 Rear flasher lamp — L.H.
120 Heated back-light
121 Temperature transmitter
122 Battery condition indicator

COLOUR CODE

B	Black	N	Brown	S	Slate
G	Green	O	Orange	U	Blue
K	Pink	P	Purple	W	White
LG	Light green	R	Red	Y	Yellow

WIRING DIAGRAM — TR7 — 1980 USA MARKET

FUEL INJECTION VEHICLES

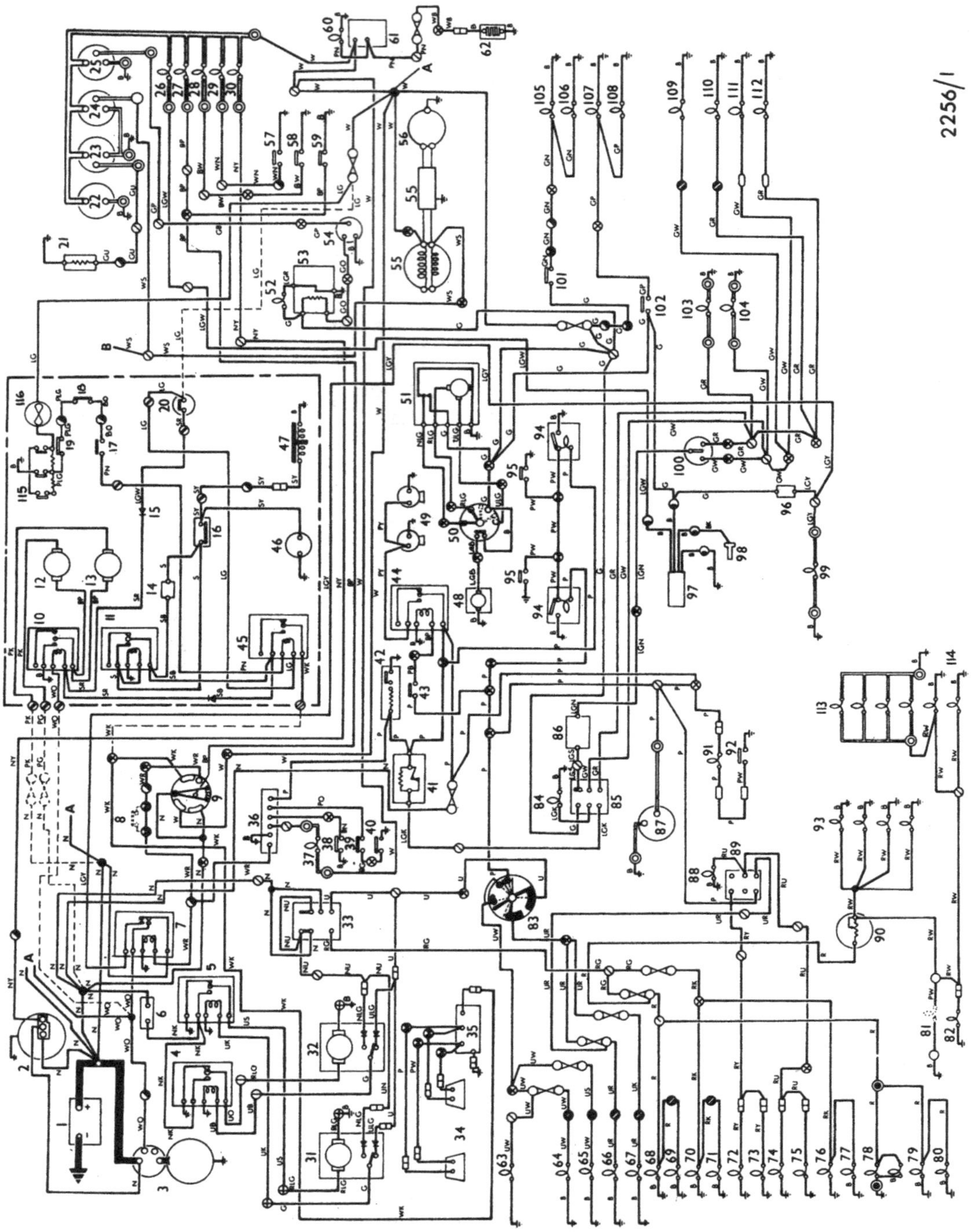

AKM3079B

WIRING DIAGRAM KEY—TR7—1980 USA MARKET—FUEL INJECTION VEHICLES

1 Battery
2 Alternator
3 Starter motor
4 Headlamp run/stop relay — R.H.
5 Headlamp run/stop relay — L.H.
6 Headlamp motor circuit breaker
7 Starter relay
8 Starter inhibitor switch (automatic transmission only)
9 Ignition/starter switch

Air conditioning:
10 Condenser fan relay
11 Clutch delay relay
12 Condenser fan — L.H.
13 Condenser fan — R.H.
14 Delay unit
15 Diode
16 Ranco valve high pressure cut-out
17 Manual cut-out switch
18 Full throttle cut-out switch
19 Cold thermostat
20 Radiator thermostat

21 Temperature transmitter
22 Battery condition indicator
23 Temperature indicator
24 Tachometer
25 Fuel indicator
26 Coolant low warning light
27 Brake warning light
28 Handbrake warning light
29 Oil pressure warning light
30 Ignition warning light
31 Headlamp actuator — L.H.
32 Headlamp actuator — R.H.
33 Master light switch
34 Speakers
35 Radio
36 Seat belt timer/buzzer unit
37 Seat belt warning light
38 Belt switch
39 Door switch
40 Key switch
41 Hazard flasher unit
42 Cigarette lighter
43 Horn-push
44 Horn relay

Air conditioning:
45 Master relay
46 Clutch
47 Throttle jack

48 Windscreen washer pump
49 Horn
50 Windscreen wash/wipe switch
51 Windscreen wiper motor
52 Low fuel warning light
53 Low fuel delay unit
54 Tank unit
55 Ignition coil and electronic module assembly
56 Ignition distributor
57 Oil pressure switch
58 Handbrake switch
59 Brake pressure differential switch
60 Heated back-light warning light
61 Heated back-light switch
62 Heated back-light
63 Main beam warning light
64 Main beam — R.H.
65 Main beam — L.H.
66 Dip beam — R.H.
67 Dip beam — L.H.
68 Front marker lamp — R.H.
69 Front parking lamp — R.H.
70 Front marker lamp — L.H.
71 Front parking lamp — L.H.
72 Front fog lamp
73 Front fog lamp
74 Rear guard lamp
75 Rear guard lamp
76 Tail lamp — L.H.
77 Rear marker lamp — L.H.
78 Plate illumination lamp
79 Rear marker lamp — R.H.
80 Tail lamp — R.H.
81 Selector panel illumination (automatic transmission only)
82 Cigarette lighter illumination
83 Column light switch
84 Hazard warning light
85 Hazard switch
86 Turn signal flasher unit
87 Clock
88 Rear guard lamp warning light
89 Fog switch
90 Panel rheostat
91 Luggage boot lamp
92 Luggage boot lamp switch
93 Heater control illumination
94 Map/courtesy lamp
95 Door switch
96 Service interval counter
97 Coolant low indicator unit
98 Coolant low sensor
99 Oxygen sensor warning light
100 Turn signal switch
101 Reverse lamp switch
102 Stop lamp switch
103 Turn signal warning light — L.H.
104 Turn signal warning light — R.H.
105 Reverse lamp
106 Reverse lamp
107 Stop lamp
108 Stop lamp
109 Front flasher lamp — R.H.
110 Front flasher lamp — L.H.
111 Rear flasher lamp — R.H.
112 Rear flasher lamp — L.H.
113 Instrument illumination
114 Switch panel illumination

Air conditioning:
115 Blower motor switch
116 Blower motor

COLOUR CODE

B	Black	N	Brown	S	Slate
G	Green	O	Orange	U	Blue
K	Pink	P	Purple	W	White
LG	Light green	R	Red	Y	Yellow

WIRING DIAGRAM—TR7—1980 FUEL INJECTION

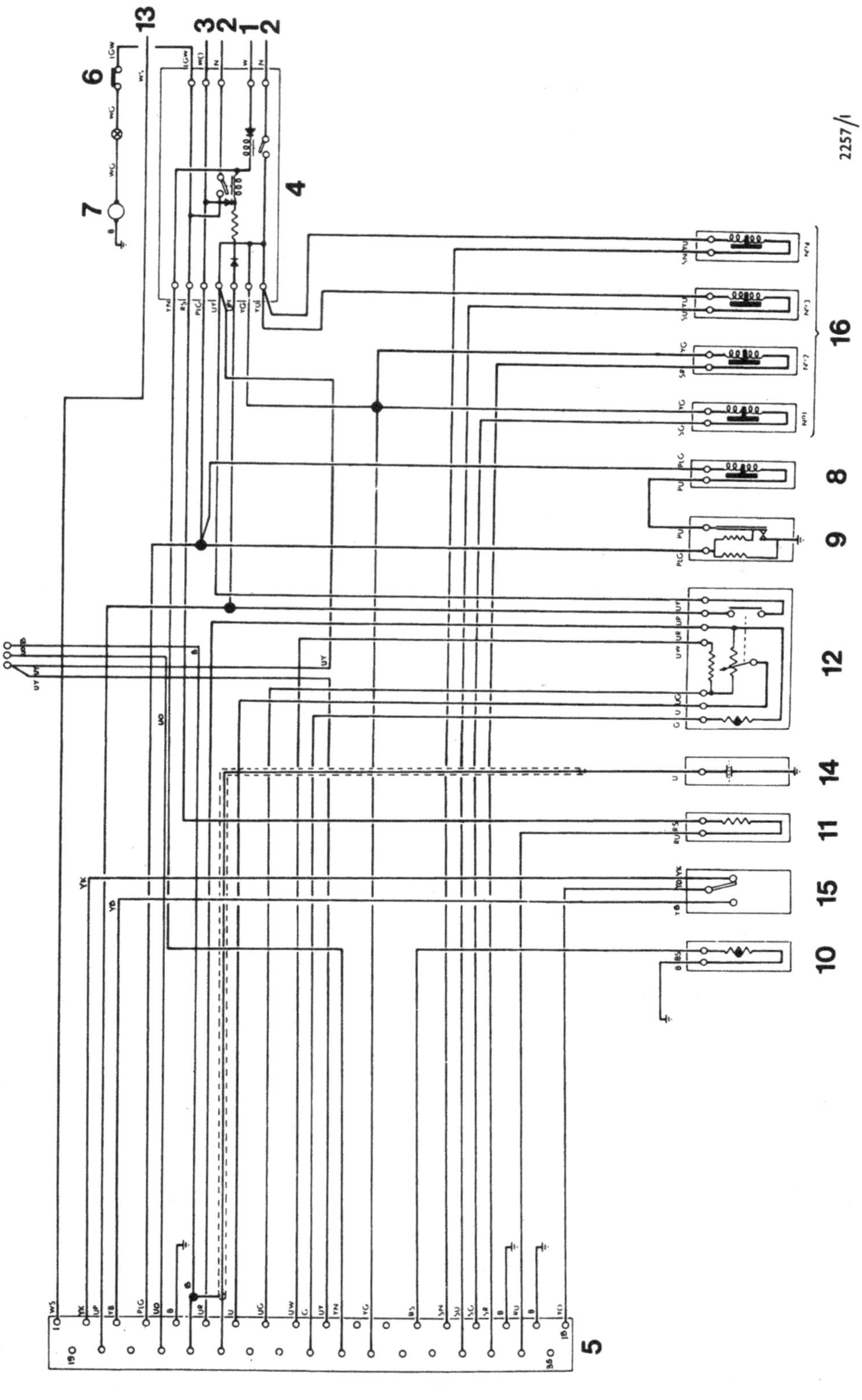

WIRING DIAGRAM KEY—TR7—1980 FUEL INJECTION

1 Ignition control input

2 Battery input

3 Start input

4 Relay

5 Electronic control unit — E.C.U.

6 Inertia switch

7 Fuel pump

8 Cold start injector

9 Thermotime switch

10 Water temperature sensor

11 Extra air valve

12 Air flow meter

13 Engine speed — input from ignition coil negative

14 Lambda sensor

15 Throttle switch

16 Injector — 4 off

17 Test facility — feedback monitor plug

COLOUR CODE

B	Black	N	Brown	S	Slate
G	Green	O	Orange	U	Blue
K	Pink	P	Purple	W	White
LG	Light green	R	Red	Y	Yellow

WIRING DIAGRAM — 1981 USA MARKET
FUEL INJECTION VEHICLES

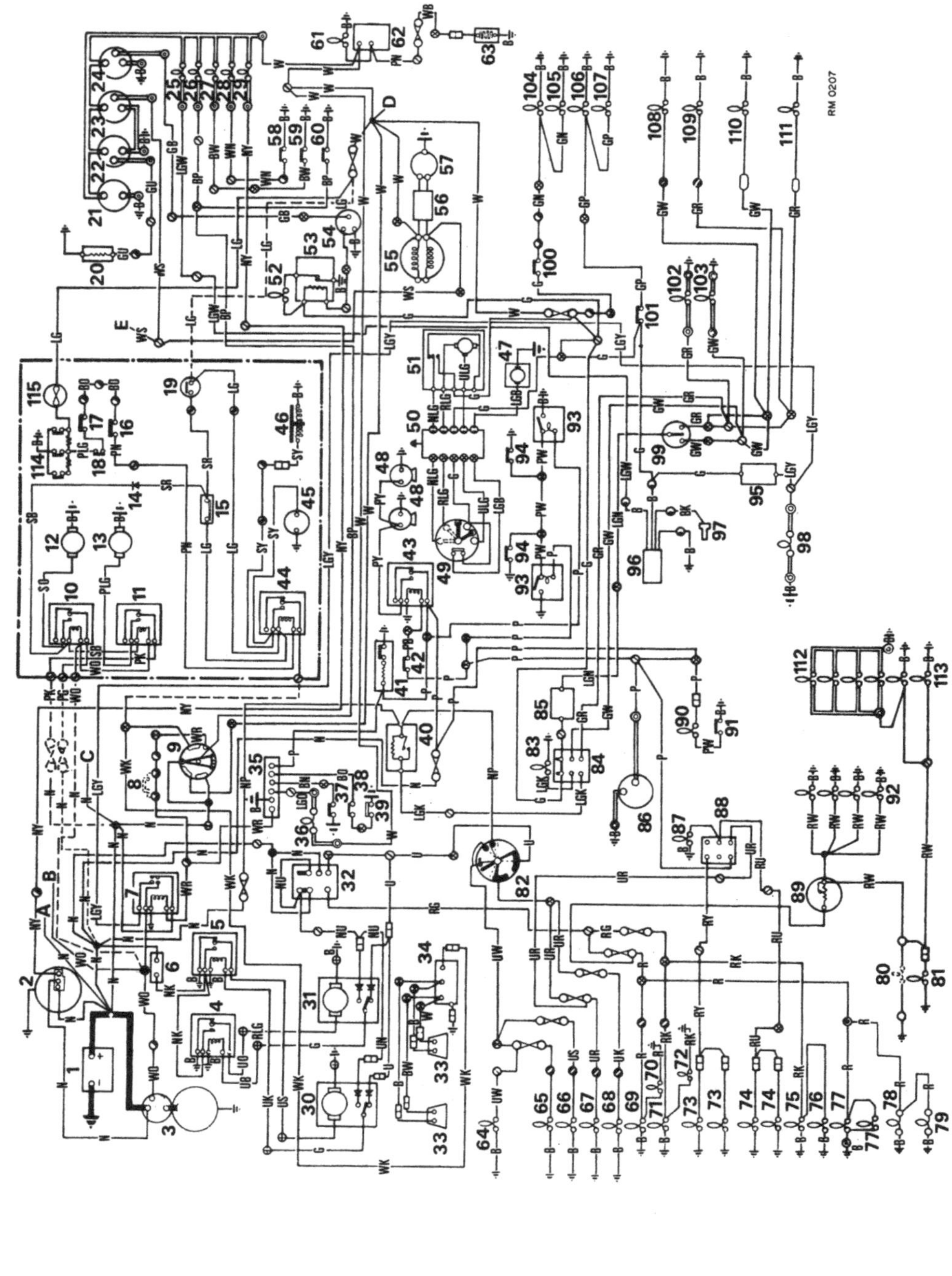

WIRING DIAGRAM KEY—1981 USA MARKET—FUEL INJECTION VEHICLES

1 Battery
2 Alternator
3 Starter motor
4 Headlamp relay — R.H.
5 Headlamp relay — L.H.
6 Headlamp motor circuit breaker
7 Starter relay
8 Starter inhibitor switch (automatic transmission only)
9 Ignition/starter switch
10 Condenser fan relay — L.H.
11 Condenser fan relay — R.H.
12 Condenser fan — L.H.
13 Condenser fan — R.H.
14 Diode
15 Pressure cut in switch
16 Air conditioning cut-out switch
17 Full throttle cut-out switch
18 Thermostat
19 Radiator thermostat
20 Temperature transmitter
21 Voltmeter
22 Temperature gauge
23 Tachometer
24 Fuel gauge
25 Coolant low warning light
26 Brake warning light
27 Handbrake warning light
28 Oil pressure warning light
29 Ignition warning light
30 Headlamp motor — L.H.
31 Headlamp motor — R.H.
32 Master light switch
33 Speaker
34 Radio
35 Seat belt timer/buzzer unit
36 Seat belt warning light
37 Belt switch
38 Door switch
39 Audible warning switch
40 Hazard flasher unit
41 Cigarette lighter
42 Horn-push
43 Horn relay
44 Clutch relay
45 Clutch
46 Engine air solenoid valve
47 Windscreen washer motor
48 Horn
49 Windscreen wash/wipe switch
50 Wiper delay unit
51 Windscreen wiper motor
52 Low fuel warning light
53 Low fuel delay unit
54 Tank unit
55 Ignition coil
56 Amplifier
57 Ignition distributor
58 Oil pressure switch
59 Handbrake switch
60 Brake pressure differential switch
61 Heated back-light warning light
62 Heated back-light switch
63 Heated back-light
64 Main beam warning light
65 Main beam — R.H.
66 Main beam — L.H.
67 Dip beam — R.H.
68 Dip beam — L.H.
69 Front marker lamp — R.H.
70 Front parking lamp — R.H.
71 Front marker lamp — L.H.
72 Front parking lamp — L.H.
73 Front fog lamp
74 Rear fog guard lamp
75 Tail lamp — L.H.
76 Rear marker lamp — L.H.
77 Number plate illumination lamp
78 Rear marker lamp — R.H.
79 Tail lamp — R.H.
80 Selector panel illumination (automatic transmission only)
81 Cigarette lighter illumination
82 Main/dip/flash switch
83 Hazard warning light
84 Hazard switch
85 Turn signal flasher unit
86 Clock
87 Rear fog guard warning light
88 Fog switch
89 Panel rheostat
90 Luggage boot lamp
91 Luggage boot lamp switch
92 Heater control illumination
93 Map/courtesy lamp
94 Door switch
95 Service interval counter
96 Coolant low indicator unit
97 Coolant low sensor
98 Oxygen sensor warning light
99 Turn signal switch
100 Reverse lamp switch
101 Stop lamp switch
102 Turn signal warning light — L.H.
103 Turn signal warning light — R.H.
104 Reverse lamp — R.H.
105 Reverse lamp — L.H.
106 Stop lamp — R.H.
107 Stop lamp — L.H.
108 Front flasher lamp — R.H.
109 Front flasher lamp — L.H.
110 Rear flasher lamp — R.H.
111 Rear flasher lamp — L.H.
112 Instrument illumination
113 Switch panel illumination
114 Blower motor switch
115 Blower motor

A To injector and ECU relay
B To diode resistor pack
C To fuel pump relay
D To diode resistor pack
E Fuel injection timing signal

(A–E: See also Fuel Injection Wiring Diagram)

COLOUR CODE

B	Black	N	Brown	S	Slate
G	Green	O	Orange	U	Blue
K	Pink	P	Purple	W	White
LG	Light green	R	Red	Y	Yellow

WIRING DIAGRAM
FUEL INJECTION — 1981

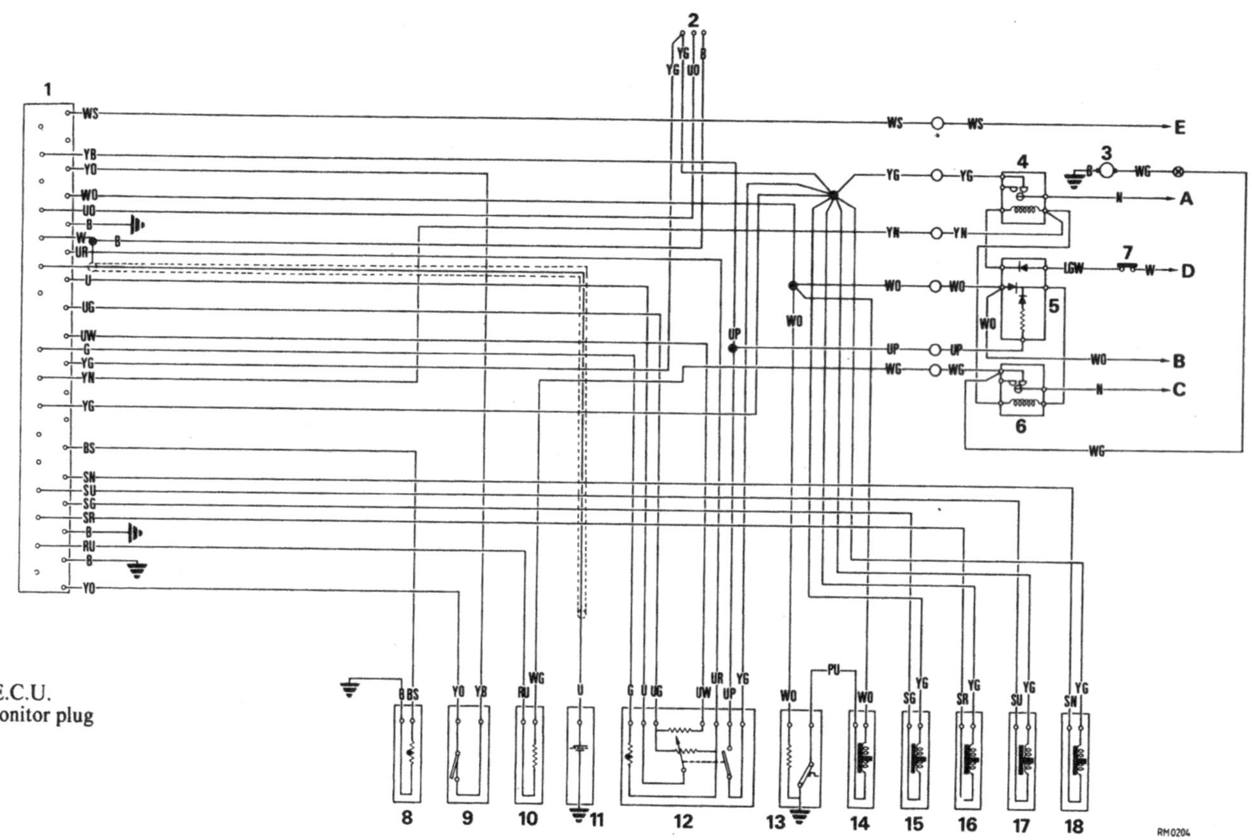

1 Electronic control unit — E.C.U.
2 Test facility — feedback monitor plug
3 Fuel pump
4 Injector and E.C.U. relay
5 Diode resistor pack
6 Fuel pump relay
7 Inertia switch
8 Water temperature sensor
9 Throttle switch
10 Extra air valve
11 Lambda sensor
12 Air flow meter
13 Thermotime switch
14 Cold start injector
15 No. 1 injector
16 No. 2 injector
17 No. 3 injector
18 No. 4 injector

COLOUR CODE

B	Black	N	Brown	S	Slate
G	Green	O	Orange	U	Blue
K	Pink	P	Purple	W	White
LG	Light green	R	Red	Y	Yellow

ALTERNATOR DATA CHART 86.10.00

CAUTION: The alternator contains polarity sensitive components that may be irreparably damaged if subjected to incorrect polarity.

Do not connect or disconnect any part of the charging circuit – including the battery leads – while the engine is running. Run the alternator with all connections made or with the unit disconnected.

Manufacturer / Type	Lucas 17 ACR		Lucas 20 ACR Dual sensed–battery sensed with machine sensed safety control European terminations	
	Lucas Part No.	**Stanpart No.**	**Lucas Part No.**	**Triumph Part No.**
Part numbers–				
assembly	–	TKC 0863	54021510	TKC 0905
–comprising –				
alternator	23818	TKC 0864	23817	TKC 0906
fan	54217652	147990	54216943	157069
pulley	54217767	154334	54201395	157083
Polarity	Negative earth only		Negative earth only	
Brush length				
– new	0.5 in (12.70 mm)		0.5 in (12.70 mm)	
– renew if less than	0.2 in (5.00 mm) protrudes from brushbox when free		0.2 in (5.00 mm) protrudes from brushbox when free	
Brush spring pressure	9 to 13 oz (255 to 370 g) at face flush with brushbox		9 to 13 oz (255 to 370 g)	
Rectifier pack				
– output rectification	6 diodes (3 live side and 3 earth side		6 diodes (3 live side and 3 earth side)	
– field winding supply rectification	3 diodes		3 diodes	
Stator windings	Three phase–star connected		Three phase–delta connected	
Field winding rotor				
– poles				
– maximum permissible speed	15,000 rev/min		15,000 rev/min	
– shaft thread	9/16 in–18 U.N.F.		M16–1.5	
Field winding resistance at 20°C			3.2 ± 5% ohm	
Control	Dual sensed–battery sensed with machine sensed safety control		–	
Regulation–type	14 TR		14 TR	
Terminations	European		–	
Adjustment bolt tapping	Metric M8–1.25		–	
Nominal output				
– condition	Hot		Hot	
– alternator speed	6000 rev/min		6000 rev/min	
– engine speed	2540 rev/min		2540 rev/min	
– control voltage	14 volt		14 volt	
– amp.	36 amp		66 amp	

ALTERNATOR DATA CHART [Lucas 17 ACR and 20 ACR]
[ALTERNATOR DATA CHART [Lucas 23 ACR and 25 ACR

CAUTION: The alternator contains polarity sensitive components that may be irreparably damaged if subjected to incorrect polarity.

Do not connect or disconnect any part of the charging circuit—including the battery leads—while the engine is running. Run the alternator with all connections made or with the unit disconnected.

Manufacturer / Type	Lucas 23 ACR	Lucas 25 ACR	
Part numbers—	**Lucas Part No.**	**Lucas Part No.**	**Triumph Part No.**
assembly	——	23859	RKC 2803
—comprising—			
alternator			RKC 2804
fan			
pulley			154334
Polarity	Negative earth only	Negative earth only	
Brush length			
—new	0.5 in (12.70 mm)	0.5 in (12.70 mm)	
—renew if less than	0.2 in (5.00 mm) protrudes from brushbox when free	0.2 in (5.00 mm) protrudes from brushbox when free	
Brush spring pressure	9 to 13 oz (255 to 370 g) at face flush with brushbox	9 to 13 oz (255 to 370 g) at face flush with brushbox	
Rectifier pack			
—output rectification	6 diodes (3 live side and 3 earth side)	6 diodes (3 live side and 3 earth side)	
—field winding supply rectification	3 diodes	3 diodes	
Stator windings	Three phase—delta connected	Three phase—delta connected	
Field winding rotor			
—poles	12	12	
—maximum permissible speed	15,000 rev/min	15,000rev/min	
Field winding resistance at 20°C	3.201 ohm	3.201 ohm	
Stator winding resistance at 20°C	.202 ohm	.152 ohm	
Regulating voltage	13.6–14.4 (at battery)	13.6–14.4 (at battery)	
Fan diameter	5.5 in (139.7 mm)	6.0 in (152.4 mm)	
Weight	10.5 lb (4.8 kg)	11 lb (5.1 kg)	
Nominal output			
—condition	Hot	Hot	
—alternator speed	6000 rev/min	6000 rev/min	
—engine speed	2540 rev/min	2540 rev/min	
—control voltage	14.2 volt	14.2 volt	
—amp	55 amp	65 amp]	

ALTERNATOR WIRING DIAGRAM

Lucas 17ACR Alternator with 14TR Regulator

Lucas Part No. 23796

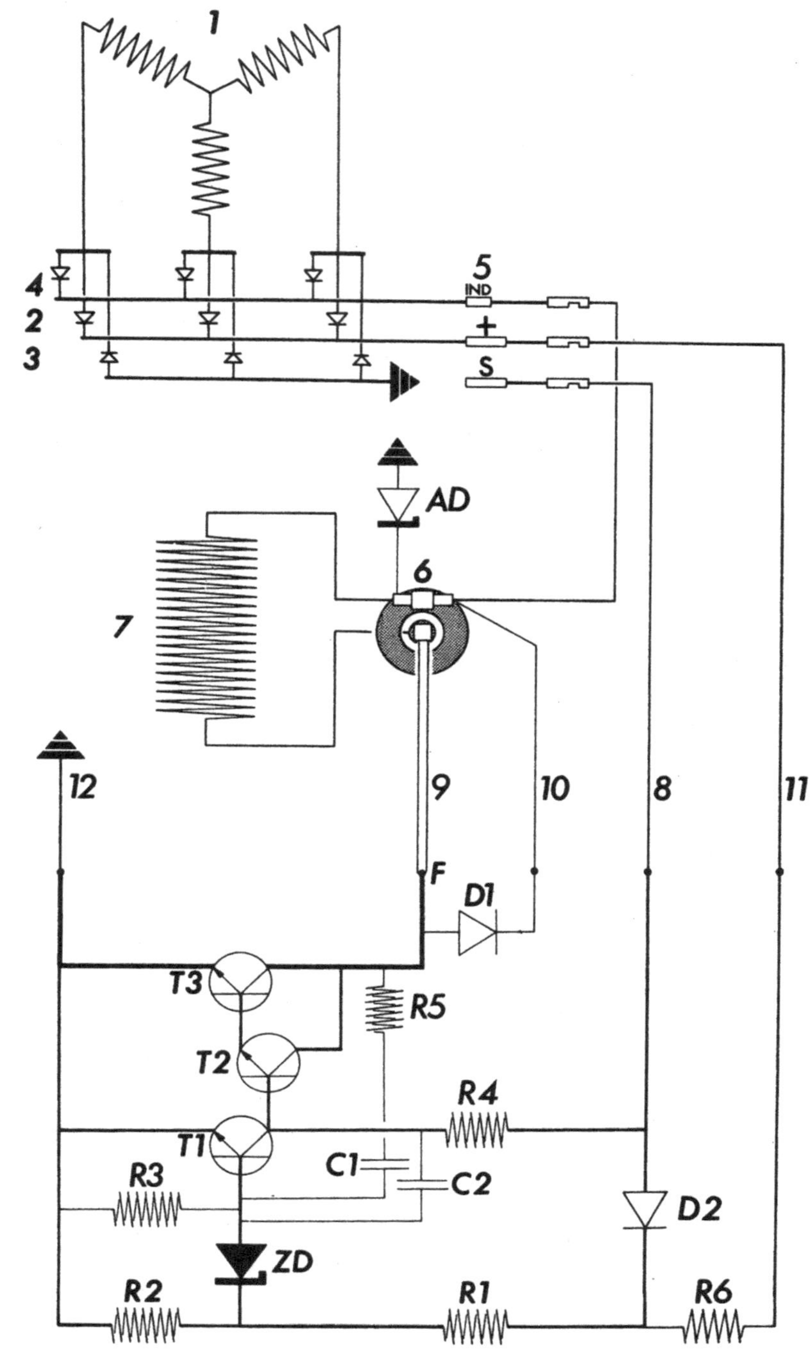

KEY TO ALTERNATOR WIRING DIAGRAM

Lucas 17ACR Alternator with 14TR Regulator

Lucas Part No. 23796

1	Stator windings	
2	Live side output diodes	
3	Earth side output diodes	
4	Field winding supply diodes	
5	European terminations	IND Ignition warning light + Main to battery S Sense to battery
6	Brushes to slip rings	
7	Field winding	
8	Battery sensed lead—white	
R4	Resistor	Restricts T2 base current supplied from 'battery sensed lead'
T2	Intermediate transistor	Controls T3 base current direct
9	Metal connector link	
T3	Output Transistor	Controls field winding earth return circuit
R1 and R2	Resistors	Potential divider—used in normal operation. Senses battery reference voltage
ZD	Zener diode	Voltage sensitive component. Opposes passage of current until breakdown voltage—approximately 8 volts—is reached. Controls T1 base current direct
T1	Input transistor	Controls T2 base current by diverting current passing through R4 to earth when ZD is conducting
C1 and R5	Capacitor and resistor	Prevents transistor overheating by providing positive feedback circuit to ensure quick switching of transistors from 'fully on' to 'fully off'
R3	Resistor	Path for small leakage current which may pass through ZD at high temperatures
10	Surge lead—yellow	
D1	Surge quench diode	Connected across field winding. Protects T3 from field winding high induced voltage surge and smooths field winding current
C2	Condenser	Radio interferance suppression
11	Machine sensed lead—red	
R6	Resistor	Fail-safe component. If main alternator to battery wire disconnects battery sensed lead voltage will fall. Alternator runway voltage is clamped by high voltage on machine sensed lead. Modified potential divider R1, R2 and addition R6
D2	Diode	Fail-safe component. If battery sensed lead disconnects alternator shuts down as supply via R4 to transistors T2 and T3 stops. D2 blocks feed into this circuit from machine sensed lead via R6
AD	Avalanche diode	Protects output transistor from high transient voltage which may occur from faulty charging circuit connectors
12	Earth lead—black	Regulator earth

ALTERNATOR – LUCAS TYPES 17ACR and 20ACR

Remove and refit **86.10.02**

Remove

1 Isolate the battery.
2 Slide the harness plug lock to the unlock position–inboard and down.
3 Disconnect the harness plug.
4 Slacken the adjustment bolt.
5 Slacken the support bracket bolt.
6 Slacken the main mounting bolt.
7 Push the alternator towards the engine and remove the drive belt from the pulley.
8 Remove the adjustment bolt and washer.
9 Remove the nut and washer.
10 Support the weight of the alternator. Withdraw the main mounting bolt and washer. Lift the alternator from the vehicle.

Refit

11 Position the alternator to the mounting bracket. Insert the main mounting bolt and washer.
12 Fit the nut and washer finger tight.
13 Fit the adjustment bolt and washer finger tight.
14 Push the alternator towards the engine and fit the drive belt to the pulley.
15 Perform 86.10.05 operations 4 to 7.
16 Connect the harness plug.
17 Slide the harness plug lock to the lock position–outboard and up.
18 Connect the battery.

ALTERNATOR

Drive belt – adjust **86.10.05**

1 Slacken the adjustment bolt.
2 Slacken the support bracket bolt.
3 Slaken the main mounting bolt.
4 Carefully lever the alternator away from the engine to tension the belt. Tighten the adjustment bolt.
CAUTION: To prevent bearing damage when tensioning the belt use a lever of soft material–preferably wood–applied to the alternator drive end bracket. Do not lever on any other part of the alternator.
5 Check the belt tension. Total movement should be 0.75 to 1.00 in (20 to 25 mm) at the mid-point of the longest run.
6 Tighten the support bracket bolt.
7 Tighten the main mounting bolt.

ALTERNATOR – LUCAS TYPE [16 ACR, 17 ACR, 23 ACR]

Overhaul **86.10.08**

This overhaul instruction is specific to alternator Lucas part number 23818. Other units may differ slightly.

Disassemble

1 Remove the moulded cover.
2 Before disturbing any wires note the wire positions and colours.
3 Remove the brushbox, regulator and surge protection diode assembly as follows:
Remove screw to release surge protection diode. Disconnect four Lucar connectors from the rectifier pack. Remove two screws to release the brushbox. Lift away the assembly.
4 If required, the regulator may be detached from the assembly as follows:
Remove screw to release one wire eyelet. Remove screw to release the regulator. Disengage two lugs and lift away the regulator. Collect up the spacer.
5 Remove screw to release the radio capacitor.
6 Note the position of the three stator wires on the rectifier pack.
7 Unsolder the three stator wire connections. Do not overheat the diodes or bend the diode pins. Solder quickly and provide a heat sink by gripping the diode pin with pliers.
8 Remove screw to release the rectifier pack earth strip. Slacken the nut and withdraw the rectifier pack.
9 Remove the through-bolts.
10 Provide an extractor tool as shown.

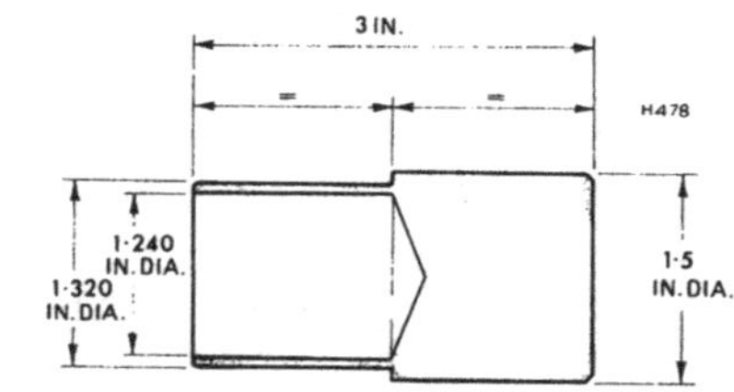

11 To remove the slip-ring end bracket, position the extractor tool to engage with the outer journal of the slip-ring end bearing. Employ a second operator to support the slip-ring end bracket by hand. Carefully tap the extractor tool to drive the bearing from the housing.
NOTE: It may be necessary to carefully file away surplus solder from the two field winding connections on the slip-ring moulding if the exterior tool will not pass over the moulding.
12 The rubber 'O' ring fitted in the slip-ring end bracket bearing housing may remain *in situ* unless replacement is contemplated.
13 Remove the stator windings from the drive end bracket.
14 Prevent the rotor turning by wrapping a scrap fan belt round the pulley and retaining by hand or vice. Remove the nut, spring washer, pulley and fan. If necessary, use a suitable extractor.
15 Remove the key.
16 Using a suitable press, remove the rotor from the drive end bracket.
CAUTION: Do not attempt to remove the rotor by applying hammer blows to the shaft end. Such action may burr over and damage the thread.
17 Collect up the thick spacer.
18 Remove the thin spacer from the rotor shaft.

Assemble

19 Fit the thin spacer to the rotor shaft.
20 Using a suitable press, the thick spacer and a suitable tube, fit the rotor to the drive end bracket by applying pressure to the bearing inner journal.
CAUTION: Do not use the drive-end bracket as a support while fitting the rotor. If the spacer is not employed, the felt ring may be damaged.
21 Fit the key.
22 Fit the fan, pulley, spring washer and nut. Prevent the rotor turning by wrapping a scrap fan belt round the pulley and retaining by hand or vice. Torque load the nut to 25 to 30 lbf ft (3.46 to 4.15 kgf m).
23 Observe the relationship of the stator windings to the drive-end bracket determined by the stator wire connections, the rectifier pack position on the slip-ring end bracket, the alignment of the mounting lugs on the end brackets and the through-bolt clearances on the stator windings.
24 Position the stator windings to the drive-end bracket.
25 Ensure that the rubber 'O' ring is fitted correctly in the slip-ring end bracket bearing housing.
26 Fit the slip-ring end bracket by carefully pushing the bearing into the housing.
27 Fit the through-bolts, tightening evenly.
28. Position the rectifier pack. Fit screw to secure the rectifier pack earth strip. Tighten the nut to secure the rectifier pack.
29 Position three stator wires on the rectifier pack as noted in operation 6.
30 Solder three stator wire connections. Note the precautions stated in operation 7 and use 'M' grade 45-55 resin core solder.
31 Position the radio capacitor. Fit the screw to secure.
32 If required, attach the regulator to the brushbox, regulator and surge protection diode assembly as follows: Position the spacer. Position the regulator. Fit screw to secure the regulator. Fit screw to secure one wire eyelet.
33 Fit the brushbox, regulator and surge protection diode assembly as follows: Position the assembly. Fit two screws to secure the brushbox, include one earth wire eyelet under one screw head. Connect four Lucar connectors to the rectifier pack. Fit screw to secure surge protection diode.
34 Fit the moulded cover.

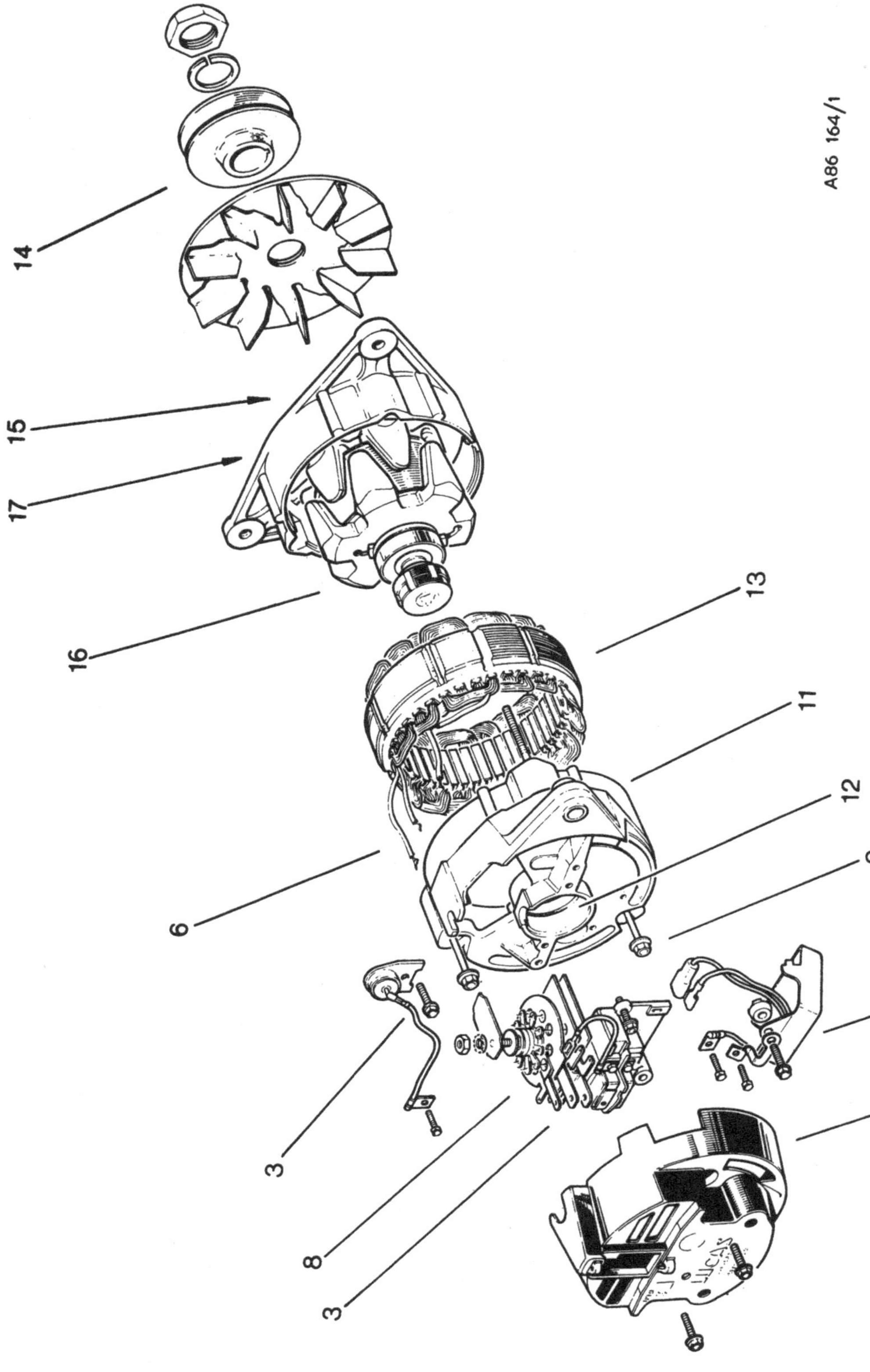
A86 164/1
14
15
17
16
13
11
12
9
6
4
1
3
8
3

ALTERNATOR – LUCAS TYPE 20ACR [and 25 ACR]

Overhaul **86.10.08**

Disassemble

1 Remove two screws, washers and spring washers. Remove the cover.
2 Note the wire positions and colour codes.
3 Disconnect the red wire small Lucar connector.
4 Disconnect the white wire small Lucar connector.
5 Disconnect the red wire large Lucar connector.
6 Remove the screw, washer and spring washer.
7 Remove the screw and spring washer.
8 Remove the screw, washer and spring washer.
9 Lift out the brushbox, regulator and surge protection diode assembly.
10 If necessary detach the regulator from the brushbox as follows–
Remove the screw to release one wire eyelet. Remove the screw to release the regulator. Bend up the contact strap. Disengage two lugs and lift away the regulator. Collect up the spacer.
11 Remove three pozidriv screws.
12 Remove the earth strap screw and spring washer.
13 Remove three screws, washers and spring washers.
14 Lift out the rectifier pack.
15 Scribe a line across the slip ring end bracket, stator winding and drive end bracket.

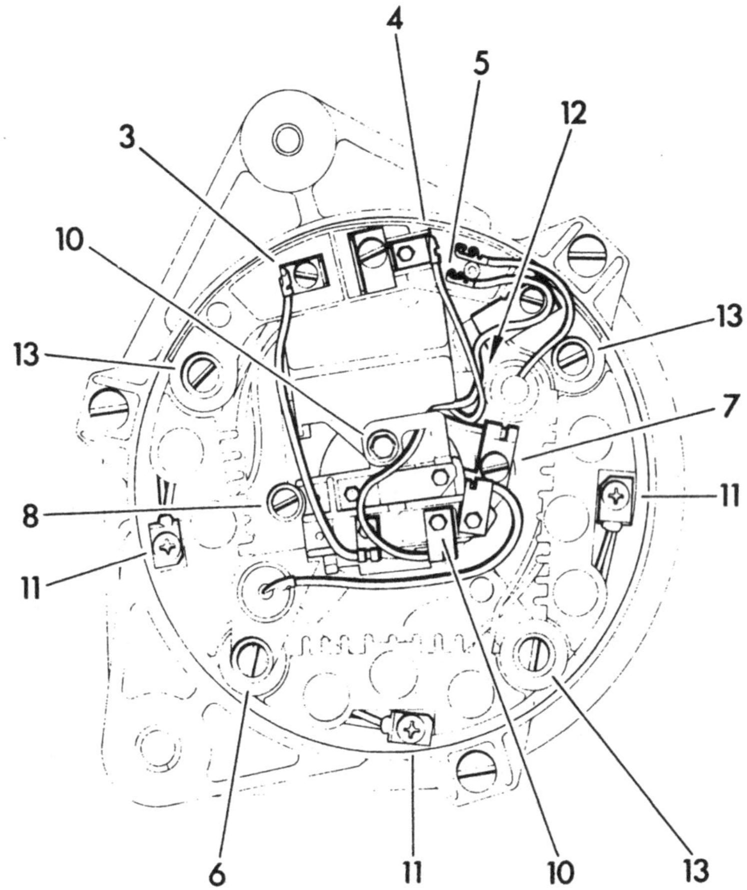

A86 068

16 Remove three side bolts and spring washers.
17 Collect up three nuts.
18 Provide an extractor tool as shown.
19 To remove the slip ring end bracket, position the extractor tool to engage with the outer journal of the slip ring end bearing. Employ a second operator to support the slip ring end bracket by hand. Carefully tap the extractor tool to drive the bearing from the housing.
NOTE: It may be necessary to carefully file away surplus solder from the two field winding connections on the slip ring moulding if the extractor tool will not pass over the moulding.
20 The rubber 'O' ring fitted in the slip ring end bracket bearing housing may remain in situ unless replacement is contemplated.
21 Remove the stator windings from the drive end bracket.
22 Remove the nut and spring washer. Prevent the rotor turning by wrapping a scrap fan belt round the pulley and retaining by hand or vice.
23 Remove the pulley.
24 Remove the fan.
25 Remove the flanged spacer.
26 Remove the key.
27 Unsolder two field winding connections. Pull the slip ring moulding from the shaft. This operation is necessary to prevent damage to the slip ring moulding during remove and refit of the rotor to the drive end bracket.
28 Using a suitable press remove the rotor from the drive end bracket.
CAUTION: Do not attempt to remove the rotor by applying hammer blows to the shaft end. Such action may burr over and damage the thread.
29 Remove the thin spacer from the rotor shaft.
30 Remove three screws. Remove the plate.
31 Remove the rubber ring.
32 Press the bearing from the drive end bracket.

Assemble

33 Press the bearing into the drive and end bracket.

34 Position the rubber ring.

35 Position the plate with the countersinks uppermost. Secure with three screws.

36 Fit the thin spacer to the rotor shaft.

37 Using a suitable press, the flanged spacer and a suitable tube, fit the rotor to the drive end bracket by applying pressure to the bearing inner journal. **CAUTION**: Do not use the drive end bracket as a support while fitting the rotor.

38 Push the slip ring moulind onto the shaft. Ensure that the moulding tongue engages correctly in the shaft slot. Solder two field winding connections.

39 Fit the key.

40 Fit the flanged spacer with the reduced diameter against the bearing inner journal.

41 Fit the fan with the blades adjacent to the drive end bracket.

42 Fit the pulley with the boss against the fan.

43 Fit the spring washer and nut. Prevent the rotor turning by wrapping a scrap fan belt round the pulley and retaining by hand or vice. Torque load the nut to 30 to 36 lbf ft (4.2 to 4.9 kgf m).

44 Position the stator windings to the drive end bracket. Align the scribe lines made at operation 15 above.

45 Ensure that the rubber 'O' ring is fitted correctly in the slip ring end bracket bearing housing.

46 Fit the slip ring end bracket by carefully pushing the bearing into the housing. Align the scribe lines made at operation 15 above.

47 Fit three side bolts with spring washers under the bolt heads. Secure with three nuts. Tighten evenly.

48 Position the rectifier pack.

49 Fit three screws, washers and spring washers.

50 Fit the earth strap screw and spring washer.

51 Fit three pozidrive screws.

52 If necessary attach the regulator to the brushbox as follows–
Position the spacer. Position the regulator. Bend down the contact strap. Fit the screw to secure the regulator. Fit the screw to secure one wire eyelet.

53 Position the brushbox, regulator and surge protection diode assembly.

54 Fit the screw, washer and spring washer.

55 Fit the screw and spring washer. Include the capacitor lug and the regulator earth wire eyelet in the assembly.

56 Fit the screw, washer and spring washer. Include the surge protection diode lug in the assembly.

57 Connect the red wire large Lucar connector.

58 Connect the white wire small Lucar connector.

59 Connect the red wire small Lucar connector.

60 Position the cover. Secure with two screws, washers and spring washers.

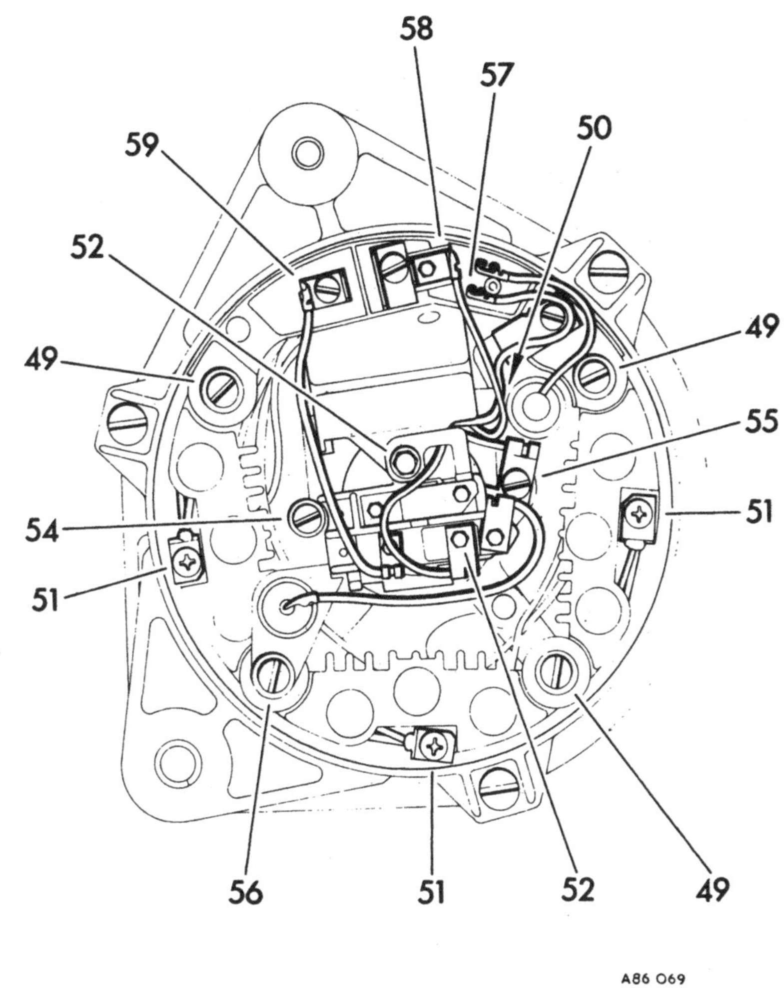

ALTERNATOR – LUCAS 25ACR

Overhaul **86.10.08**

Dismantling

1 Remove two bolts. Remove the cover.
2 Note the wire positions and the colour codes.
3 Disconnect the yellow wire small Lucar connector from the rectifier pack.
4 Remove the screw securing the avalanche diode.
5 Disconnect the white wire Lucar connector from the terminal block.
6 Disconnect the yellow wire Lucar connector from the terminal block.
7 Remove the nut, spring washer and two red wire tags from the terminal block.
8 Remove the two screws. Lift out the brushbox, regulator and avalanche diode assembly.
9 If necessary detach the regulator from the brushbox as follows:
Remove the screw to release one wire eyelet. Remove the screw to release the regulator. Bend up the contact strap. Disengage two lugs and lift away the regulator.
10 Remove screw and radio capacitor.
11 Remove three through bolts.
12 Provide an extractor tool as shown.

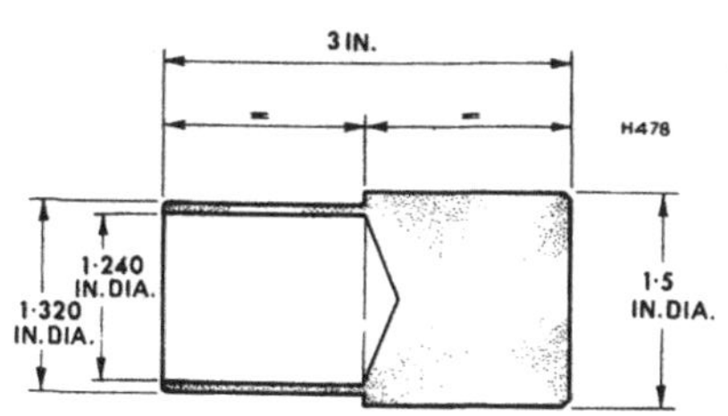

13 To remove the slip-ring end bracket, stator, rectifier and terminal block assembly position the extractor tool to engage with the outer journal of the slip-ring end bearing.
Employ a second operator to support the slip-ring end bracket and stator by hand. Carefully tap the extractor tool to drive the bearing from the housing.
NOTE: It may be necessary to carefully file away surplus solder from the two field winding connections on the slip-ring moulding if the extractor tool will not pass over the moulding.
14 The rubber 'O' ring fitted in the slip-ring end bracket bearing housing may remain *in situ* unless replacement is contemplated.
15 If necessary detach the stator, rectifier and terminal block from the slip-ring end bracket as follows:
Unsolder three wires. At each apply solder iron, bend out centre limb and disconnect wire. Remove stator. Slacken nut and withdraw rectifier. Remove two screws and terminal block.
16 Prevent the rotor turning by wrapping a scrap fan belt round the pulley and retaining by hand or vice. Remove the nut and spring washer.
17 Remove the pulley.
18 Remove the fan.
19 Remove the key.
20 Remove the outer spacer.
21 Using a suitable press remove the rotor from the drive end bracket.
CAUTION: Do not attempt to remove the rotor by applying hammer blows to the shaft end. Such action may burr over and damage the thread.
22 Remove the inner spacer.
23 Remove three screws. Remove the plate.
24 Press the bearing from the drive end bracket.
25 Collect up the metal retainer and felt washer.

Reassembling

26 Position the felt washer and metal retainer.
27 Press the bearing into the drive end bracket.
28 Position the plate. Secure with three screws.
29 Fit the inner spacer.
30 Using a suitable press, the outer spacer and a slave tube – fit the drive end bracket to the rotor by applying pressure to the bearing inner journal.
CAUTION: Do not use the drive end bracket as a support while fitting the rotor.
31 Fit the outer spacer.
32 Fit the key.
33 Fit the fan.
34 Fit the pulley.
35 Fit the spring washer and nut. Retain as at instruction 16 and torque load nut to 30 lbf ft (41 Nm).
36 If necessary, assemble the stator, rectifier and terminal block to the slip-ring end bracket as follows:
Position terminal block and secure with two screws. Insert rectifier and tighten nut. Position stator. Solder three wires using standard resin-cored solder.
37 Ensure that rubber 'O' ring is fitted in the slip-ring end bracket bearing housing.
38 Fit the slip-ring end bracket, stator, rectifier and terminal block assembly by carefully pushing the bearing into the housing. Align the main mounting lugs.
39 Fit three through bolts. Tighten evenly. Torque load to 5 lbf ft (7 Nm).
40 Position radio capacitor. Secure with screw.
41 If necessary, assemble the regulator to the brushbox as follows:
Position the regulator. Bend down the contact strap. Fit the screw to secure the regulator. Fit the screw to secure yellow wire eyelet.
42 Position the brushbox, regulator and avalanche diode assembly. Secure with two screws. Include the regulator earth wire eyelet under one screw head.
43 Position two red wire tags to the terminal block. Secure with spring washer and nut.
44 Connect the yellow wire Lucar connector to the terminal block.
45 Connect the white wire Lucar connector to the terminal block.
46 Fit the screw to secure the avalanche diode.
47 Connect the yellow wire small Lucar connector to the rectifier pack.
48 Position the cover. Secure with two bolts.

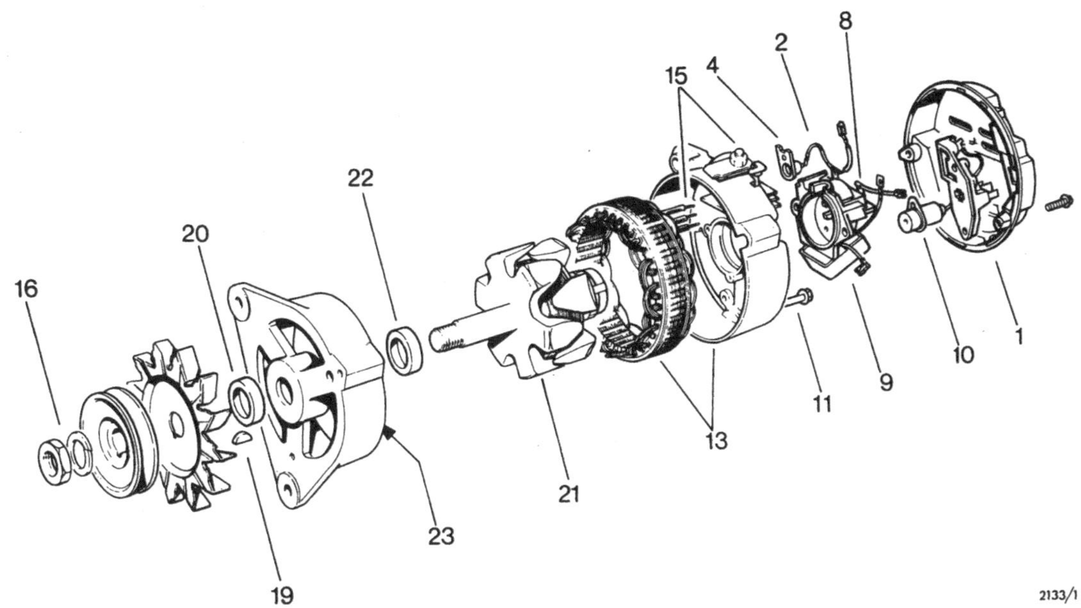

2133/1

BATTERY

Remove and refit **86.15.01**

Remove

1 Remove the battery leads.
2 Slacken two nuts. Swing the retaining assembly down.
3 Lift the battery from the tray.

Refit

4 Ensure that the retaining assembly is assembled in the swung down position with the hooks engaged in the vehicle body apertures.
5 Ensure that all leads and harness wires are outside the area of the retaining assembly.
6 Ensure that the rubber mat is correctly positioned.
7 Lift the battery onto the tray.
8 Swing the retaining assembly up. Tighten two nuts.
9 Fit the battery leads. Do not hammer the terminals to the terminal posts.
10 Coat the terminals with petroleum jelly 'Vaseline' to prevent corrosion.

IGNITION DISTRIBUTOR APPLICABILITY, IGNITION TIMING AND FEATURES CHART [86.35.00] 1975 to 1979

	UK AND EUROPE 1976/77/78/79	U.S.A. MARKET			
		Federal		California	
		1975/76	1977/78/79	1975/76	1977/78/79
Triumph part No.	RKC 0066	TKC 1204	TKC 3328	TKC 1689	TKC 3330
Lucas part No.	X	41601	41701	41603	41700
Delco Remy part No.	7992715	X	X	X	X
Ignition timing — static	10 degrees B.T.D.C.	10 degrees B.T.D.C.	10 degrees B.T.D.C.	10 degrees B.T.D.C.	10 degrees B.T.D.C.
Ignition timing — at idle	See table 86.35.15 Advance pipe disconnected	2 degrees A.T.D.C. Retard pipe connected	10 degrees B.T.D.C.	2 degrees A.T.D.C. Retard pipe connected	2 degrees A.T.D.C. Retard pipe connected
Idle speed	800 rev/min	800 rev/min	800 rev/min	800 rev/min	800 rev/min
Conventional points ignition	✓	X	X	X	X
Electronic ignition	X	✓	✓	✓	✓
Centrifugal advance	✓	✓	✓	✓	✓
Vacuum advance	✓	X	X	X	X
Retard unit	X	✓	X	✓	✓
Micrometer adjustment nut	X	X	X	X	X

IGNITION DISTRIBUTOR APPLICABILITY, IGNITION TIMING AND FEATURES CHART 1980

	UK AND EUROPE	U.S.A. MARKET				CANADA AND AUSTRALIA
		Federal			California	
	Carburetter	Carburetter	Carburetter	Fuel Injection	Fuel Injection	Carburetter
	1980	Cosmetic 1980	1980	1980	1980	1980
Triumph part No.	RKC 0066	TKC 3330	PKC 1333	RKC 3773	RKC 3773	RKC 4722
Lucas part No.	X	41700	X	X	X	X
Delco Remy part No.	7992715	X	9977215	9977120	9977120	9977254
Ignition timing — static	10 degrees B.T.D.C.	10 degrees B.T.D.C.	10 degrees B.T.D.C.	10 degrees B.T.D.C.	10 degrees B.T.D.C.	10 degrees B.T.D.C
Ignition timing — at idle	See table 86.35.15 Advance pipe disconnected	2 degrees A.T.D.C. Retard pipe connected	2 degrees A.T.D.C. Retard pipe connected	2 degrees A.T.D.C. Retard pipe connected	2 degrees A.T.D.C. Retard pipe connected	2 degrees A.T.D.C. Retard pipe connected
Idle speed	800 rev/min	800 rev/min	800 rev/min	800 rev/min	800 rev/min	800 rev/min
Conventional points ignition	✓	X	X	X	X	X
Electronic ignition	X	✓	✓	✓	✓	✓
Centrifugal advance	✓	✓	✓	✓	✓	✓
Vacuum advance	✓	X	X	X	X	X
Retard unit	X	✓	✓	✓	✓	✓
Micrometer adjustment nut	X	X	X	X	X	X

IGNITION DISTRIBUTOR APPLICABILITY, IGNITION TIMING AND FEATURES CHART 1981

	LOW COMPRESSION ENGINE	CANADA
	Carburetter	Carburetter
	1981	1981
Triumph Part No.	RKC 5525	RKC 4722
Lucas Part No.	X	X
Delco Remy Part No.	9977370	9977254
Ignition timing — static	10° B.T.D.C.	10 degrees B.T.D.C.
Ignition timing — at idle	10° B.T.D.C	2 degrees A.T.D.C. Retard pipe connected
Idle speed	800 rev/min	800 rev/min
Conventional points ignition	✓	X
Electronic Ignition	X	✓
Centrifugal Advance	✓	✓
Vacuum Advance	✓	X
Retard Unit	X	✓

IGNITION COIL AND BALLAST RESISTOR – ALL MODELS

Data

Ignition coil

Manufacturer	Lucas
Type	15C6
Lucas part number	45266
Triumph part number	UKC 2982
Primary winding resistance	1.3 to 1.45

Ballast resistor wire

Resistance	1.3 to 1.5 ohm
Wire length	63 in (1600 mm)
Wire colour code	Pink/white

Description

This system is designed to assist engine starting under adverse conditions. A ballast resistor wire built into the harness is electrically positioned in series in the normal supply to the ignition coil. This wire causes a voltage drop in the circuit so that the 12 volt supply from the ignition switch may be employed to power the nominally rated 6 volt ignition coil.

During engine start the resistor is by-passed and the battery voltage (reduced from 12 volt by the starter motor load) is applied to the coil direct from the starter motor relay. This slight voltage overload provides an increased high tension voltage at the spark plugs.

IGNITION DISTRIBUTOR – UK AND EUROPE—1976-1980

Data

Manufacturer	AC Delco
Series	D302
Delco Remy part no.	7992715
Stanpart No.	RKC 0066
Contact gap	0.014 to 0.016 in
Rotation – viewed on rotor	Anti-clockwise
Firing angles	90 ± 1 degree
Dwell angle	39 ± 1 degree
Open angle	51 ± 1 degree
Moving contact spring tension	19 to 24 oz
Condenser capacity	0.18 to 0.23 mfd
Engine firing order	1–3–4–2

Centrifugal advance

Distributor r.p.m.	Degs. distributor advance		Crankshaft r.p.m.	Degs. crankshaft advance	
	Minimum	Maximum		Minimum	Maximum
300	No advance to occur		600	No advance to occur	
500	0	1.00	1000	0	2.0
900	1.85	3.85	1800	3.7	7.7
1200	4.00	6.00	2400	8.0	12.0
1600	5.00	7.00	3200	10.0	14.0
2400	7.00	9.00	4800	14.0	18.0
3000	9.00	9.00	6000	18.0	18.0

Vacuum advance

Ins. of mercury vacuum	Degs. distributor advance		Degs. crankshaft advance	
	Minimum	Maximum	Minimum	Maximum
3		No advance to occur		
5	1.30	4.67	2.60	9.34
7	6.00	8.00	12.00	16.00
10	6.00	8.00	12.00	16:00
20	8.00	8.00	16.00	16.00

IGNITION DISTRIBUTOR UK AND EUROPE—1981

Data

Manufacturer	AC Delco
Series	D308
Delco Remy part no.	9977370
Stanpart No.	RKC 5525
Contact gap	0.014 to 0.016 in
Rotation — viewed on rotor	Anti-clockwise
Firing angles	90 ± 1 degree
Dwell angle	39 ± 1 degree
Open angle	51 ± 1 degree
Moving contact spring tension	19 to 24 ozf
Condenser capacity	0.18 to 0.23 mfd
Engine firing order	1–3–4–2

Centrifugal advance

Distributor r.p.m.	Degs. distributor advance		Crankshaft r.p.m.	Degrees crankshaft advance	
	Minimum	Maximum		Minimum	Maximum
300	No advance to occur		600	No advance to occur	
550	5	6.5	1100	10	13
1000	9.5	11	2000	19	22
1500	12	14	3000	24	28

Vacuum advance

Ins. of mercury vacuum	Degs. distributor advance		Degs. crankshaft advance	
	Minimum	Maximum	Minimum	Maximum
3		No advance to occur		
5	0	4.5	0	9
7	4	7.5	8	15
8	6.5	7.5	13	15

IGNITION COMPONENTS WITH DELCO REMY CONVENTIONAL CONTACT SYSTEM

BALLAST RESISTOR WIRE

Data

Resistance	1.3 to 1.5 ohm
Wire length	63 in (1600 mm)
Wire colour code	Pink/white

Description

This system is designed to assist engine starting under adverse conditions. A ballast resistor wire built into the harness is electrically positioned in series in the normal supply to the ignition coil. This wire causes a voltage drop in the circuit so that the 12 volt supply from the ignition switch may be employed to power the nominally rated 6 volt ignition coil.

During engine start the resistor is by-passed and the battery voltage (reduced from 12 volt by the starter motor load) is applied to the coil direct from the starter motor relay. This slight voltage overload provides an increased high tension voltage at the spark plugs.

IGNITION COIL

Data

Manufacturer	Lucas
Type	15C6
Lucas part number	45266
Triumph part number	UKC 2982
Primary winding resistance	1.3 to 1.45 ohm

IGNITION DISTRIBUTOR –

Contact assembly – remove and refit **86.35.13**

Remove

1 Remove the fresh air duct, see 80.15.31.
2 Pull the 'king' high tension lead from the distributor cover.
3 Remove the distributor cover and swing to rest across the camshaft cover.
4 Remove two screws and spring washers. Lift off the rotor.
5 Push the moving contact spring away from the terminal post and slip two wire eyelets along the post to release.
6 If necessary rotate the crankshaft to position the weights for the best access to the cross headed screws.
NOTE: The screws are cross headed to provide a choice of screwdriver position.
7 Remove two cross headed screws and spring washers.
8 Lift out the contact assembly.

Refit

9 Wipe preservative from the new contact faces.
10 Reverse 5 to 8.
11 Adjust contact gap, see 86.35.14.

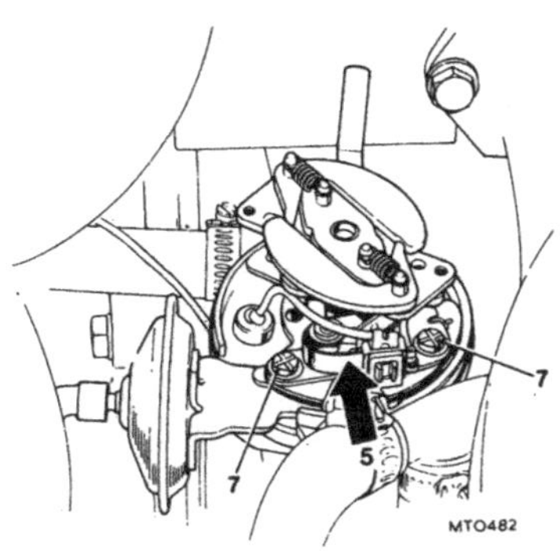

IGNITION DISTRIBUTOR –

Contact gap – adjust **86.35.14**

1 Remove the fresh air duct, see 80.15.31.
2 Pull the 'king' high tension lead from the distributor cover.
3 Remove the distributor cover and swing to rest across the camshaft cover.
4 Remove two screws and spring washers. Lift off the rotor.
5 Rotate the crankshaft to position the contact heel on a cam peak and position the weights for the best access to the contacts.
6 If the contact gap is correct a 0.015 in (0.38 mm) feeler gauge will fit between the contacts.
7 If the gap is correct operations 8 to 11 may be ignored.
8 Slacken two cross headed screws.
9 Move the fixed contact about the pivot to adjust the gap. This may be facilitated by inserting a screwdriver in the slots and twisting to position the fixed contact.
10 Tighten two cross headed screws.
11 Check that the correct gap has been maintained.
12 Reverse 1 to 4.

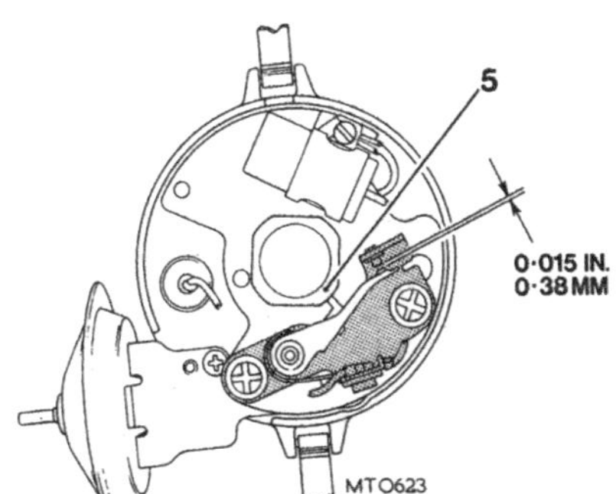

IGNITION DISTRIBUTOR –

Ignition timing – adjust **86.35.15**

If the engine can be run adjust dynamically as follows. If the engine cannot be run adjust statically as follows:

Dynamic

1 Remove the fresh air duct, see 80.15.31.
2 Pull off the advance unit pipe.
3 Connect a tachometer to the engine as instructed by the manufacturer.
NOTE: The vehicle instrument panel tachometer may be used if no other instrument is available.
4 Connect a timing light as instructed by the manufacturer. The engine is timed on number one cylinder which is at the front of the engine.
5 Run the engine.
6 Position the timing light to illuminate the crankshaft pulley and timing cover scale.
7 Hold the speed at each of the engine r.p.m. indicated. The ignition timing should be as stated.

Engine r.p.m.	Static timing	Centrifugal advance –mean	Ignition timing B.T.D.C.
1000		1.0	11.0
1800	10 degs. B.T.D.C.	5.7	15.7
2400		10.0	20.0

8 Stop the engine.
9 If the ignition timing is correct operations 10 to 13 may be ignored.
10 Use service tool S349 to slacken two distributor mounting bolts.
11 Rotate the distributor body slightly clockwise to advance the timing or anti-clockwise to retard the timing.
12 Tighten two distributor mounting bolts.
13 Repeat operation 5 onwards.
14 Remove the timing light.
15 Remove the tachometer.
16 Push on the advance unit pipe.
17 Fit the fresh air duct, see 80.15.31.

continued

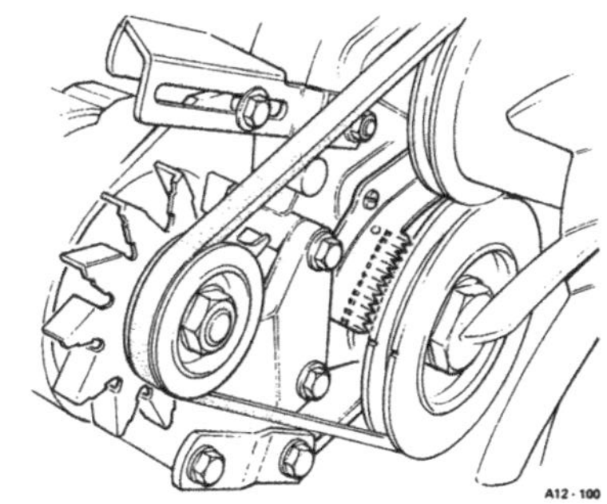

Static

18 Isolate the battery.

19 Disconnect the distributor low tension fly lead from the harness.

20 Provide a test lamp circuit as shown.

21 Rotate the crankshaft in the engine run direction to bring the crankshaft pulley notch to the start of the timing cover scale. The test lamp should now be illuminated.

22 Carefully rotate the crankshaft further until the lamp just goes out.

23 If the timing is correct the pulley notch will be aligned with the 10 degree BEFORE on the scale.

24 If the ignition timing is correct operations 25 to 30 may be ignored.

25 Remove the fresh air duct, see 80.15.31.

26 Use service tool S349 to slacken two distributor mounting bolts.

27 Rotate the crankshaft in the engine run direction to align the pulley notch with the 10 degree BEFORE on the scale.

28 Rotate the distributor body anti-clockwise past the test lamp illumination position. Carefully rotate clockwise until the lamp just goes out.

29 Tighten two distributor mounting bolts.

30 Repeat operation 21 onwards.

31 Remove the test lamp circuit.

32 Connect the distributor low tension fly lead to the harness white/slate wire.

33 Fit the fresh air duct, see 80.15.31.

34 Connect the battery.

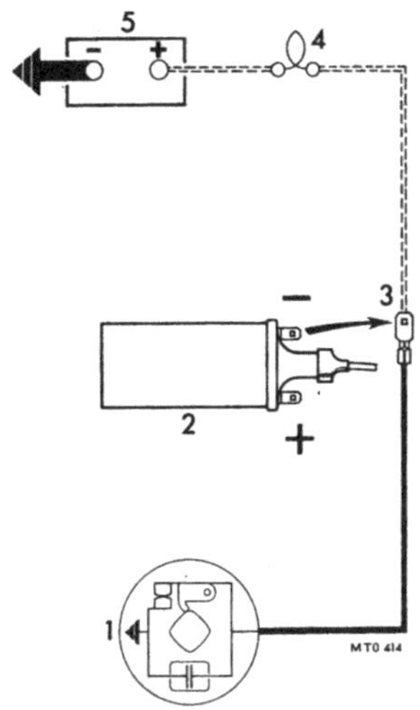

1 Distributor – diagrammatic layout.
2 Ignition coil.
3 Distributor fly lead removed from coil.
4 Test lamp – 12 volt.
5 Vehicle battery.

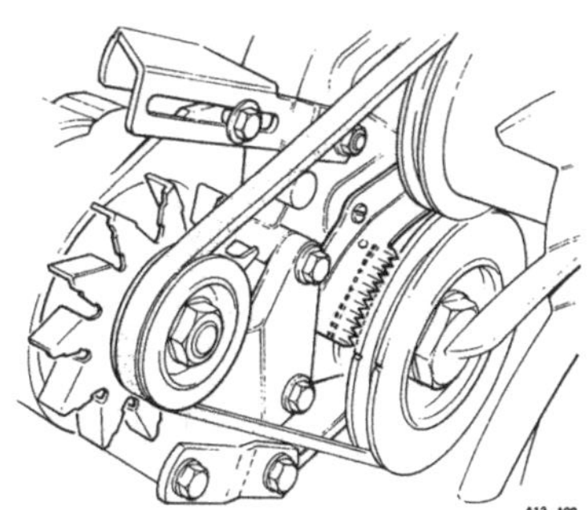

IGNITION DISTRIBUTOR – DELCO REMY

Lubrication **86.35.18**

1 Remove the fresh air duct, see 80.15.31.

2 Pull the 'king' high tension lead from the distributor cover.

3 Remove the distributor cover and swing to rest across the camshaft cover.

4 Remove two screws and spring washers. Lift off the rotor.

5 *Vehicles fitted with cam lubrication sponge only:*
Perform instructions 6 to 10.

6 Squeeze the two sides of the cam lubrication post together and withdraw the post.

7 Remove the sponge from the post.

8 Work on amount of Shell Alvania No. 2 grease or equivalent into the spong.

9 Fit the sponge to the post.

10 Squeeze the two sides of the post together and insert the post. Ensure that the curved base surface faces outwards.

11 *Vehicles not fitted with cam lubrication sponge only:*
Lightly grease the cam with Shell Alvania No. 2 grease or equivalent.

12 Inject a few drops of engine oil through the OIL hole to lubricate the top bearing.

13 Apply one drop of engine oil to each weight pivot post and each 'cam action' position.

14 Reverse instructions 1 to 4.

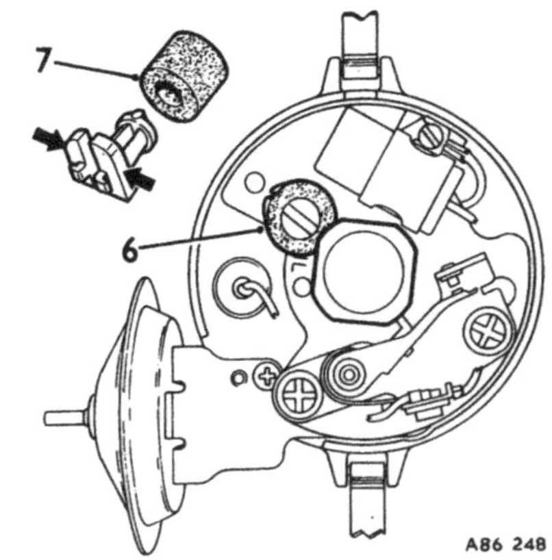

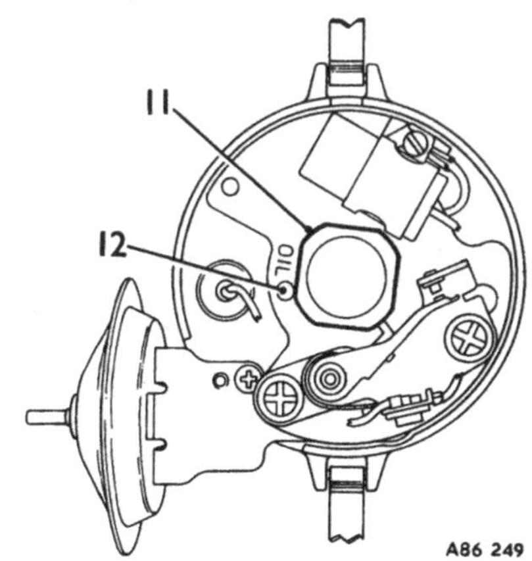

IGNITION DISTRIBUTOR – U.K. AND EUROPE

Remove and refit 86.35.20

Remove
1 Remove the fresh air duct, see 80.15.31.
2 Pull the 'king' high tension lead from the distributor cover.
3 Remove the distributor cover and swing to rest across the camshaft cover.
4 To assist the refit rotate the crankshaft in the engine run direction to bring number one piston to T.D.C. on the firing stroke. This is indicated when the rotor is pointing approximately towards the rear manifold mounting bolt as shown and the mark on the pulley is aligned with the 0 degree on the timing cover scale.
5 Pull off the advance unit pipe.
6 Disconnect the distributor low tension fly lead from the harness.
7 Use service tool S349 to remove two distributor mounting bolts, spring washers and washers.
8 Carefully withdraw and manoeuvre the distributor from the block.

Refit
9 Ensure that number one piston is at T.D.C. on the firing stroke and the mark on the pulley is aligned with the 0 degree on the timing cover scale.
10 Carefully manoeuvre and insert the distributor into the block with the vacuum unit facing exactly rearwards. Engage the drive gear so that the rotor is finally pointing approximately towards the rear manifold mounting bolt as shown.
11 Fit two distributor mounting bolts, spring washers and washers.
12 Connect the distributor low tension fly lead to the harness white/slate wire.
13 Push on the advance unit pipe.
14 Fit the distributor cover.
15 Push the 'king' high tension lead onto the distributor cover.
16 Adjust ignition timing, see 86.35.15.
17 Fit the fresh air duct, see 80.15.31.

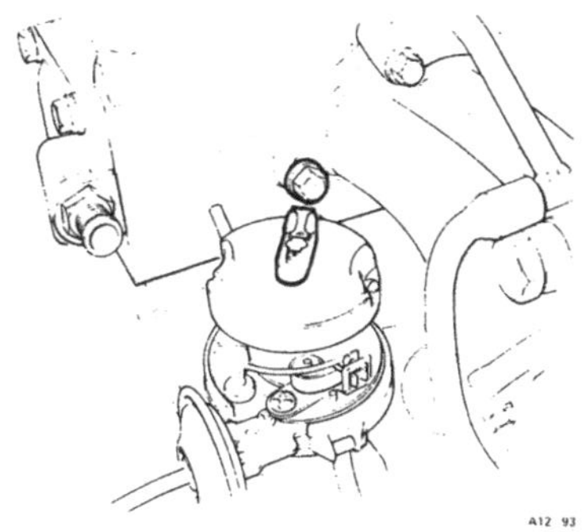

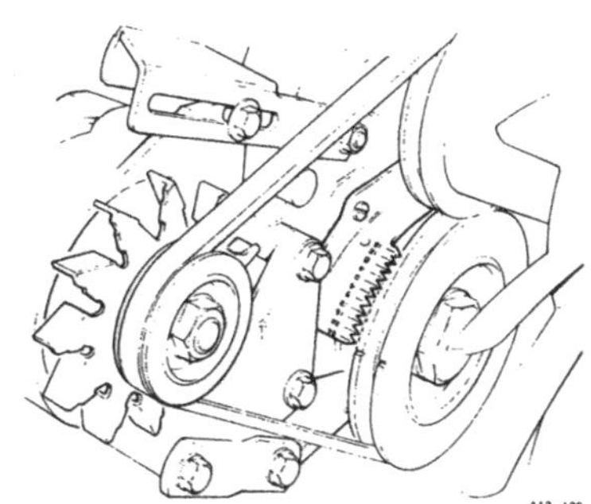

IGNITION DISTRIBUTOR – U.K. AND EUROPE

Overhaul 86.35.26

NOTE: Instructions 8, 9, 28, 29 and 30 apply to distributors fitted with cam lubrication sponge only.

Disassemble
1 Remove the cover.
2 Remove two screws and spring washers. Lift off the rotor.
3 Note the position of the two wire eyelets on the terminal post.
4 Push the moving contact spring away from the terminal post and slip two wire eyelets along the post to release.
5 Rotate the shaft to position the weights for the best access to the cross-headed screws. The screws are cross-headed to provide a choice of screwdriver position.
6 Remove two cross-headed screws and spring washers. Lift out the contact assembly.
7 Remove the screw and lift out the condenser.
8 Squeeze the two sides of the cam lubrication post together and withdraw the post.
9 Remove the sponge from the post.
10 Tap out the roll pin. Remove the drive gear and thrust washer.
11 Ensure that the shaft is burr-free. Withdraw the assembly.
12 Remove the special circlip and lift off the plate.
13 Remove two screws and lift off the earth lead and vacuum advance unit.
14 Carefully prise out the felt washer.
15 Remove the rubber grommet upwards and withdraw the wire assembly.
16 Remove the control springs, exercising great care not to distort the springs.
17 Lift off the weights.
18 Remove the cam unit by withdrawing down the shaft.

Assemble
19 Lubricate the shaft with engine oil. Fit the cam unit to the shaft.
20 Position the weights.
21 Fit the control springs, exercising great care not to distort the springs.
22 Insert the wire assembly downwards and fit the rubber grommet.
23 Saturate the felt washer with engine oil and position in the base recess.
24 Fit the vacuum advance unit. Include the earth lead tag under the appropriate screw head.
25 Lubricate the plate bearing with engine oil. Fit the plate and secure it with the special circlip.
26 Ensure that the shaft is adequately lubricated with engine oil. Insert the shaft assembly.
27 Fit the thrust washer and drive gear. Secure with the roll pin.
28 Work an amount of Shell 'Alvania No. 2' grease or equivalent into the sponge.
29 Fit the sponge to the post.
30 Squeeze the two sides of the post together and insert the post. Ensure that the curved base surface faces outwards.
31 Fit the condenser. Include the earth lead tag under the screw head.
32 Position the contact assembly. Fit two cross-headed screws and spring washers finger tight.
33 Push the moving contact spring away from the terminal post and slip two wire eyelets along the post to secure in the positions noted at operation 3 above.
34 Rotate the shaft to position the contact heel on a cam peak and position the weights for the best access to the contacts.
35 Move the fixed contact about the pivot to adjust the contact gap to 0.015 in (0.38 mm). This may be facilitated by inserting a screwdriver in the slots and twisting to position the fixed contact.
36 Tighten two cross-headed screws.
37 Check the correct gap has been maintained.
38 Apply one drop of engine oil to each weight pivot post and each 'cam action' position.
39 Position the rotor with correct location of the keying. Secure with two screws and spring washers.
40 Fit the cover.

illustration on next page—

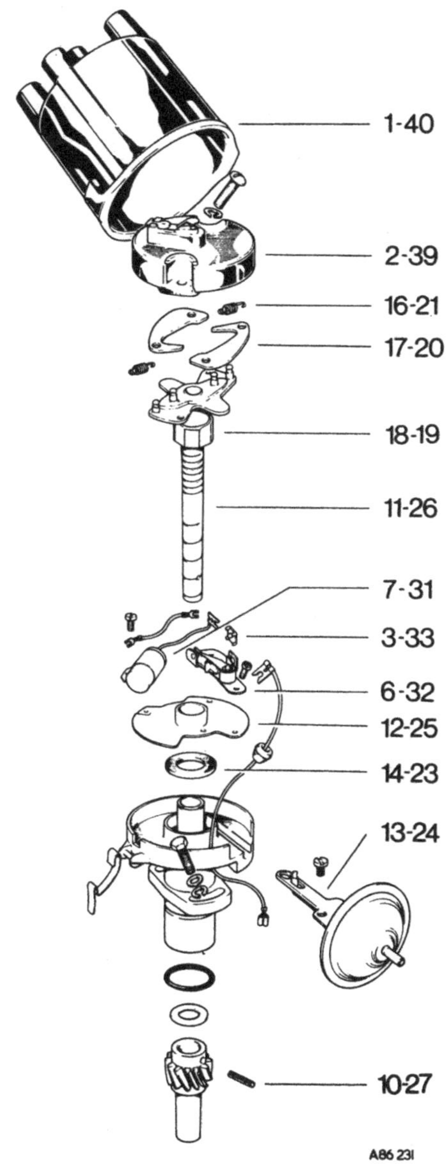

DRIVE RESISTOR – U.S.A. MARKETS [86.35.00]

Data

Manufacturer	Lucas
Lucas part number	54427556
Triumph part number	UKC 3908
Resistance	10 ± 5% ohm

Description

This unit may be considered as a detached component of the ignition distributor amplifier. Its function is associated with the transistors in the amplifier. It is independently mounted due to its size and heat dissipating requirements.

IGNITION COIL

Data

Manufacturer	**Lucas**
Type	**15C6**
Lucas part number	**45266**
Triumph part number	**UKC 2982**
Primary winding resistance	**1.3 to 1.45 ohm**

'OPUS ELECTRONIC IN BUILT' SYSTEM – U.S.A. MARKETS

Description

The 'Opus electronic in built' system consists of a conventional ignition coil and high tension circuit. The distributor contains a power transistor which manages the current flow through the ignition coil primary winding.

Distributor – deleted are the conventional cam, contacts and capacitor. These are replaced by an oscillator, timing rotor, pick up and amplifier. A conventional Lucas body, mechanical centrifugal advance and vacuum retard unit are retained.

Electronic circuit – the encapsulated oscillator supplies pulses to the pick up. The majority of these pulses are 'lost'. The timing rotor is carried on the rotating shaft and contains four ferrite rods. When one of the ferrite rods aligns with the pick up a pulse is 'caught' by the pick up and applied to the input of the encapsulated amplifier. The result is that the amplifier output switches the power transistor to the off condition.

Ignition coil and high tension circuit – the off condition of the power transistor collapses the ignition coil primary winding current. A spark occurs at the appropriate spark plug in the conventional manner.

Retard unit – the pick up is mounted on the moving plate. This assembly may be rotated through a limited angle by the vacuum retard unit. The ignition timing is thus modified by the vacuum retard unit by amending the relationship between the timing rotor and the pick up.

Drive resistor – this unit may be considered as a detached component of the amplifier. Its function is associated with the transistors in the amplifier. It is independently mounted due to its size and heat dissipating requirements.

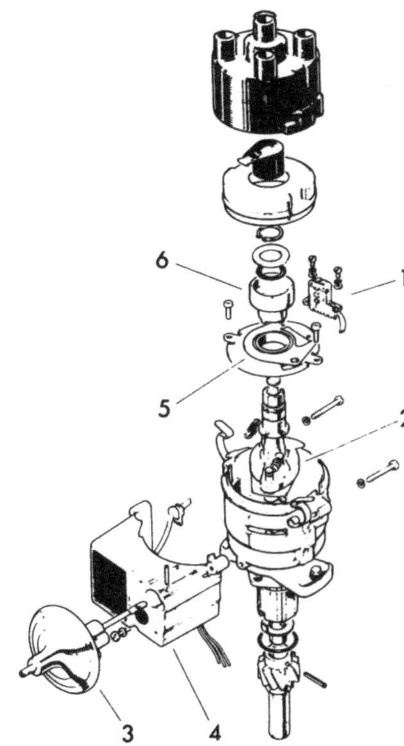

1 Pick up
2 Centrifugal advance
3 Retard unit
4 Oscillator and amplifier
5 Moving plate
6 Timing rotor

IGNITION DISTRIBUTOR – LUCAS ELECTRONIC – Triumph Part No. TKC 1204

Data

Manufacturer	Lucas
Type	47DE4
Lucas part number	41601
Triumph part number	TKC 1204
System	Opus electronic in-built
Polarity	Negative earth only
Pick-up air gap	0.014 to 0.016 in (0.35 to 0.40 mm)
Rotation–viewed on rotor	Anti-clockwise
Firing angle	90 ± 1 degree
Engine firing order	1 – 3 – 4 – 2
To be used in conjunction with an external drive resistor of	10 ± 5% ohm

Centrifugal advance

Check at increasing speeds

Distributor rev/min	Degrees distributor advance		Crankshaft rev/min	Degrees crankshaft advance	
	Minumum	Maximum		Minimum	Maximum
400	No advance to occur		800	No advance to occur	
500	0	1.0	1000	0	2
750	1.0	3.0	1500	2	6
950	2.5	4.5	1900	5	9
1100	3.5	5.5	2200	7	11
1750	6.0	8.0	3500	12	16
2100	7.0	9.0	4200	14	18
2750	7.0	9.0	5500	14	18

Retard unit

Check first with 11 inHg. Increase to 15 inHg. Finally decrease to 3 inHg.

Inches of mercury vacuum	Degrees distributor retard		Degrees crankshaft retard	
	Minimum	Maximum	Minimum	Maximum
3	No retard to occur			
11	5	7	10	14
15	5	7	10	14

IGNITION DISTRIBUTOR – LUCAS ELECTRONIC – Triumph Part No. TKC 3328

Data

Manufacturer	Lucas
Type	47DE4
Lucas part number	41701
Triumph part number	TKC 3328
System	Opus electronic in built
Polarity	Negative earth only
Pick up air gap	0.014 to 0.016 in (0.35 to 0.40 mm)
Rotation-viewed on rotor	Anti-clockwise
Firing angle	90 ± 1 degree
Dwell angle	70–80 degrees
Engine firing order	1–3–4–2
To be used in conjunction with an external drive resistor of	10 ± 5% ohm

Centrifugal advance

Check at decelerating speeds

Distributor rev/min	Degs. distributor advance		Crankshaft rev/min	Degs. crankshaft advance	
	Minimum	Maximum		Minimum	Maximum
400	No advance to occur		800	No advance to occur	
650	1.0	4.0	1300	2	8
950	4.0	6.0	1900	8	12
1400	6.5	8.5	2800	13	17
1700	7.0	9.0	3400	14	18
2500	7.0	9.0	5000	14	18

IGNITION DISTRIBUTOR – LUCAS ELECTRONIC – Triumph Part No. TKC 1689

Data

Manufacturer . . . . Lucas
Type. 47DE4
Lucas Part No. 41603
Triumph Part No. . . . . TKC 1689

System . . . . Opus electronic in-built
Polarity . . . . Negative earth only
Pick-up air gap . . . . 0.014 to 0.016 in (0.35 to 0.40 mm)
Rotation – viewed on rotor. . . . . Anti-clockwise
Firing angle . . . . 90 ± 1 degree
Engine firing order. . . . . 1 – 3 – 4 – 2
To be used in conjuction with an external drive resistor of . . . . 10 ± 5% ohm

Centrifugal advance

Check at increasing speeds

Distributor rev/min	Degrees distributor advance		Crankshaft rev/min	Degrees crankshaft advance	
	Minimum	Maximum		Minimum	Maximum
440	No advance to occur		880	No advance to occur	
650	1.5	3.5	1300	3	7
750	3.0	5.0	1500	6	10
1050	4.5	6.5	2100	9	13
1900	8.5	10.5	3800	17	21
2250	10.0	12.0	4500	20	24
3000	10.0	12.0	6000	20	24

Retard unit

Check first with 11 inHg. Increase to 15 inHg. Finally decrease to 3 inHg.

Inches of mercury vacuum	Degrees distributor retard		Degrees crankshaft retard	
	Minimum	Maximum	Minimum	Maximum
3	No retard to occur			
11	5	7	10	14
15	5	7	10	14

IGNITION DISTRIBUTOR – LUCAS ELECTRONIC – Triumph Part No. TKC 3330

Data

Manufacturer . . . . Lucas
Type 47 DE4
Lucas part number 41700
Triumph part number . . . . TKC 3330

System . . . . Opus electronic in built
Polarity . . . . Negative earth only
Pick-up air gap . . . . 0.014 to 0.016 in (0.35 to 0.40 mm)
Rotation-viewed on rotor . . . . Anti-clockwise
Firing angle . . . . 90 ± 1 degree
Dwell angle . . . . 70–80 degrees
Engine firing order 1–3–4–2
To be used in conjunction with an external drive resistor of . . . . 10 ± 5% ohm

Centrifugal advance

Check at increasing speeds

Distributor rev/min	Degs. distributor advance		Crankshaft rev/min	Degrees crankshaft advance	
	Minimum	Maximum		Minimum	Maximum
400	No advance to occur		800	No advance to occur	
650	1.0	4.0	1300	2	8
950	4.0	6.0	1900	8	12
1400	6.5	8.5	2800	13	17
1700	7.0	9.0	3400	14	18
2500	7.0	9.0	5000	14	18

Retard Unit

Ins. of mercury vacuum	Degs. distributor retard		Degs. crankshaft retard	
	Minimum	Maximum	Minimum	Maximum
3	No retard to occur			
11	5	7	10	14
15	5	7	10	14

Check first with 11 inHg. Increase to 15 inHg.
Finally decrease to 3 inHg.

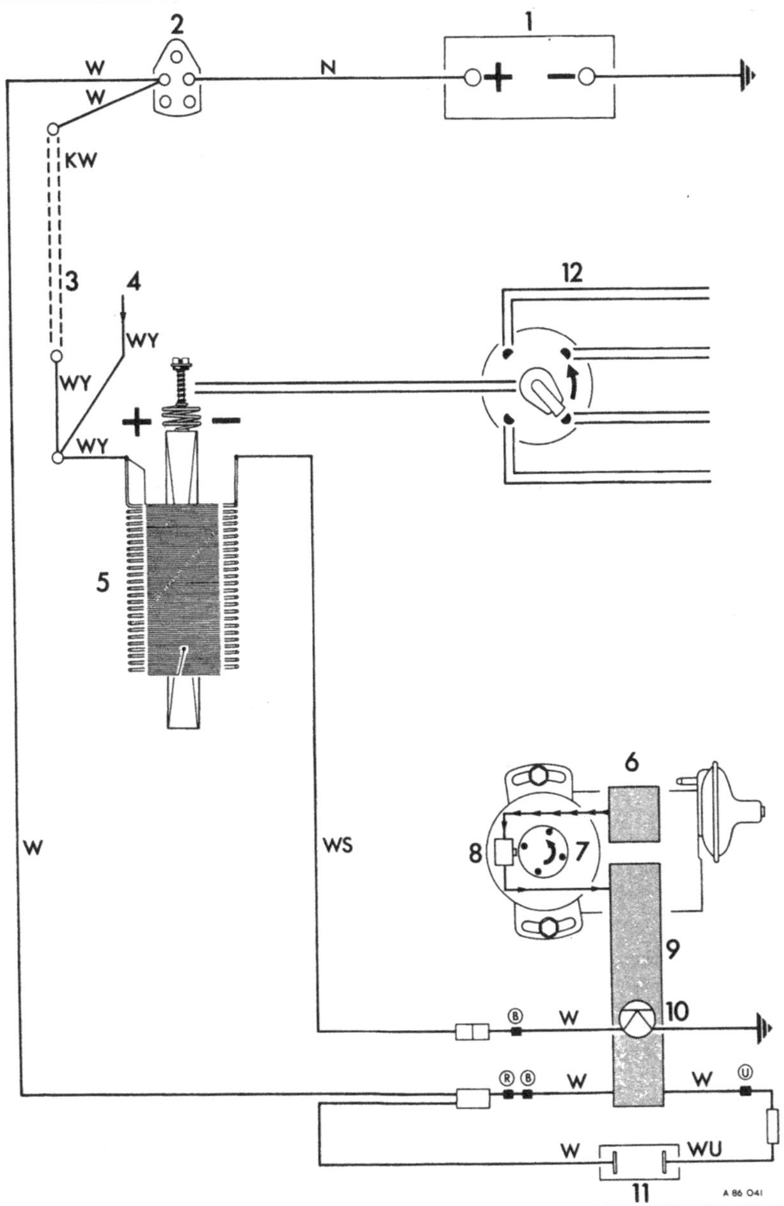

1 Battery
2 Ignition/starter switch
3 Ballast resistor wire
4 Start ignition supply from interlock starter motor relay
5 Ignition coil
6 Oscillator
7 Timing rotor
8 Pick up
9 Amplifier
10 Power transistor
11 Drive resistor
12 High tension circuit

Colour Code

N Brown
W White
KW Pink/White
WY White/Yellow
WS White/Slate
WU White/Blue
Ⓑ Black ident
Ⓡ Red ident
Ⓤ Blue ident

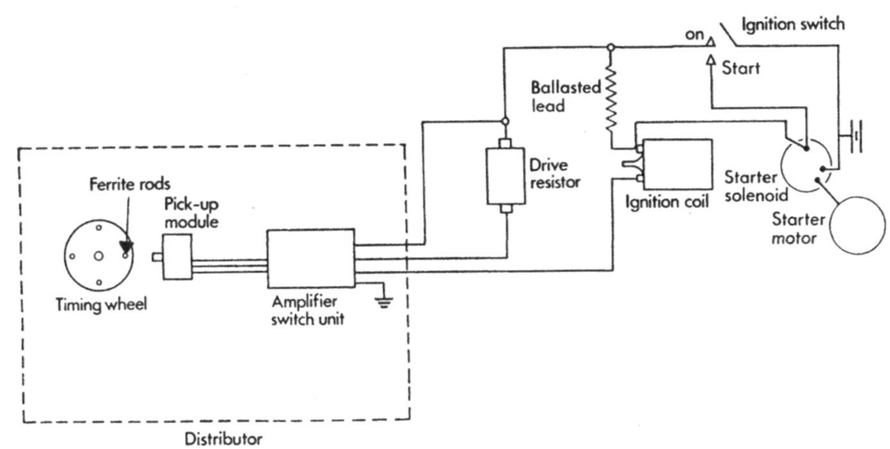

Simplified System Diagram of Electronic Ignition

IGNITION DISTRIBUTOR

Ignition timing – adjust **86.35.15**

The 'Opus electronic in built' ignition system provides a beneficial long dwell period with a short open period. For this reason a static adjustment using a test lamp is not practical. The test lamp would only flicker and would not provide an exact firing position.

If the engine can be run adjust dynamically as follows. If the engine cannot be rung adjust statically as follows.

Dynamic

1 Connect a tachometer to the engine as instructed by the manufacturer.
NOTE: The vehicle instrument panel tachometer may be used if no other instrument is available.
2 Connect a timing light as instructed by the manufacturer. The engine is timed on number one cylinder which is at the front of the engine.
3 Ensure that the retard unit pipe is connected. – Canadian and California specification vehicles only.
4 Run the engine.
5 Position the timing light to illuminate the crankshaft pulley and timing cover scale.

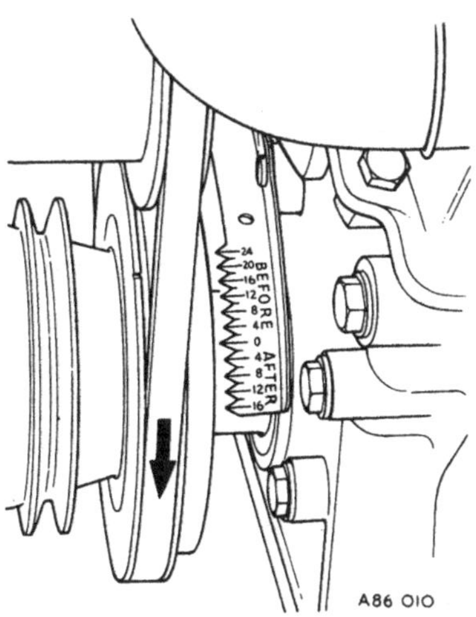

6 Hold the speed at 700 to 900 r.p.m. Nominal 800 r.p.m.
7 The ignition timing should now be 2 degrees A.T.D.C. – Canadian and California spec. vehicles only and 10° b.t.d.c. Federal specification vehicles.
8 If a correction is required perform operations 9 to 13.
9 Stop the engine.
10 Use Service tool S349 to slacken two distributor mounting bolts.
11 Rotate the distributor body slightly clockwise to advance the timing or anti-clockwise to retard the timing.
12 Tighten two distributor mounting bolts.
13 Repeat operation 4 onwards.

Static

14 Isolate the battery.
15 Rotate the crankshaft in the engine run direction to align the mark on the pulley with the 10 degree B.T.D.C. on the timing cover scale.
16 Remove the distributor cover.
17 Pull off the rotor.
18 Remove the plastic anti flash cover.
19 Inspect the relationship of the pick up to the nearest ferrite rod on the timing rotor. This should be positioned as shown.

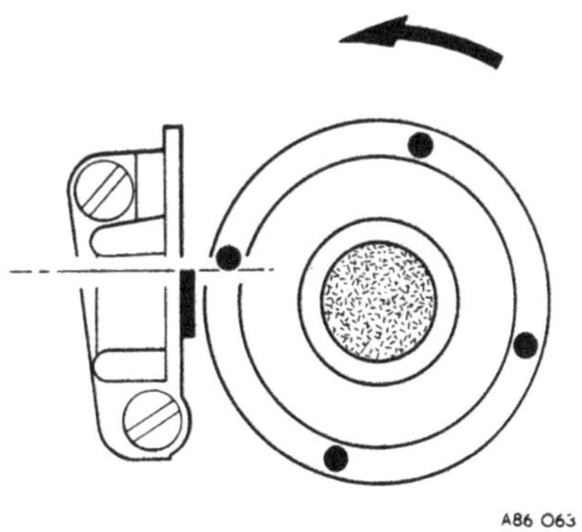

20 If a correction is required perform operations 21 to 23.
21 Use Service tool S349 to slacken two distributor mounting bolts.
22 Rotate the distributor body as required to achieve the position shown.
23 Tighten two distributor mounting bolts.
24 Fit the plastic anti flash cover with the two recesses positioned adjacent to the clips.
25 Push on the rotor.
26 Fit the cover.
27 Connect the battery.
28 When the engine can be run adjust dynamically as detailed above.

IGNITION DISTRIBUTOR

Lubrication **86.35.18**

1 Remove the fresh air duct. 80.15.31.
2 Remove the cover.
3 Pull off the rotor.
4 Remove the plastic anti flash cover.
5 Apply a few drops of engine oil to the felt pad to lubricate the rotor carrier bearing.
6 Inject a few drops of engine oil through the apertures to lubricate the centrifugal timing control.
7 Apply one drop of engine oil to each of the two lubrication apertures of the moving plate bearing.

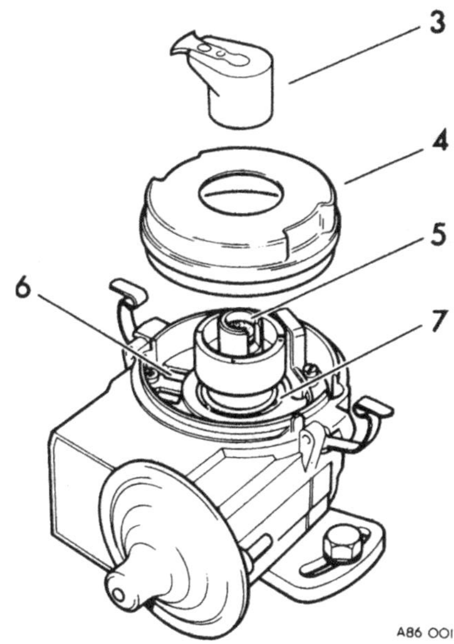

IGNITION DISTRIBUTOR

Remove and refit **86.35.20**

CAUTION: Do not connect the ignition distributor wires listed below direct to the battery positive supply –
White wire with Blue indent.
White wire with Black indent.

Remove

1 Isolate the battery.
2 Remove the fresh air duct. 80.15.31.
3 Remove the distributor cover.
4 Pull off the retard unit pipe. – Canadian and California specification vehicles only.
5 Disconnect three wire connectors.
6 Use Service tool S349 to remove two distributor mounting bolts, spring washers and washers.
7 Carefully withdraw and manoeuvre the distributor from the block.

Refit

8 Ensure that the mounting plate is correctly fitted to the block. The mounting plate is symmetrical and may be fitted either way round.
9 Rotate the crankshaft in the engine run direction to bring number one piston to T.D.C. on the firing stroke and align the mark on the pulley with the 0 degree on the timing cover scale.
10 **Distributor with retard unit only: Carefully manoeuvre and insert the distributor into the block with the retard unit facing exactly rearwards. Engage the drive gear so that the rotor is finally pointing approximately towards the rear manifold mounting bolt as shown.**
11 **Distributor with no retard unit only: Carefully manouvre and insert the distributor into the block with the oscillator and amplifier unit facing exactly rearwards. Engage the drive gear so that the rotor is finally pointing approximately towards the rear manifold mounting bolt as shown.**

continued

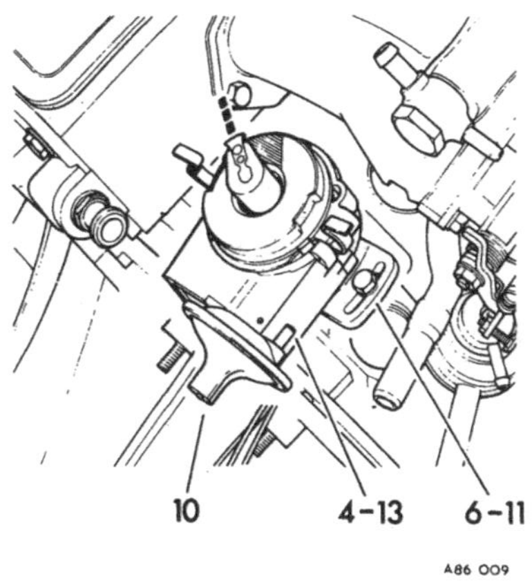

12 Fit two distributor mounting bolts, spring washers and washers.
13 Connect three wire connections as shown on illustration 86.35.00.
14 Distributor with retard unit only: Push on the retard unit pipe.
15 Fit the distributor cover.
16 Connect the battery.
17 Adjust ignition timing see, 86.35.16.
18 Fit the fresh air duct, see 80.15.31.

IGNITION DISTRIBUTOR

Overhaul 86.35.26

Disassemble

1 Remove the cover.
2 Pull off the rotor.
3 Remove the plastic anti flash cover.
4 Withdraw the felt pad.
5 Carefully remove two screws, lock washers and washers to release the pick up.
6 Remove two long screws and lock washers.
7 Hold the amplifier module in position. Remove one short screw, lock washer and washer.
8 Hold the distributor body in one hand and the amplifier module in the other hand. Carefully manoeuvre to unhook the retard unit link from the moving plate pin. This operation will not be visible.
9 Hold the distributor body and amplifier module slighlty apart. Withdraw two clips.
10 Pull out the wire grommet. Remove the amplifier module and pick up joined together by the wires.
11 **Distributor with retard unit only: Tap out the retard unit pin. Withdraw the retard unit.**
12 **Using a suitable pair of small circlip pliers remove the circlip.**
13 **Remove the plain washer.**
14 **Remove the rubber 'O' ring.**
15 **Carefully withdraw the timing rotor.**
16 **Remove two screws. Lift out the base plate.**
17 **Tap out the drive gear pin. Remove the drive gear and thrust washer. Ensure that the shaft is burr free and withdraw it.**
18 **Remove the metal distance collar.**
19 **Remove the control springs, exercising care not to distort the springs.**

NOTE: Do not attempt to dismantle the shaft and mechanism further.

continued

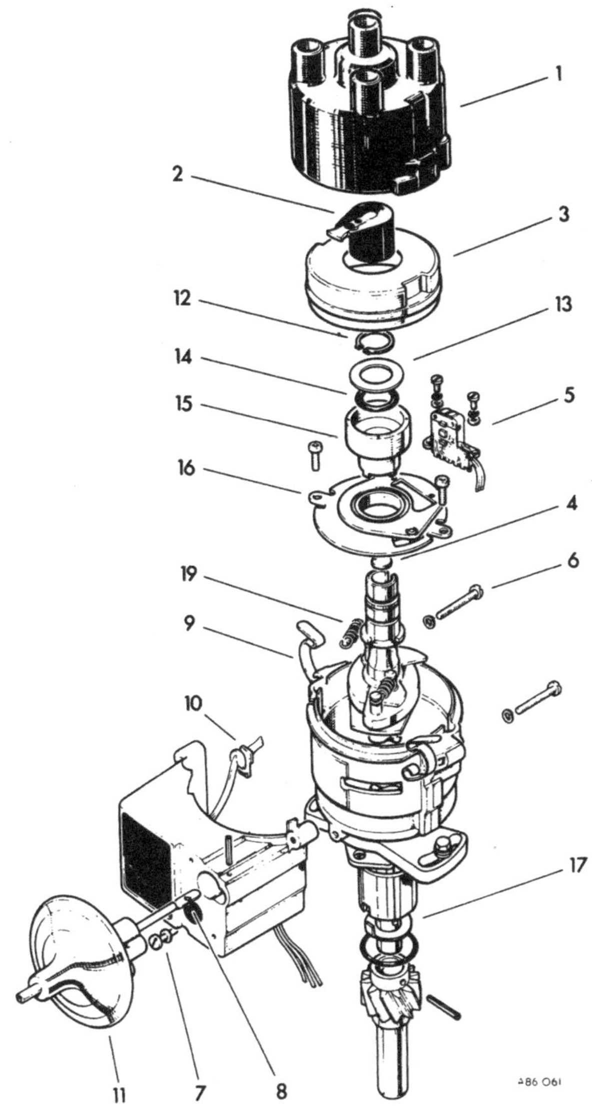

Assemble

20 Lubricate the weight assembly working surfaces with Rocol 'Moly pad'.
21 Fit the control springs, exercising care not to distort the springs.
22 Fit the metal distance collar.
23 Lubricate the shaft with Rocol 'Moly pad' and insert it into the body. Fit the thrust washer and drive gear. Secure with the drive gear pin.
24 Distributor with retard unit only: Lubricate the moving plate pin with Rocol 'Moly Pad'. Position the base plate so that the moving plate pin is correctly positioned for the retard unit link. Secure with two screws.
25 Distributor with no retard unit only: Position the base plate so that the pick-up mounting is away from the grommet cut-out. Secure with two screws.
26 Carefully insert the timing rotor. Ensure that the master projection locates correctly in the master slot.
27 Fit the rubber 'O' ring.
28 Fit the plain washer.
29 Fit the circlip.
30 Distributor with retard unit only: Insert the retard unit. Secure with the retard unit pin.
31 Hold the distributor body and amplifier module slightly apart. Push in the wire grommet.
32 Insert two clips.
33 Distributor with retard unit only: Hold the distributor body in one hand and the amplifier module in the other hand. Carefully manouvre and slightly rotate the moving plate to hook the retard unit link to the moving plate pin. This operation will not be visible.
34 Hold the amplifier module in position. Fit one short screw, lock washer and washer finger-tight.
35 Fit two long screws and lock washer finger tight.
36 Ensure that the amplifier module and two wire grommets are correctly seated. Tighten three screws evenly.
37 Position the pick-up. Carefully fit two screws, lock washers and washers finger tight.
38 Adjust the pick-up air gap, see 86.35.31.
39 Insert the felt pad.
40 Lubricate, see 86.35.18.
41 Fit the plastic anti-flash cover with the two recesses positioned adjacent to the clips.
42 Push on the rotor.
43 Fit the cover.

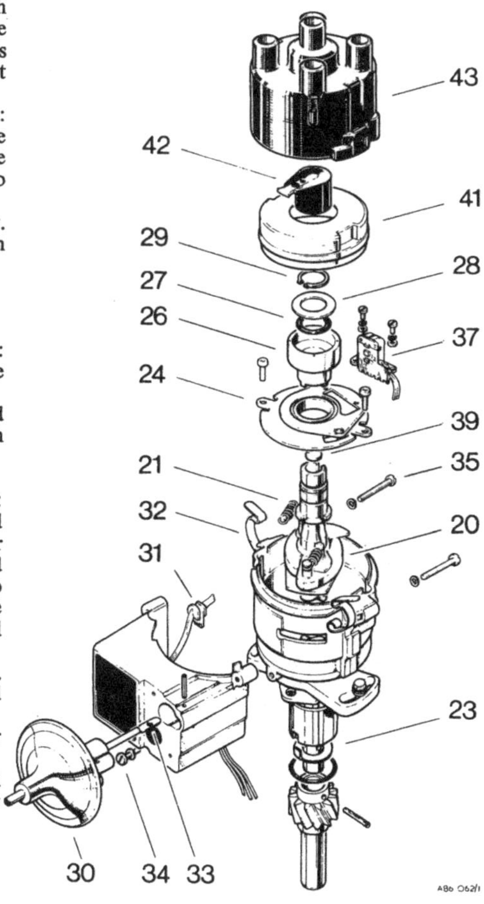

seated. Tighten three screws evenly.
36 Position the pick up. Carefully fit two screws, lock washers and washers finger tight.
37 Adjust the pick up air gap. 86.35.31.
38 Insert the felt pad.
39 Lubricate. 86.35.18.
40 Fit the plastic anti flash cover with the two recesses positioned adjacent to the clips.
41 Push on the rotor.
42 Fit the cover.

IGNITION DISTRIBUTOR

Pick up air gap – adjust **86.35.31**

CAUTION: Do not insert a feeler gauge into the pick up air gap when the ignition circuit is energised.

1 Isolate the battery.
2 remove the fresh air duct. 80.15.31.
3 Remove the cover.
4 Pull off the rotor.
5 Remove the plastic anti flash cover.
6 If the pick up air gap is correct a 0.014 to 0.016 in (0.35 to 0.40 mm) feeler gauge will just slide between the contacts.

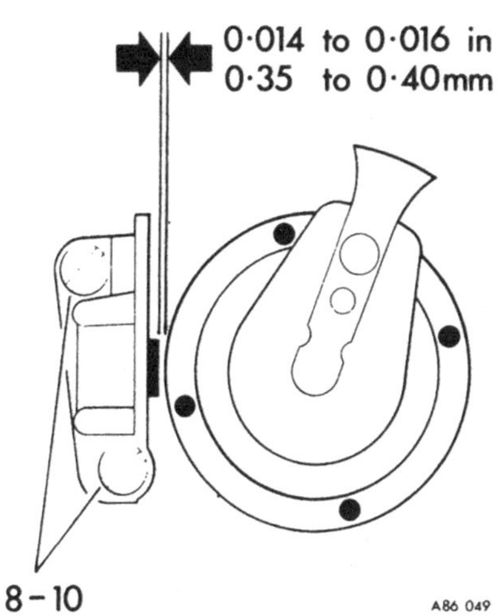

7 When the pick up air gap is correct operations 8 to 11 may be ignored.
8 If a correction is required slacken two screws.
9 Move the pick up about the pivot screw to adjust the gap.
10 Tighten two screws.
11 Check that the correct gap has been maintained.
CAUTION: Ensure that this operation is performed as the gap may change substantially when tightening the screws.
12 Connect the battery.

IGNITION COIL – all models

Remove and refit **86.35.32**

Remove

1 Drive the vehicle onto a ramp.
2 Raise the ramp.
3 Locate the ignition coil mounted in the engine bay low down on the right hand side of the bulkhead as shown.
4 Pull up the protective cover.
5 Pull off the high tension lead.
6 Disconnect two Lucar connectors.
7 Remove two nuts, spring washers and washers.
8 Remove the drive resistor lug from the body stud. – U.S.A. spec. vehicles.
9 Remove the ignition coil from two body studs.

Refit

10 Reverse 1 to 9.

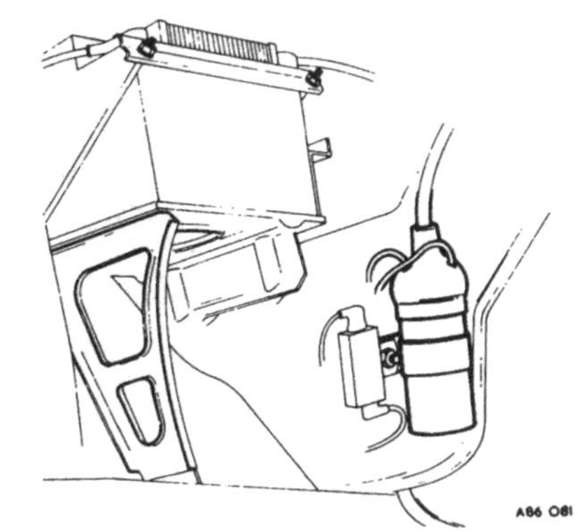

DRIVE RESISTOR – U.S.A. specification vehicles

Remove and refit **86.35.37**

Remove

1 Drive the vehicle onto a ramp.
2 Raise the ramp.
3 Locate the drive resistor mounted in the engine bay low down on the right hand side of the bulkhead as shown.
4 Disconnect two Lucar connectors.
5 Remove single nut, spring washer and washer.
6 Remove the drive resistor from the body stud.

Refit

7 Reverse 1 to 6. The Lucar connectors may be fitted either way round.

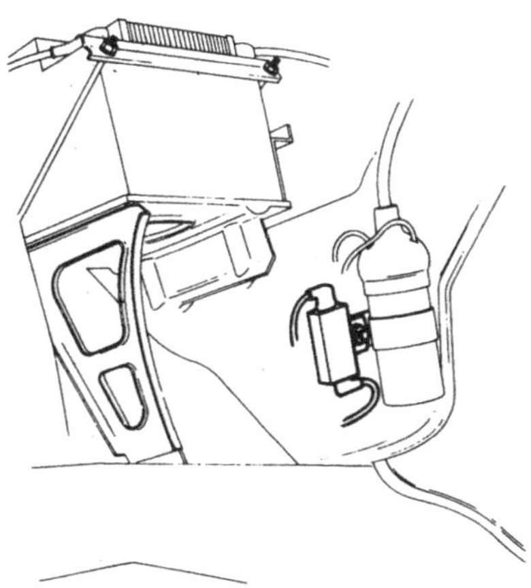

A86 082

IGNITION DISTRIBUTOR — DELCO REMY ELECTRONIC

Description **86.35.00**

The Delco Remy electronic—current controlled ignition system consists of two units. A pulse generator ignition distributor and an ignition coil and electronic module assembly.

Distributor—an angle transducer consisting of a timing rotor, static component, permanent magnet and winding supply timing signal pulses to the electronic module assembly. A conventional AC Delco body, mechanical centrifugal advance and vacuum retard unit are retained.

Retard unit—the static permanent magnet and winding may be rotated through a limited angle by the vacuum retard unit. The ignition timing is thus modified by amending the relationship between the timing rotor and the static assembly.

1 High tension rotor arm
2 Centrifugal advance mechanism
3 Angle transducer — timing rotor
4 — static component
5 — permanent magnet
6 — winding
7 Body
8 Drive gear
9 Retard unit
10 Plug to electronic module assembly

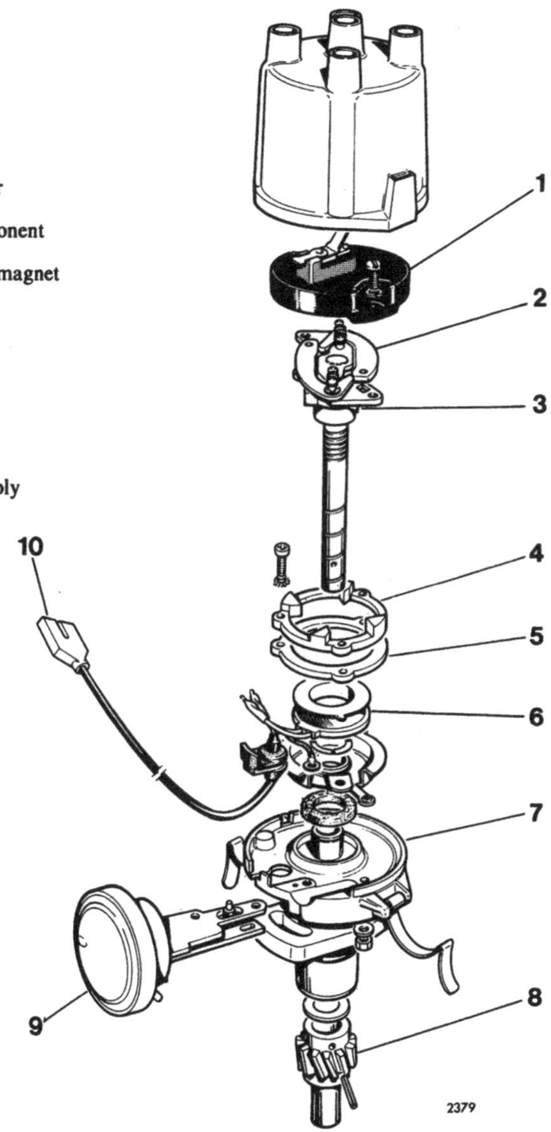

IGNITION COMPONENTS WITH
DELCO REMY—ELECTRONIC SYSTEM

IGNITION COIL AND ELECTRONIC MODULE ASSEMBLY

Data

Manufacturer	AC Delco — France
AC Delco part number	3474214
Triumph part number	PKC 0478
System	Electronic — current controlled ignition system
Polarity	Negative earth only
Primary winding:	
Maximum current controlled at	5.5 amp
Resistance	0.75 ohm

Description

The unit consists of an ignition coil winding and an electronic module assembly mounted on a light alloy die casting.

The electronic module assembly receives timing pulses from the ignition distributor. The module amplifies the signal to control the on–off switching of a power transistor.

The 'off' condition of the power transistor collapses the ignition coil primary winding current. A spark occurs at the appropriate spark plug in the conventional manner.

A feature is that the module contains a current limiting circuit, the maximum current being controlled at 5.5 amp. The ignition coil primary winding is not responsible for current control and can therefore be of low resistance — 0.75 ohm.

This design allows the ignition coil to operate efficiently with a high kilovolt output at all engine speeds.

IGNITION COIL AND ELECTRONIC MODULE ASSEMBLY

Remove and refit **86.35.37**

Removing
1 Depress the claw and pull off the pulse lead plug.
2 Pull off the high tension lead.
3 Disconnect the harness plug.
4 Remove the front bolt assembly — the nut is positioned under the front wing.
5 Remove the rear bolt — a cage nut is fitted to the mounting bracket.
6 Remove the unit from the vehicle.

Refitting
7 Reverse 1 to 6. Include the earth lead tag under the forward bolt head.

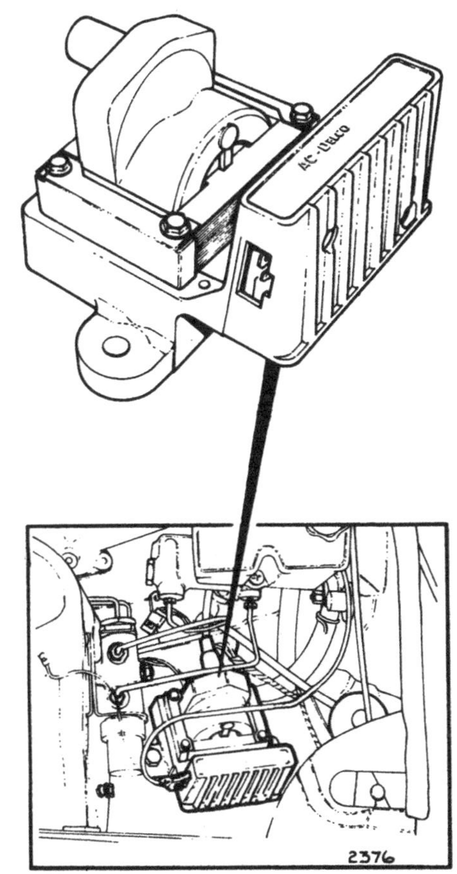

AKM3079B

HEADLAMP

Data **86.40.00**

Actuator

Manufacturer	Lucas
Type	15W
Lucas part number	75857
Triumph part number	RKC 0822
Polarity	Negative earth only
Running current – after 60 seconds from cold with link rod disconnected from crank	1.5 amp.
Running speed – crank after 60 seconds from cold with link rod disconnected from crank	46 to 52 rev/min
Armature end-float	0.002 to 0.008 in (0.05 to 0.20 mm)
Brush length – new	0.250 in (6.35 mm)
renew if less than	0.187 in (4.75 mm)
Brush spring pressure – when compressed so brush bottom is aligned with brush-box slot end	5 to 7 ozf (140 to 200 gf)

Description **86.40.00**

A base bracket is bolted rigidly to the vehicle body. A substantial light alloy member is hinged to the base bracket. The light alloy member is traversed from one position to the other by an electric actuator. Mounted to the light alloy member is a 'body colour' box casting which contains the headlamp.

Actuator – this consists of a permanent magnet motor and a gearbox unit which drives a single direction rotating crank arm. The gearbox unit includes a limit switch.

Action – rotation of the actuator crank arm is converted to an up and down motion of the light alloy member by a link rod. The weight of the assembly is counter-balanced by a spring.

Box casting – with the headlamp in the close position the top surface of the box casting provides the visible panel which follows the vehicle body contours. In the open position the box casting forms the visible fairing. The headlamp is a conventional Lucas seven inch housing and light unit assembly.

Hand knob – the actuator is provided with a hand knob. This may be used to open or close a headlamp during service or if an actuator should fail. Rotate only in the direction shown.

WARNING: Exercise care when using the hand knob. If the battery is not isolated slight rotation of the hand knob may initiate a full cycle movement. Keep all limbs clear of the mechanism to avoid personal injury.

Design – to assist manufacture and service the base bracket, light alloy member, actuator and link rod assembly is not handed. To follow the vehicle body contours the box casting is handed.

Service – special precautions must be taken to avoid scratching the visible paint surfaces of the box casting. If the headlamp assembly is removed from the vehicle onto a bench, the bench should be covered with a thick protective cloth.

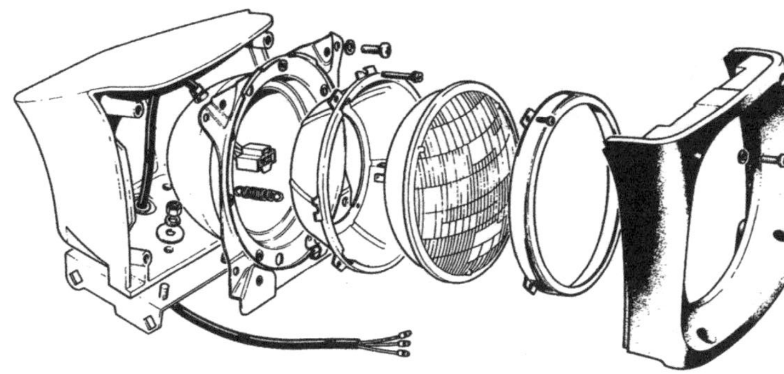

A86 006

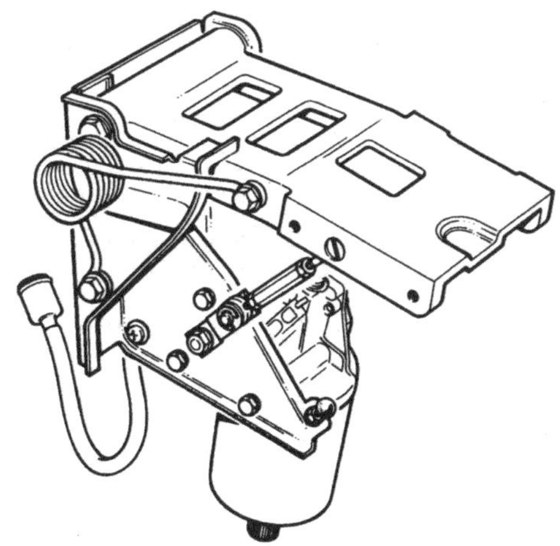

A86 045

AKM3079B

HEADLAMP ELECTRICAL CIRCUIT – U.K. and European

The circuit may be considered in two sections, the control circuit and the motor power circuit.

Control circuit – to manage the two run/stop relays for three operations.

Normal up selection – move the master light switch to headlamps. Supply to main/dip/flash switch for main or dip beam. Also supply via delay unit – terminal 2, diode D4 and terminal 1 to blue/purple up line. Circuit continues across each limit switch to appropriate run/stop relay winding. When a headlamp reaches the up position the circuit is broken by the limit switch.

Normal down selection – move the master light switch to parking lamps or off. Supply via normally closed contacts of flash relay to blue/brown down line. Circuit continues across each limit switch to appropriate run/stop relay winding. When a headlamp reaches the down position the circuit is broken by the limit switch.

Head flash selection – move the main/dip flash switch to head flash. Supply to main beams. Also signal on blue/white wire to delay unit – terminal 6 – to turn on transistor. Delay unit capacitor C1 will charge for later use. Flash relay winding will now pull in to energize blue/purple up line. Transistor base circuit earth and flash relay winding earth is via terminal 2, dip beam contacts – always closed when head flash is selected – and across dip beam to earth. When the switch is released capacitor C1 will discharge slowly through the transistor base circuit to hold the condition for approximately 4 seconds. This prevents excessive full cycle movements of the headlamps if the driver should flash the headlamps more than once.

Motor power circuit – supply through single circuit breaker. Each motor runs in a single direction only. An up selection causes the crank arm to rotate through 216 degrees. A down selection causes the crank arm to rotate through 144 degrees. When the control circuit de-energizes a relay the contacts drop to connect the motor to earth. This action causes regenerative braking of the armature which maintains consistent stopping of the crank arm.

Circuit breaker – if the motion of either headlamp is obstructed the motor will draw an excess current. This condition will be tolerated for a short time to allow the operational headlamp to complete its movement. A single Otter current/heat sensitive circuit breaker will then actuate to protect the motor power circuit. The circuit breaker is self resetting and has a quick recovery time.

Diodes – two diodes included in each limit switch prevent electrical interaction between the left-hand and right-hand side control circuits. During a motor run both the 'open stop' and the 'close stop' contacts are made. If the headlamp movements should get slightly out of phase–and the diodes were not included–the action would not be terminated. Also a selection to parking lamps or off – if the diodes were not included – would not extinguish the light units until both headlamps close.

illustration next page—

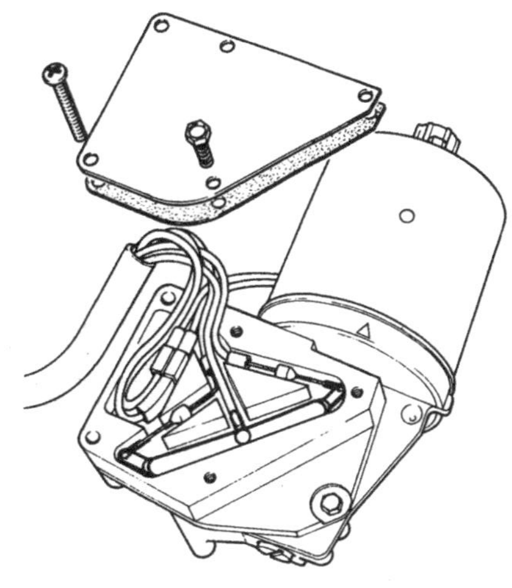

Limit switch

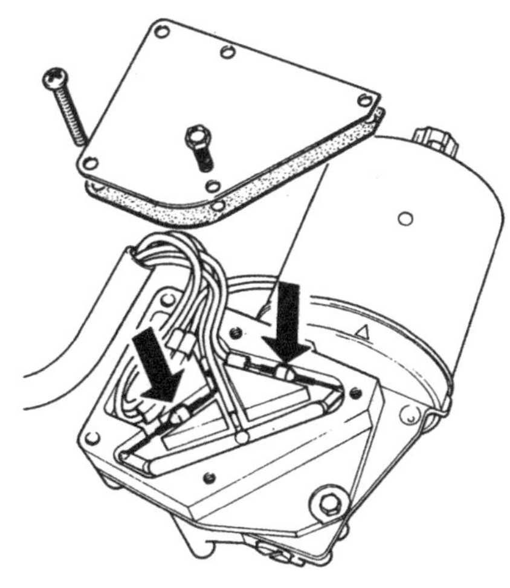

Diodes

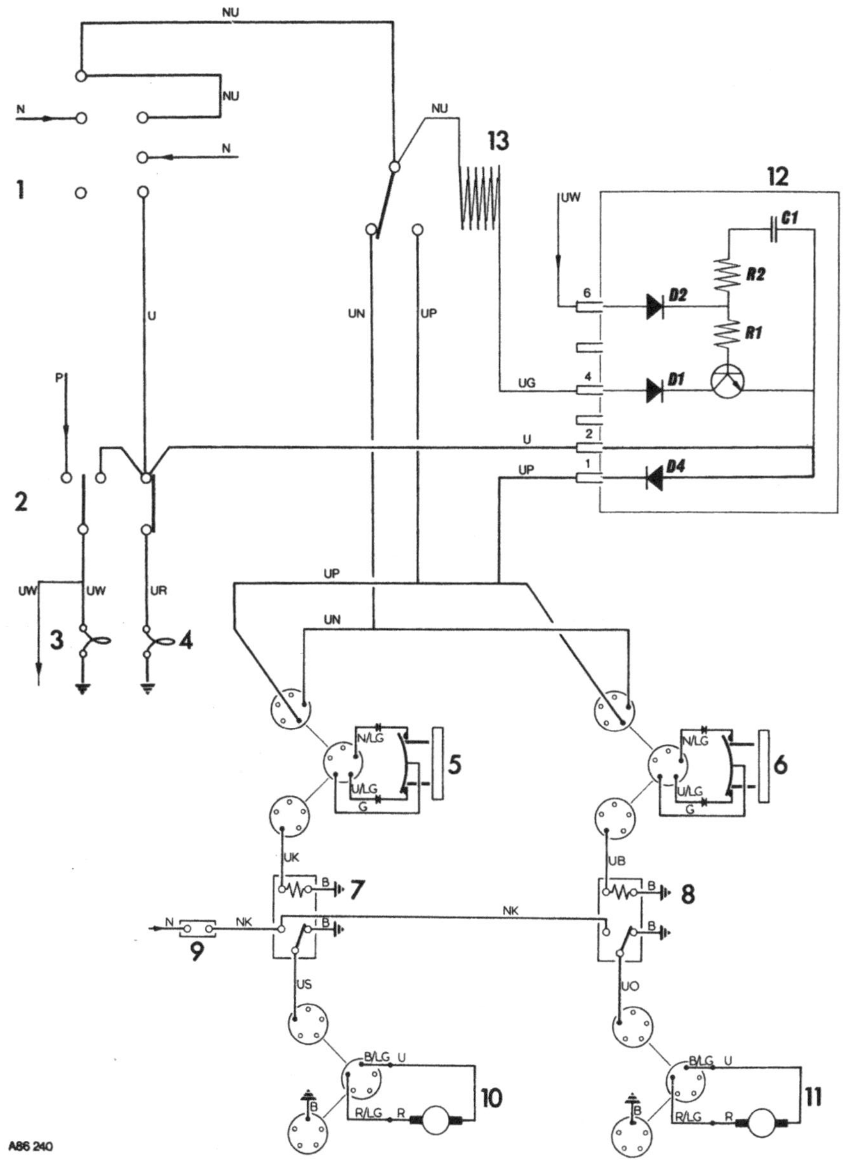

1 Master light switch
2 Main/dip/flash switch
3 Main beam
4 Dip beam
5 L.H. actuator – limit switch
6 R.H. actuator – limit witch
7 L.H. run/stop relay
8 R.H. run/stop relay
9 Circuit reaker
10 L.H. actuator – motor
11 R.H. actuator – motor
12 Delay unit
13 Flash relay

Colour code

B	Black
B/LG	Black/light green
G	Green
N	Brown
NK	Brown/pink
N/LG	Brown/light green
NU	Brown/blue
P	Purple
R	Red
R/LG	Red/light green
U	Blue
UB	Blue/black
UG	Blue/green
UK	Blue/pink
U/LG	Blue/light gree
UN	Blue/brown
UO	Blue/orange
UP	Blue/purple
UR	Blue/red
US	Blue/slate
UW	Blue/white

HEADLAMP ELECTRICAL CIRCUIT – U.S.A.

The circuit may be considered in two sections, the control circuit and the motor power circuit.

Control circuit – to manage the two run/stop relays for two operations.

Normal up selection – move the master light switch to headlamps. Supply to main/dip switch for main or dip beam. Also supply to blue up line. Circuit continues across each limit switch to appropriate run/stop relay winding. When a headlamp reaches the up position the circuit is broken by the limit switch.

Normal down selection – move the master light switch to parking lamps or off. Supply to blue/brown down line. Circuit continues across each limit switch to appropriate run/stop relay winding. When a headlamp reaches the down position the circuit is broken by the limit switch.

Motor power circuit – supply through single circuit breaker. Each motor runs in a single direction only. An up selection causes the crank arm to rotate through 216 degrees. A down selection causes the crank arm to rotate through 144 degrees. When the control circuit de-energizes a relay the contacts drop to connect the motor to earth. This action causes regenerative braking of the armature which maintains consistent stopping of the crankarm.

Circuit breaker – if the motion of either headlamp is obstructed the motor will draw an excess current. This condition will be tolerated for a short time to allow the operational headlamp to complete its movement. A single Otter current/heat-sensitive circuit breaker will then actuate to protect both left-hand and right-hand side motor power circuits. The circuit breaker is self resetting and has a quick recovery time.

Diodes – two diodes included in each limit switch prevent electrical interaction between the left-hand and right-hand side control circuits. During a motor run both the 'open stop' and the 'close stop' contacts are made. If the headlamp movement should get slightly out of phase – and the diodes were not included – the action would not be terminated. Also a selection to side or off – if the diodes were not included – would not extinguish the light units until both headlamps close.

illustration next page

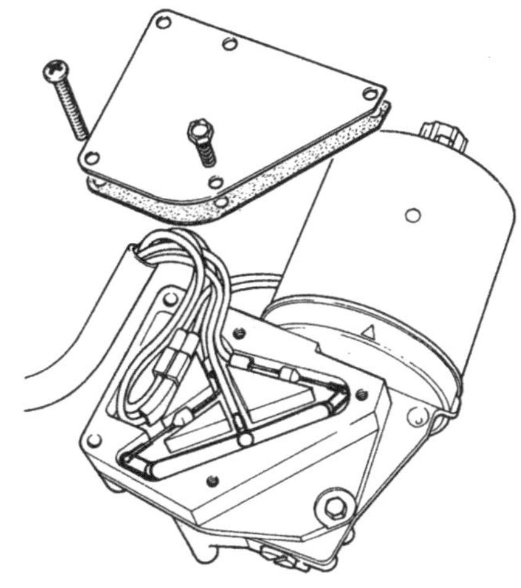

Limit switch

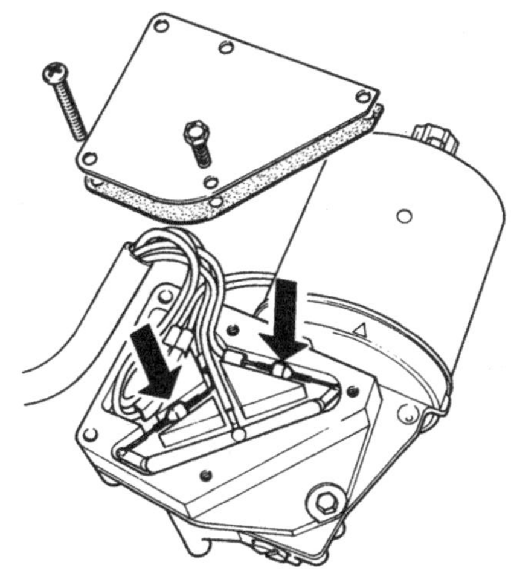

Diodes

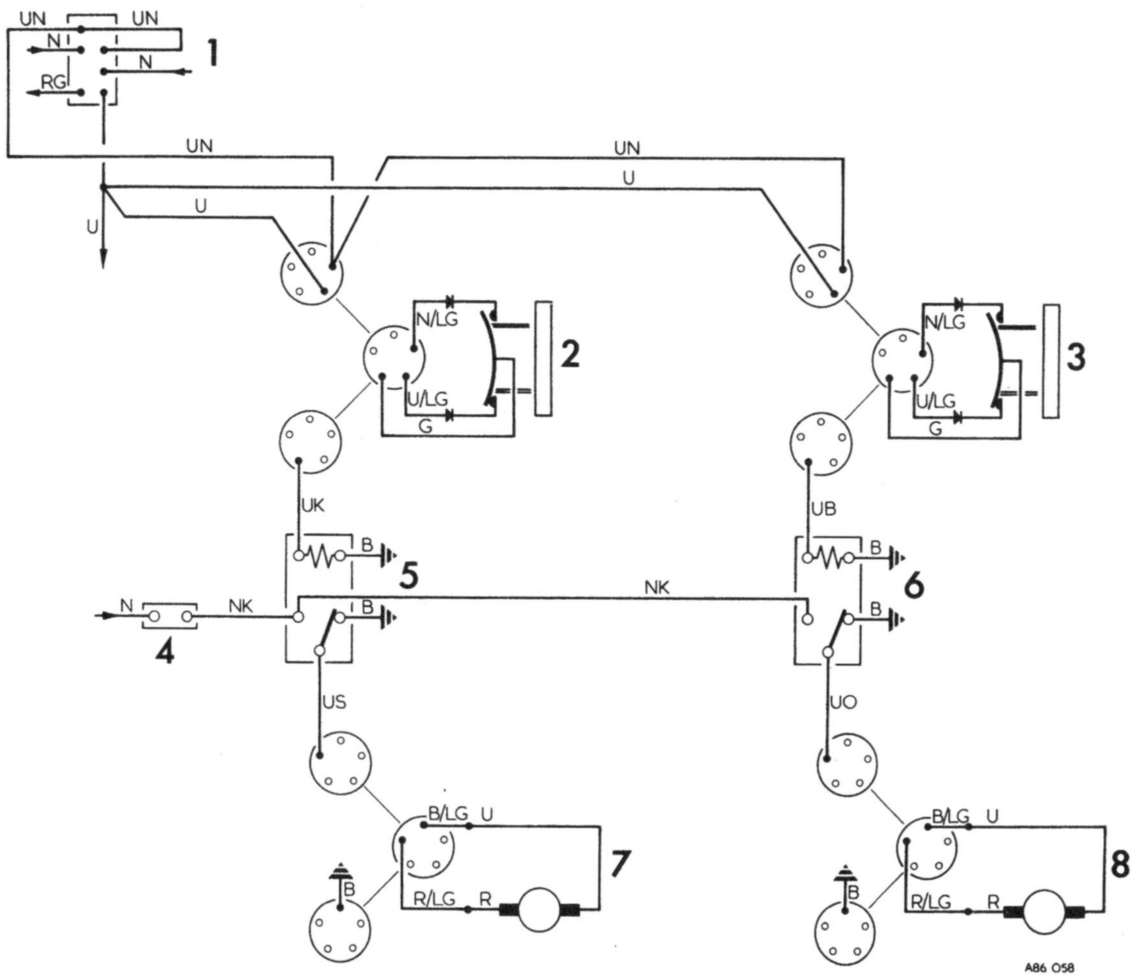

1 Master light switch
2 L.H. actuator – limit switch
3 R.H. actuator – limit switch
4 Circuit breaker
5 L.H. run/stop relay
6 R.H. run/stop relay
7 L.H. actuator – motor
8 R.H. actuator – motor

Colour code

B	Black
B/LG	Black/light green
G	Green
N	Brown
NK	Brown/pink
N/LG	Brown/light green
R	Red
RG	Red/green
R/LG	Red/light green
U	Blue
UB	Blue/black
UK	Blue/pink
U/LG	Blue/light green
UN	Blue/brown
UO	Blue/orange
US	Blue/slate

HEADLAMP

Rubber bezel – remove and refit 86.40.01

Remove

1 Select the master light switch to raise the headlamps.
2 Isolate the battery to extinguish the headlamps.
3 Unscrew four Pozidriv screws and lock washers.
4 Pull the rubber bezel from the box casting.
5 Manoeuvre the rubber bezel downwards into the headlamp cavity. If necessary manoeuvre further downwards to remove from the vehicle. The rubber bezel may be distorted slightly to facilitate this operation.
6 Collect up four Pozidriv screws and lock washers.

Refit

7 Reverse 1 to 5.

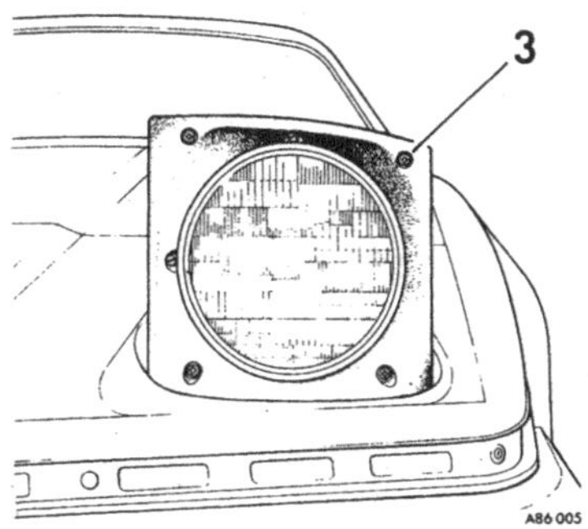

HEADLAMP

Assembly – remove and refit 86.40.02

Remove

1 Remove the rubber bezel, see 86.40.01.
2 To assist refitting, note the actuator harness and headlamp wire runs relative to the components.
3 Disconnect the actuator harness plug.
4 Disconnect three headlamp snap connectors.
5 Support the weight of the headlamp assembly. Remove four nuts, lock washers and washers. Carefully manoeuvre the headlamp assembly upwards through the body aperture.

Refit

6 Carefully manoeuvre the headlamp assembly downwards through the body aperture. Fit four washers, lock washers and nuts.
7 Connect three headlamp snap connectors. Ensure that the wire runs are as noted at operation 2 above. Connect the wires as follows:
Blue/red wire to blue/red wire
Blue/white wire to blue/white wire
Black wire to black wire.
8 Connect the actuator harness plug. Ensure that the harness run is as noted at operation 2 above.
9 Inspect the actuator harness and headlamp wire runs to ensure that no foul occurs while using the actuator manual thumb screw to traverse the mechanism over its full travel.
10 Adjust the headlamp assembly, see 86.40.16.
11 Refit the rubber bezel, see 86.40.01.

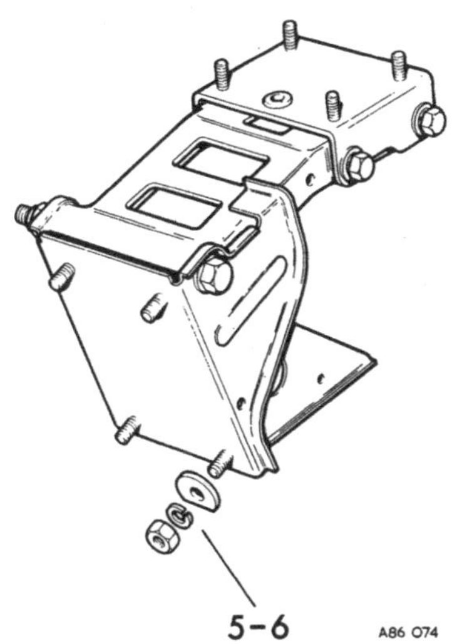

HEADLAMP

Light unit – remove and refit 86.40.09

Remove

1 Remove the rubber bezel, see 86.40.01.
2 Remove three screws to release the retaining rim and light unit.
3 Pull the connector block from the light unit.

Refit

4 Reverse 1 to 3.

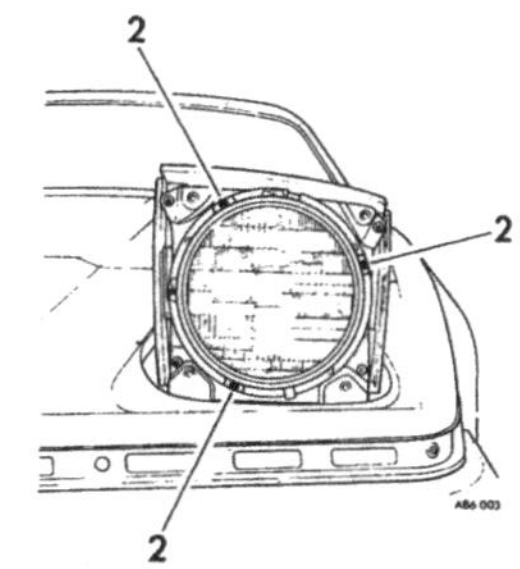

HEADLAMP

Link rod – remove and refit 86.40.12

The upper end of the link rod contains a rubber bush. This provides a measure of compliance to the link rod alignment and to the primary and secondary adjustments.

After an extended period of service wear of the upper rubber bush and the lower Oilite bush may dictate renewal of the link rod.

Remove

1 Remove the headlamp assembly, see 86.40.02.
2 Perform 86.40.15 operations 2 to 11.

Refit

3 Perform 86.40.15 operations 39 to 49.
4 Refit the headlamp assembly, see 86.40.02.

HEADLAMP

Actuator – remove and refit 86.40.13

Remove

1 Remove the headlamp assembly, see 86.40.02.
2 Safely retain the spring by rotating the actuator manual thumb screw to position the light alloy member against the metal up stop on the base bracket.
3 Remove the circlip and washer.
4 Remove the nut, lock washer, washer and screw. Remove the harness 'P' clip.
5 Remove three screws and lock washers.
6 Manoeuvre the link rod from the bush.
7 Remove the bush from the crank arm.
8 Manoeuvre the crank arm through the aperture.

Refit

9 Manoeuvre the crank arm through the aperture.
10 Lubricate the bush working surfaces with engine oil.
11 Fit the bush to the crank arm.
12 Position the link rod ends in the turnbuckle so that an equal number of threads are visible at each end.
13 Manoeuvre the link rod to the bush. It may be necessary to rotate the manual thumb screw to position the crank arm.
14 Position the actuator. It may be necessary to rotate the manual thumb screw to position the crank arm. Ensure that the actuator harness is looped up between the actuator and the base bracket. Secure with three screws and lock washers.
15 Position the harness 'P' clip. Secure with the screw, washer, lock washer and nut.
16 Fit the washer and circlip.
17 With the headlamp assembly removed from the vehicle perform the primary adjustment and the secondary adjustment, see 86.40.16.
18 Refit the headlamp assembly, see 86.40.02.

HEADLAMP

Actuator – overhaul **86.40.14**

CAUTION: The actuator operates in a hostile environment. After overhaul or disturbing any joint the unit must be sealed as detailed in operation 33 below.

Disassemble

1 Remove three hexagon screws and two Pozidriv screws. Lift off the cover and gasket.
2 Disconnect two Lucar connectors.
3 Remove one hexagon screw. Lift off the limit switch and harness.
4 Remove the nut. Withdraw the crank arm and washer.
5 Ensure that the shaft is burr-free and withdraw it. Remove the dished washer.
6 Using a large screwdriver blade carefully prise the hand knob from the armature shaft.
7 Remove the thrust screw or the thrust screw and locknut as fitted.
8 Remove the through bolts.
9 Carefully withdraw the cover and armature about 0.2 in (5 mm). Continue withdrawal, allowing the brushes to drop clear of the commutator. Ensure that the two brushes are not contaminated with grease.
10 Pull the armature from the cover against the action of the permanent magnet.
11 Remove two thrust washers.
NOTE: These may be on the armature shaft or in the cover bearing recess.
12 Remove three screws to release the brush assembly. Break the wire slot seal. Lift the assembly from the recess.

Assemble

NOTE: The following lubricants are required during assembly:

Molybdenum di-sulphide oil.
Shell Turbo 41 oil.
Ragosine Listate grease.

13 Position the brush assembly. Insert the wires in the slot. Secure with three screws.
14 Lubricate two thrust washers with Molybdenum di-sulphide oil.
15 Fit two thrust washers into the cover bearing recess.
16 Lubricate the cover bearing with Shell Turbo 41 oil
17 Position the armature to the cover against the action of the permanent magnet. Ensure that the two thrust washers remain in position during this operation.
18 Lubricate the self-aligning bearing with Shell Turbo 41 oil.

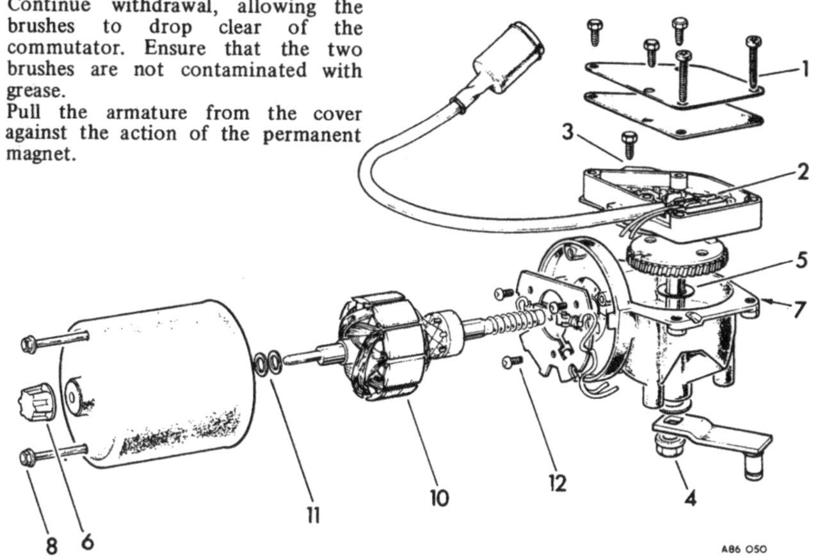

19 Carefully insert the armature shaft through the bearing. Ensure that the brushes are not contaminated with lubricant. Push the two brushes back to clear the commutator.
20 Seat the cover against the gearbox. Turn the cover to align the marks shown. Fit the through bolts.
21 Fit the thrust screw or the thrust screw and locknut as fitted.
22 If a non-adjustable thrust screw is fitted check the armature end-float as follows:
Position a feeler gauge between the armature shaft and the thrust screw. Push the armature towards the cover. End-float should be 0.002 to 0.008 in. In the unlikely event of adjustment being required end-float may be increased by fitting shim washer/washers under the thrust screw head or reduced by mounting the thrust screw in a lathe and removing metal from the underside of the head.
23 If an adjustable thrust screw and locknut is fitted adjust the armature end-float as follows:
Slacken the locknut. Screw the thrust screw in until resistance is felt. Screw the thrust screw out a quarter of a turn – maintain in this position and tighten the locknut.
24 Push the hand knob onto the armature shaft.
25 Lubricate the final gear bushes with Shell Turbo 41 oil.
26 Fit the dished washer with its concave surface facing the final gear. Insert the shaft.
27 Fit the washer. Fit the crank arm. Ensure that the relationship of the crank arm to the final gear cam is as shown. Secure with the nut.

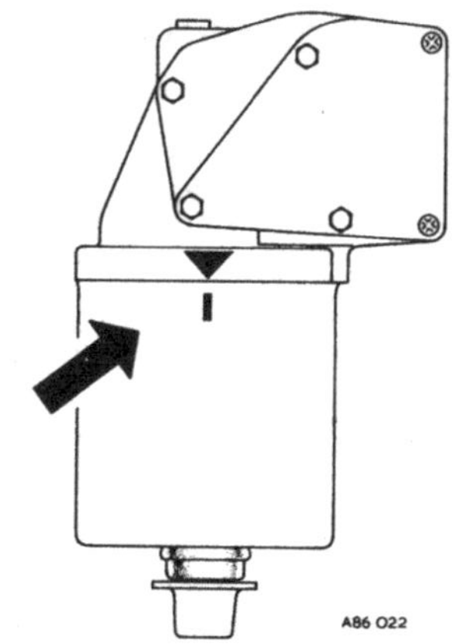

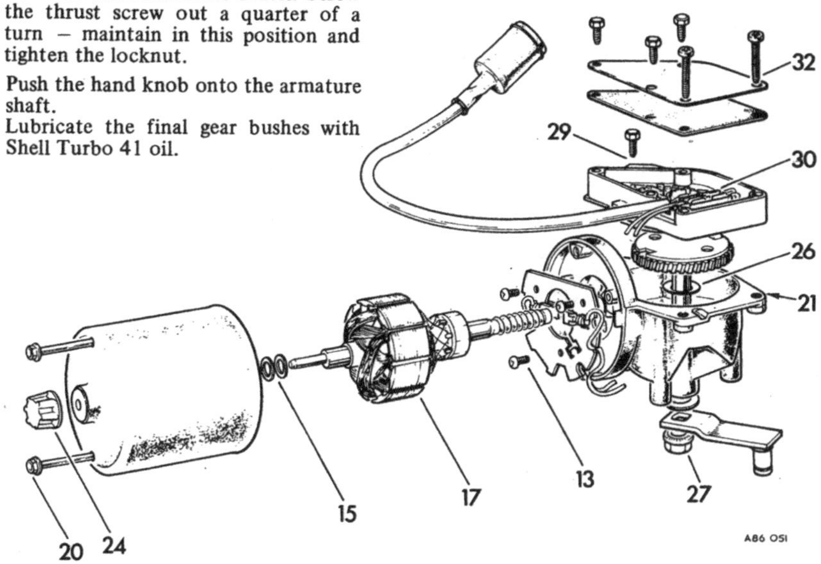

28 Lubricate the final gear cam with Ragosine Listate grease.
29 Position the limit switch and harness. Secure with one hexagon screw.
30 Connect two Lucar connectors as follows:
Red/light green wire to red wire.
Black/light green wire to blue wire.
31 Pack the wires into the limit switch recesses as shown.
32 Position the gasket and cover. Secure with three hexagon screws and two Pozidriv screws.
33 Seal the exterior of the unit by applying a rubberised sealing compound such as vehicle body underseal. Special attention must be given to disturbed joints and the wire slot seal.

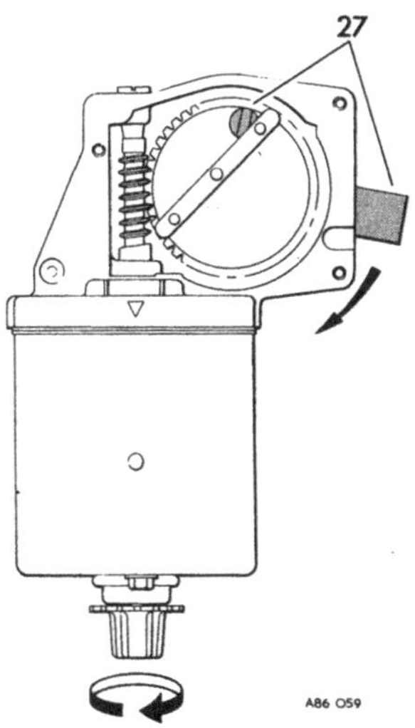

A86 059

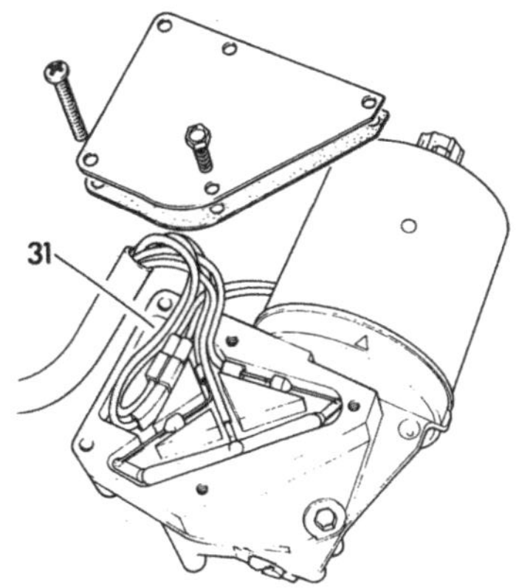

A86 056

HEADLAMP

Assembly – overhaul **86.40.15**

Disassemble

1 Remove the headlamp assembly, see 86.40.02.
2 Remove the nut, lock washer, washer and screw. Remove the harness 'P' clip.
3 To assist refitting, scribe the outline of the mounting bracket sides onto the light alloy member.
4 Remove four bolts, lock washers and washers. Separate the two assemblies.
5 Withdraw the headlamp harness from the light alloy member.
6 Safely restrain the spring by rotating the actuator hand knob to position the light alloy member against the metal up stop on the base bracket.
7 Remove the circlip and washer.
8 Remove the nut, lock washer, washer and special screw.
9 Slide the upper end of the link rod from the light alloy member.
10 Remove the link rod from the crank arm.
11 Remove the bush from the crank arm.
12 Remove three screws and lock washers. Manoeuvre the crank arm through the aperture and remove the actuator.
13 Ensure that the spring is safely restrained and extended to its weakest working position by checking that the light alloy member is against the metal up stop on the base bracket.
14 Retain the spring by strong hand pressure. Carefully remove the nut, lock washer, small washer, large washer and special bolt.
15 Remove the nut, lock washer, small washer, large washer and special bolt. Lift away the spring.
16 Remove the nut, washer, washer and bolt. Withdraw the light alloy member.
17 Withdraw the hinge pin.
18 Slacken the locknut. Remove the rubber down stop including the locknut and washer.
19 Remove three screws to release the retaining rim and light unit.
20 Pull the connector block from the light unit.
21 Remove four screws and lock washers. Withdraw the housing and plate.
22 If necessary drill out four 'pop rivets' to separate the housing plate.
23 To assist refitting, scribe the outline of the mounting bracket onto the box casting.
24 Remove four nuts, lock washers and washers. Remove the mounting bracket.
25 Remove the grommet from the mounting bracket to release the headlamp harness.

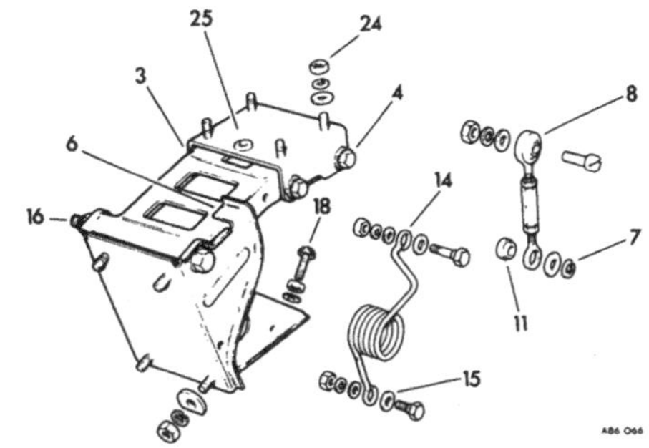

A86 066

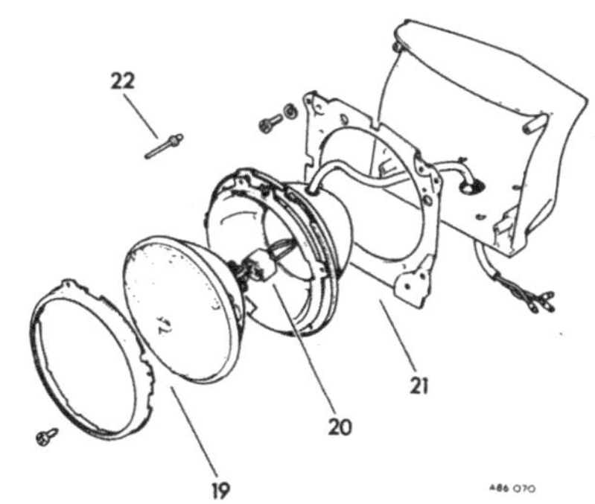

A86 070

continued

Assemble

26 Position the headlamp harness and fit the grommet to the mounting bracket.
27 Position the mounting bracket to the scribe lines on the box casting. Secure with four nuts, lock washers and washers.
28 If necessary, fit four 'pop rivets' to unite the housing and plate.
29 Insert the housing and plate. Secure with four screws and lock washers.
30 Push the connector block on to the light unit.
31 Position the retaining rim and light unit. Secure with three screws.
32 Fit the rubber down stop including the locknut and washer.
33 Lubricate the hinge pin working surfaces with P.B.C. (Poly Butyl Cuprysil) grease.
34 Insert the hinge pin into the light alloy member.
35 Insert the light alloy member into the base bracket. Secure with the bolt, washer, washer and nut.
36 Position the spring. Secure with the special bolt, large washer, small washer, lock washer and nut.
37 Retain the spring by strong hand pressure. Secure with the special bolt, large washer, small washer, lock washer and nut.
38 Manoeuvre the crank arm through the aperture and position the actuator. Ensure that the actuator harness is looped up between the actuator and the base bracket. Secure with three screws and lock washers.
39 Lubricate the bush working surfaces with engine oil.
40 Fit the bush to the crank arm.
41 Position the link rod ends in the turnbuckle so that an equal number of threads are visible at each end.
42 Fit the link rod to the bush.
43 Slide the upper end of the link rod to the light alloy member.
44 Fit the special screw, washer, lock washer and nut.
45 Fit the washer and circlip.
46 With the headlamp assembly removed from the vehicle perform the primary and secondary adjustment, see 86.40.16.
47 Insert the headlamp harness through the light alloy member.
48 Position the mounting bracket to the scribe lines on the light alloy member. Secure with four bolts, lock washers and washers.
49 Position the harness 'P' clip. Secure with the screw, washer, lock washer and nut.
50 Fit the headlamp assembly, see 86.40.02.

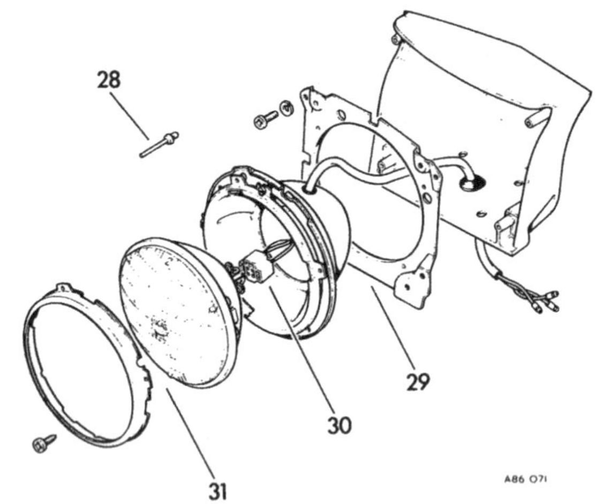

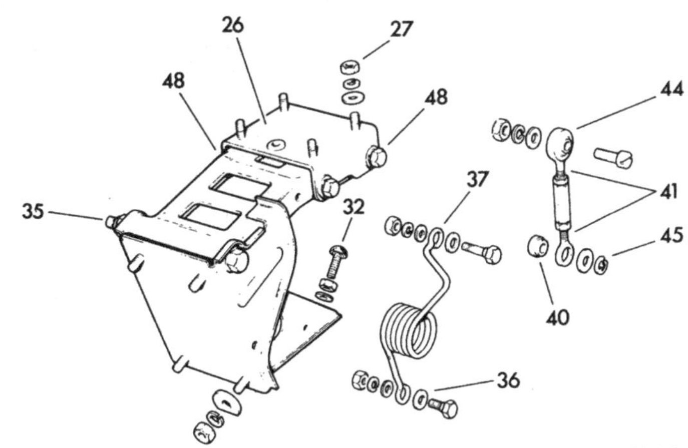

HEADLAMP

Assembly – adjust **86.40.16**

Primary adjustment

The primary adjustment is to ensure that the light alloy member is forced hard against the metal up stop on the base bracket when the crank arm stops for the up position. This condition is achieved by adjusting the length of the link rod.

This adjustment may be made with the headlamp assembly in situ or removed from the vehicle.

1 *Headlamp assembly in situ only:* Select the master light switch to raise the headlamps.
2 *Headlamp assembly removed from the vehicle only:* Using the actuator hand knob rotate the crank arm to obtain the T.D.C. position as shown.
3 Slacken two link rod lock nuts. Note that the upper lock nut has a left-hand thread.
4 Rotate the turnbuckle by hand – do not use a spanner – until slight resistance is felt as the light alloy member contacts the metal up stop on the base bracket. This is the datum point.
5 Rotate a further 4½ flats (270 degrees) using a spanner.
NOTE: This will provide a 'crowd' of 1.20 to 1.40 mm to the length of the link rod. Correct pre-load will then be applied to the up stop.
6 Hold the turnbuckle in this position and tighten two lock nuts.

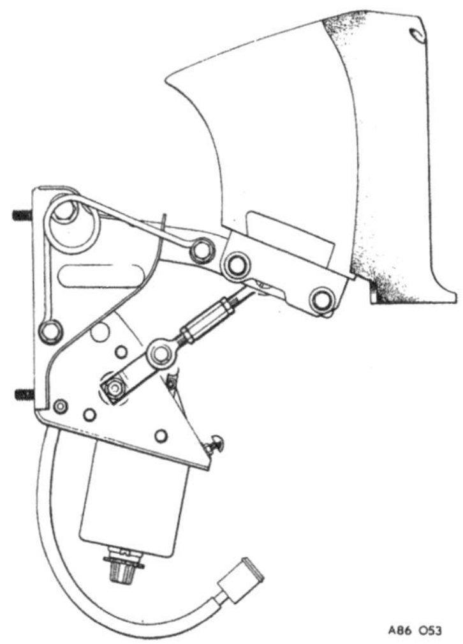

A86 053

Secondary adjustment

The secondary adjustment is to ensure that the rubber down stop mounted on the base bracket is suitably compressed by the light alloy member when the crank arm stops for the down position. This condition is achieved by adjusting the position of the rubber down stop.

This adjustment may be made with the headlamp assembly in situ or removed from the vehicle.

7 Perform the primary adjustment, see 86.40.16.
CAUTION: The position the light alloy member assumes in the down position is dependent on the length of the link rod. A correct secondary adjustment cannot be achieved until the primary adjustment is correctly set.
8 Slacken the rubber down stop lock nut. Screw the rubber down stop to its lowest position.
9 *Headlamp assembly in situ only:* Select the master light switch to retract the headlamps.
10 *Headlamp assembly removed from vehicle only:* Using the actuator hand knob rotate the crank arm to obtain B.D.C. position as shown.
11 Screw the rubber down stop out by hand until it just contacts the light alloy member. This is the datum point.
12 Screw out a further 1⅓ turns (480 degrees).
NOTE: This will provide a 'crowd' of 1.00 to 1.20 mm to the downstop.
13 Hold the rubber down stop in this position and tighten the lock nut.

continued

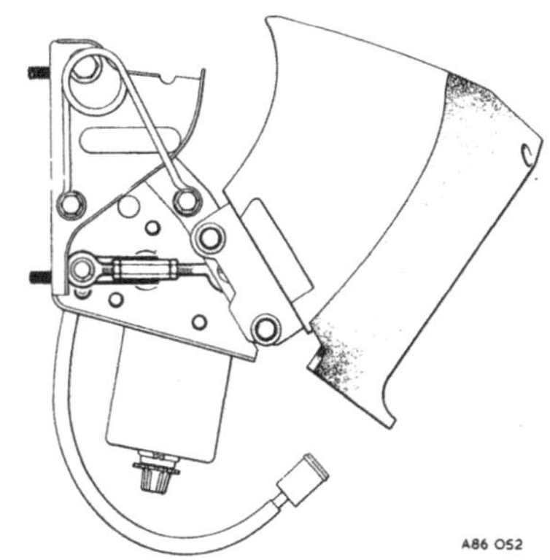

A86 052

Box casting adjustment

The box casting adjustment is to ensure that the top surface of the box casting is correctly aligned with the vehicle body contours when the headlamp is retracted. This condition is achieved by adjusting the position of components.

This adjustment may only be made with the headlamp assembly in situ.

14 Perform the primary and secondary adjustment, see 86.40.16.
CAUTION: The position the light alloy member assumes in the down position is dependent on the length of the link rod. A correct box casting adjustment cannot be achieved until the primary adjustment is correctly set.
15 Ensure that the headlamps are in the retracted position.
16 Slacken four nuts securing the base bracket to the vehicle body.
17 Move the complete headlamp assembly to align the top surface of the box casting with the vehicle body contours.
18 Hold the headlamp assembly in this position and tighten four nuts.
19 If sufficient adjustment is not available to achieve alignment perform operations 20 to 22.
20 Slacken four bolts securing the mounting bracket to the light alloy member.
21 Adjust the relationship of the components to facilitate the operation.
22 Hold the components in position and tighten four bolts.
23 Repeat operation 15 onwards.

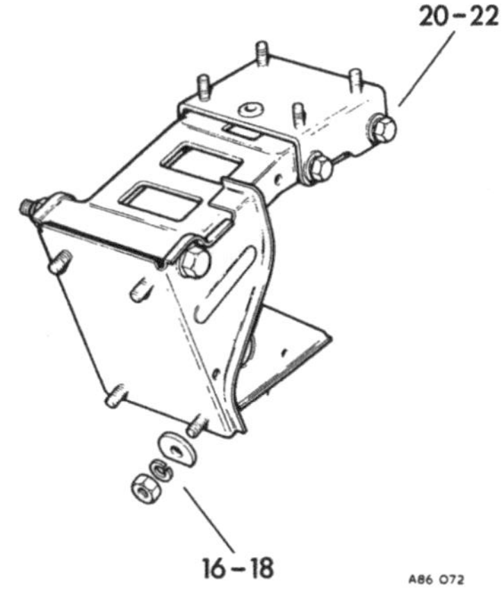

HEADLAMP

Beam aiming **86.40.17**

Beam aiming can best be accomplished using equipment such as Lucas 'Beamsetter', 'Lev-L-Lite' or 'Beam tester'. This service is available at Triumph distributors or dealers and will ensure maximum road illumination with minimum discomfort to other road users.

With the headlamps in the up position the adjustment is achieved in the same way as for conventional solid mounted Lucas headlamp units. One screw positions the beam in the horizontal plane while a second screw controls beam height.

1 Select the master light switch to raise the headlamps.
2 Ensure that the headlamp assembly is correctly adjusted, see 86.40.16.
3 Gain access to the beam aiming screws by inserting a screwdriver through the 'cut outs' provided in the rubber bezel.
4 Screw 'A' positions the beam in the horizontal plane.
5 Screw 'B' controls beam height.

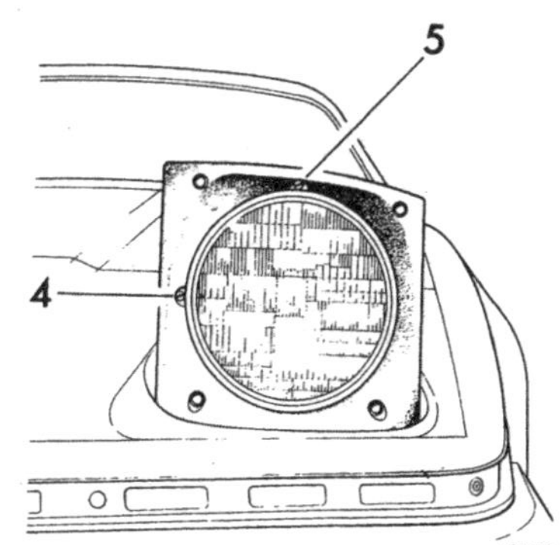

HEADLAMP

Circuit breaker – remove and refit **86.40.19**

Remove
1 Open the cubby box lid.
2 Remove the two Pozidriv screws. Withdraw the lower panel to obtain access to the component mounting plate.
3 Locate the circuit breaker attached to the harness using the appropriate illustration.
4 Pull off two electrical connectors to detach the circuit breaker from the vehicle.

Refit
5 Push on two electrical connectors. The connectors may be fitted either way round.
6 Stow the circuit breaker in a suitable position.
7 Reverse 1 to 2.

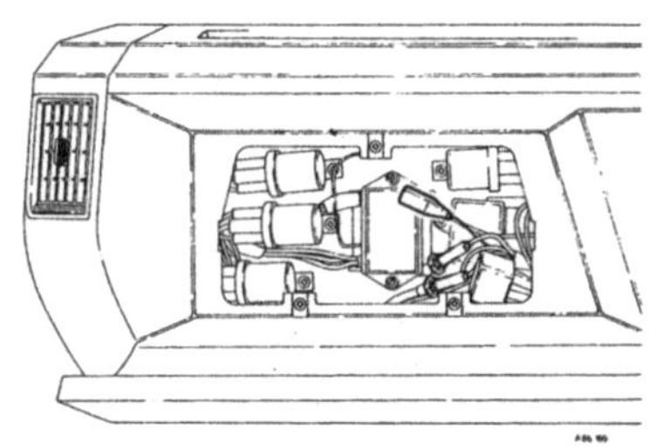

LAMPS

Front parking and flasher lamp – remove and refit 86.40.26

Remove

1 Remove the lamp harness from the body clip.
2 Disconnect three snap connectors.
3 Remove two Pozidriv screws and washers. Withdraw the clear lens and the amber plate lens.
4 Remove the lens gasket.
5 Remove two bulbs from the bayonet fittings.
6 Remove two nuts, spring washers and washers.
7 Withdraw the lamp base and gasket.

Refit

8 Insert the lamp base and gasket. Ensure that it is the correct way up with the lens securing tappings horizontal.
9 Fit two nuts, spring washers and washers.
10 Fit two bulbs to the bayonet fittings.
11 Fit the lens gasket.
12 Position the amber plate lens to the clear lens. Ensure that the assembly is the correct way up with the screw holes horizontal and the amber plate lens the correct end.
13 Insert the lens assembly. Secure with two Pozidriv screws and washers.
14 Connect three snap connectors as follows:
Left-hand lamp only –
Red wire to red wire.
Green/red wire to green wire.
Black wire to black wire.
Right-hand lamp only –
Red wire to red wire.
Green/white wire to green wire.
Black wire to black wire.
15 Fit the lamp harness to the body clip.

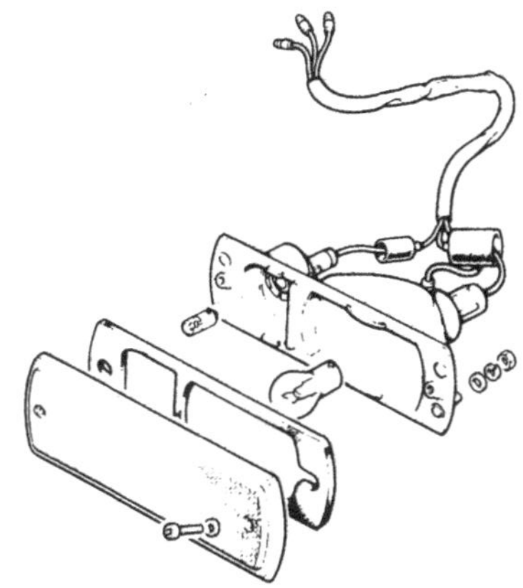

LAMPS

Front flasher repeater lamp – remove and refit 86.40.53

Remove

1 The lamp unit is attached by a groove in the rubber component of the lamp locating in an aperture provided in the vehicle body. Carefully manoeuvre the rubber to allow the complete lamp unit to be withdrawn outwards.
2 Pull off the metal bezel.
3 Remove the lens.
4 Remove the bulb from the bayonet fitting.
5 *Early vehicles only:*
Locate the terminations of the green wire and the black wire below the headlamp assembly.
6 *Later vehicles only:*
Locate the terminations of the green wire and the black wire in the engine bay behind the headlamp mounting position.
7 **Disconnect two electrical connections.**
8 **Provide a length of slave cord about 2 metres long.**

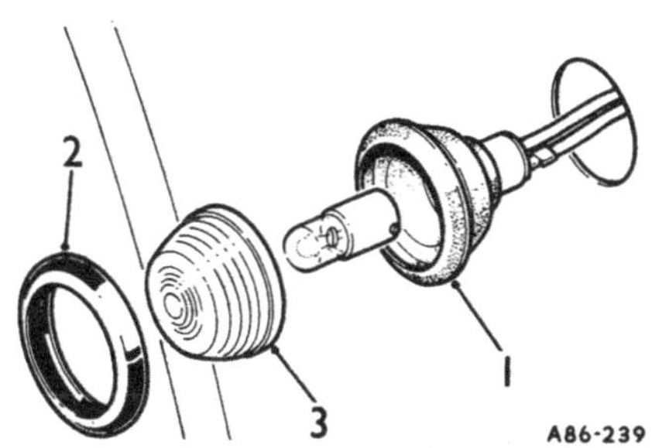

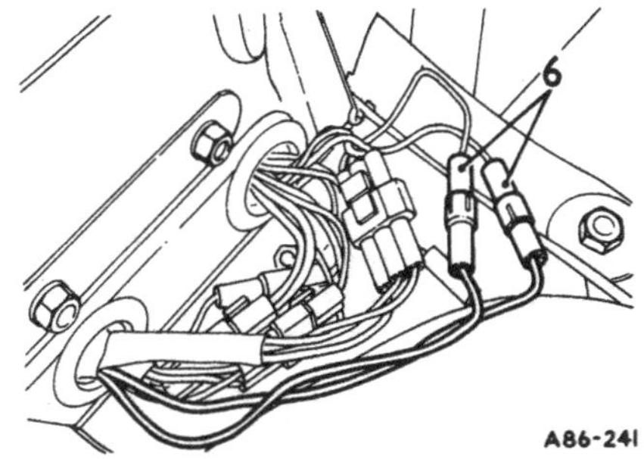

9 **Attach the cord to the two wire ends.**
10 **Attach the other end of the cord to the vehicle body to prevent it being pulled completely through.**
11 **Carefully withdraw the lamp wires from the vehicle wing cavity.**
12 **Detach the cord from the two wire ends.**

Refit

13 Reverse 1 to 12. Connect two wires as follows:
Left-hand only –
Green wire to black wire to black wire.
Right-hand lamp only –
Green wire to green/white wire.
Black wire to black wire.

LAMPS – U.S.A. specification vehicles

Front marker lamp – remove and refit 86.40.59

Remove

1 Select the master light switch to raise the headlamps.
2 Isolate the battery to extinguish the headlamps.
3 Pull the bulb holder from the lamp base.
4 Pull the bulb from the bulb holder.
5 Disconnect two Lucar connectors.
6 Remove two nuts, spring washers and washers.
7 Withdraw the lamp assembly. Do not attempt to disassemble the lens from the lamp base.

Refit

8 Ensure that the rubber gasket is secured to vehicle body.
9 Reverse 1 to 7.

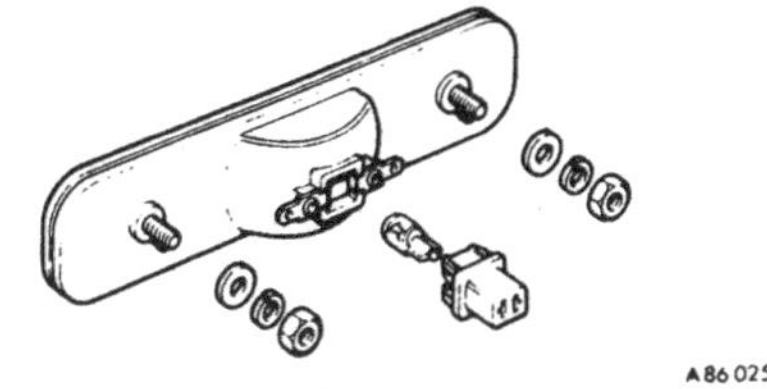

LAMPS – U.S.A. specification vehicles

Rear marker lamp – remove and refit 86.40.64

Remove

1 Open the luggage boot lid.
2 Remove two nuts, spring washers and washers.
3 Remove the clamp cover.
4 Pull the bulb holder from the lamp base.
5 Pull the bulb from the bulb holder.
6 Disconnect two Lucar connectors.
7 Withdraw the lamp assembly. Do not attempt to disassemble the lens from the lamp base.

Refit

8 Ensure that the rubber gasket is secured to the vehicle body.
9 Reverse 1 to 7.

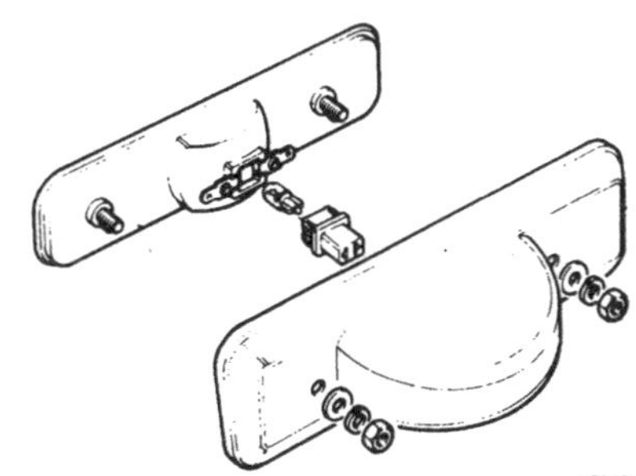

LAMPS

Rear lamp assembly – remove and refit **86.40.70**

Remove

1 Open the luggage boot lid.
2 Turn back the floor mat on the appropriate side.
3 Remove two trim panel top longer Pozidriv screws.
4 Remove two trim panel bottom shorter Pozidriv screws.
5 Remove the trim panel.
6 Rotate four bulb holders anti-clockwise and remove from the bayonet fittings.
7 Remove four bulbs from the bayonet fittings.
8 Remove one nut and spring washer. Remove the harness earth tag.
9 Remove five nuts, spring washers and washers.
10 Withdraw the lens and lens gasket. The lens is a one piece assembly. The lens section may not be replaced individually.
11 Withdraw the lamp base and lamp base gasket.

Refit

12 Reverse 1 to 11. Fit the four bulb holders so that the wire colour codes are as the left-hand assembly shown. On a right-hand assembly the last wire is GW Green/White.

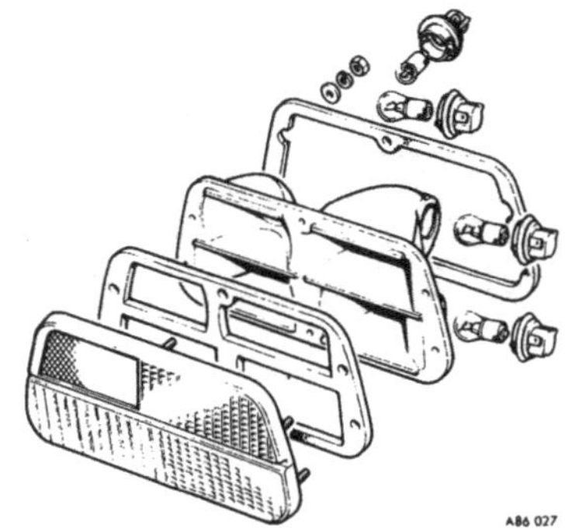

A86 027

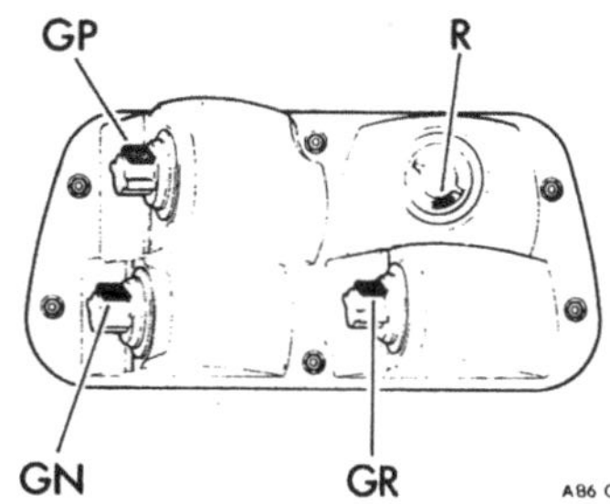

A86 030

GN	Green/Brown	GP	Green/Purple
GR	Green/Red	R	Red

LAMPS

Plate illumination lamp – remove and refit **86.40.86**

Remove

1 Open the luggage boot lid to obtain improved access.
2 Remove two Pozidriv screws.
3 Manoeuvre the lamp from the body panel aperture. Take care not to break the festoon bulb.
4 Carefully remove the festoon bulb.
5 Remove two Lucar connectors.

Refit

6 Reverse 1 to 5. The two Lucar connectors may be fitted either way round.

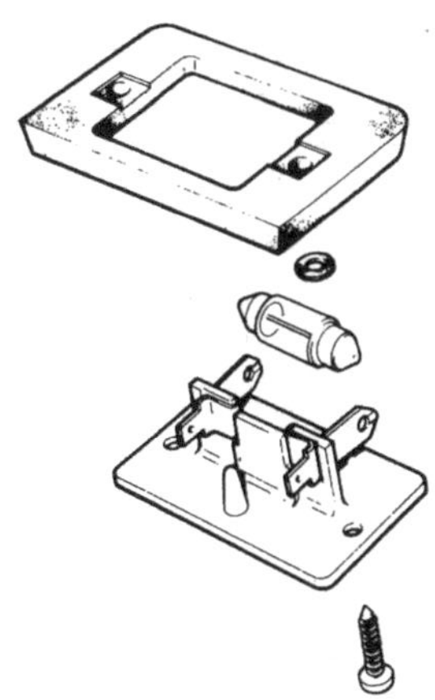

A86 028

REAR FOG LAMP ASSEMBLY

Remove and refit **86.40.99**

Removing

1 From inside the luggage compartment disconnect the lamp leads. Pass the leads throught the body grommet to hang loose below the lamp.
2 Remove two bolts, nuts and washers securing the lamp to the bracket.

Refitting

3 Reverse instructions 1 and 2

ROOF LAMP

Remove and refit **86.45.02**

Early vehicles only

Removing

1 Isolate the battery.
2 Gently squeeze the lens adjacent to the clip projections and remove the lens.
3 Carefully remove the festoon bulb.
4 Remove two Pozidriv screws.
5 Withdraw the lamp base.
6 Note the wire colour codes and positions.
7 Disconnect two terminal ends.

Refitting

8 Reverse 1 to 7. Insert the lamp base with the switch facing forward. Ensure that the right-hand screw makes a good earth connection to the vehicle body.

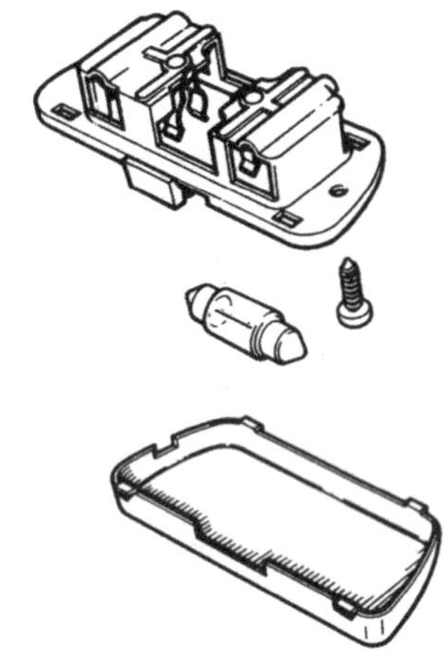

A86 036

MAP/COURTESY LAMP

Remove and refit **86.45.10**

Later vehicles only:

Removing

1 Isolate the battery.
2 Carefully prise the rearward edge of the lamp outwards to free it from the door trim pad.
3 Carefully remove the festoon bulb.
4 Note the wire colour codes and positions.
5 Remove the three Lucar connectors.

Refitting

6 Reverse 1 to 5.

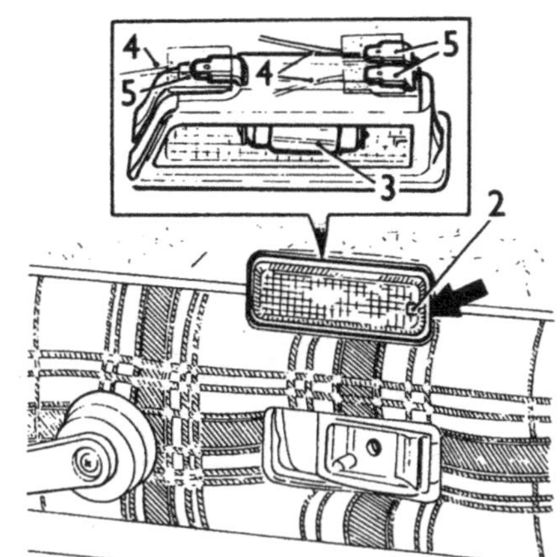

A 86230

LAMPS

Instrument illumination – remove and refit 86.45.32

The four instrument illumination asemblies may be considered as a composite part of the instrument panel. Each instrument illumination assembly consists of three parts. The bulb and bulb holder which may best be removed from the concealed side of the instrument panel. The green dome cover which will not normally require to be removed.

Remove

All operations:

1. Isolate the battery.
2. Remove the fascia centre grille, see 76.55.14.
3. Remove the fascia instrument cowl, see 76.46.17.
4. Read the remainder of this operation and decide the best approach for the specific operation that is to be performed.

Any bulb and bulb holder may be removed as follows:

5. At the concealed side of the instrument panel proceed as follows: Rotate the appropriate bulb holder anti-clockwise and carefully withdraw from the bayonet fitting.
6. If satisfactory access is not available at the concealed side of the instrument panel proceed as follows: Perform 88.20.01 operations 3 to 9. Withdraw the instrument panel as shown. Rotate the appropriate bulb holder anti-clockwise and carefully withdraw from the bayonet fitting.
7. Pull the bulb from the bulb holder.

Any green dome cover may be removed as follows:

8. Remove three Pozidriv screws and washers.
9. Carefully slide the lens upwards following the path of its natural arc.
10. Lift out the face panel.
11. Remove the speedometer, see 88.30.01 or the tachometer, see 88.30.21 to obtain access to the appropriate green dome cover.
12. Remove the green dome cover from the housing claws. Take care not to break a claw.

Refit

13. Reverse all operations performed.

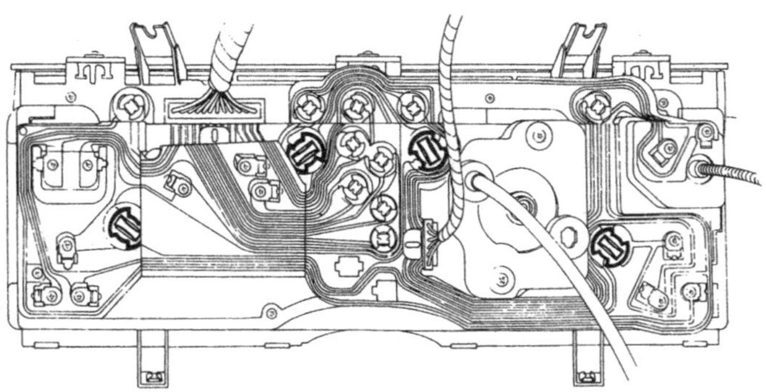

A86 073

LAMPS

Heater control illumination – remove and refit 86.45.44

Remove

1. Remove two Pozidriv screws.
2. Lower the panel downwards.
3. Any of the four bulb holders may now be removed by tilting the panel and sliding out a bulb holder.
4. Remove the bulb from the bayonet fitting.

Refit

5. Reverse 1 to 4.

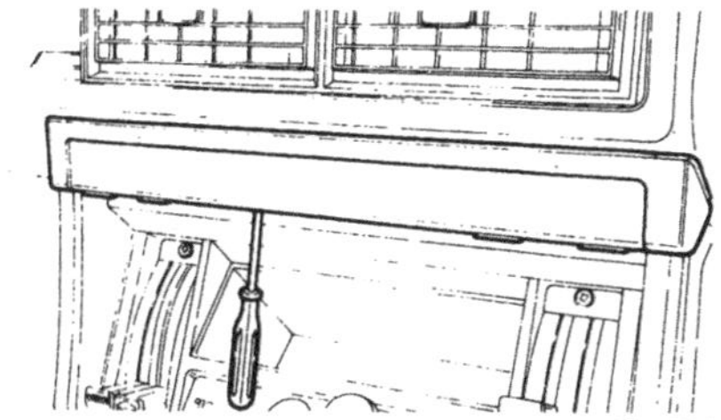

A86 254

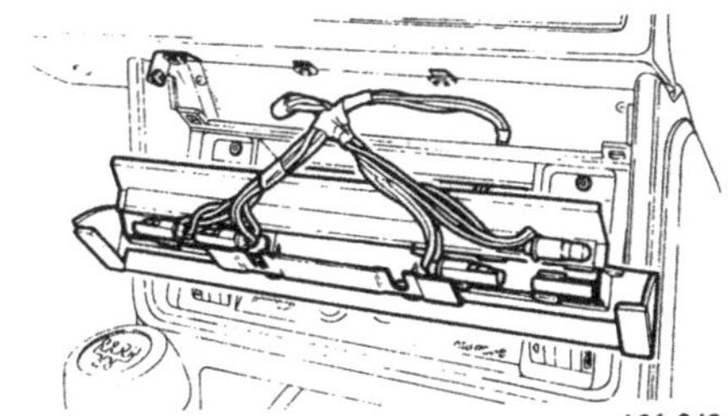

A86 242

LAMPS

Fascia switch panel illumination – remove and refit 86.45.47

Remove

1. Isolate the battery.
2. Remove two Pozidriv screws and washers.
3. Withdraw the panel.
4. Either of the two bulb holders may now be removed by carefully pulling from the housing.
5. Pull the bulb from the bulb holder.

Refit

6. Reverse 1 to 5.

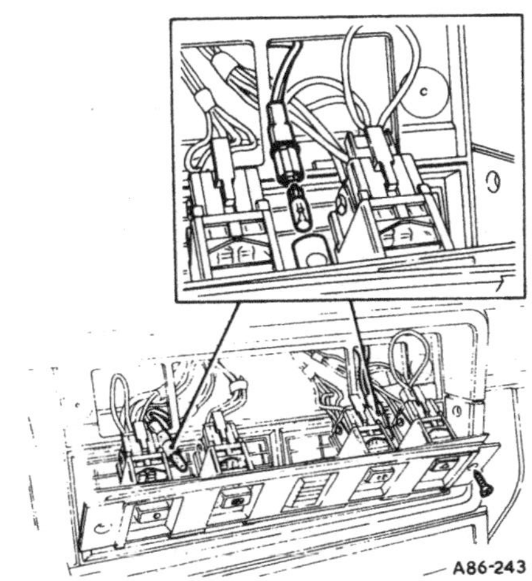

A86-243

LAMPS

Warning light – remove and refit **86.45.60**

The warning light assemblies may be considered as a composite part of the instrument panel. Each warning light assembly consists of three parts. The bulb holder which may only be removed from the concealed side of the instrument panel. The bulb which may be removed from either side of the instrument panel. The annotated lens which may only be removed from the exposed side of the instrument panel.

Remove

All operations:

1 Isolate the battery.
2 Remove the fascia centre grille, see 76.55.14.
3 Remove the fascia instrument cowl, see 76.46.17.
4 Read the remainder of this operation and decide the best approach for the specific operation that is to be performed.

Any bulb holder may be removed as follows:

5 At the concealed side of the instrument panel proceed as follows: Rotate the appropriate bulb holder anti-clockwise and carefully withdraw from the bayonet fitting.
6 If satisfactory access is not available at the concealed side of the instrument panel proceed as follows:
Perform 88.20.01 operations 3 to 9. Withdraw the instrument panel as shown. Rotate the appropriate bulb holder anti-clockwise and carefully withdraw from the bayonet fitting.
7 Pull the bulb from the bulb holder.

Any bulb or annotated lens may be removed as follows:

8 Remove three Pozidriv screws and washers.
9 Carefully slide the lens upwards following the path of its natural arc.
10 Lift out the face panel.
11 Lift out the appropriate annotated lens.
12 Pull the bulb from the bulb holder.

Refit

13 Reverse all operations performed.

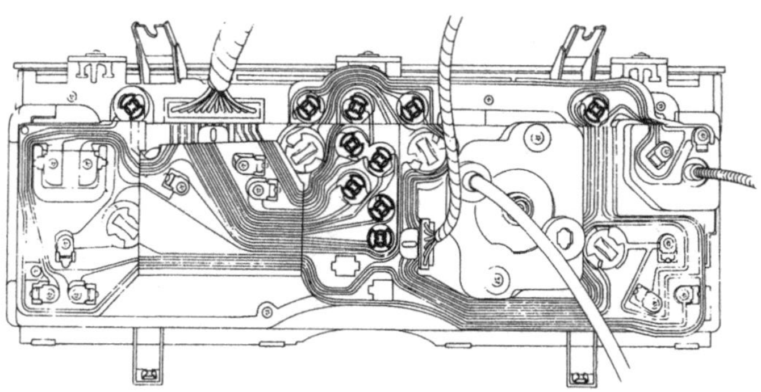

A86 064

LAMPS

Hazard warning light – remove and refit **86.45.76**

Remove

1 Isolate the battery.
2 Remove two Pozidriv screws and washers.
3 Withdraw the panel.
4 Disconnect the hazard switch harness plug.
5 Pull the bulb from the harness plug.

Refit

6 Reverse 1 to 5.

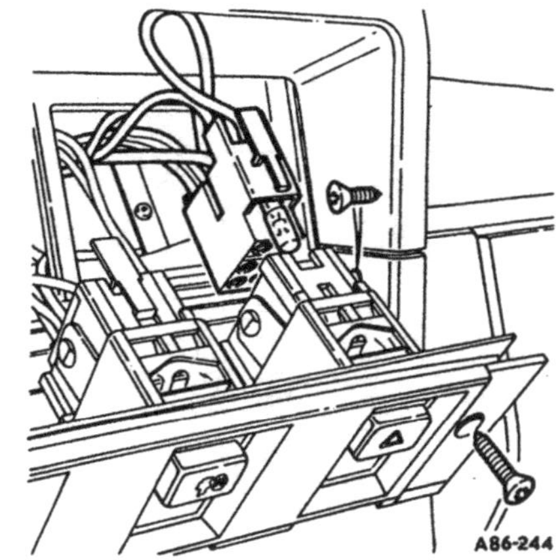

A86-244

RADIO – U.S.A. specification vehicles

Door speaker– remove and refit **86.50.13**

Removing

1 Remove the door trim pad. 76.34.01.
2 Remove the four Pozidrive screws and spring clips securing the speaker to the trim pad.
3 Withdraw the speaker.

Refitting

4 Reverse instructions 1 to 3.

COMPONENT MOUNTING PLATE
– U.K. – R.H. Steering

Early vehicles [86.55.00]

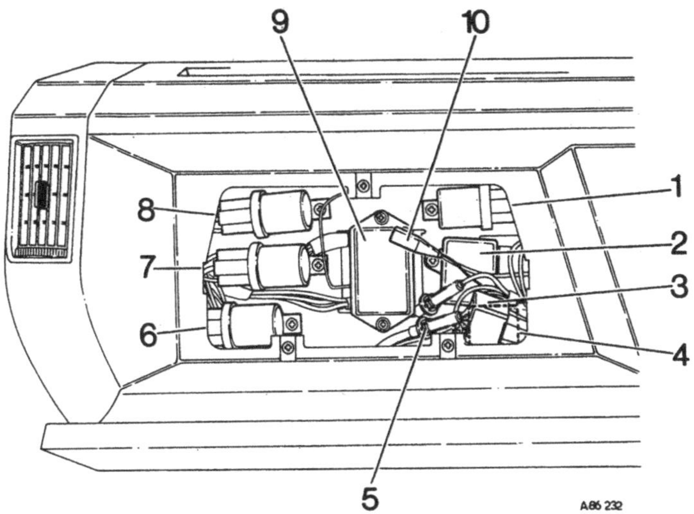

1 Starter motor relay
2 Headlamp – delay unit
3 Hazard flasher unit
4 Headlamp – R.H. run/stop relay
5 Fuses – in line
6 Headlamp – L.H. run/stop relay
7 Horn relay
8 Headlamp – flash relay
9 Fuse box
10 Headlamp – circuit breaker

Later vehicles

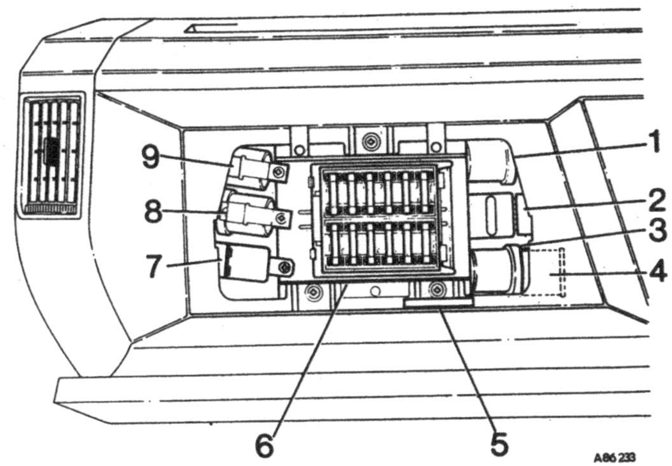

1 Headlamp – run/stop relay
2 Headlamp – delay unit
3 Headlamp – flash relay
4 Hazard flasher unit
5 Headlamp – circuit breaker
6 Fuse box
7 Horn relay
8 Starter motor relay
9 Headlamp – run/stop relay

COMPONENT MOUNTING PLATE
– European – L.H. Steering

Early vehicles

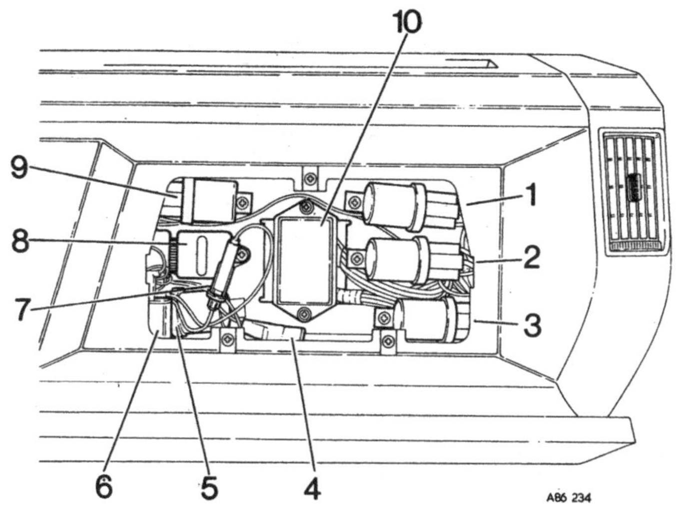

1 Headlamp – flash relay
2 Horn relay
3 Headlamp – R.H. run/stop relay
4 Headlamp – circuit breaker
5 Headlamp – L.H. run/stop relay
6 Hazard flasher unit
7 Fuses – in line
8 Headlamp – delay unit
9 Starter motor relay

Later vehicles

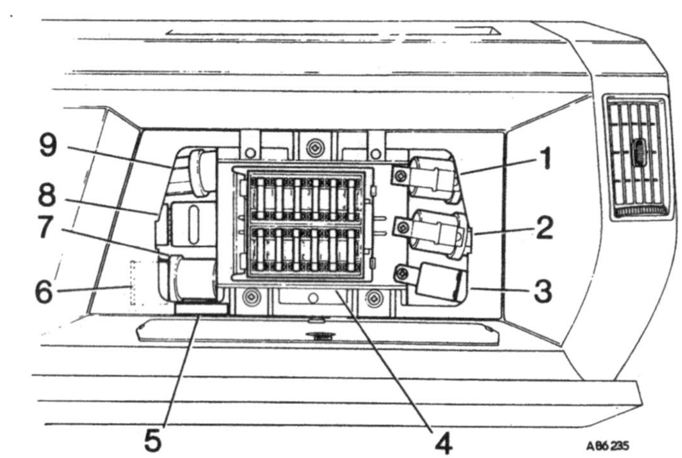

1 Headlamp – R.H. run/stop relay
2 Starter motor relay
3 Horn relay
4 Fuse box
5 Headlamp – circuit breaker
6 Hazard flasher unit
7 Headlamp – flash relay
8 Headlamp – delay unit
9 Headlamp – L.H. run/stop relay

RELAYS – U.K. AND EUROPE

RELAY

Starter motor relay – remove and refit 86.55.05

Remove
1 Open the cubby box lid.
2 Remove two Pozidriv screws. Withdraw the lower panel to obtain access to the component mounting plate.
3 Locate the starter motor relay using the appropriate illustration.
4 Remove the Pozidriv screw.
5 Pull the relay from the harness plug.

Refit
6 Reverse 1 to 5.

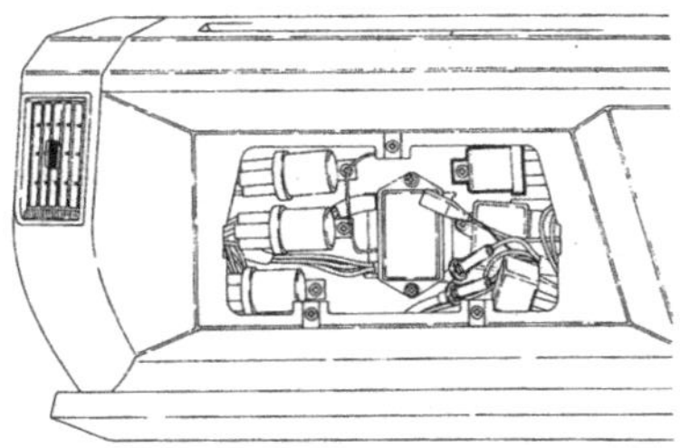

RELAY

Horn relay – remove and refit 86.55.09

Remove
1 Open the cubby box lid.
2 Remove two Pozidriv screws. Withdraw the lower panel to obtain access to the component mounting plate.
3 Locate the horn relay using the appropriate illustration.
4 Remove the Pozidriv screw.
5 Pull the relay from the harness plug.

Refit
6 Reverse 1 to 5.

FLASHER UNIT

Turn signal flasher unit – remove and refit 86.55.11

Remove
1 Locate the flasher unit mounted in a clip attached to the fascia support rail forward of the choke and bonnet release controls.
2 Pull the flasher unit from the clip.
3 Disconnect two Lucar connectors.

Refit
4 Reverse 1 to 3. Connect two Lucar connectors as follows:
Light green/slate wire to terminal B.
Light green/brown wire to terminal L.

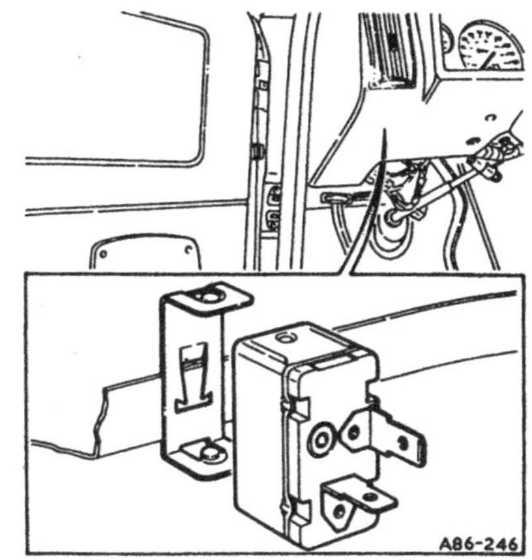

FLASHER UNIT

Hazard flasher unit – remove and refit 86.55.12

Remove
1 Locate the flasher unit mounted in a clip attached to the component mounting plate by placing the head near the floor under the glovebox and looking upwards.
NOTE: The illustration shows the flasher unit position looking through the glovebox.
2 Pull the clip from the component mounting plate.
3 Disconnect two Lucar connectors.
4 Remove the flasher unit from the clip.

Refit
5 Reverse 1 to 4. Connect two Lucar connectors as follows:
Purple wire to terminal B.
Light green/pink wire to terminal L.

RELAY

Headlamp – run/stop relay – remove and refit 86.55.20

Remove
1 Open the cubby box lid.
2 Remove two Pozidriv screws. Withdraw the lower panel to obtain access to the component mounting plate.
3 Locate the left-hand headlamp run/stop relay or the right-hand headlamp run/stop relay using the appropriate illustration.
4 Remove the Pozidriv screw.
5 Pull the relay from the harness plug.

Refit
6 Reverse 1 to 5.

RELAY

Air conditioning – delay circuit relay – remove and refit **86.55.30**

Remove
1 Open the bonnet.
2 Locate the relay attached to a forward engine bay panel.
3 Remove the Pozidriv screw.
4 Disconnect the harness plug.

Refit
5 Reverse 1 to 4. Ensure that the harness earth tag is included in the screw assembly.

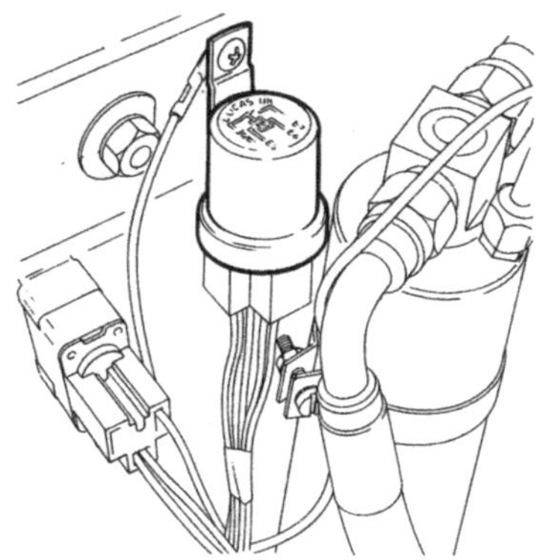

A86 087

RELAY

Headlamp – flash relay – remove and refit **86.55.37**

Remove
1 Open the cubby box lid.
2 Remove two Pozidriv screws. Withdraw the lower panel to obtain access to the component mounting plate.
3 Locate the headlamp flash relay using the appropriate illustration.
4 Remove the Pozidriv screw.
5 Pull the relay from the harness plug.

Refit
6 Reverse 1 to 5.

FLASHER UNIT

Air conditioning – delay circuit flasher unit – remove and refit **86.55.31**

Remove
1 Open the bonnet.
2 Locate the flasher unit mounted in a clip attached to a forward engine bay panel.
3 Pull the flasher unit from the clip.
4 Disconnect the harness plug.

Refit
5 Connect the harness plug. Ensure that the flasher unit Lucar blades are fitted into the harness plug operational sockets.
6 Reverse 1 to 3.

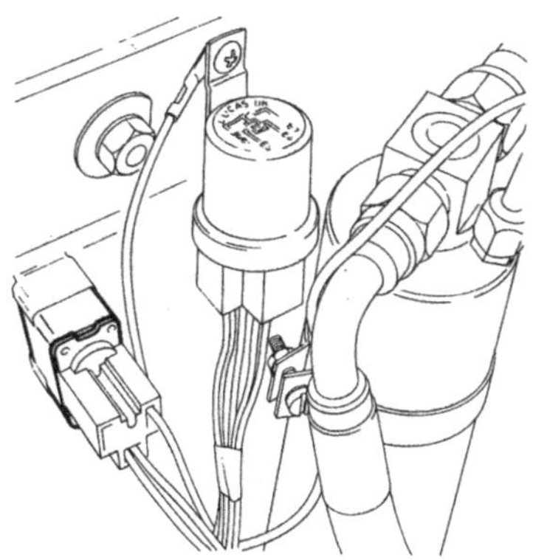

A86 088

CONTROL UNIT

Headlamp – flash control unit – remove and refit **86.55.38**

Remove
1 Open the cubby box lid.
2 Remove two Pozidriv screws. Withdraw the lower panel to obtain access to the component mounting plate.
3 Locate the headlamp flash control unit using the appropriate illustration.
4 Remove the Pozidriv screw.
5 Depress the claw lever and pull the control unit from the harness multi-contact plug.

Refit
6 Reverse 1 to 5.

illustration next page—

COMPONENT MOUNTING PLATE – U.S.A. [86.55.00]

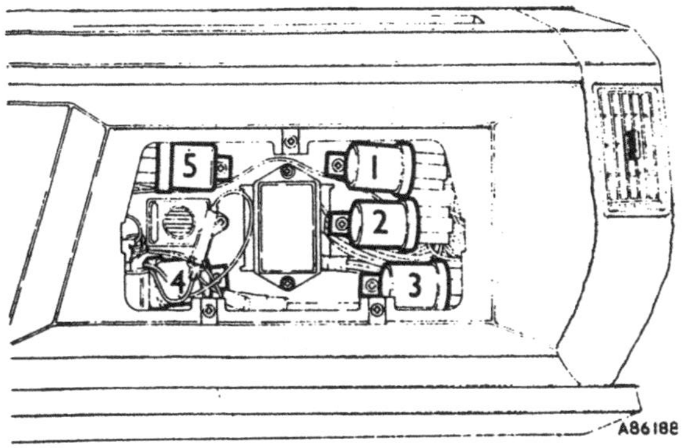

1. Starter motor relay (blue) Lucas part no. 33365, Triumph part no. UKC 7219.
2. Air conditioning—control relay (blue) Lucas part no. 33365, Triumph part no. UKC 7219.
3. Headlamp—R. H. run/stop relay (black) Lucas part no. 3362, Triumph part no. UKC 2416.
4. Headlamp—L. H. run/stop relay (black) Lucas part no. 33362, Triumph part no. UKC 2416.
5. Horn relay (some vehicles) (black) Lucas part no. 3362, Triumph part no. UKC 2416. *mounted in engine bay; see 86.55.30*—air conditioning—delay circuit relay (black) Lucas part no. 33362, Triumph part no. UKC 2416.

Early vehicles

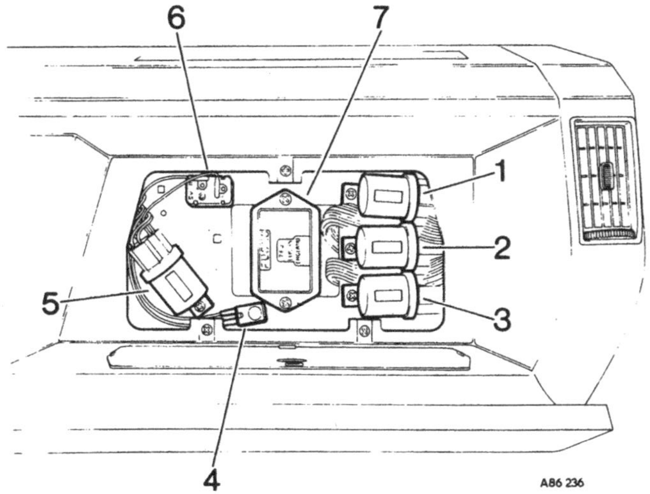

1 Interlock starter motor relay
2 Air conditioning – control relay
3 Headlamp – R.H. run/stop relay
4 Headlamp – circuit breaker
5 Headlamp – L.H. run/stop relay
6 Hazard flasher unit
7 Fuse box

Intermediate vehicles

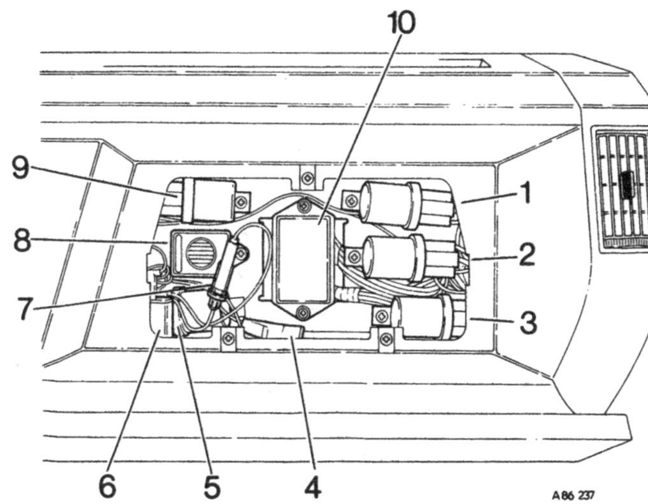

1 Starter motor relay
2 Air conditioning – control relay
3 Headlamp – R.H. run/stop relay
4 Headlamp – circuit breaker
5 Headlamp – L.H. run/stop relay
6 Hazard flasher unit
7 Fuse – in line
8 Seat belt – timer/buzzer module
9 Horn relay
10 Fuse box

COMPONENT MOUNTING PLATE – U.S.A.

Later vehicles

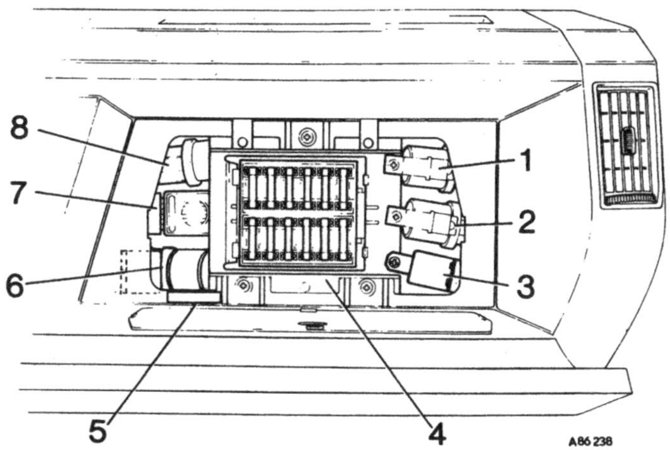

1 Headlamp – R.H. run/stop relay
2 Starter motor relay
3 Horn relay
4 Fuse box
5 Headlamp – circuit breaker
6 Hazard flasher unit
7 Seat belt – timer/buzzer module
8 Headlamp – L.H. run/stop relay

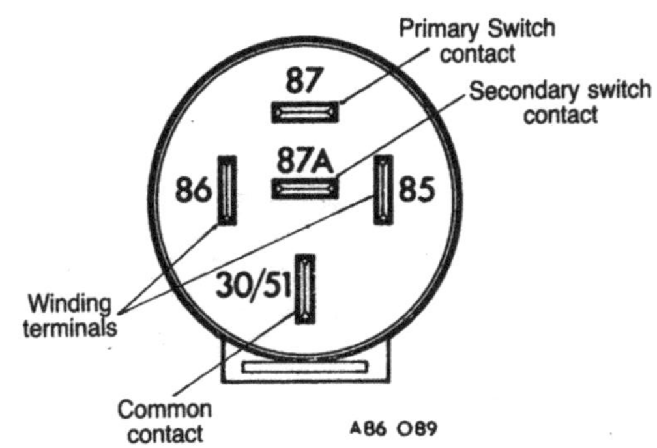

RELAYS – U.S.A. specification vehicles 86.55.00

[1977 with four fuses]

Data—[Typical Relay]

Manufacturer	Lucas
Type	26RA

RELAY

**Starter motor relay—
remove and refit** 86.55.05

Remove
1 Open the cubby box lid.
2 Withdraw the lower panel to obtain access to the component mounting plate.
3 Locate the starter motor relay using the appropriate illustration.
4 Remove the Pozidriv screw.
5 Pull the relay from the harness plug.

Refit
6 Reverse 1 to 5.

RELAY – some vehicles

**Horn relay—
remove and refit** 86.55.09

Remove
1 Open the cubby box lid.
2 Withdraw the lower panel to obtain access to the component mounting plate.
3 Locate the horn relay using the appropriate illustration.
4 Remove the Pozidriv screw.
5 Pull the relay from the harness plug.

Refit
6 Reverse 1 to 5.
NOTE – Usually found in engine bay adjacent to wiper motor.

RELAY

**Air conditioning – control relay –
remove and refit** 86.55.10

Remove
1 Open the cubby box lid.
2 Withdraw the lower panel to obtain access to the component mounting plate.
3 Locate the air conditioning – control relay using the appropriate illustration.
4 Remove the Pozidriv screw.
5 Pull the relay from the harness plug.

Refit
6 Reverse 1 to 5.

FLASHER UNIT

**Turn signal flasher unit –
remove and refit** 86.55.11

Remove
1 Locate the flasher unit mounted in a clip which is clipped to the facia rail at the extreme left hand end.
2 Push the flasher unit upwards to disengage the clip from the facia rail.
3 Disconnect two Lucar connectors.
4 Remove the flasher unit from the clip.

Refit
5 Reverse 1 to 4. Connect two Lucar connectors as follows:

Light green/slate wire to terminal B.
Light green/brown wire to terminal L.

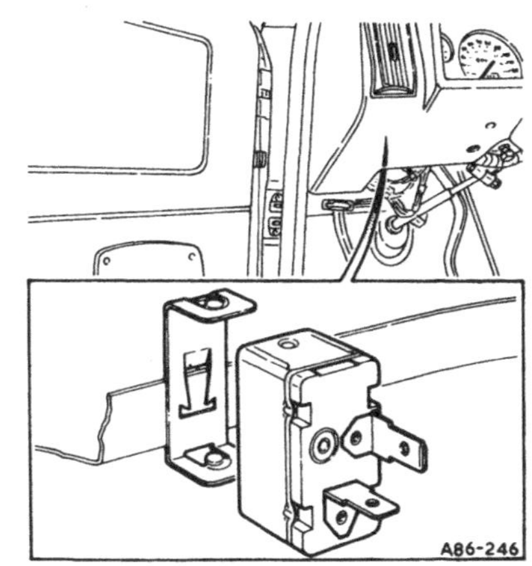

FLASHER UNIT

Hazard flasher unit – remove and refit 86.55.12

Remove
1 Open the cubby box lid.
2 Pull out the lower panel to obtain access to the component mounting plate.
3 Locate the flasher unit mounted in a clip attached to the component mounting plate.
4 Pull the flasher unit from the clip.
5 Disconnect two Lucar connectors.

Refit
6 Connect two Lucar connectors as follows:

Purple wire to terminal B.
Light green/pink wire to terminal L.

7 Reverse 1 to 4.

RELAY

Headlamp – run/stop relay – remove and refit 86.55.20

Remove
1 Open the cubby box lid.
2 Withdraw the lower panel to obtain access to the component mounting plate.
3 Locate the left-hand headlamp run/stop relay or the right-hand headlamp run/stop relay using the appropriate illustration.
4 Remove the Pozidriv screw.
5 Pull the relay from the harness plug.

Refit
6 Reverse 1 to 5.

AIR CONDITIONING— DELAY CIRCUIT RELAY

Remove and refit 86.55.30

Early vehicles only

Removing
1 Open the bonnet.
2 Locate the relay attached to a forward engine bay panel.
3 Remove the Pozidriv screw.
4 Disconnect the harness plug.

Refitting
5 Reverse 1 to 4. Ensure that the harness earth tag is included in the screw assembly.

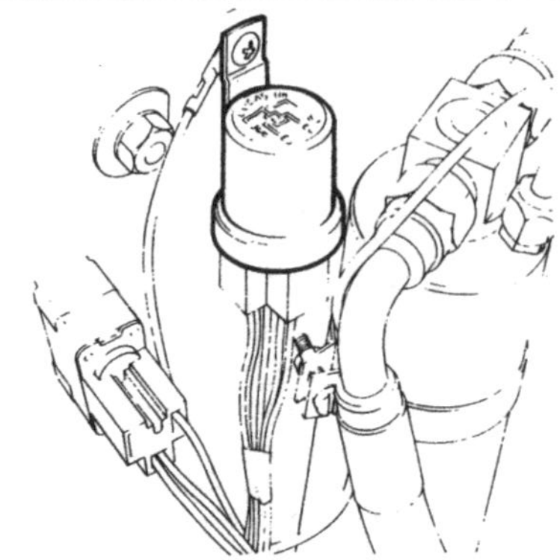

A86 087

AIR CONDITIONING—DELAY CIRCUIT FLASHER UNIT

Remove and refit 86.55.31

Early vehicles only

Removing
1 Open the bonnet.
2 Locate the flasher unit mounted in a clip attached to a forward engine bay panel.
3 Pull the flasher unit from the clip.
4 Disconnect the harness plug.

Refitting
5 Connect the harness plug. Ensure that the flasher unit Lucar blades are fitted into the harness plug operational sockets.
6 Reverse 1 to 3.

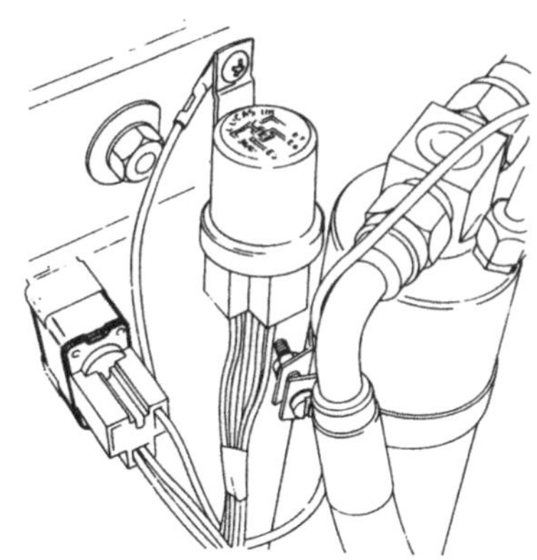

A86 088

INTERLOCK MODULE

Remove and refit **86.57.07**

Early U.S.A. vehicles only

Removing
1 Locate the interlock module clipped to the fascia rail at the extreme right-hand end.
2 Wriggle the interlock module forwards and sideways to disengage the two clips.
3 Disconnect the multi-pin harness plug.

Refitting
4 Reverse instructions 1 to 3. The multi-pin harness plug is keyed and may only be fitted in the correct position.

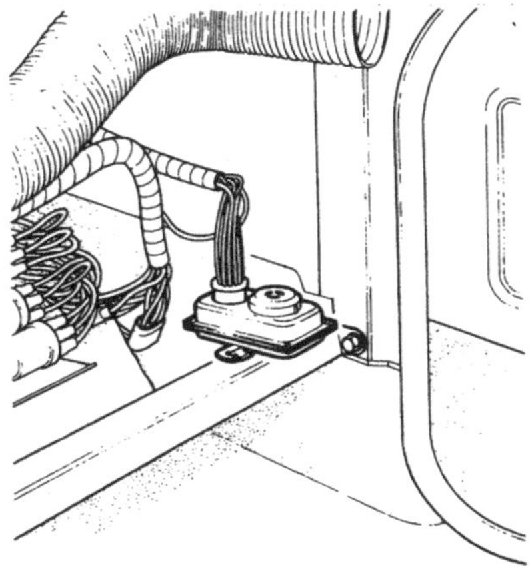

A86 032

SEAT BELT – TIMER/BUZZER MODULE

Remove and refit 86.57.08

Removing
1 Open the cubby box lid.
2 Withdraw the lower panel to obtain access to the component mounting plate.
3 Locate the timer/buzzer module.
4 Remove the Pozidriv screw.
5 Withdraw the module from the harness multi-contact plug.

Refit
6 Reverse 1 to 5.

STARTER MOTOR

86.60.00

Data	
Manufacturer	Lucas
Type	2M100 PE
Lucas part number	25703
Triumph part number	TKC 0020
Motor	
Yoke diameter	4 in (101.60 mm)
Light running – speed	6,000 rev/min
current	40 amp.
torque	Not stated
Load running – speed	1,000 rev/min
current	300 amp.
torque	7.3 lbf ft (1.01 kgf m)
Locked – speed	Nil
current	463 amp.
torque	14.4 lbf ft (1.99 kgf m)
Commutator minimum skimming thickness	0.140 in (3.56 mm)
Brush length – new	0.710 in (18.03 mm)
renew if less than.	0.375 in (9.53 mm)
Brush spring tension	36 ozf (1000 gf)
Shaft end-float: maximum between bush and Spire retaining ring	0.010 in (0.25 mm)
Bearing renewal mandrel diameter:	
Commutator end cover bearing	0.4377 in (11.118 mm)
Drive end bracket bearing	0.4729 in (12.012 mm)
Solenoid	
Pull-in winding resistance – measured between unmarked 'WR wire' connector and 'STA' terminal	0.25 to 0.27 ohm
Hold-in winding resistance – measured between unmarked 'WR wire' connector and unit body	0.76 to 0.80 ohm

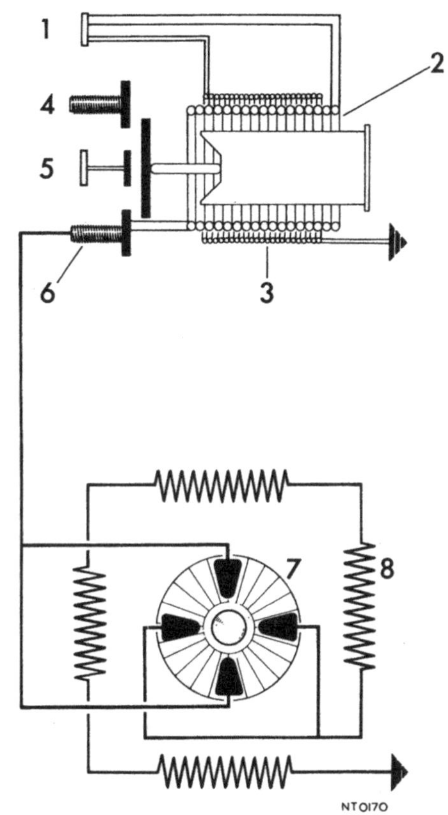

1. Unmarked 'WR wire' connector
2. Pull-in winding
3. Hold-in winding
4. Solenoid battery terminal
5. IGN connector for ballast ignition
6. Solenoid motor terminal
7. Brushes and commutator
8. Field windings

STARTER MOTOR

Remove and refit **86.60.01**

Remove

1 Drive the vehicle onto a ramp.
2 Isolate the battery.
3 Raise the ramp.
4 Lower the complete exhaust system. It is not necessary to remove it from the vehicle.
 a Remove two bolts, spring washers and washers at the gearbox bracket.
 b Remove two nuts, spring washers, washers and bolts at the bell-housing bracket.
 c Unhook two rubber rings at the front silencer hanger.
 d Unhook two rubber rings at the tail pipe hanger.
 e Remove three nuts, spring washers and washers at the manifold flange.
5 Unclip the heat shield from the starter solenoid.
6 Swing the front of the exhaust pipe under the engine and temporarily secure under the right-hand side of the vehicle.
7 Disconnect the Lucar connector.
8 Remove the nut and spring washer. Disconnect the battery lead from the solenoid.
9 Slacken one bolt indicated at the bell-housing exhaust bracket.
10 Remove the lower mounting bolt, spring washer and nut.
11 Remove the middle mounting bolt, cable clip, spring washer and nut.
12 Remove the upper mounting bolt, spring washer and nut.
NOTE: This bolt may be removed with the starter motor tilted.
13 Manoeuvre the starter motor downwards from the vehicle.

Refit

14 Position the starter motor upwards to the vehicle.
15 Fit the upper mounting bolt, spring washer and nut finger-tight.
NOTE: This bolt may be inserted with the starter motor tilted.
16 Fit the middle mounting bolt, cable clip, spring washer and nut finger-tight.
17 Fit the lower mounting bolt, spring washer and nut finger-tight. Include the exhaust bell-housing bracket in the assembly.
18 Tighten the upper bolt assembly.
19 Tighten the middle bolt assembly.
20 Tighten the lower bolt assembly.
21 Tighten one bolt indicated at the bell-housing exhaust bracket.
22 Connect the battery lead to the solenoid. Fit the spring washer and nut.
23 Connect the Lucar connector.
24 Swing the front of the exhaust pipe under the engine and position to the manifold.
25 Clip the heat shield to the starter solenoid.
26 Reverse operation 4. Fit a new exhaust manifold gasket.
27 Lower the ramp.
28 Ensure that the electrical leads are stowed along the inside edge of the heat shield.
29 Connect the battery.

STARTER MOTOR

Roller clutch drive – remove and refit **86.60.07**

Remove

1 Disassemble the starter motor, see 86.60.13.
2 Provide a special punch as shown.
3 Position the special punch over the shaft end and tap the thrust collar from the jump ring towards the roller clutch drive.
4 Prise the jump ring from the shaft groove.
5 Remove the thrust collar.
6 Remove the roller clutch drive.

Refit

7 Lubricate the splines and pinion bearing with grease. Shell SB 2628 for home market and cool climates. Shell Retinax A for hot climates.
8 Fit the roller clutch drive with the engaging lever convex surface facing the solenoid.
9 Fit the thrust collar with the open side facing the shaft end as shown.
10 Prise the jump ring into the shaft groove.
11 Force the thrust collar over the jump ring.

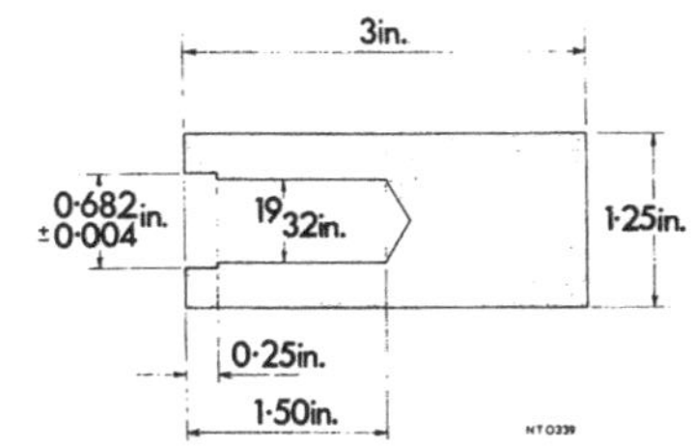

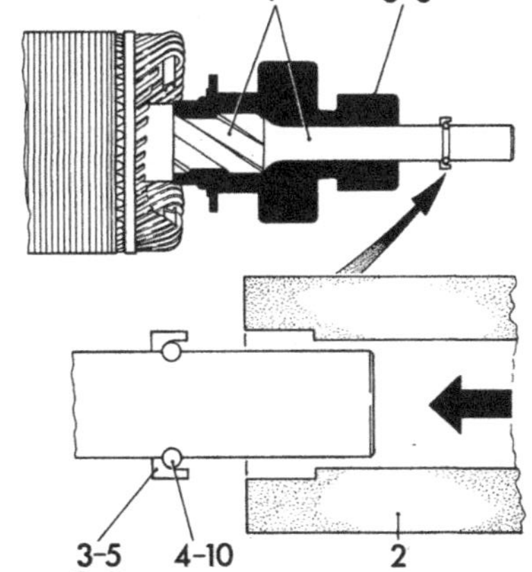

STARTER MOTOR

Overhaul **86.60.13**

Dismantling

NOTE: Dismantling of the starter motor will necessitate the destruction of one Spire retaining ring. Ensure that a new Spire retaining ring for the armature shaft is available before proceeding further. This item is included in the sundry parts kit, Lucas part no. 54246438 or Triumph part no. 520466.

1 Remove the nut, spring washer and washer to free the connector link from the solenoid.
2 Carefully prise off the metal end cap.
3 Use a small chisel to cut a number of claws and remove the Spire retaining ring. Do not prise off the Spire retaining ring without cutting a number of claws as such action may damage the bearing end face, armature shaft and bearing surface when the shaft is withdrawn.
4 Remove two through bolts and spring washers.
5 Withdraw the yoke and commutator end cover assembly.
6 Remove the thrust washer.
7 Remove the rubber seal block.
8 Employ a second operator to support the yoke by hand. Use a length of wood of approximately 0.75 in (20 mm) diameter to tap the commutator end cover from the yoke.
9 Lift out two field winding brushes from the brush box to separate the commutator end cover from the yoke.
10 Tap out the pivot pin. The staking will fold inwards during withdrawal.
11 Remove two bolts and spring washers. Withdraw the solenoid leaving the plunger attached to the engaging lever.
12 Remove the return spring.
13 Withdraw the armature, roller clutch drive and plunger assembly from the drive end bracket.
14 Unhook the plunger complete with the solenoid seal moulding from the engaging lever.

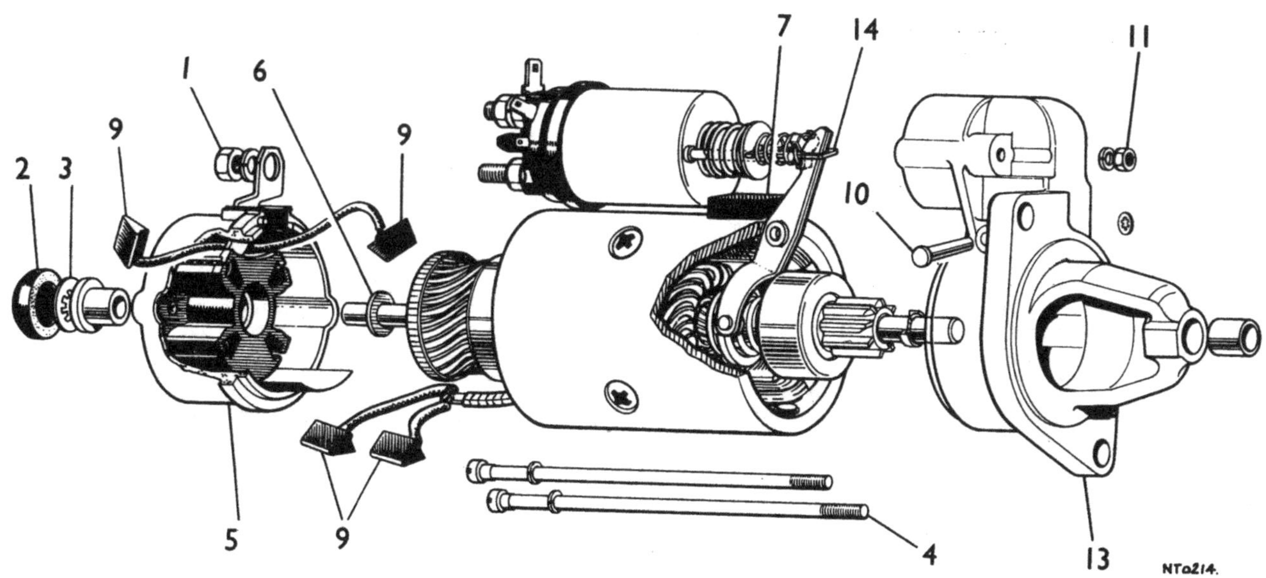

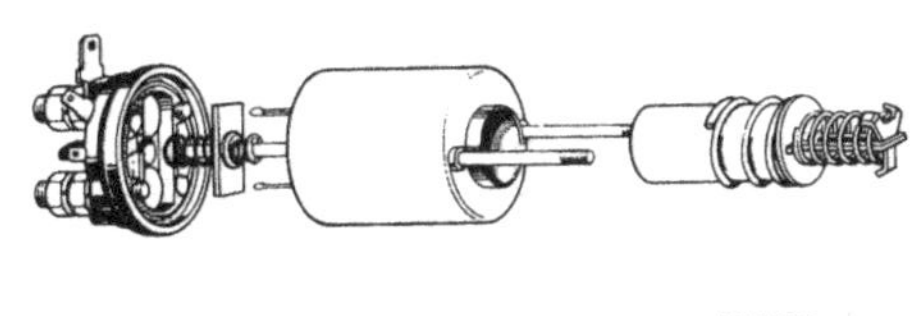

NTO150

Bearings

15 Inspect the porous bronze bearing bushes for wear.
16 If necessary renew either bush as follows:
Extract the bush using a suitable press and mandrel. Prepare the porous bronze bush by immersing it in thin engine oil for 24 hours or thin engine oil heated to 100° C for two hours. Allow the bush to cool in the oil. Fit the bush using a suitable press and a highly polished, shouldered mandrel of the appropriate dimension given in Data. Do not ream the bush after fitting or its porosity may be impaired.

Brushes

17 Clean the brushes and brush box with a petrol-moistened cloth.
18 Check that the brushes move freely in the brush box.
19 Check the brush spring pressure as shown. Position a new brush so that the top protrudes 0.060 in (1.50 mm) above the brush box. Brush spring pressure should be as given in Data. Repeat for the remaining three springs. If the pressure is low renew the commutator end bracket assembly.
20 Check the brush length. Renew the brushes if less than the length given in Data.
21 If necessary renew the commutator end cover brushes. Brushes are supplied attached to a new connector link. Withdraw two brushes from the brush box. Withdraw the connector link. Position new brushes as shown. Retain the longer flexible under the flap.
22 If necessary renew the field winding brushes. Brushes are supplied attached to a common flexible. Cut the old flexibles 0.250 in (6 mm) from the joint. Solder the new flexible to the ends of the old flexible. Do not attempt to solder direct to the field winding strip as the strip may be produced from aluminium.

Commutator

23 Clean the commutator with a petrol-moistened cloth. If the commutator is in good condition it will be smooth and free from pits or burned spots.

24 If necessary polish the commutator with fine glass-paper.

25 If necessary skim the commutator. Separate the armature from the roller clutch drive by performing 86.60.07. Mount the armature in a lathe and rotate at high speed. Using a very sharp tool, take a light cut. Polish with fine glass-paper. Do not cut below the minimum skimming thickness given in Data. Do not undercut insulators between segments.

Roller clutch drive

26 Do not wash the roller clutch in petrol as such action would remove lubricant from the sealed unit. It may be cleaned by wiping with a petrol-moistened cloth.

27 Check that the clutch locks in one direction and rotates smoothly in the other. The unit should move freely round and along the armature shaft splines.

28 The roller clutch is a sealed unit. If the above conditions are not met repair by replacement of the roller clutch unit.

Solenoid

29 Assembly of the starter solenoid involves soldering and sealing complications. It is therefore not advisable to attempt to service this unit. If the solenoid operation is suspect, repair by replacement of the solenoid unit.

30 The plunger is matched with the solenoid body. The spares unit of purchase is a matched solenoid and plunger and the box also contains a return spring. All three items should be fitted as a set.

Reassembling

31 Ensure that the bearing surfaces on the armature shaft are burr-free.

32 Hook the plunger complete with the solenoid seal moulding onto the engaging lever.

33 Insert the armature, roller clutch drive and plunger assembly into the drive end bracket.

34 Fit the return spring to the solenoid.

35 Fit the solenoid so the 'STA' terminal is adjacent to the yoke. Secure with two bolts and spring washers. Ensure that the plunger does not unhook during this operation.

36 Lightly grease the pivot pin. Align the holes and insert the pivot pin. Secure by staking. Ensure that the plunger does not unhook during this operation.

37 Insert two field winding brushes into the brush box with flexibles positioned as shown.

38 Position the commutator end cover to the yoke.

39 Position the rubber seal block.

40 Fit the thrust washer.

41 Holding the commutator end cover firmly to the yoke, insert the assembly.

42 Fit two through bolts and spring washers.

43 If necessary adjust the position of the rubber seal block.

44 Fit a new Spire retaining ring to the armature shaft.

45 Fit the metal end cap.

46 Fit washer, spring washer and nut to secure the connector link to the solenoid.

SWITCHES

Data **86.65.00**

Ignition/starter switch

Normal vehicles only

View on switch harness plug:

Position 0	Off	No connections
Position 1	Auxiliary	N to WK
Position 2	Ignition	N to WK to W
Position 3	Start	N to W to WR

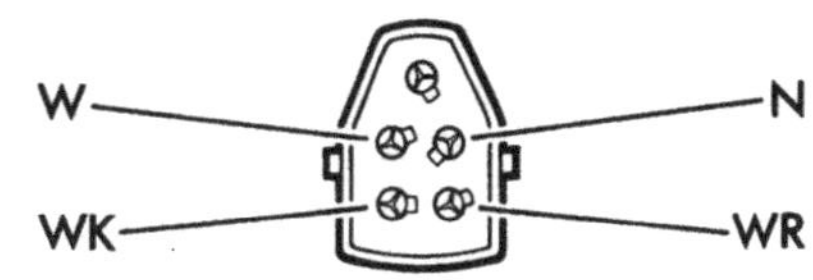

Ignition starter switch

Key warning system vehicles only

View on switch harness plug:

Position 0	Lock	N to NR
Position 1	Auxiliary	N to NR to WK
Position 2	Ignition	N to WK to W
Position 3	Start	N to W to WR

The black/green wire included in the switch harness is associated with the key warning system. A single Lucar connects to the key switch built into the steering column lock assembly.

Steering column multi-purpose switch—headlamp/turn signal/horn unit.

View on switch harness plug:

Position	Head main	U to UW
Position	Head dip	U to UR
Position	Head flash	P left hand to UW
Position	L.H. turn signal	LG/N to GR
Position	R.H. turn signal	LG/N to GW
Position	Horn	P right hand to PB

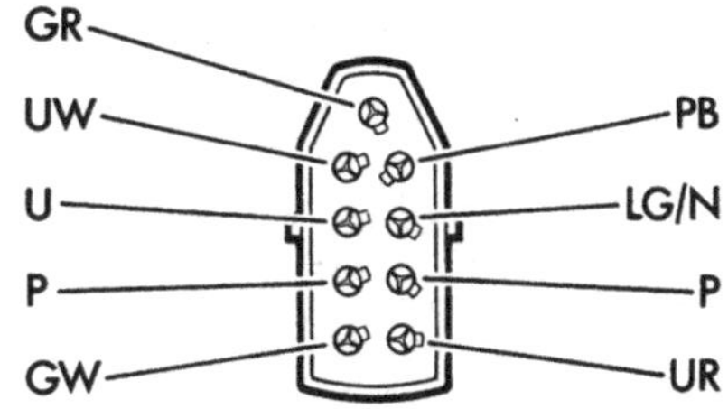

Steering column multi-purpose switch—washer/wiper unit

View on switch harness plug:

Position 0	Off	N/LG to R/LG
Position 1	Normal speed	G to R/LG
Position 2	High speed	G to U/LG
Push knob	Washers	G to L/GB

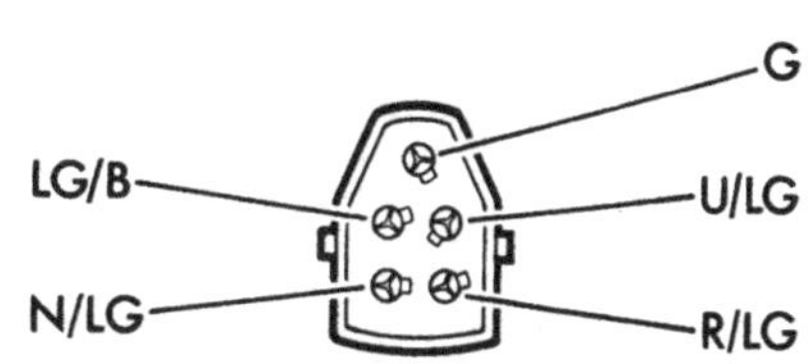

A86 046

Master light switch

Position	Off	N left hand to UN left hand
Position	Side	N left hand to RG and N right hand to UN right hand
Position	Head	N left hand to RG and N right hand to U

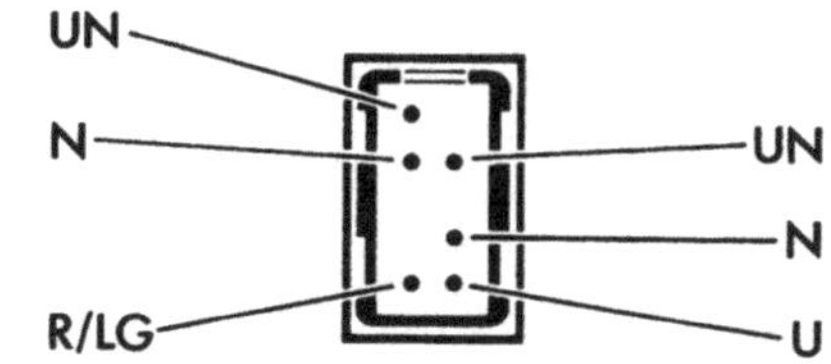

Hazard switch

Position	Off	G to LG/S
Position	Hazard	LG/K to GR to GW to LG/K

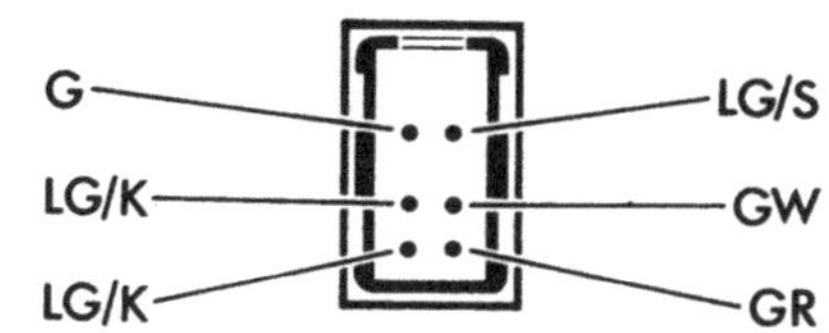

A86 047

Heated backlight switch

Position	Off	No connections
Position	On	G to WB

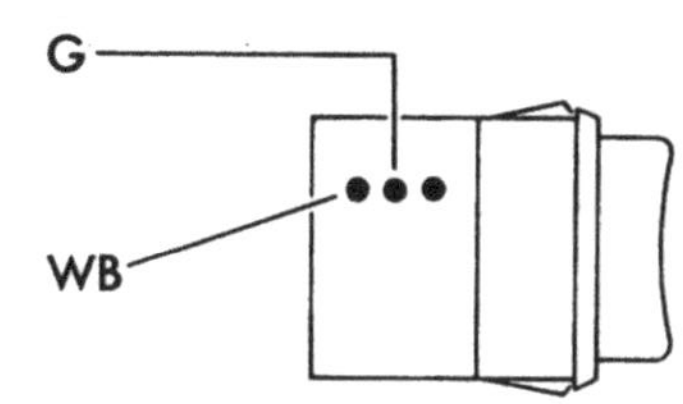

Oil pressure switch

Emission control – anti run on system vehicles only

Position	Engine static	WN to B
Position	Engine run	WP to B

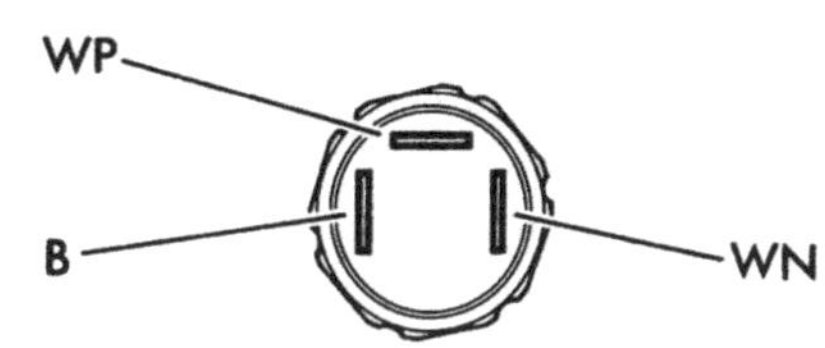

A86 048

IGNITION/STARTER SWITCH

Remove and refit 86.65.02

Removing
1 Isolate the battery.
2 Unscrew two long Pozidriv screws and remove the nacelle upper and lower halves.
NOTE: The screws have a long threaded length of approximately 20 mm. A long 'unscrew time' should be expected.
3 Note the switch harness wire run.
4 Disconnect one harness plug.
5 Remove two small Pozidriv screws.
6 Remove the switch complete with its harness from the vehicle.

Refitting
7 Position the switch harness wire run as noted at operation 3 above. Insert the switch into the steering column lock assembly. Note the keyway and ensure that the lock shaft and switch shaft align correctly.
8 Fit two small Pozidriv screws.
9 Connect one harness plug.
10 Position the nacelle upper and lower halves. Secure with two long Pozidriv screws.
11 Connect the battery.

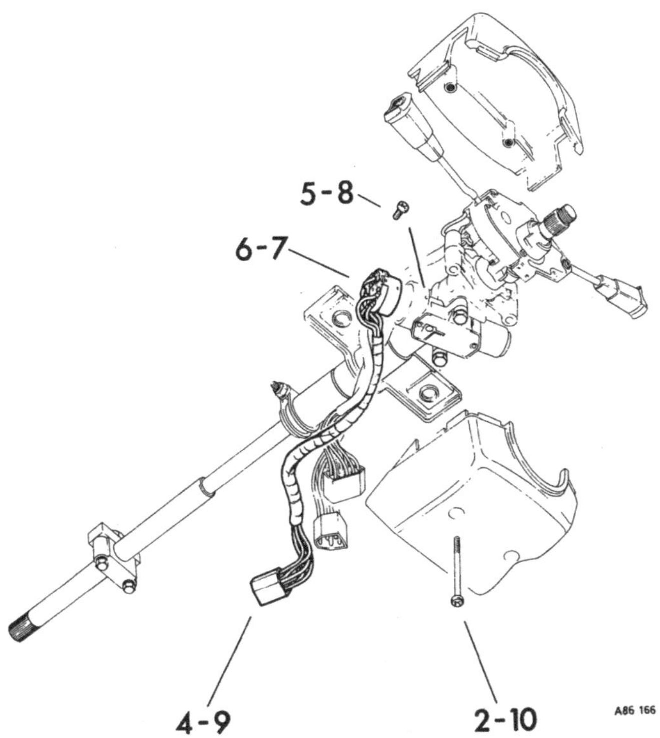

MASTER LIGHT SWITCH

– Remove and refit 86.65.09

Removing
1 Remove the fascia switch panel, see 86.65.66.
2 Using a small screwdriver carefully prise off two Spire clips.
3 Remove the face panel and the switch identification strip assembly.
4 Push inwards one plastic clip on the switch and withdraw the switch from the panel.

Refitting
5 Reverse 1 to 4. Insert the switch so that the 'symbol' is at the lower edge.

DOOR SWITCH

Remove and refit 86.65.14

Removing
1 Open the appropriate door.
2 Remove the single screw.
3 Withdraw the switch.
4 Disconnect one Lucar connector.

Refitting
5 Reverse 1 to 4.

PANEL RHEOSTAT

Remove and refit 86.65.12

Removing
1 Isolate the battery.
2 Pull out the centre console tray.
3 Locate the hole in the knob. Insert a suitable probe into the hole and while depressing pull the knob from the shaft.
4 Unscrew the bezel.
5 Withdraw the panel rheostat downwards from the centre console panel.
6 Remove the spring washer.
7 Note the wire colour codes and positions.
8 Disconnect two Lucar connectors.

Refitting
9 Reverse 1 to 8.

FRONT FOG LAMP SWITCH

Remove and refit 86.65.19

When fitted

Removing
1 Remove the fascia switch panel, see 86.65.66.
2 Using a small screwdriver carefully prise off two Spire clips.
3 Remove the face panel and the switch identification strip assembly.
4 Push inwards one plastic clip on the switch and withdraw the switch from the panel.

Refitting
5 Reverse 1 to 4. Insert the switch so that the 'symbol' is at the lower edge.

REVERSE LAMP SWITCH

Remove and refit 86.65.20

Four speed gearbox only

Removing

1 Drive the vehicle onto a ramp.
2 Raise the ramp.
3 Disconnect two snap connectors.
4 Unscrew the switch from the gearbox extension.
5 Collect up the distance washer.

Refitting

6 Fit the distance washer to the switch.
7 Screw the switch into the gearbox extension and tighten lightly.
NOTE: No further adjustment is required.
8 Connect two snap connectors. The connectors may be fitted either way round.
9 Perform a function test as follows: Switch on ignition. Select reverse. The reverse lamps should now be illuminated. Select any other gear. The reverse lamps should now be off.

Five speed gearbox only

Removing

10 Drive the vehicle onto a ramp.
11 Unscrew the gear lever knob.
12 Lift off the gear lever gaiter.
13 Remove four Pozidriv screws and washers. Lift out the draught excluder and plate.
14 Raise the ramp.
15 Disconnect two snap connectors.
16 Lower the ramp.
17 Slacken the locknut.
18 Unscrew the switch from the gearbox extension.

Refitting

19 Ensure that the locknut is fitted to the switch.
20 Screw the switch into the gearbox extension.
21 Adjust the switch position as follows: Provide a test lamp circuit as shown. Select reverse gear. Screw the switch inwards until the lamp just illuminates. This is the datum position. Screw the switch inwards a further 3 flats (180 degrees). Hold the switch in this position and tighten the locknut.
22 Raise the ramp.
23 Connect two snap connectors. The connectors may be fitted either way round.
24 Lower the ramp.
25 Perform a function test as follows: Switch on ignition. Select reverse. The reverse lamps should now be illuminated. Select any other gear. The reverse lamps should now be off.
26 Reverse 11 to 13.

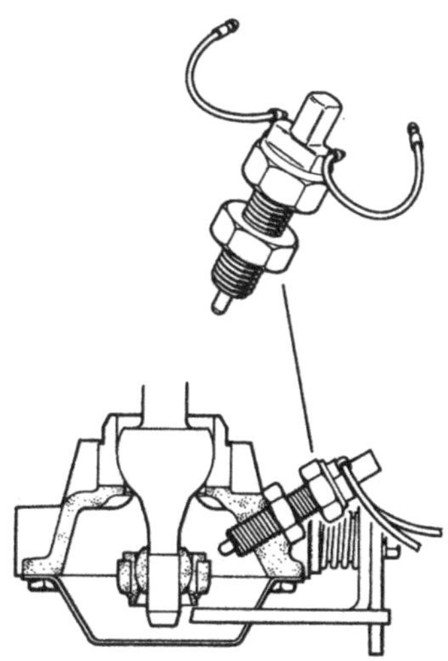

A86 141

INTERLOCK GEARBOX SWITCH

Remove and refit 86.65.28

Early U.S.A. vehicles only

Removing

1 Drive the vehicle onto a ramp.
2 Raise the ramp.
3 Locate the appropriate switch on the gearbox extension.
4 Disconnect two Lucar connectors.
5 Slacken the locknut.
6 Unscrew the switch from the tabbed bracket and locknut.
7 Collect up the locknut.

Refitting

8 Screw the switch partially into the tapped bracket.
9 Screw the locknut onto the switch.
10 Adjust the switch position as follows: Select any gear other than neutral. Screw the switch inwards until the switch plunger just contacts the selector shaft. This is the datum point. Screw the switch inwards a further 6 flats (360 degrees). Hold the switch in this position and tighten the locknut.
11 Connect two Lucar connectors. The connectors may be fitted either way round.
12 Perform a function test as follows: Sit in either the driver's or passenger's seat. Do not fasten the belt. Switch on ignition. Select any gear. The Fasten Belts warning light and buzzer should now be on. Select neutral: The visual and audible warnings should now be off.

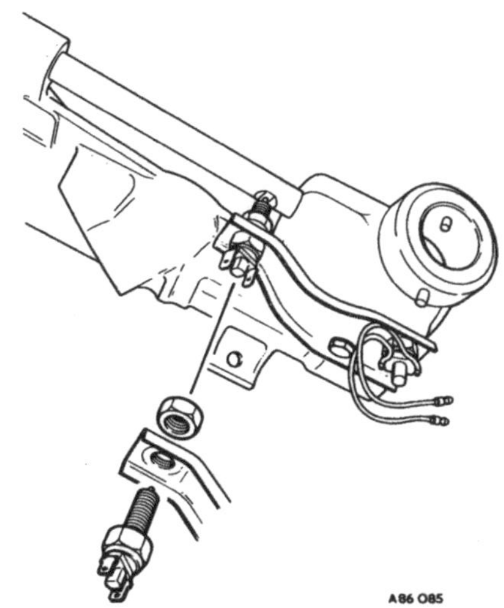

A86 085

SWITCHES

Passenger's seat switch – remove and refit 86.65.29

Remove
1. Remove the passenger's seat, see 76.70.05.
2. Unhook two front diaphragm attachment clips.
3. Note the switch position and wire run.
4. Withdraw the wires through the diaphragm hole.
5. Withdraw the switch.

Refit
6. Reverse 1 to 5.

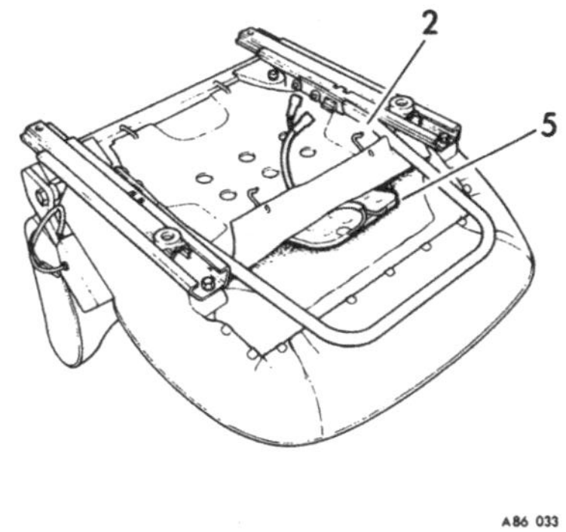

SWITCHES – U.K. AND EUROPE

Oil pressure switch – remove and refit 86.65.30

Remove
1. Locate the switch on the right-hand side of the engine adjacent to the dipstick tube.
2. Disconnect the Lucar connector.
3. Using a spanner, unscrew the switch from the oil transfer adaptor.

Refit
4. Screw the switch into the oil transfer adaptor. The thread is tapered. Do not overtighten.
5. Connect the Lucar connector.

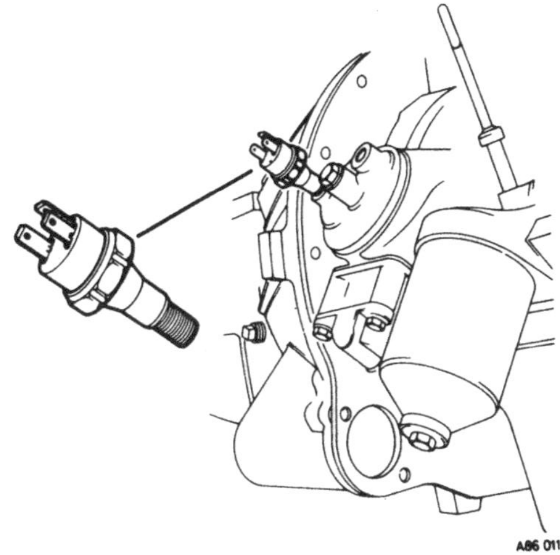

SWITCHES – U.S.A. specification vehicles

Oil pressure switch – remove and refit 86.65.30

Remove
1. Locate the switch on the right hand side of the engine adjacent to the dipstick.
2. *Normal vehicles only–*
   Disconnect one Lucar connector.
3. *Emission control–anti run on system vehicles only–*
   Disconnect three Lucar connectors.
4. Using a spanner, unscrew the switch from the oil transfer adaptor.

Refit
5. Screw the switch into the oil transfer adaptor. The thread is tapered. Do not overtighten.
6. *Normal vehicles only–*
   Connect one Lucar connector.
7. *Emission control–anti run on system vehicles only–*
   Connect three Lucar connectors as shown.

NOTE: Later vehicles have a moulded plug which can be fitted one way only. Wire colours may not conform to illustration.

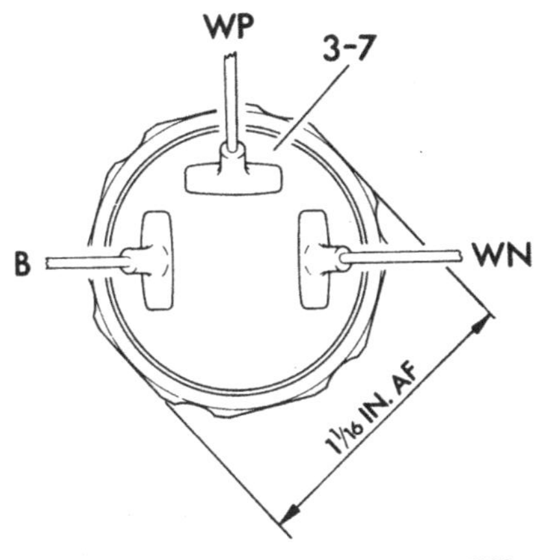

SWITCHES

Driver's belt switch – remove and refit 86.65.31

Passenger's belt switch – remove and refit 86.65.32

Remove
1. Remove the driver's seat, see 76.70.04 or the passenger's seat, see 76.70.05.
2. Remove single bolt to release the integral buckle and switch unit.
3. Collect up the wavy washer, flange bush and distance piece.

Refit
4. Reverse 1 to 3.

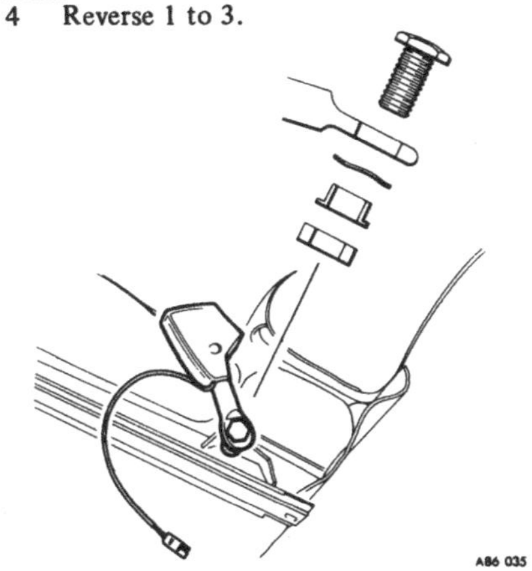

SWITCHES

Heated back light switch – remove and refit 86.65.36

Remove
1. Remove the fascia switch panel, see 86.65.66.
2. Using a small screwdriver carefully prise off two Spire clips.
3. Remove the face panel and the switch identification strip assembly.
4. Push inwards two spring clips on the switch and withdraw the switch from the panel.

Refit
5. Reverse 1 to 4.

SWITCHES

Handbrake switch – remove and refit **86.65.45**

Remove
1 Remove the console assembly, see 76.25.01.
2 Pull apart the Velcro 'touch and close' fastener strips along the top edge of the hand brake gauntlet.
3 Remove the single Pozidriv screw and detach the switch.
4 Disconnect the Lucar connector.

Refit
5 Reverse 1 to 4.

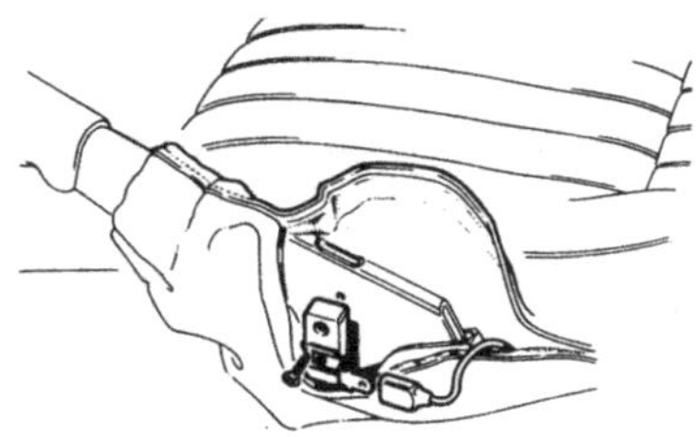

A86 031

SWITCHES

Hazard switch – remove and refit **86.65.50**

Remove
1 Remove the fascia switch panel, see 86.65.66.
2 Using a small screwdriver carefully prise off two Spire clips.
3 Remove the face panel and the switch identification strip assembly.
4 Push inwards one plastic clip on the switch and withdraw the switch from the panel.

Refit
5 Reverse 1 to 4. Insert the switch so that the 'symbol' is at the lower edge.

NOTE: The hazard warning light built into the switch has a bulb housed in the switch harness plug.

SWITCHES

Brake line failure switch – remove and refit **86.65.47**

Remove
1 Locate the brake line failure switch on the brake master cylinder.
2 Release the harness plug claws and disconnect the harness plug from the switch.
3 Using a spanner on the nylon switch body carefully unscrew the switch.

Refit
4 Carefully screw the switch to the brake master cylinder. Do not overtighten. Torque load to only 15 lbf in (0,17 kgf m).
5 Connect the harness plug. The harness plug may be fitted either way round.

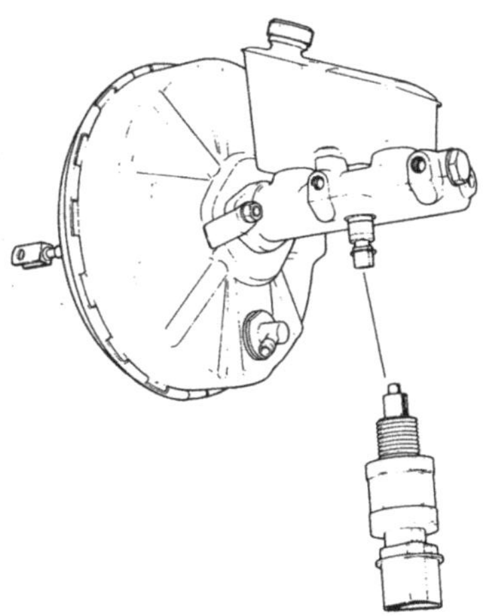

SWITCHES

Stop lamp switch – remove and refit **86.65.51**

Remove
1 Locate the switch adjacent to the brake pedal arm attachment to the brake servo.
2 Disconnect two Lucar connectors.
3 Slacken the large hexagon nut.
4 Push the brake pedal forward and remove the nut and washer/washers. **CAUTION**: Do not attempt to rotate the switch in the bracket as the switch has a locating flat.
5 Remove the switch from the bracket.

Refit
6 Position the switch to the bracket.
7 Push the brake pedal forward and fit the washer/washers and nut finger-tight.
8 Tighten the nut lightly. Do not overtighten the nut on the plastic threads or the switch may be damaged.
9 Connect two Lucar connectors. The connectors may be fitted either way round.

SWITCHES

Choke switch – remove and refit **86.65.53**

Remove

1 Locate the choke switch attached to the choke cable adjacent to the hand knob.
2 Using a small screwdriver remove the small screw.
3 Slide the clip from the switch and remove both items.
4 Disconnect one Lucar connector.

Refit

5 Connect one Lucar connector.
6 Position the switch and clip over the reduced diameter of the cable. Slide the clip onto the switch.
7 Position the assembly so that the switch plunger is located in the hole provided in the outer cable housing. Secure with the small screw.

CIGARETTE LIGHTER

Remove and refit **86.65.60**

Remove

1 Isolate the battery.
2 Pull out the centre console tray.
3 Withdraw the cigarette lighter heating unit.
4 Pull the purple wire 3 mm connector from the centre terminal.
5 Carefully insert a pair of long nosed pliers into the inner well to locate on the stronger cross piece as shown. Holding the outer well, unscrew the inner well from the outer well.
6 Remove the illumination ring.
7 Disconnect the black earth wire Lucar connector.
8 Disconnect the red/white wire single pin harness plug.
9 If necessary renew the bulb as follows: Squeeze the sides of the bulb cowl and withdraw. Unclip the bulb cowl from the bulb holder. Remove the bulb from the bayonet fitting.

Refit

10 Reverse 1 to 9. Fit the cigarette lighter so that the bulb cowl is suitably positioned.

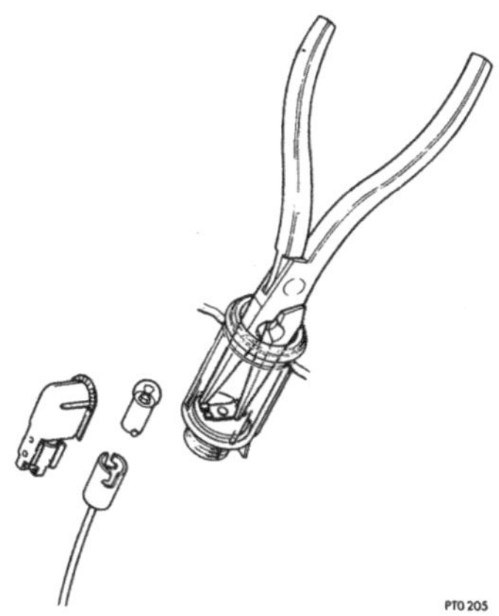

PTO 205

SWITCHES

Steering column multi-purpose switch – remove and refit **86.65.64**

Remove

1 Isolate the battery.
2 Unscrew two long Pozidriv screws and remove the nacelle upper and lower halves.
NOTE: The screws have a long threaded length of approximately 20 mm. A long 'unscrew time' should be expected.
3 Remove the steering wheel, see 57.60.01.
4 Note the switch harness wire runs.
5 Remove the Insuloid harness clip securing the switch harnesses to the fascia rail support strut. The clip is released by squeezing the projection as shown.
6 Disconnect two harness plugs.
7 Slacken the switch clamp screw.
8 Withdraw the switch complete with its harnesses from the column.

Individual switch renew

The multi-purpose switch is initially fitted as a complete unit. In service either half of the switch may be renewed as follows:

9 Do not slacken or remove two hexagon headed screws.
10 Drill out two rivets.
11 Remove one hexagon headed screw and washer.
12 Discard the defective switch.
13 Position the new switch.
14 Fit two bolts and nuts supplied with the new switch.
15 Fit one hexagon headed screw and washer. Do not fully tighten as the screw head is required to move with the switch arm.

continued

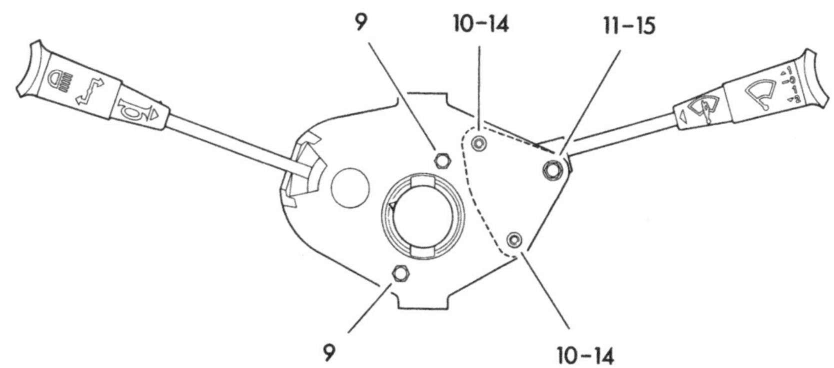

A86 165

Refit

16 Insert the switch harnesses into position. Position the switch to the column with the switch tongue located into the outer tube assembly slot.
17 Push the switch against the outer tube assembly and tighten the switch clamp screw.
18 Position the switch harness wire runs as noted at operation 4 above.
19 Connect two harness plugs.
20 Fit the Insuloid harness clip to secure the switch harnesses to the fascia rail support strut.
21 Fit the steering wheel, see 57.60.01. **NOTE:** The turn signal cancelling collar must be correctly aligned at this stage. Refer to 57.60.01.
22 Position the nacelle upper and lower halves. Secure with two long Pozidriv screws.
23 Connect the battery.

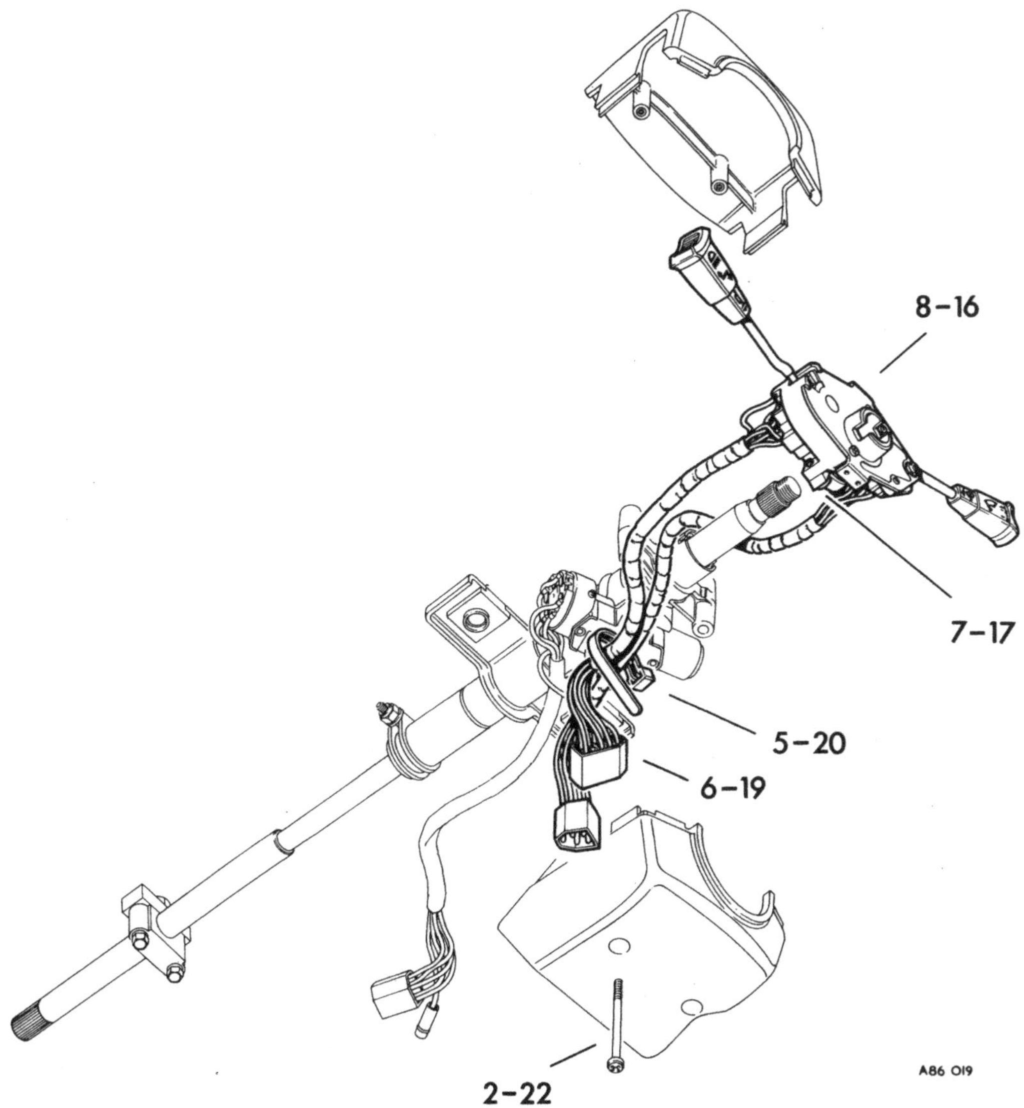

SWITCHES

Rear fog lamp switch – remove and refit **86.65.65**

When fitted

Remove

1 Remove the fascia switch panel, see 86.65.66.
2 Using a small screwdriver carefully prise off two Spire clips.
3 Remove the face panel and the switch identification strip assembly.
4 Push inwards two spring clips on the switch and withdraw the switch from the panel.

Refit

5 Reverse 1 to 4.

SWITCHES

Fascia switch panel – remove and refit **86.65.66**

Remove

1 Remove two Pozidriv screws and washers. Withdraw the fascia switch panel.
2 Pull out two switch identification strip bulb holders.
3 Disconnect all switch harness plugs.
4 Remove the fascia switch panel from the vehicle.

Refit

5 Reverse 1 to 4.

WINDTONE HORN

Adjust **[86.65.67]**

The "Windtone" horn can be adjusted as follows if it has a poor tone or fails to sound.

Adjusting

1 Turn the serrated screw counterclockwise until the horn just fails to sound. (If the horn already fails to sound, it may be necessary to first turn the screw clockwise until a sound is heard.)
2 From the "no sound" position, turn the screw clockwise about one-quarter turn.

CAUTION: Do not turn the adjusting screw while the horn is sounding or while electrical current is being delivered to the horn. Turn the screw in small increments, then test the horn between adjustments by applying current or depressing the horn control. The horn can be damaged if these procedures are not followed.

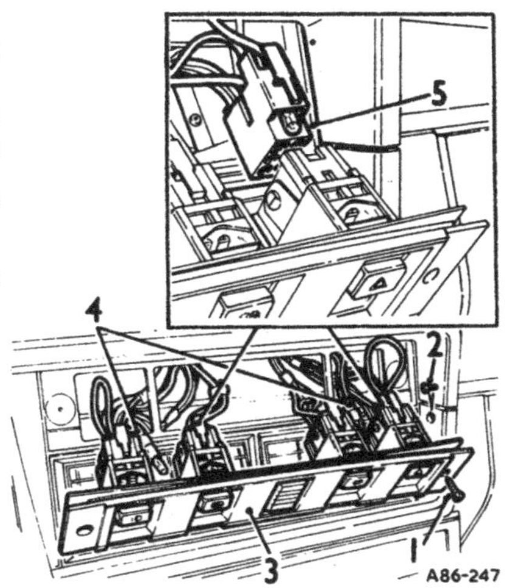

FUSE

Remove and refit **86.70.02**

Fuse box

Remove

1 Open the cubby box lid.
2 Remove two Pozidriv screws. Withdraw the lower panel to obtain access to the component mounting plate.
3 Locate the fuse box attached to the component mounting plate.
4 Pull off the plastic cover.
5 Identify the defective fuse.
6 Carefully lever the fuse from the contacts.

Refit

7 Reverse 1 to 6.

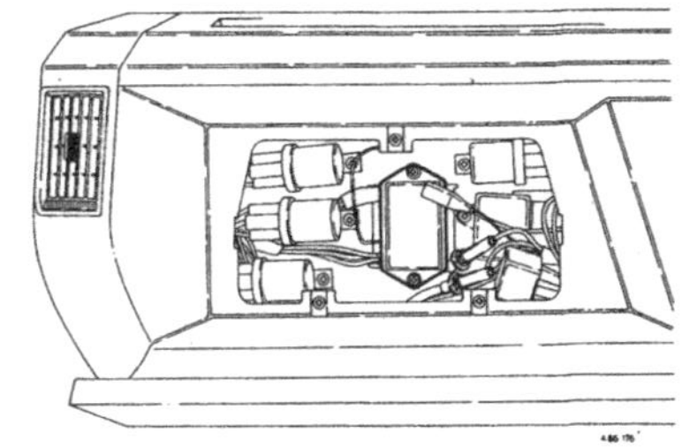

In line

Remove

1 Open the cubby box lid.
2 Remove two Pozidriv screws. Withdraw the lower panel to obtain access to the component mounting plate.
3 Locate the two in line fuses using the appropriate illustration.
4 Identify the two in line fuses as follows:
Front parking lamp and cockpit illuminations to red/green and red wires. Heated backlight to white and white/black wires.
5 Release the bayonet type fitting to separate the two parts of the fuse holder.
6 Remove the fuse.

Refit

7 Reverse 1 to 6.

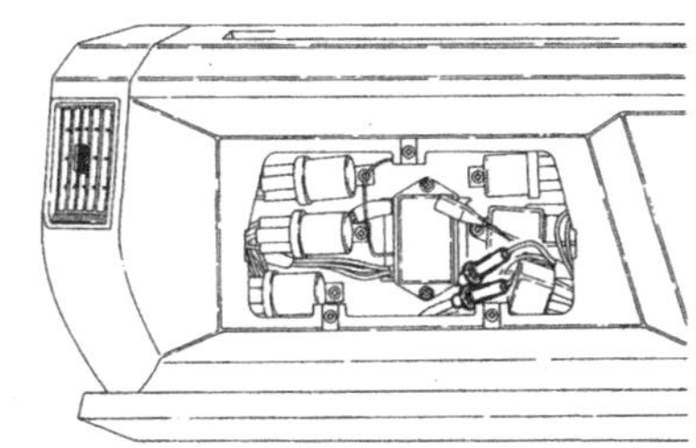

INSTRUMENTS OPERATIONS

BATTERY CONDITION INDICATION — 88.10.00

Data

Manufacturer	Smith's
Type	Bi-metal resistance
Smith's part number	BV 8115/00
Triumph part number	UKC 0512
Illumination	Remote by instrument panel illumination

Description

The Smith's 'bi-metal resistance' battery condition indicator measures the system voltage. Instrument response is slow to damp out short period variations. The indicator produces no radio interference and no suppression equipment is required.

The battery condition indicator contains a bi-metal strip surrounded by a heater series winding. Current flow through the series winding heats the bi-metal strip which consequently distorts. The pointer is suspended between the moving end of the bi-metal strip and a spring blade. This arrangement causes the pointer to take up a position over the scale that is related to current flow through the indicator. Current flow through the indicator is related to system voltage.

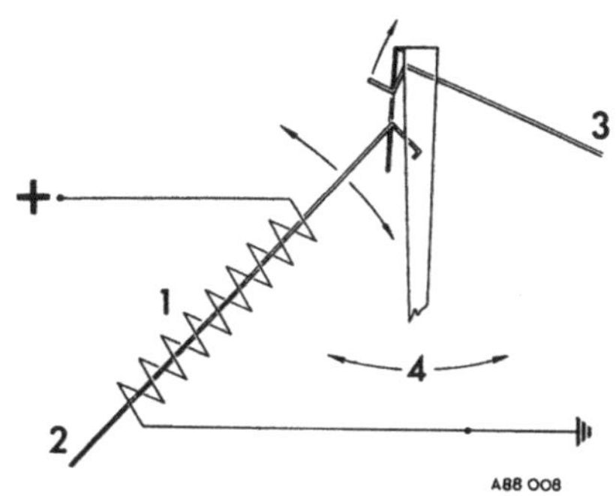

† Supply voltage
1 Heater series winding
2 Bimetal strip
3 Spring blade
4 Pointer

BATTERY CONDITION INDICATOR

Remove and refit 88.10.07

Remove

1 Remove the lens, see 88.20.17.
2 Lift away the face panel.
3 Remove one Pozidriv screw.
4 Carefully withdraw the battery condition indicator.

Refit

5 Reverse 1 to 4..

CLOCK — 88.15.00

Data

Manufacturer	Smith's–made inFrance
Type	Transistor
Smith's part number	CTJ 3702/00
Triumph part number	UKC 2408

CLOCK

Adjust 88.15.04

Set the hands to the correct time–

Locate the adjustment knob below the fascia and to the left of the speedometer trip reset knob. Push the clock adjustment knob upwards and rotate to set the hands.

Regulate the clock–

The clock is precision regulated during manufacture and no further adjustment is possible in service. If the clock is gaining or losing less than five minutes per week the clock should be periodically reset to the correct time as detailed above. If the clock is gaining or losing more than five minutes per week the clock should be replaced.

CLOCK

Remove and refit 88.15.07

Remove

1 Remove the lens, see 88.20.17.
2 Lift away the face panel.
3 Unscrew the knurled nut. Withdraw the clock cable.
4 Remove two Pozidriv screws.
5 Withdraw the clock.

Refit

6 Reverse 1 to 5.

INSTRUMENT PANEL

Description **88.20.00**

The instrument panel is an independent unit which contains a set of instruments and a comprehensive set of warning lights. These are enclosed in a plastic housing. The exposed side of the instrument panel is finished with a face panel and a one-piece curved lens produced from ICI Acrylic. The individual instruments therefore require no bodies or glasses.

Electrical connections from the vehicle harness to the printed circuit attached to the concealed side of the instrument panel is by two multi-contact harness plugs.

The mechanism of the individual instruments are screw mounted into the housing. Electrical input to the instruments is via the instrument pillars. The electrical path from the printed circuit to the instrument pillars is by metal clips screwed down against exposed sections of the printed circuit. This design allows any instrument to be removed and refitted without making any conscious electrical connections.

The warning light bulbs are retained in bulb holders which are fitted into the concealed side of the instrument panel by bayonet fittings. The warning light annotated lens blocks are positioned by housing dowels and are retained by being sandwiched between the housing and the face panel. The warning light electrical connections are made automatically by two bulb holder strips making direct contact against exposed sections of the printed circuit.

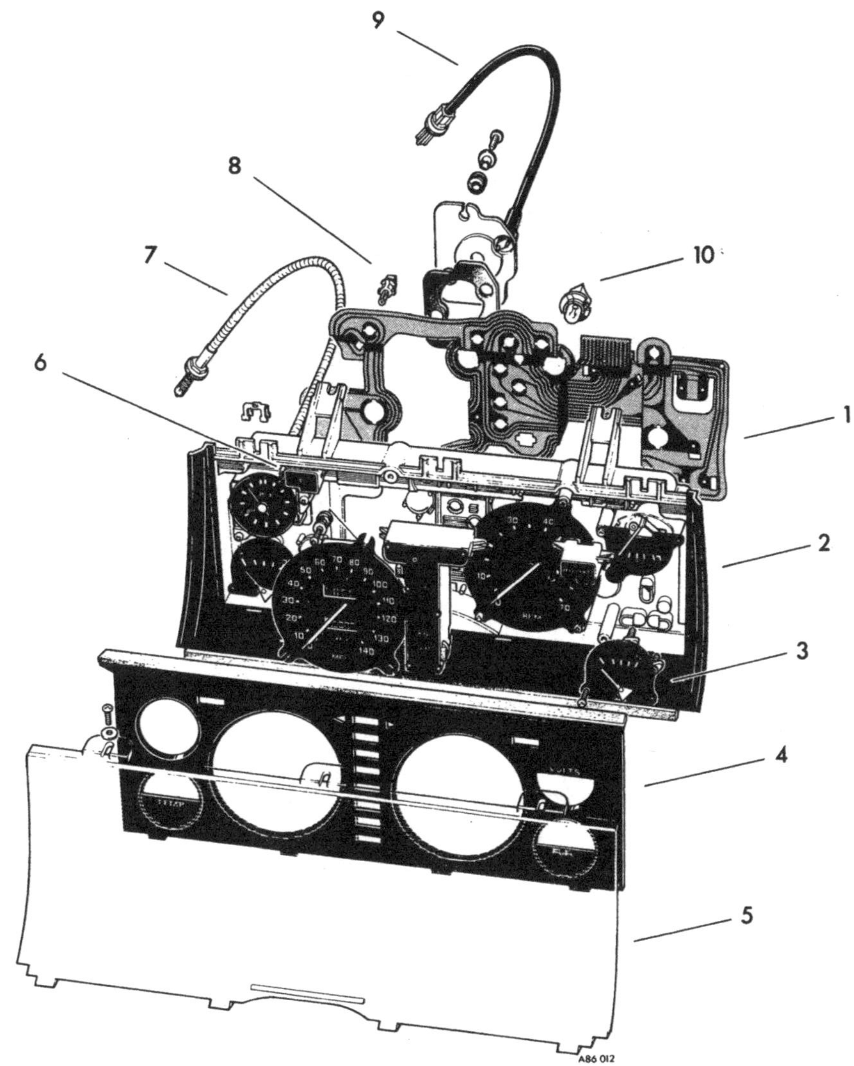

1 Printed circuit
2 Housing
3 Instrument
4 Face panel
5 Lens
6 Annotated lens block
7 Clock cable
8 Warning light
9 Speedometer trip reset cable
10 Instrument illumination

INSTRUMENT PANEL

Remove and refit 88.20.01

Remove

1 Remove the fascia centre grille, see 76.55.14.
2 Remove the fascia instrument cowl, see 76.46.17.
3 Unscrew two long Pozidriv screws and remove the nacelle upper and lower halves.
NOTE: The screws have a long threaded length of approximately 20 mm. A long 'unscrew time' should be expected.
4 Slacken the speedometer trip reset cable knurled nut. Withdraw the cable from the bracket slot.
5 Slacken the clock cable knurled nut. Withdraw the cable from the bracket slot.
6 Depress the lever to release the catch from the annular groove in the boss. Pull the speedometer cable from the instrument.
7 Remove two upper Pozidriv screws and washers.
8 Remove two lower Pozidriv screws and washers.
9 Withdraw the instrument panel slightly.
10 Disconnect two harness multi-contact plugs.
11 Remove the instrument panel from the vehicle.

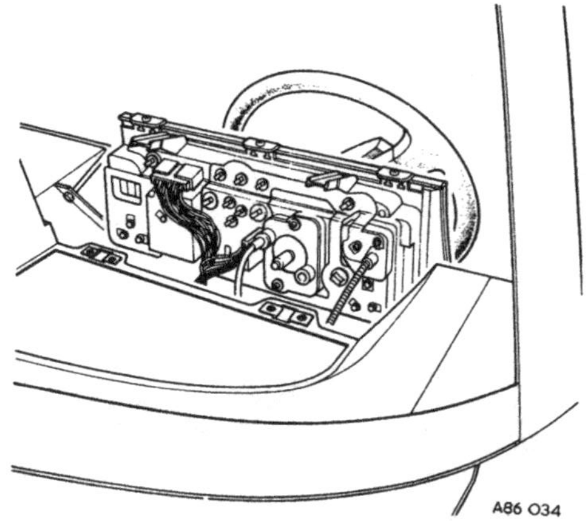

A86 034

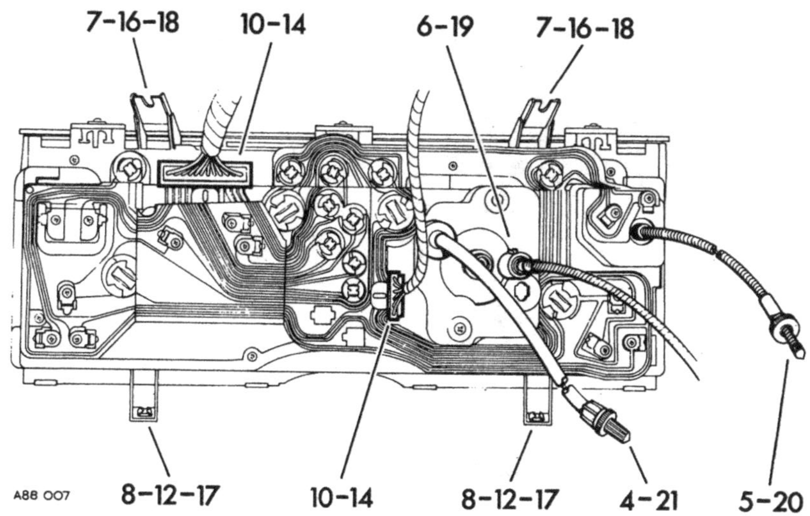

A88 007

Refit

12 Ensure that two lower Spire nuts are correctly fitted.
13 Position the instrument panel with the clock cable and the speedometer trip reset cable inserted either side of the fascia rail support strut and harnesses.
14 Connect two harness multi-contact plugs.
15 Insert the instrument panel to its correct position.
16 To help establish the correct position, fit two upper Pozidriv screws and washers finger-tight.
17 Fit two lower Pozidriv screws and washers.
18 Tighten two upper Pozidriv screws and washers.
19 Push the speedometer cable onto the instrument. Ensure that the catch engages into the annular groove in the boss.
20 Insert the clock cable into the appropriate bracket slot. Hand tighten the knurled nut.
21 Insert the speedometer trip reset cable into the appropriate bracket slot. Hand tighten the knurled nut.
22 Position the nacelle upper and lower halves. Secure with two long Pozidriv screws.
23 Fit the fascia instrument cowl, see 76.46.17.
24 Fit the fascia centre grille, see 76.55.14.

INSTRUMENT PANEL

Lens – remove and refit 88.20.17

Remove
1 Remove the fascia centre grille, see 76.55.14.
2 Remove the fascia instrument cowl, see 76.46.17.
3 Remove three Pozidriv screws and washers.
4 Carefully slide the lens upwards following the path of its natural arc.

Refit
5 Reverse 1 to 4. If difficulty is experienced to engage the four lens tongues, slacken two lower Pozidriv screws and slightly reposition the instrument panel.

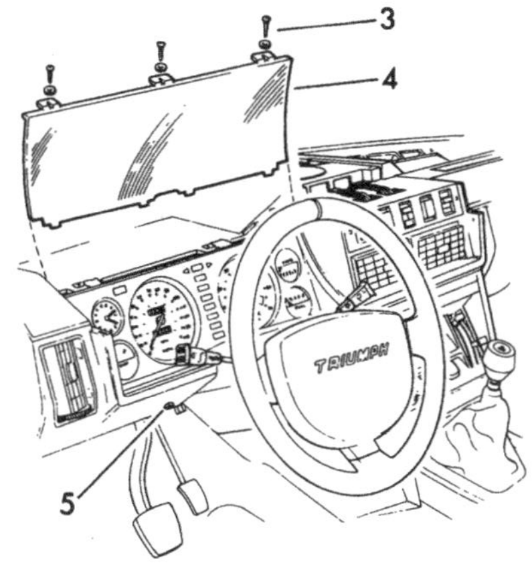

A88 003

INSTRUMENT PANEL

Printed circuit – remove and refit 88.20.19

Remove
1 Remove the instrument panel, see 88.20.01.
2 Remove a number of small Pozidriv screws.
3 Lift out a number of contact clips.
4 Rotate four large bulb holders anti-clockwise and lift from the bayonet fittings.
5 Rotate a number of small bulb holders anti-clockwise and lift from the bayonet fittings.
6 Use a small screwdriver to carefully prise up two plastic press clips. Take care not to damage the printed circuit.
7 Release any sticky tape.
8 Lift away the printed circuit.

Refit
9 Reverse 1 to 8.

FUEL INDICATION

TEMPERATURE INDICATION

Data	**88.25.00**
Fuel indicator	
Manufacturer	Smith's
Type	Air cored
Smith's part number	ACF 8103/00
Triumph part number	UKC 0511
Body	None–assembled into instrument panel
Illumination	Remote by instrument panel illumination
Fuel tank unit	
Manufacturer	Smith's
Smith's part number	TBS 5232/000EC
Triumph part number	TKC 0147
Temperature indicator	
Manufacturer	Smith's
Type	Air cored
Smith's part number	ACT 8102/00
Triumph part number	UKC 0510
Body	None–assembled into instrument panel
Illumination	Remote by instrument panel illumination
Temperature transmitter	
Manufacturer	Smith's
Type	Semi-conductor
Smith's part number	TT 4803/00A
Triumph part number	150843
Colour code	Black
Indication range	50 to 140 deg. centigrade
Thread	⅝ in–18 UNF

FUEL INDICATION

TEMPERATURE INDICATION

Description

The Smith's 'air cored' indication system is claimed to be suitable for rugged environments and able to withstand high levels of vibration. Instrument response is rapid. The system is self-compensating for variation in voltage supply and no voltage stabilizer is required. The system produces no radio interference and no suppression equipment is required.

The 'air cored' indicator consists of three individual coil windings positioned relative to each other as shown. A central shaft carries a magnet bar which is free to swing in the area enclosed by the coils. The end of the shaft carries the pointer.

The magnetic field direction produced by any one coil is fixed. The magnetic field strength produced by any one coil may be varied by adjusting the current through the appropriate coil. The final magnetic field direction is the resultant of the currents through all three coils.

The design is that two of the coils receive a constant current. The third coil receives a current managed by a variable resistor which is the appropriate system sensor. The sensor therefore controls the final magnetic field direction. The magnet bar aligns itself with this magnetic field direction and the pointer is positioned to indicate the correct dial reading.

Switch off – when the circuit supply is switched off a small permanent magnet located in the assembly influences the magnet bar and the pointer returns to the zero position. The effect of this 'pull off' magnet is allowed for when calibrating the instrument during manufacture.

The circuit diagram for the fuel indication system and the temperature indication system is shown. Coils 'A' and 'B' receive an almost constant current. In this circuit the current through coils 'A' and 'B' will vary but to a small amount compared to the variation through coil 'C'. Coil 'C' receives a current managed by the fuel tank unit or the temperature transmitter. A ballast resistor and calibration resistor are included in the circuit as shown.

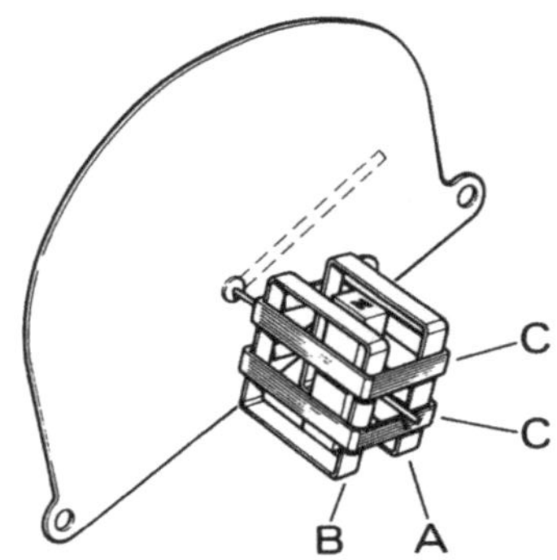

A88 005

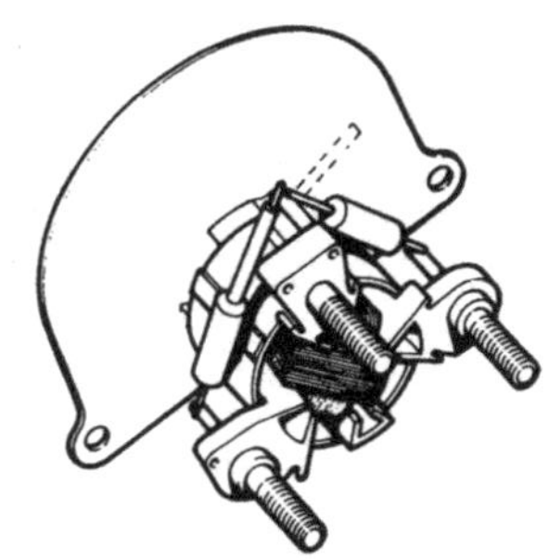

A88 004

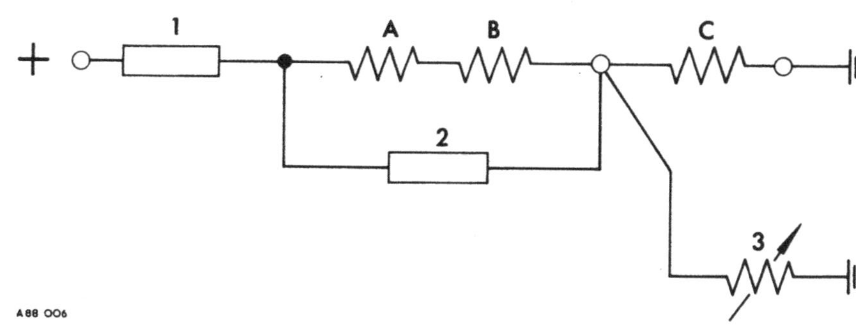

1 Ballast resistor
2 Calibration resistor
A Coil 'A'
B Coil 'B'
C Coil 'C'
3 System sensor

A88 006

TEMPERATURE INDICATION

Temperature indicator – remove and refit **88.25.14**

Remove
1 Remove the lens, see 88.20.17.
2 Lift away the face panel.
3 Remove one Pozidriv screw.
4 Carefully withdraw the temperature indicator.

Refit
5 Reverse 1 to 4.

TEMPERATURE INDICATION

Temperature transmitter – remove and refit **88.25.20**

Remove
1 Drain part of the coolant, see 26.10.01.
2 Locate the transmitter on the inlet manifold.
3 Disconnect the Lucar connector.
4 Unscrew the transmitter from the inlet manifold.

Refit
5 Reverse 1 to 4. Fit a new sealing washer if available.

FUEL INDICATION

Fuel indicator – remove and refit **88.25.26**

Remove
1 Remove the lens, see 88.20.17.
2 Lift away the face panel.
3 Remove one Pozidriv screw.
4 Carefully withdraw the fuel indicator.

Refit
5 Reverse 1 to 4.

FUEL INDICATION

Fuel tank unit – remove and refit **88.25.32**

Remove
1 Drive the vehicle onto a ramp.
2 Estimate the quantity of fuel in the fuel tank.
3 Provide a suitable container to receive the quantity.
4 Raise the ramp.
5 Disconnect three Lucar connectors.
6 Prepare for fuel spillage. Pull off the rubber fuel outlet pipe and drain the tank contents into the container.

[**WARNING:** Fuel and fuel vapors will issue from the tank as you remove the outlet pipe and the fuel tank unit. Do not smoke or work near heaters or other fire hazards. Do not conduct electrical tests near the uncovered tank, as doing this may create an electrical spark and cause an explosion. Keep a fire extinguisher handy.]

7 Release the locking ring by tapping anti-clockwise. Remove the locking ring.
NOTE: This operation may be facilitated by employing a service tool – Churchill number 18G 1001.
8 Carefully withdraw the tank unit.
9 Remove the sealing washer.

Refit
10 Reverse 1 to 9. Fit a new sealing washer if available. Connect the Lucar connectors as follows:
Green/black wire to terminal T.
Green/orange wire to terminal W.
Black wire to earth terminal.

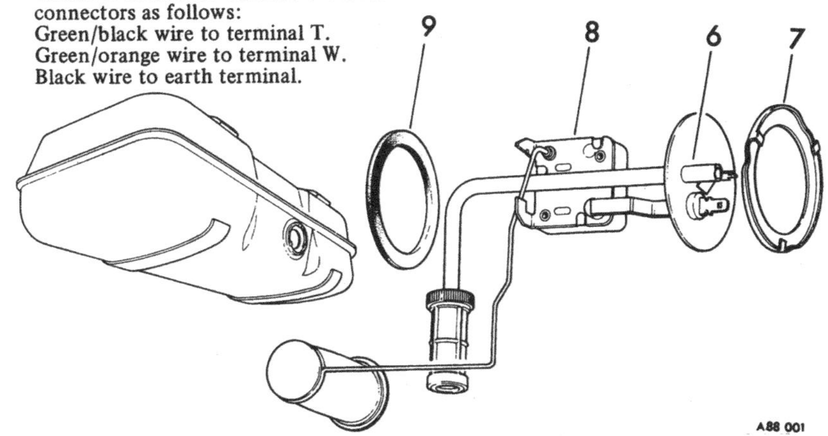

SPEEDOMETER

Remove and refit **88.30.01**

Remove
1 Remove the lens, see 88.20.17.
2 Lift away the face panel.
3 Depress the lever to release the catch from the annular groove in the boss. Pull the speedometer cable from the instrument.
4 Remove three Pozidriv screws.
5 Withdraw the speedometer.

Refit
6 To assist engagement of the speedometer trip reset cable rotate the control to pre-set the alignment.
7 Insert the speedometer.
8 Fit three Pozidriv screws.
9 Push the speedometer cable onto the instrument. Ensure that the catch engages into the annular groove in the boss.
10 Position the face panel.
11 Fit the lens, see 88.20.17.

SPEEDOMETER

Trip reset cable – remove and refit **88.30.02**

Remove
1 Remove the speedometer, see 88.30.01.
2 Note the run of the trip reset cable.
3 Slacken the trip reset cable knurled nut. Withdraw the cable from the bracket slot.
4 To assist withdrawal of the trip reset cable rotate the control to pre-set the alignment.
5 Using a small screwdriver release the trip reset cable claws. Withdraw the trip reset cable attachment from the instrument panel housing.

Refit
6 To assist insertion of the trip reset cable rotate the control to pre-set the alignment.
7 Insert the trip reset cable attachment into the instrument panel housing. Ensure that the claws engage correctly.
8 Ensure that the trip reset cable run is as noted at operation 2 above.
9 Insert the trip reset cable into the bracket slot. Hand tighten the knurled nut.
10 Fit the speedometer, see 88.30.01.

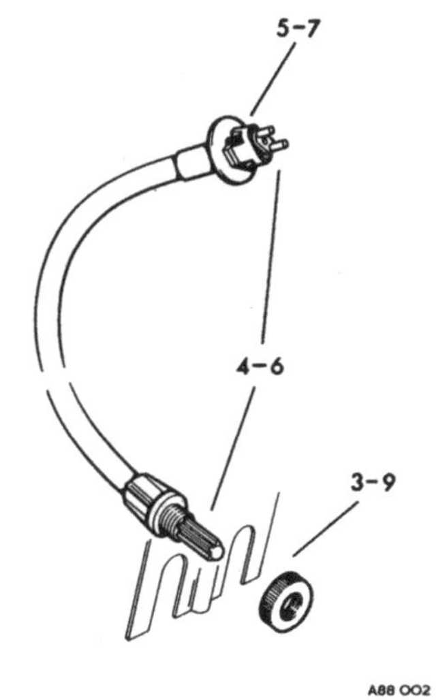

TACHOMETER

Remove and refit **88.30.21**

Remove
1 Remove the lens, see 88.20.17.
2 Lift away the face panel.
3 Remove three Pozidriv screws.
4 Carefully withdraw the tachometer.

Refit
5 Reverse 1 to 4.

All Service Tools mentioned in this Manual must be obtained direct from the manufacturers:

Messrs. V. L. Churchill & Co. Ltd.
P.O. Box No. 3
London Road
Daventry, Northants.
England NN11 4NF

Tool No.	Description
18G 47-1	Layshaft cluster bearings remover adaptor.
18G 47-3	Constant pinion remover/replacer adaptor.
18G 47 BD	Differential carrier bearings remover adaptor.
18G 47 BP	Mainshaft bearing and speedo gear remover/replacer adaptor.
18G 55 A (38U 3)	Piston ring compressor.
18G 106	Valve spring compressor.
18G 134 (550)	Bearing and oil seal replacer.
18G 134 DH	Differential bearings replacer adaptor.
18G 191	Bevel pinion setting gauge.
18G 191 M	Dummy pinion.
18G 207 (S 98A)	Preload gauge.
18G 284	Impulse extractor – U.N.F.
18G 284 AAA (S 4235-A2)	First motion shaft remover adaptor.
18G 304	Front and rear hub remover.
18G 1197	Mainshaft bearing and speedometer drive gear replacer.
18G 1198 (S 145)	Gearbox mainshaft circlip replacer.
18G 1198A	Gearbox mainshaft circlip replacer adaptor sleeve.
18G 1199 (S 114A)	Gearbox mainshaft circlip remover.
18G 1205 (S 337 RG 421)	Propeller shaft flange wrench.
18G 1208	Dummy layshaft.
18G 1256 A	Torx screw wrench.
18G 1261	Steering rack and pinion retainer spanner.
18G 1269	First motion shaft oil seal protector sleeve.

Tool No.	Description
18G 1311	Crankshaft spigot bearing remover.
18G 1312	Crankshaft spigot needle roller bearing.
47(S 4221A)	Multi-purpose hand press.
60A	Valve guide remover/replacer.
S 60A-2A	Valve guide remover/replacer adaptor.
S 60A-7	Valve guide replacer adaptor.
S 60A-8	Valve guide replacer adaptor.
S 60A-9	Valve guide replacer adaptor.
MS 68	Torx screw wrench.
S 101	Differential case spreader.
S 101-1	Adaptor.
S 304	Rear hub bearing replacer.
S 314	Mainshaft ball bearing replacer.
S 348	Water pump overhaul kit.
S 349	Distributor wrench.
S 350	Cylinder head stud remover/replacer.
S 352	Valve spring compressor adaptor.
S 353	Carburetter adjusting tool.
S 356C	Front and rear hub remover.
S 356C-5	Retaining nuts.
S 357A	Plug spanner.
RTR 359	Front suspension closure nut wrench.
RTR 360	Retaining washer holding wrench.
3072 (S 4235A)	Slide hammer.
S 4221A-10	Differential bearing remover adaptor.
S 4221A-17	Pinion head bearing remover/replacer adaptor.
S 4235A-1	Axle shaft remover.
S 4235A-10	Water pump assembly remover.
P 5045 (18G 1276)	Adjustable coil spring retainers.

Tool No.	Description
	5 SPEED TRANSMISSION
18G 47 AJ	Pinion bearing remover/replacer adaptor.
18G 284 AAH	Mainshaft pilot bearing outer track remover adaptor.
18G 705-1	Fifth speed gear remover adaptor.
18G 1271	Axle shaft inner oil seal remover.
18G 1272	Pinion nut wrench and preload adaptor.
18G 1273	Pinion oil seal protector sleeve.
S 4221A-16	Outer taper bearing remover/replacer adaptor.

Tool No.	Description
	AUTOMATIC TRANSMISSION
CBW 1C (18G 502)	Hydraulic test equipment.
CBW 1C-2	Pressure test adaptor.
18G 1307	Rear servo adjuster and special locknut front adaptor.
CBW 33	Mainshaft end-float gauge.
CBW 37A (18G 1016)	Clutch spring compressor.
CBW 41A (18G 702)	Rear clutch piston replacer.
CBW 42A (18G 1107)	Front clutch piston replacer.
CBW 60	Bench cradle.
CBW 62	Throttle cable mounting seal remover.
CBW 547A-50	Tension wrench.
CBW 547A-50-2A (18G 537B)	Rear servo adjuster adaptor.
CBW 548 (18G 681)	Torque screwdriver.

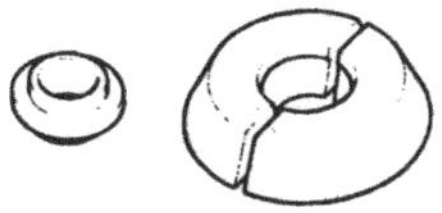

18G 47-1

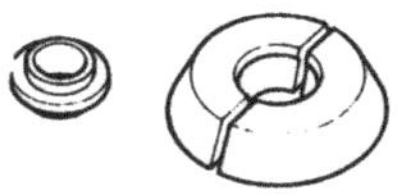

18G 47-3

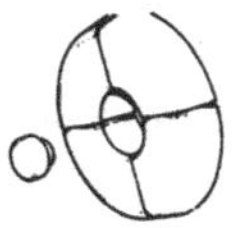

18G 47 BD

18G 47 BP

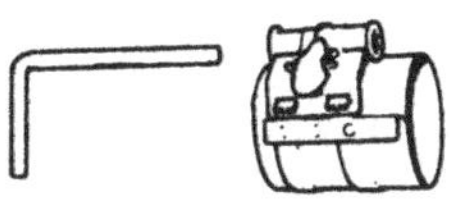

18G 55 A (38U 3)

18G 106

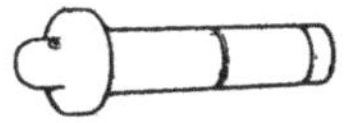

18G 134 (550)

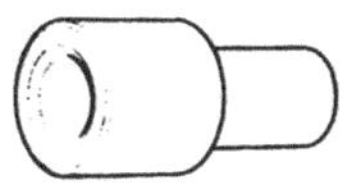

18G 134 DH

18G 191

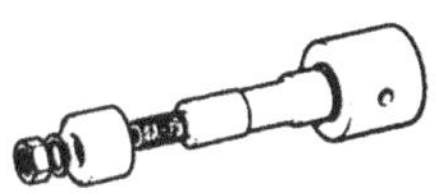

18G 191 M

18G 207 (S 98A)

18G 284

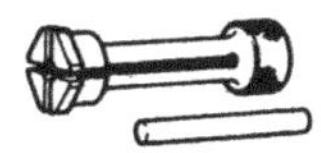

18G 284 AAA (S 4235-A2)

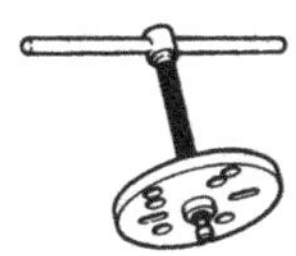

18G 304

18G 1197

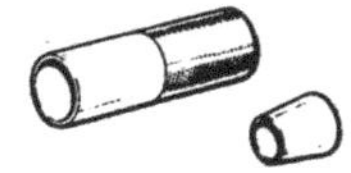

18G 1198 (S 145)

18G 1198 A

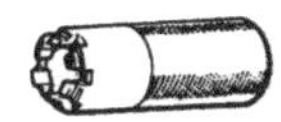

18G 1199 (S 144A)

18G 1205 (S 337)
(RG 421)

18G 1208

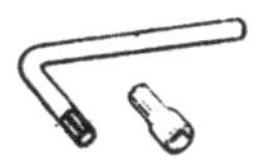

18G 1256 A

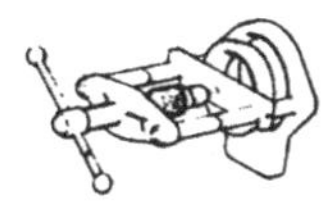

47 (S 4221A)

S 60A-9

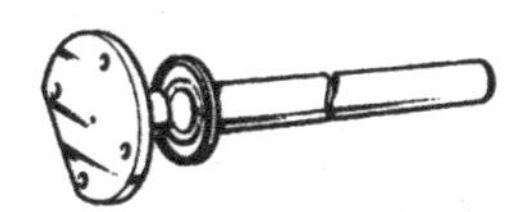

S 314

18G 1261

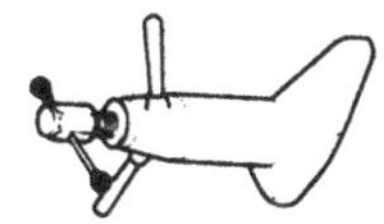

60 A

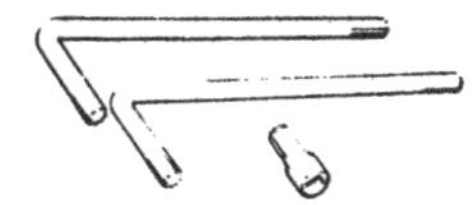

MS 68

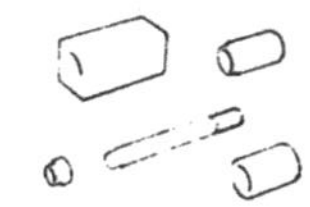

S 348

18G 1269

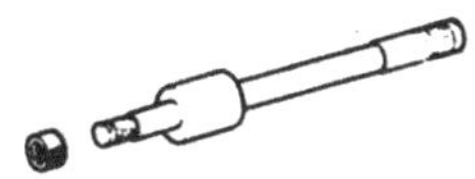

S 60A-2A

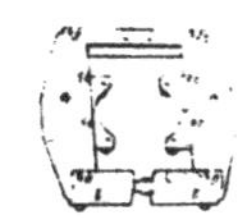

S 101

S 349

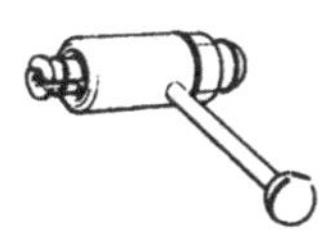

18G 1311

S 60A-7

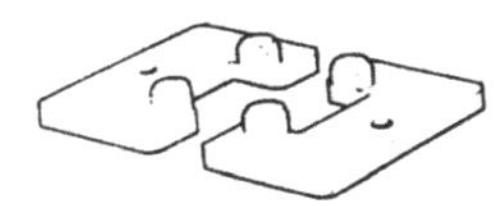

S 101-1

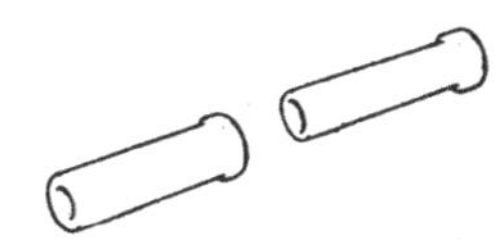

S 350

18G 1312

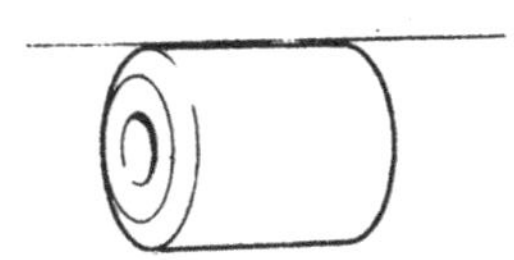

S 60A-8

S 304

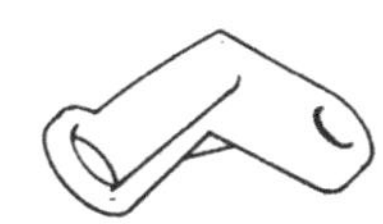

S 352

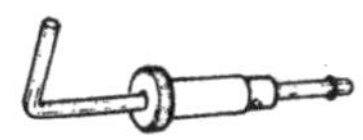
S 353

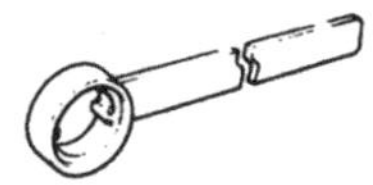
RTR 360

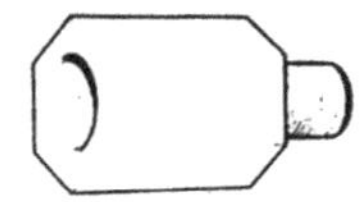
S 4235A-10

S 356C

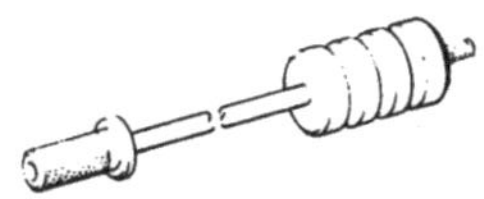
3072 (S 4235A)

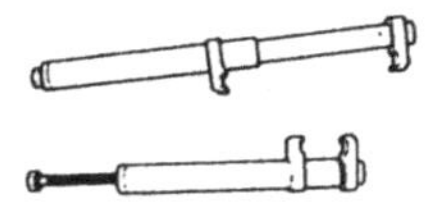
P 5045 (18G 1276)

S 356C-5

S 4221A-10

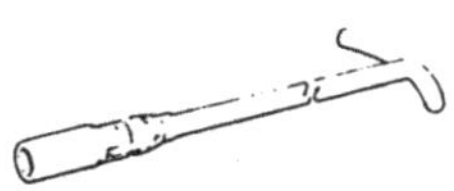
S 357A

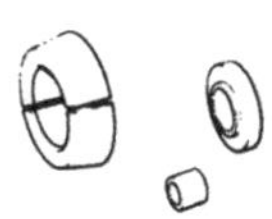
S 4221A-17

RTR 359

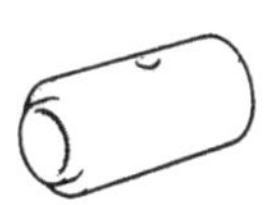
S 4235A-1

5 SPEED TRANSMISSION

18G 47 AJ

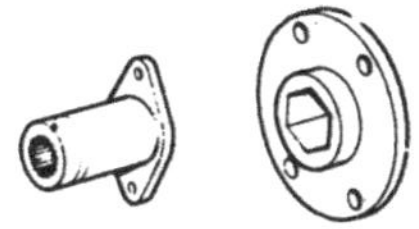

18G 1272

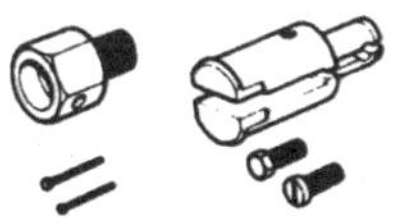

18G 284 AAH

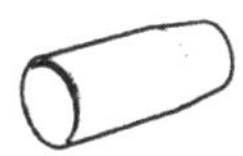

18G 1273

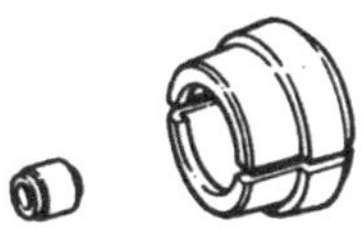

18G 705-1

S 4221A-16

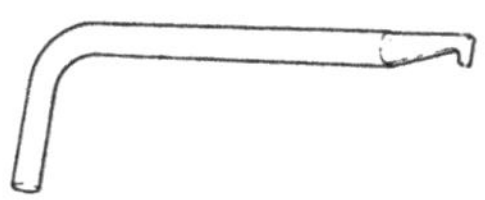

18G 1271

AUTOMATIC TRANSMISSION

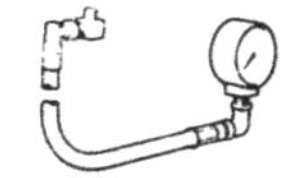

CBW 1C (18G 502)

CBW 42A (18G 110)

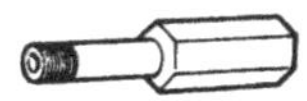

CBW 1C-2

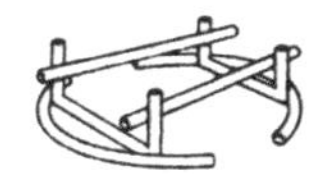

CBW 60

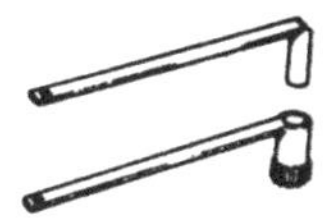

18G 1307

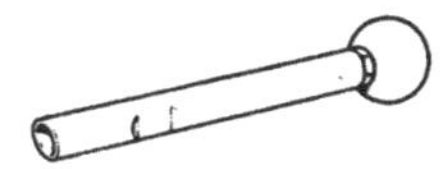

CBW 62

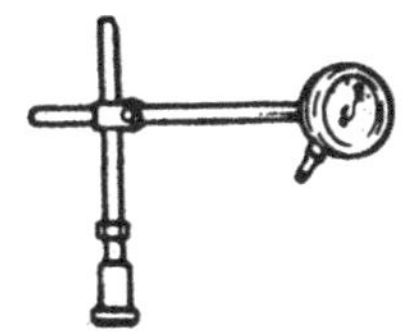

CBW 33

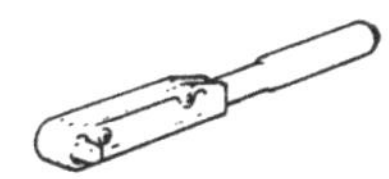

CBW 547A-50

CBW 37A (18G 1016)

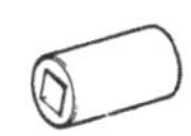

CBW 547A-50-2A (18G 537B)

CBW 41A (18G 702)

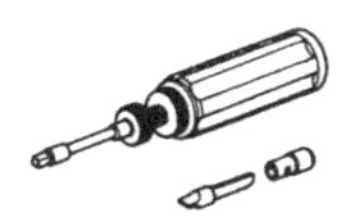

CBW 548 (18G 681)

202604R002